MW00720832

Technical Mathematics with Calculus

Linda Davis

Technical Mathematics with Calculus

Merrill Publishing Company
A Bell & Howell Information Company
Columbus Toronto London Melbourne

To Carlos, whose unending support, encouragement, and assistance made this textbook possible.

Published by Merrill Publishing Company
A Bell & Howell Information Company
Columbus, Ohio 43216

This book was set in Times Roman

Administrative Editor: Steve Helba
Developmental Editor: Monica Ohlinger
Production Editor: Mary Harlan
Art Coordinator: James Hubbard
Cover Designer: Cathy Watterson
Text Designer: Cynthia Brunk
Photo Editor: Terry Tietz

Photo Credits: All photos copyrighted by the individuals or companies listed. Jeff Greenberg: pp. 1, 41, 261, 289, 447, 525, 575, 633, 791, 879. Cobalt Productions/ Merrill: p. 85. Arlen Pennell/Riverside Methodist Hospital: p. 319. NASA: p. 349. B & K Precision: p. 413. Morning News Tribune, Tacoma, WA: 487. Larry Hamill: p. 609.

Library of Congress Catalog Card Number: 89–63150
International Standard Book Number: 0–675–20965–x
Printed in the United States of America
1 2 3 4 5 6 7 8 9—93 92 91 90

Preface

Approach

After years of watching technology students struggle with mathematics, my objective was to write a student-oriented textbook that would make learning mathematics easier. Of course, any textbook that makes learning easier for the student will also make teaching easier for the instructor. The features that make this book student oriented include

- The easy-to-follow writing style
- A spiral approach to learning
- A balance between pure math and technical applications
- The boxed step-by-step procedures
- The functional use of second color and graphics to support student learning

WRITING STYLE Experienced teachers know what approach works best in their classes for getting concepts across to students. In this book, I have used the approaches that have worked best for me in promoting student understanding. I have employed a conversational style to communicate the material at a reading level appropriate to the intended audience and to make it easy for students to read and understand. Also, I use examples to illustrate and explain every key concept presented. Further, though offering many examples is vital to the success of students in mathematics, more vital still is that they have at their disposal all of the explanation needed to grasp the concepts when they are doing their homework and the instructor is not around. So I have incorporated into the text verbal nudges similar to those a teacher might use in class or a tutoring session:

- ☐ **Learning Hints** give students practical study tips
- ☐ **Notes** point out trouble spots and help students overcome them
- ☐ **Cautions** alert students to common errors and ways to avoid them
- ☐ **Remember** notes reinforce previously learned concepts
- ☐ **Calculator symbols** in the examples and exercises let students know that they should use their calculators for those particular problems
- ☐ Application problems from the technologies are keyed to their specific area by labels next to the problem number:

  E Electronics
  M Mechanical
  I Industrial
  C Civil

SPIRAL APPROACH In the spiral approach, a concept is first introduced, then discussed again in subsequent chapters. For example, equations involving radicals are introduced in Chapter 4 and discussed again in Chapters 5, 6, and 7 in relation to material on quadratic equations, systems of equations, and solving higher degree equations, respectively. The spiral approach used in this text enhances student retention by

- ☐ building on what already has been learned
- ☐ challenging students to apply to new problems concepts that have been mastered
- ☐ periodically and methodically reviewing what has gone before

BALANCED COVERAGE In a technical mathematics textbook, the balance between pure mathematics and technology is very important. It is essential that students not only understand the mathematical concepts but also be able to apply them to technical problems. Thus, my aim has been to stress topics required for a career in technology without sacrificing the mathematics on which these topics are based. Applications are an integral part of the chapters and presume little prior technical knowledge. This integration affords students the opportunity to learn about technology as they learn math. When math can be shown to relate to the ''real world,'' students' motivation to understand it increases.

STEP-BY-STEP PROCEDURES Students progress through chapters in a consistent format:

- ☐ For key concepts, students are presented a clearly marked **Procedure Box** that gives them a step-by-step method for solving problems
- ☐ Within the chapters, there is a natural progression from simple to more difficult examples
- ☐ Each example includes a step-by-step procedure with accompanying explanation
- ☐ Rules and formulas are also featured in highlight boxes

- ☐ Within sections, subtopics are clearly labeled in the margins, making it easier for students to find those subjects they need to review

This consistent method for problem solving significantly aids students in mastering the subject of technical mathematics.

FUNCTIONAL USE OF SECOND COLOR AND GRAPHICS We have taken great care to use color to aid student learning and understanding. Additionally, we have included graphical representations of concepts where appropriate and where they will help students retain the material.

Organization

The organization of the text is also centered around a consistent format built on several key pedagogical features:

- ☐ Each chapter opens with an application problem related to the chapter material
- ☐ A detailed solution to the application problem is given at the end of the chapter
- ☐ Chapters also begin with a list of learning objectives that are keyed to sections in which relevant topics are covered
- ☐ End-of-chapter tests are, in turn, keyed to the learning objectives—incorrect answers on the self-tests can be referenced to objectives and then to chapter sections for additional study
- ☐ Every chapter concludes with a summary that serves as a handy guide for reviewing terms (page references provided) and formulas
- ☐ Chapters also end with a review that provides a sufficient number of problems for students to strengthen their knowledge of the chapter material and that separates topics by section so students can pinpoint those problem sections

Accuracy

Nothing is more frustrating to both teachers and students than inaccuracies in a mathematics textbook. Our goal has been to eliminate that frustration by putting the text—the examples, the answers in the back of the text, and the solutions manual—through rigorous accuracy checking. All told, the book went through six rounds of accuracy reviews by fifteen professors at crucial stages of its development.

The textbook includes three appendixes:

- ☐ Because Appendix A includes example problems and exercises, it can serve as Chapter 0 to review topics in arithmetic and algebra, including
- ☐ real numbers, laws of operations of real numbers, zero and orders of operations, exponents, scientific notation, roots and radicals, addition and
- ☐ subtraction of algebraic expressions, multiplication of polynomials, polynomial division, and percentages.
- ☐ Appendix B provides a review of basic geometric terms and formulas.

FIGURE 1
Examples of student-oriented features

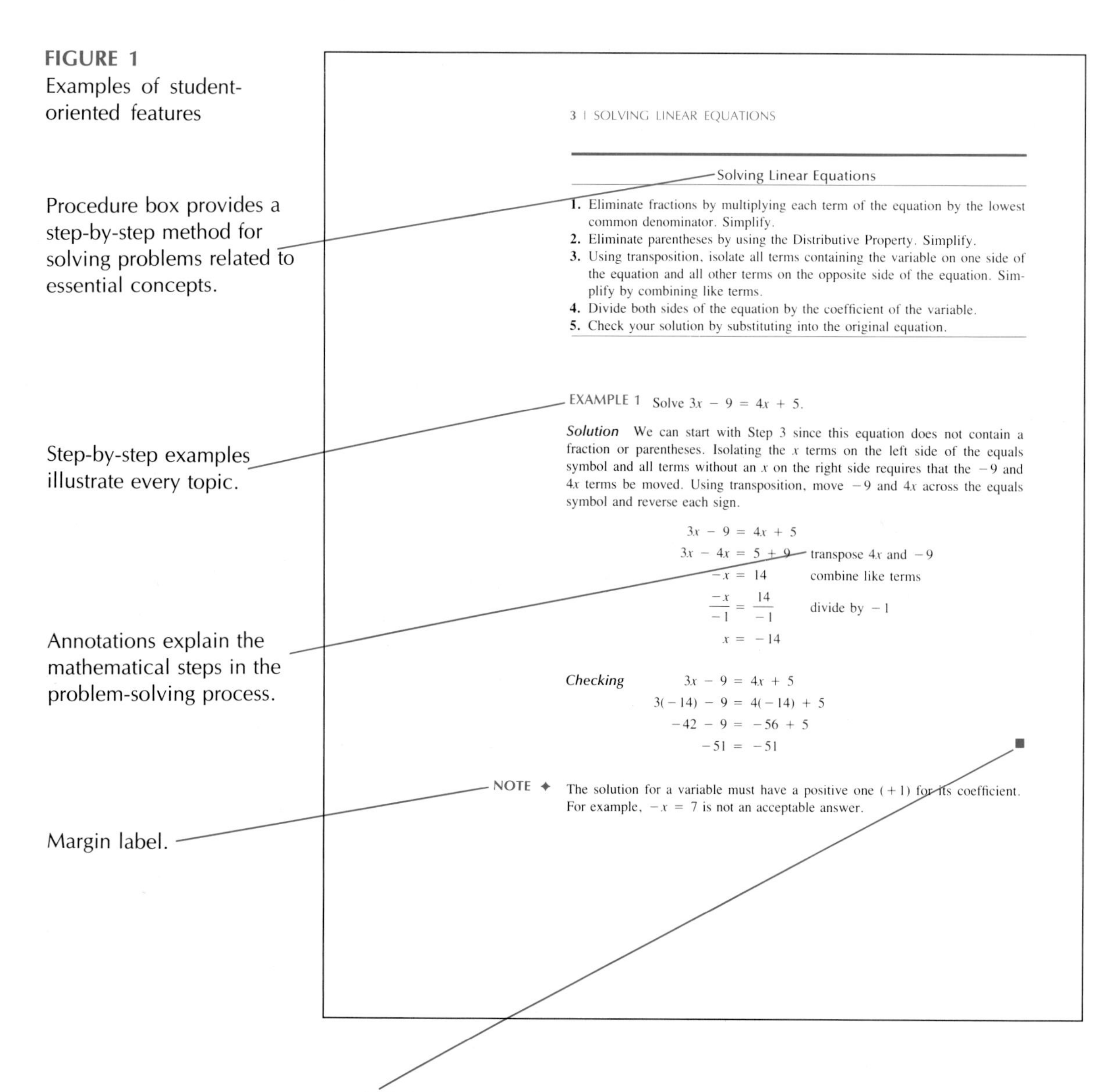

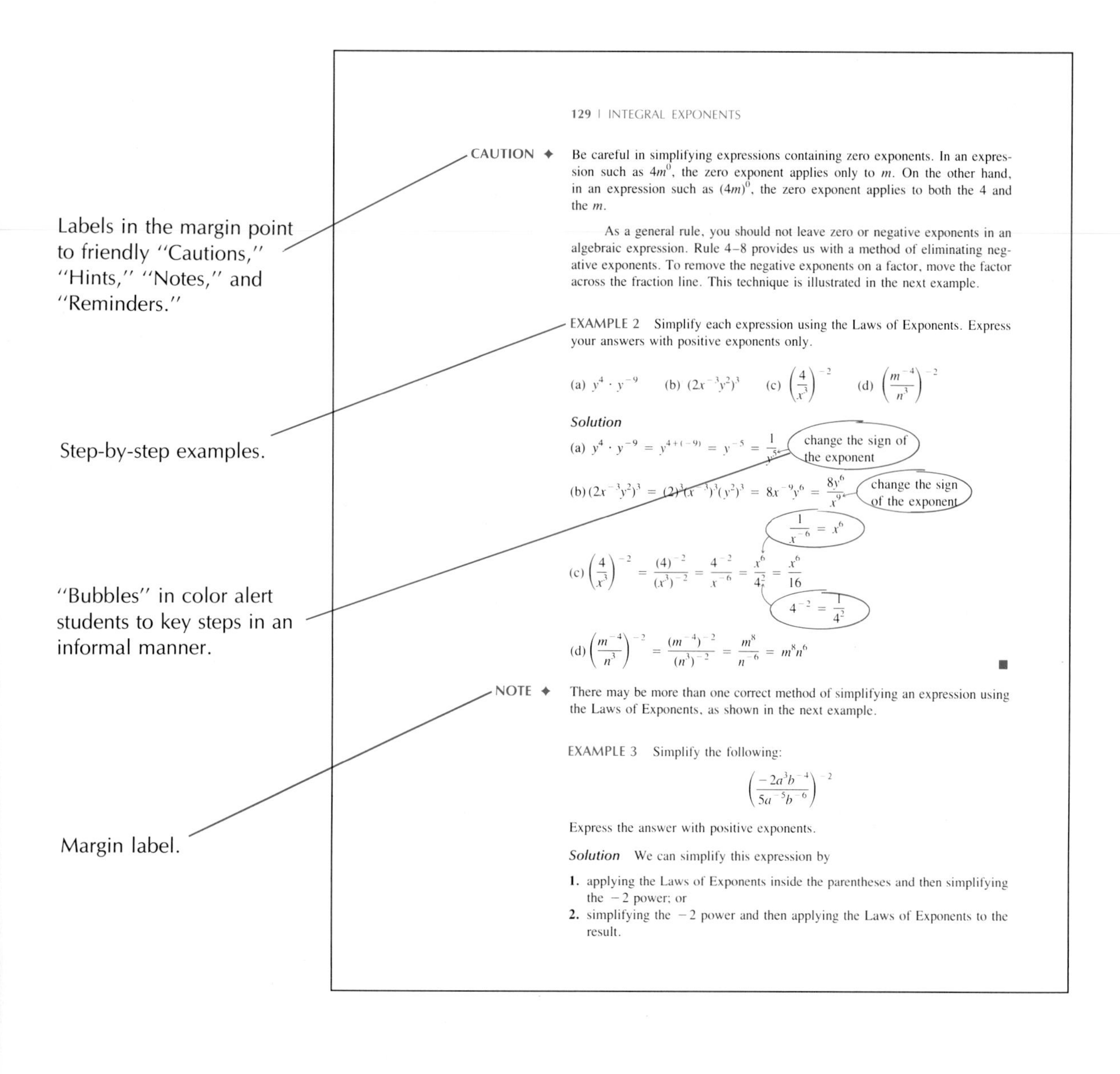

129 | INTEGRAL EXPONENTS

CAUTION ✦ Be careful in simplifying expressions containing zero exponents. In an expression such as $4m^0$, the zero exponent applies only to m. On the other hand, in an expression such as $(4m)^0$, the zero exponent applies to both the 4 and the m.

As a general rule, you should not leave zero or negative exponents in an algebraic expression. Rule 4–8 provides us with a method of eliminating negative exponents. To remove the negative exponents on a factor, move the factor across the fraction line. This technique is illustrated in the next example.

EXAMPLE 2 Simplify each expression using the Laws of Exponents. Express your answers with positive exponents only.

(a) $y^4 \cdot y^{-9}$ (b) $(2x^{-3}y^2)^3$ (c) $\left(\frac{4}{x^3}\right)^{-2}$ (d) $\left(\frac{m^{-4}}{n^3}\right)^{-2}$

Solution

(a) $y^4 \cdot y^{-9} = y^{4+(-9)} = y^{-5} = \frac{1}{y^5}$ (change the sign of the exponent)

(b) $(2x^{-3}y^2)^3 = (2)^3(x^{-3})^3(y^2)^3 = 8x^{-9}y^6 = \frac{8y^6}{x^9}$ (change the sign of the exponent)

(c) $\left(\frac{4}{x^3}\right)^{-2} = \frac{(4)^{-2}}{(x^3)^{-2}} = \frac{4^{-2}}{x^{-6}} = \frac{x^6}{4^2} = \frac{x^6}{16}$ ($\frac{1}{x^{-6}} = x^6$; $4^{-2} = \frac{1}{4^2}$)

(d) $\left(\frac{m^{-4}}{n^3}\right)^{-2} = \frac{(m^{-4})^{-2}}{(n^3)^{-2}} = \frac{m^8}{n^{-6}} = m^8n^6$ ■

NOTE ✦ There may be more than one correct method of simplifying an expression using the Laws of Exponents, as shown in the next example.

EXAMPLE 3 Simplify the following:

$$\left(\frac{-2a^3b^{-4}}{5a^{-5}b^{-6}}\right)^{-2}$$

Express the answer with positive exponents.

Solution We can simplify this expression by

1. applying the Laws of Exponents inside the parentheses and then simplifying the -2 power; or
2. simplifying the -2 power and then applying the Laws of Exponents to the result.

- ☐ Appendix C contains tables of integrals, metric units and conversions, trigonometric functions, logarithms, and powers and roots, which can be used for quick reference.

Suggestions for Use

Technical Mathematics with Calculus can be used in an introductory algebra, trigonometry, and calculus course for students in an engineering technology program. I have organized the text by grouping separately the algebra topics, the trigonometry topics, and the calculus topics. This sequencing greatly contributes to the text's flexibility, as the chapters can be easily integrated. Tables 1 through 6 indicate suggested course schedules for two-quarter and two-semester sequences of algebra and trigonometry, and Tables 7 and 8 cover calculus. Since some schools teach algebra and trigonometry as separate courses, I have included a schedule for each. I have also provided a schedule that includes coverage of Chapter 16, ''Introduction to Statistics and Empirical Curve Fitting'' and Chapter 17, ''Sequences, Series, and the Binomial Theorem.'' In preparing these schedules I have made these assumptions:

- ☐ the quarter courses are 5 hours per week for 10 weeks
- ☐ the semester courses are 3 hours per week for 18 weeks
- ☐ a test is given after each chapter, unless otherwise noted
- ☐ one day is spent on each section, unless otherwise noted

TABLE 1
Two Quarters (Algebra/Trig, excluding Chapters 16 & 17)

First Quarter (Algebra)		Second Quarter (Trig)	
Chapter	Number of days	Chapter	Number of days
1	5	9	5**
2	6	10	5**
3	6	11	6**
4	5	12	6
5	5	13	6**
6	6*	14	6**
7	5	15	9**
8	4		
*Combine Sections 6–5 and 6–6.		**Spend 2 days each on Sections 9–4, 10–4, 11–3, 11–4, 13–4, 14–4, 15–4, 15–5, 15–6.	

TABLE 2
Two Quarters (Algebra/Trig, including Chapters 16 & 17)

First Quarter (Algebra)		Second Quarter (Trig)	
Chapter	Number of days	Chapter	Number of days
1	5	9	4
2	6	10	4
3	6	11	4**
4	5	12	6
5	5	13	5
6	6*	14	6
7	5	15	6
8	4	16	4
		17	4**
*Combine Sections 6–5 and 6–6.		**Combine Chapters 10 and 11 and Chapters 16 and 17 into one chapter test.	

Ancillaries

The entire package—textbook and supplementary materials—has been designed to give instructors all the support they need to teach technical mathematics efficiently and effectively:

- ☐ The **Instructor's Resource Manual** includes solutions to all even-numbered in-text problems, and contains additional worked-out example problems for instructors to use in class—these examples can be made into transparencies.
- ☐ The **Test Bank** provides over 1700 multiple-choice questions, which are organized by chapter and section.
- ☐ The **Computerized Test Bank** (hardware requirements—IBM PC, AT, XT, PS/2, or compatible; 256K; 2 disk drives; DOS 2.0 or higher) utilizes the full 1700 questions of the printed test bank, is formatted for IBM or Apple, and provides full graphics and editing capabilities.
- ☐ The **Student Study Guide** contains worked-out solutions to every other odd-numbered problem in the text, gives students helpful study tips and additional explanation of material, and provides additional exercises and application problems.
- ☐ The **Study Disk** (hardware requirements—IBM PC, AT, XT, or compatible; 256K; 1 disk drive; CGA graphics) can be used in a lab setting as computer-aided instruction, provides students additional exercises and examples in an IBM format, and utilizes additional features, such as hints, study tips, and an on-screen calculator. (Disk is free upon adoption.)

TABLE 3
Two Semesters (Algebra/Trig, including Chapters 16 & 17)

First Semester (Algebra)		Second Semester (Trig)	
Chapter	Number of days	Chapter	Number of days
1	5	9	5**
2	6	10	4
3	8*	11	6**
4	5	12	6
5	6*	13	5
6	7	14	5
7	5	15	6
8	4	16	4
		17	4
*Spend 2 days each on Sections 3–4, 3–5, and 5–5.		**Spend 2 days each on Sections 9–4, 11–3, and 11–4.	

TABLE 4
Two Quarters (Integrated, excluding Chapters 16 & 17)

First Quarter		Second Quarter	
Chapter	Number of days	Chapter	Number of days
1	5	11	6**
2	6	12	6
9	5*	7	5
3	6	8	4
4	5	14	7**
5	5	13	6**
6	6*	15	9**
10	4		
*Combine Sections 6–5 and 6–6. Spend 2 days on Section 9–4.		**Spend 2 days each on Sections 11–3, 11–4, 13–4, 14–4, 14–5, 15–4, 15–5, 15–6.	

TABLE 5
Two Quarters (Integrated, including Chapters 16 & 17)

First Quarter		Second Quarter	
Chapter	Number of days	Chapter	Number of days
1	5	11	6**
2	6	12	6
9	5*	7	5
3	6	8	4
4	5	14	5
5	5	13	5
6	6*	15	6
10	4	16	4
		17	4
*Combine Sections 6–5 and 6–6. Spend 2 days on Section 9–4.		**Spend 2 days each on Sections 11–3 and 11–4.	

TABLE 6
Two Semesters (Integrated, including Chapters 16 & 17)

First Semester		Second Semester	
Chapter	Number of days	Chapter	Number of days
1	6*	11	6
2	6	12	6
9	5*	7	5
3	7*	8	4
4	5	14	5
5	6*	13	5
6	7	15	6
10	4	16	4
		17	4
*Combine Sections 6.5 and 6.6. Spend 2 days each on Sections 1–5, 9–4, 3–6, and 5–5.			

TABLE 7
Two Quarters (Calculus)

Calculus I		Calculus II	
Chapter	Number of days	Chapter	Number of days
18	15	21	20
19	17	22	13
20	18	23	17
Spend 2–3 days per section and schedule a test every 5–6 days			

TABLE 8
Two Semesters (Calculus)

Calculus I		Calculus II	
Chapter	Number of days	Chapter	Number of days
18	15	21	20
19	19	22	15
20	20	23	19
Spend 2–3 days per section and schedule a test every 5–6 days			

Acknowledgments

Any textbook from its inception to completion involves a great many people willing to give of their time and talent. This textbook is no exception. I would like to acknowledge the contribution of the following colleagues in the development of this textbook.

Sandra Beken, Horry Georgetown Technical College
Glenn R. Boston, Catawba Valley Community College
Robert J. Campbell, North Shore Community College
Ray E. Collings, Tri-County Technical College
Sabah Al-Hadad, California Polytechnic State University–San Luis Obispo
Harold Hauser, Mt. Hood Community College
Patricia L. Hirschy, Delaware Technical and Community College
Baback A. Izadi, DeVry Institute of Technology–Columbus
Wendell Johnson, University of Akron
Dimitrios Kyriazopolous, DeVry Institute of Technology–Chicago
Michael Latina, Community College of Rhode Island
Lynn Mack, Piedmont Technical College
David L. Potter, ITT Technical Institute–Dayton
Larry Richard, DeVry Institute of Technology–Lombard
Ursula Rhoden, Nashville State Technical Institute
William Van Alstine, Aiken Technical College
Kelly P. Wyatt, Umpqua Community College
Robert J. Zabek, Norwalk State Technical College

Accuracy checkers for the textbook and solutions manual:

John Bailey, Clark State Community College
Doug Cook, Owens Technical College
John L. Crawford, Lee College
Jane Downey, St. Cloud State University
Allan Edwards, West Virginia University–Parkersburg
J. Vernon Gwaltney, John Tyler Community College
John Heublein, Kansas Institute of Technology
Margaret Kimbell, Texas State Technical Institute–Waco
George Marketos, University of Cincinnati
Kylene Norman, Clark State Community College
Samuel F. Robinson, New Hampshire Technical College
John A. Seaton, Prince George's Community College
Molly Sumner, Pikes Peak Community College
Ed Turner, University of Cincinnati
Andrew S. Wargo, Bucks County Community College

In addition, I would especially like to thank Sandra Beken for her work on the test bank; Lynn Mack for her work on Appendixes A and B; Kylene Norman and John Bailey for their work on the solutions manual; and John

Heublein for his work on the accuracy check of the textbook. I would like to express my undying gratitude to Doug Cook for his work on the accuracy check of the textbook. His attention to detail and perseverance are greatly appreciated.

I would like to acknowledge the assistance of Monica Ohlinger, Mary Harlan, and Steve Helba at Merrill Publishing. As Development Editor, Monica was always willing to help and her cheerful attitude really made my job easier. As Production Editor, Mary Harlan was great in coordinating corrections from accuracy checks, multitudinous pieces of art, and answering my many questions. I would also like to thank Steve Helba for his assistance and support.

In conclusion, I would like to acknowledge the unending support and assistance of my husband, Carlos. His encouragement and support kept me going through many tight schedules. Also, his assistance in proofreading, accuracy checking, typing, and technical support was essential to the production of this textbook.

Contents

APPENDIXES

Solving equations or inequalities of various types is one of the most important applications of algebra. For example, a chemist is mixing a 35% hydrochloric acid solution with an 18% hydrochloric acid solution to yield 50 ml of a solution that is 26% hydrochloric acid. How much of each solution should the chemist use? (The solution to this problem is given at the end of the chapter.)

Technical problems can be stated in the form of an equation or formula. However, many technical problems are stated verbally, as the one above, and must be translated into an equation or inequality before they can be solved. In this chapter we will begin our study of solving equations and inequalities with the linear type.

Learning Objectives

After you complete this chapter, you should be able to

1. Use the Principles of Equality to solve linear equations in one variable (Section 1–1).
2. Solve an equation involving absolute value (Section 1–1).
3. Solve literal equations for a specified variable (Section 1–2).
4. Solve linear inequalities in one variable and graph the solution (Section 1–3).
5. Solve an inequality involving absolute value and graph the solution (Section 1–3).
6. Transform a verbal statement of variation into a formula (Section 1–4).
7. Solve applied problems using variation (Section 1–4).
8. Apply the techniques of solving equations and inequalities learned in this chapter to technical problems (Section 1–5).

Chapter 1

Solving Linear Equations and Inequalities

1–1

SOLVING LINEAR EQUATIONS

An **equation** is a statement that two expressions are equivalent or equal. In a **linear equation** in one variable, *the variable has an exponent of one and does not appear in the denominator of any term.* (A *term* is an expression separated by an addition or subtraction symbol.) For example, both of the following are linear equations in one variable:

$$2x + 6(3x - 1) = 7$$

and

$$8x = 7 + 5x - 10$$

However, neither of the following is a linear equation:

$$6x^2 + 5x = 10$$

and

$$\frac{7}{x} + 3x - 8 = 0$$

An equation or **open statement** may be true for some values of the variable and false for other values. However, a **solution** or **root** of an equation is a number that produces a true statement when substituted for the variable. Two is the solution or root for the equation $x - 6 = -4$ because substituting two for x results in a true statement: $2 - 6 = -4$.

Since not all linear equations are easy to solve by trial-and-error substitutions, we must develop a systematic method of solving them. Before discussing a systematic procedure, however, let us look at the Principles of Equality, which serve as the basis for solving equations. The **Principles of Equality** state that if the same operation is performed on each side of an equation, an equivalent equation results. Thus, if you add, subtract, multiply, or divide both sides of an equation by the same quantity, you do not change the solution to the equation. Another useful property in solving equations, called **transposition,** states that moving any *term* from one side of the equation to the opposite side and reversing its sign does not alter the equality. For example, $x + 4 = 0$ is equivalent to $x = -4$. Transposition produces the same result as adding or subtracting the same quantity to both sides of an equation, but transposition is shorter and faster. A procedure for solving linear equations is given in the following box. An explanation of each step involved in this process is given in later examples.

Solving Linear Equations

1. Eliminate fractions by multiplying each term of the equation by the lowest common denominator. Simplify.
2. Eliminate parentheses by using the Distributive Property. Simplify.
3. Using transposition, isolate all terms containing the variable on one side of the equation and all other terms on the opposite side of the equation. Simplify by combining like terms.
4. Divide both sides of the equation by the coefficient of the variable.
5. Check your solution by substituting into the original equation.

EXAMPLE 1 Solve $3x - 9 = 4x + 5$.

Solution We can start with Step 3 since this equation does not contain a fraction or parentheses. Isolating the x terms on the left side of the equals symbol and all terms without an x on the right side requires that the -9 and $4x$ terms be moved. Using transposition, move -9 and $4x$ across the equals symbol and reverse each sign.

$$3x - 9 = 4x + 5$$

$$3x - 4x = 5 + 9 \qquad \text{transpose } 4x \text{ and } -9$$

$$-x = 14 \qquad \text{combine like terms}$$

$$\frac{-x}{-1} = \frac{14}{-1} \qquad \text{divide by } -1$$

$$x = -14$$

Checking

$$3x - 9 = 4x + 5$$

$$3(-14) - 9 = 4(-14) + 5$$

$$-42 - 9 = -56 + 5$$

$$-51 = -51$$

■

NOTE ✦ The solution for a variable must have a positive one $(+1)$ for its coefficient. For example, $-x = 7$ is not an acceptable answer.

EXAMPLE 2 Solve $2(3x + 4) = 6 - (2x - 5)$.

Solution Since this equation does not contain fractions, we begin by using the Distributive Property to remove the parentheses.

$$
\begin{aligned}
2(3x + 4) &= 6 - (2x - 5) && \\
6x + 8 &= 6 - 2x + 5 && \text{eliminate parentheses} \\
6x + 8 &= 11 - 2x && \text{combine like terms} \\
6x + 2x &= 11 - 8 && \text{transpose } -2x \text{ and } 8 \\
8x &= 3 && \text{combine like terms} \\
\frac{8x}{8} &= \frac{3}{8} && \text{divide by } 8
\end{aligned}
$$

After checking, the solution is $x = 3/8$. ■

EXAMPLE 3 Solve $2.1x + 6 = 7.3x - 4.8$.

Solution Since this equation does not contain fractions or parentheses, we begin with Step 3. To isolate the variable on the left side of the equation, transpose 6 and $7.3x$.

$$
\begin{aligned}
2.1x + 6 &= 7.3x - 4.8 && \\
2.1x - 7.3x &= -4.8 - 6 && \text{transpose } 7.3x \text{ and } 6 \\
-5.2x &= -10.8 && \text{combine like terms} \\
x &\approx 2.1 && \text{divide by } -5.2
\end{aligned}
$$

10.8 [+/−] [÷] 5.2 [+/−] [=] ⟶ 2.076923077 ■

NOTE ✦ The [+/−] key is used to change the sign of the previous entry.

EXAMPLE 4 Solve $\frac{3}{4}x - 2 = \frac{1}{3}x + \frac{1}{2}$.

Solution Begin by removing the fractions; that is, multiply each term by the lowest common denominator, 12.

$$\overset{3}{\cancel{12}}\left(\frac{3}{\cancel{4}}x\right) - 12(2) = \overset{4}{\cancel{12}}\left(\frac{1}{\cancel{3}}x\right) + \overset{6}{\cancel{12}}\left(\frac{1}{\cancel{2}}\right)$$

$$
\begin{aligned}
9x - 24 &= 4x + 6 && \text{simplify} \\
9x - 4x &= 6 + 24 && \text{transpose } 4x \text{ and } -24 \\
5x &= 30 && \text{combine like terms} \\
x &= 6 && \text{divide by } 5
\end{aligned}
$$

■

EXAMPLE 5 Solve $\dfrac{2(x-3)}{5} = x - \dfrac{3}{5}$.

Solution First, we remove fractions by multiplying each term of the equation by the common denominator, 5.

$$\cancel{5}\left(\frac{2(x-3)}{\cancel{5}}\right) = 5(x) - \cancel{5}\left(\frac{3}{\cancel{5}}\right)$$

$$\begin{aligned} 2(x-3) &= 5x - 3 && \text{simplify} \\ 2x - 6 &= 5x - 3 && \text{eliminate parentheses} \\ 2x - 5x &= -3 + 6 && \text{transpose } 5x \text{ and } -6 \\ -3x &= 3 && \text{combine like terms} \\ x &= -1 && \text{divide by } -3 \end{aligned}$$

Checking

$$\frac{2(-1-3)}{5} = -1 - \frac{3}{5}$$

$$\frac{-8}{5} = \frac{-8}{5}$$

The solution to the equation is $x = -1$. ■

CAUTION ✦ Be sure that you transpose only *terms*. For example, in $2x - 8 = 10$, you can transpose $2x$, but you cannot transpose 2 because it is not a term but a factor.

Absolute Value Equations

Absolute value, denoted by the symbol $|\ \ |$, is defined to be the distance between a number and zero on the number line. Therefore, the statement $|x| = 4$ means that x could be 4 or -4. (See Figure 1–1.) Solving an absolute value equation may require several additional steps which are given below.

FIGURE 1–1

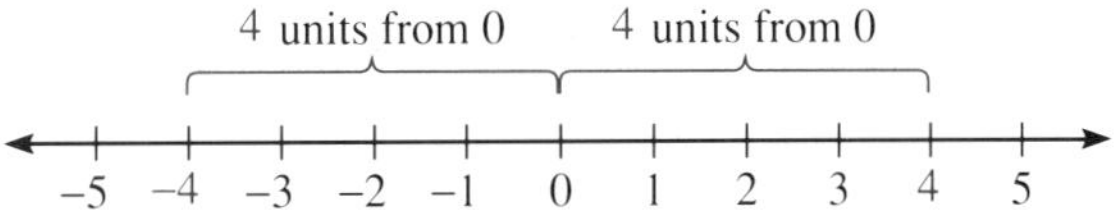

4 units from 0   4 units from 0

−5 −4 −3 −2 −1 0 1 2 3 4 5

Solving Absolute Value Equations

1. Isolate the absolute value quantity on one side of the equation.
2. Write two equations using the definition of absolute value.
3. Solve each linear equation using the steps given previously.
4. Check the solution.

EXAMPLE 6 Solve $2|4x - 5| + 3 = 15$.

Solution Isolating the absolute value quantity on one side of the equation gives

$$\begin{aligned} 2|4x - 5| &= 15 - 3 && \text{transpose 3} \\ 2|4x - 5| &= 12 && \text{combine like terms} \\ |4x - 5| &= 6 && \text{divide by 2} \end{aligned}$$

Using the definition of absolute value, the following two statements are true:

$$\begin{aligned} +(4x - 5) &= 6 & \text{or} \quad (4x - 5) &= 6 \\ 4x &= 11 & -4x + 5 &= 6 \\ x &= 11/4 & -4x &= 1 \\ & & x &= -1/4 \end{aligned}$$

Checking

$$\begin{aligned} 2|4(11/4) - 5| + 3 &= 15 & 2|4(-1/4) - 5| + 3 &= 15 \\ 2|6| + 3 &= 15 & 2|-6| + 3 &= 15 \\ 2 \cdot 6 + 3 &= 15 & 2 \cdot 6 + 3 &= 15 \\ 15 &= 15 & 15 &= 15 \end{aligned}$$

The solutions are $x = 11/4$ and $x = -1/4$. ■

EXAMPLE 7 Solve the following equation: $4|2x - 3| + 6 = 11$.

Solution First, we isolate the absolute value quantity by transposing 6.

$$\begin{aligned} 4|2x - 3| &= 11 - 6 \\ 4|2x - 3| &= 5 \end{aligned}$$

Normally, we would divide both sides of the equation by 4. However, in this case, a fraction results on the right side of the equation. Therefore, at this point, we will apply the definition of absolute value.

$$\begin{aligned} 4(2x - 3) &= 5 & \text{or} \quad -[4(2x - 3)] &= 5 \\ 8x - 12 &= 5 & -8x + 12 &= 5 \\ 8x &= 17 & -8x &= -7 \\ x &= 17/8 & x &= 7/8 \end{aligned}$$

The solutions are $x = 17/8$ and $x = 7/8$. ■

Application

EXAMPLE 8 The equivalent resistance R of two resistors R_1 and R_2 connected in parallel is given by

$$\frac{1}{R} = \frac{1}{R_1} + \frac{1}{R_2}$$

If $R_1 = 5$ ohms (Ω) and $R = 3\ \Omega$, find R_2 by substituting the given values for R_1 and R.

Solution Substituting values for R and R_2 gives the following equation:

$$\frac{1}{3} = \frac{1}{5} + \frac{1}{R_2}$$

To solve for R_2, multiply by the common denominator, $15R_2$, which gives

$$15R_2\left(\frac{1}{3}\right) = 15R_2\left(\frac{1}{5}\right) + 15R_2\left(\frac{1}{R_2}\right)$$

$$5R_2 = 3R_2 + 15$$

$$5R_2 - 3R_2 = 15$$

$$2R_2 = 15$$

$$R_2 = 7.5$$

The value of R_2 is 7.5 Ω. ■

LEARNING HINT ✦ If you have difficulty in solving linear equations, write the steps for solution given in this section on an index card. Then use the index card when doing homework, and go through these steps one at a time.

1–1 EXERCISES

Solve each of the following equations.

1. $x + 6 = -11$

2. $x + 10 = 8$

3. $2x - 6 = -14$

4. $-3x + 4 = 19$

5. $6x - 7 + 4x = 8 - 5x + 10 + 5$

6. $10x - 5 + 6 - 2x = 3x - 9$

7. $3.7x + 8.6 = 10.5$

8. $4.2 - 3.5x = 14.4$

9. $2(x - 6) = 6$

10. $3(x - 1) = 10$

11. $4(3x - 6) - 10x = 12$

12. $-2(4x - 7) + 5x = 7$

13. $\frac{x}{2} - 5 = 3$

14. $8 - \frac{x}{3} = 2$

15. $8.1x + 6 = 3.2x - 16.4$

16. $9.7x + 4.3 = 15.8 - 2.4x$

17. $\frac{3(-2x + 4)}{5} = 7$

18. $\frac{-5(x - 4)}{7} = -1$

19. $3(x - 1) - (2x + 6) = 10$

20. $4(2x - 1) - (3x + 2) = 8$

21. $7x - 5(2x + 3) = 4 - (2x + 6)$

22. $5x - 7(2x - 1) = 9 + 2(3 - x)$

23. $\frac{2(x - 2)}{3} + 5x = 10$

24. $\frac{4(3 - x)}{5} - 3x = -6$

25. $\frac{8x}{9} - 2 = \frac{-5x}{6} + \frac{1}{3}$

26. $\frac{x}{6} + \frac{3x}{4} + 1 = \frac{2x}{3} - \frac{5}{6}$

27. $\frac{x}{3} - \frac{1}{3} = \frac{x}{2} + 3$

28. $\frac{-2x}{3} + 2 = \frac{5x}{6} - \frac{3x}{2}$

29. $4(3x - 6) - 7x + 6 = -2(x - 6) + 4x$

30. $-9(x - 2) + 6 + 6x = 7x - 2(3x + 4)$

31. $-5(2x - 1) + 6 = 4(8 - x) - (3x + 2) + 5$

32. $3(2x - 4) + 16 = -5(x + 6) + (2 - 5x)$

33. $1.8x - 4.3 = 2.4(x - 3)$

34. $5.1 + 6.8x = 7.3(2x - 3.9)$

35. $\frac{3(x - 5)}{7} = 6 + 2x$

36. $\dfrac{-2(3x+4)}{5} = -7 + 3x$

37. $\dfrac{x-3}{4} + 6 = \dfrac{2x+9}{3}$

38. $\dfrac{2x+5}{6} + 1 = \dfrac{x-3}{2}$

39. $\dfrac{3(4x-6)}{2} - \dfrac{3}{4} = \dfrac{x+6}{3}$

40. $\dfrac{2(2x+3)}{3} - 2 = \dfrac{-(x+4)}{2}$

41. $|4x - 6| = 10$ **42.** $|3x - 8| = 16$

43. $5|2x - 1| + 4 = 11$ **44.** $2|3x - 6| - 2 = 15$

45. $2|x - 5| + 7 = 14$ **46.** $4|3x + 1| - 7 = 16$

47. If the efficiency of an engine is 0.56 when the absolute initial temperature is 800 kelvins (800 K), the resulting equation is

$$0.56 = 1 - \frac{T}{800}$$

Find the absolute final temperature T.

48. In finding the resultant of two forces, a scientist must solve the following equation:

$$0.94A - (0.64)200 = 100$$

Find A.

E **49.** Kirchhoff's current law states that the sum of the currents flowing into a junction equals the sum of currents flowing out of the junction. For a particular circuit the following equation results:

$$2 + \frac{v}{5} = 3$$

Solve for v.

50. A formula from physics is $v = v_0 + at$. Determine the value for t when $v = 73$ ft/s, $v_0 = 22$ ft/s, and $a = 8$ ft/s^2.

E **51.** The voltage V in a certain circuit is given by the formula $V = IR$ where I is the current and R is the resistance. Determine the current in a circuit where the voltage is 110 V and the resistance is 15 Ω.

52. The frequency of sound f' in hertz (Hz) is given by $f' = f\left(1 + \dfrac{v}{s}\right)$. Solve for f' when $f = 400$ Hz, $v = 45$ mi/h, and $s = 760$ mi/h.

53. The total surface area A of a cylinder is given by $A = 2\pi rh + 2\pi r^2$. Solve for h if $r = 36$ ft and $A = 6{,}192\pi$.

C **54.** The maximum length of a line on a map S is given by $S = \sqrt{8RE - E^2}$. Determine the maximum length of the line when $R = 0.38$ mm and $E = 0.18$ mm.

E **55.** The total resistance R_T of two resistors connected in parallel is given by

$$\frac{1}{R_T} = \frac{1}{R_1} + \frac{1}{R_2}$$

Find R_T when $R_1 = 7.5$ Ω and $R_2 = 5.6$ Ω.

56. The distance d a rocket travels is given by $d = rt$ where r = rate and t = time. How far will a rocket traveling at 8.73 km/s travel in 9 sec?

M **57.** The kinetic energy of a moving body is given by K.E. $= mv^2/2$ where m is the mass and v is the velocity. Determine the kinetic energy of a truck whose mass is 200 slugs and whose velocity is 35 ft/s. $\left(\text{Note: slugs} = \dfrac{\text{lb}}{\text{ft/s}^2}.\right)$

58. The volume V of a sphere is given by $V = 4\pi r^3/3$. Find the volume of a sphere if $r = 18$ ft.

59. The temperature in degrees Fahrenheit is given by $F = \dfrac{9}{5}C + 32$. Find the temperature Fahrenheit corresponding to 35°C.

1–2

SOLVING LITERAL EQUATIONS

One of the most important applications of solving equations is evaluating and solving formulas, or literal equations. Many times in a science or technology application, you will need to rearrange a given formula to solve for a different variable and then evaluate that formula. In evaluating a formula, we substitute a numerical value for variables and calculate the resulting numerical value of another variable. For example, the perimeter of an equilateral triangle is given by $P = 3s$. We can evaluate the perimeter of different triangles by substituting the length of the sides for s. Solving a formula requires rearranging an equation in a different form. This section discusses both of these concepts.

Notation

A common practice in algebra is to *use the first letter of the word* to represent the *variable*. For example, we could use b to represent the base of a rectangle. However, in the case of the area of a trapezoid, there are two bases of different lengths, and the letter b cannot be used to represent both bases. To alleviate this problem, we can use subscripts. A **subscript** is a number or letter written below and to the right of a variable. In the case of the area of a trapezoid, the subscripts 1 and 2 could be used on the letter b to denote the two bases. Thus, the bases of the trapezoid could be represented by b_1 and b_2. Another method used to denote different quantities with the same letter is to use uppercase and lowercase letters. The bases of the trapezoid could also be represented as b and B.

Solving Literal Equations

In order to solve a formula or literal equation for a specific variable, you follow the same steps that were given in the previous section for linear equations. These steps are repeated below for your convenience.

Solving Literal Equations

1. Eliminate fractions by multiplying each term of the equation by the lowest common denominator. Simplify.
2. Eliminate parentheses in the equation by using the Distributive Property. Simplify.
3. Using transposition, isolate all terms containing the variable on one side of the equation. Simplify.
4. Divide both sides of the equation by the coefficient of the variable.
5. Check the solution.

EXAMPLE 1 Solve the following literal equation for h:

$$A = \frac{bh}{2} \quad \text{area of a triangle}$$

Solution First, we eliminate the fraction by multiplying each term by the common denominator, 2.

$$2(A) = 2\left(\frac{bh}{2}\right) \quad \text{eliminate fractions}$$

$$2A = bh \quad \text{simplify by removing parentheses}$$

Since the term containing h is isolated, we divide both sides of the equation by the coefficient of h, which is b.

$$\frac{2A}{b} = \frac{bh}{b} \quad \text{divide by } b$$

$$h = \frac{2A}{b}$$

Checking Substitute into the equation just as we did in the previous section.

$$A = \frac{bh}{2}$$

$$A = \frac{b(2A/b)}{2}$$

$$A = A$$

■

EXAMPLE 2 Solve the following for v_2:

$$m(v_2 - v_1) = Ft \quad \text{change in momentum from physics}$$

Solution We begin by using the Distributive Property to remove the parentheses.

$$m(v_2 - v_1) = Ft$$

$$mv_2 - mv_1 = Ft \quad \text{eliminate parentheses}$$

$$mv_2 = Ft + mv_1 \quad \text{transpose}$$

$$v_2 = \frac{Ft + mv_1}{m} \quad \text{divide by } m$$

Checking

$$m\left[\left(\frac{Ft + mv_1}{m}\right) - v_1\right] = Ft$$

$$Ft = Ft$$

■

CAUTION ✦ Remember that like terms must have the same subscript and exponent. Therefore, we cannot combine v_1 and v_2 in Example 2.

EXAMPLE 3 Solve the following formula for b_2:

$$A = \frac{h(b_1 + b_2)}{2} \quad \text{area of a trapezoid}$$

Solution

$$2(A) = 2\left(\frac{h(b_1 + b_2)}{2}\right) \quad \text{eliminate fractions}$$

$$2A = h(b_1 + b_2) \quad \text{simplify}$$

$$2A = hb_1 + hb_2 \quad \text{eliminate parentheses}$$

$$2A - hb_1 = hb_2 \quad \text{transpose}$$

$$\frac{2A - hb_1}{h} = b_2 \quad \text{divide by } h$$

$$b_2 = \frac{2A - hb_1}{h}$$

■

CAUTION ✦ Be careful in simplifying the last expression in Example 3. You can divide away factors contained in both the numerator and the denominator of a fraction, but not terms. We cannot cancel the h's in the formula given in Example 3 because the h in the numerator is part of a term.

Evaluating Formulas

In evaluating a formula, we calculate the value of a specific variable, given numerical values for the remaining variables. In some instances, we may need to rearrange the formula and solve for a different variable. This process is illustrated in the next three examples.

EXAMPLE 4 The distance s an object travels in time t is given by the formula

$$s = vt + \frac{1}{2}at^2$$

where v is initial velocity, t is time, and a is acceleration. Find the acceleration of an object that travels 100 m in 8 s if the initial velocity is 10 m/s. Solve for a, then substitute numerical values.

Solution We rearrange the formula to solve for a and substitute the numerical values for s, v, and t.

$$2(s) = 2(vt) + 2\left(\frac{1}{2}at^2\right) \quad \text{eliminate fractions}$$

$$2s = 2vt + at^2 \quad \text{simplify}$$

$$2s - 2vt = at^2 \quad \text{transpose}$$

$$\frac{2s - 2vt}{t^2} = a \quad \text{divide by } t^2$$

Substituting $s = 100$, $v = 10$, and $t = 8$ into the formula gives

$$a = \frac{2(100) - 2(10)(8)}{(8)^2}$$

$$a = 0.625 \text{ m/s}^2$$

2 [×] 100 [−] 2 [×] 10 [×] 8 [=] [÷]

8 [x^2] [=] ⟶ 0.625

■

EXAMPLE 5 The formula for converting temperature in degrees Fahrenheit to degrees Celsius is

$$C = \frac{5}{9}(F - 32)$$

If the temperature is 50°C, find the corresponding temperature in degrees Fahrenheit by solving the formula for F and then substituting.

Solution First, we rearrange the formula to solve for F.

$$9(C) = 9\left(\frac{5}{9}(F - 32)\right) \quad \text{eliminate fractions}$$

$$9C = 5(F - 32) \quad \text{simplify}$$

$$9C = 5F - 160 \quad \text{eliminate parentheses}$$

$$9C + 160 = 5F \quad \text{transpose } -160$$

$$\frac{9C + 160}{5} = F \quad \text{divide by 5}$$

Then we substitute $C = 50$ into the formula.

$$F = \frac{9(50) + 160}{5}$$

$$F = \frac{610}{5}$$

$$F = 122$$

The temperature that corresponds to 50°C is 122°F.

9 [×] 50 [+] 160 [=] [÷] 5 [=] ⟶ 122 ■

EXAMPLE 6 For two resistors r and R connected in series in a dc circuit of voltage V, current I is given by

$$I = \frac{V}{(r + R)}$$

Find r if $I = 2$ amperes (A), $V = 12$ V, and $R = 5\ \Omega$, the unit of resistance. Solve and substitute.

Solution First, we solve the formula for r. We eliminate the fraction by multiplying each term by $(r + R)$.

$$(r + R)I = (r + R)\left(\frac{V}{r + R}\right)$$

$$I(r + R) = V \qquad \text{simplify}$$

$$Ir + IR = V \qquad \text{eliminate parentheses}$$

$$Ir = V - IR \qquad \text{transpose}$$

$$r = \frac{V - IR}{I} \qquad \text{divide by } I$$

Then we substitute $V = 12$, $I = 2$, and $R = 5$ into the formula.

$$r = \frac{12 - 2(5)}{2}$$

$$r = 1\ \Omega$$

12 [−] 2 [×] 5 [=] [÷] 2 [=] ⟶ 1 ■

1–2 EXERCISES

Solve each formula for the specified variable.

1. $A = Lwh$ for h

2. $P = 2L + 2w$ for L

3. $S = 2\pi rh + 2\pi r^2$ for h

4. $L = \frac{1}{2}Ps$ for P

5. $E = mgh$ for m

6. $V = \frac{4}{3}\pi r^3$ for π

7. $I = kL(T - t)$ for t

8. $R = \frac{rL}{D^2}$ for L

9. $v^2 = v_0^2 + 2as$ for a

10. $P = \frac{fs}{t}$ for f

11. $T = \frac{D - d}{2}$ for D

12. $A = \frac{hb}{2}$ for b

13. $E = \frac{1}{2}mv^2$ for m

14. $V = \frac{1}{3}\pi r^2 h$ for h

15. $L = 2\pi rh$ for r

16. $d = \frac{fL}{f + w}$ for w

17. $F = \frac{9}{5}C + 32$ for C

18. $A = \frac{h(B + b)}{2}$ for B

19. $a = \frac{v - v_0}{t}$ for v_0

20. $s = vt + \frac{1}{2}at^2$ for v

21. $v_{av} = \frac{v_f + v_0}{2}$ for v_f

22. $P = \frac{N + 2}{D_0}$ for N

23. $L = 3.14(r_1 + r_2) + 2d$ for r_1

24. $P = \frac{1}{3}Nmv^2$ for m

25. $a = \frac{w_t - w_0}{t}$ for w_t

26. $mv_2 - mv_1 = Ft$ for v_1

27. $F = \frac{kq_1q_2}{r^2}$ for q_1

28. $M = \frac{P(C + L)}{T}$ for C

29. $y = mx + b$ for m

30. $h = \frac{k(t_2 - t_1)aT}{d}$ for t_2

31. $f = \frac{f_s u}{u + v_s}$ for f_s

32. $h = \frac{D - d}{2}$ for d

33. $a = \frac{2T}{(d_1 - d_2)g}$ for d_1

34. $\frac{P_1V_1}{T_1} = \frac{P_2V_2}{T_2}$ for T_2

35. $I = \frac{E}{R + r}$ for r

36. $s = \frac{H}{m(t_2 - t_1)}$ for t_1

37. $P = \frac{V_1(V_2 - V_1)}{gJ}$ for V_2

38. $m = \frac{y_2 - y_1}{x_2 - x_1}$ for y_1

39. $r = \frac{ab}{a + b + c}$ for c

40. $y - y_0 = m(x - x_0)$ for x

Solve for the required quantity by rearranging the formula (if necessary) and then substituting values.

41. The distance d that a free-falling object falls in t seconds is $d = gt^2/2$ where $g = 9.8$ m/s^2, the acceleration due to gravity. How far will a cement block fall in 16 seconds?

E **42.** The power P in watts (W) in a dc circuit is the product of the voltage V and the current I. What power is consumed by a radio requiring 110 V and 0.15 A?

43. A golfer's handicap H is calculated using the formula $H = 0.85(A - 72)$ where A is the golfer's three-round average. If the golfer's three-round scores are 75, 87, and 80, find her handicap.

E **44.** When two resistors, R_1 and R_2, are connected in parallel, the equivalent resistance R equals their product divided by their sum. If the two resistors are 15 Ω and 9 Ω, find the equivalent resistance.

45. The living room of a house requires 120 ft^2 of carpeting. Find the length of the room if the width is 8 ft. (*Note:* $A = LW$.)

I **46.** The interest I earned on an investment equals the product of the amount invested P, the number of years invested t, and the interest rate r, expressed as a decimal. Mr. Jones earns $270 interest on an investment of $1,500 at 9%. How long did it take Mr. Jones to earn this interest?

47. A body is moving with an acceleration of 2 ft/s^2. How much time is required for its velocity to increase from 10 ft/s to 16 ft/s? (*Note:* $a = (v - v_0)/t$.)

48. A cubic foot contains 7.48 gal. What is the capacity in gallons of oil for a cylindrical tank 8 ft high with a radius of 3 ft? (*Note:* $V = \pi r^2 h$.)

49. A living room requires 20 yd^2 of carpeting. Find the width of the living room (in feet) if the length is 15 ft. ($A = LW$. Note the difference in the units of measure.)

50. You are a weatherperson responsible for converting the current temperature in degrees Fahrenheit to degrees Celsius. The formula is $F = 9C/5 + 32$. What temperature in degrees Celsius corresponds to 78°F?

1–3

SOLVING LINEAR INEQUALITIES

An **inequality** is a statement that a certain quantity is not equal to a second quantity. The **solution** to an inequality consists of all real numbers that result in a true statement when substituted for the variable. Unlike the solution to a linear equation in one variable, the solution to a linear inequality is an infinite range of numbers. For example, the solution to the linear equation $x = 3$ means that three is the only number that can be substituted for x and yield a true statement. However, the solution to the inequality $x \geq 3$ means that substituting three or any number greater than three results in a true statement. Therefore, the solution to this inequality is the infinite range of numbers starting with three.

Properties of Inequalities

Properties similar to the Principles of Equality apply to inequalities. These properties can be divided into three categories: addition and subtraction, multiplication and division by a positive number, and multiplication and division by a negative number. These properties are summarized below.

Properties of Inequalities

Addition and Subtraction

If the same quantity is added to or subtracted from both sides of an inequality, an equivalent inequality results, thus leading to the same solution. For example,

$$x > 1$$

is equivalent to

$$x - 8 > 1 - 8$$
$$x + 8 > 1 + 8$$

Multiplication and Division: Positive Number

If a positive number multiplies or divides both sides of an inequality, an equivalent inequality results. For example,

$$x > 1$$

is equivalent to

$$3x > 3$$

and to

$$\frac{x}{3} > \frac{1}{3}$$

Multiplication and Division: Negative Number

If a negative number multiplies or divides both sides of an inequality, the direction of the inequality symbol must be reversed to produce an equivalent inequality. For example,

$$x > 1$$

is equivalent to

$$-3x < -3$$

and also to

$$-\frac{x}{3} < -\frac{1}{3}$$

The process of solving a linear inequality uses the same steps that have been used throughout this chapter. The only difference in solving an inequality is that the direction of the inequality symbol must be reversed if you multiply or divide both sides of the inequality by a negative number. These steps are restated here for completeness.

Solving Linear Inequalities

1. Eliminate fractions by multiplying each term of the inequality by the lowest common denominator. Simplify.
2. Eliminate parentheses by using the Distributive Property. Simplify.
3. Using transposition, isolate all terms containing the variable on one side of the inequality. Simplify.
4. Divide both sides of the inequality by the coefficient of the variable.
5. Check the solution.

EXAMPLE 1 Solve the inequality $10 - 5x < 2x - 11$.

Solution Since this inequality does not contain fractions or parentheses, we begin by isolating the variable.

$$\begin{aligned} 10 - 5x &< 2x - 11 \\ -5x - 2x &< -11 - 10 && \text{transpose } 2x \text{ and } 10 \\ -7x &< -21 && \text{combine like terms} \\ x &> 3 && \text{divide by } -7 \text{ and simplify} \end{aligned}$$

Checking Substitute any number from the solution set into the original inequality. Substituting $x = 6$ gives

$$\begin{aligned} 10 - 5(6) &< 2(6) - 11 \\ -20 &< 1 \qquad \text{(true)} \end{aligned}$$

■

CAUTION ✦ The most common error in solving inequalities is neglecting to reverse the inequality symbol when multiplying or dividing by a negative. Every time you multiply or divide, check the sign of the number to determine if reversing the inequality symbol is necessary.

EXAMPLE 2 A student must have at least an 80 average to make a B in a math course. If the student has made 76, 84, and 73 on the first three tests, what must he make on the last test to make a B, assuming all tests are equally weighted?

Solution To calculate the student's average, we add the grades and divide by the number of grades. If x = the grade on the last test, the average is given by

$$\frac{76 + 84 + 73 + x}{4}$$

Since the average must be 80 or above, the inequality becomes

$$\frac{76 + 84 + 73 + x}{4} \geq 80$$

Solving this inequality gives

$76 + 84 + 73 + x \geq 320$	eliminate fractions
$233 + x \geq 320$	combine like terms
$x \geq 320 - 233$	transpose
$x \geq 87$	combine like terms

The student must make an 87 or above on the last test to make a B in the course. ■

Graphing the Solution

In solving linear inequalities, it is often helpful to graph the solution to gain a visual representation on the number line. The graphical solution shows the starting point and the direction of the solution. To show that the starting number is included in the solution, place a dot at the number; to show that the starting number is not part of the solution, place a circle at the number. To show the direction of the solution, draw a line with an arrow pointing in the proper direction. Figure 1–2 illustrates this information.

FIGURE 1–2

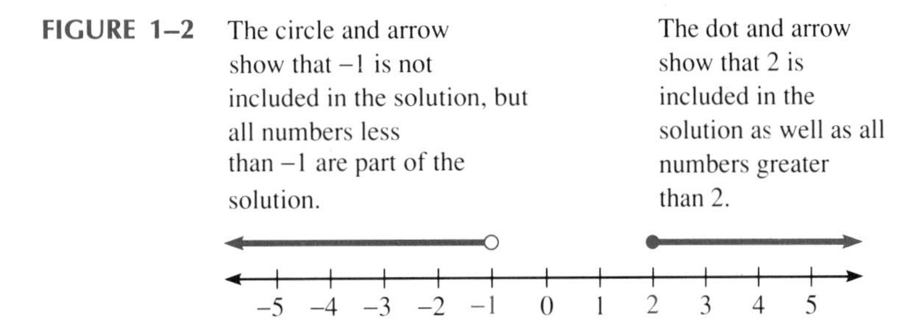

LEARNING HINT ✦ In graphing the solution, you may confuse greater than and less than. To avoid this error, always isolate the variable on the left side of the inequality. Then the inequality symbol points in the direction of the arrow on the number line.

EXAMPLE 3 Solve the following inequality and graph the solution:

$$-3(2x - 3) + 15 \geq 2x + 4(x - 6)$$

Solution

$-3(2x - 3) + 15 \geq 2x + 4(x - 6)$	
$-6x + 9 + 15 \geq 2x + 4x - 24$	eliminate parentheses
$-6x + 24 \geq 6x - 24$	combine like terms
$-6x - 6x \geq -24 - 24$	transpose
$-12x \geq -48$	combine like terms
$x \leq 4$	divide by -12

Checking Substitute a number from the range into the inequality. We substitute 3.

$$-3[2(3) - 3] + 15 \geq 2(3) + 4[(3) - 6]$$
$$-3(3) + 15 \geq 6 + 4(-3)$$
$$6 \geq -6 \qquad \text{(true)}$$

The solution is $x \leq 4$. The graph of the solution set is a dot at 4 and a line with an arrow pointing to the left of 4. The graph of the solution is shown in Figure 1–3.

FIGURE 1–3

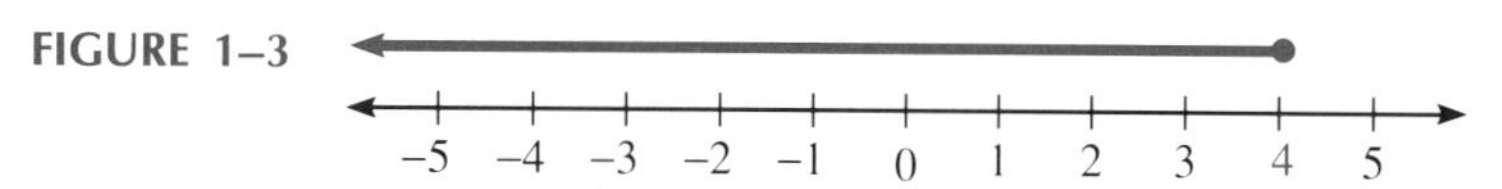

■

EXAMPLE 4 Solve the following inequality and graph the solution set:

$$\frac{2(5x - 3)}{5} < \frac{1}{3} + \frac{x}{5}$$

Solution First, we eliminate the fractions by multiplying by the common denominator of 15.

$$15\left(\frac{2(5x - 3)}{5}\right) < 15\left(\frac{1}{3}\right) + 15\left(\frac{x}{5}\right)$$

$6(5x - 3) < 5(1) + 3(x)$	simplify
$30x - 18 < 5 + 3x$	eliminate parentheses
$30x - 3x < 5 + 18$	transpose $3x$ and -18
$27x < 23$	combine like terms
$x < \frac{23}{27}$	divide by 27

The solution is $x < 23/27$. The graph of the solution set consists of a circle at 23/27 and a line with an arrow pointing to the left of 23/27. The solution is shown in Figure 1–4.

FIGURE 1–4

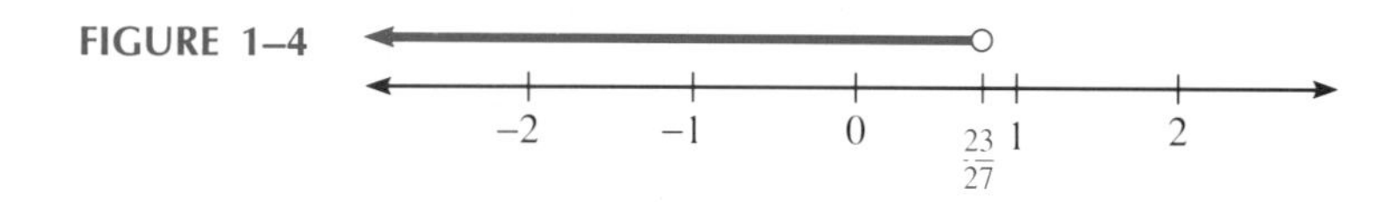

Application

EXAMPLE 5 The strength s of a sheet of material sufficient to hold a certain weight is given by $s + 7 \geq 2s + 4$, where s is measured in lb/in.^{2} Find the range of s.

Solution Solving the inequality for s gives

$$s + 7 \geq 2s + 4$$
$$s - 2s \geq 4 - 7$$
$$-s \geq -3$$
$$s \leq 3$$

The material is able to withstand no more than 3 lb/in.^{2}

Absolute Value Inequalities

Earlier in this chapter we discussed solving equations involving absolute value. Now we extend those methods to solving inequalities involving absolute value. Recall from our previous discussion that absolute value results in two equations. An absolute value inequality also results in two inequalities. When a statement consists of two or more inequalities connected by "and" or "or," the statement is called a **compound inequality.** The expression $x > 2$ or $x < -1$ is an example of a compound inequality. Inequalities involving absolute value result in a compound inequality.

Let's examine the inequality $|x| > 4$. The solution consists of all numbers more than four units from zero on the number line. This set of numbers is shown in Figure 1–5. The solution set can be written as $x > 4$ or $x < -4$. Therefore, an absolute value inequality containing the $>$ or $\geq$ symbol consists of two solutions connected by the word "or."

$$\text{solution:} \quad x < -4 \quad \text{or} \quad x > 4$$

FIGURE 1–5

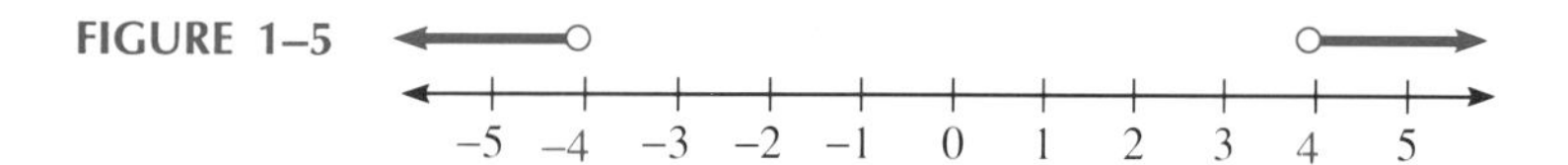

On the other hand, the solution to an absolute value inequality such as $|x| < 2$ consists of all numbers less than two units from zero on the number line. This solution set is shown in Figure 1–6. The solution consists of all numbers $x < 2$ and $x > -2$. Therefore, an absolute value inequality containing the $<$ or $\leq$ symbol consists of two inequalities connected by the word "and." This type of solution can be expressed in a simpler notation as $-2 < x < 2$.

When the solution consists of a line segment, the solution is written

$$\text{smallest number} < x < \text{largest number}$$
$$\text{or smallest number} \leq x \leq \text{largest number}$$
$$\text{solution:} \quad -2 < x < 2$$

FIGURE 1–6

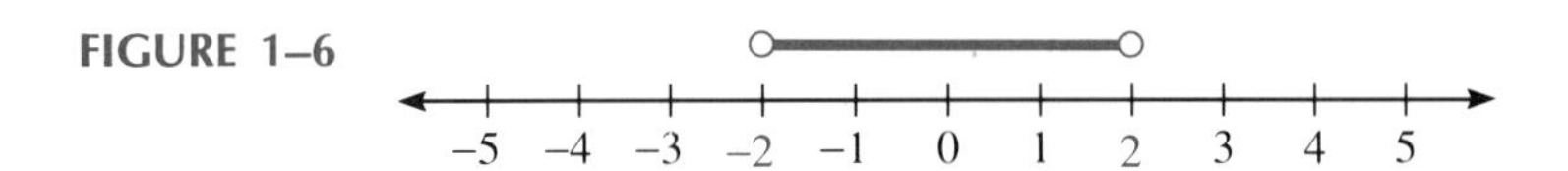

The next two examples illustrate solving absolute value inequalities.

EXAMPLE 6 Solve the inequality $|2x - 1| \leq 9$.

Solution To solve an absolute value inequality, we follow the same steps used in solving equations involving absolute value, remembering that multiplication or division of the entire inequality by a negative number reverses the direction of the inequality. Since the absolute value quantity is isolated, we write the two inequalities that result from the definition of absolute value. Remember that the $\leq$ symbol results in two inequalities connected by ''and.''

$$\begin{aligned} 2x - 1 &\leq 9 & \text{and} \quad -(2x - 1) &\leq 9 \\ 2x &\leq 9 + 1 & -2x + 1 &\leq 9 \\ 2x &\leq 10 & -2x &\leq 9 - 1 \\ x &\leq 5 & -2x &\leq 8 \\ & & x &\geq -4 \end{aligned}$$

Checking Substituting -2 into the original inequality gives

$$\begin{aligned} |2(-2) - 1| &\leq 9 \\ |-5| &\leq 9 \\ 5 &\leq 9 \qquad \text{(true)} \end{aligned}$$

The solution is the set of numbers such that $x \geq -4$ and $x \leq 5$. The solution is written

$$\text{smallest number} \leq x \leq \text{largest number}$$
$$-4 \leq x \leq 5$$

and is shown in Figure 1–7.

FIGURE 1–7

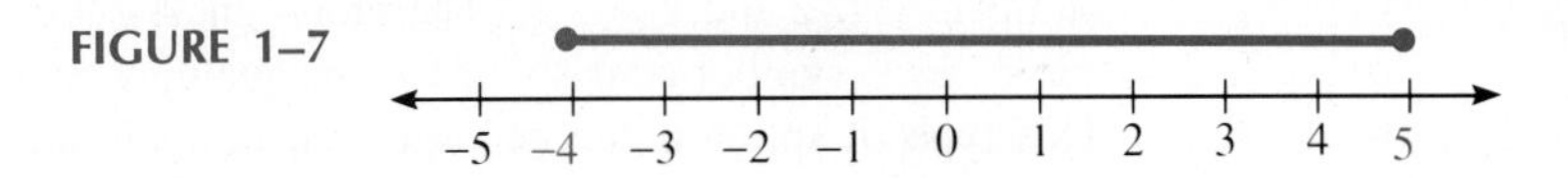

■

EXAMPLE 7 Solve $|5x + 3| > 7$.

Solution Since this absolute value inequality contains the > symbol, we write two inequalities connected by "or."

$$\begin{aligned} 5x + 3 &> 7 & \text{or} \quad -(5x + 3) &> 7 \\ 5x &> 7 - 3 & -5x - 3 &> 7 \\ 5x &> 4 & -5x &> 10 \\ x &> 4/5 & x &< -2 \end{aligned}$$

Checking To check the solution, we substitute a number from the solution set into the original inequality. Since the solution set consists of two rays, we substitute -3 and 2 into the inequality.

$$\begin{aligned} |5(-3) + 3| &> 7 & |5(2) + 3| &> 7 \\ 12 &> 7 \quad \text{(true)} & 13 &> 7 \quad \text{(true)} \end{aligned}$$

The solution set is written $x > 4/5$ or $x < -2$ and is shown in Figure 1–8.

FIGURE 1–8

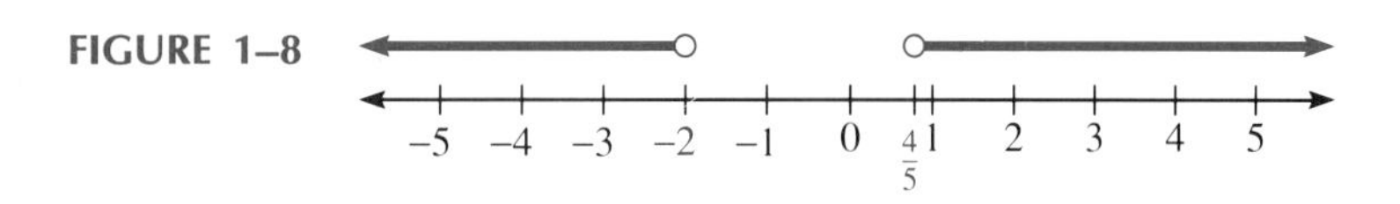

■

1–3 EXERCISES

Solve each of the following inequalities and graph the solution set on the number line.

1. $x + 5 \geq -8$

2. $7 \leq 3 + x$

3. $2x - 6 \leq 8$

4. $4x - 10 \leq -14$

5. $2(-3x + 6) > -10$

6. $-5(3 - 2x) < 15$

7. $\frac{x}{3} > -4$

8. $\frac{3x}{5} < -6$

9. $2x + 3 - 7x < 9 - x + 6$

10. $8 - x + 4 > 6x - 10 + 1$

11. $5x - 3 \geq 7x + 7$

12. $3x + 17 < 5x + 3$

13. $-6 - 4x > -9x + 24$

14. $-7x \geq 35$

15. $-(2x + 6) + 8 < 2(2x - 5)$

16. $-4x + 17 + 8x < -5 - x + 2$

17. $\frac{3x}{7} + \frac{1}{4} < 1 + \frac{x}{2}$

18. $\frac{5x}{6} - \frac{2}{3} + x > \frac{1}{2} + \frac{2x}{3}$

19. $4(3 - 2x + 1) + 3x + 4 \geq 0$

20. $-2(5 + x) - 7 + 4x < -2x - 1$

21. $\frac{2x + 1}{4} < \frac{1}{2}$

22. $\frac{3x - 5}{6} > \frac{1}{3}$

23. $3x - \frac{1}{3} > \frac{x}{2} + 6$

24. $\frac{3}{4} - 5x + 2 < \frac{1}{2} - x + \frac{1}{4}$

25. $-8x + 7 - 3(x + 1) + 17 \leq 0$

26. $\frac{2(x + 1)}{3} > 6 - \frac{1}{4} + 3x$

27. $3\left(2x - \frac{1}{3}\right) \leq 0$

28. $\frac{7x + 1}{3} \geq \frac{2x - 1}{2}$

29. $\frac{x}{4} - \frac{x}{6} + \frac{x}{3} > \frac{x}{2}$

30. $\frac{2x}{3} - \frac{x}{9} + \frac{5x}{6} > \frac{x}{2} - \frac{4x}{9}$

31. $\frac{3(2x - 4)}{5} + \frac{1}{3} > \frac{-8x}{3}$

32. $\frac{-2(5 - 6x)}{2} + 3 < \frac{-7}{2} + 14x$

33. $2x - 8 + 5x \leq 3(x - 2) + 6$

34. $8(-2x + 1) - 6x > -7 + (2x - 1) - 4$

35. $\frac{3(4 - 2x)}{4} + 1 > \frac{-(x + 1)}{3} - \frac{5}{6}$

36. $\frac{2 + 5(x - 1)}{3} < \frac{3}{4}$

37. $\frac{4(x - 3) + (2x + 1)}{2} \geq \frac{1}{2}$

38. $\frac{5(x - 6) - (2x + 3)}{3} \geq 0$

39. $-3(2x - 1) + 6 + 2x < 7 - (2x - 1)$

40. $9 - 3(x - 5) + 6x \leq -4 + 2(4x - 7) + 6x$

41. $|3x + 4| < 8$

42. $|2x - 5| \geq 7$

43. $|4x - 1| + 3 > 16$

44. $5|x - 2| + 6 \leq 25$

45. $8|x - 3| < 21$

46. $3|7x - 2| - 5 \geq 19$

47. Maria has scored 88, 93, 96, and 84 on her first four math tests. What is the lowest she can score on the fifth test to have at least a 90 average?

48. A total of 500 tickets are available to Six Flags. If adult tickets sell for $14 and children's tickets are $5, how many of each must be sold to total at least $5,200?

49. A car rental agency rents cars for $14 a day and $0.50 per mile. If Joe rents a car for two days, what is the maximum distance he can drive and keep his charges under $37?

50. Jane has 18 coins consisting of nickels and dimes. If she has at least $1.15, what is the maximum number of nickels?

M 51. The measurement m of an automobile engine part must satisfy a tolerance given by $|m - 6| \leq 0.05$, where m is measured in millimeters. Solve for m to determine the range of tolerance.

52. The in-flight time t of a projectile is given by $|4t - 9| < 9$. Find the interval of time in seconds that the object is in flight.

1–4

VARIATION

Direct Variation

Variation describes the relationship between variables. When two quantities are related so that their ratio is a constant, the variables are said to **vary directly** or to be **proportional.** If a car is traveling at 55 miles per hour, at the end of one hour the car has traveled 55 miles, at the end of two hours it has traveled 110 miles, and at the end of three hours it has traveled 165 miles. The equation $d = 55t$ describes this relationship between the distance traveled d and the time t. If we rearrange the equation, then $d/t = 55$. In other words, the ratio of the variables is a constant, called the **constant of variation.** We can describe this relationship by saying that distance varies directly with time. In general, if we let x and y represent variables and k represent the constant of variation, then the formula for direct variation could be written as $y/x = k$ or as follows:

$$y = kx \qquad \text{direct variation}$$

EXAMPLE 1 Hooke's law states that the force F required to stretch a spring is directly proportional to the change of length x of the spring. Write a formula for this relationship.

Solution Using k as the constant of variation, the formula is

$$F = kx$$

■

NOTE ✦ A variable may also vary with a second variable that has been raised to some power. In this case, the appropriate exponent is attached to the second variable.

EXAMPLE 2 When a ball rolls down an inclined plane, the distance d it travels is directly proportional to the square of the time t. Write a formula to represent this relationship.

Solution This relation is a statement of direct variation where the distance varies with the square of the time. The formula representing this relationship is

$$d = kt^2$$

■

NOTE ✦ At this point we do not know the value of k, the constant of variation, but it will be discussed later in this section.

Inverse Variation

Let us consider the possible dimensions for a rectangle when its area is fixed at 100 square feet. Several possibilities for the dimensions are 2 ft by 50 ft, 1 ft by 100 ft, and 4 ft by 25 ft. This relationship can be expressed as $LW = 100$. Two variables are said to **vary inversely** or to be **inversely proportional** if the product of the variables is a constant. In our example, L and W vary inversely because their product is a constant, and in this case 100 is the constant of variation. If x and y represent the variables and k represents the constant of variation, the formula for inverse variation could be written as $y \cdot x = k$ or as follows:

$$y = \frac{k}{x} \qquad \text{inverse variation}$$

EXAMPLE 3 Write a formula for the following relationship: The pressure of a compressed gas varies inversely with the volume.

Solution If P represents the pressure and V represents the volume, the formula is

$$P = \frac{k}{V}$$

■

One variable can also be inversely proportional to the power of a second variable, as shown in the next example.

EXAMPLE 4 The illumination provided by a light source is inversely proportional to the square of the distance from the source. Write a formula to express this relationship.

Solution If the letter I represents the illumination and d represents the distance, the formula for the inverse variation is

$$I = \frac{k}{d^2}$$

■

Joint Variation

The volume of a cylinder changes with both the radius and the height of the cylinder. This example describes one variable whose value depends upon the value of two other variables. **Joint variation** occurs when one variable varies directly with several other variables. The general formula for joint variation with y varying jointly with x and z is as follows:

$$y = kxz \qquad \text{joint variation}$$

EXAMPLE 5 The volume V of a cone varies jointly as the square of the radius r of the base and the altitude h. Write a formula to represent this relationship.

Solution Since the volume varies jointly with the square of the radius and the altitude, the formula is

$$V = kr^2h$$

■

Combined Variation

The same variable can vary directly with one variable while varying inversely or jointly with other variables. We can combine several types of variation into a single relationship by using one constant of variation. If y varies directly with x and inversely with z, the formula for **combined variation** is as follows:

$$y = \frac{kx}{z} \qquad \text{combined variation}$$

EXAMPLE 6 The safe load L for a horizontal beam supported at the ends varies jointly with the width w and the square of the depth d and inversely with the distance s between the supports. Write a formula to represent this relationship.

Solution This is a combined variation problem since it involves both joint and inverse variation. The formula for the expression "L varies jointly with w and d^2" would be written as

$$L = kwd^2$$

Similarly, the formula for "L varies inversely as s" would be written

$$L = \frac{K}{s}$$

Combining these two formulas using only one constant of variation, H, gives the formula for the verbal statement as

$$L = \frac{Hwd^2}{s}$$

■

Solving Variation Problems

Variation is used to extrapolate or go beyond given information. For example, the perimeter of a regular polygon varies directly with the length of the side. If the perimeter of a polygon is 20 ft when the length of a side is 4 ft, then what is the perimeter when the length of a side is 7 ft? Solving a problem of this type requires a procedure such as the one outlined below.

Solving Variation Problems

1. Write a variation formula from the verbal statement.
2. Solve for k, the constant of variation, by substituting the complete set of values given for the variables.
3. Substitute the value of k into the variation formula.
4. Using the value of k and the remaining variables, solve for the value of the specified variable.

EXAMPLE 7 Boyle's law states that the volume V of a gas varies inversely as the pressure P of the gas. If the volume of the gas at a certain temperature is 75 in.^{3} when the pressure is 30 lb/in.^{2}, find the volume when the pressure is 25 lb/in.^{2}

Solution We begin by writing a formula for the inverse variation described.

$$V = \frac{k}{P}$$

To solve for k, we must know values for V and P. We can have only one unknown to yield a unique solution, but the volume is 75 when the pressure is 30. Therefore, substituting in values for $V = 75$ and $P = 30$ gives

$$75 = \frac{k}{30}$$

$$2{,}250 = k$$

Next, we substitute 2,250 for k into the variation formula.

$$V = \frac{2{,}250}{P}$$

Finally, the verbal statement requires finding the volume when the pressure is 25. Substituting in the value of $P = 25$ and solving for V gives

$$V = \frac{2{,}250}{25}$$

$$V = 90$$

Thus, the volume is 90 in.^{3} when the pressure is 25 lb/in.^{2} ■

EXAMPLE 8 The power in a circuit varies jointly as the resistance and the square of the current. If the power in a 10-Ω resistor is 3.6 W when the current is 0.6 A, what is the power in the resistor when the current is 0.8 A?

Solution Using P for power, R for resistance, and I for current, the formula is given by

$$P = kRI^2$$

To solve for k, substitute $P = 3.6$, $R = 10$, and $I = 0.6$.

$$3.6 = k(10)(0.6)^2$$

$$k = 1$$

$$P = RI^2$$

To find the power when the resistance is 10 Ω and the current is 0.8 A, substitute $R = 10$ and $I = 0.8$.

$$P = (10)(0.8)^2$$

Therefore, the power is 6.4 W. ■

1–4 EXERCISES

Write a formula to express the following statements.

1. x varies directly with z.
2. m varies inversely with p^2.
3. a varies jointly with b and c.
4. m varies directly with p^2.
5. x varies directly with y and inversely with s.
6. d is inversely proportional to t^2 and directly proportional to s.
7. m varies jointly with n and p^3 and inversely with r.
8. x varies directly with y and inversely with z^3.
9. s varies jointly with t^2 and v.
10. a varies inversely with b^2 and jointly with c and d^3.

Set up a formula and solve for k, the constant of variation.

11. x varies directly as y^2, and $x = 10$ when $y = 2$.
12. m varies inversely with n, and $m = 5$ when $n = 2$.
13. a varies jointly with b and c^2, and $a = 30$ when $b = 5$ and $c = 1$.
14. s varies directly with a and inversely with t^2, and $s = 81$ when $a = 4$ and $t = 2$.
15. x varies directly with y^2 and inversely with z, and $x = 32$ when $y = 4$ and $z = 2$.

Solve the following variation problems.

16. x varies inversely with y^2. If $x = 2$ when $y = 3$, find x when $y = 4$.

17. y varies directly with t. If $y = 14$ when $t = 4$, then find y when $t = 16$.

18. m varies jointly with n and p. If $m = 30$ when $n = 1$ and $p = 3$, find m when $n = 3$ and $p = 4$.

19. a varies directly with b^2 and inversely with c. If $a = 4$ when $b = 4$ and $c = 7$, then find a when $b = 6$ and $c = 2$.

20. s varies inversely with t. If $s = 3$ when $t = 9$, find s when $t = 4$.

21. r varies directly as the square of s. If $r = 40$ when $s = 2$, then find r when $s = 3$.

22. z varies inversely as the square root of y. If $z = 1$ when $y = 9$, then find z when $y = 25$.

23. a varies jointly with b and the square of c. If $a = 72$ when $b = 9$ and $c = 2$, find b when $a = 36$ and $c = 3$.

24. d varies directly with t. If $d = 30$ when $t = 6$, then find t when $d = 55$.

25. m varies directly as n and inversely as p. If m $= 2$ when $n = 16$ and $p = 16$, then find p when $m = 16$ and $n = 25$.

26. The intensity of light varies inversely as the square of the distance from the source. If the intensity is 10 foot-candles at a distance of 2 ft, find the intensity when the distance is 7 ft.

27. Hooke's law states that the force needed to stretch a spring is directly proportional to the length of the stretch. Use Hooke's law to determine the length that a spring is stretched when a force of 5 newtons (N) is applied, if a force of 10 N stretched the spring 2.7 m.

E **28.** The power in a resistance circuit varies jointly as the resistance and the square of the current. If a power of 100 W results from a 25-Ω resistance with a 2-A current, find the resistance when the power is 75 W and the current is 6 A.

29. The area of a circle varies directly with the square of the radius. The radius of a circle whose area is 12.56 cm^2 is 2 cm. Find the area of a circle whose radius is 5 cm.

30. The volume of a gas varies directly with the temperature and inversely with the pressure. If a volume of 500 cm^3 results from a temperature of 273 K and pressure of 76 N/cm^2, find the volume if the temperature is 353 K and the pressure is 90 N/cm^2.

M **31.** The strength of a rectangular beam is directly proportional to the square of the thickness. If a 2-in.-thick beam supports 1,000 lb, how much would a 5-in. beam support?

32. The acceleration a of an object is directly proportional to the difference between the final velocity v_f and the initial velocity v_0 and inversely proportional to the elapsed time t. Write a formula to express this relationship.

E **33.** Ohm's law states that current in a circuit varies directly as the voltage and inversely as the resistance. If the current is 2 A, the voltage is 10 V, and the resistance is 5 Ω, what resistance is caused by a current of 7 A and a voltage of 3 V?

34. The volume of a cone varies jointly with the height and the square of the radius. If a volume of 12.56 m^3 results from a cone of height 3 m and radius 2 m, what is the volume of a cone whose height is 2 m and radius is 5 m?

35. The potential energy of a stationary object is jointly proportional to the mass and elevation of the object. If the potential energy of a 5-kg object whose elevation is 3 m is 147 J, then find the elevation of an object if the potential energy is 100 J and its mass is 6 kg.

36. The kinetic energy of an object varies jointly with its mass and the square of its velocity. If the kinetic energy of a 10-kg object whose velocity is 8 m/s is 320 J, find the kinetic energy of an object whose mass is 20 kg and whose velocity is 6 m/s.

37. The period of a pendulum is directly proportional to the square root of its length and inversely proportional to the square root of the acceleration due to gravity. If the period is 1.57, the length is 2 ft, and the acceleration due to gravity is 32 ft/s^2, find the period of a pendulum 5 ft long.

E **38.** The resistance of a wire varies directly as the length and inversely as the cross-sectional area of the wire. If the resistance of a wire 14 ft long with a cross-sectional area of 0.01 in.^{2} is 0.5 Ω, find the resistance of the same wire 20 ft long.

39. The distance an object falls is jointly proportional to the acceleration due to gravity and the square of time. If the acceleration due to gravity is 9.8 m/s^2, the time is 4 s, and the distance it falls is 78.4 m, how far does the object fall in 6 s?

40. The surface area of a sphere varies directly as the square of the radius. The surface area of a sphere with a radius of 8 in. is 804 in^2. Find the surface area of a sphere with a radius of 10 in.

1–5 APPLICATIONS

So far in this chapter, we have primarily solved equations and inequalities that are explicitly stated. However, most problems encountered in mathematics, science, and engineering job situations are stated verbally. Therefore, it is necessary to transform a verbal statement into an equation that can be solved. Solving problems of this type is essential to a technician, but often most difficult to master. The following steps should help in formulating an equation from a verbal statement.

Solving Verbal Problems

1. Read the entire verbal statement carefully. If possible, draw a diagram or chart.
2. Select a letter to represent one of the unknown quantities. Many times the question at the end of the verbal statement will give a clue as to the unknown quantity.
3. If there are several unknown quantities, they must all be expressed in terms of the letter selected in Step 2.
4. Write an equation from the verbal statement.
5. Solve the equation and relate the solution to the variable specified in the verbal statement.
6. Check the solution to make sure the conditions of the verbal statement are satisfied.

EXAMPLE 1 A technician earns $12.40 an hour for the first 40 hours and $18.60 for each hour over 40. How many hours of overtime must the technician work to earn $719.20 in a week?

Solution Since the verbal statement asks for overtime hours, let x = number of overtime hours. To write the equation, we know that weekly pay is the sum of salary from regular hours and salary from overtime.

$$\text{pay for regular hours} + \text{pay for overtime hours} = \text{total salary}$$
$$12.40(40) + 18.60x = 719.20$$

Solving the equation gives

$$\begin{aligned} 496.00 + 18.60x &= 719.20 \\ 18.60x &= 719.20 - 496.00 \\ 18.60x &= 223.20 \\ x &= 12 \end{aligned}$$

The technician must work 12 hours of overtime to earn \$719.20.

$$719.20 \boxed{-} 496.00 \boxed{=} \boxed{\div} 18.60 \boxed{=} \longrightarrow 12$$

Checking

$$\begin{aligned} \text{technician's salary for 40 hours} &= \$12.40(40) = \$496.00 \\ \text{technician's salary for 12 hours overtime} &= \$18.60(12) = \$223.20 \\ \text{technician's total salary} &= \$496.00 + \$223.20 = \$719.20 \end{aligned}$$

■

EXAMPLE 2 The third side of an isosceles triangle is 5 ft less than twice the length of the two equal sides. Find the length of the sides of the triangle if the perimeter is 35 ft.

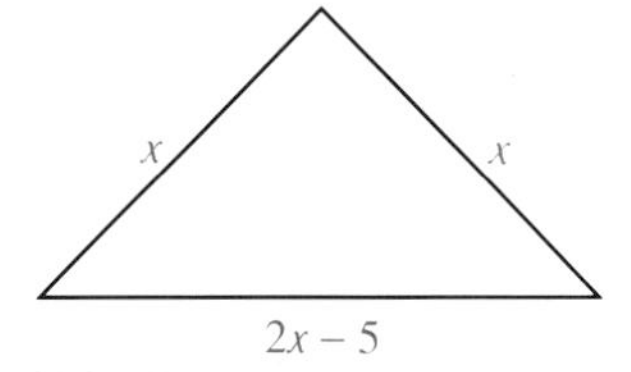

FIGURE 1–9

Solution Since the verbal statement asks for the length of the three sides, let x represent the two equal sides. Therefore, as shown in Figure 1–9,

$$\begin{aligned} x &= \text{the length of each of the two equal sides of the triangle} \\ (2x - 5) &= \text{the length of the third side of the triangle} \end{aligned}$$

Since the perimeter of a triangle is the sum of the three sides, the equation is

$$x + x + (2x - 5) = 35$$

Solving for x gives

$$\begin{aligned} 4x - 5 &= 35 \\ 4x &= 35 + 5 \\ 4x &= 40 \\ x &= 10 \end{aligned}$$

Since x represents one of the two equal sides and $2x - 5$ represents the third side, the dimensions of the triangle are 10 ft, 10 ft, and 15 ft.

Checking

$$\begin{aligned} 15 &= 2(10) - 5 \\ \text{perimeter} &= 10 + 10 + 15 = 35 \end{aligned}$$

■

Ratio and Proportion

A **ratio** is a comparison of any two (or more) quantities. For example, the ratio of Jeff's monthly income of $1,750 to Greg's monthly salary of $2,050 is 35/41.

EXAMPLE 3 Mr. Jones invested $15,000 in two savings accounts paying 7.5% and 9% interest. He invested the money in the ratio of 2 to 3, respectively. How much money did he invest in each account?

Solution If x = one share, then

$$2x = \text{the amount invested at } 7.5\%$$
$$3x = \text{the amount invested at } 9\%$$

Because the total investment is $15,000, the equation is

$$2x + 3x = 15{,}000$$
$$5x = 15{,}000$$
$$x = 3{,}000$$

Mr. Jones invested $2x$ or $6,000 at 7.5% and $3x$ or $9,000 at 9%. ■

A **proportion** is a statement of equality between two ratios. A proportion can be written in fraction form as $a/b = c/d$ or in ratio form as $a:b = c:d$ (read "a is to b as c is to d"). In the proportion

$$\frac{a}{b} = \frac{c}{d}$$

multiplying each side of the equation by bd gives

$$bd\left(\frac{a}{b}\right) = \left(\frac{c}{d}\right)bd$$
$$ad = cb$$

This result is called the **cross-product rule** because multiplying diagonally across the fractions gives the same result, as shown below.

$$\frac{a}{b} = \frac{c}{d}$$
$$ad = cb$$

Cross-Product Rule

If

$$\frac{a}{b} = \frac{c}{d}$$

then

$$ad = bc$$

where $b \neq 0$ and $d \neq 0$.

Proportions are equal as long as the resulting cross-products are equal. The following proportions are all equivalent because the cross-products of each are equal.

$$\left.\begin{aligned} \frac{a}{b} &= \frac{c}{d} \\ \frac{b}{a} &= \frac{d}{c} \\ \frac{a}{c} &= \frac{b}{d} \\ \frac{c}{a} &= \frac{d}{b} \end{aligned}\right\} ad = bc$$

The application of proportion and the cross-product rule is illustrated in the next example.

EXAMPLE 4 A typist can complete 20 pages in 3.5 hours. How many hours would it take him to type 27 pages?

Solution To solve the problem, set up a proportion letting x equal the number of hours it would take him to type 27 pages. The proportion can be set up in a number of ways, all of which result in the same cross-product. The following list gives several equivalent proportions for this problem:

$$\frac{\text{pages}}{\text{time}} = \frac{\text{pages}}{\text{time}} \qquad \frac{\text{time}}{\text{pages}} = \frac{\text{time}}{\text{pages}} \qquad \frac{\text{pages}}{\text{pages}} = \frac{\text{time}}{\text{time}}$$

Using $\frac{\text{pages}}{\text{time}} = \frac{\text{pages}}{\text{time}}$ gives the proportion

$$\frac{20}{3.5} = \frac{27}{x}$$

Solving for x gives

$$\begin{aligned} 20x &= 27(3.5) \qquad \text{(cross product rule)} \\ 20x &= 94.5 \\ x &\approx 4.7 \text{ hours} \end{aligned}$$

It would take the typist approximately 4.7 hours to type 27 pages.

27 [×] 3.5 [÷] 20 [=] ⟶ 4.725 ■

Motion

EXAMPLE 5 Two cars 600 miles apart travel towards each other. The rate of one car is 15 mi/h more than the rate of the other. Find the rate of each car if they meet in 5 hours.

Solution This problem is typical of "distance" problems. Figure 1–10 and the accompanying chart will help you organize the information. Set up the variables as follows:

$$x = \text{the rate of the slower car}$$
$$(x + 15) = \text{the rate of the faster car}$$

These values are placed in the rate column of the chart. Since the cars start at the same time, place 5 in the column for time. Since distance is the product of rate and time, fill in this column as the product of the rate and time columns.

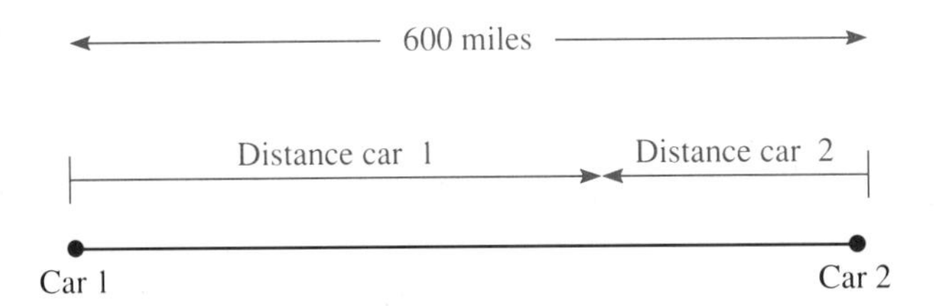

FIGURE 1–10

	rate × time = distance		
	Rate	Time	Distance
Car 1	x	5	$5x$
Car 2	$(x + 15)$	5	$5(x + 15)$

We use the last column from the chart to write the equation. Therefore, we must read the verbal statement to find the relationship concerning distance. This relationship is

$$\text{total distance} = 600$$

Therefore, the equation is

$$\begin{array}{ccccc} \text{distance car 1} & + & \text{distance car 2} & = & \text{total distance} \\ \uparrow & & \uparrow & & \uparrow \\ 5x & + & 5(x + 15) & = & 600 \end{array}$$

Solving for x gives

$$\begin{aligned} 5x + 5x + 75 &= 600 \\ 10x + 75 &= 600 \\ 10x &= 600 - 75 \\ 10x &= 525 \\ x &= 52.5 \end{aligned}$$

Since x represents the speed of the slower car, its speed is 52.5 mi/h, and the speed of the faster car is $x + 15$ or 67.5 mi/h.

Checking

$$\text{distance car 1} = 5(52.5) = 262.5$$
$$\text{distance car 2} = 5(67.5) = 337.5$$
$$\text{total distance} = 262.5 + 337.5 = 600$$

■

Mixture

EXAMPLE 6 A chemist wants to make a 37% acid solution using 70 ml of a 48% acid solution and a 22% acid solution. How much of the 22% solution should she use?

Solution This problem is typical of "mixture" problems. Again, a chart can help us organize the information. We let x represent the amount of the 22% acid solution. Since the chemist uses 70 ml of the 48% solution, the amount of the mixture is $70 + x$. The amount of each solution in the mixture is the product of the strength and amount columns.

strength × amount = amount in solution		
Strength	Amount	Amount in solution
48% acid 22% acid	70 x	0.48(70) $0.22x$
37% mixture	$70 + x$	$0.37(70 + x)$

As in the case of "motion problems," the equation uses the last column of the chart. The equation is

48% in solution + 22% in solution = amount of 37% in solution

$$\uparrow \qquad\qquad \uparrow \qquad\qquad \uparrow$$

$$0.48(70) + 0.22x = 0.37(70 + x)$$
$$33.6 + 0.22x = 25.9 + 0.37x$$
$$7.7 = 0.15x$$
$$51.3 \approx x$$

Checking

amount of 48% acid in the solution = 33.6
amount of 22% acid in the solution = 11.3
amount of 37% acid (mixture) = 44.9

■

Finance

EXAMPLE 7 Jim Parker invested a certain amount of money at 10% and \$1,000 more than three times this amount at 8%. The total annual interest from the two investments is \$760. How much did Jim invest at each rate?

Solution This problem is typical of "finance" problems. As in the previous two examples, a chart is helpful. The variables are

$$x = \text{the amount invested at } 10\%$$
$$(3x + 1{,}000) = \text{the amount invested at } 8\%$$

These values are placed in the amount column of the chart. Since interest is the product of rate and amount, we multiply the contents of the first two columns and place the result in the interest column.

rate ×	amount =	interest
Rate	Amount	Interest
10%	x	$0.10(x)$
8%	$(3x + 1{,}000)$	$0.08(3x + 1{,}000)$

In calculating the interest, we must express the rate in decimal form. Since the total interest is $760, the equation is

$$\begin{aligned}
\text{interest at } 10\% + \text{interest at } 8\% &= \text{total interest}\\
\uparrow \qquad\qquad\quad \uparrow \qquad &\qquad \uparrow\\
0.10(x) \quad + \quad 0.08(3x + 1{,}000) &= 760\\
0.10x + 0.24x + 80 &= 760\\
0.34x + 80 &= 760\\
0.34x &= 760 - 80\\
0.34x &= 680\\
x &= 2{,}000\\
3x + 1{,}000 = 3(2{,}000) + 1{,}000 &= 7{,}000
\end{aligned}$$

Jim invested $2,000 at 10% interest and $7,000 at 8% interest.

Checking

$$\begin{aligned}
\text{interest on } 10\% \text{ investment} &= 0.10(2000) = \$200\\
\text{interest on } 8\% \text{ investment} &= 0.08(7000) = \$560\\
\text{total interest} &= \$760
\end{aligned}$$

■

In this section we have discussed solving applied problems that give a linear equation. This concept is difficult to master because each problem is different, but lots of practice does make it easier.

1–5 EXERCISES

1. A farmer wishes to fence in a triangular tract of land. One side is 60 rods, another is 40 rods, and the third side is 25 rods. How much fence is needed?

E 2. The resistance in one resistor is 3 Ω more than that in a second resistor. The total resistance of the two is 16 Ω. Find the resistance in each resistor.

3. The length of a rectangular tract of land is 2 ft more than the width. If it takes 480 ft of wire to enclose the tract, find the length of the land.

4. A 6-ft board is to be cut into two pieces such that one piece is 2 ft longer than the other piece. Find the length of each piece.

M 5. The transmission ratio is the ratio of the engine speed to the drive shaft speed. The engine speed of a car is 4,000 revolutions per minute (r/min), and the drive shaft speed is 1,000 r/min. What is the transmission ratio?

C 6. In constructing a building, the architect must know the bearing capacity of the soil, which is the number of pounds the soil can support. The total weight of a building divided by the bearing capacity of the soil determines the area of footing. What area of footing is required if the bearing capacity is 15,000 lb/ft^2 and the weight of the building is 75,000 tons?

7. The diagonal of a square is 1.414 times the length of a side of the square. Find the length of the diagonal for a square whose sides are 40 in.

M 8. A service station owner wants to be sure his hydraulic lift will support a truck. The pressure on the hydraulic fluid is the force in pounds divided by the area. What pressure is required to support 8,000 lb if the area is 400 in^2?

9. The ratio of milk to cream in ice cream is 4 quarts to 3 quarts, respectively. How many quarts of milk and cream are used in 5 gallons of ice cream? (Be careful of the units of measure.)

10. It took Bob 2 hours to mow a lawn that is 17,500 ft^2. How long would it take him to mow an area of 30,000 ft^2?

11. Two cars start in Chicago and travel in opposite directions. At the end of 3 hours, they are 330 miles apart. If one car travels 10 mi/h faster than the other, find the speed of the slower car.

12. Sally saves dimes and quarters. She has three times as many quarters as dimes and a total of $12.75. How many dimes and quarters does she have?

I 13. Express the ratio of gross sales to profit if a profit of $750 was made on gross sales of $1,100.

14. If a computer can print 900 lines in 180 seconds, express the ratio of lines per minute.

C 15. Cement, sand, and gravel are mixed in a 1:3:4 proportion to make concrete. How much sand would be used to make 64 yd^3 of concrete?

16. A 50-ft building casts a shadow 30 ft long. At the same time and place, how long a shadow would a 20-ft tree cast?

17. How many liters of a 20% alcohol solution must be mixed with 40 liters of a 35% solution to get a 28% alcohol solution?

18. It takes 850 gallons of insecticide for a crop duster to spray 20 acres. Find the ratio of application in gallons per acre.

19. A video tape recorder moves 900 ft of tape past the recording head in 2 hours. Express the ratio of the amount of tape per minute.

20. It takes 7 people 12 hours to complete a job. If they worked at the same rate, how many people would it take to complete the job in 16 hours?

21. The sum of three consecutive integers is 75. Find the integers. (Represent the integers as x, $x + 1$, and $x + 2$.)

22. The Nielsen rating service found that for every 13 people who liked a new television show, 7 people did not like it. You conduct a random sample survey in your school and find that 156 people liked the television show. Based on the Nielsen rating, how many people in your school did not like the show?

23. A bar sells a 24-oz mug of beer for $1.00, while the grocery next door sells a six-pack of 12-oz cans for $3.29. Which is a better buy?

24. Two truckers drive the same route from Miami to Atlanta. One trucker travels at 60 mi/h, while the other trucker travels at 65 mi/h. The slower truck takes an hour longer. How long does it take each driver to complete the trip?

25. The power of one engine is 3 hp more than that of a second engine. A third engine is 5 hp less than twice the power of the second engine. If the total horsepower of the three engines is 34, find the horsepower of each engine.

26. A carpenter needs to cut an 8-ft board into two pieces so that one piece is 6 in. less than twice the length of the other piece. Find the length in feet of each piece.

27. A cylinder contains 50 liters of a 60% chemical solution. How much of this solution should be drained off and replaced with a 40% solution to obtain a final strength of 46%?

E 28. One inductance has a value 2/3 that of a second inductance. If the smaller inductance is increased by 10, the sum is increased to twice the original value. Find the values of the inductances.

29. A TV repairperson charges $40 per hour to repair black-and-white sets and $60 per hour for color sets. If the repairperson earns $620 for working 12 hours, how many hours were spent repairing color sets?

30. An investment counselor advises you to invest in two stocks. If one stock returns 10% and the other stock returns 13% on your investment, and you invest $1,500 in each stock for a year, how much do you earn in interest?

31. Oil tank A has a capacity twice that of tank B. Oil tank C has a capacity 20 gallons less than that of tank A. The total capacity of the tanks is 1,080 gallons. Find the capacity of tank C.

32. The perimeter of an isosceles triangle is 30 in. If the third side is 3 in. more than the length of the other sides, find the length of each side.

33. A resistor costs one-tenth as much as a transistor. If three resistors and seven transistors cost $2.28, how much does one transistor cost?

34. On a trip Jerry drove a steady speed for 3 hours. An accident slowed his speed by 30 mi/h for the last part of the trip. If the 190-mile trip took 4 h, what was his speed during the first part of the trip?

CHAPTER SUMMARY

Summary of Terms

absolute value (p. 5)
combined variation (p. 24)
compound inequality (p. 19)
constant of variation (p. 22)
cross-product rule (p. 30)
equation (p. 2)
inequality (p. 14)
inversely proportional (p. 23)
joint variation (p. 24)
linear equation (p. 2)
open statement (p. 2)
Principles of Equality (p. 2)
proportion (p. 30)
proportional (p. 22)
ratio (p. 30)
root (p. 2)
solution (pp. 2, 14)
subscript (p. 9)
transposition (p. 2)
vary directly (p. 22)
vary inversely (p. 23)

Summary of Formulas

$y = kx$ direct variation

$y = \frac{k}{x}$ inverse variation

$y = kxz$ joint variation

$y = \frac{kx}{z}$ combined variation

CHAPTER REVIEW

Section 1–1

Solve the following equations.

1. $2x - 8 = 14$

2. $7x + 12 = 9$

3. $4(2x - 3) + 6 = 7 - (10 - 2x)$

4. $2x - 5 + 6x + x = 3(2 + 2x + 1)$

5. $2.3x - 4(x + 1.2) = 6.7 - 4.8x$

6. $7.3(x - 2) + 8.6 = 3 - 6.3x$

7. $\frac{2x - 6}{4} = \frac{1}{2} + \frac{1 - x}{4}$

8. $\frac{2x - 6}{5} = \frac{4 + 3x}{4}$

9. $3|2x - 1| + 7 = 14$

10. $|3x + 4| = 8$

Section 1–2

Solve for the specified variable.

11. $a = \frac{v - v_0}{t}$ for v

12. $V = \frac{1}{3}BH$ for B

13. $y - y_0 = m(x - x_0)$ for x

14. $y = mx + b$ for x

15. $I = \dfrac{E}{R + r}$ for E

16. $Q = wc(T_1 - T_2)$ for T_1

17. $Q = \dfrac{I_2Rt}{J}$ for R

18. $A = ab + \dfrac{d}{2}(a + c)$ for c

19. $L = a + (n - 1)d$ for n

20. $S = \dfrac{n}{2}(a + 1)$ for a

21. The acceleration of an object equals the ratio of the force applied and the mass of the object. Find the force in newtons when the acceleration is 16 m/s^2 and the mass is 4 kg. [*Note:* newton = (m·kg)/s^2.]

22. Calculate a golfer's handicap if his three-round scores are 70, 85, and 83. [$H = 0.85(A - 72)$ where A is the three-round average.]

C 23. Determine the area of the footing for a building weighing 120,000,000 lb, if the bearing weight of the soil is 15,000 lb/ft^2. (Footing area is the quotient of building weight and soil-bearing capacity.)

Section 1–3

Solve the following inequalities and graph the solution.

24. $4x - 8 + x - 14 > 8x - 1$

25. $-2x + 7 - 5x < 7(x + 2) - 13$

26. $3(2x - 4) \leq -2(4 + x - 5)$

27. $2.5(3x + 1.8) \geq 7.9 - (1.6x - 5.6)$

28. $6.1(x - 3.7) - (8.5x - 5) > 7 + 3.2x$

29. $|8x + 6| > 34$

30. $|4x + 6| \leq 7$

31. $|5x + 2| \geq 18$

32. $\dfrac{3x}{5} + \dfrac{1}{2} \geq \dfrac{x}{4} - \dfrac{2}{5}$

33. $\dfrac{7x}{4} + \dfrac{7}{8} \leq \dfrac{x + 2}{2}$

Section 1–4

34. Write an equation to represent the following statement: a varies jointly with the square of b and the cube of c.

35. Write an equation for the following statement: x varies directly with the square root of y and inversely with the square of z.

36. Write an equation to represent the following statement: The volume of a pyramid varies jointly with the area of the base and the height.

37. a varies inversely with the square root of b. If $a = 3$ when $b = 4$, find a when $b = 16$.

38. m varies jointly with n and the square of p. If $m = 36$ when $n = 3$ and $p = 2$, find n when $m = 240$ and $p = 4$.

39. x varies directly as the cube root of y. If $x = 64$ when $y = 8$, find x when y is 27.

E 40. The voltage across a series circuit varies directly as the current. If the voltage is 30 V when the current is 5 A, find the current when the voltage is 66 V.

I 41. The cost of labor at a factory varies jointly with the number of workers and the number of hours worked. If 8 men work 320 hours to earn $11,648, how many men must work for 460 hours to earn $23,023?

42. The volume of a gas varies directly as the absolute temperature in rankine degrees (°R) and inversely with the pressure. If a gas at 639° R has a pressure of 12 lb/in.^{2} and a volume of 106 in.^{3}, find the volume when the temperature is 600° R and the pressure is 144 lb/in.^{2}

Section 1–5

43. Distance is the product of rate and time. If a car travels 486 miles at a rate of 56 mi/h, how many hours does the trip require?

44. A technician earns $11.40 an hour for the first 40 hours and $17.10 for each hour over 40. How many overtime hours must the technician work to earn $609.90 a week?

45. Pressure is the quotient of force and area. If the pressure is 60 lb/in.^{2} and the area is 20 in.^{2}, find the force exerted.

46. The resistance of one resistor is three less than four times the resistance of a second resistor. If the total resistance of the two resistors is 147 Ω, find the resistance of each resistor.

47. The current in one electrical component is six less than three times the current in another component. If the total current is 54 A, find the current in each component.

48. The perimeter of a triangle is 48 cm. If the first side is three more than the second side and the second side is twice the third side, find the length of each of the sides.

I 49. The cost of producting one computer is \$300 more than the cost of producing a second computer. The total cost of producing one of the first type and three of the second type is \$8,300. Find the cost of producing each type of computer.

50. Peanuts and cashews are combined to make 20 lb of a mixture that will sell for \$3.03/lb. If peanuts cost \$2.40/lb and cashews cost \$4.50/lb, how much of each should be used in the mixture?

51. A person pays social security tax, state income tax, and federal income tax on his salary in a 2 : 3 : 5 proportion. If the total deductions for these items if \$400, how much is paid in social security tax?

52. Maria invests \$800 in two stocks, one paying a return of 9.6% and the other paying 4.8%. If her total income from the two stocks is \$66.00, how much did she invest in each stock?

53. A computer line printer can print 350 characters in 2 seconds. How many seconds are required to print 1,400 characters?

54. Two planes traveling in opposite directions leave the same airport at the same time. If they are 2,550 miles apart at the end of 3 hours and one flies 50 mi/h faster than the other, how fast is the slower plane traveling?

CHAPTER TEST

The number in parentheses refers to the appropriate learning objective given at the beginning of the chapter.

1. Solve $6x - 2(4x + 5) = 3(x - 6) + 2$. (1)

2. Solve the following literal equation for a: (3)

$$s = v + \frac{1}{2}at^2$$

3. Bob invests \$6,000 in two accounts, one paying 7.3% and the other paying 9.5%. If his total interest is \$521.60, how much did he invest in each account? (8)

4. Solve the following inequality and graph the solution: (4)

$$-6x + 12 \leq 2x + 20$$

5. The area of a triangle is 1/2 the product of the base and height. Find the base of a triangle whose area is 56 ft^2 and whose height is 8 ft. Solve and then substitute. (3, 8)

6. Write an equation to represent the following relationship: d varies directly with the cube of b and inversely with c. (6)

7. Solve $3|2x + 9| - 2 = 18$. (2)

8. Solve the following inequality and graph the solution: (4)

$$\frac{1}{2}x - 3 > \frac{5x}{3}$$

E 9. Resistors connected in series have voltage drops proportional to their resistances. If the voltage drop across a 30-Ω resistor is 75 V, what is the voltage drop across a 10-Ω resistor? (7)

10. The volume of a cone varies jointly with the square of the radius and the height. If the volume is 31.0 when the radius is 2.3 and the height is 5.6, find the volume when the radius is 6.8 and the height is 3.9. (7)

11. Solve the following inequality and graph the solution: $|7x - 2| \geq 12$. (4)

12. Express the ratio of \$2.25 to \$0.75 in lowest terms. (8)

13. Fred drives from Atlanta to Orlando in 8 hours. Part of the trip he drives 55 mi/h and the remainder at 60 mi/h. If the distance from Atlanta to Orlando is 450 miles, how long did he drive at each speed? (8)

14. Solve $V = \pi r^2 h$ for h. (3)

15. Solve the following: (1)

$$\frac{3x}{4} - \frac{1}{2} = \frac{2}{3} + \frac{5x}{6}$$

16. Mr. Green mixes antifreeze and water in a ratio of 2:5. How many gallons of antifreeze are used in mixing 21 quarts? (8)

17. Solve the following literal equation for P: $a = V(k - PV)$. (3)

18. Solve $2.9x - (7.2 - 6.8x) = 4 + 3.1x$ (1)

19. Truck A can carry twice as much as truck B, and truck C carries 100 lb less than three times as much as truck A. The total capacity for the three trucks is 8,000 lb. Find the capacity of truck A. (8)

20. Solve the following and graph the solution: $3|2x - 5| < 11$. (5)

SOLUTION TO CHAPTER INTRODUCTION

To determine the amount of 35% and 18% hydrochloric acid to use in the solution, we fill in the accompanying table.

Strength	Amount	Amount in solution
35%	x	$0.35x$
18%	$50 - x$	$0.18(50 - x)$
26%	50	0.26(50)

Next we solve the following equation:

$$\text{amount in solution of 35\%} + \text{amount in solution of 18\%} = \text{total amount in solution}$$

$$\begin{aligned} 0.35x + 0.18(50 - x) &= 0.26(50) \\ 0.35x + 9 - 0.18x &= 13 \\ 0.17x &= 4 \\ x &\approx 24 \\ 50 - x &\approx 26 \end{aligned}$$

The chemist should use approximately 24 ml of the 35% solution and 26 ml of the 18% hydrochloric acid solution.

Graphs provide a visual representation of the relationship between two variables. They can be used to determine the break-even point for a company. For example, imagine that you own a company that makes and sells widgets for $2.50. The fixed cost of renting the building, paying utilities, and monthly salaries is $1,500 per month. The variable cost of raw materials, hourly salaries, and transportation is $1.25 per unit. Determine the quantity of widgets to manufacture and the resulting revenue in order to break even. (The answer to this problem is given at the end of the chapter.)

We deal with functions every day without realizing it. For example, the distance you can drive on a full tank of gas is a function of the type of car, the speed you drive, and road conditions. The velocity with which a ball strikes the ground is a function of its initial velocity, acceleration, and elapsed time.

Learning Objectives

After completing this chapter, you should be able to

1. Express one variable as a function of a second variable (Section 2–1).
2. Identify the domain of a function (Section 2–1).
3. Evaluate $f(x)$ for a given value of x (Section 2–1).
4. Plot points on the rectangular coordinate system (Section 2–2).
5. Graph a given function on the rectangular coordinate system (Section 2–3).
6. Determine the slope of a line given two points (Section 2–4).
7. Determine the distance between two points and the midpoint of two points (Section 2–4).
8. Find the solution to a given inequality or system of two inequalities on the rectangular coordinate system (Section 2–5).
9. Apply these concepts to solve an equation, find the break-even point of a business, and graph empirical data (Section 2–6).

Chapter 2

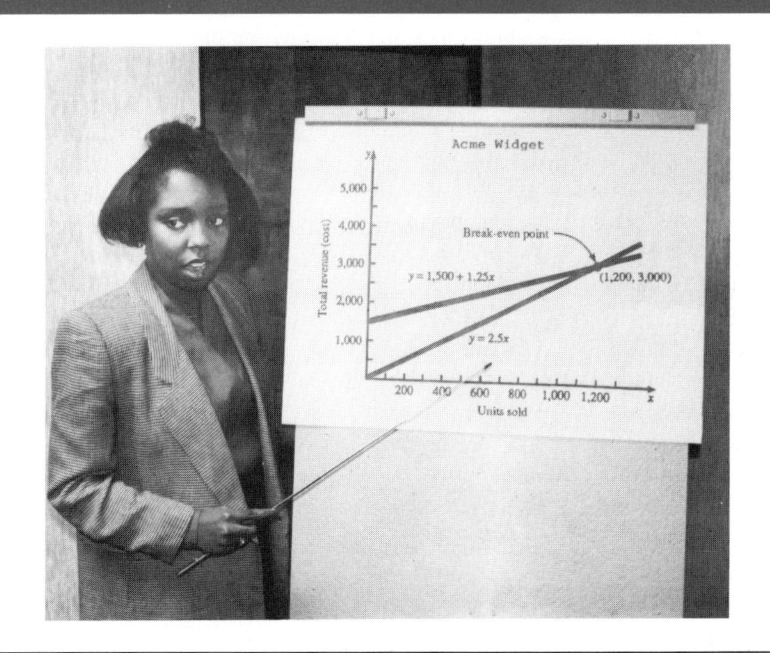

Functions and Graphs

2–1

FUNCTIONS

Definitions

A **function** describes the relationship between variables. In other words, a function is a rule, usually in the form of an equation, that generates a unique number for the dependent variable for each chosen value of the independent variable. The variable whose numerical value is arbitrarily chosen is called the **independent variable,** and the variable whose value is determined from this choice is called the **dependent variable.** In an equation involving x and y, x is usually the independent variable, and y is usually the dependent variable. With variables other than x and y, the physical constraints determine the independent variable. For example, the distance s that an object falls is given by $s = 16t^2$, where t represents elapsed time. Since the distance that the object has fallen depends on the elapsed time, time is the independent variable, and distance is the dependent variable. Although we will discuss functions of two variables, it is also possible for a function to have several independent variables and one dependent variable, such as the volume of a pyramid, given by $V = (\pi r^2 h)/3$. A function can also consist of only a dependent variable, such as $y = 3$.

Functional Notation

Returning to the example of the distance that an object falls, given by $s = 16t^2$, we can say the distance that an object falls is a function of the elapsed time. To denote this functional relationship, the equation $s = 16t^2$ could be written $f(t) = 16t^2$or $s(t) = 16t$. The notation $f(t)$ is called **functional notation** and is read "f of t" to denote that the relationship is a function of t, the independent variable. In the case of the variables x and y, the equation $y = 2x - 7$ could also be represented as $f(x) = 2x - 7$ or $y(x) = 2x - 7$ since y is a function of the independent variable x.

EXAMPLE 1 Write the indicated function using functional notation

(a) Express $m = 6p^2 - p + 5$ as a function of p.
(b) Express the area A of a triangle whose height is 10 ft as a function of the base b.
(c) Express the perimeter P of a rectangle of width 8 m as a function of length L.

Solution

(a) Since m is a function of p, one functional notation is $f(p) = 6p^2 - p + 5$. To denote m as the dependent variable, you could also write $m(p) = 6p^2 - p + 5$.
(b) The area of a triangle whose height is 10 ft is given by the formula

$$A = \frac{10b}{2}$$

$$A = 5b$$

The functional notation could be $f(b) = 5b$ or $A(b) = 5b$.

(c) The perimeter of a rectangle of width 8 m is given by

$$P = 2(8) + 2L$$
$$P = 16 + 2L$$

The functional notation could be written as $f(L) = 16 + 2L$ or $P(L) = 16 + 2L$. ■

Functional notation is useful in naming a specific value of the independent variable. For example, for the function $f(x) = 2x - 7$, the notation $f(3)$ represents the value of the dependent variable when the independent variable equals 3. To determine $f(3)$, we substitute 3 for x as follows:

$$f(x) = 2x - 7$$
$$f(3) = 2(3) - 7 \quad \text{substitute } x = 3$$
$$f(3) = -1$$

EXAMPLE 2 Given the function $f(x) = 3x^2 - 2x + 1$, find the following:

(a) $f(0)$ (b) $f(-2)$ (c) $f(4)$

Solution

(a) To find $f(0)$, we substitute 0 for x in the function.

$$f(x) = 3x^2 - 2x + 1$$
$$f(0) = 3(0)^2 - 2(0) + 1 \quad \text{substitute } x = 0$$
$$f(0) = 0 - 0 + 1$$
$$f(0) = 1$$

(b)

$$f(-2) = 3(-2)^2 - 2(-2) + 1 \quad \text{substitute } x = -2$$
$$f(-2) = 12 + 4 + 1$$
$$f(-2) = 17$$

(c)

$$f(4) = 3(4)^2 - 2(4) + 1 \quad \text{substitute } x = 4$$
$$f(4) = 48 - 8 + 1$$
$$f(4) = 41$$

■

A method similar to that used in Examples 1 and 2 is used to evaluate a function when a literal expression is substituted for the independent variable, as shown in the next example.

EXAMPLE 3 If $f(x) = x^2 - 4x$, find $f(x - 3)$.

Solution To find $f(x - 3)$, we substitute $x - 3$ for x in the function and simplify the result.

$$\begin{aligned} f(x) &= x^2 - 4x & \\ f(x-3) &= (x-3)^2 - 4(x-3) & \text{substitute} \\ f(x-3) &= x^2 - 6x + 9 - 4x + 12 & \text{multiply} \\ f(x-3) &= x^2 - 10x + 21 & \text{simplify} \end{aligned}$$

■

EXAMPLE 4 Evaluate the following functions

(a) Find $m(2)$ for $m(x) = \sqrt{4x - 5}$.

(b) Find $p(1.8)$ for $p(n) = \dfrac{n^3 - 3n}{9 - 2n}$.

(c) The displacement of a free-falling object with an initial velocity of 9 ft/s as a function of time is given by $s(t) = 9t + 16t^2$. Find $s(4) + s(t)$.

Solution

(a) To find $m(2)$, substitute $x = 2$ into the function and simplify.

$$\begin{aligned} m(x) &= \sqrt{4x - 5} \\ m(2) &= \sqrt{4(2) - 5} \\ m(2) &= \sqrt{3} \end{aligned}$$

(b) To determine $p(1.8)$, substitute $n = 1.8$ into the function.

$$\begin{aligned} p(n) &= \frac{n^3 - 3n}{9 - 2n} \\ p(1.8) &= \frac{(1.8)^3 - 3(1.8)}{9 - 2(1.8)} \\ p(1.8) &\approx 0.08 \end{aligned}$$

1.8 [y^x] 3 [−] 3 [×] 1.8 [=] [÷] [(] 9 [−] 2 [×] 1.8 [)] [=] → 0.08

(c)

$$s(t) = 9t + 16t^2$$

$$\begin{aligned} s(4) + s(t) &= \underbrace{[9(4) + 16(4)^2]}_{s(4)} + \underbrace{[9t + 16t^2]}_{s(t)} \\ &= (36 + 256) + (9t + 16t^2) \\ &= 292 + 9t + 16t^2 \end{aligned}$$

■

Restriction on the Domain

The **domain** of a function is the set of permissible values for the independent variable. The **range** of a function is the set of possible values for the dependent variable resulting from the domain. The domain can be called the *input values,* and the range can be called the *output values*. Unless otherwise noted, the domain and range of a function are assumed to be the set of all real numbers.

Since the range of a function is limited to real numbers, certain functions may require a restriction of the domain. The primary restrictions result from avoiding any attempt to divide by zero or take the square root of a negative number. These restrictions are illustrated in the following two examples.

EXAMPLE 5 Determine the domain for the following function:

$$f(x) = \frac{1}{x} + 3x$$

Solution The domain must exclude zero because it leads to division by zero. The domain is all real numbers except zero. ■

EXAMPLE 6 Determine the domain for the following functions:

(a) $f(x) = \sqrt{x - 2}$

(b) $f(p) = \dfrac{1}{p} + \dfrac{3}{p - 4}$

(c) $y(x) = \dfrac{\sqrt{9 - x}}{2x - 12}$

Solution

(a) The square root of a negative number does not result in a real number. Therefore, the quantity under the radical must be positive or zero: $x - 2 \geq 0$, $x \geq 2$. The domain is all real numbers greater than or equal to 2.

(b) Since division by zero is not allowed, we determine the values of p in the denominator that would result in division by zero.

$$p \neq 0 \quad \text{and} \quad p - 4 \neq 0$$
$$p \neq 4$$

The domain is all real numbers except 0 and 4.

(c) Since the quantity under the radical cannot be negative and the quantity in the denominator cannot equal zero, the domain is

$$9 - x \geq 0 \qquad 2x - 12 \neq 0$$
$$-x \geq -9 \qquad 2x \neq 12$$
$$x \leq 9 \qquad x \neq 6$$

or all real numbers less than or equal to 9 except 6. ■

2–1 EXERCISES

Express each function using functional notation.

1. Give the perimeter P of a square as a function of the sides s.
2. Give the perimeter P of an equilateral (3 equal length sides) triangle as a function of the sides s.
3. If the length of a rectangle is 5 ft, express the area A as a function of the width w.
4. Express the radius r of a circle as a function of its circumference C.
5. The total cost C of a taxi ride is \$0.60 for the first mile and \$0.40 for each additional mile m. Express the total cost as a function of m.
6. The temperature in degrees Celsius C is 17.8° less than the product of 5/9 and the temperature in degrees Fahrenheit F. Express C as a function of F.

E 7. Voltage V is the product of current I and resistance R. Express current as a function of voltage in a circuit containing a 10-Ω resistor.

8. Express the width w of a rectangle as a function of its area A if the length is 30 ft.

Find the function values requested.

9. $f(x) = 5x^2 + 6$ **(a)** $f(2)$ **(b)** $f(-3)$ **(c)** $f(-1)$
10. $y(x) = 4x - 6$ **(a)** $y(-2)$ **(b)** $y(0)$ **(c)** $y(4)$
11. $f(x) = -x^2 + 3x - 1$ **(a)** $f(6)$ **(b)** $f(1)$ **(c)** $f(-3)$
12. $m(p) = 2\sqrt{x} - 4$ **(a)** $m(6)$ **(b)** $m(9.7)$ **(c)** $m(2)$
13. $g(a) = \frac{3}{a} + 2a$ **(a)** $g(2.8)$ **(b)** $g(-3)$ **(c)** $g(-2.1)$
14. $b(c) = \sqrt{4c} - 15$ **(a)** $b(0)$ **(b)** $b(1.4)$ **(c)** $b(4)$
15. $g(x) = \sqrt{x} - \frac{3}{x + 1}$ **(a)** $g(0)$ **(b)** $g(7)$ **(c)** $g(3)$
16. $f(h) = \frac{3h}{h^2 - 2}$ **(a)** $f(3.7)$ **(b)** $f(-4)$ **(c)** $f(0)$
17. $s(t) = 3t^2 - 4t + 9$ **(a)** $s(-3)$ **(b)** $s(1)$ **(c)** $s(6)$
18. If $f(x) = 4x^2 - 7$, find $f(m)$ and $f(x - 3)$.
19. If $m(s) = 3s - 14$, find $m(x - 8)$ and $m(2s + 5)$.
20. If $g(x) = x^2 - 7x + 9$, find $g(t^3)$ and $g(x - 4)$.
21. If $p(n) = \frac{3n + 8}{n - 1}$, find $p(2n)$ and $p(n + 1) - p(n)$.
22. If $f(y) = 2y + 9$, find $f(3y - 4)$ and $f(y^2)$.
23. If $g(y) = 3y^2 - 7y$, find $g(y - 1) + g(3y)$.

If the range is restricted to real numbers, determine the domain of the following functions.

24. $f(x) = \frac{2}{x} + 3$
25. $g(y) = \sqrt{y + 7}$
26. $p(a) = \frac{a^2 + 6}{a - 2}$
27. $m(t) = \frac{3 + t}{2t - 1} + \sqrt{3t}$
28. $y(x) = \frac{\sqrt{x^2 - 9}}{4x - 20}$
29. $h(x) = \frac{3}{x} + \frac{4x}{x + 6}$
30. $f(p) = \frac{\sqrt{p + 4} + \sqrt{p - 5}}{3p + 9}$

Solve for the requested variable.

31. The average acceleration a of an object starting from rest and reaching 30 mi/h in elapsed time t is given by the function

$$a(t) = \frac{30}{t}$$

Find the average acceleration if the time required is 6 hours.

M 32. Horsepower (hp) is a function of force, distance, and elapsed time t. If a force of 50 lb moves an object 6 ft in time t seconds, the equivalent horsepower is

$$\text{hp} = \frac{300}{550t}$$

Find the horsepower if $t = 8$ s.

I 33. Tax rate R is a function of the levied tax L and the assessed value V of the property and is given by

$$R = \frac{L}{V}$$

If the levied tax is \$800, determine the tax rate when the property is assessed at \$1,300.

34. The surface area s of a cube as a function of the length of an edge e is given by $s = 6e^2$. Find the surface area of a cube whose edge is 8 in.
35. The volume V of a pyramid with a square base of 6 m as a function of height h is given by $V = 12h$. Find the volume of a pyramid 7 m high.

2–2

THE RECTANGULAR COORDINATE SYSTEM

In discussing functions such as $y = 3x - 2$, it is helpful to have a visual representation. The solution to a function of two variables contains a value for each variable. Since a pair of numbers cannot be graphed on the number line, in this section we will discuss a system on which to graph a function of two variables.

Definitions

A function in two variables can be graphed on the **rectangular coordinate system,** also called the **Cartesian coordinate system** named after the French philospher René Descartes. The rectangular coordinate system consists of two perpendicular number lines, called **axes.** The horizontal number line, usually called the ***x* axis,** represents the independent variable and is positive to the right of zero. The vertical number line, usually called the ***y* axis,** represents the dependent variable and is positive above zero. The intersection of the axes is called the **origin** because it represents zero for each variable. The axes divide the plane into four equal regions, called **quadrants.** The quadrants are numbered I, II, III, and IV in a counterclockwise direction, with quadrant I in the upper right. Refer to Figure 2–1 for an illustration of this terminology.

FIGURE 2–1

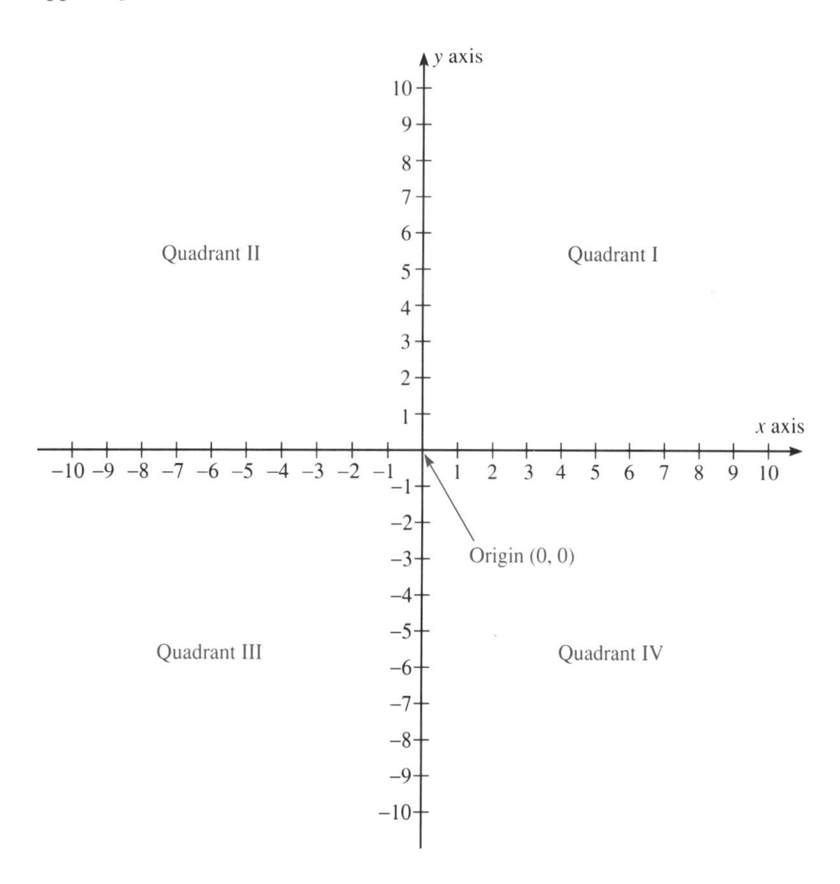

The solution to a function of two variables consists of ordered pairs (x, y) where the first element in the ordered pair is called the **x coordinate,** or abscissa, and the second element is called the **y coordinate,** or ordinate. Furthermore, there is a **one-to-one correspondence** between an ordered pair and a point on the rectangular coordinate system. That is, each ordered pair results in exactly one point on the rectangular coordinate system, and each point results in exactly one ordered pair. The process of locating points on the rectangular coordinate system is called *plotting the points*.

Plotting Points

To Plot an Ordered Pair

1. Start at the origin and move x units left or right, depending on the sign.
2. From the x location, move y units up or down, depending on the sign. Place a dot at the point.

EXAMPLE 1 Plot the point corresponding to $(4, -3)$.

Solution We start at the origin and, since the x coordinate is positive, move to the right four units. From this location move down three units because the y coordinate is negative. Place a dot. This point is shown in Figure 2–2(a). ■

EXAMPLE 2 Plot the point corresponding to $(-1, 4)$.

Solution We start at the origin and move left one unit. Then from this location move up four units and place a dot. This point is shown in Figure 2–2(b). ■

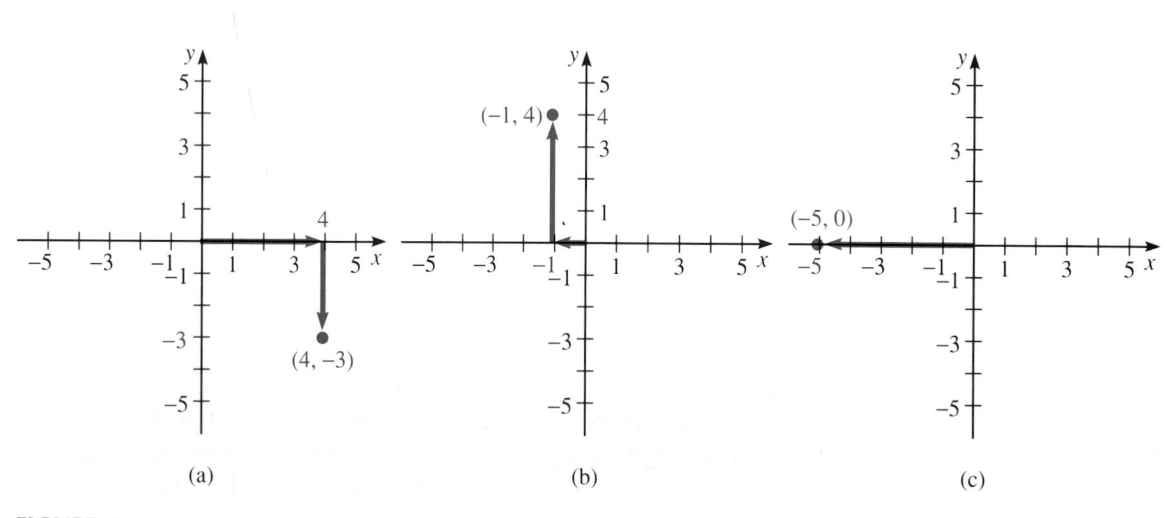

FIGURE 2–2

EXAMPLE 3 Plot the point corresponding to $(-5, 0)$.

Solution Start at the origin and move left five units because the x coordinate is negative. Place a dot at this location because the y coordinate is zero. This point is shown in Figure 2–2(c). ■

CAUTION ✦ Be very careful in plotting points. The x coordinate always appears first in the ordered pair. You may have a tendency to confuse the order of the coordinates in plotting points. When plotting a point, always move left or right first and then up or down.

EXAMPLE 4 Identify the coordinates of points A, B, and C in Figure 2–3(a), (b), and (c), respectively.

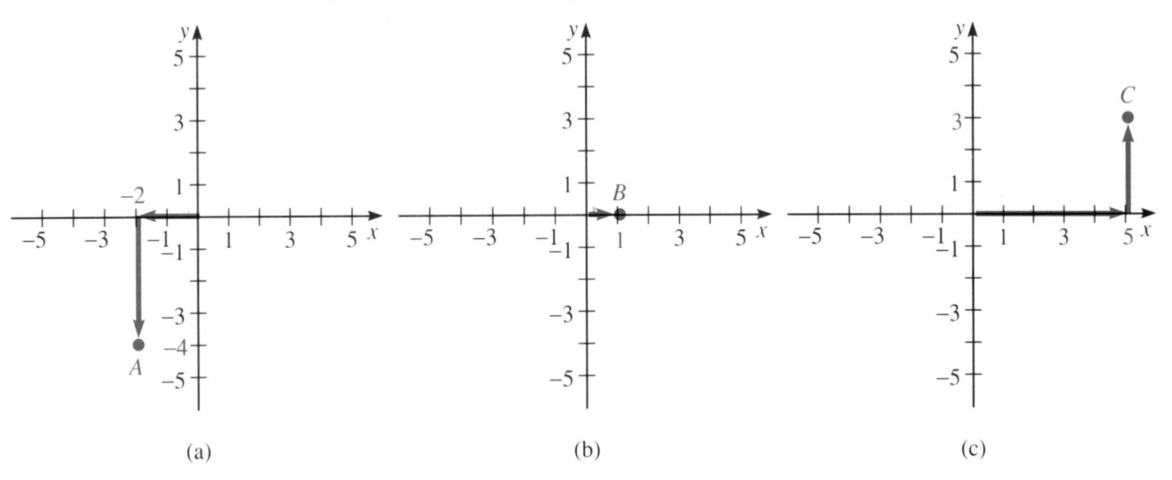

FIGURE 2–3

Solution

(a) Point A is two units to the left of the origin and four units below the x axis. The coordinates of point A are $(-2, -4)$.
(b) Point B is one unit to the right of the y axis and zero units above or below the x axis. The coordinates of point B are $(1, 0)$.
(c) Point C is five units to the right of the y axis and three units above the x axis; therefore, the coordinates of point C are $(5, 3)$. ■

EXAMPLE 5 Three vertices of a rectangle are given by $(-3, 4)$, $(2, 4)$, and $(-3, 2)$. Determine the coordinates of the fourth vertex.

Solution To find the fourth vertex, it may be helpful to plot the known vertices and join them with line segments as shown in Figure 2–4. Since opposite sides of a rectangle are equal, we can count down vertically from the upper right vertex a length equal to the opposite side, two. The coordinates of the fourth vertex are $(2, 2)$. ■

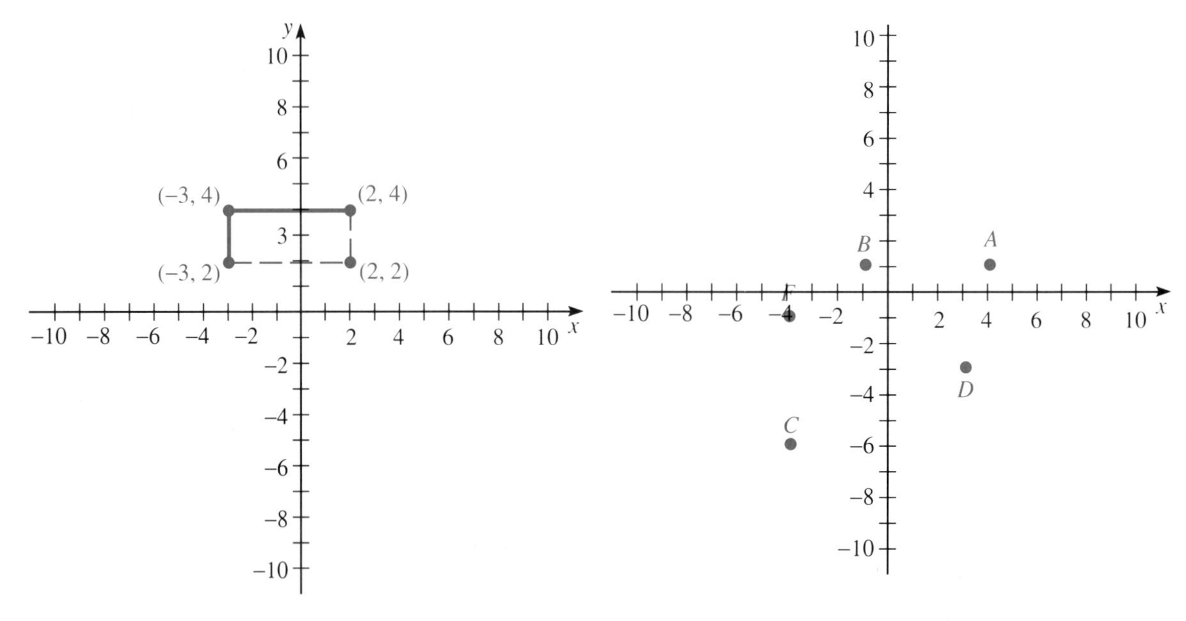

FIGURE 2–4

FIGURE 2–5

2–2 EXERCISES

Plot the following points on the rectangular coordinate system.

1. $A(3, -2)$, $B(-6, 8)$, $C(5, -4)$
2. $A(0, 4)$, $B(-3, -3)$, $C(-2, -5)$
3. $A(4, 0)$, $B(-1, 5)$, $C(5, -9)$
4. $A(4, -7)$, $B(4, 8)$, $C(8, 0)$

Find the coordinates of the points given on the graph shown in Figure 2–5.

5. A, B, C
6. D, E, F
7. In what quadrant are the points located where x is negative and y is positive?
8. In what quadrant are the points located where x is positive and y is negative?
9. In what quadrant are the points located where both x and y are negative?
10. In what quadrant are the points located where both x and y are positive?
11. Where are the points $(0, y)$ located?
12. Where are the points $(x, 0)$ located?
13. What is the y coordinate of any point located on the x axis?
14. What is the x coordinate of any point located on the y axis?
15. In what quadrant(s) is the ratio y/x positive?
16. In what quadrant(s) is the ratio y/x negative?
17. A square has three vertices at (0, 0), (5, 0), and (5, 5). Find the coordinates of the fourth vertex.
18. Three vertices of a rectangle are (4, 3), (10, 6), and (10, 3). What are the coordinates of the fourth vertex?
19. Two vertices of an equilateral triangle are $(-5, 4)$ and $(4, 4)$. What is the x coordinate of the third vertex?
20. The vertices of the base of an isosceles triangle are $(-7, 6)$ and $(6, 6)$. Find the x coordinate of the vertex.

2–3

GRAPHING A FUNCTION

In this section we will integrate the concept of a function with graphing in the rectangular coordinate system to obtain a visual representation of a function that will be used later in this chapter.

The **graph** of a function consists of all ordered pairs that satisfy the functional rule. However, an equation in two variables may have an infinite number of ordered pairs that satisfy the functional rule. Since it is impossible to plot all these ordered pairs, we plot a representative sample and draw a smooth curve through the points.

Equations of the form $ax + by = c$ where a and b are constants and not both zero are called **linear equations in two variables.** Since linear equations have graphs that are lines, it is necessary to plot only two ordered pairs that satisfy the functional rule. However, you will have fewer careless errors if you graph at least three ordered pairs.

Graphing a Function

EXAMPLE 1 Graph the function $y = f(x) = 3x - 4$.

Solution To graph the line that represents this function, we need three ordered pairs that satisfy the equation. To determine these ordered pairs, we arbitrarily choose three values for x and substitute them into the functional rule. Then we plot the points and draw a line through them. Calculating the ordered pairs gives

$$x = -2 \qquad y = f(x) = 3(-2) - 4 = -10$$
$$x = 0 \qquad y = f(x) = 3(0) - 4 = -4$$
$$x = 1 \qquad y = f(x) = 3(1) - 4 = -1$$

The ordered pairs that we plot are $(-2, -10)$, $(0, -4)$, and $(1, -1)$. The graph of $y = 3x - 4$ is shown in Figure 2–6. ■

To obtain the ordered pairs, it is often helpful to solve the equation for the dependent variable. Then, to organize the ordered pairs, fill in a table of values with columns for the independent and dependent variables. If the equation is nonlinear, you will need to obtain enough ordered pairs to draw a smooth curve. Depending on the size of the numbers required on the graph, it may also be necessary to scale one or both axes. To determine the scale for each axis, subtract the smallest value in the table of values from the largest to find the range of values required. Then count the total number of units available on the graph. Divide the number of values required by the total number of units available. You may want to round off this number since the scale should be convenient to use.

FIGURE 2–6

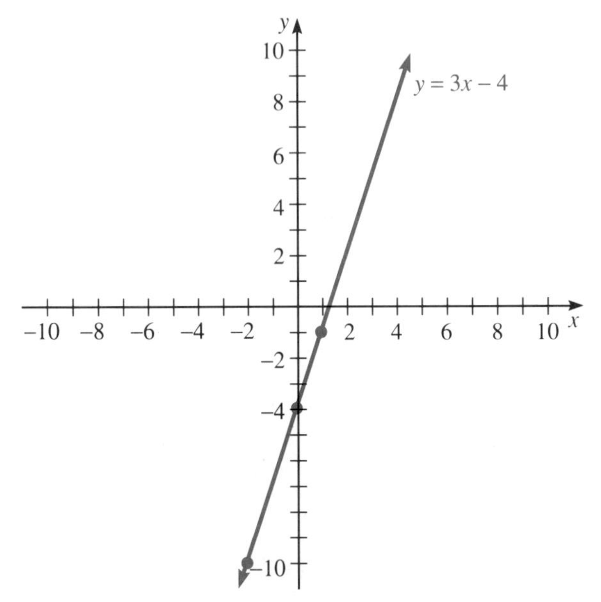

The following box summarizes the steps required to graph a function, and Example 2 illustrates scaling the axes.

To Graph a Function

1. Determine the dependent variable and solve the equation for it.
2. Determine any restrictions on the domain.
3. Construct a table of values reflecting the domain restrictions, if they exist.
4. Determine an appropriate scale for each axis. The numbers for one or both of the variables may be too large or too small to fit on the graph.
5. Plot the ordered pairs and draw a line or smooth curve through and extending beyond the points.

EXAMPLE 2 Graph $f(x) = 2x^2 + 6$.

Solution The independent variable is x, and the dependent variable is $f(x)$. Since this equation does not satisfy the definition of a linear equation, it is nonlinear and we must choose more than three values for x. The accompanying calculations and table of values are given below.

In choosing the scale for this graph, we want the graph to be large enough to show the necessary details, but not so large as to lose the relationship between the variables. Our x values range from -3 to 4, a total of 8 units, and the total number of units available on the graph is 16. Therefore, we let each unit along the x axis equal 0.5, or 8/16. Choosing a scale for the y axis

x	y	
-3	24	$2(-3)^2 + 6 = 24$
-2	14	$2(-2)^2 + 6 = 14$
-1	8	$2(-1)^2 + 6 = 8$
0	6	$2(0)^2 + 6 = 6$
1	8	$2(1)^2 + 6 = 8$
2	14	$2(2)^2 + 6 = 14$
4	38	$2(4)^2 + 6 = 38$

FIGURE 2–7

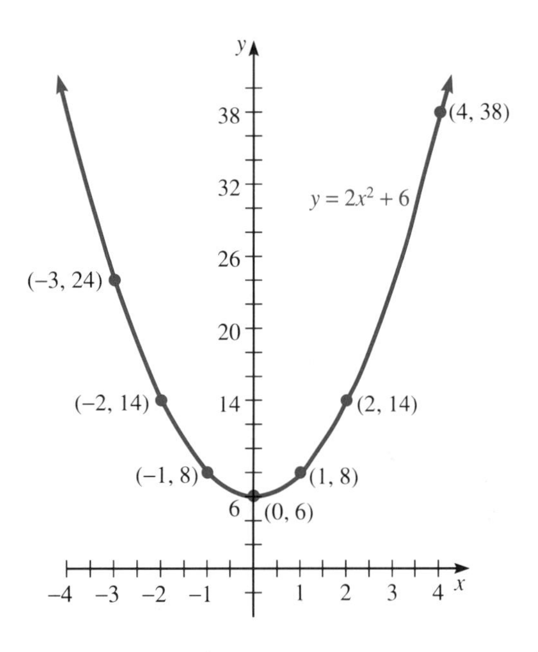

is not quite as easy in this case. The values in the table of values range from 6 to 38, but we want to include the origin in the graph. Therefore, we will use 0 to 38 for the range of values required. Since we have 19 units available on the graph, let each unit on the y axis equal 2, or 38/19. After plotting the points, we join them with a smooth curve as shown in Figure 2–7. ■

The U-shaped graph shown in Figure 2–7 is called a **parabola.** Its shape is characteristic of equations where one variable is quadratic (degree two) and the other variable is linear. The turning point of the curve (0, 6) is called the **vertex** of the parabola. Be sure that you graph the vertex of the parabola by selecting x values or by estimating its location. We will discuss quadratic equations and the parabola in more detail later in this chapter.

Restrictions on the Domain

Recall from Section 2–1 that the domain of a function may need to be restricted because the range is real numbers. The next two examples illustrate the effect that domain restrictions have on its graph. Since the table of values should contain only permissible values of the independent variable, we must determine domain restrictions before completing the table of values.

EXAMPLE 3 Graph $y = \sqrt{3x - 1}$.

Solution The independent and dependent variables are x and y, respectively. Before completing a table of values, we must determine any required restrictions to the domain. Recall from a previous section of this chapter that the value under the radical symbol must be nonnegative for the range to be only real numbers. Therefore,

$$3x - 1 \geq 0$$
$$x \geq 1/3$$

The domain is restricted to values greater than or equal to one-third, as shown in the accompanying table of values. The scale selected for the y axis is 0.5. The graph of this function is shown in Figure 2–8. The shape of the graph is characteristic of the square root function.

x	y
1/3	0
1	1.4
2	2.2
3	2.8
4	3.3
5	3.7
6	4.1

FIGURE 2–8

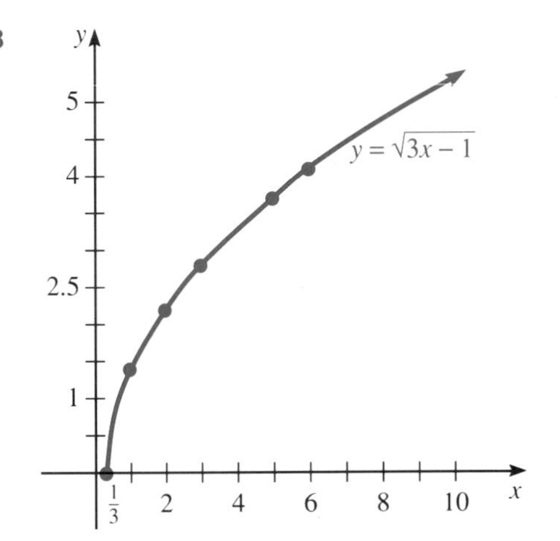

To calculate $\sqrt{3x - 1}$ for selected values of x on the calculator, follow these steps:

3 [×] 1 [÷] 3 [−] 1 [=] [√] ⟶ 0

3 [×] 2 [−] 1 [=] [√] ⟶ 2.2

3 [×] 5 [−] 1 [=] [√] ⟶ 3.7 ■

EXAMPLE 4 Graph the following function:

$$f(x) = \frac{2}{x + 1} + 3$$

Solution The domain of this function is restricted to avoid division by zero. Since the denominator of the fraction cannot equal zero,

$$x + 1 \neq 0$$
$$x \neq -1$$

The domain of the function is all real numbers except -1. The table of values is shown below, and the resulting graph is shown in Figure 2–9.

FIGURE 2–9

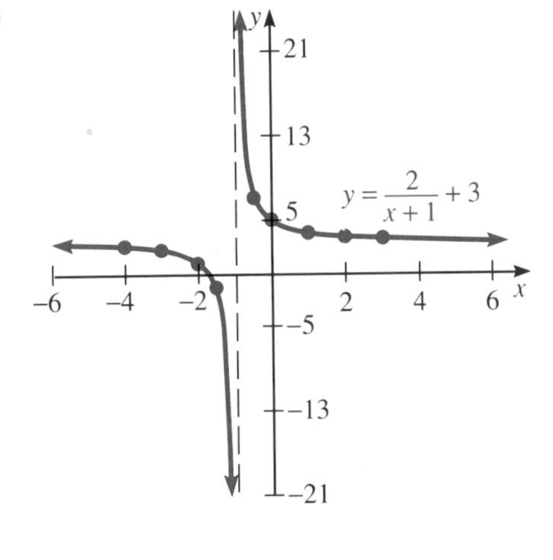

x	y
−4	2.3
−3	2
−2	1
−1.5	−1
−0.5	7
0	5
1	4
2	3.6
3	3.5

■

The graph shown in Figure 2–9 is called a **hyperbola.** The restriction of the domain ($x \neq -1$) results in two branches of the graph: one on either side of the line $x = -1$. The line $x = -1$ is called an **asymptote** because the two branches of the graph approach the line but never touch it. In graphing a function such as the one in Example 4, be sure to take some large and some small values of x.

The domain or range of a function can also be restricted by physical constraints. Certain values of a variable may be unreasonable, as illustrated in the next example.

EXAMPLE 5 The amount of work W (ft-lb) done in moving an object is the product of the force F (lb) and the distance d (ft). Plot the relationship between work and force in moving a piano a distance of 10 ft.

Solution The equation relating work, force, and distance is $W = Fd$. If the piano is moved 10 ft, the equation becomes $W = 10F$. The independent variable F is placed on the horizontal axis, and the dependent variable W is placed on the vertical axis. The table of values is shown below, and the graph for this function is given in Figure 2–10. A scale of 5 lb is chosen for the horizontal axis, and a scale of 50 ft-lb is chosen for the vertical axis. Negative values of force are not chosen because a negative force is not reasonable in this situation.

FIGURE 2–10

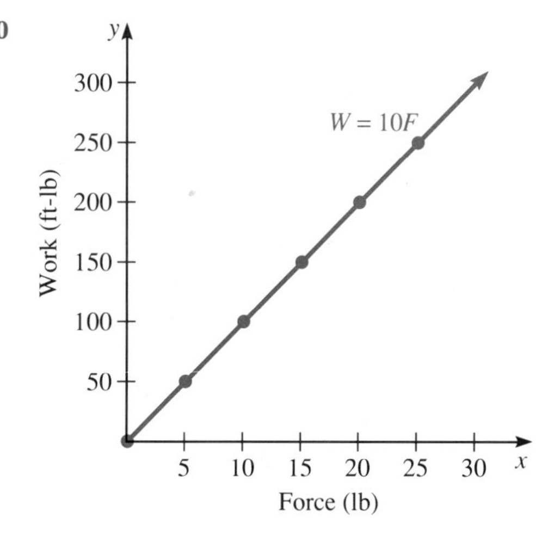

F	W
0	0
5	50
10	100
15	150
20	200
25	250

■

2–3 EXERCISES

Graph the following functions on the rectangular coordinate plane.

1. $4x - y = 8$
2. $3x - 1 = 2y$
3. $3x - y = 5$
4. $2y - 8 = 3x$
5. $5x - 3y = 9$
6. $x - 4y = 12$
7. $y = 7x - 1$
8. $y = 3x + 5$
9. $y = 9 + 2x$
10. $5x - 7y = 10$
11. $2x = 8y - 3$
12. $2x = 3y$
13. $5 - 6x + 9y = 0$
14. $6x - 10y = 2$
15. $y = x^2$
16. $y = 2x^2$
17. $y = x^2 - 3$
18. $y = 4x^2$
19. $y = 3x^2$
20. $y = 2x^2 - 3$
21. $y = 3x^2 - 5$
22. $y = 4x^2 + 3$
23. $y = -2x^2 + x - 3$
24. $y = 3(x^2 - 1)$
25. $y = 2(x^2 + 3)$
26. $y = 4x^2 - 3x + 1$
27. $y = -3x^2 - 5x + 4$
28. $y = -x^2 + 8$
29. $y = \dfrac{3}{x}$
30. $y = \dfrac{6}{x}$
31. $y = 1 - \dfrac{1}{x}$
32. $y = \dfrac{4}{x} - 3$
33. $y = \sqrt{4 - 3x}$
34. $y = \sqrt{5 - 2x}$
35. $y = \sqrt{2x + 1}$
36. $y = \sqrt{4x + 9}$

37. The velocity v (ft/s) of an object thrown upward under the influence of gravity is given by the equation $v = v_0 + at$. If an object is thrown upward with an initial velocity v_0 of 20 ft/s with the acceleration due to gravity a of -32 ft/s^2, the equation becomes $v = 20 - 32t$. Graph velocity as a function of time.

M 38. The normal stress σ (lowercase Greek letter sigma) on a bar is the load P divided by the cross-sectional area a ($\sigma = P/a$). Graph the relationship between stress and load if the cross-sectional area is 10 m^2.

39. Graph temperature in degrees Celsius as a function of temperature in degrees Fahrenheit if $C = (5F - 160)/9$.

C 40. In construction, the area of footing A of a building is the ratio of the total weight W of the building to the bearing capacity B of the soil. If a building weighs 20 tons, graph area as a function of bearing capacity.

M 41. An equation in fluid mechanics is $P = Dhg$ where P is the pressure, D is the density of the liquid, h is the height of the column of liquid, and g is the gravitational acceleration. If $g = 32$ ft/s^2 and $D = 0.8$ slugs/ft^3, graph pressure as a function of height.

2–4 DISTANCE AND SLOPE

In this section we will discuss the distance between two points, the midpoint of a line segment joining two points, and the slope of a line. We will discuss distance first.

Distance

Let us develop the formula for the distance between two points. We take two points P and Q represented by the ordered pairs (x_1, y_1) and (x_2, y_2), respectively. The distance between P and Q is represented by line segment d shown in Figure 2–11. If we draw a line through P parallel to the x axis and another line through Q parallel to the y axis, a right triangle is formed. The lengths of the sides of the triangle are shown in the figure. Since a right triangle is formed, we can use the Pythagorean theorem to solve for d, obtaining

$$d^2 = (x_2 - x_1)^2 + (y_2 - y_1)^2$$
$$d = \sqrt{(x_2 - x_1)^2 + (y_2 - y_1)^2}$$

FIGURE 2–11

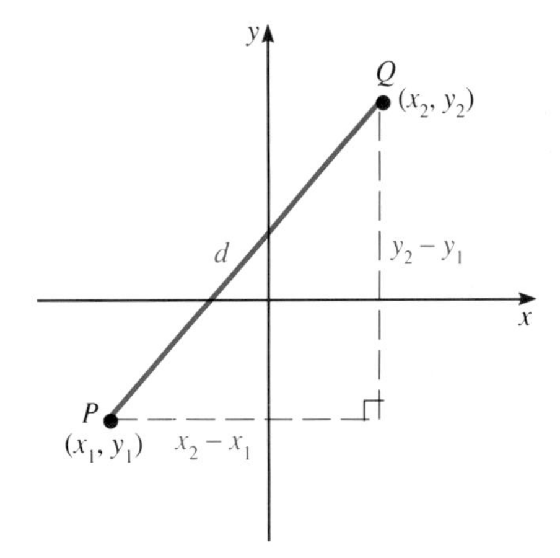

Distance Formula

Given two points represented by the ordered pairs (x_1, y_1) and (x_2, y_2), the distance d between the points is given by

$$d = \sqrt{(x_2 - x_1)^2 + (y_2 - y_1)^2}$$

EXAMPLE 1 Find the distance between $(-6, 2)$ and $(8, -5)$.

Solution First, choose one ordered pair to serve as (x_1, y_1), and the other to serve as (x_2, y_2).

$$\underset{(x_1,\, y_1)}{\underset{\uparrow}{(-6, 2)}} \qquad \underset{(x_2,\, y_2)}{\underset{\uparrow}{(8, -5)}}$$

Then we substitute the appropriate members into the distance formula and simplify the result.

$$d = \sqrt{(x_2 - x_1)^2 + (y_2 - y_1)^2}$$
$$d = \sqrt{(8 - (-6))^2 + (-5 - 2)^2}$$
$$d = \sqrt{196 + 49}$$
$$d = 7\sqrt{5} \approx 15.7$$

The line segment joining these two points is shown in Figure 2–12.

FIGURE 2–12

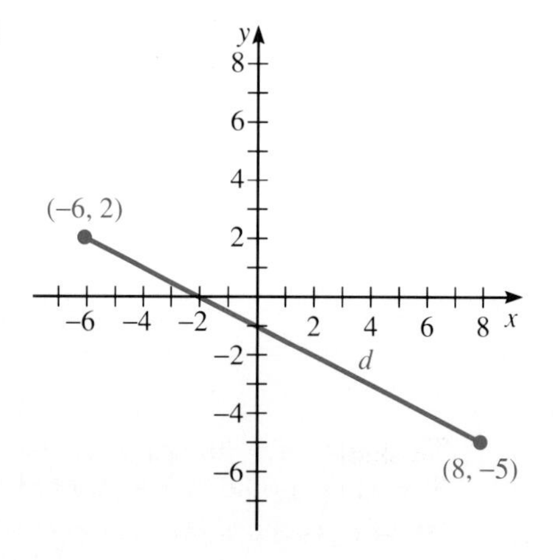

■

Midpoint

The **midpoint** of a line segment is the point halfway between the endpoints. The midpoint is given by the following formula.

Midpoint of a Line Segment

If (x_1, y_1) and (x_2, y_2) represent the endpoints of a line segment, the midpoint is represented by

$$\left(\frac{x_1 + x_2}{2}, \frac{y_1 + y_2}{2}\right)$$

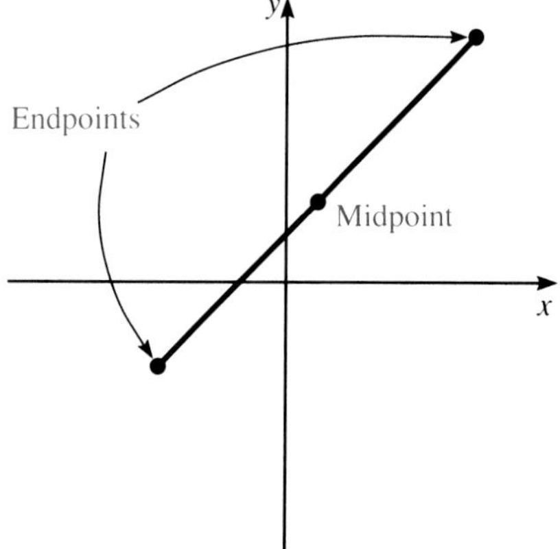

EXAMPLE 2 Find the midpoint of the line segment joining $\left(-5, \frac{3}{4}\right)$ and $\left(\frac{1}{2}, 3\right)$.

Solution First, choose the ordered pairs to serve as (x_1, y_1) and (x_2, y_2).

$$\begin{matrix} (x_1, y_1) & (x_2, y_2) \\ \uparrow & \uparrow \\ \left(-5, \frac{3}{4}\right) & \left(\frac{1}{2}, 3\right) \end{matrix}$$

The midpoint is given by

$$\left(\frac{-5 + \frac{1}{2}}{2}, \frac{\frac{3}{4} + 3}{2}\right) = \left(-\frac{9}{4}, \frac{15}{8}\right)$$ ■

Slope

The **slope** of a line refers to the direction and steepness of the line's slant. Slope can also be expressed as the ratio of rise over run. If we choose any two points (x_1, y_1) and (x_2, y_2) on the given line, and if we construct a right triangle using these points, the rise is the length of the vertical side and the run is the length of the horizontal side of the triangle. The slope formula is given below.

Slope of a Line

Given line L, choose any two points, (x_1, y_1) and (x_2, y_2), on this line. Then the slope m of line L is given by

$$m = \frac{\text{rise}}{\text{run}} = \frac{y_2 - y_1}{x_2 - x_1}$$

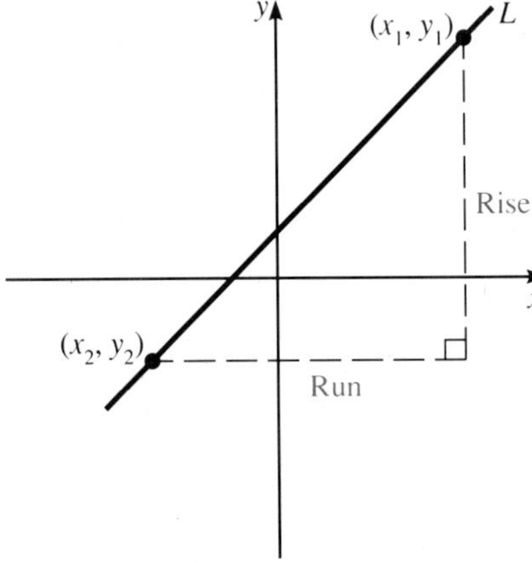

EXAMPLE 3 Find the slope of the line passing through the points (6, −5) and (8, 7).

FIGURE 2–13

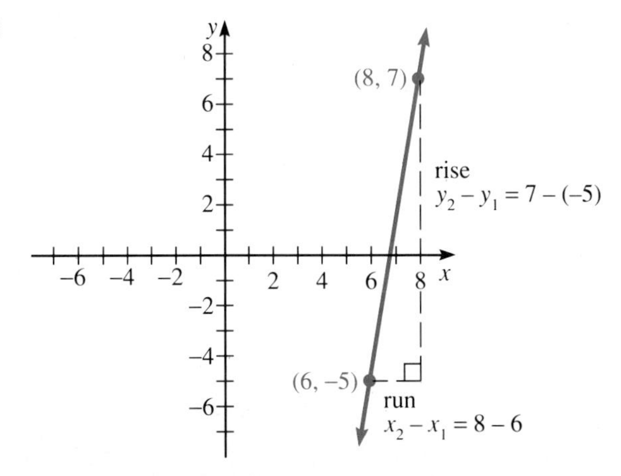

Solution Figure 2–13 shows the line passing through these points, the right triangle that results, and the rise and run. To find the slope of the line, let

$$\begin{array}{cc} (x_1, y_1) & (x_2, y_2) \\ \uparrow & \uparrow \\ (6, -5) & (8, 7) \end{array}$$

Then substitute into the slope formula.

$$m = \frac{y_2 - y_1}{x_2 - x_1}$$

$$m = \frac{7 - (-5)}{8 - 6}$$

$$m = 6$$

■

NOTE ✦ The sign of the slope denotes the direction of slant. A line that slants from lower left to upper right has a positive slope; a line that slants from upper left to lower right has a negative slope.

EXAMPLE 4 Find the slope of the line passing through the following points:

(a) (3, 6) and (3, −4) (b) (−2, 5) and (3, 5)

Solution Let L_1 represent the line through the points in (a) and L_2 represent the line through the points in (b). From Figure 2–14 you can see that L_1 is a vertical line and L_2 is a horizontal line. The slope of L_1 is

$$m = \frac{6 - (-4)}{3 - 3} = \text{undefined}$$

The slope of L_2 is

$$m = \frac{5 - 5}{3 - (-2)} = 0$$

FIGURE 2–14

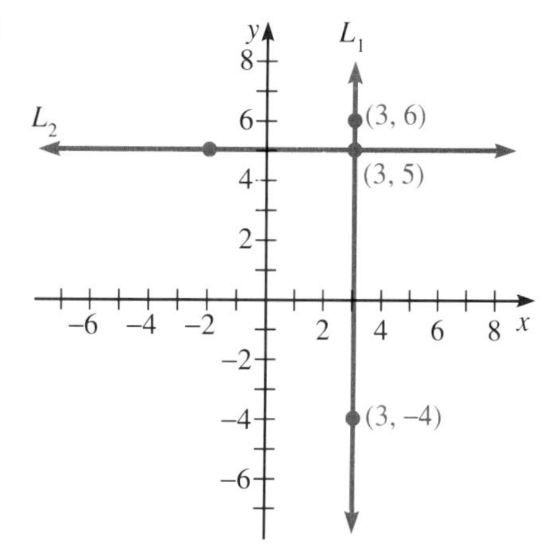

■

NOTE ✦ The slope of any vertical line is undefined, and the slope of any horizontal line is 0.

EXAMPLE 5 Find the slope of the line passing through the following points:

(a) (1, 6) and (3, 2) (b) (2, −9) and (−3, 1)

Solution Let L_1 represent the line through the points in (a) and L_2 represent the line through the points in (b). Figure 2–15 shows that L_1 and L_2 appear to be parallel. Using the slope formula to find the slope of L_1 gives

$$m = \frac{6 - 2}{1 - 3} = \frac{4}{-2} = -2$$

The slope of L_2 is

$$m = \frac{-9 - 1}{2 - (-3)} = \frac{-10}{5} = -2$$

FIGURE 2–15

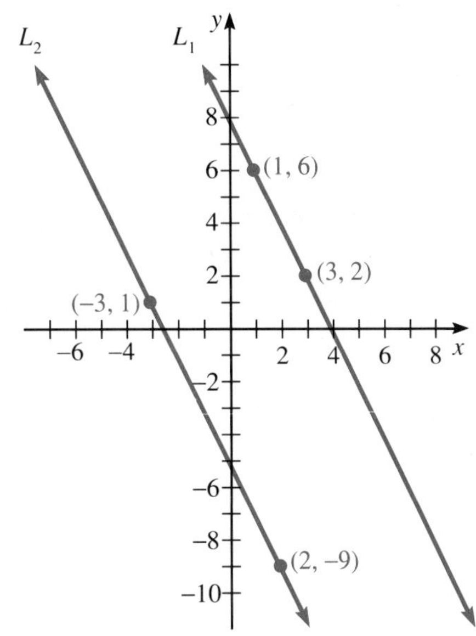

■

NOTE ✦ Parallel lines have the same slope.

EXAMPLE 6 Find the slope of the line passing through the following points:

(a) $(0, -8)$ and $(3, 1)$ (b) $(-3, 3)$ and $(6, 0)$

Solution Again, let L_1 represent the line passing through the points in (a) and L_2 represent the line through the points in (b). Figure 2–16 suggests that lines L_1 and L_2 may be perpendicular. The slope of L_1 is

$$m = \frac{-8 - 1}{0 - 3} = 3$$

The slope of L_2 is

$$m = \frac{3 - 0}{-3 - 6} = -\frac{1}{3}$$

FIGURE 2–16

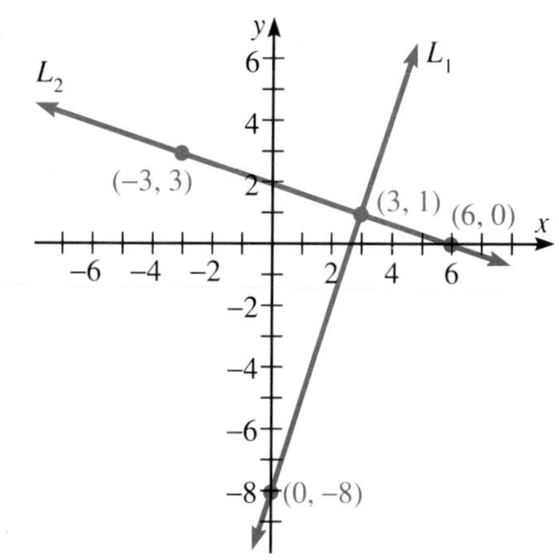

■

NOTE ✦ The slope of perpendicular lines are negative reciprocals of each other. Thus, the product of the slopes of perpendicular lines is -1 (unless they are a vertical line and a horizontal line).

Summary of Slopes

Negative slope

Positive slope

Horizontal line, slope = 0

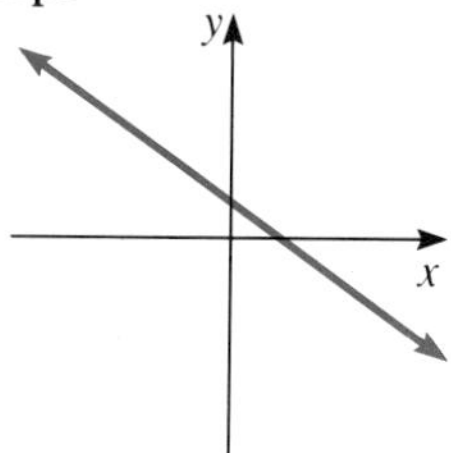

Vertical line, slope = undefined

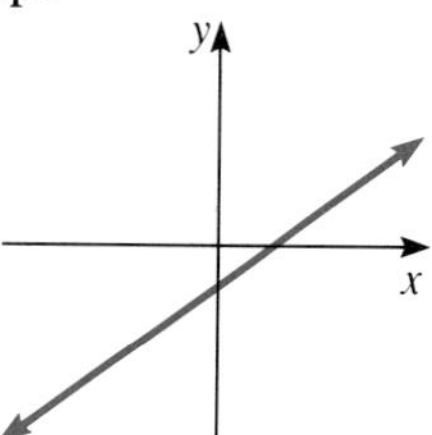

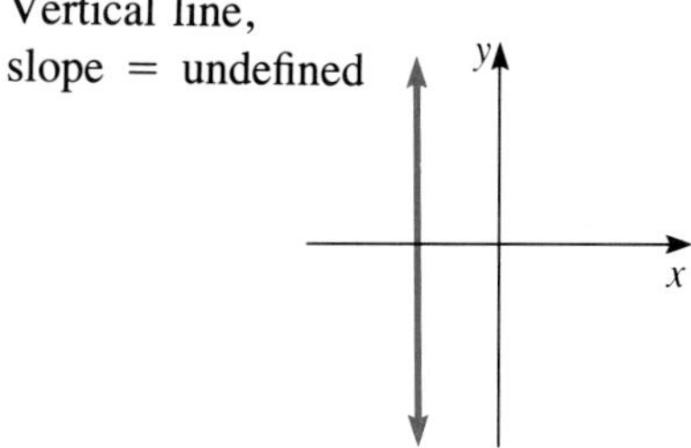

Parallel lines, same slope

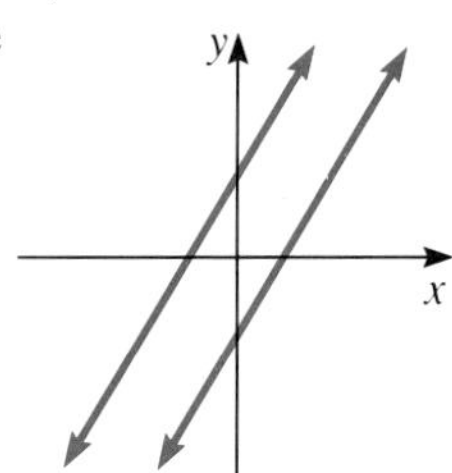

Perpendicular lines, negative reciprocal slopes

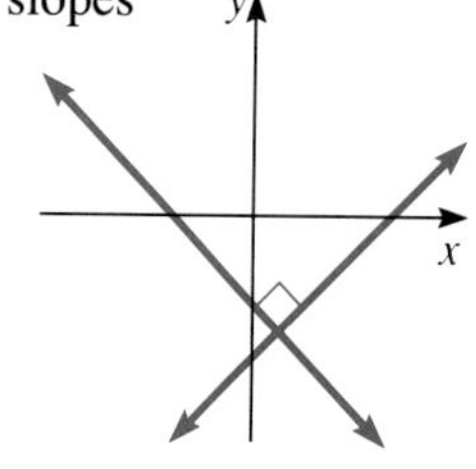

EXAMPLE 7 Show that the following represent vertices of a right triangle:

$$C: (-2, 0) \quad A: (-4, 2) \quad B: (1, 3)$$

Solution Figure 2–17 shows the triangle ABC formed by joining points A, B, and C. To prove that ABC is a right triangle, we show that the slope of CB is the negative reciprocal of the slope of AC.

$$CB = \frac{3 - 0}{1 - (-2)} = \frac{3 - 0}{1 + 2} = 1$$

$$AC = \frac{2 - 0}{-4 - (-2)} = \frac{2 - 0}{-4 + 2} = -1$$

Since the slopes are negative reciprocals, ABC is a right triangle.

FIGURE 2–17

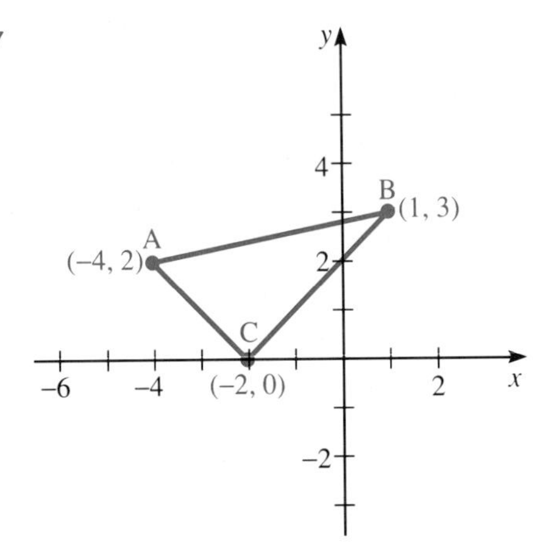

■

2–4 EXERCISES

Find the distance between the following ordered pairs.

1. $(-2, 8)$ and $(3, 1)$

2. $(-6, -4)$ and $(3, -7)$

3. $(6, -5)$ and $(3, -8)$

4. $(5, 12)$ and $(-7, -8)$

5. $\left(\frac{1}{3}, \frac{3}{5}\right)$ and $(2, 4)$

6. $\left(-\frac{1}{2}, \frac{3}{4}\right)$ and $(3, -5)$

7. $(2, 6)$ and $(-1, 6)$

8. $(-7, 10)$ and $(-7, -4)$

9. $(-5, -2)$ and $(-5, 4)$

10. $(3, -9)$ and $(-2, -9)$

Find the midpoint of the line segment joining the given ordered pairs.

11. $(-2, 0)$ and $(6, -3)$

12. $(0, 5)$ and $(-3, 10)$

13. $(3, 7)$ and $(-8, 5)$

14. $(-2, -7)$ and $(-8, -3)$

15. $(1.7, -5)$ and $(3, -2.3)$

16. $(3.2, -1.8)$ and $(6.7, 8.1)$

17. $\left(\frac{2}{3}, -10\right)$ and $(3, 5)$

18. $\left(\frac{3}{7}, 2\right)$ and $\left(-3, \frac{1}{3}\right)$

19. $\left(\frac{1}{6}, \frac{7}{8}\right)$ and $\left(\frac{15}{2}, \frac{11}{4}\right)$

20. $\left(\frac{3}{4}, -\frac{8}{3}\right)$ and $\left(\frac{15}{2}, -\frac{3}{2}\right)$

Find the slope of the line passing through the given points.

21. $(-7, 5)$ and $(4, -8)$ **22.** $(-3, 11)$ and $(7, -8)$

23. $(4, -2)$ and $(4, 10)$ **24.** $(9, 6)$ and $(12, 5)$

25. $(-3, -10)$ and $(-1, 5)$

26. $(7, -0.5)$ and $(-8, -0.5)$

27. $(-6.1, 8)$ and $(3.4, 8)$ **28.** $(-14, 3)$ and $(6, -9)$

29. $(14, -8)$ and $(-7, 4)$ **30.** $(-1, 9)$ and $(-1, 0)$

Determine if the lines passing through the given points are parallel, perpendicular, or neither.

31. L_1: $(-1, 8)$ and $(3, -1)$
L_2: $(-2, -3)$ and $(1, -9)$

32. L_1: $(-3, 6)$ and $(4, -7)$
L_2: $(2, -2)$ and $(-1, 8)$

33. L_1: $(4, 4)$ and $(-8, -5)$
L_2: $(-3, -1)$ and $(0, -5)$

34. L_1: $(-2, 13)$ and $(3, -2)$
L_2: $(0, -1)$ and $(-2, 5)$

35. L_1 $(-6, 0)$ and $(4, 3)$
L_2: $(5, 4)$ and $(-3, -2)$

36. Show that $(0.5, 5)$, $(-2, -2)$, and $(3, -2)$ represent the vertices of an isosceles triangle.

37. Show that $(0, 2)$, $(4, -2)$, and $(3, -1)$ represent the vertices of a right triangle.

38. Show that $(-2, 3)$, $(2, -2)$, $(4, 3)$, and $(-4, -2)$ represent the vertices of a parallelogram.

39. Show that $(-1, 5)$, $(2, 2)$, $(-1, 2)$, and $(2, 5)$ represent the vertices of a square.

2–5

GRAPHING AN INEQUALITY

In Chapter 1, we discussed solving linear inequalities in one variable. In this section we will discuss solving inequalities in two variables. Recall from Chapter 1 that the solution to an inequality in one variable may be represented on a number line. On the other hand, the solution to an inequality in two variables must be represented on the rectangular coordinated plane. Just as an equation in two variables may have an infinite number of ordered-pair solutions, so may an inequality in two variables. The graphical solution to an equation may be represented by a line or curve, but the solution to an inequality could be a region of the plane. The method used to graph an inequality is developed in the next example.

Inequalities in Two Variables

EXAMPLE 1 Graph the following inequality:

$$2x - 5y \geq 10$$

Solution We begin by solving the inequality for the dependent variable.

$$\begin{aligned} 2x - 5y &\geq 10 \\ -5y &\geq 10 - 2x \\ y &\leq \frac{-10 + 2x}{5} \end{aligned}$$

Next, we obtain the boundary of the solution set by graphing the equation

$$y = \frac{-10 + 2x}{5}$$

The table of values for this equation is given below. We plot the points and draw a line through them. The solution to this linear inequality is a half-plane bounded by this line. To determine which half-plane to shade, we choose a test point (which may be any point not on the boundary). Let us use the point $(7, -4)$. Substituting $(7, -4)$ into the inequality gives

$$\begin{aligned} 2(7) - 5(-4) &\geq 10 \\ 14 + 20 &\geq 10 \\ 34 &\geq 10 \qquad \text{(true)} \end{aligned}$$

Since this is a true statement, shade the half of the plane containing the test point—the lower half in this case. Since the inequality includes the ''equal to'' symbol, the line is part of the solution set. The graph of $2x - 5y \geq 10$ is shown in Figure 2–18.

FIGURE 2–18

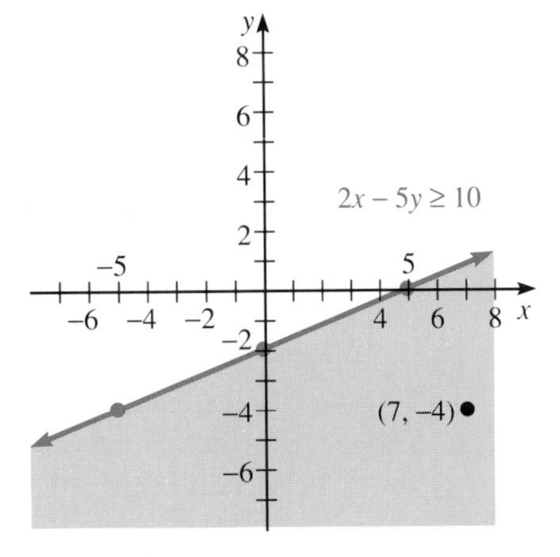

x	y
-5	-4
0	-2
5	0

■

Graphing an Inequality

1. Solve the inequality for the dependent variable. Transpose the dependent variable to the left side of the inequality.
2. Replace the inequality symbol with the equals symbol, and graph the resulting equation.
3. Join the points with a dotted curve if the inequality is less than or greater than, or with a solid curve if the inequality contains the equals symbol.
4. Shade the graph by choosing a test point and substituting it into the original inequality. If the inequality is true, shade the portion of the plane containing the test point. Otherwise, shade the other portion of the plane.

EXAMPLE 2 Graph $y < x^2 - 3x + 2$.

Solution Since the inequality is solved for y, we graph the equation $y = x^2 - 3x + 2$. The table of values is given below. Notice in the table of values that it was necessary to choose an x value between 1 and 2 because they resulted in the same y value. After connecting the points with a smooth, dotted curve, we choose a test point (0, 0) and substitute it into the original inequality.

$$0 < (0)^2 - 3(0) + 2$$
$$0 < 2 \qquad \text{(true)}$$

Therefore, we shade the portion of the plane containing (0, 0). The graph is shown in Figure 2–19. The curve is not part of the solution set.

FIGURE 2–19

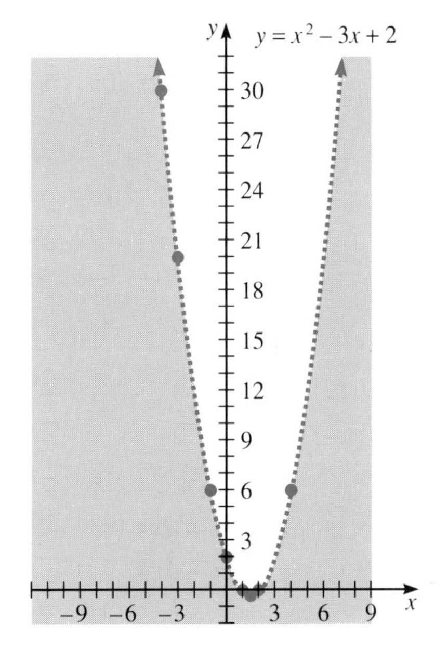

x	y
−4	30
−3	20
−1	6
0	2
1	0
2	0
4	6
1.5	−1/4

■

CAUTION ✦ Always check your solution by substituting a point from the shaded region. One of the most frequent errors in graphing inequalities is shading the wrong area.

Systems of Inequalities

In a later chapter, we will discuss solving systems of equations using both graphical and algebraic techniques. However, we will discuss solving systems of inequalities in this section because it uses the techniques we have just developed.

A **system of linear inequalities** contains more than one linear inequality in two variables. The solution to a system of linear inequalities is the region of the plane contained in the solution set of both inequalities. The remainder of this section is devoted to solving systems of linear inequalities.

EXAMPLE 3 Find the solution to the following system of inequalities:

$$x - 5y > 12$$
$$3x + 4y \leq 6$$

Solution We graph each inequality using the same methods used in the previous examples. The table of values is given below, and the graph is given in Figure 2–20. Part (a) of the figure shows the solution to the inequality $x - 5y > 12$; Part (b) shows the solution to $3x + 4y \leq 6$; and Part (c) shows the area common to both graphs. We can check the solution by choosing a point in the shaded region and substituting into both inequalities. Substituting $(3, -5)$ into each inequality gives

$$\begin{aligned} x - 5y &> 12 \\ 3 - 5(-5) &> 12 \\ 28 &> 12 \quad \text{(true)} \end{aligned} \qquad \begin{aligned} 3x + 4y &\leq 6 \\ 3(3) + 4(-5) &\leq 6 \\ -11 &\leq 6 \quad \text{(true)} \end{aligned}$$

$x - 5y = 12$	
x	y
-2	$-14/5$
0	$-12/5$
1	$-11/5$

$3x + 4y = 6$	
x	y
-2	3
0	3/2
2	0

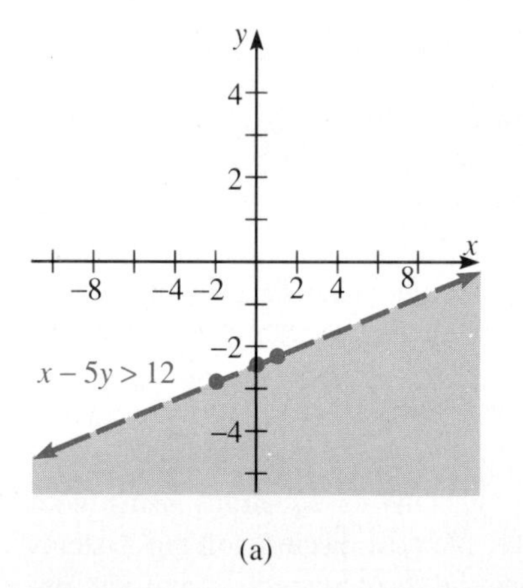

(a)

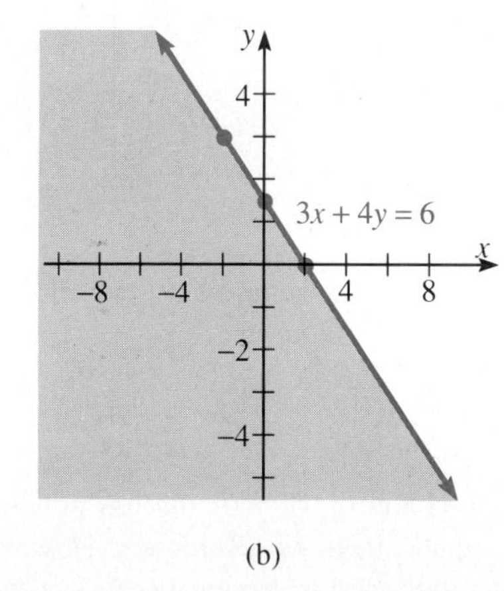

(b)

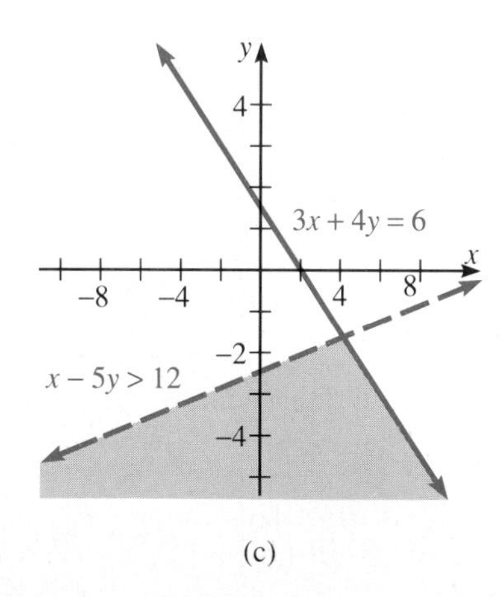

(c)

FIGURE 2–20

■

EXAMPLE 4 The manager of a donut shop pays \$1.10 a dozen for cinnamon donuts and \$1.40 a dozen for chocolate donuts. The manager wants to stock at least 52 dozen donuts but cannot spend more than \$120. Find the solution set.

Solution We must set up a system of inequalities to solve this problem. Let x represent the number of dozen of cinnamon donuts and y represent the number of dozen of chocolate donuts. If the manager wants to have at least 52 dozen donuts, then

$$x + y \geq 52$$

If the manager can spend no more than \$120 for the donuts, then

$$1.10x + 1.40y \leq 120$$

The system of inequalities to be solved graphically is

$$x + y \geq 52$$
$$1.10x + 1.40y \leq 120$$

The table of values for each equation is given below. Notice that negative values for x and y are not reasonable in this case. The solution set is shown in Figure 2–21. A scale of 5 was used for each axis. From the graph, if the manager chooses any ordered pair in the shaded region, he will satisfy the constraints of both stock and money. For example, the ordered pair (30, 50) lies in the shaded region.

$$30 + 50 > 52 \quad \text{and} \quad 33 + 70 \leq 120$$

Therefore, the manager could buy 30 dozen cinnamon donuts and 50 dozen chocolate donuts. He could also choose any other combination of dozens of donuts from the shaded region.

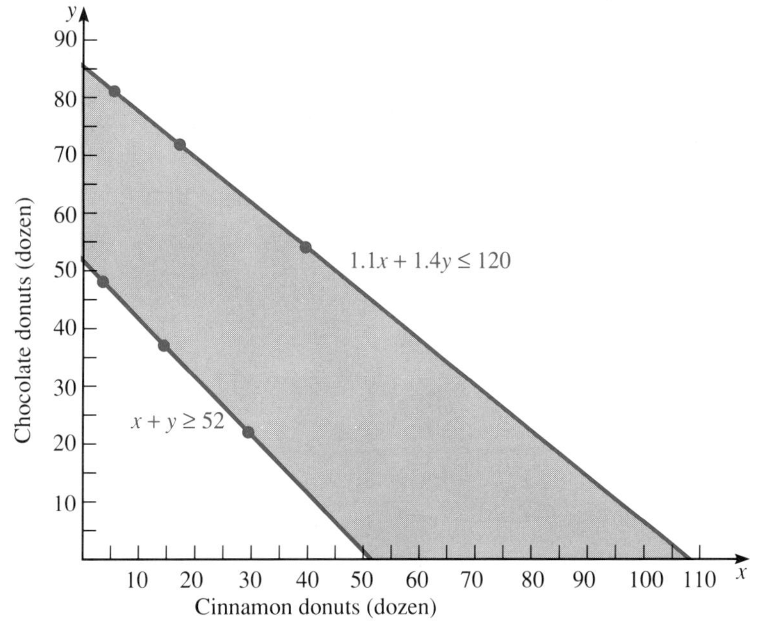

$x + y = 52$	
x	y
4	48
15	37
30	22

$1.1x + 1.4y = 120$	
x	y
6	81
18	72
40	54

FIGURE 2–21

■

Linear Programming

One of the most important applications of systems of inequalities is linear programming. Many problems in business and industry require the most cost-effective solution. **Linear programming** uses systems of inequalities to maximize or minimize such quantities as cost, profit, and allocation of raw materials subject to certain conditions. The system of inequalities is often large and complicated and is usually solved using the simplex algorithm implemented on a computer. However, in this section we will limit the discussion to the graphical solution of simple systems.

From previous experience a business derives an equation representing profit and a set of inequalities representing the allocation of raw materials. After deriving the system of inequalities, we plot each inequality on the same rectangular coordinate system and find the solution set, which should be a polygon. Then we find the coordinates of the vertices of this polygon from the graph because the solution to a linear programming problem is always found at one of these vertices. Last, we find the solution to the linear programming problem by substituting the coordinates of the polygon vertices into the profit or cost equation to find the maximum or minimum, respectively.

EXAMPLE 5 A small computer company manufactures two models of a computer, one for home use and one for business use. In the manufacturing process, a certain number of man-hours are allocated to assembling the computer components, and a certain number of man-hours are allocated to installing the computer into its case. The home computer requires 1 man-hour to be assembled and 2 man-hours to be put into its case. The business computer requires 3 man-hours to be assembled and 1 man-hour to be put into its case. Workers have a maximum of 9 man-hours per day for assembling the computer and a maximum of 8 man-hours per day for installing the computer into its case. If the company makes a profit of \$100 on each home computer and \$150 on each business computer, how many of each computer should the company manufacture to maximize its profit?

Solution The information on the assembly and installation for the home and business computers is summarized in the following table:

	Assembly (hours)	Installation (hours)
Home	1	2
Business	3	1
Maximum	9	8

The first task is to transfer the verbal statement into a series of equations and inequalities. Let

x represent the number of business computers manufactured

y represent the number of home computers manufactured

Since negative numbers are not valid for either x or y, then $x \geq 0$ and $y \geq 0$. The restriction on production relates to the number of hours available for assembling the computer components and then installing the computer into its case. These restrictions translate into the following inequalities:

$$3x + y \leq 9 \quad \text{(assembling components)}$$
$$x + 2y \leq 8 \quad \text{(installing computer in its case)}$$

We are attempting to maximize the profit represented by

$$P = 150x + 100y$$

In order to solve this problem, we find the point of intersection of the two lines. The region of solution is shown in Figure 2–22. From the graph we can determine that the coordinates for the vertices of the polygon are (0, 0), (0, 4), (2, 3), and (3, 0). Substituting each point into the profit equation gives

$$P = 150x + 100y$$
$$P = 150(0) + 100(0) = 0$$
$$P = 150(0) + 100(4) = 400$$
$$P = 150(2) + 100(3) = 600$$
$$P = 150(3) + 100(0) = 450$$

Since the point (2, 3) results in the largest profit, it is the solution to the linear programming problem. The computer company should produce 2 business computers and 3 home computers per day for a profit of $600.

FIGURE 2–22

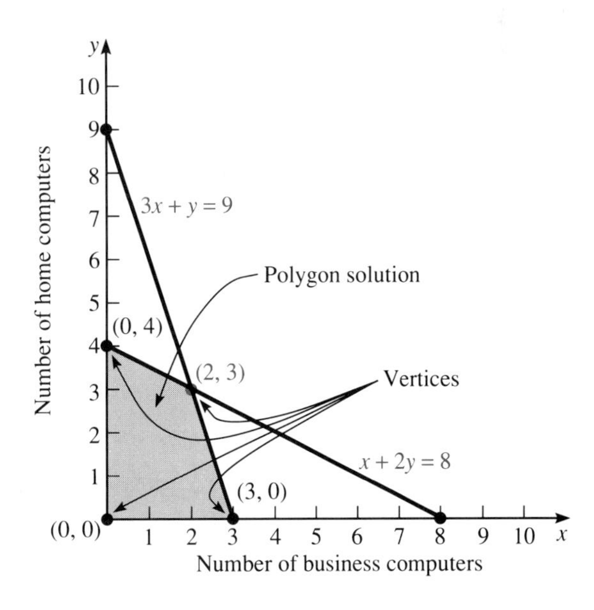

2–5 EXERCISES

Graph each of the following inequalities on the rectangular coordinate system.

1. $y > 5x - 6$ **2.** $y < 2x + 1$

3. $2x - y \le 3$ **4.** $y \ge 3(x - 1) + 6$

5. $x \ge -5$ **6.** $4x - 5y \le 8$

7. $2x < 5y + 6$ **8.** $y \le -3$

9. $3y \ge 4x - 9$ **10.** $7x \le 2y - 3$

11. $y < 1$ **12.** $x > -6$

13. $y \le 2x^2 - 5$ **14.** $y > 3x^2 - 7$

15. $y > -x^2 + 4$ **16.** $y \le -x^2 - 3$

17. $y \ge -3x^2 + 1$ **18.** $y < -2x^2 + 5$

19. $y < 4x^2 - 7$ **20.** $y \ge x^2 + 2$

21. $y > x^2 - 10$ **22.** $y > 3x^2 - 8$

Solve the following systems of inequalities.

23. $x \ge y$, $2x + y > 3$

24. $5x - y > 10$, $x - 3y > 4$

25. $2x + y < 6$, $x - y > 1$

26. $4x \le 3y$, $x \ge 2y + 1$

27. $y \ge 3x - 5$, $x < -3$ **28.** $y \le 4$, $x > 5y - 4$

29. Given the inequalities $2x - 3y \le 4$, $x + 4y \le 20$, $x \ge 0$, $y \ge 0$, graph these inequalities, find the corner points, and find the point that gives the maximum profit if the profit is given by $P = 25x + 16y$.

I 30. If the profit for a company is given by $P = 30x + 85y$, find the values for x and y that give the maximum profit subject to the following constraints: $x \ge 0$, $y \ge 0$, $x + 2y \le 11$, and $7x + y \le 12$.

I 31. A company produces two kinds of sewing machines, an economy model and a deluxe model. The economy model requires 1 man-hour to be assembled and 3 man-hours to be put into its case. The deluxe machine requires 2 man-hours to be assembled and 4 man-hours to be put into its case. Workers have a maximum of 15 hours per day to assemble the machines and 24 hours per day to put the machines into their cases. If the company makes a profit of \$50 on the economy machine and \$110 on the deluxe machine, how many of each sewing machine should the company make to maximize profit?

I 32. A watchmaker makes two types of watches. One watch requires 1 man-hour to prepare the case and 4 man-hours to assemble the parts. The second watch requires 1 man-hour to prepare the case and 2 man-hours to assemble the parts. She has a maximum of 5 hours to prepare the cases and 12 hours to assemble the parts. If she makes a profit of \$60 on the first watch and \$40 on the second watch, how many of each watch should she make to maximize profit?

I 33. A company manufactures two types of integrated circuits. The first integrated circuit requires 2 hours to manufacture and 1.25 hours to test per thousand. The second type requires 1.5 hours to manufacture and 1.75 hours to test per thousand. The company has a maximum of 7 hours to manufacture the integrated circuits and a maximum of 7.875 hours to test them. It makes a profit of \$2 on the first integrated circuit and \$2.50 on the second one. How many (in thousands) should the company manufacture to maximize its profit?

I 34. KLM Corporation manufactures two circuit analyzers. The first analyzer requires 28 man-hours to design and 18 man-hours to produce. The second analyzer requires 16 man-hours to design and 12 man-hours to produce. The company has a maximum of 8,436 hours to design the analyzers and 5,622 hours to produce them. It makes a profit of \$75 on the first analyzer and \$125 on the second analyzer. How many of each should the company manufacture?

I 35. ABC Company manufactures two portable computers. The first computer requires 265 man-hours to design and 170 man-hours to manufacture. The second computer requires 190 man-hours to design and 150 man-hours to manufacture. The company has 2,845 hours available for design and 2,050 hours available for manufacture. It makes a profit of \$200 on the first computer and \$350 on the second computer. How many of each computer should the company manufacture?

2–6

APPLICATIONS

In this section we will discuss several applications of graphing. Although graphs are not as accurate as algebraic solutions, we can use them to solve equations. We can also use them to determine the break-even point, that is, the point at which a business neither makes a profit nor suffers a loss on its product. Another application is in graphing data gained from an experiment or observations. For example, if you stand on a street corner and count the number of cars passing every hour, you can graph these data and use them to develop a traffic plan. Each of these applications will be discussed in this section.

Solving Equations Graphically

The **solution** or **root** of an equation is the x coordinate of the point where the graph intersects the x axis. This point is also called the $\boldsymbol{x}$ **intercept** of the graph. Since the equation is represented as a function, this same point is also called a **zero** of the function. To determine the solution to an equation or the zero of a function, we must find the x intercept. If the graph does not touch the x axis, the equation does not have a real number solution.

To Solve an Equation Graphically

1. Collect all terms on one side of the equation.
2. Write the equation as a function, equal to y or $f(x)$.
3. Graph the equation.
4. The point(s) where the graph intersects the x axis is the solution to the equation.

EXAMPLE 1 Solve $4x = 3(x - 1) + 6$ graphically.

Solution First, collect all terms on one side of the equation. Transposing all terms to the left side of the equation gives

$$4x - 3(x - 1) - 6 = 0$$

Setting this equation equal to y or $f(x)$ and simplifying gives

$$\begin{aligned} y = f(x) &= 4x - 3(x - 1) - 6 \\ &= x - 3 \end{aligned}$$

Next, we graph the equation using a table of values. The table of values is given below, and the resulting graph is shown in Figure 2–23. Note that the graph intersects the x axis at 3. It may be necessary to estimate the solution from the graph if the solution is not an integer. Thus, the root or solution to the equation is $x = 3$. We can verify this solution by solving the linear equation algebraically.

$$
\begin{aligned}
4x &= 3(x - 1) + 6 && \\
4x &= 3x - 3 + 6 && \text{eliminate parentheses} \\
4x &= 3x + 3 && \text{combine like terms} \\
4x - 3x &= 3 && \text{transpose and combine like terms} \\
x &= 3 && \text{divide}
\end{aligned}
$$

FIGURE 2–23

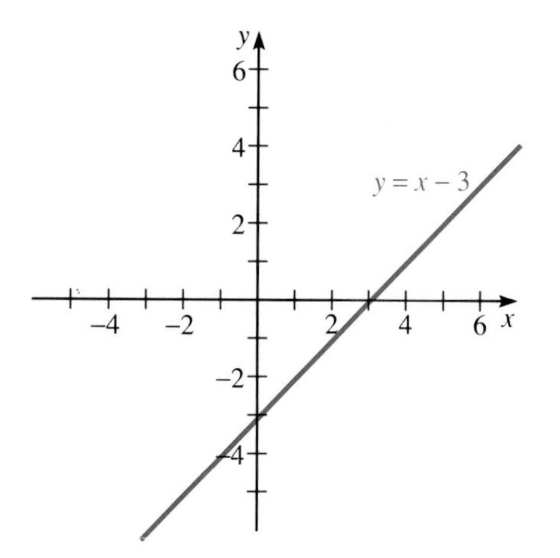

x	y
−1	−4
0	−3
2	−1

■

EXAMPLE 2 Solve $x^2 + x = 12$ graphically.

Solution First, we transpose all terms in the original equation to the left and set the result equal to y.

$$y = x^2 + x - 12$$

Then we fill in the table of values given below. As discussed earlier in this chapter, a quadratic equation for the form $y = ax^2 + bx + c$ results in a parabola. To graph the parabola, we must determine the coordinates of the vertex. They are given by $(-b/2a, f(-b/2a))$. For this equation, $y = x^2 + x - 12$, $a = 1$, and $b = 1$. Therefore, the coordinates of the vertex are

$$\frac{-b}{2a} = -\frac{1}{2}$$

$$f\left(\frac{-b}{2a}\right) = f\left(\frac{-1}{2}\right) = -12\frac{1}{4}$$

The vertex is $(-1/2, -12\frac{1}{4})$. The graph is given in Figure 2–24. From the graph, the zeros or solutions are $x = -4$ and $x = 3$. If the solution is not an integer, estimate the solution from the graph.

FIGURE 2–24

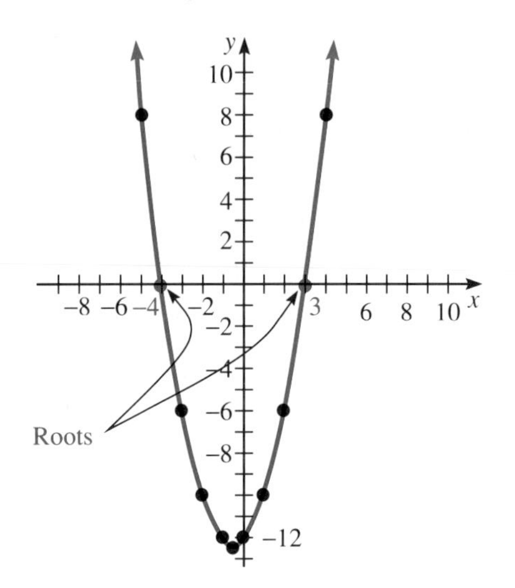

x	y
−3	−6
−2	−10
−1	−12
0	−12
1	−10
2	−6
3	0
4	8

■

Break-Even Point

In business the **break-even point** is the point in production where total cost equals total revenue. It occurs when the company neither makes a profit nor suffers a loss. Using principles of business, we can determine one equation to represent the total cost and a second equation to represent total revenue. Graphically, the point of intersection of the cost equation and the revenue equation represents the break-even quantity and the break-even revenue. Thus, in linear equations, any production above the break-even quantity usually results in a profit, while any production less than the break-even quantity usually results in a loss.

To Determine the Break-Even Point

1. Determine the equation for total cost and the equation for total revenue using x to represent the number of units to be manufactured and y to represent the total cost.
2. Graph each equation and find the point of intersection.
3. The x coordinate of the break-even point denotes the number of units of the product to manufacture, and the y coordinate denotes the resulting revenue.

EXAMPLE 3 A manufacturer sells his product for $5 per unit and sells all that he produces. The fixed cost of production is $3,500 per month, and the variable cost per unit is $1.50. Find the break-even quantity and revenue per month.

Solution We begin by determining equations to represent total cost and total revenue. Let y represent total revenue and total cost (in dollars), and let x represent the number of units of the product sold. The principles of business state that

$$\text{total cost} = \text{fixed cost} + \text{variable cost}$$
$$y = 3{,}500 + 1.50x$$

and

$$\text{total revenue} = (\text{price/unit})(\text{number of units sold})$$
$$y = 5x$$

We graph the equations $y = 5x$ and $y = 3{,}500 + 1.50x$ as shown in Figure 2–25. Locating the break-even point from the graph, we find that the coordinates are (1,000, 5,000). Therefore, the company should produce 1,000 units to break even and would receive and spend a revenue of $5,000.

FIGURE 2–25

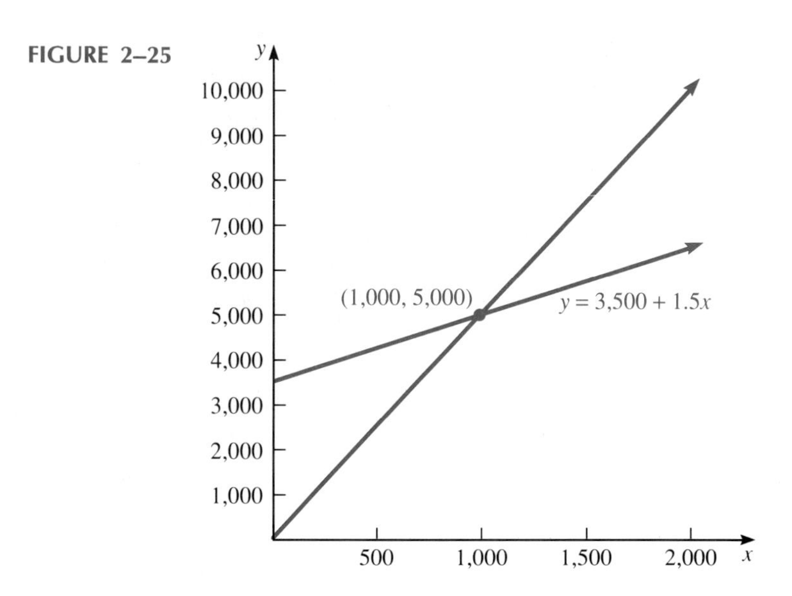

■

Empirical Data

Graphing is also used in analyzing empirical data, or data that are obtained by experiment or observation. Since measurements in a laboratory are subject to human and experimental errors, the data may not follow a functional equation. Therefore, in graphing empirical data, we graph the points and try to connect them with a smooth curve. However, all of the points may not fit on the curve. Therefore, we average the curve through the data points to obtain the **line of best fit.**

Graphing Empirical Data

1. **1.** Make a table of values from the data, if necessary.
2. **2.** Determine a scale for each axis by scanning the range of values.
3. **3.** Construct and label the axes with the scale and the units of measure.
4. **4.** Plot the points. Draw a circle around each point.
5. **5.** Draw a smooth curve by averaging the plotted points. To average the plotted points, draw the curve so that it passes midway between two points such that the vertical distance from the curve to points above and below the curve are equal. Figure 2–26 in the following example illustrates.

EXAMPLE 4 The following empirical data were gathered concerning the pressure on the bottom of a cylindrical tank exerted by a certain force:

Force (lb)	0	5	12	25	42	65
Pressure (lb/in^2)	0	0.4	1.5	2.7	4	6.5

Solution The values for force range from 0 to 65, while the values for pressure range from 0 to 6.5. The independent variable, force, is placed on the horizontal axis in increments of 5 lb. The dependent variable, pressure, is placed on the vertical axis in increments of 0.5 lb/in^2. We draw a line through the data points so that the distance from the line to data points above and below the line is equal. The graph is shown in Figure 2–26. The portion of the line of best fit beyond the data set given is denoted by a dashed line.

FIGURE 2–26

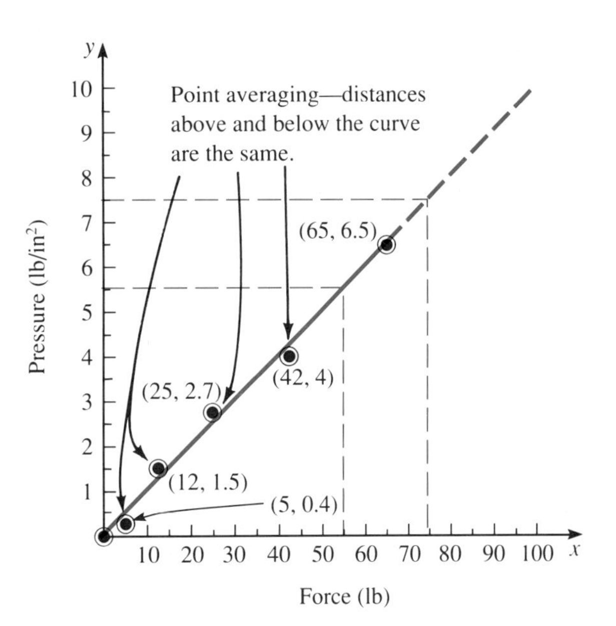

You can use a graph to determine additional information about a function. The process of estimating values between data points on a graph is called **interpolation.** In Example 4, we can estimate the pressure exerted by a particular force by locating the force on the horizontal axis, moving vertically to the graph and horizontally across to the pressure axis, and then reading the value. For example, a 55-lb force results in a 5.5 lb/in^2 pressure as read from the graph (see the dotted line in Figure 2–26). **Extrapolation** is the process of estimating the value of a function beyond the known data. We read data from the graph in exactly the same manner as for interpolation. For example, a 75-lb force results in a pressure of 7.5 lb/in^2 as read from the graph (see the dotted line in Figure 2–26). Both interpolation and extrapolation should be used cautiously because they both depend on the accuracy of the data and the graph.

2–6 EXERCISES

Find the solution to the following equations graphically.

1. $6(x - 2) + 3 = 2x - 1$

2. $3x - 5 + 2x = 8x + 10$

3. $7x - 3 + 4(x - 1) = 4$

4. $4x - (x - 5) = 2x + 7$

5. $2x^2 + 10x + 8 = 0$

6. $(9 - x) + 3x + 5 = 2(x - 3) + x - 10$

7. $3x^2 = 3x + 6$

8. $3x^2 - 3x = 18$

9. $x^2 - x = 20$

10. $x^2 + 3x = -2$

11. $-3x + 10 = 6$

12. $2x + 14 = 6$

13. $4x + (8 - 2x) = 3(x + 1)$

14. $5x - (x + 9) = 3x + 19$

15. $x^2 + 6x = -5$

16. $x^2 = 3x + 18$

I **17.** A manufacturer sells his product for \$3 per unit and sells all that he produces. The fixed cost of production is \$8,000, and the variable cost is \$1.00 per unit. Find the break-even point.

I **18.** A manufacturer sells his product for \$8 per unit and sells all that he produces. The fixed cost of production is \$12,000, and the variable cost is \$2 per unit. Find the break-even quantity and revenue.

I **19.** A manufacturer sells his product for \$14 per unit, selling all that he produces. If the fixed cost of production is \$18,000 and the variable cost is \$8 per unit, find the break-even quantity.

I **20.** A manufacturer sells his product for \$3 per unit and has fixed costs of \$2,000 and a variable cost of \$1 per unit. Under normal conditions, would the manufacturer make a profit or loss if he produces 1,200 units?

I **21.** A manufacturer sells his product for \$12 per unit and has fixed costs of \$15,000 and variable costs of \$7 per unit. Does the manufacturer make a profit if he produces 1,000 units? Why or why not?

E **22.** Graph the following empirical data relating the voltage and current of a small light bulb. Use the graph to approximate the current at 0.9 V.

Voltage (V)	0.1	0.4	0.7	1	2	3
Current (mA)	30	50	60	70	100	135

23. Graph the relationship between the time of day and the corresponding temperature (in degrees Fahrenheit).

Time	6:00 A.M.	9:00	10:00	11:00	1:00 P.M.	3:00	4:00	7:00	10:00
Temp. (°F)	35	40	43	50	54	67	65	58	50

E 24. Graph the following empirical data relating the voltage and current of a light-emitting diode (LED):

Voltage (V)	1.4	1.43	1.48	1.5	1.52	1.55	1.6	1.64
Current (mA)	0.01	0.08	0.4	0.6	1.2	3	10	30

25. Graph the relationship between time and the distance an object travels. Determine the distance of an object at 4.5 s.

Time (s)	1	2	3	4	5	6	7	8
Distance (m)	1.2	4.9	11.03	19.6	30.63	44.1	60.03	78.4

26. Graph the relationship between mass and weight.

Mass (slugs)	2	3	5	6	9	10	13	14
Weight (lb)	64	96	160	192	288	320	416	448

E 27. Graph the following empirical data relating the voltage and current of a fast switching diode. Approximate the voltage of a current of 4 mA.

Voltage (V)	0.4	0.5	0.53	0.64	0.69	0.75	0.79
Current (mA)	0.01	0.05	0.08	0.5	1.2	3	6

I 28. In business, demand (amount purchased) for a product depends on the price of the product. Graph the following data and determine the price at a demand of 27.

Price	70	60	40	30	20	8
Demand	5	10	20	24	30	36

E 29. Graph the following empirical data relating the voltage and current of a diode. Estimate the current for a voltage of 0.80 V.

Voltage (V)	0.2	0.27	0.33	0.39	0.44	0.48	0.52	0.58	0.68	0.72
Current (mA)	0.1	0.7	3	8	15	25	40	70	150	200

30. Using the following data, graph the relationship between the radius and area of a circle.

Radius (ft)	1	2	3	4	5	6
Area (ft^2)	3.14	12.56	28.3	50.3	78.5	113.1

CHAPTER SUMMARY

Summary of Terms

asymptote (p. 55)
axes (p. 47)
break-even point (p. 75)
Cartesian coordinate system (p. 47)
dependent variable (p. 42)
domain (p. 45)
extrapolation (p. 78)
function (p. 42)
functional notation (p. 42)
graph (p. 51)
hyperbola (p. 55)
independent variable (p. 42)
interpolation (p. 78)
linear equation in two variables (p. 51)
linear programming (p. 70)
line of best fit (p. 76)
midpoint (p. 59)
one-to-one correspondence (p. 48)
origin (p. 47)
parabola (p. 53)
quadrants (p. 47)
range (p. 45)
rectangular coordinate system (p. 47)
root (p. 73)
slope (p. 59)
solution (p. 73)
system of linear inequalities (p. 67)
vertex (p. 53)
x axis (p. 47)
x coordinate (p. 48)
x intercept (p. 73)
y axis (p. 47)
y coordinate (p. 48)
zero (p. 73)

Summary of Formulas

$$d = \sqrt{(x_2 - x_1)^2 + (y_2 - y_1)^2} \quad \text{distance}$$

$$\left(\frac{(x_1 + x_2)}{2}, \frac{(y_1 + y_2)}{2}\right) \quad \text{midpoint}$$

$$m = \frac{\text{rise}}{\text{run}} = \frac{y_2 - y_1}{x_2 - x_1} \quad \text{slope}$$

CHAPTER REVIEW

Section 2–1

Express each function using functional notation.

1. Give the perimeter *P* of an equilateral triangle as a function of the sides *s*.

2. Give the perimeter *P* of an isosceles triangle as a function of the equal sides *s* if the base is three more than the other sides.

3. Express the area of a triangle as a function of the base if the height is three more than the base.

4. Express the radius r of a sphere as a function of the volume V.

Determine the function values.

5. $f(x) = 3x + 6$. Find $f(-4)$ and $f(2)$.
6. $g(y) = 4y^2 - 5$. Find $g(-2)$ and $g(m)$.
7. $s(t) = t^2 - 5t + 7$. Find $s(-3)$ and $s(4)$.
8. $m(p) = 6p^2 - 1$. Find $m(p - 3)$.
9. $f(y) = 3y - 4$. Find $f(y - 6) + f(7y)$.
10. $g(x) = x^2 + 8x$. Find $g(x - 2) + g(3)$.

If the range is restricted to real numbers, determine the domain of the following functions.

11. $f(x) = \sqrt{3x + 8}$
12. $g(y) = \dfrac{2y + 7}{y - 1}$
13. $s(t) = \sqrt{t - 4}$
14. $f(h) = \dfrac{7}{h} + \dfrac{h + 1}{h - 3} - \dfrac{6}{2h}$
15. $p(y) = \dfrac{3}{y} - \dfrac{2y - 9}{3 - y}$
16. $h(m) = \sqrt{5m + 15}$

Section 2–2

Plot the following points on the rectangular coordinate system.

17. $A(6, 2)$, $B(-1, 0)$, and $C(3, -5)$
18. $D(8, -1)$, $E(-5, -2)$, and $F(0, -4)$
19. $G(7, 2)$, $H(6, 0)$, and $I(-9, 8)$
20. $J(-3, -5)$, $K(4, -8)$, and $L(0, 7)$
21. $M(-6, 4)$, $N(-3, -3)$, and $O(8, 4)$
22. Three vertices of a rectangle are given by $(-2, 4)$, $(3, 4)$, and $(-2, -2)$. Find the coordinate of the fourth vertex.

Section 2–3

Graph the following equations on the rectangular coordinate system.

23. $y = 3(x - 1) + 4$
24. $y = \dfrac{3}{x}$
25. $4x^2 - y = 10$
26. $f(x) = x^2 + 3x - 4$
27. $y = \sqrt{4x - 1}$
28. $y = \sqrt{2x - 3}$
29. The area of a square is given by $A = s^2$ where s is the length of a side. Graph the relationship between the area and length of a side.
30. The surface area of a sphere is $s = 4\pi r^2$. Graph the relationship between the surface area and radius. (Use $\pi = 3.14$.)

Section 2–4

Determine the distance between the points given, the slope of the line passing through them, and the midpoint of the line segment joining them.

31. $(-2, 6)$ and $(8, 1)$
32. $(-3, 5)$ and $(-1, -7)$
33. $(-6, -5)$ and $(1, -10)$
34. $(2, 8)$ and $(-3, -9)$
35. $(7, -3)$ and $(-8, -3)$
36. $(4, 3)$ and $(3, 10)$
37. $\left(\frac{3}{5}, -4\right)$ and $\left(-\frac{1}{3}, 3\right)$
38. $(-2, 4)$ and $(-2, 7)$
39. $(1.7, -3.6)$ and $(2.9, 8.1)$
40. $\left(-\frac{1}{3}, \frac{3}{4}\right)$ and $\left(\frac{1}{2}, -\frac{1}{5}\right)$

Section 2–5

Graph the following inequalities on the rectangular coordinate system.

41. $3x + y > 6$
42. $y < 3x^2$
43. $y > 5x - 1$
44. $2y \geq -5x + 1$
45. $y \leq (x - 1)^2$
46. $y \leq 3(2x + 5) - 7$
47. Find the solution for the following system of inequalities: $x > y + 1$ and $y \geq x$.
48. Find the solution for the following system: $4x - 3y \leq 12$ and $y \geq 2x^2 - 1$.

Section 2–6

Solve the following equations graphically.

49. $4(x - 1) + 3 = 2x - (5x + 6)$
50. $2x + 3(x - 1) = -(x - 5) + 8$
51. $8x - 2(3x + 5) = 3x - 9$
52. $5x - 3 + 7x = 4(2x + 1)$

53. $x^2 + x - 6 = 0$

54. $x^2 + 3x = 0$

55. $3(x - 1) - 5 = 0$

56. $6 - 4(x - 1) = 0$

I **57.** A company sells its product for $8 per unit and sells all that it produces. If the fixed cost of production is $12,000 and the variable cost is $2 per unit, find the break-even point.

I **58.** A company sells its product for $12 per unit and sells all that it produces. If the fixed cost is $20,000 and the variable cost is $8 per unit, find the break-even cost.

59. Graph the following empirical data on the distance traveled versus speed. Is this relationship linear?

Speed (mi/h)	15	30	45	50	60	65	70
Distance (mi)	30	60	90	100	120	130	140

CHAPTER TEST

The number in parentheses refers to the appropriate learning objective given at the beginning of the chapter.

1. If $f(x) = -(3x - 2) + 6$, find $f(-2)$. (3)

2. Graph $y = 2x^2 - 1$. (5)

3. Solve $2x^2 + 4x = 6$ graphically. (9)

4. Graph $5x - 6 \geq y$. (8)

5. Plot the following points on the rectangular coordinate system: (4)

$$(-1, 6), \quad (-1, -1), \quad (7, 6), \quad (7, -1)$$

Join the points with lines and describe the resulting figure.

6. A company sells its product for $3 per unit, selling all that it manufactures. If the fixed cost is $6,000 and variable cost is $1 per unit, find the break-even point. (9)

7. Find the solution for $3x + y \geq 6$ and $y \geq x$. (8)

8. Graph $f(x) = 2x - 15$. (5)

9. Solve $5x - 8 = 2(3x + 6) - 2x$ graphically. (9)

10. Determine the slope of the line passing through $(-3, -8)$ and $(6, -4)$. (6)

11. If the range is restricted to real numbers, find the domain for (2)

$$f(x) = \sqrt{4x - 20}$$

12. If $f(x) = 3x - 5$ and $g(x) = 5x^2 + 1$, find $2f(x) - 3g(x)$. (3)

13. Graph the following empirical data relating the current and voltage of a diode: (9)

Voltage (V)	0.3	0.4	0.5	0.6	0.7	0.8
Current (mA)	0.002	0.008	0.125	1.2	14	300

14. Determine the distance between the points $(-6, 7)$ and $(4, -10)$ and the midpoint of the line segment joining them. **(7)**

15. Graph the relationship between displacement in meters and time in seconds from the following data. Is this relationship linear? **(9)**

Displacement (m)	4	36	100	144	400
Time (s)	1	3	5	6	10

16. If $f(x) = 2x^2 + 1$, find $f(3m) - f(m)$. **(3)**

17. Plot the points $A(-6, 2)$, $B(0, 3)$, and $C(-8, 0)$. **(4)**

18. Express interest rate as a function of principal and time. **(1)**

SOLUTION TO CHAPTER INTRODUCTION

To determine the break-even point, we graph the following equations:

$$\begin{aligned} \text{total cost} &= \text{fixed cost} + \text{variable cost} \\ y &= 1{,}500 + 1.25x \\ \text{total revenue} &= (\text{price per unit})(\text{number of units}) \\ y &= 2.5x \end{aligned}$$

The graph of these equations is given in Figure 2–27. From the graph, we see that the intersection of the lines occurs at the point (1,200, 3,000). The company should produce 1,200 widgets and receive a total revenue (and cost) of \$3,000.

FIGURE 2–27

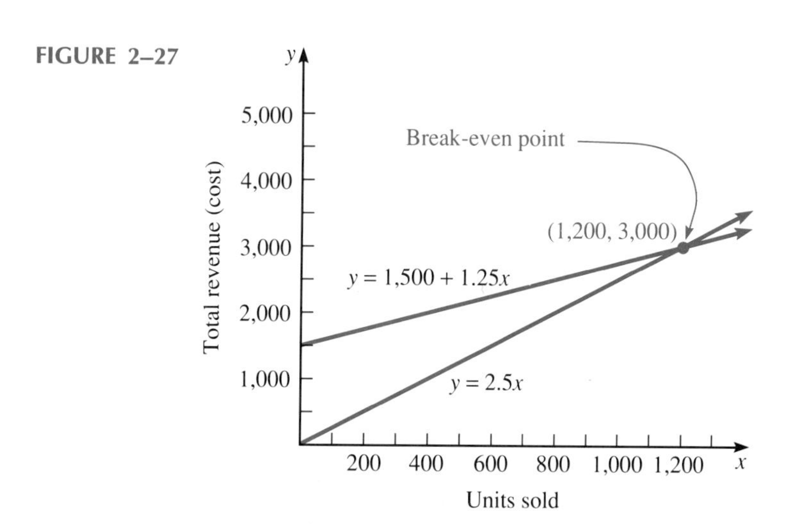

You are an accountant responsible for calculating the current worth of a computer system that your company bought for $9,600 two years ago. You know that the annual depreciation of the computer is $430. (The solution to this problem is given at the end of the chapter.)

Many technical problems, including the one given above, result in fractional expressions. For example, the measurement of parallax in surveying, the resistivity of metal conductors, and the reactance of a capacitor and inductor connected in series also involve fractional equations.

In this chapter, we will first discuss factoring because it is a tool used in simplifying fractional expressions. Then we will discuss equivalent fractions and operations with algebraic fractions. We will end the chapter with a discussion of solving equations containing algebraic fractions.

Learning Objectives

After completing this chapter, you should be able to

1. Factor a polynomial using the greatest common factor (Section 3–1).
2. Factor a binomial using the difference of two squares (Section 3–1).
3. Factor a trinomial using a perfect square trinomial (Section 3–2).
4. Factor a general trinomial using the *ac* method (Section 3–2).
5. Reduce fractions to lowest terms (Section 3–3).
6. Multiply and divide algebraic fractions (Section 3–4).
7. Add and subtract algebraic fractions (Section 3–5).
8. Simplify a complex fraction (Section 3–5).
9. Solve an equation containing fractions (Section 3–6).

Chapter 3

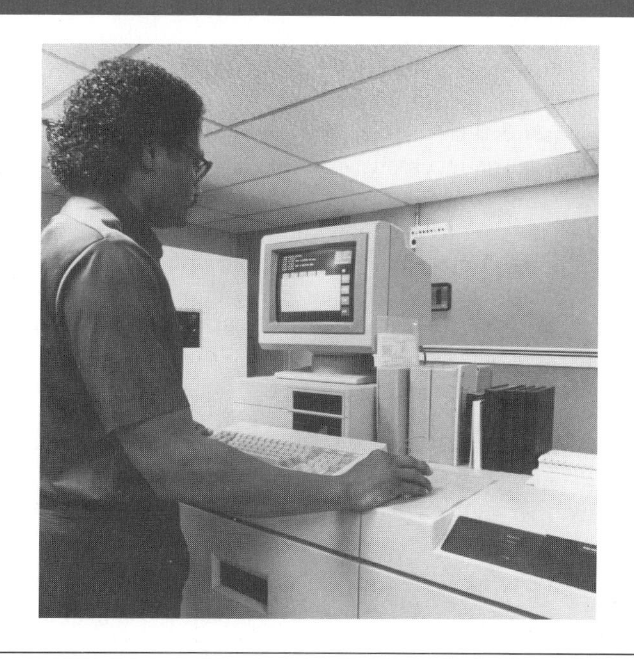

Factors and Fractions

3–1

FACTORING: GREATEST COMMON FACTOR AND DIFFERENCE OF SQUARES

In job situations you will rarely see a factoring problem alone. However, factoring an algebraic expression is used extensively in operations with algebraic fractions and in solving quadratic equations. The importance of factoring results from its usefulness as a tool to do other things. In the first two sections of this chapter, we will discuss four methods of factoring a polynomial: greatest common factor, difference of two squares, perfect square, and the *ac* method.

Definitions

The process of writing a polynomial as the product of other ''simpler'' polynomials is called **factoring.** A polynomial or a factor is called **prime** if its only factors are 1 and itself. A polynomial is said to be **factored completely** if each factor in its product is prime.

Greatest Common Factor

Factoring is the opposite of multiplication, and, specifically, factoring the greatest common factor is opposite to multiplying a monomial and a polynomial using the Distributive Property. For example, we can multiply $2a^2(3ab + 5c)$ to obtain the product $6a^3b + 10a^2c$. In factoring a greatest common factor, we reverse this process because $6a^3b + 10a^2c$ is factored to obtain the expression $2a^2(3ab + 5c)$.

Greatest Common Factor

Recognition

There is no specific method of recognizing when this technique of factoring applies. You must look for a greatest common factor in every polynomial.

Factoring

1. Determine the greatest common factor by finding the largest number that divides into each numerical coefficient and by finding the smallest exponent of any variable or binomial factor that appears in each term.
2. Divide the common factor from Step 1 into the original polynomial term by term.
3. Write the factored answer as the product of the common factor from Step 1 and the result of the division from Step 2.
4. Check your answer by multiplying the factored answer.

EXAMPLE 1 Factor $4m + 12mn$.

Solution First, we determine the common factor. Since the number 4 divides into both 4 and 12 and the variable m appears in each term, the common factor is $4m$. Note that the variable n is not part of the common factor because it does not appear in both terms. Second, we divide $4m$ into the polynomial $4m + 12mn$. This gives

$$\frac{4m}{4m} + \frac{12mn}{4m} = 1 + 3n$$

Third, the factored answer is the product of the common factor and the result from division. The factored expression is

greatest common factor   result from division

$$4m(1 + 3n)$$

Multiplication should be used to check the factoring. It does not guarantee a prime factorization, but it does check the method of factoring used.

Checking

$$4m(1 + 3n)$$
$$4m + 12mn$$

■

CAUTION ✦ When dividing $4m$ into $4m$, remember that a nonzero quantity divided by itself equals 1. You may forget to place the 1 inside the parentheses in the factored answer.

EXAMPLE 2 Factor $18a^2b^3c^4 + 36a^4c^3 - 27a^3b^2c^3$.

Solution Nine divides into each numerical coefficient; the variables a and c appear in each term, and the smallest exponents on a and c are 2 and 3, respectively. Therefore, the greatest common factor is $9a^2c^3$. Dividing $9a^2c^3$ into the polynomial gives

$$\frac{18a^2b^3c^4}{9a^2c^3} + \frac{36a^4c^3}{9a^2c^3} - \frac{27a^3b^2c^3}{9a^2c^3}$$

$$2 \cdot 1 \cdot b^3 \cdot c \qquad 4 \cdot a^2 \cdot 1 \qquad 3 \cdot a \cdot b^2 \cdot 1$$

$$2b^3c + 4a^2 - 3ab^2$$

The factored expression is

$$9a^2c^3(2b^3c + 4a^2 - 3ab^2)$$

Checking

$$9a^2c^3(2b^3c + 4a^2 - 3ab^2)$$
$$18a^2b^3c^4 + 36a^4c^3 - 27a^3b^2c^3$$

■

As illustrated in the next example, a binomial may also be a common factor.

EXAMPLE 3 An expression relating volume, pressure, and temperature of a gas is $P_1V_1T_2 - P_2V_2T_1 = 0$. If the pressure and temperature are fixed and temperature is given in kelvins, this expression becomes

$$PV_1(C + 273) - PV_2(C + 273) = 0$$

Factor the left side of this equation.

Solution The greatest common factor is $P(C + 273)$. Note that P and the binomial $(C + 273)$ appear in each term, and the largest exponent on each is 1. Dividing the polynomial by the common factor gives

$$\frac{\cancel{P}V_1\cancel{(C + 273)}}{\cancel{P}\cancel{(C + 273)}} - \frac{\cancel{P}V_2\cancel{(C + 273)}}{\cancel{P}\cancel{(C + 273)}}$$
$$V_1 - V_2$$

The factored expression is

$$P(C + 273)(V_1 - V_2)$$ ■

Difference of Two Squares

An expression such as $x^2 - y^2$ in which one square is subtracted from another is called the **difference of two squares.** Factoring the difference of two squares is the reverse process of multiplying the sum and difference of a binomial: $(x + y)(x - y)$. The form and method of factoring the difference of two squares are given below.

Difference of Two Squares

Recognition

1. The difference of two squares must have exactly two terms (a binomial).
2. Each term must be a perfect square (variables with even-numbered exponents are perfect squares).
3. The terms must be separated by subtraction.

Factoring

The binomial $x^2 - y^2$ can be written and factored as follows:

$$(x)^2 - (y)^2$$
$$(x + y)(x - y)$$

sum (points to +) difference (points to −)

EXAMPLE 4 Factor $16x^2 - 9y^2$.

Solution Since the two terms $16x^2$ and $9y^2$ are perfect squares and are separated by subtraction, the binomial is the difference of two squares. The binomial can be written and factored as follows:

$$16x^2 - 9y^2$$
$$(4x)^2 - (3y)^2$$
$$(4x + 3y)(4x - 3y)$$

sum (↑ at +) difference (↑ at −)

Checking

$$(4x + 3y)(4x - 3y)$$
$$16x^2 - 12xy + 12xy - 9y^2 = 16x^2 - 9y^2$$ ■

EXAMPLE 5 Factor the following:

(a) $25x^2 + 36y^2$ (b) $x^4 - 81$

Solution

(a) The two terms $25x^2$ and $36y^2$ are perfect squares, but the squared terms are not separated by subtraction. The binomial is prime since there is no common factor.
(b) The binomial $x^4 - 81$ is the difference of two squares. It can be written and factored as follows:

$$x^4 - 81$$
$$(x^2)^2 - (9)^2$$
$$(x^2 + 9)(x^2 - 9)$$

sum (↑ at +) difference (↑ at −)

Checking each factor shows that $x^2 - 9$ will factor again as the difference of squares. The factor $x^2 + 9$ does not factor.

$$(x^2 + 9)[(x)^2 - (3)^2]$$
$$(x^2 + 9)(x + 3)(x - 3)$$

The factored expression is

$$(x^2 + 9)(x + 3)(x - 3)$$ ■

NOTE ✦ To obtain the exponent to be placed inside the parentheses in factoring the difference of two squares, divide the original exponent by 2.

Combined Factoring

Frequently, polynomials must be factored more than once. The next two examples illustrate this process.

EXAMPLE 6 Factor $64x^5 - 16xy^6$.

Solution The polynomial contains a greatest common factor of $16x$. Factoring the common factor gives

$$16x(4x^4 - y^6)$$

Next we must attempt to factor the binomial $4x^4 - y^6$. The two terms $4x^4$ and y^6 are perfect squares separated by subtraction. Thus, $4x^4 - y^6$ factors as the difference of two squares as follows:

$$(2x^2)^2 - (y^3)^2$$
$$(2x^2 - y^3)(2x^2 + y^3)$$

Since $(2x^2 - y^3)$ and $(2x^2 + y^3)$ will not factor, the factored expression is

$$16x(2x^2 - y^3)(2x^2 + y^3)$$ ■

CAUTION ✦ When you factor out a common factor and then factor the result again, remember to write the common factor in your answer. For example,

$$8x^3 - 32xy^2 = 8x(x^2 - 4y^2) = 8x(x - 2y)(x + 2y)$$

Also, when some factors cannot be factored again, be sure to include them in the answer. For example,

$$x^4 - 81y^4 = (x^2 - 9y^2)(x^2 + 9y^2) = (x - 3y)(x + 3y)(x^2 + 9y^2)$$

EXAMPLE 7 The difference in the volumes of two cylinders with the same height but different radii is given by

$$\pi r_1^2 h - \pi r_2^2 h$$

Factor this expression.

Solution The polynomial contains a common factor of πh. Factoring the common factor gives

$$\pi h(r_1^2 - r_2^2)$$

The binomial will factor as the difference of two squares. The factored expression is

$$\pi h(r_1 - r_2)(r_1 + r_2)$$ ■

3–1 EXERCISES

Factor each of the following polynomials completely.

1. $3a^2b + 9a^3$ **2.** $32a^2b - 28a^3b^4$

3. $7x^2y - 14x + 35x^3y^2$ **4.** $17m^3n^2 - 34m^4n$

5. $49x^2 - 25y^2$

6. $15x^2y^4 + 45x^3y^3 - 20x^2y^3$

7. $30x^5 - 40x^2 + 100x^3$ **8.** $81s^2 - 100t^2$

9. $25p^2 - 64n^2$ **10.** $36x^7 - 72x^{10} + 18x^5$

11. $9y^3 + 7y^6 + 10y^4x^2$ **12.** $9p^2 - 25n^2$

13. $36m^2 - 121n^2$ **14.** $64x^2 - 9y^2$

15. $32s^2t - 40s^3t^3 + 8st$

16. $27a^3b^2 - 81a^4b - 63a^2b^3$

17. $18m^3n^4 - 9m^2n^5 + 3m^4n^3$

18. $16a^2 - 81b^2$

19. $25x^5 + 30x^2 - 15x^6$

20. $17x^3y^2 + 29x^5y - 46x^4y^3$

21. $144s^2 - 49t^2$ **22.** $225h^2 - 169k^2$

23. $38m^4n^3 + 57m^3n^4 - 19m^3$

24. $56a^5b^2c^9 + 70b^5c^4 - 35a^6b^8$

25. $16m^4 - 25n^4$ **26.** $256f^2 - 49g^2$

27. $64m^8 + 32m^4 + 48m^{15}$

28. $18x^4 - 24x^3 + 14x^2 - 36x^5$

29. $169c^2 - 36d^2$ **30.** $9m^2 - 100n^2$

31. $63a^7b^5c^{10} - 49a^5c^{10}$ **32.** $9ax + 9a^2x - 18a^3x^2$

33. $4x^2 - 49y^2$ **34.** $p^2 - 81q^2$

35. $54x^3y^2z - 63x^4y^5z^3 + 81x^3z$

36. $169s^4 - 64t^4$

37. $a^4 - 36b^4$ **38.** $9x^4y^4 - 81$

39. $46s^9t^6 + 23s^{12}t^4 + 69s^8t^9$

40. $40m^3n - 12m^{10}n^8 + 28m^8$

41. $2x(x - 3) + 3(x - 3)$

42. $x(2x + 1) - 3(2x + 1)$

43. $3(x - 5) - 5x(x - 5)$

44. $6x(2x - 3) + 1(2x - 3)$

45. The expression for the difference in the surface area of a sphere with radius r_1 and the surface area of a second sphere with radius r_2 is given by $4\pi r_2^2 - 4\pi r_1^2$. Factor this expression.

46. If an object is dropped from a building 64 ft high, the expression for the height at time t is $64 - 16t^2$. Factor this expression.

I 47. In economics the break-even point occurs when demand equals supply. If the demand is represented by the expression $1{,}125 - 4p$, and the supply is represented by $125p^2 - 4p$, then the equation for the break-even point becomes $125p^2 - 1{,}125 = 0$. Factor the left side of the equation.

E 48. The change in resistance in a circuit with a temperature change is given by $R_1 + R_1at - R_1at_1$. Factor this expression.

49. The momentum M of a mass m that has a change in velocity from v_1 to v_2 is given by the equation $M = mv_2 - mv_1$. Factor the right side of the equation.

50. The displacement s of an object is given by the equation

$$s = v_0t + \frac{1}{2}at^2$$

Factor the right side of the equation.

51. The difference in the volume of two cylinders with the same radius and heights h_1 and h_2 is given by $\pi r^2h_1 - \pi r^2h_2$. Factor this expression.

52. The surface area of a rectangular solid is given by $2hw + 2hl + 2wl$. Factor this expression.

I 53. The amount of money in a savings account at a given time is given by $P + Prt$ where P is principal, r is simple interest rate, and t is time. Factor this expression.

54. The coefficient of linear expansion, α, is the ratio of change in length per degree Celsius to the length at 0°C. The new length is given by $L_0 + L_0\alpha\Delta t$. Factor this expression.

3–2

FACTORING TRINOMIALS

The number of terms in a polynomial is the key to determining which methods of factoring to consider. After removing any common factors, try to factor a trinomial (a polynomial with three terms) as a perfect square, and then try the *ac* method. Each of these methods of factoring trinomials is discussed in this section. Since factoring a perfect square is easier, it is discussed first.

Perfect Square Trinomial

The product of two identical binomials is called a **perfect square trinomial.**

$$(x + y)(x + y) = x^2 + 2xy + y^2$$
$$(x - y)(x - y) = x^2 - 2xy + y^2$$

Since recognition is the key to factoring a perfect square trinomial, the following information summarizes how to recognize and factor one.

Perfect Square Trinomial

Recognition

1. A perfect square trinomial must have exactly three terms.
2. The polynomial can be arranged so that the first and last terms are perfect squares (with positive signs for both).
3. The magnitude (neglect the sign) of the middle term must be the product of 2 and the square root of each squared term.

Factoring

To factor a perfect square trinomial, fill in the following formula:

$$(\sqrt{\text{first term}} \pm \sqrt{\text{third term}})^2$$

Choose + or − depending on the middle sign of the polynomial.

EXAMPLE 1 Factor $9x^2 - 12xy + 4y^2$.

Solution First, we check the criteria for a perfect square trinomial:

1. The polynomial contains three terms: $9x^2$, $12xy$, and $4y^2$.
2. $9x^2$ and $4y^2$ are perfect squares.
3. The magnitude of the middle term is

$$2\sqrt{9x^2}\sqrt{4y^2} = 2(3x)(2y) = 12xy$$

Therefore, this polynomial meets the requirements for a perfect square trinomial.

Next, to factor the polynomial, fill in the formula as follows:

$$(\sqrt{\text{first term}} - \sqrt{\text{third term}})^2$$

$$\uparrow \qquad\qquad \uparrow$$

$$(\sqrt{9x^2} - \sqrt{4y^2})^2$$

$$(3x - 2y)^2$$

Checking

$$(3x - 2y)(3x - 2y)$$

$$9x^2 - 6xy - 6xy + 4y^2$$

$$9x^2 - 12xy + 4y^2$$

The factored expression is $(3x - 2y)^2$. ■

EXAMPLE 2 Factor $81x^2 - 72xy - 16y^2$.

Solution We check to see if this polynomial fits the requirements of a perfect square trinomial:

1. The polynomial contains three terms: $81x^2$, $72xy$, and $16y^2$ (ignoring signs).
2. The first term is a square, $81x^2$, but the third term, $-16y^2$, is not a square because a square must be positive.

Therefore, this trinomial is not a perfect square trinomial and does not factor using any of the other methods of factoring. ■

The *ac* Method

If a trinomial does not have a greatest common factor and is not a perfect square trinomial, then we attempt to factor it using the *ac* method. If we have a trinomial of the form $ax^2 + bx + c$, from reversing the FOIL process we know that the sum of the product of factors a and c must combine to yield b. The *ac* method is explained in detail in the next examples.

EXAMPLE 3 Factor $x^2 + 6x + 8$.

Solution Since this polynomial does not contain a common factor and does not fit the criteria for a perfect square trinomial, we try factoring by the *ac* method. By examining the polynomial, we see that $a = 1$, $c = 8$, and $ac = 8$, and factors of 8 must combine to yield b, which is 6. The factors of 8 are $\pm 8 \cdot \pm 1$ and $\pm 2 \cdot \pm 4$. The factors of 8 that combine to give 6 are $+2$ and $+4$. Therefore, we can write the polynomial as

$$x^2 + 6x + 8$$

$$x^2 + 2x + 4x + 8$$

Then grouping the terms and factoring a common factor from each group gives the following result.

$$(x^2 + 2x) + (4x + 8)$$
$$(x)(x + 2) \oplus (4)(x + 2)$$
$$(x + 4)(x + 2)$$

Checking

$$(x + 4)(x + 2)$$
$$x^2 + 2x + 4x + 8$$
$$x^2 + 6x + 8$$

■

EXAMPLE 4 Factor $x^2 + 3x - 28$.

Solution Since the trinomial does not meet the requirements for a perfect square trinomial, we use the ac method of factoring. In this polynomial, $a = 1$, $c = -28$, and $ac = -28$. Therefore, we want to determine factors of -28 that combine to yield b, or 3. The factors of -28 are $\pm 1 \cdot \pm 28$, $\pm 4 \cdot \pm 7$, and $\pm 2 \cdot \pm 14$. The factors that combine to yield 3 are $+7 \cdot -4$. We write the polynomial as

$$x^2 + 7x - 4x - 28$$

Then we group the terms and factor.

$$(x^2 + 7x) - (4x + 28)$$
$$x(x + 7) - 4(x + 7)$$
$$(x + 7)(x - 4)$$

Checking

$$(x - 4)(x + 7)$$
$$x^2 + 3x - 28$$

■

EXAMPLE 5 Factor $3x^2 + 11x - 4$.

Solution We use the ac method to factor this trinomial also. In this polynomial, $a = 3$, $c = -4$, and $ac = -12$. We must determine factors of -12 that combine to yield 11. The factors of -12 are $\pm 1 \cdot \pm 12$, $\pm 3 \cdot \pm 4$, and $\pm 2 \cdot \pm 6$. The factors of 12 that combine to yield 11 are $+12 \cdot -1$. Thus, we can write the polynomial as

$$3x^2 + 11x - 4$$
$$3x^2 + 12x - x - 4$$

Factoring gives

$$3x(x + 4) - 1(x + 4)$$
$$(x + 4)(3x - 1)$$

Checking

$$(3x - 1)(x + 4)$$
$$3x^2 + 11x - 4$$

■

EXAMPLE 6 Factor $8x^2 - 14xy + 3y^2$.

Solution Since $a = 8$, $c = 3$, and $ac = 24$, we must determine factors of 24 that combine to give -14. The factors that yield 24 are $\pm 1 \cdot \pm 24$, $\pm 2 \cdot \pm 12$, $\pm 3 \cdot \pm 8$, and $\pm 4 \cdot \pm 6$. The factors that combine to give -14 are -12 and -2. The polynomial can be written as

$$8x^2 - 12xy - 2xy + 3y^2$$

Factoring gives

$$4x(2x - 3y) - y(2x - 3y)$$
$$(4x - y)(2x - 3y)$$

Checking

$$(2x - 3y)(4x - y)$$
$$8x^2 - 14xy + 3y^2$$

■

Combined Factoring

EXAMPLE 7 Factor $6x^2 + 33x - 18$.

Solution This polynomial contains a common factor of 3.

$$3(2x^2 + 11x - 6)$$

Next, we try to factor $2x^2 + 11x - 6$ using the ac method. Since $a = 2$, $c = -6$, $ac = -12$, and $b = 11$, the factors of -12 that sum to 11 are $+12 \cdot -1$. Then we can write the polynomial and factor as follows:

$$2x^2 + 11x - 6$$
$$2x^2 + 12x - 1x - 6$$
$$2x(x + 6) - 1(x + 6)$$
$$(2x - 1)(x + 6)$$

The factored answer is $3(2x - 1)(x + 6)$. ■

CAUTION ✦ Remember to include the common factor in the factored answer.

EXAMPLE 8 An object is thrown upward with an initial velocity of 74 ft/s. To find the time it takes to reach a height of 40 ft, we must solve the equation $16t^2 - 74t + 40 = 0$. Factor the left side of this equation.

Solution First, we factor a greatest common factor of 2.

$$2(8t^2 - 37t + 20) = 0$$

Then factoring the trinomial inside the parentheses using the *ac* method gives the factored expression

$$2(8t - 5)(t - 4) = 0$$ ■

3–2 EXERCISES

Factor each of the following polynomials completely.

1. $a^2 + 8ab - 33b^2$
2. $b^2 + 8b + 16$
3. $64x^2 + 208xz + 169z^2$
4. $m^2 + 7mn + 12n^2$
5. $x^2 + xy - 30y^2$
6. $a^2 + 6ab - 16b^2$
7. $100m^2 + 180mn + 81n^2$
8. $144x^2 - 168xy + 49y^2$
9. $121x^2 - 66xy + 9y^2$
10. $c^2 + 8cd + 12d^2$
11. $m^2 - 2mn - 24n^2$
12. $m^2 + 21mn + 11n^2$
13. $4n^2 - 10n - 6$
14. $5p^2 - 9p - 2$
15. $6 + y - 12y^2$
16. $12x^2 + 5xy - 28y^2$
17. $144a^2 + 168a + 49$
18. $169p^2 + 286np + 121n^2$
19. $35k^2 + 9kg - 18g^2$
20. $40m^2 + 61mn + 18n^2$
21. $64n^2 - 208np + 169p^2$
22. $49x^2y^2 - 154xyz + 121z^2$
23. $5x^4 + x^2y^2 - 18y^4$
24. $25x^2 + 90xy - 36y^2$
25. $4m^4 + 28m^2n^2 + 49n^4$
26. $16s^4 - 40s^2t^2 + 25t^4$
27. $6x^3y + 3x^2y^2 - 3xy^3$
28. $2x^4 - 16x^2y^2 + 30y^4$
29. $21x^4 - 13x^2y^2 - 18y^4$
30. $27m^4 - 33m^2n^2 - 20n^4$
31. $64a^2 - 48ab + 9b^2$
32. $100m^2 - 140mn + 49n^2$
33. $81n^2 + 90np + 25p^2$
34. $9a^2 + 30ab + 25b^2$
35. $36x^4 - 60x^2y^2 + 25y^4$
36. $9s^4t^4 - 24s^2t^2 + 16$

Factor the following, using any of the methods discussed thus far.

37. $8x^3y^2 - 64x^4y - 80xy^4$
38. $36x^2 + 108xy + 81y^2$
39. $49x^2y + 70xy^2 + 25y^3$
40. $16x^3y - 4xy$
41. $17xyz + 34x^2z + 16y^3z$
42. $16x^2 + 64xy + 64y^2$
43. $9x^2 - 54xz + 81z^2$
44. $98x^3y - 112x^2y^2 + 32xy^3$
45. $16x^2z - 144xyz + 324y^2z$
46. $81x^4 - 6xy^4$
47. $5x^2 + 30xy + 45y^2$
48. $4x^2 - 12xy + 9y^2$
49. $12x^2 - 6xy - 90y^2$
50. $15x^2y^2 - 10x^2y + 20x^2$
51. $144x^2 - 49y^2$
52. $100m^4 - 180m^2 + 81$
53. To find the dimensions of a rectangle whose length is 2 more than twice the width and whose area is 60, you must factor the expression $2x^2 + 2x - 60$. Factor this expression.
54. If an object is projected upward from the ground a distance of 128 ft using an initial velocity of 144 ft/s, the expression for time of rise, t, becomes $128 - 144t + 16t^2 = 0$. Factor the left side of the equation.
55. The position of a projectile moving in a straight line is given by $s = 2t^2 - 5t$. If the displacement, s, is 12 m, the equation becomes $2t^2 - 5t - 12 = 0$. Factor the left side of the equation.

56. To find the length and width of a rectangle whose length is 7 m more than its width and whose area is 120 m^2, you must solve the equation $x^2 + 7x - 120 = 0$. Factor the left side of the equation.

57. A piece of cardboard 8 m square is to be made into a box by cutting a square from each corner and folding. If the area of the bottom of the box is to be 16 m^2, you must solve the equation $4x^2 - 32x + 48 = 0$. Factor the left side of the equation.

E 58. The resistance of a conductor varies with the temperature. Under certain conditions the resistance is given by $30 + 236t - 32t^2$. Factor this expression.

3–3

EQUIVALENT FRACTIONS

In the remainder of this chapter, we will discuss operations with algebraic fractions as we continue to add to our basic algebra skills. Many technical formulas and problems require simplifying expressions containing algebraic fractions. The process of factoring is essential to operations with algebraic fractions, and you must understand it thoroughly.

Definitions

If a and b ($b \neq 0$) represent algebraic expressions, then a/b represents an **algebraic fraction.** A fraction is said to be in **lowest terms** or **reduced** if the numerator and denominator do not contain common factors. Two fractions are said to be **equivalent** if they can both be reduced to the same fraction. We can produce an equivalent fraction by multiplying or dividing the numerator and denominator of the fraction by the same quantity.

Equivalent Fractions

EXAMPLE 1 Change $\frac{4xy}{7z}$ to an equivalent fraction with a denominator of $21z^2$.

Solution To change the denominator of $7z$ to $21z^2$, we must multiply by $3z$. In order to produce an equivalent fraction, we must multiply the numerator and denominator by $3z$ as follows:

$$\frac{4xy}{7z} \cdot \frac{3z}{3z} = \frac{12xyz}{21z^2}$$

■

EXAMPLE 2 Change $\frac{5}{6x - 10}$ to an equivalent fraction with a denominator of $6x^2 - 4x - 10$.

Solution We must determine what factor to multiply the denominator $6x - 10$ by to yield $6x^2 - 4x - 10$. This process is more difficult than in the previous example. Let us factor each denominator.

$$\begin{array}{cc} 6x - 10 & 6x^2 - 4x - 10 \\ 2(3x - 5) & 2(3x - 5)(x + 1) \end{array}$$

From the analysis of the denominators, we can see that we must multiply the numerator and denominator of the fraction by $x + 1$. The equivalent fraction is

$$\frac{5}{6x - 10} \cdot \frac{x + 1}{x + 1} = \frac{5x + 5}{6x^2 - 4x - 10}$$ ■

In producing an equivalent fraction in the two previous examples, we multiplied the numerator and denominator by the same quantity. In reducing a fraction to lowest terms, we divide the numerator and denominator by the same quantity.

Reducing Fractions

Reducing algebraic fractions to lowest terms uses the same procedure as that used in arithmetic. For example, if you were asked to reduce 27/36 to lowest terms, you would find the largest factor contained in both 27 and 36 and divide it out.

$$\frac{27}{36} = \frac{\not{9} \cdot 3}{\not{9} \cdot 4} = \frac{3}{4}$$

More commonly, we say that we divide away factors common to the numerator and denominator.

EXAMPLE 3 Reduce the following to lowest terms:

$$\frac{16x^4y^2z^8}{32y^5z^6}$$

Solution We must divide out the greatest common factor (GCF) contained in both the numerator and denominator.

$$\frac{\cancel{16}\,x^4\,\cancel{y^2}\,\overset{z^2}{\cancel{z^8}}}{\underset{2}{\cancel{32}}\;\underset{y^3}{\cancel{y^5}}\,\cancel{z^6}}$$

The reduced fraction is

$$\frac{x^4z^2}{2y^3}$$ ■

The concept presented in Example 3 also applies to reducing polynomial fractions. We divide out the greatest common factor contained in both the numerator and denominator. However, you must factor polynomials before canceling. Study the next examples carefully.

EXAMPLE 4 Reduce the following fraction to lowest terms:

$$\frac{8x^2 - 10x + 3}{6x^2 - x - 1}$$

Solution To reduce polynomial fractions, factor the numerator and denominator, cancel or divide out the greatest common factors, and write the product of the factors that remain. Factoring the numerator and denominator of this fraction gives

$$\frac{(4x - 3)(2x - 1)}{(2x - 1)(3x + 1)}$$

Canceling the greatest common factor contained in both the numerator and the denominator, $(2x - 1)$, gives

$$\frac{(4x - 3)\cancel{(2x - 1)}}{\cancel{(2x - 1)}(3x + 1)}$$

The reduced fraction is

$$\frac{4x - 3}{3x + 1}$$

■

EXAMPLE 5 Reduce the following fraction to lowest terms:

$$\frac{6x^2 + 16x + 10}{8x^2 - 12x - 20}$$

Solution We begin by completely factoring the numerator and denominator of the fraction.

$$\frac{2(3x^2 + 8x + 5)}{4(2x^2 - 3x - 5)}$$

$$\frac{2(3x + 5)(x + 1)}{4(2x - 5)(x + 1)}$$

Canceling the greatest common factor gives

$$\frac{\cancel{2}(3x + 5)\cancel{(x + 1)}}{\underset{2}{\cancel{4}}(2x - 5)\cancel{(x + 1)}}$$

The reduced fraction is

$$\frac{3x + 5}{2(2x - 5)}$$

■

CAUTION ✦ Be sure to cancel only common *factors,* not terms, when reducing a fraction to lowest terms.

In reducing fractions to lowest terms, you may find some factors that differ only in their signs, such as $x - y$ and $y - x$. In this instance, it is helpful to remember that $-x = (-1)x$. Therefore, the expression $y - x$ can be written as $(-1)(x - y)$. Reducing the fraction $(x - y)/(y - x)$ using this technique gives

$$\frac{x - y}{y - x} = \frac{\cancel{(x - y)}}{-1\cancel{(x - y)}} = -1$$

Factors that differ only in their signs in all terms cancel and yield -1. These factors are considered opposites like a and $-a$. This concept is illustrated in the next two examples.

EXAMPLE 6 Reduce the following to lowest terms:

$$\frac{16 - x^2}{x - 4}$$

Solution First, we completely factor the numerator of the fraction.

$$\frac{(4 - x)(4 + x)}{(x - 4)}$$

Then we cancel the greatest common factor. Notice that $(4 - x)$ and $(x - 4)$ differ only in signs. From the previous discussion, we can write the factors as

$$\frac{-1(x - 4)(4 + x)}{(x - 4)}$$

Canceling common factors gives the reduced fraction as

$$-(4 + x)$$

■

EXAMPLE 7 Reduce the following fraction to lowest terms:

$$\frac{x^2 - 2x - 15}{25 - x^2}$$

Solution To reduce the fraction to lowest terms, factor the numerator and denominator, cancel the greatest common factor, and write the product of the factors that remain. Performing these steps gives

$$\frac{(x + 3)(x - 5)}{(5 - x)(5 + x)}$$

Notice that $(x - 5)$ and $(5 - x)$ differ only in signs. The cancellation of these factors gives -1. After the greatest common factor is canceled, the reduced fraction is

$$-\frac{x + 3}{5 + x}$$

■

CAUTION ✦ The factors $x - a$ and $x + a$ are not opposites and do not cancel, but $x - a$ and $a - x$ or $x + a$ and $-x - a$ are opposites and divide to produce -1.

3–3 EXERCISES

Change each fraction to an equivalent fraction with the indicated denominator.

1. $\frac{x}{5y}$ to $\frac{?}{15y^2z}$

2. $\frac{2x}{7}$ to $\frac{?}{14y}$

3. $\frac{xy}{3z}$ to $\frac{?}{12z^3}$

4. $\frac{5ab}{9c}$ to $\frac{?}{45c^3}$

5. $\frac{(a + b)}{2ab}$ to $\frac{?}{8a^2b}$

6. $\frac{(3x - y)}{mn}$ to $\frac{?}{3m^2n^3}$

7. $\frac{6m}{2(m - 3n)}$ to $\frac{?}{6m(m - 3n)}$

8. $\frac{9xy}{4(x - y)}$ to $\frac{?}{8x^2(x - y)}$

9. $\frac{x - 4}{2x - 5}$ to $\frac{?}{6xy - 15y}$

10. $\frac{3x}{4x - 7}$ to $\frac{?}{12x^3 - 21x^2}$

11. $\frac{3x}{x - 5y}$ to $\frac{?}{2x^2 - 9xy - 5y^2}$

12. $\frac{2x - 1}{2x + 1}$ to $\frac{?}{4x^2 - 1}$

Reduce each fraction to lowest terms. Some fractions may be written in lowest terms.

13. $\frac{9x^4y}{27y^4}$

14. $\frac{12m^4n^2}{20m^7}$

15. $\frac{54a^4bc^8}{18b^5}$

16. $\frac{40s^{10}t^9}{75s^2t^{11}}$

17. $\frac{8x^3y^2z^5}{72xy^2z^8}$

18. $\frac{62k^5p^8m^{10}}{3l^3m^{12}}$

19. $\frac{24mn^8p^3}{30m^6n^5p}$

20. $\frac{17a^{17}b^3}{51a^{10}b^3}$

21. $\frac{84a^3cd^5}{12ac^8d^2}$

22. $\frac{48s^{13}t^3w}{64s^{10}t^8}$

23. $\frac{6ab(2x - y)}{3a^2(2x - y)}$

24. $\frac{5s^3t(3x - 2)}{10t(3x - 2)}$

25. $\frac{14x^2(x - 1)}{16x}$

26. $\frac{36m^4n^2(2m - n)}{8m^6n}$

27. $\frac{(2x - y)(3 - y)}{(y - 3)(x + 2y)}$

28. $\frac{(3x - 1)(x + 7)}{(x - 7)(1 - 3x)}$

29. $\frac{(x - 5)(x + 5)}{(x - 5)(5 - x)}$

30. $\frac{10x^3(2x - 3y)}{-5x^4(3y - 2x)}$

31. $\frac{14a^3b(a - 2b)}{7b^3(a - 2b)}$

32. $\frac{6x^2 - 7x + 2}{8 - 2x - 3x^2}$

33. $\frac{4x^2 - 8x}{8x - 16}$

34. $\frac{25x^2 - 36y^2}{15x^2 + 43xy + 30y^2}$

35. $\frac{2a^2b^2 - 6a^2b}{6ab^5 - 18ab^4}$

36. $\frac{x^2 + 4xy + 4y^2}{4y^2 - x^2}$

37. $\frac{7x^2 + 41xy - 6y^2}{2x^2 + 15xy + 18y^2}$

38. $\frac{7x^2 + 20xy - 3y^2}{7x^2 + 6xy - y^2}$

3–4

MULTIPLICATION AND DIVISION OF FRACTIONS

Multiplying Fractions

The rules for multiplying algebraic fractions are the same as those used in arithmetic:

multiply numerators

$$\frac{a}{b} \cdot \frac{c}{d} = \frac{ac}{bd}$$

multiply denominators

However, the resulting fraction may need to be reduced to lowest terms. Thus, it is better to divide out any common factors prior to multiplying. In summary, when you multiply fractions in arithmetic, you divide out common factors and then multiply the remaining factors. For example, to multiply

$$\frac{16}{27} \cdot \frac{30}{48}$$

we remove common factors

$$\frac{\overset{1}{\cancel{16}}}{\underset{9}{\cancel{27}}} \cdot \frac{\overset{10}{\cancel{30}}}{\underset{3}{\cancel{48}}}$$

and multiply the remaining factors to obtain

$$\frac{10}{27}$$

We perform these same steps in multiplying algebraic fractions. A summary of this process follows.

Multiplying Algebraic Fractions

1. Factor each polynomial.
2. Cancel the greatest common factor contained in both the numerator and denominator.
3. Write the fraction as the product of the resulting factors.

EXAMPLE 1 Multiply the following:

$$\frac{8a^3b^4}{54a^2c^4} \cdot \frac{16c^2}{20ab^5}$$

Solution To multiply these fractions, we remove common factors.

$$\frac{\overset{2}{\cancel{8}}a^3b^4}{\underset{27}{\cancel{54}}a^2c^4} \cdot \frac{\overset{8}{\cancel{16}}c^2}{\underset{5}{\cancel{20}}ab^5} \quad \text{reducing numerical coefficients}$$

$$\frac{2\overset{a^2}{\cancel{a^3}}b^4}{27\underset{1}{\cancel{a^2}}c^4} \cdot \frac{8c^2}{5\cancel{a}b^5} \quad \text{reducing the } a\text{'s}$$

$$\frac{2\cancel{b^4}}{27c^4} \cdot \frac{8c^2}{5\underset{b}{\cancel{b^5}}} \quad \text{reducing the } b\text{'s}$$

$$\frac{2}{27\underset{c^2}{\cancel{c^4}}} \cdot \frac{8\cancel{c^2}}{5b} \quad \text{reducing the } c\text{'s}$$

$$\frac{2}{27c^2} \cdot \frac{8}{5b}$$

Multiplying the remaining numerator and denominator gives the product of the fraction as

$$\frac{16}{135c^2b}$$

■

CAUTION ✦ If you divide out the greatest common factor, the resulting fraction should be reduced to lowest terms; however, you should always check the answer just in case you have overlooked a common factor prior to multiplying.

EXAMPLE 2 Multiply the following fractions:

$$\frac{6x^2y}{4x^2 - 4x} \cdot \frac{5x - 5}{15x^3}$$

Solution To multiply these fractions, we begin by factoring the binomials.

$$\frac{6x^2y}{4x(x - 1)} \cdot \frac{5(x - 1)}{15x^3}$$

Then cancel common factors.

$$\frac{\overset{\cancel{2}\ \cancel{x}}{\cancel{6}\cancel{x^2}}y}{\underset{2}{\cancel{4}}\cancel{x}\cancel{(x - 1)}} \cdot \frac{\cancel{5}\cancel{(x - 1)}}{\underset{\cancel{3}\ x^2}{\cancel{15}\cancel{x^3}}}$$

The resulting product is

$$\frac{y}{2x^2}$$

■

EXAMPLE 3 Multiply the following:

$$\frac{4x^2 - 25y^2}{3x^2 + 11xy + 10y^2} \cdot \frac{3x^2 + 2xy - 5y^2}{25y - 10x}$$

Solution First, we factor each polynomial.

$$\frac{(2x - 5y)(2x + 5y)}{(3x + 5y)(x + 2y)} \cdot \frac{(3x + 5y)(x - y)}{5(5y - 2x)}$$

Then we cancel the greatest common factor.

$$\frac{\overset{-1}{\cancel{(2x - 5y)}}(2x + 5y)}{\cancel{(3x + 5y)}(x + 2y)} \cdot \frac{\cancel{(3x + 5y)}(x - y)}{5\cancel{(5y - 2x)}}$$

The product is

$$\frac{-(2x + 5y)(x - y)}{5(x + 2y)}$$

■

NOTE ✦ Although it is permissible to multiply the products in the answer, it is preferable to leave the fraction in factored form.

EXAMPLE 4 In converting the units on the rate of flow of a liquid from 150 in.^{3}/s to ft^3/min, it is necessary to multiply in.^{3}/s by the following conversion factors:

$$\frac{1 \text{ ft}^3}{1{,}728 \text{ in.}^3} \quad \text{and} \quad \frac{60 \text{ s}}{1 \text{ min}}$$

Perform this conversion.

Solution To convert 150 in.^{3}/s to ft^3/min, we multiply the three fractions, canceling units just as we have canceled algebraic factors.

$$\frac{150 \cancel{\text{in.}^3}}{\cancel{\text{s}}} \cdot \frac{1 \text{ ft}^3}{1{,}728 \cancel{\text{in.}^3}} \cdot \frac{60 \cancel{\text{s}}}{1 \text{ min}}$$

Canceling units and performing the arithmetic with the coefficients gives

5.2 ft^3/min

150 [×] 60 [÷] 1728 [=] ⟶ 5.2083

■

Dividing Fractions

Recall from arithmetic that you divide fractions by inverting the divisor and multiplying the resulting fractions.

invert divisor

$$\frac{a}{b} \div \frac{c}{d} = \frac{a}{b} \cdot \frac{d}{c} = \frac{ad}{bc}$$

multiply

Applying this procedure to an arithmetic example gives

invert divisor

$$\frac{2}{3} \div \frac{5}{9} = \frac{2}{3} \cdot \frac{9}{5} = \frac{2}{\cancel{3}_1} \cdot \frac{\cancel{9}^3}{5} = \frac{6}{5}$$

multiply

This same process applies to dividing algebraic fractions and is summarized below.

Dividing Fractions

1. Invert the divisor.
2. Multiply the resulting fractions by following the procedure given for multiplying fractions.

EXAMPLE 5 Divide the following fractions:

$$\frac{9m^2 - 18m^3}{16m^2n^5} \div \frac{3 - 6m}{18m^3}$$

Solution To divide the fractions, take the reciprocal of the divisor and multiply.

$$\frac{9m^2 - 18m^3}{16m^2n^5} \cdot \frac{18m^3}{3 - 6m}$$

To multiply the fractions, factor each polynomial, cancel the greatest common factors, and write the product of the factors that remain.

$$\frac{\overset{3}{\cancel{9}}\cancel{m^2}\cancel{(1-2m)}}{\underset{8}{\cancel{16}}\cancel{m^2}n^5} \cdot \frac{\overset{9}{\cancel{18}}m^3}{\cancel{3}\cancel{(1-2m)}}$$

The quotient of the fractions is

$$\frac{27m^3}{8n^5}$$

■

CAUTION ✦ To avoid errors in dividing fractions, always invert the divisor first and then factor and cancel as necessary. Do *not* factor or cancel first.

EXAMPLE 6 The length of an arc of a circle is given by $\frac{2n\pi r}{360}$, while the area of a sector is given by $\frac{n\pi r^2}{360}$, where n is the measure of the central angle in degrees and r is the radius. Express the ratio of the length of the arc to the area of the sector.

Solution The term *ratio* means division. Setting up the ratio gives

$$\frac{\dfrac{2n\pi r}{360}}{\dfrac{n\pi r^2}{360}} \quad \begin{matrix}\text{length of arc}\\[2ex] \text{area of sector}\end{matrix}$$

Converting to the use of a division symbol instead of the fraction bar gives

$$\frac{2n\pi r}{360} \div \frac{n\pi r^2}{360}$$

Dividing gives

$$\frac{2n\pi r}{360} \cdot \frac{360}{n\pi r^2}$$

Canceling common factors gives the ratio

$$\frac{2}{r}$$

■

3–4 EXERCISES

Perform the indicated operations.

1. $\dfrac{4x^2y^3}{2x^3} \cdot \dfrac{16x^4}{8y^3}$

2. $\dfrac{35a^3b}{6b^7} \cdot \dfrac{22b^4}{55a^5b}$

3. $\dfrac{5xy^3}{7x^3y} \cdot \dfrac{6x^2y}{30x^4y^3}$

4. $\dfrac{14m^4n^2}{42mn^5} \cdot \dfrac{m^2}{7m^5}$

5. $\dfrac{4x(2x + 3)}{(x - 1)} \cdot \dfrac{(x - 1)}{8x^2}$

6. $\dfrac{13ab^3}{2a^4(3 - 2x)} \cdot \dfrac{(2x - 3)}{26a^4}$

7. $\dfrac{(7x - 1)}{x(x + 2)} \cdot \dfrac{5x^2(x + 2)}{(7x + 3)}$

8. $\dfrac{(3x + 4)(x - 1)}{(1 - x)(2x + 3)} \cdot \dfrac{(2x + 3)}{3x}$

9. $\dfrac{8x^2y}{3y} \div \dfrac{24x}{27xy}$

10. $\dfrac{18m^4n^2}{4mn} \div \dfrac{6m^3n}{12mn}$

11. $\dfrac{2ab^2c}{9a} \div \dfrac{10b^3}{3a^4c^3}$

12. $\dfrac{48a(2 - b)}{27a^3} \div \dfrac{16(b - 2)}{9a^4}$

13. $\dfrac{8m^2(a + 2b)}{32m} \div \dfrac{16(a - 2b)}{64m^2}$

14. $\dfrac{(x - 3)(2x - 7)}{3x} \div \dfrac{(2x - 7)}{(x - 3)}$

15. $\dfrac{14x^2(x + y)}{3x^3(2x - y)} \div \dfrac{28(x + y)}{18(y - 2x)}$

16. $\dfrac{12a^3(a + 2b)}{(a + 2b)} \div \dfrac{6ab^3}{2b}$

17. $\dfrac{6x^2 + x - 2}{16x^3} \cdot \dfrac{4x^2 + 12x}{2x^2 + 5x - 3}$

18. $\dfrac{8m^2 - 4m}{20m^3} \cdot \dfrac{2m + 1}{4m^2 - 1}$

19. $\dfrac{24x^2 + 2x - 2}{16x^2 - 1} \cdot \dfrac{4x^2 - 19x - 5}{x - 5}$

20. $\dfrac{3a^2 + 5ab - 2b^2}{5a^2 + 6ab - 8b^2} \cdot \dfrac{5a^2 + ab - 4b^2}{2a^2 + 7ab + 5b^2}$

21. $\dfrac{a^2 - b^2}{a^2 + b^2} \cdot \dfrac{2a^2 + ab - b^2}{a^2 + 2ab + b^2}$

22. $\dfrac{7s^2 + 8st - 12t^2}{21s^2 - 25st + 6t^2} \cdot \dfrac{3s - t}{4s^2 + 17st + 18t^2}$

23. $\dfrac{7m^2 + 19m - 6}{2m^2 + m - 15} \cdot \dfrac{3m^2 + 5m - 8}{14m^2 + 17m - 6}$

24. $\dfrac{12x^2 + 23xy - 2y^2}{6x^2 + 13xy - 5y^2} \cdot \dfrac{4x^2 + 20xy + 25y^2}{x^2 + 3xy + 2y^2}$

25. $\dfrac{72s^2 - 42st + 5t^2}{18s^2 - 9st + t^2} \cdot \dfrac{9s^2 - t^2}{36s^2 - 3st - 5t^2}$

26. $\dfrac{10m^2 - 17mn + 3n^2}{m + 5n} \cdot \dfrac{7m^2 + 36mn + 5n^2}{14m^2 - 19mn - 3n^2}$

27. $\dfrac{14a^2 + 5ab - b^2}{2a + b} \div \dfrac{3a^2 - 11ab - 4b^2}{15a^2 + 2ab - b^2}$

28. $\dfrac{4s^2 - 11st - 3t^2}{8s^2 + 14st + 3t^2} \div \dfrac{5s^2 - 6st + t^2}{2s^2 + st - 3t^2}$

29. $\dfrac{9x^2 - 10xy + y^2}{4x^2 - 3xy - y^2} \div \dfrac{18x^2 + 7xy - y^2}{6x^2 + 13xy + 5y^2}$

30. $\dfrac{2a^2 + 11ab + 5b^2}{2a^2 - 3ab + b^2} \div \dfrac{3a^2 - 4ab - 15b^2}{2a^2 - 7ab + 3b^2}$

31. $\dfrac{x^2 - 9y^2}{x^2 + 4xy + 4y^2} \div \dfrac{x^2 + 2xy - 3y^2}{2x^2 - 5xy - 3y^2}$

32. $\dfrac{6x^2 - 7xy + y^2}{6x - 6y} \div \dfrac{36x^2 - y^2}{12x^2 + 8xy + y^2}$

33. $\dfrac{5s^2 + 11st + 6t^2}{3s^2 - 4st + t^2} \div \dfrac{s^2 - t^2}{6s^2 - 23st + 7t^2}$

34. $\dfrac{7m^2 - 21mn}{2m^2 - 5mn - 3n^2} \div \dfrac{28m + 14n}{6m^2 + 5mn + n^2}$

35. $\dfrac{9m^2 + 12mn + 4n^2}{3m^2 + 10mn - 8n^2} \div \dfrac{3m^2 - mn - 2n^2}{m^2 + 3mn - 4n^2}$

36. $\dfrac{6s^2t - 3st^2}{2s^2 + st - t^2} \div \dfrac{3s + 4t}{s^2 - t^2}$

37. $\dfrac{x^2 - 16}{x^2 + 6x + 9} \div \dfrac{4 - x}{x^2 - 9}$

38. $\dfrac{(5 - x)(x + 1)}{(x - 1)} \cdot \dfrac{10x^2 - 10}{x^2 - 25}$

39. The centripetal force F of an object traveling in a circular path is its mass m times its centripetal acceleration a. If $m = w/g$ and $a = v^2/r$, find the expression for F.

40. Find the ratio of the area of a circle (πr^2) and the circumference of a circle ($2\pi r$).

M 41. Average power is defined to be the time rate of doing work. If the formula for power is solved for time, the units in the metric system become

$$\frac{N \cdot m}{N\left(\frac{m}{s}\right)}$$

Simplify this expression to obtain the correct units for time.

42. The area A of a regular polygon is one-half the product of the apothem a and the perimeter p ($A = ap/2$). If the apothem is given by $as/2$, and the perimeter is $4s^2/a$, find the area of the polygon in terms of s.

43. Convert 250 mi/h to ft/s by multiplying 5,280 ft/1 mi and 1 h/3,600 s.

44. Convert 30 yd³/s to ft³/min by multiplying by 27 ft³/1 yd³ and 60 s/1 min.

3–5

ADDITION AND SUBTRACTION OF FRACTIONS

So far, each operation with algebraic fractions has followed the same procedure as the respective operation with arithmetic fractions. The operations of addition and subtraction of algebraic fractions are no exception. To add and subtract both algebraic fractions and arithmetic fractions, we must find a common denominator. Before actually adding and subtracting fractions, however, we will examine how to find the **lowest common denominator (LCD),** which is the smallest quantity that is divisible by each denominator.

To find the lowest common denominator in arithmetic, first break each denominator into prime factors. Then form the common denominator by finding the product of the primes to the highest exponent in which the primes occur in any given denominator. For example, to find a common denominator for $\frac{7}{42} + \frac{5}{18}$, we break the denominators into prime factors as $42 = 7 \cdot 3 \cdot 2$ and $18 = 3^2 \cdot 2$. The common denominator consists of each factor raised to is largest power. Thus, $7 \cdot 3^2 \cdot 2 = 126$ is the lowest common denominator for 42 and 18. As so many times in the past, the process of finding a lowest common denominator for an algebraic fraction is the same as that used in arithmetic.

EXAMPLE 1 Find the lowest common denominator for the following fractions:

$$\frac{8}{15x^2y}, \quad \frac{15}{40xy^4}, \quad \frac{18}{x^3y^2}$$

Solution First, we factor each denominator.

$$15x^2y = 5 \cdot 3 \cdot x^2 \cdot y$$
$$40xy^4 = 5 \cdot 2^3 \cdot x \cdot y^4$$
$$x^3y^2 = x^3 \cdot y^2$$

Choose the largest exponent for each factor: $5 \cdot 3 \cdot 2^3 \cdot x^3 \cdot y^4$

The lowest common denominator is the product of these factors, or $120x^3y^4$. ■

EXAMPLE 2 Find the lowest common denominator for the following fractions:

$$\frac{6x}{2x - 4}, \quad \frac{2x - 3}{x^2 - 4x + 4}, \text{ and } \frac{7}{x^2 - 4}$$

Solution First, we factor each denominator.

$$\begin{aligned} 2x - 4 &= 2 \cdot (x - 2) \\ x^2 - 4x + 4 &= (x - 2)^2 \\ x^2 - 4 &= (x - 2) \cdot (x + 2) \end{aligned}$$

Choose the largest exponent for each factor:

$$2 \cdot (x - 2)^2 \cdot (x + 2)$$

The lowest common denominator is $2(x - 2)^2(x + 2)$. ■

To begin the discussion of adding and subtracting fractions, let us look once again at the arithmetic example. To add

$$\frac{7}{42} + \frac{5}{18}$$

we first find the common denominator of 126, as done previously. Then we change each fraction to an equivalent fraction with 126 as the denominator (refer to Section 3–3 on equivalent fractions). Thus,

$$\frac{7}{42} = \frac{7 \cdot 3}{42 \cdot 3} = \frac{21}{126}$$

$$\frac{5}{18} = \frac{5 \cdot 7}{18 \cdot 7} = \frac{35}{126}$$

Next, we combine the fractions with the common denominator.

$$\frac{21}{126} + \frac{35}{126} = \frac{21 + 35}{126} = \frac{56}{126}$$

The final step is to reduce the fraction to lowest terms, if necessary.

$$\frac{56}{126} = \frac{28}{63} = \frac{4}{9}$$

The sum of the two fractions is 4/9. We follow this same procedure in adding and subtracting algebraic fractions.

Addition or Subtraction of Algebraic Fractions

1. Find the lowest common denominator.
2. Change each fraction to an equivalent fraction with the common denominator as the denominator.
3. Simplify the numerator by removing parentheses and combining like terms.
4. Reduce the fraction to lowest terms, if possible.

EXAMPLE 3 Perform the indicated operation:

$$\frac{2x}{3x - 1} - \frac{6}{2x + 5}$$

Solution The lowest common denominator is $(3x - 1)(2x + 5)$. Change each fraction to an equivalent fraction with $(3x - 1)(2x + 5)$ as the denominator.

$$\frac{2x}{3x - 1} \cdot \frac{2x + 5}{2x + 5} = \frac{2x(2x + 5)}{(3x - 1)(2x + 5)}$$

$$\frac{6}{2x + 5} \cdot \frac{3x - 1}{3x - 1} = \frac{6(3x - 1)}{(3x - 1)(2x + 5)}$$

Next, we combine the two fractions over the common denominator.

$$\frac{2x(2x + 5) - 6(3x - 1)}{(3x - 1)(2x + 5)}$$

Then we remove parentheses and combine like terms.

$$\frac{4x^2 + 10x - 18x + 6}{(3x - 1)(2x + 5)}$$

$$\frac{4x^2 - 8x + 6}{(3x - 1)(2x + 5)}$$

Last, we try to reduce the fraction to lowest terms.

$$\frac{2(2x^2 - 4x + 3)}{(3x - 1)(2x + 5)}$$

Since factoring the numerator reveals no common factors, the fraction is reduced and either form is acceptable. ■

EXAMPLE 4 Subtract the following fractions:

$$\frac{x - 3}{x^2 + 7x + 10} - \frac{2x + 1}{2x^2 + x - 6}$$

Solution First, we *determine the common denominator.*

$$x^2 + 7x + 10 = (x + 5)(x + 2)$$
$$2x^2 + x - 6 = \quad (x + 2)(2x - 3)$$

The lowest common denominator is $(x + 5)(x + 2)(2x - 3)$. Next, we *change each fraction to an equivalent fraction* with the lowest common denominator as the denominator.

$$\frac{x - 3}{(x + 5)(x + 2)} \cdot \frac{2x - 3}{2x - 3} = \frac{(x - 3)(2x - 3)}{(x + 2)(x + 5)(2x - 3)}$$
$$\frac{2x + 1}{(2x - 3)(x + 2)} \cdot \frac{x + 5}{x + 5} = \frac{(2x + 1)(x + 5)}{(x + 2)(x + 5)(2x - 3)}$$

Then we place the numerator of the fraction over the common denominator.

$$\frac{(x - 3)(2x - 3) - (2x + 1)(x + 5)}{(x + 2)(x + 5)(2x - 3)}$$

We *simplify the numerator* by removing parentheses and combining like terms.

$$\frac{2x^2 - 3x - 6x + 9 - 2x^2 - 10x - x - 5}{(x + 2)(x + 5)(2x - 3)}$$
$$\frac{-20x + 4}{(x + 2)(x + 5)(2x - 3)}$$

Last, we try to *reduce the fraction to lowest terms* by factoring the numerator.

$$\frac{-4(5x - 1)}{(x + 2)(x + 5)(2x - 3)}$$

Since factoring the numerator does not reveal a common factor, either of the last two fractions is an acceptable answer. ■

EXAMPLE 5 The equivalent capacitance C_e of three capacitors in series is given by

$$\frac{1}{C_e} = \frac{1}{C_1} + \frac{1}{C_2} + \frac{1}{C_3}$$

Solve for C_e by combining the fractions on the right side of the equation and inverting both sides of the equation.

Solution To add the fractions on the right side of the equation, we must obtain a common denominator, $C_1C_2C_3$, and write each fraction as an equivalent fraction with the common denominator.

$$\frac{1}{C_e} = \frac{C_2C_3 + C_1C_3 + C_1C_2}{C_1C_2C_3}$$

Taking the reciprocal of each side of the equation gives

$$C_e = \frac{C_1C_2C_3}{C_2C_3 + C_1C_3 + C_1C_2}$$

■

Complex Fractions

As you will see in the next chapter, removing negative exponents in an algebraic expression can lead to a **complex fraction,** which is a fraction whose numerator and/or denominator contains one or more fraction. Simplifying a complex fraction can involve addition, subtraction, multiplication, and division of fractions.

Simplifying a Complex Fraction

1. Simplify the numerator and denominator of the complex fraction by adding or subtracting the fractions, as necessary.
2. Divide the fractions by inverting the divisor and applying the rules for multiplication.

EXAMPLE 6 Simplify the following complex fraction:

$$\frac{\dfrac{x + 2}{x} - \dfrac{3}{2}}{\dfrac{7}{x} + \dfrac{5}{4x^2}}$$

Solution First, we must *obtain a single fraction* in the numerator and denominator by using addition or subtraction.

$$\frac{\dfrac{2(x + 2) - 3x}{2x}}{\dfrac{7(4x) + 5(1)}{4x^2}} = \frac{\dfrac{-x + 4}{2x}}{\dfrac{28x + 5}{4x^2}}$$

Then we *divide the complex fraction* by inverting the divisor and multiplying.

$$\frac{-x + 4}{2x} \cdot \frac{\overset{2x}{4x^2}}{28x + 5}$$

$$\frac{2x(-x + 4)}{28x + 5}$$

■

EXAMPLE 7 Simplify the following complex fraction:

$$\frac{\dfrac{2x + 1}{x - 3} + \dfrac{5}{x + 3}}{\dfrac{6x - 8}{x^2 - 9}}$$

Solution First, we *add the fractions in the numerator*.

$$\frac{\dfrac{(2x + 1)(x + 3) + 5(x - 3)}{(x - 3)(x + 3)}}{\dfrac{6x - 8}{x^2 - 9}}$$

$$\frac{\dfrac{2x^2 + 7x + 3 + 5x - 15}{(x - 3)(x + 3)}}{\dfrac{6x - 8}{x^2 - 9}}$$

$$\frac{\dfrac{2x^2 + 12x - 12}{(x - 3)(x + 3)}}{\dfrac{6x - 8}{x^2 - 9}}$$

Last, we *divide the fraction* by inverting the divisor and multiplying.

$$\frac{2x^2 + 12x - 12}{(x - 3)(x + 3)} \cdot \frac{x^2 - 9}{6x - 8}$$

$$\frac{\cancel{2}(x^2 + 6x - 6)}{\cancel{(x - 3)(x + 3)}} \cdot \frac{\cancel{(x + 3)(x - 3)}}{\cancel{2}(3x - 4)}$$

$$\frac{x^2 + 6x - 6}{3x - 4}$$

■

3–5 EXERCISES

Add or subtract the following fractions.

1. $\dfrac{2x}{8x} + \dfrac{3}{6x^2} - \dfrac{2x^2}{4x^3}$

2. $\dfrac{4}{7y} + \dfrac{8}{2y^3} + \dfrac{6x}{14}$

3. $\dfrac{5x - 1}{2} + \dfrac{6x}{3x} - \dfrac{1}{4x^2}$

4. $\dfrac{3x}{4x^4} - \dfrac{7x + 4}{6x} - \dfrac{2}{3x^3}$

5. $\dfrac{2m + 3}{6m^2} - \dfrac{5m + 1}{5m^3}$

6. $\dfrac{4x - 9}{7x^4} - \dfrac{3x - 1}{2x}$

7. $\dfrac{4x - 1}{x + 1} - \dfrac{6x}{x^2}$

8. $\dfrac{8}{3n^3} + \dfrac{7n}{2n - 1}$

9. $\dfrac{4y - 3}{7y + 2} - \dfrac{6y}{3y^2}$

10. $\dfrac{x}{3x + 5} - \dfrac{5x - 6}{4x}$

11. $\dfrac{2x}{2x^2 - x} + \dfrac{x + 1}{3x}$

12. $\dfrac{3x - 4}{4x - 8} + \dfrac{6}{8x}$

13. $\dfrac{a + 1}{3a - 6} - \dfrac{5a}{a - 2}$

14. $\dfrac{7m}{7m - 14} + \dfrac{4}{21m^3}$

15. $\dfrac{4x + 1}{5x^2 - 15x} + \dfrac{8x}{10x}$

16. $\dfrac{5x - 2}{6x^2 - 18x} - \dfrac{3x}{3x^2}$

17. $\dfrac{3y}{2y - 1} + \dfrac{4y}{3y + 2}$

18. $\dfrac{8x}{4x + 5} - \dfrac{7x}{6x - 1}$

19. $\dfrac{9}{3x + 6} - \dfrac{4}{2x - 5}$

20. $\dfrac{3}{4c^2 - c} + \dfrac{c}{5c + 7}$

21. $\dfrac{3z + 1}{7z^2 - 14z} + \dfrac{z - 1}{2z + 3}$

22. $\dfrac{2x}{5x - 1} + \dfrac{3x - 1}{2x + 7}$

23. $\dfrac{2x}{8x + 7} + \dfrac{x - 1}{3x - 8}$

24. $\dfrac{d - 7}{4d + 1} - \dfrac{d^2 - 6}{7d + 3}$

25. $\dfrac{b - 3}{2b + 1} - \dfrac{2b + 9}{5b + 3}$

26. $\dfrac{x + 3}{4x + 5} + \dfrac{2x - 4}{5 - 3x}$

27. $\dfrac{9x - 2}{7x - 4} + \dfrac{3x + 6}{4x + 3}$

28. $\dfrac{x^2}{2x + 11} - \dfrac{3x + 1}{2x + 7}$

29. $\dfrac{4n}{n - 4} + \dfrac{2}{4 - n}$

30. $\dfrac{7x}{6x - 1} - \dfrac{x^2}{1 - 6x}$

31. $\dfrac{2k - 1}{3k^2 + 5k - 2} + \dfrac{5k}{6k^2 + k - 1}$

32. $\dfrac{y + 3}{y^2 - 16} - \dfrac{6y}{2y^2 - 11y + 12}$

33. $\dfrac{10x + 7}{12x^2 - 40x + 12} - \dfrac{5x + 2}{6x^2 + 16x - 6}$

34. $\dfrac{x^2 + 5x}{x^2 - x - 12} + \dfrac{4 + 6x}{3x^2 - 7x - 20}$

35. $\dfrac{2y - 4}{9 - y^2} + \dfrac{6y + 5}{y - 3}$

36. $\dfrac{7b + 3}{25 - b^2} - \dfrac{b - 9}{b - 5}$

37. $\dfrac{2x}{x^2 - 3x - 4} - \dfrac{x}{3x^2 + 5x + 2}$

38. $\dfrac{4c}{c^2 - 25} + \dfrac{3}{c^2 - 10c + 25}$

39. $\dfrac{2x + 1}{x^2 - x - 6} + \dfrac{3x}{x^2 + 3x - 4}$

40. $\dfrac{3x - 4}{x^2 + 5x + 6} + \dfrac{2x + 5}{x^2 + 4x + 3}$

Simplify the following complex fractions.

41. $\dfrac{\dfrac{5}{x} - \dfrac{3x}{y}}{\dfrac{6}{x^2} + 5}$

42. $\dfrac{\dfrac{7x + 1}{x^2} + \dfrac{4}{x}}{\dfrac{22x + 2}{5x}}$

43. $\dfrac{\dfrac{4x}{x + 1} - \dfrac{2}{x - 1}}{\dfrac{3}{x - 1} + \dfrac{6x}{x + 1}}$

44. $\dfrac{\dfrac{7}{x} - \dfrac{4x}{x - 1}}{\dfrac{3}{x + 4} + \dfrac{2x}{x}}$

45. $\dfrac{\dfrac{7}{2x + 1} - \dfrac{3x}{x - 1}}{\dfrac{4x + 5}{2x^2 - x - 1}}$

46. $\dfrac{\dfrac{8x}{5x^2} + \dfrac{4}{x^2 - x}}{\dfrac{4x + 3}{10x^3 - 10x^2}}$

47. $\dfrac{\dfrac{6}{x - 4} + \dfrac{2x}{2x + 3}}{\dfrac{2x - 5}{3} - \dfrac{4}{x}}$

48. $\dfrac{\dfrac{3x - 2}{x^2 - 1} + \dfrac{6x}{x + 1}}{\dfrac{3x - 9}{x^2 - 2x - 3}}$

49. $\dfrac{\dfrac{4a}{2a - b} + \dfrac{6b}{2a + b}}{\dfrac{4b}{6a + 3b} - \dfrac{3}{a}}$

50. $\dfrac{\dfrac{7x - 1}{2x - 4} + \dfrac{3x}{4x - 8}}{\dfrac{3x}{x^2 - 4} + \dfrac{8}{x + 2}}$

E 51. The reciprocal of the total resistance in a parallel circuit is the sum of the reciprocals of the individual resistances. If the individual resistances are represented by $8x$, $16x^3$, and $8x^2 - 32x$, respectively, find a simplified expression for the reciprocal of the total resistance.

52. Find a simplified expression for the sum of the reciprocals of two consecutive even integers.

53. Find a simplified expression for the sum of the reciprocals of two consecutive integers.

E 54. The reactance X of an inductor and capacitor in series is given by

$$X = \omega L - \frac{1}{\omega C}$$

Find a simplified expression for the reactance.

55. The distance s that a ball falls, neglecting friction, is given by

$$s = vt + \frac{at^2}{2}$$

Simplify the expression on the right side of the equation.

56. An expression that results from Einstein's theory of relativity relating the velocity of two objects approaching each other is

$$\frac{v_1 - v_2}{1 - \dfrac{v_1 v_2}{c^2}}$$

where c is the speed of light. Simplify this expression.

57. A car travels a distance d_1 at a rate v_1; then it travels another distance d_2 at another rate v_2. The average speed for the entire trip is given by

$$\frac{d_1 + d_2}{\dfrac{d_1}{v_1} + \dfrac{d_2}{v_2}}$$

Simplify this expression.

M 58. An equation in fluids is $P = Dhg$ where P is the pressure, D is the density of the liquid, h is the height of the column of liquid, and g is the gravitational acceleration. If this equation is solved for h, we have $h = P/Dg$. Substituting in the proper units gives

$$\frac{\dfrac{\text{kg} \cdot \text{m/s}^2}{\text{m}^2}}{\dfrac{\text{kg}}{\text{m}^3} \cdot \dfrac{\text{m}}{\text{s}^2}}$$

Simplify this expression to get the correct units for the height of the column of liquid.

3–6

FRACTIONAL EQUATIONS AND APPLICATIONS

Thus far in this chapter, we have discussed operations with algebraic fractions. The technique learned in this chapter as well as in Chapter 1 allows us to add another type of equation to the list that we can solve: equations with algebraic fractions. We follow the same procedure to solve these equations as we used in Chapter 1. These steps are repeated below.

Solving Equations with Fractions

1. Eliminate fractions by multiplying each term of the equation by the lowest common denominator (LCD). Find the LCD as outlined in Section 3–5.
2. Eliminate parentheses.
3. Isolate the variable using the Transposition Property.
4. Divide both sides of the equation by the coefficient of the variable.
5. Check your answer by substituting into the original equation.

EXAMPLE 1 Solve the following equation:

$$\frac{4}{x} - \frac{7}{x^2} = \frac{5}{x}$$

Solution First, we *eliminate the fractions* by multiplying each term by the LCD, x^2.

$$\textcircled{x^2}\left(\frac{4}{x}\right) - \textcircled{x^2}\left(\frac{7}{x^2}\right) = \textcircled{x^2}\left(\frac{5}{x}\right)$$

$$4x - 7 = 5x$$

Then we *isolate the variable* by transposing $4x$.

$$-7 = 5x - 4x$$
$$-7 = x$$

Checking

$$\frac{4}{-7} - \frac{7}{(-7)^2} = \frac{5}{-7}$$
$$-\frac{4}{7} - \frac{1}{7} = \frac{-5}{7}$$

Since checking results in a true statement, the solution is $x = -7$. ■

Extraneous Roots

In solving equations that contain fractions with the variable in the denominator, always check for extraneous solutions that result in division by zero.

EXAMPLE 2 Solve the following:

$$\frac{1}{x-3} - \frac{1}{x+3} = \frac{2x}{x^2-9}$$

Solution First, we eliminate the fractions by multiplying each term by the lowest common denominator, $(x - 3)(x + 3)$.

$$\frac{\cancel{(x-3)}(x+3)(1)}{\cancel{(x-3)}} - \frac{\cancel{(x+3)}(x-3)(1)}{\cancel{(x+3)}} = \frac{2x\cancel{(x+3)}\cancel{(x-3)}}{\cancel{(x+3)}\cancel{(x-3)}}$$

$$1(x + 3) - 1(x - 3) = 2x$$

Then we remove parentheses and solve for x.

$$x + 3 - x + 3 = 2x \quad \text{remove parentheses}$$
$$6 = 2x \quad \text{combine like terms}$$
$$3 = x \quad \text{divide}$$

Checking

$$\frac{1}{3-3} - \frac{1}{3+3} = \frac{6}{9-9}$$

Since $x = 3$ leads to division by zero, $x = 3$ is an extraneous solution, and this equation has no solution. ■

Solving Formulas

As noted in Chapter 1, formulas must often be rearranged and solved for a different variable. The next two examples illustrate solving a formula containing fractions.

EXAMPLE 3 In business, straight-line depreciation of equipment is given by the equation

$$d = \frac{C - C_L}{L}$$

where C = cost of new equipment, d = annual amount of depreciation, and C_L = value of the equipment at L years. Solve the formula for C_L and find the value of a typewriter after 5 years if the depreciation is \$125/year and the typewriter cost \$1,000 originally.

Solution To solve the formula for C_L, we begin by removing fractions.

$$L(d) = L\left(\frac{C - C_L}{L}\right) \quad \text{multiply by LCD}$$

$$dL = C - C_L \quad \text{eliminate fractions}$$

$$dL - C = -C_L \quad \text{transpose}$$

$$C_L = C - dL \quad \text{divide by } -1$$

Substitute the numerical values $d = 125$, $L = 5$, and $C = 1{,}000$.

$$C_L = 1{,}000 - 125(5)$$

$$C_L = 1{,}000 - 625$$

$$C_L = 375$$

The typewriter is worth \$375 at the end of 5 years. ■

EXAMPLE 4 In surveying, parallax p must be considered in aerial photos. Parallax is an apparent change in the direction of an object caused by a change in the observer's position. The relationship is given by

$$\frac{H - h}{f} = \frac{B}{p}$$

Solve the formula for p and find the value for p if $H = 3{,}000$ ft, $h = 653$ ft, $f = 96$ ft, and $B = 300{,}000$ ft.

Solution

$$fp\left(\frac{H - h}{f}\right) = fp\left(\frac{B}{p}\right) \quad \text{multiply by LCD}$$

$$p(H - h) = Bf \quad \text{eliminate fraction}$$

$$p = \frac{Bf}{H - h} \quad \text{divide}$$

Substituting the numerical value gives

$$p = \frac{(300{,}000)(96)}{3{,}000 - 653}$$

$$p = 12{,}271 \text{ ft}$$

■

Solving Rate Problems

One application that results in a fractional equation is called a *rate problem*. In these problems you are asked to find the time it takes for two people or machines to complete a task working at constant rates. The next problem illustrates a rate problem.

EXAMPLE 5 One draftsman can complete a drawing in 3 hours, while a second draftsman takes 5 hours to complete the same job. How long would it take to complete the job if the two draftsmen worked together?

Solution The amount of work completed is the product of time and the rate of work. Therefore, the work done in one unit of time is 1 divided by the rate of work. If the first draftsman can do the entire job in 3 hours, then he can complete 1/3 of the job in one hour. Similarly, the second draftsman will have completed 1/5 of the job in one hour; and if x represents the time the job will require if they work together, then $1/x$ of the job will be done in one hour. Since their combined rate is equal to the sum of their individual rates, the equation is

$$\frac{1}{3} + \frac{1}{5} = \frac{1}{x}$$

First, multiply by the common denominator, $15x$.

$$15x\left(\frac{1}{3}\right) + 15x\left(\frac{1}{5}\right) = 15x\left(\frac{1}{x}\right)$$

Then solve the resulting equation for x.

$$5x + 3x = 15$$

$$8x = 15$$

$$x = \frac{15}{8}$$

Therefore, it takes 1.875 hours to complete the drawing if the draftsmen work together.

■

Solving Motion Problems

In this section we will extend our knowledge of motion problems to include fractional equations. The method of "setting up" and solving these equations is the same as discussed in Chapter 1.

EXAMPLE 6 Fred can go 10 miles upstream in the same time that it takes to go 15 miles downstream. If the speed of the current is 5 mi/h, what is the speed of Fred's boat in still water?

Solution We will fill in a chart as we did in Chapter 1. The distance upstream is 10 miles, while the distance downstream is 15 miles. Let $x =$ the speed of the boat in still water. Then $x - 5$ will represent the speed of the boat upstream (against the current), and $x + 5$ will represent the speed of the boat downstream (with the current). Time is the quotient of distance and rate. These values are entered in the accompanying chart.

	distance ÷	rate =	time
	Distance	Rate	Time
Upstream	10	$x - 5$	$\frac{10}{x - 5}$
Downstream	15	$x + 5$	$\frac{15}{x + 5}$

Since the amount of time required to go upstream equals the time required to go downstream, the equation is

$$\frac{10}{x - 5} = \frac{15}{x + 5}$$

To solve the equation, multiply by the common denominator, $(x - 5)(x + 5)$.

$$(x + 5)\cancel{(x - 5)}\left(\frac{10}{\cancel{x - 5}}\right) = \cancel{(x + 5)}(x - 5)\left(\frac{15}{\cancel{x + 5}}\right)$$

$$10(x + 5) = 15(x - 5)$$

Eliminate parentheses and solve the equation.

$$\begin{aligned} 10x + 50 &= 15x - 75 \\ 10x - 15x &= -75 - 50 \\ -5x &= -125 \\ x &= 25 \end{aligned}$$

Fred's boat goes 25 mi/h in still water. ■

3–6 EXERCISES

Solve the equation for the variable.

1. $\frac{x}{3} - 1 = \frac{2}{5} + x$

2. $4 - x + \frac{1}{7} = \frac{1}{3}$

3. $\frac{8}{9} + \frac{x-2}{3} = 2$

4. $\frac{5}{6} + \frac{x}{4} = \frac{2x-1}{2}$

5. $\frac{8}{x} + \frac{1}{3} = \frac{4}{x}$

6. $\frac{3}{x} + 8 = \frac{3x+18}{x}$

7. $\frac{9}{x} = \frac{5}{x} - \frac{3}{4}$

8. $\frac{2x+6}{x} = \frac{4}{3}$

9. $\frac{7}{2x^2} = \frac{4}{x^2} - \frac{3}{4x}$

10. $\frac{8}{3x-1} = \frac{7x}{3x-1} - 2$

11. $\frac{1}{x} - \frac{4}{x^2} = \frac{1}{3x}$

12. $\frac{15}{x} - 7 = \frac{1}{x}$

13. $\frac{7}{3x^3} + \frac{1}{x^2} = \frac{5}{2x^3}$

14. $\frac{7}{x^4} + \frac{6}{x^3} = \frac{-11}{x^4}$

15. $\frac{2}{x^2} - \frac{9}{x^3} = \frac{5}{x^2}$

16. $\frac{2}{2x+5} = \frac{6}{3x-3}$

17. $\frac{3}{x-4} + \frac{1}{x+4} = \frac{5x+4}{x^2-16}$

18. $\frac{8}{x-2} + \frac{6}{x+2} = \frac{32}{x^2-4}$

19. $\frac{2x+3}{x-1} = \frac{4x}{2x+1}$

20. $\frac{9x}{x^2-9} = \frac{6}{x+3}$

21. $\frac{7}{3x-2} + \frac{5}{3x+5} = 0$

22. $\frac{10}{4x-3} + \frac{7}{2x+5} = 0$

23. $\frac{5}{8x+1} = \frac{2}{x-4}$

24. $\frac{3}{x+5} - \frac{4}{x-5} = \frac{30}{x^2-25}$

25. $\frac{3}{2x-1} + \frac{6}{2x+1} = \frac{15}{4x^2-1}$

26. $\frac{7}{x+3} - \frac{5}{x-4} = \frac{9}{x^2-x-12}$

27. $\frac{2x}{x-1} - \frac{3}{x+1} = \frac{2(x^2+2)}{x^2-1}$

28. $\frac{5}{2x+1} + \frac{2}{x-3} = \frac{14}{2x^2-5x-3}$

29. $\frac{1}{x+2} + \frac{3x}{x-5} = \frac{3x^2+30}{x^2-3x-10}$

30. $\frac{7x}{3x+5} - \frac{3}{x-1} = \frac{7x^2-3x-8}{3x^2+2x-5}$

31. The harmonic mean m of two numbers x and y is given by

$$m = \frac{2xy}{x+y}$$

Solve for y.

32. The equation for displacement s of an object thrown upward is given by

$$s = v_0t + \frac{1}{2}at^2$$

Solve for a.

M 33. An equation from hydrodynamics is

$$\frac{p}{d} = \frac{M}{d} - \frac{m^2}{2\pi^2r^2}$$

Solve for p.

I 34. In straight-line depreciation, solve for C if d = \$45, C_L = \$200, and L = 3 yr given that

$$d = \frac{C - C_L}{L}$$

35. In physics, the density of a substance is the mass of the substance divided by its volume. Find the volume of a substance if its density is 9.6 kg/m^3 and its mass is 4 kg.

M 36. The mechanical advantage of a mechanism is the quotient of the output force and input force. If the mechanical advantage is 830 and the output force is 25 lb, find the input force.

E 37. The power required by an electrical appliance is the product of voltage and current (watts = volts · amperes). What current is required for an iron when the power is 660 watts and the voltage is 110 volts.

E 38. Five electric light bulbs, each with a rating of 60 W and 110 V, are being operated by a 110-V dc power source. What is the value of the total load resistance? Use $P_T = V^2/R_T$ where P_T is total power, V is voltage, and R_T is total resistance.

E 39. A 50-m coil of copper wire has a resistance of 1.05 Ω. If 15 m of wire is cut from the coil, what is the resistance of the remaining 35 m of wire? (*Note:* $R_1/R_2 = L_1/L_2$.)

40. The formula relating the temperature in degrees Celsius C to the temperature in degrees Fahrenheit F is

$$C = \frac{5F - 160}{9}$$

Find the temperature Fahrenheit that corresponds to 20°C.

E 41. Conductors of different metals are compared by their resistivity. The formula is $\sigma = RA/L$ where σ is resistivity in ohm · meters; R is resistance in ohms; A is cross-sectional area in square meters; and L is length in meters. Determine the resistance of a nichrome wire if $\sigma = 0.000001\ \Omega$, $A = 0.000000128\ \text{m}^2$, and $L = 2\text{m}$.

C 42. In construction, the area of footing of a building is the ratio of the total weight of the building to the bearing capacity of the soil. If the area of footing is 24 ft^2 and the weight of the building is 60,000 lb, what is the bearing capacity of the soil?

E 43. The reactance of an inductor and capacitor in series is given by

$$X = \omega L - \frac{1}{\omega C}$$

If $L = 6$ henries (H), $W = 3$ rad/s, and $X = 3.5\ \Omega$, find C in microfarads (μF).

44. Find two consecutive integers such that four times the reciprocal of the larger integer is twice the reciprocal of the smaller integer.

45. If a number is added to the numerator and subtracted from the denominator of 7/9, the result is 3. Find the number.

46. Find three consecutive odd integers such that the quotient of the second and third integers is 3.

47. Find three consecutive even integers such that the quotient of the third and first integers is 2.

48. If twice a number is added to the numerator and three times the number is subtracted from the denominator of $-9/5$, the result is 5. Find the number.

49. One biology lab assistant can clean the monkey cages in 3 hours, while the other lab assistant can clean the same cages in 2 hours. How long does it take if they work together?

50. A carpenter can complete a certain job in 5 days, while his apprentice takes 8 days to do the same job. How long will it take if they work together?

51. A swimming pool can be filled by an inlet pipe in 2 hours, while it takes 5 hours to drain the pool. Starting with an empty pool, how long will it take to fill the pool if both inlet and drain are open? (Set this problem up as a rate problem, except subtract the two times since they work against each other.)

52. A battery can be charged in 4 hours, while the load will discharge it in 9 hours. Assuming the battery is dead and being charged and discharged at the same time, how long will it take to recharge the battery?

CHAPTER SUMMARY

Summary of Terms

algebraic fraction (p. 97)

complex fraction (p. 112)

difference of two squares (p. 88)

equivalent fraction (p. 97)

factored completely (p. 86)

factoring (p. 86)

lowest common denominator (LCD) (p. 108)

lowest terms (p. 97)

perfect square trinomial (p. 92)

prime (p. 86)

reduced fraction (p. 97)

Summary of Formulas

$x^2 - y^2 = (x + y)(x - y)$ difference of two squares

$$\left.\begin{array}{l} x^2 + 2xy + y^2 = (x + y)^2 \\ x^2 - 2xy + y^2 = (x - y)^2 \end{array}\right\} \text{perfect square trinomial}$$

CHAPTER REVIEW

Section 3–1

Factor using greatest common factor.

1. $7x^3 + 38x^8 - 28x^2$
2. $27x^5 - 18x^{10} + 3x^2$
3. $15a^2b + 40b^2c - 25ac^3$
4. $48yz^5 + 32y^4z^3$
5. $32m^8 + 8m - 24m^4$
6. $90a^5b^4 + 27a^3b^8 - 63a^6b^5$
7. $18x^3y^2 + 36x^2z - 72x^5y^3z^2$
8. $56x^3y^5 - 72x^8y^3$
9. $7x(2a + b) - 6y(2a + b)$
10. $4s^2(3x - 4y) - 3t(3x - 4y)$
11. $12x^2(m - 3n) + 6(m - 3n)$
12. $8a(2c - d) + 4(2c - d)$
13. $3r^2(2s - t) - 3r^2(7s + 5t)$
14. $r^3(5x - 4y) - r^3(2x + y)$
15. $8r^3(3x - y) + 4r^2(y + 4z)$
16. $6m^3(p - 4s) - 3m(m - p)$

Factor as the difference of two squares, if possible.

17. $16x^2 - 25y^2$
18. $64m^7 - 25n$
19. $9x^2 + 49y^2$
20. $121r^{16} - 81s^{14}$
21. $169s^2 - 121t^2$
22. $144x^2 - 49y^2$
23. $25c^4 - 36d^4$
24. $81z^4 - 16a^8$
25. The difference in the area of two cones with the same height but different radii r_1 and r_2 is $\frac{1}{3}\pi r_1^2 - \frac{1}{3}\pi r_2^2$. Factor this expression.
26. The difference in the volume of two cylinders with the same radius but different heights is $\pi r^2 h_1 - \pi r^2 h_2$. Factor this expression.
27. I The total weekly revenue of a company is represented by $400p - 8p^2$ where p is the price in dollars of the product. Factor this expression.

Section 3–2

Factor as a perfect square trinomial, if possible.

28. $64x^2 - 144xy + 81y^2$
29. $9p^2 - 6np + 4n^2$
30. $4x^2y^2 + 4xyz + z^2$
31. $169a^2 + 130a + 25$
32. $9a^2 - 6a + 1$
33. $49m^2 - 112mn + 64n^2$
34. $16m^4 - 40m^2n^2 - 25n^4$
35. $100s^4 - 260s^2t^2 + 169t^4$
36. The area of a square is given by $x^2 + 6x + 9$. Factor this expression.

Factor by the *ac* method, if possible.

37. $10x^2 + 23xy - 42y^2$
38. $72a^2 - ab - 56b^2$
39. $2a^2 - 7ab - 99b^2$
40. $28r^2 + rs - 45s^2$
41. $m^2 + 16m + 39$
42. $4x^2 - 7x + 3$
43. $32s^2 + 12st - 35t^2$
44. $b^2 + 9b + 8$
45. $72x^2 - 19xy - 40y^2$
46. $54s^4 + 69s^2 + 20$
47. $18s^4 - 73s^2y^2 + 35y^4$
48. $20a^2 - 58ac + 42c^2$
49. An object thrown upward with an initial velocity v_0 of 48 ft/s travels a distance s of 64 ft. The expression for the time t is $16t^2 - 48t + 64$. Factor the left side of this expression.

Section 3–3

Reduce each fraction to lowest terms.

50. $\dfrac{40x^2y^3}{45x^4y}$
51. $\dfrac{27m^4n^3}{54mn^5}$
52. $\dfrac{8x^3}{4x^5 - 16x^2}$
53. $\dfrac{6x^2y}{3x^3y^4 - 9xy^2}$
54. $\dfrac{16x^2 - 1}{12x^2 - 5x - 2}$
55. $\dfrac{8x^2 - 8}{5x^2 + x - 6}$
56. $\dfrac{6x^2 + 13x - 5}{1 - 9x^2}$
57. $\dfrac{4x^2 - x - 5}{25 - 16x^2}$

Section 3–4

Perform the indicated operation.

58. $\dfrac{7x^2y^3}{36x} \cdot \dfrac{3xy^4}{14y^9}$
59. $\dfrac{36x^5y^2}{18y^3} \cdot \dfrac{14y^3}{48x^8y^4}$
60. $\dfrac{4x - 4}{16x} \div \dfrac{9x^3 - 9x^2}{5x^5}$
61. $\dfrac{25a^5c^3}{36b^4c} \div \dfrac{30a^2b}{24a^4b^2c^5}$
62. $\dfrac{4x - 8}{3x - 5} \cdot \dfrac{25 - 9x^2}{3x^2 - x - 10}$
63. $\dfrac{14x^2 - 6x}{6x^2 + 7x - 20} \cdot \dfrac{2x^2 + 3x - 5}{6x^3 - 6x^2}$
64. $\dfrac{4x^2 - 4xy + y^2}{4x^2 - y^2} \div \dfrac{x^2 + 2xy + y^2}{x^2 - y^2}$

65. $\dfrac{15x^2 + 7xy - 2y^2}{3x^2 + 13xy + 4y^2} \cdot \dfrac{3x^2 - 5xy + 2y^2}{9x^2 - 4y^2}$

66. $\dfrac{15x^2y - 5xy^2}{14x^2 + 9xy + y^2} \div \dfrac{6x^2 + xy - y^2}{7x^3y + x^2y^2}$

67. $\dfrac{18x^2 - 24xy}{6x^2 - 10xy - 24y^2} \div \dfrac{9x^2 - 16y^2}{18x^3 - 54x^2y}$

Section 3–5

Perform the indicated operation.

68. $\dfrac{8x}{y^2} - \dfrac{3}{xy} + \dfrac{5}{x^3}$

69. $\dfrac{14}{5x} + \dfrac{9}{3x^2y} - \dfrac{5}{2y^3}$

70. $\dfrac{7x}{4x - 4} + \dfrac{5}{16x^2}$

71. $\dfrac{3x - 1}{2x - 8} + \dfrac{4}{x - 4}$

72. $\dfrac{3x}{2x + 1} - \dfrac{3x - 1}{4x^2 - 1}$

73. $\dfrac{6x}{x^2 - x - 2} + \dfrac{4x}{x^2 - 2x - 3}$

74. $\dfrac{x - 1}{x^2 - x - 12} + \dfrac{2x + 1}{x^2 - 3x - 4}$

75. $\dfrac{5x - 3}{9x^2 - 16} - \dfrac{2x + 7}{6x^2 + 5x - 4}$

76. $\dfrac{3x + 4}{4x^2 - 1} - \dfrac{x - 7}{2x^2 + 9x - 5}$

77. $\dfrac{9x + 2}{3x^2 + 11x - 20} + \dfrac{2x - 6}{x^2 + 4x - 5}$

78. $\dfrac{\dfrac{x + 2}{6}}{\dfrac{x^2 - 4}{3x^3}}$

79. $\dfrac{\dfrac{2x + 4}{3xy^2}}{\dfrac{8x + 16}{5x}}$

80. $\dfrac{\dfrac{7x - 14}{5x^2y^3}}{\dfrac{3x - 6}{x^2 - xy}}$

81. $\dfrac{\dfrac{9x^2 - 1}{6xy^3}}{\dfrac{3x + 1}{4y}}$

82. $\dfrac{\dfrac{x}{6} - \dfrac{2y^2}{3x}}{\dfrac{6x^2 - 12xy}{x + 2y}}$

83. $\dfrac{\dfrac{x}{8} - \dfrac{2}{4x}}{\dfrac{x - 2}{6x^3}}$

84. $\dfrac{\dfrac{7x}{x - 1} + \dfrac{5}{x + 1}}{\dfrac{3}{x} - \dfrac{2}{x - 1}}$

85. $\dfrac{\dfrac{9}{3x - 1} + \dfrac{6x}{x + 2}}{\dfrac{3}{x} - \dfrac{4}{21x^2}}$

86. $\dfrac{\dfrac{2x + 1}{3x - 2} + \dfrac{6x}{x - 1}}{\dfrac{4x + 3}{x - 3} + \dfrac{7}{x - 1}}$

87. $\dfrac{\dfrac{5x - 3}{x - 3} - \dfrac{2x + 3}{x + 2}}{\dfrac{2x}{5x + 4} + \dfrac{x + 1}{6x - 5}}$

Section 3–6

Solve the following equations.

88. $\dfrac{7}{x^2} + \dfrac{5}{x} = \dfrac{-3}{x^2}$

89. $\dfrac{9}{x} = \dfrac{-1}{2}$

90. $\dfrac{6}{x + 1} - \dfrac{3}{x - 1} = \dfrac{3}{x^2 - 1}$

91. $\dfrac{3}{x - 2} - \dfrac{2}{x + 2} = \dfrac{8}{x^2 - 4}$

92. $\dfrac{9}{x - 3} + \dfrac{7}{x - 4} = \dfrac{5x}{x^2 - 7x + 12}$

93. $\dfrac{5}{x + 1} + \dfrac{7}{x - 2} = \dfrac{10x}{x^2 - x - 2}$

94. $\dfrac{1}{3x} - \dfrac{4}{x - 1} = \dfrac{21}{3x^2 - 3x}$

95. $\dfrac{6x}{8x - 16} + \dfrac{3}{4} = \dfrac{3x}{x - 2}$

96. $\dfrac{-2}{x + 1} + \dfrac{4}{x - 3} = \dfrac{8x}{x^2 - 2x - 3}$

97. $\dfrac{7}{x + 4} - \dfrac{3}{x - 5} = \dfrac{-7}{x^2 - x - 20}$

CHAPTER TEST

The number in parentheses refers to the appropriate learning objective given at the beginning of the chapter.

1. Reduce the following to lowest terms: (5)

$$\frac{-18x^3y^4z}{48x^4yz}$$

2. Factor completely $36x^3y - 27x^2y^3 + 18x^2y$. **(1)**

3. Subtract the following: **(7)**

$$\frac{x - 4}{x - 3} - \frac{2x + 1}{2x - 3}$$

4. Factor completely $70x^3 - 115x^2y + 15xy^2$. **(1, 4)**

5. If 7 is added to two-fifths of a number, the result is 9. Find the number. **(9)**

6. Multiply the following: **(6)**

$$\frac{6x^2 - 5x - 4}{-3x^2 + 19x - 20} \cdot \frac{5x^2 - 9x - 18}{10x^2 + 17x + 6}$$

7. Factor completely $8m^3 - 28m^2n - 60mn^2$. **(1, 4)**

8. Simplify the following: **(8)**

$$\frac{\dfrac{4x + 2}{x}}{\dfrac{6x + 3}{x^2}}$$

9. Factor completely $28x^2 - 51xy - 27y^2$. **(4)**

10. Divide the following: **(6)**

$$\frac{7m^3n^2(n + 1)}{8m^5n^6} \div \frac{14n^4(n - 1)}{16mn^8}$$

11. The net rate of radiation is the difference between the rate of energy emission and the rate of energy absorption. This relationship is given by $e\pi T_1^4 - e\pi T_2^4$. Factor this expression. **(1, 2)**

12. One draftsman can complete a drawing in 4 hours, while it takes another draftsman 6 hours to do the same drawing. How long will it take them working together? **(9)**

13. Solve the following for x: **(9)**

$$\frac{3x - 1}{x - 1} - \frac{6x}{x + 2} = \frac{-3x^2}{x^2 + x - 2}$$

14. Factor completely $25x^2 - 20xy + 4y^2$. **(3)**

15. The ratio of yield per acre is given as $(16x^2 + 24x - 40):(x - 1)$. Simplify this ratio and calculate the yield per acre when $x = 5$. **(5)**

16. Solve the following for x: **(9)**

$$\frac{8x - 1}{x^2} + \frac{3}{x} = \frac{2}{x}$$

17. Work is defined to be the product of force and displacement. If 50 ft-lb of work is done in lifting an 8-lb object, how far was the object moved? **(9)**

SOLUTION TO CHAPTER INTRODUCTION

The formula for straight-line depreciation is given by

$$d = \frac{C - C_L}{L}$$

where d = annual depreciation, C = cost of new equipment, and C_L = value of the equipment after L years. You are asked to determine the value of the \$9,600 computer system after 2 years if the annual depreciation is \$430. Before substituting the numerical values, we solve the formula for C_L.

$$\begin{aligned} L(d) &= \left(\frac{C - C_L}{L}\right) L && \text{eliminate fractions} \\ Ld &= C - C_L && \text{simplify} \\ Ld - C &= -C_L && \text{transpose} \\ C - Ld &= C_L && \text{divide by } -1 \\ 9{,}600 - 2(430) &= C_L && \text{substitute numerical values} \\ 8{,}740 &= C_L && \end{aligned}$$

The computer system is currently worth \$8,740.

You are asked to find the inductance of a circuit containing an inductor and a capacitor given that the resonant frequency is 1,200 Hz and the capacitance is 1.4×10^{-6} H. (The solution to this problem is given at the end of the chapter.)

In this chapter we will discuss integral, fractional, and negative exponents. Since fractional exponents can be expressed in radical form, we will also discuss simplifying radicals and operations with radicals. We will conclude the chapter with a discussion of solving equations containing radicals, such as the one resulting from the problem above.

Learning Objectives

After completing this chapter, you should be able to

1. Simplify an algebraic expression with integral exponents (Section 4–1).
2. Simplify an algebraic expression with fractional exponents (Section 4–2).
3. Simplify radical expressions to simplest form (Section 4–3).
4. Add, subtract, multiply, and divide radical expressions (Section 4–4).
5. Solve equations involving radicals (Section 4–5).

Chapter 4

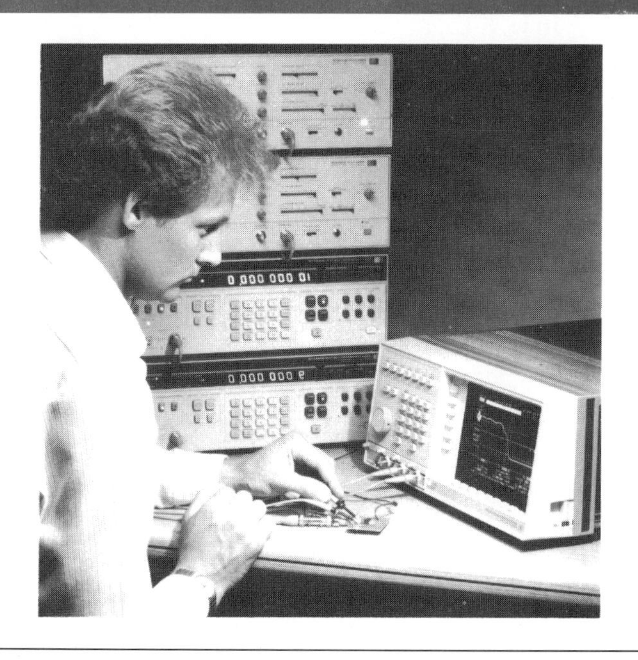

Exponents and Radicals

4–1

INTEGRAL EXPONENTS

In this section we will briefly review the Laws of Exponents and discuss negative and zero exponents. The Laws of Exponents are summarized below.

Laws of Exponents

For any real numbers a and b and integers m and n, the following rules apply:

Rule 4–1. $a^m \cdot a^n = a^{m+n}$
Rule 4–2. $(ab)^m = a^m b^m$
Rule 4–3. $(a^n)^m = a^{nm}$
Rule 4–4. $\dfrac{a^m}{a^n} = a^{m-n}$ if $m > n$, $a \neq 0$
Rule 4–5. $\dfrac{a^m}{a^n} = \dfrac{1}{a^{n-m}}$ if $m < n$, $a \neq 0$
Rule 4–6. $\left(\dfrac{a}{b}\right)^m = \dfrac{a^m}{b^m}$ $(b \neq 0)$
Rule 4–7. $a^0 = 1$ $(a \neq 0)$
Rule 4–8. $a^{-n} = \dfrac{1}{a^n}$ $(a \neq 0)$

You should be familiar with operations concerning positive, integer ex ponents. Example 1 reviews the use of Rules 4–1 through 4–7. Then we wil discuss in greater detail how to eliminate negative exponents using Rule 4–8.

EXAMPLE 1 Simplify each expression by applying the Laws of Exponents Your answer may contain negative exponents.

(a) $x^5 \cdot x^{-1}$ (b) $(a^2b^{-3})^{-4}$ (c) $\dfrac{x^8}{x^4}$ (d) $\dfrac{y^2}{y^8}$ (e) $\left(\dfrac{x^{-2}}{y^3}\right)^{-4}$
(f) $4m^0$ (g) $(4m)^0$

Solution

(a) $x^5 \cdot x^{-1} = x^{5+(-1)} = x^4$ — Rule 4–1

(b) $(a^2b^{-3})^{-4} = (a^2)^{-4}(b^{-3})^{-4} = a^{-8}b^{12}$ — Rules 4–2 and 4–3

(c) $\dfrac{x^8}{x^4} = x^{8-4} = x^4$ — Rule 4–4

(d) $\dfrac{y^2}{y^8} = \dfrac{1}{y^{8-2}} = \dfrac{1}{y^6}$ — Rule 4–5

(e) $\left(\dfrac{x^{-2}}{y^3}\right)^{-4} = \dfrac{(x^{-2})^{-4}}{(y^3)^{-4}} = \dfrac{x^8}{y^{-12}}$ — Rules 4–6 and 4–3

(f) $4m^0 = 4(1) = 4$ — Rule 4–7

(g) $(4m)^0 = 4^0m^0 = 1$ — Rules 4–2 and 4–7 ■

CAUTION ✦ Be careful in simplifying expressions containing zero exponents. In an expression such as $4m^0$, the zero exponent applies only to m. On the other hand, in an expression such as $(4m)^0$, the zero exponent applies to both the 4 and the m.

As a general rule, you should not leave zero or negative exponents in an algebraic expression. Rule 4–8 provides us with a method of eliminating negative exponents. To remove the negative exponents on a factor, move the factor across the fraction line. This technique is illustrated in the next example.

EXAMPLE 2 Simplify each expression using the Laws of Exponents. Express your answers with positive exponents only.

(a) $y^4 \cdot y^{-9}$ (b) $(2x^{-3}y^2)^3$ (c) $\left(\frac{4}{x^3}\right)^{-2}$ (d) $\left(\frac{m^{-4}}{n^3}\right)^{-2}$

Solution

(a) $y^4 \cdot y^{-9} = y^{4+(-9)} = y^{-5} = \frac{1}{y^5}$ ← change the sign of the exponent

(b) $(2x^{-3}y^2)^3 = (2)^3(x^{-3})^3(y^2)^3 = 8x^{-9}y^6 = \frac{8y^6}{x^9}$ ← change the sign of the exponent

(c) $\left(\frac{4}{x^3}\right)^{-2} = \frac{(4)^{-2}}{(x^3)^{-2}} = \frac{4^{-2}}{x^{-6}} = \frac{x^6}{4^2} = \frac{x^6}{16}$

$\frac{1}{x^{-6}} = x^6$

$4^{-2} = \frac{1}{4^2}$

(d) $\left(\frac{m^{-4}}{n^3}\right)^{-2} = \frac{(m^{-4})^{-2}}{(n^3)^{-2}} = \frac{m^8}{n^{-6}} = m^8n^6$ ■

NOTE ✦ There may be more than one correct method of simplifying an expression using the Laws of Exponents, as shown in the next example.

EXAMPLE 3 Simplify the following:

$$\left(\frac{-2a^3b^{-4}}{5a^{-5}b^{-6}}\right)^{-2}$$

Express the answer with positive exponents.

Solution We can simplify this expression by

1. applying the Laws of Exponents inside the parentheses and then simplifying the -2 power; or
2. simplifying the -2 power and then applying the Laws of Exponents to the result.

Both methods are shown below.

Applying the first method gives

$$\left(\frac{-2a^3b^{-4}}{5a^{-5}b^{-6}}\right)^{-2}$$

$$\left(\frac{-2a^3b^6a^5}{5b^4}\right)^{-2}$$ eliminate negative exponents inside parentheses (Rule 4–8)

$$\left(\frac{-2a^8b^2}{5}\right)^{-2}$$ simplify inside parentheses (Rules 4–1 and 4–4)

$$\frac{(-2)^{-2}(a^8)^{-2}(b^2)^{-2}}{(5)^{-2}}$$ eliminate parentheses (Rule 4–3)

$$\frac{(5)^2}{(-2)^2(a^8)^2(b^2)^2}$$ eliminate negative exponents (Rule 4–8)

$$\frac{25}{4a^{16}b^4}$$ simplify (Rule 4–3)

Applying the second method of removing the negative power on the parentheses gives

$$\frac{(-2)^{-2}(a^3)^{-2}(b^{-4})^{-2}}{(5)^{-2}(a^{-5})^{-2}(b^{-6})^{-2}}$$ Rule 4–2

$$\frac{(-2)^{-2}a^{-6}b^8}{(5)^{-2}a^{10}b^{12}}$$ Rule 4–3

$$\frac{(5)^2b^8}{(-2)^2a^{10}b^{12}a^6}$$ Rule 4–8

$$\frac{25}{4a^{16}b^4}$$ Rules 4–1, 4–3, and 4–5 ■

NOTE ✦ When you have a negative exponent on a fraction, you can eliminate the negative exponent by inverting the fraction inside the parentheses. Then simplify the result as we did it in the previous example. For example,

$$\left(\frac{-2a^3b^{-4}}{5a^{-5}b^{-6}}\right)^{-2} = \left(\frac{5a^{-5}b^{-6}}{-2a^3b^{-4}}\right)^2$$

Multinomial Expressions

The Laws of Exponents also apply to simplifying a multinomial expression. However, we must be careful to distinguish between terms and factors when we simplify negative exponents.

EXAMPLE 4 Simplify $a^{-2} + b - c^{-1}$.

Solution Each of the previous examples contained a single term. However, this example contains three terms: a^{-2}, b, and, ignoring the signs, c^{-1}. We must remove negative exponents within each *term* by moving *factors* with negative exponents across the fraction line. Removing the negative exponents within each term gives

$$a^{-2} + b - c^{-1}$$

$$\frac{1}{a^2} + b - \frac{1}{c}$$

After the negative exponents have been removed, no further simplification of exponents is necessary. However, we must simplify this fractional expression by adding the fractions using a common denominator. The simplified expression is

$$\frac{c + a^2bc - a^2}{a^2c}$$

■

EXAMPLE 5 Simplify $(3x + y^{-2})^{-3}$.

Solution To remove the negative exponent outside the parentheses, we take the factor across the fraction line.

$$\frac{1}{(3x + y^{-2})^3}$$ remove negative exponent outside the parentheses (Rule 4–8)

$$\frac{1}{\left(3x + \frac{1}{y^2}\right)^3}$$ remove negative exponent within the parentheses (Rule 4–8)

Notice that eliminating the negative exponents has led to a complex fraction. We simplify the complex fraction as we did in Chapter 3.

$$\frac{1}{\left(\frac{3xy^2 + 1}{y^2}\right)^3}$$ add the fractions within the parentheses

$$\frac{1}{\frac{(3xy^2 + 1)^3}{y^6}}$$ eliminate parentheses (Rule 4–6)

$$\frac{y^6}{(3xy^2 + 1)^3}$$ divide the fraction

We do not need to expand the denominator of the fraction. ■

EXAMPLE 6 The current I in a circuit with an external resistance R, a cell of electromotive force E, and internal resistance r can be written $I = Er^{-1}(Rr^{-1} + 1)^{-1}$. Simplify this expression.

Solution First, we remove the negative exponents outside the parentheses by taking the factors across the fraction line.

$$I = \frac{E}{r(Rr^{-1} + 1)}$$

Second, we remove the negative exponent in the denominator.

$$I = \frac{E}{r\left(\frac{R}{r} + 1\right)}$$

Last, we simplify the complex fraction.

$$I = \frac{E}{r\left(\frac{R + r}{r}\right)}$$

$$I = \frac{E}{R + r}$$

The next example illustrates a type of expression that you will need to simplify in calculus. ■

EXAMPLE 7 Simplify the following:

$$4(x - 2)^2(4x - 1)^{-2} + 7(x - 2)(4x - 1)^{-1}$$

Solution First, remove the negative exponents (which are both outside the parentheses) within each term by moving that factor with a negative exponent across the fraction line.

$$\frac{4(x - 2)^2}{(4x - 1)^2} + \frac{7(x - 2)}{(4x - 1)}$$

Next, add the fractions and simplify.

$$\frac{4(x - 2)^2 + 7(x - 2)(4x - 1)}{(4x - 1)^2}$$ find LCD ($(4x - 1)^2$), determine equivalent fractions, and express as one fraction

$$\frac{4(x^2 - 4x + 4) + 7(4x^2 - 9x + 2)}{(4x - 1)^2}$$ multiply binomials in numerator

$$\frac{4x^2 - 16x + 16 + 28x^2 - 63x + 14}{(4x - 1)^2}$$ remove parentheses in numerator

$$\frac{32x^2 - 79x + 30}{(4x - 1)^2}$$ add like terms in numerator

$$\frac{(32x - 15)(x - 2)}{(4x - 1)^2}$$ factor numerator ■

4–1 EXERCISES

Simplify and express all answers with positive exponents.

1. $b^9 \cdot b^{-4}$
2. $y^{-6} \cdot y^2$
3. $m^{-10} \cdot m^6$
4. $x^{12} \cdot x^3$
5. $(3x^{-2}y^4)^{-3}$
6. $(2ab^{-3})^{-2}$
7. $\left(\frac{2x^{-1}}{y}\right)^2$
8. $\left(\frac{4m}{n^{-3}}\right)^2$
9. $\left(\frac{6a^{-4}}{c^{-3}}\right)^{-2}$
10. $\left(\frac{5x}{y^{-4}}\right)^{-3}$
11. $(2a + b)^0$
12. $(14a + 8b)^0$
13. $m^0 + n^0$
14. $m + 3n^0$
15. $3x^0 + y$
16. $7x^0 + 8y^0$
17. $\left(\frac{3a}{2b^2}\right)^0$
18. $\left(\frac{17x^3}{54y^2}\right)^0$
19. 18^0
20. $(46^2)^0$
21. $\frac{x^{-4}y^3z^3}{x^6y^{-4}z^{-5}}$
22. $\frac{(a^0b^{-1})^{-3}}{(-a)^{-2}b^5}$
23. $\frac{(-ab^{-2})^0c^3}{a^{-3}}$
24. $\frac{x^4y^{-3}z^0}{-2x^{-3}z^{-4}}$
25. $\frac{m^{-3}n^{-2}p^3}{m^0p^{-3}}$
26. $\frac{m^{-2}n^{-3}}{m^4n^{-6}}$
27. $\left(\frac{a^{-3}b^4}{c^{-3}a}\right)^{-2}$
28. $\left(\frac{3x^{-2}y^4}{9x^3y^2}\right)^{-2}$
29. $\left(\frac{x^{-4}y^3z^0}{z^4y^{-5}}\right)^3$
30. $\left(\frac{a^{-4}b^2c}{ab^{-6}c^3}\right)^4$
31. $\left(\frac{2x^2y^{-3}}{4x^{-5}y}\right)^{-3}$
32. $\left(\frac{9m^{-6}n^3}{81m^{-7}n^{-4}}\right)^{-1}$
33. $3(x + 4)^2(2x + 5)^{-2} + (x + 4)(2x + 5)^{-1}$
34. $2(x + 1)^2(3x - 1)^{-2} - (x + 1)(3x - 1)^{-1}$
35. $a^{-1} + b^2 - c^0$
36. $x^{-2} + y^{-2}$
37. $x^2 + y^{-3}$
38. $(4x - 5)(x + 2)^{-2} + 3(4x - 5)(x + 2)^{-1}$
39. $m^4 + n^{-3} + p^0$
40. $(a^0 + b)^{-2} + c^2$
41. $(x - 6)^2(x + 1)^{-2} - 2(x - 6)(x + 1)^{-1}$
42. $m^{-2} + n^{-1} - p^2$
43. $(a + b)^{-2} + c$
44. $x^2 + y^2 - z^{-3}$
45. $x^{-2} + y^{-1}$
46. $(a + b^{-2})^{-3}$
47. $(2a^{-1} + b)^{-2}$
48. $(3x^{-1} + 2y^{-2})^{-1}$

49. The following equation gives the magnetic field H at a distance r cm from the center of the magnet whose length is $2L$ and whose pole strengths are at $+m$ and $-m$:

$$H = \frac{Lr(r^2 - L^2)^{-2}}{(4m)^{-1}}$$

Simplify this expression.

E 50. The total resistance R_T of two conductors in parallel is given by

$$R_T = \frac{1}{r_2^{-1} + r_1^{-1}}$$

Simplify this expression.

51. The focal length F of two lenses with focal lengths f_1 and f_2 separated by a distance d is given by

$$F = \frac{1}{f_2^{-1} + f_1^{-1} - d(f_1f_2)^{-1}}$$

Simplify this expression.

52. The units of measure for the height of a liquid under pressure are given by

$$\frac{(\text{kg} \cdot \text{m})(\text{s} \cdot \text{m})^{-2}}{\text{kg} \cdot \text{m}^{-2} \cdot \text{s}^{-2}}$$

Remove negative exponents and then simplify.

I 53. The following equation gives the amount A of money you would need to invest to yield y dollars at the end of x years at n rate of interest compounded annually:

$$A = y(1 + n)^{-x}$$

Write this expression without negative exponents.

4–2

RATIONAL EXPONENTS

All Laws of Exponents discussed in the previous section apply to rational (fractional) exponents as well as integral exponents. However, an additional rule is used to convert a fractional exponent into a radical expression. You can convert fractional exponents into a radical expression by using the following rule.

Fractional Exponents

If $\sqrt[n]{a}$ exists as a real number, then

$$a^{m/n} = \sqrt[n]{a^m} = (\sqrt[n]{a})^m$$

(power: m; root: n)

where m and n are integers and n is positive.

Given the conditions stated above, the expressions $\sqrt[n]{a^m}$ and $(\sqrt[n]{a})^m$ are equivalent. But the second expression is easier to evaluate because we are taking the root of a smaller number.

So far we have discussed positive radicands. If the radicand is negative and the index is an odd number, the root is real and negative. On the other hand, if the radicand is negative and the index is an even number, the root is an imaginary number. We will discuss imaginary and complex numbers in more detail later. For now, to evaluate a radical that results in an imaginary number, take the root of the number and place a j behind it.

EXAMPLE 1 Evaluate the following expressions:

(a) $27^{2/3}$ (b) $(-32)^{1/5}$ (c) $(-16)^{1/2}$

Solution

(a) From the rule of fractional exponents,

$$27^{2/3} = (\sqrt[3]{27})^2$$

Since $(3)^3 = 27$, $\sqrt[3]{27} = 3$ and the expression becomes $(3)^2 = 9$.

$$27^{2/3} = 9$$

(b) From the rule of fractional exponents,

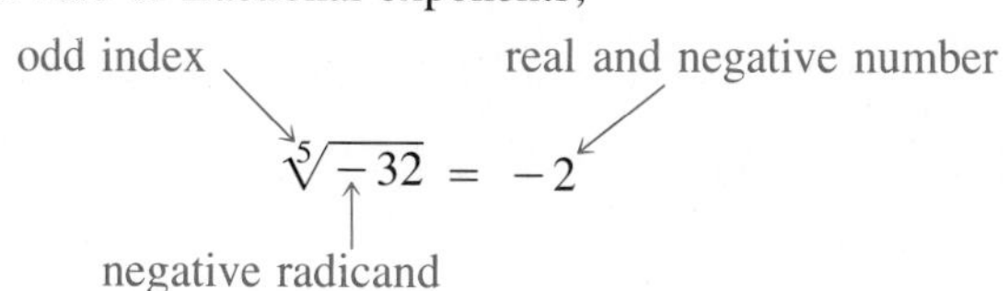

(c) We convert from a fractional exponent to a radical.

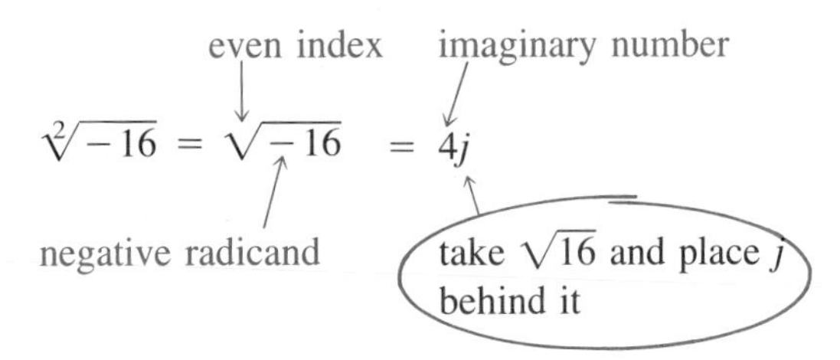

EXAMPLE 2 Evaluate $64^{-5/6}$.

Solution Before converting the fractional exponent to a radical, remove the negative exponent.

$\dfrac{1}{64^{5/6}}$	eliminate negative exponent
$\dfrac{1}{(\sqrt[6]{64})^5}$	convert fractional exponent to radical
$\dfrac{1}{(2)^5}$	evaluate the radical ($\sqrt[6]{64} = 2$)
$\dfrac{1}{32}$	evaluate the exponent ($2^5 = 32$)

EXAMPLE 3 The instantaneous current in a capacitor is given by

$$I = \frac{V}{R} e^{-t/(RC)}$$

where C = capacitance, V = terminal voltage of the battery, t = time, R = resistance, and e is the irrational number approximately equal to 2.71828. Simplify this expression by removing negative exponents and converting the fractional exponent to a radical.

Solution First, we remove the negative exponent by moving the factor across the fraction line.

$$I = \frac{V}{Re^{t/(RC)}}$$

Next, we convert the fractional exponent into a radical expression.

$$I = \frac{V}{R\sqrt[RC]{e^t}}$$

Fractional Exponents with a Calculator

Some expressions require a calculator to evaluate roots because the answer is not an integer. Most algebraic calculators have a $\boxed{y^x}$ button for this purpose. However, you must express the fractional exponent in decimal form to use this button.

EXAMPLE 4 Evaluate the following:

(a) $54^{3/5}$ (b) $9^{3/4}$

Solution

(a) 54 $\boxed{y^x}$ $\boxed{(}$ 3 $\boxed{\div}$ 5 $\boxed{)}$ $\boxed{=}$ $\longrightarrow$ 10.950558

(b) 9 $\boxed{y^x}$ $\boxed{(}$ 3 $\boxed{\div}$ 4 $\boxed{)}$ $\boxed{=}$ $\longrightarrow$ 5.196152 ■

EXAMPLE 5 The total yield y from an investment is given by

$$y = a(1 + n)^x$$

where a = initial investment, n = interest rate, and x = time in years. Find the yield on an initial investment of \$5,000 at a rate of 9% compounded annually for 1½ years.

Solution Substituting into the formula $a = 5{,}000$, $n = 0.09$, and $x = 3/2$ gives

$$y = 5{,}000(1 + 0.09)^{3/2}$$
$$y = 5{,}000(1.09)^{3/2}$$

The yield is \$5,689.97.

5000 $\boxed{\times}$ 1.09 $\boxed{y^x}$ $\boxed{(}$ 3 $\boxed{\div}$ 2 $\boxed{)}$ $\boxed{=}$ $\rightarrow$ 5689.97 ■

Operations with Fractional Exponents

The Laws of Exponents discussed in Section 4–1 also apply to fractional exponents. The next two examples illustrate their use.

EXAMPLE 6 Simplify each expression and give all answers with positive exponents.

(a) $4^{3/5} \cdot 4^{2/5}$ (b) $(27^{2/3})^{1/2}$ (c) $(16a^3)^{3/4}$ (d) $\left(\dfrac{m^{2/3}}{m}\right)^{3/4}$

Solution

(a) $4^{3/5} \cdot 4^{2/5} = 4^{(3/5)+(2/5)} = 4^{5/5} = 4^1 = 4$

(b) $(27^{2/3})^{1/2} = 27^{(2/3)\cdot(1/2)} = 27^{1/3} = 3$

(c) $(16a^3)^{3/4} = 16^{3/4}(a^3)^{3/4} = 8a^{9/4}$

(d) $\left(\dfrac{m^{2/3}}{m}\right)^{3/4} = \dfrac{(m^{2/3})^{3/4}}{m^{3/4}} = \dfrac{m^{1/2}}{m^{3/4}} = \dfrac{1}{m^{3/4-1/2}} = \dfrac{1}{m^{3/4-2/4}} = \dfrac{1}{m^{1/4}}$ ■

EXAMPLE 7 Simplify each expression and give all answers with positive exponents.

(a) $\left(\dfrac{9x^{-2}y^{1/3}}{4x^{1/2}y^{-3/4}}\right)^{3/2}$ (b) $(2x - 1)(x + 6)^{-1/3} + (x + 6)^{2/3}$

Solution

(a)

$$\left(\frac{9x^{-2}y^{1/3}}{4x^{1/2}y^{-3/4}}\right)^{3/2}$$

$$\left(\frac{9y^{13/12}}{4x^{5/2}}\right)^{3/2} \quad \text{Rules 4–1 and 4–8}$$

$$\frac{9^{3/2}(y^{13/12})^{3/2}}{4^{3/2}(x^{5/2})^{3/2}} \quad \text{Rule 4–2}$$

$$\frac{27y^{13/8}}{8x^{15/4}} \quad \text{Rule 4–3}$$

(b) $(2x - 1)(x + 6)^{-1/3} + (x + 6)^{2/3}$

$$\frac{2x - 1}{(x + 6)^{1/3}} + (x + 6)^{2/3} \quad \text{Rule 4–8}$$

$$\frac{2x - 1 + (x + 6)^{1/3}(x + 6)^{2/3}}{(x + 6)^{1/3}} \quad \text{add fractions}$$

$(x + 6)^{(1/3)+(2/3)} = (x + 6)$

$$\frac{2x - 1 + (x + 6)}{(x + 6)^{1/3}} \quad \text{multiply}$$

$$\frac{3x + 5}{(x + 6)^{1/3}} \quad \text{combine like terms}$$

■

4–2 EXERCISES

Evaluate the given expression.

1. $8^{2/3}$
2. $32^{2/5}$
3. $64^{1/3}$
4. $256^{1/4}$
5. $216^{1/3}$
6. $243^{3/5}$
7. $125^{2/3} - 81^{3/4}$
8. $9^{1/2} + 125^{2/3}$
9. $8^{2/3} + 16^{1/4}$
10. $16^{3/4} - 27^{2/3}$
11. $\dfrac{25^{1/2} \cdot 8^{-2/3}}{81^{-1/3}}$
12. $49^{1/2} \cdot 36^{-1/2}$
13. $\dfrac{9^{1/2} \cdot 16^{-3/4}}{8^{2/3}}$
14. $\dfrac{64^{-2/3} \cdot 81^{1/4}}{27^{-1/3}}$
15. $4^{-1/2} + 27^{1/3}$
16. $16^{-3/4} + 8^{-2/3}$

Calculate and round to hundredths.

17. $143^{2/3}$
18. $67^{3/5}$
19. $\dfrac{8^{-1/5}}{8^{3/4}}$
20. $6^{1.7} \cdot 14^{1/3}$
21. $2^{1/4} \cdot 3^{-3/4}$
22. $\dfrac{11^{2/3}}{11^{1/2}}$
23. $(7^{3/4})^{-1/6}$
24. $(18^{2/3})^{-2/5}$

Simplify each expression and give all answers with positive, rational exponents.

25. $\dfrac{x^{-2/3}}{x^{3/4}}$ **26.** $\dfrac{y^{3/5}}{y^{-1/3}}$

27. $a^{4/7} \cdot a^{3/5}$ **28.** $m^{3/7} \cdot m^{1/3}$

29. $(m^{-2/3}n^2)^{-3/4}$ **30.** $(x^{1/2}y^{-3/4})^{-1}$

31. $y^{1/8}y^{-3/4}$ **32.** $b^{2/9}b^{1/3}$

33. $\dfrac{b^{-2}}{b^{1/3}}$ **34.** $\dfrac{x^{-3/5}}{x^{-3/4}}$

35. $\dfrac{x^{1/2}x^{3/4}}{x^{1/8}}$ **36.** $\dfrac{a^{2/7}a^{-1/3}}{a^{-1/2}}$

37. $\dfrac{a^{-3}a^{3/7}}{a^{-1/3}}$ **38.** $\dfrac{x^{1/2}y^{-1/4}}{(xy)^{3/8}}$

39. $\dfrac{x^{-2/3}y^2}{x^{-1/4}y^{1/4}}$ **40.** $\dfrac{(m^{-2}n^{1/3})^2}{m^{1/3}n^{-1}}$

41. $9^{1/3} \cdot 9^{5/3}$ **42.** $8^{3/4} \cdot 8^{1/4}$

43. $\dfrac{5^{1/2}}{5^{3/2}}$ **44.** $\dfrac{10^{1/2}}{10^{-3/2}}$

45. $\dfrac{6^{1/6}}{6^{-5/6}}$ **46.** $\dfrac{8^{-3/4}}{8^{1/4}}$

47. $(2x + 5)(x - 1)^{-1/4} + (x - 1)^{3/4}$

48. $(x + 7)(3x + 5)^{-1/2} - (3x + 5)^{1/2}$

49. $(3y + 4)(y + 2)^{-3/5} - 8(y + 2)^{2/5}$

50. $(4m - 1)(m + 4)^{-2/3} + 6(m + 4)^{1/3}$

51. The magnetic field is given by

$$H = \frac{2mL}{(r^2 + L^2)^{3/2}}$$

Convert the fractional exponent into a radical expression.

52. The emissive power of a blackbody at wavelength λ (the Greek letter lambda) can be written

$$E = \frac{c\lambda^{-5}}{e^{c/\lambda t} - 1}$$

Remove negative exponents and convert the fractional exponent into a radical.

I 53. Find the yield on an initial investment of $a =$ \$8,000 at $n = 8\%$ for $x = 3\frac{1}{2}$ years. (Refer to Example 5.)

M 54. An expression used to estimate the shear stress in pounds per square inch which must be transferred between the immediate stiffeners and web during shear transfer is

$$f = h\left(\frac{F}{340}\right)^{3/2}$$

Write this expression in radical form.

55. The number N of radioactive nuclei left after undergoing a decay process is given by $N = N_0e^{-Dt}$, where N_0 is the initial number of unstable nuclei and D is the disintegration constant for the particular nuclei. The half-life of the nuclei is defined as the length of time required for half the initial number of nuclei to decay. This leads to $N = N_0e^{0.69t/h}$, where h is the half-life of the nuclei. The fraction of nuclei remaining after time t can be written as $N = (0.5)^{t/h}$ where h and t have the same units. Find the fractional amount of Strontium 90 left after 8 years if its half-life is 28 years.

M 56. An approximation of engine efficiency E is given by $E = 100(1 - R^{-2/5})$ where R is the compression ratio. Find the efficiency of an engine whose compression ratio is 240/35.

4–3

SIMPLEST RADICAL FORM

So far in this chapter, we have discussed expressions written in exponential form. However, in the previous section we discussed the relationship between fractional exponents and radical form. In the remainder of this chapter we will discuss algebraic expressions written in radical form. Namely, in this section we will discuss converting a radical to its simplest form.

Laws of Exponents in Radical Form

Several of the Laws of Exponents can be written in radical form. These rules, which will form the basis for simplifying radicals, are summarized below.

Radical Form of Laws of Exponents

For positive integers m and n, the following rules apply:

Rule 4–9. $\sqrt[n]{a^n} = a \quad \text{for } a \geq 0$

Rule 4–10. $\sqrt[n]{ab} = \sqrt[n]{a}\sqrt[n]{b} \quad \text{for } a,b \geq 0$

Rule 4–11. $\sqrt[m]{\sqrt[n]{a}} = \sqrt[mn]{a}$

Rule 4–12. $\sqrt[n]{\dfrac{a}{b}} = \dfrac{\sqrt[n]{a}}{\sqrt[n]{b}} \quad \text{where } b \neq 0$

Criteria for Simplest Radical Form

A simplified radical implies certain requirements of the answer. These requirements are as follows:

- ☐ An nth-root radical does not contain any nth factors.
- ☐ There are no radicals contained inside other radicals.
- ☐ The denominator of any fraction does not contain a radical.
- ☐ The order of the radical is reduced.

The following examples show how to simplify a radical to meet each of these requirements.

Removing nth-Root Factors

EXAMPLE 1 Simplify the radical expression $\sqrt{72}$.

Solution In simplifying $\sqrt{72}$, we are trying to remove any squares contained in 72. If you know that $72 = 36 \cdot 2$ and $36 = 6^2$, then you can immediately simplify $\sqrt{72} = \sqrt{36}\sqrt{2} = 6\sqrt{2}$. However, if you have difficulty determining the squares contained in 72, break 72 into prime factors. Then write these prime factors in terms of exponents of 2 because the index of the radical is 2. Breaking 72 into prime factors gives

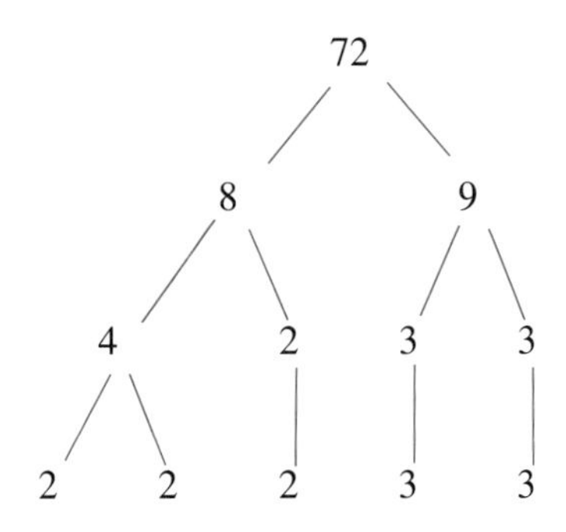

NOTE ✦ When the index of a radical is omitted, it is understood to be 2.

Writing the factors in terms of an exponent of 2 gives

$$72 = 2 \cdot 2 \cdot 2 \cdot 3 \cdot 3 = 2^2 \cdot 3^2 \cdot 2$$

Therefore,

$$\sqrt{72} = \sqrt{2^2 \cdot 3^2 \cdot 2}$$

Using Rule 4–9 for simplifying radicals gives

$$2 \cdot 3\sqrt{2} = 6\sqrt{2}$$

■

CAUTION ✦ Only the base is removed from the radical when the index and the exponent are equal $\left(\sqrt[n]{a^n} = a\right)$.

The same process of removing nth-root factors applies to variable expressions also.

EXAMPLE 2 Simplify the radical expression $\sqrt[4]{64a^7b^8}$.

Solution To simplify this radical, we must remove from the radical any 4th-root factors. Therefore, *each factor under the radical should be written in terms of exponents of 4.*

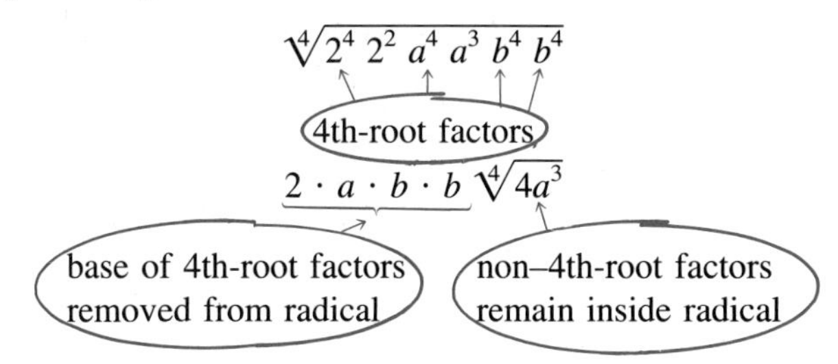

The simplified radical is $2ab^2\ \sqrt[4]{4a^3}$. ■

Radicals within Radicals

According to Rule 4–11, you can simplify a radical inside another radical to a single radical by multiplying indices, as shown in the next example.

EXAMPLE 3 Simplify the following:

(a) $\sqrt[3]{\sqrt[5]{4x}}$ (b) $\sqrt{3\sqrt{m}}$

Solution

(a) $\sqrt[3]{\sqrt[5]{4x}} = \sqrt[(3 \cdot 5)]{4x} = \sqrt[15]{4x}$ (b) $\sqrt{3\sqrt{m}} = \sqrt{\sqrt{9m}} = \sqrt[4]{9m}$

Since 3 cannot be removed from the outer radical, you must include it in the inner radical by squaring it.

■

Rationalizing the Denominator

Simplifying a radical also involves eliminating radicals from the denominator. This process is called **rationalizing the denominator** because we multiply the numerator and denominator by a quantity that results in a rational denominator.

EXAMPLE 4 Simplify the radical expression $\sqrt{\frac{5a}{7}}$.

Solution First, let us apply Rule 4–12 and write the expression as two radicals.

$$\frac{\sqrt{5a}}{\sqrt{7}}$$

To remove the radical in the denominator, we must have a square under the radical ($\sqrt[2]{a^2} = a$, from the more general statement $\sqrt[n]{a^n} = a$). Therefore, we must multiply the numerator and denominator by $\sqrt{7}$.

$$\frac{\sqrt{5a}}{\sqrt{7}} \cdot \quad \text{rationalize denominator}$$

$$\frac{\sqrt{35a}}{\sqrt{49}} \quad \text{multiply radicals}$$

$$\frac{\sqrt{35a}}{7} \quad \text{simplify denominator}$$

■

EXAMPLE 5 Simplify $\sqrt[3]{\frac{5a^4}{4b^3}}$.

Solution First, simplify the radical by removing any cubes contained under the radical.

$$\frac{\sqrt[3]{5a^3a}}{\sqrt[3]{4b^3}}$$

$$\frac{a\sqrt[3]{5a}}{b\sqrt[3]{4}}$$

Before we can eliminate the radical in the denominator, each factor under the radical must have an exponent of 3. To determine the factor for rationalizing, write each factor in terms of exponents.

$$\frac{a\sqrt[3]{5a}}{b\sqrt[3]{2^2}}$$

From this information, we can see that we rationalize by multiplying the numerator and denominator by $\sqrt[3]{2}$.

$$\frac{a\sqrt[3]{5a}}{b\sqrt[3]{2^2}} \cdot \frac{\sqrt[3]{2}}{\sqrt[3]{2}} \qquad \text{rationalize denominator}$$

$$\frac{a\sqrt[3]{10a}}{b\sqrt[3]{2^3}} \qquad \text{multiply radicals}$$

The simplified radical is

$$\frac{a\sqrt[3]{10a}}{2b}$$

■

NOTE ✦ Always simplify a radical by removing nth-root factors before rationalizing the denominator. In some instances, rationalizing may not be necessary after this operation has been performed.

Reducing the Index of a Radical

The final requirement for simplifying a radical is that the order of the radical be reduced. Fractional exponents make this operation easier.

NOTE ✦ Before you can convert several fractional exponents to a radical, the denominator of the fractions must be the same.

$$a^{m/n}b^{p/n} = \sqrt[n]{a^m b^p}$$

In some instances, you may have to obtain a common denominator, as shown in the following example.

EXAMPLE 6 Simplify the following:

(a) $\sqrt[6]{4x^2}$ (b) $\sqrt[6]{81y^2}$

Solution

(a) First, *write the radicand in exponential form.*

$$\sqrt[6]{2^2x^2}$$

Then *express the radical expression with fractional exponents.*

$$(2^2x^2)^{1/6}$$

$$(2^2)^{1/6}(x^2)^{1/6} \qquad \text{Rule 4–2}$$

$$2^{2/6}x^{2/6} \qquad \text{Rule 4–3}$$

Last, *reduce the fractional exponent,* if possible, and convert back to radical form.

$$2^{1/3}x^{1/3}$$
$$\sqrt[3]{2x}$$

(b)

$$\sqrt[6]{81y^2}$$

$\sqrt[6]{3^4y^2}$	convert to exponential form
$(3^4y^2)^{1/6}$	convert to fractional exponent
$3^{4/6}y^{2/6}$	simplify exponents
$3^{2/3}y^{1/3}$	reduce fractional exponent
$\sqrt[3]{9y}$	convert to radical form

■

Polynomial Radicands

Polynomial radicands must also meet the requirements for simplest form.

EXAMPLE 7 Simplify the following radical expression:

$$\sqrt{3x^2 - 2x + \frac{1}{3}}$$

Solution

$\sqrt{\dfrac{9x^2 - 6x + 1}{3}}$	combine the fractions
$\sqrt{\dfrac{(3x - 1)^2}{3}}$	factor the numerator
$(3x - 1)\sqrt{\dfrac{1}{3}}$	remove the square from the radical
$(3x - 1)\dfrac{\sqrt{1}}{\sqrt{3}} \cdot \dfrac{\sqrt{3}}{\sqrt{3}}$	rationalize the denominator
$\dfrac{(3x - 1)\sqrt{3}}{3}$	

■

CAUTION ✦ Only *n*th-root *factors* can be removed from the radical. It is necessary to combine terms before simplifying a radical.

4–3 EXERCISES

Express each of the following radicals in simplest radical form.

1. $\sqrt{80a^4b^3}$
2. $\sqrt{98m^4}$
3. $\sqrt{150x^2y^5}$
4. $\sqrt{104a^5b}$
5. $\sqrt[3]{32x^5y^5}$
6. $\sqrt[4]{64x^6}$
7. $\sqrt{162m^3n^5y^8}$
8. $\sqrt{250p^5m^8}$
9. $\sqrt{363ab^5c^{12}}$
10. $\sqrt{375k^3n^8}$

11. $\sqrt[5]{96x^5y^8z^2}$
12. $\sqrt[4]{81a^6b^{12}}$
13. $\sqrt{\sqrt[3]{y}}$
14. $\sqrt[3]{4\sqrt{a}}$
15. $\sqrt{\sqrt[3]{4x}}$
16. $\sqrt[4]{\sqrt{6m}}$
17. $\sqrt{\dfrac{72a^7}{9b^6}}$
18. $\sqrt[3]{\dfrac{128m^5}{27p^6}}$
19. $\sqrt[3]{192x^5y^2z^8}$
20. $\sqrt[3]{324a^6b^{10}c^3}$
21. $\sqrt[4]{80m^4n^7p^{15}}$
22. $\sqrt{300m^7n^{13}p^4}$
23. $\sqrt[5]{64k^5m^{12}}$
24. $\sqrt[4]{256x^8y^5z^{15}}$

Simplify by rationalizing the denominator.

25. $\sqrt{\dfrac{17}{6}}$
26. $\sqrt{\dfrac{18}{5}}$
27. $\sqrt{\dfrac{21}{5}}$
28. $\sqrt{\dfrac{19}{11}}$
29. $\sqrt{\dfrac{46a^2b}{3a}}$
30. $\sqrt{\dfrac{40}{13a}}$
31. $\sqrt{\dfrac{7a^3}{8}}$
32. $\sqrt{\dfrac{13n^2}{54m^3}}$
33. $\sqrt{\dfrac{81m^4}{7n}}$
34. $\sqrt{\dfrac{36y^3}{27x}}$
35. $\sqrt{\dfrac{5a}{20}}$
36. $\sqrt{\dfrac{4b^2}{18a^3}}$
37. $\dfrac{4b}{\sqrt{32b}}$
38. $\dfrac{\sqrt{60m^3}}{\sqrt{20p^2}}$
39. $\sqrt{\dfrac{40m^2}{7n^3}}$
40. $\sqrt{\dfrac{108a^5b^3}{27c^5}}$
41. $\sqrt[3]{\dfrac{6m^3}{4}}$
42. $\sqrt[3]{\dfrac{16y^6}{81x^4}}$
43. $\sqrt[3]{\dfrac{32m^5}{9n^4}}$
44. $\sqrt[3]{\dfrac{x^3}{4y}}$
45. $\sqrt[4]{\dfrac{81a^2}{4b^2}}$
46. $\sqrt{\dfrac{64a^6}{27b^3}}$
47. $\sqrt{\dfrac{72x^2y^3}{5z^3}}$
48. $\sqrt[4]{\dfrac{32y^8}{9x^2}}$
49. $\dfrac{6m^3}{\sqrt[3]{n^5}}$
50. $\dfrac{14m}{\sqrt[4]{8n^2}}$
51. $\dfrac{\sqrt{x^2 - 4}}{x}$
52. $\dfrac{\sqrt{y^2 - 8}}{y}$
53. $\sqrt{x^2 + 8x + 16}$
54. $\sqrt{m^2 + 2m + 1}$
55. $\sqrt{y^2 + 5y + \dfrac{25}{4}}$
56. $\sqrt{3x^2 + 8x + \dfrac{16}{3}}$
57. $\sqrt{8m^2 - 4m + \dfrac{1}{2}}$
58. $\sqrt{5a^2 - 4a + \dfrac{4}{5}}$

59. The angular velocity ω after movement through the arc θ is

$$\omega = \sqrt{\omega_0^2 + 2\alpha\theta}$$

where ω_0 is the initial angular velocity and α (alpha) is the angular acceleration. Find the angular velocity if $\omega_0 = 30$ rad/s, $\theta = 1.5$ rad, and $\alpha = 60$ rad/s^2. Leave your answer in simplest radical form.

60. The fundamental frequency n of a vibrating string is given by

$$n = \frac{1}{2L}\sqrt{\frac{T}{m}}$$

where L = length, T = tension, and m = mass/unit length. A string 20 cm long with a mass per unit length of 0.04 kg/m is under a tension of 450 N. Find the fundamental frequency. Round to hundredths.

61. When the map of a region is available, you can find its area by dividing the region into geometric shapes. When the three sides of a triangle have been scaled from the map, the area of a triangle with sides a, b, and c is found by

$$\sqrt{s(s - a)(s - b)(s - c)}$$

where $s = (a + b + c)/2$. Find the area of a triangle with sides 40 m, 25 m, and 35 m. Round to hundredths.

62. The period T of a pendulum is given by

$$T = 2\pi\sqrt{\frac{L}{g}}$$

where T = time in seconds, L = length in centimeters, and g = acceleration due to gravity. Find the period of a pendulum 40 cm long if $g = 980$ cm/s^2. Leave your answer in simplified radical form.

4 – 4

OPERATIONS WITH RADICALS

In this section we will discuss adding, subtracting, multiplying, and dividing expressions containing a radical. To perform these operations with radicals, we use many of the same techniques as we did with the corresponding operation with polynomials.

Adding and Subtracting Radicals

To add and subtract expressions containing a radical, combine like terms, just as with polynomials. However, we must define like radical terms. **Like radicals** have the same index and radicand.

Adding and Subtracting Radicals

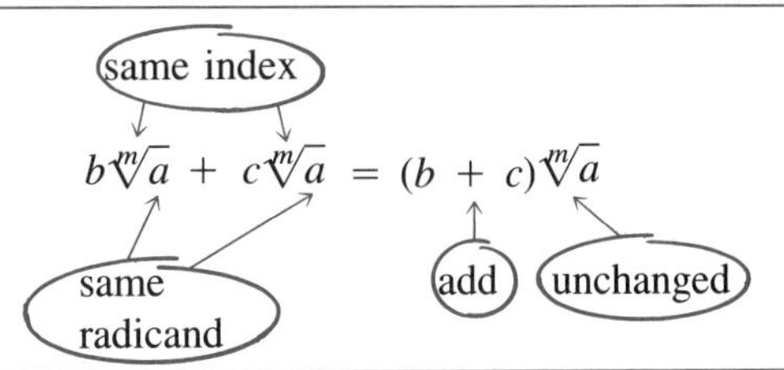

We cannot always identify like radicals at a glance. Usually the radicals must be simplified first.

EXAMPLE 1 Add the following radical expressions:

$$4\sqrt{8} + 5\sqrt{18}$$

Solution First, we simplify each radical by removing any squares.

$$4\sqrt{8} + 5\sqrt{18}$$
$$4\sqrt{2^2 \cdot 2} + 5\sqrt{3^2 \cdot 2}$$
$$8\sqrt{2} + 15\sqrt{2}$$

same index

$$8\sqrt{2} + 15\sqrt{2}$$

same radicand

Second, we combine the two like terms.

$$(8 + 15)\sqrt{2}$$

$$23\sqrt{2}$$

■

EXAMPLE 2 Simplify $2a\sqrt[3]{16} + \sqrt[3]{81a^3} - 7a\sqrt[3]{2}$.

Solution First, we simplify each radical by removing any cubes contained under the radical.

$$2a\sqrt[3]{2^3 \cdot 2} + \sqrt[3]{3^3 \cdot 3 \cdot a^3} - 7a\sqrt[3]{2}$$

$$4a\sqrt[3]{2} + 3a\sqrt[3]{3} - 7a\sqrt[3]{2}$$

Next, we combine like terms. The terms $4a\sqrt[3]{2}$ and $-7a\sqrt[3]{2}$ contain like radicals, but $3a\sqrt[3]{3}$ is not like the other two terms because it has a different radicand. Since only the first and last terms are like, we can combine only these two terms. The result is

$$-3a\sqrt[3]{2} + 3a\sqrt[3]{3}$$

■

EXAMPLE 3 Simplify the following:

$$\sqrt{\frac{a^3}{12}} + \sqrt{27a^3} - \sqrt{\frac{27}{a^3}}$$

Solution First, we simplify each radical by removing any squares.

$$\frac{a}{2}\sqrt{\frac{a}{3}} + 3a\sqrt{3a} + \frac{3}{a}\sqrt{\frac{3}{a}}$$

Second, we rationalize the denominator, where necessary.

$$\frac{a\sqrt{a}\sqrt{3}}{2\sqrt{3}\sqrt{3}} + 3a\sqrt{3a} + \frac{3\sqrt{3}\sqrt{a}}{a\sqrt{a}\sqrt{a}}$$

$$\frac{a}{2}\sqrt{\frac{3a}{3^2}} + 3a\sqrt{3a} + \frac{3}{a}\sqrt{\frac{3a}{a^2}}$$

$$\frac{a}{6}\sqrt{3a} + 3a\sqrt{3a} + \frac{3}{a^2}\sqrt{3a}$$

Finally, we combine like terms. Since all three terms contain like radicals, we add as follows:

$$\left(\frac{a}{6} + 3a + \frac{3}{a^2}\right)\sqrt{3a}$$

Adding the coefficients using a common denominator of $6a^2$ gives

$$\left(\frac{a^3 + 18a^3 + 18}{6a^2}\right)\sqrt{3a}$$

$$\left(\frac{19a^3 + 18}{6a^2}\right)\sqrt{3a}$$

■

Multiplying Radicals

We multiply radicals using the same techniques used to multiply polynomials: the Distributive Property or FOIL. By restating Rule 4–10,

$$\sqrt[n]{a} \cdot \sqrt[n]{b} = \sqrt[n]{ab}$$

we can see that to multiply radicals of the same order, we multiply their radicands. Therefore,

same index

$$\sqrt[3]{6} \cdot \sqrt[3]{2} = \sqrt[3]{12}$$

multiply radicands

In addition to multiplying the radicals, we must also multiply all factors outside the radical separately.

multiply radicals

$$4\sqrt[3]{6} \cdot 5\sqrt[3]{2} = 20\sqrt[3]{12}$$

multiply factors outside the radical

NOTE ✦ Once the multiplication has been performed, each radical must be simplified and like radicals combined.

EXAMPLE 4 Multiply the following radical expressions:

$$2\sqrt{3}(\sqrt{5} - 6\sqrt{7})$$

Solution To multiply this monomial and binomial, we use the Distributive Property.

$2\sqrt{3}(\sqrt{5} - 6\sqrt{7})$	
$2\sqrt{3}(\sqrt{5}) - 2\sqrt{3}(6\sqrt{7})$	Distributive Property
$2 \cdot 1\sqrt{3 \cdot 5} - 2 \cdot 6\sqrt{3 \cdot 7}$	multiply
$2\sqrt{15} - 12\sqrt{21}$	simplify

Since each radical is simplified and there are no like radicals, the product is $2\sqrt{15} - 12\sqrt{21}$. ■

EXAMPLE 5 Multiply $(4\sqrt{3} - \sqrt{6})(3\sqrt{3} + 5\sqrt{6})$.

Solution We multiply these two binomials by using FOIL.

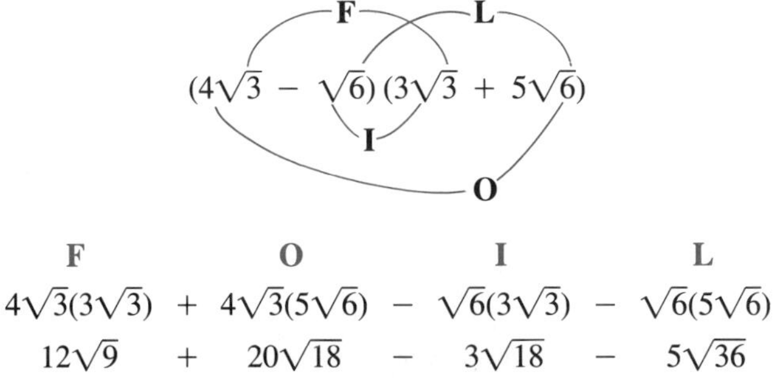

$$\begin{array}{ccccccc} \text{F} & & \text{O} & & \text{I} & & \text{L} \\ 4\sqrt{3}(3\sqrt{3}) & + & 4\sqrt{3}(5\sqrt{6}) & - & \sqrt{6}(3\sqrt{3}) & - & \sqrt{6}(5\sqrt{6}) \\ 12\sqrt{9} & + & 20\sqrt{18} & - & 3\sqrt{18} & - & 5\sqrt{36} \end{array}$$

Next, we simplify the radicals and combine like radicals.

$$36 + 60\sqrt{2} - 9\sqrt{2} - 30$$
$$6 + 51\sqrt{2}$$

■

Multiplying Radicals of Different Order

In the previous discussion on multiplication, only radicals of the same order could be multiplied. However, if we convert to fractional exponents, we can multiply radicals of different order. The next example illustrates this process.

EXAMPLE 6 Multiply $\sqrt[3]{6} \cdot \sqrt{4x}$.

Solution First, *write each radical in fractional exponent form.*

$$(6)^{1/3}(4x)^{1/2}$$

Next, *change the fractional exponents to equivalent fractions* with a common denominator.

$$(6)^{2/6}(4x)^{3/6}$$
$$[(6)^2(4x)^3]^{1/6}$$

Then *change back to radical form and simplify.*

$$\sqrt[6]{6^2 \cdot (4x)^3}$$

$$\sqrt[6]{6^2 \cdot 4^3 \cdot x^3}$$

$$\sqrt[6]{(2 \cdot 3)^2 \cdot (2^2)^3 \cdot x^3}$$

$$\sqrt[6]{2^2 \cdot 3^2 \cdot 2^6 \cdot x^3}$$

$$2\sqrt[6]{36x^3}$$

■

Dividing Radicals

To divide radicals, we use Rule 4–12, repeated here for convenience.

$$\frac{\sqrt[n]{a}}{\sqrt[n]{b}} = \sqrt[n]{\frac{a}{b}} \qquad \text{where } b \neq 0$$

Then we simplify the result as necessary.

EXAMPLE 7 Divide the following:

(a) $\dfrac{\sqrt{42}}{\sqrt{7}}$ (b) $\dfrac{4\sqrt{10x}}{3\sqrt{2x^2}}$

Solution

(a) $$\frac{\sqrt{42}}{\sqrt{7}} = \sqrt{\frac{42}{7}} = \sqrt{6}$$

(b) $$\frac{4\sqrt{10x}}{3\sqrt{2x^2}} = \frac{4}{3}\sqrt{\frac{10x}{2x^2}} = \frac{4}{3}\sqrt{\frac{5}{x}} = \frac{4\sqrt{5}\sqrt{x}}{3\sqrt{x}\sqrt{x}} = \frac{4\sqrt{5x}}{3x}$$ ■

Binomial Denominators

Rationalizing a binomial denominator requires multiplying by the conjugate of the denominator. Two binomials, for example, $(a + b)(a - b)$, whose product is the difference of two squares are called **conjugates.** For example, $(\sqrt{2} - \sqrt{3})$ is the conjugate of $(\sqrt{2} + \sqrt{3})$.

EXAMPLE 8 Simplify the following:

$$\frac{(\sqrt{x} - 3\sqrt{y})}{(\sqrt{x} - \sqrt{y})}$$

Solution The conjugate of $\sqrt{x} - \sqrt{y}$ is $\sqrt{x} + \sqrt{y}$. We will multiply the numerator and denominator of the fraction by the conjugate of the denominator and simplify the result.

$$\frac{(\sqrt{x} - 3\sqrt{y})}{(\sqrt{x} - \sqrt{y})} \cdot \frac{(\sqrt{x} + \sqrt{y})}{(\sqrt{x} + \sqrt{y})}$$

multiply using FOIL

$$\frac{\overbrace{\sqrt{x^2} + \sqrt{xy} - 3\sqrt{xy} - 3\sqrt{y^2}}}{\underbrace{(\sqrt{x})^2 - (\sqrt{y})^2}}$$

difference of squares

$$\frac{x - 2\sqrt{xy} - 3y}{x - y}$$ ■

4–4 EXERCISES

Perform the indicated operation and leave the answer in simplest form.

1. $\sqrt{50} - \sqrt{8} + \sqrt{18}$

2. $\sqrt{27} - \sqrt{75} + \sqrt{48}$

3. $2\sqrt{80} + \sqrt{45} - 2\sqrt{125}$

4. $\sqrt{28} + \sqrt{63} + \sqrt{175}$

5. $3\sqrt{75} - 4\sqrt{20} - \sqrt{80}$

6. $-\sqrt{24} + 2\sqrt{216} - 5\sqrt{54}$

7. $\sqrt{8} - 2\sqrt{27} + 5\sqrt{18}$

8. $3\sqrt{40a^3} - 2\sqrt{90a^3} - 5a\sqrt{250a}$

9. $a\sqrt{40a^3} + 3\sqrt{10a^5} + 6a\sqrt{90}$

10. $3x\sqrt{18x^3} + \sqrt{2x^5} + x^2\sqrt{32x}$

11. $\sqrt{16a^3b^4} - \sqrt{36a^5b^2}$

12. $\sqrt{48m^4n^7} + \sqrt{192m^4n^7}$

13. $\sqrt[3]{16} - \sqrt[3]{54}$

14. $\sqrt[3]{81} + \sqrt[3]{648}$

15. $7\sqrt[3]{81} - 2\sqrt[3]{24}$

16. $5\sqrt[3]{16x^4} - 9\sqrt[3]{128x^4}$

17. $6\sqrt{32x^2} + 4\sqrt{162x^2}$

18. $8m\sqrt[3]{24m^3} + 4m\sqrt[3]{54m^3} - m\sqrt[3]{375m^3}$

19. $\sqrt{\frac{8}{a^2}} + \frac{3}{a}\sqrt{18}$

20. $\sqrt{9a} + \sqrt{\frac{4}{a}}$

21. $3\sqrt{\frac{2}{a^3}} - \sqrt{8a} + 4\sqrt{\frac{9}{a}}$

22. $6x\sqrt{\frac{27}{x}} - x\sqrt{\frac{12}{x}} + \frac{1}{x}\sqrt{3x}$

23. $\sqrt{3}(\sqrt{7} - 5)$

24. $\sqrt{6}(\sqrt{7} - 2\sqrt{10})$

25. $2\sqrt{6}(\sqrt{8} - 4\sqrt{5})$

26. $7\sqrt{3}(\sqrt{3} + 2\sqrt{5})$

27. $(\sqrt{3} + \sqrt{6})(\sqrt{2} - \sqrt{5})$

28. $(\sqrt{7} - \sqrt{2})(\sqrt{4} + \sqrt{3})$

29. $(\sqrt[3]{2x} + 7)(\sqrt[3]{10x} + 6)$

30. $(\sqrt[5]{8x} - 4)(\sqrt[5]{2x} + 7)$

31. $(2\sqrt{3} - 6\sqrt{7})(4\sqrt{5} + \sqrt{7})$

32. $(8\sqrt{3a} + \sqrt{6})(5\sqrt{3a} + \sqrt{2})$

33. $(7\sqrt{3} + 8)(6\sqrt{3} - 5)$

34. $(7\sqrt{12} - \sqrt{5})(2\sqrt{2} + \sqrt{5})$

35. $(3\sqrt[4]{8} + 7\sqrt[4]{6})(2\sqrt[4]{5} - 4\sqrt[4]{12})$

36. $(9\sqrt[3]{40} + \sqrt[3]{48})(6\sqrt[3]{4} + \sqrt[3]{32})$

37. $(8\sqrt{7m} + 6\sqrt{10n})(\sqrt{7m} + 2\sqrt{10n})$

38. $(\sqrt{6} + \sqrt{10})^2$

39. $(5\sqrt{8} - 2\sqrt{10})^2$

40. $(3\sqrt{5} - 7\sqrt{6})^2$

41. $(6\sqrt{3} - 7)^2$

42. $(\sqrt{7} + 4\sqrt{12})(\sqrt{7} - 4\sqrt{12})$

43. $(4\sqrt{5} + \sqrt{6})(4\sqrt{5} - \sqrt{6})$

44. $\frac{\sqrt{18x^3}}{\sqrt{x^2}}$

45. $\frac{\sqrt{3m^2}}{\sqrt{21m^2}}$

46. $\frac{4\sqrt{3}}{2\sqrt{5}}$

47. $\frac{12\sqrt{6}}{6\sqrt{5}}$

48. $\frac{\sqrt{6}}{3\sqrt{5} - \sqrt{3}}$

49. $\frac{\sqrt{7}}{6 - 9\sqrt{2}}$

50. $\frac{2\sqrt{x} - \sqrt{y}}{\sqrt{x} + 2\sqrt{y}}$

51. $\frac{3\sqrt{a} + 2\sqrt{b}}{4\sqrt{a} + b}$

52. The effective mass of the electron has been determined experimentally to be

$$m = \frac{m_0}{\sqrt{1 - (v/c)^2}}$$

where m = effective mass at velocity v, m_0 = rest mass, v = velocity of the electron, and c = velocity of light. Simplify this expression.

4–5

EQUATIONS WITH RADICALS

Many technical formulas contain radical expressions. In rearranging these formulas, you must often solve an equation containing radicals. Equations containing radicals may be solved using the **Power Rule,** which states that *an equivalent equation results from raising both sides of an equation to the same* *power*. We can use the Power Rule to remove the radicals, thus making it easier to solve the equation. However, raising each side of an equation to a power can lead to an extraneous solution. An **extraneous solution** appears to be a solution but does not satisfy the original equation. For this reason, all solutions to equations containing a radical must be checked and extraneous roots discarded.

EXAMPLE 1 Solve $\sqrt{3x - 4} - 5 = 0$.

Solution First, we *isolate the radical on one side of the equation.*

$$\sqrt{3x - 4} = 5$$

Next, we *square both sides of the equation.*

$$(\sqrt{3x - 4})^2 = (5)^2$$
$$3x - 4 = 25$$

Then we *solve the resulting equation for x.*

$$3x = 29$$
$$x = 29/3$$

Checking

$$\sqrt{3(29/3) - 4} - 5 = 0$$
$$\sqrt{25} - 5 = 0$$
$$5 - 5 = 0$$
$$0 = 0$$

Remember that we use the principal or positive square root in evaluating these radicals. The solution is 29/3. ■

Solving an Equation with Radicals

1. Isolate the radical term on one side of the equation. If there are two expressions containing a radical, place one radical expression on each side of the equation.
2. Raise each side of the equation to a power equal to the index of the radical.
3. Combine like terms.
4. If the equation still contains a radical, repeat Steps 1, 2, and 3.
5. Solve the equation for the variable.
6. Check the solution in the original equation. Discard any extraneous roots.

EXAMPLE 2 Solve the following equation for x:

$$\sqrt[3]{2x - 1} = 3$$

Solution

$$\begin{aligned} \sqrt[3]{2x - 1} &= 3 \\ (\sqrt[3]{2x - 1})^3 &= (3)^3 && \text{cube both sides} \\ 2x - 1 &= 27 && \text{simplify} \\ 2x &= 28 && \text{transpose and add} \\ x &= 14 && \text{divide} \end{aligned}$$

Checking

$$\begin{aligned} \sqrt[3]{2(14) - 1} &= 3 \\ \sqrt[3]{27} &= 3 \\ 3 &= 3 \end{aligned}$$

Since checking results in a true statement, the solution is $x = 14$. ■

EXAMPLE 3 Solve $\sqrt{y - 1} + 2 = \sqrt{y - 3}$.

Solution Since the two radicals are on opposite sides of the equation, we begin by squaring both sides of the equation.

$$\begin{aligned} (\sqrt{y - 1} + 2)^2 &= (\sqrt{y - 3})^2 && \text{square both sides} \\ y - 1 + 4\sqrt{y - 1} + 4 &= y - 3 && \text{simplify} \\ 4\sqrt{y - 1} &= -6 && \text{combine like terms} \\ (4\sqrt{y - 1})^2 &= (-6)^2 && \text{square both sides} \\ 16(y - 1) &= 36 && \text{simplify} \\ 16y - 16 &= 36 && \text{eliminate parentheses} \\ 16y &= 52 && \text{transpose and add} \\ y &= \frac{13}{4} && \text{divide} \end{aligned}$$

Checking

$$\begin{aligned} \sqrt{\frac{13}{4} - 1} + 2 &= \sqrt{\frac{13}{4} - 3} \\ \frac{3}{2} + 2 &\neq \frac{1}{2} \end{aligned}$$

Since $13/4$ is an extraneous solution, this equation has no solution. ■

CAUTION ✦ In squaring a binomial such as $(\sqrt{y - 1} + 2)$, remember the product of the outside and inside terms:

$$(a + b)^2 \neq a^2 + b^2$$

EXAMPLE 4 Solve $\sqrt{x^2 - 4x} - 5 = x - 2$.

Solution Isolating the radical on one side of the equation gives

$$\begin{aligned} \sqrt{x^2 - 4x} &= x + 3 && \text{transpose and add} \\ (\sqrt{x^2 - 4x})^2 &= (x + 3)^2 && \text{square both sides} \\ x^2 - 4x &= x^2 + 6x + 9 && \text{simplify} \\ x^2 - 4x - x^2 - 6x &= 9 && \text{transpose} \\ -10x &= 9 && \text{combine like terms} \\ x &= \frac{-9}{10} && \text{divide} \end{aligned}$$

Checking

$$\begin{aligned} \sqrt{\left(-\frac{9}{10}\right)^2 - 4\left(-\frac{9}{10}\right)} - 5 &= -\frac{9}{10} - 2 \\ \sqrt{\frac{81}{100} + \frac{36}{10}} - 5 &= -\frac{29}{10} \\ \sqrt{\frac{441}{100}} - 5 &= -\frac{29}{10} \\ \frac{21}{10} - 5 &= -\frac{29}{10} \\ -\frac{29}{10} &= -\frac{29}{10} \end{aligned}$$

Since checking results in a true statement, the solution is

$$x = \frac{-9}{10}$$

■

Applications

EXAMPLE 5 The velocity of an object moving in a horizontal circle is

$$V = \sqrt{\frac{Fr}{m}}$$

where F = centripetal force, r = radius of the circle, and m = mass of the object. Find the radius of the circle if an object has a centripetal force of 6.8 N, a velocity of 4.1 m/s, and a mass of 1.7 kg. Solve for r and then substitute.

Solution First, square both sides of the equation.

$$\begin{aligned} (V)^2 &= \left(\sqrt{\frac{Fr}{m}}\right)^2 && \text{square both sides} \\ V^2 &= \frac{Fr}{m} && \text{simplify} \end{aligned}$$

Then solve for r.

$$V^2m = Fr \quad \text{eliminate fractions}$$
$$\frac{V^2m}{F} = r \quad \text{divide by } F$$

Last, substitute the values $F = 6.8$ N, $V = 4.1$ m/s, and $m = 1.7$ kg $\left(\text{N} = \frac{\text{kg} \cdot \text{m}}{\text{s}^2}\right)$.

$$\frac{(4.1 \text{ m/s})^2(1.7 \text{ kg})}{6.8 \text{ N}} = r$$
$$4.2 \text{ m} = r$$

The radius of the circle is 4.2 m.

4.1 [x^2] [×] 1.7 [÷] 6.8 [=] → 4.2025 ■

EXAMPLE 6 The formula to determine the range d of a ship-to-shore radio message is given by

$$d = \sqrt{2rh + h^2}$$

Solve this literal equation for r.

Solution

$$d = \sqrt{2rh + h^2}$$
$$(d)^2 = (\sqrt{2rh + h^2})^2 \quad \text{square both sides}$$
$$d^2 = 2rh + h^2 \quad \text{simplify}$$
$$d^2 - h^2 = 2rh \quad \text{transpose}$$
$$\frac{d^2 - h^2}{2h} = r \quad \text{divide by } 2h$$

■

4–5 EXERCISES

Solve the following equations for the variable.

1. $\sqrt{3y + 1} - 5 = 0$ **2.** $2\sqrt{y} + 8 = 0$
3. $\sqrt{4y - 7} = 3$ **4.** $\sqrt{5x - 1} = 3$
5. $\sqrt{5a - 8} = 3$ **6.** $\sqrt{7m - 6} = 6$
7. $\sqrt{2m + 10} = -4$ **8.** $\sqrt{4a - 2} = 6$
9. $\sqrt{6x - 9} = 9$ **10.** $\sqrt[3]{5a - 1} = 3$
11. $\sqrt[3]{7a - 1} = 3$ **12.** $\sqrt[3]{6b - 4} = -4$
13. $\sqrt[3]{4y + 12} = -2$ **14.** $\sqrt[3]{2y - 5} = 3$
15. $\sqrt[3]{7m - 8} = 3$ **16.** $2\sqrt{x - 1} = 5$
17. $\sqrt{3y + 7} + 5 = 8$ **18.** $3\sqrt{y - 2} = 9$
19. $\sqrt{6a - 1} + 9 = 5$ **20.** $\sqrt{3m + 12} - 5 = 4$
21. $\sqrt{8b + 4} - 2 = 8$ **22.** $6 - \sqrt{8m - 1} = 10$

23. $\sqrt{2y + 1} + 2 = \sqrt{2y - 5}$

24. $\sqrt{5x - 2} + 3 = \sqrt{7 + 5x}$

25. $\sqrt{3x - 1} + 10 = \sqrt{3x + 8}$

26. $\sqrt[3]{9a - 6} = \sqrt[3]{3a + 18}$

27. $\sqrt[4]{5m - 1} = \sqrt[4]{2m + 8}$

28. $\sqrt[4]{5x - 4} - \sqrt[4]{2x + 5} = 0$

29. $\sqrt{7p - 8} = \sqrt{2p + 5}$

30. $\sqrt{10m + 6} = \sqrt{2m + 22}$

31. $\sqrt[3]{3x + 6} - \sqrt[3]{9x + 22} = 0$

32. $\sqrt[3]{11x - 5} = \sqrt[3]{8x + 4}$

33. $\sqrt{3a + 4} - \sqrt{2a + 4} = 0$

34. $\sqrt{a^2 - 3a + 5} = a - 7$

35. $\sqrt{m^2 + 8m - 17} = m - 5$

36. $\sqrt{4b^2 - 6b - 2} = 2b - 10$

37. $\sqrt{1 - 2p + p^2} = p + 3$

38. $\sqrt{16m^2 - 9} = 4m - 1$

39. $\sqrt{y^2 + 5y + 9} = y + 3$

40. $3x + 2 = \sqrt{9x^2 + 5x}$

41. $\sqrt{9a^2 + 4a} - 3a = 2$

42. $\sqrt{5 - 2a + 16a^2} = 4a - 1$

43. $\sqrt{a^2 + 2a + 4} = a + 4$

44. $\sqrt{a^2 - 16} = 4 + a$

45. $\sqrt{4a^2 + 8a - 5} = 2a + 3$

46. $a + 6 = \sqrt{a^2 + 7a - 4}$

47. The average velocity of molecules is given by

$$v = \sqrt{\frac{3kT}{m}}$$

where T is temperature, m is mass, and k is a constant involving the universal gas constant. Solve this literal equation for T.

E **48.** The resonance frequency of a circuit containing inductance L and capacitance C is

$$f = \frac{1}{2\pi\sqrt{LC}}$$

Solve for C.

49. The radius of a sphere in terms of its volume is given by

$$r = \sqrt[3]{\frac{3V}{4\pi}}$$

Solve for V.

50. The velocity V of an object in simple harmonic motion is given by

$$V = \frac{2\pi}{T}\sqrt{r^2 - x^2}$$

where r = amplitude, T = period, and x = displacement of the vibrating body from its position of rest. Solve for x.

51. The velocity v of an object with initial velocity v_0, displacement s, and acceleration a is given by

$$v = \sqrt{v_0^2 + 2as}$$

Solve this literal equation for s.

52. The lateral surface area S of a cone as a function of the radius of the base r and height h is given by

$$S = \pi r\sqrt{r^2 + h^2}$$

Solve this formula for h.

CHAPTER SUMMARY

Summary of Terms

conjugates (p. 149)

extraneous solution (p. 151)

like radicals (p. 145)

Power Rule (p. 150)

rationalizing the denominator (p. 141)

CHAPTER REVIEW

Section 4–1

Simplify by removing zero and negative exponents and applying the rules for exponents.

1. $\dfrac{2x^{-3}y^2}{4^{-1}y^{-6}}$
2. $\dfrac{6a^{-4}b^3c^0}{9a^3b^{-5}c^3}$
3. $\dfrac{16m^{-6}n^3}{24m^{-4}n^{-8}}$
4. $\dfrac{15x^{-8}y^0z^4}{20x^{10}y^{-4}z^{-7}}$
5. $\dfrac{3m^{-1}p^3}{mnp^{-6}}$
6. $\dfrac{3xy^{-2}z^4}{9x^{-4}y^{-3}}$
7. $\left(\dfrac{a^{-4}b^3c^{-1}}{a^0c^{-6}}\right)^{-3}$
8. $\left(\dfrac{3x^3y^{-4}}{x^{-6}y^{-3}}\right)^2$
9. $x^{-1} + 3y^{-2}$
10. $4m^{-1} - 6n^{-2}$

Section 4–2

Evaluate.

11. $81^{3/4}$
12. $343^{-2/3}$
13. $16^{-1/4}$
14. $64^{5/6}$
15. $8^{-2/3} - 125^{1/3}$
16. $16^{3/4} + 216^{-2/3}$

Evaluate and round to hundredths.

17. $6^{1/3} \cdot 6^{1/2}$
18. $8^{3/4} \cdot 9^{2/3}$
19. $(16^{3/4})^{1/2}$
20. $(7^{3/5})^{-2}$

Simplify.

21. $\dfrac{m^{3/2}n^{2/5}}{m^{-1/3}n^2}$
22. $\dfrac{a^3b^{-1/2}}{a^{-2/3}b^0}$
23. $\dfrac{(x^{-2}y^{1/3})^2}{x^{-3/4}z^{-2}}$
24. $\dfrac{(m^{-3/5}n^2)^{-5}}{n^{-8}}$

Section 4–3

Express each answer in simplest radical form.

25. $\sqrt{180x^2y^5}$
26. $\sqrt{108m^5n^8}$
27. $\sqrt{54x^5y^8}$
28. $\sqrt{64x^9y^3}$
29. $\sqrt{\dfrac{8x^2y^3}{z^6}}$
30. $\sqrt{\dfrac{64b^7c^3}{a^8}}$
31. $\sqrt{\dfrac{27x^3y^5}{16z^5}}$
32. $\sqrt{\dfrac{81ab^4}{c^7}}$
33. $\sqrt{\dfrac{a^5b^9}{4c}}$
34. $\sqrt{\dfrac{9m^8n^3}{8p}}$
35. $\sqrt{2x^2 + 10x + \dfrac{25}{2}}$
36. $\sqrt{5y^2 - 2y + \dfrac{1}{5}}$

Section 4–4

Perform the indicated operation. Leave your answer in simplest radical form.

37. $3\sqrt{50} - 5\sqrt{32} + \sqrt{8}$
38. $3\sqrt{8} + 5\sqrt{18} - 3\sqrt{72}$
39. $\sqrt{27} + 2\sqrt{72} - 3\sqrt{48}$
40. $\sqrt{125} - \sqrt{20} - \sqrt{45}$
41. $2\sqrt{16} - \sqrt{54} + 3\sqrt{128}$
42. $\sqrt{81} - 2\sqrt{24} + 5\sqrt{3}$
43. $3\sqrt{2}(\sqrt{5} + 4\sqrt{3})$
44. $\sqrt{7}(2\sqrt{3} - \sqrt{5})$
45. $(\sqrt{3} + \sqrt{8})(\sqrt{3} - 3\sqrt{5})$
46. $(3\sqrt{6} + \sqrt{5})(\sqrt{6} - 2\sqrt{5})$
47. $(2\sqrt{3} + 6\sqrt{5})(\sqrt{3} - 3\sqrt{5})$
48. $(\sqrt{8} - \sqrt{6})(2\sqrt{8} + 3\sqrt{6})$
49. $\dfrac{\sqrt{20x^3}}{\sqrt{5x^5}}$
50. $\dfrac{\sqrt{60y^5}}{\sqrt{4y}}$
51. $\dfrac{6\sqrt{3}}{2\sqrt{7}}$
52. $\dfrac{4\sqrt{5}}{16\sqrt{3}}$
53. $\dfrac{3\sqrt{x} - 4}{\sqrt{x} + 2}$
54. $\dfrac{6 - \sqrt{y}}{5 + \sqrt{y}}$
55. $\dfrac{2\sqrt{m} - \sqrt{n}}{\sqrt{m} + \sqrt{n}}$
56. $\dfrac{\sqrt{a} + 3\sqrt{b}}{\sqrt{a} - 3\sqrt{b}}$

Section 4–5

Solve the following equations.

57. $3\sqrt{y} = 9$
58. $6\sqrt{x} = 12$
59. $\sqrt{2y + 3} = 5$
60. $\sqrt{3y + 4} = 4$
61. $\sqrt{3z + 1} + 5 = 8$
62. $\sqrt{4x - 1} + 5 = 12$
63. $6 - \sqrt{x + 3} = 7$
64. $7 + \sqrt{2y - 1} = -1$
65. $3\sqrt{a - 4} = \sqrt{2a + 9}$
66. $4\sqrt{m - 5} = 20$
67. $\sqrt{x + 5} = \sqrt{x - 1} + 6$
68. $\sqrt{y + 3} - 6 = \sqrt{y - 7}$
69. $\sqrt{x - 3} + 6 = \sqrt{x - 8}$

CHAPTER TEST

The number in parentheses refers to the appropriate learning objective at the beginning of the chapter.

1. Combine $2\sqrt{27} - 7\sqrt{24} + 5\sqrt{3}$. (4)

2. Simplify $\sqrt[3]{108x^5y^8z}$. (3)

3. Evaluate the following expression: $8^{-2/3} + 16^{3/4}$ (2)

4. Simplify the following and rationalize the denominator: (4)

$$\sqrt[3]{\frac{16x^4y^3}{z^4}}$$

5. Pressure has units of force per unit area. This gives (1)

$$\frac{(M)(L)(T)^{-2}}{(L)^2}$$

Remove negative exponents and simplify.

6. Solve the following for x: (5)

$$\sqrt{3x - 1} + 8 = 10$$

7. Apply the Laws of Exponents to simplify the following: (1)

$$\frac{9x^{-1}y^{-3}z^0}{15x^4y^{-8}z^4}$$

8. Multiply the following binomials: (4)

$$(3\sqrt{2} + \sqrt{5})(\sqrt{2} - 4\sqrt{5})$$

9. Simplify the radical $\sqrt{216x^5y^3z}$. (3)

I **10.** A sinking fund is an annuity established to produce an amount of money at some future date. If amount A is deposited at the end of each period and interest is compounded at rate i per period, the total P will be produced at the end of n periods. (1)

$$P = A\left[\frac{(1 + i)^n - 1}{i}\right]$$

Evaluate this expression if A = \$5,000, i = 8.5%, and n = 2.5 yr.

11. Simplify (4)

$$\sqrt{\frac{x^9z^2}{18y^5}}$$

12. Multiply the following radicals: (4)

$$6\sqrt{3}(\sqrt{8} - 3\sqrt{6})$$

13. Simplify **(1)**

$$\left(\frac{4x^{-1}y^3}{18x^3y^{-4}}\right)^{-2}$$

14. Evaluate $81^{-3/4}$. **(2)**

15. The frequency n of a stretched string of length L, tension T, density d, and radius r is given by **(3)**

$$n = (2rL)^{-1}\sqrt{\frac{T}{\pi d}}$$

Simplify this expression when $T = 21$ N (2.1×10^6 g·cm/s^2), $d = 8$ g/cm^3, $r = 0.05$ cm, and $L = 20$ cm.

16. Simplify the following: **(2)**

$$\frac{9^{3/4} \cdot 9^{-1/4}}{9}$$

17. Solve the equation $\sqrt{4x + 1} + 6 = \sqrt{4x + 2}$. **(5)**

SOLUTION TO CHAPTER INTRODUCTION

The formula relating resonant frequency f, inductance L, and capacitance C is

$$f = \frac{1}{2\pi\sqrt{LC}}$$

To determine the value of the inductance in the given circuit, we solve the equation for L and substitute the given values.

$$f^2 = \left(\frac{1}{2\pi\sqrt{LC}}\right)^2 \qquad \text{eliminate radical}$$

$$f^2 = \frac{1}{4\pi^2LC} \qquad \text{simplify}$$

$$4\pi^2f^2LC = 1 \qquad \text{eliminate fraction}$$

$$L = \frac{1}{4\pi^2f^2C} \qquad \text{divide}$$

$$L = \frac{1}{4\pi^2(1{,}200)^2(1.4 \times 10^{-6})}$$

$$L = 1.3 \times 10^{-2}\ \text{F}$$

You are an architect designing a rectangular swimming pool addition for a client. Because of the size of the lot, the pool area is 40 ft by 30 ft. What are the dimensions of a walkway of uniform width around the pool if the area of the pool is 704 ft^2? (The solution to this problem is given at the end of the chapter.)

Solving problems of this type results in a quadratic equation. In this chapter we will discuss three methods for solving quadratic equations: factoring, completing the square, and the quadratic formula. Also, we will discuss solving nonlinear inequalities and the application of these concepts to technical problems.

Learning Objectives

After completing this chapter, you should be able to

1. Solve a quadratic equation by factoring (Section 5–1).
2. Solve a quadratic equation by completing the square (Section 5–2).
3. Solve a quadratic equation by using the quadratic formula (Section 5–3).
4. Using the discriminant, describe the number and type of roots in a quadratic equation (Section 5–3).
5. Solve a quadratic inequality (Section 5–4).
6. Apply the techniques for solving quadratic equations to technical problems (Section 5–5).

Chapter 5

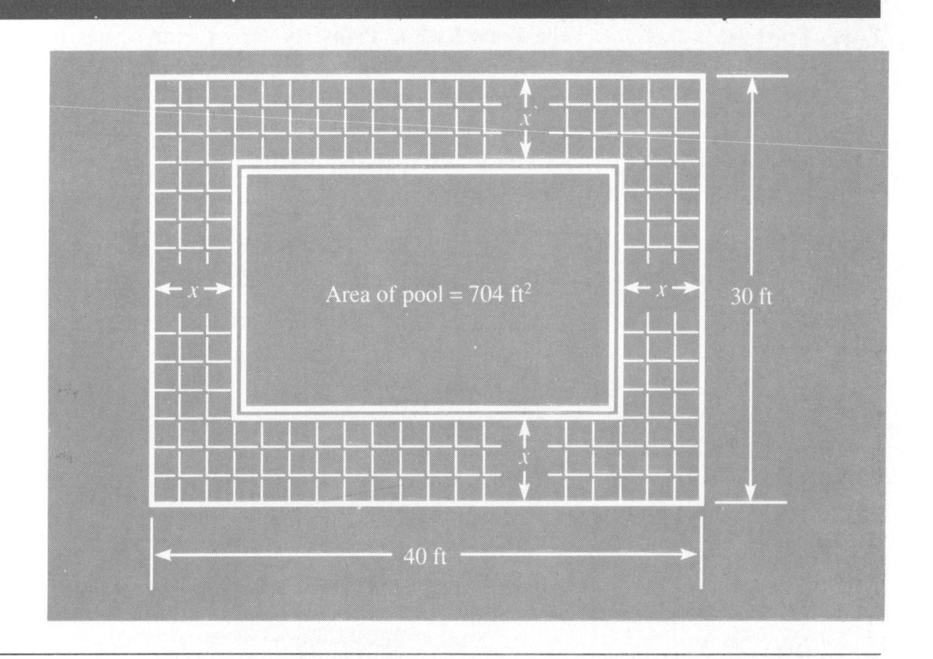

Quadratic Equations

5–1

SOLUTION BY FACTORING

Definitions

An equation of the form $ax^2 + bx + c = 0$ where a, b, and c are constants and $a \neq 0$ is called a **quadratic equation in standard form.** A **solution** or **root** of a quadratic equation is any value of the variable that results in a true statement. The solution may be two real numbers or a single real number or a pair of complex numbers. We begin the discussion of solving quadratic equations by factoring with the Zero Factor Property.

Zero Factor Property

The **Zero Factor Property** serves as the basis for solving quadratic equations by factoring. It states that *if the product of two or more factors equals zero, then at least one of the factors must equal zero*. For example, if $(x + 3)(x - 4) = 0$, then by the Zero Factor Property, $x + 3 = 0$ or $x - 4 = 0$. As a result of applying the Zero Factor Property, we can say that the solutions to the equation are $x = -3$ and $x = 4$ since both numbers result in a true statement.

Solution by Factoring

The next example illustrates the use of the Zero Factor Property in solving quadratic equations by factoring.

EXAMPLE 1 Solve $6x^2 + 7x - 20 = 0$.

Solution First, we factor the left side of the equation.

$$(3x - 4)(2x + 5) = 0$$

Next, we apply the Zero Factor Property.

$$3x - 4 = 0 \qquad 2x + 5 = 0$$

Then we solve each equation for x.

$$\begin{aligned} 3x - 4 &= 0 \\ 3x &= 4 \\ x &= 4/3 \end{aligned} \qquad \begin{aligned} 2x + 5 &= 0 \\ 2x &= -5 \\ x &= -5/2 \end{aligned}$$

Checking $x = 4/3$:

$$\begin{aligned} 6(4/3)^2 + 7(4/3) - 20 &= 0 \\ 32/3 + 28/3 - 20 &= 0 \\ 0 &= 0 \end{aligned}$$

$x = -5/2$:

$$\begin{aligned} 6(-5/2)^2 + 7(-5/2) - 20 &= 0 \\ 75/2 - 35/2 - 20 &= 0 \\ 0 &= 0 \end{aligned}$$

Since both answers result in a true statement, the solutions are $x = 4/3$ and $x = -5/2$. ■

Solution by Factoring

1. Arrange the quadratic equation in standard form, $ax^2 + bx + c = 0$.
2. Factor $ax^2 + bx + c$.
3. Apply the Zero Factor Property and solve for the variable.
4. Check the answer.

NOTE ✦ Since this method of solving quadratic equations is based on factoring, you may want to review the methods of factoring discussed in Chapter 3.

EXAMPLE 2 Solve $2m^2 = 23m - 63$.

Solution First, we arrange the quadratic equation in standard form.

$$2m^2 - 23m + 63 = 0$$

Then we factor the left side of the equation.

$$(2m - 9)(m - 7) = 0$$

Next, we apply the Zero Factor Property and solve for m.

$$\begin{aligned} 2m - 9 &= 0 & \qquad m - 7 &= 0 \\ 2m &= 9 & & \\ m &= 9/2 & m &= 7 \end{aligned}$$

Since both answers check, the solutions are $m = 9/2$ and $m = 7$. ■

CAUTION ✦ Be sure to arrange the quadratic equation in standard form before you begin to factor. The Zero Factor Property applies only to factors whose product equals zero. You may have a tendency to apply the Zero Factor Property to equations like $3(x + 4)(x - 8) = 10$, but the Zero Factor Property cannot be applied to this equation in its current form because it does not equal zero.

EXAMPLE 3 Solve $9x^2 = 10x$.

Solution Arranging the quadratic equation in standard form gives

$$9x^2 - 10x = 0$$

Notice that the equation does not contain a constant term; however, this does not affect the way we solve the equation. Next, we factor the left side of the equation using the greatest common factor technique of factoring.

$$x(9x - 10) = 0$$

Last, we apply the Zero Factor Property and solve for x.

$$x = 0 \qquad 9x - 10 = 0$$
$$9x = 10$$
$$x = 10/9$$

The solutions are $x = 0$ and $x = 10/9$. ■

EXAMPLE 4 Solve $9y^2 = 4$.

Solution First, we must arrange the equation in $ax^2 + bx + c = 0$ form.

$$9y^2 - 4 = 0$$

Next, we factor the left side of the equation using the difference of two squares method of factoring.

$$(3y - 2)(3y + 2) = 0$$

Finally, we apply the Zero Factor Property and solve for y.

$$3y - 2 = 0 \qquad 3y + 2 = 0$$
$$3y = 2 \qquad 3y = -2$$
$$y = 2/3 \qquad y = -2/3$$

The solutions are $y = 2/3$ and $y = -2/3$. ■

EXAMPLE 5 Solve $4x(x - 2) + 7 = 8(x - 1)$.

Solution First, we must arrange this equation in standard form.

$$4x^2 - 8x + 7 = 8x - 8 \quad \text{eliminate parentheses}$$
$$4x^2 - 8x + 7 - 8x + 8 = 0 \quad \text{transpose } 8x \text{ and } -8$$
$$4x^2 - 16x + 15 = 0 \quad \text{combine like terms}$$

Next, we factor the equation.

$$(2x - 3)(2x - 5) = 0$$

Finally, we apply the Zero Factor Property and solve for x.

$$2x - 3 = 0 \qquad 2x - 5 = 0$$
$$2x = 3 \qquad 2x = 5$$
$$x = 3/2 \qquad x = 5/2$$

The solutions are $x = 3/2$ and $x = 5/2$. ■

Fractional Equations

Like linear equations, quadratic equations are easier to solve if fractions are eliminated. Some equations do not appear to be quadratic until the fractions are actually eliminated, as shown in the next example.

EXAMPLE 6 Solve the following equation:

$$\frac{x + 2}{x} = \frac{6x}{x + 1}$$

Solution First, arrange the equation in standard form.

$\cancel{x}(x + 1)\left(\dfrac{x + 2}{\cancel{x}}\right) = x(\cancel{x + 1})\left(\dfrac{6x}{\cancel{x + 1}}\right)$	multiply by LCD
$(x + 1)(x + 2) = x(6x)$	eliminate fractions
$x^2 + 2x + x + 2 = 6x^2$	eliminate parentheses
$-5x^2 + 3x + 2 = 0$	transpose and simplify
$5x^2 - 3x - 2 = 0$	multiply by -1

Next, factor the left side of the equation.

$$(5x + 2)(x - 1) = 0$$

Finally, apply the Zero Factor Property and solve for x.

$$5x + 2 = 0 \qquad x - 1 = 0$$
$$5x = -2$$
$$x = -2/5 \qquad x = 1$$

After checking, the solutions are $x = -2/5$ and $x = 1$. ■

CAUTION ✦ Remember to check for division by zero when the original equation contains a fraction.

Word Problems

EXAMPLE 7 Two computers networked together can complete a job in 2 hours. It takes one computer 3 hours longer than the other to complete the job alone. How long does it take the slower computer to complete the job alone?

Solution From the previous discussion of rate problems, recall that we want to represent the portion of the job completed in one unit of time. Let

x = amount of time it takes the slow computer alone

$(x - 3)$ = amount of time it takes the faster computer alone

2 = amount of time it takes the two computers together

Thus, the slower computer completes $1/x$ of the job in one hour; the faster computer completes $1/(x - 3)$ of the job in one hour; and together they complete 1/2 of the job in one hour. The equation is

$$\frac{1}{x} + \frac{1}{x - 3} = \frac{1}{2}$$

The equation does not appear to be quadratic, but it leads to a quadratic equation after simplification.

$$\begin{aligned}
2\cancel{x}(x - 3)\left(\frac{1}{\cancel{x}}\right) + 2x\cancel{(x - 3)}\left(\frac{1}{\cancel{x - 3}}\right) &= \cancel{2}x(x - 3)\left(\frac{1}{\cancel{2}}\right) && \text{multiply by LCD} \\
2(x - 3) + 2x &= x(x - 3) && \text{eliminate fractions} \\
2x - 6 + 2x &= x^2 - 3x && \text{eliminate parentheses} \\
4x - 6 &= x^2 - 3x && \text{combine like terms} \\
-x^2 + 4x - 6 + 3x &= 0 && \text{transpose} \\
-x^2 + 7x - 6 &= 0 && \text{combine like terms} \\
x^2 - 7x + 6 &= 0 && \text{multiply by } -1
\end{aligned}$$

Factoring the equation and applying the Zero Factor Property to solve for x gives

$$(x - 6)(x - 1) = 0$$

$$x - 6 = 0 \qquad x - 1 = 0$$

$$x = 6 \qquad x = 1$$

Relating this answer to the problem statement gives

slower computer's time:	$x = 6$	or	$x = 1$
faster computer's time:	$x - 3 = 3$		$x - 3 = -2$

Note that when the slower computer's time is 1 hour, then the faster computer's time is -2 hours. Since time is not measured in negative numbers, the only solution is $x = 6$. Therefore, it takes the slower computer 6 hours to complete the job alone. ■

Equations with Radicals

Some equations containing radicals do not appear to be quadratic, but simplification leads to a quadratic equation, as seen in the next example.

EXAMPLE 8 Solve the following:

$$a = \sqrt{\frac{6 - 13a}{5}}$$

Solution This equation also results in a quadratic equation after simplification.

$$a^2 = \frac{6 - 13a}{5} \quad \text{eliminate radical by squaring}$$

$$5a^2 = 6 - 13a \quad \text{eliminate fractions}$$

$$5a^2 + 13a - 6 = 0 \quad \text{transpose}$$

$$(5a - 2)(a + 3) = 0 \quad \text{factor}$$

$$5a - 2 = 0 \qquad a + 3 = 0$$

$$a = 2/5 \qquad a = -3$$

Since -3 does not check, the solution is $a = 2/5$. ■

CAUTION ✦ Squaring both sides of an equation to remove the radical may lead to extraneous roots. Always check your answer.

5–1 EXERCISES

Solve the following quadratic equations by factoring.

1. $x^2 + 3x - 18 = 0$

2. $x^2 - 8x + 15 = 0$

3. $a^2 - 9a + 20 = 0$

4. $m^2 + 5m - 36 = 0$

5. $2x^2 + 11x - 21 = 0$

6. $18x - 14x^2 = 0$

7. $4x^2 = 18x$

8. $6y^2 + 13y - 5 = 0$

9. $5b^2 + 22b - 15 = 0$

10. $6r^2 - 18r = 0$

11. $81 = 4x^2$

12. $9x^2 = 1$

13. $15x^2 + 13x = -2$

14. $15d^2 = 24 - 2d$

15. $t = 72 - t^2$

16. $-16x = 35 - 3x^2$

17. $25x^2 - 15 = 23x - 3x^2$

18. $2s^2 = 70 - 4s$

19. $2f^2 = 8f$

20. $x^2 = 64$

21. $16x^2 - 25 = 0$

22. $3m(m - 2) = 2(m + 5) - 9m$

23. $6z^2 + 3(z + 5) = -16z$

24. $(x - 1)(2x + 1) = 3(x + 2) + 9x$

25. $(y + 3)(y - 1) = 2y(y + 3)$

26. $(a - 3)(a + 3) = 6a - 18$

27. $(4x - 1)(x + 2) = x(2x - 3) - 10$

28. $(3x + 4)(x - 2) = 2x(x + 1) + 4$

29. $\dfrac{3c}{c - 2} + \dfrac{6}{c + 2} = \dfrac{24}{c^2 - 4}$

30. $\dfrac{5x}{x + 1} + \dfrac{7x}{x - 1} = \dfrac{3x + 1}{x^2 - 1}$

31. $\dfrac{7}{x} - \dfrac{1}{x^2} = 6$

32. $\dfrac{10}{m + 3} + \dfrac{10}{m - 3} = \dfrac{7}{2}$

33. $\dfrac{1}{y} + \dfrac{1}{y + 6} = \dfrac{1}{4}$

34. $\dfrac{60}{b} + 5 = \dfrac{60}{b - 1}$

35. $\dfrac{1}{a} + \dfrac{1}{a + 3} = \dfrac{1}{2}$

36. $\dfrac{2x}{x - 1} - \dfrac{3}{x - 2} = 0$

37. $\dfrac{1}{x} + \dfrac{1}{x + 9} = \dfrac{1}{6}$

38. $\dfrac{6}{6 - m} + \dfrac{6}{6 + m} = \dfrac{32}{15}$

39. $y = \sqrt{\dfrac{5y + 12}{3}}$

40. $3z = \sqrt{8 - 6z}$

41. $x = -\sqrt{3x + 10}$

42. $x = -\sqrt{\dfrac{6x + 8}{5}}$

43. $a = \sqrt{\dfrac{-3a + 9}{2}}$

44. $y = \sqrt{\dfrac{5y - 6}{6}}$

45. $2x = \sqrt{-3x + 10}$

46. If three times the square of a number is added to ten times the number, the result is 25. Find the number.

47. Working together, Fred and Larry can paint the exterior of a house in 8 hours, while it takes Larry 12 hours longer than Fred if each works alone. How long does it take Larry to paint the house alone?

48. When an object is dropped from a 64-ft-tall building, the resulting equation for its height is $h = 64 - 16t^2$. Find the time, t, in seconds, required for the object to strike the ground ($h = 0$).

E49. The power dissipated in a 25-Ω resistor is 49 W. The equation to find the current passing through the resistor is $49 = 25I^2$. Solve for the current, I.

50. A tract of land is triangular. If the area of the tract is 1,200 m² and the height is 40 m more than the base, find the dimensions of the tract.

5–2

SOLUTION BY COMPLETING THE SQUARE

The factoring method of solving quadratic equations is limited in its usefulness. A second method for solving quadratic equations, called *completing the square,* can be used to solve any quadratic equation.

Before actually discussing completing the square, however, we must solve a particular type of quadratic equation: equations of the form $(ax + b)^2 = c$. Solving this type of equation is based on the **Square Root Property,** which states that *if m and n are real numbers and $m^2 = n^2$, then $m = n$ or $m = -n$.* In other words, taking the square root of both sides of an equation produces an equivalent equation.

EXAMPLE 1 Solve $(x + 3)^2 = 25$.

Solution First, we *apply the square root property.*

$$\sqrt{(x + 3)^2} = \pm\sqrt{25}$$
$$x + 3 = \pm 5$$

The notation ± 5 is a shorthand method of writing both $+5$ and -5. Then we *solve the resulting equation for x.*

$$x = -3 \pm 5$$

The following solutions result from solving the two equations:

$$x = -3 + 5 \qquad x = -3 - 5$$
$$x = 2 \qquad x = -8$$

The solutions are $x = 2$ and $x = -8$. ■

EXAMPLE 2 Solve $(3x + 4)^2 = 48$.

Solution By the square root property, we know that

$$\sqrt{(3x + 4)^2} = \pm\sqrt{48}$$
$$3x + 4 = \pm\sqrt{48}$$

Next, simplify the radical using the techniques from the previous chapter (that is, $\sqrt{48} = \sqrt{16 \cdot 3} = 4\sqrt{3}$).

$$3x + 4 = \pm 4\sqrt{3}$$

Then solve the linear equation for x.

$$3x = -4 \pm 4\sqrt{3}$$

$$x = \frac{-4 \pm 4\sqrt{3}}{3}$$

■

The equations in Examples 1 and 2 were easy to solve because the expression containing the variable was a perfect square. The process of completing the square transforms an equation to this form.

Solution by Completing the Square

1. Arrange the equation in $ax^2 + bx = c$ form.
2. If $a \neq 1$, divide each term of the equation by a, giving

$$x^2 + \frac{b}{a}x = \frac{c}{a}$$

3. Complete the square by *adding* the magnitude (omit the sign) of $\left(\frac{b}{2a}\right)^2$ to both sides of the equation.
4. Write the left-hand side of the equation as a perfect square:

$$(\sqrt{\text{first term}} \pm \sqrt{\text{third term}})^2$$

5. Apply the Square Root Property.
6. Solve the resulting linear equation for the variable.

EXAMPLE 3 Solve $x^2 + 6x = 24$.

Solution Since the equation is arranged in the required form and $a = 1$, we begin with Step 3 by *completing the square*. We must add the magnitude of

$$\left(\frac{b}{2a}\right)^2 = \left(\frac{6}{2}\right)^2 = 9$$

to both sides of the equation.

$$x^2 + 6x \textcircled{+ 9} = 24 \textcircled{+ 9}$$

Next, we *write the left side of the equation as a perfect square* and *simplify* the right side.

$$(x + 3)^2 = 33$$

Then we *apply the Square Root Property*.

$$\sqrt{(x + 3)^2} = \pm\sqrt{33}$$
$$x + 3 = \pm\sqrt{33}$$

Finally, we solve the linear equation for x.

$$x = -3 \pm \sqrt{33}$$

■

EXAMPLE 4 Solve $2x^2 + 3x - 8 = 0$. Express the solution in decimal form rounded to thousandths.

Solution First, we arrange the equation in the required form.

$$2x^2 + 3x = 8$$

Since $a \neq 1$, we divide both sides of the equation by the value of a, which is 2.

$$\frac{2x^2}{2} + \frac{3x}{2} = \frac{8}{2}$$
$$x^2 + \frac{3}{2}x = 4$$

Next, we complete the square by adding

$$\left(\frac{b}{2a}\right)^2 = \left(\frac{3}{4}\right)^2 = \frac{9}{16}$$

to both sides of the equation.

$$x^2 + \frac{3}{2}x + \frac{9}{16} = 4 + \frac{9}{16}$$

Then we write the left side of the equation as a perfect square and simplify the right-hand side.

$$\left(x + \frac{3}{4}\right)^2 = \frac{73}{16}$$

Take the square root of both sides of the equation.

$$x + \frac{3}{4} = \pm\frac{\sqrt{73}}{4}$$

Solve for x.

$$x = \frac{-3 \pm \sqrt{73}}{4}$$

The roots are approximately 1.386 and -2.886.

First root: 3 [+/−] [+] 73 [√] [=] [÷] 4 [=] → 1.386

Second root: 3 [+/−] [−] 73 [√] [=] [÷] 4 [=] → −2.886 ■

REMEMBER ✦ The square root of a negative number is an imaginary number, as we discussed in Chapter 4 and as illustrated in the following example.

EXAMPLE 5 Solve the following quadratic equation by completing the square:

$$3x^2 + 10 + 2x = 0$$

Solution First, arrange the equation in the required order.

$$3x^2 + 2x = -10$$

Since $a \neq 1$, we divide each term by the value of a, which is 3.

$$x^2 + \frac{2}{3}x = \frac{-10}{3}$$

Next, we complete the square by adding

$$\left(\frac{b}{2a}\right)^2 = \left(\frac{2}{6}\right)^2 = \frac{1}{9}$$

to each side of the equation.

$$x^2 + \frac{2}{3}x + \frac{1}{9} = \frac{-10}{3} + \frac{1}{9}$$

Write the left side of the equation as a perfect square and combine like terms on the right side.

$$\left(x + \frac{1}{3}\right)^2 = \frac{-29}{9}$$

Take the square root of both sides of the equation.

$$x + \frac{1}{3} = \pm j\frac{\sqrt{29}}{3}$$

(Remember, $\sqrt{-29} = \sqrt{29}\sqrt{-1} = j\sqrt{29}$, since $j = \sqrt{-1}$.) The solution is the pair of complex numbers

$$x = \frac{-1 \pm j\sqrt{29}}{3}$$

■

Pythagorean Theorem

In a right triangle, the two perpendicular sides are called the **legs,** and the side opposite the right angle is called the **hypotenuse.** A very important relationship involving the sides of a right triangle, called the **Pythagorean theorem,** states that *the square of the length of the hypotenuse equals the sum of the squares of the length of the legs.* In other words,

$$(\text{hypotenuse})^2 = (\text{leg}_1)^2 + (\text{leg}_2)^2$$

We use the Pythagorean theorem to solve the problem in the next example.

EXAMPLE 6 One leg of a right triangle is 21 m longer than the other leg, and the hypotenuse is 39 m. Find the length of the legs of the triangle.

FIGURE 5–1

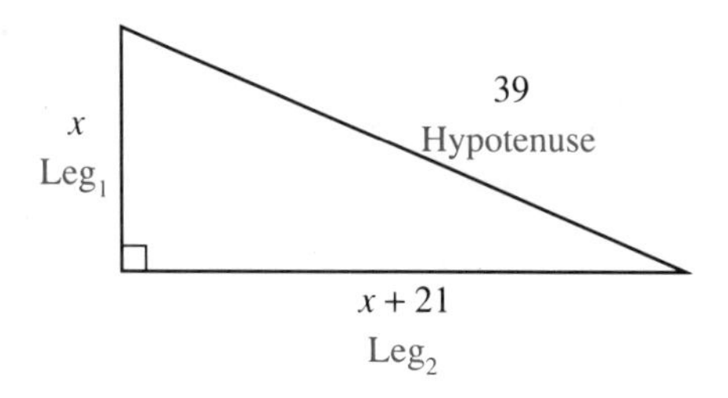

Solution As shown in Figure 5–1, let

$$x = \text{length of the shorter leg} = \text{leg}_1$$
$$(x + 21) = \text{length of the longer leg} = \text{leg}_2$$

Using the Pythagorean theorem, we obtain the equation

$$\begin{array}{ccccc} (\text{hypotenuse})^2 & = & (\text{leg}_1)^2 & + & (\text{leg}_2)^2 \\ \uparrow & & \uparrow & & \uparrow \\ (39)^2 & = & x^2 & + & (x + 21)^2 \end{array}$$

We simplify the equation and arrange it in $ax^2 + bx + c = 0$ form.

$$\begin{aligned} 1{,}521 &= x^2 + x^2 + 42x + 441 && \text{eliminate parentheses} \\ 1{,}521 &= 2x^2 + 42x + 441 && \text{combine like terms} \\ 2x^2 + 42x &= 1{,}080 && \text{transpose} \\ x^2 + 21x &= 540 && \text{divide by 2} \end{aligned}$$

Complete the square by adding $(21/2)^2 = 441/4$ to both sides of the equation.

$$x^2 + 21x + \frac{441}{4} = 540 + \frac{441}{4}$$

Simplify each side of the equation.

$$\left(\sqrt{x^2} + \sqrt{\left(\frac{21}{2}\right)^2}\right)^2 = \frac{2{,}601}{4}$$

Take the square root of both sides of the equation.

$$x + \frac{21}{2} = \pm\frac{51}{2}$$

Solve the resulting equation for x.

$$x = \frac{-21}{2} \pm \frac{51}{2}$$

$$x = 15 \quad \text{and} \quad x = -36$$

Since the lengths of the sides are not measured in negative numbers, $x = 15$ and $x + 21 = 36$. The lengths of the legs are 15 m and 36 m. ■

5–2 EXERCISES

Solve the following quadratic equations by completing the square.

1. $x^2 - 4x = 20$

2. $a^2 + 6a = 12$

3. $m^2 + 6m = 14$

4. $x^2 - 8x = 15$

5. $x^2 - 2x = 3$

6. $x^2 + 5x = 10$

7. $x^2 + 5x - 8 = 0$

8. $z^2 - 3z = 7$

9. $k^2 - 9k - 8 = 0$

10. $3x^2 + 6x - 17 = 0$

11. $2x^2 + 8x = 13$

12. $5x^2 - 15x = 23$

13. $4d^2 - 12d + 19 = 0$

14. $2y^2 + 5y + 10 = 0$

15. $3x^2 - 9x = 14$

16. $4x^2 + 10x = 31$

17. $20m = 24 - 5m^2$

18. $7x = 24 - 3x^2$

19. $7x^2 - 21x + 17 = 0$

20. $6z^2 = 25 - 3z$

21. $x^2 + 5x + 1 = 0$

22. $5x^2 = 3x - 14$

23. $-9c = -2 - 2c^2$

24. $3a^2 + 4a = -8$

25. $3x^2 + 7x + 6 = 0$

26. $8x^2 + 10x = 17$

27. $17y - 4y^2 = 20$

28. $2y^2 = 8 - 7y$

29. $7x^2 - 14 = 6x$

30. $15 - 3x^2 = 5x$

E **31.** The power dissipated in a circuit with two currents flowing through a resistor is given by

$$P = (I_1 + I_2)^2 R$$

Find I_1 if $I_2 = 0.3$ A, $R = 100\ \Omega$, and $P = 11$ W. Round your answer to hundredths.

32. The sum of two numbers is 10 and their product is 24. Find the numbers.

33. One leg of a right triangle is 4 more than the smaller leg, and the hypotenuse is 8 more than the smaller leg. Find the length of the sides of the right triangle.

34. Find two consecutive integers such that the sum of their squares is 613.

I **35.** Total revenue R is the product of price and quantity. If the price is a function of the quantity sold q, the equation is $R = (1{,}000 - q)q$. Determine the production quantity to earn \$187,500 in revenue.

5–3

SOLUTION BY THE QUADRATIC FORMULA

A third method of solving quadratic equations is the quadratic formula. You can use the quadratic formula, as well as completing the square, to solve any quadratic equation. However, you will find that the quadratic formula is easier to use. The derivation of the quadratic formula follows.

Deriving the Quadratic Formula

We derive the quadratic formula by taking the standard form of a quadratic equation and completing the square.

$ax^2 + bx + c = 0$	
$ax^2 + bx = -c$	transpose c
$x^2 + \frac{b}{a}x = -\frac{c}{a}$	divide by a
$x^2 + \frac{b}{a}x + \left(\frac{b}{2a}\right)^2 = -\frac{c}{a} + \left(\frac{b}{2a}\right)^2$	add the square of half the linear term's coefficient
$x^2 + \frac{b}{a}x + \frac{b^2}{4a^2} = -\frac{c}{a} + \frac{b^2}{4a^2}$	simplify
$\left(x + \frac{b}{2a}\right)^2 = \frac{-4ac + b^2}{4a^2}$	factor left side and simplify right side
$x + \frac{b}{2a} = \pm\sqrt{\frac{b^2 - 4ac}{4a^2}}$	take square root of both sides
$x = \frac{-b}{2a} \pm \sqrt{\frac{b^2 - 4ac}{4a^2}}$	solve for x
$x = \frac{-b \pm \sqrt{b^2 - 4ac}}{2a}$	simplify and add fractions

The quadratic formula is summarized below.

Quadratic Formula

If $ax^2 + bx + c = 0$, where $a \neq 0$, then

$$x = \frac{-b \pm \sqrt{b^2 - 4ac}}{2a}$$

Notice that the quadratic formula results in the following two solutions:

$$\frac{-b + \sqrt{b^2 - 4ac}}{2a} \quad \text{and} \quad \frac{-b - \sqrt{b^2 - 4ac}}{2a}$$

Solution by the Quadratic Formula

To solve a quadratic equation using the formula, go through the following steps:

1. Arrange the equation in standard form,

$$ax^2 + bx + c = 0$$

2. Identify a, b, and c. The coefficient of the squared term is a; the coefficient of the linear term is b; and the constant term is c.
3. Substitute the values of a, b, and c into the formula.
4. Simplify the expression.
Note: The coefficients a, b, and c include their signs.

The process of solving several types of quadratic equations is illustrated in the next three examples.

EXAMPLE 1 Solve $5x^2 - 6x = 3$.

Solution First, we arrange the equation in standard form and identify the values of a, b, and c.

$$\underset{\substack{\uparrow \\ a = 5}}{5x^2} \quad \underset{\substack{\uparrow \\ b = -6}}{-6x} \quad \underset{\substack{\uparrow \\ c = -3}}{-3} = 0$$

Then we substitute values for a, b, and c into the quadratic formula.

$$x = \frac{-b \pm \sqrt{b^2 - 4ac}}{2a}$$

$$x = \frac{-(-6) \pm \sqrt{(-6)^2 - 4(5)(-3)}}{2(5)}$$

$$x = \frac{6 \pm \sqrt{36 + 60}}{10}$$

$$x = \frac{6 \pm \sqrt{96}}{10}$$

Next, we simplify the radical as follows: $\sqrt{96} = \sqrt{16 \cdot 6} = 4\sqrt{6}$ (remove any squares just as we did in Chapter 4).

$$x = \frac{6 \pm 4\sqrt{6}}{10}$$

Last, we reduce the fractional expression to lowest terms.

$$x = \frac{2(3 \pm 2\sqrt{6})}{10}$$

$$x = \frac{3 \pm 2\sqrt{6}}{5}$$

■

EXAMPLE 2 Solve the following:

$$\frac{y + 1}{y - 2} - \frac{6}{y^2 - 5y + 6} = \frac{2y}{y - 3}$$

Solution First, we must *arrange the equation in standard form*. We *eliminate fractions* by multiplying by the common denominator, $(y - 3)(y - 2)$.

$$\begin{aligned}
(y + 1)(y - 3) - 6 &= 2y(y - 2) && \text{eliminate fractions}\\
y^2 - 2y - 3 - 6 &= 2y^2 - 4y && \text{eliminate parentheses}\\
-y^2 + 2y - 9 &= 0 && \text{transpose and}\\
y^2 \quad -2y \quad +9 &= 0 && \text{combine like terms}\\
\uparrow \qquad \uparrow \qquad \uparrow & && \text{multiply by } -1\\
a = 1 \quad b = -2 \quad c = 9 & &&
\end{aligned}$$

Substituting into the quadratic formula and *simplifying* the result gives

$$y = \frac{-(-2) \pm \sqrt{(-2)^2 - 4(1)(9)}}{2(1)}$$

$$y = \frac{2 \pm \sqrt{4 - 4(1)(9)}}{2(1)}$$

$$y = \frac{2 \pm \sqrt{-32}}{2}$$

$$y = \frac{2 \pm 4j\sqrt{2}}{2} \qquad \text{simplify radical}$$

$$y = \frac{2(1 \pm 2j\sqrt{2})}{2} \qquad \text{factor numerator}$$

$$y = 1 \pm 2j\sqrt{2} \qquad \text{reduce fraction}$$

■

EXAMPLE 3 Solve $(3x + 2)(5x - 1) - 1 = 3x(x - 1)$.

Solution First, we eliminate parentheses and arrange the equation in standard form.

$$\begin{aligned}
(3x + 2)(5x - 1) - 1 &= 3x(x - 1) &&\\
15x^2 - 3x + 10x - 2 - 1 &= 3x^2 - 3x && \text{eliminate parentheses}\\
15x^2 + 7x - 3 - 3x^2 + 3x &= 0 && \text{add and transpose}\\
12x^2 + 10x - 3 &= 0 && \text{combine like terms}
\end{aligned}$$

Next, we substitute $a = 12$, $b = 10$, and $c = -3$ into the quadratic formula and simplify.

$$x = \frac{-10 \pm \sqrt{100 - 4(12)(-3)}}{2(12)}$$

$$x = \frac{-10 \pm \sqrt{100 + 144}}{24}$$

$$x = \frac{-10 \pm \sqrt{244}}{24}$$

$$x = \frac{-10 \pm 2\sqrt{61}}{24}$$

$$x = \frac{2(-5 \pm \sqrt{61}}{24}$$

Reducing the fraction gives the solutions as

$$x = \frac{-5 \pm \sqrt{61}}{12}$$

■

Equations with Radicals

The following example illustrates solving an equation which contains radicals and which must be squared twice to eliminate the radical. The resulting equation is quadratic and can be solved using the quadratic formula.

EXAMPLE 4 Solve $\sqrt{y+1} - 8 = \sqrt{3y-4}$. Express the answer to the nearest thousandths.

Solution First, we must arrange the equation in standard form.

$$\begin{aligned} \sqrt{y+1} - 8 &= \sqrt{3y-4} && \\ y + 1 - 16\sqrt{y+1} + 64 &= 3y - 4 && \text{square both sides} \\ y + 65 - 16\sqrt{y+1} &= 3y - 4 && \text{combine like terms} \\ -16\sqrt{y+1} &= 3y - 4 - y - 65 && \text{transpose} \\ -16\sqrt{y+1} &= 2y - 69 && \text{combine like terms} \\ 256(y+1) &= 4y^2 - 276y + 4{,}761 && \text{square both sides} \\ 256y + 256 &= 4y^2 - 276y + 4{,}761 && \text{eliminate parentheses} \\ -4y^2 + 532y - 4{,}505 &= 0 && \text{transpose and simplify} \end{aligned}$$

Next, we substitute $a = -4$, $b = 532$, and $c = -4{,}505$ into the quadratic formula and simplify the result.

$$y = \frac{-532 \pm \sqrt{(532)^2 - 4(-4)(-4{,}505)}}{2(-4)}$$

$$y = \frac{-532 \pm \sqrt{210{,}944}}{-8}$$

$$y = 9.089 \quad \text{and} \quad y = 123.911$$

To calculate $\sqrt{\ }$: 532 [x^2] [−] 4 [×] 4 [+/−] [×] 4505 [+/−] [=] [$\sqrt{\ }$] [STO] → 459.2864030

To determine y: 532 [+/−] [+] [RCL] [=] [÷] 8 [+/−] [=] → 9.089

To determine y: 532 [+/−] [−] [RCL] [=] [÷] 8 [+/−] [=] → 123.911

Checking $y = 9.089$: $\sqrt{9.089 + 1} - 8 = \sqrt{3(9.089) - 4}$

$$-4.824 \neq 4.824$$

$y = 123.911$: $\sqrt{123.911 + 1} - 8 = \sqrt{3(123.911) - 4}$

$$3.176 \neq 19.176$$

Since neither of the answers checks, this equation has no solution. ■

CAUTION ✦ When you square both sides of an equation, remember to check the solution for extraneous roots.

Quadratic Literal Equations

You can also use the quadratic formula to solve a formula in which the coefficients are literal, rather than numerical.

EXAMPLE 5 The distance fallen, s, of a free-falling object is given by

$$s = vt + \frac{gt^2}{2}$$

where v = initial velocity, t = time, and g = acceleration due to gravity. Solve this formula for t.

Solution This formula is quadratic in the variable t, so we must use one of the three methods for solving a quadratic equation. Using the quadratic formula, we must arrange the equation in standard form.

$$2(s) = 2(vt) + 2\left(\frac{gt^2}{2}\right) \quad \text{eliminate fractions}$$

$$2s = 2vt + gt^2 \quad \text{simplify}$$

$$-gt^2 - 2vt + 2s = 0 \quad \text{transpose}$$

Remember that a represents the coefficient of the squared variable, b represents the coefficient of the linear variable, and c represents the constant term. Next, we substitute $a = -g$, $b = -2v$, and $c = 2s$ into the quadratic formula.

$$t = \frac{-(-2v) \pm \sqrt{(-2v)^2 - 4(-g)(2s)}}{2(-g)}$$

$$t = \frac{2v \pm \sqrt{4v^2 - 4(-g)(2s)}}{-2g}$$

$$t = \frac{2v \pm \sqrt{4v^2 + 8gs}}{-2g}$$

Then we simplify the radical expression.

$$t = \frac{2v \pm \sqrt{4(v^2 + 2gs)}}{-2g}$$

$$t = \frac{2v \pm 2\sqrt{v^2 + 2gs}}{-2g}$$

Reducing the fraction to lowest terms gives the following solutions:

$$t = \frac{v \pm \sqrt{v^2 + 2gs}}{-g}$$

■

EXAMPLE 6 The deflection y of a beam is given by $y = x^2 + xL - 1.5L^2$ where L = length and x = distance from the end. Find the position, x, where the deflection is 0.5 ft if the length, L, is 6 ft.

Solution Since the deflection is 0.5, the equation becomes

$$0.5 = x^2 + xL - 1.5L^2$$

Substituting $L = 6$ and arranging the equation in standard form gives

$$x^2 + 6x - 54.5 = 0$$

Then substitute $a = 1$, $b = 6$, and $c = -54.5$ into the quadratic formula.

$$x = \frac{-6 \pm \sqrt{36 - 4(1)(-54.5)}}{2}$$

$$x = \frac{-6 \pm \sqrt{254}}{2}$$

$$x = \frac{-6 \pm 15.93737745}{2}$$

$$x = 4.9687 \text{ and } -10.9687$$

Since position is not measured in negative numbers, the position we want is 4.9687 ft.

6 [±] [+] 254 [√‾] [=] [÷] 2 [=] → 4.968688725

6 [±] [−] 254 [√‾] [=] [÷] 2 [=] → −10.96868872 ■

Discriminant

The **discriminant** is the expression that appears under the radical in the quadratic formula.

$$b^2 - 4ac$$

You can use the sign of the discriminant to determine the number and type of solutions to a quadratic equation. This information is summarized below.

Number and Type of Solutions	
Value of the Discriminant	Number and Type of Solutions
Zero (0)	One real solution
Negative (−)	Two complex solutions
Positive (+)	Two real solutions

CAUTION ✦ You must arrange the quadratic equation in standard form before determining the value of the discriminant.

EXAMPLE 7 Using the discriminant, describe the number and type of solutions for the quadratic equation $2x^2 - 6x + 10 = 0$.

Solution Since the quadratic equation is arranged in standard form, $a = 2$, $b = -6$, and $c = 10$. Substituting into the discriminant, we have

$$\begin{aligned} b^2 - 4ac &= (-6)^2 - 4(2)(10) \\ &= 36 - 80 \\ &= -44 \end{aligned}$$

Since the discriminant is negative, the quadratic equation has two complex solutions of the form $a + bj$. ■

NOTE ✦ The discriminant does not give the actual solution to a quadratic equation, but rather it gives information about the number and type of solutions. When imaginary numbers are not a feasible solution, it would not be necessary to solve a quadratic equation whose discriminant is a negative number.

EXAMPLE 8 Using the discriminant, determine the number and type of solutions for $3x^2 - 6 = -5x$.

Solution First, we arrange the equation in standard form and identify a, b, and c.

$$3x^2 + 5x - 6 = 0$$

Then we substitute $a = 3$, $b = 5$, and $c = -6$ into the discriminant.

$$\begin{aligned} b^2 - 4ac &= (5)^2 - 4(3)(-6) \\ &= 25 + 72 \\ &= 97 \end{aligned}$$

Since the discriminant is positive, this quadratic equation has two real number solutions. ■

5–3 EXERCISES

Solve the following equations using the quadratic formula.

1. $4x^2 + 5x - 6 = 0$
2. $3x^2 + 8x - 1 = 0$
3. $8z - 2z^2 = 10$
4. $7 - 5p = 4p^2$
5. $12 = 9y^2 - 5y$
6. $15x - 4 = x^2$
7. $4x^2 + 8 = 9x$
8. $6n - 9 + 7n^2 = 0$
9. $10x^2 - x + 11 = 0$
10. $13k + 2k^2 - 8 = 0$
11. $7 - 2z^2 + 5z = 14 - 2z$
12. $x^2 + 3x = 6 + 9x - 4x^2$
13. $11x - 14 = 6x - 2x^2 + 7$
14. $3y^2 + 10 = 14y - 2$
15. $8m^2 - 9 = 0$
16. $5x^2 - 7x = 0$
17. $8x^2 = 15x$
18. $3z^2 - 14 = 0$
19. $-3x^2 - 8 = 2x(x - 3)$
20. $4(3x + 2) + 5x(x - 1) = 8x$
21. $4d(d - 2) = (2d - 1)(d + 3)$
22. $(2x + 1)(x - 5) = 7(3x + 2)$
23. $(2m - 3)m + 6(m^2 - 4m) = 8$
24. $9t(t - 2) = 17 - 3(5t + 6)$
25. $(x - 4)(x + 3) - 5(2x - 3) = 0$
26. $8(3x + 3) = 9x(2x + 5)$
27. $3(t - 5) + 4t(2t - 1) = 38$
28. $(y + 3)(4y - 1) = 6y(y - 2) + 13$
29. $\frac{6}{x} + 9 = \frac{5}{x^2}$
30. $7 = \frac{2}{x} - \frac{11}{x^2}$
31. $\frac{2s}{s - 1} + \frac{6}{s + 3} = \frac{4s}{s^2 + 2s - 3}$
32. $\frac{5x}{2x + 3} - \frac{16}{2x^2 - x - 6} = \frac{x}{x - 2}$
33. $\frac{1}{z} + \frac{1}{z - 2} = \frac{1}{7}$
34. $\frac{1}{y} + \frac{1}{y + 2} = \frac{1}{12}$
35. $\frac{1}{x} + \frac{1}{x - 1} = \frac{1}{4}$
36. $\frac{1}{x} + \frac{1}{x - 3} = \frac{1}{12}$
37. $y = \sqrt{\frac{3y - 4}{7}}$
38. $x = \sqrt{\frac{6x - 9}{5}}$
39. $\sqrt{x^2 - 9x} = \sqrt{7 - 3x^2}$
40. $\sqrt{2x - 3} = \sqrt{4x - 7x^2}$

Using the discriminant, describe the number and type of solutions for the following equations.

41. $3x^2 - 2x = 7$
42. $9 - y^2 = 6y$
43. $6y^2 = 5y$
44. $10x^2 - 7x = 0$

45. $9x^2 = 6x$

46. $8z^2 - 11 = 0$

47. $10z^2 = 25 - 4z^2$

48. $5x^2 + 13x + 25 = 0$

49. $18x = -25 - 4x^2$

50. $14n - 17 = 3n^2$

51. $\frac{d^2}{6} + \frac{5d}{3} = 2$

52. $\frac{7x^2}{2} = 10 - \frac{5x}{3}$

53. $\frac{t^2}{4} - \frac{3}{5} = 2t$

54. $\frac{4s}{5} - 3 = \frac{s^2}{2}$

55. $\frac{2x^2}{7} - \frac{x}{3} = -2$

56. The number of diagonals N of a polygon given the number of sides n is

$$N = \frac{n(n - 3)}{2}$$

How many sides does a polygon with 35 diagonals ($N = 35$) have?

57. The formula for the surface area S of a cylinder is $S = 2\pi r^2 + 2\pi rh$ where r is the radius and h is the height. Solve the formula for r.

58. The formula for the area A of a sector of a circle is

$$A = \frac{r^2\theta}{2}$$

where r is the radius and θ is the measure of the central angle in radians. Solve the formula for r.

I 59. Total revenue is the product of price and quantity. If q represents quantity and price is a function of quantity given by $1{,}500 - 3q$, find the quantity that would produce a total revenue of \$5,000.

60. The formula to determine the range d of a ship's radio communication is given by

$$d = \sqrt{2rh + h^2}$$

Solve the formula for h.

5–4

NONLINEAR INEQUALITIES

In Chapter 1, we solved linear equations and inequalities by using much the same technique for both. In this section, we will discuss solving nonlinear inequalities. The technique is based on factoring, just as with quadratic equations, but it differs from the techniques used to solve quadratic equations in the previous sections of this chapter. Solving nonlinear inequalities is based on the concept of critical values.

Critical Values

The zeros of a function (roots of an equation), called the **critical values,** *divide the number line into intervals over which the function is always positive or always negative*. For example, the critical values of the equation $x^2 - 2x - 3 = 0$ are $x = 3$ and $x = -1$. These two critical values divide the number line into three regions where the function can be positive or negative, as shown in the following figure:

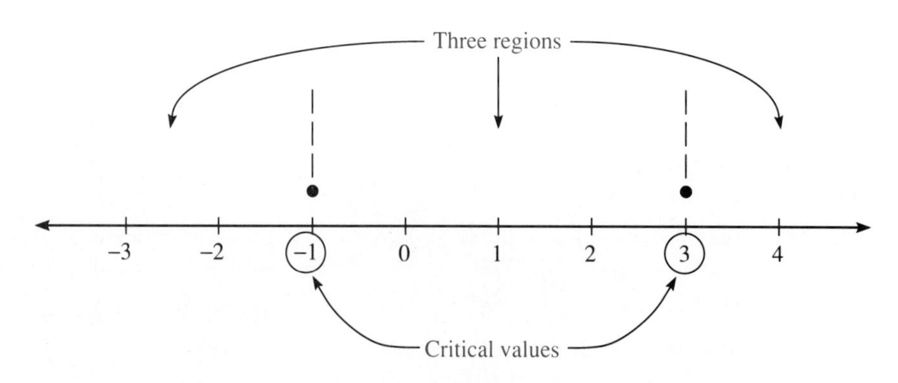

Quadratic Inequalities

In the next example, we illustrate how to use the critical values to solve a quadratic inequality.

EXAMPLE 1 Solve $x^2 + 3x - 4 > 0$.

Solution First, we must *find the critical values*. Since the critical values occur where the function $x^2 + 3x - 4$ equals zero, we solve the following equation:

$$x^2 + 3x - 4 = 0$$

$$(x + 4)(x - 1) = 0$$

$$x + 4 = 0 \qquad x - 1 = 0$$

$$x = -4 \qquad x = 1$$

The critical values are -4 and 1. Since the original inequality is strictly less than, place circles at -4 and 1 on the number line. Notice that these critical values divide the number line into three regions, as shown below:

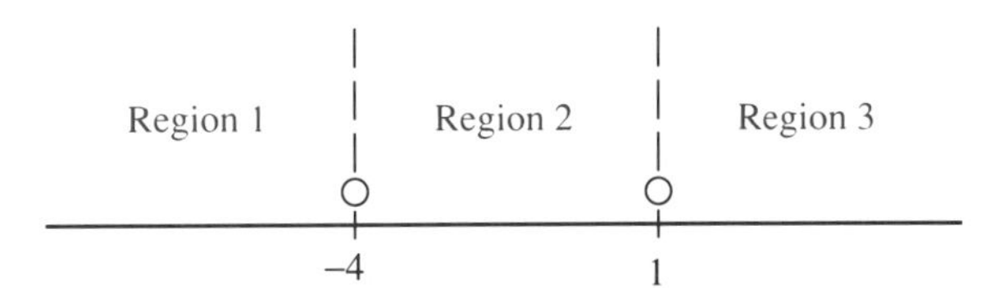

Next, we must *determine which of these regions satisfy the original inequality,* $x^2 + 3x - 4 > 0$. To do so, we will determine the sign of each factor $(x + 4)$ and $(x - 1)$ in each region. Since the sign of the inequality, and thus each factor, does not change throughout a region, we can choose any number in a given region to substitute in the factors. This process is shown below:

	Region 1		Region 2		Region 3	
Factor	Substitute $x = -6$	Sign of factor	Substitute $x = 0$	Sign of factor	Substitute $x = 3$	Sign of factor
$x + 4$	$-6 + 4 = -2$	$-$	$0 + 4 = 4$	$+$	$3 + 4 = 7$	$+$
$x - 1$	$-6 - 1 = -7$	$-$	$0 + 1 = -1$	$-$	$3 - 1 = 2$	$+$

–4 1

Since $x^2 + 3x - 4 = (x + 4)(x - 1)$, the sign of the function in each region is as given in the following figure:

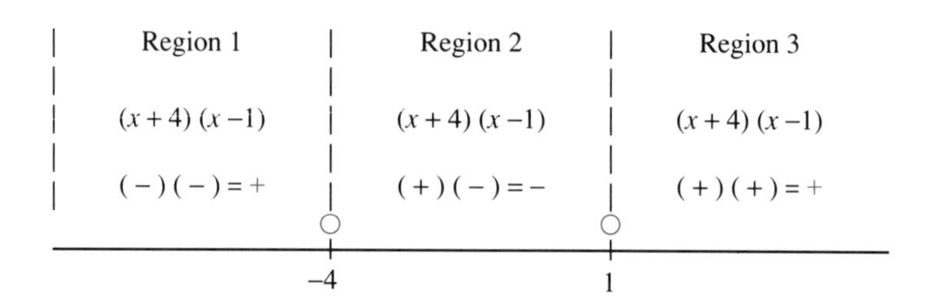

Since the inequality is greater than zero, the solution consists of those regions where the sign of the result is positive, or regions 1 and 3. Since the original inequality does not contain the "equal to" symbol ($>$ rather than $\geq$), the critical values are not included in the solution; that is, $x < -4$ or $x > 1$. ■

Solving Quadratic Inequalities

1. Arrange the inequality in $ax^2 + bx + c \lesseqgtr 0$ form. Change to an equation.
2. Factor and determine the critical values.
3. Draw a number line and mark each critical value. Determine the sign of each factor in each region.
4. Determine the sign of the inequality in each region and the solution.

EXAMPLE 2 Solve the following inequality:

$$2x^2 \leq 15 - x$$

Solution First, arrange the inequality in $ax^2 + bx + c \leq 0$ form, and then change it to an equation.

$$2x^2 + x - 15 \leq 0$$
$$2x^2 + x - 15 = 0$$

Second, factor and determine the critical values.

$$(2x - 5)(x + 3) = 0$$

$$2x - 5 = 0 \qquad x + 3 = 0$$
$$x = 5/2 \qquad x = -3$$

Third, place dots at 5/2 and -3 on the number line, and mark the resulting regions. To determine the sign of each factor in each region, substitute a number from the region into the factor. Fourth, determine the sign of the factored inequality in each region as shown below:

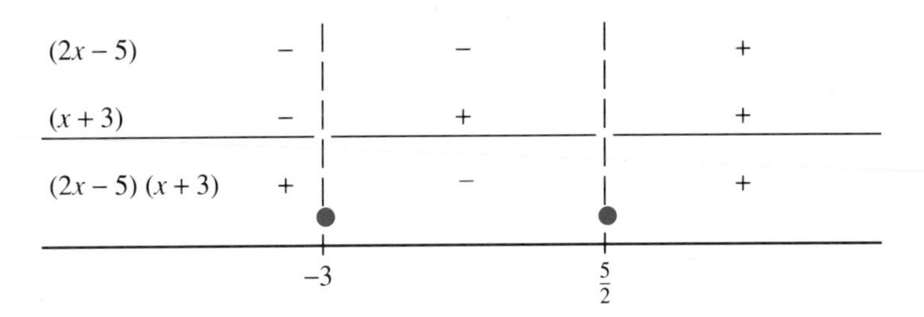

The solution is $-3 \le x \le 5/2$. The critical values are included in the solution because the original inequality contains the "equal to" symbol. ■

Rational Inequalities

In a rational expression, the critical values occur where the numerator or denominator equals zero. A rational expression changes signs only at its critical values. In solving a rational inequality, remember domain restrictions necessary to avoid division by zero.

EXAMPLE 3 Solve the following rational inequality:

$$\frac{(x - 5)^2(x + 3)}{(x - 1)} \ge 0$$

Solution Since the critical values occur where the numerator or denominator equals zero, the critical values are $x = 5, -3$, and 1. On the number line we place a point at 5 and -3, but we must place a circle at 1 because $x = 1$ leads to division by zero. The critical values, the resulting regions, the sign of each factor in each region, and the sign of the original fraction in each region are as follows:

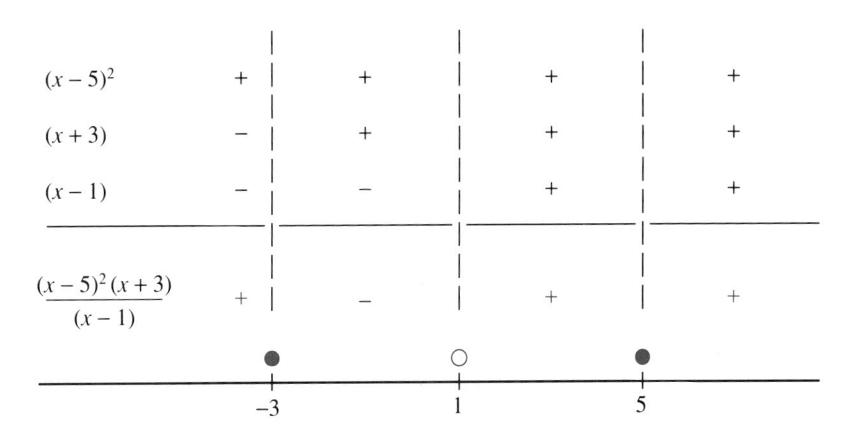

Since the original inequality is greater than or equal to zero, the solution includes those regions where the fraction is positive. The solution is $x \leq -3$ or $x > 1$. ■

Application

EXAMPLE 4 The formula for displacement (the number of feet fallen), s, of a free-falling object is given by

$$s = v_0 t + \frac{gt^2}{2}$$

where v_0 = initial velocity = -32 ft/s, t = time, and g = acceleration due to gravity = -32 ft/s^2. An object thrown from a 250-ft building has an initial velocity of 32 ft/s. How many seconds after release will the object be more than 122 ft above the ground?

Solution Substituting the given information into the formula gives the inequality ($250 - 122 = 128$)

$$-128 < -32t - 16t^2$$

The signs are negative due to the downward direction of movement. Finding the critical values gives

$$16t^2 + 32t - 128 < 0 \quad \text{transpose}$$
$$t^2 + 2t - 8 < 0 \quad \text{divide by 16}$$
$$(t + 4)(t - 2) < 0 \quad \text{factor}$$

Since the critical values are $t = -4$ and 2, we place circles at -4 and 2 on the number line. The sign of each factor and the product in each region are given in the following figure:

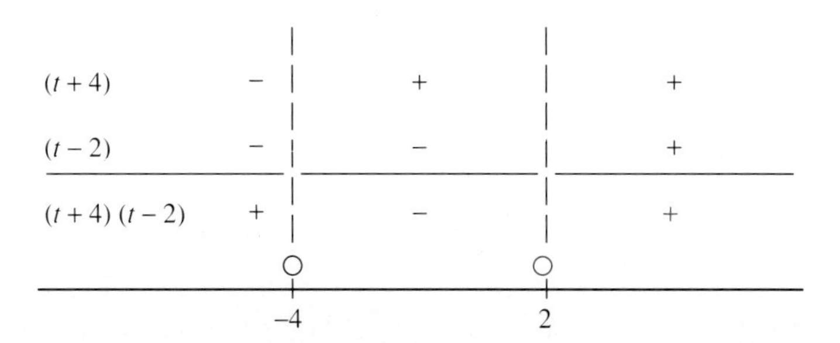

Since the inequality is less than zero, the solution includes the region where the product is negative. The solution is $-4 < t < 2$. However, from the physical constraints of this problem, time is not negative. Therefore, the solution is $0 < t < 2$. The object is more than 122 ft above the ground up to the first 2 seconds after release. ■

5–4 EXERCISES

Solve the following nonlinear inequalities.

1. $2x^2 > 12 - 2x$
2. $x^2 + 2x \leq 35$
3. $3x^2 + 19x + 6 \leq 0$
4. $18x^2 + 34x > 4$
5. $4x^2 - 12x \geq 0$
6. $x^2 - x - 30 \leq 0$
7. $21x^2 - 10 > -29x$
8. $16x^2 > 9$
9. $49x^2 \geq 81$
10. $36x^2 - 25 \leq 0$
11. $6x^2 + 43x - 40 < 0$
12. $12x^2 + 28x \geq 5$
13. $x^2 + 11x \leq -28$
14. $4x^2 - 20x + 24 > 0$
15. $x(2x + 1)(x - 5) \geq 0$
16. $6x(x - 3)(5x + 6) < 0$
17. $2x^3 - 8x \leq 0$
18. $3x^3 - 27x \geq 0$
19. $\dfrac{2x - 1}{x + 4} > 0$
20. $\dfrac{3x - 9}{x - 5} < 0$
21. $\dfrac{x^2 + 8x - 9}{x + 3} \geq 0$
22. $\dfrac{x^2 - 5x - 36}{x + 1} > 0$
23. $\dfrac{x(x + 5)^2(x + 2)}{x - 3} < 0$
24. $\dfrac{2x(x - 1)^2(x + 4)}{x + 7} \leq 0$
25. $\dfrac{(3x - 4)(x + 6)}{x + 1} \leq 0$
26. $\dfrac{(x - 5)(2x + 6)}{x - 3} > 0$
27. The velocity v of an object can be found by $v^2 = v_0^2 + 2as$ where v_0 is initial velocity, a is acceleration, and s is displacement. Find the value for the initial velocity if the velocity must be greater than 150 ft/s and $a = 2.5$ ft/s^2 when $s = 5$ ft.
28. Find the values of the radius r that make the surface area s of a cylinder greater than 32π m^2 if the height h is 6 m. (*Note:* $s = 2\pi r^2 + 2\pi rh$.)
29. For what values of the length of a rectangle is the area less than 40 m^2 if the length of the rectangle is 3 m more than the width?

5–5

APPLICATIONS

In the previous sections of this chapter, the method of solving a quadratic equation was specified. However, in a job situation where solving a quadratic equation is necessary, your supervisor will not come to you and say, "Here is a problem, use a quadratic equation to model the problem, and solve the quadratic equation by factoring."

Since we have three methods for solving a quadratic equation, we need a strategy for selecting an appropriate method for a given situation. You can use both completing the square and the quadratic formula to solve any quadratic equation. However, factoring can be used only for equations that have rational roots. One strategy is to first attempt to solve the equation by factoring. If you do not easily see how to factor the equation, then use the quadratic formula.

EXAMPLE 1 The formula for the displacement, s, of a free-falling object is given by

$$s = v_0 t + \frac{gt^2}{2}$$

where v_0 = initial velocity, t = time, and g = acceleration due to gravity. In the English system of measurement, where $g = 32$ ft/sec^2 the displacement is given by $s = v_0t + 16t^2$. If a baseball thrown straight down from the top of a 100-ft building has an initial velocity of 20 ft/s, how long does it take the ball to strike the ground? Substitute and then solve for t.

Solution We substitute $s = 100$ and $v_0 = 20$ into the formula.

$$100 = 20t + 16t^2$$

Arranging the equation in standard form gives

$$-16t^2 - 20t + 100 = 0 \quad \text{transpose}$$
$$4t^2 + 5t - 25 = 0 \quad \text{divide by } -4$$

This equation does not appear to factor, so we will solve for t using the quadratic formula.

$$a = 4 \qquad b = 5 \qquad c = -25$$

$$t = \frac{-5 \pm \sqrt{25 - 4(4)(-25)}}{2(4)}$$

$$t = \frac{-5 \pm \sqrt{25 + 400}}{8}$$

$$t = \frac{-5 \pm \sqrt{425}}{8}$$

$$t = \frac{-5 \pm 20.62}{8}$$

The solutions are $t = -3.20$ s and $t = 1.95$ s. Since time is not measured in negative numbers, it will take 1.95 s for the baseball to strike the ground.

To calculate $\sqrt{}$: 25 [−] 4 [×] 4 [×] 25 [+/−] [=] [√‾] [STO]

Solutions: 5 [+/−] [−] [RCL] [=] [÷] 8 [=] → −3.20

5 [+/−] [+] [RCL] [=] [÷] 8 [=] → 1.95 ■

EXAMPLE 2 In business, the **equilibrium point** is the point where the demand for a product equals the supply. If a company knows that the demand for a product is given by the expression $(500 - x^2)$/day and the supply is given by the expression $(10x + 300)$/day, find the equilibrium point.

Solution Because the equilibrium point is the point where the demand equals the supply, the equation is

$$500 - x^2 = 10x + 300$$

We arrange the equation in standard form and solve by factoring.

$$-x^2 - 10x + 200 = 0 \quad \text{transpose and add}$$
$$x^2 + 10x - 200 = 0 \quad \text{multiply by } -1$$
$$(x + 20)(x - 10) = 0 \quad \text{factor}$$

$$x + 20 = 0 \qquad x - 10 = 0 \quad \text{Zero Factor Property}$$
$$x = -20 \qquad x = 10$$

Since the manufacturer cannot produce -20 items, the supply equals the demand when 10 items are produced per day. Thus, the manufacturer is underproducing if he produces fewer than 10 items, and he is overproducing if he produces more than 10 items. ■

EXAMPLE 3 Diane drives from Atlanta to Orlando, a distance of 360 miles, and back along the same route to Atlanta. Her average speed from Atlanta to Orlando is 15 mi/h less than her speed on the return trip. If the total driving time is 14 hours, what is her average speed from Atlanta to Orlando?

Solution If we let

$$x = \text{average speed from Atlanta to Orlando}$$

then

$$x + 15 = \text{average speed from Orlando to Atlanta}$$

Since Diane drives the same route both to and from Orlando, the distance for each leg of the trip is 360 miles. This information is included in the accompanying chart. Since we have expressions for both distance and rate, we can use the relationship time = distance/rate to fill in an expression for time in the chart.

distance ÷ rate = time			
	Distance	Rate	Time
Atlanta to Orlando	360	x	$\frac{360}{x}$
Orlando to Atlanta	360	$(x + 15)$	$\frac{360}{(x + 15)}$

Since the equation is set up using the last column, we will need to read the problem again to find a relationship for time. The problem states that the total time for the trip is 14 hours. The equation is as follows:

$$\begin{array}{ccccc} \text{time (Atlanta to Orlando)} & + & \text{time (Orlando to Atlanta)} & = & \text{total time} \\ \uparrow & & \uparrow & & \uparrow \\ \dfrac{360}{x} & + & \dfrac{360}{(x+15)} & = & 14 \end{array}$$

First, eliminate the fractions by multiplying by the common denominator, $x(x + 15)$.

$$\begin{aligned} 360(x + 15) + 360x &= 14x(x + 15) && \text{eliminate fractions} \\ 360x + 5{,}400 + 360x &= 14x^2 + 210x && \text{eliminate parentheses} \\ -14x^2 + 510x + 5{,}400 &= 0 && \text{transpose} \\ 7x^2 - 255x - 2{,}700 &= 0 && \text{divide by } -2 \end{aligned}$$

Using the quadratic formula to solve, we have

$$x = \frac{255 \pm \sqrt{(-255)^2 - 4(7)(-2700)}}{2(7)}$$

$$x = 45 \quad \text{or} \quad x \approx -8.57$$

Since speed is not measured in negative numbers, the solution is $x = 45$. Diane's average speed from Atlanta to Orlando is 45 mi/h.

To calculate $\sqrt{}$: 255 [x^2] [±] [−] 4 [×] 7 [×] 2700 [±] [=] [$\sqrt{}$] [STO]

Solution: 255 [+] [RCL] [=] [÷] 14 [=] → 45

255 [−] [RCL] [=] [÷] 14 [=] → −8.57 ■

EXAMPLE 4 A landscape architect wants to surround a rectangular swimming pool with a brick walk of uniform width. If the area of the brick walk is 405 ft^2 and the swimming pool is 15 ft by 20 ft, how wide should the walk be?

Solution If we let

$$x = \text{width of the walkway}$$

then the dimensions of both the pool and the walkway are $(15 + 2x)$ and $(20 + 2x)$ (the dimensions of the pool are 15 by 20). As shown in Figure 5–2, the area of the walkway is the difference in the area of the larger rectangle and the area of the small rectangle. Since the area of a rectangle is the product of length and width, the area of the large rectangle is $(20 + 2x)(15 + 2x)$, and the area of the small rectangle is 20(15). The equation is

FIGURE 5–2

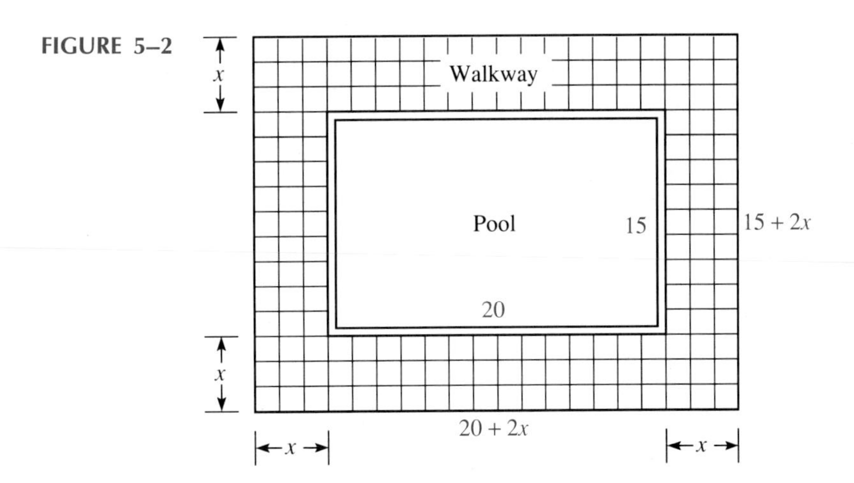

$$\begin{array}{ccccc} \text{area of large rectangle} & - & \text{area of small rectangle} & = & 405 \\ \uparrow & & \uparrow & & \uparrow \\ (20 + 2x)(15 + 2x) & - & 20(15) & = & 405 \end{array}$$

First, we arrange the equation in standard form.

$$300 + 40x + 30x + 4x^2 - 300 = 405$$
$$4x^2 + 70x - 405 = 0$$

Then we solve the equation by the quadratic formula.

$$a = 4 \qquad b = 70 \qquad c = -405$$

$$x = \frac{-70 \pm \sqrt{4{,}900 - 4(4)(-405)}}{2(4)}$$

$$x = \frac{-70 \pm \sqrt{4{,}900 + 6{,}480}}{8}$$

$$x = \frac{-70 \pm \sqrt{11{,}380}}{8}$$

$$x \approx 4.58 \quad \text{or} \quad x \approx -22.08$$

The walkway should be 4.58 ft wide.

To calculate $\sqrt{}$: 70 [x^2] [−] 4 [×] 4 [×] 405 [+/−] [=] [$\sqrt{}$] [STO]

Solution: 70 [+/−] [+] [RCL] [=] [÷] 8 [=] → 4.58

70 [+/−] [−] [RCL] [=] [÷] 8 [=] → −22.08 ■

EXAMPLE 5 The total surface area of a cylinder with a top and a bottom is given by the formula $A = 2\pi r^2 + 2\pi rh$. If a cylinder has a surface area of 64π cm^2 and a height of 6 cm, find the radius. Substitute and solve.

Solution First, substitute $A = 64\pi$ and $h = 6$ into the formula.

$$64\pi = 2\pi r^2 + 12\pi r$$

$$2\pi r^2 + 12\pi r - 64\pi = 0 \qquad \text{transpose and multiply all terms by } -1$$

$$r^2 + 6r - 32 = 0 \qquad \text{divide by } 2\pi$$

$$a = 1 \qquad b = 6 \qquad c = -32$$

$$r = \frac{-6 \pm \sqrt{36 - 4(1)(-32)}}{2}$$

$$r \approx 3.40 \quad \text{or} \quad r \approx -9.40$$

The radius of the cylinder is 3.4 cm. ■

Maximum and Minimum

The graph of a quadratic function is a parabola. The "tip" of the parabola is called the **vertex.** For a quadratic equation written in $y = ax^2 + bx + c$ form, the *parabola opens upward and the vertex is a minimum if $a > 0$. If $a < 0$, the parabola opens downward and the vertex is a maximum*. The x coordinate of the vertex, given by $-b/(2a)$, defines where the maximum or minimum value occurs. (We will discuss the parabola in greater detail in Chapter 15.) Figure 5–3 illustrates the maximum and minimum.

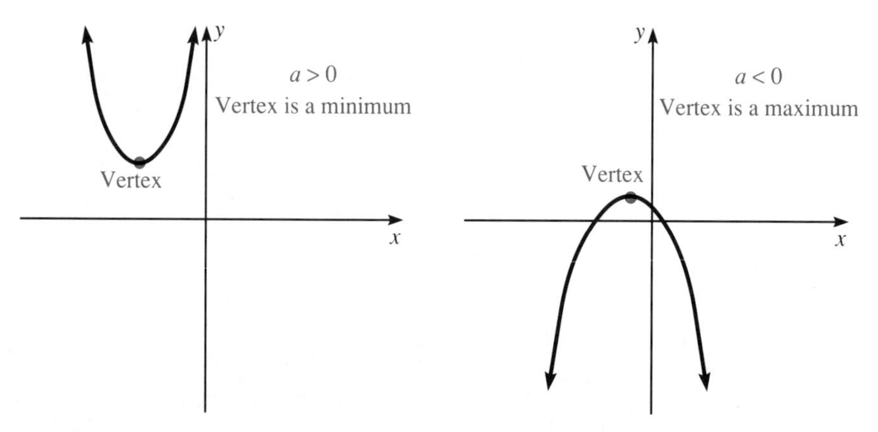

FIGURE 5–3

EXAMPLE 6 A rectangular dog run is to be fenced with 40 yd of barbed wire. Find the maximum area that can be fenced if one width is not enclosed. See Figure 5–4.

FIGURE 5–4

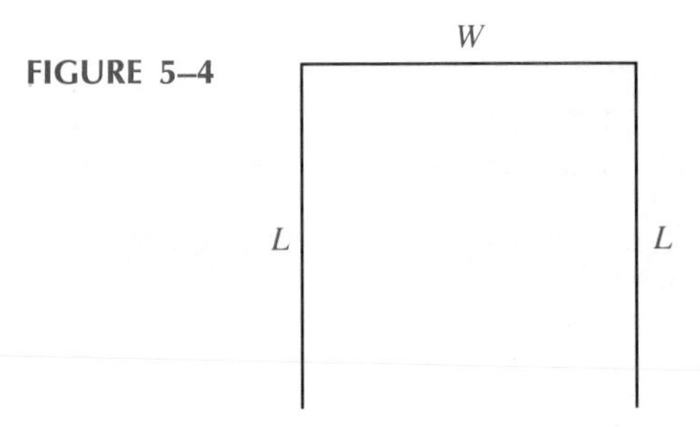

Solution Since we are fencing three sides of the tract, the amount of barbed wire is given by

$$2L + W = 40$$

If we solve for W, we obtain

$$W = 40 - 2L$$

However, we want to fence the tract so that the area is a maximum. The area of the rectangular tract is given by

$$A = LW$$

If we substitute for W in the area formula, we have

$$A = L(40 - 2L)$$
$$A = 40L - 2L^2$$

The equation in standard form is $y = -2L^2 + 40L - A$. Since the coefficient of L^2 is negative, the function does have a maximum, and that maximum occurs at the x coordinate of the vertex given by $-b/(2a)$.

$$\frac{-b}{2a} = \frac{-40}{2(-2)} = 10$$

The maximum area occurs when $L = 10$. The maximum area is

$$A = 40L - 2L^2$$
$$A = 40(10) - 2(10)^2$$
$$A = 200$$

The maximum rectangular area that can be fenced with 40 yd of barbed wire is 200 yd^2. ■

EXAMPLE 7 A company sells 1,000 widgets each week at the price of $2.00. However, the company needs to raise the price. Market research shows that for each $0.10 increase in price, the company will sell 20 fewer widgets. If the company wants to receive $2,450 in revenue each week, what price should it charge?

Solution Let n = the number of \$0.10 price increases. Each increase in price is represented by $0.10n$, and this results in $20n$ fewer units sold. From business principles, we know that

$$\begin{array}{ccc} \text{total revenue} = & (\text{price/unit}) & \cdot\ (\text{number of units sold}) \\ \uparrow & \uparrow & \uparrow \\ 2{,}450 = & (2 + 0.10n) & \cdot\ (1{,}000 - 20n) \end{array}$$

Solving the equation gives

$$\begin{aligned} 2{,}450 &= 2{,}000 - 40n + 100n - 2n^2 && \text{eliminate parentheses} \\ 2n^2 - 60n + 450 &= 0 && \text{transpose and add} \\ n^2 - 30n + 225 &= 0 && \text{divide by 2} \\ (n - 15)(n - 15) &= 0 && \text{factor} \\ n &= 15 \end{aligned}$$

Since the price/unit is given by $2 + 0.10n$, the company should charge

$$\begin{gathered} 2 + 0.10(15) \\ 2 + 1.50 \\ \$3.50 \end{gathered}$$

for each widget. ■

5–5 EXERCISES

1. If a ball thrown from the top of a 120-ft building has an initial velocity of 8 ft/s, how long will it take the ball to strike the ground? (Solve $120 = 8t + 16t^2$.)

I **2.** If the demand equation for an integrated circuit is given by $p = -600 - x^2$ and the supply equation is given by $p = 400 - 3x$, find the equilibrium quantity.

3. The city decides to put a brick walkway of uniform width around a rectangular flower garden that is 40 ft by 20 ft. If the area of the walkway is half the area of the garden, find the width of the walkway.

4. $N = n(n - 3)/2$ is the formula for the number of diagonals N of a polygon with n sides. Find the number of sides for a polygon with five diagonals.

5. The sum of six times the number and the square of the number is -8. Find the number.

6. The length of the base of a triangle is 3 cm less than the length of the height. Find the length of the base if the area of the triangle is 40 cm^2. (The area of a triangle = 1/2 · base · height.)

7. Jeff can survey a tract of land in three days less than Marvin. If it takes eight days for them to survey the land together, how long does it take Marvin alone? Compute the answer to two decimal places.

E **8.** When two dc currents flow in the same direction in a common resistor, the power dissipated in the resistor is given by $P = R(I_1 + I_2)^2$. If $P = 147$ W, $R = 7$ Ω, and $I_1 = 4$ A, find I_2 by substituting and solving. (Compute the answer to two decimal places.)

I **9.** A company sells its product for \$3 and sells 1,500 per week at this price. To increase its revenue, the company needs to increase its price; however, for each \$0.40 increase in price, the company will sell 100 fewer items. If the company wants to earn a total revenue of \$5,060 per week, what price should it charge? (See Example 7.)

10. Find two consecutive integers whose product is 132. Let x and $x + 1$ represent the consecutive integers.

11. Lucy wants to mat a picture that is 10 cm by 14 cm. If the area of the matting is 28 cm^2, find the width of the matting.

12. Jose drove from Los Angeles to the Mexican border and back, a total distance of 240 miles. His average speed from Los Angeles to the border was 15 mi/h more than his speed on the return trip. If the total driving time was 6 hours, what was Jose's average speed from the border back to Los Angeles? (Compute the answer to two decimal places.)

M 13. The area of a washer needs to be 3π cm^2. If the radius of the outside circle is 2 cm, find the radius of the inner circle. (*Note:* $A = \pi r^2$, area of a circle.)

14. A workman drops a hammer from the top of a 90-ft building. How long does it take the hammer to strike the ground if the initial velocity is zero? (Solve $90 = 16t^2$. Compute your answer to two decimal places.)

15. If the product of two consecutive odd integers is increased by six, the result is nine. Find the integers. (Let x and $x + 2$ represent the consecutive odd integers.)

16. Two inlet pipes can fill a swimming pool in 23½ hours. If the smaller pipe alone takes 12 h more to fill the pool, how much time is required for the larger pipe to fill the pool alone?

17. In order to conserve space, an oil dealer decides to tear down two oil storage tanks and replace them with one tank equal in capacity to the two tanks. If the radii of the two tanks are 18 ft and 20 ft and the height of each is 15 ft, what should be the radius of the new tank if its height is 25 ft? (The volume of a cylinder is $\pi r^2 h$. Give your answer to two decimal places.)

E 18. An electrical circuit contains two resistors. The sum of their resistances is 46 Ω, and their product is 465 Ω. Find the two resistances.

I 19. If the demand equation is given by $p = 8x - 5x^2$, and the supply equation is $p = 18x - 40$, find the equilibrium quantity, where supply equals demand. (Refer to Example 2.)

E 20. The power in the load P_L of a series circuit is given by the equation $P_L = -I_L^2 R_L + V_s I_L$. Find I_L when $P_L = 5$ W, $R_L = 5\ \Omega$, and $V_s = 10$ V.

21. Use the fact that the vertex results in a maximum or minimum to find the maximum height s an object will reach if the equation of flight is given by $s = 20t - 16t^2$ (thrown upward with initial velocity of 20 ft/sec).

22. Find the maximum area of a rectangle whose perimeter is 36 ft. (Refer to Example 6.)

23. Divide 20 into two parts such that the product will be a maximum.

I 24. A company sells a product for \$8 and sells 1,200 per week at this price. The company finds that for each \$0.50 increase in price, 40 fewer of the product will be sold. If the company wants to make \$10,260 per week, what price should it charge? (Refer to Example 7.)

25. The base of a triangle is 9 m longer than its height. Find the height if the area of the triangle is 90 m^2.

CHAPTER SUMMARY

Summary of Terms

critical values (p. 182)
discriminant (p. 180)
equilibrium point (p. 188)
hypotenuse (p. 172)
legs (p. 172)
Pythagorean theorem (p. 172)
quadratic equation in standard form (p. 162)
root (p. 162)
solution (p. 162)
Square Root Property (p. 168)
vertex (p. 192)
Zero Factor Property (p. 162)

Summary of Formulas

$(\text{hypotenuse})^2 = (\text{leg}_1)^2 + (\text{leg}_2)^2$ Pythagorean theorem

$x = \dfrac{-b \pm \sqrt{b^2 - 4ac}}{2a}$ quadratic formula

$b^2 - 4ac$ discriminant

CHAPTER REVIEW

Section 5–1 Solve by factoring.

1. $3x^2 + 16x = 12$ **2.** $8x = 20 - x^2$

3. $x^2 = 16$ **4.** $5x^2 - 21 = 8x$

5. $x(2x + 4) = -(7x + 5)$

6. $9x^2 = 81$

7. $10x = 14x^2$

8. $7x^2 + 18x - 12 = 2x^2 + 7x$

9. $4x(x + 1) - 6(2x + 3) = 3x + 2$

10. $5x(x + 4) - 10 = -2(x - 19)$

11. $\frac{x}{x - 5} + \frac{4}{x + 5} = \frac{x}{x^2 - 25}$

12. $\frac{6}{x^2} + 8 = \frac{14}{x}$

13. $\frac{x}{x + 3} + \frac{x}{x + 1} = \frac{3 - x}{x^2 + 4x + 3}$

14. $\frac{3}{x + 2} - \frac{4}{x - 2} = \frac{-2 - x^2}{x^2 - 4}$

15. $24x^2 = 31x + 15$ **16.** $x^2 = \frac{11x + 3}{4}$

Section 5–2

Solve by completing the square.

17. $x^2 + 6x = 12$ **18.** $x^2 + 10x - 15 = 0$

19. $x^2 = 5x$ **20.** $7x(x + 2) = 25$

21. $2x^2 - 6x = 14$ **22.** $9x^2 + 18x = 30$

23. $3x^2 + 7x + 17 = 0$ **24.** $4x^2 + 11x - 20 = 0$

25. $5x(x - 1) = 11 - 10x$

26. $8x^2 + 7x = -21$

27. $\frac{3x^2}{2} + x = \frac{1}{3}$ **28.** $\frac{4}{x} - \frac{1}{x^2} = 5$

29. $\frac{x - 1}{x + 2} + \frac{x}{x - 3} = \frac{2}{x^2 - x - 6}$

30. $\frac{6}{2x - 1} + \frac{2x}{x + 3} = \frac{3x^2}{2x^2 + 5x - 3}$

Section 5–3

Solve by the quadratic formula.

31. $14x^2 - 8 = 0$ **32.** $7x^2 - 3x + 10 = 0$

33. $5x^2 - 12x + 13 = 0$ **34.** $5x^2 - x = 7$

35. $9x^2 + 8x = 0$ **36.** $36x^2 - 1 = 0$

37. $x^2 + 3(x - 4) = 7 - 2x(x - 1)$

38. $7x^2 - 14x = 0$

39. $2x(x - 3) + 6 = 4(x^2 + 5)$

40. $6x(x + 2) - 5 = (x + 1)(x - 3)$

41. $x^2 = 18$

42. $11x + 15x^2 - 5 = 8x^2 + 5x$

43. $\frac{2x - 3}{x + 1} + \frac{3x}{x - 6} = \frac{10}{x^2 - 5x - 6}$

44. $\frac{13x}{2x + 3} + \frac{5}{x - 5} = \frac{6}{2x^2 - 7x - 15}$

45. $\frac{4x^2}{9} - \frac{1}{3} = \frac{2x - 1}{6}$ **46.** $\frac{9}{x^2} = 16 + \frac{8}{x}$

Using the discriminant, describe the number and type of solutions.

47. $4x - x(8 - x) = 20$

48. $14x^2 - 8x = 15 + 6x + 2x^2$

49. $\frac{7}{x^2} + 6 = \frac{1}{x}$

50. $\frac{9x}{x - 1} + \frac{7}{x + 1} = \frac{6}{x^2 - 1}$

51. $5x(x + 6) = (x - 3)(x - 7)$

52. $15x^2 - 9x = 6(x - 3) + 12x(x - 1)$

53. $17x^2 + 6x - 3(x + 1) = 15 + 4x(x + 2)$

54. $13x^2 - 10x + 6 = 2x + 12 + 6x^2$

Section 5–4

Solve the following inequalities.

55. $4x^2 \geq 3 - x$

56. $3x^2 + 12x < -(x + 4)$

57. $2x^3 + 7x^2 - 15x < 0$

58. $2x^3 + 8x^2 \geq 5x^2 + 20x$

59. $\frac{x - 3}{x + 1} < 0$ **60.** $\frac{2x - 1}{x + 5} \geq 0$

61. $\frac{(x - 1)(x + 3)}{(x - 2)} \geq 0$ **62.** $\frac{(x + 4)(x - 2)}{x^2} < 0$

Section 5–5

63. Find two consecutive integers whose product is 342.

64. The length of a rectangle is five more than the width. Find the dimensions of the rectangle if its area is 104 in.^{2}

E **65.** The equivalent resistance, R_T, of two resistors connected in parallel is given by the relationship

$$\frac{1}{R_T} = \frac{1}{R_1} + \frac{1}{R_2}$$

while two resistors connected in series have an equivalent resistance given by $R_T = R_1 + R_2$. Find two resistances that will give an equivalent resistance of 25 Ω in series and 4 Ω in parallel.

E **66.** The power P_L is given by $P_L = EI - RI^2$ where E = voltage, I = current, and R = resistance. Find the current in a circuit where $P_L = 9$ W, $E = 12$ V, and $R = 4$ Ω.

67. One computer can run a program in 3 hours less than a second computer. If the computers are networked together, they can run the program in 8 hours. How long does it take to run the program on each computer alone? (Round your answer to hundredths.)

CHAPTER TEST

The number in parentheses refers to the appropriate learning objective at the beginning of the chapter.

1. Solve the following by factoring: **(1)**

$$7x(x - 2) - 10 = 2 + 3x$$

2. Using the discriminant, describe the number and kind of solutions for $3x^2 + 9(x - 3) = 4(2x + 5)$. **(4)**

3. Solve the following: **(1, 2, 3)**

$$\frac{6x}{(2x + 5)} - \frac{3}{(2x^2 + 3x - 5)} = \frac{4x}{(x - 1)}$$

4. Find two consecutive even integers whose product is 120. **(6)**

5. Solve the following by factoring: **(1)**

$$10x^2 - 29x + 21 = 0$$

6. Solve the following by completing the square: **(2)**

$$2x^2 + 6x - 13 = 0$$

7. Solve the following by using the quadratic formula: **(3)**

$$(8x - 3)(2x + 5) + 8(x - 1) = 5(3 - 2x) + 9$$

8. The length of a rectangle is 7 ft more than the width. Find the width of the rectangle if its area is 120 ft^2. **(6)**

9. Solve the following quadratic equation: **(1, 2, 3)**

$$9x^2 - 14 = 0$$

10. Using the discriminant, determine the number and type of solutions for $7x^2 - 8x + 6 = 4x(x - 3)$. **(4)**

11. Solve the following quadratic equation: **(1, 2, 3)**

$$3x^2 - 8x + 5 = 0$$

12. One computer can execute a program in 4 minutes less than a second computer. If it takes 12 minutes for the two computers together to complete the job, how long does it take for the faster computer to complete the job alone? (Round your answer to hundredths.) **(6)**

13. Solve the following by the quadratic formula: **(3)**

$$\frac{7x^2}{2} + \frac{5x}{3} = 4$$

14. Solve the following quadratic equation: **(1, 2, 3)**

$$\frac{4x}{x - 1} + \frac{6}{x + 3} = \frac{-24}{x^2 + 2x - 3}$$

15. Solve the following quadratic equation: **(1, 2, 3)**

$$\frac{2x}{x - 3} + \frac{7x}{2x - 1} = \frac{3x(2x - 5)}{2x^2 - 7x + 3}$$

16. A stone weighing 2 pounds thrown down from the top of a 20-ft house has an initial velocity of 8 ft/s. How long will it take the stone to strike the ground? (Solve $20 = 8t + 16t^2$.) Round your answer to hundredths. **(6)**

17. Solve the following quadratic equation: **(1, 2, 3)**

$$8x^2 + 28x = 6x + 21$$

18. Use the discriminant to describe the number and type of solutions for the following: **(4)**

$$4x(x - 1) + 6x = (x - 2)(3x + 6)$$

19. Solve by completing the square: **(2)**

$$\frac{1}{2}x^2 + 8x - 9 = 0$$

20. Solve the following quadratic inequality: **(5)**

$$4x^2 + 6x \leq 18$$

SOLUTION TO CHAPTER INTRODUCTION

To determine the width of the walkway, we draw Figure 5–5. Let x represent the width of the walkway. From the information given, the area of the pool is 704 ft^2; therefore, the equation is

$$\text{(length of pool)(width of pool)} = \text{area of pool}$$
$$(40 - 2x) \cdot (30 - 2x) = 704$$

FIGURE 5–5

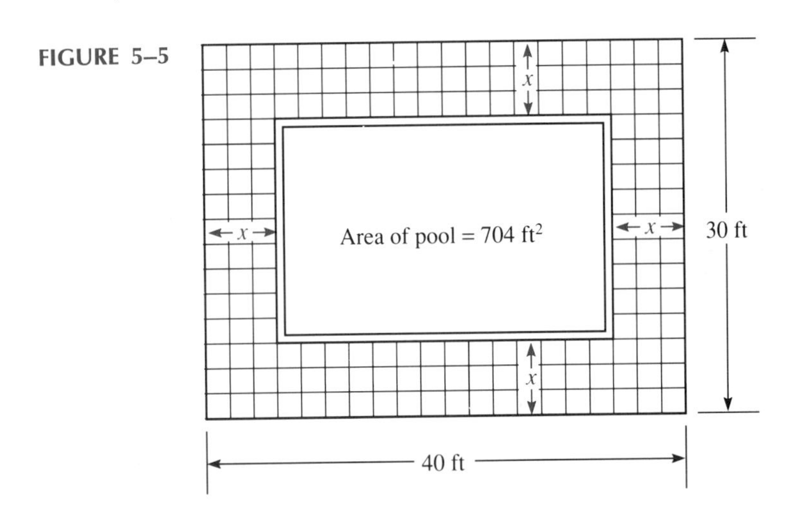

Simplifying this equation gives

$$1{,}200 - 80x - 60x + 4x^2 = 704$$
$$4x^2 - 140x + 496 = 0$$
$$x^2 - 35x + 124 = 0$$

Solving this quadratic equation by factoring gives

$$(x - 31)(x - 4) = 0$$

$$x = 31 \quad \text{and} \quad x = 4$$

Since 31 ft is more than the width of the pool enclosure, the walkway should be 4 ft wide.

As an electronics technician, you are to calculate the currents I_1, I_2, and I_3 flowing through the circuit shown on the facing page. Since you are finding three variables, you will use a system of three equations. (The solution to this problem is given at the end of the chapter.)

So far, we have discussed solving linear and quadratic equations of various types, but all the equations discussed have been similar in that they contained only one variable. However, in this chapter we will discuss solving systems of equations that contain two or more variables. We will discuss solving systems graphically, algebraically, and using determinants. We will also discuss solving systems of nonlinear equations using algebraic techniques.

Learning Objectives

After completing this chapter, you should be able to

1. Determine whether an ordered pair is the solution to a system of equations (Section 6–1).
2. Solve a system of linear equations graphically (Section 6–1).
3. Solve a system of two or three linear equations algebraically (Sections 6–2 and 6–3).
4. Solve a nonlinear system of equations (Section 6–4).
5. Evaluate a determinant (Section 6–5).
6. Solve a system of linear equations by using Cramer's rule (Section 6–6).
7. Apply the concepts of solving a system of equations to technical situations (Section 6–7).

Chapter 6

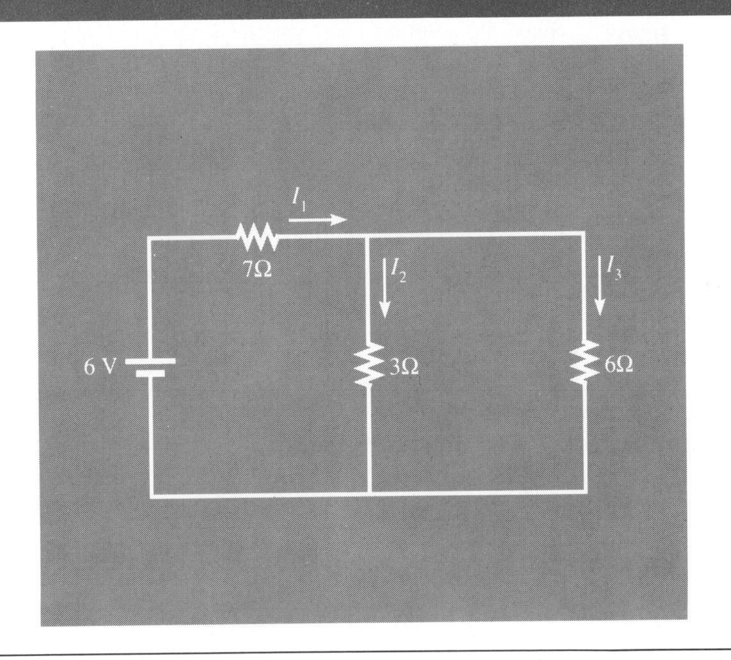

Systems of Equations

6–1

GRAPHICAL SOLUTION OF SYSTEMS OF LINEAR EQUATIONS

In Chapter 2, we discussed linear equations in two variables, such as $4x - 7y = 10$. Two equations of this form are called a **system of linear equations.** The **solution** to the system is the ordered pair (x, y) that satisfies both equations simultaneously.

Checking Solutions

Since the solution to a system of equations must satisfy both equations simultaneously, we can *check* a possible solution *by substituting the ordered pair into both equations*. The next example illustrates this process.

EXAMPLE 1 For each of the following systems, determine whether the given ordered pair is a solution:

(a) $\left.\begin{aligned} 3x - y &= 8 \\ x + 5y &= 24 \end{aligned}\right\} (4, 4)$ (b) $\left.\begin{aligned} 4x - 3y &= 10 \\ x + 2y &= 0 \end{aligned}\right\} (2, -1)$

Solution

(a) We substitute the ordered pair (4, 4) into each equation.

$$\begin{aligned} 3x - y &= 8 \\ 3(4) - (4) &= 8 \\ 8 &= 8 \quad \text{(true)} \end{aligned} \qquad \begin{aligned} x + 5y &= 24 \\ (4) + 5(4) &= 24 \\ 24 &= 24 \quad \text{(true)} \end{aligned}$$

Since both substitutions result in a true statement, the ordered pair is the solution to this system.

(b)

$$\begin{aligned} 4x - 3y &= 10 \\ 4(2) - 3(-1) &= 10 \\ 8 + 3 &= 10 \quad \text{(false)} \end{aligned} \qquad \begin{aligned} x + 2y &= 0 \\ (2) + 2(-1) &= 0 \\ 0 &= 0 \quad \text{(true)} \end{aligned}$$

Since $(2, -1)$ does not satisfy the equation $4x - 3y = 10$, it cannot be the solution to this system of equations. ■

Solving Graphically

Recall from Chapter 2 that a linear equation in two variables is represented graphically by a line. Therefore, the graph of a system of two linear equations contains two lines. The **graphical solution** to such a system is *the point of intersection of the lines* since that point simultaneously satisfies both equations.

EXAMPLE 2 Solve the following system of equations graphically:

$$2x - y = 6$$
$$3x + y = 4$$

Solution To graph each equation, we can fill in a table of values just as we did in Chapter 2.

$2x - y = 6$	
x	y
-1	-8
0	-6
2	-2

$3x + y = 4$	
x	y
-2	10
0	4
1	1

The graph of these two equations is given in Figure 6–1. As shown in the graph, the point of intersection (solution) is the ordered pair $(2, -2)$.

FIGURE 6–1

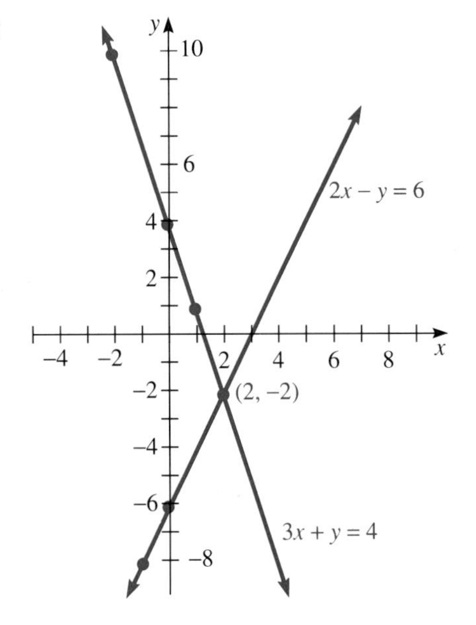

■

Graphing Using the Intercepts

An alternate method of graphing linear equations is to use the x and y intercepts. The **x intercept** is the x coordinate of the point where the graph crosses the x axis, and the **y intercept** is the y coordinate of the point where the graph crosses the y axis. To calculate the x intercept, substitute $y = 0$ into the equation; to find the y intercept, substitute $x = 0$.

EXAMPLE 3 Solve the following system graphically using intercepts:

$$3x + 2y = -6$$
$$4x - y = 8$$

Solution To determine the x intercept for each equation, substitute $y = 0$ into the equation.

x intercept ($y = 0$)	
$3x + 2y = -6$	$4x - y = 8$
$3x + 2(0) = -6$	$4x - (0) = 8$
$x = -2$	$x = 2$

To determine the y intercept for each equation, substitute $x = 0$ into the equation.

y intercept ($x = 0$)	
$3x + 2y = -6$	$4x - y = 8$
$3(0) + 2y = -6$	$4(0) - y = 8$
$y = -3$	$y = -8$

The x intercept for the graph of $3x + 2y = -6$ is -2, and the y intercept is -3. The x intercept for the graph of $4x - y = 8$ is 2, and the y intercept is -8. Graphing both lines as shown in Figure 6–2, we can see that the intersection does not occur at integer values. Therefore, it is necessary to estimate the solution to be $(0.9, -4.3)$.

FIGURE 6–2

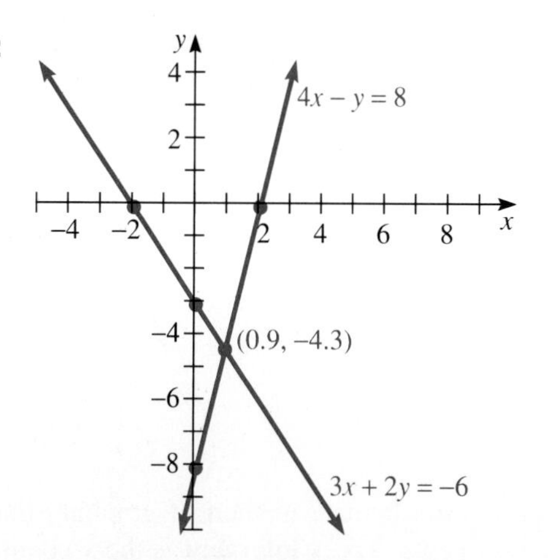

Checking

$$3x + 2y = -6 \qquad\qquad 4x - y = 8$$
$$3(0.9) + 2(-4.3) = -6 \qquad\qquad 4(0.9) - (-4.3) = 8$$
$$-5.9 \approx -6 \qquad\qquad 7.9 \approx 8$$

■

Application

EXAMPLE 4 A 12-ft beam weighing 50 lb is supported on each end. The uniform beam supports a 75-lb pile of bricks 4 ft from the right end and a 160-lb man 3 ft from the left end (see Figure 6–3). A system of equations that results from this situation is

$$A + B = 285$$
$$6A - 6B = 330$$

where A represents the force supported by the left end of the beam and B represents the force supported by the right end. Use a graph to find how much weight each end supports.

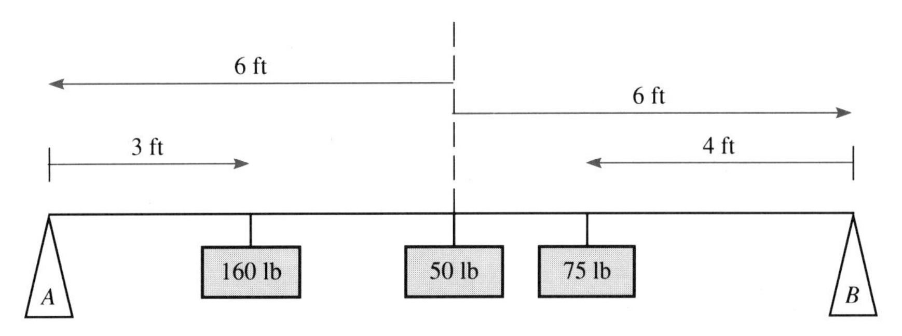

FIGURE 6–3

Solution Placing A on the horizontal axis and B on the vertical axis, we find that the x and y intercepts for the graph of $A + B = 285$ are 285 and 285, respectively. The x and y intercepts of $6A - 6B = 330$ are 55 and -55, respectively. From Figure 6–4, we can see that the *point of intersection* is estimated to be $A = 170$ and $B = 115$. Therefore, the left end supports 170 lb and the right end supports 115 lb. Large numbers also make graphing difficult.

FIGURE 6–4

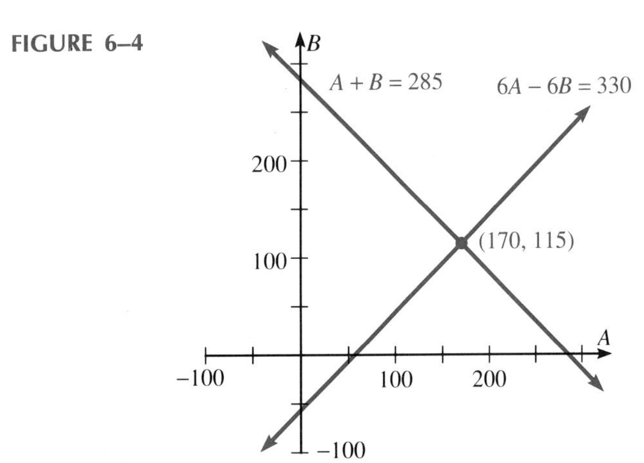

■

Inconsistent and Dependent Systems

The systems of equations solved thus far are called *consistent* and *independent* because the graphs intersect in exactly one point and the system has exactly one solution. However, not every system of equations has one solution. Since parallel lines never intersect, a *system of equations whose graph consists of parallel lines* has no solution and is called **inconsistent.** If the *graph of a system of two linear equations produces one line,* every point on the line is a solution. A system of this type is called **dependent.** Inconsistent and dependent systems are illustrated in the next example.

EXAMPLE 5 Solve the following systems of equations graphically:

(a) $5x - 3y = 11$ (b) $4x - y = 3$

$10x - 6y = 7$ $8x - 2y = 6$

Solution

(a) The x and y intercepts of the graph for $5x - 3y = 11$ are $11/5$ and $-11/3$, respectively, while the intercepts of $10x - 6y = 7$ are $7/10$ and $-7/6$, respectively. The graph of these equations is shown in Figure 6–5(a). Since the lines have no point of intersection, this system of equations is inconsistent and has no solution.

(b) The x and y intercepts of the graph for both $4x - y = 3$ and $8x - 2y = 6$ are $3/4$ and -3, respectively. This system of equations is dependent, as shown in Figure 6–5 (b), and the solutions are all the points on the line.

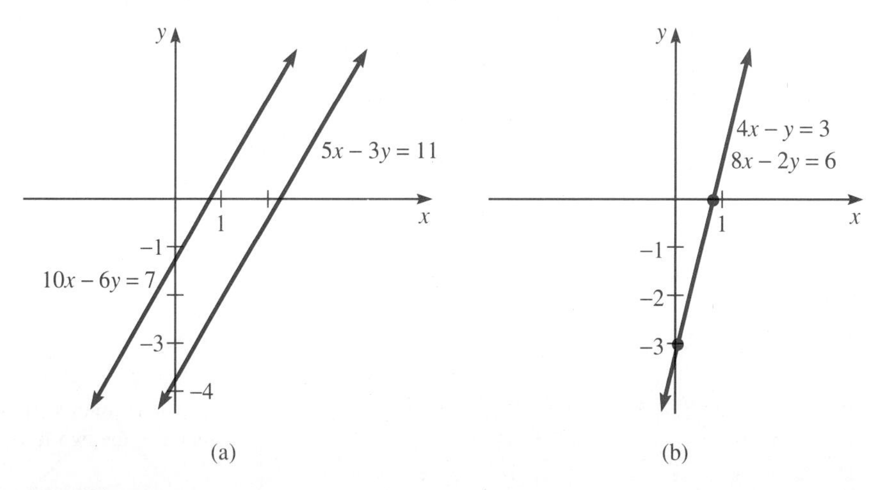

FIGURE 6–5 ■

6–1 EXERCISES

Determine whether the given ordered pair is the solution to the system of equations.

1. $\left.\begin{aligned} 7x - y &= 8 \\ 3x + 2y &= 6 \end{aligned}\right\}(0, 1)$

2. $\left.\begin{aligned} x + 4y &= -5 \\ y &= 2x + 10 \end{aligned}\right\}(-5, 0)$

3. $\left.\begin{aligned} 4x - y &= 5 \\ x &= 3y - 7 \end{aligned}\right\}(2, 3)$

4. $\left.\begin{aligned} 5x - y &= 7 \\ 2x + 9y &= 22 \end{aligned}\right\}(2, 2)$

5. $\left.\begin{aligned} 6x + 7y &= 13 \\ 5x + y &= -1 \end{aligned}\right\}(-1, 4)$

6. $\left.\begin{aligned} x - y &= 15 \\ 2x - 3y &= -3 \end{aligned}\right\}(7, 8)$

7. $\left.\begin{aligned} 2x - 4y &= 14 \\ 2y &= -13 - x \end{aligned}\right\}(-3, -5)$

8. $\left.\begin{aligned} 3x - 5y &= 7 \\ x - 3y &= 1 \end{aligned}\right\}(4, 1)$

9. $\left.\begin{aligned} 3x + y &= 1 \\ 8x + 2y &= -2 \end{aligned}\right\}(-2, 7)$

10. $\left.\begin{aligned} x + 3y &= -3 \\ 2x - y &= 15 \end{aligned}\right\}(6, -3)$

Solve the following systems of equations graphically. Estimate the answer to the nearest tenth, if necessary.

11. $2x + 3y = -6$
$x - 4y = -3$

12. $2x - 3y = 1$
$5x - y = -17$

13. $x + y = 6$
$3x - y = 10$

14. $5x - 4y = 12$
$9x + 2y = -6$

15. $7x - 2y = 11$
$3y = x - 7$

16. $3x + 2y = 13$
$x - 5y = 10$

17. $x + 3y = 10$
$2x - y = 8$

18. $8x - 3y = 12$
$2x + 3y = 8$

19. $2x + 3y = 8$
$4x + 6y = 10$

20. $10x = 3y + 4$
$2x - y = 6$

21. $5x + 7y = 13$
$y = 2x - 9$

22. $3x - 7y = 14$
$9x - 21y = 42$

23. $3x - y = 7$
$12x - 4y = 28$

24. $7x + y = 10$
$x - 5y = 8$

25. $4x - y = 10$
$3x + 5y = 8$

26. $8x - y = 15$
$16x - 2y = 10$

27. A uniform 100-ft bridge weighing 200 tons is supported on each end. If a truck weighing 14 tons is 10 ft from the west end and a car weighing 3 tons is 30 ft from the east end, the system of equations could be written as

$$A + B = 217$$
$$50A + 500 = 50B$$

where A and B represent the weight supported by the west and east ends, respectively. Solve this system for A and B.

E 28. Applying Ohm's law to a particular circuit results in the following system of equations:

$$E - 6I = 0$$
$$E + 10I = 8$$

Solve this system graphically.

C 29. The perimeter of a rectangular plot of land is 28 rods. If the length L of the tract is 4 rods more than the width W, the system of equations is

$$2L + 2W = 28$$
$$L = W + 4$$

Solve this system graphically to find the dimensions of the rectangle.

30. The sum of two numbers is 25, and their difference is 9. If x and y represent the numbers, the system of equations is

$$x + y = 25$$
$$x - y = 9$$

Find the two numbers by finding the point of intersection of the two lines represented by the equations.

6–2

ALGEBRAIC SOLUTION OF TWO LINEAR EQUATIONS

As shown in Section 6–1, graphing may lead to an approximate solution to a system of linear equations. For this reason, we will now discuss two methods of algebraic solution: substitution and addition.

Solving by Substitution

In order to solve a system of two linear equations, we must eliminate one of the variables to obtain one linear equation with one variable. One such method of eliminating a variable is called **substitution.** The process used in substitution is developed in the next example.

EXAMPLE 1 Solve the following system of equations by substitution:

$$2x - y = 8$$
$$3x + 4y = 10$$

Solution First, we solve one of the two equations for a variable. Solving the first equation for y gives

$$2x - y = 8$$
$$-y = 8 - 2x$$
$$y = -8 + 2x$$

Then we substitute this quantity for y into the second equation.

$$3x + 4y = 10$$
$$3x + 4(-8 + 2x) = 10$$

The resulting equation is a linear equation in one variable that we now solve.

$$3x - 32 + 8x = 10 \quad \text{eliminate parentheses}$$
$$11x = 10 + 32 \quad \text{transpose}$$
$$x = 42/11 \quad \text{divide}$$

To solve for y, we substitute the value determined for x into the expression previously determined for y.

$$y = -8 + 2x$$
$$y = -8 + 2(42/11)$$
$$y = -4/11$$

The solution is $\left(\frac{42}{11}, \frac{-4}{11}\right)$. ■

Solution by Substitution

1. Solve one of the equations for a variable. (Choose a variable with a numerical coefficient of positive or negative one ($+1$ or -1), if possible, to make the problem easier to solve.)
2. Substitute for the variable from Step 1 into the remaining equation.
3. Solve the resulting equation for the remaining variable.
4. Solve for the other variable by substituting the value of the variable from Step 3 into the equation from Step 1.
5. Check your answer by substituting into both equations.

Note: The solution to a system of equations is not dependent on which variable you solve for in Step 1.

In some instances, the substitution method is simple and straightforward, as in Example 2. However, as you will see in Example 3, the substitution method can be long and tedious.

EXAMPLE 2 Solve the following system of equations using substitution:

$$4x + y = 10$$
$$6x - 3y = 6$$

Solution First, we solve the first equation for y because it has a coefficient of $+1$.

$$4x + y = 10$$
$$y = 10 - 4x$$

Second, we substitute this expression for y in the second equation.

$$6x - 3y = 6$$
$$6x - 3(10 - 4x) = 6$$

Third, we solve the resulting equation for x.

$$6x - 30 + 12x = 6$$
$$18x = 6 + 30$$
$$18x = 36$$
$$x = 2$$

Fourth, we solve for y by substituting $x = 2$ into the equation from Step 1.

$$y = 10 - 4x$$
$$y = 10 - 4(2)$$
$$y = 2$$

Checking $\quad 4(2) + (2) = 10 \qquad 6(2) - 3(2) = 6$

$$10 = 10 \qquad 6 = 6$$

Since $x = 2$ and $y = 2$ checks in both equations, the solution is the ordered pair (2, 2). ■

CAUTION ✦ In Step 2, be sure to substitute into the equation that you did not use in Step 1.

EXAMPLE 3 Solve the following system of equations by substitution:

$$7x = 3y + 9$$
$$2x + 5y = 12$$

Solution Since none of the variables has a coefficient of 1, we solve the second equation for x.

$$2x + 5y = 12$$
$$2x = 12 - 5y$$
$$x = \frac{12 - 5y}{2}$$

Substituting the quantity for x from the second equation into the first equation gives

$$7x = 3y + 9$$
$$7\left(\frac{12 - 5y}{2}\right) = 3y + 9$$

Then we solve the resulting equation.

$$2\left[7\left(\frac{12 - 5y}{2}\right)\right] = 2(3y) + 2(9) \quad \text{multiply by LCD}$$
$$7(12 - 5y) = 6y + 18 \quad \text{eliminate fractions}$$
$$84 - 35y = 6y + 18 \quad \text{eliminate parentheses}$$
$$84 - 18 = 6y + 35y \quad \text{transpose}$$
$$66 = 41y \quad \text{combine like terms}$$
$$\frac{66}{41} = y \quad \text{divide}$$

Next, solve for x by substituting 66/41 for y.

$$x = \frac{12 - 5(66/41)}{2}$$

$$x = \frac{12 - (330/41)}{2}$$

$$x = \frac{162}{82}$$

$$x = \frac{81}{41}$$

The solution is the ordered pair $\left(\frac{81}{41}, \frac{66}{41}\right)$.

Checking

$$7x = 3y + 9 \qquad\qquad 2x + 5y = 12$$

$$7\left(\frac{81}{41}\right) = 3\left(\frac{66}{41}\right) + 9 \qquad\qquad 2\left(\frac{81}{41}\right) + 5\left(\frac{66}{41}\right) = 12$$

$$\frac{567}{41} = \frac{198}{41} + \frac{369}{41} \qquad\qquad \frac{162}{41} + \frac{330}{41} = 12$$

$$\frac{567}{41} = \frac{567}{41} \qquad\qquad 12 = 12$$

■

CAUTION ✦ In an ordered pair solution, x is always written first. You may have a tendency to write your first answer first in the ordered pair whether it is x or not.

Solution by Addition

As you can see from Example 3, some systems of equations are difficult to solve by substitution. Such systems are generally solved more easily by the **addition method.** Like the substitution method, the addition method of algebraic solution eliminates one of the variables. However, unlike substitution, the elimination is accomplished by addition of the equations. The process used in the addition method is illustrated in the next example.

EXAMPLE 4 Solve the following system of equations by addition:

$$3x - y = 10$$
$$2x + y = 15$$

Solution Notice that in these two equations, y has opposite coefficients. Therefore, *we can eliminate the variable y by adding the two equations.* (This can be done because equals added to equals are equal.)

$$\begin{array}{r} 3x - y = 10 \\ \underline{2x + y = 15} \\ 5x \quad\;\; = 25 \end{array}$$

Solving the resulting linear equation in one variable gives

$$\begin{aligned} 5x &= 25 \\ x &= 5 \end{aligned}$$

To solve for y, substitute $x = 5$ into either of the original equations in the system.

$$\begin{aligned} 3x - y &= 10 \\ 3(5) - y &= 10 \\ 15 - y &= 10 \\ -y &= 10 - 15 \\ y &= 5 \end{aligned}$$

Checking

$$\begin{aligned} 3x - y &= 10 \qquad & 2x + y &= 15 \\ 3(5) - (5) &= 10 \qquad & 2(5) + (5) &= 15 \\ 10 &= 10 \qquad & 15 &= 15 \end{aligned}$$

Since the ordered pair satisfies both equations, the solution is (5, 5). ■

Solution by Addition

1. Arrange the equations in $ax + by = c$ form, if necessary.
2. Arrange opposite coefficients on one of the variables. To do so, you may have to multiply one or both equations by a constant.
3. Add the resulting equations and solve for the other variable.
4. Solve for the remaining variable by substituting the value from Step 3 into either equation in the original system.
5. Check your solution in each equation.

EXAMPLE 5 Solve the following system of equations by addition:

$$\begin{aligned} 4x - 2y &= -4 \\ 2x + y &= 10 \end{aligned}$$

Solution First, arrange both equations in $ax + by = c$ form. Second, obtain opposite coefficients on the variable y by multiplying the second equation by 2 as follows:

$$\begin{aligned} 4x - 2y &= -4 \longrightarrow & 4x - 2y &= -4 \\ 2(2x) + 2(y) &= 2(10) \longrightarrow & 4x + 2y &= 20 \end{aligned}$$

Third, add the equations and solve for x.

$$\begin{aligned} 4x - 2y &= -4 \\ 4x + 2y &= 20 \\ \hline 8x \quad &= 16 \\ x &= 2 \end{aligned}$$

Fourth, substitute $x = 2$ into one of the original equations.

$$\begin{aligned} 2x + y &= 10 \\ 2(2) + y &= 10 \\ 4 + y &= 10 \\ y &= 6 \end{aligned}$$

Since this ordered pair checks in both equations, the solution is (2, 6). ■

NOTE ✦ The choice of which variable to eliminate is completely arbitrary. The solution is the same ordered pair whether you initially eliminate x or y.

EXAMPLE 6 Solve the following system of equations by addition:

$$\frac{1}{2}x + \frac{1}{3}y = 1$$

$$\frac{1}{3}x + \frac{1}{5}y = 2$$

Solution Solving this system by addition will be easier if we eliminate the fractions first.

$$6\left(\frac{1}{2}x\right) + 6\left(\frac{1}{3}y\right) = 6(1) \qquad \text{multiply by LCD of 6}$$

$$15\left(\frac{1}{3}x\right) + 15\left(\frac{1}{5}y\right) = 15(2) \qquad \text{multiply by LCD of 15}$$

$$\begin{aligned} 3x + 2y &= 6 \\ 5x + 3y &= 30 \end{aligned}$$

Notice that if we elect to eliminate x, we can obtain opposite coefficients on x by multiplying the first equation by -5 and the second equation by 3. (We could also eliminate x by multiplying by 5 and -3, respectively.)

$$\begin{aligned} -5(3x) + -5(2y) &= -5(6) \longrightarrow & -15x - 10y &= -30 \\ 3(5x) + \quad 3(3y) &= 3(30) \longrightarrow & 15x + \quad 9y &= 90 \\ & & -y &= 60 \\ & & y &= -60 \end{aligned}$$

Substitute $y = -60$ into one of the original equations and solve for x.

$$\frac{1}{2}x + \frac{1}{3}(-60) = 1$$
$$\frac{1}{2}x - 20 = 1$$
$$\frac{1}{2}x = 21$$
$$x = 42$$

The solution is the ordered pair $(42, -60)$. ■

CAUTION ✦ Be sure that the variable you wish to eliminate has opposite coefficients; that is, you must have opposite signs.

Inconsistent and Dependent Systems

In the previous section, we discussed the graphical solution to inconsistent and dependent systems of equations. These systems also occur in algebraic solutions, and we must be able to identify them.

- ☐ If *both variables are eliminated* and a *true* statement results, the system is *dependent* (same line).
- ☐ If *both variables are eliminated* and a *false* statement results, the system is *inconsistent* (parallel lines).

Both inconsistent and dependent systems of equations are illustrated in Example 7.

EXAMPLE 7 Solve the following systems of equations:

(a)
$$\begin{aligned} 3x - 2y &= 8 \\ -9x + 6y &= 20 \end{aligned}$$

(b)
$$\begin{aligned} 4x - 5y &= -10 \\ -8x + 10y &= 20 \end{aligned}$$

Solution

(a) Solving by addition gives

$$\begin{array}{rcr} 3(3x) - 3(2y) = 3(8) & \longrightarrow & 9x - 6y = 24 \\ -9x + 6y = 20 & \longrightarrow & -9x + 6y = 20 \\ \hline & & 0 = 44 \quad \text{(false)} \end{array}$$

Since *both variables are eliminated* and a *false* statement, $0 = 44$, resulted, this system of equations is *inconsistent*.

(b) Solving by addition gives

$$\begin{array}{rcl} 2(4x) - 2(5y) = 2(-10) & \longrightarrow & 8x - 10y = -20 \\ -8x + 10y = 20 & \longrightarrow & -8x + 10y = 20 \\ \hline & & 0 = 0 \quad \text{(true)} \end{array}$$

Since *both variables are eliminated* and a *true* statement, $0 = 0$, resulted, this system of equations is *dependent*. ■

Applications

EXAMPLE 8 A chemist wants to make 10 gallons of a 25% solution of sulfuric acid using a 15% solution and a 30% solution. How much of each solution should the chemist use?

Solution We have discussed similar mixture problems in previous chapters. However, in this section we can use two variables to set up the problem. Let

$$x = \text{amount of the 30\% solution used}$$
$$y = \text{amount of the 15\% solution used}$$

With this information we fill in the table just as we did in Chapter 1.

% × amount = amount in mixture		
%	Amount	Amount in Mixture
30% 15%	x y	$0.30x$ $0.15y$
25%	10	0.25(10)

Unlike the previous treatment of mixture problems, we have used two variables. Therefore, we must write two equations. First, the total amount of the mixture is 10 gallons and is represented by the equation

$$x + y = 10$$

Second, the total amount of each solution in the mixture is given by the equation

$$\begin{array}{ccccc} \text{amount of 30\%} & + & \text{amount of 15\%} & = & \text{amount of 25\%} \\ \uparrow & & \uparrow & & \uparrow \\ 0.30x & + & 0.15y & = & 0.25(10) \end{array}$$

The system of equations is

$$\begin{aligned} x + y &= 10 \\ 0.30x + 0.15y &= 0.25(10) \end{aligned}$$

Using substitution, we solve the first equation for y and substitute into the second equation.

$$\begin{aligned} y &= 10 - x & \\ 0.30x + 0.15y &= 2.50 & \\ 0.30x + 0.15(10 - x) &= 2.50 & \text{substitute} \\ 0.30x + 1.50 - 0.15x &= 2.50 & \text{eliminate parentheses} \\ 0.15x &= 1 & \text{transpose and add} \\ x &= 6.67 & \text{divide} \end{aligned}$$

2.5 [−] 1.5 [=] [÷] [(] 0.30 [−] 0.15 [)] [=] → 6.67

Then substitute $x = 6.67$ to find y.

$$\begin{aligned} y &= 10 - x \\ y &= 10 - 6.67 \\ y &= 3.33 \end{aligned}$$

The chemist should use 6.67 gal of the 30% solution and 3.33 gal of the 15% solution. ■

6–2 EXERCISES

Solve the following systems of equations by addition.

1. $3x - y = 10$, $2x + y = 5$

2. $x + y = 8$, $3x - y = 8$

3. $4x + 3y = -4$, $8x - 2y = -24$

4. $x - 4y = 8$, $2x + 6y = 9$

5. $6x - 8y = -14$, $2y = 5x + 14$

6. $3x - y = 8$, $4y = 12x + 10$

7. $\frac{1}{2}x - \frac{1}{3}y = 2$, $\frac{3}{4}x + 2y = -12$

8. $\frac{1}{2}x + \frac{2}{3}y = \frac{5}{9}$, $4x - \frac{1}{2}y = \frac{25}{3}$

Solve the following systems of equations by substitution.

9. $y = 2x + 6$, $3x + 4y = 13$

10. $x = 2y + 3$, $7y + 3x = 9$

11. $x = y + 10$, $3x - 2y = -10$

12. $3x + y = -2$, $y = 4x + 5$

13. $6x - y = 14$, $3x + 4y = -2$

14. $4x - 5y = 13$, $y - 2x = 1$

15. $\frac{1}{2}x - \frac{2}{3}y = 1$, $\frac{1}{4}x + \frac{1}{6}y = -1$

16. $\frac{2}{5}x - \frac{1}{3}y = \frac{7}{3}$, $\frac{1}{2}x + \frac{3}{4}y = \frac{7}{4}$

17. $3x + 4y = 10$, $7x - 2y = 14$

18. $6x - 3y = 8$, $4x = 18 - 7y$

Solve the following systems of equations.

19. $x + 3y = 20$, $x - 4y = -22$

20. $-2x + 7y = 10$, $x - 9y = -5$

21. $2x - 3y = 15$, $4y = 3x - 19$

22. $4x - 6y = -3$, $x + 4y = 2$

23. $2x + y = 10$, $6x + 3y = 15$

24. $x + 3y = -2$, $5x + 7y = 6$

25. $x - \frac{1}{3}y = \frac{5}{6}$, $\frac{1}{3}x - \frac{1}{4}y = \frac{5}{12}$

26. $4x - 9y = 20$, $2x - 7y = 8$

27. $x - 2y = 7$
$3x + 4y = 1$

28. $3x + 2y = 11$
$5x + y = 9$

29. $x + 5y = 10$
$3x - 4y = 12$

30. $3x - y = 6$
$4x + 5y = 20$

31. Find the equation of the line of the form $ax + by = 10$ that passes through the points (1, 6) and (2, 4). (*Hint:* Each point must satisfy the equation. Substitute the ordered pairs for x and y into this equation to obtain a system of two linear equations that you then solve for a and b.)

32. Find two numbers whose sum is 30 if the larger number is 6 less than twice the smaller number.

33. Mrs. Smith buys apples and nectarines at the market. She buys one more pound of apples at \$0.69 a pound than nectarines at \$0.89 a pound. If the total cost is \$3.85, how many pounds of each did she buy?

34. Two cars leave the same town and travel in opposite directions. At the end of 8 hours, the cars are 896 miles apart. Using a system of equations, find the rate of each car if the rate of one car is 4 mi/h more than the rate of the other.

35. Betsy invests \$13,000 in two stocks, one yielding 7% and the other yielding 8.3%. If her total yield for a year is \$1,000, how much did she invest in each stock?

36. Find two numbers whose sum is 32 and whose difference is 14.

6–3

ALGEBRAIC SOLUTION OF THREE LINEAR EQUATIONS

Addition and substitution can be extended to solve a system of three linear equations using those techniques presented in the previous section. We eliminate one variable to obtain a system of two equations in two variables and solve this system just as we did in the last section. Then we obtain the solution to the system with three variables.

EXAMPLE 1 Solve the following system of equations:

$$x + y - z = 8 \qquad (1)$$
$$2x - y + z = 10 \qquad (2)$$
$$x + 3y + z = 6 \qquad (3)$$

Solution First, we reduce this system of three equations to a system of two equations containing the same two variables. Thus, we must eliminate one of the variables from two equations. We can eliminate z by adding equations (1) and (2), obtaining

$$\begin{array}{rcll} x + y - z &=& 8 & (1)\\ 2x - y + z &=& 10 & (2)\\ \hline 3x &=& 18 & \end{array}$$

Then we can also eliminate z by adding equations (1) and (3).

$$\begin{array}{rcll} x + y - z &=& 8 & (1)\\ x + 3y + z &=& 6 & (3)\\ \hline 2x + 4y &=& 14 & \end{array}$$

Now, we solve the resulting system of two equations.

$$\begin{aligned} 3x &= 18 \\ 2x + 4y &= 14 \\ \hline 3x &= 18 \\ x &= 6 \end{aligned}$$

$$\begin{aligned} 2(6) + 4y &= 14 \quad \text{solve for } y \\ 12 + 4y &= 14 \\ 4y &= 2 \\ y &= 1/2 \end{aligned}$$

At this point, we know that $x = 6$ and $y = 1/2$. To complete the solution of the original system of three equations, we must solve for z. Substituting the numerical values for x and y into equation (1) gives

$$\begin{aligned} x + y - z &= 8 \qquad \textbf{(1)} \\ (6) + (1/2) - z &= 8 \\ 13/2 - z &= 8 \\ -z &= 8 - 13/2 \\ -z &= 3/2 \\ z &= -3/2 \end{aligned}$$

To check, we must substitute the values for x, y, and z into all three equations. The solution is written as the ordered triple $(6, 1/2, -3/2)$. ■

Solving a System of Three Equations

1. Eliminate one variable from two equations.
2. Solve the resulting system of two equations by using substitution or elimination by addition.
3. Solve for the remaining variable by substituting into one of the original equations.
4. Check the solution.

EXAMPLE 2 Solve the following system of equations:

$$\begin{aligned} x - y + z &= -4 \qquad \textbf{(1)} \\ 2x + y + 2z &= -5 \qquad \textbf{(2)} \\ 3x - y - z &= -6 \qquad \textbf{(3)} \end{aligned}$$

Solution The variable we choose to eliminate is completely arbitrary, but one variable may be easier to eliminate than another.

Step 1: Eliminate One Variable. In this system, let us eliminate y from two equations by using addition. Adding equations (1) and (2) gives

$$\begin{array}{rcl} x - y + z &=& -4 \quad (1)\\ 2x + y + 2z &=& -5 \quad (2)\\ \hline 3x \qquad + 3z &=& -9 \end{array}$$

Adding equations (2) and (3) gives

$$\begin{array}{rcl} 2x + y + 2z &=& -5 \quad (2)\\ 3x - y - z &=& -6 \quad (3)\\ \hline 5x \qquad + z &=& -11 \end{array}$$

The resulting system of two equations is

$$\begin{aligned} 3x + 3z &= -9\\ 5x + z &= -11 \end{aligned}$$

Step 2: Solve the System of Two Variables. We solve this system using addition, and we eliminate z by multiplying the second equation by -3.

$$\begin{array}{rcl} 3x + 3z = -9 & \longrightarrow & 3x + 3z = -9\\ -3(5x) + (-3)(z) = -3(-11) & \longrightarrow & -15x - 3z = 33\\ \hline & & -12x \qquad = 24\\ & & x = -2 \end{array}$$

Solving for z, we substitute $x = -2$ into $5x + z = -11$.

$$\begin{aligned} 5(-2) + z &= -11\\ z &= -11 + 10\\ z &= -1 \end{aligned}$$

Step 3: Solve for y. Next, we solve for y by substituting $x = -2$ and $z = -1$ into equation (1).

$$\begin{aligned} x - y + z &= -4 \quad (1)\\ (-2) - y + (-1) &= -4\\ -3 - y &= -4\\ -3 + 4 &= y\\ 1 &= y \end{aligned}$$

Step 4: Check.

$$\begin{array}{ccc} x - y + z = -4 & 2x + y + 2z = -5 & 3x - y - z = -6\\ -2 - 1 - 1 = -4 & -4 + 1 - 2 = -5 & -6 - 1 + 1 = -6\\ -4 = -4 & -5 = -5 & -6 = -6 \end{array}$$

Since the answer checks, the solution is the ordered triple $(-2, 1, -1)$. ■

EXAMPLE 3 Solve the following system of equations:

$$x - 3y + 2z = 13 \quad (1)$$
$$2x + y - 3z = -9 \quad (2)$$
$$4x - 2y + z = 8 \quad (3)$$

Solution *Step 1: Eliminate One Variable*. We can eliminate x by solving equation (1) for x and substituting that quantity into equations (2) and (3).

$$x = 13 + 3y - 2z \quad (1)$$

$$\begin{aligned} 2x + y - 3z &= -9 \quad (2) \\ 2(13 + 3y - 2z) + y - 3z &= -9 \\ 26 + 6y - 4z + y - 3z &= -9 \\ 7y - 7z &= -35 \end{aligned}$$

$$\begin{aligned} 4x - 2y + z &= 8 \quad (3) \\ 4(13 + 3y - 2z) - 2y + z &= 8 \\ 52 + 12y - 8z - 2y + z &= 8 \\ 10y - 7z &= -44 \end{aligned}$$

The resulting system of equations is

$$\begin{aligned} 7y - 7z &= -35 \\ 10y - 7z &= -44 \end{aligned}$$

Step 2: Solve the System of Two Variables. To solve this system of equations by addition, we can eliminate z by multiplying the second equation by -1.

$$\begin{aligned} 7y - 7z = -35 &\longrightarrow \quad 7y - 7z = -35 \\ -1(10y - 7z = -44) &\longrightarrow \quad \underline{-10y + 7z = 44} \\ & \qquad\quad -3y \qquad = 9 \\ & \qquad\qquad\qquad y = -3 \end{aligned}$$

Substituting $y = -3$ into the equation $7y - 7z = -35$ and solving for z gives

$$\begin{aligned} 7(-3) - 7z &= -35 \\ -21 - 7z &= -35 \\ -7z &= -14 \\ z &= 2 \end{aligned}$$

Step 3: Solve for x. We solve for x by substituting $y = -3$ and $z = 2$ into the expression for x in Step 1.

$$x = 13 + 3y - 2z$$
$$x = 13 + 3(-3) - 2(2)$$
$$x = 13 - 9 - 4$$
$$x = 0$$

The solution is the ordered triple $(0, -3, 2)$. ■

EXAMPLE 4 The general form for the equation of a parabola is $y = ax^2 + bx + c$. Find the equation of the parabola that passes through the points $(1, 9)$, $(-1, 3)$, and $(-2, 6)$.

Solution We must determine the value of the constants a, b, and c. Since the parabola passes through the three points, each must satisfy the equation $y = ax^2 + bx + c$. Substituting each ordered pair into the equation gives the following system of three equations in three variables:

$$9 = a(1)^2 + b(1) + c$$
$$3 = a(-1)^2 + b(-1) + c$$
$$6 = a(-2)^2 + b(-2) + c$$

Step 1: Eliminate One Variable.

$$9 = a + b + c \qquad (1)$$
$$3 = a - b + c \qquad (2)$$
$$6 = 4a - 2b + c \qquad (3)$$

We can eliminate b by adding equations (1) and (2).

$$\begin{array}{rl} 9 = & a + b + c \quad (1) \\ 3 = & a - b + c \quad (2) \\ \hline 12 = & 2a \qquad + 2c \end{array}$$

Then we can eliminate b by multiplying equation (1) by 2 and adding the result to equation (3).

$$\begin{array}{rcrl} 2(9 = a + b + c) & \longrightarrow & 18 = & 2a + 2b + 2c \\ 6 = 4a - 2b + c & \longrightarrow & 6 = & 4a - 2b + c \\ \hline & & 24 = & 6a \qquad + 3c \end{array}$$

The resulting system of two equations is

$$12 = 2a + 2c$$
$$24 = 6a + 3c$$

Step 2: Solve the System of Two Equations. Using substitution, we solve the first equation for a.

$$12 - 2c = 2a$$
$$6 - c = a$$

We substitute that quantity for a into the second equation.

$$24 = 6a + 3c$$
$$24 = 6(6 - c) + 3c$$
$$24 = 36 - 6c + 3c$$
$$-12 = -3c$$
$$4 = \mathrm{c}$$

Then we solve for a.

$$a = 6 - c$$
$$a = 6 - 4$$
$$a = 2$$

Step 3: Solve for b. Substituting $a = 2$ and $c = 4$ into equation (1) gives

$$9 = a + b + c \qquad \textbf{(1)}$$
$$9 = (2) + b + (4)$$
$$3 = b$$

Since $a = 2$, $b = 3$, and $c = 4$, the equation of the parabola is $y = 2x^2 + 3x + 4$. ■

EXAMPLE 5 The perimeter of a triangle is 26 cm. The largest side is 2 cm less than the sum of the other two sides. Also, the largest side is 4 cm more than the middle-sized side. Find the length of the sides of the triangle.

Solution Since we are trying to find the three sides of the triangle, let

L = length of the largest side
M = length of the middle-sized side
S = length of the smallest side

Since we have three variables, we must have at least three equations in order to obtain a unique solution. From the verbal statement, the three equations are as follows:

The perimeter is 26.

$$L + M + S = 26 \tag{1}$$

The largest side is 2 less than the sum of the other two sides.

$$L = M + S - 2 \tag{2}$$

The largest side is 4 more than the middle-sized side.

$$L = M + 4 \tag{3}$$

Step 1: Eliminate One Variable. We can eliminate the variable L from the system by substituting equation (3), which is solved for L, into equations (1) and (2).

$$(M + 4) + M + S = 26 \tag{1}$$

$$2M + S = 22$$

$$(M + 4) = M + S - 2 \tag{2}$$

$$6 = S$$

Step 2: Solve the System of Two Equations. The system of two equations resulting from Step 1 is

$$2M + S = 22$$

$$S = 6$$

If $S = 6$, then

$$2M + S = 22$$

$$2M + 6 = 22$$

$$2M = 16$$

$$M = 8$$

Step 3: Solve for L. If $S = 6$ and $M = 8$, then

$$L = M + 4$$

$$L = 8 + 4$$

$$L = 12$$

The sides of the triangle are 12 cm, 8 cm, and 6 cm. ■

6–3 EXERCISES

Solve the following systems of equations.

1. $x - y + 2z = -5$
   $2x + y + z = -1$
   $x + y - z = 3$

2. $x + y - z = 0$
   $x - y - z = 4$
   $x + 2y + z = -4$

3. $2x - y + 2z = 8$
   $x + 2y - z = -1$
   $x + y + z = 1$

4. $x - 2y + 2z = -3$
   $x + y - 2z = 4$
   $3x - y + z = 1$

5. $3x + y - z = 9$
   $x + 3y + 2z = 1$
   $2x - y + z = 1$

6. $4x + 2y - z = -7$
   $2x + y + z = -2$
   $x - y - 2z = -5$

7. $2x - y + 3z = 9$
   $x + 2y + 2z = 5$
   $3x - 2y + z = 10$

8. $x + 3y - z = -3$
   $2x - 3y + 2z = 6$
   $x + y + z = 3$

9. $3x + 2y + z = 4$
   $2x - 3y + 4z = -6$
   $x - 2y - z = -4$

10. $2x - y + 3z = 13$
    $x + 3y - 2z = -4$
    $4x + 2y - z = 12$

11. $4x - 3y + z = -6$
    $2x + 2y - 4z = 6$
    $3x + y + 3z = -9$

12. $2x + y + 4z = 10$
    $x - 3y + 2z = 12$
    $3x - 2y + z = 17$

13. $x - 2z = 6$
    $2y - 3z = 12$
    $2x + 3y = 13$

14. $3x + 4y = 9$
    $-2y + z = -7$
    $2x - 3z = 1$

15. $2y - 3z = -7$
    $x + y = 1$
    $3x + 2z = 6$

16. $2x + 3y = 8$
    $x + 4z = -6$
    $2y - z = 9$

17. $y + 4z = 9$
    $2x + 3z = 2$
    $x - 3y = -5$

18. $3x - y = -17$
    $x + 3z = -5$
    $2y - z = 4$

19. $2x + y = -7$
    $3z - x = -5$
    $2y + z = -1$

20. $y - 4z = -6$
    $2x - 3y = 18$
    $x + 2y = 2$

21. $y - 3x = 9$
    $2x + 3z = -6$
    $4y + 2z = 8$

22. $4x + 2y = -6$
    $x - 5z = -5$
    $6z - 2y = 0$

23. $6x - 4y = -8$
    $z - 3x = 6$
    $2y + z = -2$

24. $z + x = 4$
    $3y + 4z = 1$
    $2y - x = -5$

25. $2x - 3y = 2$
    $4z + 6y = -2$
    $5z + 4x = 2$

26. $4x - 3z = -4$
    $2y + 5z = 5$
    $y - 8x = 2$

27. $5y + 2z = -3$
    $4x - z = 2$
    $2x - 10y = -2$

28. $3x + 2y = 2$
    $4x - 3z = 2$
    $4y - 6z = -6$

E 29. When Kirchhoff's laws are applied to the circuit shown in Figure 6–6, the following equations result:

$$9 - 6I_1 - 5I_2 = 0$$
$$9 - 6I_1 - 10I_3 = 0$$
$$I_1 - I_2 - I_3 = 0$$

Solve this system of equations for the currents I_1, I_2, and I_3.

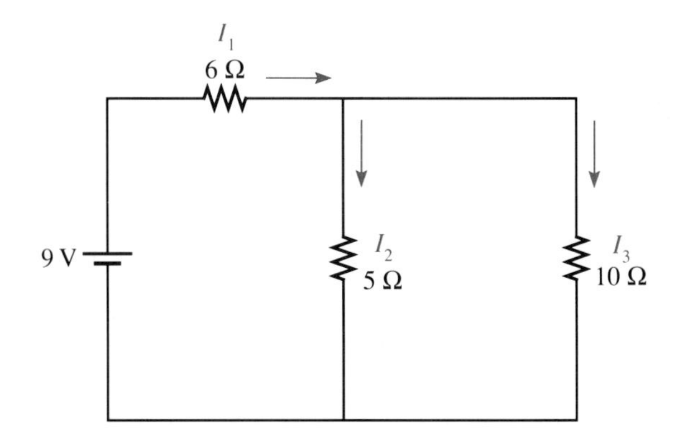

FIGURE 6–6

30. The perimeter of a triangle is 72 yd. The longest side is 10 yd more than twice the shortest side, and twice the middle side is 2 yd more than the longest side. Find the length of the sides of the triangle.

31. Find three numbers whose sum is 77 if the largest number is 2 more than twice the smallest number, the largest number is 12 less than twice the middle number.

32. Jane has a collection of 48 coins consisting of nickels, dimes, and quarters valued at $6.20. How many of each coin does she have if the number of quarters is eight less than twice the number of dimes?

33. Find the equation of the parabola, $y = ax^2 + bx + c$, that passes through the points $(0, 5)$, $(-2, 21)$, and $(1, 6)$. (Refer to Example 4.)

34. Find the equation of the parabola passing through the points $(0, 6)$, $(1, 5)$, and $(-3, -15)$. Refer to Example 4.)

I 35. A company manufactures three types of stereos, types 1, 2, and 3. The company manufactures four less type 2 than twice type 1, and seven more type 2 than the sum of type 1 and type 3. If the total number of stereo units manufactured per day is 51, how many of each type are manufactured?

6 – 4

SOLVING NONLINEAR SYSTEMS

In this section, we will discuss nonlinear systems of two equations. The graphical solution to a nonlinear system is the point of intersection of the two figures. For example, if the two equations represent a circle and a parabola, they may intersect from zero to four times, depending on their position. Figure 6–7 illustrates the possible intersection of the circle and parabola. As you can see from this graphical example, nonlinear systems may have more than one solution. The emphasis in this section is on algebraic solutions to nonlinear systems. Graphs are used only to verify the algebraic solution.

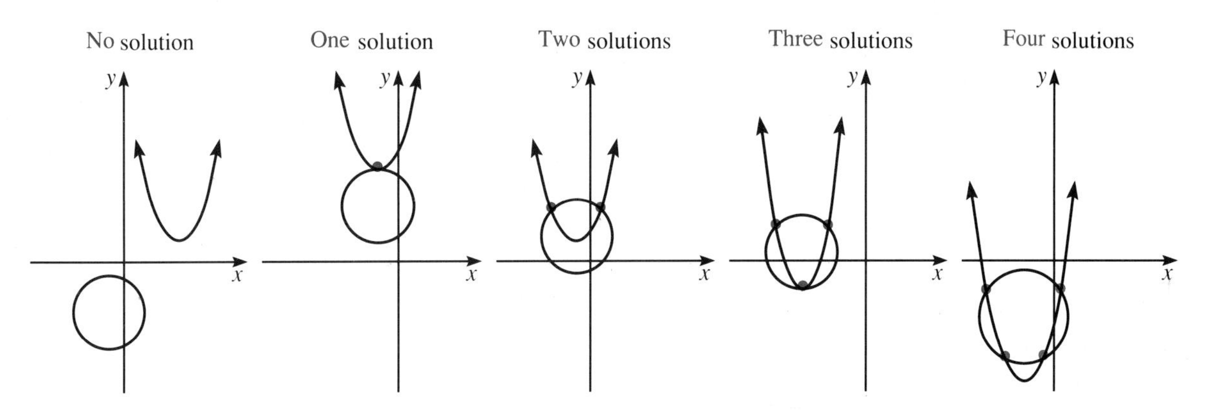

FIGURE 6–7

EXAMPLE 1 Solve the following system of equations for all real number solutions:

$$x - 4y = 8 \qquad \textbf{(1)}$$
$$x^2 + y^2 = 4 \qquad \textbf{(2)}$$

Solution The substitution method is easier to use in a nonlinear system where one of the equations is linear. We solve equation (1) for x.

$$x = 8 + 4y \qquad \textbf{(1)}$$

Then we substitute into equation (2).

$$x^2 + y^2 = 4 \qquad \textbf{(2)}$$
$$(8 + 4y)^2 + y^2 = 4$$
$$64 + 64y + 16y^2 + y^2 = 4$$
$$17y^2 + 64y + 60 = 0$$

We solve this quadratic equation for y by factoring.

$$(17y + 30)(y + 2) = 0$$

$$17y = -30 \qquad\qquad y = -2$$

$$y = \frac{-30}{17} \approx -1.76$$

Substituting each of these values into the expression for x gives

$$x = 8 + 4y$$

$$x = 8 + 4(-30/17) \qquad x = 8 + 4(-2)$$

$$x \approx 0.941 \qquad\qquad x = 0$$

The ordered pair solutions are $(0.941, -1.76)$ and $(0, -2)$. Figure 6–8 shows the graphical solution to this nonlinear system.

FIGURE 6–8

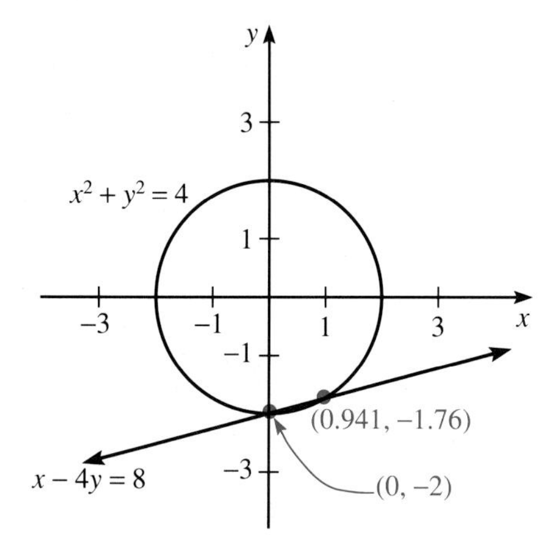

■

EXAMPLE 2 Find the real number solutions to the following nonlinear system of equations:

$$x + y = 5 \qquad \textbf{(1)}$$

$$2x + y^2 = 3 \qquad \textbf{(2)}$$

Solution We solve this nonlinear system by substitution. Solve equation (1) for x.

$$x = 5 - y \qquad \textbf{(1)}$$

Substitute that quantity for x into equation (2).

$$2x + y^2 = 3 \qquad \textbf{(2)}$$

$$2(5 - y) + y^2 = 3$$

$$10 - 2y + y^2 = 3$$

$$y^2 - 2y + 7 = 0$$

Since the quadratic equation does not factor, we use the quadratic formula where $a = 1$, $b = -2$, and $c = 7$.

$$y = \frac{2 \pm \sqrt{4 - 4(1)(7)}}{2} = \frac{2 \pm 2j\sqrt{6}}{2} = \frac{2(1 \pm j\sqrt{6})}{2}$$

$$y = 1 \pm j\sqrt{6}$$

This nonlinear system has no real number solution. Figure 6–9 verifies this result since the line and the parabola do not intersect.

FIGURE 6–9

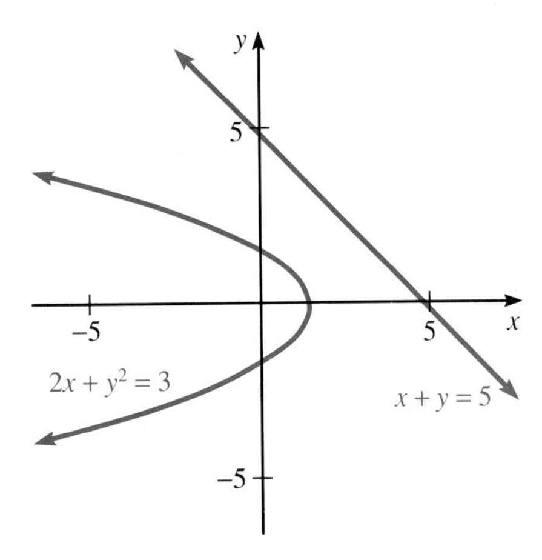

EXAMPLE 3 Find the real number solution to the following nonlinear system of equations:

$$x^2 + y^2 = 25$$
$$4x^2 - 9y^2 = 36$$

Solution Since both equations are quadratic, the addition method is easier to use.

$$x^2 + y^2 = 25 \qquad \textbf{(1)}$$
$$4x^2 - 9y^2 = 36 \qquad \textbf{(2)}$$

We can eliminate y by multiplying equation (1) by 9 and adding the result to equation (2).

$$9(x^2 + y^2 = 25) \longrightarrow \quad 9x^2 + 9y^2 = 225$$
$$4x^2 - 9y^2 = 36 \longrightarrow \quad 4x^2 - 9y^2 = 36$$
$$13x^2 = 261$$
$$x^2 = \frac{261}{13} \approx 20.08$$
$$x \approx \pm\sqrt{20.08} \approx \pm 4.48$$

Substituting into equation (1) to determine y gives

$$(\pm 4.48)^2 + y^2 = 25$$
$$20.08 + y^2 = 25$$
$$y^2 = 25 - 20.08$$
$$y^2 = 4.92$$
$$y \approx \pm\sqrt{4.92} \approx \pm 2.22$$

The ordered pairs are (4.48, 2.22), (−4.48, 2.22), (−4.48, −2.22), and (4.48, −2.22). Do not forget the plus-and-minus signs before the radical when solving an equation such as $y^2 = 4.92$. Figure 6–10 shows the four points of intersection of the circle and the hyperbolas.

FIGURE 6–10

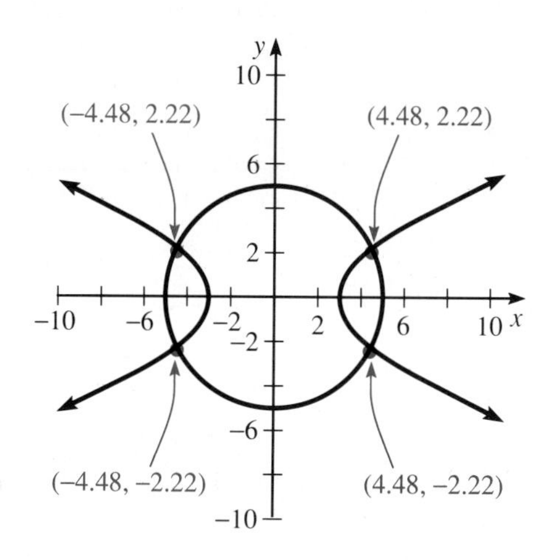

■

CAUTION ✦ Remember that when you solve an equation, you want all solutions to the equation. Therefore, an equation such as $x^2 = 20.08$ leads to two solutions, $x = \pm 4.48$.

Applications

EXAMPLE 4 Find the dimensions of a rectangle if the area is 24 m² and the perimeter is 20 m.

Solution We can solve this problem by using one variable as we did in Chapter 1; however, it is easier to set up the problem if we use two variables. Let

L = length of the rectangle

W = width of the rectangle

Since the area of the rectangle is 24, the equation is

$$LW = 24$$

Since the perimeter is 20, the equation is

$$2L + 2W = 20$$

The system of equations that we must solve to find the dimensions of the rectangle is

$$LW = 24 \qquad \textbf{(1)}$$

$$2L + 2W = 20 \qquad \textbf{(2)}$$

Solve the second equation for W, substitute that quantity for W into equation (1), and solve.

$$\begin{aligned} 2W &= 20 - 2L && \textbf{(1)} \\ W &= 10 - L && \\ LW &= 24 && \textbf{(2)} \end{aligned}$$

$$\begin{aligned} L(10 - L) &= 24 && \\ 10L - L^2 &= 24 && \text{eliminate parentheses} \\ -L^2 + 10L - 24 &= 0 && \text{transpose} \\ L^2 - 10L + 24 &= 0 && \text{multiply by } -1 \\ (L - 4)(L - 6) &= 0 && \text{factor} \\ L = 4 \qquad L &= 6 && \text{solve for } L \end{aligned}$$

Then solve for W given that $L = 4$ and $L = 6$.

$$W = 10 - L$$

$$\begin{aligned} W &= 10 - 4 \qquad & W &= 10 - 6 \\ W &= 6 & W &= 4 \end{aligned}$$

Both substitutions result in the same dimensions. The rectangle is 4 m by 6 m. ■

EXAMPLE 5 In business, market equilibrium is the point where demand for a product equals its supply. If p represents the unit price of the product and q represents the quantity demanded at price p, the demand curve for a particular product may be given by

$$4p^2 + 3q = 45$$

and the supply equation may be given by

$$8p^2 + 20p - 2q = 24$$

Find the market equilibrium quantity and the equilibrium price.

Solution To find the market equilibrium point, solve the following system of equations:

$$4p^2 + 3q = 45 \quad \textbf{(1)}$$
$$8p^2 + 20p - 2q = 24 \quad \textbf{(2)}$$

Solve the first equation for q, substitute that quantity for q into equation (2), and solve.

$$4p^2 + 3q = 45 \quad \textbf{(1)}$$
$$3q = 45 - 4p^2$$
$$q = \frac{45 - 4p^2}{3}$$

$$8p^2 + 20p - 2\left(\frac{45 - 4p^2}{3}\right) = 24 \quad \textbf{(2)}$$
$$24p^2 + 60p - 90 + 8p^2 = 72 \quad \text{eliminate fractions and parentheses}$$
$$32p^2 + 60p - 162 = 0 \quad \text{add}$$
$$16p^2 + 30p - 81 = 0 \quad \text{divide by 2}$$
$$(8p + 27)(2p - 3) = 0 \quad \text{factor}$$

$$8p + 27 = 0 \qquad 2p - 3 = 0$$
$$p = -3.375 \qquad p = 1.5$$

Since the unit price of the product cannot be a negative number, we substitute $p = 1.5$ to determine the value of q.

$$q = \frac{45 - 4p^2}{3}$$
$$q = \frac{45 - 4(1.5)^2}{3}$$
$$q = 12$$

45 [−] 4 [×] 1.5 [x^2] [=] [÷] 3 [=] → 12

The point of intersection is $p = 1.5$ and $q = 12$. The equilibrium price is \$1.50, and the equilibrium quantity is 12. The two equations are graphed in Figure 6–11. Notice that in the graph, only nonnegative values of p and q are considered.

FIGURE 6–11

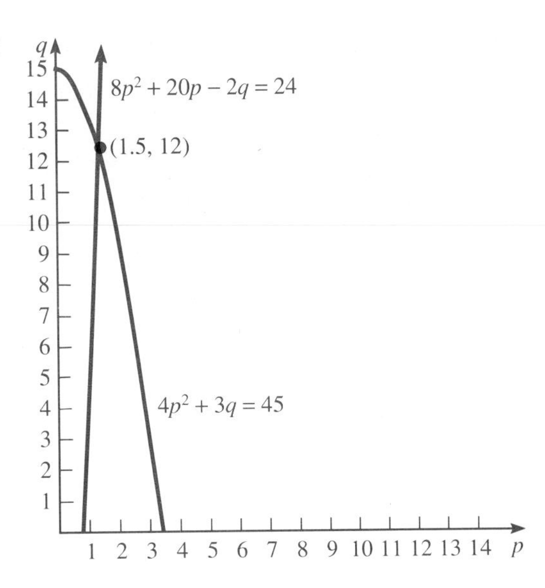

6–4 EXERCISES

Solve the following nonlinear systems for real number solutions.

1. $5x + y = -7$
$y^2 = -4x$

2. $y = x^2$
$x - y = -2$

3. $y = 4 - x$
$4x^2 + y = -5$

4. $x^2 + 16y^2 = 144$
$x + 4y + 12 = 0$

5. $x = 1$
$4x^2 + 4y^2 = 16$

6. $x^2 + y^2 = 9$
$y = 1$

7. $x^2 = 8y$
$x^2 + y^2 = 20$

8. $4x^2 + y^2 = 4$
$y^2 = -4(x - 1)$

9. $x^2 + y^2 = 9$
$4x^2 + 9y^2 = 27$

10. $x^2 + y^2 = 4$
$x^2 = 3y$

11. $x^2 + y^2 = 25$
$3x^2 + 9y^2 = 27$

12. $x^2 + y^2 = 16$
$3x^2 - 9y^2 = 27$

13. $3x^2 + 3y^2 = 12$
$x^2 - y^2 = 1$

14. $x^2 + y^2 = 4$
$4y^2 - 9x^2 = 36$

15. $x^2 - y^2 - 4 = 0$
$x^2 + y^2 = 1$

16. $x^2 - y^2 = 3$
$x^2 + y^2 = 7$

17. $x^2 + 4y^2 = 16$
$3x^2 - 4y^2 = 8$

18. $3x^2 - 5y^2 = 15$
$x^2 + y^2 = 5$

19. $x^2 + 3y^2 = 4$
$2x^2 - 9y^2 = 18$

20. $x^2 + y^2 = 8$
$x^2 - 2y^2 = 2$

21. Find two numbers such that the sum of their squares is 73 and the difference between the numbers is 5.

22. Use a system of equations to find the dimensions of a rectangle whose length is 4 yd more than its width and whose area is 192 yd^2.

E 23. The total resistance R of two resistors connected in series is 8 Ω and the total resistance for the same resistors connected in parallel is 1.875 Ω. Find the two resistances. [*Note:* total resistance in series is given by $R = R_1 + R_2$, and in parallel by $R = R_1R_2/(R_1 + R_2)$.]

24. Find two numbers such that the sum of their squares is 157 and the difference of their squares is 85.

I 25. Find the market equilibrium quantity in hundreds and the price in dollars if the demand curve is given by $2p^2 + 3q = 72$ and the supply equation is given by $4p^2 - 3p + 6q = 135$.

C 26. Find the dimensions of a rectangular yard whose area is 960 m^2 if it requires 128 m of fencing to enclose the yard.

6 – 5

DETERMINANTS

A **matrix** is a rectangular array of numbers consisting of **elements** arranged in rows and columns. A matrix is enclosed in brackets, and its size is given as the number of rows and the number of columns. The following matrix is a two-by-three (also written 2 × 3) matrix, that is, two rows by three columns:

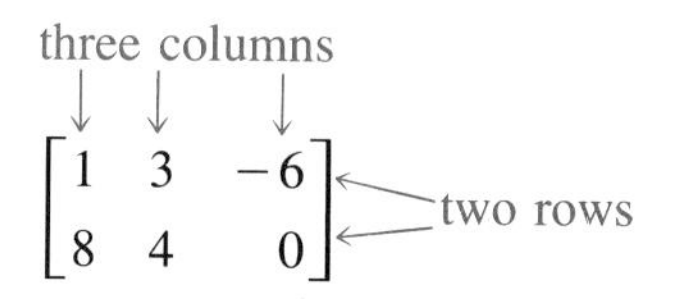

$$\begin{bmatrix} 1 & 3 & -6 \\ 8 & 4 & 0 \end{bmatrix}$$

Each square matrix (that is, one having the same number of rows and columns) has a unique number associated with that matrix. That number, called its **determinant,** is written in the following form:

$$\begin{vmatrix} a & b \\ c & d \end{vmatrix}$$

Two-by-Two Determinant

To find the value of a 2 × 2 determinant, take the difference in the diagonal products as follows:

$$\begin{vmatrix} a & b \\ c & d \end{vmatrix} = ad - bc$$

The reasoning behind this method is explained in the next section.

EXAMPLE 1 Evaluate the following determinant:

$$\begin{vmatrix} -6 & 2 \\ -1 & 5 \end{vmatrix}$$

Solution

$$\begin{vmatrix} -6 & 2 \\ -1 & 5 \end{vmatrix} = (-6)(5) - (-1)(2) = -28$$

The value of the determinant is −28. ■

Three-by-Three Determinant

Evaluating a 3 × 3 determinant is a little more difficult. First, we will discuss a method of evaluation that works *only* for a 3 × 3 determinant. Then we will discuss expansion by cofactors, a method that can be used for all higher-order determinants. Probably the easier method of evaluating a 3 × 3 determinant is as follows:

☐ Rewrite the first two columns of the determinant to the right of the existing determinant.

- ☐ Multiply the six diagonals, three in the downward (left to right) direction and three in the upward direction (left to right). Diagonals in the downward direction are added, and diagonals in the upward direction are subtracted. The following determinant illustrates:

$$\begin{vmatrix} a & b & c \\ d & e & f \\ g & h & i \end{vmatrix} \begin{matrix} a & b \\ d & e \\ g & h \end{matrix} = aei + bfg + cdh - gec - hfa - idb$$

EXAMPLE 2 Evaluate the following determinant:

$$\begin{vmatrix} 1 & 3 & -2 \\ 4 & -1 & -3 \\ 2 & 5 & 2 \end{vmatrix}$$

Solution First, we rewrite the first two columns of the determinant to the right of the existing determinant.

$$\begin{vmatrix} 1 & 3 & -2 \\ 4 & -1 & -3 \\ 2 & 5 & 2 \end{vmatrix} \begin{matrix} 1 & 3 \\ 4 & -1 \\ 2 & 5 \end{matrix}$$

Then we multiply the six diagonals.

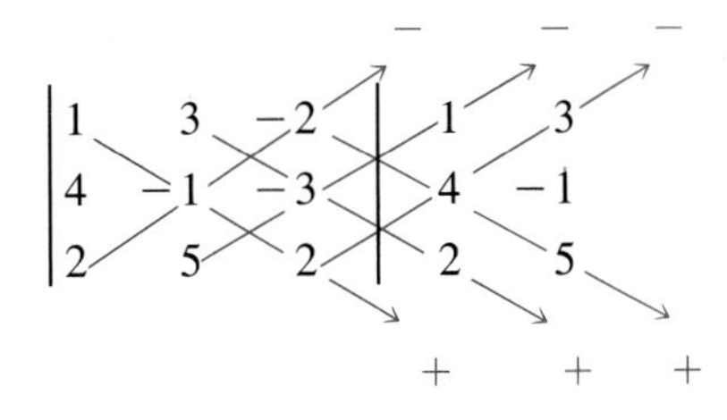

$$\overbrace{(-2) + (-18) + (-40)}^{\text{add}} - \overbrace{(4) - (-15) - (24)}^{\text{subtract}}$$

$$-2 - 18 - 40 - 4 + 15 - 24$$

$$-73$$

The value of the determinant is -73. ■

LEARNING HINT ✦ In multiplying the diagonals of a 3×3 determinant, draw the diagonals one at a time as you multiply.

EXAMPLE 3 Evaluate the following determinant:

$$\begin{vmatrix} 3 & 0 & -2 \\ -1 & 6 & 4 \\ 5 & -3 & 1 \end{vmatrix}$$

Solution We rewrite the first two columns and multiply the diagonals.

$$\left|\begin{matrix} 3 & 0 & -2 \\ -1 & 6 & 4 \\ 5 & -3 & 1 \end{matrix}\right| \begin{matrix} 3 & 0 \\ -1 & 6 \\ 5 & -3 \end{matrix}$$

$$(18) + (0) + (-6) - (-60) - (-36) - (0)$$
$$18 - 6 + 60 + 36$$
$$108$$

■

Minors

Before evaluating a determinant using expansion by cofactors, we must define *minor*. The **minor** of a particular element is *the determinant that remains after the row and the column that contain the element* have been deleted. For example, in the following determinant, the minor of e is the determinant remaining after we have deleted the row and the column that contain the element e:

delete this column ↓

$$\begin{vmatrix} a & b & c \\ d & e & f \\ g & h & i \end{vmatrix} \leftarrow \text{delete this row}$$

The minor for the element e is the determinant

$$\begin{vmatrix} a & c \\ g & i \end{vmatrix}$$

EXAMPLE 4 Determine the minor for the element k in the following determinant:

$$\begin{vmatrix} a & f & m \\ t & d & x \\ c & k & g \end{vmatrix}$$

Solution If we delete the row and the column containing k, we have

$$\begin{vmatrix} a & f & m \\ t & d & x \\ c & k & g \end{vmatrix}$$

The minor of the element k is

$$\begin{vmatrix} a & m \\ t & x \end{vmatrix}$$

■

Cofactors

The **cofactor** of an element is *its minor with a sign attached*. To obtain the sign diagram, start in the upper left corner of the determinant with a plus sign, and alternate signs in both the vertical and the horizontal direction (*not* diagonal). The following sign diagram is for a 4 × 4 determinant:

$$\begin{vmatrix} + & - & + & - \\ - & + & - & + \\ + & - & + & - \\ - & + & - & + \end{vmatrix}$$

You can expand the sign diagram to any size by continuing to alternate signs.

Expansion by Cofactors

To **expand a determinant by cofactors,** *choose any column or row* about which to expand. The determinant is the sum of the product of each element in the chosen row or column and its cofactor.

EXAMPLE 5 Evaluate the following determinant by using expansion by cofactors:

$$\begin{vmatrix} 2 & -4 & 3 \\ -1 & 5 & -2 \\ 7 & -8 & 1 \end{vmatrix}$$

Solution The value of the determinant will be the same no matter which row or column we choose. Thus, if a row or column has smaller numbers or zero elements, choose it because the calculations are easier. Let us expand this determinant using the third column. Expanding about the third column gives

column elements

$$+3\begin{vmatrix} -1 & 5 \\ 7 & -8 \end{vmatrix} - (-2)\begin{vmatrix} 2 & -4 \\ 7 & -8 \end{vmatrix} + 1\begin{vmatrix} 2 & -4 \\ -1 & 5 \end{vmatrix}$$

cofactor signs

$$+3(8 - 35) + 2[-16 - (-28)] + 1(10 - 4)$$
$$3(-27) + 2(12) + 1(6)$$
$$-81 + 24 + 6$$
$$-51$$

■

EXAMPLE 6 Evaluate the following determinant by using expansion by cofactors:

$$\begin{vmatrix} -1 & 2 & 5 & 3 \\ 0 & -4 & 0 & 1 \\ 2 & 1 & -2 & -3 \\ -3 & 0 & -1 & 4 \end{vmatrix}$$

Solution Since row 2 has the most zero elements, we use it to expand the determinant.

$$+(-4)\begin{vmatrix} -1 & 5 & 3 \\ 2 & -2 & -3 \\ -3 & -1 & 4 \end{vmatrix} + 1\begin{vmatrix} -1 & 2 & 5 \\ 2 & 1 & -2 \\ -3 & 0 & -1 \end{vmatrix}$$

We evaluate the 3×3 determinants by writing the first two columns again and multiplying diagonals.

$$-4\begin{vmatrix} -1 & 5 & 3 \\ 2 & -2 & -3 \\ -3 & -1 & 4 \end{vmatrix}\begin{matrix} -1 & 5 \\ 2 & -2 \\ -3 & -1 \end{matrix} + 1\begin{vmatrix} -1 & 2 & 5 \\ 2 & 1 & -2 \\ -3 & 0 & -1 \end{vmatrix}\begin{matrix} -1 & 2 \\ 2 & 1 \\ -3 & 0 \end{matrix}$$

$$-4[8 + 45 + (-6) - 18 - (-3) - 40] +$$
$$1[1 + 12 + 0 - (-15) - 0 - (-4)]$$
$$-4(-8) + 1(32)$$
$$32 + 32$$
$$64$$

■

NOTE ✦ Remember that to evaluate a determinant using expansion by cofactors, find the product of a chosen element and its minor. Compare the expansion in Examples 5 and 6 and notice the difference in the number of cofactors. This difference results from the fact that two of the elements in Example 6 are zero, resulting in a zero product.

6–5 EXERCISES

Evaluate the following determinants.

1. $\begin{vmatrix} 2 & -3 \\ 6 & -4 \end{vmatrix}$

2. $\begin{vmatrix} 0 & -5 \\ -6 & -8 \end{vmatrix}$

3. $\begin{vmatrix} 9 & 2 \\ 5 & 8 \end{vmatrix}$

4. $\begin{vmatrix} 1 & 3 \\ -6 & 7 \end{vmatrix}$

5. $\begin{vmatrix} 0 & 1 \\ 1 & 0 \end{vmatrix}$

6. $\begin{vmatrix} 4.7 & 3.8 \\ 5.9 & 2.4 \end{vmatrix}$

7. $\begin{vmatrix} -7.4 & 0 \\ 18.5 & 2.3 \end{vmatrix}$

8. $\begin{vmatrix} -6 & -4 \\ 2 & -5 \end{vmatrix}$

9. $\begin{vmatrix} 0 & -3 \\ -7 & -10 \end{vmatrix}$

10. $\begin{vmatrix} 2 & -1 & 6 \\ -3 & 0 & 4 \\ 1 & -2 & 1 \end{vmatrix}$

11. $\begin{vmatrix} 1 & 0 & 6 \\ -4 & 2 & 5 \\ -1 & 1 & 3 \end{vmatrix}$

12. $\begin{vmatrix} 1 & 4 & 2 \\ 0 & -1 & 3 \\ -3 & -2 & 0 \end{vmatrix}$

13. $\begin{vmatrix} 0 & 2 & -1 \\ 3 & 0 & 1 \\ 4 & -2 & 0 \end{vmatrix}$

14. $\begin{vmatrix} 6.1 & -5.6 & 0 \\ 4.6 & 1.4 & 0 \\ -1 & 2.3 & 0 \end{vmatrix}$

15. $\begin{vmatrix} 0 & 0 & 0 \\ 1 & 3 & 2 \\ 5 & -2 & -1 \end{vmatrix}$

16. $\begin{vmatrix} -1 & 3 & 0 \\ 0 & 4 & 2 \\ -3 & 0 & -2 \end{vmatrix}$

17. $\begin{vmatrix} 4.9 & 0 & 0 \\ 0 & -1.6 & 3.7 \\ -2.8 & 1 & 2.3 \end{vmatrix}$

18. $\begin{vmatrix} 0 & -5 & 2 \\ -1 & 6 & 0 \\ 0 & 1 & -3 \end{vmatrix}$

Evaluate the following determinants by using expansion by cofactors.

19. $\begin{vmatrix} 0 & -1 & 2 \\ -3 & 4 & 0 \\ -2 & 0 & 1 \end{vmatrix}$

20. $\begin{vmatrix} -3 & 0 & 2 \\ 4 & 1 & 0 \\ -2 & 0 & -1 \end{vmatrix}$

21. $\begin{vmatrix} -4 & 1 & -3 \\ 1 & 2 & -1 \\ 0 & 6 & 0 \end{vmatrix}$

22. $\begin{vmatrix} 0 & 3 & 2 \\ -4 & -1 & 5 \\ 1 & -6 & 1 \end{vmatrix}$

23. $\begin{vmatrix} 1 & 0 & -5 \\ -6 & 2 & 0 \\ 4 & -1 & 3 \end{vmatrix}$

24. $\begin{vmatrix} 7 & 2 & -9 \\ 0 & -1 & -3 \\ 3 & 8 & 1 \end{vmatrix}$

25. $\begin{vmatrix} -5 & 2 & -1 \\ 1 & 0 & 0 \\ 4 & -2 & -3 \end{vmatrix}$

26. $\begin{vmatrix} 1 & 0 & -2 & 0 \\ 2 & 1 & 3 & -4 \\ 0 & 5 & -1 & 4 \\ -1 & 0 & -2 & 1 \end{vmatrix}$

27. $\begin{vmatrix} 1 & 2 & 2 & 1 \\ 0 & -1 & 0 & -3 \\ 3 & 1 & 1 & -4 \\ 4 & 3 & -3 & 0 \end{vmatrix}$

28. $\begin{vmatrix} 5 & 1 & -1 & 2 \\ 3 & -2 & 0 & 1 \\ 3 & 0 & -3 & 0 \\ 0 & 2 & -1 & 3 \end{vmatrix}$

29. $\begin{vmatrix} 0 & 4 & -3 & 2 \\ 1 & 2 & 2 & -3 \\ 1 & -4 & 2 & -1 \\ 0 & 1 & -3 & 2 \end{vmatrix}$

30. $\begin{vmatrix} -1 & 3 & 2 & -4 \\ 2 & 1 & -5 & -2 \\ 6 & 1 & 0 & -2 \\ -2 & 4 & -1 & 1 \end{vmatrix}$

6–6

SOLUTION OF LINEAR SYSTEMS BY DETERMINANTS

Deriving Cramer's Rule

In this section, we will discuss a third method of solving a system of linear equations. The method is called **Cramer's rule.** To derive Cramer's rule, we take a system of equations such as

$$ax + by = c$$
$$dx + ey = f$$

and solve it by the addition method.

$$x = \frac{ce - bf}{ae - bd} \qquad y = \frac{af - cd}{ae - bd}$$

Converting the numerator and denominator into determinants gives Cramer's rule.

Cramer's Rule

$$x = \frac{\begin{vmatrix} c & b \\ f & e \end{vmatrix}}{\begin{vmatrix} a & b \\ d & e \end{vmatrix}} = \frac{D_x}{D} \qquad y = \frac{\begin{vmatrix} a & c \\ d & f \end{vmatrix}}{\begin{vmatrix} a & b \\ d & e \end{vmatrix}} = \frac{D_y}{D}$$

where $D \neq 0$

Rather than memorize the determinants used in Cramer's rule, notice the following:

☐ The determinant D is the same for both variables and consists of the coefficients of x and y from the original system.

$$\begin{aligned} ax + by &= c \\ dx + ey &= f \end{aligned} \Rightarrow \begin{vmatrix} a & b \\ d & e \end{vmatrix} = D$$

☐ To find the determinant in the numerator, replace the column of coefficients of the subscripted variable with the constants

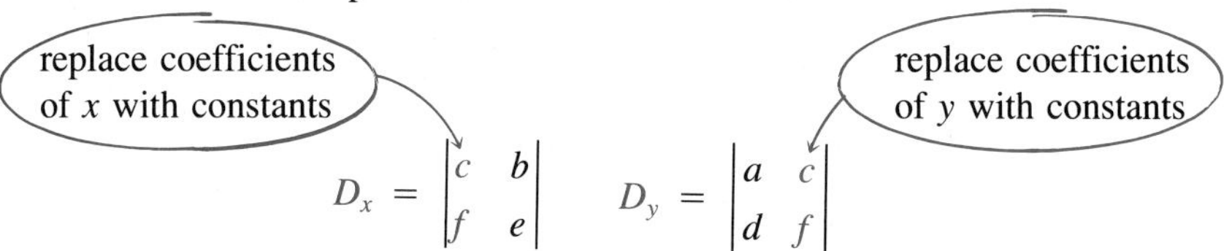

$$D_x = \begin{vmatrix} c & b \\ f & e \end{vmatrix} \qquad D_y = \begin{vmatrix} a & c \\ d & f \end{vmatrix}$$

EXAMPLE 1 Solve the following system of equations using Cramer's rule:

$$\begin{aligned} 4x - 5y &= 10 \\ 3x + 7y &= 14 \end{aligned}$$

Solution To solve this system for x and y, we must calculate the determinants D, D_x, and D_y.

$$D = \begin{vmatrix} 4 & -5 \\ 3 & 7 \end{vmatrix} = 28 + 15 \qquad \text{coefficients of } x \text{ and } y$$

$$D = 43$$

$$D_x = \begin{vmatrix} 10 & -5 \\ 14 & 7 \end{vmatrix} = 70 + 70 \qquad \text{replace coefficients of } x \text{ with constants}$$

$$D_x = 140$$

$$D_y = \begin{vmatrix} 4 & 10 \\ 3 & 14 \end{vmatrix} = 56 - 30 \qquad \text{replace coefficients of } y \text{ with constants}$$

$$D_y = 26$$

The solution is

$$x = \frac{D_x}{D} = \frac{140}{43}$$

$$y = \frac{D_y}{D} = \frac{26}{43}$$

The solution is the ordered pair $\left(\frac{140}{43}, \frac{26}{43}\right)$. ■

CAUTION ✦ When applying Cramer's rule, be sure that the equations are arranged in standard form, $ax + by = c$.

EXAMPLE 2 Use Cramer's rule to solve the following system of equations:

$$\begin{aligned} x + y - z &= 5 \\ 2x - 3y + 6z &= -6 \\ x + 4y - 3z &= 9 \end{aligned}$$

Solution Since these equations are in standard form, we need to calculate D, D_x, D_y, and D_z,

$$D = \left|\begin{matrix} 1 & 1 & -1 \\ 2 & -3 & 6 \\ 1 & 4 & -3 \end{matrix}\right| \begin{matrix} 1 & 1 \\ 2 & -3 \\ 1 & 4 \end{matrix}$$

$$D = 9 + 6 - 8 - 3 - 24 + 6$$
$$D = -14$$

$$D_x = \left|\begin{matrix} 5 & 1 & -1 \\ -6 & -3 & 6 \\ 9 & 4 & -3 \end{matrix}\right| \begin{matrix} 5 & 1 \\ -6 & -3 \\ 9 & 4 \end{matrix}$$

$$D_x = 45 + 54 + 24 - 27 - 120 - 18$$

$$D_x = -42$$

$$D_y = \left|\begin{matrix} 1 & 5 & -1 \\ 2 & -6 & 6 \\ 1 & 9 & -3 \end{matrix}\right| \begin{matrix} 1 & 5 \\ 2 & -6 \\ 1 & 9 \end{matrix}$$

$$D_y = 18 + 30 - 18 - 6 - 54 + 30$$

$$D_y = 0$$

$$D_z = \left|\begin{matrix} 1 & 1 & 5 \\ 2 & -3 & -6 \\ 1 & 4 & 9 \end{matrix}\right| \begin{matrix} 1 & 1 \\ 2 & -3 \\ 1 & 4 \end{matrix}$$

$$D_z = -27 - 6 + 40 + 15 + 24 - 18$$

$$D_z = 28$$

The solution is given by

$$x = \frac{D_x}{D} = \frac{-42}{-14} = 3$$

$$y = \frac{D_y}{D} = \frac{0}{-14} = 0$$

$$z = \frac{D_z}{D} = \frac{28}{-14} = -2$$

The solution is the ordered triple $(3, 0, -2)$. ■

Applications

EXAMPLE 3 Jack has a collection of 25 coins consisting of dimes and nickels. The coins are worth \$2.00. Use a system of equations to determine the number of each coin.

Solution Let

$$n = \text{number of nickels}$$
$$d = \text{number of dimes}$$

coin value × number = total value		
	Number	Total Value
Nickels	n	$0.05n$
Dimes	d	$0.10d$
Total	25	2.00

The equations are

$$\text{total number of coins} = 25$$

$$n + d = 25$$

$$\text{total value of nickels} + \text{total value of dimes} = \$2.00$$

$$\uparrow \qquad\qquad \uparrow \qquad\qquad \uparrow$$

$$0.05n + 0.10d = 2.00$$

Since the equations are in standard form, we calculate D, D_n, and D_d.

$$D = \begin{vmatrix} 1 & 1 \\ 0.05 & 0.10 \end{vmatrix} = 0.10 - 0.05$$

$$D = 0.05$$

$$D_n = \begin{vmatrix} 25 & 1 \\ 2.00 & 0.10 \end{vmatrix} = 2.50 - 2.00$$

$$D_n = 0.50$$

$$D_d = \begin{vmatrix} 1 & 25 \\ 0.05 & 2.00 \end{vmatrix} = 2.00 - 1.25$$

$$D_d = 0.75$$

The solution is

$$n = \frac{D_n}{D} = \frac{0.50}{0.05} = 10$$

$$d = \frac{D_d}{D} = \frac{0.75}{0.05} = 15$$

Jack has 15 dimes and 10 nickels. ■

EXAMPLE 4 The sum of the angles of any triangle is 180°. In a particular triangle, the largest angle is 5° more than three times the smallest angle. The second largest angle is 25° more than the smallest angle. Using a system of equations, find the angles of the triangle.

Solution Let

a = largest angle

b = second largest angle

c = smallest angle

The verbal statement gives the following system of equations:

The sum of the angles is 180°.

$$a + b + c = 180$$

The largest angle is 5° more than 3 times the smallest angle.

$$a = 3c + 5$$

The second largest angle is 25° more than the smallest angle.

$$b = c + 25$$

Arranging the system of equations in standard form gives

$$\begin{aligned} a + b + c &= 180 \\ a - 3c &= 5 \\ b - c &= 25 \end{aligned}$$

Evaluating the determinants D, D_a, D_b, and D_c gives

$$D = \begin{vmatrix} 1 & 1 & 1 \\ 1 & 0 & -3 \\ 0 & 1 & -1 \end{vmatrix} \begin{matrix} 1 & 1 \\ 1 & 0 \\ 0 & 1 \end{matrix}$$

$$D = 0 + 0 + 1 - 0 + 3 + 1$$

$$D = 5$$

$$D_a = \begin{vmatrix} 180 & 1 & 1 \\ 5 & 0 & -3 \\ 25 & 1 & -1 \end{vmatrix} \begin{matrix} 180 & 1 \\ 5 & 0 \\ 25 & 1 \end{matrix}$$

$$D_a = 0 - 75 + 5 + 0 + 540 + 5$$

$$D_a = 475$$

$$D_b = \begin{vmatrix} 1 & 180 & 1 \\ 1 & 5 & -3 \\ 0 & 25 & -1 \end{vmatrix} \begin{matrix} 1 & 180 \\ 1 & 5 \\ 0 & 25 \end{matrix}$$

$$D_b = -5 + 0 + 25 - 0 + 75 + 180$$

$$D_b = 275$$

$$D_c = \begin{vmatrix} 1 & 1 & 180 \\ 1 & 0 & 5 \\ 0 & 1 & 25 \end{vmatrix} \begin{matrix} 1 & 1 \\ 1 & 0 \\ 0 & 1 \end{matrix}$$

$$D_c = 0 + 0 + 180 - 0 - 5 - 25$$

$$D_c = 150$$

The solution is

$$a = \frac{D_a}{D} = \frac{475}{5} = 95$$

$$b = \frac{D_b}{D} = \frac{275}{5} = 55$$

$$c = \frac{D_c}{D} = \frac{150}{5} = 30$$

The angles of the triangle are 95°, 55°, and 30°. ■

CAUTION ✦ When a particular variable does not appear in an equation, remember to fill in a zero for its position in the determinant.

Special Cases

Inconsistent and dependent systems of equations also occur when Cramer's rule is applied. The next example illustrates how to recognize these systems.

EXAMPLE 5 Solve the following systems of equations by using Cramer's rule:

(a) $2x - 3y = 7$
$-4x + 6y = 10$

(b) $x + y = -2$
$-3x - 3y = 6$

Solution

$$\text{(a)}\ x = \frac{\begin{vmatrix} 7 & -3 \\ 10 & 6 \end{vmatrix}}{\begin{vmatrix} 2 & -3 \\ -4 & 6 \end{vmatrix}} = \frac{42 + 30}{12 - 12} = \frac{72}{0} = \text{undefined (inconsistent)}$$

$$y = \frac{\begin{vmatrix} 2 & 7 \\ -4 & 10 \end{vmatrix}}{\begin{vmatrix} 2 & -3 \\ -4 & 6 \end{vmatrix}} = \frac{20 + 28}{12 - 12} = \frac{48}{0} = \text{undefined (inconsistent)}$$

If *only the determinant in the denominator equals zero,* the system of equations is *inconsistent* and has *no solution.*

$$\text{(b)}\ x = \frac{\begin{vmatrix} -2 & 1 \\ 6 & -3 \end{vmatrix}}{\begin{vmatrix} 1 & 1 \\ -3 & -3 \end{vmatrix}} = \frac{6 - 6}{3 - 3} = \frac{0}{0}\ \text{(dependent)}$$

$$y = \frac{\begin{vmatrix} 1 & -2 \\ -3 & 6 \end{vmatrix}}{\begin{vmatrix} 1 & 1 \\ -3 & -3 \end{vmatrix}} = \frac{6 - 6}{0} = \frac{0}{0}\ \text{(dependent)}$$

If the *determinant in both the numerator and the denominator equals zero,* the system of equations is *dependent.* ■

6–6 EXERCISES

Use determinants to solve the following systems of equations.

1. $3x - 2y = 8$
$2x + y = -4$

2. $3x + 2y = 3$
$4x + 5y = 11$

3. $x - 2y = 8$
$3x + 4y = -6$

4. $4x - 2y = -4$
$3x + y = -8$

5. $2x - 5y = 12$
$x + 7y = 6$

6. $2.9x - 5.1y = 1.3$
$3.6x + 4.5y = 13.2$

7. $4.1x + 7.3y = -6.8$
$3.7x + 8.5y = 1.7$

8. $3x + 2y = 13$
$6x - 5y = -19$

9. $6x - 8y = 0$
$7x + 9y = 0$

10. $2x + 7y = 6$
$3x + 4y = -4$

11. $3x - y = 3$
$4x - 2y = 0$

12. $6x + 4y = 1$
$12x + 2y = -1$

13. $3x = 4y + 3$
$6y = 9x - 6$

14. $8x = -3y$
$4x - 5y = 13$

15. $5x - 12y = 0$
$10x + 8y = 8$

16. $3x - 2y + 4z = -4$
$x + 7y - 3z = 2$
$4x + 5y + z = -2$

17. $4x + 2y - z = 10$
$3x - y + 2z = -7$
$2x - 3y - 5z = 1$

18. $x + 3y - 2z = 5$
$2x - y - 4z = 3$
$3x + 2y + x = -13$

19. $4.7x + 3.6y - 5.1z = 10$
$3.5x + y + 7.2z = -2.9$
$x + 6.8y - 4.5z = 0$

20. $3x - y - 2z = 2$
$x + 4y + 2z = 7$
$2x + 7y - 4z = -17$

21. $5x + 7y - 6z = 6$
$3x - 4y + 2z = -2$
$2x + 5y - 4z = 4$

22. $-2.7x + 5.2y + 8.3z = -1$
$6.1x + 4.5y = -6$
$-3.3y + 2.7z = -7$

23. $2x - 4z = 16$
$3y - 5x = -6$
$3z + y = 0$

24. $4x + 3z = -1$
$y - 2x = 0$
$2y + 3z = -1$

25. $2x + 3y - z = -3$
$4y + 3z = 7$
$x - 3y = 10$

26. $2x - 4y - z = 3$
$3y + 7z = 1$
$8z - 3x = -10$

27. $4x - 6y = -2$
$8z + 3y = -6$
$2x - 3y - 4z = 3$

28. $3z + 8x = 1$
$2y + 6z = 2$
$12x + y + 3z = 4$

29. $5x + y - 4z = 0$
$3y - 8x = 9$
$8z + 5x = 6$

30. $7y - 4z = 2$
$6x + 9y - 2z = -3$
$5y + 9x = -6$

31. Use Cramer's rule to find the width of a rectangle if the perimeter is 120 ft and the length is 9 ft less than twice the width.

32. Use Cramer's rule to find out the smaller acute angle of a right triangle if the angles of a triangle sum to 180° and twice the smallest angle is 21° more than the other acute angle.

33. Use Cramer's rule to find the largest of three numbers if the sum of the three numbers is 87 and the largest number is three times the smallest number. The second largest number is nine less than twice the smallest number.

34. Ray has invested $20,000 in investments paying 8%, 9.5%, and 11%. If the total annual interest is $2,005 and the sum of the amounts at 9.5% and 11% is four times the amount at 8%, how much does Ray have invested at 9.5%?

35. A theater owner sold 245 tickets, and receipts totaled $895.00. How many children's tickets did he sell if a child's ticket is $2 and an adult's ticket is $4.25?

6–7

APPLICATIONS

In this section, we will apply the methods of solving a system of equations to technical problems. The first application involves **torque,** which is defined as the tendency of a force to produce a change in rotational motion. Mathematically, torque is the product of a force and its lever arm, defined to be the perpendicular distance from the pivot point to the line of action of the force. For example, the torque produced by the 100-lb force in Figure 6–12 is

$$\text{torque} = (100 \text{ lb})(8 \text{ ft}) = 800 \text{ ft-lb}$$

EXAMPLE 1 A 100-ft uniform bridge weighing 100 tons is supported at each end by a pillar. A 15-ton truck is 25 ft from one end, and a car weighing 1.5 tons is 40 ft from the opposite end of the bridge. Under these conditions, how much weight does each pillar support?

FIGURE 6–12

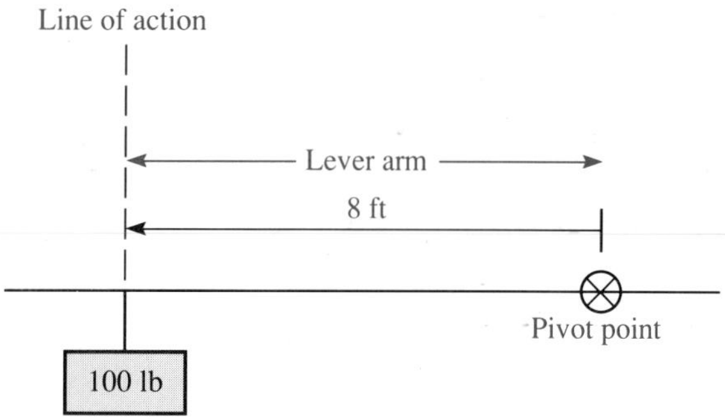

Line of action

Lever arm

8 ft

Pivot point

100 lb

Solution First, we draw a figure to represent this problem, shown in Figure 6–13. Let

$$A = \text{weight supported by the left pillar}$$
$$B = \text{weight supported by the right pillar}$$

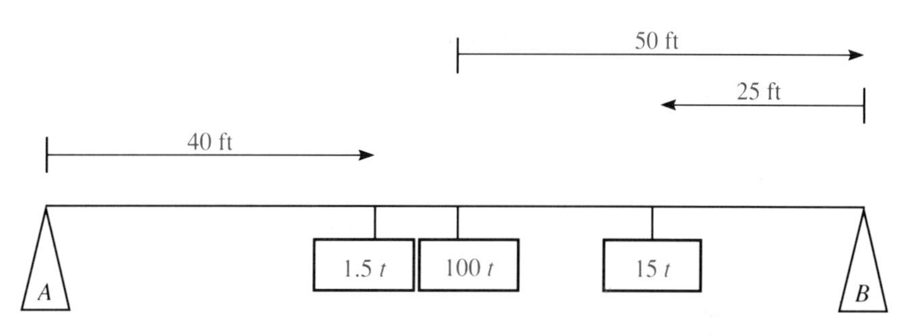

FIGURE 6–13

Next, we draw a force diagram showing the direction in which each force acts on the bridge. This force diagram is shown in Figure 6–14. Notice that the weight of the bridge, the truck, and the car act downward on the bridge, and the supports act upward on the bridge.

FIGURE 6–14

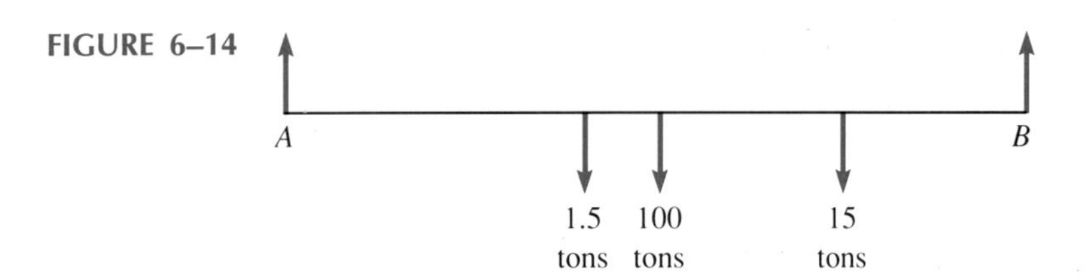

The First Condition of Equilibrium states that the sum of the forces acting upward on an object equals the sum of the forces acting downward on the object. Using the First Condition of Equilibrium and the force diagram, we can write the following equation:

$$\overbrace{A + B}^{\text{upward}} = \overbrace{1.5 + 100 + 15}^{\text{downward}}$$
$$A + B = 116.5$$

The Second Condition of Equilibrium states that the sum of the torques that tend to cause clockwise (CW) rotation equals the sum of the torques that tend to cause counterclockwise (CCW) rotation. To arrive at an equation, we must choose a pivot point, and we must determine the torque produced by each force and the direction of the force. Let us choose the center of the bridge as the pivot point (the choice of a pivot point is arbitrary). Figure 6–15 shows the pivot point, the lever arm for each force, and the direction of the resulting torque.

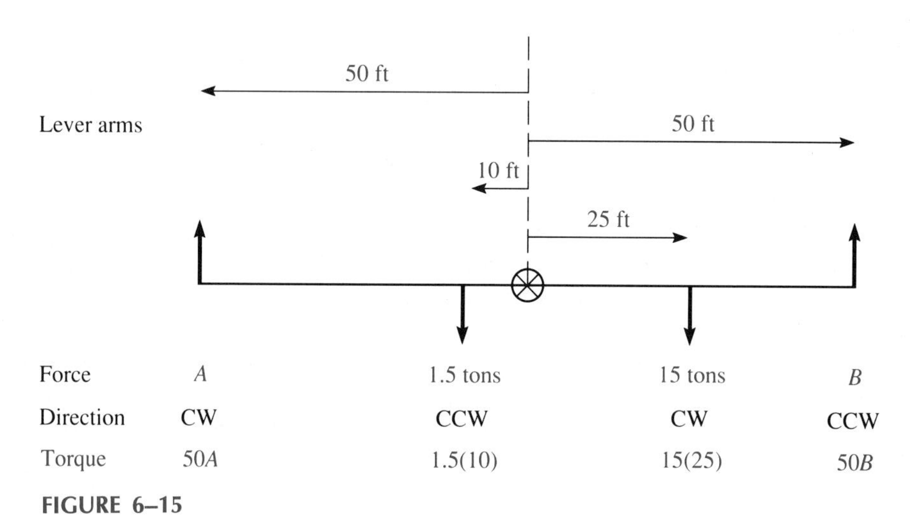

FIGURE 6–15

Note that the 100-ton force produces no torque since its lever arm is zero. Since torque is the product of a force and its lever arm, applying the Second Condition of Equilibrium gives the following equation:

$$\overbrace{50(A) + 15(25)}^{\text{clockwise}} = \overbrace{1.5(10) + 50(B)}^{\text{counterclockwise}}$$
$$50A - 50B = -360$$

We must solve the following system to determine A and B:

$$A + B = 116.5 \quad \textbf{(1)}$$
$$50A - 50B = -360 \quad \textbf{(2)}$$

We use Cramer's rule to solve for A.

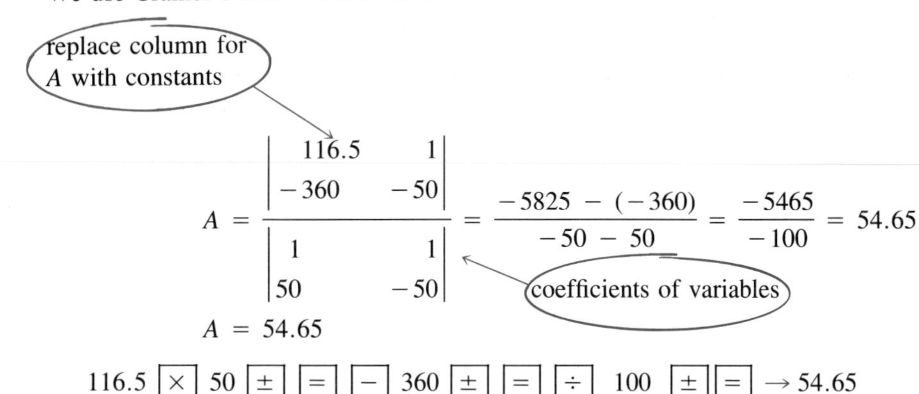

$$A = \frac{\begin{vmatrix} 116.5 & 1 \\ -360 & -50 \end{vmatrix}}{\begin{vmatrix} 1 & 1 \\ 50 & -50 \end{vmatrix}} = \frac{-5825 - (-360)}{-50 - 50} = \frac{-5465}{-100} = 54.65$$

$$A = 54.65$$

116.5 [×] 50 [±] [=] [−] 360 [±] [=] [÷] 100 [±] [=] → 54.65

Solve for B by substituting $A = 54.65$ into equation (1).

$$A + B = 116.5 \qquad \textbf{(1)}$$
$$54.65 + B = 116.5$$
$$B = 61.85$$

One pillar supports 61.85 tons, and the other pillar supports 54.65 tons. ■

LEARNING HINT ✦ To determine the direction of rotation of torque, hold one end of your pencil fixed at the pivot point and rotate the other end of the pencil in the direction of the arrow for the force.

EXAMPLE 2 From the plat of a triangular tract of land, a surveyor knows that the perimeter is 70 m. The lengths on the plat have been blurred, but the surveyor knows that the length of the longest side is 10 m less than the sum of the other two sides, and the longest side is 2 m less than twice the shortest side. Find the length of the three sides of the triangular tract.

Solution Let

x = length of the longest side
y = length of the middle side
z = length of the shortest side

The system of equations from the verbal statement is

The perimeter is 70.

$$x + y + z = 70 \qquad \textbf{(1)}$$

The longest side is 10 less than the sum of the other two sides.

$$x = y + z - 10 \qquad \textbf{(2)}$$

The longest side is 2 less than twice the shortest side.

$$x = 2z - 2 \tag{3}$$

We use the substitution method to reduce this system of three equations to two. Substituting the quantity for x from equation (3) into equations (1) and (2) gives

$$(2z - 2) + y + z = 70 \tag{1}$$
$$3z + y = 72$$

$$(2z - 2) = y + z - 10 \tag{2}$$
$$z - y = -8$$

Then we solve the resulting system.

$$\begin{aligned} 3z + y &= 72 \\ z - y &= -8 \\ \hline 4z \quad &= 64 \\ z \quad &= 16 \end{aligned}$$

We substitute $z = 16$ into the equation and solve for y.

$$\begin{aligned} z - y &= -8 \\ 16 - y &= -8 \\ y &= 24 \end{aligned}$$

We substitute $z = 16$ into equation (3) to solve for x.

$$\begin{aligned} x &= 2z - 2 \\ x &= 2(16) - 2 \\ x &= 30 \end{aligned}$$

The triangular tract is 30 m by 24 m by 16 m. ■

EXAMPLE 3 A small textile company uses two different spinning machines, A and B, in the manufacture of cotton thread and wool yarn. Each unit of cotton thread requires 2 hours on machine A and 1 hour on machine B, while wool yarn requres 1 hour on machine A and 1 hour on machine B. Machine A can run a maximum of 22 hours per day, while machine B can run a maximum of 16 hours per day. How many units of cotton thread and wool yarn should the company make to keep the machines running at maximum capacity?

Solution Let

x = amount of cotton thread produced

y = amount of wool yarn produced

The system of equations is

$$\text{Machine } A\text{:}\quad 2x + y = 22 \qquad (1)$$
$$\text{Machine } B\text{:}\quad x + y = 16 \qquad (2)$$

Using the addition method to solve the system, we multiply equation (2) by -1.

$$\begin{array}{rcr} 2x + y = 22 & \longrightarrow & 2x + y = 22 \\ -1(x + y = 16) & \longrightarrow & -x - y = -16 \\ \hline & & x \qquad = 6 \end{array}$$

Solve for y by substituting $x = 6$ into equation (2).

$$x + y = 16 \qquad (2)$$
$$6 + y = 16$$
$$y = 10$$

Therefore, the textile company should produce 6 units of cotton thread and 10 units of wool yarn per day to keep the machines running at maximum capacity. ■

Kirchhoff's Laws

Kirchhoff's laws state the following:

1. The sum of the voltage rises and drops around a closed loop is zero.
2. The sum of the currents entering a node equals the sum of currents leaving the node.

A **closed loop** is any path that touches a point exactly once and returns to the starting point. A **node** is a point where three or more conductors (lines) are joined.

EXAMPLE 4 Find currents I_1, I_2, I_3 in the circuit given in Figure 6–16.

Solution We begin by applying Kirchhoff's current law. We must apply this law at enough nodes so that each current appears in at least one equation. There are nodes at b and e in Figure 6–16. From Kirchhoff's current law, the equation at node b is

$$I_1 = I_2 + I_3$$
$$I_1 - I_2 - I_3 = 0$$

Since each current appears in the previous equation, we do not need to write the equation for node e.

FIGURE 6–16

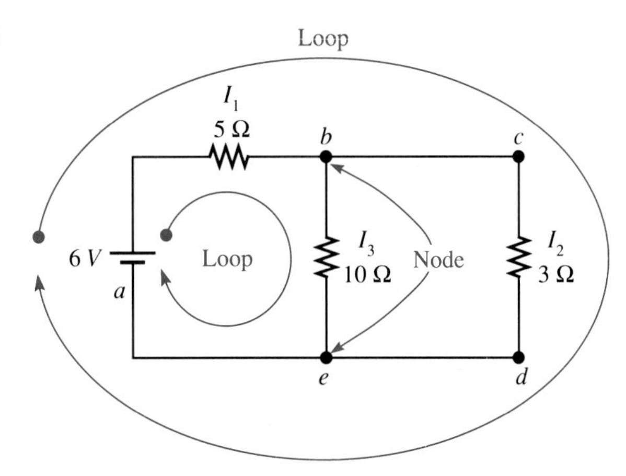

Next, we apply Kirchhoff's voltage law to closed loops until every current is included in at least one equation. Figure 6–16 has two closed loops: *abea* and *abcdea*. Applying the voltage law ($V = IR$) for both loops gives the following equations, respectively:

$$6 - 5I_1 - 10I_3 = 0$$
$$6 - 5I_1 - 3I_2 = 0$$

We must solve the following system of equations:

$$I_1 - I_2 - I_3 = 0$$
$$-5I_1 - 10I_3 = -6$$
$$-5I_1 - 3I_2 = -6$$

We solve this system by using Cramer's rule.

$$D = \left|\begin{array}{ccc|cc} 1 & -1 & -1 & 1 & -1 \\ -5 & 0 & -10 & -5 & 0 \\ -5 & -3 & 0 & -5 & -3 \end{array}\right.$$

$$D = 0 - 50 - 15 - 0 - 30 - 0$$
$$D = -95$$

$$I_1 = \frac{\left|\begin{array}{ccc|cc} 0 & -1 & -1 & 0 & -1 \\ -6 & 0 & -10 & -6 & 0 \\ -6 & -3 & 0 & -6 & -3 \end{array}\right.}{-95}$$

$$I_1 = -78/-95 \approx 0.82 \text{ A}$$

$$I_2 = \frac{\begin{vmatrix} 1 & 0 & -1 \\ -5 & -6 & -10 \\ -5 & -6 & 0 \end{vmatrix} \begin{matrix} 1 & 0 \\ -5 & -6 \\ -5 & -6 \end{matrix}}{-95}$$

$$I_2 = -60/-95 \approx 0.63 \text{ A}$$

$$I_3 = \frac{\begin{vmatrix} 1 & -1 & 0 \\ -5 & 0 & -6 \\ -5 & -3 & -6 \end{vmatrix} \begin{matrix} 1 & -1 \\ -5 & 0 \\ -5 & -3 \end{matrix}}{-95}$$

$$I_3 = -18/-95 \approx 0.19 \text{ A}$$

■

Partial Fractions

In calculus, it is sometimes necessary to break an algebraic fraction into partial fractions in order to integrate it. In Chapter 3, we combined several algebraic fractions into a single fraction. For example, if we add two fractions such as $\frac{4}{x+1}$ and $\frac{3}{x+3}$, we get

$$\frac{4}{x+1} + \frac{3}{x+3} = \frac{4(x+3) + 3(x+1)}{(x+1)(x+3)} = \frac{7x+15}{(x+1)(x+3)}$$

Reversing this process to *transform a single fraction into several separate fractions* is called **decomposing into partial fractions.**

EXAMPLE 5 Decompose the following into partial fractions:

$$\frac{-4x + 23}{(2x - 1)(x + 3)}$$

Solution To decompose this fraction, we must write the fraction as a partial fraction of the form

$$\frac{\text{constant}}{\text{linear factor}}$$

Since there are two linear factors in the denominator of this fraction, we have

$$\frac{-4x + 23}{(2x - 1)(x + 3)} = \frac{A}{2x - 1} + \frac{B}{x + 3}$$

To simplify this equation, we eliminate fractions by multiplying the equation by the lowest common denominator, $(2x - 1)(x + 3)$.

$$-4x + 23 = A(x + 3) + B(2x - 1)$$
$$-4x + 23 = Ax + 3A + 2Bx - B \quad \text{eliminate parentheses}$$
$$-4x + 23 = (A + 2B)x + (3A - B) \quad \text{combine like terms}$$

In order for this equation to be satisfied, the coefficient of the same variable must be equal. Therefore,

$$-4x + 23 = (A + 2B)x + (3A - B)$$
$$A + 2B = -4$$
$$3A - B = 23$$

To decompose the original fraction into partial fractions, we solve the resulting system of equations. We solve this system by the addition method.

$$\begin{array}{lcl} A + 2B = -4 & \longrightarrow & A + 2B = -4 \\ 2(3A - B = 23) & \longrightarrow & 6A - 2B = 46 \\ \hline & & 7A \quad = 42 \\ & & A \quad = 6 \end{array}$$

$$A + 2B = -4$$
$$6 + 2B = -4$$
$$2B = -10$$
$$B = -5$$

Decomposing the fraction into partial fractions gives

$$\frac{-4x + 23}{(2x - 1)(x + 3)} = \frac{6}{2x - 1} - \frac{5}{x + 3}$$ ■

6–7 EXERCISES

I

1. On an assembly line it takes 2 hours to complete two tasks. One task takes 2/3 hour more than the other. Use a system of equations to find how long each task requires.
2. A pile of quarters and nickels has a value of \$9.00. Twice the number of quarters is four more than three times the number of nickels. How many of each coin are there?
3. A chemist wants to make 400 ml of a 20% sulfuric acid solution using a 5% solution and a 30% solution. How much of each solution should she use?

M

4. A 12-ft scaffold of negligible weight is supported at each end. Two brick masons weighing 180 lb and 155 lb stand 3 ft and 5 ft, respectively, from opposite ends. If a pile of bricks weighing 100 lb is posi-

tioned at the center of the scaffold, how much weight does each end support?

5. Decompose the following into partial fractions:

$$\frac{16x + 3}{(2x - 2)(x + 3)}$$

6. How much solder containing 60% tin should be mixed with a solder containing 40% tin to make 65 lb of solder containing 52% tin?

C 7. A surveyor is trying to find the dimensions of a rectangular tract. The surveyor knows that the perimeter of the tract is 122 yd and the length is 7 yd more than the width. Find the dimensions of the tract.

8. Helen invested $25,000, part at 10% and part at 12%, yielding an annual income of $2,640. How much did Helen invest at each rate?

M 9. Weights of 100 N, 150 N, and 75 N are placed on a board resting on two supports as shown in Figure 6–17. Neglecting the weight of the board, what are the forces exerted by the supports?

10. Decompose the following into partial fractions:

$$\frac{-22x + 47}{(3x - 5)(2x + 7)}$$

C 11. A builder wants to combine a concrete mixture that is 25% cement with a concrete mixture that is 40% cement to make a 100-yd^3 concrete mixture that is 35% cement. How much of each concrete mixture should he use?

E 12. Find currents I_1 and I_2 shown in Figure 6–18.

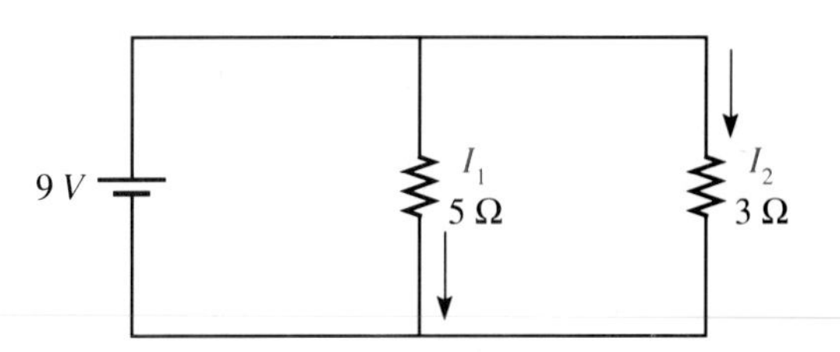

FIGURE 6–18

13. A plane traveled 1,200 miles with the wind in 3 hours and made the return trip against the wind in 4 hours. Determine the speed of the plane and the speed of the wind.

14. Decompose the following into partial fractions:

$$\frac{37x - 35}{(x + 1)(5x - 7)}$$

I 15. The cost of manufacturing a product consists of a fixed charge plus a variable charge for each item. If it costs $1,370 to produce 650 units and $1,640 to produce 800 units, what are the fixed and variable charges?

16. The perimeter of a rectangle is 130 yd. Find the length and width of the rectangle if the length is five more than twice the width.

17. Decompose the following into partial fractions:

$$\frac{-2x - 42}{(x + 3)(x - 1)}$$

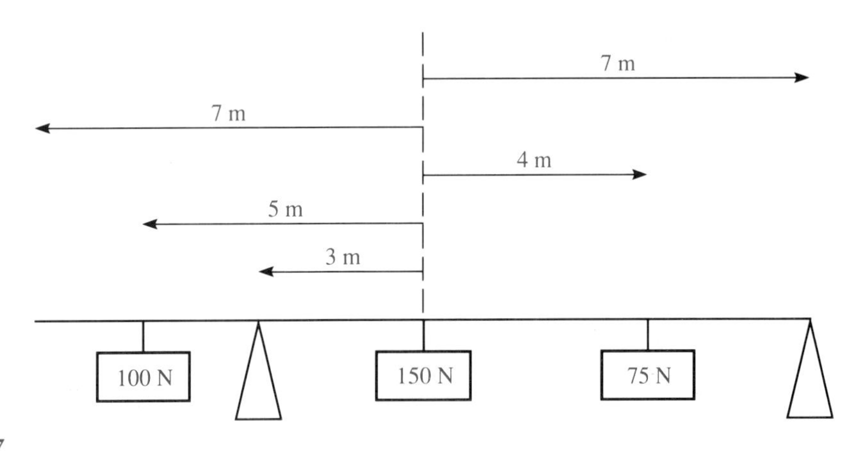

FIGURE 6–17

M 18. A 50-ft uniform bridge weighing 100 tons is supported on each end. Two trucks weighing 7 tons and 5 tons are 10 ft and 18 ft, respectively, from opposite ends. How much weight does each end support?

E 19. Find currents I_1, I_2, and I_3 shown in Figure 6–19.

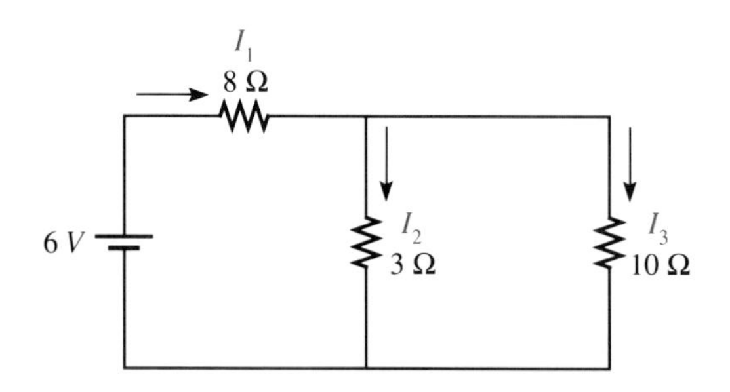

FIGURE 6–19

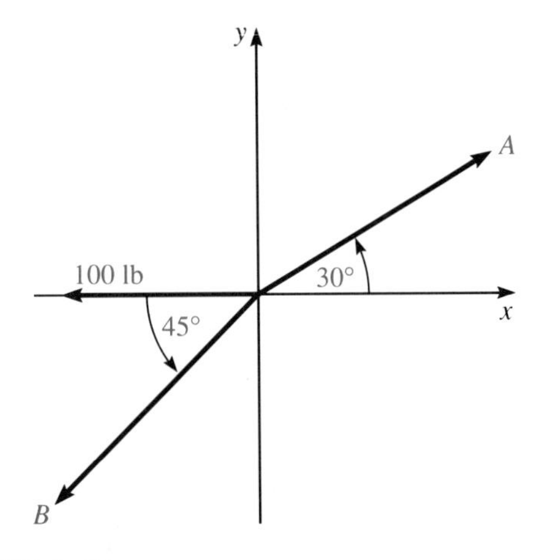

FIGURE 6–20

M 20. A system of equations for forces A and B shown in Figure 6–20 is

$$0.5A = 0.71B$$
$$0.87A = 100 + 0.71B$$

Find forces A and B.

E 21. Find currents I_1, I_2, and I_3, in amperes, in Figure 6–21. Round your answer to tenths.

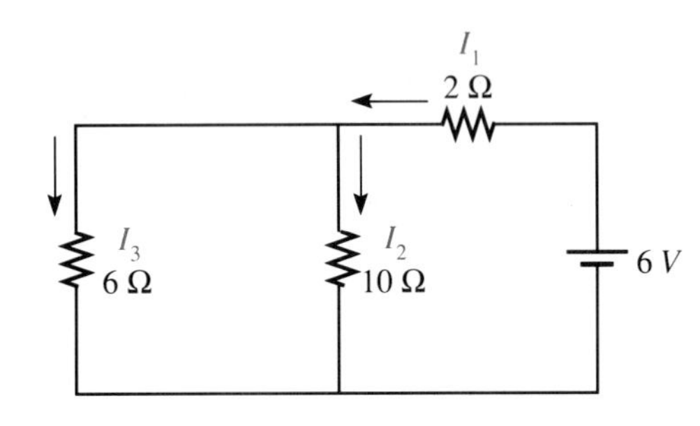

FIGURE 6–21

22. Find three numbers whose sum is 42. The largest number is two less than the sum of the other two numbers. Twice the smallest number is four less than the largest number.

23. Bill has a collection of dimes and quarters. If he has 30 coins worth $5.70, how many dimes and quarters does he have?

CHAPTER SUMMARY

Summary of Terms

addition method (p. 211)
closed loop (p. 249)
cofactor (p. 235)
Cramer's rule (p. 237)
decomposing into partial fractions (p. 251)
dependent system (p. 206)
determinant (p. 232)
elements (p. 232)
expanding a determinant by cofactors (p. 235)
graphical solution (p. 202)
inconsistent system (p. 206)
matrix (p. 232)
minor (p. 234)
node (p. 249)
solution (p. 202)
substitution method (p. 208)
system of linear equations (p. 202)
torque (p. 244)
x intercept (p. 203)
y intercept (p. 203)

CHAPTER REVIEW

Section 6–1

1. Is $(-2, 3)$ the solution to the following system of equations:

$$2x + y = -1$$
$$x - 4y = 8$$

Why or why not?

2. Is $(-4, 0)$ the solution to the following system of equations:

$$2x + 3y = -8$$
$$y = 3x + 12$$

Why or why not?

Solve the following systems of equations graphically.

3. $x - 6y = 6$; $4x + y = -1$
4. $x - 3y = -7$; $2x + y = 7$
5. $3x - 5y = -7$; $y = 2x + 7$
6. $x = 2y - 11$; $3x + y = 2$
7. $4x - y = 6$; $3x + 4y = 8$
8. $x - 2y = 6$; $5x + y = 8$

Section 6–2

Solve the following systems of equations by using the addition method.

9. $2x + y = 0$; $3x + 2y = 3$
10. $3x + y = 11$; $4x - 3y = -7$
11. $3x - 2y = -2$; $3y = -5x - 1$
12. $3x + 7y = 8$; $-9x = -21y + 10$

Solve the following systems of equations by using the substitution method.

13. $y = 3x - 4$; $3x - 2y = 2$
14. $x = 4y - 1$; $2x - 5y = 4$
15. $x - 8y = 5$; $16y = 2x - 10$
16. $3x + 4y = 8$; $2x + 3y = 6$
17. Mr. Roberts invests \$15,000 in two savings accounts, one at 8.5% and the other at 9.75%. If he wants to earn \$1,387.50 in interest per year, how much should he invest at each rate?
18. Find two numbers such that the first number increased by 25 is 3 more than the second number, and twice the first number is 14 more than the second number.

Section 6–3

Solve each of the following systems by using addition or substitution.

19. $x - 3y + z = -2$; $2x + y - z = -6$; $3x - y + 2z = -1$
20. $2x + 3y - z = 0$; $x - 3y + 4z = -9$; $-x + y + z = 5$
21. $4x - 3y + 2z = -12$; $x + 5y - 3z = 20$; $2x + 2y - z = 8$
22. $x + 2y + 3z = 3$; $3x - y - 2z = -6$; $2x + y + z = -1$
23. $x - 2y = -5$; $3z = 9$; $2x + 4z = 10$
24. $z - 3x = -14$; $4y + z = -4$; $2x - y = 4$
25. $4x - z = -1$; $6y + 7z = -11$; $2x - 3y = 9$
26. Fred has a collection of five-, ten-, and twenty-dollar bills. If the total value of the bills is \$175, and there are twice as many tens as fives and two more twenties than fives, how many of each bill does Fred have?
27. The sum of the angles of any triangle is 180°. In a certain triangle, the largest angle is 9° more than three times the smallest angle, and the middle-sized angle is 11° more than the smallest angle. Find the measure of each angle.

Section 6–4

Solve the following nonlinear systems of equations.

28. $x + y = 6$; $x^2 + y^2 = 9$
29. $2x - y = 8$; $y = x^2 - 8$
30. $x^2 + y^2 = 16$; $x^2 + y^2 = 4$
31. $3x^2 + 4y = 12$; $y = x^2 - 4$
32. $5x^2 - 4y^2 = 20$; $x^2 + y^2 = 25$
33. $2x^2 + 3y^2 = 6$; $x^2 + y^2 = 4$

Section 6–5

Evaluate the following determinants.

34. $\begin{vmatrix} -6 & 2 \\ 4 & -1 \end{vmatrix}$
35. $\begin{vmatrix} a & c \\ 3 & -2 \end{vmatrix}$

36. $\begin{vmatrix} 2 & 0 & 5 \\ -1 & 3 & 1 \\ 4 & -2 & 0 \end{vmatrix}$

37. $\begin{vmatrix} 0 & -1 & 3 \\ 4 & -2 & 2 \\ 1 & 1 & 5 \end{vmatrix}$

38. $\begin{vmatrix} 8 & -6 & 2 \\ 0 & 3 & 4 \\ -1 & -2 & 1 \end{vmatrix}$

39. $\begin{vmatrix} -3 & 2 & 1 \\ 0 & 4 & -2 \\ -1 & 2 & -1 \end{vmatrix}$

Using expansion by cofactors, evaluate the following determinants.

40. $\begin{vmatrix} -1 & 2 & 0 & 3 \\ 0 & 4 & -2 & 6 \\ 1 & 1 & 0 & 2 \\ 3 & -2 & 1 & 4 \end{vmatrix}$

41. $\begin{vmatrix} 3 & 2 & 0 & -4 \\ 1 & 0 & 1 & 6 \\ -1 & 6 & -2 & -3 \\ -2 & 3 & 0 & 2 \end{vmatrix}$

Section 6–6

Use Cramer's rule to solve the following systems of equations.

42. $3x - 7y = -27$
$4x + 9y = 19$

43. $2x - 11y = 10$
$x + 9y = 5$

44. $4x - 5y = 7$
$10x + 8y = -3$

45. $9x + 2y = 2$
$3x - 4y = -18$

46. $2x - y + 3z = -4$
$x + 4y - 2z = 8$
$3x - 2y + z = -11$

47. $-2x + y - 3z = -17$
$x + 5y + 6z = 1$
$3x - 4y + z = 26$

Section 6–7

M **48.** Find the forces shown in Figure 6–22 if the resulting system of equations is

$$0.64A - 0.5B = 0$$
$$0.77A + 0.87B = 200$$

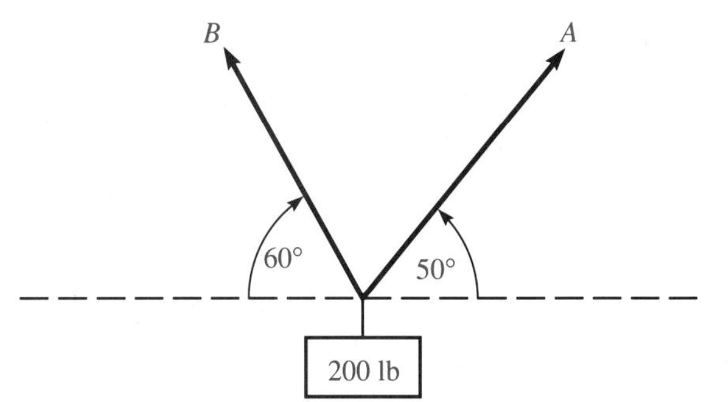

FIGURE 6–22

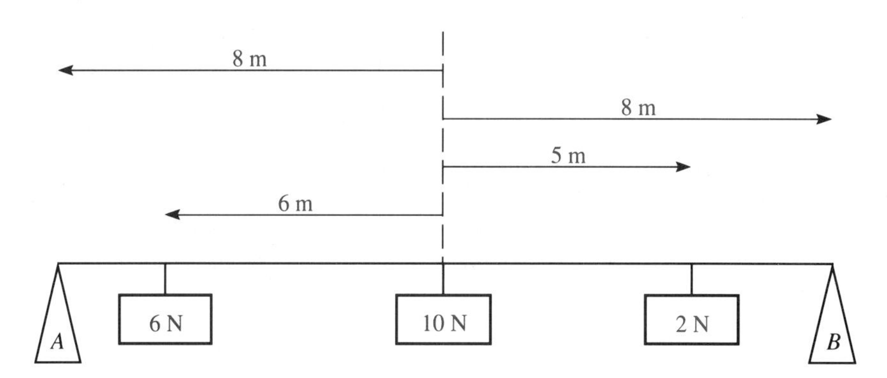

FIGURE 6–23

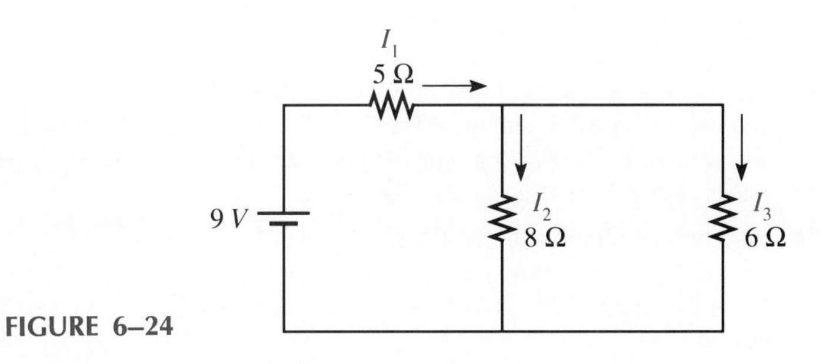

FIGURE 6–24

49. Find two numbers whose sum is 55 if the larger number is one more than twice the smaller number.

M 50. Find the weight supported by each end, A and B, shown in Figure 6–23.

E 51. Find currents I_1, I_2, and I_3 shown in Figure 6–24.

52. Jim has a collection of nickels, dimes, and quarters. He has a total of 24 coins worth \$3.90. Find how many of each coin Jim has if he has the same number of nickels and dimes.

53. Decompose the following into partial fractions:

$$\frac{34x - 18}{8x^2 - 2x - 3}$$

Factor the denominator first.

CHAPTER TEST

The number in parentheses refers to the appropriate learning objective given at the beginning of the chapter.

1. Use the addition method to solve the following system: **(3)**

$$3x + 2y = 4$$
$$4x + 5y = -11$$

2. Evaluate the following determinant: **(5)**

$$\begin{vmatrix} -1 & 3 & -2 \\ 2 & 0 & 4 \\ -3 & 1 & -1 \end{vmatrix}$$

3. Is the ordered pair $(-1, 0)$ the solution to the following system of equations: **(1)**

$$-6x + y = 6$$
$$3x + 7x = -3$$

Why or why not?

4. The sum of the angles of any triangle is 180°. In a certain triangle, the two smallest angles are of equal measure and the largest angle is 20° more than the sum of the other two angles. Find the angles of the triangle. **(7)**

5. Solve the following system of equations graphically: **(2)**

$$2x + 3y = -6$$
$$x - 4y = 13$$

6. Solve the following system of equations algebraically: **(3)**

$$2x + y - 3z = -5$$
$$x - 3y + 2z = -6$$
$$3x + 2y - z = 1$$

7. Solve the following system of equations by using the substitution method: **(3)**

$$6x + y = 2$$
$$2y = 10 - 6x$$

8. A grocer wants to mix coffee costing \$1.99/lb with coffee costing \$3.49/lb to make 50 lb of a coffee blend selling for \$2.80/lb. How many pounds of each coffee should he use? (7)

9. Evaluate the following determinant by using expansion by cofactors: (5)

$$\begin{vmatrix} -1 & 3 & 0 & 5 \\ 0 & 2 & -1 & 0 \\ 3 & 0 & -2 & 6 \\ 0 & 1 & 3 & 2 \end{vmatrix}$$

10. Use Cramer's rule to solve the following for x: (6)

$$3x - 5y = 13$$
$$4x + 7y = 10$$

11. Solve the following nonlinear system of equations: (4)

$$x^2 + y^2 = 16$$
$$4x^2 + y^2 = 16$$

12. Find forces A and B in Figure 6–25. (7)

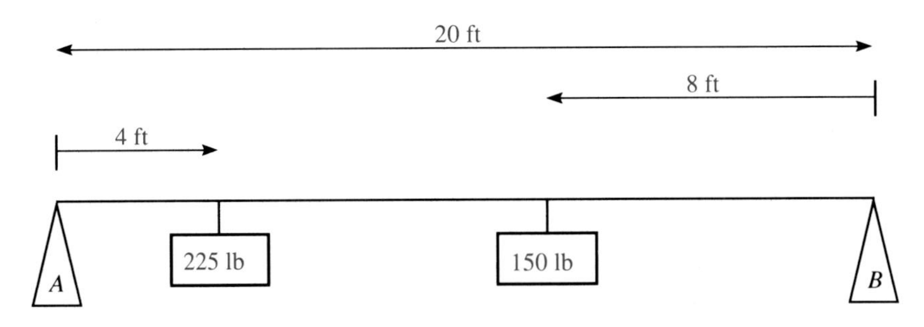

FIGURE 6–25

13. Solve the following system of equations for x and y: (3)

$$2x + y = 7$$
$$6x + 3y = 14$$

14. Solve the following system of equations graphically: (2)

$$3x - 5y = -11$$
$$4y = 5x + 14$$

15. Solve the following nonlinear system of equations: (4)

$$x^2 - y^2 = 9$$
$$2x^2 - 5y^2 = 6$$

SOLUTION TO CHAPTER INTRODUCTION

Using Kirchhoff's current law, we obtain the following equation:

$$I_1 = I_2 + I_3$$

Applying Kirchhoff's voltage law to the closed loops, we obtain the following equations:

$$6 - 7I_1 - 3I_2 = 0$$
$$6 - 7I_1 - 6I_3 = 0$$

The system of equations that you must solve is

$$\begin{aligned} I_1 - I_2 - I_3 &= 0 \\ 7I_1 + 3I_2 &= 6 \\ 7I_1 + 6I_3 &= 6 \end{aligned}$$

Using determinants to solve this system gives

$$D = \begin{vmatrix} 1 & -1 & -1 \\ 7 & 3 & 0 \\ 7 & 0 & 6 \end{vmatrix}$$

$$D = 81$$

$$D_1 = \begin{vmatrix} 0 & -1 & -1 \\ 6 & 3 & 0 \\ 6 & 0 & 6 \end{vmatrix}$$

$$D_1 = 54$$

$$D_2 = \begin{vmatrix} 1 & 0 & -1 \\ 7 & 6 & 0 \\ 7 & 6 & 6 \end{vmatrix}$$

$$D_2 = 36$$

$$D_3 = \begin{vmatrix} 1 & -1 & 0 \\ 7 & 3 & 6 \\ 7 & 0 & 6 \end{vmatrix}$$

$$D_3 = 18$$

The currents are $I_1 \approx 0.67$ A, $I_2 \approx 0.44$ A, and $I_3 \approx 0.22$ A.

You are responsible for designing an open box, with no waste, from a rectangular piece of cardboard 16 in. by 24 in. The box is to have a square base and a volume of 640 in.^{3} What should the dimensions of the box be? The cardboard comes in certain sizes, and you want to maximize the volume of the box to minimize waste. (The solution to this problem is given at the end of the chapter.)

To solve a problem of this type, you must know how to solve a higher-degree equation. In this chapter we will develop a technique for solving problems that result in equations of degree three or more.

Learning Objectives

After completing this chapter, you should be able to

1. Apply the Remainder Theorem (Section 7–1).
2. Use the Factor Theorem to determine whether a quantity is a factor of a given polynomial, a zero of a given function, or the root of a given equation (Section 7–1).
3. Perform synthetic division and use it to determine whether a quantity is a root, a zero, or a factor of a second quantity (Section 7–2).
4. Find the roots of a higher-degree equation (Sections 7–3 and 7–4).
5. Use Descartes' Rule of Signs to determine the maximum number of positive and negative roots of an equation (Section 7–4).
6. List all possible rational roots of a polynomial equation (Section 7–4).
7. Find irrational roots of a polynomial equation by linear approximation (Section 7–5).

Chapter 7

Solving Higher-Degree Equations

7–1

THE REMAINDER THEOREM AND THE FACTOR THEOREM

In previous chapters we discussed both linear and quadratic equations. Both types of equations are special cases of polynomial equations. A **polynomial equation** is of the form

$$f(x) = a_0x^n + a_1x^{n-1} + \cdots + a_n$$

where a_0, a_1, $\cdots$, a_n are constant, real numbers that serve as coefficients of the equation, and n is a nonnegative integer that determines the degree of the equation. In this chapter we will extend the techniques of solving equations to include equations of degree greater than two. Division will play a major role in solving these polynomial equations.

In polynomial division, the polynomial we are dividing by is called the **divisor;** the polynomial we divide into is called the **dividend;** and the answer is called the **quotient.** However, some polynomial division results in a **remainder.** Mathematically, these definitions can be written as follows:

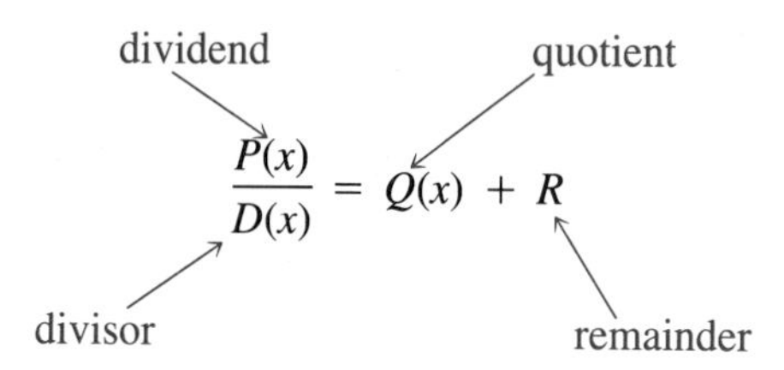

Remainder Theorem

The **Remainder Theorem** states that *if a polynomial P(x) is divided by the linear factor $x - r$, then $f(r) = R$*. For example, if we divide $3x^2 - 2x + 5$ by $x - 3$, the remainder is 26. Similarly, if we evaluate $f(3)$, the answer is 26. Thus, $f(3)$ equals the remainder resulting from division by $x - 3$.

NOTE ✦ Recall from Chapter 2 that $f(r)$ is functional notation denoting the value of the function when r is substituted for the variable.

EXAMPLE 1 Verify the Remainder Theorem by dividing $x^2 - 6x - 8$ by $x + 1$ and then evaluating $f(-1)$.

Solution First, dividing $x^2 - 6x - 8$ by $x + 1$ gives

$$\begin{array}{r} x - 7 \\ x + 1\overline{)x^2 - 6x - 8} \\ \ominus \quad \ominus \quad\quad \\ \underline{x^2 + x} \quad\quad \\ -7x - 8 \\ \oplus \quad \oplus \\ \underline{-7x - 7} \\ -1 \end{array}$$

The remainder resulting from this division is -1. Evaluating $f(-1)$ (as we did in Chapter 2) gives

$$\begin{aligned} f(x) &= x^2 - 6x - 8 \\ f(-1) &= (-1)^2 - 6(-1) - 8 \\ f(-1) &= 1 + 6 - 8 \\ f(-1) &= -1 \end{aligned}$$

Since the remainder from polynomial division and $f(r)$ are equal, the Remainder Theorem is verified for this example. ■

CAUTION ✦ For the Remainder Theorem to apply, the divisor must be in the form $x - r$.

EXAMPLE 2 Using the Remainder Theorem, determine the remainder when $2x^3 + 5x^2 - 6x + 8$ is divided by $x - 3$.

Solution We will find the remainder by evaluating $f(r)$. First, we must determine the value of r. Since $x - r = x - 3$, $r = 3$, and we evaluate $f(3)$.

$$\begin{aligned} f(x) &= 2x^3 + 5x^2 - 6x + 8 \\ f(3) &= 2(3)^3 + 5(3)^2 - 6(3) + 8 \\ f(3) &= 54 + 45 - 18 + 8 \\ f(3) &= 89 \end{aligned}$$

Since $f(3) = 89$, then by the Remainder Theorem, the remainder is also 89. ■

Factor Theorem

An important special case of the Remainder Theorem is the **Factor Theorem,** which states that if $f(r) = R = 0$, then $(x - r)$ is a factor of the polynomial. This theorem is very important in solving higher-degree equations because it states that if the remainder is zero,

- ☐ $(x - r)$ is a factor of the polynomial;
- ☐ $x = r$ is a root (solution) of the polynomial equation; and
- ☐ r is a zero (the x intercept of the graph) of the polynomial function.

EXAMPLE 3 Determine if $x + 4$ is a factor of the following polynomial:

$$f(x) = 2x^3 - 6x + 8$$

Solution According to the Factor Theorem, $x + 4$ is a factor of $2x^3 - 6x + 8$ if and only if the remainder is zero. To determine the remainder, we evaluate $f(-4)$.

$$\begin{aligned} f(x) &= 2x^3 - 6x + 8 \\ f(-4) &= 2(-4)^3 - 6(-4) + 8 \\ f(-4) &= -128 + 24 + 8 \\ f(-4) &= -96 \end{aligned}$$

Since the remainder is *not* zero, then $x + 4$ is *not* a factor of $2x^3 - 6x + 8$. Furthermore, -4 is not a zero of the polynomial function, and -4 is not a root of the equation $2x^3 - 6x + 8 = 0$. ■

CAUTION ✦ Be sure to distinguish between the r in the expression $x - r$ and $f(r)$. When determining zeros, roots, remainders, and factors by evaluating $f(r)$, always be sure that you have solved for r.

EXAMPLE 4 Determine if -1 is a root of the following polynomial equation:

$$6x^4 + 5x^3 - x^2 + 3 = 0$$

Solution To determine if -1 is a root of the polynomial equation, we evaluate $f(-1)$ and determine if the result is zero.

$$f(-1) = 6(-1)^4 + 5(-1)^3 - (-1)^2 + 3$$
$$f(-1) = 6 - 5 - 1 + 3$$
$$f(-1) = 3$$

Since $f(-1) \neq 0$, -1 is not a root of the polynomial equation. ■

EXAMPLE 5 Determine if 2 is a zero of the following polynomial function:

$$3x^3 - 8x^2 + 5x - 2$$

Solution We must find $f(2)$ and determine if the result is zero.

$$f(2) = 3(2)^3 - 8(2)^2 + 5(2) - 2$$
$$f(2) = 24 - 32 + 10 - 2$$
$$f(2) = 0$$

Since $f(2) = 0$, 2 is a zero of the polynomial function. In addition, we also know that $x - 2$ is a factor of the polynomial $3x^3 - 8x^2 + 5x - 2$ and 2 is a root of the polynomial equation $3x^3 - 8x^2 + 5x - 2 = 0$. ■

7–1 EXERCISES

Verify the Remainder Theorem by using long division and evaluating $f(r)$.

1. $(2x^3 + x^2 - 3x + 1) \div (x - 2)$

2. $(5x^3 + 6x^2 - 3x + 1) \div (x - 3)$

3. $(6x^3 + 8x - 3) \div (x + 3)$

4. $(2x^3 - 7x + 3) \div (x + 4)$

5. $(12x^3 - 7x^2 + 16x + 12) \div (3x - 1)$

6. $(3x^3 + 4x^2 - x + 5) \div (x - 1)$

7. $(x^3 + 5x^2 - 3) \div (x - 1)$

8. $(10x^3 + 28x^2 - 41x + 21) \div (5x - 1)$

9. $(x^4 - x^3 + 7x - 8) \div (x + 3)$

10. $(3x^4 + 8x^2 - 7x + 5) \div (x + 1)$

11. $(2x^4 + 8x^2 - 3x + 1) \div (x + 2)$

12. $(x^4 + 6x^3 - 4x^2 + 8) \div (x - 7)$

13. $(15x^4 + 10x^3 + 21x^2 + 5x - 8) \div (3x + 2)$

14. $(6x^4 - 5x^3 - 12x^2 + 28x + 7) \div (6x - 5)$

Using the Remainder Theorem, find the remainder R.

15. $(3x^3 + 6x^2 - 4x + 1) \div (x + 1)$

16. $(7x^3 - 4x^2 + 6x - 2) \div (x + 1)$

17. $(x^3 + x^2 + x - 6) \div (x - 2)$

18. $(4x^4 + 5x^2 - 9x + 8) \div (x - 4)$

19. $(2x^3 - 7x^2 + 5x + 1) \div (x - 4)$

20. $(x^3 - 3x^2 + 9) \div (x - 3)$

21. $(6x^3 + 8x - 3) \div (x + 5)$

22. $(2x^3 - 6x + 14) \div (x + 7)$

23. $(2x^4 - 5x^2 + 7x + 9) \div (x + 2)$

24. $(3x^4 + 2x^3 - x + 6) \div (x + 3)$

25. $(x^4 + 6x^3 - 2x + 10) \div (x - 1)$

26. $(5x^4 - 4x^2 + 8) \div (x + 4)$

27. $(x^3 + 3x^2 - 8) \div (2x - 1)$

28. $(x^3 + 5x - 7) \div (2x + 3)$

29. $(3x^3 + 2x - 6) \div (3x - 2)$

30. $(2x^3 + 7x^2 + 9) \div (4x - 1)$

Use the Factor Theorem to determine if the second expression is a factor of the first.

31. $7x^3 - 6x^2 + 9x - 1;\ x + 1$

32. $x^3 - 3x^2 - 4x + 12;\ x + 2$

33. $3x^3 + 18x^2 - 47x + 8;\ x + 8$

34. $x^3 + 8x - 5;\ x - 1$

35. $9x^3 + 18x^2 - 13x + 2;\ 3x - 1$
(Hint: Let $r = 1/3$.)

36. $2x^3 - 6x^2 + 3;\ x + 3$

37. $x^4 - 2x^3 - 5x^2 + 17x + 14;\ x - 2$

38. $x^4 + 5x^3 + 3x^2 - 6x - 8;\ x + 4$

39. $2x^4 - 6x^3 + 8x^2 + 5x - 15;\ x - 3$

40. $3x^4 + x^3 - 5x + 8;\ x - 2$

41. $6x^3 - 3x^2 + 14x - 10;\ 2x - 1$

42. $2x^3 - 7x^2 - 13x + 3;\ 2x + 3$

43. $3x^3 - 13x^2 - 7x + 2;\ 3x + 2$

44. $4x^3 - 3x^2 - 6x + 18;\ 4x - 3$

Determine if the given number is a zero of the function $f(x)$.

45. $2x^3 - 7x + 8;\ -2$

46. $2x^3 + x^2 - 14x + 3;\ -3$

47. $x^3 + 3x^2 - 19x + 3;\ 3$

48. $6x^3 - 7;\ -2$

49. $3x^3 - 2x^2 + 2x + 7;\ -1$

50. $x^3 + 6x^2 + 5x - 12;\ -4$

51. $x^4 + x^3 - 30x^2 - 4x + 20;\ 5$

52. $x^4 - 1;\ -1$

53. $x^4 + 8x^2 - 7x + 6;\ 2$

54. $3x^4 - 7x^2 + 8x - 1;\ 1$

55. $6x^3 + x^2 - 22x + 7;\ 1/3$

56. $2x^4 - 5x^3 - 8x^2 + 24x - 10;\ 5/2$

57. $3x^4 - 5x^3 + 9x - 6;\ 1/2$

58. $x^3 + 2x^2 - 3x + 6;\ -1/3$

Determine if the given number is a root of the equation.

59. $5x^3 + 15x^2 - 6x - 18 = 0;\ -3$

60. $3x^3 + 5x^2 - 7 = 0;\ -1$

61. $2x^3 + 13x^2 + x - 30 = 0;\ -6$

62. $3x^3 + 5x^2 - 10x + 2 = 0;\ 1$

63. $x^3 - 4x^2 - 22x + 7 = 0;\ -7$

64. $2x^3 - 4x^2 + 7x - 2 = 0;\ 2$

65. $x^3 - 6x^2 + 8 = 0;\ -4$

66. $5x^3 + 9x - 11 = 0;\ -3$

67. $x^4 + 3x^3 - 4x^2 - 32 = 0;\ 2$

68. $3x^3 + 22x^2 + 18x - 8 = 0;\ -4/3$

69. $2x^3 - 5x^2 + 6x - 9 = 0;\ -1/2$

70. $2x^3 + 3x^2 - 11x + 3 = 0;\ 3/2$

7–2

SYNTHETIC DIVISION

In this section we will discuss a shortcut to polynomial long division called **synthetic division.** With a higher-degree polynomial, the computation to determine $f(r)$ is difficult. Generally, synthetic division is much easier than evaluating $f(r)$ or performing long division. We develop the method of synthetic division in the next example.

EXAMPLE 1 Divide $3x^3 + 5x^2 - x + 4$ by $x + 2$.

Solution

$$
\begin{array}{r|rrrr}
 & 3x^2 & - & x & + 1 \\
\hline
x + 2 & 3x^3 + 5x^2 & - & x & + 4 \\
 & \ominus \quad \ominus & & & \\
 & 3x^3 + 6x^2 & & & \\
\cline{2-2}
 & -x^2 & - & x & \\
 & \oplus & & \oplus & \\
 & -x^2 & - & 2x & \\
\cline{2-4}
 & & & x & + 4 \\
 & & & \ominus & \ominus \\
 & & & x & + 2 \\
\cline{4-5}
 & & & & 2
\end{array}
$$

In performing this division, we need to consider only the coefficients, and many of them are repeated. Let us rewrite the long division from above, omitting the variables and any repeated coefficients.

$$
\begin{array}{r|rrrr}
 & & 3 & -1 & 1 \\
\hline
2 & 3 & 5 & -1 & 4 \\
 & & 6 & & \\
\cline{2-3}
 & & -1 & & \\
 & & & -2 & \\
\cline{3-4}
 & & & 1 & \\
 & & & & 2 \\
\cline{4-5}
 & & & & 2
\end{array}
$$

Then let us compress this problem by moving all the numbers below the dividend into two rows.

$$
\begin{array}{r|rrrr}
 & & \textcircled{3} & \textcircled{-1} & \textcircled{1} \\
\hline
2 & 3 & 5 & -1 & 4 \\
 & & 6 & -2 & 2 \\
\hline
 & & -1 & 1 & \textcircled{2}
\end{array}
$$

All the coefficients of the quotient and the remainder appear in the bottom line except for the first one. Therefore, we repeat the first coefficient, 3, in the bottom line and omit the quotient above the dividend. Also, to avoid subtraction, we change the divisor of 2 to -2, the actual value of r, and change each sign in the second row.

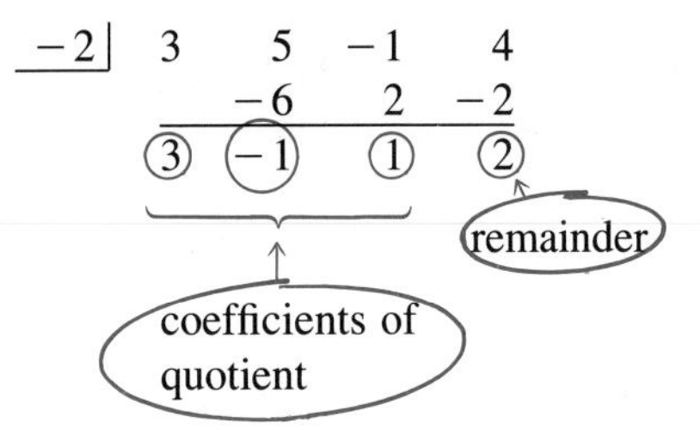

The first three numbers in the bottom row represent the coefficients of the quotient, and the last number represents the remainder. To write the quotient in polynomial form, we write the variable in descending powers, starting with an exponent one less than the degree of the dividend. In this polynomial, the degree of the dividend is three; therefore, the degree of the quotient is two. Filling in descending powers of x starting with 2 gives the quotient as $3x^2 - x + 1$ with a remainder of 2. ■

Example 2 shows the step-by-step procedure for synthetic division.

EXAMPLE 2 Use synthetic division to divide $2x^4 - 6x^3 + x^2 + 7x - 5$ by $x - 1$.

Solution First, write the dividend coefficients in descending powers and find the value of r.

$$\underline{1}| \quad 2 \quad -6 \quad 1 \quad 7 \quad -5$$

Bring down the first dividend coefficient.

$$\begin{array}{r|rrrrr} 1 & 2 & -6 & 1 & 7 & -5 \\ & \downarrow & & & & \\ \hline & 2 & & & & \end{array}$$

Then multiply this coefficient by r, placing the product under the next dividend coefficient.

$$\begin{array}{r|rrrrr} 1 & 2 & -6 & 1 & 7 & -5 \\ & & 2 & & & \\ \hline & 2 & & & & \end{array}$$

multiply

Add the result.

$$\begin{array}{r|rrrrr} 1 & 2 & -6 & 1 & 7 & -5 \\ & & 2 & & & \\ \hline & 2 & -4 & \text{add} & & \end{array}$$

Then continue this process until all the dividend coefficients have been used.

$$\begin{array}{r|rrrrr} 1 & 2 & -6 & 1 & 7 & -5 \\ & & 2 & -4 & -3 & 4 \\ \hline & 2 & -4 & -3 & 4 & -1 \end{array}$$

Write the quotient using the coefficients in the bottom row. Insert powers of x, starting with a power one less than the degree of the original polynomial.

$$\begin{array}{ccccc} 2 & -4 & -3 & 4 & -1 \\ \downarrow & \downarrow & \downarrow & \downarrow & \downarrow \\ x^3 & x^2 & x & x^0 & \text{remainder} \end{array}$$

The quotient is $2x^3 - 4x^2 - 3x + 4$ with a remainder of -1. ■

Synthetic Division

1. Arrange the coefficients of the dividend in descending powers, placing a zero for any missing terms. Determine r and place it to the left of the dividend coefficients.
2. Bring down the first coefficient of the dividend.
3. Multiply r by this coefficient and place the product under the next dividend coefficient. Add the result.
4. Repeat Step 3 until all dividend coefficients have been used.

Function Values

Since the Remainder Theorem equates $f(r)$ and R, we can determine a function value by using synthetic division to determine the remainder.

EXAMPLE 3 Find $f(-3)$ if $f(x) = 7x^4 + 3x^3 - 5x + 6$.

Solution We arrange the dividend coefficients in descending powers and insert a 0 for the missing x^2 term.

$$7 \quad 3 \quad 0 \quad -5 \quad 6$$

Dividing gives

$$\begin{array}{r|rrrrr} -3 & 7 & 3 & 0 & -5 & 6 \\ & & -21 & 54 & -162 & 501 \\ \hline & 7 & -18 & 54 & -167 & 507 \end{array}$$

Because the remainder is 507, $f(-3) = 507$. You can check this result by evaluating $f(-3) = 7(-3)^4 + 3(-3)^3 - 5(-3) + 6$. ■

NOTE ✦ Graphically, $f(-3) = 507$ translates into the ordered pair $(-3, 507)$. Therefore, we could also use synthetic division to fill in a table of values for graphing.

Testing Factors

The Factor Theorem states that $x - r$ is a factor of a polynomial if the remainder is zero. We can also use synthetic division to test factors.

EXAMPLE 4 Use synthetic division to determine if $4x + 1$ is a factor of the following polynomial function:

$$12x^4 + 11x^3 - 2x^2 + 11x + 3$$

Solution The Factor Theorem is valid only for factors of the form $x - r$. To test a factor such as $ax - b$, write the factor as $a\left(x - \frac{b}{a}\right)$ and use synthetic division to test $\left(x - \frac{b}{a}\right)$. If the remainder is zero and the quotient is divisible by a, then $ax - b$ is a factor of the polynomial. We write $4x + 1$ as $4\left(x + \frac{1}{4}\right)$ and divide by $-1/4$.

$$\begin{array}{r|rrrrr} -\frac{1}{4} & 12 & 11 & -2 & 11 & 3 \\ & & -3 & -2 & 1 & -3 \\ \hline & 12 & 8 & -4 & 12 & 0 \end{array}$$

Since the remainder is zero and the quotient is divisible by 4, $4x + 1$ is a factor of $12x^4 + 11x^3 - 2x^2 + 11x + 3$. ■

Testing Zeros

From the previous section, we know that r is a zero of a polynomial function if the remainder is zero. We can determine the remainder by synthetic division.

EXAMPLE 5 Determine if $\frac{1}{2}$ is a zero of the function $6x^3 + 13x^2 - 26x + 9$.

Solution To determine if $\frac{1}{2}$ is a zero, we perform synthetic division using $r = \frac{1}{2}$ and look at the remainder.

$$\begin{array}{r|rrrr} \frac{1}{2} & 6 & 13 & -26 & 9 \\ & & 3 & 8 & -9 \\ \hline & 6 & 16 & -18 & 0 \end{array}$$

Since the remainder is zero, $\frac{1}{2}$ is a zero of the function $6x^3 + 13x^2 - 26x + 9$. ■

7–2 EXERCISES

Divide by using synthetic division.

1. $(x^3 + 6x^2 - x + 9) \div (x - 3)$
2. $(x^3 + 5x^2 - 3x + 8) \div (x + 2)$
3. $(2x^3 - 5x^2 + 3x + 8) \div (x + 1)$
4. $(6x^3 - x^2 + 2x - 7) \div (x - 4)$
5. $(3x^3 - 7x^2 + x - 9) \div (x - 3)$
6. $(x^3 + 9x^2 + 7x - 11) \div (x - 2)$
7. $(4x^3 - x^2 + 6x - 7) \div (x - 1)$
8. $(7x^3 + x^2 - 8x - 12) \div (x + 1)$
9. $(x^3 + 3x^2 - 10) \div (x - 2)$
10. $(2x^3 + 7x^2 - 9) \div (x + 3)$
11. $(9x^3 - 8x + 11) \div (x - 4)$
12. $(9x^3 + 11 \quad 5) \div (x - 4)$
13. $(3x^4 - 2x^3 + 7x^2 - x + 9) \div (x - 5)$
14. $(x^4 - 6x^3 + 2x^2 + 11x - 8) \div (x - 1)$
15. $(x^4 - 7x^2 + 8) \div (x + 2)$
16. $(6x^4 - 7x + 10) \div (x - 6)$
17. $(x^4 + x^3 - 3x^2 + 2x - 6) \div (x + 3)$
18. $(2x^4 + 9x^3 + 5x - 3) \div (x + 5)$
19. $(2x^5 - 6x^3 - x + 5) \div (x - 1)$
20. $(3x^5 - 8x^2 + 4) \div (x + 2)$
21. $(x^5 + x^4 - 8x^2 + 6) \div (x + 4)$
22. $(x^6 + 4x^4 - x^3 + 8) \div (x + 4)$
23. $(3x^6 - 3x^3 + 8) \div (x - 2)$
24. $(2x^6 - 2x^2 + 9) \div (x + 1)$

Using synthetic division, determine if the second expression is a factor of the first.

25. $x^3 + 3x^2 - 16x + 12; x + 6$
26. $2x^3 - 9x^2 - 8x + 15; x - 5$
27. $3x^3 - 10x^2 - 9x + 4; x - 4$
28. $2x^3 + 7x^2 - 6x - 3; x + 3$
29. $x^3 + 6x^2 + 8x - 7; x + 7$
30. $x^3 - x^2 - 4x + 4; x - 1$
31. $4x^3 + x^2 - x + 2; x + 1$
32. $3x^3 + 7x^2 - 8x + 5; x - 3$
33. $2x^3 - 9x + 15; x - 5$
34. $x^3 + 8x - 9; x + 1$
35. $x^4 + 8x^3 - x^2 + 3x + 10; x + 2$
36. $x^4 + 2x^2 - 24; x - 2$
37. $x^4 + 3x^3 - 3x - 9; x + 3$
38. $x^4 + 6x^2 - 12; x - 4$
39. $2x^4 + 3x^3 - 2x^2 - x - 6; 2x + 3$
40. $6x^4 - 2x^3 - 7x + 7; 3x - 1$
41. $3x^3 + 2x^2 - 7x + 2; 3x - 1$
42. $2x^4 - x^3 - 2x + 1; 2x - 1$

Using synthetic division, determine if the given number is a zero of the given function $f(x)$.

43. $3x^3 + 8x - 9; 3$
44. $5x^3 + 9x^2 + 7x + 18; -2$
45. $x^3 - 4x^2 - x + 4; 4$
46. $x^3 + 8x^2 - 4x + 11; -1$
47. $3x^4 - 2x^3 - 3x^2 + 8x - 4; 2/3$
48. $2x^4 - 7x^3 + 9x - 5; -1/2$
49. $2x^4 + 5x^3 - 2x - 3; -3/2$
50. $5x^4 - 2x^3 - 5x^2 + 7x - 2; 2/5$

7–3

ROOTS OF AN EQUATION

In this section we will discuss solving equations of degree three or more. In solving higher-degree equations, you will find it advantageous to know as much as possible about the number and type of roots. We will discuss several rules that will prove helpful in this area.

Number of Roots

The first rule, called the **Fundamental Theorem of Algebra,** *states that every polynomial equation has at least one root,* which may be a real or a complex number. By this theorem, the equation $f(x) = 0$ of degree n has at least one root r_1, and $(x - r_1)$ is a factor of $f(x)$. Thus, we can write

$$f(x) = (x - r_1)q_1(x) = 0$$

where $q_1(x)$ is a polynomial of degree $n - 1$. However, by the Fundamental Theorem, $q_1(x)$ has at least one root r_2. Thus, $f(x)$ can be written

$$f(x) = (x - r_1)(x - r_2)q_2(x) = 0$$

This process continues until $f(x)$ can be written in terms of only linear factors. As a result of the Fundamental Theorem of Algebra, we know the following:

- If $f(x)$ is of degree n, there will be n linear factors.
- Every polynomial equation of degree n has exactly n roots.

For example, this theorem states that the equation $6x^4 - 3x^2 + 5x - 1 = 0$ has exactly 4 linear factors and 4 roots. However, this theorem does not state that all n roots are different. When a root occurs m times, the root is said to have a **multiplicity** of m. For example, if $f(x)$ can be written as

$$f(x) = (x - 1)(x + 5)^2(x - 8)^3 = 0$$

then −[illegible] a root of multiplicity two and 8 is a root of multiplicity three.

Reducing Equations to Quadratic Form

If we know all the roots but two in solving a higher-degree equation, we can use the methods of solving quadratic equations to find the remaining roots.

EXAMPLE 1 Solve the equation $x^3 - 2x^2 - 5x + 6 = 0$ if 3 is one of the roots.

Solution Since the degree of this equation is three, there are three roots, one of which is the number 3. Performing synthetic division by 3 gives

$$\begin{array}{r|rrrr} 3 & 1 & -2 & -5 & 6 \\ & & 3 & 3 & -6 \\ \hline & 1 & 1 & -2 & 0 \end{array}$$

Then we can write the original equation as the product of the following factors:

$$x^3 - 2x^2 - 5x + 6 = (x - 3)(x^2 + x - 2)$$

The remaining two solutions are roots of the quadratic equation $x^2 + x - 2 = 0$. We can solve this quadratic equation by factoring.

$$x^2 + x - 2 = 0$$
$$(x + 2)(x - 1) = 0$$

The original equation can be written in terms of the following linear factors:

$$(x - 3)(x + 2)(x - 1) = 0$$

The roots of the equation are 3, -2, and 1. ■

EXAMPLE 2 Solve the equation $x^4 + 4x^3 - 7x^2 - 4x + 6 = 0$ given that -1 and 1 are roots. Round the solution to hundredths.

Solution Since -1 and 1 are both roots of the equation, we ause synthetic division twice to yield a quadratic equation.

$$\begin{array}{r|rrrrr} -1 & 1 & 4 & -7 & -4 & 6 \\ & & -1 & -3 & 10 & -6 \\ \hline 1 & 1 & 3 & -10 & 6 & 0 \\ & & 1 & 4 & -6 & \\ \hline & 1 & 4 & -6 & 0 & \end{array}$$

The resulting quadratic equation is $x^2 + 4x - 6 = 0$ which we can solve by using the quadratic formula.

$$x = \frac{-4 \pm \sqrt{16 - 4(1)(-6)}}{2}$$

$$x = \frac{-4 \pm 2\sqrt{10}}{2}$$

$$x = -2 \pm \sqrt{10}$$

2 [+/−] [+] 10 [√] [=] → 1.162

2 [+/−] [−] 10 [√] [=] → −5.162

The four roots of the equation are 1, -1, $-2 + \sqrt{10} \approx 1.16$, and $-2 - \sqrt{10} \approx -5.16$. ■

Complex Roots

The Fundamental Theorem of Algebra states that the roots of an equation may be complex numbers. *Complex roots will always occur in conjugate pairs*. For example, if $a + bj$ is a root, then $a - bj$ is also a root of the equation.

EXAMPLE 3 Solve the equation $2x^4 + 7x^3 + 6x^2 + 4x + 8 = 0$ if -2 is a root of multiplicity two.

Solution Using synthetic division twice gives

$$\begin{array}{r|rrrrr} -2 & 2 & 7 & 6 & 4 & 8 \\ & & -4 & -6 & 0 & -8 \\ \hline -2 & 2 & 3 & 0 & 4 & 0 \\ & & -4 & 2 & -4 & \\ \hline & 2 & -1 & 2 & 0 & \end{array}$$

The resulting quadratic equation is $2x^2 - x + 2 = 0$ which we solve by the quadratic formula.

$$x = \frac{1 \pm \sqrt{1 - 4(2)(2)}}{4}$$

$$x = \frac{1 \pm \sqrt{-15}}{4}$$

$$x = \frac{1 \pm j\sqrt{15}}{4}$$

The roots of the equation are -2, -2, $\frac{1 + j\sqrt{15}}{4}$, and $\frac{1 - j\sqrt{15}}{4}$. ■

EXAMPLE 4 Solve the equation $x^4 - 5x^3 + 12x^2 - 20x + 32 = 0$ given that $2j$ is a root of the equation.

Solution Since $2j$ is a root, $-2j$ is also a root of the equation. We now use synthetic division twice, dividing by $2j$ and then $-2j$, to reduce to a quadratic equation. (Remember from Chapter 4 that $j = \sqrt{-1}$. Squaring both sides of this equation gives $j^2 = -1$.)

$$\begin{array}{r|rrrrr} 2j & 1 & -5 & +12 & -20 & +32 \\ & & 0 + 2j & -4 - 10j & 20 + 16j & -32 \\ \hline -2j & 1 & -5 + 2j & 8 - 10j & 0 + 16j & 0 \\ & & 0 - 2j & 0 + 10j & 0 - 16j & \\ \hline & 1 & -5 & 8 & 0 & \end{array}$$

The resulting quadratic equation is $x^2 - 5x + 8 = 0$ which we solve by the quadratic formula.

$$x = \frac{5 \pm \sqrt{25 - 4(1)(8)}}{2}$$

$$x = \frac{5 \pm \sqrt{-7}}{2}$$

$$x = \frac{5 \pm j\sqrt{7}}{2}$$

The roots of the equation are $2j$, $-2j$, $\frac{5 + j\sqrt{7}}{2}$, and $\frac{5 - j\sqrt{7}}{2}$. ■

7–3 EXERCISES

Solve each equation given the roots indicated.

1. $x^3 + 2x^2 - 10x - 56 = 0$; 4
2. $x^3 + 6x^2 + 9x + 200 = 0$; -8
3. $2x^3 + 3x^2 - 2x + 21 = 0$; -3
4. $2x^3 + 11x^2 - x + 30 = 0$; -6
5. $4x^3 - x^2 - 11x + 8 = 0$; 1
6. $3x^3 + 16x^2 + 6x + 5 = 0$; -5
7. $6x^3 + 11x^2 + 3x + 10 = 0$; -2
8. $2x^3 - 2x^2 - 13x + 3 = 0$; 3
9. $x^3 - 12x^2 + 44x - 45 = 0$; 5
10. $4x^3 - 7x^2 + 8x - 5 = 0$; 1
11. $2x^3 - 11x^2 + 25x - 22 = 0$; 2
12. $x^3 + 9x^2 + 17x - 12 = 0$; -4
13. $x^4 + 4x^3 - 7x^2 - 34x - 24 = 0$; -2, -1
14. $x^4 - 2x^3 - 8x^2 + 19x - 6 = 0$; -3, 2
15. $x^4 + 2x^3 - 12x^2 - 40x - 32 = 0$; -2 is a root of multiplicity two.
16. $x^4 - 7x^3 + 17x^2 - 17x + 6 = 0$; 1 is a root of multiplicity two.
17. $x^4 + 2x^3 - 32x^2 + 30x + 63 = 0$; 3 is a root of multiplicity two.
18. $x^4 + x^3 - 46x^2 - 4x + 168 = 0$; 6, -7
19. $x^4 - 4x^3 - 13x^2 + 28x + 60 = 0$; 5, 3
20. $x^4 - 6x^3 + 5x^2 - 8x + 80 = 0$; 4 is a root of multiplicity two.
21. $x^4 + 5x^2 - 36 = 0$; $3j$
22. $x^4 + 3x^3 + 6x^2 + 3x + 5 = 0$; $-j$
23. $x^4 - x^3 - 9x^2 - 11x - 4 = 0$; 4, -1
24. $x^4 + 2x^3 - 14x^2 + 7x + 10 = 0$; 2, -5
25. $x^4 + 2x^3 - 10x^2 - 17x + 6 = 0$; -2, 3
26. $x^4 + 4x^3 - 53x^2 - 60x + 108 = 0$; 6, -2
27. $x^3 - 7x^2 + 12x - 10 = 0$; $1 + j$
28. $x^4 - 3x^3 + 6x^2 - 3x + 5 = 0$; j
29. $x^4 + x^3 + x^2 + 4x - 12 = 0$; $2j$
30. $x^4 + 7x^3 + 19x^2 + 63x + 90 = 0$; $3j$

7–4 RATIONAL ROOTS

In the previous section we solved higher-degree equations when we knew enough roots to obtain a quadratic equation. However, in reality we seldom initially know any of the roots of a higher-degree equation. Finding the rational roots of a higher-degree equation by trial and error can be a lengthy process, but in this section we will develop some guidelines to shorten this process. We begin the discussion with a rule.

RULE ✦ If the coefficients of the equation

$$f(x) = a_0x^n + a_1x^{n-1} + a_2x^{n-2} + \cdots + a_n = 0$$

are integers, then any rational roots are factors of a_n divided by factors of a_0.

This rule does not guarantee the existence of rational roots, but it does limit the number of trials we must make in finding rational roots. Moreover, if $a_0 = 1$, we need try only the factors of a_n, the constant term.

EXAMPLE 1 List all possible rational roots of the following equation:

$$x^3 - x^2 - 14x + 24 = 0$$

Solution Since the coefficient of the highest-degree term is 1, any rational roots must be factors of the constant term, 24. The possible rational roots are ± 1, ± 2, ± 3, ± 4, ± 6, ± 8, ± 12, and ± 24. We must include the plus-and-minus sign before each factor because, for example, both $+2$ and -2 could be roots of the equation. ■

EXAMPLE 2 List all possible rational roots of the following equation:

$$4x^4 - 7x^3 + 9x^2 - x + 10 = 0$$

Solution The possible rational roots are all combinations of *factors of 10 divided by factors of 4*. The factors of 10 are ± 1, ± 2, ± 5, and ± 10, and the factors of 4 are ± 1, ± 2, and ± 4. The possible rational roots are

$$\pm 1, \quad \pm 2, \quad \pm 5, \quad \pm 10, \quad \pm 1/2, \quad \pm 5/2, \quad \pm 1/4, \quad \pm 5/4$$ ■

Descartes' Rule of Signs

To further aid in limiting the number of possible rational roots, we can use **Descartes' Rule of Signs,** which states that the number of positive real roots of $f(x) = 0$ cannot exceed the number of sign changes in the terms of $f(x)$, and the number of negative roots cannot exceed the number of sign changes in the terms of $f(-x)$.

EXAMPLE 3 Determine the maximum number of positive and negative roots for the following equation:

$$3x^3 + 4x^2 - 6x + 8 = 0$$

Solution To determine the maximum number of positive, real roots, we write the sign associated with each term of $f(x)$ as follows:

$$f(x) = 3x^3 + 4x^2 - 6x + 8$$

+ + − +
(1) (2)

Since the sign changes twice, the number of positive, rational roots cannot exceed two. To determine the maximum number of negative roots, we determine the number of sign changes in $f(-x)$. To do so, we replace x with $-x$ in the equation.

$$f(x) = 3x^3 + 4x^2 - 6x + 8$$
$$f(-x) = 3(-x)^3 + 4(-x)^2 - 6(-x) + 8$$
$$f(-x) = -3x^3 + 4x^2 + 6x + 8$$

The signs are

$$- \quad + \quad + \quad +$$
(1)

Since there is one sign change, there is only one negative, rational root. ■

Maximum and Minimum Values

One additional rule will help restrict the number of rational numbers that we must consider in finding the roots of $f(x) = 0$. If we divide $f(x)$ by a number and the quotient has only positive signs, that number is a maximum of the roots. Similarly, if we divide $f(x)$ by a number and the signs on the quotient alternate, that number is a minimum of the roots.

EXAMPLE 4 Determine the roots of the following equation:

$$2x^4 + x^3 - 35x^2 - 28x + 96 = 0$$

Solution First, we determine the maximum number of positive and negative roots.

$f(x)$ $+ \quad + \quad - \quad - \quad +$ (1) (2) — a maximum number of two positive roots

$f(-x)$ $+ \quad - \quad - \quad + \quad +$ (1) (2) — a maximum number of two negative roots

Second, we list the possible factors, the quotient of factors of 96 and factors of 2. The possible factors are

$$\pm 1, \quad \pm 2, \quad \pm 3, \quad \pm 4, \quad \pm 6, \quad \pm 8, \quad \pm 12, \quad \pm 16,$$
$$\pm 24, \quad \pm 32, \quad \pm 48, \quad \pm 96, \quad \pm 1/2, \quad \pm 3/2$$

Third, we test the possibilities until we find all but two roots. We begin by testing positive roots. With an extensive list of possibilities, let us choose a number in the middle of the list, 6.

6	2	1	−35	−28	96
		12	78	258	1,380
	+2	+13	+43	+230	+1,476

Although 6 is not a root, the quotient contains only positive signs; therefore, 6 is a maximum value for the roots. Next, let us test 4.

$$\begin{array}{r|rrrrr} 4 & 2 & 1 & -35 & -28 & 96 \\ & & 8 & 36 & 4 & -96 \\ \hline & 2 & 9 & 1 & -24 & 0 \end{array}$$

We could continue to test another positive root, but let us test some negative values since there could be two negative roots. Again, let us test a number in the middle of the list, -6.

$$\begin{array}{r|rrrr} -6 & 2 & 9 & 1 & -24 \\ & & -12 & 18 & -114 \\ \hline & +2 & -3 & +19 & -138 \end{array}$$

Although -6 is not a root, it is a minimum value for the roots because the signs on the quotient alternate. Next, let us test -3

$$\begin{array}{r|rrrr} -3 & 2 & 9 & 1 & -24 \\ & & -6 & -9 & 24 \\ \hline & 2 & 3 & -8 & 0 \end{array}$$

There are two roots remaining represented by the quadratic equation $2x^2 + 3x - 8 = 0$, which we solve by using the quadratic formula.

$$x = \frac{-3 \pm \sqrt{9 - 4(2)(-8)}}{4} = \frac{-3 \pm \sqrt{73}}{4}$$

$$x \approx 1.39 \quad \text{and} \quad x \approx -2.89$$

The roots of the equation are 4, -3, $\dfrac{-3 + \sqrt{73}}{4} \approx 1.39$, and $\dfrac{-3 - \sqrt{73}}{4} \approx -2.89$. ■

EXAMPLE 5 A rectangular piece of plastic 12 cm by 13 cm is to be molded into an open box, the same thickness as the original piece of plastic, with no material wasted. The box is to have a square base and a volume of 180 cm^3. What are the possible dimensions of the box?

Solution Let

L = length and width of the square base
H = height of the box

Then the *volume* of the box is represented by

$$L^2H = 180$$

and the *surface area* of the box is represented by

$$L^2 + 4LH = 156$$

The surface area of the box equals the area of the original piece of plastic: (12 cm)(13 cm). We solve this nonlinear system of equations by substituting $H = 180/L^2$ into the second equation for H.

$$L^2 + 4L\left(\frac{180}{L^2}\right) = 156$$

This equation simplifies to yield

$$L^3 + 720 - 156L = 0$$
$$L^3 - 156L + 720 = 0$$

There are a large number of factors of 720, but let us test 6.

$$\begin{array}{r|rrrr} 6 & 1 & 0 & -156 & 720 \\ & & 6 & 36 & -720 \\ \hline & 1 & 6 & -120 & 0 \end{array}$$

One possibility for the length is 6. The resulting quotient is $L^2 + 6L - 120 = 0$. Solving by the quadratic formula gives

$$L = \frac{-6 \pm \sqrt{36 - 4(1)(-120)}}{2} = \frac{-6 \pm \sqrt{516}}{2}$$

$$L \approx 8.36 \quad \text{and} \quad L \approx -14.36$$

Since the length of the box is not measured in negative numbers, the only reasonable answer from the quadratic formula is 8.36. Thus,

$$L = 6 \quad \text{or} \quad L \approx 8.36$$

$$H = (180/L^2) = 5 \quad \text{or} \quad H \approx 2.58$$

Therefore, the box is 6 cm by 6 cm by 5 cm, or 8.36 cm by 8.36 cm by 2.58 cm. ■

7–4 EXERCISES

List the possible rational roots for the equations given.

1. $x^3 + 8x^2 - 3x + 10 = 0$
2. $x^3 + 8x^2 - 16 = 0$
3. $x^4 + 8x^3 + 7x - 24 = 0$
4. $x^4 - 7x^3 + 8x + 48 = 0$
5. $x^3 - 5x + 36 = 0$
6. $x^5 + 3x^4 - 5x^2 + 6x - 50 = 0$
7. $3x^3 + 8x^2 + 5x - 18 = 0$
8. $6x^3 + 12x - 35 = 0$
9. $2x^4 - 6x^2 + 9x + 14 = 0$
10. $9x^3 - 7x^2 + 6x + 36 = 0$
11. $5x^3 + 6x - 12 = 0$
12. $4x^4 + 11x^2 + 12 = 0$
13. $6x^4 - 7x^3 + 9x + 18 = 0$

14. $7x^4 - 9x^3 + 6x - 24 = 0$
15. $4x^5 + 7x^3 - 3x^2 + 20 = 0$
16. $3x^5 - 10x^4 + 7x^2 + 18 = 0$

Determine the maximum number of positive and negative roots for the equations given.

17. $5x^4 - 6x^3 - 8x^2 + x - 5 = 0$
18. $8x^4 + 7x^3 - x^2 - x + 6 = 0$
19. $2x^3 + x^2 - 10 = 0$
20. $9x^3 - 6x^2 - 8x - 5 = 0$
21. $4x^5 + x^3 - 8x^2 + 6x - 1 = 0$
22. $6x^7 + x^5 - 4x^2 + 9 = 0$
23. $x^7 - 6x^4 + 3x^3 + x^2 - 4x + 8 = 0$
24. $3x^8 + 7x^5 - 3x^3 + 6x + 10 = 0$
25. $8x^6 + x^5 + 7x^2 + 10 = 0$
26. $5x^6 + 4x^3 + 7x^2 - 3 = 0$
27. $-x^9 + 6x^7 + x^6 - 3x^2 + x + 5 = 0$
28. $x^{10} - 4x^6 + 7x^3 + 8x - 5 = 0$

Solve the following equations.

29. $2x^3 + 3x^2 - 8x + 3 = 0$
30. $3x^3 + x^2 - 12x - 4 = 0$
31. $3x^3 - 4x^2 - 28x - 16 = 0$
32. $4x^3 - 19x^2 - 8x + 15 = 0$
33. $x^3 + 10x^2 + 29x + 24 = 0$
34. $6x^3 - 5x^2 - 8x + 3 = 0$
35. $x^3 + 2x^2 - 21x + 18 = 0$
36. $2x^3 + 7x^2 - 10x - 50 = 0$
37. $x^3 + 8x^2 + 5x - 14 = 0$
38. $6x^3 + 7x^2 - 44x + 32 = 0$
39. $2x^4 + x^3 - x^2 + x - 3 = 0$
40. $2x^4 + 9x^3 - 40x^2 - 54x - 7 = 0$
41. $x^4 + 3x^3 - 21x^2 - 43x + 60 = 0$
42. $x^4 + 8x^3 - x^2 - 68x + 60 = 0$
43. $x^5 - 6x^4 + 7x^3 + 14x^2 - 48x + 32 = 0$
44. $2x^4 + 5x^3 - 19x^2 - 45x + 9 = 0$
45. $x^5 - 6x^4 + 7x^3 - 9x^2 - 8x + 15 = 0$
46. $x^5 + 7x^4 - 18x^3 - 72x^2 + 80x + 128 = 0$
47. A rectangular piece of plastic 5 in. by 37 in. is to be molded into an open box with no material wasted. If the box is to have a square base, the same thickness as the original piece of plastic, and a volume of 200 in^3, what should the dimensions of the box be? (See Example 5.)
48. A piece of plastic 16 cm by 10 cm is to be molded into an open box with no material wasted. The box is to have the same thickness as the original piece of plastic, a square base, and a volume of 144 cm^3. Find the dimensions of the box.
49. Under certain conditions the horizontal displacement of an object as a function of time is given by $s(t) = t^3 - 3t^2 + 4$. Find the time t when the displacement equals zero i.e., when $s(t) = 0$.
50. The radius of a cylindrical tank is 5 m less than its height. If the volume of the cylinder is $3{,}042\pi\ \text{m}^3$, find the dimensions of the tank.
51. You can make a rectangular box from a piece of cardboard 10 cm by 7 cm, by cutting a square from each corner and bending up the sides. What are the resulting dimensions of the box if its volume is 36 cm^3?

7–5

IRRATIONAL ROOTS BY LINEAR APPROXIMATION

In this section we will develop a method of solving for irrational roots. When we cannot reduce a polynomial equation to a quadratic equation by finding all but two rational roots, we need another method to find irrational roots.

Locating Roots Graphically

In Chapter 2, we solved equations graphically by locating the x intercept of the graph. This procedure is illustrated in Example 1.

EXAMPLE 1 Locate the real roots of the equation $x^3 + 2x^2 - x + 1 = 0$ graphically.

Solution First, we fill in the accompanying table of values. As shown in the graph in Figure 7–1, the intercept and root occur between -3 and -2. The table of values also indicates that the sign of $f(x)$ changes between $x = -3$ and $x = -2$.

x	-3	-2	-1.5	-1	0	1	2
$f(x)$	-5	3	3.625	3	1	3	15

sign change in $f(x)$

FIGURE 7–1

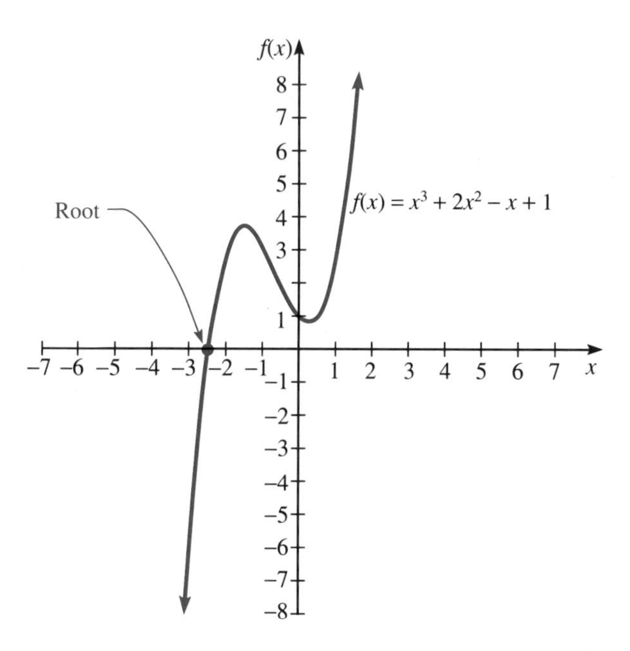

■

Linear Interpolation

One method of *approximating irrational roots* is called **linear interpolation.** This method is based on the assumption that if two points are sufficiently close to each other, then the straight line joining the points is a good approximation of the actual curve between the points. To find irrational roots, we begin by finding an interval on which a root exists, and then we successively narrow the interval until we find the root to the desired degree of accuracy. We know that when a curve crosses the x axis, the x value is a root of the equation. Therefore, finding an interval on which a root exists is based on the fact that the value of y or $f(x)$ changes sign whenever the graph crosses the x axis. This translates to a sign change on the remainder from synthetic division.

EXAMPLE 2 Using linear interpolation, find the root, to the nearest hundredth, between -2 and -1 of the equation $x^3 - 6x^2 + 10 = 0$.

Solution We know that since $f(-2) = -22$ and $f(-1) = 3$, the graph crosses the x axis and a root does exist between -2 and -1. Since $f(-1)$ is closer to zero, the root is closer to $x = -1$. Let us test $x = -1.2$ by using synthetic division.

-1.2	1	-6	0	10
		-1.2	8.64	-10.37
	1	-7.2	8.64	-0.37

Since $f(-1.2) \approx -0.37$ and $f(-1) = 3$, the root is between -1 and -1.2. Next, we will test -1.1.

-1.1	1	-6	0	10
		-1.1	7.81	-8.59
	1	-7.1	7.81	1.41

Since $f(-1.2) \approx -0.37$ and $f(-1.1) \approx 1.41$, the root is closer to $x = -1.2$. Let us test $x = -1.18$.

-1.18	1	-6	0	10	
		-1.18	8.47	-10	(rounded)
	1	-7.18	8.47	0	(approximately)

Since the remainder is approximately zero, the root rounded to hundredths is -1.18. ■

CAUTION ✦ When $f(x)$ changes signs between $x = a$ and $x = b$, we know only that an odd number of roots occur in the interval. If $f(x)$ does not change signs on an interval, there may be no real roots or an even number of real roots in the interval.

EXAMPLE 3 By linear interpolation, find the root, to the nearest hundredth, between 1 and 2 of the equation $x^4 - 2x^3 + 6x - 6 = 0$.

Solution Since $f(1) = -1$ and $f(2) = 6$, a root does exist in the interval. Next, let us approximate the root by constructing a graph of the straight line joining the points (1, −1) and (2, 6) as shown in Figure 7–2. The graph illustrates that the root is closer to 1; therefore, let us test $x = 1.2$ and $x = 1.3$ by synthetic division.

1.2	1	−2	0	6	−6
		1.2	−0.96	−1.15	5.82
	1	−0.8	−0.96	4.85	−0.18

1.3	1	−2	0	6	−6
		−1.3	−0.91	−1.18	6.26
	1	−0.7	−0.91	4.82	0.26

FIGURE 7–2

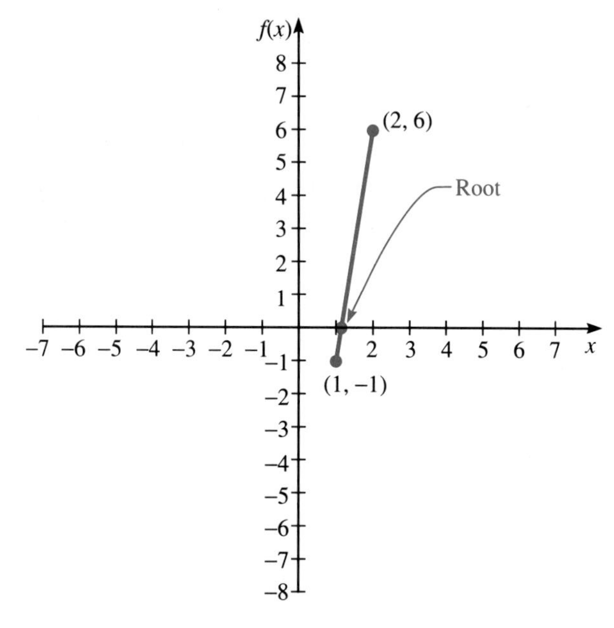

Since the sign on the remainder changes, a root exists in the interval between 1.2 and 1.3. Since the remainder is closer to zero for 1.2, let us test 1.24 and 1.25.

1.25	1	−2	0	6	−6
		1.25	−0.94	−1.17	6.04
	1	−0.75	−0.94	4.83	0.04

1.24	1	−2	0	6	−6
		1.24	−0.94	−1.17	5.99
	1	−0.76	−0.94	4.83	−0.01

The approximate root in the given interval is 1.24 ■

EXAMPLE 4 Find all the real roots, to the nearest tenth, of the equation $6x^4 - 3x^2 + 5x - 4 = 0$.

Solution First, we complete the accompanying table of values and graph the equation so that we can approximately locate roots. The graph is shown in Figure 7–3.

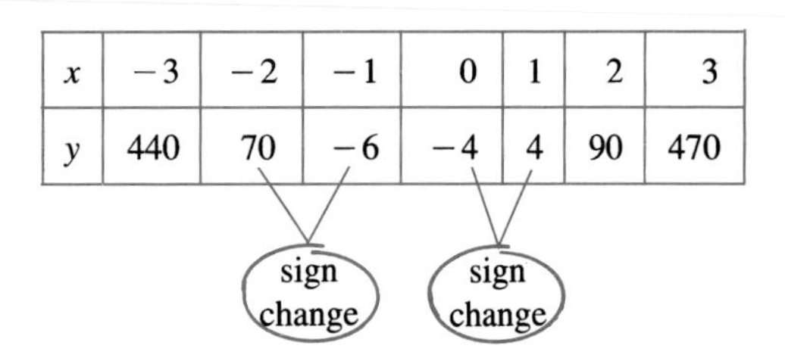

x	-3	-2	-1	0	1	2	3
y	440	70	-6	-4	4	90	470

FIGURE 7–3

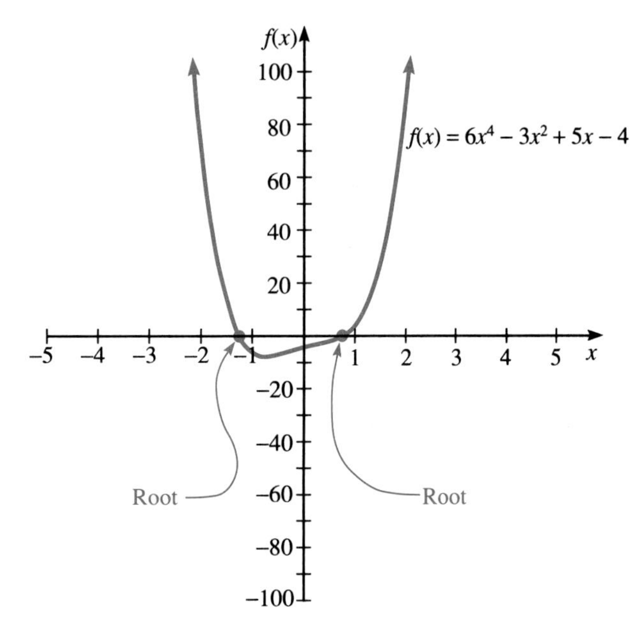

The graph crosses the x axis between -2 and -1 and again between 0 and 1. Since the graph crosses the x axis closer to -1 than -2, let us test -1.1 and -1.2.

$$\begin{array}{r|rrrrr} -1.1 & 6 & 0 & -3 & 5 & -4 \\ & & -6.6 & 7.3 & -4.7 & -0.3 \\ \hline & 6 & -6.6 & 4.3 & 0.3 & -4.3 \end{array}$$

$$\begin{array}{r|rrrrr} -1.2 & 6 & 0 & -3 & 5 & -4 \\ & & -7.2 & 8.6 & -6.7 & 2.0 \\ \hline & 6 & -7.2 & 5.6 & -1.7 & -2.0 \end{array}$$

Since the remainder is still negative, let us test $x = -1.3$.

-1.3	6	0	-3	5	-4
		-7.8	10.1	-9.3	5.6
	6	-7.8	7.1	-4.3	1.6

Since the sign changes and the remainder of 1.6 is the smaller, the approximate root is $x = -1.3$. Similarly, we approximate the root between 0 and 1, testing 0.7 and 0.8.

0.7	6	0	-3	5	-4
		4.2	2.9	-0.1	3.4
	6	4.2	-0.1	4.9	-0.6

0.8	6	0	-3	5	-4
		4.8	3.8	0.6	4.5
	6	4.8	0.8	5.6	0.5

Since $f(0.8) \approx 0.5$, the approximate root is 0.8. The approximate real roots of this equation rounded to tenths are 0.8 and -1.3. Because there are only two real roots, there are also two complex number roots. ■

7–5 EXERCISES

Find the real roots, to the nearest tenth, of the following equations.

1. $2x^3 - 8x^2 + x - 12 = 0$

2. $x^3 + 6x^2 - 5x + 8 = 0$

3. $2x^4 - 5x^3 + 3x - 9 = 0$

4. $x^3 + 2x^2 - 3x + 12 = 0$

5. $3x^3 + 7x - 9 = 0$

6. $4x^3 - 3x^2 + 2x - 5 = 0$

7. $x^3 - 2x^2 + x - 3 = 0$

8. $2x^4 - 3x^2 - 6 = 0$

9. $x^4 + 6x^3 - 5x + 9 = 0$

10. $x^4 - x^3 + 2x^2 + x - 8 = 0$

11. $x^4 + 6x^3 + x^2 - 3x + 10 = 0$

12. $5x^4 + 6x - 8 = 0$

13. $6x^4 - 2x^3 + x^2 + 3x - 6 = 0$

14. $3x^4 + 8x^2 - 14 = 0$

15. $2x^4 - 3x^3 + x^2 + 5x - 4 = 0$

16. $x^5 - x^3 + 6x^2 + x - 10 = 0$

17. $2x^5 + x^4 - 3x^3 - 4x^2 + x + 8 = 0$

18. $x^5 + x^4 - x^3 + 2x^2 + x - 8 = 0$

19. $2x^5 - 4x^4 + 6x^2 - x + 9 = 0$

20. $x^5 + 2x^3 - 2 = 0$

21. The dimensions of a rectangular box are 5, 8, and 6 cm. If each dimension of the box is increased by a fixed amount, say x, the volume is 792 cm^3. Find the amount by which to increase each dimension.

22. A cylindrical storage tank 10 ft high holds 1,131 ft^3. Determine the thickness (to hundredths) of the tank if the outside radius is 6.2 ft.

23. The position x of a particle that moves in a line as a function of time t is given by $x = t^3 + 4t^2 - 10t + 20$. Find the time when the particle has moved 195 ft.

CHAPTER SUMMARY

Summary of Terms

Descartes' Rule of Signs (p. 275)
dividend (p. 262)
divisor (p. 262)
Factor Theorem (p. 263)
Fundamental Theorem of Algebra (p. 271)
linear interpolation (p. 280)
multiplicity (p. 271)
polynomial equation (p. 262)
quotient (p. 262)
remainder (p. 262)
Remainder Theorem (p. 262)
synthetic division (p. 266)

CHAPTER REVIEW

Section 7–1

Using the Remainder Theorem, find the remainder.

1. $(4x^3 - 7x^2 + 6x - 9) \div (x + 3)$
2. $(2x^4 + 3x^2 + 7) \div (x - 1)$
3. $(5x^4 + 7x^3 + 3x - 8) \div (x + 2)$
4. $(9x^3 + 8x^2 + x - 10) \div (x + 5)$

Use the Factor Theorem to determine if the second expression is a factor of the first.

5. $2x^3 - 6x^2 + 9; x - 3$
6. $x^3 - 5x^2 + 7x - 3; x - 1$
7. $3x^3 - 15x^2 - 11x + 5; 3x - 1$
8. $4x^3 - x^2 + 8; 2x + 1$
9. $x^4 + 6x^3 - 8x + 8; x + 5$
10. $2x^4 - x^2 + 3x - 144; x + 3$

Section 7–2

Using synthetic division, find the quotient and remainder.

11. $6x^3 - 5x^2 + 3x - 1; x + 1$
12. $9x^3 + 7x^2 - x + 6; x - 2$
13. $x^4 - 3x^3 + 7x - 8; x + 2$
14. $2x^4 - 7x^2 + 6x - 10; x + 5$
15. $4x^4 - x^3 + 2x^2 - 9; x - 3$
16. $4x^3 - 9x^2 + 3x - 5; x - 1$

Use synthetic division to determine if the given number is a zero of the function.

17. $6x^3 - 11x^2 + 9x - 2; 1/3$
18. $2x^3 + x^2 + 5x - 6; -3$
19. $5x^4 + 6x^3 - 8x + 7; -1$
20. $7x^5 + 6x^3 + x^2 - 6x + 14; 2$
21. $x^4 + 6x^3 + 8x^2 + 43x - 30; -6$
22. $4x^4 + 27x^3 - 7x^2 - 20x + 5; 1/4$

Section 7–3

Find the remaining roots of the equation given the root(s) indicated.

23. $2x^3 + 3x^2 - 8x + 3 = 0; -3$
24. $x^3 + 5x^2 + 3x + 54 = 0; -6$
25. $2x^3 - 3x^2 + 4x + 9 = 0; -1$
26. $6x^3 - 37x^2 + 47x + 20 = 0; 5/2$
27. $2x^4 - 5x^3 + 3x^2 - 15x - 9 = 0; -1/2$ and 3
28. $2x^4 + x^3 - 24x^2 - 20x + 21 = 0; 7/2$ and -3
29. $x^4 - 8x^3 + 25x^2 - 58x + 40 = 0; 5$ and 1
30. $x^4 - 4x^3 - x^2 + 16x - 12 = 0; 2$ and 3

Section 7–4

List all possible rational roots for the equations given.

31. $4x^3 + 7x^2 - 3x + 2 = 0$
32. $2x^3 + 6x^2 - 9x + 10 = 0$
33. $3x^4 + 7x^3 - 5x^2 + 12 = 0$
34. $8x^3 - 7x^2 + x + 6 = 0$
35. $5x^3 + 8x^2 - 10x + 15 = 0$
36. $3x^4 - 5x^2 + 7x - 8 = 0$

Using Descartes' Rule of Signs, list the maximum number of positive and negative roots for the given equation.

37. $8x^3 + 7x^2 - 6x + 10 = 0$

38. $2x^3 - 7x^2 - 5x - 3 = 0$

39. $4x^4 - 5x^3 - 6x^2 + 7x - 3 = 0$

40. $3x^4 + x^3 - 2x^2 - x + 6 = 0$

41. $x^5 - 4x^3 - 2x^2 - x + 9 = 0$

42. $2x^5 - x^4 + 2x^3 + 4x^2 - 3x - 11 = 0$

Solve the following equations.

43. $2x^3 - 9x^2 - 2x + 24 = 0$

44. $3x^3 - 4x^2 - 16x + 3 = 0$

45. $4x^3 - 28x^2 - 69x - 27 = 0$

46. $2x^3 - 11x^2 - 41x + 140 = 0$

47. $x^3 + x^2 - 14x - 24 = 0$

48. $x^4 + x^3 - 14x^2 + 21x - 45 = 0$

49. $2x^4 + 13x^3 - 14x^2 - 43x + 42 = 0$

50. $2x^4 - 4x^3 - 25x^2 - 3x + 30 = 0$

Section 7–5

Use linear approximation to find the irrational roots, to the nearest tenth, of the following equations.

51. $3x^3 + 6x - 10 = 0$ **52.** $x^3 + x^2 - 6x + 7 = 0$

53. $x^3 + 2x^2 + 3x + 4 = 0$

54. $x^3 - 5x^2 + x - 8 = 0$

CHAPTER TEST

The number in parentheses refers to the appropriate learning objective given at the beginning of the chapter.

1. Use synthetic division to determine if -2 is a zero of the function $x^4 - 3x^2 - x - 6$. Explain why or why not. (3)
2. Determine the rational roots of the following equation: (7)

$$x^3 + x^2 - 2x - 2 = 0$$

3. Find the remaining roots of the equation $6x^3 - 35x^2 + 21x + 20 = 0$ if 5 is one of the roots. (4)
4. Solve the following equation: (4)

$$x^3 - 14x^2 + 58x - 80 = 0$$

5. Use the Factor Theorem to determine if $2x + 5$ is a factor of $4x^3 + 8x^2 - x - 15$. (2)
6. List all possible rational roots for $2x^4 + 6x^3 - x^2 + 5x - 6 = 0$. (6)
7. Solve the equation $2x^3 - 7x^2 - 28x - 12 = 0$. (4)
8. Apply the Remainder Theorem to find the remainder when $8x^4 + 7x^3 + 5x - 10$ is divided by $x + 2$. (1)
9. Using Descartes' Rule of Signs, determine the maximum number of positive and negative roots for $2x^4 - 6x^3 - x^2 + 7x - 10 = 0$. (5)
10. You can make a rectangular box from a piece of cardboard by cutting a square from each corner and bending up the sides. What are the resulting dimensions of a box made from a 16-cm-by-17-cm piece of cardboard if the volume of the box is to be 312 cm^3? (4)
11. Use synthetic division to determine the quotient and remainder when $7x^4 + 8x^3 - 6x^2 + 10$ is divided by $x + 2$. (3)
12. Use linear approximation to find the irrational root(s) of the following equation: (7)

$$x^4 + 6x^3 - 7x^2 + x + 8 = 0$$

13. Use Descartes' Rule of Signs to list the maximum number of positive and negative roots for $3x^5 - 4x^4 + 7x^2 + 8x + 5 = 0$. **(5)**

14. Solve the equation $6x^3 + 43x^2 + 60x - 25 = 0$. **(4)**

15. Using synthetic division, find the quotient and remainder resulting from the division of $x^5 + 6x^4 - 7x^2 + 8$ by $x - 2$. **(3)**

16. List all possible rational roots of $3x^3 + 5x^2 - 10x + 16 = 0$. **(6)**

SOLUTION TO CHAPTER INTRODUCTION

Let

$$L = \text{length and width of the square base}$$
$$H = \text{the height of the box}$$

The volume is given by

$$L^2H = 640$$

and the surface area is given by

$$L^2 + 4LH = 384$$

Solving the first equation for H and substituting into the second equation gives

$$\begin{aligned} L^2 + 4L\left(\frac{640}{L^2}\right) &= 384 \\ L^3 + 2{,}560 &= 384L \\ L^3 - 384L + 2{,}560 &= 0 \end{aligned}$$

Last, solve this higher-degree equation by using the technique developed in this chapter.

$$\begin{array}{r|rrrr} 8 & 1 & 0 & -384 & 2{,}560 \\ & & 8 & 64 & -2{,}560 \\ \hline & 1 & 8 & -320 & 0 \end{array}$$

Therefore, $L = 8$ is one measure.

Using the quadratic formula to solve the resulting quadratic equation gives

$$L = \frac{-8 \pm \sqrt{64 - 4(1)(-320)}}{2} = \frac{-8 \pm \sqrt{1344}}{2}$$

$$L \approx \frac{-8 \pm 36.66}{2}$$

$$L \approx 14.33 \quad \text{or} \quad L \approx -22.33$$

The box could be 8 in. by 8 in. by 10 in., or 14.33 in. by 14.33 in. by 3.12 in.

You are the traffic planner for a city. Based on its current population of 56,180 and its current growth rate of 4.8% annually, you must determine the population 15 years from now, in order to assure adequate public facilities to serve the population. (The solution to this problem is given at the end of the chapter.)

Population growth is one example of an exponential function. The exponential function and its inverse, the logarithmic function, are very important to many areas of technology. For example, bacteria in a petri dish and the world's population both grow at an exponential rate. Energy loss and absorption, radioactive decay, and compound interest are also examples of exponential functions. The logarithmic and exponential functions and their applications are discussed in this chapter.

Learning Objectives

After completing this chapter, you should be able to

1. Graph exponential and logarithmic functions (Sections 8–1 and 8–2).
2. Evaluate exponential expressions using a calculator (Section 8–1).
3. Convert a logarithmic expression to an exponential expression and vice versa (Section 8–2).
4. Solve simple logarithmic equations by converting them to exponential form (Section 8–2).
5. Apply the properties of logarithms to write logarithmic expressions in different forms, and evaluate the resulting expressions (Section 8–3).
6. Solve exponential and logarithmic equations (Section 8–4).

Chapter 8

Exponential and Logarithmic Functions

8–1

THE EXPONENTIAL FUNCTION

An **exponential function** is a function y or $f(x)$ in which *the base is a constant b and the exponent contains the independent variable x.* The exponential function is defined by the following expression.

Definitions

The Exponential Function

$$y = b^x$$

where

b = any positive real number other than 1
x = any real number
y = any positive real number

The exponential function differs from the exponential expressions discussed in Chapter 4, such as $3x^3y^4$, because the independent variable appears in the exponent. Some examples of exponential functions are

$$y = 3^x \qquad y = 6^{2x-1} \qquad y = 5e^{3x}$$

Graphs of Exponential Functions

To graph an exponential function, we follow the procedure outlined in Chapter 2. We obtain ordered pairs by substituting a chosen value for x into the equation and calculating the resulting value of y. Since the graph of an exponential function is a curved line, we must choose a sufficient number of values for x to sketch a smooth curve through the plotted points.

EXAMPLE 1 Sketch the graph of $y = 3^x$.

Solution First, we prepare a table of values by substituting integral values of x from -4 to 4. Then we plot the points and connect them with a smooth curve as shown in Figure 8–1.

x	-4	-3	-2	-1	0	1	2	3	4
y	1/81	1/27	1/9	1/3	1	3	9	27	81

■

NOTE ✦ Notice that in Figure 8–1, the value of y increases without bound as x increases, and y approaches zero as x decreases. This relation is characteristic of an **exponential growth function.**

FIGURE 8–1

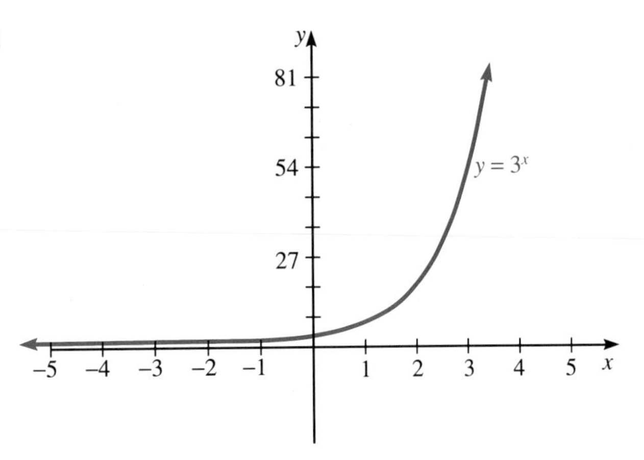

EXAMPLE 2 Graph $y = 2^{-x}$.

Solution To graph this function, we fill in a table of values just as we did in the previous example. Remember from Chapter 4 that 2^{-x} is equivalent to $\left(\frac{1}{2}\right)^x$. The graph of $y = 2^{-x}$ is shown in Figure 8–2.

x	−4	−3	−2	−1	0	1	2	3	4
y	16	8	4	2	1	1/2	1/4	1/8	1/16

FIGURE 8–2

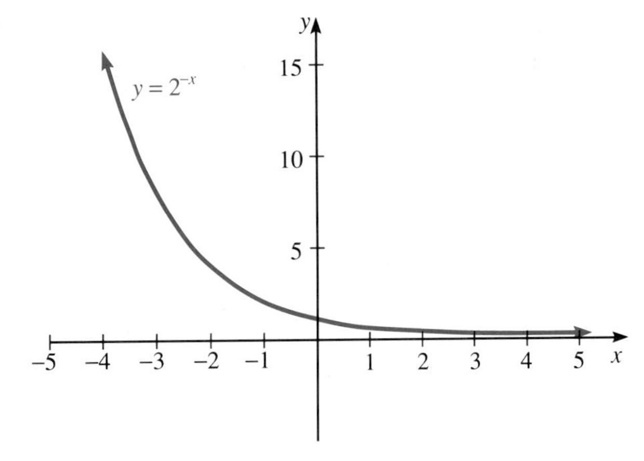

■

NOTE ✦ Notice the difference between the graph of $y = 2^{-x}$ (Figure 8–2) and the graph of $y = 3^x$ (Figure 8–1). Figure 8–2 represents an **exponential decay function** where y approaches zero as x increases and infinity as x decreases.

From the graphs shown in Figures 8–1 and 8–2, we can summarize the following properties of exponential functions.

Properties of Exponential Functions

- ☐ For $b > 1$, an exponential growth function results as shown in Figure 8–1.
- ☐ For $0 < b < 1$ and $x > 0$, an exponential decay function results as shown in Figure 8–2.
- ☐ The graph of an exponential function of the form $y = b^x$ is above the x axis but approaches the x axis as an asymptote.
- ☐ Exponential functions of the form $y = b^x$ intersect the y axis at $(0, 1)$.

Application

EXAMPLE 3 Compound interest is given by the formula

$$y = P\left(1 + \frac{r}{m}\right)^{mt}$$

where y = accumulated value, P = principal, r = rate of interest, m = frequency of compounding, and t = time. Graph the relationship between the time and accumulated value if $P = \$500$, $r = 10\%$, and $m = 12$ (interest is compounded monthly). Then use the graph to approximate how long it would take to have an accumulated value of \$1,000.

Solution Substituting the given values into the formula, we have

$$y = 500\left(1 + \frac{0.10}{12}\right)^{12t}$$

$$y = 500(1.0083)^{12t}$$

Substituting positive integral values of t gives the following table of values:

t	0	1	2	3	4	5	6
y	500	552.36	610.20	674.09	744.68	822.65	908.80

The graph of this function is given in Figure 8–3. To determine how long it takes for the accumulated value to reach \$1,000, we find \$1,000 along the y axis; then we move horizontally to the graph and vertically down to the t axis. As shown by the graph, it takes about 7 years for \$500 to have an accumulated value of \$1,000. ■

FIGURE 8–3

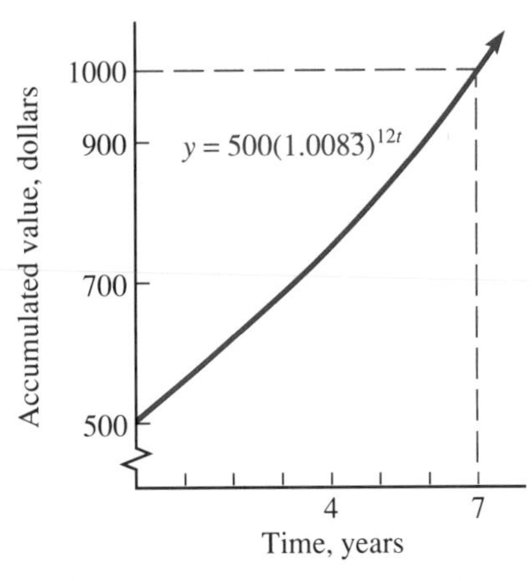

Base e

The base of an exponential function can be any positive real number other than 1. In the previous discussion, the base has been a rational number, but irrational numbers are also possible. The number $e = 2.71828\ldots$ is irrational, just as π and $\sqrt{2}$ are. The expression $(1 + (1/n))^n$ approaches the value of e as n approaches infinity. The number e, named after Swiss mathematician Leonhard Euler, is used extensively in applications of exponential growth and decay. You can find the value of e^x on the calculator using an $\boxed{e^x}$ key or $\boxed{\text{inv}}$ $\boxed{\ln x}$ keys, depending on your calculator. In this text, calculator keystrokes use the $\boxed{e^x}$ key.

Graphs of Exponential Functions

EXAMPLE 4 Graph the function $y = e^x$.

Solution We fill in a table of values using the calculator. The graph of $y = e^x$ is given in Figure 8–4. As you can see by comparing the graphs in Figures 8–1 and 8–4, this graph also represents exponential growth. The equation $y = ae^{nx}$ represents exponential growth with a base of e, if n is positive.

x	y	
-4	0.02	4 $\boxed{+/-}$ $\boxed{e^x}$ $\rightarrow$ 0.018
-3	0.05	3 $\boxed{+/-}$ $\boxed{e^x}$ $\rightarrow$ 0.050
-2	0.14	2 $\boxed{+/-}$ $\boxed{e^x}$ $\rightarrow$ 0.14
-1	0.37	1 $\boxed{+/-}$ $\boxed{e^x}$ $\rightarrow$ 0.37
0	1	0 $\boxed{e^x}$ $\rightarrow$ 1
1	2.7	1 $\boxed{e^x}$ $\rightarrow$ 2.7
2	7.4	2 $\boxed{e^x}$ $\rightarrow$ 7.4
3	20.1	3 $\boxed{e^x}$ $\rightarrow$ 20.1
4	54.6	4 $\boxed{e^x}$ $\rightarrow$ 54.6

FIGURE 8–4

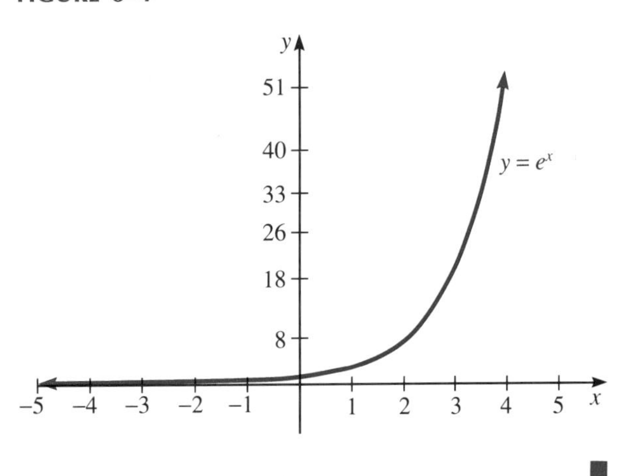

■

EXAMPLE 5 The circuit shown in Figure 8–5 consists of a resistance R, a capacitance C, a constant voltage V, and a switch. The instantaneous current I at any time t is given by

$$I = \frac{V}{R}e^{-t/(RC)}$$

Graph I as a function of t when $V = 120$ V, $R = 500\ \Omega$, and $C = 0.0002$ F.

Solution Substituting in the given values, we obtain the following equation:

$$I = \frac{120}{500}e^{-(t/[500(0.0002)])}$$
$$I = 0.24e^{-t/0.10}$$
$$I = 0.24e^{-10t}$$

The table of values is given below. The graph of this function is shown in Figure 8–6 and is characteristic of the exponential decay curve. The general equation $y = ae^{-nx}$ represents exponential decay with a base of e if n is positive.

t	0	0.05	0.10	0.15	0.20	0.25	0.30
I	0.24	0.15	0.09	0.05	0.03	0.02	0.012

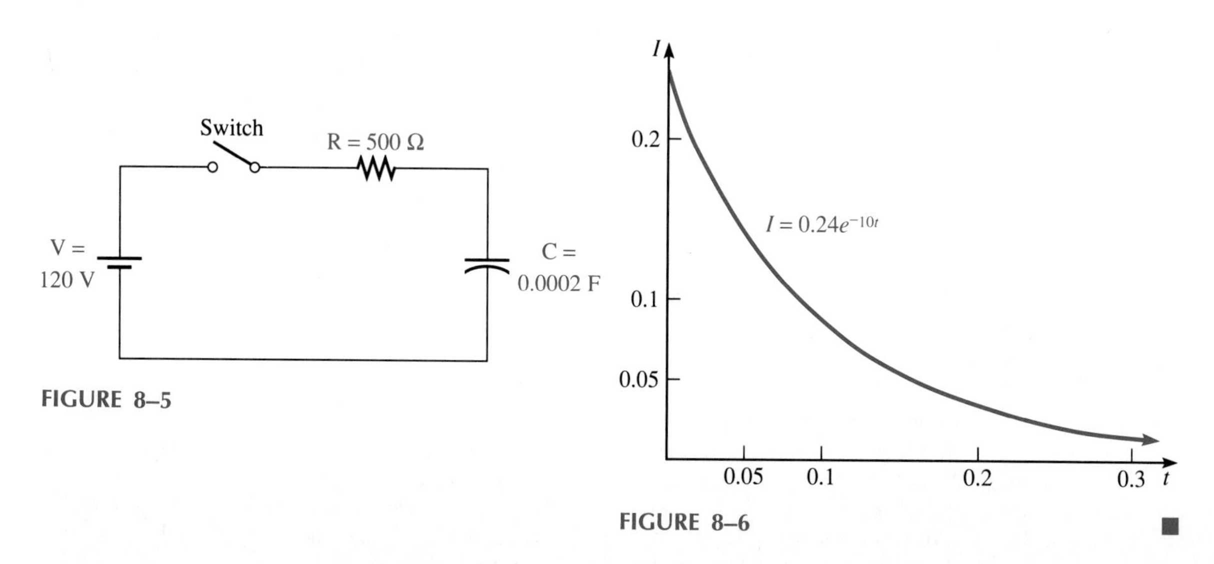

FIGURE 8–5

FIGURE 8–6

Application

EXAMPLE 6 A biologist finds that bacteria in a petri dish grow at a rate of 1.8% each hour. If the biologist starts the culture with 750 cells, how many cells of the culture will she have at the end of 8 hours? (Use a base of e.)

Solution The general equation for exponential growth with a base of e is

$$y = ae^{nx}$$

where a = initial amount, n = growth rate, and x = time. Filling in the values $a = 750$, $n = 0.018$, and $x = 8$, we obtain the following equation:

$$y = 750e^{(0.018)(8)}$$

Performing the calculation yields

$$y = 866 \text{ cells}$$

The biologist will have 866 cells of the culture at the end of 8 hours.

750 [×] [(] 0.018 [×] 8 [)] [e^x] [=] → 866.16 ■

8–1 EXERCISES

Graph the following functions.

1. $y = 4^x$
2. $y = 5^{-x}$
3. $y = 3^{-x}$
4. $y = 2^{3x}$
5. $y = \left(\frac{1}{2}\right)^{3x}$
6. $y = \left(\frac{1}{4}\right)^{-x}$
7. $y = \left(\frac{1}{3}\right)^{-x}$
8. $y = \left(\frac{1}{3}\right)^{2x}$
9. $y = e^x$
10. $y = e^{-x}$
11. $y = e^{2x}$
12. $y = e^{4x}$
13. $y = 4e^{-x}$
14. $y = 2e^{-x}$
15. $y = 3e^{-2x}$
16. $y = 4e^{-3x}$

17. The growth of a bacteria is given by the equation $N = N_0 3^t$ where N_0 = initial number of bacteria and t = elapsed time. Graph the relationship between N and t if $N_0 = 100$. Using the graph, find the number of bacteria at the end of 4 hours. Verify your answer using a calculator.

18. Population growth P is given by the equation $P = P_0 e^{kt}$ where P_0 = initial population, t = time, and k = rate of growth. Graph the yearly population growth of the city of Glenwood from 1976 to 1983 if its population in 1976 was 250 and $k = 0.05$. Use the graph to predict the population of Glenwood in 1989.

I 19. On his daughter's sixteenth birthday, a father promises to give her $30,000 on her twenty-fifth birthday. If interest is compounded continuously at an annual rate of 9%, how much money should the father put into the account initially? (*Note:* $P = Se^{-rt}$ where P = initial investment, S = yield at t years, and r = rate of interest.)

20. A radioisotope decays at a rate equal to $N = N_0(1/2)^x$ where N_0 = initial number of nuclei present after x half-lives. How many nuclei remain after 3 half-lives if the initial number is 1,000,000?

I 21. A *continuous* annuity is an annuity in which R dollars are dispersed annually by uniform payments that are payable continuously. The present value y of a continuous annuity for x years is

$$y = R\left(\frac{1 - e^{-rx}}{r}\right)$$

where r = annual interest rate compounded continuously. Find the present value of an annuity in which $3,000 is invested annually for 10 years at 12% annual interest.

22. A yeast manufacturer finds that yeast grows exponentially at a rate of 20% per hour. How many pounds of yeast will the manufacturer have at the end of 12 hours if 100 pounds is used initially? (Use $y = ae^{nx}$. See Example 6.)

E 23. The circuit shown in Figure 8–7 consists of a voltage source V, a resistance R, a capacitance C, and a switch. The current is given by $I = (V/R)e^{-t/(RC)}$. Find the current after 2 seconds if $V = 100$ V, $R = 350\ \Omega$, and $C = 0.1$ mF.

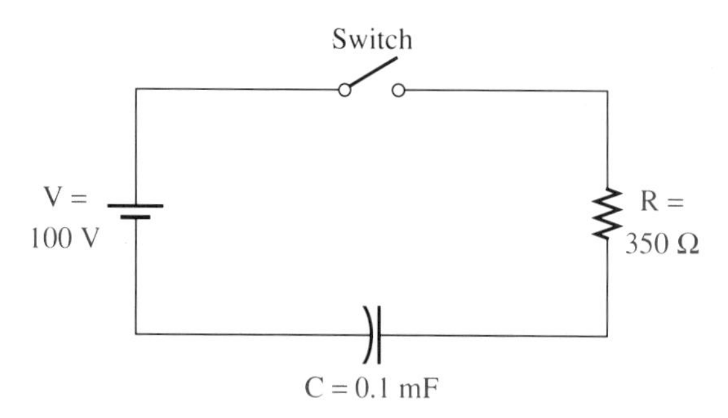

FIGURE 8–7

E 24. In the circuit shown in Figure 8–8, determine the instantaneous voltage across the capacitor 2 seconds after the switch is closed. Use the relationship $V_c = V(1 - e^{-t/(RC)})$ when $V = 110$ V, $R = 50\ \Omega$, and $C = 0.05$ F.

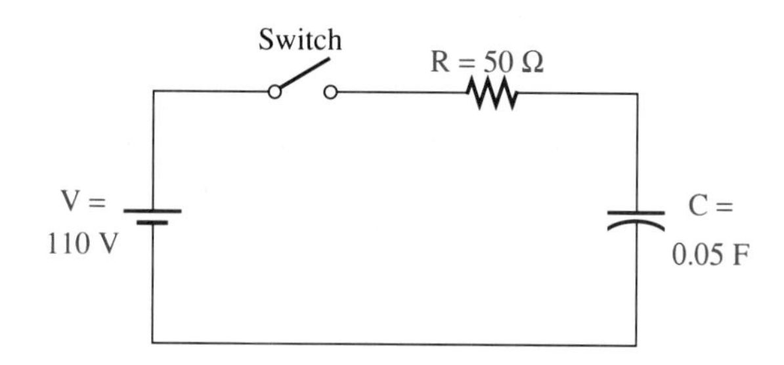

FIGURE 8–8

E 25. Plot the relationship between I and t in the formula $I = (V/R)(1 - e^{-Rt/L})$ where $V = 9$ V, $R = 1.5\ \Omega$, and $L = 10$ H.

I 26. An amount of money R invested every year for x years at i rate of interest will accumulate to y dollars. The formula relating these factors is given by

$$y = R\left(\frac{(1 + i)^x - 1}{i}\right)$$

If Jose pays \$2,000 each year into a retirement plan that earns interest at 12% compounded annually, what will be the value of his annuity after 30 years?

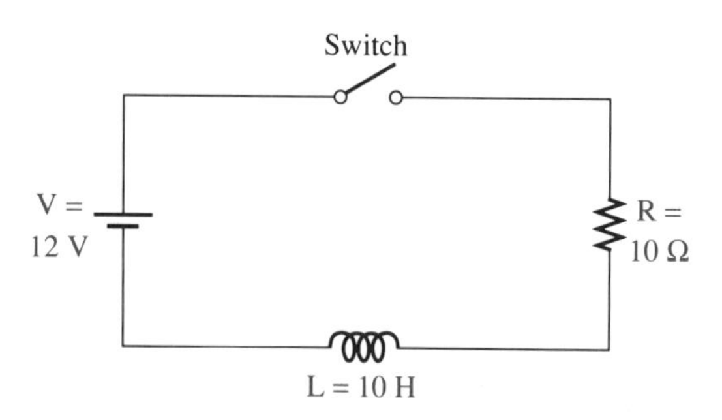

FIGURE 8–9

E 27. Determine the instantaneous current 3 seconds after the switch is closed in the circuit shown in Figure 8–9. Use the relationship $I = (V/R)(1 - e^{-Rt/L})$ when $V = 12$ V, $R = 10\ \Omega$, and $L = 10$ H.

E 28. When the switch in Figure 8–10 is closed, the current I grows exponentially according to the formula $I = (V/R)(1 - e^{-(R/L)t})$ where L is inductance in henries and R is resistance in ohms. Find the current at 0.05 s if $R = 500\ \Omega$, $L = 100$ H, and $V = 120$ V.

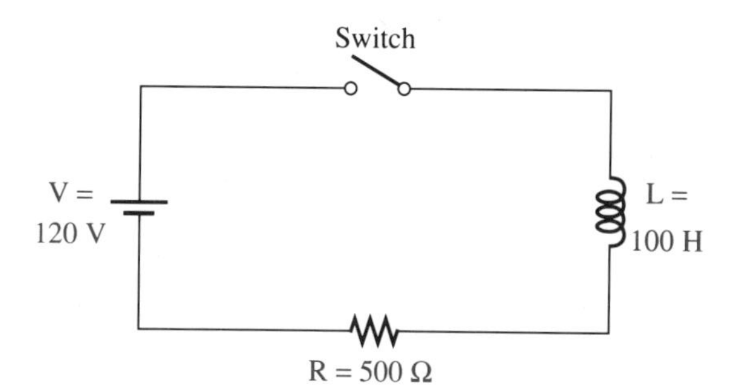

FIGURE 8–10

I 29. The capital recovery factor is the annual payment y required to completely pay off some present amount P over x years at i rate of interest. The capital recovery factor is given by

$$y = P\left(\frac{i(1 + i)^x}{(1 + i)^x - 1}\right)$$

If a businessman wants to pay off a \$100,000-debt over 10 years at an interest rate of 8% compounded annually, what annual payment is required?

8–2

THE LOGARITHMIC FUNCTION

When working with the exponential function $y = b^x$, often we must solve for x. To do so, we must define a new type of function called a *logarithm*.

Definition

Definition of Logarithm

$$y = b^x$$

if and only if

$$x = \log_b y$$

where

x = any real number
b = any positive real number, $b \neq 1$
y = any positive real number

NOTE ✦ Since y must be a positive number, the logarithm of zero and negative numbers is undefined.

The definition of a logarithm provides a mechanism of solving an exponential function for the exponent. Therefore, we must be able to translate from exponential form to logarithmic form. This relationship is as follows:

$y = b^x$ exponential form
$x = \log_b y$ logarithmic form

EXAMPLE 1 Express each of the following equations in exponential form:

(a) $\log_4 16 = 2$ (b) $\log_2 8 = 3$ (c) $\log_e x = 3$
(d) $\log_{25} 5 = 1/2$ (e) $\log_b 1 = 0$

Solution

(a) $4^2 = 16$ (b) $2^3 = 8$ (c) $e^3 = x$
(d) $25^{1/2} = 5$ (e) $b^0 = 1$

In part (e), since any nonzero base raised to the zero power equals one, the logarithm of one is always zero. ■

EXAMPLE 2 Express each of the following equations in logarithmic form.

(a) $3^4 = 81$ (b) $5^3 = 125$
(c) $b^8 = 246$ (d) $e^4 = d$

Solution

(a) $\log_3 81 = 4$ (b) $\log_5 125 = 3$
(c) $\log_b 246 = 8$ (d) $\log_e d = 4$ ■

The logarithmic bases that are used most frequently are 10 and e. Logarithms with a base of 10 are called **common logarithms** and are written as $\log x$. If the base of a logarithm is not written, it is understood to be ten. Likewise, logarithms with a base of e are called **natural logarithms** and are written as $\ln x$.

Solving Logarithmic Equations

To solve a logarithmic equation, convert the equation to exponential form. This procedure is developed in the next example.

EXAMPLE 3 Solve the following equations for the variable:

(a) $\log_x 27 = 3$ (b) $\log_2 16 = x$ (c) $\log_4 x = 3$
(d) $\log_9(1/81) = x$ (e) $\log_3 81^{-1} = x$

Solution

(a) $\log_x 27 = 3$ is equivalent to $x^3 = 27$, so $x = 3$.
(b) $\log_2 16 = x$ is equivalent to $2^x = 16$, so $x = 4$.
(c) $\log_4 x = 3$ is equivalent to $4^3 = x$, so $x = 64$.
(d) $\log_9(1/81) = x$ is equivalent to $9^x = 1/81$. Since $81 = 9^2$, it will be easier if we express both sides of the equation in terms of the same base.

$$9^x = \frac{1}{81}$$

$$9^x = \frac{1}{9^2}$$

$$9^x = 9^{-2}$$

$$x = -2$$

(e) $\log_3 81^{-1} = x$ is equivalent to $3^x = 81^{-1}$ or $3^x = (3^4)^{-1}$, so $x = -4$. ■

Solving Exponential Equations

To solve an exponential equation, convert the equation to a logarithmic expression.

EXAMPLE 4 The formula to evaluate the initial investment P needed to accumulate an amount S at a rate r compounded continuously in a certain time t is

$$P = Se^{-rt}$$

Solve the equation for t and determine the time required for a \$2,000 investment to accumulate to \$3,500 at an interest rate of 8% compounded continuously.

Solution To solve the equation for t, we divide both sides by S.

$$\frac{P}{S} = e^{-rt}$$

Using the definition of logarithms, we convert to the following equation:

$$\ln\left(\frac{P}{S}\right) = -rt$$

Dividing by $-r$ gives the solution for t.

$$\frac{\ln\left(\frac{P}{S}\right)}{-r} = t$$

Substituting the values $P = 2{,}000$, $S = 3{,}500$, and $r = 0.08$ gives

$$\frac{\ln(0.5714)}{-0.08} = t$$

$$7 \approx t$$

2000 [÷] 3500 [=] [lnx] [÷] 0.08 [+/−] [=] → 6.995 ■

Logarithmic Functions

The logarithmic relationship is also a function. However, in writing a function, we normally label x the independent variable and y the dependent variable. By interchanging x and y in the previous definition, we can write the logarithmic function in the traditional manner as follows.

Logarithmic Function

$$y = \log_b x$$

where $x > 0$, $b > 0$, and $b \neq 1$

Graphing Logarithmic Functions

The graphs in the next two examples illustrate some of the properties of logarithms.

EXAMPLE 5 Graph $y = 3 \log x$.

Solution The domain of the logarithmic function is restricted. Remember that the function is undefined for $x \leq 0$. The table of values is given below, and the graph is shown in Figure 8–11.

x	1/2	1	2	4	5	8	10	15	20
y	−0.9	0	0.9	1.8	2.1	2.7	3	3.5	3.9

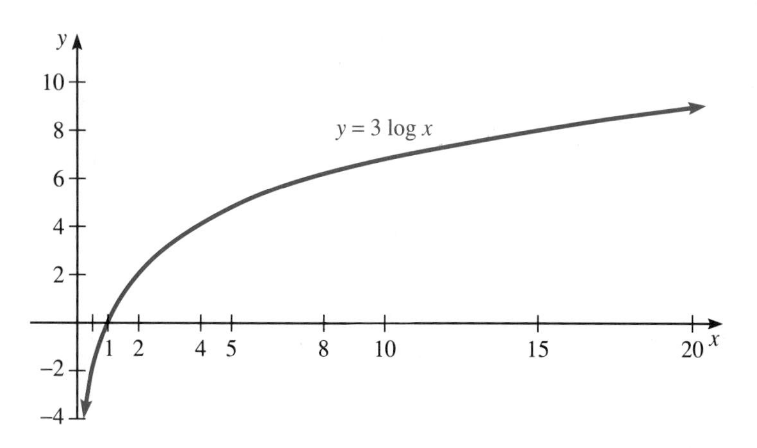

FIGURE 8–11

■

EXAMPLE 6 Graph $y = 3 \log_{0.4}x$.

Solution Since the calculator performs only logarithmic operations with base 10 or e, we must convert this equation to exponential form as follows:

$$y = 3 \log_{0.4}x$$

$$\frac{y}{3} = \log_{0.4}x$$

$$0.4^{y/3} = x$$

Then we fill in a table of values by choosing values for y and calculating x. The graph of $y = 3 \log_{0.4}x$ is given in Figure 8–12.

x	y	
3.39	−4	0.4 [y^x] [(] 4 [+/−] [÷] 3 [)] [=] → 3.39
1.84	−2	0.4 [y^x] [(] 2 [+/−] [÷] 3 [)] [=] → 1.84
1.36	−1	0.4 [y^x] [(] 1 [+/−] [÷] 3 [)] [=] → 1.36
1	0	0.4 [y^x] [(] 0 [÷] 3 [)] [=] → 1
0.74	1	0.4 [y^x] [(] 1 [÷] 3 [)] [=] → 0.74
0.40	3	0.4 [y^x] [(] 3 [÷] 3 [)] [=] → 0.40
0.29	4	0.4 [y^x] [(] 4 [÷] 3 [)] [=] → 0.29

FIGURE 8–12

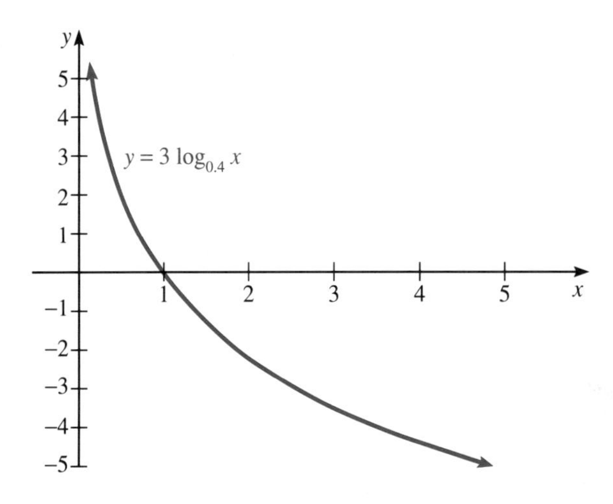

Examples 5 and 6 demonstrate the following properties of the logarithmic function $y = \log_b x$:

- ☐ If $0 < b < 1$, the function is decreasing; that is, y decreases as x increases (Example 6).
- ☐ If $b > 1$, the function is increasing (Example 5).
- ☐ The y axis is an asymptote of the curve.
- ☐ The x intercept is the point (1, 0).
- ☐ $\text{Log}_b x$ is undefined for $x \leq 0$.

8–2 EXERCISES

Convert the following exponential expressions into logarithmic form.

1. $6^3 = 216$
2. $3^4 = 81$
3. $2^5 = 32$
4. $4^3 = 64$
5. $3^{-4} = 1/81$
6. $2^{-4} = 1/16$
7. $(1/3)^3 = 1/27$
8. $e^4 = 54.6$
9. $e^2 = 7.39$
10. $27^{2/3} = 9$
11. $16^{-3/4} = 1/8$
12. $81^{-3/4} = 1/27$

13. $e^{-1} = 0.37$

14. $e^{-3} = 0.05$

15. $8^{-2/3} = 1/4$

16. $125^{-1/3} = 1/5$

Convert the following logarithmic expressions into exponential form.

17. $\log_3 27 = 3$

18. $\log_4 64 = 3$

19. $\log_2 16 = 4$

20. $\log_6 216 = 3$

21. $\log_5 125 = 3$

22. $\log_2 128 = 7$

23. $\ln 20.1 = 3$

24. $\log_{343} 49 = 2/3$

25. $\log_6 1/36 = -2$

26. $\ln 2.72 = 1$

27. $\ln 7.39 = 2$

28. $\ln 148.41 = 5$

29. $\log_{8/27} 9/4 = -2/3$

30. $\log_{16/81}(27/8) = -3/4$

Graph the following functions.

31. $y = \log_2 x$

32. $y = \log_4 x$

33. $y = \log_{0.05} x$

34. $y = 2 \log x$

35. $y = \log x$

36. $y = \log_{0.7}(x - 1)$

37. $y = 4 \log(x + 1)$

38. $y = \log(-x)$

39. $y = -2 \log_{0.4} x$

40. $y = 3 \log_{0.8} 2x$

41. $y = \ln 2x$

42. $y = \ln(x + 3)$

43. $y = -3 \ln x$

44. $y = 4 \ln x$

45. $y = 4 \ln(x - 2)$

46. $y = 5 \ln(2x + 1)$

Without using a calculator, solve the following equations for the variable.

47. $\log_2 64 = x$

48. $\log_4 y = 4$

49. $\log_8 y = 3$

50. $\log_{1/2} 4 = x$

51. $\log_b 16/9 = -2$

52. $\log_b 8 = -3$

53. $\log 10^8 = x$

54. $\log_b 12^{-3} = -3$

55. $\ln e^x = 6$

56. $\ln e^4 = x$

57. $\ln e = x$

58. $\ln(1/e^2) = x$

59. $\ln(1/e) = x$

60. $\ln e^x = 8$

61. The equation for population growth P is given by $P = P_0 e^{kt}$ where P_0 = initial population, t = time, and k = rate of growth. Solve for t.

62. A certain radioisotope decays at a rate equal to $N = N_0(1/2)^x$ where N_0 is initial number of nuclei present after x half-lives. Solve for x.

E 63. A circuit consists of a voltage source V, a resistance R, a capacitance C, and a switch. The current is given by $I = (V/R)e^{-t/(RC)}$. Solve for t.

64. According to Lambert's law, the intensity I of light after passing through a thickness x of a material whose absorption coefficient is k and original intensity I_0 is given by $I = I_0 e^{-kx}$. Solve for x.

65. Yeast grows at a rate given by $y = 100e^{0.2x}$. Solve for x in terms of y.

8–3

PROPERTIES OF LOGARITHMS

Since a logarithm represents an exponential expression, the properties of logarithms are based on the Laws of Exponents. The following Laws of Exponents from Chapter 4 are of greatest importance.

Laws of Exponents

$$b^x \cdot b^y = b^{x+y} \quad \text{multiplication}$$

$$\frac{b^x}{b^y} = b^{x-y} \quad \text{division, } b \neq 0$$

$$(b^x)^p = b^{xp} \quad \text{power}$$

Next, we will relate each of these rules to properties of logarithms. If we let $m = \log_b x$ and $n = \log_b y$, then converting these expressions to exponential form gives $b^m = x$ and $b^n = y$. Applying the multiplication rule for exponents gives

$$xy = b^m b^n$$
$$xy = b^{m+n}$$

Converting this expression to logarithmic form gives

$$\log_b(xy) = m + n$$

Substituting the assigned values $m = \log_b x$ and $n = \log_b y$ gives the following property of logarithms.

Logarithm of a Product

$$\log_b(xy) = \log_b x + \log_b y \quad \text{Rule 8–1}$$

This property of logarithms is based on the multiplication law of exponents.

$$b^x \cdot b^y = b^{x+y}$$

Verbally, this rule states that the logarithm of a product is the sum of the logarithm of each factor.

Similarly, using the same definition for m and n, we have

$$\frac{x}{y} = \frac{b^m}{b^n} = b^{(m-n)}$$

Writing this equation in logarithmic form gives

$$\log_b\left(\frac{x}{y}\right) = m - n$$

Substituting $m = \log_b x$ and $n = \log_b y$ gives the following property of logarithms.

Logarithm of a Quotient

$$\log_b\left(\frac{x}{y}\right) = \log_b x - \log_b y \quad \text{Rule 8–2}$$

This property of logarithms is based on the division law of exponents:

$$\frac{b^x}{b^y} = b^{x-y}$$

Verbally, this rule states that the logarithm of a quotient is the logarithm of the numerator minus the logarithm of the denominator.

Let $m = \log_b x$. Then converting to an exponential expression gives

$$b^m = x$$

Raising both sides of the equation to the p power gives

$$(b^m)^p = x^p$$
$$b^{mp} = x^p$$

Converting this exponential expression to a logarithmic form gives

$$\log_b(x^p) = mp$$

Substituting for m gives the following logarithmic expression.

Logarithm of a Power

$$\log_b(x^p) = p \log_b x \quad \text{Rule 8–3}$$

This property of logarithms is based on the power law of exponents.

$$(b^x)^p = b^{xp}$$

Verbally, this rule states that the logarithm of a quantity raised to a power equals the power times the logarithm of the quantity.

Other properties of logarithms that result from the three properties are listed below.

Other Properties of Logarithms

$$\log_b(b^n) = n \quad \text{Rule 8–4}$$
$$\log_b b = 1 \quad \text{Rule 8–5}$$
$$\log_b 1 = 0 \quad \text{Rule 8–6}$$

CAUTION ✦ Be very careful in using the properties of logarithms. There is no rule for simplifying such expressions as $\log_b(x + y)$ or $\log_b(x - y)$ because there are no laws of exponents for $b^x + b^y$ or $b^x - b^y$. Keep relating the properties of logarithms back to the rules for exponents.

The following examples illustrate the properties of logarithms.

EXAMPLE 1 Express each of the following as a sum, difference, or multiple of logarithms:

(a) $\log_2(5x)$ (b) $\log_7\left(\frac{6}{x}\right)$ (c) $\log_3(2x^4)$

Solution

(a) $\log_2(5x) = \log_2 5 + \log_2 x$ Rule 8–1

(b) $\log_7\left(\frac{6}{x}\right) = \log_7 6 - \log_7 x$ Rule 8–2

(c) $\log_3(2x^4) = \log_3 2 + 4\log_3 x$ Rules 8–1 and 8–3 ■

EXAMPLE 2 Express the following as a sum, difference, or multiple of logarithms:

$$\log_b\left(\frac{64c^4}{a^2}\right)$$

Solution

$\log_b(64c^4) - \log_b a^2$ Rule 8–2

$\log_b 2^6 + \log_b c^4 - \log_b a^2$ Rule 8–1

$6\log_b 2 + 4\log_b c - 2\log_b a$ Rule 8–3 ■

EXAMPLE 3 Find the exact value of each of the following expressions by applying the properties of logarithms:

(a) $\log_3(9 \cdot 81)$ (b) $\log_3(3^4 \cdot 243)^5$

Solution

(a)

$\log_3 9 + \log_3 81$ Rule 8–1

$\log_3 3^2 + \log_3 3^4$ Rule 8–4

$2 + 4$ ($3^2 = 9$ and $3^4 = 81$)

6

(b)

$5\log_3(3^4 \cdot 243)$ Rule 8–3

$5[\log_3 3^4 + \log_3 243]$ Rule 8–1

$5(4 + 5)$ Rule 8–4, $3^5 = 243$

45 ■

EXAMPLE 4 Express the following as a single logarithm and simplify:

(a) $2\ln 6 - 3\ln 3 + \ln 9$ (b) $\log_b(x^2 - 1) + \frac{1}{2}\log_b x - \log_b(x + 1)$

Solution

(a)

$$\ln 6^2 - \ln 3^3 + \ln 9 \quad \text{Rule 8–3}$$

$$\ln\left(\frac{6^2}{3^3}\right) + \ln 9 \quad \text{Rule 8–2}$$

$$\ln\left(\frac{6^2 \cdot 9}{3^3}\right) \quad \text{Rule 8–1}$$

$$\ln 12$$

(b)

$$\log_b(x^2 - 1) + \log_b\sqrt{x} - \log_b(x + 1) \quad \text{Rule 8–3}$$

(Remember, $1/2 \log x = \log x^{1/2} = \log \sqrt{x}$.)

$$\log_b(x^2 - 1)(\sqrt{x}) - \log_b(x + 1) \quad \text{Rule 8–1}$$

$$\log_b\left(\frac{(x^2 - 1)(\sqrt{x})}{(x + 1)}\right) \quad \text{Rule 8–2}$$

$$\log_b [\sqrt{x}(x - 1)] \quad \text{simplify}$$

■

EXAMPLE 5 An expression for the current I with voltage source V, resistance R, capacitance C, and a switch is

$$\ln I + \ln R - \ln V = \frac{-t}{RC}$$

Solve for I.

Solution Let us begin by simplifying the left side of the equation into a single logarithmic expression.

$$\ln(I \cdot R) - \ln V = \frac{-t}{RC} \quad \text{Rule 8–1}$$

$$\ln\left(\frac{IR}{V}\right) = \frac{-t}{RC} \quad \text{Rule 8–2}$$

Then we convert from logarithmic form to exponential form.

$$e^{-t/(RC)} = \frac{IR}{V}$$

Solving the equation for I gives

$$Ve^{-t/(RC)} = \left(\frac{IR}{V}\right)V$$

$$Ve^{-t/(RC)} = IR$$

$$\frac{V}{R}e^{-t/(RC)} = I$$

■

Change of Base

Sometimes it is necessary to change from one logarithmic base to another. The formula for changing bases is as follows.

Change of Base

$$\log_b x = \frac{\log_a x}{\log_a b}$$

where a can be any base but usually 10 or e.

EXAMPLE 6 Find $\log_4 280$ by changing to base 10.

Solution By comparing $\log_4 280$ with the formula for change of base, we find that $b, = 4$, $x = 280$, and $a = 10$.

$$\log_4 280 = \frac{\log_{10} 280}{\log_{10} 4}$$

$$\log_4 280 \approx 4.0646$$

280 [log] [÷] 4 [log] [=] → 4.0646

8–3 EXERCISES

Evaluate the logarithms by using the properties of logarithms.

1. $\log_2(16 \cdot 8)$
2. $\log_3(81 \cdot 27)$
3. $\log_3 81^2$
4. $\log_4\left(\frac{256}{16}\right)$
5. $\log_2\left(\frac{4 \cdot 64}{16}\right)$
6. $\log_2 16^3$
7. $\log_2(8 \cdot 32)^3$
8. $\log_3(27 \cdot 81)^{1/2}$
9. $\log_3\sqrt{81 \cdot 3 \cdot 27}$
10. $\log_4\sqrt{\frac{64}{16}}$
11. $\log_2\sqrt[3]{\frac{64}{8 \cdot 2}}$
12. $\log_2(16\sqrt{8})$

Express as a single logarithmic expression and simplify, if possible.

13. $\log x - 2\log(x - 3)$
14. $3\log x - 2\log(y + 1)$
15. $3\log(x + 1) + \log(x - 3) - \log(x + 1)$
16. $\log(3x - 1) + 2\log x$
17. $\log x + 3\log y - \frac{1}{2}\log(x - 5)$
18. $\log x - 2\log y + \log(x + 1) - \log(y + 3)$
19. $\log(5y) + \frac{1}{3}\log(2y + 5) - 2\log y$
20. $\frac{1}{2}\log(x + 6) + 4\log x - \log(2x)$
21. $2\log x + \log(x + 1) - \log x - \frac{1}{3}\log x$
22. $\log(y + 3) - 2\log(y + 3) - \log y$
23. $3\log(x - 1) - (\log x + 8\log x)$
24. $\log(2x - 3) - 2\log(x + 1) - \log x$
25. $\log\left(\frac{x}{y}\right) + 3\log\left(\frac{x}{z}\right)$
26. $2\log\left(\frac{x}{a}\right) + 3\log\left(\frac{y}{b}\right)$

Given that $\log 2 = 0.3010$, $\log 3 = 0.4771$, and $\log 5 = 0.6990$, find the following logarithms by using the properties of logarithms.

27. $\log 6$

28. $\log 12$

29. $\log\left(\frac{5}{2}\right)$

30. $\log 45$

31. $\log \sqrt{15}$

32. $\log \sqrt{8}$

33. $\log 24$

34. $\log\left(\frac{16}{3}\right)$

35. $\log 75$

36. $\log 200$

37. $\log 90$

38. $\log \sqrt{36}$

39. $\log\left(\frac{75}{4}\right)$

40. $\log\left(\frac{81}{25}\right)$

Use the formula for changing bases to evaluate the following expressions.

41. $\log_6 203$

42. $\log_2 0.0068$

43. $\log_4(16 \cdot 84)$

44. $\log_3(287 \cdot 5)$

45. $\log_8 576$

46. Graph $y = 2 \log x$ and $y = \log x^2$ on the same set of axes. How do the two graphs compare?

47. Graph $y = \log(x + 3)$ and $y = \log x + \log 3$ on the same set of axes, and compare the results.

48. The human ear hears in a logarithmic manner given by the equation

$$\frac{N}{10} = \log\left(\frac{P(\text{out})}{P(\text{in})}\right)$$

where N = gain or loss, $P(\text{in})$ = input power, and $P(\text{out})$ = output power. Express the right side of this equation as a sum, difference, or power of logarithms.

49. The atmospheric pressure P at h miles above sea level leads to the equation

$$\ln P - \ln 14.7 \approx -0.21h$$

Solve this equation for P.

8–4

EXPONENTIAL AND LOGARITHMIC EQUATIONS

Exponential Equations

As defined earlier, an exponential function is a function in which the independent variable appears as an exponent. As we discussed earlier, one technique for solving an exponential equation is to transform each side into an expression with the same base, set the exponents equal, and solve for the variable. However, this technique is not always easy to apply. Therefore, the most applicable method of solving an exponential equation is to take the logarithm of both sides of the equation and then solve. This technique is illustrated in the next four examples.

EXAMPLE 1 Solve the equation $3^x = 2^{(x+1)}$.

Solution It would be difficult to write each side of this equation with the same base. Therefore, *we take the logarithm of each side* of the equation.

$$\log 3^x = \log 2^{(x+1)}$$

Applying the properties of logarithms, we obtain

$$x \log 3 = (x + 1)\log 2$$

Then we solve the equation for x.

$$\begin{aligned} x \log 3 &= x \log 2 + \log 2 && \text{eliminate parentheses} \\ x \log 3 - x \log 2 &= \log 2 && \text{transpose} \\ x(\log 3 - \log 2) &= \log 2 && \text{factor} \\ x &= \frac{\log 2}{(\log 3 - \log 2)} && \text{divide} \\ x &\approx 1.7 \end{aligned}$$

2 [log] [÷] [(] 3 [log] [−] 2 [log] [)] [=] → 1.7095 ■

CAUTION ✦ Always check your answer to make sure it does not result in the logarithm of a negative number.

EXAMPLE 2 Solve the following equation:

$$3(2^{(3x-4)}) = 12^{(x+1)}$$

Solution Taking the logarithm of each side of the equation gives

$$\log[3(2^{(3x-4)})] = \log 12^{(x+1)}$$

Next, apply the properties of logarithms.

$$\log 3 + (3x - 4)\log 2 = (x + 1)\log 12$$

Then solve for x.

$$\begin{aligned} \log 3 + 3x \log 2 - 4 \log 2 &= x \log 12 + \log 12 \\ 3x \log 2 - x \log 12 &= \log 12 - \log 3 + 4 \log 2 \\ x(3 \log 2 - \log 12) &= \log 12 - \log 3 + 4 \log 2 \\ x &= \frac{\log 12 - \log 3 + 4 \log 2}{(3 \log 2 - \log 12)} \\ x &\approx -10.3 \end{aligned}$$

12 [log] [−] 3 [log] [+] 2 [log] [×] 4 [=] [÷]
[(] 2 [log] [×] 3 [−] 12 [log] [)] [=] → −10.2571

Although x is a negative number, this answer does not result in taking the logarithms of a negative number; therefore, $x \approx -10.3$. ■

Applications

One application of exponential equations is exponential growth or decay, discussed earlier in this chapter. The next two examples illustrate exponential growth.

EXAMPLE 3 The equation for population growth is given by $P = P_0e^{kt}$. How long will it take Smalltown to achieve a population of 775 if the current population is 490, and its growth rate is 3% per year?

Solution Substituting the given figures into the formula gives

$$775 = 490e^{0.03t}$$

Then we employ the method used in the previous examples to solve for t. Since the base is e, we take the natural logarithm of both sides of the equation.

$$\ln 775 = \ln(490e^{0.03t})$$
$$\ln 775 = \ln 490 + \ln e^{0.03t}$$

Recall that one of the properties of logarithms is $\log_b b^n = n$. Therefore,

$$\ln e^{0.03t} = 0.03t$$
$$\ln 775 = \ln 490 + 0.03t$$

Solving for t gives

$$\ln 775 - \ln 490 = 0.03t$$
$$\frac{\ln 775 - \ln 490}{0.03} = t$$
$$15.3 \approx t$$

It will require 15.3 years for Smalltown to grow from a population of 490 to 775 people if it continues its current rate of growth.

775 [ln x] [−] 490 [ln x] [=] [÷] 0.03 [=] → 15.28 ■

EXAMPLE 4 Tom wants to invest \$15,000 at an interest rate of 10% compounded annually and accumulate this principal to \$25,000. How long will this take given the formula $S = P(1 + i)^n$?

Solution We substitute and solve the equation for n.

$$S = P(1 + i)^n$$
$$25{,}000 = 15{,}000(1 + 0.10)^n$$
$$\log 25{,}000 = \log[15{,}000(1 + 0.10)^n]$$
$$\log 25{,}000 = \log 15{,}000 + n \log 1.10$$
$$\log 25{,}000 - \log 15{,}000 = n \log 1.10$$
$$\frac{\log 25{,}000 - \log 15{,}000}{\log 1.10} = n$$
$$5.4 \approx n$$

25000 [log] [−] 15000 [log] [=] [÷] 1.10 [log] [=] → 5.4

Tom's \$15,000 will accumulate to \$25,000 in 5.4 years if it is invested at 10% compounded annually and if he does not make any deposits or withdrawals. ■

Logarithmic Equations

A logarithmic equation is an equation that contains the logarithm of the variable. To solve a logarithmic equation, apply the properties of logarithms to *obtain a single logarithmic expression;* then *convert to exponential form*. This technique is illustrated in the next four examples.

EXAMPLE 5 Solve the following equation:

$$\log(2x - 3) - 2 = \log(4x - 1)$$

Solution Isolating the logarithmic terms on one side of the equation gives

$$\log(2x - 3) - \log(4x - 1) = 2$$

Converting to a single logarithmic expression gives

$$\log\left(\frac{2x - 3}{4x - 1}\right) = 2$$

Converting from a logarithmic to an exponential expression gives

$\dfrac{2x - 3}{4x - 1} = 10^2$	base of logarithm is 10
$\dfrac{2x - 3}{4x - 1} = 100$	square 10
$400x - 100 = 2x - 3$	eliminate fractions
$398x = 97$	transpose and add
$x \approx 0.244$	divide

Since $\log(2x - 3)$ and $\log(4x - 1)$ are logarithms of negative numbers when $x = 0.244$, this equation has no solution. ■

EXAMPLE 6 Solve the following equation:

$$2 \log x - 1 = \log(x - 1)$$

Solution Isolating the logarithmic terms and applying the properties of logarithms gives

$2 \log x - \log(x - 1) = 1$	isolate logarithms
$\log\left(\dfrac{x^2}{x - 1}\right) = 1$	convert to single logarithmic expression
$\dfrac{x^2}{x - 1} = 10$	convert to exponential form
$x^2 = 10x - 10$	eliminate fraction
$x^2 - 10x + 10 = 0$	transpose

Use the quadratic formula to solve for x.

$$x = \frac{10 \pm \sqrt{100 - 4(1)(10)}}{2}$$

$$x = \frac{10 \pm \sqrt{60}}{2}$$

$$x \approx 8.87 \quad \text{and} \quad x \approx 1.13$$

Since neither answer results in taking the logarithm of a negative number, the solutions are $x \approx 8.87$ and $x \approx 1.13$. ■

EXAMPLE 7 Solve the equation $\ln(2x + 3) - 3 = \ln(x - 5)$.

Solution We use the same process to solve an equation with natural logarithms. Isolating the logarithmic terms gives

$$\ln(2x + 3) - \ln(x - 5) = 3$$

Then we convert to a single logarithmic expression.

$$\ln\left(\frac{2x + 3}{x - 5}\right) = 3$$

Last, we convert to exponential form and solve for x. Note that the base of the logarithm is e instead of 10.

$$e^3 = \frac{2x + 3}{x - 5}$$

$$20 = \frac{2x + 3}{x - 5}$$

$$20x - 100 = 2x + 3$$

$$18x = 103$$

$$x \approx 5.72$$

■

EXAMPLE 8 In a circuit with a voltage source V, a resistance R, a capacitance C, and a switch, the current I at any time t is given by

$$\ln R + \ln I - \ln V = \frac{-t}{RC}$$

Find the current in a circuit where $R = 500\ \Omega$, $V = 120$ V, $C = 0.0002$ F, and $t = 0.025$ s.

Solution To provide extra practice, let us solve the equation for I and then substitute the given values. Converting to a single logarithmic expression and solving for I gives

$$\ln\left(\frac{RI}{V}\right) = \frac{-t}{RC}$$

$$e^{-t/(RC)} = \frac{RI}{V} \qquad \text{convert to exponential form}$$

$$\frac{V}{R}e^{-t/(RC)} = I \qquad \text{divide}$$

Substituting the given values yields

$$\frac{120}{500}e^{-0.025/[500(0.0002)]} = I$$

$$0.19 \text{ A} \approx I$$

The current is 0.19 A. ■

8–4 EXERCISES

Solve the following exponential equations.

1. $3^{2x} = 5^{x+1}$
2. $4^{5x} = 6^{x+1}$
3. $2^{3x+1} = 6^x$
4. $7^{3x} = 21$
5. $4^{3x} = 7^{x+1}$
6. $5^{3x+1} = 8^{2x}$
7. $6 \cdot 3^{x+5} = 8$
8. $2 \cdot 4^{3x} = 10^x$
9. $3 \cdot 2^{4x} = 6^{x-3}$
10. $6 \cdot 3^x = 7^{x-2}$
11. $e^{3x} = 5$
12. $e^{5x} = 2$
13. $e^{x+1} = 8$
14. $2e^{x+1} = 9$
15. $3e^{2x-1} = 7$
16. $7e^x = 15$

Solve the following logarithmic equations.

17. $\log(3x + 1) - 1 = \log 2x$
18. $\log 2x + \log 3 = \log 12$
19. $2 + \log(5x - 3) = \log x$
20. $\log(6x + 2) = 1 + \log 2$
21. $\log(x^2 - 4) - 3 = \log(x + 2)$
22. $\log(x^2 + 2x - 8) = 2 + \log(x + 4)$
23. $\log(x^2 - 9) - 1 = \log(x - 3)$
24. $\log(x^2 - 25) - \log(x + 5) = 1$
25. $\log x + \log(x + 1) = \log(2x - 1)$
26. $\log(x + 1) = 2 - \log x$
27. $2 \log x + 1 = \log x$
28. $\log(x^2 + 3) - 1 = 2 \log x$
29. $\log(x - 1) = 2 + 2 \log x$
30. $\ln(2x + 1) - \ln x = 3$
31. $\ln x - \ln(x - 1) = 1$
32. $\ln 3x + \ln x = \ln 9$
33. $\ln(3x + 2) - 2 = \ln(x + 1)$
34. $\ln(3x - 2) = \ln(x + 1)$
35. The logarithmic mean temperature difference ΔT_m is given by

$$\Delta T_m = \frac{\Delta T_a - \Delta T_b}{\ln(\Delta T_a/\Delta T_b)}$$

where ΔT_a = change in temperature of the hot liquid and ΔT_b = change in temperature of the cooling liquid. A heat exchanger is operating with a change in the temperature of the hot liquid of 100°, and the change in the temperature of the cooling liquid of 210°. Find the logarithmic mean temperature difference.

36. The growth A of a certain bacteria is given by

$$A = me^{nx}$$

where m = initial amount of bacteria, n = rate of growth, and x = time. If the biologist starts with 500 cells of the bacteria and they grow at a rate of 3% per hour, how long will it take to produce 750 cells of the bacteria?

I **37.** An amount y that must be invested now to produce an amount S in x years at rate i compounded annually is given by

$$y = S\frac{1}{(1+i)^x} \quad \text{or} \quad S(1+i)^{-x}$$

If Jenny invests $2,000 at 12% compounded annually, how long will it take her to reach $7,000?

38. The amount of time it takes for a radioisotope to decay to half the initial amount of radioactivity is called the *half-life* and is expressed by the equation $A = A_0e^{-kt}$. Determine the half-life of a radioisotope by letting $A = A_0/2$ and solving for t.

39. Determine the power gain N of an amplifier in decibels when an input power of 2 W produces an output power of 150 W if $N = 10 \log(P_{out}/P_{in})$.

40. In chemistry the pH of a solution is given by

$$\text{pH} = -\log(\text{H}^+)$$

where H^+ is hydrogen ion concentration. If the pH of a solution is 8.763, find the hydrogen ion concentration.

CHAPTER SUMMARY

Summary of Terms

common logarithm (p. 298)

exponential decay function (p. 291)

exponential function (p. 290)

exponential growth function (p. 290)

natural logarithms (p. 298)

Summary of Formulas

$y = b^x$ exponential function

$x = \log_b y$ if $y = b^x$ logarithm

$y = \log_b x$ logarithmic function

$\log_b(xy) = \log_b x + \log_b y$ logarithm of a product

$\log_b\left(\frac{x}{y}\right) = \log_b x - \log_b y$ logarithm of a quotient

$\log_b(x^p) = p \log_b x$ logarithm of a power

$$\left.\begin{aligned}\log_b(b^n) &= n\\ \log_b b &= 1\\ \log_b 1 &= 0\end{aligned}\right\} \text{other properties of logarithms}$$

$\log_b x = \dfrac{\log_a x}{\log_a b}$ change of base

CHAPTER REVIEW

Section 8–1

Graph the following exponential functions.

1. $y = 3^x$

2. $y = 6^{-x}$

3. $y = (1/5)^x$

4. $y = 4^{2x}$

5. $y = 2e^x$

6. $y = e^{2x}$

7. $y = e^{-3x}$

8. $y = 3e^x$

I **9.** Compound interest is given by $A = P(1 + i)^t$ where P = principal, i = interest rate, t = time in years, and A = accumulated amount. If P = \$1,000 and i = 10% compounded annually, plot the relationship between t and A.

E 10. In the circuit shown in Figure 8–13, the instantaneous voltage across the capacitor, V_c, is given by

$$V_c = V(1 - e^{-t/(RC)})$$

Find V_c when V = 120 V, t = 2 s, R = 50 Ω, and C = 0.003 F.

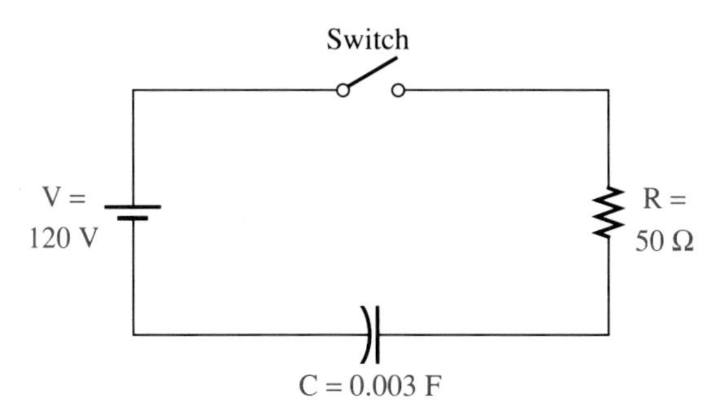

FIGURE 8–13

I 11. Depreciation of business equipment is given by the formula

$$S = C(1 - r)^n$$

where S = scrap value, C = original cost, r = rate of depreciation, and n = useful life in years. Find the scrap value of a machine costing \$8,000 and having a useful life of 10 years and a depreciation rate of 25%.

Section 8–2

Convert the exponential expressions to logarithmic form.

12. $8^3 = 512$
13. $4^{-2} = 1/16$
14. $(1/2)^{-2} = 4$
15. $(81)^{3/4} = 27$
16. $e^0 = 1$
17. $e^3 \approx 20.1$
18. $e^{-4} \approx 0.018$
19. $e^{-2} \approx 0.135$

Convert the logarithmic expressions to exponential form.

20. $\log_{10}100 = 2$
21. $\log_3 9 = 2$
22. $\log_5(1/125) = -3$
23. $\log_2 32 = 5$
24. $\ln 1.65 \approx 1/2$
25. $\ln 0.25 \approx -1.4$
26. $\ln 0.607 \approx -1/2$
27. $\ln 20.1 \approx 3$

Solve for x (without using a calculator) by converting to exponential form.

28. $\log_x 49 = 2$
29. $\log_8 2 = x$
30. $\log_3 x = -3$
31. $\log_x 6^5 = 5$
32. $\ln e^{-4} = x$
33. $\ln(1/e^4) = x$

Section 8–3

Use the properties of logarithms to evaluate the following.

34. $\log_3 81^4$
35. $\log_5(125 \cdot 25)$
36. $\log_4(256 \div 16)$
37. $\log_2(16 \cdot 32 \div 8)$
38. $\log_2(8^3 \cdot 4)$
39. $\log_3(27^3 \cdot 243)$

Write as a single logarithmic expression and simplify, if possible.

40. $\log(x^2 + 1) - \log x + 3 \log x$
41. $3 \log(x + 1) - \log(x + 1)$
42. $\log(x^2 - 16) + \log(x + 4) - \log(x - 8)$
43. $4 \log x + 6 \log(x + 1) - \log x$
44. $\log(2x + 1) - (\log x + 3 \log y)$
45. $\log\left(\frac{x}{y}\right) + 2 \log\left(\frac{y}{z}\right)$

Section 8–4

Solve the following equations.

46. $2^x = 3^{2x-1}$
47. $7^{3x} = 6^{2x+1}$
48. $4(3^{5x-1}) = 18$
49. $7(4^{x+1}) = 6^{2x}$
50. $e^{5x-1} = 28$
51. $e^{2x} = 90$
52. $\log x - \log 3 = \log 8$
53. $\log(x + 1) + 1 = \log(3x + 7)$
54. $\log x + \log(x + 2) = -1$
55. $2 \log x = \log(x + 2)$
56. $\ln x - \ln(3x + 2) = 2$
57. $\ln(x + 3) - \ln x = 4$

I 58. Given the equation $A = P(1 + i)^t$, determine how long it would take \$750 to grow into \$1,200 at an interest rate of 10.75% compounded annually.

59. Given the equation $P = P_0e^{kt}$, determine how long it would take a town of 1,200 people (P_0) to grow to 2,000 people (P) if the annual rate of growth k is 8%.

CHAPTER TEST

The number in parentheses refers to the appropriate learning objective at the beginning of the chapter.

1. Graph the logarithmic function $y = 3\log(x + 1)$. (1)
2. Evaluate the following by using the properties of logarithms: (5)

$$\log_3(81)^8$$

3. Depreciation is given by the formula $S = C(1 - r)^n$. What is the useful life n of a computer whose original cost C is \$10,000, depreciation rate r is 15%, and scrap value S is \$2,500? (6)
4. Graph the exponential function $y = 5e^{-x}$. (1)
5. The formula for business depreciation can be written as (3, 5)

$$\log S - \log C = \log(1 - r)^n$$

Solve for S.

6. Solve the following for x: (6)

$$e^{x+1} = 40$$

7. Convert $\left(\frac{1}{8}\right)^{1/3} = \frac{1}{2}$ to logarithmic form. (3)
8. Graph $x = e^y$ and $\ln x = y$ on the same set of axes. How do the graphs compare? (1)
9. The intensity (in lumens) of a light, after passing through a thickness x (in cm) of a medium having an absorption coefficient of 0.3, is given by (2)

$$I = 1{,}500e^{-0.3x}$$

What is the intensity of a light beam that passes through 8 cm of this medium?

10. Write the following as a single logarithmic expression: (5)

$$\log(x + 1) - 2\log(x - 1) + 3\log x - \log(3x + 4)$$

11. Solve the following for x: (6)

$$\log(x + 3) - \log 2 = \log 6x$$

12. Convert $\ln 0.37 \approx -1$ into an exponential expression. (3)
13. Given that $\log 5 \approx 0.6990$ and $\log 11 \approx 1.0414$, find $\log(50 \cdot 11^2)$ using the properties of logarithms. (Do not use a calculator.) (5)
14. Solve the following for x: (6)

$$3^{x+1} = 7^{2x-1}$$

15. Solve the following for x by converting to exponential form: (4)

$$\log_6 216 = x$$

16. Newton's law of cooling states that when a heated object is immersed in a cooling solution, the temperature T of the object at any time t after immersion is given by (2)

$$T = T_0 e^{-0.4t}$$

where T_0 = initial temperature of the cooling solution. Find T when $T_0 = 230°$ and $t = 5$ s.

17. Given the equation $\text{pH} = -\log[H^+]$, find the hydrogen ion concentration H^+ of a solution whose pH is 5.73. (6)

SOLUTION TO CHAPTER INTRODUCTION

The population of the city is an exponential growth function given by $y = ae^{nx}$ where a = initial population, n = rate of growth, and x = number of years. Using this information, we compute the population as follows:

$$y = 56{,}180e^{0.048(15)}$$

$$y \approx 115{,}418$$

The city will have a population of 115,418 in 15 years if its annual rate of increase is 4.8%.

You are a radiologist treating a tumor 8.76 cm below the surface of the skin. However, to avoid the lungs, the radiation source must be aimed at an angle. The radiation source is moved 13.49 cm horizontally. What is the angle, relative to the patient's skin, at which you must aim the radiation source in order to hit the tumor, and how far does the beam travel through the body before reaching the tumor? (The answer to this problem is given at the end of the chapter.)

Ancient Greek engineers used the field of trigonometry to measure objects that could not be measured directly. Trigonometry has evolved over the years to its current status as an invaluable tool in almost every area of science and technology. In this chapter we discuss right angle trigonometry.

Learning Objectives

After completing this chapter, you should be able to

1. Convert an angle measured in degrees, minutes, and seconds into decimal parts of a degree, and vice versa (Section 9–1).
2. Draw an angle and determine angles coterminal with it (Section 9–1).
3. Given a point on the terminal side of an angle in standard position, determine any of the six trigonometric functions of that angle (Section 9–2).
4. Given the value of one trigonometric function, determine the value of any of the remaining functions of that angle (Section 9–2).
5. Using a calculator, find the value of a given trigonometric function (Section 9–3).
6. Using a calculator, find the first quadrant angle for a given trigonometric value (Section 9–3).
7. Solve a right triangle from given information (Section 9–4).
8. Apply the methods of solving a right triangle to technical problems (Section 9–4).

Chapter 9

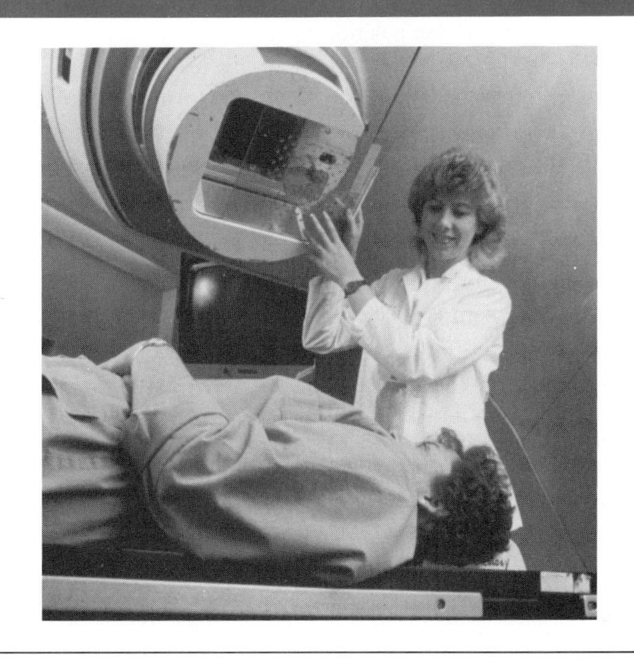

Right Angle Trigonometry

9–1

ANGLES

This chapter begins the study of the branch of mathematics called **trigonometry,** which literally means *triangle measurement*. Trigonometry deals primarily with six ratios called **trigonometric functions,** which are used extensively in physics and engineering. However, before discussing the trigonometric functions, we must define the terminology used.

Definitions

An **angle** is defined as the amount of rotation required to move a ray (a half-line) from one position to another. *The original position of the ray* is called the **initial side** of the angle, and *the final position of the ray* is called the **terminal side** of the angle. *The point about which the rotation occurs and at which the initial and terminal sides of the angle intersect* is called the **vertex.** The measure of the angle itself is the amount of rotation between the initial and terminal sides. The sign of an angle is determined by the direction of rotation of the initial side. *If the rotation from the initial to terminal side is in a counterclockwise direction,* the angle measure is said to be *positive,* and if *the rotation is in a clockwise direction,* the angle measure is said to be *negative*. To specify an angle, we need the initial side, the terminal side, and a curved arrow extending from the initial to the terminal side to show the direction of rotation. Figure 9–1 illustrates this terminology, with angle 1 a positive angle and angle 2 a negative angle.

FIGURE 9–1

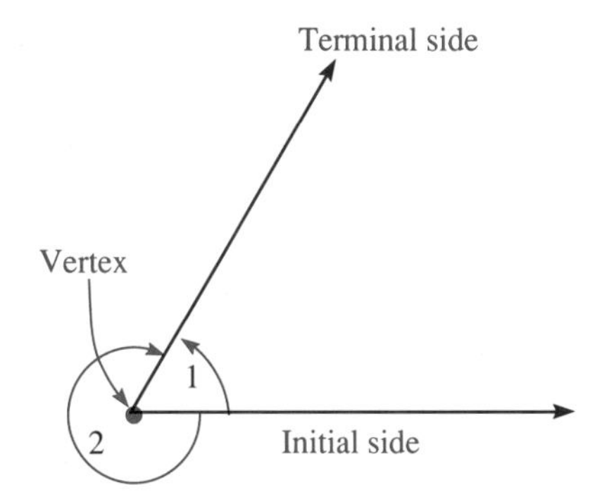

Angular Measure

Angles can be measured in degrees, radians, or grads. One rotation measures 360 degrees, 2π radians, or 400 grads. The discussion in this chapter will concentrate on degree measure, and radians will be discussed in a later chapter. *The degree can be further divided into 60 equal parts* called **minutes,** and *each minute can be divided into 60 equal parts* called **seconds.** The symbols °, ′, and ″ are used to denote degrees, minutes, and seconds, respectively. Calculators usually use decimal parts of a degree for angular measure rather than degrees, minutes, and seconds. Therefore, in order to use a calculator, you must convert angles in degrees, minutes, and seconds to degrees. The next two examples illustrate this conversion process.

EXAMPLE 1 Convert an angle whose measure is 47°39′53″ to a decimal part of a degree. Round to hundredths.

Solution First, we convert 53 seconds into a decimal part of a minute by dividing by 60, obtaining 0.88. At this point we have converted 47°39′53″ to 47°39.88′. Then we must change 39.88 minutes into a decimal portion of a degree by dividing by 60. Thus, 47°39′53″ is equivalent to 47.66°.

53 [÷] 60 [=] $0.88\overline{3}$ [+] 39 [=] [÷] 60 [=] [+] 47 [=] → 47.66 ■

EXAMPLE 2 Convert an angle whose measure is 158.48° to an angle measured to the nearest minute.

Solution We must convert 0.48° to minutes by multiplying 0.48° by 60′. The angle 158.48° is equivalent to 158°29′. ■

NOTE ✦ Some calculators have a button to convert degrees, minutes, and seconds to a decimal part of a degree, and vice versa. Look in your owner's manual for details on the procedure.

Coterminal Angles

Two angles are **coterminal** if they have the *same initial and terminal sides*. In Figure 9–2, angles A, B, and C are all coterminal angles. To find angles coterminal to a given angle, add or subtract multiples of 360° to it.

FIGURE 9–2

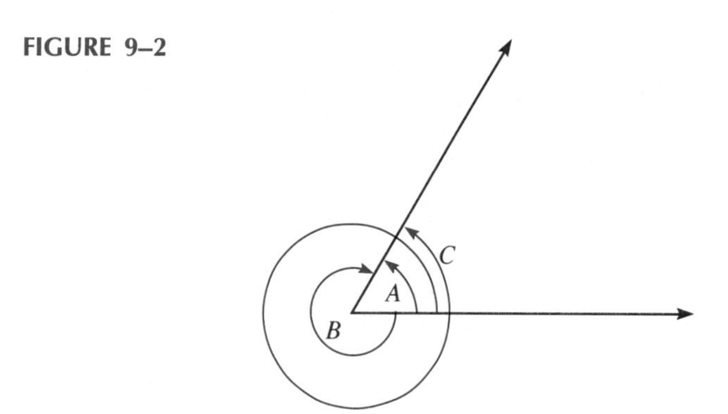

EXAMPLE 3 Find two angles that are coterminal with an angle of 112°45′.

Solution Since there are 360° in one complete rotation, you can find one angle coterminal with a 112°45′ angle by adding 360° to 112°45′, which gives 472°45′. To find a second angle coterminal with 112°45′, find the negative angle with the same initial and terminal sides. To find the measure of this angle, subtract 360° from 112°45′. To subtract these angles, you must subtract like units: degrees from degrees and minutes from minutes. Thus, subtract 359°60′ from 112°45′ to obtain −247°15′. These angles are illustrated in Figure 9–3. ■

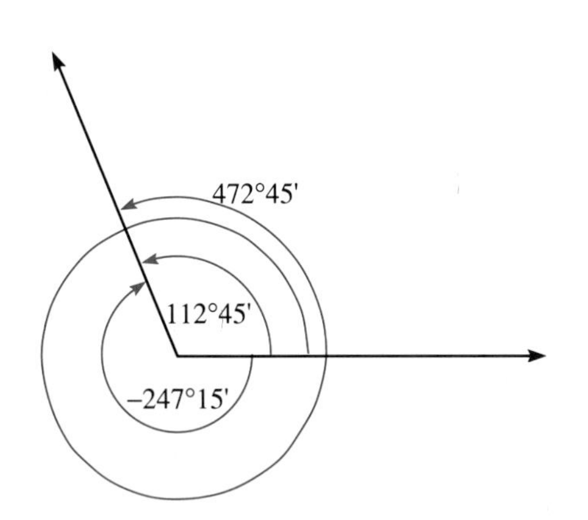

FIGURE 9–3

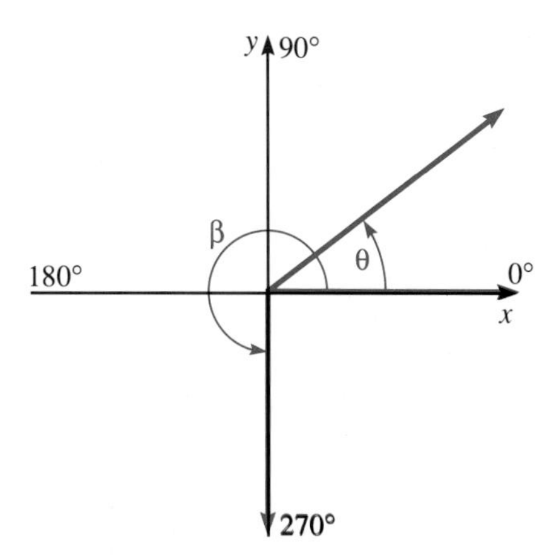

FIGURE 9–4

Angles and Rectangular Coordinates

Often we will draw angles on the rectangular coordinate plane. If an angle has *its initial side on the positive x axis and its vertex at the origin,* the angle is said to be in **standard position.** Therefore, the quadrant of an angle in standard position, disregarding the direction of rotation, is determined by the location of the terminal side. An angle whose *terminal side is located on one of the axes* is called a **quadrantal angle.** Figure 9–4 shows angle θ in standard position and angle β a quadrantal angle of 270°. The figure also gives the measure of the other quadrantal angles.

EXAMPLE 4 Draw an angle in standard position whose measure is 160°.

Solution We start by positioning the initial side of the angle on the positive x axis. Then we rotate that ray in a counterclockwise direction 160°, which places the terminal side of the angle in the second quadrant. This angle is shown in Figure 9–5. This is a second quadrant angle. ■

An angle in standard position whose terminal side is in the first, second, third, or fourth quadrant is called a *first, second, third,* or *fourth quadrant angle,* respectively. An angle can be specified by angular measure, as in Example 4, or by a point on the terminal side, as in the next example.

EXAMPLE 5 Draw an angle in standard position whose terminal side passes through the point $(-6, 4)$.

Solution To draw this angle, we place the initial side of the angle on the positive x axis. Then we plot the point $(-6, 4)$ in the coordinate plane and draw a line from the origin through the point. This angle is a second quadrant angle and is shown in Figure 9–6. ■

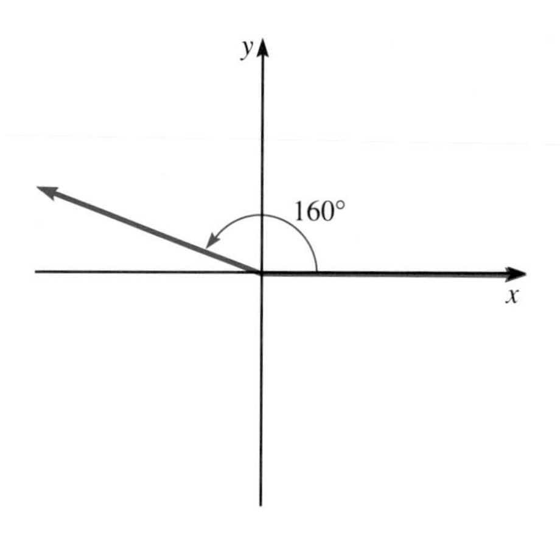

FIGURE 9–5

FIGURE 9–6

9–1 EXERCISES

Convert the following angles measured in degrees, minutes, and seconds to angles measured to the nearest hundredths of a degree.

1. 34°46′23″ **2.** 158°37′48″
3. 318°13′53″ **4.** 98°26′38″
5. 148°17′36″ **6.** 229°26′15″
7. 39°18′40″ **8.** 346°12′56″
9. 256°25′37″

Convert the following angles measured in degrees to angles measured to the nearest minute.

10. 35.68° **11.** 274.14°
12. 57.20° **13.** 158.34°
14. 63.56° **15.** 314.73°
16. 263.8° **17.** 195.71°
18. 208.43°

Draw the following angles and find two angles coterminal with it.

19. 35°47′ **20.** 167.46°
21. 318°34′45″ **22.** 193.27°
23. 56°29′41″ **24.** 75.8°
25. 18°56′ **26.** 347°
27. 253°38′45″

Draw the following angles in standard position and determine whether the angle is a first, second, third, or fourth quadrant angle.

28. 156.7° **29.** 93°26′
30. 395° **31.** 43°51′28″
32. 126.67° **33.** 169°45′
34. 216°36′49″ **35.** 184.6°
36. 283°45′

Draw an angle in standard position such that the terminal side passes through the point given.

37. (3, 8) **38.** (−4, 6)
39. (3, −6) **40.** (6, 9)
41. (−5, −2) **42.** (−5, 8)
43. (6, −1) **44.** (−8, −5)
45. (7, −4)

9–2

DEFINING THE TRIGONOMETRIC FUNCTIONS

Trigonometry is based on six ratios which we will now develop. If we place a first quadrant angle in standard position and drop a perpendicular line from a point (x, y) on the terminal side of the angle to the x axis, we form a right triangle. As illustrated in Figure 9–7, *the side opposite the right angle* is called the **hypotenuse** or the **radius,** r. *The side opposite angle* θ is called the **opposite side,** and its length can be denoted by the y coordinate of the point (x, y). *The side next to angle* θ is called the **adjacent side,** and its length can be denoted by the x coordinate of the point (x, y). Given this right triangle, the trigonometric functions are defined as the ratio of the sides of the triangle. The **trigonometric functions** are sine (sin), cosine (cos), tangent (tan), secant (sec), cosecant (csc), and cotangent (cot). These functions are defined in the following box.

FIGURE 9–7

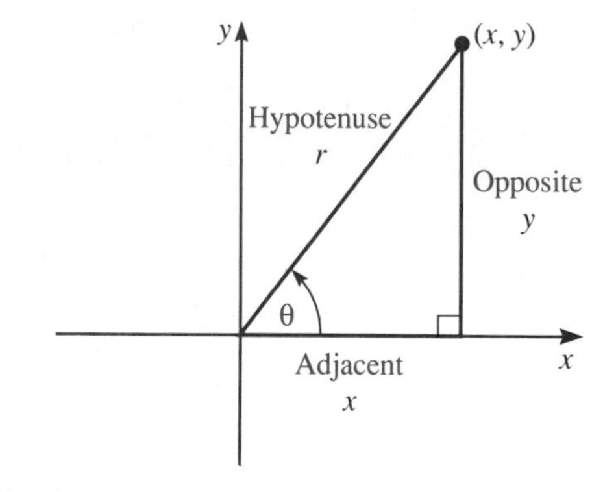

The Trigonometric Functions

Given the right triangle defined in Figure 9–7,

$$\sin\theta = \frac{\text{opposite}}{\text{hypotenuse}} = \frac{y}{r}$$

$$\cos\theta = \frac{\text{adjacent}}{\text{hypotenuse}} = \frac{x}{r}$$

$$\tan\theta = \frac{\text{opposite}}{\text{adjacent}} = \frac{y}{x}$$

$$\sec\theta = \frac{\text{hypotenuse}}{\text{adjacent}} = \frac{r}{x}$$

$$\csc\theta = \frac{\text{hypotenuse}}{\text{opposite}} = \frac{r}{y}$$

$$\cot\theta = \frac{\text{adjacent}}{\text{opposite}} = \frac{x}{y}$$

NOTE ✦ In determining the trigonometric functions for a given angle, it does not matter which point on the terminal side you choose because the resulting triangles are similar. Since the ratio of the sides of similar triangles are equal, an equivalent trigonometric ratio would result from any point chosen on the terminal side.

Reciprocal Trigonometric Ratios

Note from the definition of the six trigonometric functions that three of the functions are the reciprocals of the other three. The sine and cosecant, the cosine and secant, and the tangent and cotangent are reciprocals. These reciprocal relationships are very important to remember, especially when you are using a calculator. They are summarized in the following box.

Reciprocal Trigonometric Functions

$\sin\theta = \dfrac{y}{r}$	$\csc\theta = \dfrac{r}{y}$	$\csc\theta = \dfrac{1}{\sin\theta}$
$\cos\theta = \dfrac{x}{r}$	$\sec\theta = \dfrac{r}{x}$	$\sec\theta = \dfrac{1}{\cos\theta}$
$\tan\theta = \dfrac{y}{x}$	$\cot\theta = \dfrac{x}{y}$	$\cot\theta = \dfrac{1}{\tan\theta}$

Finding Trigonometric Values

Given the sides of a right triangle, we can determine the value of all six trigonometric functions.

EXAMPLE 1 For the triangle given in Figure 9–8, find the six trigonometric functions. Round to hundredths.

FIGURE 9–8

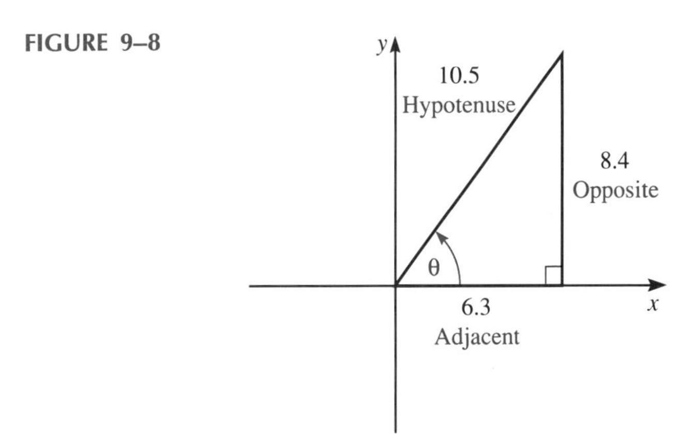

Solution For angle θ, 8.4 is the length of the opposite side, or y; 6.3 is the length of the adjacent side, or x; and 10.5 is the length of the hypotenuse, or r. Therefore, the trigonometric functions are as follows:

$$\sin\theta = \frac{y}{r} = \frac{\text{opposite}}{\text{hypotenuse}} = \frac{8.4}{10.5} = 0.80$$

$$\cos\theta = \frac{x}{r} = \frac{\text{adjacent}}{\text{hypotenuse}} = \frac{6.3}{10.5} = 0.60$$

$$\tan\theta = \frac{y}{x} = \frac{\text{opposite}}{\text{adjacent}} = \frac{8.4}{6.3} = 1.33$$

$$\csc\theta = \frac{r}{y} = \frac{\text{hypotenuse}}{\text{opposite}} = \frac{10.5}{8.4} = 1.25$$

$$\sec\theta = \frac{r}{x} = \frac{\text{hypotenuse}}{\text{adjacent}} = \frac{10.5}{6.3} = 1.67$$

$$\cot\theta = \frac{x}{y} = \frac{\text{adjacent}}{\text{opposite}} = \frac{6.3}{8.4} = 0.75$$

■

EXAMPLE 2 The terminal side of angle β passes through the point (5, 8). Assuming that β is in standard position, determine the trigonometric functions of this angle to four decimal places.

FIGURE 9–9

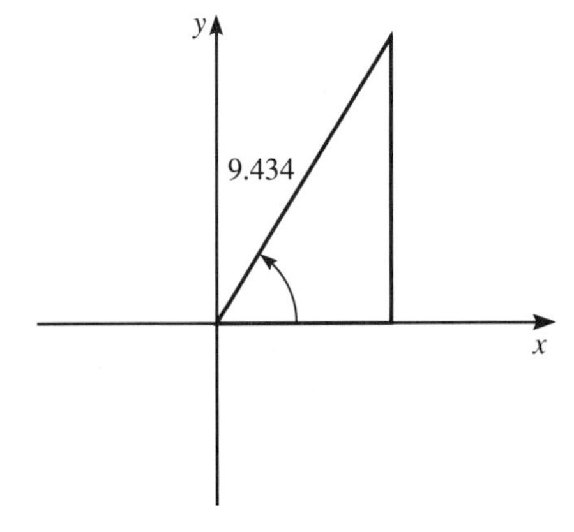

Solution To determine the value of the trigonometric functions, we must know the value of the three sides of the right triangle, which is shown in Figure 9–9. Therefore, we calculate r by using the Pythagorean theorem.

$$\text{hypotenuse}^2 = \text{leg}_1^2 + \text{leg}_2^2$$

$$r^2 = 8^2 + 5^2$$

$$r = \sqrt{8^2 + 5^2}$$

$$r = 9.4340$$

8 $\boxed{x^2}$ $\boxed{+}$ 5 $\boxed{x^2}$ $\boxed{=}$ $\boxed{\sqrt{\ }}$ → 9.4340

$$\sin \beta = \frac{\text{opposite}}{\text{hypotenuse}} = \frac{8}{9.434} = 0.8480$$

$$\cos \beta = \frac{\text{adjacent}}{\text{hypotenuse}} = \frac{5}{9.434} = 0.5300$$

$$\tan \beta = \frac{\text{opposite}}{\text{adjacent}} = \frac{8}{5} = 1.6000$$

$$\csc \beta = \frac{\text{hypotenuse}}{\text{opposite}} = \frac{9.434}{8} = 1.1792$$

$$\sec \beta = \frac{\text{hypotenuse}}{\text{adjacent}} = \frac{9.434}{5} = 1.8868$$

$$\cot \beta = \frac{\text{adjacent}}{\text{opposite}} = \frac{5}{8} = 0.6250$$

■

Determining Values from a Given Value

Given the value of one of the trigonometric functions, we can find the value of the remaining functions.

EXAMPLE 3 Given that $\cos \alpha = 6/11$, find the missing trigonometric functions of α to four decimal places.

Solution Since the cosine is defined to be the ratio of the adjacent side and the hypotenuse, we can label the appropriate sides of the right triangle as shown in Figure 9–10. To find the missing side, we apply the Pythagorean theorem.

$$11^2 = 6^2 + y^2$$
$$y = 9.2195$$

11 [x^2] [−] 6 [x^2] [=] [$\sqrt{\ }$] → 9.2195

FIGURE 9–10

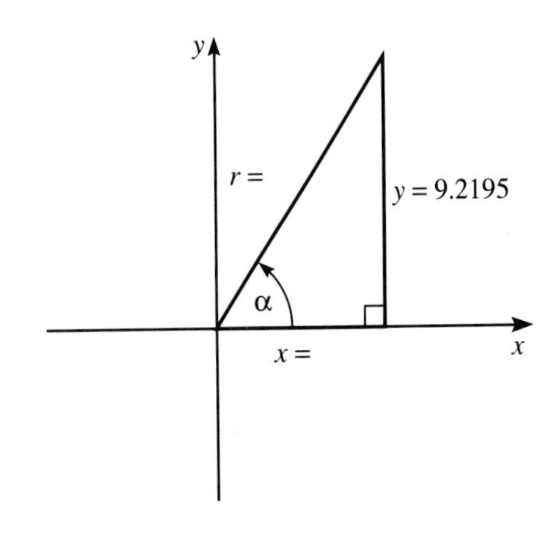

The remaining trigonometric values are calculated as follows:

$$\sin\alpha = \frac{9.2195}{11} = 0.8381 \qquad \csc\alpha = \frac{11}{9.2195} = 1.1931$$

$$\tan\alpha = \frac{9.2195}{6} = 1.5366 \qquad \sec\alpha = \frac{11}{6} = 1.8333$$

$$\cot\alpha = \frac{6}{9.2195} = 0.6508$$

■

EXAMPLE 4 Given that $\cot\theta = 0.6669$, find $\sin\theta$ and $\sec\theta$.

Solution First, we write the cotangent value as a ratio in order to determine x and y.

$$\cot\theta = \frac{x}{y} = \frac{0.6669}{1} \quad \begin{matrix} \to & x = 0.6669 \\ \to & y = 1 \end{matrix}$$

Next, we use the Pythagorean theorem to find r ($r > 0$).

$$r^2 = (0.6669)^2 + 1^2$$
$$r^2 \approx 1.4448$$
$$r \approx 1.2020$$

0.6669 $\boxed{x^2}$ $\boxed{+}$ 1 $\boxed{x^2}$ $\boxed{=}$ $\boxed{\sqrt{\ }}$ $\to$ 1.2020

Then we determine $\sin\theta$ and $\sec\theta$ by using the definition and the triangle shown in Figure 9–11.

$$\sin\theta = \frac{y}{r} = \frac{1}{1.2020} = 0.8319$$

$$\sec\theta = \frac{r}{x} = \frac{1.2020}{0.6669} = 1.8024$$

FIGURE 9–11

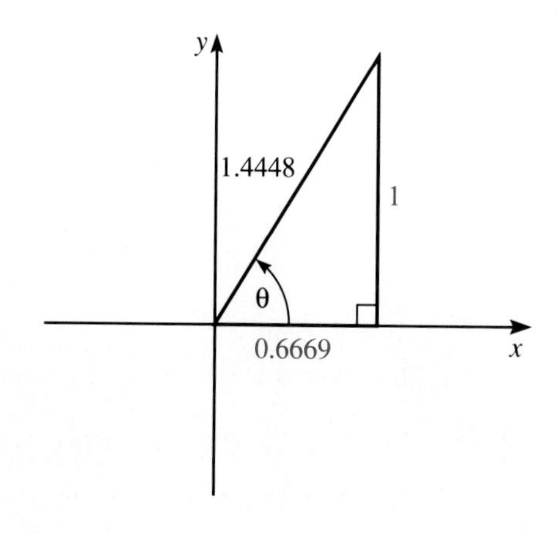

■

9–2 EXERCISES

Each of the following is a point on the terminal side of an angle in standard position. Determine the value to four decimal places of all six trigonometric functions of the angle.

1. (3, 2)
2. $(4, \sqrt{5})$
3. (6, 13)
4. (15, 3)
5. (9, 8)
6. (10, 7)
7. $(\sqrt{3}, 1)$
8. (1, 6)
9. (1.7, 3.6)
10. (7.4, 11.8)
11. (16, 12)
12. (9.1, 6.5)

Use the given trigonometric function to find the value of the indicated trigonometric function. Give your answer to four decimal places.

13. $\sin\theta = \frac{6}{7}$; find $\tan\theta$ and $\csc\theta$.

14. $\cot\theta = \frac{14}{5}$; find $\sin\theta$ and $\cos\theta$.

15. $\sec\theta = \frac{11}{7}$; find $\sin\theta$ and $\cot\theta$.

16. $\cos\theta = \frac{5}{8}$; find $\tan\theta$ and $\csc\theta$.

17. $\tan\theta = 3$; find $\cos\theta$ and $\csc\theta$.

18. $\sin\theta = 0.7458$; find $\tan\theta$ and $\sec\theta$.

19. $\csc\theta = 1.6587$; find $\sec\theta$ and $\cot\theta$.

20. $\cot\theta = 6.8431$; find $\tan\theta$ and $\sec\theta$.

21. $\tan\theta = 1.6300$; find $\cot\theta$ and $\cos\theta$.

22. $\sec\theta = 2.7568$; find $\tan\theta$ and $\sin\theta$.

9–3

VALUES OF THE TRIGONOMETRIC FUNCTIONS

In Section 9–2, we discussed finding the value of the trigonometric functions given the length of the sides of the right triangle. However, in reality we are usually trying to find the value of a trigonometric function given the measure of the angle. Later in this section we will discuss finding the value of trigonometric functions using a calculator. We will begin our discussion with some special angles.

Functions of 45°

If we construct an isosceles (two sides of equal length) right triangle with legs each measuring one unit, the resulting angles of the triangle are 45°, 45°, and 90°. Remember that the sum of the angles of a triangle is 180°. Using the Pythagorean theorem, we find that the hypotenuse of this right triangle is

$$\text{leg}_1^2 + \text{leg}_2^2 = \text{hypotenuse}^2$$

$$\sqrt{\text{leg}_1^2 + \text{leg}_2^2} = \text{hypotenuse}$$

$$\sqrt{1^2 + 1^2} = \sqrt{2}$$

The sides of a 45°, 45°, 90° triangle are 1, 1, and $\sqrt{2}$. The basic trigonometric functions for 45° are given in the following box.

REMEMBER ✦ The longest side of a triangle is always opposite the largest angle, and the shortest side is opposite the smallest angle.

Functions of 45°

$$\sin 45^\circ = \frac{1}{\sqrt{2}} = \frac{\sqrt{2}}{2}$$

$$\cos 45^\circ = \frac{1}{\sqrt{2}} = \frac{\sqrt{2}}{2}$$

$$\tan 45^\circ = \frac{1}{1} = 1$$

NOTE ✦ Remember that we obtain $\frac{\sqrt{2}}{2}$ from $\frac{1}{\sqrt{2}}$ by rationalizing the denominator.

Functions of 30° and 60°

If we construct an equilateral (three sides of equal length) triangle with sides each two units long, each of the resulting angles is 60° (180°/3 = 60°). If we construct an altitude to the base of the triangle, it bisects both the angle and the base as shown in Figure 9–12. Using the Pythagorean theorem to determine the length of the altitude gives

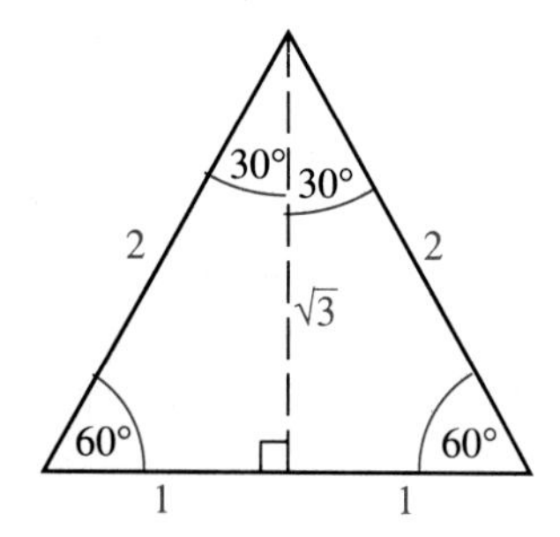

FIGURE 9–12

$$\text{leg}_1^2 + \text{leg}_2^2 = \text{hypotenuse}^2$$

$$\sqrt{\text{hypotenuse}^2 - \text{leg}_1^2} = \text{leg}_2$$

$$\sqrt{2^2 - 1^2} = \sqrt{3}$$

The sides of a 30°, 60°, 90° triangle are 1, $\sqrt{3}$, and 2, respectively. Using this triangle, we can calculate the following trigonometric functions.

Functions of 30°

$$\sin 30^\circ = \frac{1}{2}$$

$$\cos 30^\circ = \frac{\sqrt{3}}{2}$$

$$\tan 30^\circ = \frac{\sqrt{3}}{3}$$

Functions of 60°

$$\sin 60^\circ = \frac{\sqrt{3}}{2}$$

$$\cos 60^\circ = \frac{1}{2}$$

$$\tan 60^\circ = \frac{\sqrt{3}}{1} = \sqrt{3}$$

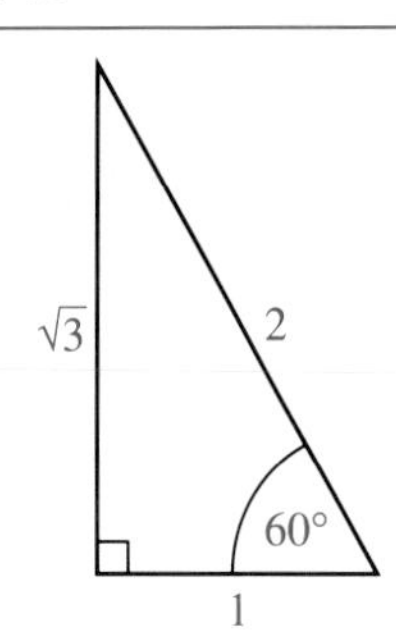

Functions of 0° and 90°

For the quadrantal angle 0° with $r = 1$, we can find the following trigonometric functions.

Functions of 0°

$$\sin 0^\circ = \frac{y}{r} = \frac{0}{1} = 0$$

$$\cos 0^\circ = \frac{x}{r} = \frac{1}{1} = 1$$

$$\tan 0^\circ = \frac{y}{x} = \frac{0}{1} = 0$$

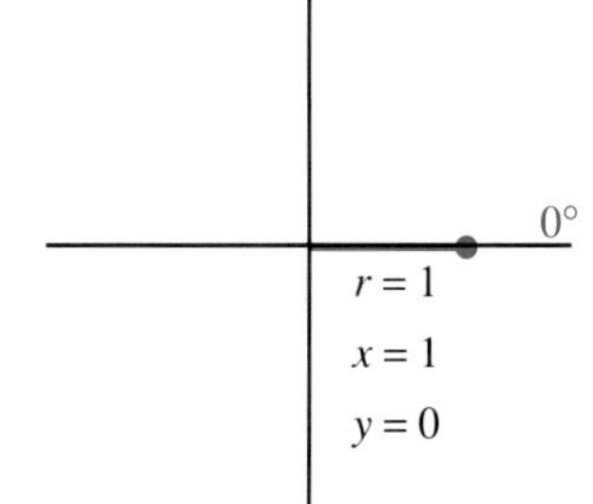

Similarly, for the quadrantal angle 90° with $r = 1$, we can find the following trigonometric functions.

Functions of 90°

$$\sin 90^\circ = \frac{y}{r} = \frac{1}{1} = 1$$

$$\cos 90^\circ = \frac{x}{r} = \frac{0}{1} = 0$$

$$\tan 90^\circ = \frac{y}{x} = \frac{1}{0} = \text{undefined}$$

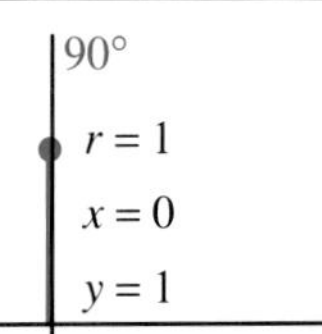

Table 9–1 summarizes the trigonometric functions of the angles discussed in this section. The values of these trigonometric functions are used often enough that you should memorize them. Also remember that the secant, cosecant, and cotangent functions are reciprocals of the cosine, sine, and tangent functions, respectively, and each can be found from the value of its reciprocal.

TABLE 9–1

	Special Angles				
Function	0°	30°	45°	60°	90°
sin	0	$\frac{1}{2}$	$\frac{\sqrt{2}}{2}$	$\frac{\sqrt{3}}{2}$	1
cos	1	$\frac{\sqrt{3}}{2}$	$\frac{\sqrt{2}}{2}$	$\frac{1}{2}$	0
tan	0	$\frac{\sqrt{3}}{3}$	1	$\sqrt{3}$	undef.
cot	undef.	$\sqrt{3}$	1	$\frac{\sqrt{3}}{3}$	0
sec	1	$\frac{2\sqrt{3}}{3}$	$\sqrt{2}$	2	undef.
csc	undef.	2	$\sqrt{2}$	$\frac{2\sqrt{3}}{3}$	1

Evaluating Trigonometric Functions

Since the geometric methods used to find the trigonometric value of angles such as 30° or 45° are difficult for other angles, we use a calculator to find the trigonometric functions for most acute angles. To find the value of a trigonometric function with an algebraic calculator, enter the angle and then press the desired trigonometric function key. For example, to find sin 48.6°, make sure your calculator is in the degree mode and your angle is in decimal degrees. Press 48.6 and then the [sin] key.

EXAMPLE 1 Using your calculator, find the following trigonometric functions rounded to three decimal places:

(a) sin 63° (b) cos 38.7° (c) tan 57.49° (d) sin 36°42′

Solution Be sure your calculator is in the degree mode when performing this operation.

(a) 63 [sin] → 0.891 (b) 38.7 [cos] → 0.780 (c) 57.49 [tan] → 1.569

(d) The calculator does not recognize an angle measured in degrees and minutes. Therefore, it is necessary to change 42 minutes into a decimal portion of a degree by dividing it by 60. Thus, $36°42'$ becomes $36.7°$.

36.7 [sin] → 0.598 ■

If you look at your calculator, you will not find keys to represent the secant, cosecant, or cotangent functions. In order to find the value of these trigonometric functions on the calculator, we must use the reciprocal relationships of the cosine, sine, and tangent, respectively. We use the [$1/x$] key on the calculator to rapidly compute this reciprocal relationship.

EXAMPLE 2 Find each of the following to three decimal places:

(a) sec 16.75° (b) cot 43.3° (c) csc 74°17′

Solution Be sure your calculator is in the degree mode.

(a) Since the secant is the reciprocal of the cosine function, we can write

$$\sec 16.75° = \frac{1}{\cos 16.75°}$$

We calculate this expression as

16.75 [cos] [$1/x$] → 1.044

(b) Since the cotangent is the reciprocal of the tangent function,

$$\cot 43.3° = \frac{1}{\tan 43.3°}$$

43.3 [tan] [$1/x$] → 1.061

(c) Since the cosecant is the reciprocal of the sine function,

$$\csc 74°17' = \frac{1}{\sin 74°17'}$$

In calculating the expression on the right side of the equation, we must change $74°17'$ into degrees.

74 [+] 17 [÷] 60 [=] [sin] [$1/x$] → 1.039 ■

Determining the Angle from the Function Value

In applications it is sometimes necessary to find the measure of an angle from the value of its trigonometric function. To perform this operation on the calculator, use the [inv] or [2nd] key, followed by the appropriate trigonometric function. On some calculators, it's [$\sin^{-1}$], [$\cos^{-1}$], or [$\tan^{-1}$].

EXAMPLE 3 Find θ to the nearest tenth of a degree for each of the following:

(a) $\sin \theta = 0.3684$ (b) $\cos \theta = 0.7983$ (c) $\tan \theta = 2.865$

Solution Be sure your calculator is in the degree mode.

(a) 0.3684 [inv] [sin] $\rightarrow$ 21.6 $21.6° \approx \theta$

(b) 0.7983 [inv] [cos] $\rightarrow$ 37.0 $37.0° \approx \theta$

(c) 2.865 [inv] [tan] $\rightarrow$ 70.8 $70.8° \approx \theta$ ■

EXAMPLE 4 Find θ to the nearest minute for the following:

(a) $\cos \theta = 0.5436$ (b) $\tan \theta = 0.3685$

Solution

(a) 0.5436 [inv] [cos] $\rightarrow$ 57.071

Then change to minutes by pressing the following keys:

[−] 57 [=] [×] 60 [=] 4.26

$57°4' \approx \theta$

(b) 0.3685 [inv] [tan] $\rightarrow$ 20.23

Write down 20 and then change to minutes by pressing the following keys:

[−] 20 [=] [×] 60 [=] $\rightarrow$ 13.7

$20°14' \approx \theta$ ■

REMEMBER ✦ Some calculators have a button to convert decimal degrees to degrees and minutes, making the conversion shown in Example 4 unnecessary.

EXAMPLE 5 Find θ to the nearest tenth of a degree for each of the following:

(a) $\sec \theta = 3.465$ (b) $\cot \theta = 0.1679$ (c) $\csc \theta = 2.693$

Solution

(a) $\sec\theta = \dfrac{1}{\cos\theta} = 3.465$

$\cos\theta = \dfrac{1}{3.465}$

3.465 [1/x] [inv] [cos] $\rightarrow$ 73.2 $73.2° \approx \theta$

(b) $\cot\theta = \dfrac{1}{\tan\theta} = 0.1679$

$\tan\theta = \dfrac{1}{0.1679}$

0.1679 [1/x] [inv] [tan] $\rightarrow$ 80.5 $80.5° \approx \theta$

(c) $\csc\theta = \dfrac{1}{\sin\theta} = 2.693$

$\sin\theta = \dfrac{1}{2.693}$

2.693 [1/x] [inv] [sin] $\rightarrow$ 21.8 $21.8° \approx \theta$ ■

CAUTION ✦ To avoid errors, it is best to write down the reciprocal trigonometric relationships and then perform the operation on your calculator.

EXAMPLE 6 Find θ to the nearest minute for the following:

(a) $\csc\theta = 2.684$ (b) $\cot\theta = 1.87$

Solution

(a) $\csc\theta = \dfrac{1}{\sin\theta} = 2.684$

2.684 [1/x] [inv] [sin] [−] 21 [=] [×] 60 [=] $\rightarrow$ 52 (converting to minutes)

$21°52' \approx \theta$

(b) $\cot\theta = \dfrac{1}{\tan\theta} = 1.87$

1.87 [1/x] [inv] [tan] [−] 28 [=] [×] 60 [=] $\rightarrow$ 8 (converting to minutes)

$28°8' \approx \theta$ ■

9–3 EXERCISES

Find the value of the given trigonometric function.

1. sin 56.8° **2.** tan 78.56°
3. cos 84° **4.** sec 18.6°
5. cos 63.71° **6.** sin 19.3°
7. tan 65.7° **8.** csc 27.8°
9. sec 31.56° **10.** cot 51.35°
11. sin 22.6° **12.** cos 44°
13. tan 17.4° **14.** sec 23°
15. csc 72° **16.** cot 83.76°
17. csc 43.75° **18.** cos 32.56°
19. cot 48.7° **20.** sin 67.9°
21. cos 36°18′ **22.** sec 18°43′
23. tan 73°52′ **24.** sin 58°12′
25. csc 61°39′ **26.** cot 10°24′
27. sec 76°19′ **28.** csc 29°8′

Find θ to the nearest tenth of a degree.

29. $\sin\theta = 0.763$ **30.** $\tan\theta = 2.268$
31. $\cos\theta = 0.378$ **32.** $\csc\theta = 2.631$
33. $\cot\theta = 3.76$ **34.** $\cos\theta = 0.968$
35. $\sin\theta = 0.138$ **36.** $\cot\theta = 0.67$
37. $\sec\theta = 2.87$ **38.** $\cot\theta = 1.37$
39. $\tan\theta = 5.678$ **40.** $\cos\theta = 0.421$
41. $\tan\theta = 3.617$ **42.** $\sin\theta = 0.861$
43. $\csc\theta = 4.815$ **44.** $\sec\theta = 1.973$

Find θ to the nearest minute.

45. $\tan\theta = 6.739$ **46.** $\cos\theta = 0.391$
47. $\csc\theta = 4.358$ **48.** $\sec\theta = 2.78$
49. $\sin\theta = 0.765$ **50.** $\cot\theta = 2.538$
51. $\tan\theta = 1.328$ **52.** $\cot\theta = 3.614$
53. $\csc\theta = 1.38$ **54.** $\sec\theta = 3.482$
55. $\sin\theta = 0.758$ **56.** $\cos\theta = 0.382$

M **57.** The torque required to move the load up the thread of a screw is given by

$$T = \frac{FD}{2}\left(\frac{\cos A \tan B + f}{\cos A - f \tan B}\right)$$

Find T if $F = 500$ lb, $D = 1.2$ in., $f = 0.15$, $A = 12°$, and $B = 2.5°$.

58. The axial force required for a cone brake is given by

$$F = \frac{T(\sin A + f\cos A)}{fR}$$

Find F if $T = 75$ ft-lb, $A = 12°$, $f = 0.25$, and $R = 0.4$ ft.

59. For a leading power factor, the voltage induced in an armature is given by

$$E = \sqrt{(V\cos\theta + IR)^2 + (V\sin\theta - IX)^2}$$

Find the induced voltage if $V = 500$ V, $I = 200$ A, $R = 0.1\ \Omega$, $X = 1.2\ \Omega$, and $\theta = 53°$.

60. The index of refraction is defined as

$$n = \frac{\sin i}{\sin r}$$

where i = angle of incidence and r = angle of refraction. Find the index of refraction if $r = 40°$ and $i = 34.7°$.

9–4

THE RIGHT TRIANGLE

We can model many applied problems by using a right triangle. Trigonometry is extremely useful in solving for unknown angles and unknown sides of a right triangle. Solving a right triangle means determining the measure of all sides and angles of the triangle. In this section we will discuss techniques for solving a right triangle.

EXAMPLE 1 Find angle A shown in Figure 9–13.

Solution Recall that each of the six trigonometric functions utilizes one of the acute angles and two sides of the right triangle. To find angle A, we set up a trigonometric ratio using the angle and two known sides of the triangle. As shown in Figure 9–13, we are given the side opposite the angle and the hypotenuse. Therefore, we use the sine function.

$$\sin A = \frac{\text{opposite}}{\text{hypotenuse}}$$

$$\sin A = \frac{17.2}{25.6}$$

$$\sin A \approx 0.672$$

$$A = 42.2°$$

17.2 [÷] 25.6 [=] [inv] [sin] → 42.21 ■

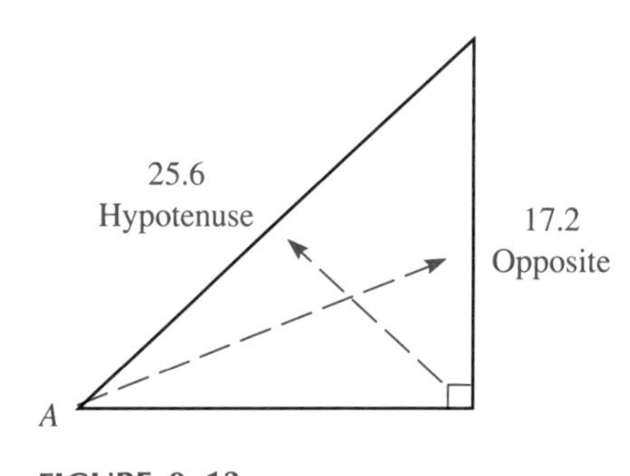

FIGURE 9–13

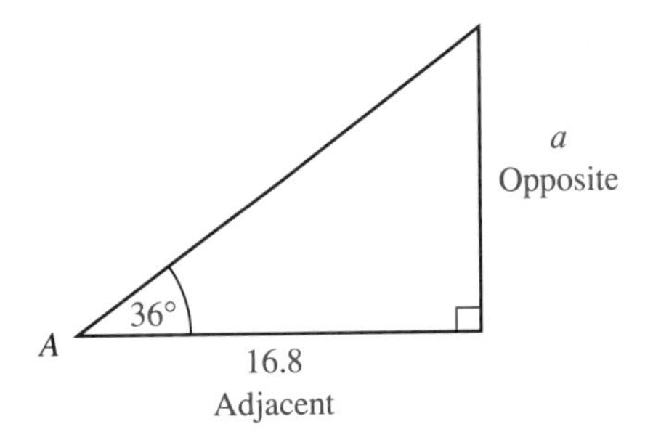

FIGURE 9–14

EXAMPLE 2 Find side a in the triangle shown in Figure 9–14 if $A = 36°$ and $b = 16.8$. Round a to have the same accuracy as b.

Solution We set up a trigonometric ratio involving side a, a known angle, and a known side.

$$\tan A = \frac{\text{opposite}}{\text{adjacent}}$$

$$\tan 36° = \frac{a}{16.8}$$

$$a = 16.8(\tan 36°)$$

$$a = 12.2$$

16.8 [×] 36 [tan] [=] → 12.2 ■

NOTE ✦ There is usually more than one correct way to solve a right triangle. For example, the cotangent function could have been used in Example 2.

Applications

We can use the trigonometric functions to find information in applied problems. The remainder of this section deals with methods of solving applied problems.

EXAMPLE 3 The angle of elevation from an observer on the ground to the top of a building is 37.4°. Find the height of the building if the observer is 240 ft from the base of the building.

Solution The **angle of elevation** is the *angle formed by the horizontal and the line of sight* of an observer. As shown in Figure 9–15, the angle of elevation is measured above the horizontal. Figure 9–16 gives a visual representation of the verbal statement.

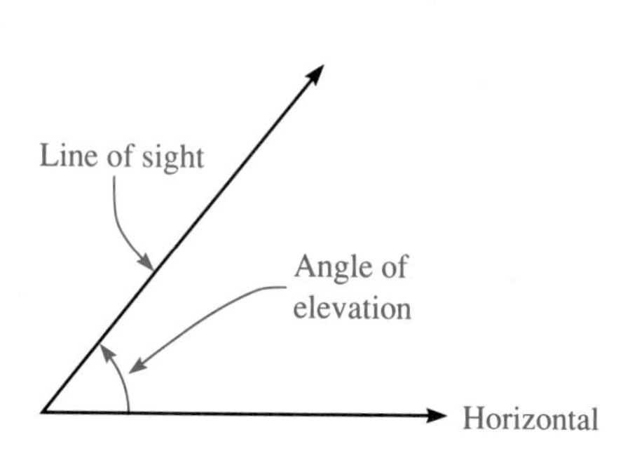

FIGURE 9–15

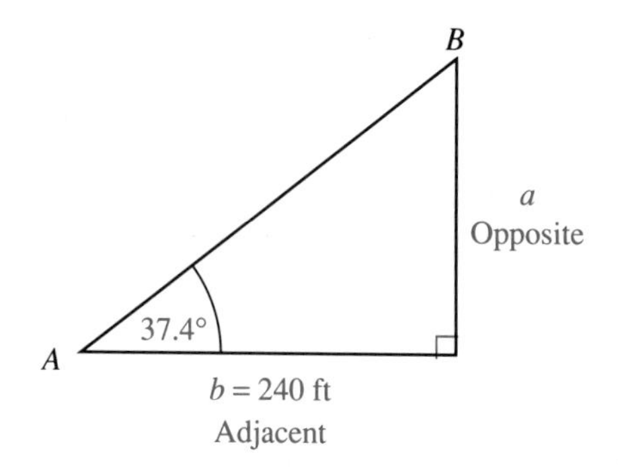

FIGURE 9–16

Next, we set up a trigonometric ratio and solve for a, the height.

$$\tan A = \frac{a}{b}$$

$$\tan 37.4° = \frac{a}{240}$$

$$a = 240(\tan 37.4°)$$

$$a = 183 \text{ ft}$$

240 [×] 37.4 [tan] [=] → 183 ■

Solving Applied Problems

1. Read the verbal statement.
2. Make a drawing to depict the verbal statement. Label all known quantities and check your figure against the verbal statement.
3. Set up a trigonometric ratio containing the unknown, and solve for the unknown.
4. Check your solution to see if it is reasonable.

EXAMPLE 4 The end of a loading ramp is 3.4 m from the base of the loading dock. Determine the length of the ramp so that it makes an angle of 18° relative to the ground.

Solution Figure 9–17 gives a drawing of the verbal statement. We set up a trigonometric ratio and solve for the unknown. From the drawing we know the acute angle of 18°, the side adjacent to the angle, and we want to calculate the hypotenuse. Therefore, we choose the cosine function. The calculation is as follows:

$$\cos 18° = \frac{3.4}{c}$$

$$c = \frac{3.4}{\cos 18°}$$

$$c = 3.6 \text{ m}$$

The ramp needs to be 3.6 m long.

3.4 [÷] 18 [cos] [=] → 3.6

FIGURE 9–17

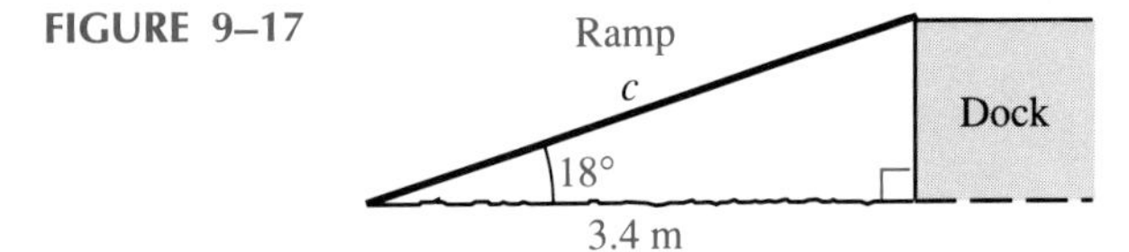

EXAMPLE 5 A man in a plane flying at an altitude of 850.0 ft sees a boat on the lake. If the angle of depression is 25°, find the straight-line distance from the plane to the boat.

Solution The **angle of depression** is measured from the horizontal down to the line of sight of an observer. As shown in Figure 9–18, the angle of depres-

FIGURE 9–18

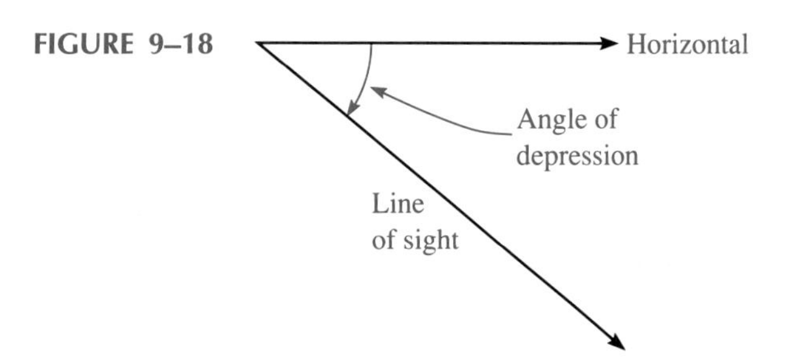

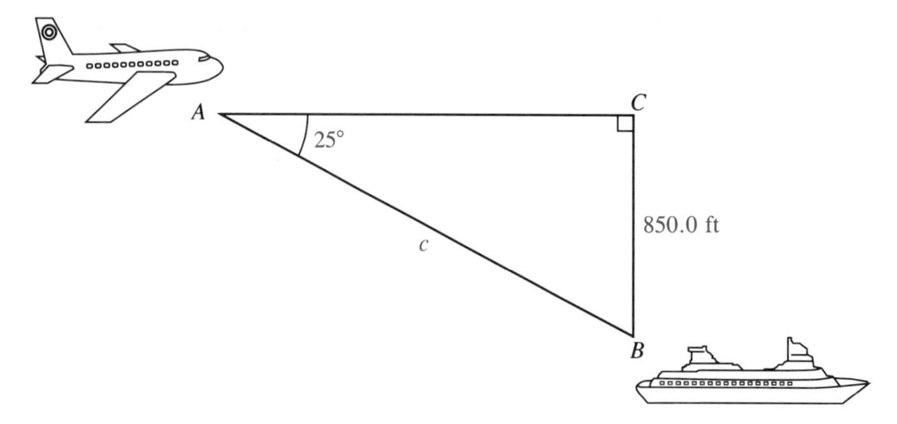

FIGURE 9–19

sion is measured below the horizontal. A drawing of the verbal statement is shown in Figure 9–19.

$$\sin 25° = \frac{850}{c}$$

$$c = \frac{850}{\sin 25°}$$

$$c = 2{,}011 \text{ ft}$$

850 [÷] 25 [sin] [=] → 2011.27 ■

EXAMPLE 6 **Inductive reactance** X is a *measure of the retardation of current flow* by the capacitance and inductance. **Impedance** Z, measured in ohms Ω, is a *measure of the retardation of current flow* by all components of a circuit. Reactance, impedance, and total resistance R form a right triangle as shown in Figure 9–20. The angle θ is called the **phase angle.** In an ac circuit with an inductive reactance of 50.0 Ω and a total resistance of 73 Ω, find the phase angle and impedance.

FIGURE 9–20

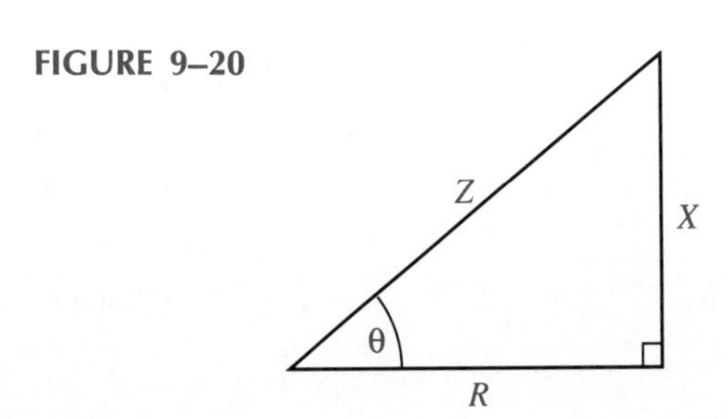

Solution Figure 9–21 gives a visual representation of the verbal statement. We can solve for impedance Z by using the Pythagorean theorem.

$$Z^2 = (50)^2 + (73)^2$$
$$Z^2 = 7{,}829$$
$$Z = 88\ \Omega$$

FIGURE 9–21

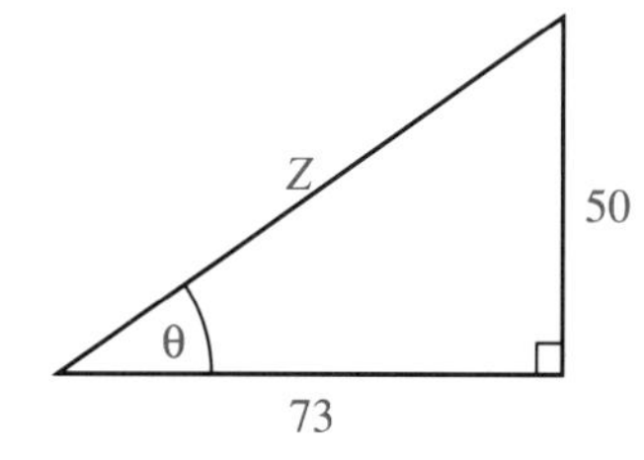

We use the tangent function to solve for the phase angle θ.

$$\tan\theta = \frac{50}{73}$$
$$34^\circ = \theta$$

50 [÷] 73 [=] [inv] [tan] → 34 ■

9–4 EXERCISES

Solve the accompanying right triangle having the given parts. Express your answer to the same degree of accuracy as that given in the problem.

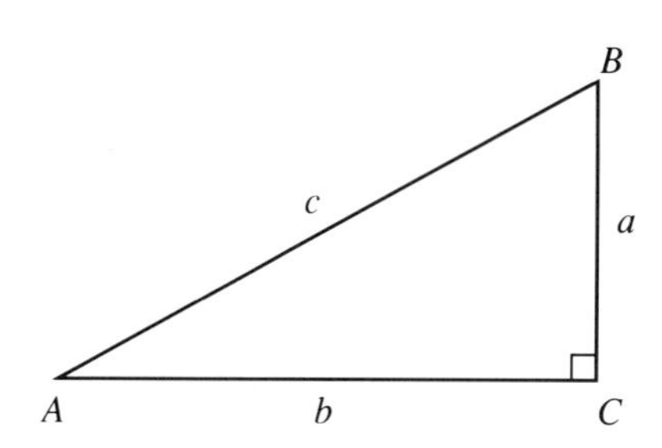

1. $a = 6.8$ m, $B = 43.8°$
2. $c = 16$ ft, $a = 9.8$ ft
3. $A = 36°43'$, $c = 39.5$ cm
4. $b = 123.6$ yd, $c = 196.83$ yd
5. $B = 14.5°$, $a = 16$ m
6. $A = 36°18'$, $c = 27.8$ ft
7. $A = 65°37'$, $b = 48.63$ cm
8. $B = 76.8°$, $a = 14.56$ yd
9. $c = 64$ ft, $a = 38$ ft
10. $A = 49.25°$, $c = 89.67$ m
11. $b = 43.3$ ft, $a = 29.8$ ft
12. $B = 57°14'$, $b = 78.8$ cm
13. From the roof of a building 80 ft high, the angle of depression to an object on the street is 58°. How far from the base of the building is the object?

E 14. The phase angle of an ac circuit is 28°, and the inductive reactance is 38 Ω. Find the impedance. (See Example 6.)

15. The average rise in elevation of a mountain road is 11.4°. What is the change in altitude (vertical) if the road length is 16.8 mi?

16. A telephone pole is to be supported by two guy wires. How much total wire is needed if the telephone pole is 40 ft high and the angle between the ground and the wire is 58°?

17. Find the length of a ramp if it makes an angle of 23° with the ground and the loading platform is 4.1 ft above the ground.

18. A house is 30 ft wide and the roof has an angle of inclination of 23°. Find the length L of the rafters. See Figure 9–22.

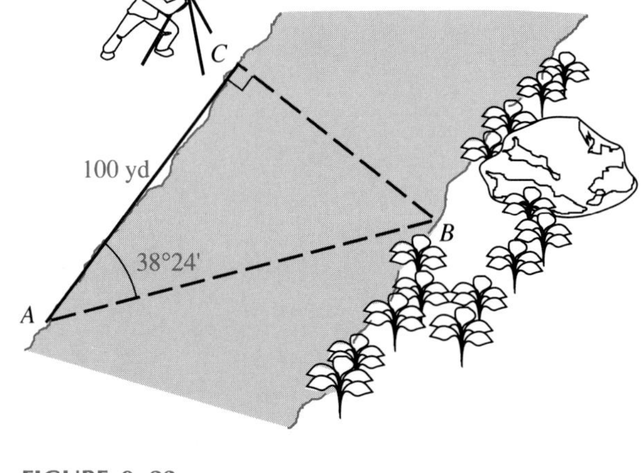

FIGURE 9–23

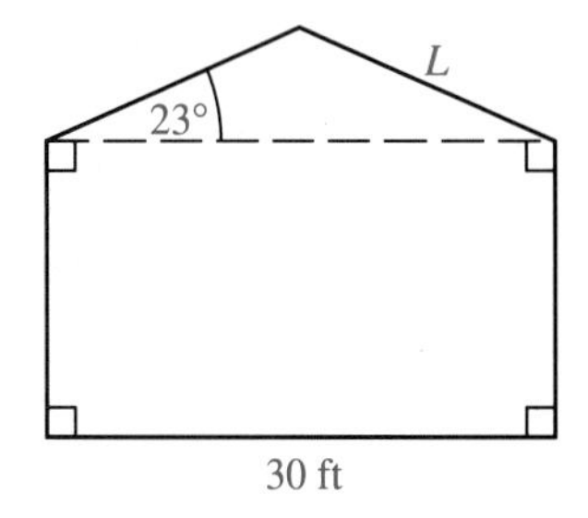

FIGURE 9–22

C 19. A surveyor wishes to determine the width of a river that is too wide to measure directly. She sights a point B on the opposite side of the river from where she stands at point C. Then she measures 100 yd from point C to point A, thus forming a right triangle. Sighting from point A to point B, she determines that angle A is 38°24′. Find the width of the river. See Figure 9–23.

20. On a blueprint the walls of a rectangular room are 5.4 cm and 8.6 cm. Find the angle between the shorter wall and the diagonal across the room.

21. A weather balloon ascends at a rate of 3.86 km/h while moving at an angle of 71.8° with the ground. At the end of an hour, what is the horizontal travel x of the balloon? Refer to Figure 9–24.

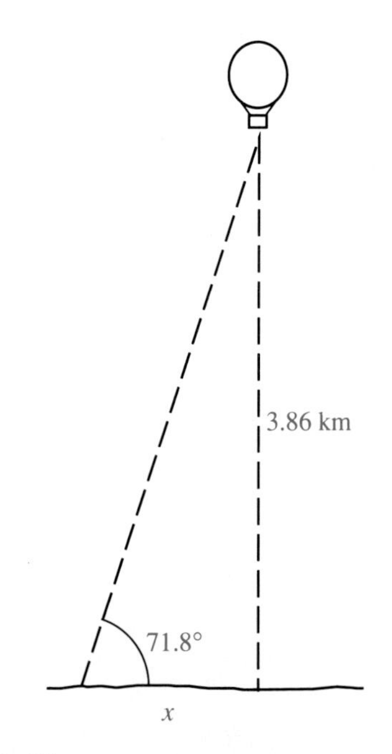

FIGURE 9–24

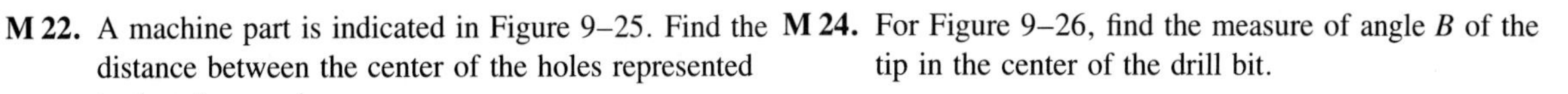

M 22. A machine part is indicated in Figure 9–25. Find the distance between the center of the holes represented by lengths x and y.

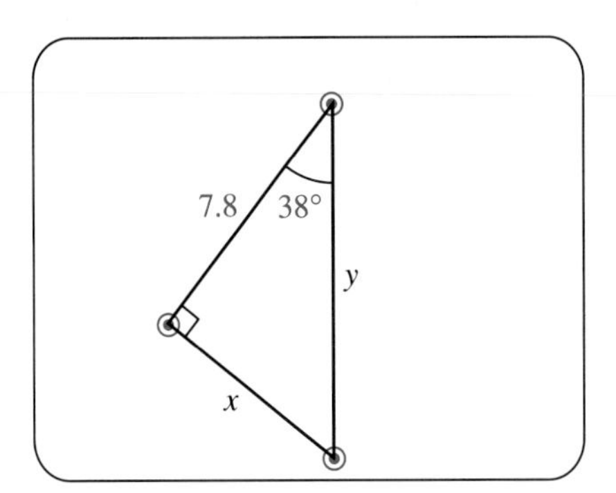

FIGURE 9–25

23. Radiation therapy is used to treat a tumor 7.62 cm below the surface of the skin. However, the radiologist must aim the radiation source at an angle to avoid the lungs, and he moves the radiation source 10.54 cm horizontally. What is the angle, relative to the patient's skin, at which the radiologist must aim the radiation source in order to hit the tumor? How far does the beam travel through the body before reaching the tumor?

M 24. For Figure 9–26, find the measure of angle B of the tip in the center of the drill bit.

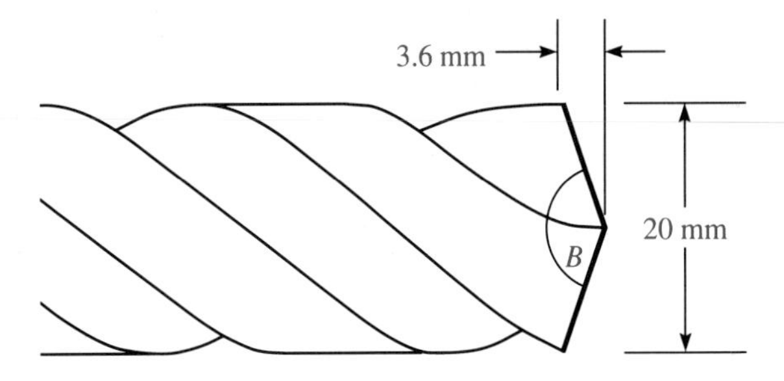

FIGURE 9–26

25. A grain conveyor belt used to store corn in a silo is 177 ft long. If the barn opening is 50 ft from the ground, at what angle relative to the ground will the conveyor belt sit?

26. The angle of elevation from point C through point B to point A is 12°18′, and the distances are as given in Figure 9–27. Find the vertical and horizontal distance from A to C.

C 27. A plot plan requires the length of a property line to be 315.6 ft; however, a tree obstructs direct measurement. From the property line, point C, the surveyor turns the transit perpendicular and measures 54.7 ft

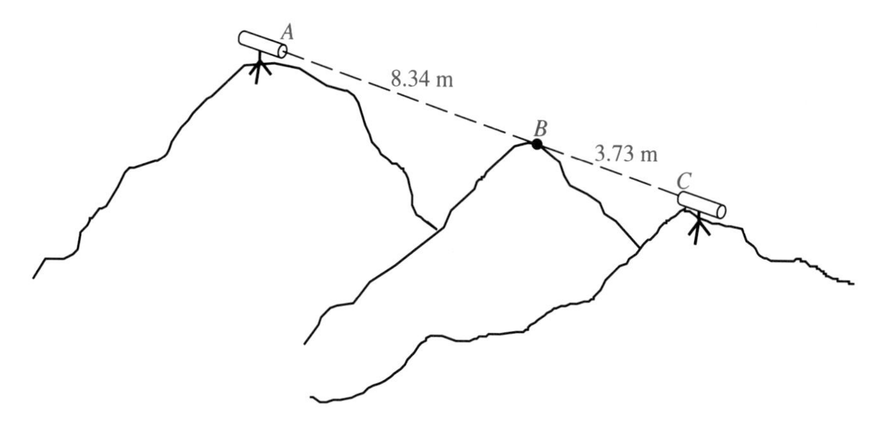

FIGURE 9–27

to point A. The surveyor sets up the transit at point A and sights back to point B. Determine the measure of angle A and the distance c to locate B correctly. Refer to Figure 9–28.

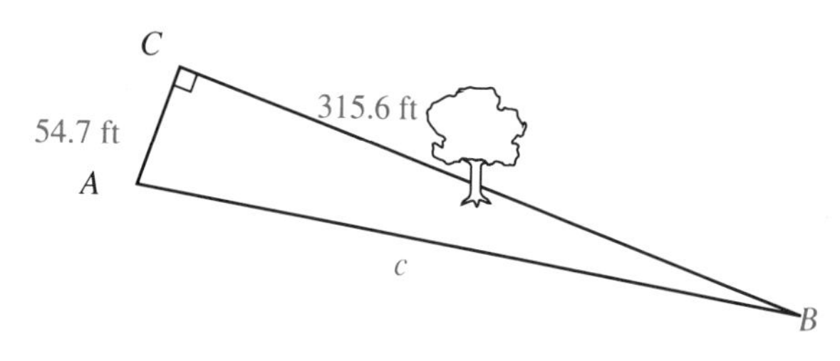

FIGURE 9–28

28. If light strikes a reflecting surface, the angle of incidence i is equal to the angle of reflection r. If a light beam has an angle of incidence equal to 42.3°, find the distance y from the surface of the plane to a point on the beam if the horizontal distance x is 16.34 in. See Figure 9–29.

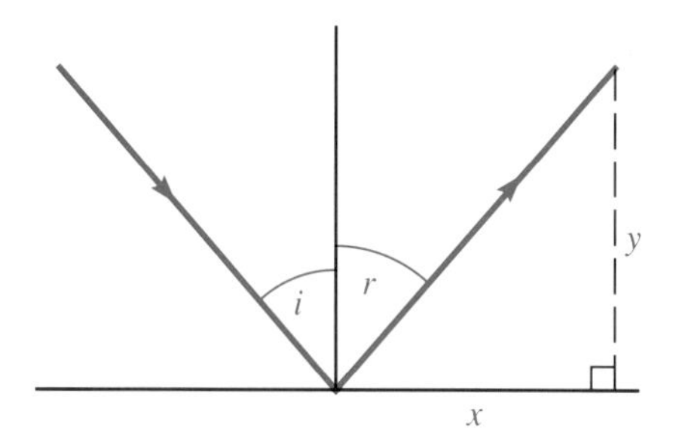

FIGURE 9–29

29. An observer in a balloon notes that the angle of depression to the top of a building is 35.8° and the angle of depression to the bottom of the building is 67.3°. If the balloon is 59.4 ft from the base of the building, find the height of the balloon.

30. From the ground the angle of elevation from an observer to the second floor is 11.3°, while the angle of elevation to the third floor is 21.8°. Find the distance between floors if the observer is 25 m from the building.

CHAPTER SUMMARY

Summary of Terms

adjacent side (p. 324)
angle (p. 320)
angle of elevation (p. 338)
angle of depression (p. 339)
coterminal (p. 321)
hypotenuse (p. 324)
impedance (p. 340)
inductive reactance (p. 340)
initial side (p. 320)
minutes (p. 320)
opposite side (p. 324)
phase angle (p. 340)
quadrantal angle (p. 322)
radius (p. 324)
seconds (p. 320)
standard position (p. 322)
terminal side (p. 320)
trigonometric functions (pp. 320 and 324)
trigonometry (p. 320)
vertex (p. 320)

Summary of Formulas

$$\sin \theta = \frac{\text{opposite}}{\text{hypotenuse}} = \frac{y}{r}$$

$$\cos \theta = \frac{\text{adjacent}}{\text{hypotenuse}} = \frac{x}{r}$$

$$\tan \theta = \frac{\text{opposite}}{\text{adjacent}} = \frac{y}{x}$$

$$\csc \theta = \frac{\text{hypotenuse}}{\text{opposite}} = \frac{r}{y}$$

$$\sec \theta = \frac{\text{hypotenuse}}{\text{adjacent}} = \frac{r}{x}$$

$$\cot \theta = \frac{\text{adjacent}}{\text{opposite}} = \frac{x}{y}$$

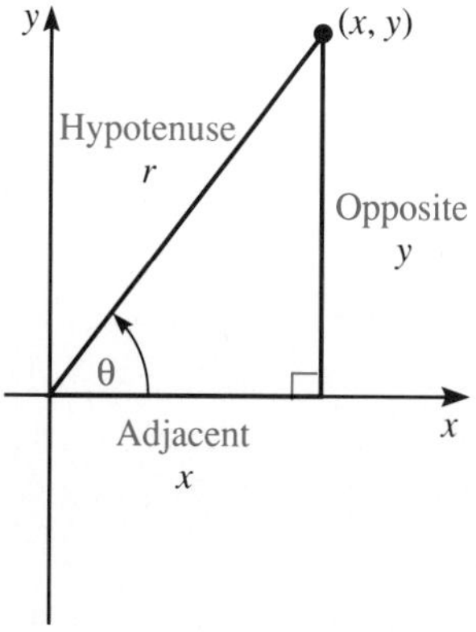

CHAPTER REVIEW

Section 9–1

Convert the following angles measured in degrees, minutes, and seconds to the nearest tenth of a degree.

1. $67°43'18''$ **2.** $134°06'27''$
3. $274°43'12''$ **4.** $324°15'23''$
5. $218°47'58''$ **6.** $157°37'10''$

Convert the following angles measured in degrees to angles measured to the nearest minute.

7. $16.78°$ **8.** $131.6°$
9. $318.15°$ **10.** $249.6°$
11. $108.43°$ **12.** $68.27°$

Draw the following angles in standard position, determine the quadrant in which the terminal side lies, and find two angles (one positive and one negative) coterminal with the given angle.

13. $147.8°$ **14.** $67°34'$
15. $290°18'12''$ **16.** $205°38'$
17. $314.63°$ **18.** $147.5°$

Section 9–2

Each of the following is a point on the terminal side of an angle in standard position. Determine the value to four decimal places of all six trigonometric functions of the angle.

19. $(6, 12)$ **20.** $(12, 3)$
21. $(10, 15)$ **22.** $(4, 7)$
23. $(\sqrt{3}, 2)$ **24.** $(4, \sqrt{5})$

Use the given trigonometric function to find the value of the indicated trigonometric function. Give your answer to four decimal places. (Do not find θ.)

25. $\sin\theta = 9/10$; find $\tan\theta$ and $\sec\theta$.
26. $\csc\theta = 15/7$; find $\sec\theta$ and $\tan\theta$.
27. $\tan\theta = 2.76$; find $\sin\theta$ and $\cos\theta$.
28. $\cos\theta = 6/13$; find $\sin\theta$ and $\cot\theta$.
29. $\sec\theta = 1.7864$; find $\cos\theta$ and $\tan\theta$.
30. $\cot\theta = 4.86$; find $\tan\theta$ and $\sin\theta$.

Section 9–3

Using a calculator, find the value of each trigonometric function rounded to three decimal places.

31. $\cos 16.84°$ **32.** $\tan 21°47'$
33. $\csc 67.8°$ **34.** $\cos 47°38'45''$
35. $\sec 72°46'$ **36.** $\sin 26.9°$

Find θ to the nearest tenth of a degree.

37. $\tan\theta = 3.7864$ **38.** $\csc\theta = 2.7684$
39. $\sec\theta = 1.6327$ **40.** $\cos\theta = 0.3798$
41. $\cot\theta = 2.5378$ **42.** $\sin\theta = 0.9156$

Section 9–4

Given right triangle *ABC* with the right angle at *C*, solve the triangle from the given information.

43. $b = 7.3$ yd, $A = 56.7°$
44. $B = 49°37'$, $c = 67.8$ ft
45. $c = 71.8$ m, $b = 56.3$ m
46. $a = 123$ cm, $c = 256$ cm
47. $A = 53°48'$, $a = 98.3$ km
48. The average rise in elevation of a mountain road is 14°. What is the change in altitude (vertical) if the road length is 4 km?
49. From the top of a 50-ft building, the angle of depression to a car on the ground is 37°. Find the horizontal distance from the base of the building to the car.
50. A cable is stretched from the top of a pole to a point 20 ft from the base of the pole. If the cable makes an angle of 60° with the ground, how tall is the pole?

CHAPTER TEST

The number in parentheses refers to the appropriate learning objective given at the beginning of the chapter.

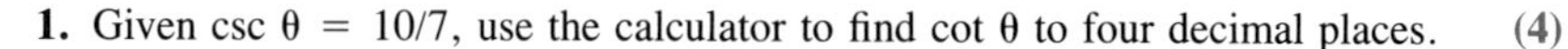

1. Given csc $\theta = 10/7$, use the calculator to find cot θ to four decimal places. (4)

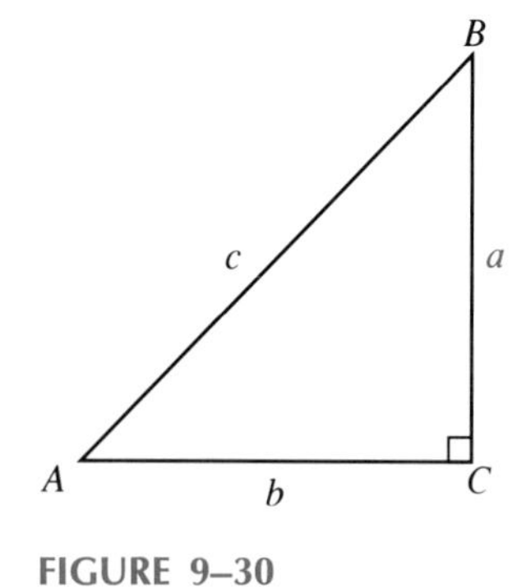

FIGURE 9–30

2. Given right triangle ABC shown in Figure 9–30, find a if $b = 56.8$ and $A = 42.5°$. (7)
3. If (4, 9) is a point on the terminal side of angle A in standard position, then find cos A and cot A using a calculator. (3)
4. Convert 127°14′48″ into an angle measured to the nearest tenth of a degree. (1)
5. Using a calculator, find angle B to the nearest tenth of a degree if cot $B =$ 2.2673. (6)
6. Using a calculator, evaluate csc 73°46′. (5)
7. Draw a 156° angle and determine the measure of two angles coterminal with it. (2)
8. A cathedral ceiling connects a wall 8 ft high with a wall 12 ft high. If the horizontal distance between the two walls is 20 ft, what is the angle of the rise of the ceiling relative to the horizontal? (8)
9. Given that (8, 5) is a point on the terminal side of angle A in standard position, find csc A using a calculator. (3)
10. Given that cot $B = 0.6544$, find sec B. (4)
11. Convert 124.83° to an angle measured in degrees and minutes. (1)
12. From a helicopter 600 ft high, the angle of depression to a house on the ground is 50°. Find the straight-line distance from the helicopter to the house. (8)
13. Find sec 24°18′46″. (5)
14. Given right triangle ABC shown in Figure 9–31, find c if $A = 38.7°$ and $a = 73.5$. (7)

FIGURE 9–31

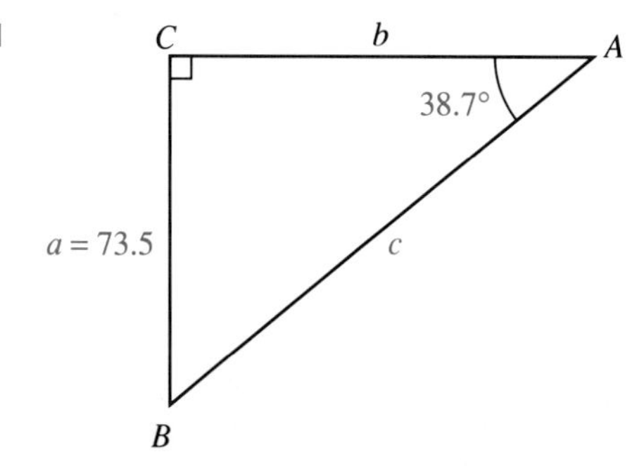

15. Calculate angle B to the nearest tenth of a degree if csc $B = 2.0371$. (6)
16. If sin $A = 0.8437$, find sec A. (4)

SOLUTION TO CHAPTER INTRODUCTION

To determine the angle and length of travel of the beam, we first draw Figure 9–32. We must find angle A and length b in this figure. The calculations follow.

FIGURE 9–32

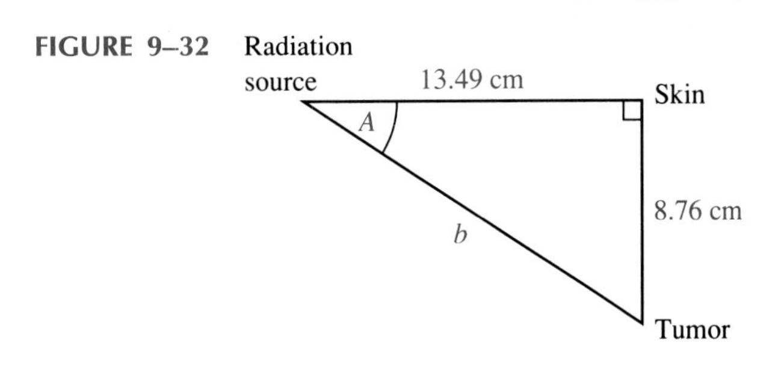

$$\tan A = \frac{8.76}{13.49}$$

$$A = 33.0°$$

$$b^2 = (13.49)^2 + (8.76)^2$$

$$b^2 = 258.72$$

$$b = 16.1 \text{ cm}$$

You should aim the radiation source at a 33.0° angle below the skin, and the beam will travel 16.1 cm through the patient's body to reach the tumor.

You are a space scientist trying to find the linear velocity of the antenna on a geosynchronous satellite that is orbiting the earth with orbital radius of 22,300 miles. At this altitude, the satellite and the earth beneath it have the same angular velocity of $\pi/12$ radians/hour. (The solution to this problem is given at the end of the chapter.)

In the previous chapter we discussed the trigonometric functions of a first quadrant angle. In this chapter, we will discuss the trigonometric functions for obtuse angles and angles measured in radians.

Learning Objectives

After completing this chapter, you should be able to

1. Determine the sign of a trigonometric function in any of the four quadrants, or if given sign(s) of trigonometric function(s), determine the quadrant(s) (Section 10–1).
2. Determine the six trigonometric functions from a point on the terminal side of an angle in any quadrant (Section 10–1).
3. Determine the reference angle for an angle whose terminal side lies in any quadrant (Section 10–2).
4. Find the value of the trigonometric functions for any given angle (Section 10–2).
5. Determine the measure of an angle in any quadrant given its trigonometric value (Section 10–2).
6. Convert angles measured in degrees to radians, and vice versa (Section 10–3).
7. Find the value of a trigonometric function given an angle measured in radians, and vice versa (Section 10–3).
8. Apply radian measure to finding the length of a circular arc, the area of a circular sector, and angular and linear velocity (Section 10–4).

Chapter 10

Trigonometric Functions of Any Angle

10–1

SIGNS OF THE TRIGONOMETRIC FUNCTIONS

Our discussion of trigonometry in Chapter 9 was limited to angles in the first quadrant. In this section we will discuss the sign of the trigonometric functions in the remaining three quadrants. Before we discuss the sign of the trigonometric functions, let us review several concepts discussed earlier.

First, we must be able to determine the quadrant in which the terminal side of an angle in standard position lies. For a given angle θ, the measurement boundaries for each quadrant are as summarized in Figure 10–1.

FIGURE 10–1

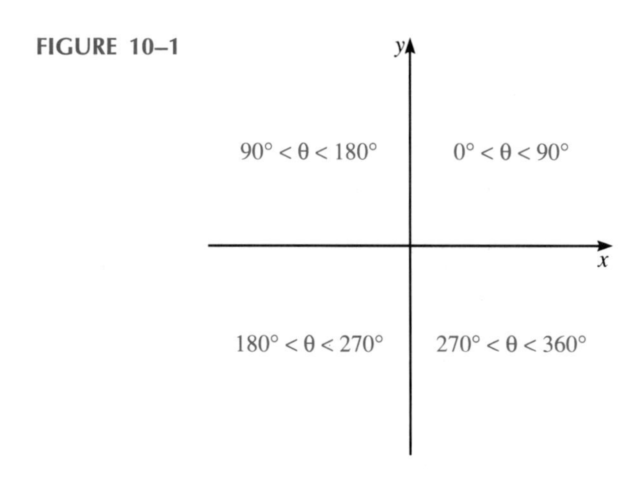

Second, we must know the definition of each trigonometric function in order to determine its sign. Therefore, the definitions of the trigonometric functions are repeated in the following box.

Definitions of the Trigonometric Functions

$$\sin\theta = \frac{y}{r} \qquad \csc\theta = \frac{r}{y}$$

$$\cos\theta = \frac{x}{r} \qquad \sec\theta = \frac{r}{x}$$

$$\tan\theta = \frac{y}{x} \qquad \cot\theta = \frac{x}{y}$$

Because r is always positive, the signs of the various trigonometric functions are determined by the sign of x and y. Recall from Chapter 9 that all the trigonometric functions are positive in quadrant I. Now, let us discuss the signs in the other quadrants.

REMEMBER ✦ A second, third, or fourth quadrant angle means that the terminal side of the angle lies in the second, third, or fourth quadrant, respectively.

Quadrant II

Figure 10–2 shows a quadrant II angle and its resulting right triangle. From the figure we can see that y and r are both positive and x is negative. From this information, we can determine that only functions that involve x will be negative in quadrant II. This information is summarized as follows:

Positive	Negative
$\sin\theta\left(\frac{y}{r}=\frac{+}{+}=+\right)$	$\cos\theta\left(\frac{x}{r}=\frac{-}{+}=-\right)$
$\csc\theta\left(\frac{r}{y}=\frac{+}{+}=+\right)$	$\sec\theta\left(\frac{r}{x}=\frac{+}{-}=-\right)$
	$\tan\theta\left(\frac{y}{x}=\frac{+}{-}=-\right)$
	$\cot\theta\left(\frac{x}{y}=\frac{-}{+}=-\right)$

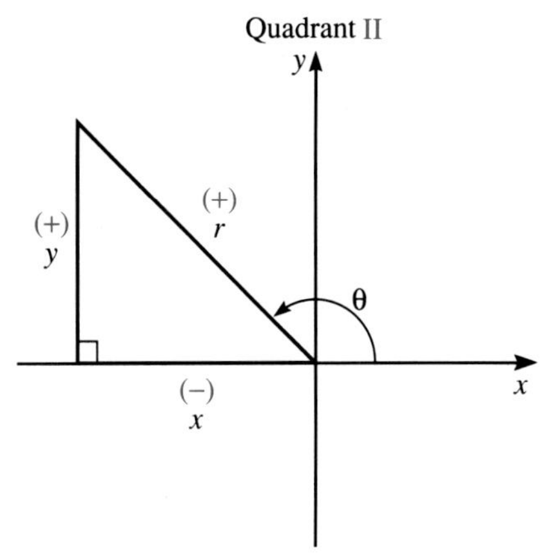

FIGURE 10–2

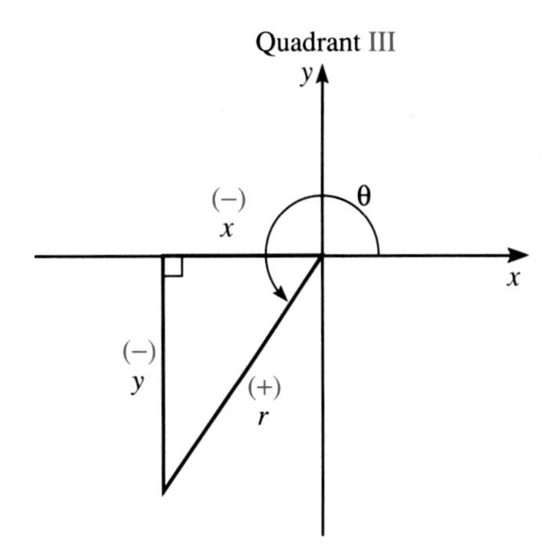

FIGURE 10–3

Quadrant III

Figure 10–3 shows a quadrant III angle and its resulting triangle. From the figure we can see that both x and y are negative and r is positive. The only positive functions in quadrant III involve both x and y. This information is summarized as follows:

Positive	Negative
$\tan\theta \left(\dfrac{y}{x} = \dfrac{-}{-} = +\right)$	$\sin\theta \left(\dfrac{y}{r} = \dfrac{-}{+} = -\right)$
$\cot\theta \left(\dfrac{x}{y} = \dfrac{-}{-} = +\right)$	$\cos\theta \left(\dfrac{x}{r} = \dfrac{-}{+} = -\right)$
	$\sec\theta \left(\dfrac{r}{x} = \dfrac{+}{-} = -\right)$
	$\csc\theta \left(\dfrac{r}{y} = \dfrac{+}{-} = -\right)$

Quadrant IV

Figure 10–4 shows a quadrant IV angle and its resulting triangle. From the figure we can see that x and r are both positive and y is negative. Therefore, the negative functions in Quadrant IV involve y. This information is summarized as follows:

Positive	Negative
$\cos\theta$	$\sin\theta$
$\sec\theta$	$\csc\theta$
	$\tan\theta$
	$\cot\theta$

FIGURE 10–4

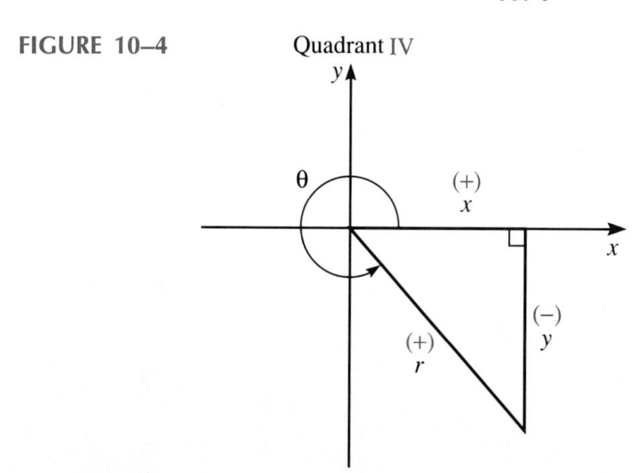

The sign of each trigonometric function in each quadrant is given in Table 10–1.

TABLE 10–1
Signs of the Trigonometric Functions

Function	Quadrant I	II	III	IV
Sine	+	+	−	−
Cosine	+	−	−	+
Tangent	+	−	+	−
Cosecant	+	+	−	−
Secant	+	−	−	+
Cotangent	+	−	+	−

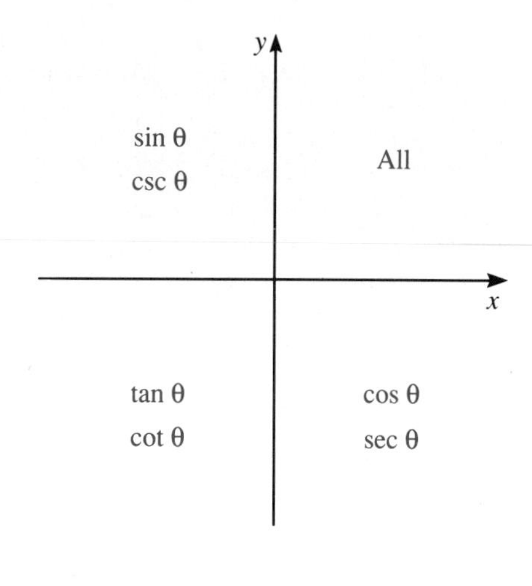

FIGURE 10–5

The graph in Figure 10–5 summarizes the quadrants where the trigonometric functions are positive.

EXAMPLE 1 Determine the sign of each of the following trigonometric functions:

(a) tan 146° (b) cos 256° (c) sec 313°
(d) sin 218° (e) csc 94° (f) cot 190°
(g) sin(−230°) (h) tan 463° (i) sec(−108°)

Solution

(a) An angle of 146° has its terminal side in quadrant II where both r and y are positive and x is negative. Therefore,

$$\tan 146^\circ = \frac{y}{x} = \frac{(+)}{(-)} = (-)$$

(b) An angle of 256° terminates in quadrant III where both x and y are negative and r is positive. Therefore,

$$\cos 256^\circ = \frac{x}{r} = \frac{(-)}{(+)} = (-)$$

(c) An angle of 313° is a quadrant IV angle where both x and r are positive and y is negative. Thus,

$$\sec 313^\circ = \frac{r}{x} = \frac{(+)}{(+)} = (+)$$

(d) $\sin 218° = \frac{y}{r} = \frac{(-)}{(+)} = (-)$

(e) $\csc 94° = \frac{r}{y} = \frac{(+)}{(+)} = (+)$

(f) $\cot 190° = \frac{x}{y} = \frac{(-)}{(-)} = (+)$

(g) Since an angle of $-230°$ is rotated clockwise from the positive x axis, $-230°$ has its terminal side in quadrant II. Therefore,

$$\sin(-230°) = \frac{y}{r} = \frac{(+)}{(+)} = (+)$$

(h) An angle of 463° involves revolving around the circle once and terminating in quadrant II at 103°. Thus,

$$\tan 463° = \tan 103° = \frac{y}{x} = \frac{(+)}{(-)} = (-)$$

(i) $\sec(-108°) = \frac{r}{x} = \frac{(+)}{(-)} = (-)$ ■

EXAMPLE 2 Determine the quadrant(s) in which the following conditions are true:

(a) $\tan \theta$ is positive.
(b) $\sin \theta$ and $\cos \theta$ are both positive.
(c) $\csc \theta$ is negative and $\cot \theta$ is positive.
(d) $\sec \theta$ and $\cos \theta$ are both positive.

Solution

(a) I and III (b) I (c) III (d) I and IV ■

EXAMPLE 3 Determine the six trigonometric functions of an angle whose terminal side passes through the point $(-3, -8)$. Round to three decimal places.

Solution Figure 10–6 represents the right triangle determined by this point. First, we use the Pythagorean theorem to determine the length of r.

$$r = \sqrt{x^2 + y^2} = \sqrt{(-3)^2 + (-8)^2} = \sqrt{73}$$
$$\approx 8.544$$

FIGURE 10–6

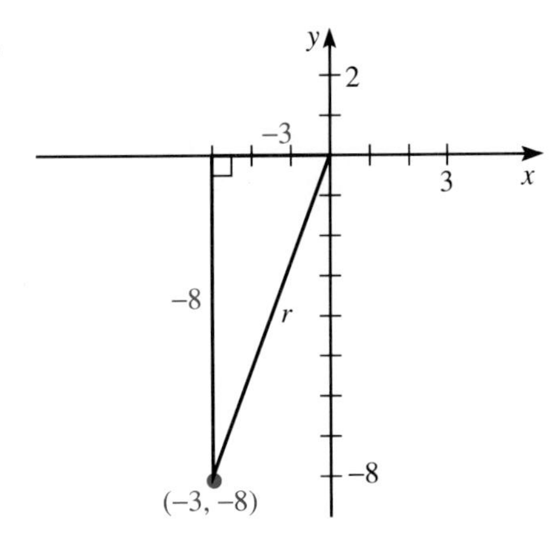

Substituting the values into the trigonometric functions gives the following values:

$$\sin\theta = \frac{y}{r} = \frac{-8}{8.544} = -0.936 \qquad \csc\theta = \frac{r}{y} = \frac{8.544}{-8} = -1.068$$

$$\cos\theta = \frac{x}{r} = \frac{-3}{8.544} = -0.351 \qquad \sec\theta = \frac{r}{x} = \frac{8.544}{-3} = -2.848$$

$$\tan\theta = \frac{y}{x} = \frac{-8}{-3} = 2.667 \qquad \cot\theta = \frac{x}{y} = \frac{-3}{-8} = 0.375$$

■

10–1 EXERCISES

Determine the sign of the following trigonometric expressions without using a calculator.

1. cot 135° **2.** sin 256°
3. csc 98° **4.** tan 318°
5. cos 235° **6.** sin 278°
7. sec 287° **8.** cot 97.3°
9. sin 115° **10.** csc 108°
11. tan 158.3° **12.** cos 263.5°
13. cot 310° **14.** sec 184.3°
15. cos 245° **16.** 4.6 csc 358°
17. $(\tan 138.6°)^3$ **18.** cos 294.8°
19. cot 196.9° **20.** sin 177.9°
21. $\sqrt{\csc 175.0°}$ **22.** $(\tan 145°)^2$
23. 3.6 sin 190.6° **24.** cos 520°
25. sec(−550°) **26.** $\dfrac{\csc 596°}{4.82}$
27. tan(−624°) **28.** $\cot^2(-195°)$
29. cos(−203°) **30.** $[\cot(-286.3°)]^{1/3}$

Find the quadrant(s) where the following conditions are true.

31. cos θ is positive. **32.** csc θ is negative.
33. cot θ is negative. **34.** sin θ is positive.
35. tan θ is negative. **36.** sec θ is negative.
37. tan θ is negative, sin θ is negative.
38. csc θ is positive, cot θ is negative.
39. cos θ is negative, tan θ is positive.
40. sec θ is negative, csc θ is negative.
41. cot θ is negative, cos θ is positive.
42. cos θ is positive, sin θ is negative.
43. sin θ is positive, csc θ is positive.
44. sin θ is positive, tan θ is negative.

45. sec θ is negative, sin θ is negative.

46. tan θ is negative, cot θ is negative.

Using a calculator, find the six trigonometric functions for an angle θ whose terminal side passes through the point given. Round your answer to four decimal places.

47. $(-4, 9)$
48. $(10, -3)$
49. $(-5, -8)$
50. $(6, -10)$
51. $(-7, -2)$
52. $(-6, 4)$
53. $(-9, -4)$
54. $(-1, 6)$
55. $(9, -7)$
56. $(-6, 9)$
57. $(7, -5)$
58. $(-3, -6)$
59. $(8, -2)$
60. $(-3, -4)$

10–2

TRIGONOMETRIC FUNCTIONS OF ANY ANGLE

In the previous section we discussed the sign of the trigonometric functions in each quadrant. In this section we will discuss finding the value of a trigonometric function in each quadrant and finding an angle given its trigonometric value. We begin the discussion with finding the trigonometric value of obtuse angles.

EXAMPLE 1 Using your calculator, find the value of the following trigonometric functions rounded to four decimal places:

(a) sin 218.6° (b) tan 108.34° (c) cos 313°16′
(d) sec 176°25′ (e) cot 346.73°

Solution Using the same calculator keystrokes as discussed earlier gives the following values:

(a) 218.6 [sin] → −0.6239

(b) 108.34 [tan] → −3.0167

(c) 16 [÷] 60 [+] 313 [=] [cos] → 0.6854

(d) 25 [÷] 60 [+] 176 [=] [cos] [1/x] → −1.0020

(e) 346.73 [tan] [1/x] → −4.2402 ■

From the results in Example 1, we can see that the calculator assigns the correct sign for the trigonometric functions in any quadrant. However, if we attempt to find an angle whose trigonometric value is given, the calculator may not give the desired angle. When we discuss the inverse trigonometric functions, you will understand why this is possible. This occurrence makes it necessary to discuss reference angles.

Reference Angle

In general, the numerical value of a trigonometric function in any of the quadrants is equivalent to the same function of a first quadrant angle, called the reference angle. The **reference angle** A of any angle θ is the positive *angle formed by the terminal side of the angle and the nearest x axis*. The reference angle A for a given angle θ in each of the quadrants is shown in Figure 10–7.

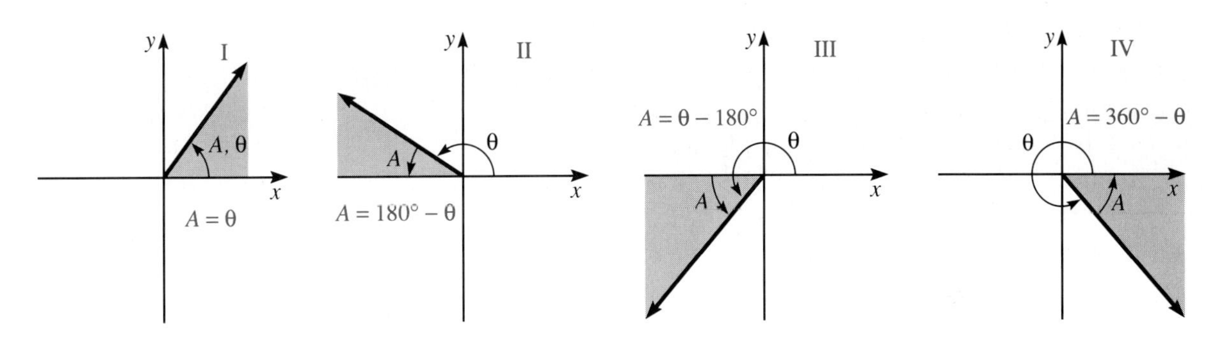

FIGURE 10–7

CAUTION ✦ The reference angle is always the positive angle measured between the terminal side of the angle and the nearest *x axis*.

A summary of how to calculate the reference angle A from a given angle θ is given in the following box.

Reference Angle A for a Given Angle θ

Quadrant I	$A = \theta$
Quadrant II	$A = 180° - \theta$
Quadrant III	$A = \theta - 180°$
Quadrant IV	$A = 360° - \theta$

EXAMPLE 2 Find the reference angle A for each of the following given angles θ:

(a) 246° (b) 136° (c) 318.4° (d) 168°37′

Solution

(a) Quadrant III

$$A = \theta - 180°$$
$$A = 246° - 180°$$
$$A = 66°$$

θ

180°

A

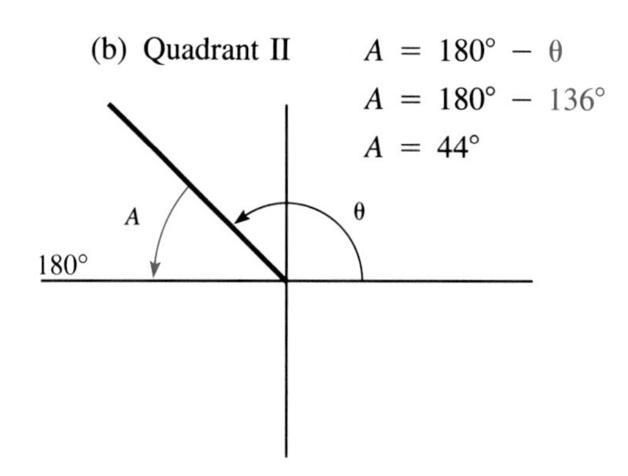

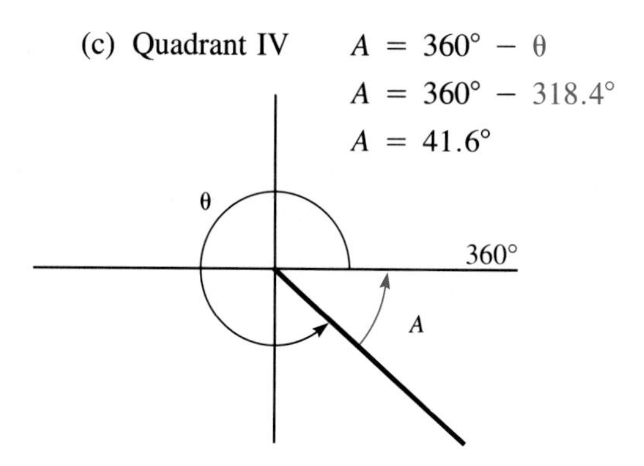

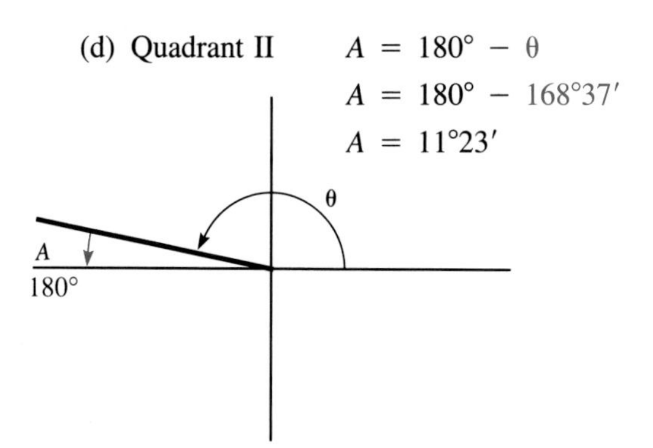

■

CAUTION ✦ Reference angles are always determined using 180° or 360°.

Finding Angles from a Trigonometric Value

Recall from the previous chapter that we used the [inv] key to find a first quadrant angle given its trigonometric value. However, when we extend this process to the remaining quadrants, there should be two answers for each angle θ. For example, if $\cos \theta = -0.6587$, then θ is both a second quadrant and a third quadrant angle because the cosine function is negative in both quadrants. Since the calculator does not give both angles, we use the absolute value of the function to yield a reference angle and then use the sign of the function to determine the two values of θ. This process is illustrated in the next examples.

EXAMPLE 3 Find all angles θ, in degrees, such that $\tan\theta = -0.5051$.

Solution First, we find A given that

$$\begin{aligned} \tan A &= |\tan\theta| \\ \tan A &= |-0.5051| \\ \tan A &= 0.5051 \\ A &= 26.8^\circ \end{aligned}$$

Second, we place the reference angle A in quadrants II and IV because the tangent is negative in both quadrants. Third, we calculate θ using the reference angle.

Quadrant II

$$\begin{aligned} \theta &= 180^\circ - A \\ \theta &= 180^\circ - 26.8^\circ \\ \theta &= 153.2^\circ \end{aligned}$$

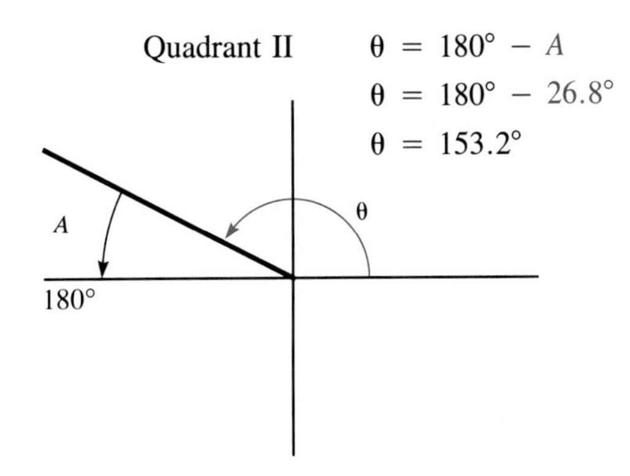

Quadrant IV

$$\begin{aligned} \theta &= 360^\circ - A \\ \theta &= 360^\circ - 26.8^\circ \\ \theta &= 333.2^\circ \end{aligned}$$

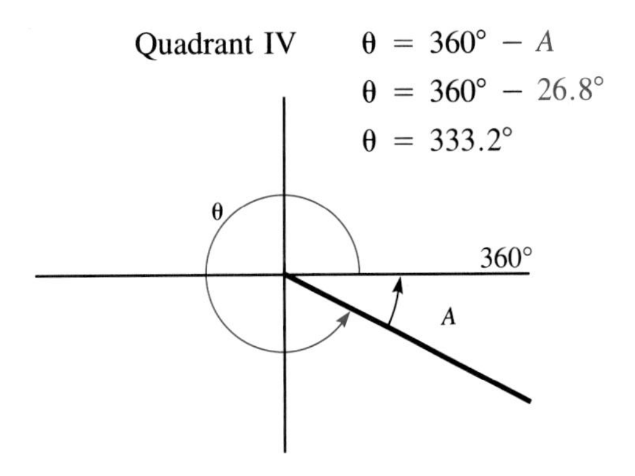

■

EXAMPLE 4 Find sec θ when $\tan\theta = -1.6255$ and $\sin\theta < 0$.

Solution Since both $\tan\theta$ and $\sin\theta$ are negative, θ must be located in quadrant IV. Finding the reference angle gives

$$\tan A = |-1.6255|$$
$$\tan A = 1.6255$$
$$A = 58.4^\circ$$

$$1.6255 \; \boxed{\text{inv}} \; \boxed{\tan} \rightarrow 58.4$$

Then we find θ in quadrant IV given the reference angle A.

$$\theta = 360^\circ - A$$
$$\theta = 360^\circ - 58.4^\circ$$
$$\theta = 301.6^\circ$$

Last, we find the secant of 301.6°.

$$\sec 301.6^\circ = \frac{1}{\cos 301.6^\circ}$$
$$\sec\theta = 1.9084$$

$$301.6 \; \boxed{\cos} \; \boxed{1/x} \rightarrow 1.9084$$

■

EXAMPLE 5 Find θ if $(-8, 3)$ is a point on the terminal side of the angle. Round to tenths of a degree.

Solution First, we find the reference angle A using the absolute value of the tangent function.

$$\tan A = \left|\frac{3}{-8}\right|$$
$$A = 20.6^\circ$$

$$3 \; \boxed{\div} \; 8 \; \boxed{=} \; \boxed{\text{inv}} \; \boxed{\tan} \rightarrow 20.6$$

Second, we find θ in quadrant II (because $x < 0$, $y > 0$) using the reference angle A.

$$A = 180^\circ - \theta$$
$$\theta = 180^\circ - A$$
$$\theta = 180^\circ - 20.6^\circ$$
$$\theta = 159.4^\circ$$

■

Quadrantal Angles

In Section 9–3, we discussed the value of the trigonometric functions for the quadrantal angles 0° and 90°. Once again, if we let $r = 1$, the values of x, y, and r at 180° and 270° are as shown in Figure 10–8. The values of the trigonometric functions for the quadrantal angles are summarized in Table 10–2.

FIGURE 10–8

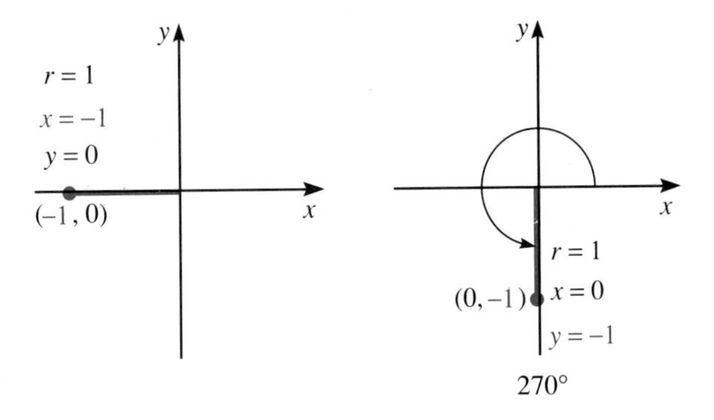

TABLE 10–2
Values of the Trigonometric Functions

	Quadrantal Angles			
Function	0°	90°	180°	270°
Sine	0	1	0	−1
Cosine	1	0	−1	0
Tangent	0	undef.	0	undef.
Cosecant	undef.	1	undef.	−1
Secant	1	undef.	−1	undef.
Cotangent	undef.	0	undef.	0

10–2 EXERCISES

Use a calculator to find the value of each of the following. Round your answer to thousandths.

1. $(\sin 218°)^3$
2. $\tan 114.6°$
3. $\csc 320.6°$
4. $\sqrt{\sec 68.5°}$
5. $\cos 325.7°$
6. $\csc 76.7°$
7. $\sin 173.9°$
8. $\cos 278.3°$
9. $\sec^3 98.4°$
10. $\sin 186.5°$
11. $\cot 254°26'$
12. $3(\tan 124°) + 5$
13. $\tan 173.65°$
14. $(\cot 295.36°)^2$
15. $\sec 85.73°$
16. $(\csc 160°53')^{1/2}$
17. $\sin 284°47'$
18. $\sqrt{\sin 139.5°}$
19. $\cos 330°18'$
20. $\cot 99.27°$

21. $\cot 786°$

22. $8 \cos 194°18'$

23. $\sin 418°39'$

24. $[\sec(-286.56°)]^{1/3}$

25. $\csc(-146°)$

26. $\sec(-684°)$

27. $\tan(-512°)$

28. $\sin 156°26'$

Determine the quadrant in which the terminal side of the given angle lies, and determine its reference angle.

29. 113.6°

30. 218.75°

31. 278.1°

32. 195.6°

33. 326.73°

34. 94.8°

35. 83.8°

36. 301.57°

37. 210.9°

38. 283.65°

39. 137.43°

40. 74°8′

41. 656°

42. 136°38′

43. 218°13′

44. 834°

45. 410°23′

46. 324°43′

47. 99°29′

48. 263°14′

Find all θ in degrees, $0° \le \theta < 360°$, such that the following statements are true. Round your answer to the nearest hundredth degree for Exercises 49–60 and to the nearest minute for Exercises 61–68.

49. $\sin \theta = 0.8151$

50. $\cos \theta = -0.4099$

51. $\tan \theta = -0.9036$

52. $\sec \theta = -1.0255$

53. $\tan \theta = 2.2148$

54. $\cot \theta = -0.9163$

55. $\sin \theta = 0.9897$

56. $\csc \theta = -1.7305$

57. $\sec \theta = -1.3809$

58. $\sin \theta = 0.4020$

59. $\csc \theta = -1.0209$

60. $\cot \theta = 1.2174$

61. $\csc \theta = -3.1516$

62. $\sec \theta = -2.1855$

63. $\cos \theta = 0.6807$

64. $\sin \theta = 0.9990$

65. $\cot \theta = 0.6720$

66. $\tan \theta = -0.5914$

67. $\sec \theta = -1.1754$

68. $\cos \theta = -0.9752$

Given the following conditions, determine the function value.

69. Find $\sin \theta$ given $\sec \theta = 3.6727$ and $\tan \theta < 0$.

70. Find $\tan \theta$ given $\sin \theta = -0.3804$ and $\cos \theta < 0$.

71. Find $\csc \theta$ given $\tan \theta = -0.0332$ and $\sec \theta > 0$.

72. Find $\cos \theta$ given $\tan \theta = 0.8973$ and $\cot \theta > 0$.

73. Find $\sec \theta$ given $\cot \theta = -0.3385$ and $\sin \theta < 0$.

74. Find $\cot \theta$ given $\cos \theta = -0.5195$ and $\csc \theta > 0$.

Find the angle, to the nearest tenth degree, whose terminal side passes through the point given.

75. $(-3, 10)$

76. $(-5, -7)$

77. $(7, -3)$

78. $(-8, -12)$

79. $(9, -4)$

80. $(-6, 1)$

81. $(4, -5)$

82. $(-1, -4)$

83. A ray of light passing through an optical wedge is shifted by amount e given by

$$e = \frac{t \sin(i - R)}{\cos R}$$

where i = angle of rotation, R = angle of reflection, and t = thickness of the wedge. Find the ray shift if $t = 9$ in., $i = 56°$, and $R = 18°$.

84. For a ray of light entering a prism, if the angle of incidence is in air, the following relationship is true:

$$n_R \sin R = \sin i$$

where i = angle of incidence, n_R = refractive index of the medium in which reflection takes place, and R = angle of refraction. Find the angle of refraction R if the angle of incidence is 66° and the refraction index is 1.375.

85. For objects projected with velocity v at an angle A above the horizontal, the time necessary to reach maximum height is given by

$$t = \frac{v \sin A}{g}$$

If $t = 2$ s, $v = 30$ m/s, and $g = 9.8$ m/s^2, find angle A.

86. In optics, the critical angle i_c is the incident angle that results in a 90° angle of refraction. This relationship is given by

$$\sin i_c = \frac{n_R}{n_i}$$

where n_R = refractive index of the refraction medium and n_i = refractive index of the incidence medium. Find the critical angle for a ray passing through a sheet of glass and refracted into air, given that $n_R = 1$ and $n_i = 1.57$.

E 87. The power P absorbed in an ac circuit is given by $P = IV \cos \theta$ where I = current, V = voltage, and θ = phase angle. Find the power P absorbed in a 2-A, 120-V circuit with a phase angle of 71°.

10–3

RADIAN MEASURE

Definitions

Until now our discussion of angles has focused on the degree as the unit of angular measure. However, some technical problems require angles to be measured in radians. Given an angle whose vertex is at the center of the circle, a **radian** is defined to be the angle that intercepts an arc equal to the radius of the circle, as shown in Figure 10–9.

FIGURE 10–9

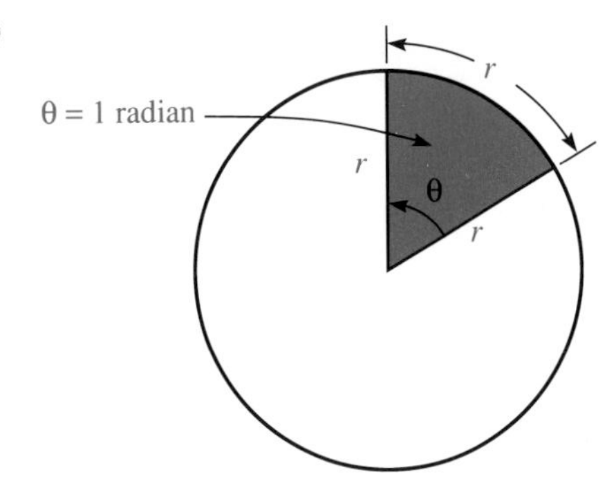

NOTE ✦ Radians are actually dimensionless units of angular measure. Therefore, an angle with the units omitted can be assumed to be measured in radians.

The quadrant boundaries in radians are given in the following box.

Quadrantal Angles

$$90° = \frac{\pi}{2} \approx 1.57 \text{ rad} \qquad 270° = \frac{3\pi}{2} \approx 4.71 \text{ rad}$$

$$180° = \pi \approx 3.14 \text{ rad} \qquad 360° = 2\pi \approx 6.28 \text{ rad}$$

EXAMPLE 1 Determine the quadrant in which the terminal side of the angle lies:

(a) $\frac{6\pi}{5}$ (b) 2.68 rad (c) $\frac{5\pi}{3}$

(d) 0.673 rad (e) $\frac{5\pi}{4}$ (f) -1.374 rad

Solution

(a) III (b) II (c) IV
(d) I (e) III (f) IV

■

Radian and Degree Conversion

To convert an angle measured in degrees to radians, or vice versa, use the following proportion:

$$\frac{\text{radians}}{\text{degrees}} = \frac{\text{radians}}{\text{degrees}}$$

Since a complete circle of 360° contains 2π radians, π radians equals 180°. Substituting this value into the proportion gives the following conversion proportion.

Converting Degrees and Radians

$$\frac{\pi}{180°} = \frac{\text{radians}}{\text{degrees}}$$

EXAMPLE 2

(a) Convert 240° to radians. Leave your answer as a rational multiple of π.

(b) Convert $\frac{7\pi}{6}$ radians to degrees.

Solution

(a) We take the basic proportion given above, substitute 240 in the degrees location and x in the radians location, and solve for x.

$$\frac{\pi}{180°} = \frac{x}{240°}$$

$$240°\pi = 180°x$$

$$\frac{240°\pi}{180°} = x$$

$$\frac{4\pi}{3} = x$$

(b) If we take the basic proportion and substitute $\frac{7\pi}{6}$ in the radian location and x in the degrees location, we have

$$\frac{\pi}{180°} = \frac{7\pi/6}{x}$$

$$\pi x = 180°\frac{7\pi}{6}$$

$$x = 210°$$

■

Finding Function Values

We find the function value of an angle measured in radians just as we did in the previous sections when using angles measured in degrees. Make sure the calculator is in the radian mode, enter the radian measure of the angle, and press the desired trigonometric function key.

EXAMPLE 3 Find each of the following trigonometric values rounded to four decimal places:

(a) $\sin 3.68$ (b) $\tan 1.25$ (c) $\sec 2.15$

(d) $\cos \frac{3\pi}{4}$ (e) $\csc \frac{8\pi}{5}$ (f) $\cot \frac{7\pi}{6}$

Solution Make sure that your calculator is in the radian mode. Then use the following keystrokes:

(a) 3.68 [sin] → −0.5128

(b) 1.25 [tan] → 3.0096

(c) 2.15 [cos] [1/x] → −1.8270

(d) 3 [×] [π] [÷] 4 [=] [cos] → −0.7071

(e) 8 [×] [π] [÷] 5 [=] [sin] [1/x] → −1.0515

(f) 7 [×] [π] [÷] 6 [=] [tan] [1/x] → 1.7321 ■

Finding an Angle from a Function Value

To find an angle given its trigonometric value, we use a reference angle just as we did in the previous section. Figure 10–10 illustrates the reference angle A for an angle θ measured in radians.

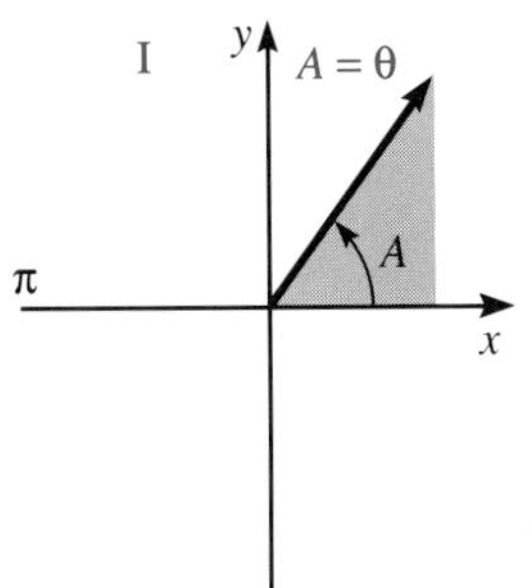

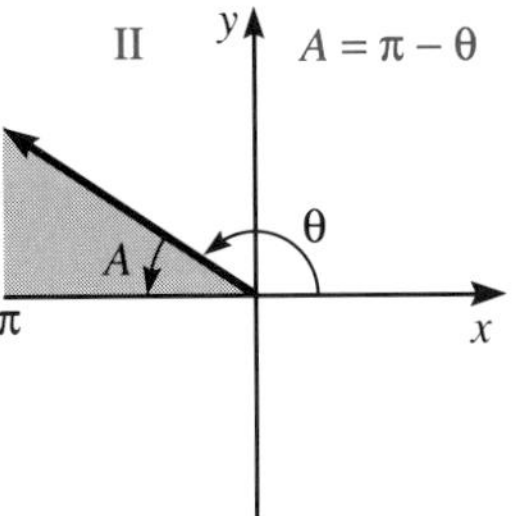

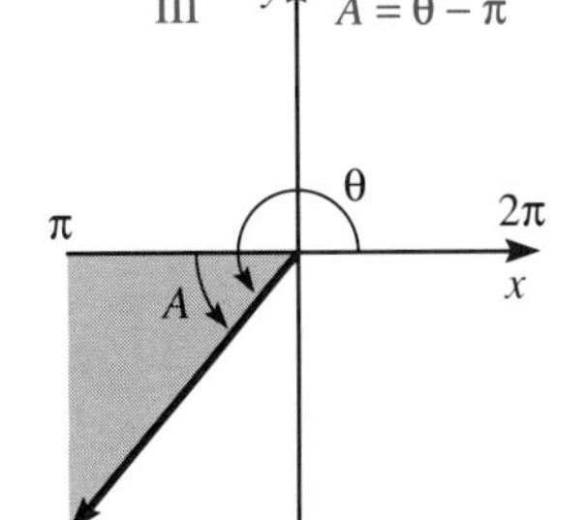

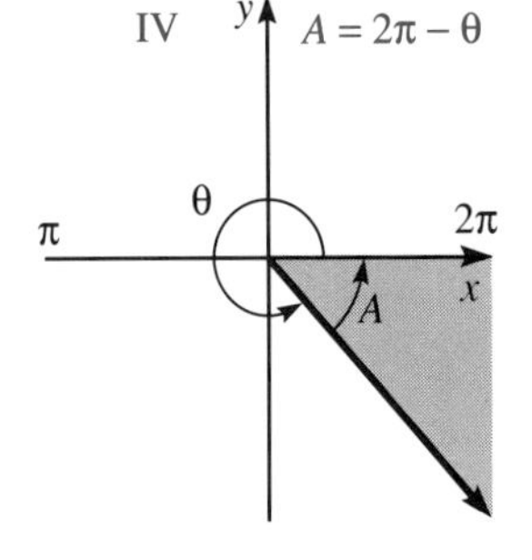

FIGURE 10–10

EXAMPLE 4 Find all values of θ in radians to hundredths such that each of the following is true:

(a) $\sin \theta = -0.8634$ (b) $\csc \theta = 3.236$

(c) $\cot \theta = 0.3839$ (d) $\cos \theta = -0.8786$

Solution

(a) First, we determine the reference angle A using the absolute value of the trigonometric function.

$$0.8634 \boxed{\text{inv}} \boxed{\sin} \rightarrow 1.04$$

Then we find θ in both quadrants III and IV where the sine function is negative.

Quadrant III	$A = \theta - \pi$	or	$\theta = A + \pi$	$\theta = 4.18$
Quadrant IV	$A = 2\pi - \theta$	or	$\theta = 2\pi - A$	$\theta = 5.24$

Therefore, $\theta = 4.18$ and $\theta = 5.24$.

(b) First, we find the reference angle A by pressing the following keys:

$$3.236 \boxed{1/x} \boxed{\text{inv}} \boxed{\sin} \rightarrow 0.31$$

Then, using the reference angle $A = 0.31$ radian, we find θ in quadrants I and II where the sine function is positive.

Quadrant I	$A = \theta$	so	$\theta = 0.31$	
Quadrant II	$A = \pi - \theta$	so	$\theta = \pi - A$	$\theta = 2.83$

Therefore, $\theta = 0.31$ and $\theta = 2.83$.

(c) To determine the reference angle A, press the following keys:

$$0.3839 \boxed{1/x} \boxed{\text{inv}} \boxed{\tan} \rightarrow 1.20$$

Next, calculate θ in quadrants I and III.

Quadrant I	$A = \theta$	so	$\theta = 1.20$	
Quadrant III	$A = \theta - \pi$	so	$\theta = A + \pi$	$\theta = 4.34$

Therefore, $\theta = 1.20$ and $\theta = 4.34$.

(d) To find the reference angle A, press the following keys:

$$0.8786 \boxed{\text{inv}} \boxed{\cos} \rightarrow 0.50$$

Since the cosine function is negative in quadrants II and III, $\theta = 2.64$ and $\theta = 3.64$. ■

CAUTION ✦ Be careful in performing operations with the cotangent, secant, and cosecant functions. Write the reciprocal relationships and solve before using the calculator. Many errors result from pressing calculator keys before analyzing the situation.

10–3 EXERCISES

Determine the quadrant in which the terminal side of each angle (measured in radians) lies.

1. 2.13 **2.** 5.34

3. $\frac{4\pi}{3}$ **4.** 0.86

5. -3.65 **6.** $\frac{\pi}{8}$

7. $-\frac{7\pi}{3}$ **8.** 3.29

Convert each angle measured in degrees to radian measure. Leave your answer as a rational multiple of π.

9. 68° **10.** 256°

11. 318° **12.** 146°

13. 330° **14.** 187°

15. $-280°$ **16.** $-159°$

Convert each angle measured in degrees to decimal radians. Round your answer to hundredths.

17. 46.8° **18.** 315.67°

19. 98°10′ **20.** 221°36′

21. 194.9° **22.** $-256°18'$

23. $-324.7°$ **24.** 136°47′

Convert each angle measured in radians to degrees. Round your answer to hundredths.

25. 1.64 **26.** 3.57

27. 5.63 **28.** $\frac{3\pi}{4}$

29. $\frac{7\pi}{3}$ **30.** 1.15

31. $-\frac{11\pi}{6}$ **32.** -6.15

Find the value of the following functions. Round your answer to thousandths.

33. sin 2.86 **34.** sec 5.34

35. $\sqrt[3]{\tan 4.37}$ **36.** csc 3.29

37. $(\cot 1.17)^{1/2}$ **38.** $\cos^2 2.17$

39. sec 0.73 **40.** sin 1.69

41. csc 4.18 **42.** sin 3.59

43. $\cot \frac{5\pi}{4}$ **44.** $\cos \frac{5\pi}{6}$

45. $\cos \frac{3\pi}{8}$ **46.** $\tan \frac{5\pi}{3}$

47. $\csc \frac{7\pi}{6}$ **48.** $\tan \frac{3\pi}{4}$

Find all values of θ, $0 \leq \theta < 2\pi$, to the nearest hundredth radian, that satisfy the following conditions.

49. $\cos \theta = 0.8345$ **50.** $\tan \theta = -2.673$

51. $\sin \theta = -0.3576$ **52.** $\cot \theta = -3.875$

53. $\csc \theta = 2.53$ **54.** $\cos \theta = 0.5641$

55. $\sec \theta = -1.635$ **56.** $\sin \theta = 0.335$

57. $\sec \theta = -2.973$ **58.** $\sin \theta = 0.8937$

59. $\cot \theta = -3.75$ **60.** $\tan \theta = 0.8634$

61. The area K of a regular polygon with n sides of length s is given by

$$K = \frac{1}{4} ns^2 \cot \frac{\pi}{n}$$

Find the area of an octagon ($n = 8$) whose sides are 3.4 m long.

62. The radius r of a circle inscribed in a regular polygon with n sides of length s is given by

$$r = \frac{s}{2} \cot \frac{\pi}{n}$$

Find the radius of a circle inscribed in a hexagon ($n = 6$) whose sides are 8 in. long.

63. A weight attached to a spring is pulled down 2 in. and released, thereby oscillating with simple harmonic motion and taking 0.12 s to return to its lowest position. This relationship is given by

$$y = 2 \cos \left(\frac{50\pi}{3} t\right)$$

Find the position y of the weight after 0.03 s.

E 64. The relationship between sinusoidal voltage V and time t is given by $V = V_0 \sin(2\pi ft)$ where V_0 is peak voltage and f is frequency. If the peak voltage is 163 V and the frequency is 60 Hz, find the voltage at time $t = 0.003$ s.

10 – 4

APPLICATIONS OF RADIAN MEASURE

In this section we will discuss three applications of radian measure: the length of a circular arc, the area of a circular sector, and angular velocity. Each of these applications requires that the angles be measured in radians.

Length of a Circular Arc

If we let s represent the length of an arc, then the length of an arc for a complete circle is given by the expression $2\pi r$. The expression 2π represents a central angle for the entire circle; however, we are attempting to find the arc length s for a given central angle θ that is less than 2π. The resulting expression for the length of a circular arc is as follows:

Length of a Circular Arc

$$s = r\theta$$

where

s = length of the arc

r = radius of the circle

θ = measure of the central angle in radians

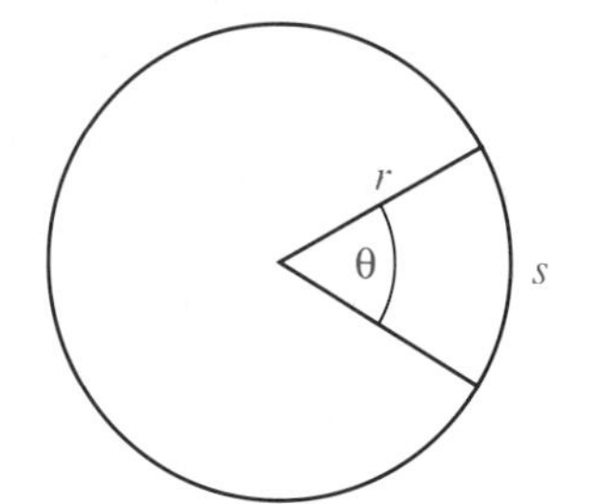

EXAMPLE 1 Find the length of the arc of a circle whose radius is 4.3 in. and central angle is $\frac{3\pi}{4}$.

Solution Using the given values of $r = 4.3$ in. and $\theta = \frac{3\pi}{4}$, we solve for s by substituting into the formula as follows:

$$s = r\theta$$
$$s = (4.3 \text{ in.})\left(\frac{3\pi}{4}\right)$$
$$s = 10 \text{ in.}$$

■

EXAMPLE 2 A pendulum bob moves through a central angle of 10° and travels a length of 6.6 in. Find the length of the pendulum bob.

Solution The path traveled by the pendulum bob is shown in Figure 10–11. From the drawing, we can see that we are asked to find the length of the

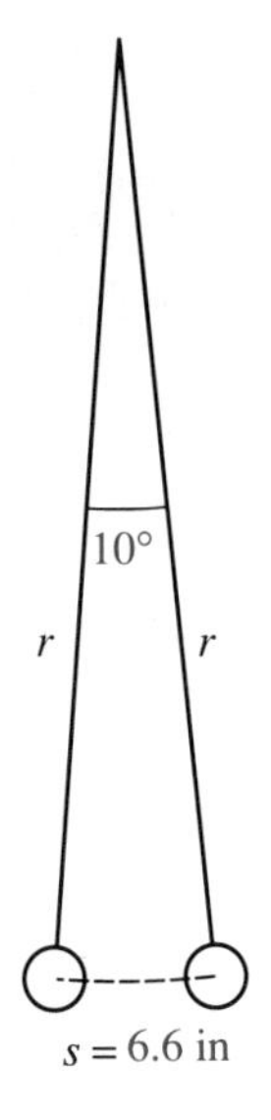

FIGURE 10–11

pendulum, labeled r in the figure. However, θ must be measured in radians, so first we convert 10° to radians.

$$\frac{\pi}{180^\circ} = \frac{\theta}{10^\circ}$$
$$10^\circ\pi = 180^\circ\theta$$
$$\frac{10^\circ\pi}{180^\circ} = \theta$$
$$\frac{\pi}{18} = \theta$$

Next, we solve the formula for r.

$$s = r\theta$$
$$r = \frac{s}{\theta}$$

Last, we substitute $\theta = \pi/18$ and $s = 6.6$ in. into the formula.

$$r = \frac{(6.6 \text{ in.})}{\pi/18}$$
$$r = 37.82 \text{ in.}$$

6.6 [÷] [(] [π] [÷] 18 [)] [=] → 37.82 ■

Area of a Sector

Radian measure is also used in finding the area of a sector of a circle. The area A of a sector is proportional to its central angle θ.

$$\frac{\text{area of sector } A}{\text{area of circle}} = \frac{\text{central angle of sector}}{\text{central angle of circle}}$$
$$\frac{A}{\pi r^2} = \frac{\theta}{2\pi}$$

Solving this equation for A gives the following formula for the area of a circular sector.

Area of a Sector

$$A = \frac{r^2\theta}{2}$$

where

A = area of the sector
r = radius of the circle
θ = measure of the central angle in radians

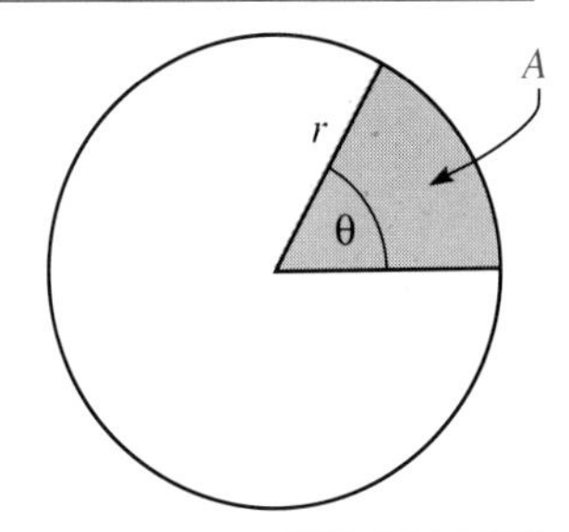

EXAMPLE 3 Find the area of the sector of a circle whose central angle is 2.34 radians and whose radius is 3.15 ft.

Solution Substituting $\theta = 2.34$ and $r = 3.15$ into the formula gives

$$A = \frac{(3.15\text{ ft})^2(2.34)}{2}$$
$$A = 11.6\text{ ft}^2$$

3.15 [x^2] [×] 2.34 [=] [÷] 2 [=] → 11.61 ■

EXAMPLE 4 Part of a baseball field and diamond is to be planted with grass seed. Determine the area of the field if the angle between the first and third baselines is 93° and the radius of the sector to be seeded is 205 ft from home plate.

Solution Figure 10–12 illustrates the verbal statement. We are asked to find the area of a sector of the circle. First, however, we must convert 93° to radians.

$$\frac{93^\circ}{\theta} = \frac{180^\circ}{\pi}$$
$$\theta \approx 1.62$$

FIGURE 10–12

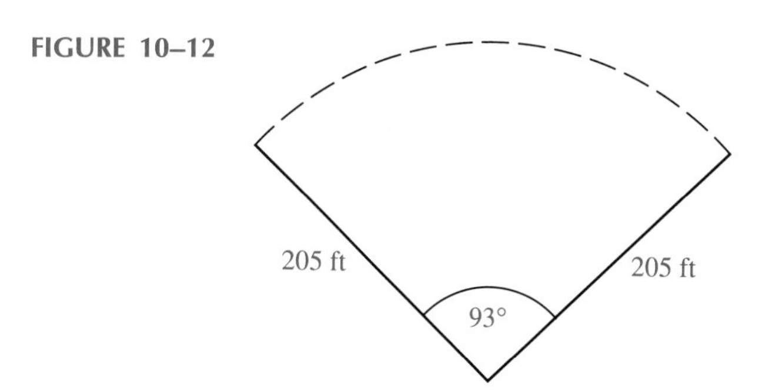

Then we find the area of the sector.

$$A = \frac{(205)^2(1.62)}{2}$$
$$A \approx 34{,}107\text{ ft}^2$$

93 [×] [π] [÷] 180 [=] [×] 205 [x^2] [=] [÷] 2 [=] → 34,106.57 ■

Linear and Angular Velocity

Average velocity v is given by the expression

$$v = \frac{s}{t}$$

where s is distance and t is elapsed time. If an object is traveling in a circular path at a constant speed, the distance traveled equals the length of the arc. If we take the formula for arc length and divide both sides by t, we have

$$s = \theta r$$

$$\frac{s}{t} = \frac{\theta}{t} r$$

$$v = \omega r$$

where v = **average linear velocity** and ω = **average angular velocity** ($\omega = \theta/t$).

Angular Motion

$$v = \omega r$$

where

ω = angular velocity measured in radians per unit of time
v = linear velocity
r = radius of the circle

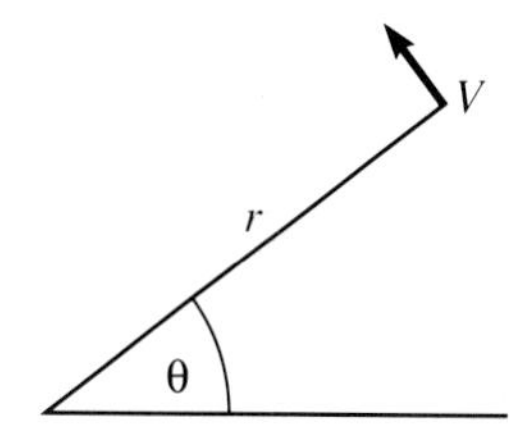

EXAMPLE 5 An object is moving along a circular path of radius 8.1 m with an angular velocity of 6.4 rad/s. Find the linear velocity of the object.

Solution Substituting the quantities $r = 8.1$ m and $\omega = 6.4$ rad/s into the formula gives

$$v = \omega r$$

$$v = (6.4 \text{ rad/s})(8.1 \text{ m})$$

$$v \approx 52 \text{ m/s}$$

■

EXAMPLE 6 A truck with 30.0-in. diameter tires is traveling at 55.0 mi/h. What is the angular velocity of the tires in rad/s?

Solution Since we are asked to find the angular velocity ω, we must rearrange the formula.

$$v = \omega r$$

$$\omega = \frac{v}{r}$$

Then we substitute $r = 15$ in. (diameter $= 30$ in.) and $v = 55$ mi/h into the formula.

$$\omega = \frac{55 \text{ mi/h}}{15 \text{ in.}}$$

However, the units of measure do not give rad/s. In order to obtain the required units of rad/s, we must convert 55 mi/h into in./s. This process is as follows:

$$\omega = \frac{\dfrac{55 \text{ mi}}{\text{h}} \cdot \dfrac{5{,}280 \text{ ft}}{1 \text{ mi}} \cdot \dfrac{12 \text{ in}}{1 \text{ ft}} \cdot \dfrac{1 \text{ h}}{3{,}600 \text{ s}}}{15 \text{ in.}}$$

$$\omega = 64.5 \text{ rad/s}$$

■

10–4 EXERCISES

For a circle with radius r, diameter d, central angle θ, and arc length s, find the unknown quantity.

1. $r = 8.31$ ft, $\theta = \frac{\pi}{3}$, $s = ?$
2. $r = 18.4$ in., $\theta = \frac{2\pi}{7}$, $s = ?$
3. $s = 10.6$ cm, $d = 24$ cm, $\theta = ?$
4. $d = 38$ in., $\theta = 110°$, $s = ?$
5. $s = 34.8$ yd, $\theta = 240°$, $r = ?$
6. $r = 21$ m, $s = 35$ m, $\theta = ?$

For a sector with area A, central angle θ, and circle with radius r, find the unknown quantity.

7. $r = 27.2$ cm, $\theta = \frac{7\pi}{8}$, $A = ?$
8. $r = 6.83$ ft, $\theta = \frac{5\pi}{6}$, $A = ?$
9. $r = 16$ m, $\theta = 220°$, $A = ?$
10. $r = 6.7$ yd, $\theta = 330°$, $A = ?$
11. $A = 486$ ft^2, $r = 18$ ft, $\theta = ?$
12. $A = 56.7$ cm^2, $r = 6.5$ ft, $\theta = ?$

13. $A = 283.5 \text{ in.}^2$, $\theta = 110°$, $r = ?$

14. $A = 687.4 \text{ m}^2$, $\theta = 240°$, $r = ?$

15. A pendulum 37.4 in. long swings through an angle θ which subtends an arc length of 23 in. Determine angle θ in degrees. (Hint: Find θ in radians, then convert.)

16. Determine the length of a pendulum if it swings through an angle of 15° and intercepts an arc of 6.8 in.

17. A flywheel rotates at an angular velocity of 4.83 rad/s. Determine the linear velocity, in feet per second, of a point on the rim if the radius of the flywheel is 1.66 ft.

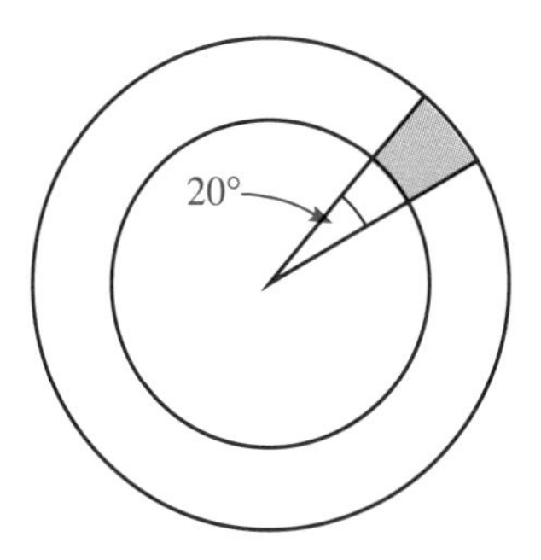

FIGURE 10–13

18. Two concentric circles have radii of 18 ft and 23 ft. Using Figure 10–13, find the shaded area.

19. An isosceles triangle is set in a circle of radius 24 ft as shown in Figure 10–14. Determine the measure of angle θ (in degrees) if the base intercepts an arc 28 ft long.

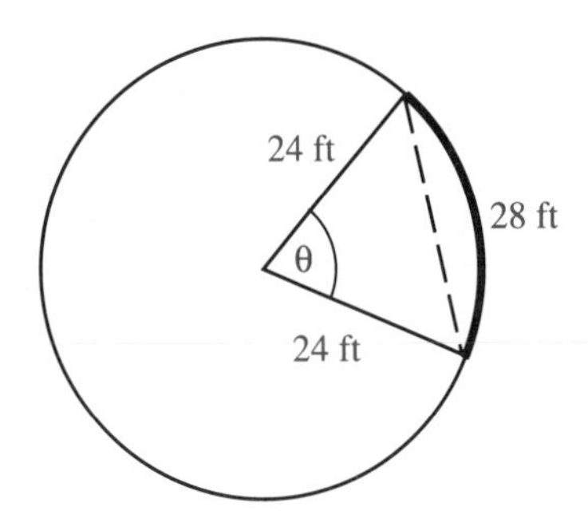

FIGURE 10–14

20. A car with 24-in. diameter tires is traveling at 35 mi/h. What is the angular velocity of the tires in rad/s? (See Example 6.)

21. A phonograph record 12 inches in diameter rotates 33 times per minute. What is the linear velocity, in ft/s, of a point on the rim?

22. In a circle of radius 12 ft, the arc length of a sector is 18 ft. What is the area of the sector?

23. A circular sector whose central angle is 190° is cut from a circular piece of sheet metal of diameter 218 mm. A cone is then formed by joining the radii of the circular sector. Find the lateral surface area of the cone. (Hint: Remember its original shape.)

24. A satellite is in a circular orbit 210 mi above the equator of the earth. How many miles must the satellite move to change its longitude by 73°? (Assume that the radius of the earth at the equator is 3,960 mi.)

25. A windmill has a blade 20 ft in diameter that rotates at 18 r/min. What is the linear velocity in ft/s of a point on the end of the blade?

CHAPTER SUMMARY

Summary of Terms

average angular velocity (p. 371)
average linear velocity (p. 371)
average velocity (p. 371)
radian (p. 363)
reference angle (p. 356)

Summary of Formulas

$s = r\theta$ length of circular arc

$A = \dfrac{r^2\theta}{2}$ area of sector

$v = \omega r$ angular motion

CHAPTER REVIEW

Section 10–1

Without using a calculator, determine the signs of the following trigonometric functions.

1. sec 218°
2. cos 112°
3. tan(−334°)
4. −3 sin(−136°)
5. $\sqrt{\cot 68.4°}$
6. $(\csc 148.67°)^3$
7. sec 286°14′
8. csc 154°46′
9. cot 73.6°

Find the quadrant(s) where the following condition(s) are true.

10. tan θ is negative, cos θ is positive.
11. sec θ is positive, cot θ is positive.
12. sec θ is negative, cot θ is positive.
13. cot θ is negative, cos θ is negative.
14. csc θ is positive, tan θ is negative.
15. sin θ is negative, cos θ is positive.

Find the six trigonometric functions for an angle θ whose terminal side passes through the point given.

16. (−8, 3)
17. (7, −6)
18. (−10, −7)
19. (18, −11)
20. (−4, −7)
21. (−9, 5)

Section 10–2

Use a calculator to find the value of the following trigonometric functions. Round your answer to four decimal places.

22. csc 174.8°
23. $\sqrt{\sec 218°46'}$
24. sin 347°26′
25. −8 sec(−248°)
26. cos(−118°)
27. $(\cot 146.8°)^3$

Find the reference angle for the following angles.

28. 131.6°
29. 247°18′
30. 456°
31. 316.7°
32. 347°23′
33. 114°43′

Find θ, $0° \leq \theta < 360°$, from the given information. Round your answer to the nearest tenth degree.

34. $\sin\theta = -0.5519$ and $\cos\theta < 0$
35. $\cot\theta = 0.3739$ and $\sin\theta > 0$
36. $\sec\theta = 5.0159$ and $\tan\theta > 0$
37. $\cos\theta = -0.2351$ and $\csc\theta < 0$
38. $\tan\theta = 0.4411$ and $\sec\theta > 0$
39. $\sin\theta = -0.5008$ and $\cot\theta < 0$

Section 10–3

Convert each angle measured in degrees to radians. Express your answer as a rational multiple of π.

40. 210°
41. 120°
42. 135°
43. 330°
44. 218°
45. 310°

Convert each angle measured in degrees to decimal radians. Round your answer to hundredths.

46. 247°54′
47. 127.56°
48. 158.73°
49. 318°27′
50. 334.3°
51. 214.5°

Convert each angle measured in radians to the nearest tenth degree.

52. $\frac{7\pi}{6}$
53. 2.756
54. 1.238
55. $\frac{7\pi}{4}$
56. −3.687
57. −5.147

Find the value of the following functions measured in radians.

58. $\cot\frac{7\pi}{4}$
59. csc 4.78
60. sec 1.73
61. $\tan\frac{3\pi}{4}$
62. cos 8.643
63. sin 9.576

Find all θ, $0 \leq \theta < 2\pi$, that satisfy the following conditions.

64. $\sin\theta = -0.7478$
65. $\cot\theta = -1.7811$
66. $\csc\theta = 1.5442$
67. $\cos\theta = -0.0733$ and $\tan\theta < 0$
68. $\cot\theta = 1.0769$ and $\sin\theta > 0$

Section 10–4

For a circle with radius r, central angle θ, and arc length s, find the unknown quantity.

69. $s = 18.6$ m, $\theta = 126°$, $r = ?$

70. $\theta = 1.74$, $s = 2.86$ km, $r = ?$

71. $r = 14$ in., $s = 31$ in., $\theta = ?$

72. $r = 36.7$ ft, $\theta = 56°$, $s = ?$

For a sector with area A, central angle θ, and radius r, find the unknown quantity.

73. $A = 684$ in.^{2}, $\theta = 5.74$, $r = ?$

74. $A = 24.7$ m^2, $r = 11.8$ m, $\theta = ?$

75. $r = 23.5$ yd, $\theta = 256°$, $A = ?$

76. $r = 15.7$ ft, $\theta = 327°$, $A = ?$

77. Two concentric circles have radii of 15 m and 29 m. Using a central angle of 84°, find the area of the sector between the two circles.

78. A circular sector whose central angle is 215° is cut from a circular piece of sheet metal of diameter 18 in. If a cone is formed from the sector by bringing the radii together, find the surface area of the cone.

79. The propeller of a plane is 3.2 m in diameter and rotates at 2,500 r/min. What is the linear velocity in m/min of a point on the end of the blade?

M 80. The armature of a dynamo is 17.4 cm in radius and rotates 1,100 r/min. What is the linear velocity of a point on the rim of the armature?

CHAPTER TEST

The number in parentheses refers to the appropriate learning objective at the beginning of the chapter.

1. Use a calculator to find csc 236°18′. **(4)**

2. The point $(-4, -7)$ is on the terminal side of angle θ. Determine $\sin\theta$ and $\cot\theta$. **(2)**

3. Find the length of the arc subtended by a circle of radius 12 ft if the central angle is 38°. **(8)**

4. Find $\csc(7\pi/6)$. **(7)**

5. Determine all θ, $0° \le \theta < 360°$, such that $\csc\theta = -1.1964$ and $\tan\theta < 0$. Give your answer to the nearest tenth degree. **(3, 5)**

6. For a circle with radius 15 in. and arc length 18 in., find the measure of the central angle to the nearest tenth degree. **(8)**

7. Determine the central angle of a pendulum 18 ft long if it swings through an arc 26.8 ft. **(8)**

8. Determine the quadrant(s) where $\cos\theta$ is negative and $\tan\theta$ is positive. **(1)**

9. Convert 3.87 radians to the nearest tenth degree. **(6)**

10. A car with 24-in. diameter tires is traveling at 45 mi/h. What is the angular velocity, in rad/s, of a point on the rim of one of the tires? **(8)**

11. In a circle of radius 20 m, the arc length of a sector is 14 m. What is the area of this sector? **(8)**

12. Determine the reference angle of 218°34′. **(3)**

13. Find the central angle of a circle of radius 14 in. if the area of the sector is 36 in^2. **(8)**

14. Find all θ, $0 \le \theta < 2\pi$, such that $\sec\theta = -1.2776$ and $\tan\theta < 0$. **(3, 7)**

15. Convert 140° to radians. Leave your answer as a rational multiple of π. **(6)**

16. Find all θ, $0 \leq \theta < 2\pi$, such that $\csc \theta = -1.2877$ and $\tan \theta < 0$. **(3, 7)**

17. A conical tent is made from a circular piece of canvas 22 ft in diameter by removing a central angle of 218°. What is the surface area of the tent made from the larger piece? **(8)**

SOLUTION TO CHAPTER INTRODUCTION

Figure 10–15 depicts the problem outlined in the chapter introduction. We must calculate the linear velocity v given that the angular velocity is $\pi/12$ rad/h and the radius is 22,300 miles. Completing the calculation gives a linear velocity of

$$v = \omega r$$
$$v = (\pi/12)(22{,}300)$$
$$v = 5{,}840 \text{ mi/h}$$

FIGURE 10–15

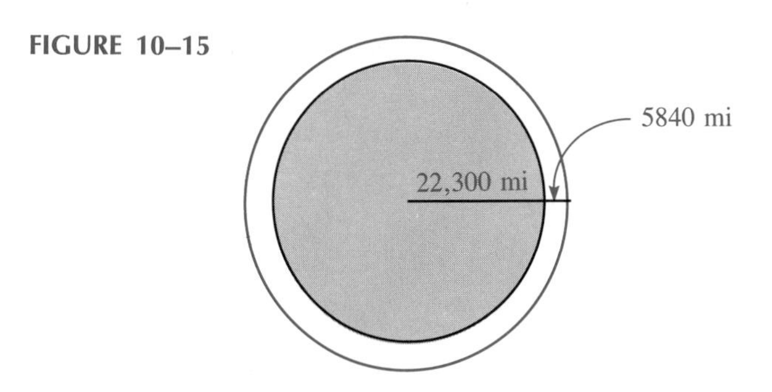

You are an airplane pilot who wants to leave Madison and fly to a destination 17° north of east of your current position. You know that the cruising speed of your plane is 350 mi/h with respect to the air, and the velocity of the wind is 40 mi/h from the west. What heading would you take to reach your destination? (The solution to this problem is given at the end of the chapter.)

A problem of this type involves the vector quantities displacement and velocity. Also, if you make a drawing of this problem, as shown on the facing page, an oblique triangle results. Therefore, to solve a problem of this kind, you must know how to use vectors to solve problems and how to solve oblique triangles; both topics are discussed in this chapter.

Learning Objectives

After completing this chapter, you should be able to

1. Find the resultant of a given set of vectors (Section 11–1).
2. Use vectors to solve problems in science and technology (Section 11–2).
3. Apply the Law of Sines to solve oblique triangles (Section 11–3).
4. Apply the Law of Cosines to solve oblique triangles (Section 11–4).
5. Apply the Law of Sines and Law of Cosines to solve technical problems (Sections 11–3 and 11–4).

Chapter 11

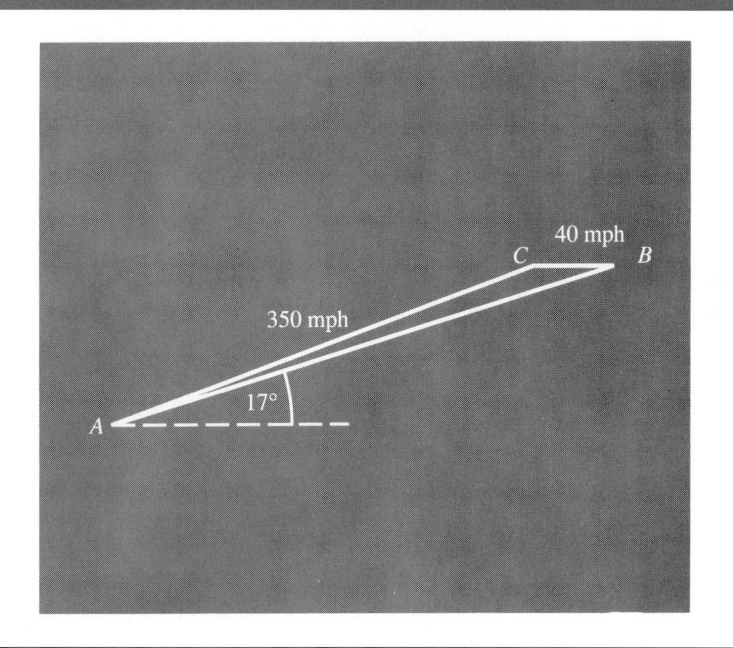

Vectors and Oblique Triangles

11–1

VECTORS

In technical areas some quantities called **scalars** can be specified using *only a number* called the **magnitude.** For example, the length of a table, the speed of a car, and the value of a trigonometric function all represent scalar quantities. However, there are other quantities, called **vectors,** that must be specified by both magnitude and direction. A car traveling 35 mi/h south is an example of the vector quantity called velocity. In the text a vector quantity will be represented in **boldface** type.

Geometric Representation

Vector quantities are represented visually in the rectangular coordinate plane by a directed line segment, as shown in Figure 11–1. The *beginning point* 0, positioned at the origin, is called the **initial point** or **tail** of the vector, and the point S is called the **terminal point** or **head** of the vector. The direction of the arrow represents the direction of the vector, and the length of the line segment represents the magnitude of the vector.

FIGURE 11–1

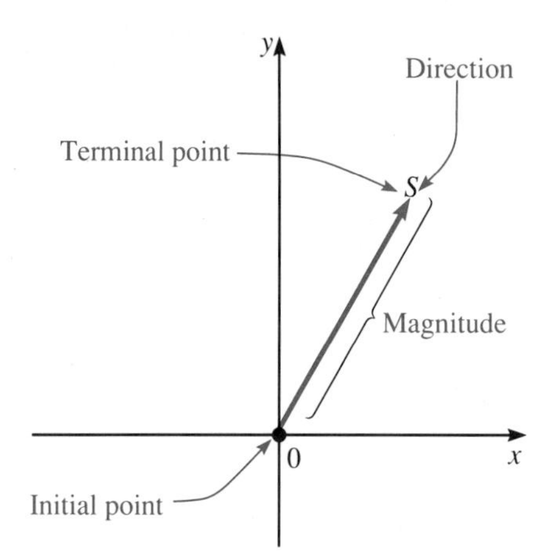

Addition of Vectors

The *sum of two or more vectors,* represented as $\mathbf{A} + \mathbf{B}$, is called their **resultant.** The resultant of several vectors is a single vector that has the same physical effect as the original vectors. Since the resultant is a vector, both its magnitude and its direction must be specified. However, before we discuss vector addition, we must discuss vector components.

Vector Components

If we draw a vector **A** as shown in Figure 11–2 and draw lines from the tail of the vector parallel to the x and y axes, the *two new vectors* are called the **components** of vector **A.** The original vector **A** is the sum of its x and y components, expressed by

$$\mathbf{A} = \mathbf{A}_x + \mathbf{A}_y$$

FIGURE 11–2

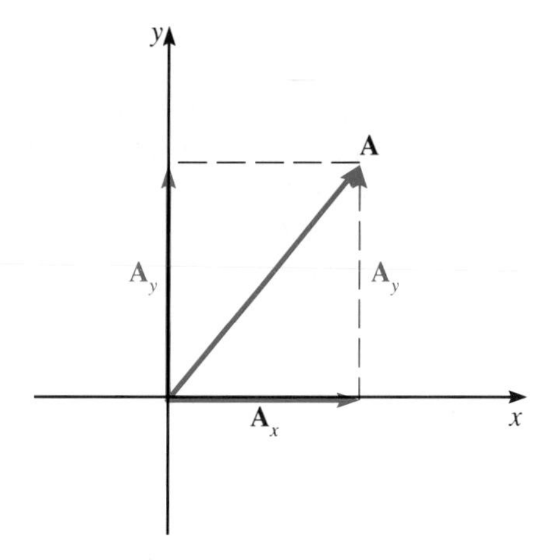

The *process of finding the components* of a vector is called **resolving the vector into its components.** If you are given the magnitude A and direction θ of vector **A,** the calculations for the components are as summarized in the following box.

Components of a Vector

Using the definition of the sine and cosine functions, we have

$$\cos \theta = \frac{A_x}{A} \quad \text{or} \quad A_x = A \cos \theta$$

$$\sin \theta = \frac{A_y}{A} \quad \text{or} \quad A_y = A \sin \theta$$

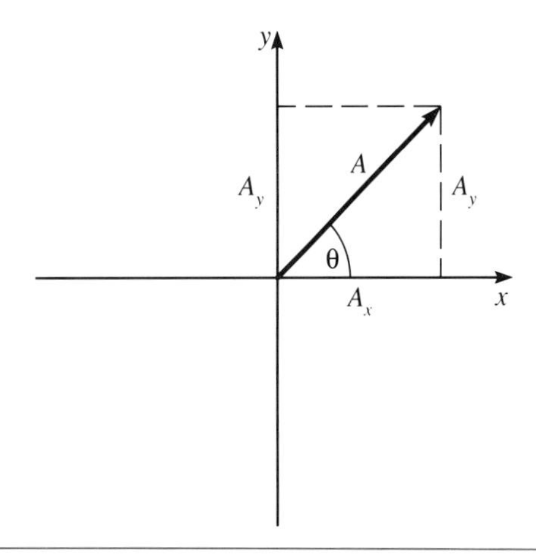

EXAMPLE 1 Resolve the following vectors into components:

(a) $s = 56$ m; $\theta = 67°$ (b) $F = 83$ lb; $\theta = 158°$

Solution

(a) $s_x = s \cos\theta = 56 \cos 67°$ 56 [×] 67 [cos] [=] → 22
$s_x = 22$ m
$s_y = s \sin\theta = 56 \sin 67°$ 56 [×] 67 [sin] [=] → 52
$s_y = 52$ m

A graphical illustration of resolving this vector into its components is given in Figure 11–3(a).

(b) $F_x = F \cos\theta = 83 \cos 158°$ 83 [×] 158 [cos] [=] → −77
$F_x = -77$ lb
$F_y = F \sin\theta = 83 \sin 158°$ 83 [×] 158 [sin] [=] → 31
$F_y = 31$ lb

A graphical illustration of resolving this vector into its components is given in Figure 11–3(b).

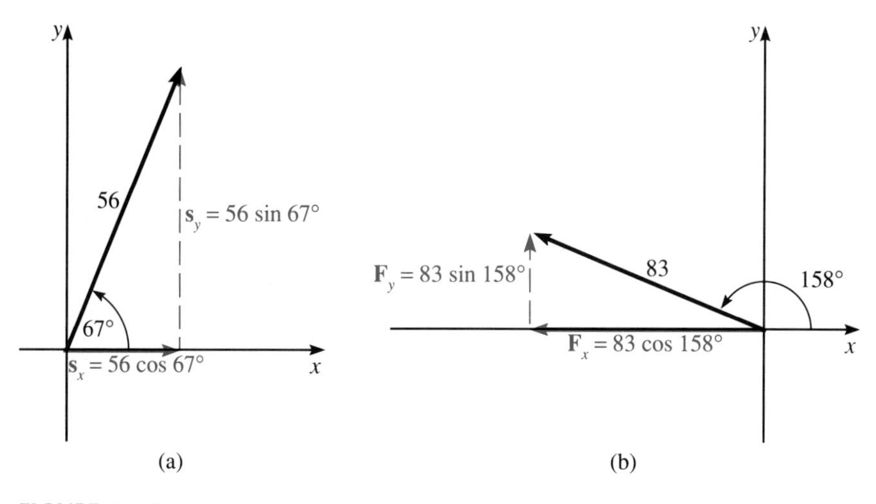

FIGURE 11–3

Vector Resultants Algebraically

Now we are in a position to add vectors algebraically. This process is illustrated in the next three examples.

EXAMPLE 2 Find the resultant of two forces, **A** and **B**. **A** = 63 lb east and **B** = 47 lb south.

Solution Vectors **A** and **B** are shown in Figure 11–4. To find the magnitude of the resultant represented by R in the figure, we can use the Pythagorean theorem.

$$R^2 = A^2 + B^2$$
$$R = \sqrt{A^2 + B^2}$$
$$R = \sqrt{(63)^2 + (47)^2}$$
$$R = 79 \text{ lb}$$

FIGURE 11–4

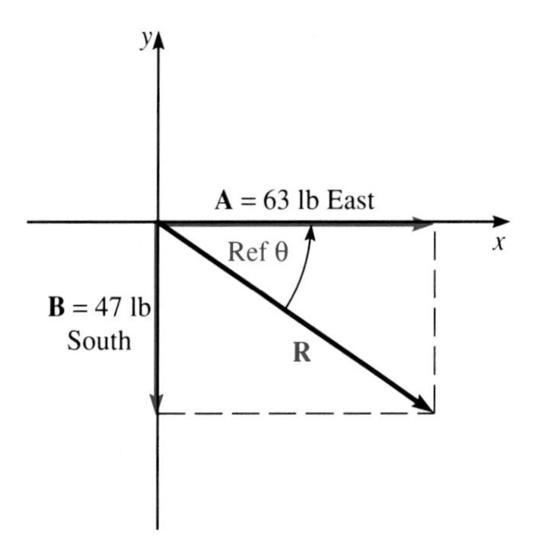

To find the direction of the resultant, we find the reference angle, labeled Ref θ in Figure 11–4, and then calculate θ.

$$\tan(\text{Ref } \theta) = \left|\frac{-47}{63}\right|$$
$$\text{Ref } \theta = 37^\circ$$

Since θ is a fourth quadrant angle, we calculate

$$\theta = 360^\circ - \text{Ref } \theta$$
$$\theta = 360^\circ - 37^\circ$$
$$\theta = 323^\circ$$

The resultant of vectors **A** and **B** is 79 lb at 323° from standard position. ■

NOTE ✦ To be sure that everyone is using the same angle, θ should always be an angle in standard position.

Vector Resultants by Components

Since the vectors in Example 2 were perpendicular, it was not necessary to resolve them into components. However, any vector that does not lie on the x or y axis must be resolved into components. The following box summarizes vector resultants, and the next example illustrates this procedure.

Resultant of Vectors

Magnitude: $$R = \sqrt{R_x^2 + R_y^2}$$

Direction: $$\tan(\text{Ref } \theta) = \left|\frac{R_y}{R_x}\right|$$

where

$$R_x = \text{sum of all } x \text{ components}$$
$$R_y = \text{sum of all } y \text{ components}$$

EXAMPLE 3 Find the resultant of two displacements, $\mathbf{A} = 48$ m at 63° and $\mathbf{B} = 73$ m at 118°.

Solution Figure 11–5 shows the two vectors and their respective components. Using the formulas to find the x and y components of the vector gives

$$A_x = 48 \cos 63° \approx 22 \qquad A_y = 48 \sin 63° \approx 43$$
$$B_x = 73 \cos 118° \approx -34 \qquad B_y = 73 \sin 118° \approx 64$$

We use the formula to calculate the magnitude and direction of the resultant.

$$R = \sqrt{R_x^2 + R_y^2}$$
$$R \approx \sqrt{(22 - 34)^2 + (43 + 64)^2}$$
$$R \approx \sqrt{(-12)^2 + (107)^2}$$
$$R \approx 110 \text{ m}$$

$$\tan(\text{Ref } \theta) = \left|\frac{R_y}{R_x}\right|$$
$$\tan(\text{Ref } \theta) \approx \left|\frac{107}{-12}\right|$$
$$\text{Ref } \theta \approx 83.4°$$

Since R_y is positive and R_x is negative, θ is a second quadrant angle. Figure 11–6 shows R_x, R_y, R, and θ.

$$\theta = 180° - \text{Ref } \theta$$
$$\theta \approx 180° - 83.4°$$
$$\theta \approx 96.6°$$

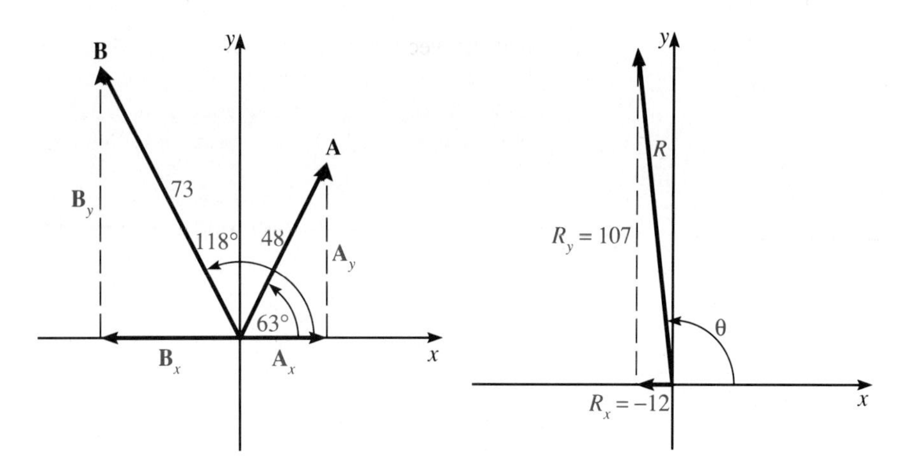

FIGURE 11–5

FIGURE 11–6

The resultant of vectors **A** and **B** is 110 m at 96.6°. ■

A table proves useful in finding the resultant of more than two vectors. The table approach is shown in the next example.

EXAMPLE 4 Find the resultant of the following vectors:

A = 38.0 N at 218° **B** = 56.0 N at 48° **C** = 23.0 N at 108°

Solution We fill in a table showing the x and y components for each vector and the sum of these components.

Vector	x Component	y Component
A	$38 \cos 218° \approx -29.9440$	$38 \sin 218° \approx -23.395$
B	$56 \cos 48° \approx 37.471$	$56 \sin 48° \approx 41.616$
C	$23 \cos 108° \approx -7.107$	$23 \sin 108° \approx 21.874$
Sum	$R_x \approx 0.420$	$R_y \approx 40.095$

The calculation for the magnitude of the resultant gives

$$R = \sqrt{R_x^2 + R_y^2}$$
$$R \approx \sqrt{(0.420)^2 + (40.095)^2}$$
$$R \approx \sqrt{1{,}607.785425}$$
$$R \approx 40.1$$

The direction of the resultant is given by

$$\tan(\text{Ref } \theta) = \left|\frac{R_y}{R_x}\right|$$

$$\tan(\text{Ref } \theta) \approx \left|\frac{40.095}{0.420}\right|$$

$$\text{Ref } \theta \approx 89°$$

Since R_x and R_y are both positive, θ is a quadrant I angle.

$$\theta \approx 89°$$

The resultant of **A, B,** and **C** is approximately 40.1 N at 89°. ■

11–1 EXERCISES

Find the resultant of the following vectors. Give the direction as an angle in standard position.

1. 18 N west; 31 N east

2. 49 lb south; 36 lb north

3. 254 m south; 189 m north

4. 158 yd east; 176 yd west

5. 53.4 ft/s north; 39.5 ft/s west

6. 17.8 mi east; 28.3 mi north

7. 59.3 lb south; 76.8 lb west

8. 7.6 m/s east; 13.4 m/s south

9. 217 N north; 309 N west

10. 9.3 m/s^2 east; 14.8 m/s^2 north

11. 18.4 mi south; 30.5 mi east

12. 156.7 lb west; 204.5 lb south

13. 55.6 ft/s north; 42.3 ft/s west

14. 37.4 m east; 26.9 m south

15. 32.6 ft/s^2 south; 21.5 ft/s^2 west

Find the x and y components of the following vectors.

16. 148 N at 35°

17. 118 ft/s at 73.5°

18. 18.8 m/s^2 at 212°

19. 56.8 lb at 136°

20. 78.3 N at 142.6°

21. 123.8 yd at 285.3°

22. 38.9 m/s at 243.6°

23. 66.9 m at 195.8°

24. 315.7 lb at 312.6°

25. 418.3 mi at 156.7°

Find the resultant of the following vectors. Give the direction as an angle in standard position.

26. **A** = 16.7 lb at 126.4°; **B** = 31.8 lb at 78°

27. **A** = 14.8 ft/s at 243.5°; **B** = 29.5 ft/s at 118.7°

28. **A** = 119.5 m at 113.8°; **B** = 79.3 m at 212.6°

29. **A** = 48.6 yd at 82°; **B** = 63.5 yd at 108.4°

30. **A** = 27.8 N at 217.5°; **B** = 40.6 N at 126.8°

31. **A** = 63 m/s^2 at 93.4°; **B** = 87 m/s^2 at 317.8°

32. **A** = 212.4 lb at 246.8°; **B** = 186.5 lb at 301.6°

33. **A** = 82.1 m at 108°; **B** = 61.8 m east; **C** = 59.2 m at 51.2°

34. **A** = 24 ft/s at 62.4°; **B** = 35.6 ft/s at 284.6°; **C** = 41.9 ft/s at 125°

35. **A** = 46 N at 136°; **B** = 73 N at −73°; **C** = 52 N at 314°

36. **A** = 18 cm at −60°; **B** = 38 cm at 98°; **C** = 59 cm at 203°

37. **A** = 384 mi at 123°; **B** = 239 mi at 207°; **C** = 315 mi at 246°

38. **A** = 14.8 m/s at 78.3°; **B** = 37.4 m/s at 163.5°; **C** = 10.9 m/s at 309.7°

39. **A** = 486 N at 197°; **B** = 217 N at 318°; **C** = 524 N at −153°

40. **A** = 316 ft at −122°; **B** = 214 ft at 326°; **C** = 190 ft at 118°

11–2

APPLICATIONS OF VECTORS

Vectors have numerous applications in science and technology. In this section we will discuss vectors as they apply to velocity, displacement, systems of forces, and inclined planes.

Velocity

Velocity is defined to be the rate at which displacement changes with respect to time. The next example shows an application of vectors to velocity.

EXAMPLE 1 A plane flies at a rate of 300 mi/h 60° north of east with respect to the air. If the velocity of the wind is 100 mi/h in a direction 23° south of east, what is the velocity of the plane with respect to the ground?

Solution As shown in Figure 11–7, we are finding the resultant of two vectors: the air velocity of the plane and the velocity of the wind. The calculations for the resultant are as given in the accompanying table.

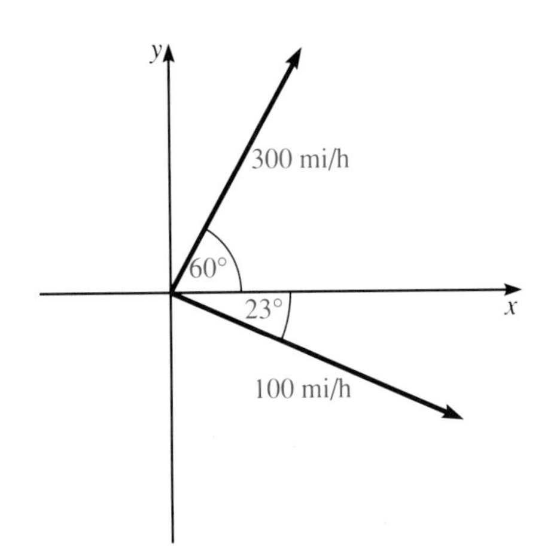

FIGURE 11–7

Vector	x Component	y Component
Plane	$300 \cos 60° = 150$	$300 \sin 60° = 259.81$
Wind	$100 \cos 23° = 92.05$	$100 \sin 23° = -39.07$
Sum	$R_x = 242.05$	$R_y = 220.74$

The x component of the resultant is 242.05, and the y component is 220.74. The calculations for the magnitude and direction of the resultant are as follows:

$$R = \sqrt{R_x^2 + R_y^2}$$
$$R \approx \sqrt{(242.05)^2 + (220.74)^2}$$
$$R \approx 328$$

$$\tan(\text{Ref } \theta) = \frac{R_y}{R_x}$$
$$\tan(\text{Ref } \theta) \approx \frac{220.74}{242.05}$$
$$\text{Ref } \theta \approx 42°$$

Since R_x and R_y are both positive, $\theta = 42°$. The ground velocity of the plane is 328 mi/h at 42° north of east. ■

Displacement

Displacement is a vector quantity that describes an object's change in position. This application is illustrated in the next example.

EXAMPLE 2 A surveyor leaves her transit and walks 20 m 23° north of east, then 10 m west, and finally 6 m 56° south of west. How far is she from her starting position?

Solution We must find the resultant of the following three vectors, shown in Figure 11–8:

A = 20 m 23° north of east **B** = 10 m west **C** = 6 m 56° south of west

The accompanying table summarizes the x and y components of these vectors.

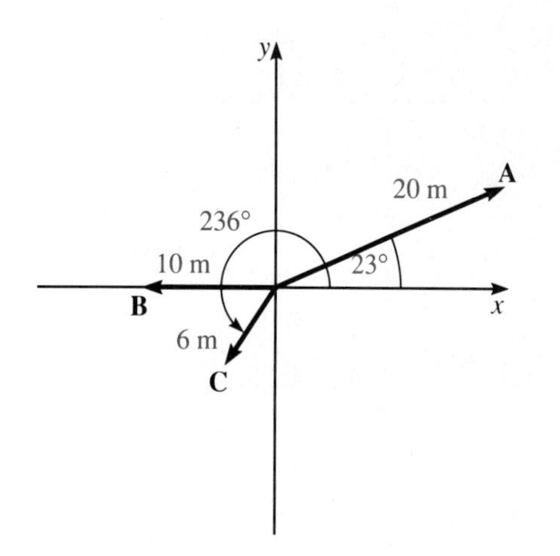

FIGURE 11–8

Vector	x Component	y Component
A	20 cos 23° = 18.4	20 sin 23° = 7.8
B	10 cos 180° = −10	10 sin 180° = 0
C	6 cos 236° = −3.4	6 sin 236° = −5.0
Sum	R_x = 5	R_y = 2.8

Next, we calculate the magnitude and direction for the resultant of the vectors.

$$R = \sqrt{R_x^2 + R_y^2}$$
$$R \approx \sqrt{(5)^2 + (2.8)^2}$$
$$R \approx 6$$

$$\tan(\text{Ref } \theta) = \left|\frac{R_y}{R_x}\right|$$

$$\tan(\text{Ref } \theta) \approx \frac{2.8}{5}$$

$$\text{Ref } \theta \approx 29°$$

Since R_y is positive and R_x is positive, θ is in quadrant I. Therefore, $\theta = 29°$. The surveyor is 6 m at 29° from her starting position. ■

Force

Force is a vector quantity (push or pull) that tends to cause motion. According to the First Condition of Equilibrium, if an object is at rest or moving at a constant velocity, the resultant of all the forces acting on that body is zero. Therefore, the sum of all vectors along the x axis is zero, and the sum of all vectors along the y axis is zero. The next two examples illustrate the application of vectors to a system of forces.

EXAMPLE 3 A sign weighing 100 lb is hung from the ceiling by cables that make 43° and 62° angles with the ceiling. Determine the tension in each of the cables.

Solution First, we draw Figure 11–9, depicting the verbal statement. Second, we resolve the vectors into components as shown in Figure 11–10. Third, we apply the First Condition of Equilibrium to yield the following equations:

$$\text{resultant of vectors along } x \text{ axis} = 0$$
$$A \cos 43° - B \cos 62° = 0$$
$$\text{resultant of vectors along } y \text{ axis} = 0$$
$$A \sin 43° + B \sin 62° + (-100) = 0$$

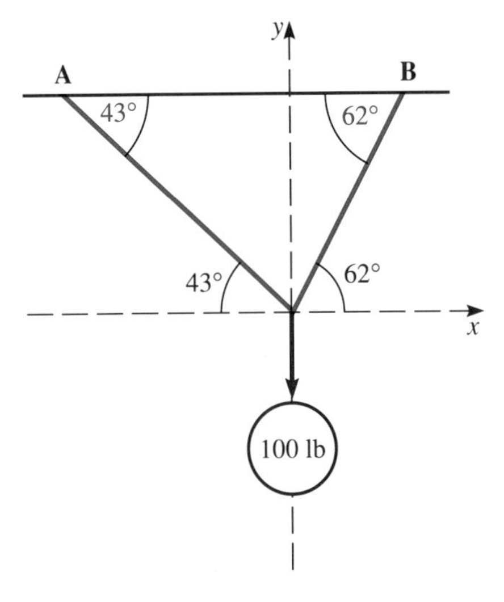

FIGURE 11–9

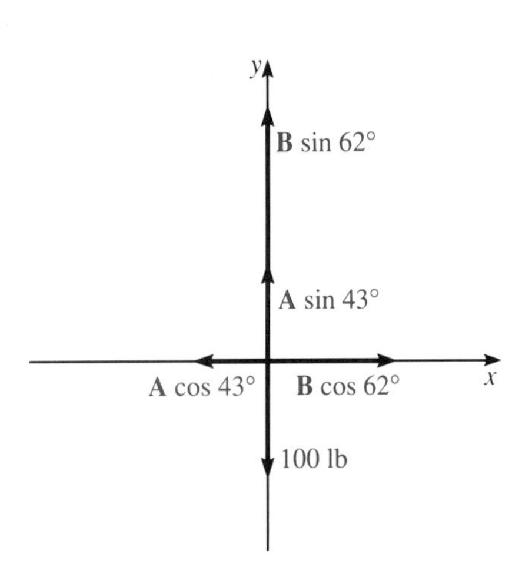

FIGURE 11–10

Using substitution to solve the system of equations for A and B gives

$$A = \frac{B \cos 62^\circ}{\cos 43^\circ}$$

$$A = 0.64B$$

$$0.64B \sin 43^\circ + B \sin 62^\circ - 100 = 0$$

$$0.44B + 0.88B = 100$$

$$1.32B = 100$$

$$B \approx 76 \text{ lb}$$

$$A = 0.64B = 0.64(76) \approx 49 \text{ lb}$$

The cable at the 43° angle has a tension of 49 lb, and the cable at the 62° angle has a tension of 76 lb. ■

Inclined Plane

A second application using a system of forces involves an object moving at a constant speed or at rest on an inclined plane. Figure 11–11 shows the forces acting in such a system. The normal force **N,** acting perpendicular to the plane, is the force exerted by the inclined plane on the object. The weight **W** of the object, acting vertically downward, is the force exerted by the object on the inclined plane. The frictional force **F** acts parallel to the inclined plane in the opposite direction of any possible movement. For ease in solving inclined plane problems, the x axis is drawn parallel to the inclined plane, and the y axis is drawn perpendicular to the inclined plane.

FIGURE 11–11

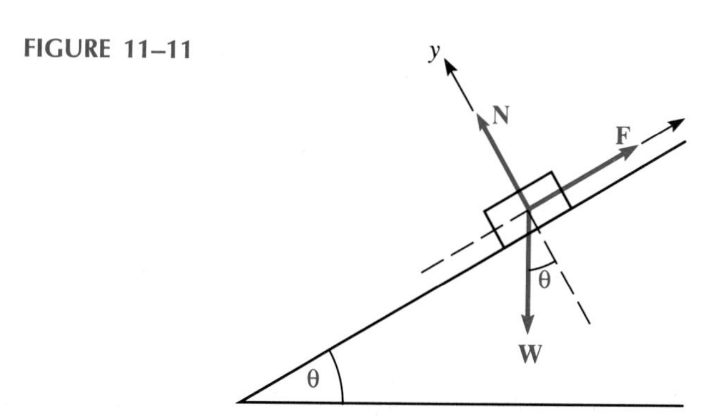

EXAMPLE 4 A 56-lb object is at rest on an inclined plane that makes an angle of 30° with the horizontal. Find the frictional and normal forces.

Solution Figure 11–12 shows the inclined plane with the pertinent information labeled. If we place the x and y axes in the manner described previously and rotate them, we obtain the system of forces shown in Figure 11–13. Using similar triangles, we find that the angle of the inclined plane equals the angle between the weight and the y axis. Resolving the vectors into components and applying the First Condition of Equilibrium gives the following equations:

$$56 \cos 60^\circ = \mathbf{F}$$
$$56 \sin 60^\circ = \mathbf{N}$$

Solving the above equations for **F** and **N** gives the frictional force as 28 lb and the normal force as 48 lb.

56 [×] 60 [cos] [=] → 28

56 [×] 60 [sin] [=] → 48 ■

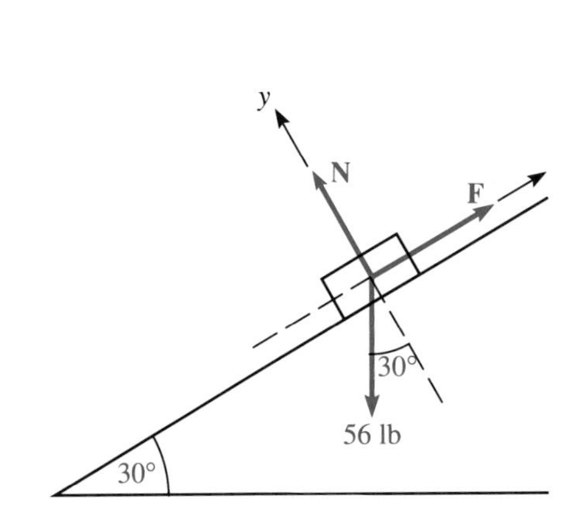

FIGURE 11–12

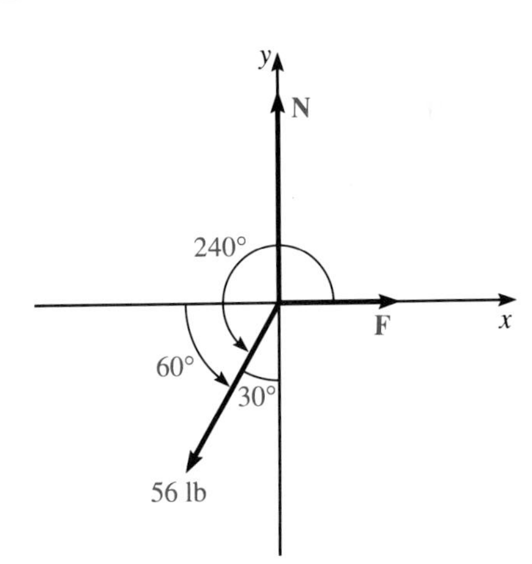

FIGURE 11–13

11–2 EXERCISES

1. Jane leaves the bakery, walks 8 blocks east, turns and walks 15 blocks at 45°, and then walks 10 blocks north. What is Jane's displacement from her initial position?
2. A boat travels horizontally across a river. If the boat travels 8 mi/h in still water and the river current travels 2 mi/h, find the velocity of the boat with respect to the water.
3. A plane is headed 37° north of east at a velocity of 457 mi/h with respect to the air. If the wind is from the northwest at 56 mi/h, what is the resultant velocity of the plane?

4. Movers are pushing a 200-lb sofa up a 20° ramp, applying a force of 150 lb at an angle of 35° with the horizontal. Determine the frictional force if the piano is moving at a constant velocity.

5. Two forces, one of 186 N and the other of 137 N, act on an object at right angles to each other. Find the magnitude of the resultant of these forces.

6. A jet flies 685 km south and then turns and flies 946 km at 58.4° west of south. What is the plane's displacement from its starting point?

7. A 20-lb ball is attached to a rope that hangs vertically from the ceiling. If the ball is pulled horizontally by a force **F** such that the rope makes a 35° angle with the vertical, find the force **F** and the tension in the rope.

8. A force of 15 lb is exerted on a bolt as shown in Figure 11–14. Determine the horizontal and vertical components of the force.

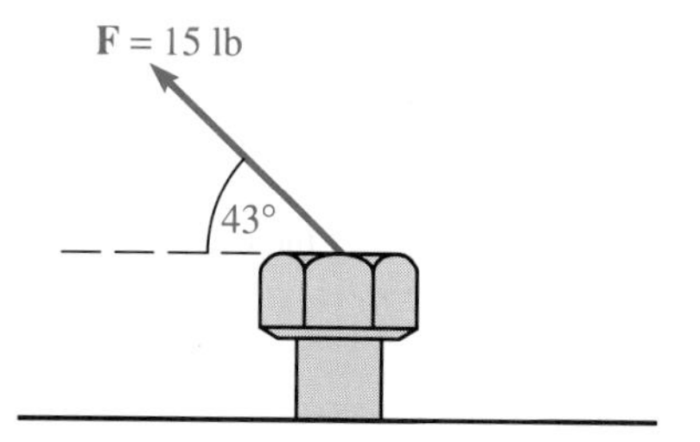

FIGURE 11–14

9. A collar that slides on a vertical rod is acted on by three forces, as shown in Figure 11–15. The direction of **F** may vary. Determine the direction of **F** so that the resultant of the forces is horizontal if the magnitude of **F** is 200 lb.

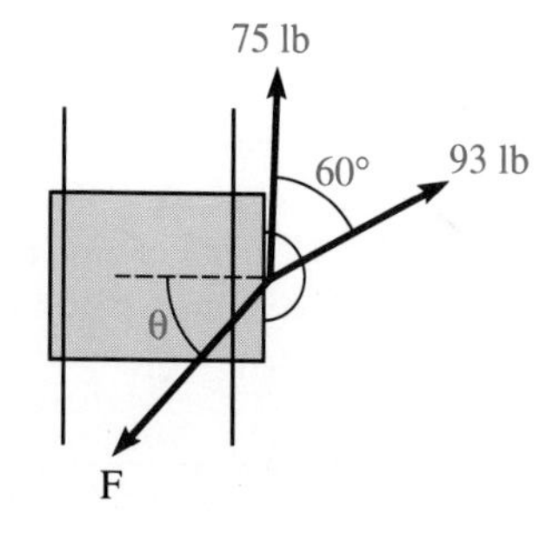

FIGURE 11–15

10. Two cables are tied together at point C and loaded as shown in Figure 11–16. Determine the tension in cables AC and BC.

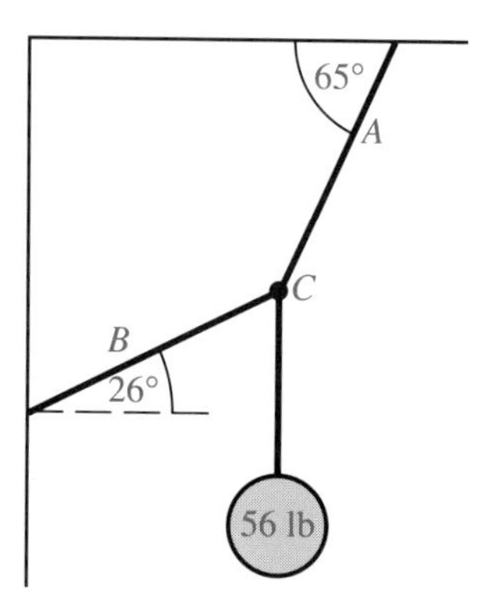

FIGURE 11–16

11. Frank wants to swim to a point directly across the river from his starting point. If he swims at a rate of 2.4 mi/h and the rate of the current is 1.5 mi/h, at what angle with respect to the shore should he swim?

12. A 4800-N crate lying between two buildings is being lifted onto a truck. The crate is supported by a vertical cable joined by two ropes A and B that pass over pulleys attached to the buildings. From Figure 11–17, determine the tension in ropes A and B.

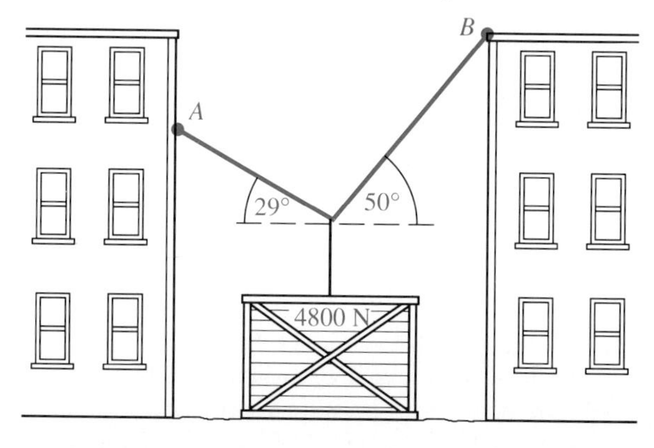

FIGURE 11–17

13. Four forces act on a screw as shown in Figure 11–18. Determine the resultant force.

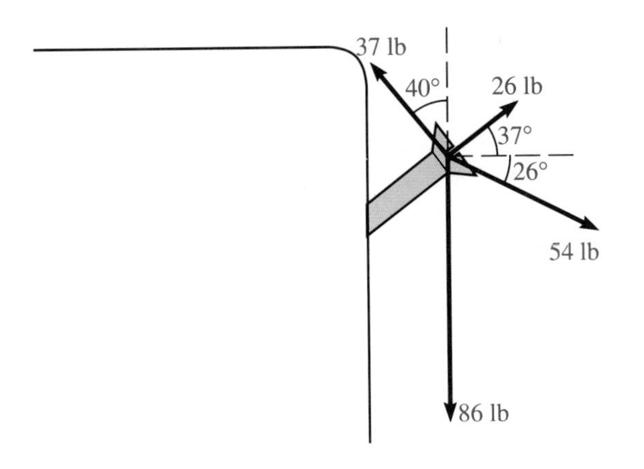

FIGURE 11–18

14. An airplane heads 37° north of west at a speed of 418 mi/h with respect to the air. If a wind blows at 38 mi/h from a direction 18° north of east, what is the velocity of the plane with respect to the ground?

15. Two cables are tied together at point C as shown in Figure 11–19. Determine the tension in cables AC and BC. (You will have to use right triangle trigonometry to determine the necessary angles.)

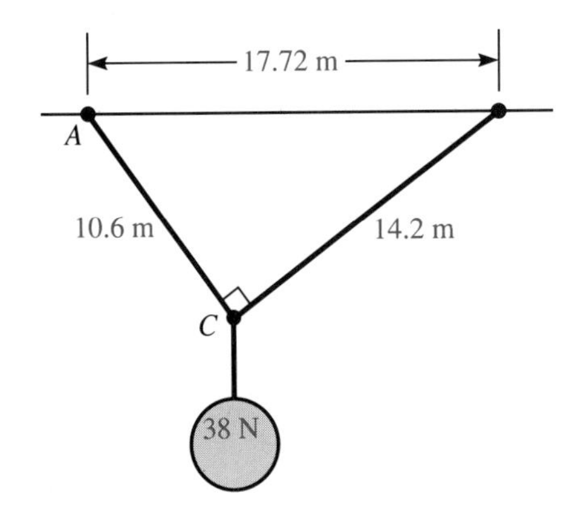

FIGURE 11–19

16. If an object weighing 75 N sits on a ramp inclined at 32° with the horizontal, what is the force of friction exerted on the object by the surface of the ramp?

17. A plane flies 27.8° south of east at a velocity of 380 mi/h with respect to the air. If the wind is blowing at 85 mi/h from a direction of 46.5° north of west, what is the velocity of the plane with respect to the ground?

18. Two ropes are tied together at point C as shown in Figure 11–20. The maximum permissible tension in each rope is 678 lb. What is the maximum force **F** that can be applied without breaking the ropes if the direction of this force is 20°?

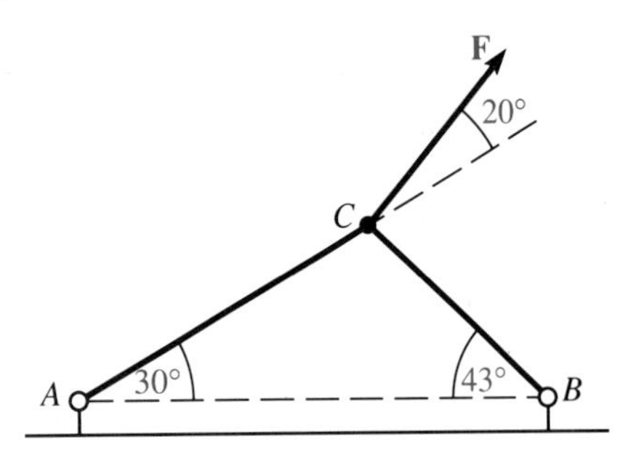

FIGURE 11–20

19. A sign weighing 250 lb is supported by two cables hung from the ceiling. One cable makes an angle of 48° with the ceiling, and the other cable makes an angle of 63° with the ceiling. Find the tension in each cable.

20. An object is dropped from a plane moving at 110 m/s at an angle of 15° below the horizontal. If the vertical velocity of the object as a function of time is given by $v_y = -110 \sin 15° - 9.8t$, what are the velocity and the direction of the object after 3 s?

21. Two cables with known tensions are attached to a telephone pole at point C. A third cable AC used as a guy wire is also attached to the pole at C. Determine the tension in AC so that the resultant of the three forces will be vertical. See Figure 11–21.

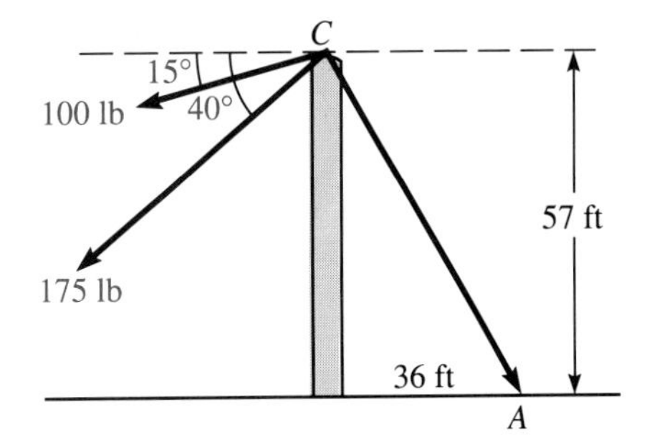

FIGURE 11–21

11–3

THE LAW OF SINES

In previous sections we have seen how to use trigonometry to solve right triangles. However, situations arise where it is necessary to solve an **oblique triangle,** a triangle without a right angle. We can use the Law of Sines and the Law of Cosines to solve oblique triangles, and the remainder of this chapter is devoted to these two topics.

Derivation of Law of Sines

We begin the discussion of solving oblique triangles by deriving the Law of Sines. Let *ABC* be an oblique triangle with sides *a, b,* and *c* opposite their respective angles, as shown in Figure 11–22. If an altitude *h* is drawn to the base, we can write the following relationships:

$$\sin A = \frac{h}{b} \qquad \sin B = \frac{h}{a}$$

$$h = b \sin A \qquad h = a \sin B$$

FIGURE 11–22

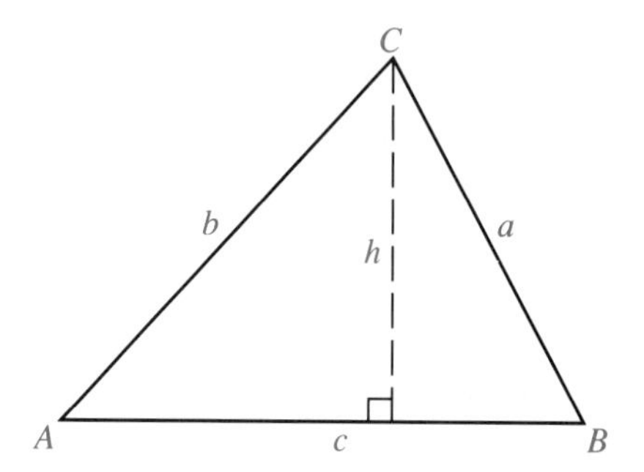

Equating the two expressions for *h* gives

$$b \sin A = a \sin B$$

Dividing both sides of the equation by $\sin A \sin B$ gives the following relationship:

$$\frac{b}{\sin B} = \frac{a}{\sin A}$$

Similarly, if we draw an altitude from angle *A* to side *a,* we can derive the following expression:

$$\frac{c}{\sin C} = \frac{b}{\sin B}$$

Combining these two results gives the Law of Sines, summarized as follows.

Law of Sines

In any triangle ABC,

$$\frac{a}{\sin A} = \frac{b}{\sin B} = \frac{c}{\sin C}$$

The length of a side is proportional to the sine of the angle opposite that side.

We can use the Law of Sines to solve the following two cases of triangles:

- ☐ *Case 1.* Two angles and one side are given.
- ☐ *Case 2.* Two sides and the angle opposite one of the sides are given.

A discussion of each of these cases follows.

Case 1

Example 1 illustrates the case where two angles and one side are given.

EXAMPLE 1 Given $A = 38°$, $B = 62°$, and $b = 16.2$, solve triangle ABC shown in Figure 11–23.

FIGURE 11–23

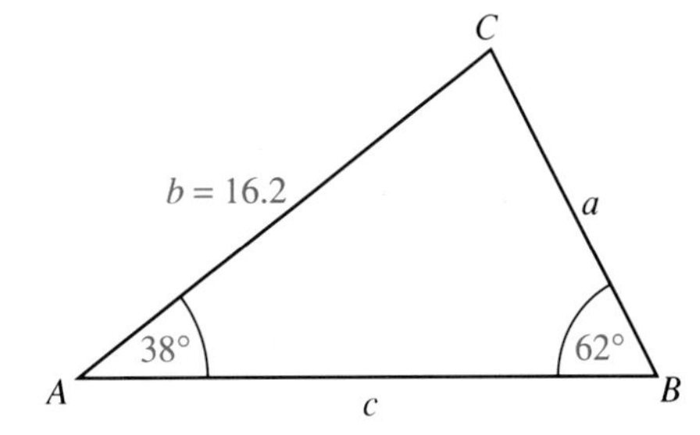

Solution From the figure, we can see that we must solve for angle C, side a, and side c. Since we know that the sum of the angles of a triangle is 180°, we can solve for C using the following equation:

$$180° = A + B + C$$
$$C = 180° - 38° - 62°$$
$$C = 80°$$

To solve for side a, we can use the Law of Sines. The calculation is as follows:

This ratio contains the unknown. $\rightarrow \dfrac{a}{\sin A} = \dfrac{b}{\sin B} \leftarrow$ Both terms of this ratio are known.

$$\frac{a}{\sin 38^\circ} = \frac{16.2}{\sin 62^\circ}$$

$$a = \frac{16.2 \sin 38^\circ}{\sin 62^\circ}$$

$$a = 11.3$$

16.2 [×] 38 [sin] [÷] 62 [sin] [=] → 11.3

To solve for side c. we can use the Law of Sines again.

$$\frac{c}{\sin C} = \frac{b}{\sin B}$$

$$\frac{c}{\sin 80^\circ} = \frac{16.2}{\sin 62^\circ}$$

$$c = \frac{16.2 \sin 80^\circ}{\sin 62^\circ}$$

$$c = 18.1$$

16.2 [×] 80 [sin] [÷] 62 [sin] [=] → 18.1 ■

CAUTION ✦ A figure is helpful in solving oblique triangles; however, the triangles are not drawn to scale. Do not rely only on the figure for information.

Case 2

When two sides and the angle opposite one of them is given, there may be no, one, or two solutions to the triangle. For this reason, Case 2 is called the **ambiguous case.** The next two examples illustrate the possibilities.

EXAMPLE 2 Given $c = 15$, $b = 23$, and $C = 42^\circ$, solve triangle ABC shown in Figure 11–24.

FIGURE 11–24

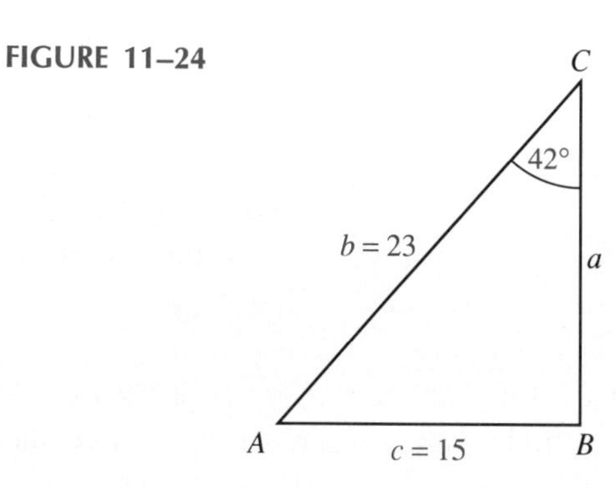

Solution From the figure, we must first solve for angle B using the Law of Sines.

$$\frac{b}{\sin B} = \frac{c}{\sin C}$$

$$\frac{23}{\sin B} = \frac{15}{\sin 42^\circ}$$

$$\sin B = \frac{23 \sin 42^\circ}{15}$$

$$\sin B = 1.03$$

Since the value of the sine of an angle is never greater than 1, no triangle can be formed with the dimensions given here. ■

EXAMPLE 3 Given $a = 21$, $b = 18$, and $A = 38^\circ$, solve triangle ABC shown in Figure 11–25.

FIGURE 11–25

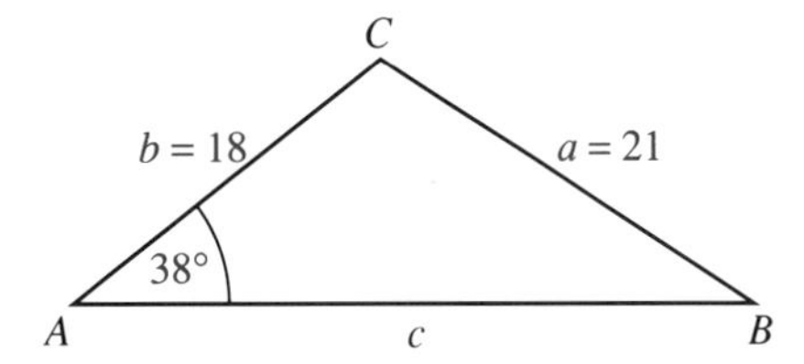

Solution From the figure, we can see that we should first solve for angle B using the Law of Sines.

$$\frac{a}{\sin A} = \frac{b}{\sin B}$$

$$\frac{21}{\sin 38^\circ} = \frac{18}{\sin B}$$

$$\sin B = \frac{18 \sin 38^\circ}{21}$$

$$\sin B \approx 0.5277$$

$$B \approx 32^\circ$$

We have found a reference angle of 32° for angle B, but the sine function is positive in the first two quadrants. Therefore, there are two possibilities for the measure of angle B: 32° or 148°. Since we cannot determine which is true, we must consider both possibilities.

If $B = 32°$,

$$C = 180° - 38° - 32°$$
$$C = 110°$$
$$\frac{a}{\sin A} = \frac{c}{\sin C}$$
$$\frac{21}{\sin 38°} = \frac{c}{\sin 110°}$$
$$c = \frac{21 \sin 110°}{\sin 38°}$$
$$c = 32$$

If $B = 148°$,

$$C = 180° - 38° - 148°$$
$$C = -6°$$

Since the angles of a triangle are not measured in negative numbers, the solution to the triangle is $B = 32°$, $C = 110°$, and $c = 32$. ■

Examples 2 and 3 show that we can determine the number of situations to the ambiguous case by working through the problem. However, the number of solutions can also be determined from the given information. These results are summarized below.

Solutions to the Ambiguous Case

Given sides a and b and angle A of oblique triangle ABC, the following solutions are possible:

- ☐ no solution if $a < b \sin A$
- ☐ right triangle solution if $a = b \sin A$
- ☐ two solutions if $b \sin A < a < b$
- ☐ one solution if $a > b$ or $A > 90°$

EXAMPLE 4 A pilot wants to fly to his destination, which is 14° north of west from his current position. If the wind is blowing from the east at 20 mi/h and the plane flies at 400 mi/h with respect to the air, what heading should the pilot take?

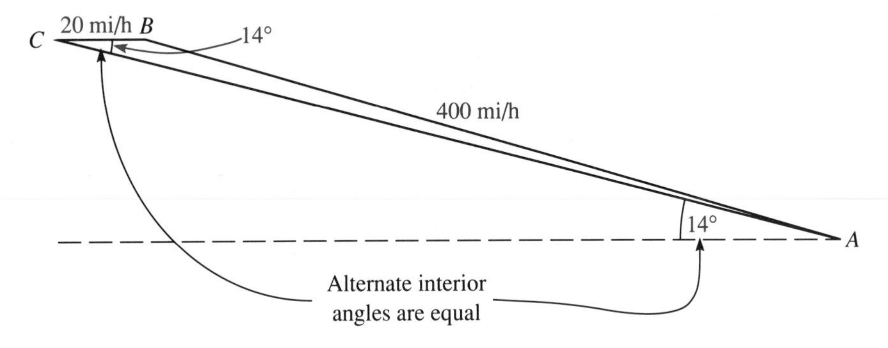

FIGURE 11–26

Solution Figure 11–26 is a diagram of the verbal statement. To find the pilot's heading, we must solve for the angle labeled A using the Law of Sines. From the information given above for the ambiguous case, there is one solution since $c > a$.

$$\frac{a}{\sin A} = \frac{c}{\sin C}$$

$$\frac{20}{\sin A} = \frac{400}{\sin 14^\circ}$$

$$\sin A = \frac{20 \sin 14^\circ}{400}$$

$$\sin A = 0.012$$

$$A = 0.69^\circ$$

The pilot's heading should be 14.69° north of west. ■

CAUTION ✦ When you are given two sides and the angle opposite one of them, remember to consider the possibility of two solutions. Neglecting to do so could be costly. For example, consider a navigator who considers only an angle of 47° and neglects the 133° angle. The plane may land at the wrong airport.

EXAMPLE 5 From the top of a building, the angle of depression to the base of a tree on the ground is 48°, while the angle of depression from a window 26 ft vertically below the top of the building is 34°. Find the distance from the top of the building to the base of the tree.

FIGURE 11–27

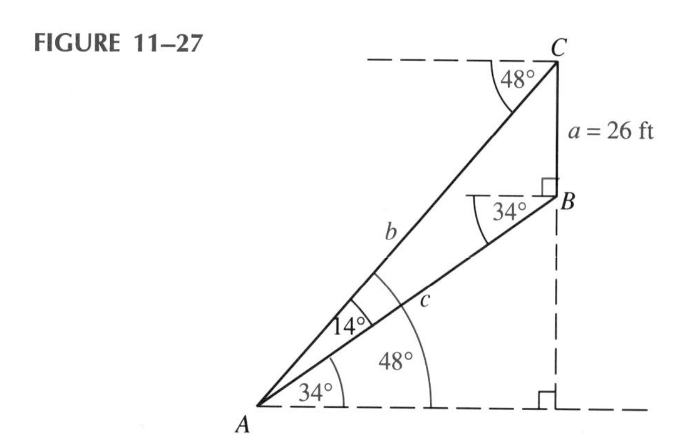

Solution Figure 11–27 gives a diagram of the resulting triangle. Since we are given two angles and a side, we do not need to consider the ambiguous case. Using the Law of Sines to solve for side b gives

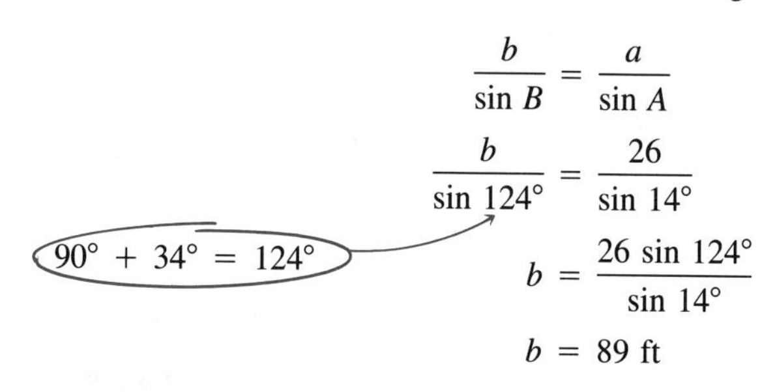

$$\frac{b}{\sin B} = \frac{a}{\sin A}$$

$$\frac{b}{\sin 124^\circ} = \frac{26}{\sin 14^\circ}$$

$$b = \frac{26 \sin 124^\circ}{\sin 14^\circ}$$

$$b = 89 \text{ ft}$$

■

11–3 EXERCISES

Solve triangle ABC given the information below.

1. $a = 15$; $C = 53^\circ$; $B = 100^\circ$

2. $a = 6.3$; $b = 4.8$; $A = 43^\circ$

3. $a = 13$; $c = 10$; $A = 43^\circ$

4. $b = 8.7$; $C = 53^\circ$; $A = 39^\circ$

5. $B = 67^\circ$; $C = 70^\circ$; $c = 23$

6. $b = 18$; $c = 25$; $C = 49^\circ$

7. $a = 7.5$; $B = 58^\circ$; $b = 9.0$

8. $A = 48^\circ$; $C = 63^\circ$; $a = 40$

9. $B = 42.8^\circ$; $c = 36.7$; $C = 57.9^\circ$

10. $C = 36.7^\circ$; $a = 56.3$; $c = 39.8$

11. $A = 95^\circ$; $C = 40^\circ$; $b = 15$

12. $a = 39.1$; $C = 95.3^\circ$; $c = 46.8$

13. $C = 82^\circ$; $a = 38$; $B = 37^\circ$

14. $A = 76.2^\circ$; $b = 49.3$; $C = 37.8^\circ$

15. $c = 14.8$; $b = 20.1$; $C = 30.2^\circ$

16. $B = 63.4^\circ$; $c = 14.8$; $A = 48.3^\circ$

17. $C = 45.3^\circ$; $b = 56.2$; $B = 72.5^\circ$

18. $a = 31$; $c = 42$; $A = 23^\circ$

19. $b = 25.3$; $C = 67.6^\circ$; $c = 16.8$

20. $b = 18.6$; $B = 68.1^\circ$; $A = 23.4^\circ$

21. A pilot wants to fly to a city 46° east of north from his starting position. If the wind is blowing from the east at 25 mi/h and the plane flies at 400 mi/h with respect to the air, find the heading the pilot should take.

22. From a helicopter flying due west, the bearing of a stationary ship is 28° west of south at 3 P.M. and 15° west of south at 3:30 P.M. If the ground speed of the helicopter is 400 km/h, find the distance from the helicopter to the ship at 3 P.M.

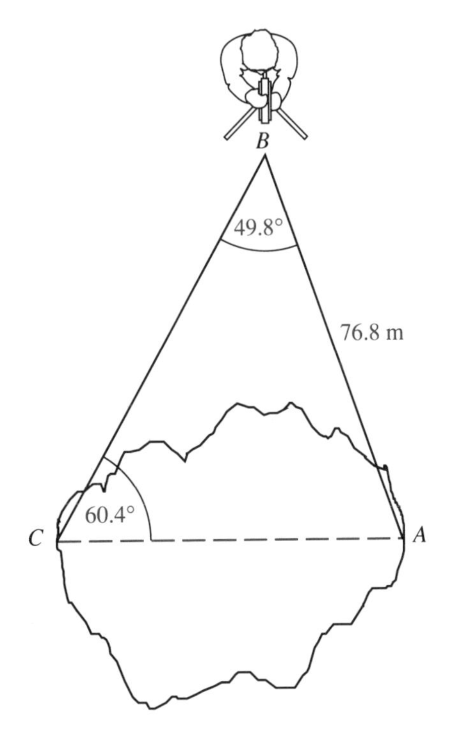

FIGURE 11–28

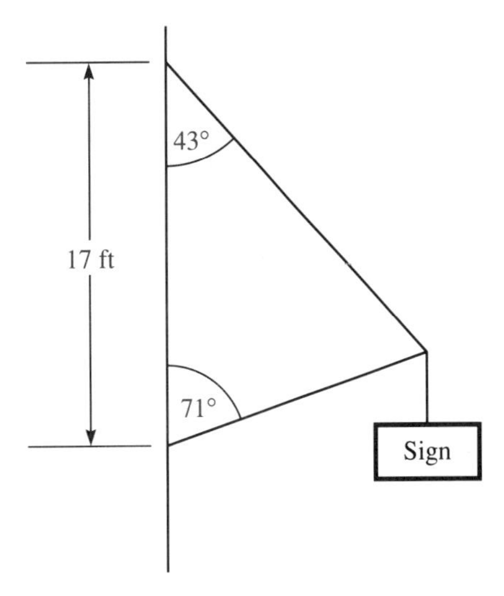

FIGURE 11–29

23. A surveyor needs to determine the distance from C to A in Figure 11–28 in order to estimate the area of the pond. Using the information from the plat, find this length.

24. A sign is supported from the side of a building by two steel struts, as shown in Figure 11–29. Find the length of the struts from the information given in the figure.

25. A ship leaves port and travels due south. At a certain point, the ship turns 24.6° north of east and travels on this course for 65.1 mi to a point 83.4 mi from port. Determine the distance from port to where the ship changed course.

26. An antenna 50 ft high is placed on the top of a building. The angle of elevation from a point on the ground to the top of the antenna is 63.7°, and the angle of elevation to the bottom of the antenna is 54.3°. Find the distance from the point on the ground to the top of the antenna.

27. An object is sighted from a helicopter at an angle of depression of 58°. After the helicopter has traveled 7 km, the angle of depression to the same side of the object is 64°. Determine the distance at that point from the helicopter to the object.

28. Town B is 28.3° south of west of town A. Determine the heading of a plane traveling from A to B if the wind is 38 mi/h from the east and the plane's speed is 400 mi/h with respect to air.

29. A telephone pole standing on level ground makes an angle of 80° with the horizontal. The pole is supported by a 30-ft prop whose base is 28 ft from the base of the pole. Find the angle made by the prop and the horizontal.

30. A 50-m-high antenna is mounted vertically on ground that slopes 13°. If two cables are attached to the top of the antenna at an angle of 53° with the horizontal, find the length of the cables.

11–4

LAW OF COSINES

The second method of solving oblique triangles is the Law of Cosines. To derive the Law of Cosines, we are given triangle ABC with altitude h, shown in Figure 11–30. The altitude divides side AB into two parts: x and $c - x$. Using the Pythagorean theorem for each triangle gives

$$a^2 = h^2 + x^2 \qquad b^2 = (c - x)^2 + h^2$$

FIGURE 11–30

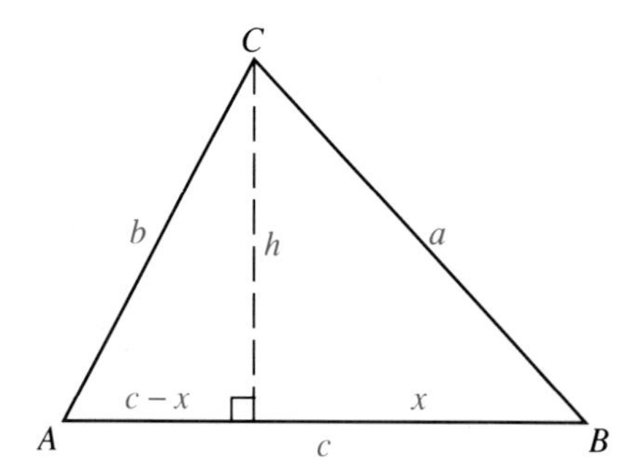

Solving each of these equations for h^2 gives

$$h^2 = a^2 - x^2 \qquad h^2 = b^2 - (c - x)^2$$

Equating the two expressions for h^2 gives

$$a^2 - x^2 = b^2 - (c - x)^2$$
$$a^2 - x^2 = b^2 - c^2 + 2cx - x^2$$

Solving the equation for b^2 gives

$$b^2 = a^2 + c^2 - 2cx$$

From Figure 11–30, we can see that the following expression is true:

$$\cos B = \frac{x}{a}$$
$$x = a \cos B$$

Substituting this expression for x gives one form of the Law of Cosines.

$$b^2 = a^2 + c^2 - 2ac \cos B$$

Using the same method and drawing altitudes to sides CB and AC gives similar results. The Law of Cosines is summarized as follows.

Law of Cosines

For oblique triangle ABC,

$$a^2 = b^2 + c^2 - 2bc \cos A$$
$$b^2 = a^2 + c^2 - 2ac \cos B$$
$$c^2 = a^2 + b^2 - 2ab \cos C$$

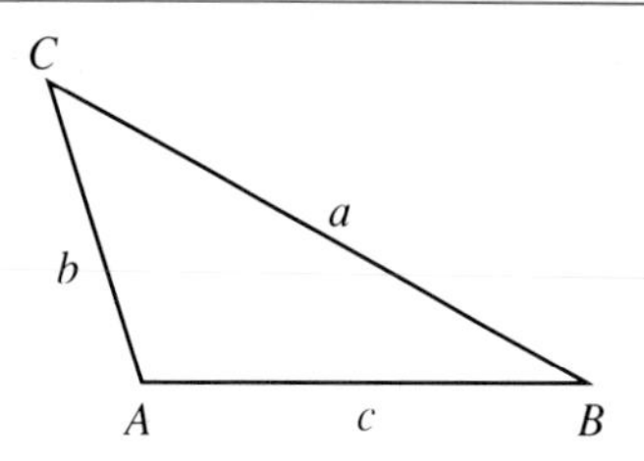

We can use the Law of Cosines to solve the following two additional cases of unknowns of an oblique triangle:

- ☐ *Case 3*. Two sides and the angle between those sides are given.
- ☐ *Case 4*. Three sides are given.

A discussion of each of these cases follows.

Case 3

The next example illustrates the case where we are given two sides and the included angle.

EXAMPLE 1 Solve triangle ABC if $a = 18.4$, $c = 26.3$, and $B = 47.9°$.

Solution First, we make a drawing of the oblique triangle as shown in Figure 11–31. We can use the Law of Cosines to solve for side b.

$$b^2 = a^2 + c^2 - 2ac \cos B$$
$$b^2 = (18.4)^2 + (26.3)^2 - 2(18.4)(26.3)\cos 47.9°$$
$$b^2 = 381.4$$
$$b = 19.5$$

FIGURE 11–31

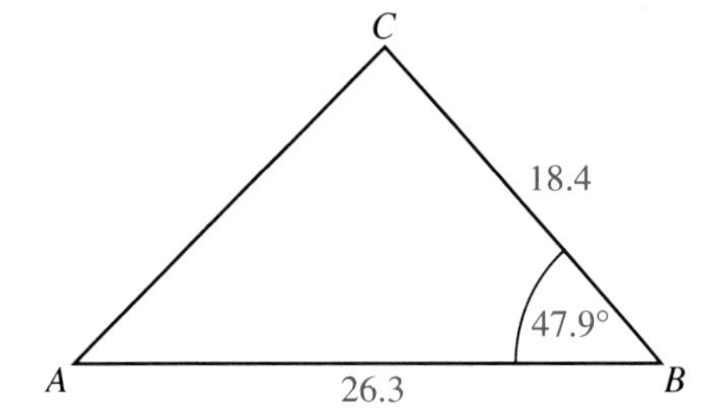

At this point, we can use either the Law of Cosines or the Law of Sines to solve for angles A and C. Since an angle is easier to find using the Law of Sines, we use it to solve for A.

$$\frac{a}{\sin A} = \frac{b}{\sin B}$$

$$\frac{18.4}{\sin A} = \frac{19.5}{\sin 47.9^\circ}$$

$$\sin A = \frac{18.4 \sin 47.9^\circ}{19.5}$$

$$\sin A = 0.7001$$

$$A = 44.4^\circ$$

Then we can solve for angle C using the fact that the sum of the angles of a triangle is 180°. This calculation gives

$$C = 180^\circ - A - B$$

$$C = 180^\circ - 44.4^\circ - 47.9^\circ$$

$$C = 87.7^\circ$$

■

Case 4

The next example illustrates solving an oblique triangle given three sides.

EXAMPLE 2 Solve triangle ABC given that $a = 17.8$, $b = 21.9$, and $c = 23.2$.

FIGURE 11–32

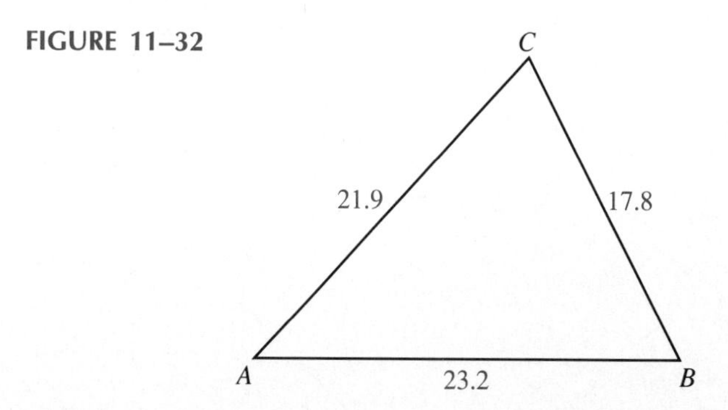

Solution First, we draw the oblique triangle ABC as shown in Figure 11–32. From the figure, we can see that we are given three sides of the triangle. We use the Law of Cosines to solve for angle A.

$$a^2 = b^2 + c^2 - 2bc \cos A$$

$$\cos A = \frac{a^2 - b^2 - c^2}{-2bc}$$

$$\cos A = \frac{(17.8)^2 - (21.9)^2 - (23.2)^2}{-2(21.9)(23.2)}$$

$$\cos A = 0.6899$$

$$A = 46.4°$$

We can use the Law of Sines to solve for angle B.

$$\frac{b}{\sin B} = \frac{a}{\sin A}$$

$$\frac{21.9}{\sin B} = \frac{17.8}{\sin 46.4°}$$

$$\sin B = 0.8910$$

$$B = 63.0°$$

Then we solve for angle C using the fact that the sum of the angles of a triangle is 180°.

$$C = 180° - A - B$$

$$C = 180° - 46.4° - 63.0°$$

$$C = 70.6°$$

■

Applications

The Law of Cosines has many uses in solving oblique triangles resulting from applied problems. The next two examples illustrate this application.

EXAMPLE 3 Two planes are 350 mi apart. One plane is 480 mi from an airport, and the second plane is 315 mi from the same airport. At what angle are the planes converging on the airport?

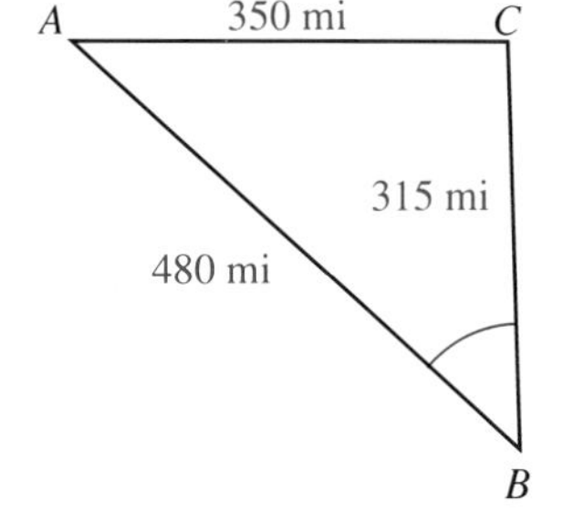

FIGURE 11–33

Solution First, we draw Figure 11–33. From the figure we are asked to find B. Since we are given three sides, we apply the Law of Cosines. Setting up the formula and solving for B gives

$$b^2 = a^2 + c^2 - 2ac \cos B$$

$$\cos B = \frac{b^2 - a^2 - c^2}{-2ac}$$

$$\cos B = \frac{(350)^2 - (315)^2 - (480)^2}{-2(315)(480)}$$

$$\cos B = 0.6849$$

$$B = 46.8°$$

The planes are converging on the airport at an angle of 46.8°. ■

EXAMPLE 4 A vertical telephone pole is to be placed on a hill that makes an angle of 8° with the horizontal. If two guy wires are attached to the pole 30.0 ft from the ground at points 15.0 ft from the base of the pole, determine the length of each of the guy wires.

FIGURE 11–34

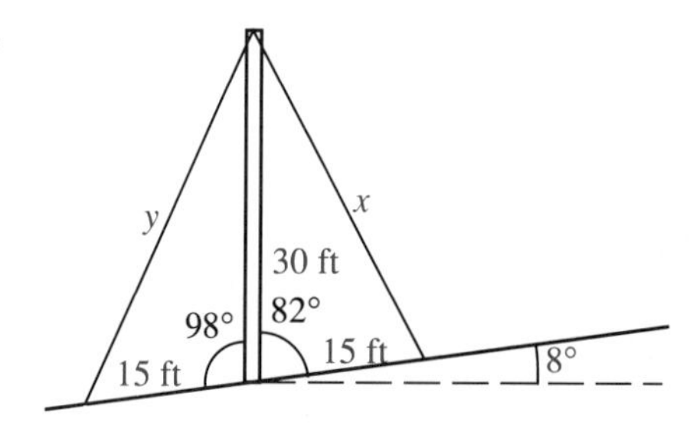

Solution We begin by drawing Figure 11–34. From the figure, we must find the length of sides x and y. We can use the Law of Cosines since we are given two sides and the included angle (Case 3).

$$x^2 = (30)^2 + (15)^2 - 2(15)(30)\cos 82^\circ$$
$$x^2 = 999.74$$
$$x = 31.6$$

Similarly, we can use the Law of Cosines to find side y.

$$y^2 = (15)^2 + (30)^2 - 2(15)(30)\cos 98^\circ$$
$$y^2 = 1{,}250.3$$
$$y = 35.4$$

The two wires are 35.4 ft and 31.6 ft long. ■

11–4 EXERCISES

Solve triangle ABC given the information below.

1. $a = 17.6$; $b = 20.3$; $c = 27.2$
2. $b = 48$; $c = 56$; $A = 72^\circ$
3. $a = 34$; $b = 26$; $C = 43^\circ$
4. $a = 28.9$; $b = 47.3$; $c = 30.6$
5. $b = 16.8$; $c = 29.4$; $A = 73.8^\circ$
6. $a = 78.6$; $c = 110.4$; $B = 56.7^\circ$
7. $a = 63$; $b = 74$; $c = 58$
8. $a = 8.1$; $b = 17.8$; $c = 24.2$
9. $a = 23.7$; $b = 16.8$; $c = 11.7$
10. $b = 16$; $c = 25$; $A = 98^\circ$
11. $a = 469$; $b = 342$; $c = 769$
12. $b = 238$; $a = 173$; $C = 83^\circ$
13. $a = 68$; $c = 49$; $B = 102^\circ$
14. $a = 14.8$; $b = 32.0$; $c = 27.5$
15. $C = 75.3^\circ$; $a = 97.5$; $b = 109.4$
16. $B = 102.3^\circ$; $c = 123.8$; $a = 175.8$
17. $a = 293$; $b = 436$; $c = 158$

18. $a = 36.8$; $b = 74.9$; $c = 44.6$

19. $A = 108°$; $b = 58$; $c = 73$

20. $a = 749$; $C = 86°$; $b = 486$

21. Find side AB in the quadrilateral in Figure 11–35.

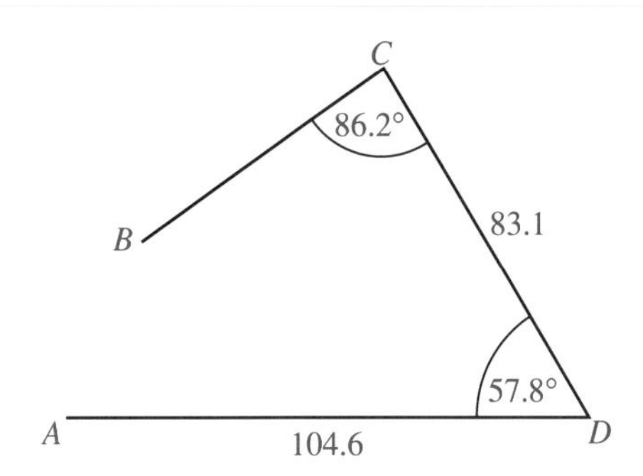

FIGURE 11–35

22. From a point on level ground directly between two telephone poles, cables are attached to the top of each pole. One cable is 74.8 ft long, and the other is 66.7 ft long. If the angle of intersection between the two cables is 103.6°, find the distance between the poles.

23. A tract of land in the shape of a parallelogram has sides 101 m and 123 m. If one diagonal is 146 m, find the measure of the angles of the parallelogram.

24. A cruise ship heads north from port for 20 mi and then turns to a heading of 18° east of south and travels another 12 mi. At this point, how far is the ship from port?

25. A triangular piece of land is bounded by 186 ft of freeway fence, 135 ft of brick wall, and 215 ft of concrete median. Find the angle between the fence and the median.

26. Two airplanes are 805 km apart. The planes are flying toward the same airport which is 322 km from one plane and 513 km from the other. Find the angle at which their paths intersect at the airport.

27. A boat moving at a speed of 8.3 knots with respect to the water heads directly southeast. If a current develops with a speed of 2.3 knots from 18° south of west, what will be the actual heading and speed of the boat?

28. A 50-ft antenna is placed on the top edge of a building. The distance from the top of the antenna to the base of a tree on the ground is 210 ft, and the distance from the bottom of the antenna to the same ground point is 185 ft. Find the angle of depression from the top of the antenna to the base of the tree.

CHAPTER SUMMARY

Summary of Terms

ambiguous case (p. 396)
components (p. 380)
head (p. 380)
initial point (p. 380)
magnitude (p. 380)
oblique triangle (p. 394)
resolving into components (p. 381)
resultant (p. 380)
scalars (p. 380)
tail (p. 380)
terminal point (p. 380)
vectors (p. 380)

Summary of Formulas

$$\left.\begin{aligned} A_x &= A\cos\theta \\ A_y &= A\sin\theta \end{aligned}\right\} \text{components of a vector}$$

$$\left.\begin{aligned} R &= \sqrt{R_x^2 + R_y^2} \quad \text{magnitude} \\ \tan(\text{Ref }\theta) &= \left|\frac{R_y}{R_x}\right| \quad \text{direction} \end{aligned}\right\} \text{resultant of vectors}$$

$$\frac{a}{\sin A} = \frac{b}{\sin B} = \frac{c}{\sin C} \quad \text{law of sines}$$

$$\left.\begin{aligned} a^2 &= b^2 + c^2 - 2bc\cos A \\ b^2 &= a^2 + c^2 - 2ac\cos B \\ c^2 &= a^2 + b^2 - 2ab\cos C \end{aligned}\right\} \text{law of cosines}$$

CHAPTER REVIEW

Section 11–1

Find the resultant of the following vectors.

1. 186 lb south; 265 lb west
2. 283 N east; 165 N south
3. 15.6 m/s at 118°; 29.3 m/s at 67°
4. 45 ft/s^2 at 193°; 72 ft/s^2 at 72°
5. 258 N at 72°; 197 N at 200°; 118 N at 156°
6. 96 yd at 283°; 149 yd at 147°; 103 yd at 61°
7. 18 lb at 136°; 38 lb at 310°; 10 lb at 66°
8. 86 m at 98°; 128 m at 172°; 59 m at 310°

Find the x and y components of the following vectors.

9. 291 lb at 221°
10. 17.84 ft at 198.3°
11. 486.7 km at 58.4°
12. 312.5 m/s at 127.6°

Section 11–2

13. Luke starts at the bicycle shop and walks 5 blocks north, then 8 blocks east, and finally 11 blocks southeast. What is Luke's final displacement from the bicycle shop?
14. A force of 218 lb is exerted on a screw as shown in Figure 11–36. Determine the horizontal and vertical components of the force.
15. A 120-lb crate is supported in the manner shown in Figure 11–37. Find the tension in cables A and B.
16. An object is dropped from a plane moving at 224 m/s at an angle of 23° above the horizontal. The vertical velocity of the object as a function of time is $v_y = 87.5 - 9.8t$. What are the velocity and the direction of the object after 12 s?
17. If an object weighing 253 lb rests on a ramp that is inclined at 18°, determine the friction force and the normal force.
18. Two ropes are tied together at point C, as shown in Figure 11–38. If the maximum tension in each rope without breaking the ropes is 756 lb, what is the maximum force **F** that can be applied? What should the direction of the force be?
19. An airplane heads 32° north of west at a velocity of 410 mi/h with respect to the air. If the wind is blowing at 63 mi/h from 16° north of east, what is the velocity of the plane with respect to the ground?
20. An 89-N object attached to a cable hangs vertically from the ceiling. A horizontal force **F** is applied to

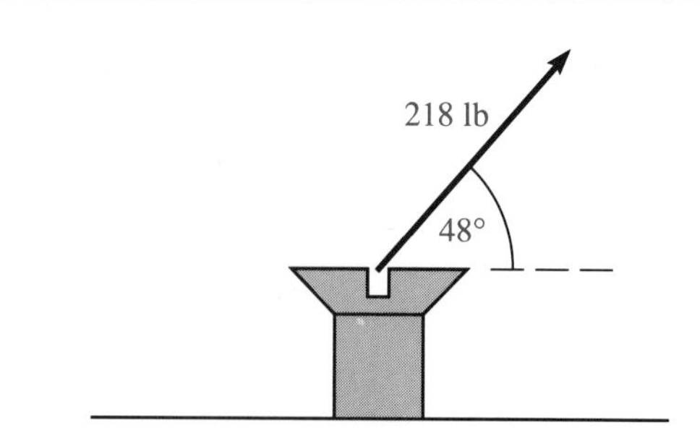

FIGURE 11–36

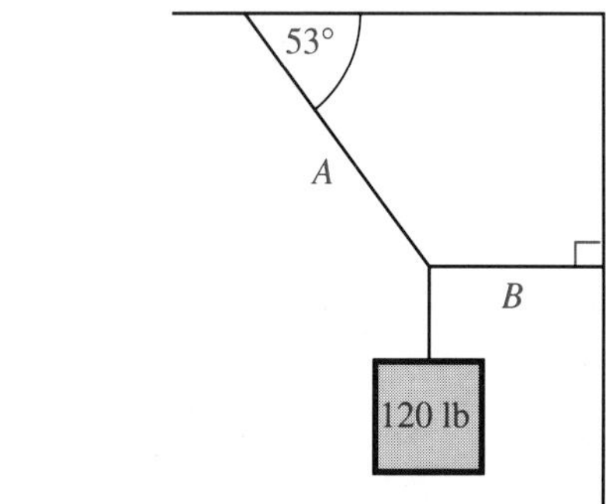

FIGURE 11–37

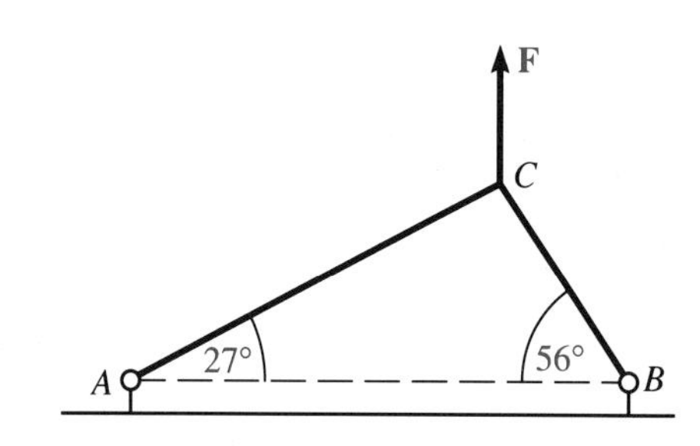

FIGURE 11–38

the object such that the angle of the cable with the vertical is 22°. Find the force **F** and the tension **T** in the cable.

Section 11–3

Solve triangle ABC given the information below.

21. $a = 73.8$; $C = 16.2°$; $B = 93.1°$

22. $B = 72.4°$; $a = 124$; $b = 230$

23. $b = 14$; $C = 57°$; $c = 26$

24. $a = 18.7$; $b = 30.4$; $A = 39.7°$

25. $a = 56.8$; $A = 67.1°$; $b = 23.7$

26. $b = 39.6$; $c = 56.7$; $C = 102.1°$

27. A pilot wants to fly to a city that is 57° south of west from his current position. If the wind is blowing from the south at 56 mi/h and the plane flies at 350 mi/h with respect to the air, what should the heading of the plane be?

28. A cruise ship leaves port and travels due east. At a certain location, the ship turns to a course 47.5° north of east and travels on this course for 125.3 mi to a point 381.4 mi from port. Find the distance from port to where the ship changed course.

29. A telephone pole standing on level ground makes an angle of 85° with the horizontal. The pole is supported by a 56-ft prop that is placed 34 ft from the base of the pole. Find the angle made by the prop and the ground.

Section 11–4

Solve triangle ABC using the information given.

30. $a = 139$; $b = 110$; $c = 76$

31. $a = 257$; $B = 39°$; $c = 310$

32. $b = 10.8$; $c = 7.3$; $A = 63.1°$

33. $a = 18.4$; $b = 12.5$; $c = 21.8$

34. $a = 57.8$; $b = 63.4$; $c = 43.4$

35. $b = 186$; $C = 53.6°$; $a = 237$

36. Two boats are 176 mi apart. The boats are sailing toward the same port which is 128 mi from one ship and 239 mi from the other ship. Find the angle at which the ships' paths intersect at the port.

37. Cables are attached to the top of two telephone poles and anchored at a point on level ground between the poles. One cable is 17 m long, and the other is 21 m long. If the angle of intersection between the two cables is 86°, find the distance between the poles.

38. A pilot flies north from an airport for 350 mi and then turns 36.1° west of north and travels another 275 mi. At this point, how far is the plane from the airport?

39. A small tract of land in the shape of a parallelogram has sides 25 yd and 27 yd long. If one diagonal is 34 yd, find the measure of the angle of the two sides of the parallelogram.

CHAPTER TEST

The number in parentheses refers to the appropriate learning objective given at the beginning of the chapter.

1. Find the resultant of each of the following vectors: (1)

125 lb at 43.1° north of east   217 lb at 56.2° south of west

2. From a position 6 ft above the ground, an observer finds the angle of elevation to the top of a building to be 87°. From the same position the angle of elevation to a window 40 ft below the top is 63°. Find the distance from the observer to the top of the building. (5)

3. In triangle ABC, $A = 16°$, $c = 24.6$, and $B = 93°$. Find side a. (3)

4. In triangle ABC, $a = 173$, $b = 158$, and $c = 163$. Find angle C. (4)

5. A force of 56 N is applied to the top of a bolt at an angle of 48° with the horizontal. Find the vertical component of this force. (2)
6. Movers are pushing a 125-lb stereo cabinet up an 18° ramp, applying a force of 80 lb at an angle of 30° with the horizontal. If the cabinet is moving at a constant velocity, what is the frictional force of the ramp on the cabinet? (5)
7. Find the resultant of each of the following vectors: (1)

   16.8 m east 37.4 m at 30° south of west 21.3 m west

8. A vertical telephone pole is placed on a hill that makes an angle of 12° with the horizontal. Guy wires are attached to the pole 21 ft from the ground at points 12 ft from the base of the pole. Determine the length of each of the guy wires. (5)
9. In triangle ABC, $a = 16.8$, $b = 23.6$, and $B = 56.1°$. Solve for side c. (3)
10. Solve triangle ABC if $a = 86.2$, $b = 118.4$, and $c = 99.3$. (3, 4)
11. In triangle ABC, $a = 176$, $B = 97°$, $c = 198$. Find side b. (4)
12. An airplane pilot wants to fly from his starting point to his destination, which is 63° north of west from his current position. If the wind is blowing from the east at 43 mi/h and the plane flies at 475 mi/h with respect to the air, what heading should the pilot take? (5)
13. A boat travels across a river. If the boat travels 10 mi/h in still water and the current flows at 5 mi/h, what is the velocity of the boat relative to the shore? (2)
14. Find the resultant of each of the following vectors: (1)

   18 m/s at 63° 31 m/s at 174° 56 m/s at −38°

15. Solve triangle ABC if $A = 46°$, $B = 93°$, and $a = 29$. (3)

SOLUTION TO CHAPTER INTRODUCTION

From the drawing given at the beginning of the chapter, we must find angle A by using the Law of Sines.

$$\frac{a}{\sin A} = \frac{b}{\sin B}$$

$$\frac{40}{\sin A} = \frac{350}{\sin 17°} \qquad (B = 17° \text{ due to alternate interior angles})$$

$$A \approx 2°$$

The pilot should head 19° north of east.

A wooden block is attached to the end of a spring. The block is pulled down 18 cm and released; as a result, it oscillates in simple harmonic motion and takes 5 s to return to its point of release. Assuming the phase angle is zero, determine the position of the block 1.75 s after release. (The solution to this problem is given at the end of the chapter.)

Sound, water, and light waves are all examples of periodic waves; that is, the shape of the wave repeats at regular intervals. For example, a water particle moving down a stretched string will trace out a pattern duplicating the graph of a sine or cosine function. Another application of trigonometric graphs occurs in the screen display of an oscilloscope, which graphically displays an electrical signal. For example, you can graph a sinusoidal voltage and a cosine voltage on the x and y axes of an oscilloscope screen to display the resulting voltage. In this chapter we will discuss the basic shape of graphs of the trigonometric functions and alterations to the graph resulting from changes in amplitude, period, and phase angle.

Learning Objectives

After completing this chapter, you should be able to

1. Identify the amplitude, period, and phase angle of a sine or cosine trigonometric function (Sections 12–1, 12–2, and 12–3).
2. Sketch the graph of a sine or cosine function using its amplitude, period, and phase angle (Sections 12–1, 12–2, and 12–3).
3. Graph the secant, cosecant, cotangent, and tangent functions (Section 12–4).
4. Apply the concept of amplitude, period, and phase shift to simple harmonic motion and ac circuits (Section 12–5).
5. Use addition of ordinates to graph composite functions (Section 12–6).
6. Graph parametric equations (Section 12–6).

Chapter 12

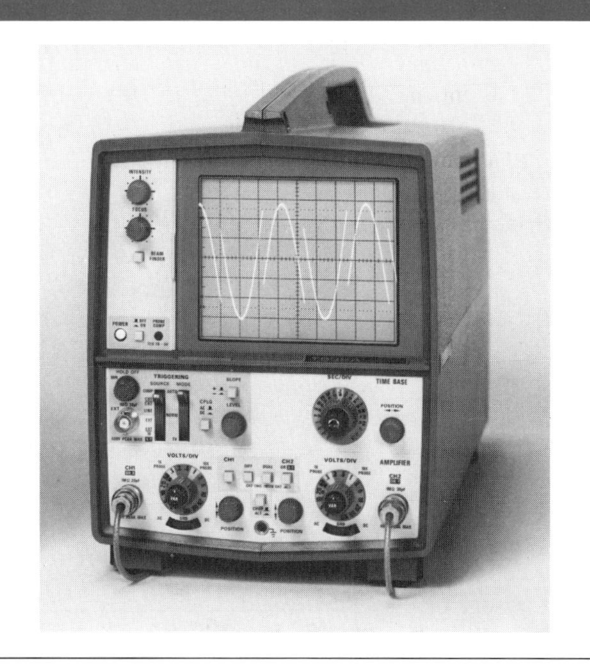

Graphs of the Trigonometric Functions

12–1

GRAPHS OF SINE AND COSINE FUNCTIONS: AMPLITUDE

In this chapter we discuss the graphs of the six trigonometric functions. In graphing a trigonometric function such as $y = \sin x$, we use the same techniques that we used in graphing an algebraic function such as $2x - 5y = 7$. Namely, we fill in a table of values by choosing values for x, substituting each value into the function, and obtaining the corresponding value for y. However, in graphing the trigonometric functions, the value chosen for x will be an angle, normally expressed in radians.

Graph of $y = \sin x$ and $y = \cos x$

EXAMPLE 1 Graph $y = \sin x$.

Solution We begin by filling in the accompanying table of values. We can use such special angles as $\pi/4$, $\pi/3$, and $\pi/6$, or the quadrantal angles, or a calculator. The graph is shown in Figure 12–1.

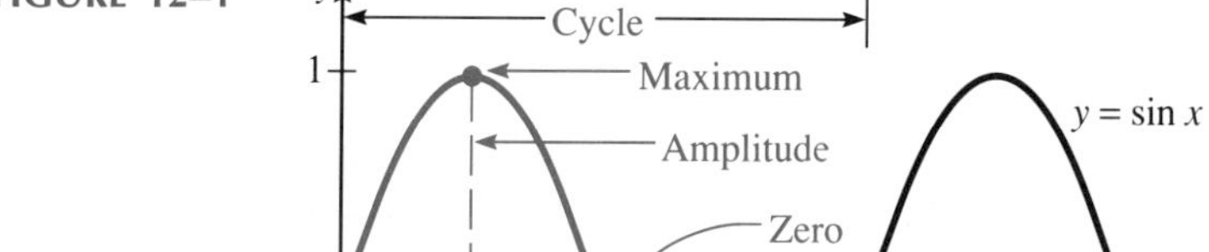

x	0	$\pi/4$	$\pi/2$	$3\pi/4$	π	$5\pi/4$	$3\pi/2$	$7\pi/4$	2π	$5\pi/2$	3π
y	0	0.7	1	0.7	0	−0.7	−1	−0.7	0	1	0

FIGURE 12–1

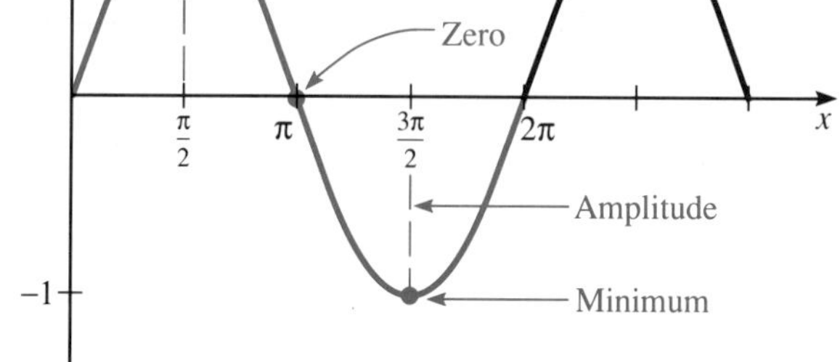

■

Two important aspects of the graph of $y = \sin x$ should be noted. First, the function is **periodic;** that is, the curve repeats itself at a regular interval. The **period** of the sine function is 2π, so the graph of the sine function looks exactly the same every 2π units, as shown by the color portion of the curve in Figure 12–1. The graph of the function through one period is called a **cycle.** Second, the amplitude of the sine function is 1. The **amplitude** represents the maximum variation of the curve from the x axis and is shown by the dashed line in Figure 12–1. Other properties of the sine function are summarized in the following box.

Properties of the Sine Function Graph

1. The graph crosses the x axis at the initial point, endpoint, and midpoint of a cycle. These points are called the **zeros of the function** (see Figure 12–1).
2. A maximum occurs midway between the first two zeros in a cycle, and a minimum occurs midway between the last two zeros. The maximum and minimum are determined by the amplitude (see Figure 12–1).

EXAMPLE 2 Graph $y = \cos x$.

Solution Using the same technique, we fill in the accompanying table of values for $y = \cos x$. The graph of $y = \cos x$ is given in Figure 12–2.

x	0	$\pi/4$	$\pi/2$	$3\pi/4$	π	$5\pi/4$	$3\pi/2$	$7\pi/4$	2π	$5\pi/2$	3π
y	1	0.7	0	-0.7	-1	-0.7	0	0.7	1	0	-1

FIGURE 12–2

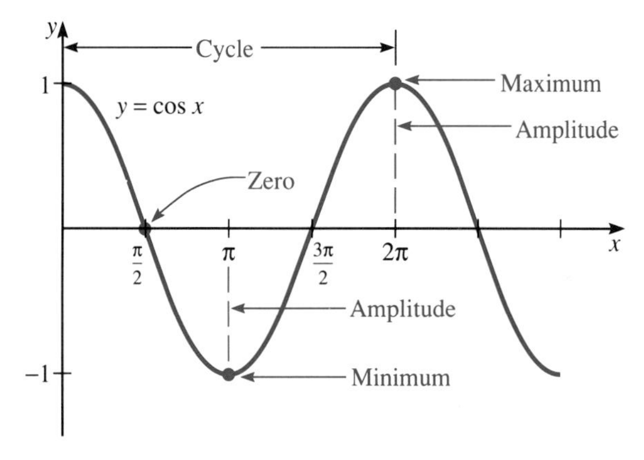

■

Notice that the graph of the cosine function in Figure 12–2 is also periodic with period 2π, and it also has an amplitude of 1. Additional properties of the cosine function are summarized below.

Properties of the Cosine Function Graph

1. A maximum occurs at the initial point and the endpoint of a cycle. A minimum occurs midway in the cycle. The maximum and minimum values are determined by the amplitude (see Figure 12–2).
2. The zeros occur midway between a maximum and a minimum (see Figure 12–2.

LEARNING HINT ✦ In the remainder of this chapter, we will discuss how changes in amplitude, period, and phase angle affect the graph of the trigonometric functions. It is extremely important that you learn the properties of the sine and cosine functions outlined in this section.

Graph of $y = a \sin x$ and $y = a \cos x$

Next, let us examine the effect of a constant multiplier on the sine and cosine functions.

EXAMPLE 3 Graph $y = 3 \sin x$ through one period.

Solution We will fill in a table of values just as we did previously, but a calculator will be helpful here. Figure 12–3 shows the graph of $y = 3 \sin x$. For comparison, the dotted line represents the graph of $y = \sin x$.

x	0	$\pi/4$	$\pi/2$	$3\pi/4$	π	$5\pi/4$	$3\pi/2$	$7\pi/4$	2π
y	0	2.1	3	2.1	0	-2.1	-3	-2.1	0

FIGURE 12–3

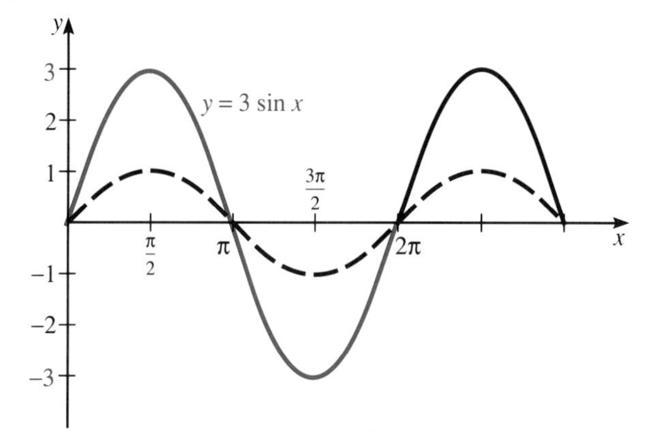

■

Notice that in the graph in Figure 12–3, the zeros of $y = 3 \sin x$ are the same as the zeros of $y = \sin x$. The difference in the two graphs occurs in their amplitude. Therefore, *multiplying the sine or cosine function by a constant changes only the amplitude*. The graphs of $y = a \sin x$ or $y = a \cos x$ will have a maximum of $|a|$ and a minimum of $-|a|$. With this knowledge we can sketch the graph of the sine and cosine functions, and we do not have to fill in a table of values. This process is illustrated in the next example.

EXAMPLE 4 Sketch the graph of $y = 4 \cos x$ through one period.

Solution We can sketch the graph of this function without a table of values because we know the basic properties of the cosine graph, and we know that the constant multiplier of 4 results in an amplitude change. To summarize, we know the following:

1. A maximum of 4 occurs at 0 and 2π, which are endpoints of the period.
2. A minimum of -4 occurs midway between the endpoints at π.
3. Zeros occur midway between each maximum and minimum at $\pi/2$ and $3\pi/2$.

The graph of $y = 4 \cos x$ is given in Figure 12–4.

FIGURE 12–4

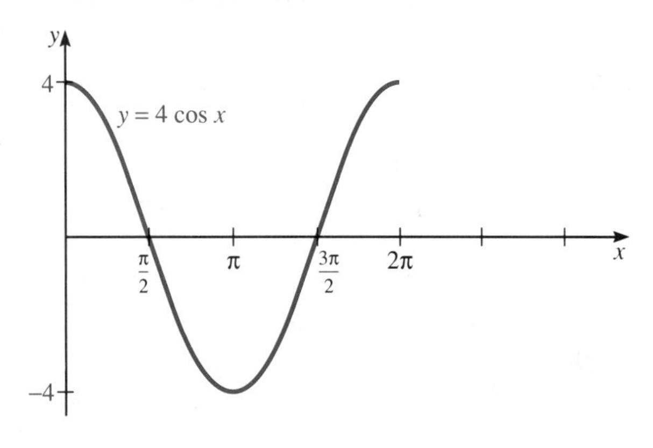

■

Graph of $y = -a \sin x$ and $y = -a \cos x$

In this section we discuss how a negative sign placed before the multiplier affects the graph.

EXAMPLE 5 Graph $y = -2 \sin x$ through one period.

Solution We know from our previous discussion that the 2 in front of the sine function results in an amplitude change that affects the maximum and minimum values. Since we do not know the effect of the negative sign on the graph, we fill in a table of values. The graph is shown in Figure 12–5. For comparison, the dotted line represents the graph of $y = 2 \sin x$.

x	0	$\pi/4$	$\pi/2$	$3\pi/4$	π	$5\pi/4$	$3\pi/2$	$7\pi/4$	2π
y	0	-1.4	-2	-1.4	0	1.4	2	1.4	0

FIGURE 12–5

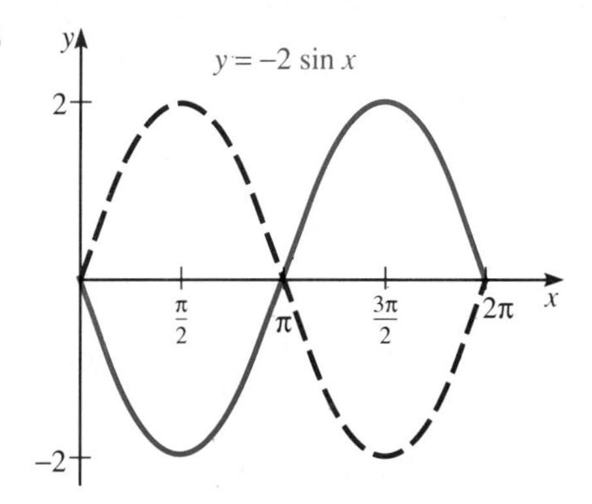

■

Notice that in Figure 12–5, the *negative sign in front of the amplitude has the effect of inverting the graph*. The locations along the x-axis where the maximum and minimum values occur are interchanged, but the zeros remain in the same location.

Summary

$$y = a \sin x$$

or

$$y = a \cos x$$

$|a|$ = amplitude.
A negative sign before a inverts the graph.

EXAMPLE 6 Sketch the graph of $y = -3 \cos x$ through one period.

Solution As in Example 4, we can sketch the graph of this function without a table of values, because we know the following:

1. The maximum of 3 and minimum of -3 occur at the endpoints and midpoint of the period, respectively.
2. The locations along the x axis where the maximum and minimum occur have been interchanged due to the negative sign.
3. Zeros occur midway between each maximum and minimum at $\pi/2$ and $3\pi/2$.

The graph of $y = -3 \cos x$ is shown in Figure 12–6.

FIGURE 12–6

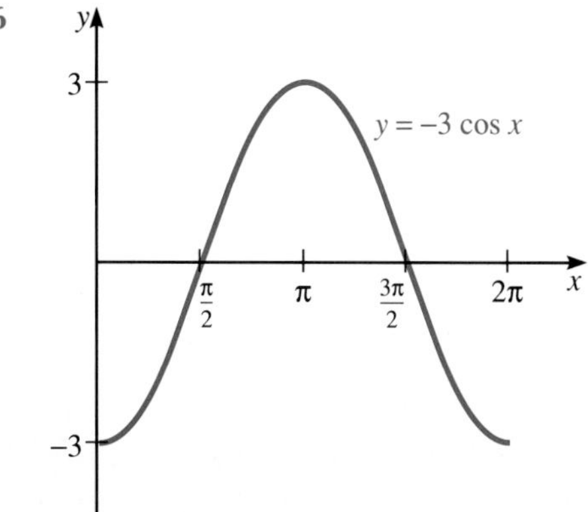

■

LEARNING HINT ✦ In sketching the graph of the sine and cosine functions, list the known information as in Example 6. Then use the calculator to check several points on the graph.

12–1 EXERCISES

Sketch the graph of each function through one period.

1. $y = 3 \cos x$
2. $y = 2 \sin x$
3. $y = -4 \sin x$
4. $y = -\cos x$
5. $y = 2 \cos x$
6. $y = 2.5 \sin x$
7. $y = -3 \sin x$
8. $y = -4.8 \cos x$
9. $y = -\sin x$
10. $y = 1.5 \sin x$
11. $y = -8 \cos x$
12. $y = 6 \sin x$
13. $y = -5 \sin x$
14. $y = 3.9 \cos x$
15. $y = 0.5 \cos x$
16. $y = -0.8 \cos x$
17. $y = -1.3 \sin x$
18. $y = 1.5 \cos x$
19. $y = 4 \sin x$

12–2

GRAPHS OF SINE AND COSINE FUNCTIONS: PERIOD

In the previous section, we found that the graph of $y = a \sin x$ or $y = a \cos x$ each have a period of 2π. However, such is not the case for all sine and cosine functions, as you will see in this section.

Graph of $y = \sin bx$ or $y = \cos bx$

EXAMPLE 1 Graph $y = \cos 2x$.

Solution Since we are uncertain about this graph, we fill in a table just as in the previous section. The graph of $y = \cos 2x$ is shown in Figure 12–7. The dotted line represents the graph of $y = \cos x$.

x	0	$\pi/4$	$\pi/2$	$3\pi/4$	π	$5\pi/4$	$3\pi/2$	$7\pi/4$	2π
y	1	0	-1	0	1	0	-1	0	1

FIGURE 12–7

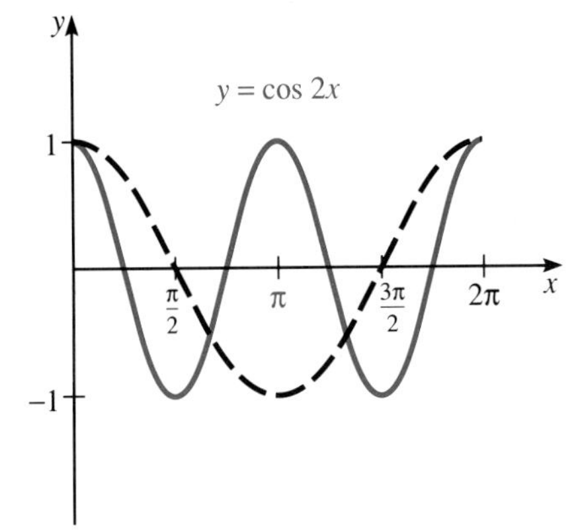

■

Comparing the two graphs in Figure 12–7 shows that the period of $y = \cos 2x$ is π instead of 2π. For graphs of the form $y = a \sin bx$ or $y = a \cos bx$, the period is given by $2\pi/b$. The zeros, minimum, and maximum occur at the same location relative to the period. For example, the maximum and minimum still occur at the endpoints and midpoint, respectively, of the period, but these locations are 0, $\pi/2$, and π for the graph of $y = \cos 2x$.

EXAMPLE 2 Sketch the graph of $y = \sin \dfrac{x}{2}$ through one period.

Solution This function undergoes a period change given by

$$\frac{2\pi}{1/2} = 4\pi$$

In graphing this function, we know the following:

1. The period is 4π.
2. The amplitude is 1.
3. The zeros occur at the endpoints and midpoint of the period, which are 0, 2π, and 4π.
4. The maximum and minimum occur midway between the zeros, at π and 3π.

The graph is shown in Figure 12–8.

FIGURE 12–8

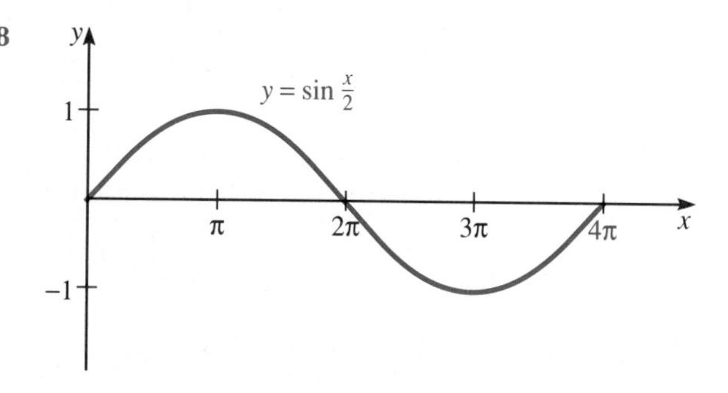

■

LEARNING HINT ✦ Sketching the graph for the sine and cosine functions will be easier if you list the known information, such as period and amplitude, before sketching the graph.

Graph of $y = a \sin bx$ or $y = a \cos bx$

EXAMPLE 3 Graph $y = 2 \cos 2\pi x$ through one period.

Solution From previous discussions, we know the following:

1. The amplitude is 2.
2. The period is $2\pi/2\pi = 1$.
3. The maximum occurs at the endpoints of the period, which are 0 and 1.
4. The minimum occurs at the midpoint of the period, 0.5.
5. The zeros occur midway between each maximum and minimum, 0.25 and 0.75.

The graph is shown in Figure 12–9. Since the period is 1 radian, the x axis is marked in numerical values of radians, rather than in units of π. ■

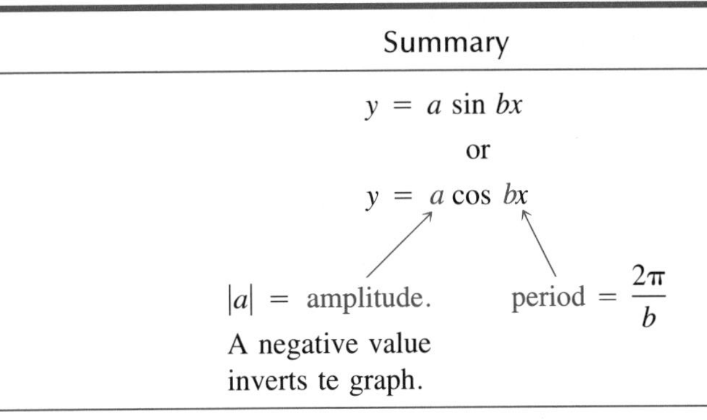

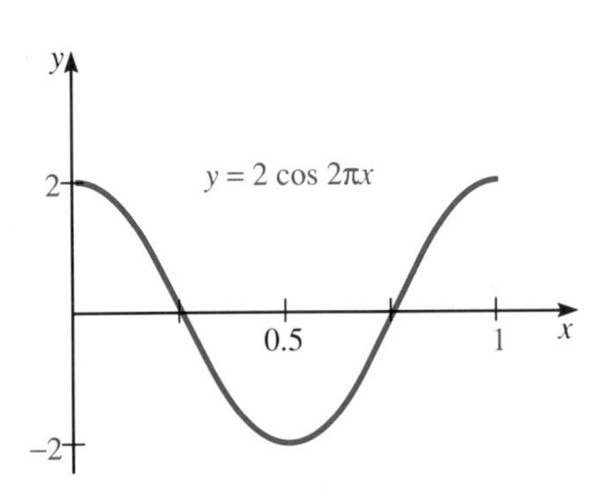

FIGURE 12–9

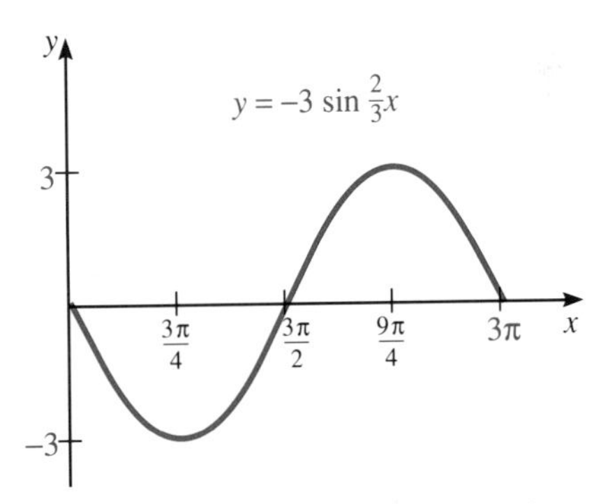

FIGURE 12–10

EXAMPLE 4 Sketch the graph of $y = -3 \sin \frac{2}{3}x$.

Solution For this graph, we know the following:

1. The amplitude is 3.
2. Due to the negative sign in front of the amplitude, the graph is inverted; the locations of the maximum and minimum are interchanged.
3. The period is $\frac{2\pi}{2/3} = 3\pi$.
4. The zeros occur at the endpoints and midpoint of the period, which are 0, 1.5π, and 3π.
5. The maximum and minimum values occur midway between the zeros at 0.75π and 2.25π, respectively.

The graph of $y = -3 \sin \frac{2}{3}x$ is shown in Figure 12–10. ■

LEARNING HINT ✦ To aid in graphing, write the properties for the sine and cosine functions, along with the summary on amplitude and period changes, on an index card. Then use this information while you do the exercises.

12–2 EXERCISES

State the amplitude and period for each function, and sketch its graph through one period.

1. $y = \sin 4x$

2. $y = \cos 3x$

3. $y = -\sin 6x$

4. $y = -\cos \frac{x}{4}$

5. $y = -\cos \frac{x}{3}$

6. $y = \sin 2x$

7. $y = 3 \sin 3x$

8. $y = 1.5 \cos 2\pi x$

9. $y = 2 \sin \pi x$

10. $y = 3.7 \sin \frac{x}{4}$

11. $y = -2 \cos 2\pi x$

12. $y = -5 \sin 6\pi x$

13. $y = 6 \sin \frac{2x}{7}$

14. $y = 1.5 \cos \frac{3x}{4}$

15. $y = -\frac{1}{2} \sin \frac{x}{2}$

16. $y = \frac{1}{2} \sin 6x$

17. $y = 4 \cos 5\pi x$

18. $y = 3 \cos \frac{4x}{3}$

19. $y = -3 \sin \frac{3x}{2}$

20. $y = -2 \cos 6\pi x$

21. Linear displacement x of an oscillating object as a function of time t can be written as

$$x = 3 \sin 2\pi t$$

Graph this function through two cycles.

E 22. The current I in an ac circuit can be given by $I = 6 \cos 120\pi t$. Sketch the graph of I as a function of t for $0 \leq t \leq 0.1$ s.

23. The displacement of an object oscillating at the end of a spring is given by $y = 8 \cos 6\pi t$. Sketch the graph of y as a function of t for $0 \leq t \leq 1$.

E 24. Under certain conditions the voltage in an ac circuit is given by $V = 100 \sin \pi t$. Graph V as a function of t for $0 \leq t \leq 0.5$ s.

12–3

GRAPHS OF SINE AND COSINE FUNCTIONS: PHASE SHIFT

If you study Figures 12–1 and 12–2, you can see that the graphs would be identical if the graph of the sine function were shifted $\pi/2$ units to the left. For this reason, the cosine function is said to *lead* the sine by $\pi/2$. This horizontal movement along the x axis is called a **phase shift.**

Graph of $y = \sin(x + c)$ and $y = \cos(x + c)$

EXAMPLE 1 Graph $y = \cos(x + \pi)$ through one period.

Solution We fill in the accompanying table of values. To calculate y when $x = \pi/2$, press the following calculator keys:

[π] [÷] 2 [+] [π] [=] [cos] → 0

We know that the period of this function is 2π and the amplitude is 1.

x	0	$\pi/2$	π	$3\pi/2$	2π
y	-1	0	1	0	-1

FIGURE 12–11

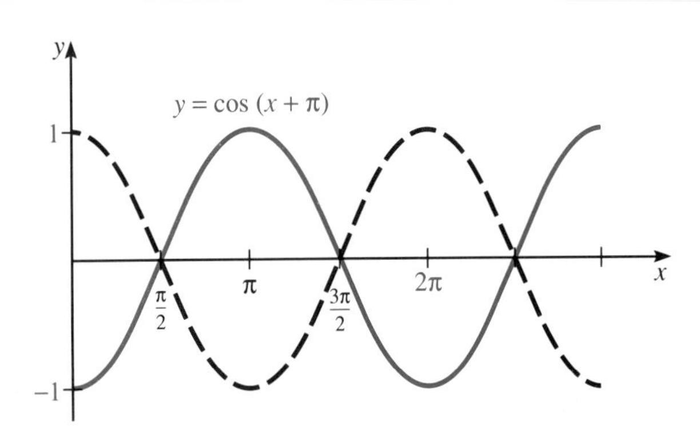

Figure 12–11 shows the graph of $y = \cos(x + \pi)$. Notice that the *graph of $y = \cos x$ has just been moved along the x axis π units to the left*. For comparison, the dotted line represents the graph of $y = \cos x$. ■

For an equation of the form $y = a \sin(bx + c)$ or $y = a \cos(bx + c)$, the **phase shift** is given by c/b, and the *graph is shifted along the x axis in the opposite direction of the sign in front of c*. The angle c is called the **phase angle.**

LEARNING HINT ✦ Students frequently have difficulty in determining the direction of the phase shift. The x axis is positive to the right of zero and negative to the left. The graph shifts along the x axis in the opposite direction of the sign in front of the phase angle. For example, $y = \sin(x - \pi)$ shifts the graph π units to the right.

Graph of $y = a \sin (bx + c)$ and $y = a \cos (bx + c)$

EXAMPLE 2 Sketch the graph of $y = 3 \cos(x - \pi)$.

Solution From our previous discussion, we know the following:

1. The amplitude is 3.
2. The period is 2π.
3. The phase shift is π units along the x axis and to the right.
4. The graph has a maximum at the endpoints and a minimum at the midpoint of its cycle.
5. The graph has a zero between each maximum and minimum.

To sketch the graph, we *begin the curve at* π and draw a cosine function with an amplitude of 3 and a period of 2π. Then we extend the graph back to the y axis. Figure 12–12 gives the graph of $y = 3 \cos(x - \pi)$.

FIGURE 12–12

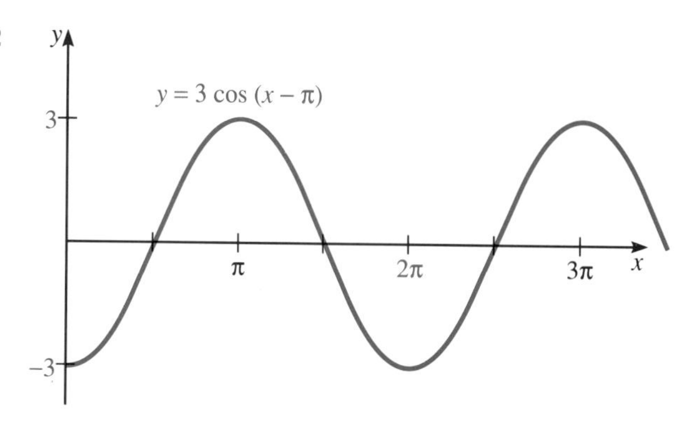

EXAMPLE 3 Sketch $y = -2\sin\left(2x + \frac{\pi}{2}\right)$ through one cycle.

Solution We know the following about the graph of this function:

1. The curve is inverted; the locations of the maximum and minimum along the x axis are reversed.
2. The amplitude is $|-2| = 2$.
3. The period is $\frac{2\pi}{b}$, or π.
4. The phase shift is $\frac{c}{b} = \frac{\pi}{4}$ units to the left since c is positive.
5. The zeros occur at the endpoints, $-\pi/4$, and $-\pi/4 + \pi = 3\pi/4$, and at the midpoint, $(-\pi/4 + 3\pi/4)/2 = \pi/4$, of the period.
6. Since the graph is inverted, the minimum is midway between the first two zeros, $-\pi/4$ and $\pi/4$ at 0, and the maximum is midway between the last two zeros, $\pi/4$ and $3\pi/4$ at $\pi/2$.

To sketch the graph, *begin at* $-\pi/4$ and extend the graph to the right so that one cycle appears to the right of the y axis. Figure 12–13 shows the graph of this function.

Summary

$$y = a\sin(bx + c)$$

or

$$y = a\cos(bx + c)$$

amplitude $= |a|$ — period $= \frac{2\pi}{b}$ — phase shift $= \frac{c}{b}$

A negative sign inverts the graph.

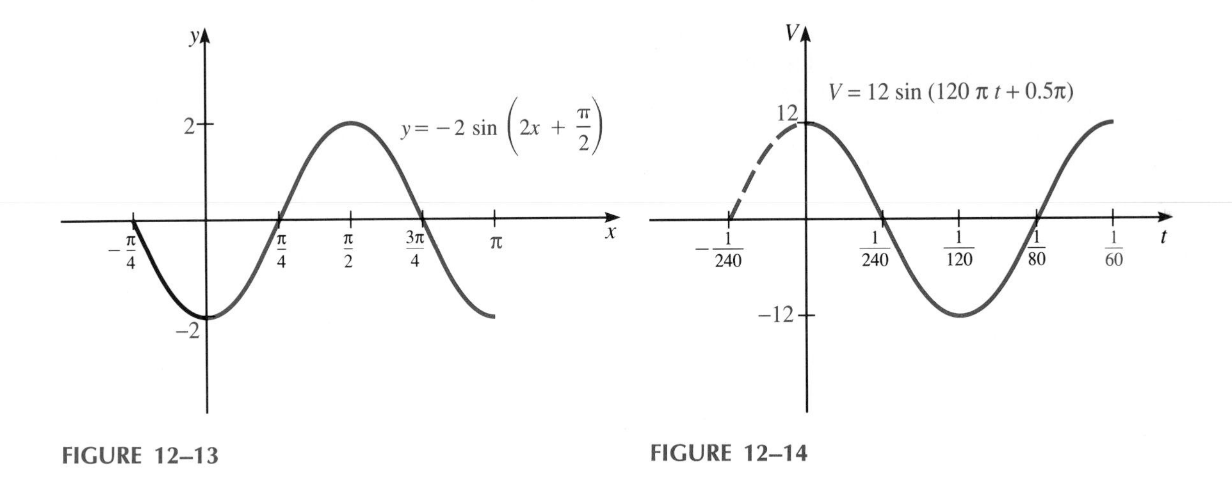

FIGURE 12–13

FIGURE 12–14

EXAMPLE 4 In an ac circuit with only a constant inductance, the voltage is given by

$$V = V_m \sin\left(2\pi f t + \frac{\pi}{2}\right)$$

where f = the frequency of the current, V_m = maximum voltage (amplitude), and t = time. In a 60-Hz system with a maximum voltage of 12 V, the equation becomes

$$V = 12\sin(120\pi t + 0.5\pi)$$

Graph the voltage as a function of time.

Solution From the equation, we know that the amplitude is 12; the period, which is the reciprocal of the frequency, is 1/60 s; and the phase shift is 1/240 s. The graph is shown in Figure 12–14. ■

Equations of Sine and Cosine Functions

EXAMPLE 5 Determine the equation of an inverted cosine function with an amplitude of 3, a period of 3π, and a phase shift of $3\pi/2$ units to the left along the x axis.

Solution To write the equation of this function, we must substitute for a, b, and c in the equation $y = a\cos(bx + c)$ where $|a|$ represents amplitude, b represents period change, and c represents the phase angle. From the given information, $|a| = 3$. To find the value of b, we know that $2\pi/b$ gives the period for the cosine function. Since the period of this function is 3π, we have

$$\frac{2\pi}{b} = 3\pi$$

$$b = \frac{2}{3}$$

To find the value of c, we know that c/b represents the phase shift and $b = 2/3$. Since the phase shift is $3\pi/2$, we have

$$\frac{c}{b} = \frac{3\pi}{2}$$

$$\frac{c}{2/3} = \frac{3\pi}{2}$$

$$c = \pi$$

Since the phase shift is left, the sign in front of c is positive. Also, since the cosine function is inverted, there is a negative sign in front of the amplitude. The equation of the function is

$$y = -3\cos\left(\frac{2}{3}x + \pi\right)$$

You can check your answer by comparing the amplitude, period, and phase shift of this function with the original information. ■

Application

EXAMPLE 6 Under certain conditions, the angular displacement s of a point at the end of an oscillating pendulum is given by

$$s = 3\cos(\pi t - \pi)$$

Determine the amplitude, period, and phase shift of this function, and sketch its graph for $0 \leq t \leq 3$.

Solution From the equation we can determine that $a = 3$, $b = \pi$, and $c = -\pi$. Using this information, we obtain

$$\text{amplitude} = 3$$

$$\text{period} = \frac{2\pi}{\pi} = 2$$

$$\text{phase shift} = \frac{c}{b} = \frac{-\pi}{\pi} = -1 \text{ unit right}$$

The graph is shown in Figure 12–15.

FIGURE 12–15

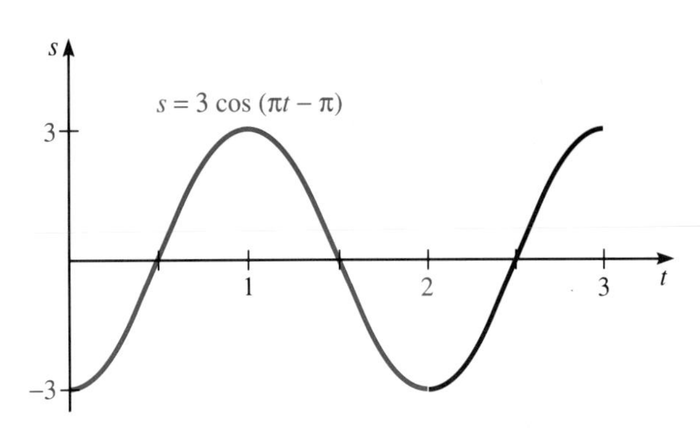

12–3 EXERCISES

State the amplitude, period, and phase shift of the following functions. Sketch the graph through one cycle.

1. $y = \cos\left(x + \frac{\pi}{4}\right)$
2. $y = \sin\left(x - \frac{3\pi}{2}\right)$
3. $y = -\sin\left(x - \frac{\pi}{3}\right)$
4. $y = \cos\left(x + \frac{3\pi}{4}\right)$
5. $y = 2\cos(x + 1)$
6. $y = -\sin(x - 2)$
7. $y = -\cos(2x - \pi)$
8. $y = \sin(3\pi x - \pi)$
9. $y = \sin(3x + 2\pi)$
10. $y = -\cos\left(2x + \frac{\pi}{3}\right)$
11. $y = 2\cos\left(x + \frac{\pi}{4}\right)$
12. $y = 3\sin(x - 2\pi)$
13. $y = -3\sin(x + 3)$
14. $y = -\cos\left(x + \frac{3\pi}{4}\right)$
15. $y = 2\sin(2\pi x + 1)$
16. $y = 6\sin(3x - 1)$
17. $y = -\cos(3\pi x - 2)$
18. $y = 3.1\sin(2x - \pi)$
19. $y = 4\sin\left(6\pi x + \frac{1}{2}\right)$
20. $y = -3\cos(3\pi x + 1)$
21. $y = -2.5\cos(3x - 2\pi)$
22. $y = 1.4\cos(4x - 3\pi)$
23. $y = \sin(2\pi x - 5\pi)$
24. $y = -5\sin(3x - 6\pi)$
25. Write the equation of a cosine function with an amplitude of 3 and period of 4.
26. Write the equation of a sine function with an amplitude of 2 and a phase shift of $\pi/3$ units to the right.
27. Write the equation of an inverted sine function with amplitude π, frequency 1/6 (frequency is the reciprocal of period), and phase shift $\pi/3$ units left.
28. Write the equation of a cosine function with amplitude 1/2, period 1/3, and phase shift 2/3 unit right.
29. Write the equation of a cosine function with amplitude 4, period π, and phase shift $\pi/3$ units right.
30. Write the equation of an inverted sine function with amplitude 6, frequency 1/2 (frequency is the reciprocal of period), and phase shift $3\pi/4$ units left.

E 31. In a certain ac circuit the current I is given by

$$I = 5.3\cos\left(120\pi t + \frac{\pi}{2}\right)$$

Graph this function through one cycle.

M 32. The angular displacement of a point at the end of a pendulum is given by

$$s = A\cos(\omega t + \theta)$$

Sketch the graph of s versus t when $A = 3.4$, $\omega = 5.6$ rad/s, and $\theta = \pi/6$ for $0 \leq t \leq 2$.

33. The equation of a certain sinusoidal wave on a string is given by

$$y = 4\sin\left(2\pi t - \frac{\pi}{2}\right)$$

Graph this function through one cycle.

M 34. The displacement y of a point on the rim of a rotating gear is given by

$$y = 8\cos\left(t + \frac{\pi}{2}\right)$$

Graph this function through one cycle.

12–4

GRAPHS OF THE OTHER TRIGONOMETRIC FUNCTIONS

Thus far in this chapter, we have discussed changes in amplitude, period, and phase angle for the sine and cosine functions only. In this section we will briefly consider the graphs of the remaining trigonometric functions. After developing the basic shape of each function, we will sketch the graph of these functions with changes in amplitude, period, and phase angle.

Graphs of the Tangent, Cotangent, Secant, and Cosecant Functions

EXAMPLE 1 Sketch the graph of $y = \tan x$.

Solution To graph the tangent function, we fill in the accompanying table of values using our knowledge of special angles and using the calculator. Note from the table of values that the tangent function is undefined at $\pi/2$ and $3\pi/2$ due to division by zero. This is represented on the graph of Figure 12–16 by dashed vertical lines at $\pi/2$ and $3\pi/2$. These dashed lines are called **asymptotes,** which are straight lines that approach but never touch a curve. Figure 12–16 gives the graph of $y = \tan x$.

x	0	$\pi/4$	$\pi/2$	$3\pi/4$	π	$5\pi/4$	$3\pi/2$	$7\pi/4$	2π
y	0	1	undef.	-1	0	1	undef.	-1	0

■

To graph the remaining trigonometric functions, we can use the reciprocal relationships.

$$\cot x = \frac{1}{\tan x} \qquad \csc x = \frac{1}{\sin x} \qquad \sec x = \frac{1}{\cos x}$$

We can fill in a table of values using values from the corresponding reciprocal function. Figure 12–17 shows the graph of $y = \cot x$. Figures 12–18 and 12–19 show the graphs of $y = \sec x$ and $y = \csc x$, respectively.

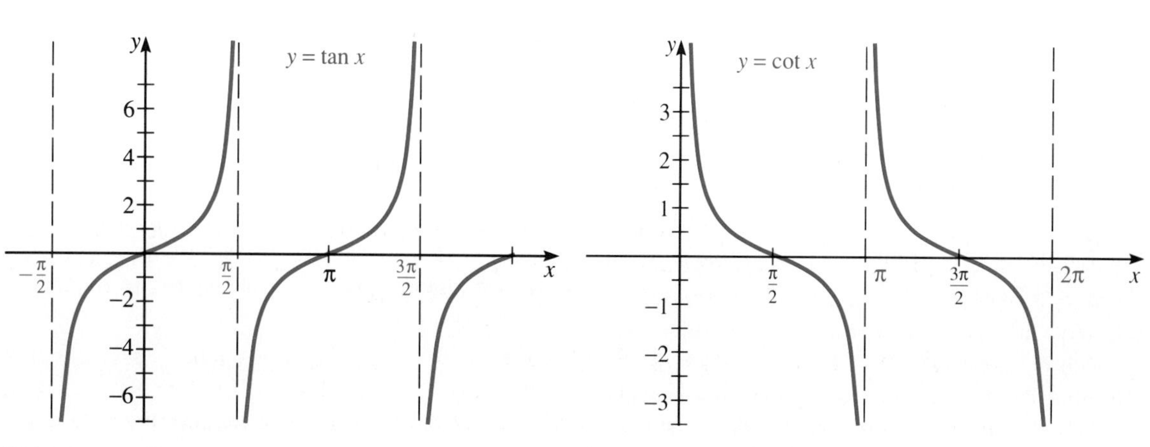

FIGURE 12–16

FIGURE 12–17

FIGURE 12–18

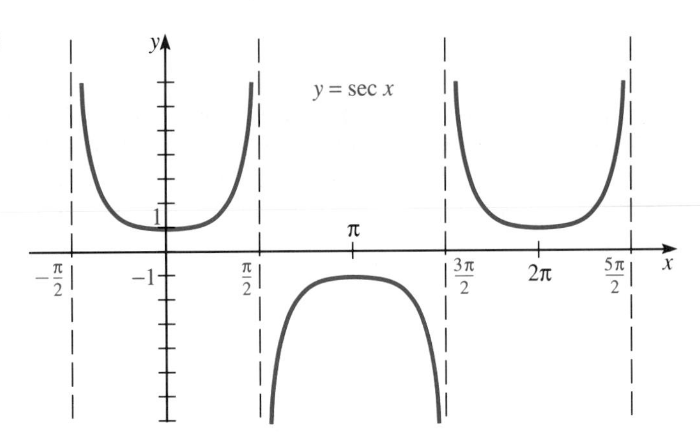

FIGURE 12–19

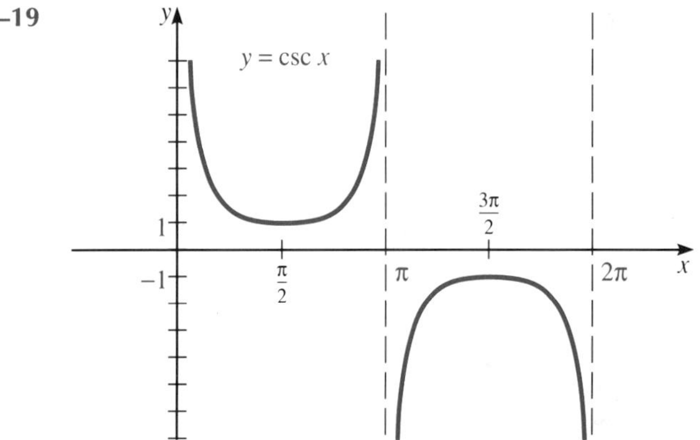

CAUTION ✦ If you use a calculator to find some of the values, you can get an error message when no error exists. For example, if you calculate cot $\pi/2$ using a calculator, you strike the following keys:

[π] [÷] 2 [=] [tan] [1/x]

However, after you strike the [tan] key, you will get an error message because the tangent of $\pi/2$ is undefined. However, cot $\pi/2 = 0$. For the quadrantal angles, it is probably best not to use a calculator.

The period of both the tangent and the cotangent functions is π, while the period of both the secant and the cosecant functions is 2π. None of these functions has an amplitude since there is no maximum distance from the x axis to the graph. Sketching the graph of the secant, cosecant, tangent, or cotangent using only amplitude, period, and phase angle is more difficult than the sine and cosine. For this reason, it is best to fill in a table of values when graphing these functions.

LEARNING HINT ✦ Be sure to learn the basic shape and location of asymptotes for these functions. It will make graphing much easier.

Changes in Amplitude, Period, and Phase Angle

EXAMPLE 2 Sketch the graph of $y = 2 \tan x$ through one cycle.

Solution Since the period of this tangent function is π, the table of values need not extend beyond π. The graph of $y = 2 \tan x$ is shown in Figure 12–20. The dotted line represents the graph of $y = \tan x$. Notice that the 2 multiplier alters the slope of the graph.

x	0	$\pi/4$	$\pi/2$	$3\pi/4$	π
y	0	2	undef.	-2	0

FIGURE 12–20

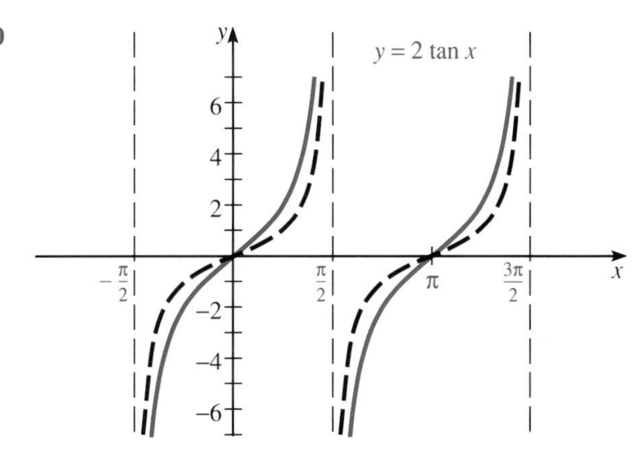

EXAMPLE 3 Graph $y = \sec(2x - \pi)$ through one cycle.

Solution Examining the equation of this function reveals a period of π and a phase shift of $\pi/2$ units to the right. We fill in the accompanying table of values using a calculator. The graph of $y = \sec(2x - \pi)$ is given in Figure 12–21.

x	0	$\pi/8$	$\pi/4$	$3\pi/8$	$\pi/2$	$5\pi/8$	$3\pi/4$	$7\pi/8$	π
y	-1	-1.41	undef.	$+1.41$	1	$+1.41$	undef.	-1.41	-1

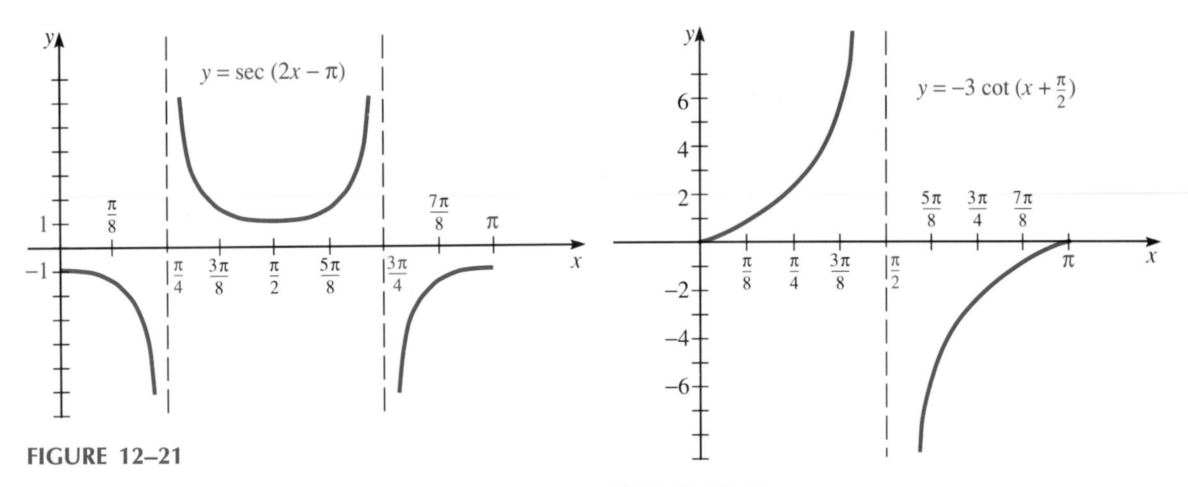

FIGURE 12–21

FIGURE 12–22

EXAMPLE 4 Graph $y = -3\cot\left(x + \frac{\pi}{2}\right)$ through one cycle.

Solution The graph of this function is inverted; the slope of the graph is altered; the period is π; and the graph is shifted $\pi/2$ units to the left. The table of values is given below, and the graph is shown in Figure 12–22.

x	0	$\pi/8$	$\pi/4$	$3\pi/8$	$\pi/2$	$5\pi/8$	$3\pi/4$	$7\pi/8$	π
y	0	7.2	3	7.2	undef.	−7.2	−3	−1.2	0

■

12–4 EXERCISES

Sketch the graph of each function through one cycle.

1. $y = 3\tan x$
2. $y = -\csc x$
3. $y = -2\sec x$
4. $y = 4\cot x$
5. $y = \tan\left(x - \frac{\pi}{2}\right)$
6. $y = \sec(x + \pi)$
7. $y = \csc(x + \pi)$
8. $y = \tan(3x - \pi)$
9. $y = 4\tan\left(x - \frac{\pi}{2}\right)$
10. $y = \csc\left(2x - \frac{\pi}{4}\right)$
11. $y = -\sec(2\pi x - \pi)$
12. $y = -\cot\frac{x}{3}$
13. $y = -\cot\left(x + \frac{\pi}{4}\right)$
14. $y = 3\sec 2x$
15. $y = 2\tan\left(\frac{x}{2} - \frac{\pi}{2}\right)$
16. $y = -2\tan(2\pi x - 3\pi)$
17. $y = -\sec\left(2x - \frac{\pi}{2}\right)$
18. $y = \csc(4x - \pi)$
19. $y = \csc\left(\frac{x}{2} + \pi\right)$
20. $y = -4\cot\left(\pi x + \frac{\pi}{2}\right)$

12–5

APPLICATIONS

Trigonometric functions are very useful in describing periodic phenomena, such as wave movement, oscillation of mechanical systems, and current flow in a circuit. In this section we will discuss two such applications: simple harmonic motion and alternating current.

Simple Harmonic Motion

When a spring with an object attached to the end is stretched and released, the object does not simply return to its equilibrium or starting position. The object actually oscillates above and below the equilibrium point. The oscillation of this system is called *simple harmonic motion*. The displacement y of an object in simple harmonic motion is given by the following equation.

Simple Harmonic Motion

$$y = a\sin(2\pi ft + \theta)$$

where

$$a = \text{amplitude of motion}$$

$$f = \text{frequency} = \frac{1}{\text{period}}$$

$$= \text{number of oscillations per unit of time}$$

$$t = \text{time}$$

$$\theta = \text{phase angle}$$

EXAMPLE 1 A system vibrating in a simple harmonic motion is given by the equation

$$y = 2\sin\left(6\pi t + \frac{\pi}{2}\right)$$

where y = displacement in inches and t = time in seconds.

(a) Find the amplitude, frequency, period, and phase shift for this function.
(b) Determine the displacement of the object 3 s after release.
(c) Determine when the object will first reach its maximum negative displacement.

Solution

(a) Determine the amplitude, frequency, period, and phase shift just as in the previous sections.

$$y = 2\sin\left(6\pi t + \frac{\pi}{2}\right)$$

$$\text{amplitude} = 2$$

$$\text{frequency} = \frac{1}{\text{period}} = 6\pi/2\pi$$
$$f = 3$$

$$\text{period} = \frac{1}{\text{frequency}} = \frac{2\pi}{6\pi} = \frac{1}{3}$$

$$\text{phase shift} = \frac{\pi/2}{6\pi} = \frac{1}{12}$$

(b) To determine the displacement y at 3 s, substitute $t = 3$ into the equation and solve for y.

$$y = 2\sin\left(6\pi(3) + \frac{\pi}{2}\right)$$
$$y = 2\sin\left(18\pi + \frac{\pi}{2}\right)$$
$$y = 2\sin\frac{37\pi}{2}$$
$$y = 2 \text{ in., its maximum height}$$

(c) The first maximum negative displacement for the regular sine function $y = a\sin(x)$ occurs at $x = 3\pi/2$. Therefore, the first time the object reaches its maximum negative displacement occurs when the angle equals $3\pi/2$. Thus,

$$6\pi t + \frac{\pi}{2} = \frac{3\pi}{2}$$

Solving for t gives

$$6\pi t = \pi$$
$$t \approx 0.17 \text{ s}$$

■

EXAMPLE 2 A wooden block attached to the end of a spring is pulled down 10 cm and released, oscillating with simple harmonic motion and taking 3 s to return to its point of release. The phase angle is zero.

(a) Determine the period, frequency, and amplitude of the motion.
(b) Write an equation to describe the block's displacement.
(c) Determine the position of the block 1.25 s after release.

Solution

(a) The maximum displacement or amplitude is 10 cm. Since the block returns to its starting point in 3 s, its period is 3 s. Frequency, which is 1/period, equals 1/3.

(b) The general form of the equation is

$$y = a \sin(2\pi f t + \theta)$$

We know that $\theta = 0$, and we just found that

$$a = 10$$
$$f = \frac{1}{3}$$

Therefore, the equation is

$$y = 10 \sin \frac{2}{3}\pi t$$

(c) We can find the position of the block 1.25 s after release by substituting $t = 1.25$ into the above equation.

$$y = 10 \sin\left(\frac{2}{3}\pi(1.25)\right)$$
$$y = 10 \sin(2.618)$$
$$y = 5 \text{ cm}$$

■

Alternating Current

Alternating current (ac) is generated when a coil of wire rotates in a magnetic field. In an ac circuit, the current I as a function of time t is given by the following equation.

Sinusoidal Current

$$I = I_m \sin(2\pi f t + \theta)$$

where

$$I_m = \text{peak current}$$
$$f = \text{frequency}$$
$$\theta = \text{phase angle}$$

Similarly, voltage V in an ac circuit is given by the following equation.

Sinusoidal Voltage

$$V = V_0 \sin(2\pi ft + \theta)$$

where

$$V_0 = \text{peak voltage}$$
$$f = \text{frequency}$$
$$\theta = \text{phase angle}$$

NOTE ✦ The applications in this section use the sine function; however, the cosine function could just as easily be used since $\sin\theta = \cos(90° - \theta)$.

EXAMPLE 3 An alternating current has a peak current of 3 A and a frequency of 50 Hz. If the phase angle is zero, write the equation of current as a function of time, find the period, and find the current at $t = 0.1$ s.

Solution To write the equation, we need values of I_m, f, and θ. From the verbal statement, we know that $I_m = 3$, $f = 50$, and $\theta = 0$. Therefore, the equation of current as a function of time is

$$I = I_m \sin(2\pi ft + \theta)$$
$$I = 3 \sin 100\pi t$$

The period is the reciprocal of the frequency; therefore,

$$\text{period} = \frac{1}{50} \text{ s}$$

We find the current at time 0.1 s by substituting $t = 0.1$ into the equation.

$$I = 3 \sin(100\pi(0.1))$$
$$I = \sin(100\pi(0.1))$$
$$I = \sin 10\pi = 0$$

■

EXAMPLE 4 A certain alternating voltage has a peak voltage of 220 V and a frequency of 60 Hz. Write an equation for the voltage V if the phase angle is $\pi/2$. Also, determine the period and the voltage when $t = 0.03$ s. Graph the function through one cycle.

Solution The general equation is given by

$$V = V_0 \sin(2\pi ft + \theta)$$

Since $V_0 = 220, f = 60$, and $\theta = \pi/2$, the equation becomes

$$V = 220 \sin\left(120\pi t + \frac{\pi}{2}\right)$$

The period is the reciprocal of the frequency; therefore,

$$\text{period} = \frac{1}{60} \text{ s}$$

The voltage when $t = 0.03$ s is given by

$$V = 220 \sin\left(120\pi(0.03) + \frac{\pi}{2}\right)$$

$$V = 68.0 \text{ V}$$

The graph through one cycle is given in Figure 12–23.

FIGURE 12–23

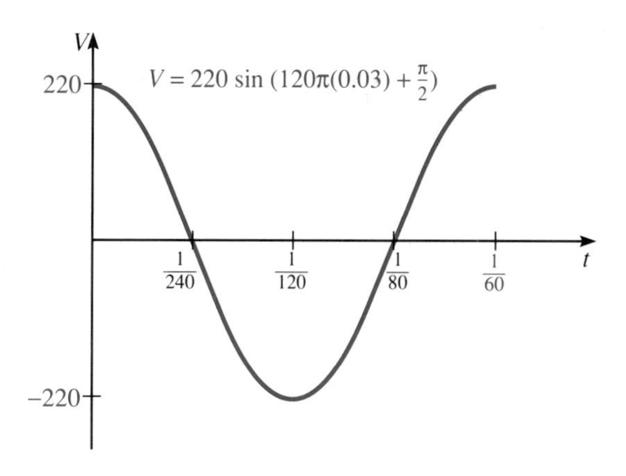

■

12–5 EXERCISES

For the equations of systems vibrating in simple harmonic motion in Exercises 1–6, do the following:

(a) Find the amplitude, period, and phase angle.
(b) Determine the displacement y of the object after 1.5 s.
(c) Determine when the object will reach its maximum positive displacement.
(d) Sketch the graph through one cycle.

1. $y = 1.4 \sin\left(6\pi t - \frac{\pi}{2}\right)$
2. $y = 3 \sin\left(2\pi t + \frac{\pi}{4}\right)$
3. $y = 2 \sin\left(3\pi t + \frac{\pi}{2}\right)$
4. $y = 2.8 \sin\left(4\pi t - \frac{\pi}{3}\right)$
5. $y = 3.1 \sin(\pi t - \pi)$
6. $y = 2.5 \sin\left(\frac{\pi t}{2} + \frac{\pi}{4}\right)$

Write an equation describing the simple harmonic motion of each system.

7. period = 2 s; amplitude = 4 in.; phase angle = $\pi/3$
8. period = 0.5 s; amplitude = 1.2 m; phase angle = $\pi/2$
9. period = 1.3 s; amplitude = 6 cm; phase angle = $-\pi/4$
10. period = 0.75 s; amplitude = 2.1 ft; phase angle = $-5\pi/6$
11. period = 4 s; amplitude = 2 yd; phase angle = π
12. period = 1.25 s; amplitude = 0.75 m; phase angle = $-\pi$

Sketch two cycles of $I = I_m \sin(2\pi ft + \theta)$ for the following.

13. $I_m = 4.6$ A; $f = 60$ Hz; $\theta = -\pi/2$
14. $I_m = 7.4$ A; $f = 50$ Hz; $\theta = \pi/6$

Sketch two cycles of $V = V_0 \sin(2\pi ft + \theta)$ for the following.

15. $V_0 = 65$ V; $f = 450$ Hz; $\theta = \pi/4$
16. $V_0 = 100$ V; $f = 60$ Hz; $\theta = -\pi$

M 17. A weight at the end of a spring vibrates in simple harmonic motion. The weight moves from its starting position to a maximum height of 18 cm and completes a cycle in 1.5 s. Assuming the phase angle is zero, write an equation to describe the position of the weight as a function of time, and find its position at 0.65 s.

E 18. The voltage in an ac circuit is given by the equation

$$V = 240 \sin(60\pi t - \pi)$$

Find the amplitude, period, and phase shift of this function, and sketch its graph through two cycles.

E 19. A certain alternating voltage has a peak voltage of 135 V and a frequency of 50 Hz. Assuming the phase angle is zero, write an equation for the voltage, find the amplitude and period, and graph the function through two cycles.

E 20. In an ac circuit the current I is given by

$$I = 4.8 \sin\left(60\pi t - \frac{\pi}{2}\right)$$

Find the amplitude, period, and phase shift of this function, and graph the function through two cycles.

12–6

COMPOSITE TRIGONOMETRIC CURVES

In the previous sections of this chapter, we dealt with graphs of one function at any given time. However, many applications deal with a combination of several functions. In this section we will discuss graphing composite functions and parametric equations.

Addition of Ordinates

A **composite function** is the sum or difference of several functions. One method of graphing composite functions is by **addition of ordinates,** whereby we graph each of the functions separately and then graph the composite function by adding the y values (ordinates) for selected points. This procedure is demonstrated in the next two examples.

EXAMPLE 1 Graph $y = x + \sin x$.

Solution First, graph the functions $y = x$ and $y = \sin x$ on the same set of coordinate axes as shown in Figure 12–24. Then locate selected points along the x axis. Any x coordinate can be used; however, let us choose points on the x axis where either function has a maximum, minimum, or zero. These locations are denoted with dotted vertical lines in Figure 12–24. Next, add the y coordinate of the two graphs at each of these points, and place a point at the new y coordinate. Finally, draw a curve through these points. The graph of $y = x + \sin x$ is shown as the color curve in Figure 12–24.

FIGURE 12–24

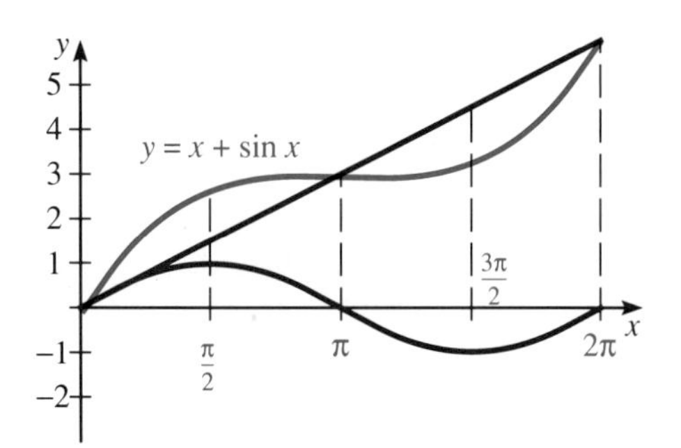

■

Graphing by Addition of Ordinates

To graph composite functions using addition of ordinates, follow these steps:

1. Graph each function on the same set of axes.
2. Locate relevant points on the x axis, such as the maximum, minimum, or zero of either function, and draw dotted vertical lines through these points.
3. Add the y coordinate of each function at the points chosen in Step 2, and place a point at the location of the new y coordinate.
4. Complete the graph by drawing a smooth curve through the points from Step 3.

EXAMPLE 2 Graph $y = \cos 2x - 3 \sin x$.

Solution First, graph the functions $y = \cos 2x$ and $y = -3 \sin x$ on the same set of axes as shown in Figure 12–25. Second, locate points on the x axis where a maximum, minimum, or zero occurs, and draw a vertical line through these points: 0, $\pi/2$, π, $3\pi/2$, and 2π (Figure 12–25). Next, place *dots* at points that represent the sum of the y coordinates at these points. Last, draw a smooth curve through these points. The graph of $y = \cos 2x - 3 \sin x$ is shown as the color curve in Figure 12–25.

FIGURE 12–25

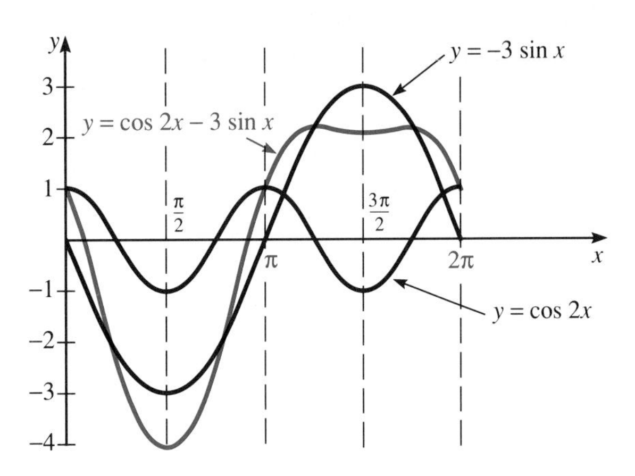

NOTE ✦ If each function is periodic, the period of the composite function is the least common multiple of the period of the individual functions. For example, $y = \cos 2x - 3 \sin x$ has a period of 2π because the period of $\cos 2x$ is π and the period of $-3 \sin x$ is 2π.

CAUTION ✦ When adding ordinates, be careful in adding negative numbers. The y coordinate resulting from adding 3 and -8 is -5.

Parametric Equations

The equations $y = 3m$ and $x = 2m + 11$ are called **parametric equations** *because the variables x and y are defined in terms of the same third variable m,* called a **parameter.** *When parametric equations describe simple harmonic motion,* the resulting graph is called a **Lissajous figure.**

One application of parametric equations is the oscilloscope. An oscilloscope converts an electrical signal into a graphic display. If we have two voltage signals represented by equations of the form

$$V = V_0 \sin(2\pi ft + \theta)$$

we can compare them by associating one signal with the x axis and the other signal with the y axis. For given values of time, we can plot values for x and y. If the two signals are equal in frequency and are in-phase, the resulting figure is a straight line. If the signals are equal in frequency but are out-of-phase, the figure will be a circle or an ellipse. A difference in frequency between the two signals results in a more complex figure.

EXAMPLE 3 Graph $x = 3 \sin t$ and $y = \cos(t - \pi)$.

Solution Since these functions are equal in frequency but are out-of-phase, we can expect the graph to be a circle or an ellipse. We fill in the accompanying table of values by choosing values for t and substituting them into each equation to find the resulting value for x and y. Then we plot the x and y coordinates and join the points with a smooth curve. The graph is shown in Figure 12–26.

t	0	$\pi/4$	$\pi/2$	$3\pi/4$	π	$5\pi/4$	$3\pi/2$	$7\pi/4$	2π
x	0	2.1	3	2.1	0	−2.1	−3	−2.1	0
y	−1	−0.7	0	0.7	1	0.7	0	−0.7	−1

FIGURE 12–26

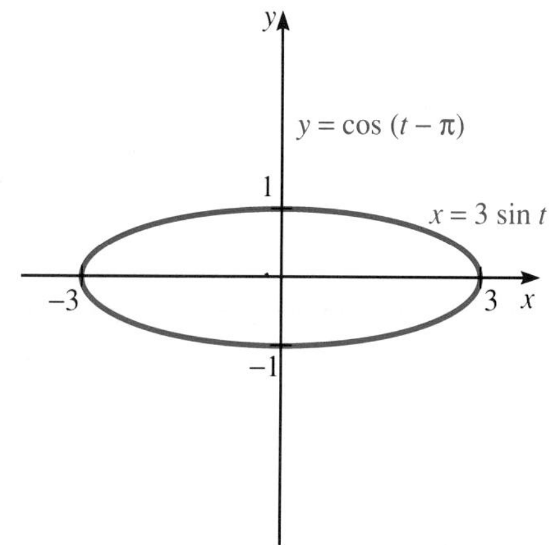

■

LEARNING HINT ✦ If you arrange the table of values in order such that t, x, and y represent the rows, you will make fewer graphing errors because the (x, y) ordered pairs are in the correct order.

EXAMPLE 4 Graph $x = 2 \sin(t - \pi/2)$ and $y = \cos(2t + \pi)$

Solution Since these functions are out-of-phase and have a different frequency, we can expect their graph to be somewhat more complex than in the previous example. However, the procedure used to graph these equations is the same as in Example 3. The table of values is given below, and the graph is shown in Figure 12–27. This graph is a portion of a parabola. Notice that the points repeat in reverse order after $t = \pi$.

t	0	$\pi/4$	$\pi/2$	$3\pi/4$	π	$5\pi/4$	$3\pi/2$	$7\pi/4$	2π
x	-2	-1.4	0	1.4	2	1.4	0	-1.4	-2
y	-1	0	1	0	-1	0	1	0	-1

FIGURE 12–27

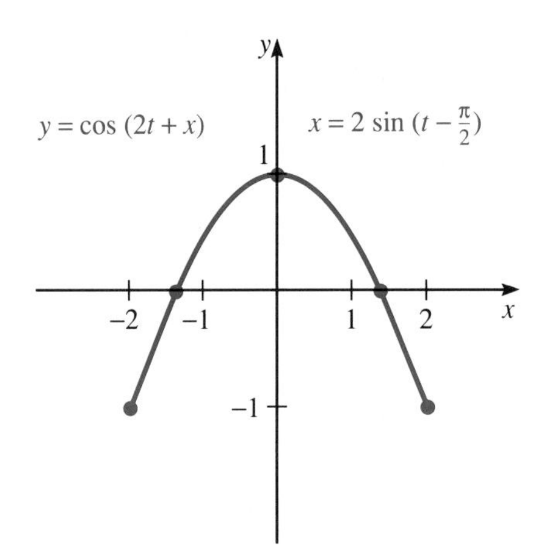

12–6 EXERCISES

Graph the following composite functions by using addition of ordinates.

1. $y = 2x + \sin x$
2. $y = \dfrac{x}{2} + \cos x$
3. $y = x - \cos x$
4. $y = -x + 2 \sin x$
5. $y = (3x - 1) + \sin 2x$
6. $y = x - 3 \cos 2x$
7. $y = \dfrac{x}{2} - 3 \sin x$
8. $y = \dfrac{x}{4} + 2 \sin x$
9. $y = -4x + \cos\left(x - \dfrac{\pi}{2}\right)$
10. $y = 3x + \sin\left(x + \dfrac{\pi}{2}\right)$
11. $y = x^2 - \sin(2x - \pi)$
12. $y = x^2 + 3 \cos\left(x - \dfrac{\pi}{4}\right)$
13. $y = 3 \cos 2x + \sin(x + \pi)$
14. $y = \cos x - 4 \sin \dfrac{x}{2}$
15. $y = \cos\left(x + \dfrac{\pi}{2}\right) + 3 \cos x$
16. $y = -2 \sin x + \sin \dfrac{x}{3}$
17. $y = \sin x - 2 \cos x$
18. $y = \cos x + \sin\left(2x - \dfrac{\pi}{4}\right)$
19. $y = 2 \sin x + 3 \cos \dfrac{x}{2}$
20. $y = \sin \dfrac{x}{2} - 2 \cos\left(x - \dfrac{\pi}{2}\right)$

Graph the following parametric equations.

21. $x = \sin t;\ y = 2\cos t$

22. $x = 2\sin t;\ y = -3\sin t$

23. $x = \sin\left(t + \frac{\pi}{4}\right);\ y = \cos\left(t - \frac{\pi}{2}\right)$

24. $x = 2\cos\frac{t}{3};\ y = -3\sin t$

25. $x = 1.5\cos 4t;\ y = \sin\left(\frac{t}{2} + \frac{\pi}{4}\right)$

26. $x = 2\sin(t + \pi);\ y = 3\sin\left(2t - \frac{\pi}{2}\right)$

27. $x = \sin\left(\frac{t}{2} - \pi\right);\ y = -\cos 3t$

28. $x = \cos 3t;\ y = 3\sin t$

E 29. The current in a certain ac circuit is given by the expression

$$i = 1.5\sin(\pi t) + 3\cos\left(t - \frac{\pi}{2}\right)$$

Graph the current as a function of time.

E 30. The voltage in a certain ac circuit is given by

$$V = 30\cos\pi t + 60\cos 40\pi t$$

Graph this relationship.

31. The x and y components of a certain object shot 43° above the horizontal are given by $x = 20t\cos 43°$ and $y = 20t\sin 43° - 16t^2$. Graph these parametric equations.

32. The first two terms in a certain Fourier series are given by

$$y = \cos\pi x + \frac{1}{9}\cos 3\pi x$$

Sketch the graph.

33. Under certain conditions, the path of a pendulum is given by $x = 4\sin\pi t$ and $y = \cos 2\pi t$. Plot the path of the pendulum.

CHAPTER SUMMARY

Summary of Terms

addition of ordinates (p. 437)
amplitude (p. 414)
asymptotes (p. 428)
composite function (p. 437)
cycle (p. 414)
Lissajous figure (p. 439)
parameter (p. 439)
parametric equation (p. 439)
period (p. 414)
periodic (p. 414)
phase angle (p. 423)
phase shift (p. 422)
zeros of a function (p. 415)

Summary of Formulas

$$y = a\sin(bx + c) \quad \text{or} \quad y = a\cos(bx + c)$$

where

$$\text{amplitude} = |a|$$

$$\text{period} = \frac{2\pi}{b}$$

$$\text{phase shift} = \frac{c}{b}$$

CHAPTER REVIEW

Section 12–1

Sketch the graph of each function through one period.

1. $y = -\sin x$ **2.** $y = 2\cos x$

3. $y = 1.8\cos x$ **4.** $y = -3\cos x$

5. $y = -5\sin x$ **6.** $y = 2.4\sin x$

Section 12–2

State the amplitude and period for each function, and sketch the graph through one period.

7. $y = \sin 2x$ **8.** $y = \cos \frac{x}{2}$

9. $y = 3\cos \frac{x}{4}$ **10.** $y = -2\sin 2\pi x$

11. $y = 6\sin 4x$ **12.** $y = 3\cos \frac{x}{3}$

13. $y = 1.6\cos 3\pi x$ **14.** $y = -4\cos 6\pi x$

15. The linear displacement x of an object oscillating at the end of a spring is given by $x = 4\cos \pi t$. Graph this function through two cycles.

16. The voltage in a certain ac circuit is given by $v = 210\sin 40\pi t$. Graph v as a function of t for $0 \le t \le 0.1$ s.

Section 12–3

Give the amplitude, period, and phase shift for the following functions, and sketch the graph through one cycle.

17. $y = 2\sin\left(x + \frac{\pi}{4}\right)$ **18.** $y = \cos\left(3x - \frac{\pi}{2}\right)$

19. $y = \cos\left(2x - \frac{\pi}{2}\right)$ **20.** $y = 1.4\sin\left(x + \frac{\pi}{4}\right)$

21. $y = -3\sin(2\pi x - 3)$

22. $y = -4\cos\left(\frac{x}{2} + \pi\right)$

23. $y = 4\cos\left(\frac{x}{3} + \frac{\pi}{2}\right)$ **24.** $y = 6\sin(\pi x - 3\pi)$

25. Find the equation of a cosine function with an amplitude of 2, a frequency of π, and a phase shift of $\pi/2$ units right.

26. Find the equation of a sine function with an amplitude of 1.4, a frequency of 4, and a phase shift of $\pi/3$ units left.

Section 12–4

Sketch the graph of the following functions through one cycle.

27. $y = -\csc 2x$ **28.** $y = 4\tan x$

29. $y = -3\cot x$ **30.** $y = -4\sec x$

31. $y = \tan\left(\frac{x}{2} - \pi\right)$ **32.** $y = \cot(\pi x - 3\pi)$

33. $y = 2\cot\left(x + \frac{\pi}{2}\right)$ **34.** $y = -2\csc(x + \pi)$

35. $y = -\sec(2\pi x - \pi)$ **36.** $y = 3\sec\left(\frac{x}{3} - \frac{\pi}{3}\right)$

Section 12–5

The equations in Problems 37–40 represent a system vibrating in simple harmonic motion.

(a) Find the amplitude, period, and phase shift.
(b) Determine the displacement y after 2.4 s.
(c) Determine when the object will reach maximum positive displacement.
(d) Sketch the graph through one cycle.

37. $y = 8\sin\left(t - \frac{\pi}{4}\right)$ **38.** $y = 6\sin(3\pi t + 2\pi)$

39. $y = 3.1\sin(2t + \pi)$ **40.** $y = 2\sin\left(\frac{t}{3} - \frac{\pi}{2}\right)$

E 41. The voltage in an ac circuit is given by $V = 60\sin(30t - \pi)$. Find the amplitude, period, and phase shift of this function, and sketch the graph through two cycles.

Section 12–6

Use addition of ordinates to graph the following functions.

42. $y = 2x + 3\cos x$ **43.** $y = x - 4\sin 2\pi x$

44. $y = x^2 - 4\sin x$

45. $y = \cos\left(\pi x + \frac{\pi}{2}\right) - x^2$

46. $y = \sin 2x + 3\cos x$ **47.** $y = 3\cos x - \sin \frac{x}{2}$

48. $y = 2\cos\left(x - \frac{\pi}{2}\right) + 3\sin(x + \pi)$

49. $y = \sin\left(\frac{x}{2} + \pi\right) + 4\cos\left(x - \frac{\pi}{4}\right)$

Graph the following parametric equations.

50. $x = 3 \sin t;\ y = \cos\left(t + \frac{\pi}{2}\right)$

51. $x = -\sin 2t;\ y = 2 \sin 2t$

52. $x = \cos\left(\frac{t}{2} - 2\pi\right);\ y = 4 \sin 3t$

53. $x = \sin\left(t - \frac{\pi}{3}\right);\ y = \cos\left(t + \frac{\pi}{2}\right)$

54. $x = \cos 4t;\ y = -4 \sin(t + 1)$

CHAPTER TEST

The number in parentheses refers to the appropriate learning objective given at the beginning of the chapter.

1. Sketch the graph of $y = -2 \cos\left(x + \frac{\pi}{2}\right)$. (2)

2. Graph $y = 2x + 2 \sin x$ by addition of ordinates. (5)

3. Write an equation describing the simple harmonic motion of an object if the period is 0.8 s, the amplitude is 2 in., and the phase angle is $\pi/3$ radians. (4)

4. For the following trigonometric function, identify the amplitude, period, and phase shift: (1)

$$y = 1.7 \sin\left(3\pi x - \frac{\pi}{3}\right)$$

5. Sketch the graph of $y = 3 \sin 4x$ through one cycle. (2)

6. Sketch the graph of $y = -2 \tan x$ through one cycle. (3)

7. Sketch the graph of $y = -\sin(x - \pi)$ through one cycle. (2)

8. An alternating current has an amplitude of 2.4 A and a frequency of 60 Hz. If the phase angle is zero, write the equation of current as a function of time, and find the current at $t = 1.8$ s. (4)

9. Sketch the graph of $y = \cot\left(x + \frac{\pi}{2}\right)$ through one cycle. (3)

10. Graph $x = \sin 2t$ and $y = \cos\left(t - \frac{\pi}{2}\right)$. (6)

11. For $y = -3 \cos(\pi x - \pi)$, identify the amplitude, period, and phase shift. (1)

12. The following two voltages are applied to the horizontal and vertical planes of an oscilloscope: (6)

$$v_1 = 60 \sin\left(100\,\pi t - \frac{\pi}{2}\right) \quad \text{and} \quad v_2 = 100 \cos 120\pi t$$

Sketch the figure that would appear on the screen.

13. Write an equation for sinusoidal voltage if the peak voltage is 115 V, the frequency is 50 Hz, and the phase angle is $-\pi/6$. Graph this function through one cycle. (2, 4)

14. Sketch the graph of $y = -2 \csc(x + \pi)$ through one cycle. (3)

SOLUTION TO CHAPTER INTRODUCTION

The general equation of an object undergoing simple harmonic motion is

$$y = a \sin(2\pi ft + \theta)$$

From the statement of the problem, $\theta = 0$, $a = 18$ cm, and $f = 1/5$. Substituting these values into the formula gives

$$y = 18 \sin \frac{2\pi t}{5}$$

To find the position of the block when $t = 1.75$, calculate

$$y = 18 \sin \frac{2\pi(1.75)}{5}$$

$$y = 14.6 \text{ cm}$$

The displacement of a piston is given by the expression $y = 4 \sin A + \sin 2A$. Determine the angle at which the displacement is zero. (The solution to this problem is given at the end of the chapter.)

Electronic signals can be monitored on an oscilloscope and described mathematically by equations using trigonometric functions. Also, sound waves from a stereo can be represented by sinusoidal functions. In this chapter we will discuss trigonometric identities, that is, formulas that can be substituted into a trigonometric expression to transform and simplify it. The trigonometric identities are useful in solving a trigonometric equation, another topic of this chapter.

Learning Objectives

After completing this chapter, you should be able to

1. Prove that a given trigonometric expression is an identity (Section 13–1).
2. Apply the formula for the sum or difference of two angles to simplify a trigonometric expression and to verify an identity (Section 13–2).
3. Apply the double- and half-angle formulas to simplify a trigonometric expression and to verify an identity (Section 13–3).
4. Write a trigonometric expression as a single term by using the appropriate sum, difference, half- or double-angle formula (Sections 13–2 and 13–3).
5. Solve a trigonometric equation (Section 13–4).
6. Use the inverse trigonometric functions to evaluate an expression (Section 13–5).

Chapter 13

Trigonometric Equations and Identities

13–1

BASIC TRIGONOMETRIC IDENTITIES

An **identity** is a relationship that is *true for all permissible values* of the variable(s). For example, the expression $3(x + 1) + 8 = x + 11 + 2x$ is an algebraic identity because any real number substituted for x results in a true statement. In this section we will discuss identities as they relate to trigonometry.

Historically, trigonometric identities were used to find the value of trigonometric functions for specific angles such as 75°. With the use of a calculator this is no longer necessary; however, the trigonometric identities are useful in other applications, such as solving trigonometric equations and Laplace transforms in calculus. We will begin the discussion with a review of some of the basic identities discussed earlier.

Basic Identities

From Chapter 9, we know that certain trigonometric functions are reciprocals: sine and cosecant, cosine and secant, and tangent and cotangent. Rearranging these relationships leads to the following identities:

$$\csc \theta = \frac{1}{\sin \theta} \quad \text{or} \quad \sin \theta = \frac{1}{\csc \theta} \quad \text{or} \quad \sin \theta \csc \theta = 1$$

$$\sec \theta = \frac{1}{\cos \theta} \quad \text{or} \quad \cos \theta = \frac{1}{\sec \theta} \quad \text{or} \quad \cos \theta \sec \theta = 1$$

$$\cot \theta = \frac{1}{\tan \theta} \quad \text{or} \quad \tan \theta = \frac{1}{\cot \theta} \quad \text{or} \quad \tan \theta \cot \theta = 1$$

LEARNING HINT ✦ Notice the different forms that these reciprocal identities can take. It is extremely important that you be able to identify the different forms of any given identity. This requires great familiarity with each one.

In Chapter 9, we also defined the trigonometric functions in terms of x, y, and r. These relationships were given as follows:

$$\sin \theta = \frac{y}{r} \qquad \csc \theta = \frac{r}{y}$$

$$\cos \theta = \frac{x}{r} \qquad \sec \theta = \frac{r}{x}$$

$$\tan \theta = \frac{y}{x} \qquad \cot \theta = \frac{x}{y}$$

If we take the definition for the tangent function and divide the numerator and denominator by r, we obtain the following additional identity:

$$\tan\theta = \frac{y}{x} = \frac{y/r}{x/r} = \frac{\sin\theta}{\cos\theta}$$

Since the cotangent function is the reciprocal of the tangent, the following relationship is also true:

$$\cot\theta = \frac{\cos\theta}{\sin\theta}$$

Several additional identities result from applying the Pythagorean theorem.

$$\begin{aligned} x^2 + y^2 &= r^2 && \text{Pythagorean theorem} \\ \frac{x^2}{r^2} + \frac{y^2}{r^2} &= \frac{r^2}{r^2} && \text{divide by } r^2 \\ \left(\frac{x}{r}\right)^2 + \left(\frac{y}{r}\right)^2 &= 1 && \text{simplify} \\ (\cos\theta)^2 + (\sin\theta)^2 &= 1 && \text{substitute} \\ \cos^2\theta + \sin^2\theta &= 1 && \text{simplify} \end{aligned}$$

NOTE ✦ The expressions $\cos^2\theta$ and $(\cos\theta)^2$ are identical; $\cos^2\theta$ is just easier to write.

Similarly, by applying the Pythagorean theorem, we can derive the following identities:

$$\begin{aligned} x^2 + y^2 &= r^2 && \text{Pythagorean theorem} \\ \frac{x^2}{y^2} + \frac{y^2}{y^2} &= \frac{r^2}{y^2} && \text{divide by } y^2 \\ \left(\frac{x}{y}\right)^2 + 1 &= \left(\frac{r}{y}\right)^2 && \text{simplify} \\ \cot^2\theta + 1 &= \csc^2\theta && \text{substitute} \end{aligned}$$

and

$$\begin{aligned} x^2 + y^2 &= r^2 && \text{Pythagorean theorem} \\ \frac{x^2}{x^2} + \frac{y^2}{x^2} &= \frac{r^2}{x^2} && \text{divide by } x^2 \\ 1 + \left(\frac{y}{x}\right)^2 &= \left(\frac{r}{x}\right)^2 && \text{simplify} \\ 1 + \tan^2\theta &= \sec^2\theta && \text{substitute} \end{aligned}$$

The eight basic identities derived so far are summarized below.

Basic Trigonometric Identities

$$\sec\theta = \frac{1}{\cos\theta} \qquad \tan\theta = \frac{\sin\theta}{\cos\theta}$$

$$\csc\theta = \frac{1}{\sin\theta} \qquad \cot\theta = \frac{\cos\theta}{\sin\theta}$$

$$\cot\theta = \frac{1}{\tan\theta} \qquad \sin^2\theta + \cos^2\theta = 1$$

$$1 + \tan^2\theta = \sec^2\theta \qquad \cot^2\theta + 1 = \csc^2\theta$$

EXAMPLE 1 Using $\theta = 45°$, verify the following identity:

$$\tan\theta = \frac{\sin\theta\cos\theta}{1 - \sin^2\theta}$$

Solution To verify the identity for a specific value, substitute the value 45° for θ and prove equality as follows:

$$\tan 45° = \frac{\sin 45° \cos 45°}{1 - \sin^2 45°}$$

$$1 = \frac{\left(\frac{\sqrt{2}}{2}\right)\left(\frac{\sqrt{2}}{2}\right)}{1 - \left(\frac{\sqrt{2}}{2}\right)^2}$$

$$1 = \frac{1/2}{1/2}$$

$$1 = 1$$

■

Proving an Identity

To prove that an expression is an identity, we must prove the equality of the equation for all permissible values of the variable. It would be impossible to substitute every permissible value to prove an identity. Therefore, we prove an identity by transforming one member of the equation into the other.

EXAMPLE 2 Prove the following identity:

$$\cot x = \frac{\sec x \cos x}{\tan x}$$

Solution To prove this identity, we must show that the two sides of the equation are identical for all permissible values of x. To do so, we choose the more complicated side of the equation, substitute trigonometric identities, and perform algebraic operations to simplify the expression until the two sides of the equation are identical. To transform the right member into $\cot x$, we can substitute 1 for $\sec x \cos x$.

$$\cot x = \frac{\sec x \cos x}{\tan x}$$

$$\cot x = \frac{1}{\tan x} \qquad \text{substitute}$$

$$\cot x = \cot x \qquad \text{substitute}$$

■

LEARNING HINT ✦ There is no set procedure for proving an identity; however, some general guidelines are given in the following box. There may be several correct ways to prove an identity, and it may take several false starts before you find a correct way. However, the more you practice, the easier the proofs become.

Guidelines for Proving Identities

1. Substitute trigonometric identities into the more complicated side of the equation until it is identical to the less complicated side (see Example 3).
2. When working with the more complicated side of an equation, always be aware of the expression you are working toward. If the less complicated side involves only one trigonometric function, then substitute in terms of that function on the other side of the equation. Another useful transformation is to multiply or divide the numerator and denominator of a fraction by the same quantity (see Example 5).
3. Perform any algebraic operations to simplify an expression. This step may involve finding a common denominator to add fractions, factoring a polynomial, or multiplying polynomials. If one side of the equation has a single term in the denominator and several terms in the numerator, write each term in the numerator over the denominator and simplify (see Examples 4 and 6).
4. When in doubt, transform the more complicated member to sine and cosine functions. This technique may not be the easiest one, but it will usually give acceptable results (See Example 3).

EXAMPLE 3 Prove the following identity:

$$\frac{\sin x \sec x}{\sin^2 x - \tan^2 x} = -\frac{\cos x}{\sin^3 x}$$

Solution The left side of the equation appears more complicated; therefore, we will simplify it. Since the right side involves only the sine and cosine functions, let us substitute the sine and cosine functions on the left side of the equation.

$$\frac{\sin x \sec x}{\sin^2 x - \tan^2 x} = -\frac{\cos x}{\sin^3 x}$$

$$\frac{\sin x \dfrac{1}{\cos x}}{\sin^2 x - \dfrac{\sin^2 x}{\cos^2 x}} = -\frac{\cos x}{\sin^3 x}$$

After substituting, we perform normal algebraic simplification by multiplying the fractions in the numerator and adding the fractions in the denominator.

$$\frac{\dfrac{\sin x}{\cos x}}{\dfrac{\sin^2 x \cos^2 x - \sin^2 x}{\cos^2 x}} = -\frac{\cos x}{\sin^3 x}$$

$$\frac{\sin x}{\cos x} \cdot \frac{\cos^2 x}{\sin^2 x(\cos^2 x - 1)} = -\frac{\cos x}{\sin^3 x} \quad \text{divide and factor}$$

$$\frac{\cos x}{\sin x(\cos^2 x - 1)} = -\frac{\cos x}{\sin^3 x} \quad \text{simplify}$$

$$\frac{\cos x}{\sin x(-\sin^2 x)} = -\frac{\cos x}{\sin^3 x} \quad \text{substitute}$$

$$-\frac{\cos x}{\sin^3 x} = -\frac{\cos x}{\sin^3 x} \quad \text{simplify}$$

■

EXAMPLE 4 Prove the following identity:

$$\frac{\sin x}{1 + \cos x} + \frac{1 + \cos x}{\sin x} = 2 \csc x$$

Solution Since the left side of the equation appears to be more complicated, we simplify by adding the fractions.

$$\frac{\sin x}{1 + \cos x} + \frac{1 + \cos x}{\sin x} = 2 \csc x$$

$$\frac{\sin^2 x + (1 + \cos x)^2}{\sin x(1 + \cos x)} = 2 \csc x \quad \text{add}$$

$$\frac{\sin^2 x + 1 + 2 \cos x + \cos^2 x}{\sin x(1 + \cos x)} = 2 \csc x \quad \text{square the binomial}$$

$$\frac{(1 - \cos^2 x) + 1 + 2\cos x + \cos^2 x}{\sin x(1 + \cos x)} = 2\csc x \quad \text{substitute}$$

$$\frac{2 + 2\cos x}{\sin x(1 + \cos x)} = 2\csc x \quad \text{simplify}$$

$$\frac{2(1 + \cos x)}{\sin x(1 + \cos x)} = 2\csc x \quad \text{factor}$$

$$\frac{2}{\sin x} = 2\csc x \quad \text{simplify (reduce)}$$

$$2\left(\frac{1}{\sin x}\right) = 2\csc x \quad \text{simplify}$$

$$2\csc x = 2\csc x \quad \text{substitute}$$

■

EXAMPLE 5 Verify the following identity:

$$\tan x = \tan x \csc^2 x - \cot x$$

Solution Since the right hand of the equation appears to be more complicated, we transform it to look like the left side. Notice that the left side of the equation contains only the tangent function; therefore, the substitutions that we make on the right side should all lead toward the tangent.

$$\tan x = \tan x \csc^2 x - \cot x$$

$$\tan x = \tan x(1 + \cot^2 x) - \cot x \quad \text{substitute}$$

$$\tan x = \tan x + \tan x \cot^2 x - \cot x \quad \text{multiply}$$

$$\tan x = \tan x + \tan x\left(\frac{1}{\tan^2 x}\right) - \frac{1}{\tan x} \quad \text{substitute}$$

$$\tan x = \tan x + \frac{1}{\tan x} - \frac{1}{\tan x} \quad \text{simplify}$$

$$\tan x = \tan x$$

■

EXAMPLE 6 Verify the following identity:

$$\sec^2 x + \tan^2 x = \sec^4 x - \tan^4 x$$

Solution Since the right side of the equation appears to be more complicated, we begin by factoring it.

$$\sec^2 x + \tan^2 x = \sec^4 x - \tan^4 x$$

$$\sec^2 x + \tan^2 x = (\sec^2 x - \tan^2 x)(\sec^2 x + \tan^2 x) \quad \text{factor}$$

$$\sec^2 x + \tan^2 x = 1(\sec^2 x + \tan^2 x) \quad \text{substitute}$$

$$\sec^2 x + \tan^2 x = \sec^2 x + \tan^2 x \quad \text{simplify}$$

■

13–1 EXERCISES

Prove the following identities.

1. $\cos x \csc x = \cot x$
2. $\dfrac{\tan x}{\sin x} = \sec x$
3. $\sin x \cos x \csc x \sec x = 1$
4. $\dfrac{\cot x}{\cos x} = \csc x$
5. $\dfrac{\sin^2 A}{1 - \cos A} = 1 + \cos A$
6. $\cos x(1 + \tan^2 x) = \sec x$
7. $\dfrac{\tan x - 1}{1 - \cot x} = \tan x$
8. $\dfrac{\cos A}{1 - \sin A} = \dfrac{1 + \sin A}{\cos A}$
9. $\sin x + \cot x \cos x = \csc x$
10. $(\cot x + \csc x)(\cot x - \csc x) = -1$
11. $\cot A(\tan A + \cot A) = \csc^2 A$
12. $(\sec x - \csc x)(\sin x + \cos x) = \tan x - \cot x$
13. $\dfrac{\cos x}{\sin^2 x - 1} = -\sec x$
14. $\dfrac{\tan^4 x - 1}{\sec^2 x} = \tan^2 x - 1$
15. $\cos x(\tan^2 x + 1) = \sec x$
16. $\csc x \cot x \cos x = \tan^2 x$
17. $\tan x = \sec x \csc x - \cot x$
18. $\sin A \tan A + \cos A = \sec A$
19. $\dfrac{2 - \sin^2 A}{\cos A} = \sec A + \cos A$
20. $\cos x(\sec x - \cos x) = \sin^2 x$
21. $\cos^2 x \csc x - \csc x = -\sin x$
22. $\dfrac{1}{\csc A - \cot A} = \csc A + \cot A$
23. $\csc A + \tan A + \cot A = \dfrac{\cos A + 1}{\sin A \cos A}$
24. $\dfrac{4 \csc^4 x - 4 \cot^4 x}{\csc^2 x + \cot^2 x} = 4$
25. $\dfrac{1 + \cot x}{1 - \cot x} = \dfrac{1 + \tan x}{\tan x - 1}$
26. $\dfrac{\cot x}{\csc x + 1} = \dfrac{\csc x - 1}{\cot x}$
27. $(\sin A - \cos A)(\sin A + \cos A) = \cos^2 A(\tan^2 A - 1)$
28. $\dfrac{\sin x - \tan x}{\sin x \tan x} = \dfrac{\cos x - 1}{\sin x}$
29. $\dfrac{\cot x + \cos x}{1 + \sin x} = \cot x$
30. $\cos^2 x - \sin^2 x = 1 - 2 \sin^2 x$
31. $\dfrac{1 + \sin y}{\cos y} + \dfrac{\cos y}{1 + \sin y} = 2 \sec y$
32. $\csc x \cot x = \dfrac{1 + \cot^2 x}{\sec x}$
33. $\dfrac{\sin x}{1 + \cos x} = \csc x - \cot x$
34. $\dfrac{\cot x - \cos x}{\cot x} = \dfrac{\cos^2 x}{1 + \sin x}$

13–2

THE SUM OR DIFFERENCE OF TWO ANGLES

The trigonometric identities in the previous section dealt with functions of one angle. In this section we will consider trigonometric functions that are the sum or difference of two angles. Let us begin by deriving an expression for $\sin (A + B)$.

Deriving the Sum and Difference Formulas

In Figure 13–1, angle A is in standard position, while angle B has its initial side as the terminal side of angle A. The angle $A + B$ is an angle in standard position. From a point P on the terminal side of angle $A + B$, perpendiculars are drawn to the x axis at point T and to the terminal side of angle A at point

R. Then perpendicular lines are drawn from R to the x axis at point U and from R to line PT at point Q. Angle TPR is equal to angle A (similar triangles). **Similar triangles** have equal corresponding angles and proportional corresponding sides.

FIGURE 13–1

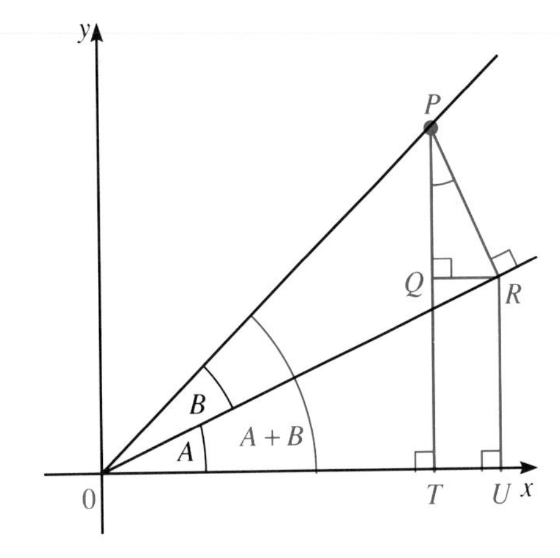

From Figure 13–1,

$$\sin(A + B) = \frac{PT}{OP} = \frac{TQ + QP}{OP} = \frac{RU}{OP} + \frac{QP}{OP}$$

From triangle ORU,

$$RU = OR \sin A$$

and from triangle PQR,

$$QP = PR \cos A$$

Substituting gives

$$\sin(A + B) = \frac{OR \sin A}{OP} + \frac{PR \cos A}{OP}$$

$$\sin(A + B) = \sin A \frac{OR}{OP} + \cos A \frac{PR}{OP}$$

But, in triangle OPR,

$$\frac{OR}{OP} = \cos B \quad \text{and} \quad \frac{PR}{OP} = \sin B$$

Substituting gives the final result.

$$\sin(A + B) = \sin A \cos B + \cos A \sin B$$

Similarly,

$$\cos(A + B) = \cos A \cos B - \sin A \sin B$$

To derive the formula for the difference of two angles, we substitute $-B$ for angle B in the sum formula as follows:

$$\sin(A + (-B)) = \sin A \cos(-B) + \cos A \sin(-B)$$

Since $\sin(-B) = -\sin B$ and $\cos(-B) = \cos B$, we have

$$\sin(A - B) = \sin A \cos B - \cos A \sin B$$

Similarly,

$$\cos(A - B) = \cos A \cos B + \sin A \sin B$$

To derive the formula for the tangent of the sum or difference of two angles, we use the fact that $\tan C = \sin C/\cos C$. Thus,

$$\tan(A \pm B) = \frac{\sin(A \pm B)}{\cos(A \pm B)} = \frac{\sin A \cos B \pm \cos A \sin B}{\cos A \cos B \mp \sin A \sin B}$$

Dividing the numerator and denominator by $\cos A \cos B$ gives

$$\tan(A \pm B) = \frac{\tan A \pm \tan B}{1 \mp \tan A \tan B}$$

The formulas for the sum and difference of two angles are summarized below.

Sum or Difference of Two Angles

$$\sin(A \pm B) = \sin A \cos B \pm \cos A \sin B$$

$$\cos(A \pm B) = \cos A \cos B \mp \sin A \sin B$$

$$\tan(A \pm B) = \frac{\tan A \pm \tan B}{1 \mp \tan A \tan B}$$

NOTE ✦ The two signs given in the formula above denote the signs for two formulas, one for sum and one for difference. In each formula, pair the upper signs together and the lower signs together.

Applying the Sum and Difference Formulas

EXAMPLE 1 Without using a calculator, find exact values for $\sin 15°$ and $\cos 75°$.

Solution To find exact values for $\sin 15°$ and $\cos 75°$, we can express $15°$ and $75°$ in terms of the special angles $0°$, $30°$, $45°$, $60°$, or $90°$, because we know exact values for these angles. We can write $\sin 15°$ as $\sin(45° - 30°)$

and cos 75° as cos(30° + 45°). Applying the appropriate formula and simplifying the result gives

$$\sin 15° = \sin(45° - 30°) = \sin 45° \cos 30° - \cos 45° \sin 30°$$

$$\sin 15° = \frac{\sqrt{2}}{2}\left(\frac{\sqrt{3}}{2}\right) - \frac{\sqrt{2}}{2}\left(\frac{1}{2}\right)$$

$$\sin 15° = \frac{\sqrt{6}}{4} - \frac{\sqrt{2}}{4}$$

$$\sin 15° = \frac{\sqrt{6} - \sqrt{2}}{4}$$

$$\cos 75° = \cos(30° + 45°) = \cos 30° \cos 45° - \sin 30° \sin 45°$$

$$\cos 75° = \frac{\sqrt{3}}{2}\left(\frac{\sqrt{2}}{2}\right) - \frac{1}{2}\left(\frac{\sqrt{2}}{2}\right)$$

$$\cos 75° = \frac{\sqrt{6} - \sqrt{2}}{4}$$

■

CAUTION ✦ Remember that $\sin(30° + 45°) \neq \sin 30° + \sin 45°$, just as $(a + b)^2 \neq a^2 + b^2$ in algebra.

EXAMPLE 2 For this problem, we know that $\sin A = 12/13$ and A is in quadrant I and that $\cos B = -4/5$ and B is in quadrant II.

(a) Find $\sin(A + B)$.
(b) Find $\cos(A + B)$.
(c) Determine the quadrant in which the terminal side of $(A + B)$ lies.

Solution In order to find $\sin(A + B)$ and $\cos(A + B)$, we need to have values for $\sin A$, $\cos A$, $\sin B$, and $\cos B$. Angles A and B are shown in Figure 13–2. From the information given, we can apply the Pythagorean theorem to

FIGURE 13–2

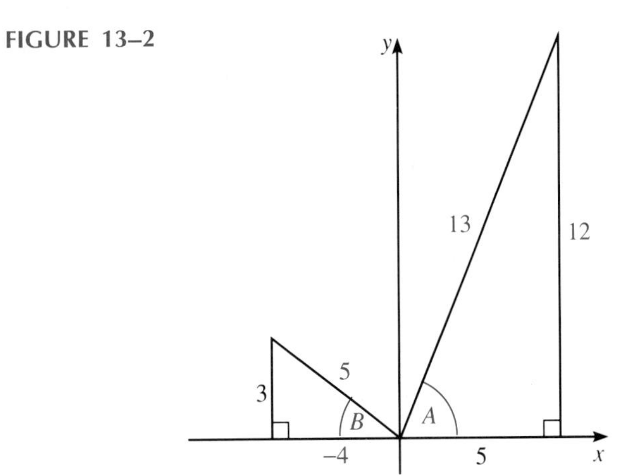

find the remaining side of each triangle and the required values for the sine and cosine. These values are

$$\sin A = \frac{12}{13} \qquad \sin B = \frac{3}{5}$$

$$\cos A = \frac{5}{13} \qquad \cos B = -\frac{4}{5}$$

We substitute into the formula to obtain the following:

(a) $\sin(A + B) = \sin A \cos B + \cos A \sin B$

$$\sin(A + B) = \frac{12}{13}\left(-\frac{4}{5}\right) + \frac{5}{13}\left(\frac{3}{5}\right)$$

$$\sin(A + B) = \frac{-48 + 15}{65}$$

$$\sin(A + B) = -\frac{33}{65}$$

(b) $\cos(A + B) = \cos A \cos B - \sin A \sin B$

$$\cos(A + B) = \frac{5}{13}\left(-\frac{4}{5}\right) - \frac{12}{13}\left(\frac{3}{5}\right)$$

$$\cos(A + B) = \frac{-20 - 36}{65}$$

$$\cos(A + B) = -\frac{56}{65}$$

(c) Finally, since $\sin(A + B)$ and $\cos(A + B)$ are both negative, the terminal side of angle $A + B$ lies in quadrant III, where the sine and cosine functions are both negative. ■

EXAMPLE 3 Reduce the following expressions to a single term:

(a) $\cos 3A \cos 2A - \sin 3A \sin 2A$

(b) $\dfrac{\tan 5A - \tan A}{1 + \tan 5A \tan A}$

Solution

(a) This expression fits the expanded form of $\cos(A + B)$. Therefore, we can write

$$\cos 3A \cos 2A - \sin 3A \sin 2A = \cos(3A + 2A) = \cos 5A$$

(b) Since this expression fits the expanded form of $\tan(A - B)$,

$$\frac{\tan 5A - \tan A}{1 + \tan 5A \tan A} = \tan(5A - A) = \tan 4A$$

■

Proving an Identity

We use the same techniques developed in the previous section to prove an identity. We have just added the formulas for the sum or difference of two angles to the list of possible trigonometric substitutions that we can make.

EXAMPLE 4 Verify the identity $\sin(A + B)\sin(A - B) = \sin^2 A - \sin^2 B$

Solution Since the left side of the equation appears to be more complicated, simplify it by expanding $\sin(A + B)$ and $\sin(A - B)$.

$$\underbrace{\sin(A + B)}_{(\sin A \cos B + \cos A \sin B)}\ \underbrace{\sin(A - B)}_{(\sin A \cos B - \cos A \sin B)} = \sin^2 A - \sin^2 B$$

$$(\sin A \cos B + \cos A \sin B)(\sin A \cos B - \cos A \sin B) = \sin^2 A - \sin^2 B$$

Because the left side is the product of the sum and difference of a binomial, multiplying gives

$$\sin^2 A \cos^2 B - \cos^2 A \sin^2 B = \sin^2 A - \sin^2 B$$

Since the right side of the equation contains only the sine function, let us substitute for $\cos^2 B$ and $\cos^2 A$.

$$\sin^2 A(1 - \sin^2 B) - (1 - \sin^2 A)\sin^2 B = \sin^2 A - \sin^2 B$$

Multiplying and simplifying the result gives

$$\sin^2 A - \sin^2 A \sin^2 B - \sin^2 B + \sin^2 A \sin^2 B = \sin^2 A - \sin^2 B$$
$$\sin^2 A - \sin^2 B = \sin^2 A - \sin^2 B$$

■

EXAMPLE 5 Prove the following identities:

(a) $\tan\left(\frac{\pi}{4} - x\right) = \frac{1 - \tan x}{1 + \tan x}$

(b) $\frac{\sin(x + y)}{\sin x \cos y} = 1 + \cot x \tan y$

Solution

(a) To prove this identity, expand the left side of the equation using the formula $\tan(A - B)$.

$$\tan\left(\frac{\pi}{4} - x\right) = \frac{1 - \tan x}{1 + \tan x}$$

$$\frac{\tan \pi/4 - \tan x}{1 + \tan \pi/4 \tan x} = \frac{1 - \tan x}{1 + \tan x}$$

Substituting $\tan \pi/4 = 1$ gives the final result.

$$\frac{1 - \tan x}{1 + \tan x} = \frac{1 - \tan x}{1 + \tan x}$$

(b) To prove this identity, let us expand the left side of the equation using the formula for $\sin(A + B)$ and simplify the result.

$$\frac{\sin(x + y)}{\sin x \cos y} = 1 + \cot x \tan y$$

$$\frac{\sin x \cos y + \cos x \sin y}{\sin x \cos y} = 1 + \cot x \tan y$$

$$\frac{\sin x \cos y}{\sin x \cos y} + \frac{\cos x \sin y}{\sin x \cos y} = 1 + \cot x \tan y$$

$$1 + \cot x \tan y = 1 + \cot x \tan y$$

■

13–2 EXERCISES

Use the sum or difference formula to find the exact value of the given function.

1. $\sin 105°$ **2.** $\tan 15°$

3. $\cos 15°$ **4.** $\sin 75°$

5. $\cos 195°$ **6.** $\tan 75°$

7. $\cos 165°$ **8.** $\tan 285°$

9. Given that $\cos A = 12/13$ and A is in quadrant I and that $\sin B = 15/17$ and B is in quadrant II, find the following:
(a) $\cos(A + B)$ **(b)** $\tan(A + B)$
(c) quadrant of $A + B$

10. Given that $\tan A = -3/4$ and A is in quadrant II and that $\cos B = 5/13$ and B is in quadrant I, find the following:
(a) $\sin(A - B)$ **(b)** $\cos(A - B)$
(c) quadrant of $A - B$

11. Given that $\sin A = 15/17$ and A is in quadrant II and that $\tan B = 12/5$ and B is in quadrant III, find the following:
(a) $\sin(A - B)$ **(b)** $\cos(A - B)$
(c) quadrant of $A - B$

12. Given that $\cos A = 7/25$ and A is in quadrant I and that $\sin B = 12/13$ and B is in quadrant II, find the following:
(a) $\tan(A + B)$ **(b)** $\sin(A + B)$
(c) quadrant of $A + B$

Reduce each expression to a single term.

13. $\sin 75° \cos 40° - \cos 75° \sin 40°$

14. $\cos 48° \cos 36° + \sin 48° \sin 36°$

15. $\cos A \cos 5A - \sin A \sin 5A$

16. $\sin 3x \cos y + \cos 3x \sin y$

17. $\sin A \sin 3A - \cos A \cos 3A$

18. $\cos(x - y) \cos x - \sin(x - y) \sin x$

19. $\dfrac{\tan(A + B) - \tan A}{1 + \tan(A + B)\tan A}$

20. $\sin(C - D) \cos D + \cos(C - D) \sin D$

Verify the following identities.

21. $\cos\left(\dfrac{\pi}{2} + x\right) = -\sin x$

22. $\sin\left(\dfrac{\pi}{3} - x\right) = \dfrac{\sqrt{3} \cos x - \sin x}{2}$

23. $\tan(\pi + x) = \tan x$

24. $\sin(30° + x) = \dfrac{\cos x + \sqrt{3} \sin x}{2}$

25. $\dfrac{\sin 5A}{\csc 2A} + \dfrac{\cos 5A}{\sec 2A} = \cos 3A$

26. $\cos(A + B) - \cos(A - B) = -2 \sin A \sin B$

27. $\cos(A + B) \cos(A - B) = \cos^2 A - \sin^2 B$

28. $\sin 2A = 2 \sin A \cos A$ (Hint: $2A = A + A$.)

29. $\sin(A + B) - \sin(A - B) = 2 \cos A \sin B$

30. $\csc(A + B) = \dfrac{\csc A \csc B}{\cot B + \cot A}$

31. $\cot(A + B) = \dfrac{\cot A \cot B - 1}{\cot A + \cot B}$

32. $\sin(A + B + C) = \sin A \cos B \cos C + \cos A \sin B \cos C + \cos A \cos B \sin C - \sin A \sin B \sin C$

33. The expression for the displacement y of an object in simple harmonic motion can be written

$$y = a \sin 2\pi ft \cos B + a \cos 2\pi ft \sin B$$

Write the expression on the right side of the equation as a single term.

34. A spring vibrating in simple harmonic motion described by

$$y_1 = A_1\cos\left(wt + \frac{\pi}{2}\right)$$

is subjected to another motion described by

$$y_2 = A_2\cos(wt - \pi)$$

Show that the resultant motion is

$$y_1 + y_2 = -A_2\cos wt - A_1\sin wt$$

35. The displacement y of a wave traveling through some liquid can be given by

$$y = 12 \sin\left(\frac{\pi t}{3} - \frac{3\pi}{2}\right)$$

Expand and simplify this expression.

36. The displacement of light passing through glass can be written as

$$d = \frac{t \sin(A - B)}{\cos B}$$

Simplify this expression.

13–3

DOUBLE-ANGLE AND HALF-ANGLE IDENTITIES

Derivation of the Double-Angle Identity

Using the formulas for the sum of two angles, we can derive the formulas for a double angle. The derivations for $\sin 2A$, $\cos 2A$, and $\tan 2A$ follow.

$$\sin 2A = \sin(A + A)$$
$$\sin 2A = \sin A \cos A + \sin A \cos A$$
$$\sin 2A = 2 \sin A \cos A$$

$$\cos 2A = \cos(A + A)$$
$$\cos 2A = \cos A \cos A - \sin A \sin A$$
$$\cos 2A = \cos^2 A - \sin^2 A$$
$$\cos 2A = (1 - \sin^2 A) - \sin^2 A$$
$$\cos 2A = 1 - 2 \sin^2 A$$
$$\cos 2A = \cos^2 A - (1 - \cos^2 A)$$
$$\cos 2A = 2 \cos^2 A - 1$$

$$\tan 2A = \tan(A + A)$$
$$\tan 2A = \frac{\tan A + \tan A}{1 - \tan A \tan A}$$
$$\tan 2A = \frac{2 \tan A}{1 - \tan^2 A}$$

The double-angle formulas are summarized in the following box.

Double-Angle Formulas

$$\sin 2A = 2 \sin A \cos A$$

$$\cos 2A = \cos^2 A - \sin^2 A$$
$$\cos 2A = 1 - 2 \sin^2 A$$
$$\cos 2A = 2 \cos^2 A - 1$$

$$\tan 2A = \frac{2 \tan A}{1 - \tan^2 A}$$

EXAMPLE 1 Use a double-angle formula to find the exact value of $\cos 2A$, $\sin 2A$, and $\tan 2A$ if $\sin A = 8/17$ and the terminal side of angle A lies in quadrant I.

FIGURE 13–3

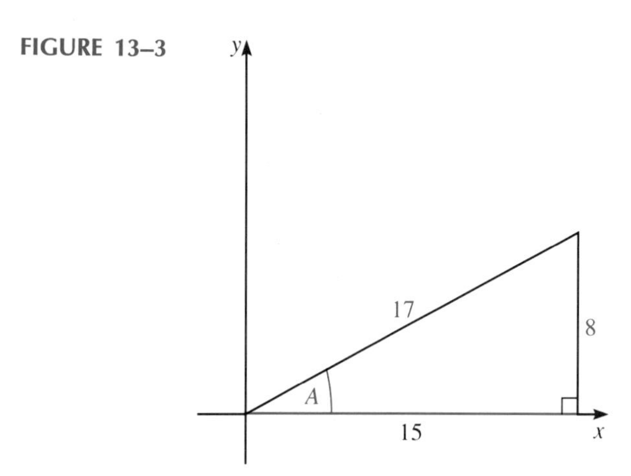

Solution In order to fill in the required double-angle formulas, we need to know the value of $\sin A$, $\cos A$, and $\tan A$. We can determine these values using the Pythagorean theorem, as shown in Figure 13–3).

$$\sin A = \frac{8}{17} \qquad \cos A = \frac{15}{17} \qquad \tan A = \frac{8}{15}$$

Calculating $\sin 2A$, $\cos 2A$, and $\tan 2A$ gives the following:

$$\sin 2A = 2 \sin A \cos A$$

$$\sin 2A = 2\left(\frac{8}{17}\right)\left(\frac{15}{17}\right)$$

$$\sin 2A = \frac{240}{289}$$

$$\cos 2A = \cos^2 A - \sin^2 A$$

$$\cos 2A = \left(\frac{15}{17}\right)^2 - \left(\frac{8}{17}\right)^2$$

$$\cos 2A = \frac{225 - 64}{289}$$

$$\cos 2A = \frac{161}{289}$$

$$\tan 2A = \frac{2 \tan A}{1 - \tan^2 A}$$

$$\tan 2A = \frac{2(8/15)}{1 - (8/15)^2}$$

$$\tan 2A = \frac{16}{15} \cdot \frac{225}{161}$$

$$\tan 2A = \frac{240}{161}$$

Since $\sin 2A$, $\cos 2A$, and $\tan 2A$ are all positive, the terminal side of angle $2A$ lies in quadrant I. We check by calculating angle A using $\sin A = 8/17$ and multiplying the resulting angle by 2.

$$\sin A = \frac{8}{17} \qquad A \approx 28° \qquad 2A \approx 56°$$

■

EXAMPLE 2 Write $4 \sin^2 2x - 2$ as a single expression.

Solution This expression appears to be one of the identities for $\cos 2A$: $1 - 2 \sin^2 A$. However, we need to arrange it in the same order as the identity.

$$-2 + 4 \sin^2 2x$$

Next, factor out -2.

$$-2(1 - 2 \sin^2 2x)$$

The expression in parentheses is equivalent to $\cos 2A$ where $A = 2x$. The expression is simplified to

$$-2 \cos 4x$$

■

EXAMPLE 3 Prove the identity $\cos x(1 - \cos 2x) = \sin x \sin 2x$.

Solution Since the left side of the equation appears to be more complicated, simplify it. Also, because the right side of the equation contains only the sine function, all substitutions on the left side will work toward that end.

$$\begin{aligned}
\cos x(1 - \cos 2x) &= \sin x \sin 2x & \\
\cos x - \cos x \cos 2x &= \sin x \sin 2x & \text{multiply} \\
\cos x - \cos x(1 - 2 \sin^2 x) &= \sin x \sin 2x & \text{substitute} \\
\cos x - \cos x + 2 \sin^2 x \cos x &= \sin x \sin 2x & \text{multiply} \\
2 \sin^2 x \cos x &= \sin x \sin 2x & \text{simplify} \\
\sin x(2 \sin x \cos x) &= \sin x \sin 2x & \text{simplify} \\
\sin x \sin 2x &= \sin x \sin 2x & \text{substitute}
\end{aligned}$$

■

Half-Angle Formulas

Using the double-angle formulas, we can derive expressions for $\sin A/2$, $\cos A/2$, and $\tan A/2$. To derive $\cos A/2$, we solve for $\cos B$ in the identity $\cos 2B = 2 \cos^2 B - 1$.

$$\begin{aligned}
\cos 2B &= 2 \cos^2 B - 1 \\
\frac{\cos 2B + 1}{2} &= \cos^2 B \\
\pm\sqrt{\frac{\cos 2B + 1}{2}} &= \cos B
\end{aligned}$$

This formula is true for all angles; therefore, substituting $A/2$ for B gives the identity

$$\cos \frac{A}{2} = \pm\sqrt{\frac{1 + \cos A}{2}}$$

The sign in front of the radical is determined by the quadrant in which the terminal side of angle $A/2$ lies. In a similar manner, use the identity $\cos 2B = 1 - \sin^2 B$, solve for $\sin B$, and substitute $B = A/2$ to obtain the half-angle formula for the sine function, as follows:

$$\begin{aligned}
\cos 2B &= 1 - 2 \sin^2 B \\
\frac{1 - \cos 2B}{2} &= \sin^2 B \\
\pm\sqrt{\frac{1 - \cos 2B}{2}} &= \sin B
\end{aligned}$$

Substituting $\dfrac{A}{2} = B$ gives

$$\sin \frac{A}{2} = \pm\sqrt{\frac{1 - \cos A}{2}}$$

We can find $\tan A/2$ using the identity

$$\tan\frac{A}{2} = \frac{\sin A/2}{\cos A/2}$$

$$\tan\frac{A}{2} = \frac{\pm\sqrt{\dfrac{1 - \cos A}{2}}}{\pm\sqrt{\dfrac{1 + \cos A}{2}}}$$

$$\tan\frac{A}{2} = \pm\sqrt{\frac{1 - \cos A}{1 + \cos A}}$$

If we multiply the numerator and denominator by $\sqrt{1 - \cos A}$ and simplify, the result is

$$\tan\frac{A}{2} = \frac{1 - \cos A}{\sin A}$$

If we multiply the numerator and denominator by $\sqrt{1 + \cos A}$, the result is

$$\tan\frac{A}{2} = \frac{\sin A}{1 + \cos A}$$

The half-angle formulas are summarized below.

Half-Angle Formulas

$$\sin\frac{A}{2} = \pm\sqrt{\frac{1 - \cos A}{2}}$$

$$\cos\frac{A}{2} = \pm\sqrt{\frac{1 + \cos A}{2}}$$

$$\tan\frac{A}{2} = \frac{1 - \cos A}{\sin A} = \frac{\sin A}{1 + \cos A}$$

Proving Identities

EXAMPLE 4 Use the half-angle formula to find an exact value of $\cos \pi/8$.

Solution Since $\pi/8$ is half of $\pi/4$, we can use the following half-angle formula:

$$\cos \frac{A}{2} = \pm\sqrt{\frac{1 + \cos A}{2}}$$

$$\cos \frac{\pi}{8} = \pm\sqrt{\frac{1 + \cos \frac{\pi}{4}}{2}}$$

$$\cos \frac{\pi}{8} = +\sqrt{\frac{1 + \frac{\sqrt{2}}{2}}{2}}$$

$$\cos \frac{\pi}{8} = \frac{\sqrt{2 + \sqrt{2}}}{2}$$

Since $\pi/8$ is in quadrant I, we use the plus sign before the radical. ■

EXAMPLE 5 Given that $\sin A = -12/13$ and A is in quadrant III, find the following:

(a) $\cos \frac{A}{2}$ (b) $\tan \frac{A}{2}$

Solution To find $\cos A/2$ and $\tan A/2$, we need to know the value of $\cos A$, which we obtain using the Pythagorean theorem.

$$\cos A = -\frac{5}{13}$$

(a)
$$\cos \frac{A}{2} = \pm\sqrt{\frac{1 + \cos A}{2}}$$

$$\cos \frac{A}{2} = \pm\sqrt{\frac{1 + \frac{-5}{13}}{2}}$$

$$\cos \frac{A}{2} = \pm\sqrt{\frac{4}{13}}$$

$$\cos \frac{A}{2} = -\frac{2}{13}\sqrt{13}$$

Since angle A is in quadrant III, $180° < A < 270°$, $A/2$ is in quadrant II, and $90° < A/2 < 135°$. Because the cosine is negative in quadrant II, we place a negative sign in front of the answer.

(b) $\tan \frac{A}{2} = \frac{1 - \cos A}{\sin A}$

$$\tan \frac{A}{2} = \frac{1 + \frac{5}{13}}{\frac{-12}{13}}$$

$$\tan \frac{A}{2} = -\frac{3}{2}$$

■

Notice that in Example 5, the sign of $\tan A/2$ is determined by the answer. Only the sine and cosine functions require determining the correct sign.

EXAMPLE 6 Verify the following identity:

$$\cos^2 \frac{B}{2} = \frac{\sec B + 1}{2 \sec B}$$

Solution To verify this identity, let us start by substituting for $\cos^2 B/2$.

$$\cos^2 \frac{B}{2} = \frac{\sec B + 1}{2 \sec B}$$

$$\left(\pm\sqrt{\frac{1 + \cos B}{2}}\right)^2 = \frac{\sec B + 1}{2 \sec B} \qquad \text{substitute}$$

$$\frac{1 + \cos B}{2} = \frac{\sec B + 1}{2 \sec B} \qquad \text{simplify}$$

If we compare the first term of the numerator on each side of the equation, we know that we must transform 1 into $\sec B$. However, $\sec B$ equals $1/\cos B$. Thus, we should divide the numerator and denominator of the left side of the equation by $\cos B$ as follows:

$$\frac{\frac{1}{\cos B} + \frac{\cos B}{\cos B}}{2\left(\frac{1}{\cos B}\right)} = \frac{\sec B + 1}{2 \sec B}$$

$$\frac{\sec B + 1}{2 \sec B} = \frac{\sec B + 1}{2 \sec B}$$

■

CAUTION ✦ Be very careful in using the double- and half-angle formulas. Determine the necessary value of A, $2A$, and $A/2$ before substituting into the formula.

13–3 EXERCISES

Use double- or half-angle formulas to find exact values for the following functions.

1. $\sin 15°$

2. $\tan 67.5°$

3. $\cos 105°$

4. $\cos 165°$

5. $\tan 120°$

6. $\sin 112.5°$

7. If $\sin A = 8/17$ and A is in quadrant I, find $\cos 2A$ and $\tan A/2$.

8. If $\tan A = -40/9$ and A is in quadrant II, find $\sin 2A$ and $\cos A/2$.

9. If $\cos A = -24/25$ and $180° < A < 270°$, find $\tan A/2$ and $\sin A/2$.

10. If $\sin A = 24/25$ and $90° < A < 180°$, find $\sin A/2$ and $\tan 2A$.

Write each expression as a single term.

11. $4 \sin 2A \cos 2A$

12. $\dfrac{2 \tan 3B}{1 - \tan^2 3B}$

13. $4 \sin^2 x \cos^2 x$

14. $16 \sin^2 A - 8$

15. $-\sqrt{\dfrac{1 + \cos 210°}{2}}$

16. $\dfrac{1 - \cos 70°}{\sin 70°}$

Prove the following identities.

17. $\cos 3x = \cos x(1 - 4 \sin^2 x)$

18. $\cos 2x = \dfrac{1 - \tan^2 x}{1 + \tan^2 x}$

19. $\tan B \sin 2B + \cos 2B = 1$

20. $\cot 2A = \dfrac{\cot^2 A - 1}{2 \cot A}$

21. $\tan \dfrac{A}{2} = \csc A - \cot A$

22. $\dfrac{\tan 2x}{\sin 2x} = \sec 2x$

23. $\sec 2C = \dfrac{\tan C + \cot C}{\cot C - \tan C}$

24. $(\sin x + \cos x)^2 = 1 + \sin 2x$

25. $\sin 2A = \dfrac{2 \tan A}{1 + \tan^2 A}$

26. $\tan \dfrac{A}{2}(\cos A + 1) = \sin A$

27. $\cos^2 A - \sin^2 A = \cos 2A$

28. $\sec \dfrac{x}{2} = \pm\sqrt{\dfrac{2}{1 + \cos x}}$

29. $\left(\cos \dfrac{x}{2} - \sin \dfrac{x}{2}\right)^2 = 1 - \sin x$

30. $\dfrac{\cos 2x - \cos x}{\cos x - 1} = 2 \cos x + 1$

31. $1 = 2 \cos^2 \dfrac{B}{2} - \cos B$

32. $\cot \dfrac{B}{2} = \csc B + \cot B$

33. The index of refraction n is given by

$$n = \frac{\sin \dfrac{(A + D)}{2}}{\sin \dfrac{A}{2}}$$

where A = angle of prism and D = angle of minimum deviation. Simplify this expression.

34. The horizontal range y of an object projected with velocity v at an angle A above the horizontal is given by $y = vt \cos A$, but

$$t = \frac{2v \sin A}{g}$$

Substitute for t and simplify the result.

13–4

TRIGONOMETRIC EQUATIONS

So far in this chapter, we have verified identities, which are equations that are true for all permissible values of the variable. In this section we will solve **conditional equations,** that is, equations that are true only for specific values of the variable. Solving a trigonometric equation means the same thing as solving an algebraic equation, namely, finding values of the variable that satisfy the equation. To solve trigonometric equations, we use the same techniques used in solving algebraic equations. In addition to these techniques, we will also simplify the equation by substituting trigonometric identities.

EXAMPLE 1 Solve the equation $2 \cos A + 1 = 0$ for $0 \le A < 2\pi$.

Solution Since this is a linear trigonometric equation, we use the techniques of solving an algebraic linear equation as follows:

$$\begin{aligned} 2 \cos A + 1 &= 0 \\ 2 \cos A &= -1 \quad \text{transpose} \\ \cos A &= -\frac{1}{2} \quad \text{divide} \end{aligned}$$

To find angle A, we must determine all angles $0 \le A < 2\pi$ whose cosine value equals 1/2. The reference angle is $\pi/3$. Since the cosine is negative in quadrants II and III,

$$A = \frac{2\pi}{3} \quad \text{and} \quad A = \frac{4\pi}{3}$$

After checking, the solutions are $2\pi/3$ and $4\pi/3$. ■

CAUTION ✦ Be sure to find *all* angles that satisfy the equation. You may have a tendency to stop with one solution.

EXAMPLE 2 Solve the equation $\sin x - 2 \sin^2 x = 0$ for $0 \le x < 2\pi$.

Solution Since this is a quadratic trigonometric equation, we solve it by factoring.

$$\begin{aligned} \sin x - 2 \sin^2 x &= 0 \\ \sin x(1 - 2 \sin x) &= 0 \end{aligned}$$

Next, we set each factor equal to zero and solve for x.

$$\sin x = 0 \qquad\qquad 1 - 2 \sin x = 0$$

$$x = 0 \quad \text{and} \quad x = \pi \qquad\qquad -2 \sin x = -1$$

$$\sin x = \frac{1}{2}$$

$$x = \frac{\pi}{6} \quad \text{and} \quad x = \frac{5\pi}{6}$$

After checking, the solutions are $x = 0, \pi/6, 5\pi/6$, and π. ■

Guidelines for Solving Trigonometric Equations

1. If the equation contains one function of a single angle, use algebraic techniques to solve for the angle (see Example 1).
2. Solve a quadratic equation containing a single function of the same angle by factoring, if possible. Otherwise, use the quadratic formula (see Example 2).
3. If the equation contains several functions of the same angle, substitute trigonometric identities to obtain a single function (see Example 3).
4. If the equation contains several angles, substitute trigonometric identities to obtain a function of a single angle (see Example 4).

EXAMPLE 3 Solve $2\sin^2 A - \cos^2 A = 0$ for $0 \le A < 2\pi$.

Solution Since this equation involves two functions of the same angle, substitute trigonometric identities to obtain one trigonometric function.

$$
\begin{aligned}
2\sin^2 A - \cos^2 A &= 0 \\
2\sin^2 A - (1 - \sin^2 A) &= 0 && \text{substitute} \\
2\sin^2 A - 1 + \sin^2 A &= 0 && \text{simplify} \\
3\sin^2 A - 1 &= 0 && \text{simplify} \\
3\sin^2 A &= 1 && \text{transpose} \\
\sin^2 A &= \frac{1}{3} && \text{divide} \\
\sin A &= \pm\frac{\sqrt{3}}{3} && \text{square root}
\end{aligned}
$$

Since A is not one of the special angles, we use the calculator to find the reference angle, $A = 0.6155$ radian. Since $\sin A$ is both positive and negative, we place this angle in all four quadrants: $A = 0.6155$, 2.5261, 3.7571, and 5.6677. After three values have been checked for A, the solutions are 0.6155, 2.5261, 3.7571, and 5.6677. ■

EXAMPLE 4 Solve the equation $\cos 2t + 3\sin t = 2$ for $0 \le t < 2\pi$.

Solution Since this equation involves more than one function and will not factor, we should make a trigonometric substitution. We want an expression in terms of a single angle, $2t$ or t, and a single trigonometric function, cosine or sine. Substituting $1 - 2\sin^2 t$ for $\cos 2t$ gives

$$\begin{aligned}
\cos 2t + 3 \sin t &= 2 \\
(1 - 2 \sin^2 t) + 3 \sin t &= 2 && \text{substitute} \\
-2 \sin^2 t + 3 \sin t - 1 &= 0 && \text{transpose} \\
2 \sin^2 t - 3 \sin t + 1 &= 0 && \text{multiply by } -1 \\
(2 \sin t - 1)(\sin t - 1) &= 0 && \text{factor}
\end{aligned}$$

$$\begin{aligned}
2 \sin t - 1 &= 0 & \sin t - 1 &= 0 && \text{set factors equal to zero} \\
2 \sin t &= 1 & \sin t &= 1 \\
\sin t &= \frac{1}{2} & t &= \frac{\pi}{2}
\end{aligned}$$

$$t = \frac{\pi}{6} \quad \text{and} \quad t = \frac{5\pi}{6}$$

After checking, the solutions are $t = \pi/6$, $\pi/2$, and $5\pi/6$. ■

EXAMPLE 5 Solve the equation $\cos 5x \cos 3x + \sin 5x \sin 3x = -1$ for $0 \le x < 2\pi$.

Solution Since the left side of this equation reduces to $\cos(5x - 3x) = \cos 2x$, this equation becomes

$$\cos 2x = -1$$

Substituting for $\cos 2x$ and solving gives

$$\begin{aligned}
2 \cos^2 x - 1 &= -1 && \text{substitute} \\
2 \cos^2 x &= 0 && \text{transpose and add} \\
\cos^2 x &= 0 && \text{divide} \\
\cos x &= 0 && \text{square root of both sides}
\end{aligned}$$

$$x = \frac{\pi}{2} \quad \text{and} \quad x = \frac{3\pi}{2}$$

■

EXAMPLE 6 Solve the equation $\sin A/2 = 1 - \cos A$ for $0 \le A < 2\pi$.

Solution First, we substitute for $\sin A/2$.

$$\pm\sqrt{\frac{1 - \cos A}{2}} = 1 - \cos A$$

Next, we remove the radical by squaring both sides of the equation.

$$\frac{1 - \cos A}{2} = (1 - \cos A)^2$$

$$\frac{1 - \cos A}{2} = 1 - 2\cos A + \cos^2 A \qquad \text{multiply}$$

$$1 - \cos A = 2 - 4\cos A + 2\cos^2 A \qquad \text{remove fractions}$$

$$0 = 1 - 3\cos A + 2\cos^2 A \qquad \text{simplify}$$

$$0 = (1 - \cos A)(1 - 2\cos A) \qquad \text{factor}$$

$$1 - \cos A = 0 \qquad 1 - 2\cos A = 0 \qquad \text{solve}$$

$$\cos A = 1 \qquad \cos A = \frac{1}{2}$$

$$A = 0 \qquad A = \frac{\pi}{3} \text{ and } A = \frac{5\pi}{3}$$

Since we squared both sides of the equation, we must check each value by substituting into the original equation. The solutions are $A = 0$, $\pi/3$, and $5\pi/3$. ■

EXAMPLE 7 Solve the equation $-\cot x + 3\tan x = 2$ for $0 \le x < 2\pi$.

Solution We begin by substituting $\cot x = 1/\tan x$.

$$-\cot x + 3\tan x = 2$$

$$-\frac{1}{\tan x} + 3\tan x = 2 \qquad \text{substitute}$$

$$-1 + 3\tan^2 x = 2\tan x \qquad \text{remove fraction}$$

$$3\tan^2 x - 2\tan x - 1 = 0 \qquad \text{transpose}$$

$$(3\tan x + 1)(\tan x - 1) = 0 \qquad \text{factor}$$

$$3\tan x + 1 = 0 \qquad \tan x - 1 = 0 \qquad \text{solve}$$

$$3\tan x = -1 \qquad \tan x = 1$$

$$\tan x = \frac{-1}{3}$$

$$x = 2.8 \text{ and } x = 6.0 \qquad x = \frac{\pi}{4} \text{ and } x = \frac{5\pi}{4}$$

After the above values have been checked for x, the solutions are $x = \pi/4$, 2.8, $5\pi/4$, and 6.0. ■

13–4 EXERCISES

Solve the following equations for $0 \leq x < 2\pi$. Use a calculator only when x is not one of the special angles.

1. $(\sqrt{3}\sin x - 1)(2\cos x + 1) = 0$
2. $(\sqrt{3}\tan x + 1)(\tan x + 1) = 0$
3. $\csc x + 1 = 0$
4. $\cot x - 1 = 0$
5. $4\sin^2 x = 1$
6. $\cot^2 x - 1 = 0$
7. $\tan^2 x + 6\tan x + 7 = 0$
8. $\sin x \csc x - \csc x = 0$
9. $\cot^2 x = \cot x$
10. $\csc^2 x - 5\cot x + 3 = 0$
11. $\sin^2 x = 1 - \cos x$
12. $\sin x = \sin \frac{x}{2}$
13. $\cot^2 x - 1 = \csc x$
14. $\cos 2x - \cos x = 0$
15. $3\csc x \sec x = \sec x$
16. $2\sin^2 x = \cos 2x - 2$
17. $\cos 2x + \sin x = 1$
18. $\dfrac{\tan 3x + \tan x}{1 - \tan 3x \tan x} = -1$
19. $5\cos^2 x = 4\cos x + 1$
20. $\sin x - 2\sin^3 x = 0$
21. $\sin 4x \cos x + \cos 4x \sin x = -1$
22. $\cos 2x - 2\cos^2 x = 0$
23. $\tan^2 x = 1 + \sec x$
24. $\dfrac{1 - \cos x}{\sin x} = \sqrt{3}$
25. $2\cos 3x = 1$
26. $2\sin 5x = \sqrt{3}$
27. $\sin 2x = \cos x$
28. $\cos x \cos 3x - \sin x \sin 3x = -1$
29. $\cot^2 x - \cot x = 0$
30. $2\cos^2 3x = 1$
31. The angle of a projectile shot at 86 m/s with a range of 500 m is given by $7{,}396 \tan A = 2450 \sec^2 A$. Find angle A.
32. The vertical displacement y of an object at the end of a spring is given by $y = 3\cos t + \sin 2t$ where t represents time. Determine the least amount of time required for the displacement to be zero.
33. The height from the water level of a point on the rim of a waterwheel is given by

$$y = 3 + 4\cos\left(t - \frac{\pi}{2}\right)$$

At what times, t, is the point at water level ($y = 0$)?

34. The displacement x of a piston is given by $x = 3\sin A + \sin 2A$. Determine the angle A where the displacement is zero.

13–5

THE INVERSE TRIGONOMETRIC FUNCTIONS

In Chapter 7, we discussed exponential and logarithmic functions. They are inverse functions because $y = b^x$ is equivalent to $x = \log_b y$. In this section we will discuss inverse functions of another type, inverse trigonometric functions. Although we did not refer to them as inverse trigonometric functions, we discussed them briefly in Chapter 9 when we solved for an angle given two sides of a right triangle. For example, to find angle A given that $\sin A = 16/19$, we pressed the [inv] and then the [sin] key on the calculator. The condition $\sin A = 16/19$ can also be written as

$$\sin^{-1}\left(\frac{16}{19}\right) = A \quad \text{or} \quad \arcsin \frac{16}{19} = A$$

where the -1 exponent or the ''arc'' prefix is used to denote an inverse trigonometric function. We will begin the discussion of the inverse trigonometric functions with their graphs.

Graphs of the Inverse Trigonometric Functions

In general, to find the inverse of any function, we interchange x and y in the equation and solve for y. We implement this procedure in graphing by interchanging the x and y coordinates of each ordered pair, as shown in the next example.

EXAMPLE 1 Graph $y = \sin x$ and $y = \arcsin x$.

Solution We know how to graph $y = \sin x$ from the previous chapter. The table of values for $y = \sin x$ is as follows:

x	$-\pi/2$	0	$\pi/2$	π	$3\pi/2$
y	-1	0	1	0	-1

To form the table of values of $y = \arcsin x$, we interchange the x and y coordinates in the table of values given above for $y = \sin x$. Thus, we obtain the following table:

x	-1	0	1	0	-1
y	$-\pi/2$	0	$\pi/2$	π	$3\pi/2$

The graphs of $y = \sin x$ and $y = \arcsin x$ are shown in Figure 13–4(a) and (b), respectively.

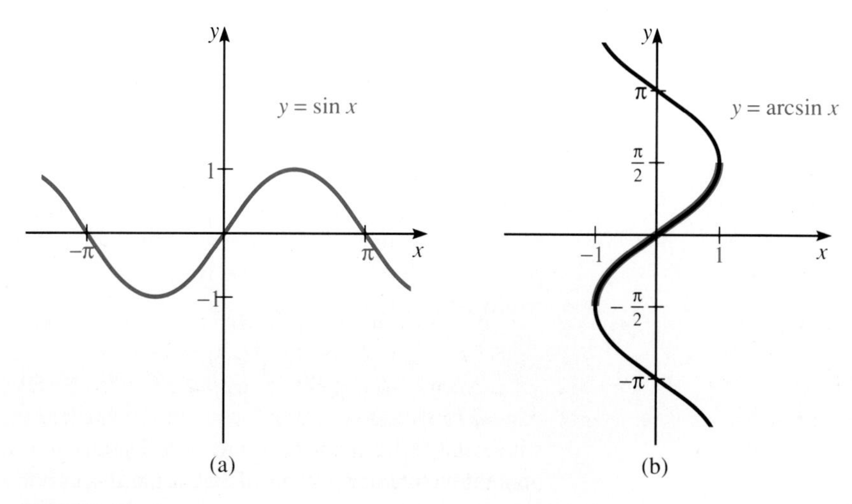

FIGURE 13–4

■

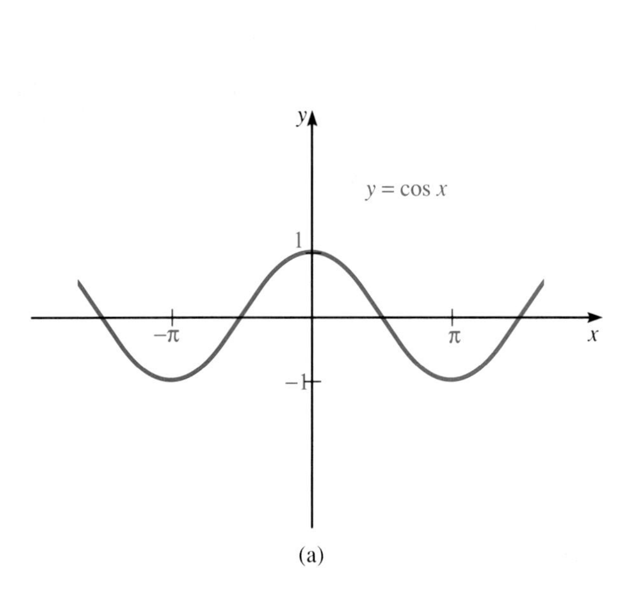

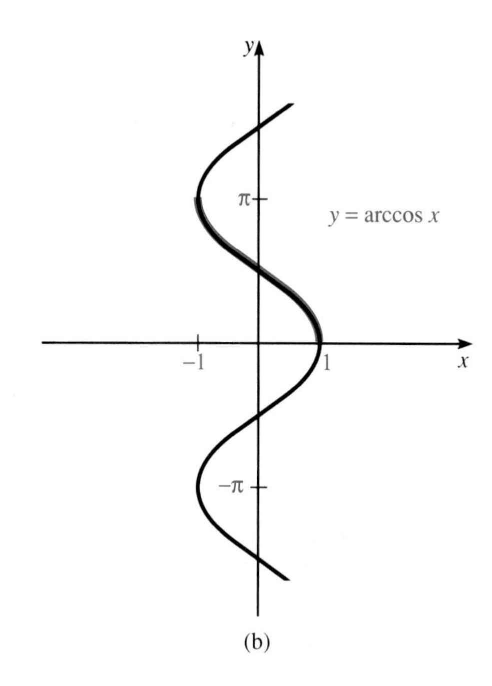

FIGURE 13–5

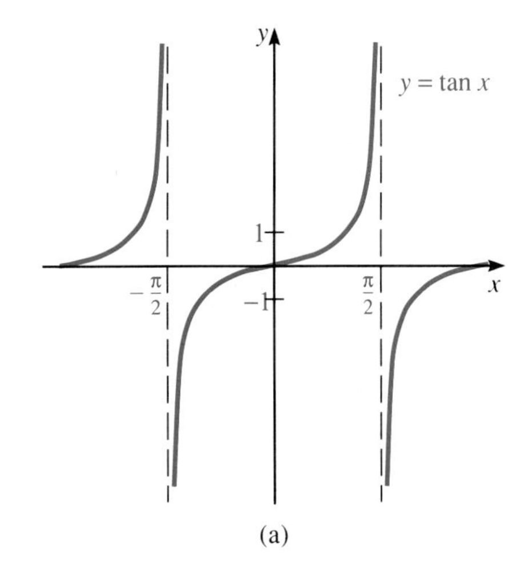

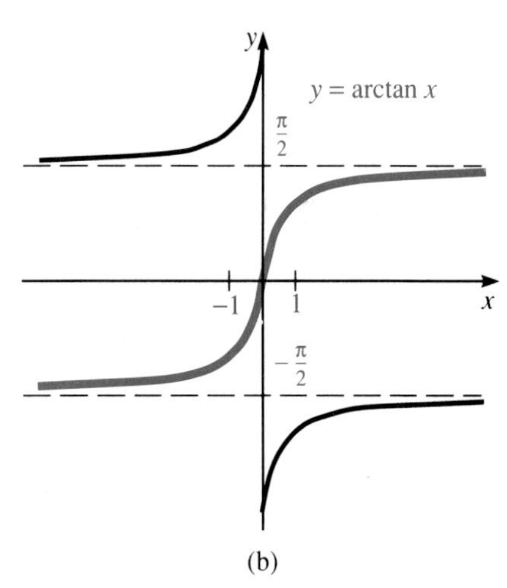

FIGURE 13–6

The graphs of $y = \cos x$ and $y = \arccos x$ are shown in Figure 13–5(a) and (b) respectively, and the graphs of $y = \tan x$ and $y = \arctan x$ are shown in Figure 13–6 (a)and (b) respectively. Notice that in Figures 13–4, 13–5, and 13–6, the graphs for arcsin, arccos, and arctan do not represent a function since a given x can be paired with more than one y coordinate. To define the inverse trigonometric functions, we must restrict the range, or the values, of y. These restrictions are denoted by the color line on the appropriate graph in

Figures 13–4, 13–5, and 13–6. The domain and range for each of the inverse trigonometric functions are summarized in Table 13–1.

TABLE 13–1

Function	Domain	Range	Quadrants
$y = \text{Arcsin } x$	$-1 \leq x \leq 1$	$-\pi/2 \leq y \leq \pi/2$	I, IV
$y = \text{Arccos } x$	$-1 \leq x \leq 1$	$0 \leq y \leq \pi$	I, II
$y = \text{Arctan } x$	all reals	$-\pi/2 < y < \pi/2$	I, IV
$y = \text{Arcsec } x$	$x \leq -1, x \geq 1$	$0 \leq y \leq \pi, y \neq \frac{\pi}{2}$	I, II
$y = \text{Arccsc } x$	$x \leq -1, x \geq 1$	$-\pi/2 \leq y \leq \pi/2, y \neq 0$	I, IV
$y = \text{Arccot } x$	all reals	$0 < y < \pi$	I, II

NOTE ✦ To denote the function form of the trigonometric inverses, we use an initial capital letter, such as Arcsin or Sin^{-1}.

EXAMPLE 2 Find the following:

(a) $y = \text{Arcsin } \frac{\sqrt{3}}{2}$ (b) $y = \text{Arctan } -\frac{\sqrt{3}}{3}$ (c) $y = \text{Arccos}(0.348)$

Solution

(a) The result of an inverse trigonometric function is an angle; therefore, the expression $y = \text{Arcsin } \frac{\sqrt{3}}{2}$ requires finding an angle y whose sine function equals $\frac{\sqrt{3}}{2}$ over the interval $\frac{-\pi}{2}$ to $\frac{\pi}{2}$. We know from previous chapters that the reference angle for y is $\frac{\pi}{3}$. Since the Arcsin function is positive in quadrant I, $y = \frac{\pi}{3}$.

(b) The expression $y = \text{Arctan } -\frac{\sqrt{3}}{3}$ requires finding an angle y whose tangent function equals $-\frac{\sqrt{3}}{3}$. The reference angle for y is $\frac{\pi}{6}$. Since the Arctangent function is negative in quadrant IV, angle y is $\frac{-\pi}{6}$.

(c) The expression $y = \text{Arccos}(0.348)$ requires that we find the angle whose cosine function equals 0.348. We will need to use the calculator because y is not one of the special angles. The calculator keystrokes are

$$0.348 \boxed{\text{inv}} \boxed{\cos} \rightarrow 1.215$$

$$y = 1.215$$

When performing this operation, be sure your calculator is in the radian mode. ■

EXAMPLE 3 Evaluate the following expressions:

(a) $\sin(\text{Arctan } \sqrt{3})$ (b) $\cos(\text{Arcsin } -1/2)$ (c) $\tan(\text{Arcsin } 0.876)$

Solution

(a) If we analyze the expression $\sin(\text{Arctan } \sqrt{3})$, the expression in parentheses, $(\text{Arctan } \sqrt{3})$, results in an angle for which we calculate the sine value.

$$\sin(\text{Arctan } \sqrt{3}) = \sin \frac{\pi}{3}$$

$$\sin(\text{Arctan } \sqrt{3}) = \frac{\sqrt{3}}{2}$$

(b) To evaluate $\cos(\text{Arcsin } -1/2)$, we determine an angle in quadrant I or IV whose sine is $-1/2$. Then we calculate the cosine value of that angle.

$$\cos\left(\text{Arcsin } \frac{-1}{2}\right) = \cos\left(-\frac{\pi}{6}\right)$$

$$\cos\left(\text{Arcsin } \frac{-1}{2}\right) = \frac{\sqrt{3}}{2}$$

(c) Begin by determining an angle whose sine value is 0.876. You will need to use the calculator since this is not one of the special angles. Using this angle, calculate the value of the tangent.

$$\tan(\text{Arcsin } 0.876) = \tan(1.0675)$$

$$\tan(\text{Arcsin } 0.876) = 1.8163$$

$$0.876 \boxed{\text{inv}} \boxed{\sin} \boxed{\tan} \rightarrow 1.8163$$ ■

EXAMPLE 4 Without using a calculator, evaluate the following expressions:

(a) $\tan\left(\text{Arcsin } \frac{3}{5}\right)$ (b) $\cos\left(\text{Arctan } \frac{-8}{15}\right)$

Solution

(a) We must find an angle A whose sine value is 3/5, and then we must calculate its tangent value. It is not necessary to know the actual measure of angle A. Figure 13–7 shows an angle A whose sine value is 3/5. Applying the Pythagorean theorem to find the length of the remaining side gives

$$\begin{aligned} 5^2 &= x^2 + 3^2 \\ 25 - 9 &= x^2 \\ 16 &= x^2 \\ \pm 4 &= x \end{aligned}$$

FIGURE 13–7

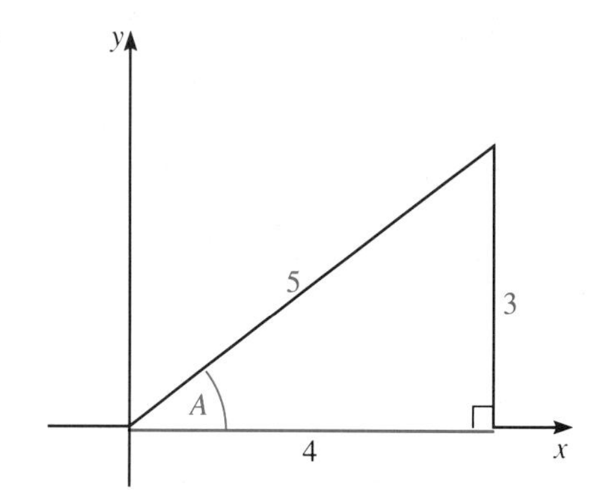

After finding the remaining side, we find that $\tan A = 3/4$. Therefore,

$$\tan\left(\text{Arcsin}\ \frac{3}{5}\right) = \frac{3}{4}$$

(b) Similarly, we can find an angle B whose tangent value is $-8/15$. Since the Arctangent function is negative in quadrant IV, angle B is in quadrant IV as shown in Figure 13–8. Using the Pythagorean theorem, we find the

FIGURE 13–8

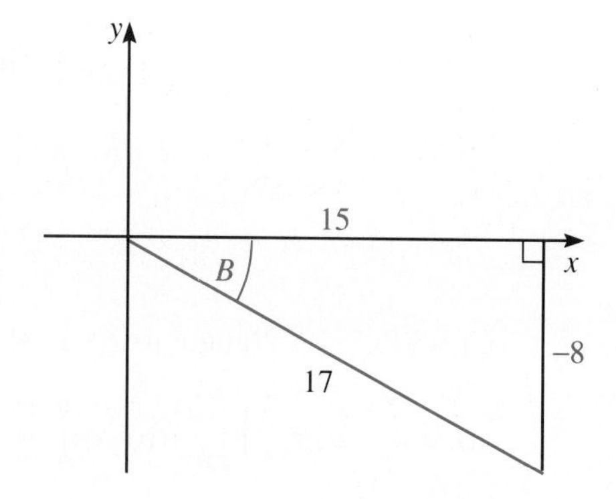

hypotenuse of the triangle to be 17. Then we find the cosine value for angle B to be 15/17.

$$\cos\left(\text{Arctan}\ \frac{-8}{15}\right) = \frac{15}{17}$$

■

EXAMPLE 5 Find an algebraic expression for sin(Arccos $3x$).

Solution First, let A be an angle whose cosine value is $3x = 3x/1$ as shown in Figure 13–9. Using the Pythagorean theorem, we find that the remaining side of the triangle is $\sqrt{1 - 9x^2}$. Then we evaluate the sine value of angle A.

$$\sin(\text{Arccos}\ 3x) = \sqrt{1 - 9x^2}$$

■

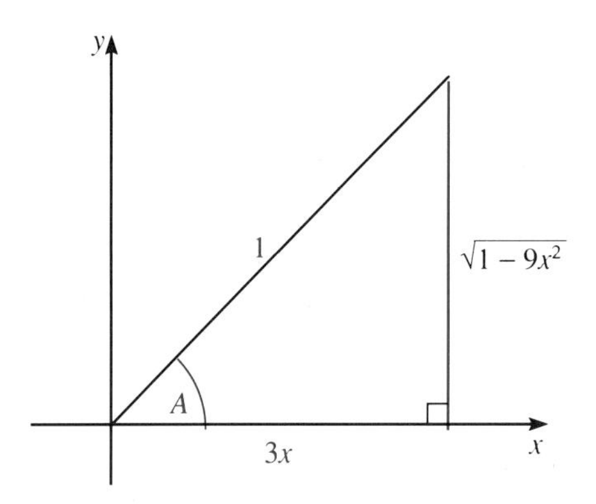

FIGURE 13–9

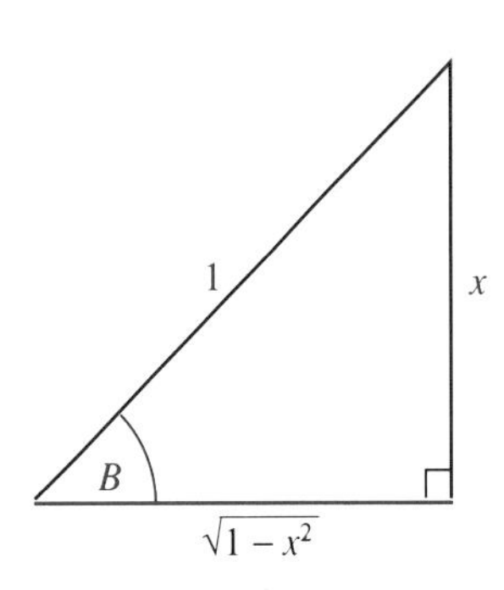

FIGURE 13–10

EXAMPLE 6 Find an algebraic expression for cos(2 Arcsin x).

Solution First, we substitute an angle for Arcsin x. Let B represent an angle whose sine value is x as shown in Figure 13–10). Making the substitution of B for Arcsin x gives

$$\cos(2\ \text{Arcsin}\ x) = \cos 2B$$

Next, we evaluate $\cos 2B$ using the formula $\cos 2B = \cos^2 B - \sin^2 B$. This step requires values for $\sin B$ and $\cos B$. Using the Pythagorean theorem to find the remaining side of the triangle in Figure 13–10, we find $\cos B = \sqrt{1 - x^2}$ and $\sin B = x$. Evaluating the expression gives

$$\cos(2\ \text{Arcsin}\ x) = \cos^2 B - \sin^2 B$$

$$\cos(2\ \text{Arcsin}\ x) = \left(\sqrt{1 - x^2}\right)^2 - (x)^2$$

$$\cos(2\ \text{Arcsin}\ x) = 1 - x^2 - x^2$$

$$\cos(2\ \text{Arcsin}\ x) = 1 - 2x^2$$

■

EXAMPLE 7 Find an algebraic expression for $\sin(\text{Arctan}(-x) + \text{Arccos } 2x)$.

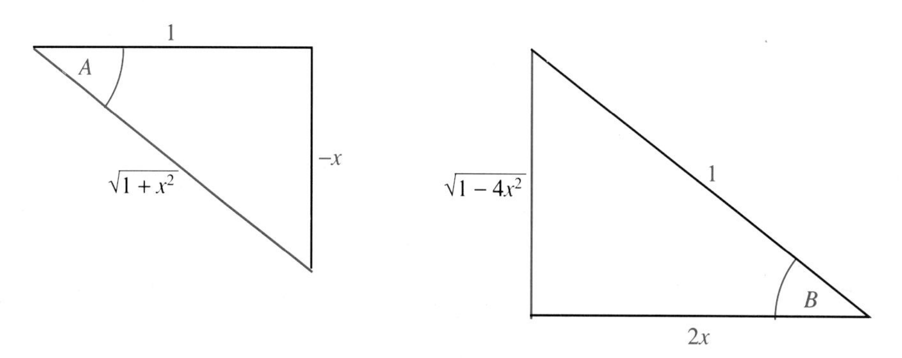

FIGURE 13–11

Solution First, we substitute angle A for $\text{Arctan}(-x)$ and angle B for Arccos $2x$ as shown in Figure 13–11. Substituting angles A and B into the expression gives

$$\sin(\underbrace{\text{Arctan}(-x)}_{A} + \underbrace{\text{Arccos } 2x}_{B}) = \sin(A + B)$$

To evaluate the expression $\sin(A + B)$, we must know the value of $\sin A$, $\sin B$, $\cos A$, and $\cos B$. Using the Pythagorean theorem to determine the remaining side of each of the triangles gives the following function values:

$$\sin A = \frac{-x}{\sqrt{1 + x^2}} \qquad \cos A = \frac{1}{\sqrt{1 + x^2}}$$

$$\sin B = \sqrt{1 - 4x^2} \qquad \cos B = 2x$$

Substituting into the formula for $\sin(A + B)$ gives

$$\underbrace{\left(\frac{-x}{\sqrt{1 + x^2}}\right)}_{\sin A}\underbrace{(2x)}_{\cos B} + \underbrace{\left(\frac{1}{\sqrt{1 + x^2}}\right)}_{\cos A}\underbrace{(\sqrt{1 - 4x^2})}_{\sin B}$$

$$\frac{\sqrt{1 - 4x^2} - 2x^2}{\sqrt{1 + x^2}}$$

■

13–5 EXERCISES

Evaluate the following without using a calculator. Express all answers in radians.

1. $\text{Arcsin}\frac{\sqrt{3}}{2}$
2. Arctan 1
3. $\text{Arccos } \frac{1}{2}$
4. Arccos(− 1)
5. Arctan 0
6. Arcsin 1
7. $\text{Arcsin } \frac{\sqrt{2}}{2}$
8. $\text{Arccos } \frac{\sqrt{3}}{2}$
9. $\text{Arcsin } \frac{1}{2}$
10. $\text{Arctan } \sqrt{3}$
11. Arcsin(− 1)
12. $\text{Arctan } \frac{1}{\sqrt{3}}$

Use a calculator to evaluate the following. Express your answer in radians.

13. Arcsin 0.7483
14. Arctan 2.68
15. Arccos(−0.67)
16. Arctan(− 1.6)
17. Arcsin(−0.8743)
18. sin(Arctan 0.378)
19. cos(Arcsin 0.586)
20. tan(Arccos 0.657)
21. sin(Arcsin 0.5)
22. cos(Arctan 1)
23. $\sin\left(\text{Arccos } \frac{\sqrt{3}}{2}\right)$
24. cos(Arcsin − 1)
25. $\tan\left(\text{Arcsin } \frac{\sqrt{2}}{2}\right)$
26. $\cos(\text{Arctan } \sqrt{3})$
27. tan(Arccos 0)
28. $\cos\left(\text{Arcsin } -\frac{15}{17}\right)$
29. $\tan\left(\text{Arccos } \frac{4}{5}\right)$
30. $\sin\left(\text{Arcsin } \frac{4}{7}\right)$
31. $\sin\left(\text{Arctan } -\frac{12}{5}\right)$
32. $\tan\left(\text{Arccos } \frac{9}{41}\right)$
33. $\sin\left(\text{Arctan } -\frac{15}{8}\right)$

Find an algebraic expression for each of the following.

34. tan(Arcsin x)
35. csc(Arccos $-x$)
36. $\cos\left(\text{Arctan } -\frac{2x}{3}\right)$
37. sin(2 Arccos x)
38. sec(2 Arcsin $2x$)
39. $\tan\left(\frac{1}{2} \text{Arcsin } x\right)$
40. $\sin\left(\frac{1}{2} \text{Arctan } \frac{1}{x}\right)$
41. sin(Arccos x − Arcsin $3x$)
42. $\tan\left(\text{Arcsin } 2x + \text{Arccos } \frac{1}{x}\right)$
43. If the angle between a magnet and its magnetic field is A, the coupling force C is given by $C = HmL \sin A$. Using an inverse trigonometric function, solve for A.
44. The maximum height of a projectile with a velocity v at an angle A above the horizontal is given by

$$h = \frac{v^2 \sin^2 A}{2g}$$

Solve for A.

45. The complete period T of a simple pendulum of period P for an arc length A is given by

$$T = \frac{P}{1 + \frac{1}{4} \sin^2 \frac{A}{4}}$$

Solve for A.

CHAPTER SUMMARY

Summary of Terms

conditional equation (page 468)

identity (p. 448)

similar triangles (p. 455)

Summary of Formulas

Sum or difference of two angles:

$$\sin(A \pm B) = \sin A \cos B \pm \cos A \sin B$$

$$\cos(A \pm B) = \cos A \cos B \mp \sin A \sin B$$

$$\tan(A \pm B) = \frac{\tan A \pm \tan B}{1 \mp \tan A \tan B}$$

Double-angle formulas:

$$\sin 2A = 2 \sin A \cos A$$

$$\cos 2A = \cos^2 A - \sin^2 A$$

$$\cos 2A = 1 - 2\sin^2 A$$

$$\cos 2A = 2\cos^2 A - 1$$

$$\tan 2A = \frac{2 \tan A}{1 - \tan^2 A}$$

Half-angle formulas:

$$\sin \frac{A}{2} = \pm\sqrt{\frac{1 - \cos A}{2}}$$

$$\cos \frac{A}{2} = \pm\sqrt{\frac{1 + \cos A}{2}}$$

$$\tan \frac{A}{2} = \frac{1 - \cos A}{\sin A} = \frac{\sin A}{1 + \cos A}$$

CHAPTER REVIEW

Section 13–1

Prove the following identities.

1. $\dfrac{\sin A}{1 - \cos A} = \dfrac{1 + \cos A}{\sin A}$

2. $\dfrac{\sec x - 1}{1 - \cos x} = \sec x$

3. $(\sin A + \cos A)(\sin A - \cos A) = \sin^2 A - \cos^2 A$

4. $\csc B + \sin B = \dfrac{2 - \cos^2 B}{\sin B}$

5. $\dfrac{1 - \cos^2 B}{1 + \cot^2 B} = \sin^4 B$

6. $\dfrac{\sin x - \tan x}{\sin x \tan x} = \dfrac{\cos x - 1}{\sin x}$

7. $\dfrac{\sin x}{1 - \cos x} - \dfrac{1}{\sin x} = \cot x$

8. $\dfrac{\sec x + \csc x}{1 + \cot x} = \sec x$

9. $\dfrac{\cot A \cos A}{\cot A - \cos A} = \dfrac{\cos A}{1 - \sin A}$

10. $\sec A \csc A - \tan A = \cot A$

11. $\dfrac{\tan x}{\sec x + 1} = \dfrac{\sec x - 1}{\tan x}$

12. $1 - \cot B \sin B \cos B = \sin^2 B$

Section 13–2

13. Given that $\tan A = -12/5$ and A is in quadrant II and that $\cos B = 9/41$ and B is in quadrant I, find the following:
(a) $\cos(A - B)$ **(b)** $\tan(A + B)$ **(c)** $\sin(A - B)$
(d) quadrant of $(A - B)$

14. Given that $\sin A = 8/17$, $\cos B = 12/13$, and both A and B are in quadrant I, find the following:
(a) $\sin(A + B)$ **(b)** $\cos(A - B)$ **(c)** $\tan(A + B)$
(d) quadrant of $(A + B)$

Reduce each expression to a single term.

15. $\sin(C - D) \cos B + \cos(C - D) \sin B$

16. $\cos 3A \cos A - \sin 3A \sin A$

17. $\dfrac{\tan 3x + \tan 2x}{1 - \tan 3x \tan 2x}$

18. $\sin 56° \cos 10° - \cos 56° \sin 10°$

Verify the following identities.

19. $\sin\left(\frac{\pi}{2} - x\right) = \cos x$

20. $\tan\left(x + \frac{\pi}{4}\right) = \dfrac{\tan x + 1}{1 - \tan x}$

21. $\sin\left(x + \frac{\pi}{2}\right) - \cos\left(x + \frac{\pi}{2}\right) = \cos x + \sin x$

22. $\dfrac{\sin(x - y)}{\sin x \cos y} = 1 - \cot x \tan y$

23. $\dfrac{\sin A}{\sin 3A} + \dfrac{\cos A}{\cos 3A} = \dfrac{\sin 4A}{\sin 3A \cos 3A}$

24. $\dfrac{\sin(A + B)}{\cos(A - B)} = \dfrac{\tan A + \tan B}{1 + \tan A \tan B}$

25. $\sin(A - \pi) + \cos(A + \pi) = -\sin A - \cos A$

Section 13–3

Write each expression as a single term.

26. $\sqrt{\dfrac{1 - \cos 9A}{2}}$

27. $8 - 16 \sin^2 B$

28. $10 \sin 6A \cos 6A$

29. $\dfrac{1 - \cos 5B}{\sin 5B}$

30. If $\sin A = 40/41$ and A is in quadrant II, find $\tan \frac{A}{2}$ and $\cos 2A$.

31. If $\tan B = -8/15$ and $90° < B < 180°$, find $\sin \frac{B}{2}$ and $\cos 2B$.

Verify the following identities.

32. $\sin \frac{A}{2} \cos \frac{A}{2} = \frac{\sin A}{2}$

33. $\sin 4x = 4 \sin x \cos x \cos 2x$

34. $\dfrac{\csc^2 x}{\csc 2x} = 2 \cot x$

35. $\sin x \sin 2x = \cos x(1 - \cos 2x)$

36. $1 - 8 \sin^2 x \cos^2 x = \cos 4x$

37. $\sin 2x = \tan x(1 + \cos 2x)$

38. $\tan\left(x + \frac{x}{2}\right) = \dfrac{\sin 2x + \sin x}{\cos 2x + \cos x}$

Section 13–4

Solve the following equations for $0 \le x < 2\pi$. Use a calculator only when x is not one of the special angles.

39. $\sqrt{3} \sin x = 1$

40. $\tan x = -1$

41. $6 \cos^2 x + \cos x = 2$

42. $\cos \frac{x}{2} = \cos x + 1$

43. $\tan^2 x - \tan x = 0$

44. $4 \sin^2 x = 3 - 8 \cos x$

45. $\sin^2 x + 5 \sin x = 6$

46. $\sin 3x \cos x - \cos 3x \sin x = 0$

47. $5 \cos^2 x - \sin^2 x = 0$

48. $3 \cos 2x + 2 = 0$

Section 13–5

Evaluate the following without using a calculator.

49. $\sin\left(\text{Arccos} - \frac{15}{17}\right)$

50. $\cos\left(\text{Arctan} \frac{4}{5}\right)$

51. $\tan\left(\text{Arcsin} - \frac{40}{41}\right)$

52. $\sec\left(\text{Arcsin} - \frac{12}{13}\right)$

Give an algebraic expression for each of the following.

53. $\sin(\text{Arctan } x)$

54. $\csc(\text{Arcsin } - 2x)$

55. $\sin(\pi - \text{Arccos } x)$

56. $\cos\left(\text{Arcsec } \frac{2}{x}\right)$

57. $\cos\left(\frac{1}{2} \text{Arcsin } x\right)$

58. $\sin(2 \text{ Arctan } x)$

CHAPTER TEST

The number in parentheses refers to the appropriate learning objective given at the beginning of the chapter.

1. If $\cos A = -9/41$ and $90° < A < 180°$, find $\tan A/2$ and $\sin 2A$. (3)

2. Solve the following equation for $0 \le x < 2\pi$: (4, 5)

$$5 \cos 2x + 3 = 0$$

3. Prove the identity (1)

$$\frac{1 + \cos^2 A}{\sin^2 A} + 1 = 2 \csc^2 A$$

4. Find an algebraic expression for the following: (6)

$$\cos\left(\frac{1}{2} \text{Arcsin } 3x\right)$$

5. Reduce the following to an expression containing a single term: (2)

$$\cos(A + B)\cos(C - D) - \sin(A + B)\sin(C - D)$$

6. Verify the following identity: (2)

$$\cos(x - y)\cos(x + y) = \cos^2 x - \sin^2 y$$

7. Evaluate the following without using a calculator. Express the answer in radians. (6)

$$\sec\left(\text{Arcsin } - \frac{12}{13}\right)$$

8. Prove the following identity: (1)

$$\frac{1 + \cot^2 x}{\cot^2 x} = \sec^2 x$$

9. Solve the following equation for $0 \le y < 2\pi$: (5)

$$6 \sin^2 A - \sin A = 1$$

10. Verify the identity (3)

$$\csc^2 \frac{A}{2} = \frac{2 \sec A}{\sec A - 1}$$

11. The angle A of a projectile shot at 25 m/s with a range of 625 m is given by $625 \tan A = 306.25 \sec^2 A$. Solve for A. (5)

12. Reduce the following to a single term: (4)

$$\sin \frac{A}{4} \cos \frac{A}{4}$$

13. Verify the identity (1)

$$\frac{\cot x \cos x}{\cot x - \cos x} = \frac{\cos x}{1 - \sin x}$$

14. Verify the following identity: (3)

$$\frac{\cot^2 x - 1}{2 \cot x} = \cot 2x$$

15. A spring vibrating in harmonic motion is subjected to two forces. The motion of each is described as (2)

$$y_1 = A_1 \sin\left(x + \frac{\pi}{4}\right) \quad \text{and} \quad y_2 = A_2 \cos\left(x + \frac{\pi}{4}\right)$$

Find a simplified expression for the resultant, $y_1 + y_2$.

SOLUTION TO THE CHAPTER INTRODUCTION

To find the angle where the displacement is zero, we must solve the equation $4 \sin A + \sin 2A = 0$ for A as follows:

$$\begin{aligned} 4 \sin A + \sin 2A &= 0 \\ 4 \sin A + 2 \sin A \cos A &= 0 \\ 2 \sin A(2 + \cos A) &= 0 \end{aligned}$$

$$2 \sin A = 0 \qquad \cos A = -2$$

$$\sin A = 0 \qquad \text{no solution}$$

$$A = 0 \text{ and } \pi$$

At 0 and π, the displacement of the piston is zero.

You are an electrical engineer responsible for designing a tuning circuit for a radio station. The circuit is to contain a variable capacitor, a 0.0020-H coil, and a 20-Ω resistor. You must calculate the capacitance necessary to tune the station to a broadcast frequency of 101 MHz. (The solution to this problem is given at the end of the chapter.)

Complex numbers, discussed in this chapter, apply to electronics. For example, one of these applications is resonant frequency as it relates to electrical systems. However, the concept of resonance also relates to mechanical systems. If mechanical impulses are applied to a system at the proper frequency, the system will resonate or vibrate. A historical note of such an occurrence is the collapse of the Tacoma Narrows Bridge in 1940 due to oscillations caused by a 42-mi/h wind. The height of the oscillations or vibrations increased to the point that the main span of the bridge broke apart and fell into Puget Sound.

Learning Objectives

After completing this chapter, you should be able to

1. Add, subtract, multiply, and divide imaginary numbers (Section 14–1).
2. Simplify powers of the imaginary number j (Section 14–1).
3. Add, subtract, multiply, and divide complex numbers in rectangular form (Section 14–2).
4. Convert complex numbers among rectangular, polar, and exponential forms (Section 14–3).
5. Use polar and exponential forms to find the product, quotient, power, and roots of complex numbers (Section 14–4).
6. Apply the concept of complex numbers to ac circuits (Section 14–5).

Chapter 14

Complex Numbers

14–1

INTRODUCTION TO COMPLEX NUMBERS

We briefly discussed complex numbers in Chapter 4 when we took the square root of a negative number. We also used complex numbers in solving quadratic equations in Chapter 5. In this chapter we will discuss operations with complex numbers in rectangular, polar, and exponential forms and applications of complex numbers to electronics.

A **complex number** is *any number that can be written as* $a + bj$ where a and b represent real numbers. The **real part** of the complex number is *represented by the letter* a, and the **imaginary part** is *represented by* bj. A number *consisting only of the imaginary part* is called a **pure imaginary number.** A complex number *written as* $a + bj$ is said to be in **rectangular form.**

Operations with Imaginary Numbers

We perform algebraic operations with imaginary numbers much the same way as in previous chapters with algebraic expressions.

EXAMPLE 1 Perform the following indicated operation:

$$(\sqrt{-6})^2$$

Solution Before squaring this number, we must write the expression in terms of j.

$$\sqrt{-6} = \sqrt{-1 \cdot 6} = \sqrt{-1} \cdot \sqrt{6} = j\sqrt{6}$$

Therefore, this expression can be written as follows:

$$(j\sqrt{6})^2$$

NOTE ✦ The j is written in front of the radical to avoid confusion as to whether or not j is under the radical.

Once we have the expression in terms of j, we square it just as we have done previously.

$$j^2(6)$$

However, $j^2 = -1$. Therefore,

$$(\sqrt{-6})^2 = -6$$ ■

CAUTION ✦ If a number is expressed as the square root of a negative, always write the number in terms of j before performing any algebraic operation.

EXAMPLE 2 Multiply the following complex numbers:

$$\sqrt{-2} \cdot \sqrt{-32}$$

Solution Before multiplying the two radicals, we must express each negative number under the radical in terms of j.

$$\sqrt{-2} \cdot \sqrt{-32}$$
$$j\sqrt{2} \cdot j\sqrt{32}$$

Then we multiply and simplify the result.

$$j^2\sqrt{64}$$
$$(-1)(8)$$
$$-8$$

■

EXAMPLE 3 Write $\sqrt{-27}$ in simplified form.

Solution First, let us write the radical in terms of j as follows:

$$j\sqrt{27}$$

Second, we simplify the radical by removing $\sqrt{9} = 3$. Thus, the simplified form is

$$3j\sqrt{3}$$

■

Powers of j

Many problems involve powers of j greater than two, making simplification of higher powers of j necessary. Powers of j are cyclic, meaning they are periodic. Look at the following powers of j and note the pattern of the answers:

$$j^1 = j \qquad j^5 = j^4 \cdot j = j$$
$$j^2 = -1 \qquad j^6 = j^4 \cdot j^2 = -1$$
$$j^3 = j^2 \cdot j = -j \qquad j^7 = j^6 \cdot j = -j$$
$$j^4 = j^2 \cdot j^2 = 1 \qquad j^8 = j^4 \cdot j^4 = 1$$

Notice that the powers of j repeat in the sequence j, -1, $-j$, and 1.

Simplifying Powers of j

To simplify a power of j, perform the following steps:

1. Because powers of j have a cycle of 4, divide the exponent by 4.
2. Pair the remainder from Step 1 with the corresponding exponent below, and find the resulting value.

$$j^0 = 1$$
$$j^1 = j$$
$$j^2 = -1$$
$$j^3 = -j$$

CAUTION ✦ In simplifying any expression involving j, never leave a power of j greater than one.

EXAMPLE 4 Simplify j^{43}.

Solution We *divide 43 by 4* and obtain a remainder of 3. The preceding box tells us that $j^3 = -j$. Therefore,

$$j^{43} = -j$$ ■

EXAMPLE 5 Simplify $6j^{24} - 5j^{18}$.

Solution First, we simplify the powers of j, and then we subtract.

$$6j^{24} - 5j^{18}$$
$$6j^{0} - 5j^{2}$$
$$6(1) - 5(-1)$$
$$6 + 5$$
$$11$$ ■

Equality of Complex Numbers

Two complex numbers are equal if their real parts are equal and their imaginary parts are equal. For example, $3 + 8j$ is equal to $x + yj$ only if $x = 3$ and $y = 8$. We can use this principle of equality to solve equations with complex numbers by equating real parts and by equating imaginary parts. The next two examples illustrate this process.

EXAMPLE 6 Determine x and y such that $2x + 5yj = 10 + 15j$.

Solution According to the definition of equality, these two complex numbers are equal if

imaginary parts are equal

$$2x + 5yj = 10 + 15j$$

real parts are equal

Therefore,

$$2x = 10 \quad \text{and} \quad 5yj = 15j$$

Thus,

$$x = 5 \quad \text{and} \quad 5y = 15$$
$$y = 3$$ ■

EXAMPLE 7 Solve the following for x and y:

$$3 - 5j + x = 2(x + 2yj) + 3j$$

Solution To determine the value of x and y, we must be able to equate the real parts and the imaginary parts of these complex numbers. Therefore, we must be able to separate the two complex numbers into their real and imaginary parts. Rearranging the complex numbers gives

$$(3 + x) - 5j = 2x + (4yj + 3)j$$

real parts

$$\overbrace{(3 + x)} - 5j = 2x + \underbrace{(4y + 3)j}$$

imaginary parts

Equating real parts and imaginary parts and solving for x and y gives

$$3 + x = 2x \quad \text{and} \quad -5j = (4y + 3)j$$

$$3 = x \qquad\qquad -5 = 4y + 3$$

$$-5 - 3 = 4y$$

$$-2 = y$$

■

Complex Conjugates

The **conjugate** of the complex number $a + bj$ is the complex number $a - bj$. To determine the complex conjugate of a number, *reverse the sign of the imaginary part*.

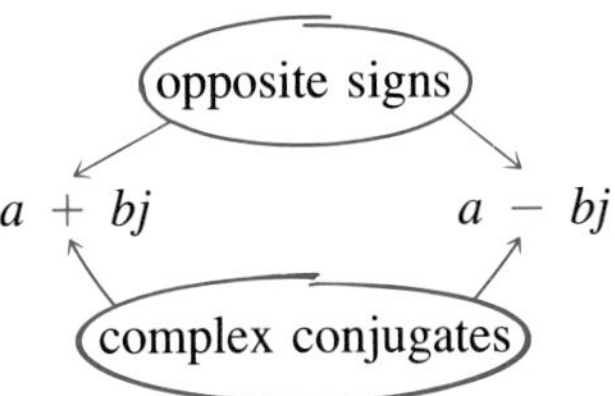

EXAMPLE 8 Determine the complex conjugate of the following complex numbers:

(a) $3 - 7j$ (b) $-5 + 11j$ (c) $8j$ (d) -5

Solution

(a) We obtain the complex conjugate of $3 - 7j$ by reversing the sign of the imaginary part: $3 + 7j$.
(b) To determine the complex conjugate of $-5 + 11j$, we reverse the sign of the imaginary part to obtain $-5 - 11j$.
(c) To find the complex conjugate of $8j$, first, we write the number in $a + bj$ form $(0 + 8j)$ and then reverse the sign of the imaginary part: $0 - 8j$. Thus, the complex conjugate of $8j$ is $-8j$.

(d) Similarly, if we write -5 in $a + bj$ form, we have $-5 + 0j$. Since the imaginary part is zero, reversing its sign does not affect the number. Therefore, the complex conjugate of -5 is -5. ■

CAUTION ✦ Be sure to reverse only the sign of the imaginary part to obtain a complex conjugate.

14–1 EXERCISES

Write each expression in terms of j.

1. $\sqrt{-9}$
2. $\sqrt{-36}$
3. $\sqrt{-8}$
4. $\sqrt{-24}$
5. $-\sqrt{-0.09}$
6. $-\sqrt{-0.16}$
7. $-\sqrt{-40}$
8. $\sqrt{-16/25}$
9. $\sqrt{-8/49}$

Perform the indicated operation and simplify.

10. $\sqrt{-8} \cdot \sqrt{-4}$
11. $\sqrt{-2} \cdot \sqrt{-6}$
12. $\sqrt{-5} \cdot \sqrt{-8}$
13. $\sqrt{-9} \cdot \sqrt{6}$
14. $\sqrt{-20} \cdot \sqrt{5}$
15. $(3\sqrt{-6})^2$
16. $(7\sqrt{-3})^2$
17. $j^9 + 6\sqrt{-25}$

Simplify the following powers of j.

18. j^{15}
19. j^{36}
20. j^{49}
21. j^{127}
22. j^{86}
23. j^{74}
24. j^{264}
25. j^{471}
26. $j^{14} + 3j^{39}$
27. $7j^{56} - j^{34}$
28. $6j^{60} - 7j^{82}$
29. $8j^{59} + 5j^{97}$

Solve for x and y.

30. $2x + 3yj = 10 + 18j$
31. $7 + 6yj = 14x - 42j$
32. $4j - 10 = 2x + yj - 8$
33. $10 + 3yj + 8j = 2(x + 11j)$
34. $3(x + 5j) = 7yj - 6 + 8yj$
35. $19 - 8yj + 2x = 15 - 2yj + 18j$
36. $x + 7yj - 4 = 5 + 21j + 2x$
37. $2x + 10j - 4yj = 8j - 16 + 2yj$
38. $4(x - yj) = 8 + 2yj + 10j$
39. $7x - 5yj + 18j = 16 - 3yj - 4j + 9x$

Determine the conjugate of the following complex numbers.

40. $2 + 7j$
41. $8 - 5j$
42. $-5 - 11j$
43. $-7 + 9j$
44. $-3j$
45. $-2j$
46. 8
47. -10

48. What type of number results from (a) the sum and (b) the product of a complex number and its conjugate?

E 49. The impedance Z in a resistance-inductance-capacitance (RLC) series ac circuit is given by

$$Z = R + jX_L - jX_C$$

Determine an expression for the impedance if $R = 44\ \Omega$, $X_L = 36\ \Omega$, and $X_C = 50\ \Omega$.

14–2

OPERATIONS IN RECTANGULAR FORM

In this section we will discuss addition, subtraction, multiplication, and division of complex numbers in rectangular form, $a + bj$. Remember that you must write the imaginary part of a complex number in terms of j before performing any operation. Also, remember that the answer should not contain any power of j greater than one.

Addition and Subtraction

To add or subtract complex numbers, combine the real parts and combine the imaginary parts.

EXAMPLE 1 Perform the following indicated operations:

(a) $(-3 + 5j) + (6 - 7j)$ (b) $(-10 + \sqrt{-64}) - (-3 - \sqrt{-49})$

Solution

(a) To add these complex numbers, we combine the real parts, -3 and 6, and we combine the imaginary parts, $5j$ and $-7j$.

$$(-3 + 5j) + (6 - 7j) = \underbrace{(-3 + 6)}_{\text{real parts}} + \underbrace{(5j - 7j)}_{\text{imaginary parts}}$$
$$= 3 - 2j$$

(b) First, we must change the complex numbers to $a + bj$ form:$(-10 + 8j) - (-3 - 7j)$. Subtracting the two complex numbers gives

$$(-10 + 8j) - (-3 - 7j) = -10 + 8j + 3 + 7j$$
$$= (-10 + 3) + (8j + 7j)$$
$$= -7 + 15j$$

■

Graphical Solution

Addition and subtraction of complex numbers in rectangular form can also be done graphically. Complex numbers are graphed in the **complex plane** where the horizontal axis represents the real part of the complex number and the vertical axis represents the imaginary part. The complex number $8 - 6j$ is represented by the ordered pair $(8, -6)$. The graph of $8 - 6j$ is shown in Figure 14–1.

FIGURE 14–1

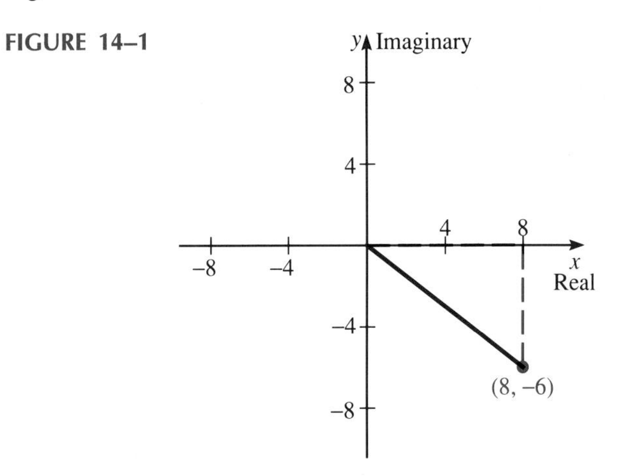

Graphical addition and subtraction of complex numbers involve the same technique as graphically finding the resultant of vectors shown in Chapter 11. The next two examples illustrate the process.

EXAMPLE 2 Find $(-2 + 3j) + (5 + 4j)$ graphically.

Solution First, we graphically represent the complex numbers $-2 + 3j$ and $5 + 4j$. Then we complete a parallelogram and read the ordered pair for the resultant as shown in Figure 14–2. The ordered pair for the resultant is (3, 7) which is equivalent to $3 + 7j$. You can check this result by adding the two complex numbers algebraically.

$$(-2 + 3j) + (5 + 4j) = (-2 + 5) + (3 + 4)j = 3 + 7j$$

FIGURE 14–2

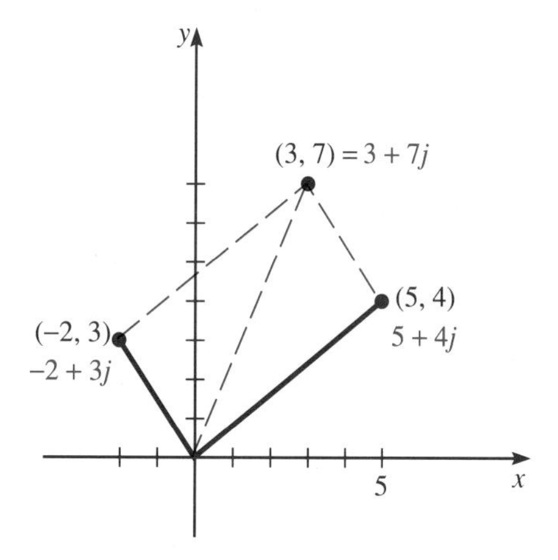

■

EXAMPLE 3 Subtract $6 - 5j$ from $4 + 9j$ graphically.

Solution Subtracting $6 - 5j$ is equivalent to adding $-6 + 5j$. Following the same procedure as Example 2, we add $4 + 9j$ and $-6 + 5j$. The ordered pair for the difference is $(-2, 14)$ which is equivalent to the complex number $-2 + 14j$ as shown in Figure 14–3.

FIGURE 14–3

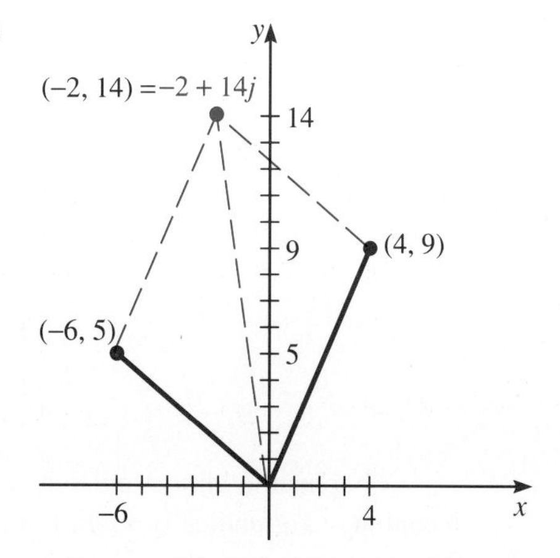

■

Multiplication

After converting to $a + bj$ form, we multiply complex numbers using the same techniques used to multiply polynomials. Then we simplify powers of j and combine like terms.

EXAMPLE 4 Multiply the following:

(a) $(3 + 5j)(8 - 7j)$ (b) $(2 + 3\sqrt{-3})(5 - \sqrt{-3})$

Solution

(a) Since the numbers are written in $a + bj$ form, we *multiply the binomials.*

$$(3 + 5j)(8 - 7j) = 24 - 21j + 40j - 35j^2$$

Making the substitution $j^2 = -1$ and *combining like terms* gives

$$\begin{aligned}(3 + 5j)(8 - 7j) &= 24 - 21j + 40j - 35(-1)\\ &= 24 - 21j + 40j + 35\\ &= 59 + 19j\end{aligned}$$

(b) First, we must convert the complex numbers to $a + bj$ form.

$$(2 + 3\sqrt{-3})(5 - \sqrt{-3}) = (2 + 3j\sqrt{3})(5 - j\sqrt{3})$$

Next, we multiply.

$$(2 + 3j\sqrt{3})(5 - j\sqrt{3}) = 10 - 2j\sqrt{3} + 15j\sqrt{3} - 3j^2\sqrt{9}$$

Last, we substitute $j^2 = -1$, simplify radicals, and combine like terms.

$$\begin{aligned}(2 + 3j\sqrt{3})(5 - j\sqrt{3}) &= 10 - 2j\sqrt{3} + 15j\sqrt{3} - 3(-1)(3)\\ &= 19 + 13j\sqrt{3}\end{aligned}$$

■

Division

Dividing complex numbers is very similar to dividing radical expressions (Chapter 4) because j is a radical. Divide complex numbers by rationalizing the denominator. If you remember that $j = \sqrt{-1}$, you can understand why rationalizing the denominator is necessary.

EXAMPLE 5 Divide the following complex numbers:

(a) $\dfrac{4 + \sqrt{-6}}{\sqrt{-1}}$ (b) $\dfrac{3 - 2j}{5 + 7j}$

Solution

(a) First, we *convert the complex number to $a + bj$ form.*

$$\frac{4 + \sqrt{-6}}{\sqrt{-1}} = \frac{4 + j\sqrt{6}}{j}$$

Then we *rationalize the denominator* by multiplying the numerator and denominator by the conjugate of j, which is $-j$.

$$\frac{4 + \sqrt{-6}}{\sqrt{-1}} = \frac{4 + j\sqrt{6}}{j} \cdot \frac{-j}{-j}$$
$$= \frac{-4j - j^2\sqrt{6}}{-j^2}$$

Finally, we substitute $j^2 = -1$ and *simplify* the result.

$$\frac{4 + \sqrt{-6}}{\sqrt{-1}} = \frac{-4j - (-1)\sqrt{6}}{-(-1)}$$
$$= \frac{\sqrt{6} - 4j}{1}$$
$$= \sqrt{6} - 4j$$

(b) Since this complex number is given in $a + bj$ form, we multiply the numerator and denominator by the conjugate of $5 + 7j$, which is $5 - 7j$.

$$\frac{3 - 2j}{5 + 7j} = \frac{(3 - 2j)}{(5 + 7j)} \cdot \frac{(5 - 7j)}{(5 - 7j)}$$

Multiplying gives

$$\frac{3 - 2j}{5 + 7j} = \frac{15 - 21j - 10j + 14j^2}{25 - 49j^2}$$

Substituting $j^2 = -1$ and combining like terms gives

$$\frac{3 - 2j}{5 + 7j} = \frac{15 - 21j - 10j + 14(-1)}{25 - 49(-1)}$$
$$= \frac{1 - 31j}{74}$$

■

14–2 EXERCISES

Perform the indicated operations. Leave your answer in $a + bj$ form.

1. $(7 - 5j) + (-4 - 2j)$
2. $(8 - 16j) + (7 + 3j)$
3. $(-8 + 11j) - (6 - 5j)$
4. $(14 + 8j) + (17 - 9j)$
5. $(9j + 2) + (4 - j)$
6. $(-12 + j) - (19j + 6)$
7. $(3 + 6j) - (8j - 5) + (-2 + 3j)$
8. $(2 + j) + (7j - 8) - (16 + 5j)$
9. $(1 - 4j) + (-9 + 4j) - (15 - 9j)$
10. $(j - 2) + (-5 + 7j) + (8j - 13)$
11. $(\sqrt{-16} - 6) + (-8 - \sqrt{-9})$
12. $(\sqrt{-4} + 8) - (16 + \sqrt{-25})$
13. $(-5 + \sqrt{-36}) - (3 + \sqrt{-9})$
14. $(-9 + \sqrt{-49}) + (16 - \sqrt{-100})$
15. $(4 + \sqrt{-49}) + (18 - \sqrt{-81})$
16. $(18 - \sqrt{-36}) - (\sqrt{-64} + 10)$
17. $\sqrt{-9}(\sqrt{-4} + \sqrt{-81})$
18. $\sqrt{-64}(\sqrt{25} - \sqrt{-81})$
19. $5j(3 - 6j)$
20. $-3j(7 + 10j)$
21. $(3 - \sqrt{-4})(9 + \sqrt{-16})$
22. $(4 + \sqrt{-1})(\sqrt{-9} + 6)$
23. $(\sqrt{-25} + 11)(3 + \sqrt{-64})$
24. $(\sqrt{-16} + 5)(9 - \sqrt{-25})$
25. $(8 - 3j)(7 + 4j)$
26. $(8 - 7j)(4 - 3j)$
27. $(2 + 5j)^2$
28. $(7 - 3j)^2$
29. $(7 - 2j)(7 + 2j)$
30. $(8 - 5j)(8 + 5j)$
31. $(-6 - 3j)^2$
32. $(-7 + 9j)^2$
33. $(2 - 3j)^3$
34. $(1 - 2j)^3$
35. $\dfrac{6 + 7j}{2j}$
36. $\dfrac{5 - 6j}{4j}$
37. $\dfrac{3 - \sqrt{-4}}{\sqrt{-9}}$
38. $\dfrac{\sqrt{-9} + 6}{\sqrt{-16}}$
39. $\dfrac{2 + \sqrt{-36}}{\sqrt{-25}}$
40. $\dfrac{9 + \sqrt{-25}}{\sqrt{-4}}$
41. $\dfrac{8 - 5j}{j}$
42. $\dfrac{9 + 11j}{2j}$
43. $\dfrac{2 - 7j}{1 - j}$
44. $\dfrac{6 - 10j}{j + 5}$
45. $\dfrac{-3 + 9j}{2 - 5j}$
46. $\dfrac{-8 + 6j}{3j - 10}$
47. $\dfrac{9 - j}{2j - 3}$
48. $\dfrac{9j - 5}{4j - 1}$
49. $\dfrac{-8 + 5j}{6 - j}$
50. $\dfrac{1 + j}{1 - j}$

Perform the indicated operations graphically.

51. $(-3 + 2j) + (-5j - 8)$
52. $(3 + 5j) + (-6 + 4j)$
53. $(4j - 3) - (7 + 2j)$
54. $(7 - 8j) - (-4 - 3j)$
55. $(5 + 6j) + (5 - 6j)$
56. $(-2 + 5j) + (-2 - 5j)$

14–3

POLAR FORM AND EXPONENTIAL FORM

In the previous two sections, we discussed complex numbers written in rectangular form: $a + bj$. However, in this section we will discuss two additional forms of complex numbers: polar and exponential. Certain operations, such as finding roots, are performed more easily when the complex number is written in exponential or polar form.

Polar Form

In the previous section we briefly discussed graphing complex numbers. If we take the complex number $x + yj$, plot it on the complex plane, and drop a perpendicular line to the real axis, we obtain a right triangle as shown in Figure 14–4. Recall that this procedure is the same one used in Chapter 11 for vectors. By forming the right triangle, we have the hypotenuse of the triangle, labeled r, and an angle θ in standard position. From our previous knowledge of trigonometry, we know that

$$x = r\cos\theta \qquad \tan\theta = \frac{y}{x}$$

$$y = r\sin\theta \qquad r = \sqrt{x^2 + y^2}$$

FIGURE 14–4

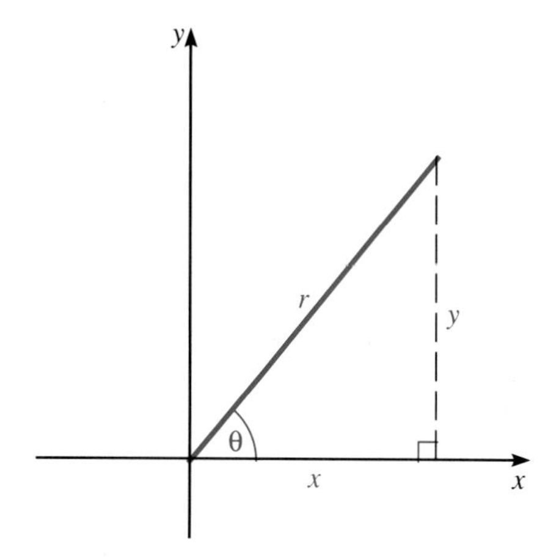

If we take the rectangular form of the complex number, substitute for x and y, and simplify, we obtain the polar form of the complex number.

$$\begin{aligned} x + yj &= (r\cos\theta) + (r\sin\theta)j && \text{substitute} \\ &= r(\cos\theta + j\sin\theta) && \text{factor} \end{aligned}$$

NOTE ✦ The j is placed in front of $\sin\theta$ so as not to confuse what is multiplied. It is j times $\sin\theta$, not sine of the product of θ and j.

Polar Form of a Complex Number

The polar form of the complex number $x + yj$ is

$$r(\cos\theta + j\sin\theta)$$

where

$$r = \sqrt{x^2 + y^2} \qquad x = r\cos\theta$$

$$\tan\theta = \frac{y}{x} \qquad y = r\sin\theta$$

Other notations for polar form are $r\angle\theta$ and r cis θ. The length r is called the **magnitude** of the complex number, and the angle θ is called the **argument.**

EXAMPLE 1 Represent $6 + 7j$ graphically and determine its polar form.

Solution We plot the ordered pair (6, 7) on the complex plane and draw a vector from the origin to the point as shown in Figure 14–5. To convert $6 + 7j$ to polar form, we must determine r and θ.

$$r = \sqrt{x^2 + y^2} \qquad \tan\theta = \frac{y}{x}$$

$$r = \sqrt{36 + 49} \qquad \tan\theta = \frac{7}{6}$$

$$r \approx 9.22 \qquad \theta \approx 49.4^\circ$$

The polar form of $6 + 7j$ is $9.22(\cos 49.4^\circ + j \sin 49.4^\circ)$.

6 [x^2] [+] 7 [x^2] [=] [$\sqrt{\ }$] → 9.22

7 [÷] 6 [=] [inv] [tan] → 49.4 ■

NOTE ✦ Be sure that your calculator is in the degree mode.

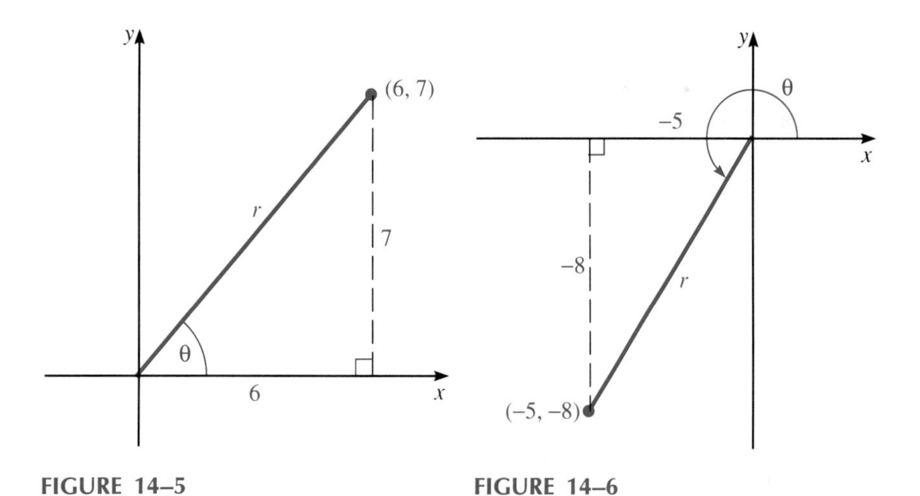

FIGURE 14–5

FIGURE 14–6

EXAMPLE 2 Represent $-5 - 8j$ graphically and find its polar form.

Solution To represent $-5 - 8j$ graphically, plot the ordered pair $(-5, -8)$ and draw a vector from the origin to the point as shown in Figure 14–6. To find its polar form, we must determine r and θ. We can find r just as we did in Example 1; however, we should find a reference angle A to determine θ because it is not a first quadrant angle.

$$r = \sqrt{(-5)^2 + (-8)^2} \qquad \tan A = \frac{|-8|}{|-5|}$$
$$r = \sqrt{89} \qquad A \approx 57.99°$$
$$r \approx 9.43 \qquad \theta = 180° + A = 237.99°$$

The polar form of $-5 - 8j$ is $9.43(\cos 237.99° + j \sin 237.99°)$, or $9.43\angle 237.99°$.

5 [+/−] [x^2] [+] 8 [+/−] [x^2] [=] [√] → 9.43

8 [÷] 5 [=] [inv] [tan] 57.99 [+] 180 [=] → 237.99 ■

CAUTION ✦ Do not rely on your calculator to determine the value of θ. Calculate the reference angle and then determine θ according to the quadrant in which the complex number lies.

EXAMPLE 3 Represent $7.8(\cos 136.8° + j \sin 136.8°)$ graphically and express the complex number in rectangular form.

Solution The graph of $7.8(\cos 136.8° + j \sin 136.8°)$ is given in Figure 14–7. To convert this complex number from polar to rectangular form, we must determine x and y. We know that $r = 7.8$ and $\theta = 136.8°$. Thus,

$$x = r\cos\theta \quad \text{and} \quad y = r\sin\theta$$
$$x = 7.8\cos 136.8° \qquad y = 7.8\sin 136.8°$$
$$x \approx -5.7 \qquad y \approx 5.3$$

7.8 [×] 136.8 [cos] [=] → −5.69

7.8 [×] 136.8 [sin] [=] → 5.34

FIGURE 14–7

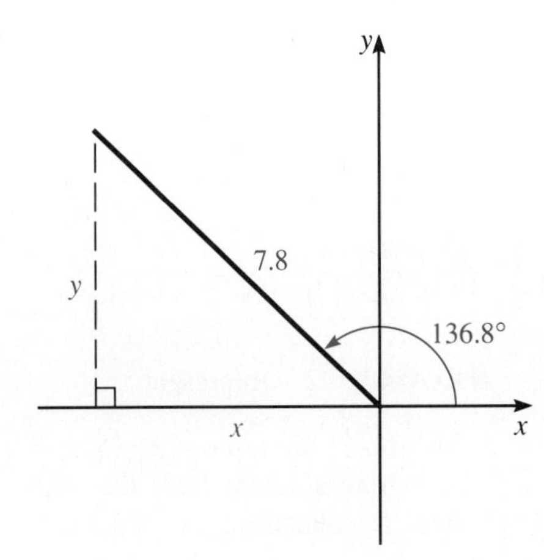

The rectangular form $x + yj$ is $-5.69 + 5.34j$. From Figure 14–7 we can verify that this answer is reasonable because the point $(-5.69, 5.34)$ would be placed in quadrant II. ■

Exponential Form

The exponential form of a complex number $x + yj$ is given below.

Exponential Form of a Complex Number

The exponential form of a complex number $x + yj$ is

$$re^{j\theta}$$

where θ is in radians and

$$r = \sqrt{x^2 + y^2} \qquad \tan\theta = \frac{y}{x}$$

Recall that e is the base for natural logarithms, discussed in Chapter 8. To convert a complex number in polar form to exponential form, convert θ from degrees to radians.

EXAMPLE 4 Convert $3 - 2j$ to polar form and exponential forms.

Solution To convert $3 - 2j$ into polar form, we must determine r and θ. Once again, we should find a reference angle A and then calculate θ.

$$r = \sqrt{(3)^2 + (-2)^2} \qquad \tan A = \frac{2}{3}$$

$$r \approx 3.61 \qquad A \approx 33.69°$$

$$\theta = 326.31°$$

3 [x^2] [+] 2 [+/−] [x^2] [=] [√] → 3.61

2 [÷] 3 [=] [inv] [tan] 33.69

[STO] 360 [−] [RCL] [=] → 326.31

The polar form of $3 - 2j$ is

$$3.61(\cos 326.31° + j \sin 326.31°)$$

To find the exponential form, we convert θ from degrees to radians ($326.31° \approx 5.70$ radians). The exponential form of $3 - 2j$ is

$$3.61e^{5.70j}$$

■

EXAMPLE 5 Convert $5.83e^{4.76j}$ to polar and rectangular forms.

Solution From $5.83e^{4.76j}$, we know that $r = 5.83$ and $\theta = 4.76$ radians. Before we convert from exponential form to polar form, θ should be in degrees. Therefore, we convert 4.76 radians to 272.73°. Some calculators have a button to convert between radians and degrees. The polar form of $5.83e^{4.76j}$ is

$$5.83(\cos 272.73° + j \sin 272.73°)$$

To convert to rectangular form, we calculate $x = r \cos \theta$ and $y = r \sin \theta$. The rectangular form of $5.83e^{4.76j}$ is

$$0.2775 - 5.823j$$

5.83 [×] 272.73 [cos] [=] → 0.2775

5.83 [×] 272.73 [sin] [=] → −5.823 ■

EXAMPLE 6 In an ac circuit, the current is represented by the complex number $2.5 - 4.8j$ amperes. Write this number in polar and exponential forms, and determine the magnitude of the current.

Solution First, convert from rectangular to polar form. For polar form, we must determine r and θ in degrees.

$$r = \sqrt{x^2 + y^2} \qquad \tan A = \left|\frac{y}{x}\right| = \frac{4.8}{2.5}$$

$$r = \sqrt{(2.5)^2 + (-4.8)^2} \qquad A \approx 62.5°$$

$$r \approx 5.4 \qquad \theta = 297.5° \quad \text{(quadrant IV)}$$

2.5 [x^2] [+] 4.8 [+/−] [x^2] [=] [$\sqrt{\ }$] → 5.4

4.8 [÷] 2.5 [=] [inv] [tan] 62.5

[STO] 360 [−] [RCL] [=] ⟶ 297.5

The polar form of $2.5 - 4.8j$ is

$$5.4(\cos 297.5° + j \sin 297.5°)$$

To obtain the exponential form of this complex number, we convert from degrees to radians. The exponential form of $2.5 - 4.8j$ is

$$5.4e^{5.2j}$$

The magnitude of the current is determined by r, which is 5.4 A. ■

The different forms of a complex number are summarized below.

Forms of a Complex Number

☐ Rectangular:

$$x + yj$$

☐ Polar:

$$r(\cos \theta + j \sin \theta)$$

where θ is usually expressed in degrees.

☐ Exponential:

$$re^{j\theta}$$

where θ is in radians and

$$x = r \cos \theta \qquad y = r \sin \theta$$

$$r = \sqrt{x^2 + y^2} \qquad \tan \theta = \frac{y}{x}$$

14–3 EXERCISES

Represent each complex number graphically and express each number in polar and exponential forms.

1. $-2 + 3j$ **2.** $5 - 6j$

3. $-7 - 10j$ **4.** $5 - 11j$

5. $6 + 9j$ **6.** $3 + 2j$

7. $6 - 13j$ **8.** $-8 + 7j$

9. $-7.5 + 10.1j$ **10.** $-6.9 - 15j$

11. $-5.7 - 9.3j$ **12.** $-1.5 + 3.8j$

13. $8j$ **14.** $-15j$

15. 15 **16.** -10

Express each complex number in rectangular form.

17. $6(\cos 300° + j \sin 300°)$

18. $10(\cos 120° + j \sin 120°)$

19. $258 \text{ cis } 100°$ **20.** $4.7 \text{ cis } 228°$

21. $7.2\angle -56.1°$ **22.** $75\angle -38°$

23. $137(\cos 180° + j \sin 180°)$

24. $9.3(\cos 201° + j \sin 201°)$

25. $7e^{1.8j}$ **26.** $4.8e^{4.1j}$

27. $163e^{2.6j}$ **28.** $0.328e^{5.4j}$

29. $28.6e^{1.31j}$ **30.** $274e^{2.88j}$

31. $0.739e^{1.1j}$ **32.** $56e^{4.9j}$

Express each complex number in polar and rectangular forms.

33. $6.75e^{3.9j}$ **34.** $11.6e^{4.1j}$

35. $175e^{1.8j}$ **36.** $238e^{5.2j}$

37. $26.9e^{2.5j}$ **38.** $67.4e^{0.75j}$

E 39. The current in an ac circuit is given by the complex number $3.6 - 5.3j$. Write this number in polar and exponential forms.

E 40. Determine the magnitude of a voltage in an ac circuit represented by $V = 108.6 + 58.7j$ volts.

14 – 4

PRODUCTS, QUOTIENTS, POWERS, AND ROOTS OF COMPLEX NUMBERS

In Section 14.2, we discussed the product and quotient of complex numbers in rectangular form. However, these operations and others are done more easily when the complex numbers are written in alternate forms. In this section we will discuss the product, quotient, power, and roots of complex numbers written in polar and exponential forms.

Multiplication

When multiplying complex numbers in exponential form, we apply the Laws of Exponents to obtain

$$r_1e^{j\theta_1} \cdot r_2e^{j\theta_2} = r_1r_2e^{j(\theta_1+\theta_2)}$$

Using this information to express the product of complex numbers in polar form, we have

$$r_1e^{j\theta_1} \cdot r_2e^{j\theta_2} = r_1(\cos\theta_1 + j\sin\theta_1)r_2(\cos\theta_2 + j\sin\theta_2)$$

If we expand the expression for the product of the exponential forms above, we have

$$r_1r_2e^{j(\theta_1+\theta_2)} = r_1r_2[(\cos(\theta_1 + \theta_2) + j\sin(\theta_1 + \theta_2)]$$

Since the two exponential expressions are equal, the two polar expressions are also equal. Thus,

$$r_1(\cos\theta_1 + j\sin\theta_1) \cdot r_2(\cos\theta_2 + j\sin\theta_2) = r_1r_2[\cos(\theta_1 + \theta_2) + j\sin(\theta_1 + \theta_2)]$$

Multiplication of Complex Numbers

☐ Exponential form:

$$r_1e^{j\theta_1} \cdot r_2e^{j\theta_2} = r_1r_2e^{j(\theta_1+\theta_2)}$$

☐ Polar form:

$$r_1(\cos\theta_1 + j\sin\theta_1) \cdot r_2(\cos\theta_2 + j\sin\theta_2) = r_1r_2[(\cos(\theta_1 + \theta_2) + j\sin(\theta_1 + \theta_2)]$$

That is, multiply magnitudes and add the angles.

EXAMPLE 1 Using polar form, multiply $2 - 7j$ and $-5 + 9j$.

Solution First, we convert each number into polar form.

For $2 - 7j$:

$$r_1 = \sqrt{2^2 + (-7)^2} \qquad \tan A = \frac{7}{2}$$
$$r_1 = \sqrt{53} \qquad A = 74.1^\circ$$
$$r_1 = 7.28 \qquad \theta_1 = 285.9^\circ$$

$$7.28(\cos 285.9^\circ + j \sin 285.9^\circ)$$

2 [x^2] [+] 7 [x^2] [$\sqrt{\ }$] → 7.28

7 [÷] 2 [=] [inv] [tan] 74.1 [STO] 360 [−] [RCL] [=] → 285.9

For $-5 + 9j$:

$$r_2 = \sqrt{(-5)^2 + 9^2} \qquad \tan A = \frac{9}{5}$$
$$r_2 = \sqrt{106} \qquad A = 60.9^\circ$$
$$r_2 = 10.3 \qquad \theta_2 = 119.1^\circ$$

$$10.3(\cos 119.1^\circ + j \sin 119.1^\circ)$$

5 [x^2] [+] 9 [x^2] [=] [$\sqrt{\ }$] → 10.3

9 [÷] 5 [=] [inv] [tan] 60.9 [STO] 180 [−] [RCL] [=] → 119.1

Multiplying the two complex numbers in polar form gives

$$7.28(\cos 285.9^\circ + j \sin 285.9^\circ) \cdot 10.3(\cos 119.1^\circ + j \sin 119.1^\circ)$$

multiply magnitudes — add angles

$$= (7.28)(10.3)[\cos(285.9^\circ + 119.1^\circ) + j \sin(285.9^\circ + 119.1^\circ)]$$
$$= 75.0(\cos 405.0^\circ + j \sin 405.0^\circ)$$

We do not leave an angular measure greater than 360° or 2π radians; therefore, the product is

$$75.0(\cos 45.0^\circ + j \sin 45.0^\circ)$$

■

EXAMPLE 2 Using exponential form, multiply $3 + 8j$ and $-6 - 5j$.

Solution First, convert each number to exponential form.

For $3 + 8j$:

$$r_1 = \sqrt{9 + 64} \qquad \tan A = \frac{8}{3}$$

$$r_1 = 8.5 \qquad A = 1.2$$

$$\theta_1 = 1.2$$

$$8.5e^{1.2j}$$

For $-6 - 5j$:

$$r_2 = \sqrt{36 + 25} \qquad \tan A = \frac{5}{6}$$

$$r_2 = 7.8 \qquad A = 0.69$$

$$\theta_2 = 3.8$$

$$7.8e^{3.8j}$$

Multiplying the complex numbers in exponential form gives

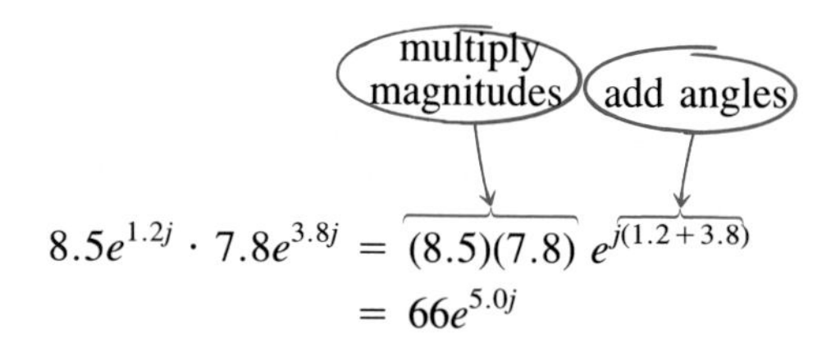

$$8.5e^{1.2j} \cdot 7.8e^{3.8j} = (8.5)(7.8)\, e^{j(1.2+3.8)}$$
$$= 66e^{5.0j}$$

■

Division

Similarly, we can derive the following formulas for division of complex numbers.

Division of Complex Numbers

☐ Exponential form:

$$r_1e^{j\theta_1} \div r_2e^{j\theta_2} = \frac{r_1}{r_2}e^{j(\theta_1 - \theta_2)}$$

☐ Polar form:

$$\frac{r_1(\cos\theta_1 + j\sin\theta_1)}{r_2(\cos\theta_2 + j\sin\theta_2)} = \frac{r_1}{r_2}[\cos(\theta_1 - \theta_2) + j\sin(\theta_1 - \theta_2)]$$

That is, divide magnitudes and subtract angles.

EXAMPLE 3 Using polar form, divide the first complex number in Example 1 by the second.

Solution From Example 1, the polar forms of these two numbers are $7.28\angle 285.9°$ and $10.3\angle 119.1°$, respectively. Dividing these numbers gives

$$\frac{7.28(\cos 285.9° + j \sin 285.9°)}{10.3(\cos 119.1° + j \sin 119.1°)}$$

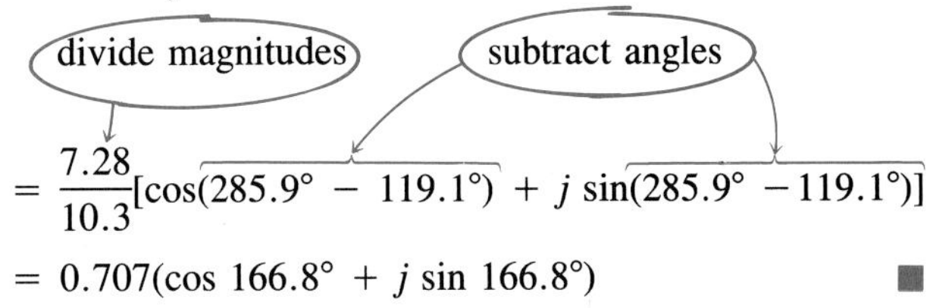

$$= \frac{7.28}{10.3}[\cos(285.9° - 119.1°) + j \sin(285.9° - 119.1°)]$$

$$= 0.707(\cos 166.8° + j \sin 166.8°)$$

■

EXAMPLE 4 Using exponential form, divide the first number in Example 2 by the second number.

Solution The exponential forms from Example 2 are $8.5e^{1.2j}$ and $7.8e^{3.8j}$. Dividing gives

$$\frac{8.5e^{1.2j}}{7.8e^{3.8j}} = \frac{8.5}{7.8}e^{j(1.2-3.8)}$$

$$= 1.1e^{-2.6j}$$

Since we want a nonnegative angle less than 2π, add -2.6 to 2π. The quotient is

$$1.1e^{3.7j}$$

■

Powers of Complex Numbers

Exponential and polar forms are helpful in raising a complex number to a power. The rule for the power of a complex number in exponential form is based on the Law of Exponents. *The formula for the nth power of a complex number in polar form* is called **DeMoivre's theorem.** The formula for the power of a complex number written in polar and exponential forms is summarized in the following box.

Powers of Complex Numbers

☐ Exponential form:

$$(re^{j\theta})^n = r^n e^{jn\theta}$$

☐ Polar form:

$$[r(\cos\theta + j\sin\theta)]^n = r^n(\cos n\theta + j \sin n\theta)$$

EXAMPLE 5 Using polar and exponential forms, find $(4 - 6j)^3$.

Solution First *converting* $4 - 6j$ *to polar form* gives

$$r \approx 7.21 \qquad \theta = 303.7°$$

$$[7.21(\cos 303.7° + j \sin 303.7°)]^3$$

Applying DeMoivre's theorem gives

$$(7.21)^3[\cos 3(303.7°) + j \sin 3(303.7°)]$$

$$375(\cos 191.1° + j \sin 191.1°$$

$$191.1° = 911.1° - 720°$$

To find $(4 - 6j)^3$ using exponential form, we must *convert* $4 - 6j$ *to exponential form.*

$$7.21e^{5.30j}$$

Applying the formula gives

$$(7.21e^{5.30j})^3 = (7.21)^3e^{3(5.30j)} = 375e^{15.9j}$$

Converting the angle to less than 2π gives

$$375e^{3.33j}$$

■

Roots of Complex Numbers

DeMoivre's theorem was originally intended for positive integral powers of *n;* however, you can also use it to find rational roots of a complex number. Every complex number has *n* distinct *n*th roots.

EXAMPLE 6 Find all possible roots for $\sqrt{-4 + 7j}$.

Solution First, we *convert* $-4 + 7j$ *to polar form.*

$$-4 + 7j = 8.1\angle 119.7°$$

$$[8.1(\cos 119.7° + j \sin 119.7°)]^{1/2}$$

Next, we *apply DeMoivre's theorem.*

$$\sqrt{8.1}\left(\cos \frac{119.7°}{2} + j \sin \frac{119.7°}{2}\right)$$

$$2.8(\cos 59.9° + j \sin 59.9°)$$

We should have two square roots. Since two complex numbers are equal if their magnitudes are equal and the angles are coterminal, we can find the second square root by adding 360° to the original angle.

$$\sqrt{8.1}\left(\cos\frac{119.7° + 360°}{2} + j\sin\frac{119.7° + 360°}{2}\right)$$

$$2.8(\cos 239.9° + j\sin 239.9°)$$

The two square roots of $-4 + 7j$ are

$$2.8(\cos 59.9° + j\sin 59.9°)$$
$$2.8(\cos 239.9° + j\sin 239.9°)$$

■

Roots of Complex Numbers

$$\sqrt[n]{x + yj} = \sqrt[n]{r}\left(\cos\frac{\theta + 360k°}{n} + j\sin\frac{\theta + 360k°}{n}\right)$$

where

$$k = 0, 1, 2, \ldots, (n - 1)$$

EXAMPLE 7 Find the exact values for the fourth root of -1.

Solution Converting -1 to polar form gives

$$1(\cos 180° + j\sin 180°)$$

Evaluating $\sqrt[4]{-1}$ gives

$$\sqrt[4]{1}\left(\cos\frac{180° + 360k°}{4} + j\sin\frac{180° + 360k°}{4}\right)$$

where $k = 0, 1, 2,$ and 3; there are four roots.

$$k = 0 \quad 1(\cos 45° + j\sin 45°) = 0.71 + 0.71j$$
$$k = 1 \quad 1(\cos 135° + j\sin 135°) = -0.71 + 0.71j$$
$$k = 2 \quad 1(\cos 225° + j\sin 225°) = -0.71 - 0.71j$$
$$k = 3 \quad 1(\cos 315° + j\sin 315°) = 0.71 - 0.71j$$

■

CAUTION ✦ When you are finding roots of a complex number, the number of answers should equal the index of the root. For example, the cube root of a complex number has three answers. Also, the complex number should be in polar form before you find powers and roots.

14–4 EXERCISES

Perform the indicated operations. Leave your answer in exponential form.

1. $4e^{2j} \cdot 7e^{3j}$
2. $2.8e^{1.3j} \cdot 3.4e^{4.6j}$
3. $8.2e^{3.2j} \cdot 6.1e^{2.9j}$
4. $7e^{3j} \cdot 6e^{j}$
5. $(5 - j)(2 + 3j)$
6. $(2 - 7j)(3 + j)$
7. $\dfrac{8e^{6j}}{2e^{j}}$
8. $\dfrac{16e^{4j}}{10e^{3j}}$
9. $\dfrac{10.4e^{5.3j}}{2.5e^{6.4j}}$
10. $\dfrac{34.8e^{2.7j}}{14.7e^{5.9j}}$
11. $\dfrac{9 + 6j}{3 - 4j}$
12. $\dfrac{11 - 8j}{7 + 4j}$
13. $(2 - 5j)^3$
14. $(1 + j)^4$

Perform the indicated operations. Leave your answer in polar form.

15. $(14\angle 146.8°)(8\angle 287.4°)$
16. $(7.8\angle 318.1°)(11.3\angle 78.4°)$
17. $16e^{4.8j} \cdot 12e^{6.1j}$
18. $5.8e^{4.5j} \cdot 2.3e^{3.2j}$
19. $\dfrac{6(\cos 120° + j \sin 120°)}{20(\cos 75° + j \sin 75°)}$
20. $\dfrac{15(\cos 216° + j \sin 216°)}{7(\cos 108° + j \sin 108°)}$
21. $\dfrac{7.8e^{1.1j}}{3.2e^{4.6j}}$
22. $\dfrac{2.5e^{6.2j}}{1.4e^{3.5j}}$
23. $(6\angle 147°)^3$
24. $(2\angle 214°)^3$
25. $(4 + 7j)^4$
26. $(8 - 5j)^2$

Find all possible roots.

27. $\sqrt[3]{1 - j}$
28. $\sqrt[4]{2 - 5j}$
29. $\sqrt{6(\cos 312° - j \sin 312°)}$
30. $\sqrt[3]{2(\cos 120° + j \sin 120°)}$
31. $\sqrt[4]{j}$
32. $\sqrt[3]{-j}$
33. $\sqrt[6]{1}$
34. $\sqrt[4]{1}$
35. $\sqrt{2 + 4j}$
36. $\sqrt{3 - 5j}$

E 37. In an ac circuit the voltage V is the product of the current I and the impedance Z. If $I = 6\angle 40°$ A and $Z = 4\angle 75°\ \Omega$, find the voltage.

E 38. In an ac circuit the impedance Z is the quotient of the voltage V and the current I. If the voltage is $12\ \angle 16.2°$ V and the current is $4.2\ \angle 14.1°$ A, find the impedance.

E 39. In an ac circuit impedance $Z_1 = 3 + 2j$ and $Z_2 = 1 + 6j$ are connected in parallel. Find the equivalent impedance Z given by

$$Z = \frac{Z_1 Z_2}{Z_1 + Z_2}$$

E 40. Find the equivalent impedance in Exercise 39 if $Z_1 = 8 + 7j$ and $Z_2 = 10 - 2j$.

14–5

APPLICATIONS

In this section we will discuss the application of complex numbers to voltage and current in an alternating current (ac) circuit. We will begin with a discussion of the basic elements of a circuit.

Resistance is the opposition to the flow of current. It is denoted by R, measured in ohms Ω, and represented by ⩘ in a circuit diagram. A **capacitor** consists of two closely spaced conductors that store an electrical charge. **Capacitance** is denoted by C, measured in farads F, and represented by ⟂⟂ in a circuit diagram. An **inductor** is a coil of wire that induces current. **Inductance** is denoted by L, measured in henries H, and represented by ⌇⌇⌇ in a circuit diagram. **Reactance,** denoted by X, is the effective resistance of any component of a circuit. Since both capacitors and inductors resist the flow of current, we can define capacitive reactance and inductive reactance by the following equations.

Capacitive Reactance

$$X_C = \frac{1}{2\pi fC}$$

Inductive Reactance

$$X_L = 2\pi fL$$

where

f = frequency of current
C = capacitance
L = inductance

In a parallel ac circuit with multiple frequencies, an inductor and a resistor together act as a "high-pass filter" to allow passage of only high-frequency currents. Low-frequency currents are diverted through the inductor. Similarly, a "low-pass filter" allows only low-frequency currents to pass.

The voltage V across any component is given by $V = IX$, where I represents current in amperes and X represents reactance in ohms. From this information, the following equations give the voltage across a resistor, a capacitor, and an inductor.

Voltage Drop across a Circuit Component

- Resistor:

$$V_R = IR$$

- Capacitor:

$$V_C = IX_C$$

- Inductor:

$$V_L = IX_L$$

EXAMPLE 1 The ac circuit shown in Figure 14–8 contains a 2.6-Ω resistor, a 0.03-H inductor, and a 0.00002-F (20-μF) capacitor. If a 9-V, 1,000-Hz power supply is connected across the circuit, determine the capacitive and inductive reactances.

FIGURE 14–8

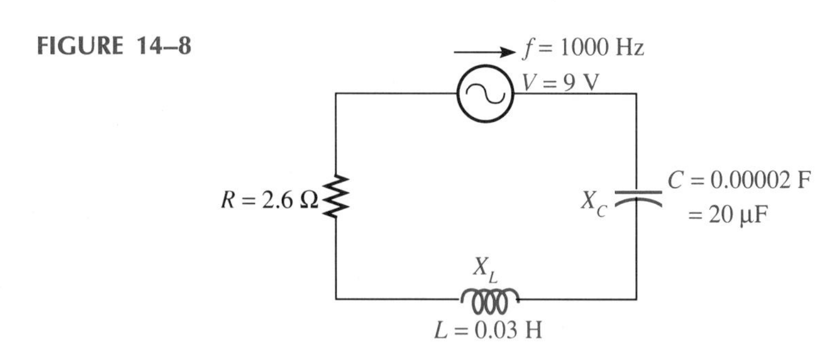

Solution From the statement of the problem, we know that $R = 2.6\ \Omega$, $L = 0.03$ H, $C = 0.00002$ F, $V = 9$ V, and $f = 1{,}000$ Hz.

Calculating the capacitive reactance X_C gives

$$X_C = \frac{1}{2\pi fC}$$

$$X_C = \frac{1}{2\pi(1{,}000)(0.00002)}$$

$$X_C = 8\ \Omega$$

2 [×] [π] [×] 1000 [×] 0.00002 [=] [1/x] → 7.96

Calculating X_L gives

$$X_L = 2\pi fL$$

$$X_L = 2\pi(1{,}000)(0.03)$$

$$X_L = 190\ \Omega$$

2 [×] [π] [×] 1000 [×] 0.03 [=] → 188.5 ■

To determine the voltage across a combination of resistors, inductors, and capacitors, we must account for the reactance and the phase of the voltage across individual components. A polar diagram shows the effective voltage across a resistor, capacitor, and inductor as vectors, with the phase differences appearing as angles. Since the voltage across a resistor is in-phase with the current, a real number R can be used to represent resistance. Graphically, this relationship is illustrated in Figure 14–9(a). Since the voltage across an inductor leads the current by 90°, the inductive reactance is represented by the positive, pure imaginary number $X_L j$ as shown in Figure 14–9(b). Because the voltage across a capacitor lags 90° behind the current, it can be represented by the negative, pure imaginary number $-X_L j$ as shown in Figure 14–9(c).

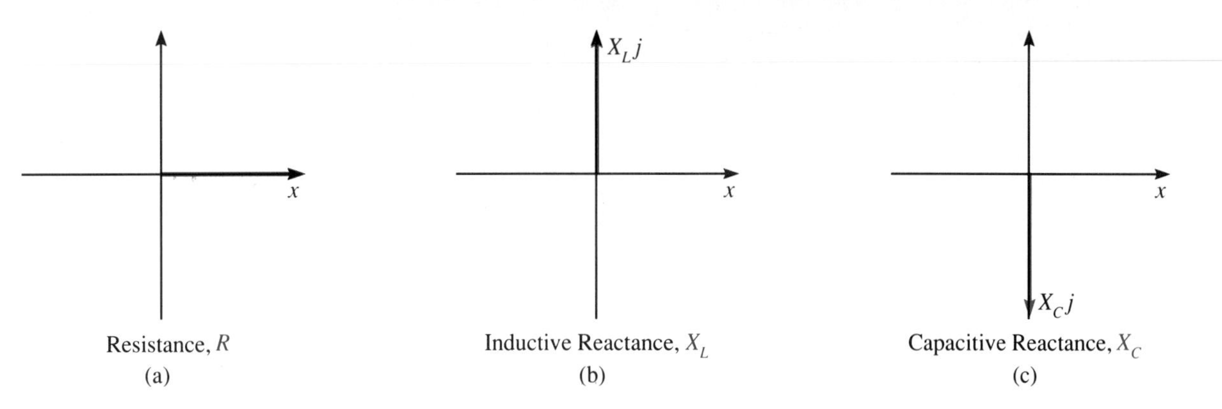

FIGURE 14–9

The total voltage V across a circuit containing a resistor, capacitor, and inductor is the sum of the individual voltages, represented by the following equation:

$$V = V_R + V_L + V_C$$
$$V = IR + IX_L j - IX_C j$$
$$V = I[R + j(X_L - X_C)]$$

Impedance Z is the effective resistance of the entire circuit and is given by $R + j(X_L - X_C)$. Making the substitution gives the following equation.

Impedance

$$V = IZ$$

where

$$Z = R + j(X_L - X_C)$$

The magnitude of Z, denoted $|Z|$, is

$$|Z| = \sqrt{R^2 + (X_L - X_C)^2}$$

and the phase angle θ between the current and voltage is given by

$$\tan \theta = \frac{X_L - X_C}{R}$$

EXAMPLE 2 For the circuit shown in Figure 14–10, assume that $X_L = 95\ \Omega$, $X_C = 103\ \Omega$, $R = 2.70\ \Omega$, and $V = 12.0$ V. Calculate the following:

(a) phase angle
(b) magnitude of the impedance
(c) current in the circuit
(d) voltage across each circuit component

FIGURE 14–10

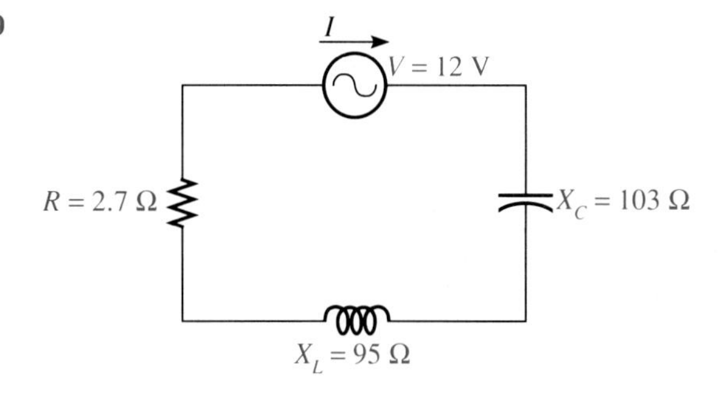

Solution

(a) The phase angle is given by

$$\tan\theta = \frac{X_L - X_C}{R} = \frac{95.0 - 103}{2.70}$$
$$\theta = -71.4°$$

The negative sign on the phase angle means that the voltage lags the current by 71.4°.

(b) The magnitude of the impedance is given by

$$|Z| = \sqrt{R^2 + (X_L - X_C)^2}$$
$$|Z| = 8.44$$

2.7 [x^2] [+] [(] 95 [−] 103 [)] [x^2] [=] [$\sqrt{\ }$] [STO] → 8.44

(c) The current is given by

$$I = \frac{V}{|Z|} = \frac{12.0}{8.44}$$
$$I \approx 1.42 \text{ A}$$

12 [÷] [RCL] [=] [STO] → 1.42

(d)

$V_R = IR = (1.42)(2.70)$ [RCL] [×] 2.7 [=] → 3.83

$V_L = IX_L = (1.42)(95.0)$ [RCL] [×] 95 [=] → 135

$V_C = IX_C = (1.42)(103)$ [RCL] [×] 103 [=] → 146 ■

EXAMPLE 3 An electric motor with an inductance of 0.020 H and a resistance of 18 Ω is connected to a 110-V, 60-Hz power source.

(a) Calculate the magnitude of the impedance of the motor.
(b) Calculate the current through the motor.
(c) If a 0.00042-F (420 μF) capacitor is added to the circuit, how much current will the circuit draw?
(d) Find the phase angle in part (c).

Solution

(a) To calculate the magnitude of the impedance, we must know R, X_L, and X_C. From the information given, $R = 18\ \Omega$, $X_C = 0$ since there is no capacitor in the circuit, and $X_L = 2\pi fL = 2\pi(60)(0.02) \approx 7.5\ \Omega$. Therefore, the magnitude of the impedance is

$$|Z| = \sqrt{R^2 + (X_L - X_C)^2} = \sqrt{(18)^2 + (7.5)^2}$$
$$|Z| = 20$$

18 [x^2] [+] 7.5 [x^2] [=] [$\sqrt{\ }$] → 19.5

(b) To calculate the current through the motor, we solve the equation $V = IZ$ for I and substitute.

$$I = \frac{V}{|Z|} = \frac{110}{20}$$
$$I = 5.5\text{ A}$$

(c) If a capacitor is added to the circuit, its capacitive reactance is

$$X_C = \frac{1}{2\pi fC} = \frac{1}{2\pi(60)(0.00042)}$$
$$X_C = 6.3\ \Omega$$

2 [×] [π] [×] 60 [×] 0.00042 [=] [1/x] → 6.3

Calculating the magnitude of the impedance gives

$$|Z| = \sqrt{R^2 + (X_L - X_C)^2}$$
$$|Z| = 18\ \Omega$$

18 [x^2] [+] [(] 7.5 [−] 6.3 [)] [x^2] [=] [$\sqrt{\ }$] → 18

The resulting current is

$$I = \frac{V}{|Z|} = \frac{110}{18}$$
$$I = 6.1\text{ A}$$

(d) The phase angle between the voltage and current is given by

$$\tan\theta = \frac{X_L - X_C}{R} = \frac{7.5 - 6.3}{18}$$

$$\theta = 3.8^\circ$$

The voltage leads the current by 3.8°. ■

Complex Current

In the discussion so far, the phase angle of the current has been taken as zero. However, if an arbitrary angle is chosen, current, voltage, and impedance must be treated as complex numbers.

EXAMPLE 4 In a particular circuit the current is given as $4 - j$ amperes, and the impedance is given as $5 + 3j$ ohms. Determine the magnitude of the voltage across the circuit.

Solution We use the formula $V = IZ$ to calculate the voltage. Multiplying gives

$$V = (4 - j)(5 + 3j) = 20 + 12j - 5j - 3j^2$$

$$V = 23 + 7j \text{ V}$$

The magnitude of the voltage is

$$|V| = \sqrt{(23)^2 + (7)^2}$$

$$|V| = 24 \text{ V}$$

■

Resonance

Since both inductive and capacitive reactance depend on the frequency of the voltage, impedance also depends on frequency. When inductive reactance equals capacitive reactance, the equation for impedance becomes $|Z| = R$. When this is true, the impedance is a minimum, and the current is a maximum. The frequency f_0 at which inductive reactance and capacitive reactance are equal is called the **resonant frequency.** The formula for f_0 given below is the result of setting $X_L = X_C$ and solving for f.

Resonant Frequency

$$f_0 = \frac{1}{2\pi\sqrt{LC}} \text{ Hz}$$

EXAMPLE 5 Find the resonant frequency of a circuit containing an inductor of 0.0300 H, a resistor of 7 Ω, a capacitor of 0.0000190 F (19 μF), and a voltage source of 9 V.

Solution The resonant frequency is given by

$$f_0 = \frac{1}{2\pi\sqrt{LC}} = \frac{1}{2\pi\sqrt{(0.03)(0.000019)}}$$

$$f_0 = 211 \text{ Hz}$$

2 [×] [π] [×] [(] 0.03 [×] 0.000019 [)] [√] [=] [1/x] → 211 ■

The resonant circuit described in Example 5 is fundamental to the operation of a variety of electrical systems. The frequency response of this circuit is shown in Figure 14–11. Radio and television receivers use this type of circuit. When the receiver is tuned to a particular station, it is set at or near the resonant frequency f_0. As a result, stations transmitting at frequencies to the left or right of f_0 do not interfere with the station's broadcast.

FIGURE 14–11

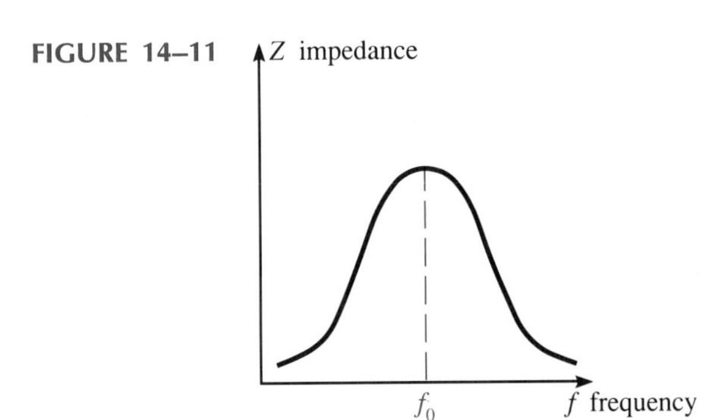

EXAMPLE 6 The tuning circuit of a television contains a variable capacitor, a 0.0030-H coil, and a 15-Ω resistor. Calculate the capacitance necessary to tune the circuit to a television station broadcasting at 60×10^6 Hz.

Solution To solve for the capacitance, use the formula for resonant frequency.

$$f_0 = \frac{1}{2\pi\sqrt{LC}}$$

Next, solve for C.

$$f_0^2 = \frac{1}{4\pi^2 LC}$$
$$4f_0^2\pi^2 LC = 1$$
$$C = \frac{1}{4f_0^2\pi^2 L}$$

Then substitute values for L and f_0.

$$C = \frac{1}{4(60 \times 10^6)^2\pi^2(0.0030)}$$
$$C = 2.3 \times 10^{-15}\ \text{F}$$

4 [×] 60000000 [x^2] [×] [π] [x^2] [×] 0.003 [=] [1/x] → 2.3 −15 ■

14–5 EXERCISES

1. An ac circuit contains a 0.036-H inductor, a 0.0000080-F capacitor, and a 7.5-Ω resistor. If a 9.0-V, 60-Hz power source is connected across the circuit, find the following:
   (a) capacitive reactance
   (b) inductive reactance
   (c) magnitude of the impedance
   (d) phase angle between the current and voltage
   (e) current
   (f) voltage across the resistor
   (g) voltage across the capacitor
   (h) voltage across the inductor

E 2. Answer parts (a)–(h) from Exercise 1 if the circuit contains a 0.0540-H inductor, a 0.00000600-F capacitor, a 12.6-Ω resistor, and a 110-V, 60.0-Hz power source.

E 3. A current of 0.15 A flows through a 0.31-H inductor that is connected to a 115-V ac source. What is the frequency of the source?

E 4. A 0.00000007-F capacitor is connected to a 9-V, 60-Hz power source. Find the following:
   (a) capacitive reactance
   (b) current flow

E 5. In a circuit, $R = 7.0\ \Omega$, $X_C = 13\ \Omega$, and $X_L = 8.0\ \Omega$. Find the magnitude of the impedance and the phase angle between the current and voltage.

E 6. Calculate the magnitude of the voltage in a circuit if the complex current is $3.00 - 4.00j$ and the complex impedance is $5.00 + 7.00j$.

E 7. If the complex current in a particular circuit is given by $2.00 + 5.00j$ and the complex impedance is given by $7.00 - 6.00j$, find the magnitude of the voltage.

E 8. The voltage across a circuit is given by $8.00 - 3.00j$, and the current is $9.00 + 5.00j$. Find the magnitude of the impedance.

E 9. The voltage across a particular circuit is given by $1.00 + 8.00j$, and the impedance is given by $7.00 - 3.00j$. Find the magnitude of the current.

E 10. A 0.0000060-F (6 μF) capacitor, a 0.060-H inductor, and an 8.0 Ω resistor are connected with a 28-V, 400-Hz power source. Find the following:
   (a) impedance of the circuit
   (b) current that flows through it

E 11. Find the resonant frequency of the circuit in Exercise 10.

E 12. A 0.0000090-F (9 μF) capacitor, a 0.090-H inductor, and a 45-Ω resistor are connected across a 115-V, 60-Hz power line. Find the following:
   (a) current in the circuit
   (b) voltage drop across each element of the circuit

E 13. The tuning circuit of a radio contains an 8.5-Ω resistor, a 0.018-H coil, and a variable capacitor. Calculate the capacitance required to tune the radio to a station broadcasting at 7.3×10^5 Hz (730 kHz).

E 14. The antenna circuit of a radio contains a 0.20-H coil, a variable capacitor, and a 48-Ω resistor. Find the capacitance required for resonance if the station transmits at 6.4×10^5 Hz (640 kHz).

E 15. The power P dissipated in an ac circuit is given by $P = IV \cos \theta$ where V = voltage, I = current, and θ = phase angle between the current and voltage. Find the power dissipated in the circuit in Exercise 10.

E 16. Find the power dissipated by the circuit described in Exercise 12.

CHAPTER SUMMARY

Summary of Terms

argument (p. 499)
capacitance (p. 510)
capacitor (p. 510)
complex number (p. 488)
complex plane (p. 493)
conjugate (p. 491)
DeMoivre's theorem (p. 507)
imaginary part (p. 488)
impedance (p. 513)
inductance (p. 510)
inductor (p. 510)
magnitude (p. 499)
pure imaginary number (p. 488)
reactance (p. 510)
real part (p. 488)
rectangular form (p. 488)
resistance (p. 510)
resonant frequency (p. 516)

Summary of Formulas

$re^{j\theta}$ where θ is in radians — exponential form of complex numbers

Multiplication of complex numbers:

$r_1e^{j\theta_1} \cdot r_2e^{j\theta_2} = r_1r_2e^{j(\theta_1+\theta_2)}$ exponential form

$r_1(\cos \theta_1 + j \sin \theta_1) \cdot r_2(\cos \theta_2 + j \sin \theta_2) = r_1r_2[(\cos(\theta_1 + \theta_2) + j \sin(\theta_1 + \theta_2)]$ polar form

Division of complex numbers:

$r_1e^{j\theta_1} \div r_2e^{j\theta_2} = \dfrac{r_1}{r_2}e^{j(\theta_1-\theta_2)}$ exponential form

$\dfrac{r_1(\cos \theta_1 + j \sin \theta_1)}{r_2(\cos \theta_2 + j \sin \theta_2)} = \dfrac{r_1}{r_2}[\cos(\theta_1 - \theta_2) + j \sin(\theta_1 - \theta_2)]$ polar form

Powers of complex numbers:

$(re^{j\theta})^n = r^ne^{jn\theta}$ exponential form

$[r(\cos \theta + j \sin \theta)]^n = r^n(\cos n\theta + j \sin n\theta)$ polar form

$$\sqrt[n]{x + yj} = \sqrt[n]{r}\left(\cos \frac{\theta + 360k^\circ}{n} + j \sin \frac{\theta + 360k^\circ}{n}\right)$$ roots of complex numbers

$r = \sqrt{x^2 + y^2}$ $\quad \tan \theta = \dfrac{y}{x}$ $\quad x = r \cos \theta$ $\quad y = r \sin \theta$

CHAPTER REVIEW

Section 14–1

Perform the indicated operation and simplify.

1. $\sqrt{-8} \cdot \sqrt{3}$
2. $\sqrt{-54} \cdot \sqrt{-9}$
3. $(6\sqrt{-5})^2$
4. $\sqrt{-16} + \sqrt{-9}$
5. $\sqrt{-4/25} + (3\sqrt{-7})^2$
6. $j^{14} + \sqrt{-49}$
7. $\sqrt{-4/9} + j^{25}$

Simplify the following powers of j.

8. j^{786}
9. j^{127}
10. j^{39}
11. j^{472}
12. $j^{34} + j^{66}$
13. $3j^{102} - j^{40}$
14. $10j^{109} + 7j^{213}$
15. $5j^{415} + j^{33}$

Section 14–2

Perform the indicated operations. Leave your answer in $a + bj$ form.

16. $(-2 + 3j) + (5j - 6)$
17. $(16 - 7j) + (-2 - 15j)$
18. $(10 - \sqrt{-25}) - (\sqrt{-16} + 7)$
19. $(\sqrt{-114} + 9) - (4 + \sqrt{-49})$
20. $(4 - 9j)(4 + 9j)$
21. $(2 - 5j)(2 + 5j)$
22. $(\sqrt{-9} + 6)(7 - \sqrt{-81})$
23. $(4 + \sqrt{-25})(11 - \sqrt{-64})$
24. $(8 + 7j)(5 - 4j)$
25. $(3 - 5j)^3$
26. $(2 + 9j)^3$
27. $\dfrac{10 - \sqrt{-49}}{-j}$
28. $\dfrac{5 - \sqrt{-18}}{j}$
29. $\dfrac{2j - 1}{9 - j}$
30. $\dfrac{8 + j}{3 - 5j}$

Section 14–3

Express each complex number in polar and exponential forms.

31. $-7 + 8j$
32. $13 - 10j$
33. $-14 - 8j$
34. $6.8 - 7.9j$
35. $8j$
36. 12
37. 14
38. $-5j$

Express each complex number in rectangular form.

39. $14\angle 276°$
40. $25\angle -62°$
41. $1.7(\cos 214° + j \sin 214°)$
42. $14.2(\cos 170° + j \sin 170°)$
43. $23e^{5.7j}$
44. $1.7e^{3.4j}$
45. $17.8e^{2j}$
46. $47.3e^{1.1j}$

Section 14–4

Perform the indicated operations. Leave your answer in exponential form.

47. $(1 - j)^2$
48. $(2 + 3j)^3$
49. $3e^{7j} \cdot 5e^{1.7j}$
50. $6.7e^{1.2j} \cdot 3e^{4.9j}$
51. $\dfrac{6.9e^{2.7j}}{2.1e^{3.4j}}$
52. $\dfrac{11.8e^{5.7j}}{6.6e^{3.9j}}$

Perform the indicated operation. Leave your answer in polar form ($0 \le \theta < 360.0°$).

53. $(6.5\angle 112°)(9.3\angle 318°)$
54. $(7.9e^{5.8j})(1.6e^{3.7j})$
55. $\dfrac{8.7e^{1.8j}}{16.4e^{2.9j}}$
56. $\dfrac{8(\cos 78.4° + j \sin 78.4°)}{10(\cos 218.1° + j \sin 218.1°)}$
57. $(5 - 6j)^2$
58. $(7 + 9j)^2$

Find all possible roots.

59. $\sqrt[3]{2 - 5j}$
60. $\sqrt{3 + 7j}$
61. $\sqrt[4]{1 + 2j}$
62. $\sqrt{16(\cos 240° + j \sin 240°)}$
63. $\sqrt[6]{j}$
64. $\sqrt[3]{1.8(\cos 78° + j \sin 78°)}$
65. $\sqrt[5]{-1}$
66. $\sqrt[4]{-1}$

Section 14–5

E For Problems 67–74, an ac circuit contains a 0.043-H inductor, a 0.0000060-F (6 μF) capacitor, and a 9.3-Ω resistor. A 12-V, 50-Hz power source is connected across the circuit.

67. Find the capacitive reactance.
68. Find the inductive reactance.

69. Find the magnitude of the impedance.

70. Find the phase angle between the current and voltage.

71. Find the current.

72. Find the voltage across the resistor.

73. Find the voltage across the capacitor.

74. Find the voltage across the inductor.

75. Calculate the magnitude of the voltage in a circuit if the complex current is $6.00 + 11.00j$ and the complex impedance is $7.00 - 7.33j$.

76. The voltage across a circuit is given by $10.00 - 2.00j$, and the current is given by $1.00 + 7.00j$. Find the magnitude of the impedance.

77. A television tuner circuit contains a 0.15 μH (1.5×10^{-7} H) coil, a variable capacitor, and a 52-Ω resistor. Find the capacitance required for resonance if the station transmits at 6.4×10^7 Hz.

78. A 0.038-H inductor is connected to a 110-V, 60-Hz power source. Find the following:
(a) inductive reactance
(b) current flow

CHAPTER TEST

The number in parentheses refers to the appropriate learning objective given at the beginning of the chapter.

1. Add $\sqrt{-16} + \sqrt{-9} + \sqrt{-81}$. (1)
2. Convert $8.6\angle 218.4°$ to exponential form. (4)
3. An ac circuit contains a 0.015-H inductor, a 4.2-Ω resistor, and a 0.0000040-F capacitor. A 115-V, 60-Hz power source is connected across the circuit. Find the inductive and capacitive reactance. (6)
4. Simplify the power of j in j^{94}. (2)
5. Find all possible roots for $\sqrt[3]{2 + 5j}$. (5)
6. Multiply $(3 - 7j)(6 + 9j)$. (3)
7. Using exponential form, divide the following and leave your answer in rectangular form: (4, 5)

$$\frac{8e^{2.7j}}{10e^{1.4j}}$$

8. Use DeMoivre's theorem to evaluate $(6 - 8j)^4$. (5)
9. Multiply and simplify $\sqrt{-18} \cdot \sqrt{-40}$. (1)
10. In a given circuit, $R = 6.3\ \Omega$, $X_C = 11.0\ \Omega$, and $X_L = 8.6\ \Omega$. Find the magnitude of the impedance and the phase angle between the current and voltage. (6)
11. Simplify the power of j in j^{986}. (2)
12. Express $7 - 10j$ in polar and exponential forms. (4)
13. Given that a complex current in a particular circuit is $6.00 - 11.00j$ and the complex impedance is $9.00 + 7.00j$, what is the magnitude of the voltage? (6)
14. Add $\sqrt{-9/16} + \sqrt{-49}$. (1)

15. Using polar form multiply the following and express your answer in exponential form: **(4, 5)**

$$[1.9(\cos 218° + j \sin 218°)][3.4(\cos 314° + j \sin 314°)]$$

16. Express $7.4e^{3.6j}$ in polar and rectangular forms. **(4)**

17. An electric motor has an inductance of 0.040 H and a resistance of 15.4 Ω. If the motor is connected to a 115-V, 60-Hz power source, find the current through the motor. **(6)**

SOLUTION TO CHAPTER INTRODUCTION

To solve for the capacitance, use the following formula, solve for C, and substitute the values given:

$$f_0 = \frac{1}{2\pi \sqrt{LC}}$$

$$f_0^2 = \frac{1}{4\pi^2 LC}$$

$$4f_0^2\pi^2 LC = 1$$

$$C = \frac{1}{4f_0^2\pi^2 L}$$

$$C = \frac{1}{4(101 \times 10^6)^2\pi^2(0.0020)}$$

$$C = 1.2 \times 10^{-15} \text{ F}$$

You are a civil engineer responsible for the construction of a semielliptical bridge. The specifications require a 15-ft minimum clearance 20 ft from the right end. If the bridge is 70 ft long, what maximum height at the center is needed to meet this specification? (The solution to this problem is given at the end of the chapter.)

A problem such as this one requires a knowledge of the equation of an ellipse, one of the conic sections to be discussed in this chapter. In addition to the ellipse, we will discuss the straight line, circle, parabola, and hyperbola.

Learning Objectives

After completing this chapter, you should be able to

1. Determine the equation of a line from given information (Section 15–1).
2. Determine the slope, x intercept, and y intercept of an equation, and use this information to graph the line (Section 15–1).
3. Find the center and radius of a circle given the equation in standard or general form (Section 15–2).
4. Find the equation of a circle given pertinent information (Section 15–2).
5. Find the focus, directrix, and vertex of a parabola from its equation; and sketch the graph of the parabola (Section 15–3).
6. Find the equation of a parabola from given information (Section 15–3).
7. Find the center, vertices, foci, and endpoints of the minor axis of an ellipse; and sketch the graph of the ellipse (Section 15–4).
8. Find the equation of an ellipse from given information (Section 15–4).
9. Find the center, vertices, foci, endpoints of the conjugate axis, and slope of the asympotes of a hyperbola; and sketch the graph of the hyperbola (Section 15–5).
10. Find the equation of a hyperbola from given information (Section 15–5).
11. Given a general second-degree equation, determine the type of conic section; determine pertinent information about the conic section; and sketch its graph (Section 15–6).
12. Apply the concepts of analytic geometry to technical problems (Section 15–6).

Chapter 15

Analytic Geometry

15–1

THE STRAIGHT LINE

In this section we will discuss how to find the equation for a line from given information, which may include such quantities as slope, x intercept, y intercept, or points through which the line passes. We can use two forms for the equation of a line: the slope–intercept form or the point–slope form. We will start the discussion with the slope–intercept form.

Slope–Intercept Form

In Chapter 2, we defined the x and y intercepts as the points at which the graph crosses the x axis and y axis, respectively. The intercepts are represented as the ordered pairs $(x, 0)$ and $(0, y)$. The **slope–intercept form** of a line uses *the slope and the y intercept to uniquely determine the equation of a line.*

Slope–Intercept Form

If m represents the slope of the line and b represents the y coordinate of the y intercept, the slope–intercept form is given by

$$y = mx + b$$

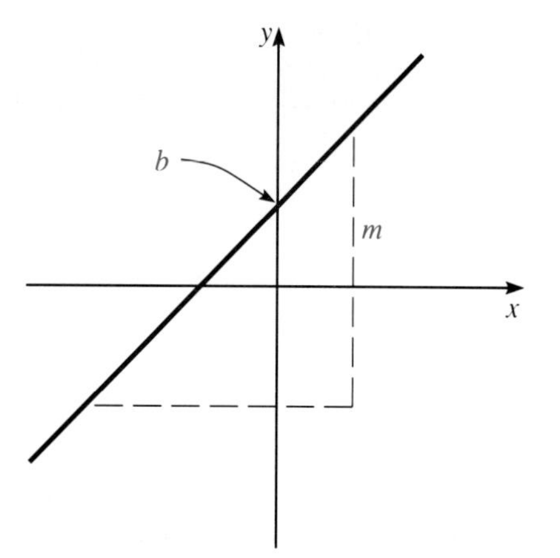

The **general form of a line** is given by

$$Ax + By + C = 0$$

where A, B, and C represent integers.

NOTE ✦ In this section you should always place the equation of the straight line in general form, which may require rearranging the equation. Remember that by algebraic convention, we usually do not leave fractional coefficients or leave the first term negative.

EXAMPLE 1 Determine the equation of the straight line with slope -3 and y intercept 5. Place your answer in general form.

Solution To use slope–intercept form of a line, we must know the slope of the line and the y intercept. Here we are given both of these quantities. Therefore, we substitute $m = -3$ and $b = 5$ into the formula.

$$y = mx + b$$
$$y = -3x + 5$$

Then we must arrange the equation in general form, $Ax + By + C = 0$.

$$3x + y - 5 = 0$$ ■

EXAMPLE 2 Determine the equation of the line having slope 2/3 and passing through the point $(0, -4)$.

Solution We are given the slope of the line and the y intercept, written in point form. Therefore, $m = 2/3$ and $b = -4$. Substituting into slope–intercept form and rearranging the equation to general form gives

$$y = mx + b$$
$$y = \frac{2}{3}x - 4$$
$$3y = 2x - 12$$
$$-2x + 3y + 12 = 0$$
$$2x - 3y - 12 = 0$$ ■

Graphs

The slope–intercept form of a line is also very useful in graphing the line, as illustrated in the next example.

EXAMPLE 3 Find the slope, y intercept, and x intercept for the equation $2x - 5y = 8$, and use this information to graph the line.

Solution To determine the slope and the y intercept from the equation, we rearrange the equation in slope–intercept form by solving the equation for y.

$$2x - 5y = 8$$
$$-5y = -2x + 8$$
$$y = \frac{-2x + 8}{-5}$$
$$y = \frac{2}{5}x - \frac{8}{5}$$

slope ($\frac{2}{5}$) y intercept ($-\frac{8}{5}$)

To find the x intercept, we substitute $y = 0$ into the equation and solve for x.

$$2x - 5y = 8$$
$$2x - 5(0) = 8$$
$$2x = 8$$
$$x = 4$$

To graph this equation using the slope and the y intercept, start by placing a point on the y axis at the intercept. Then, from that point, plot the slope to determine a second point.

$$m = \frac{\text{rise}}{\text{run}} = \frac{\text{up or down depending on sign (in } y \text{ direction)}}{\text{left or right depending on sign (in } x \text{ direction)}}$$

In this case, since the slope is 2/5, we go up 2 units and right 5 units. Place the point and draw a line. Figure 15–1 shows the graph of $2x - 5y = 8$. We can verify that this line is correct by using the x intercept.

FIGURE 15–1

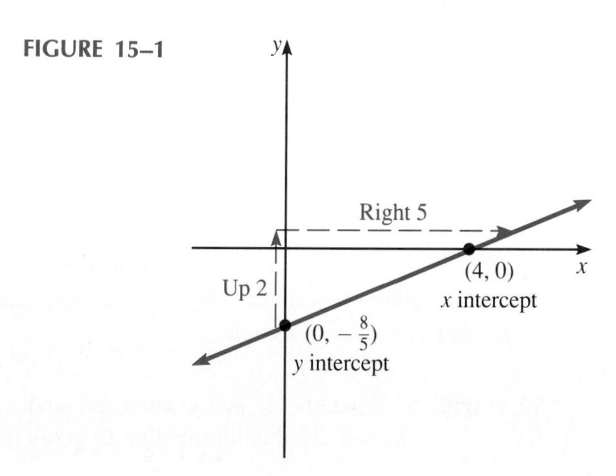

■

Point–Slope Form

The second form of a linear equation that we discuss in this section is called **point–slope,** defined as follows.

Point–Slope Form

If m is the slope of the line and (x_1, y_1) represents a point through which the line passes, the equation of the line can be written as

$$y - y_1 = m(x - x_1)$$

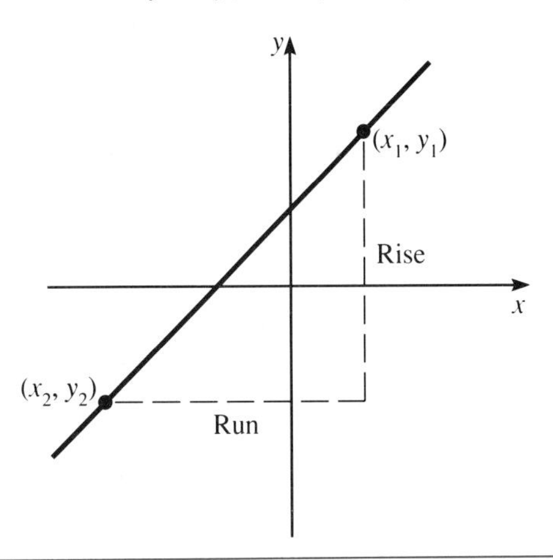

EXAMPLE 4 Determine the equation of a line having slope -3 and crossing the x axis at 4.

Solution We are given the slope of the line and the x intercept. Since we are given the x intercept, it is easier to use the point–slope form, which requires that we write the x intercept as a point: (4, 0). Substituting $m = -3$ and $(x_1, y_1) = (4, 0)$ gives

$$y - y_1 = m(x - x_1)$$
$$y - 0 = -3(x - 4)$$
$$y = -3x + 12$$

Arranging the equation in general form, $Ax + By + C = 0$, gives

$$3x + y - 12 = 0$$

■

EXAMPLE 5 Find the equation of the line passing through $(-6, 3)$ and $(5, 4)$.

Solution First, we must calculate the slope of the line. Let $(x_1, y_1) = (-6, 3)$ and $(x_2, y_2) = (5, 4)$. Then

$$m = \frac{y_2 - y_1}{x_2 - x_1}$$

$$m = \frac{4 - 3}{5 - (-6)} = \frac{1}{11}$$

Then we substitute $m = 1/11$ and use $(5, 4)$ as (x_1, y_1). ■

NOTE ✦ Using either point will result in the same equation. Substituting into the point–slope equation and rearranging gives

$$\begin{aligned} y - y_1 &= m(x - x_1) & \\ y - 4 &= \frac{1}{11}(x - 5) & \text{substitute} \\ 11y - 44 &= x - 5 & \text{eliminate fractions} \\ -x + 11y - 39 &= 0 & \text{transpose} \\ x - 11y + 39 &= 0 & \text{multiply by } -1 \end{aligned}$$

Horizontal, Vertical, and Perpendicular Lines

It is easier to write the equation for a horizontal or vertical line by inspection rather than by using the point–slope formula. The following box provides a summary to aid you in this process.

Horizontal Line

Equation: $y = b$
Slope: 0
Recognition: Ordered pairs have the same y coordinate.

$(x_1, b) \quad (x_2, b)$

Vertical Line

Equation: $x = a$
Slope: undefined
Recognition: Ordered pairs have the same x coordinate.

$(a, y_1) \quad (a, y_2)$

EXAMPLE 6 Find the equation of each line satisfying the following conditions:

(a) a line having 0 slope and passing through $(-2, 6)$ and
(b) a line passing through $(-7, 4)$ and $(-7, 2)$.

Solution

(a) Since the *slope is 0,* this is a horizontal line with equation $y = b$. From the ordered pair we can see that the y coordinate is 6. Therefore, the equation is

$$y = 6$$

(b) Since the ordered pairs have the *same x coordinate,* this is a vertical line with equation $x = a$ where a is the x coordinate of the ordered pairs. The equation is

$$x = -7$$ ■

Recall from Chapter 2 that perpendicular lines, excepting vertical and horizontal lines, have negative reciprocal slopes. The next example illustrates finding the equation of a line perpendicular to a given line.

EXAMPLE 7 Find the equation of the line perpendicular to $3x + y = 6$ and passing through $(2, -3)$.

Solution A line perpendicular to $3x + y = 6$ will have a negative reciprocal slope to the slope of this line. The slope of this line is

$$3x + y = 6$$
$$y = -3x + 6$$

slope (pointing to -3)

The slope of the perpendicular line is 1/3, and the line passes through the point $(2, -3)$. Substituting this information into point–slope form and arranging the result in general form gives

$$y - y_1 = m(x - x_1)$$
$$y - (-3) = \frac{1}{3}(x - 2)$$
$$3y + 9 = x - 2$$
$$-x + 3y + 11 = 0$$
$$x - 3y - 11 = 0$$ ■

EXAMPLE 8 The resistance of a certain electrical circuit component increases by 0.015 Ω for every increase of 1°C. The resistance at 0°C is found to be 1.5 Ω. Find the equation for resistance R in terms of temperature T, and use the equation to find the resistance when the temperature is 30°C.

Solution Since we are asked to find resistance in terms of temperature, R is the dependent variable and the equation takes the following form:

$$R = mT + b$$

This form is known and used because the change in resistance is constant for a given change in temperature. To determine the slope, recall that slope is defined to be

$$m = \frac{\text{change in } R}{\text{change in } T}$$

However, we are given that resistance changes 0.015 Ω for each 1°C change in temperature. Thus,

$$m = \frac{0.015\ \Omega}{1°\text{C}} = 0.015\ \Omega/°\text{C}$$

Then we determine the R intercept (comparable to a y intercept). We are given that the resistance is 1.5 Ω at 0°C. Therefore, $b = 1.5\ \Omega$. The resulting equation is given by

$$R = 0.015T + 1.5$$

To find the resistance at 30°C, we substitute $T = 30°$ into the equation.

$$R = 0.015T + 1.5$$
$$R = 0.015(30) + 1.5$$
$$R = 1.95$$

At 30°C, the resistance is 1.95 Ω. ■

EXAMPLE 9 The length L of a spring varies linearly with the force F applied. If a force of 2.0 lb produces a length of 18 in., and a force of 5.0 lb produces a length of 24 in., find the equation relating length and force. Leave your answer in slope–intercept form.

Solution Since force is the independent variable, we are asked to find the equation relating the ordered pairs (2, 18) and (5, 24). First, we must find the slope of the line passing through these two points.

$$m = \frac{24 - 18}{5 - 2} = 2.0\ \text{in/lb}$$

Then we use point–slope form of a line with $m = 2$ and $(F_1, L_1) = (2, 18)$. The equation is

$$L - L_1 = m(F - F_1)$$
$$L - 18 = 2(F - 2)$$
$$L = 2.0F + 14$$

■

15–1 EXERCISES

From the given information determine the equation of the line. Leave your answer in general form, $Ax + By + C = 0$, with A, B, and C integers and $A \geq 0$.

1. Slope $= -3$; y intercept $= 2/3$.
2. Slope $= -3/4$; y intercept $= -1$.
3. Slope $= 1/2$; x intercept $= -2$.
4. Slope $= -1$; passing through $(4, -2)$.
5. Passing through $(-2, 4)$ and $(3, 1)$.
6. 0 slope; passing through $(-8, -3)$.
7. Slope $= 4$; passing through $(-5, 6)$.
8. Passing through $(5, -7)$ and $(-3, -4)$.
9. Undefined slope; passing through $(4, -7)$.
10. Parallel to line $y = -4$ and passing through $(-7, 0)$.
11. Passing through $(3, -2)$ and $(-5, 7)$.
12. Slope $= -4$; x intercept $= -5$.
13. Horizontal line; passing through $(7, -3)$.
14. Passing through $(6, 9)$ and $(-4, 8)$.
15. Parallel to line $x = -2$ and passing through $(3, 1)$. (Hint: Parallel lines have equal slopes.)
16. Perpendicular to line $y = 6$ and passing through $(-8, 4)$.
17. Parallel to the line $y = 3x - 6$ and passing through $(-1, 6)$.
18. Parallel to the line $3x + 2y = 10$ and passing through $(8, -4)$.
19. Perpendicular to the line $2x - y = 7$ and passing through $(-1, -4)$.
20. Perpendicular to the line $y = 4x - 7$ and passing through $(4, 3)$.

Find the slope, y intercept, and x intercept for the following equations. Use the slope and y intercept to graph the line.

21. $3x - 4y = 12$
22. $5x + 8y = 16$
23. $x = -8$
24. $x + 6y = 10$
25. $4x - 6y + 15 = 0$
26. $y = 5$

By arranging each equation in slope–intercept form, determine whether the following lines are parallel, perpendicular, or neither.

27. $2x + y = 9$ and $2y - x = 14$
28. $4x + 3y = 6$ and $10 + 6x - 8y = 0$
29. $7x - 5y + 8 = 0$ and $y = 7x + 12$
30. $3x + 2y = 5$ and $6y = 3x + 19$
31. $3y - 2x + 8 = 0$ and $4x - 9 = 6y$
32. $x + 5y - 7 = 0$ and $5x + y + 10 = 0$

I 33. Fixed cost is the cost of operation no matter how many items are manufactured, while variable cost is the cost of manufacturing each item. Total cost is the sum of fixed and variable costs. Let x represent the number of units manufactured per day and C represent total cost. If fixed cost is \$800 per day and variable cost is \$3 per unit, write an equation relating total cost to units manufactured.

34. The length of a heated object varies with temperature. A metal rod was found to be 12.4 m at 0°C and 12.5 m at 20.0°C. Write an equation for length as a function of temperature.

35. The pressure P on an object under water varies linearly with depth D. Find the equation relating pressure and depth if the pressure at 55 ft is 38 lb/in.^{2} and the pressure at 28 ft is 26 lb/in.^{2}

E **36.** The electric resistance of a resistor increases 0.03 Ω for each 1°C increase in temperature. If its resistance at 0°C is 2 Ω, find the equation relating resistance to temperature.

37. The average velocity v of an object is the ratio of change in displacement s and change in time t. Find the equation relating displacement and time if the average velocity is 20 ft/s and $s = 15$ ft when $t = 0$ s.

38. The length of a spring varies linearly with the force applied. If a spring that is initially 12 in. long measures 16 in. when a force of 4.0 lb is applied, find the equation relating length and force.

E **39.** The resistance of a circuit component increases 0.075 Ω for each 1.0°C increase in temperature. If the resistance is 3.5 Ω at 0°C, write the equation relating resistance to temperature.

40. The acceleration a of an object is the change in velocity v divided by the change in time t. Find the equation relating velocity and time if the acceleration is 24 m/s^2 and $v = 14$ m/s when $t = 0$ s.

15–2

THE CIRCLE

The **circle** is defined as the set of points in a plane that are equidistant from a fixed point. *The fixed point* is called the **center** of the circle, and the distance from *the center to a point on the circle* is called the **radius.** If (h, k) represents the center of the circle and r represents the radius, the standard equation of a circle is as follows.

Standard Form of the Circle

If r represents the radius of a circle and (h, k) represents the center, then

$$(x - h)^2 + (y - k)^2 = r^2$$

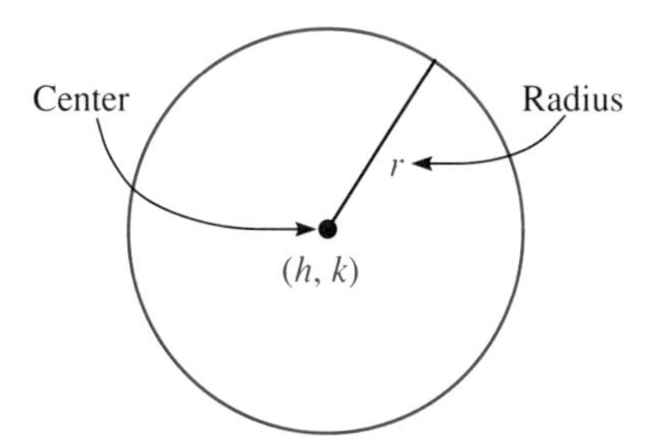

EXAMPLE 1 Determine the center, radius, and graph of the circle represented by the following equation:

$$(x - 4)^2 + (y + 3)^2 = 5$$

Solution Since the equation is in standard form, the center is $(4, -3)$ and the radius is $\sqrt{5}$. To graph the circle, place a point at $(4, -3)$ to represent the center. Then draw a circle $\sqrt{5} \approx 2.2$ units from the center. The graph is shown in Figure 15–2.

FIGURE 15–2

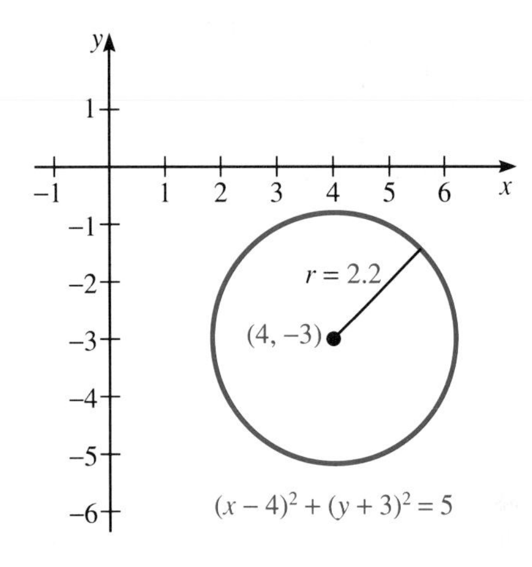

■

General Form

The **general form of the equation of a circle** results from simplifying its standard form.

General Form of the Circle

$$x^2 + y^2 + Ax + By + C = 0$$

where A, B, and C represent constants.

The next two examples illustrate the procedure for finding the center and the radius of a circle from its equation in general form.

EXAMPLE 2 Determine the center, radius, and graph for the following circle:

$$x^2 + y^2 - 2x - 6y + 6 = 0$$

Solution To determine the center and radius, we must arrange this equation in standard form by completing the square on the x and y terms. First, we group the x and y terms and transpose the constant.

$$x^2 - 2x + y^2 - 6y = -6$$

Then we complete the square by adding the square of half the linear term to both sides of the equation.

$$(x^2 - 2x + 1) + (y^2 - 6y + 9) = -6 + 1 + 9$$

Next, we simplify the result:

$$(x - 1)^2 + (y - 3)^2 = 4$$

The center is (1, 3) and the radius is 2. Figure 15–3 shows the graph of the circle.

FIGURE 15–3

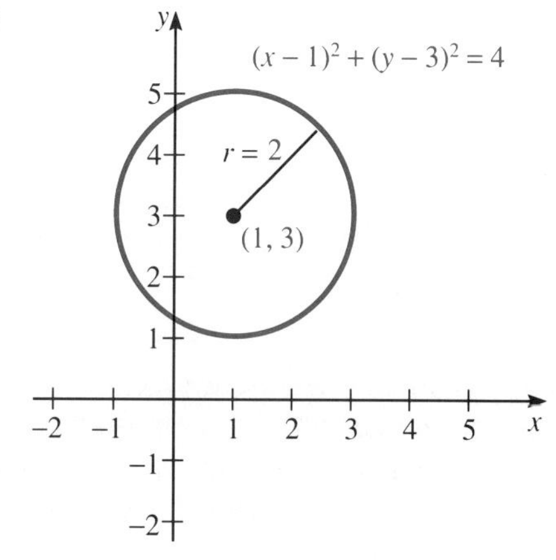

■

EXAMPLE 3 Find the center, radius, and graph of the following circle:

$$3x^2 + 3y^2 + 6x - 12y = 8$$

Solution Unlike the previous equation, the quadratic terms have a coefficient of 3. The standard form of the equation for a circle requires that the coefficient of the squared terms be 1. Therefore, before completing the square, we must divide the equation by 3.

$$3x^2 + 3y^2 + 6x - 12y = 8$$

$$x^2 + y^2 + 2x - 4y = \frac{8}{3}$$

Then we complete the square as in the previous example.

$$(x^2 + 2x + 1) + (y^2 - 4y + 4) = \frac{8}{3} + 1 + 4$$

$$(x + 1)^2 + (y - 2)^2 = \frac{23}{3}$$

The center is $(-1, 2)$ and the radius is $\sqrt{23/3} \approx 2.8$. The graph is shown in Figure 15–4.

FIGURE 15–4

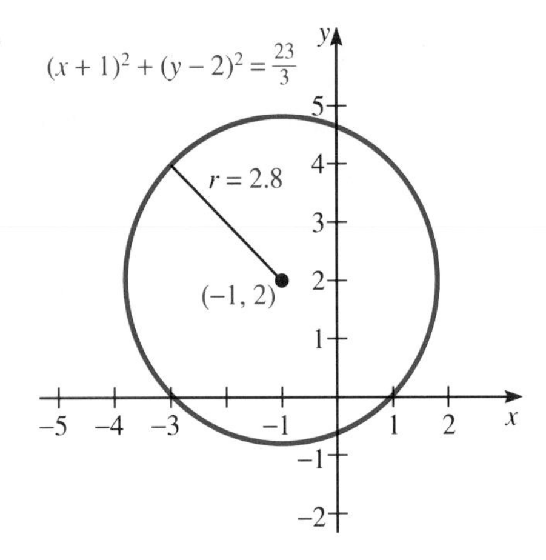

■

CAUTION ✦ When determining the center of a circle, be careful of the signs. The standard form is $(x - h)^2 + (y - k)^2 = r^2$. Thus, the coordinates for the center of the circle have signs opposite those given in the standard form.

Finding the Equation of a Circle

Until now we have found the radius and center of a circle from its equation. However, we must be able to determine the equation of a circle from its geometric information. The next two examples illustrate this process.

EXAMPLE 4 Determine the equation of a circle whose center is at $(-2, 3)$ and radius is 5. Leave your answer in standard form.

Solution To find the equation of this circle, we substitute the coordinates for the center and the value of the radius into the standard form for the equation of a circle.

$$(x - h)^2 + (y - k)^2 = r^2$$

$$h = -2 \qquad k = 3 \qquad r = 5$$

The equation is

$$(x + 2)^2 + (y - 3)^2 = 25$$

■

EXAMPLE 5 Determine the equation of a circle having center $(3, -1)$ and passing through $(-1, 2)$. Leave your answer in standard form.

Solution We are given the center but must determine the radius of the circle. The radius is defined as the distance from the center to any point on the circle represented by the distance from (3, −1) to (−1, 2). Thus,

$$r = \sqrt{(x_2 - x_1)^2 + (y_2 - y_1)^2}$$
$$r = \sqrt{(3 + 1)^2 + (-1 - 2)^2}$$
$$r = \sqrt{25}$$
$$r = 5$$

We substitute into the standard form, $(x - h)^2 + (y - k)^2 = r^2$, the values $h = 3$, $k = -1$, and $r = 5$. The equation is

$$(x - 3)^2 + (y + 1)^2 = 25$$

■

Application

EXAMPLE 6 An archway is composed of a semicircle atop a rectangle. If the rectangular portion is drawn to scale on the rectangular coordinate system, the four vertices of the rectangle are (3, 2), (9, 2), (3, 12), and (9, 12). Find the equation of the circle containing the semicircle top. Use standard form for your answer.

FIGURE 15–5

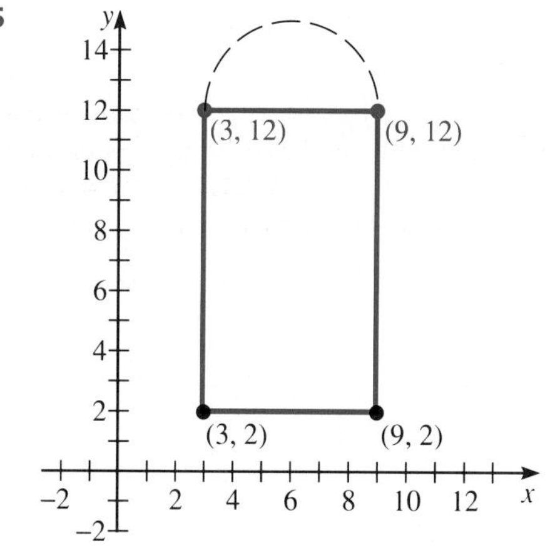

Solution Figure 15–5 shows the rectangle. The center of the circle is represented by (6, 12), which is the midpoint of the line representing the top of the rectangle.

$$x_{\text{mid}} = \frac{3 + 9}{2} = \frac{12}{2} = 6$$

$$y_{\text{mid}} = \frac{12 + 12}{2} = \frac{24}{2} = 12$$

$$\text{center} = (6, 12)$$

The radius is half the diameter represented by the distance between the vertices.

$$r = \frac{9 - 3}{2} = \frac{6}{2} = 3$$

The equation of the circle is given by

$$(x - 6)^2 + (y - 12)^2 = 9$$ ■

15–2 EXERCISES

Find the radius and center of the circle and sketch its graph.

1. $x^2 + y^2 = 9$

2. $x^2 + y^2 = 16$

3. $(x - 3)^2 + (y + 1)^2 = 6$

4. $(x + 4)^2 + (y - 3)^2 = 4$

5. $(x - 1)^2 + y^2 = \sqrt{3}$

6. $x^2 + (y + 5)^2 = \sqrt{2}$

7. $x^2 + y^2 + 8x + 2y = 5$

8. $x^2 + y^2 - 16y - 2x - 8 = 0$

9. $x^2 + y^2 - 6y + 4x = 2$

10. $x^2 + y^2 - 5x + 12y = 4$

11. $2x^2 + 2y^2 + 10x - 4y - 6 = 0$

12. $3x^2 + 3y^2 + 15y - 9x = 0$

13. $3x^2 + 3y^2 + 6x - 8y + 5 = 0$

14. $4x^2 + 4y^2 - 8y + 12x + 7 = 0$

Determine the standard form of the equation for the circle from the given information.

15. Center at (0, 0); radius of 9.

16. Center at the origin; radius of 7.

17. Center at (−3, −1); radius of 3.

18. Center at (6, 10); radius of 4.

19. Center at (0, −4); radius of $\sqrt{6}$.

20. Center at (6, 1); radius of $\sqrt{9}$.

21. Center at (−8, 5); passing through (2, 3).

22. Center at (3, 6); passing through (1, 4).

23. Center at (6, −4); passing through (3, 2).

24. Center at (−7, −6); passing through (−4, 0).

25. Center at (3, 4); tangent to x axis.

26. Center at (−5, 4); tangent to y axis.

27. A diameter is the segment from (−6, −4) to (3, 5).

28. A diameter is the segment from (7, −3) to (−4, 5).

29. Centripetal force F is given by

$$F = \frac{mv^2}{R}$$

where m = mass of an object moving with speed v in a circular path of radius R. A particle moves in the xy plane, and its center is (−4, 6). Find the equation of the circular path traveled by the particle, which has mass 0.0004 kg and moves at 3.1 m/s under a centripetal force of 1,200 N.

M 30. A draftsman is drawing a friction drive as shown in Figure 15–6. Determine the equation for each circle if the origin is at the center of the larger circle and the radii of the circles are 9 cm and 4 cm.

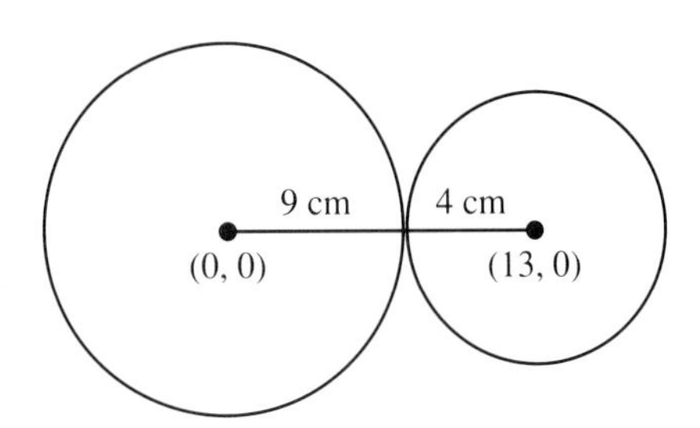

FIGURE 15–6

E 31. The impedance Z in an ac circuit is given by

$$Z^2 = R^2 + (X_L - X_C)^2$$

where R = resistance, X_L = inductive reactance, and X_C = capacitive reactance. Graph the relationship between resistance and inductive reactance if $Z = 30\ \Omega$ and $X_C = 50\ \Omega$.

15–3

THE PARABOLA

Definition

The **parabola** is defined as the set of points in a plane equidistant from a given point and a given line. *The given point* is called the **focus,** and *the given line* is called the **directrix.** *The line through the focus and perpendicular to the directrix* is called the **axis** of the parabola. *The point of intersection of the axis and the parabola* is called the **vertex.** Figure 15–7 illustrates this terminology.

FIGURE 15–7

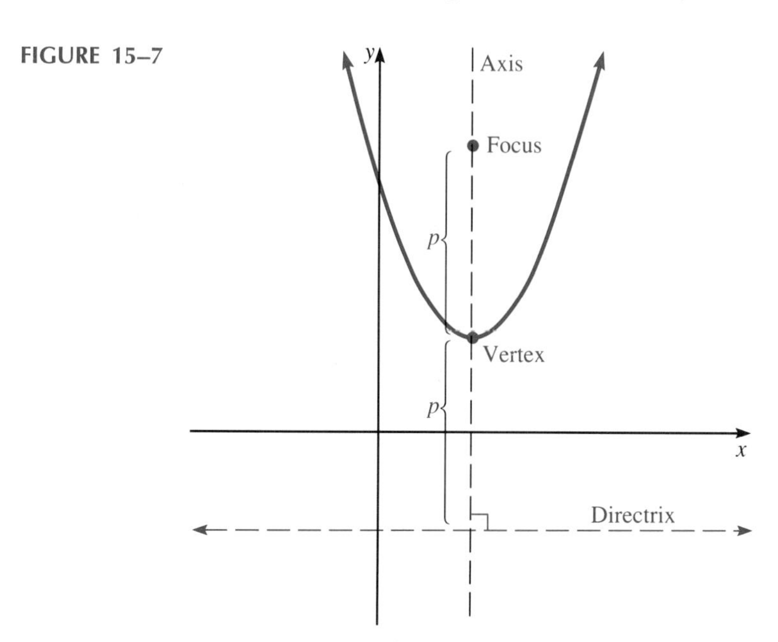

Geometry

The parabola can be placed in such a manner that its axis is a vertical line or a horizontal line. A parabola with a vertical axis is given by the following equation:

$$(x - h)^2 = 4p(y - k)$$

where (h, k) represents the coordinate of the vertex, and p represents the directed distance along the parabola's axis from the vertex to the focus or directrix. If $4p$ is a positive number, the parabola opens upward; if it is a negative value, the parabola opens downward. Figure 15–8(a) and (b) illustrates this terminology.

FIGURE 15–8

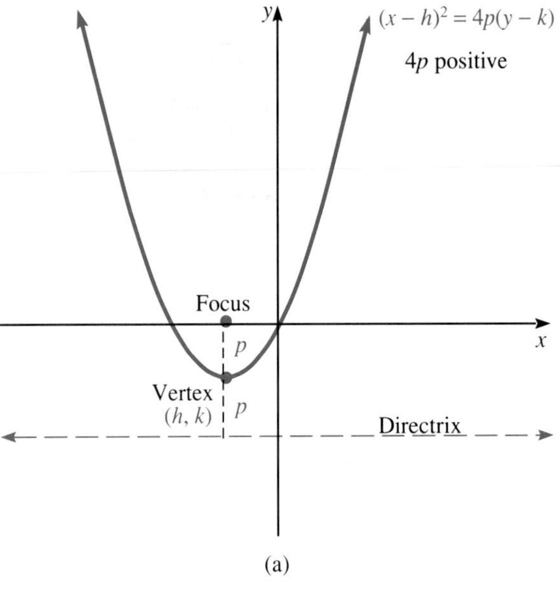

(a)

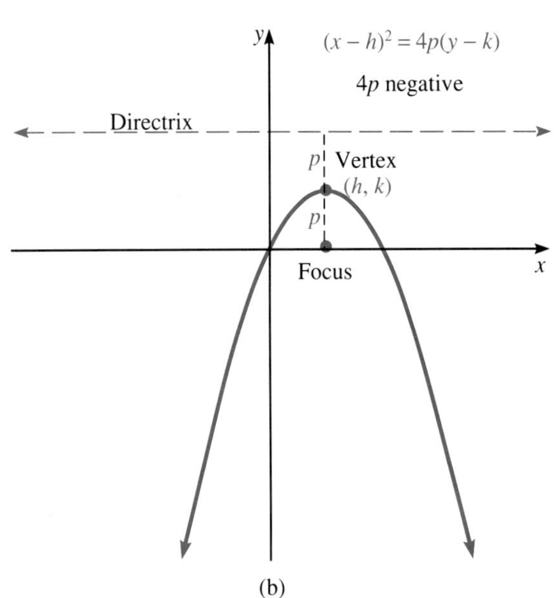

(b)

NOTE ✦ The focus is *always* located inside the parabola, and the directrix is *always* a line outside the parabola.

LEARNING HINT ✦ Rather than trying to memorize formulas, we will find p in each case and count from the vertex to determine the focus and directrix.

Similarly, a parabola whose axis is a horizontal line is represented by the following equation:

$$(y - k)^2 = 4p(x - h)$$

A negative value for $4p$ means that the parabola opens left, and a positive value means that the parabola opens right. Figure 15–9(a) and (b) illustrates this type of parabola.

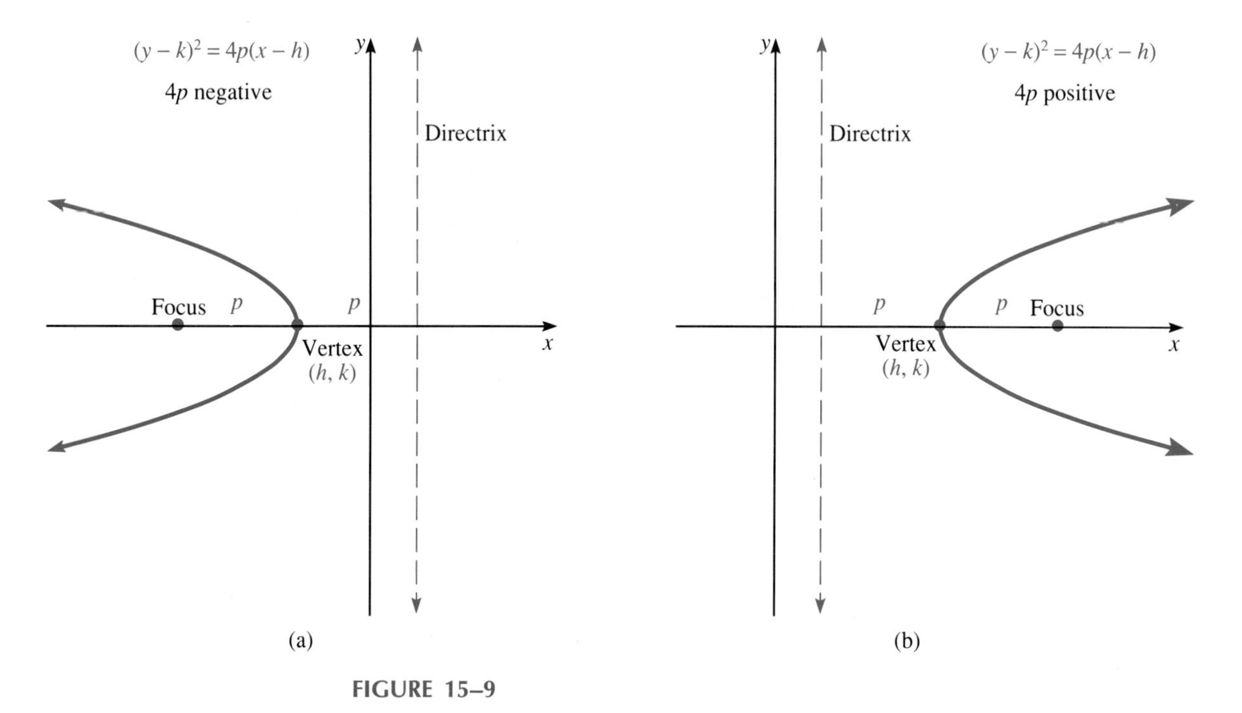

FIGURE 15–9

Finding the Focus, Directrix, and Vertex

EXAMPLE 1 Determine the vertex, directrix, focus, and graph of the parabola given by

$$x^2 = 16y$$

Solution The standard equation for a parabola of this form is

$$(x - h)^2 = 4p(y - k)$$

Writing the given equation in this form yields

$$(x - 0)^2 = 16(y - 0)$$

From this form of the equation, we can determine that the vertex is (0, 0) and $4p = 16$ or $p = 4$. Let us determine the remaining information from the graph shown in Figure 15–10. Since the value of $4p$ is positive, the graph opens upward. The focus is 4 units from the vertex inside the parabola, and the directrix is 4 units from the vertex outside the parabola. The vertex, focus, and directrix are summarized as follows:

vertex: (0, 0)
focus: (0, 4)
directrix: $y = -4$ ■

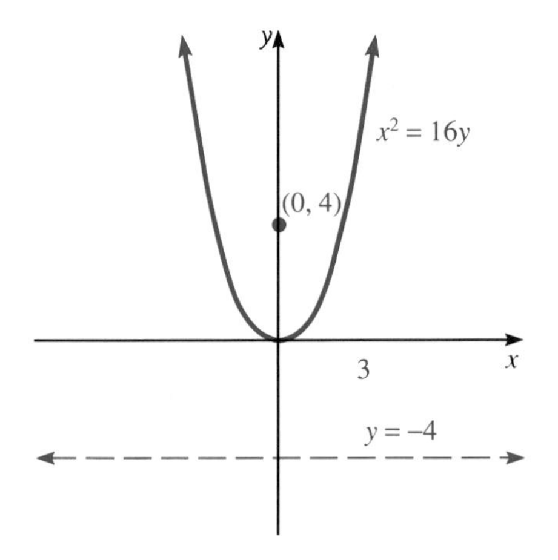

FIGURE 15–10

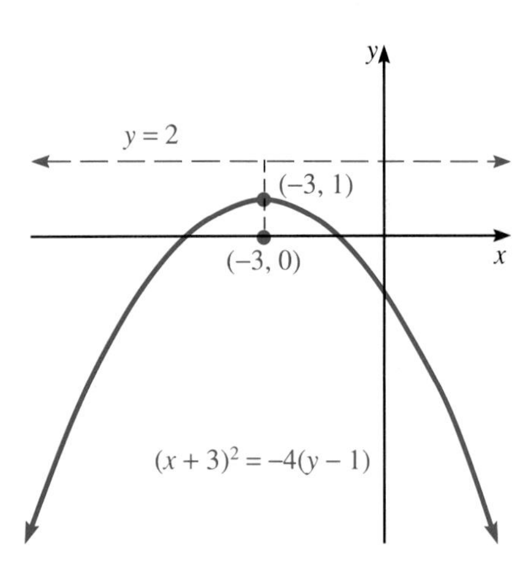

FIGURE 15–11

EXAMPLE 2 Determine the vertex, directrix, focus, and graph of the parabola given by

$$(x + 3)^2 = -4(y - 1)$$

Solution Since the equation is in standard form, the vertex is (−3, 1); the parabola opens down because $4p$ is negative; and $p = -1$. We can determine the focus and directrix from the graph as shown in Figure 15–11. The focus is 1 unit inside the parabola, and the directrix 1 unit outside the parabola.

vertex: (−3, 1)
focus: (−3, 0)
directrix: $y = 2$ ■

EXAMPLE 3 Determine the vertex, directrix, focus, and graph of the parabola given by

$$y^2 + 8y = 12x - 40$$

Solution Since this equation is not in standard form, we must complete the square.

$$
\begin{aligned}
(y^2 + 8y + 16) &= 12x - 40 + 16 && \text{complete the square} \\
(y + 4)^2 &= 12x - 24 && \text{simplify} \\
(y + 4)^2 &= 12(x - 2) && \text{factor}
\end{aligned}
$$

The vertex is $(2, -4)$, the parabola opens to the right, and $p = 3$. The focus, vertex, and directrix are as follows:

$$
\begin{aligned}
\text{vertex: } & (2, -4) \\
\text{focus: } & (5, -4) \\
\text{directrix: } & x = -1
\end{aligned}
$$

The graph is shown in Figure 15–12. ■

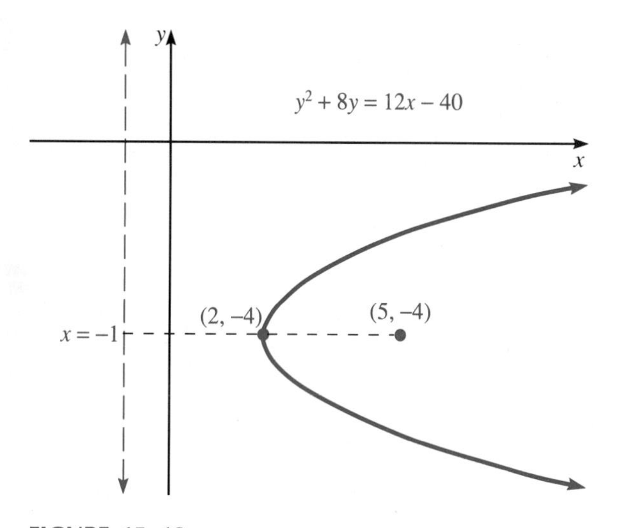

FIGURE 15–12

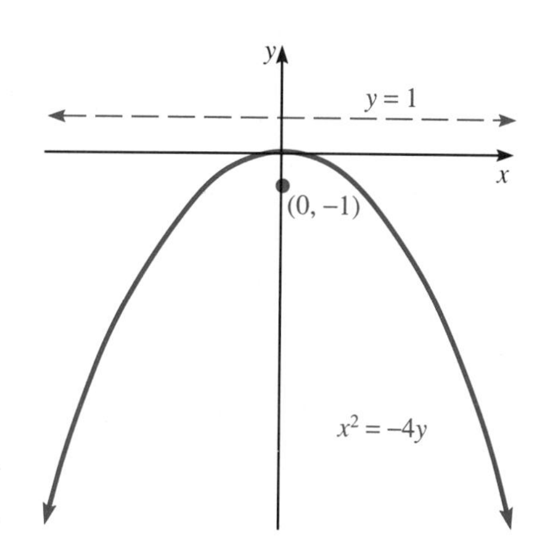

FIGURE 15–13

Finding the Equation of a Parabola

EXAMPLE 4 Find the equation of a parabola whose directrix is the line $y = 1$ and whose vertex is the point $(0, 0)$.

Solution To determine the equation of a parabola, we must fill in the value of p and (h, k). We do not know which form of the formula to use until we look at the position of the directrix. We are given the coordinate of the vertex, but a graph, shown in Figure 15–13, will help us determine the value of p. Since the directrix is outside the parabola, we know that the parabola opens downward. We also know that p is the directed distance along the parabola's axis from the vertex to the focus and directrix; therefore, $p = -1$. The equation of the parabola is

$$x^2 = -4y$$ ■

EXAMPLE 5 Find the equation of the parabola whose directrix is $x = 5$ and whose focus is $(1, -2)$.

Solution Figure 15–14 shows a graph of the parabola. We know that the focus is located on the axis of the parabola, and the vertex is located on the axis of the parabola halfway between the focus and the directrix. Therefore, the coordinates of the vertex are $(3, -2)$, and the value of p is -2. The equation of the parabola is given by

$$(y + 2)^2 = 4(-2)(x - 3)$$
$$(y + 2)^2 = -8(x - 3)$$ ■

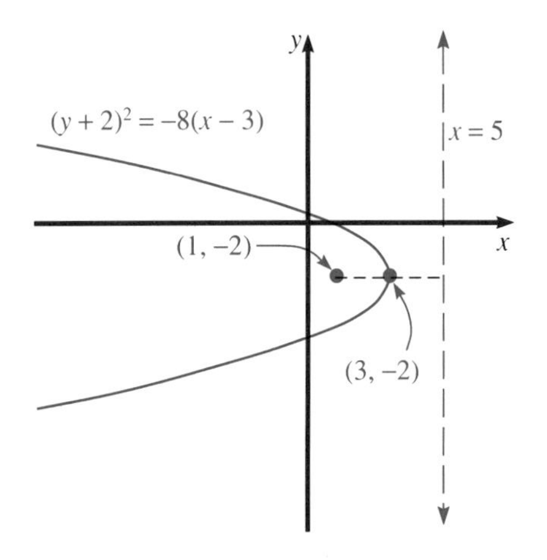

FIGURE 15–14

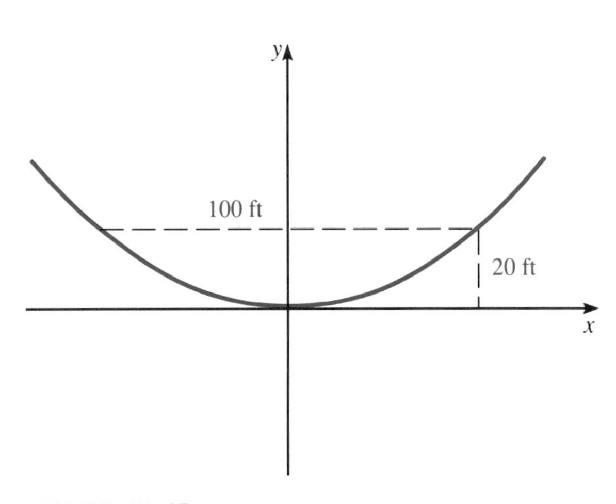

FIGURE 15–15

Application

EXAMPLE 6 A cable used to support a swinging bridge approximates the shape of a parabola. Determine the equation of the parabola if the length of the bridge is 100 ft and the vertical distance from where the cable is attached to the bridge to the lowest point of the cable is 20 ft. Assume that the origin is at the lowest point of the cable.

Solution Figure 15–15 gives a sketch of the parabola. Since we can assume that the origin is at the lowest point, the vertex is (0, 0), and the general form of the equation is $x^2 = 4py$. Next, we must determine the value of p. From the information given, the points (50, 20) and $(-50, 20)$ are on the curve. Substituting the point (50, 20) to determine p gives

$$\begin{aligned} x^2 &= 4py \\ (50)^2 &= 4p(20) \\ 2{,}500 &= 80p \\ 31.25 &= p \end{aligned}$$

The equation for the parabola is

$$\begin{aligned} x^2 &= 4py \\ x^2 &= 125y \end{aligned}$$

■

15–3 EXERCISES

Find the coordinates of the vertex and focus, determine the equation of the directrix, and sketch the graph of the parabola given by each of the following equations.

1. $y^2 = 8x$

2. $y^2 = 12x$

3. $x^2 = 12y$

4. $x^2 = 2y$

5. $y^2 = -16x$

6. $y^2 = -20x$

7. $x^2 = -10y$

8. $x^2 = -y$

9. $(y - 3)^2 = -8x$

10. $(y + 5)^2 = 6x$

11. $(x + 2)^2 = 12(y - 3)$

12. $(x - 3)^2 = -8(y - 2)$

13. $(y - 6)^2 = -2(x + 4)$

14. $(y + 2)^2 = 3(x - 5)$

15. $(x - 7)^2 = -16(y + 3)$

16. $(x + 8)^2 = 4(y + 1)$

17. $x^2 + 2x = -y + 3$

18. $x^2 - 4y + 8x = 0$

19. $y^2 + 10y = 2x - 15$

20. $y^2 - 12y + 8x - 20 = 0$

21. $3x^2 + 6x - 4y + 15 = 0$

22. $2x^2 + 8x + 4y - 20 = 0$

Find the equation of the parabola from the given geometric properties.

23. Vertex (0, 0); directrix $y = 4$.

24. Focus (1, 8); directrix $y = -2$.

25. Vertex (3, 2); focus (3, 5).

26. Directrix $x = 4$; vertex (6, 3).

27. Directrix $x = -3$; focus (1, 4).

28. Vertex $(-4, 6)$; focus $(-4, 5)$.

29. Vertex $(0, -3)$, passes through (4, 7), and directrix is parallel to the x axis.

30. Directrix $x = -3$; focus $(-1, -2)$.

31. Vertex $(3, -6)$; focus $(3, -2)$.

32. Vertex $(3, -2)$, passes through (1, 6); axis parallel to y axis.

33. The height s of a projectile is given by

$$s = v(\sin\theta)t + \frac{at^2}{2}$$

If $v = 50$ m/s, $\theta = 30°$, and $a = -9.8$ m/s^2, write the equation in standard form.

34. A cable hangs in a parabolic curve between two poles 70 ft apart. At a distance 20 ft from each pole, the cable is 4 ft above its lowest point. Find the height of the cable 45 ft from each pole.

35. The stopping distance y of a car traveling at v mi/h is given by

$$y = v + \frac{v^2}{20}$$

Arrange this equation in standard form and plot its graph.

36. If a light beam passes through the focus of a parabolic reflector, the light rays will be reflected on a line parallel to the axis of the parabola. A light beam passes through the focus of a parabolic reflector with its cross-section represented by $y^2 = 18x$ and strikes the reflector at (0.5, 3). Find the equation of the reflected line.

E 37. Heat H in a resistor R develops at a rate given by $H = Ri^2$ where i = current. Graph H versus i if $R = 10\ \Omega$.

38. A shipping lane is determined by points equidistant from a lighthouse and the opposite shoreline. Determine the equation for the shipping lane if the lighthouse is 3 mi from the shoreline.

15–4

THE ELLIPSE

Definition

The **ellipse** is defined as *the set of points in the plane the sum of whose distance from two fixed points is constant. The two fixed points* are called the **foci** (plural of **focus**) of the ellipse. The **center** of the ellipse is *the midpoint of the line segment joining the foci. The line segment whose endpoints are on the ellipse and passing through the foci* is called the **major axis.** The major axis is also the longer axis. *The endpoints of the major axis* are called the **vertices** of the ellipse. *The line segment whose endpoints are on the ellipse and passing through the center perpendicular to the major axis* is called the **minor axis.** *The endpoints of the minor axis* are simply called the **endpoints of the minor axis.** This terminology is illustrated in Figure 15–16.

FIGURE 15–16

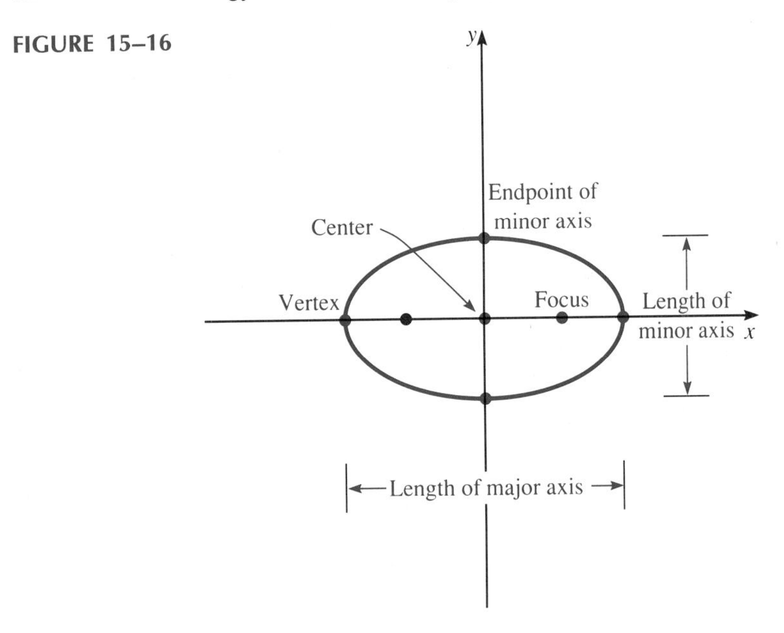

Geometry

An ellipse may have its major axis horizontal or vertical. The equation of *an ellipse elongated horizontally* is given by

$$\frac{(x - h)^2}{a^2} + \frac{(y - k)^2}{b^2} = 1 \qquad a^2 > b^2$$

where (h, k) represent the coordinates of the center, a is the distance from the center to the vertices, and b is the distance from the center to the endpoints of the minor axis. The distance from the center to either focus is represented by c where

$$c = \sqrt{a^2 - b^2}$$

Notice that the foci are always located on the major axis. Figure 15–17 illustrates the terminology for this ellipse.

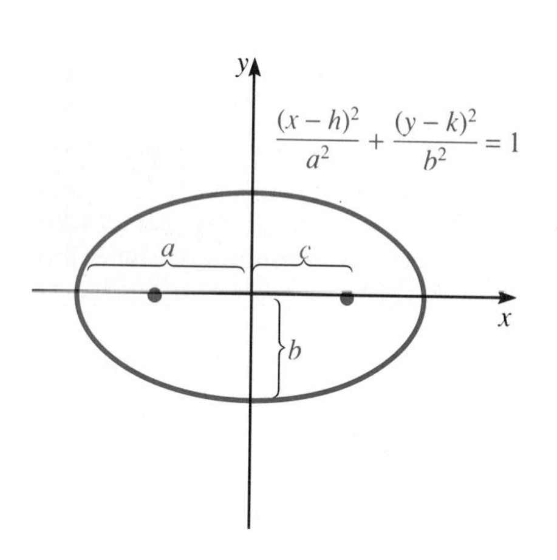

FIGURE 15–17

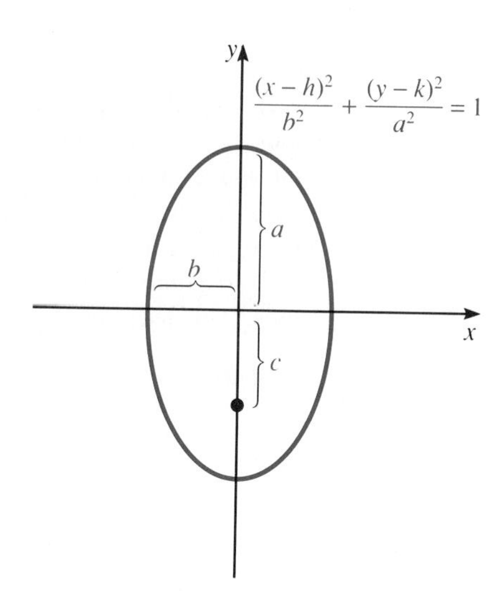

FIGURE 15–18

Similarly, *an ellipse elongated vertically* is represented by the equation

$$\frac{(x - h)^2}{b^2} + \frac{(y - k)^2}{a^2} = 1$$

where h, k, a, and b represent the same quantities as above. The major axis, vertices, and foci are located on a vertical line. Figure 15–18 illustrates the terminology for this ellipse.

LEARNING HINT ✦ For an equation of an ellipse in standard form, a^2 is the larger number in the denominator. Furthermore, the term under which a^2 appears determines the direction of elongation.

EXAMPLE 1 Find the coordinates for the center, vertices, endpoints of the minor axis, and foci of the ellipse represented by

$$16x^2 + 25y^2 = 400$$

Also sketch the graph.

Solution To begin, we place this equation in standard form by dividing the equation by 400.

$$\frac{x^2}{25} + \frac{y^2}{16} = 1 \quad \text{or} \quad \frac{(x - 0)^2}{25} + \frac{(y - 0)^2}{16} = 1$$

The center is at the origin. Since the larger denominator is under x^2, the major axis is horizontal. To find the vertices, move $\sqrt{25} = 5$ units horizontally from the center. To find the endpoints of the minor axis, move $\sqrt{16} = 4$ units vertically from the center. To find the coordinates of the foci, move $\sqrt{25 - 16} = 3$ units horizontally from the center. Figure 15–19 gives the graph. The results are summarized as follows:

center: $(0, 0)$
vertices: $(-5, 0)$ and $(5, 0)$
endpoints of minor axis: $(0, 4)$ and $(0, -4)$
foci: $(3, 0)$ and $(-3, 0)$

FIGURE 15–19

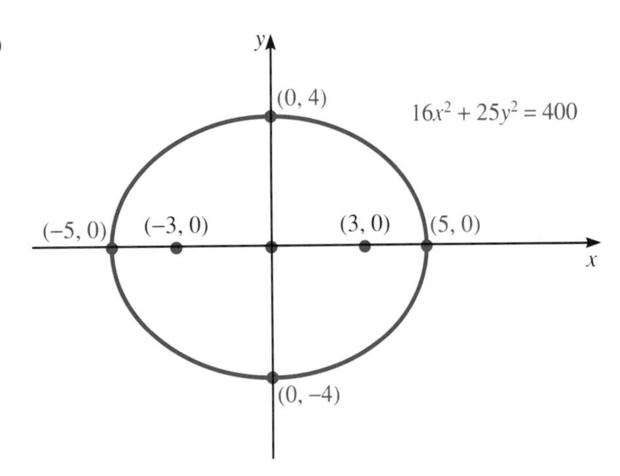

■

EXAMPLE 2 Find the coordinates for the center, vertices, endpoints of the minor axis, and foci of the ellipse represented by

$$\frac{(x - 3)^2}{4} + \frac{(y + 1)^2}{49} = 1$$

Also sketch the graph.

Solution Since the equation is in standard form, we can determine that the center is the point $(3, -1)$. Also, because the larger denominator is under the y term, the major axis is vertical. To determine the vertices, move $\sqrt{49} = 7$ units vertically from the center. To determine the endpoints of the minor axis, move $\sqrt{4} = 2$ units horizontally from the center. To determine the coordinates of the foci, move $\sqrt{49 - 4} = \sqrt{45} \approx 6.7$ units vertically from the center. The graph is shown in Figure 15–20, and the results are summarized as follows:

center: $(3, -1)$
vertices: $(3, 6)$ and $(3, -8)$
endpoints of minor axis: $(1, -1)$ and $(5, -1)$
foci: $(3, 5.7)$ and $(3, -7.7)$

FIGURE 15–20

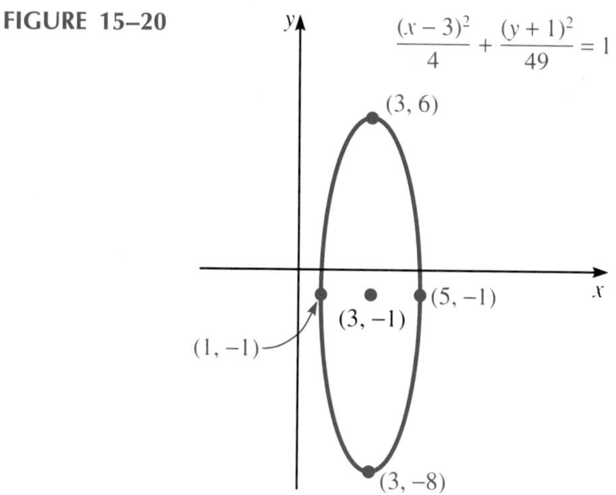

■

EXAMPLE 3 Find the coordinates for the center, vertices, endpoints of the minor axis, and foci of the ellipse represented by the equation

$$2x^2 + 3y^2 + 8x - 18y = 1$$

Also sketch the graph.

Solution First, we begin by arranging the equation in standard form and completing the square.

$$(2x^2 + 8x) + (3y^2 - 18y) = 1 \qquad \text{group terms}$$
$$2(x^2 + 4x) + 3(y^2 - 6y) = 1 \qquad \text{factor}$$
$$2(x^2 + 4x + 4) + 3(y^2 - 6y + 9) = 1 + 8 + 27 \qquad \text{complete the square}$$
$$2(x + 2)^2 + 3(y - 3)^2 = 36 \qquad \text{factor}$$
$$\frac{(x + 2)^2}{18} + \frac{(y - 3)^2}{12} = 1 \qquad \text{divide}$$

The graph is shown in Figure 15–21. Summarizing the information gives

$$a = \sqrt{18} \approx 4.2 \qquad b = \sqrt{12} \approx 3.5 \qquad c = \sqrt{18 - 12} \approx 2.4$$

center: $(-2, 3)$
vertices: $(2.2, 3)$ and $(-6.2, 3)$
endpoints of minor axis: $(-2, 6.5)$ and $(-2, -0.5)$
foci: $(0.4, 3)$ and $(-4.4, 3)$ ■

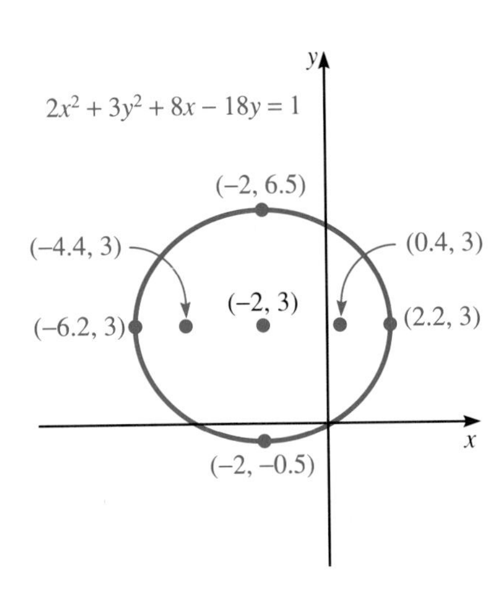

FIGURE 15–21

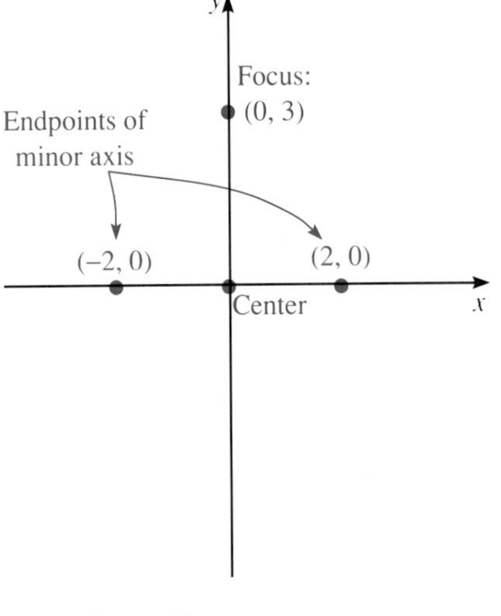

FIGURE 15–22

Finding the Equation of an Ellipse

Since we know the geometric properties of the ellipse, we can use them to find its equation. The next three examples illustrate this process.

EXAMPLE 4 Determine the equation of an ellipse centered at the origin if one focus is at (0, 3) and the endpoints of the minor axis are at (2, 0) and (−2, 0). Figure 15–22 shows this information.

Solution To determine the equation, we must know the values of h, k, a^2, and b^2. Since the center is at the origin, $h = k = 0$. Since the endpoints of the minor axis are (2, 0) and (−2, 0), $b = 2$ and $b^2 = 4$. From the location of the focus, we know that $c = 3$ and $c^2 = 9$. Then $c^2 = a^2 - b^2$ and $c^2 + b^2 = a^2$. Thus,

$$a^2 = 9 + 4$$
$$a^2 = 13$$

The major axis is in the y direction; therefore, the equation is

$$\frac{(x - \overset{h}{0})^2}{\underset{b^2}{4}} + \frac{(y - \overset{k}{0})^2}{\underset{a^2}{13}} = 1$$

■

EXAMPLE 5 Determine the equation of an ellipse if the foci are at $(-6, 3)$ and $(4, 3)$ and the length of the major axis is 14. Figure 15–23 shows the given information.

FIGURE 15–23

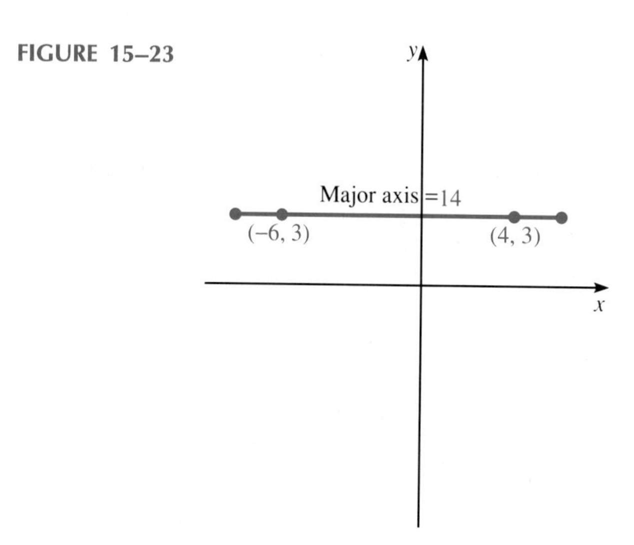

Solution Once again we must determine h, k, a^2, and b^2. The center of the ellipse is located midway between the foci. The center is

$$\left(\frac{-6 + 4}{2}, \frac{3 + 3}{2}\right) = (-1, 3)$$

Therefore, $h = -1$ and $k = 3$. Since c represents the distance from the center to each focus, $c = 5$ and $c^2 = 25$. Because the length of the major axis is 14, $2a = 14$, $a = 7$, and $a^2 = 49$. From this information we can determine b^2.

$$c^2 = a^2 - b^2$$
$$b^2 = a^2 - c^2$$
$$b^2 = 49 - 25$$
$$b^2 = 24$$

From the location of the foci, we can determine that the major axis is horizontal; therefore, the equation is

$$\frac{(x - (-1))^2}{49} + \frac{(y - 3)^2}{24} = 1$$

$$\frac{(x + 1)^2}{49} + \frac{(y - 3)^2}{24} = 1$$

■

EXAMPLE 6 A semielliptical arch is used in the construction of a shopping center. If the span of the archway is 150 ft and the maximum height is 20 ft, determine the equation of the archway.

Solution Figure 15–24 shows the archway. For convenience let us place the origin at the center of the semiellipse. From the drawing we can determine that $a = 75$ ft and $b = 20$ ft. Therefore, the equation of the ellipse is

$$\frac{x^2}{5{,}625} + \frac{y^2}{400} = 1$$

FIGURE 15–24

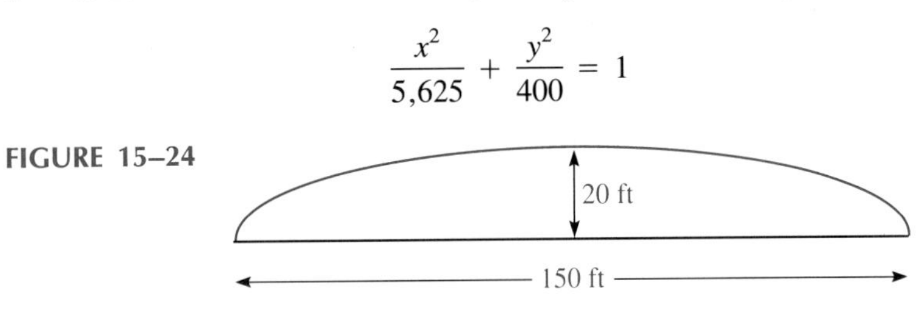

■

15–4 EXERCISES

Determine the coordinates for the center, vertices, endpoints of the minor axis, and foci of the ellipse. Round to tenths, if necessary. Sketch the graph.

1. $\frac{x^2}{25} + \frac{y^2}{9} = 1$
2. $\frac{x^2}{49} + \frac{y^2}{25} = 1$
3. $\frac{x^2}{16} + \frac{y^2}{36} = 1$
4. $\frac{x^2}{16} + \frac{y^2}{81} = 1$
5. $\frac{x^2}{10} + \frac{y^2}{6} = 1$
6. $\frac{x^2}{4} + \frac{y^2}{36} = 1$
7. $\frac{(x + 2)^2}{4} + \frac{(y - 6)^2}{25} = 1$
8. $\frac{(x - 3)^2}{64} + \frac{(y - 2)^2}{25} = 1$
9. $\frac{(x + 1)^2}{36} + \frac{y^2}{81} = 1$
10. $\frac{(y + 6)^2}{8} + \frac{(x - 4)^2}{16} = 1$
11. $\frac{(y - 1)^2}{4} + \frac{(x - 3)^2}{12} = 1$
12. $\frac{(x - 2)^2}{10} + \frac{(y + 7)^2}{36} = 1$
13. $\frac{(x + 8)^2}{30} + \frac{(y + 4)^2}{12} = 1$
14. $3x^2 + 7y^2 = 21$
15. $16x^2 + 36y^2 - 144 = 0$
16. $9x^2 + 25y^2 = 225$
17. $x^2 + 8x + 3y^2 - 12y - 1 = 0$
18. $y^2 + 5x^2 - 20x = 20$
19. $4x^2 + 7y^2 + 14y + 24x = 75$
20. $8x^2 - 32x + 2y^2 - 12y - 23 = 0$
21. $6x^2 + y^2 + 24x = 12$
22. $x^2 + 6y^2 - 24y - 40 = 0$

Find the equation of the ellipse from the given geometric properties.

23. Center at origin; endpoint of major axis (0, 7); and focus at (0, 3).

24. Center at origin; endpoint of minor axis (5, 0); and focus at $(0, \sqrt{12})$.

25. Center at (4, −3); focus at (3, −3); and length of minor axis is 4.

26. Center at (1, 5); length of minor axis is 6; length of major axis is 10; and major axis is horizontal.

27. Foci at (−7, −3) and (1, −3); length of minor axis is 6.

28. Vertices at (3, −1) and (3, 7); focus at (3, 1).

29. Vertices at (−3, −8) and (−3, 4); length of minor axis is 10.

30. Endpoints of the minor axis at (4, 10) and (4, 7); length of major axis is 16.

31. A satellite orbits the earth in an elliptical path. The maximum altitude of the satellite is 325 mi, the minimum altitude is 130 mi, and the earth's center is at one focus. Determine the equation of the satellite's path. (The radius of the earth is 3,960 mi.)

C 32. A bridge in the shape of a semiellipse spans 80 ft and is 20 ft high at its maximum. What is the maximum height of a truck that can cross the bridge at a point 30 ft from the center?

15–5

THE HYPERBOLA

Definition

The **hyperbola** is defined as *the set of points in a plane the difference of whose distance from two fixed points is a constant. The two fixed points* are called the **foci.** *The midpoint of the line segment joining the foci* is called the **center** of the hyperbola. *The line segment whose endpoints are on the hyperbola and if extended would join the foci* is called the **transverse axis.** *The points where the hyperbola intersects the transverse axis* are called the **vertices** of the hyperbola. *The line segment passing through the center and perpendicular to the transverse axis* is called the **conjugate axis.** These definitions are illustrated in Figure 15–25.

FIGURE 15–25

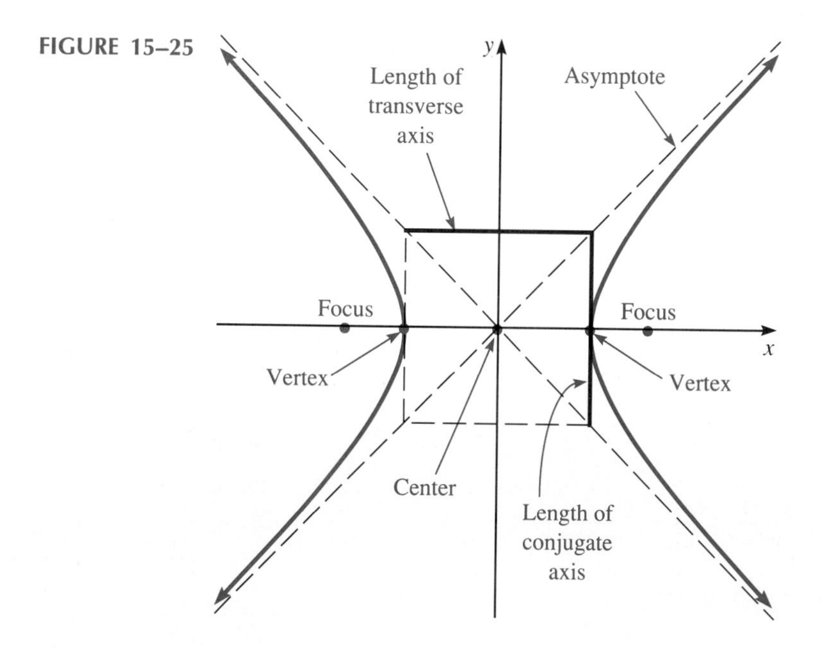

Geometry

A hyperbola can have a horizontal or a vertical transverse axis. The equation of a *hyperbola with a horizontal transverse axis* is given by

$$\frac{(x - h)^2}{a^2} - \frac{(y - k)^2}{b^2} = 1$$

where the point (h, k) is the center, a represents the distance from the center to a vertex, and b represents the distance from the center to an endpoint of the conjugate axis. The slopes of the asymptotes are given by

$$m = \pm\frac{b}{a}$$

The foci are a distance c from the center where $c = \sqrt{a^2 + b^2}$ along the transverse axis. Figure 15–26 illustrates this terminology.

FIGURE 15–26

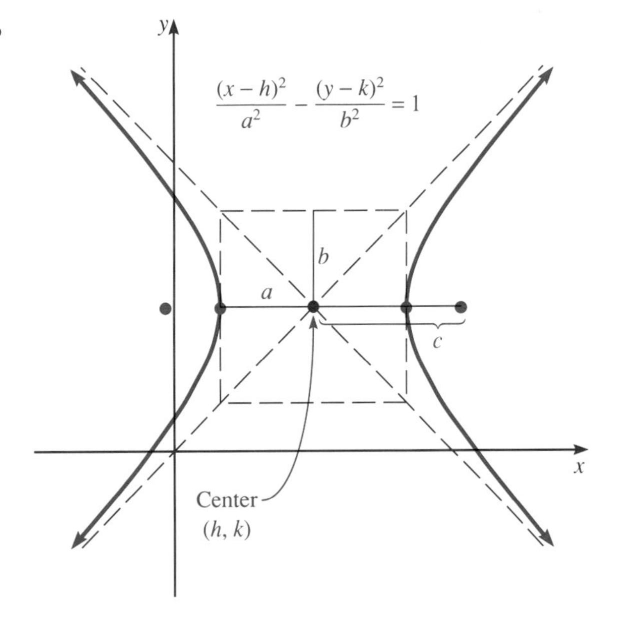

The equation of a *hyperbola whose transverse axis is vertical* is given by

$$\frac{(y - k)^2}{a^2} - \frac{(x - h)^2}{b^2} = 1$$

where h, k, a, and b represent the same quantities as above. The slopes of the asymptotes are given by

$$m = \pm\frac{a}{b}$$

Figure 15–27 illustrates this hyperbola.

FIGURE 15–27

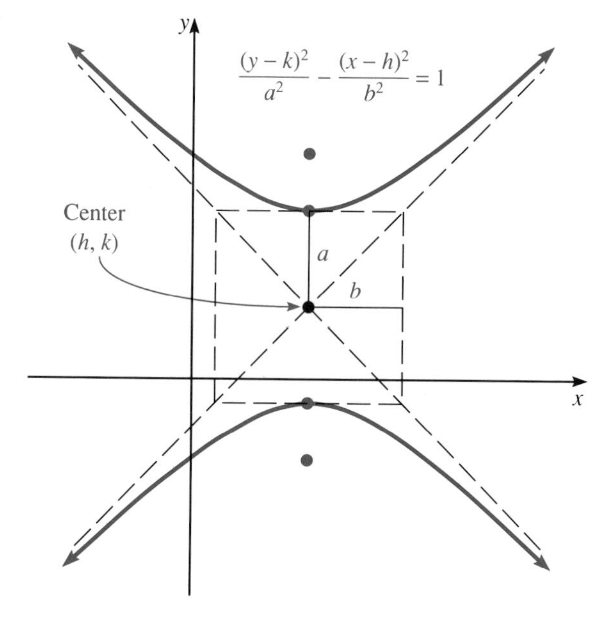

LEARNING HINT ✦ If the equation is in standard form, the direction of opening of the hyperbola is determined by which term is positive. If the y term is positive, the hyperbola opens upward and downward. If the x term is positive, the hyperbola opens left and right. The letter a^2 always appears in the denominator of the positive term (and a^2 need *not* be larger than b^2).

To graph a hyperbola, locate the center. Use the distance a to locate the vertices, and use the distance b to locate the endpoints of the conjugate axis. Draw a rectangle through the vertices and endpoints. The asymptotes pass through the center and opposite corners of this rectangle. Finally, draw the hyperbola starting at the vertices and extending toward the asymptotes.

EXAMPLE 1 Determine the foci, center, vertices, endpoints of the conjugate axis, and slopes of the asymptotes of the following hyperbola:

$$\frac{x^2}{16} - \frac{y^2}{25} = 1$$

Also sketch the graph.

Solution Since the equation is in standard form, the center is at (0, 0), $a = \sqrt{16} = 4$, $b = \sqrt{25} = 5$, and the transverse axis is horizontal. Figure 15–28 shows the graph of this hyperbola, and the results are summarized as follows:

center:	(0, 0)
vertices:	(4, 0) and (−4, 0)
endpoints of conjugate axis:	(0, 5) and (0, −5)
foci:	(6.4, 0) and (−6.4, 0)
	$(c = \sqrt{a^2 + b^2} = \sqrt{16 + 25} \approx 6.4)$
slope of asymptotes:	$m = \pm\frac{5}{4}$

FIGURE 15–28

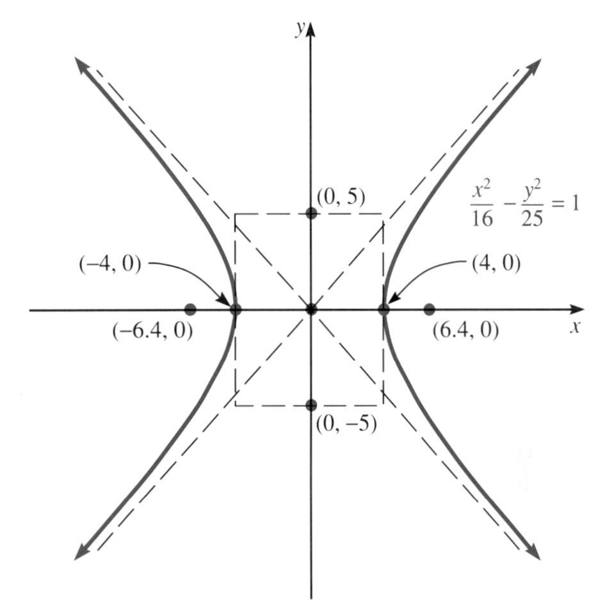

■

EXAMPLE 2 Determine the foci, center, vertices, endpoints of the conjugate axis, and slopes of the asymptotes of the following hyperbola:

$$\frac{(y + 1)^2}{36} - \frac{(x - 3)^2}{81} = 1$$

Also sketch the graph.

Solution Since this equation is in standard form, we know that the center is (3, −1), $a = 6$, $b = 9$, and the transverse axis is vertical. The distance from

FIGURE 15–29

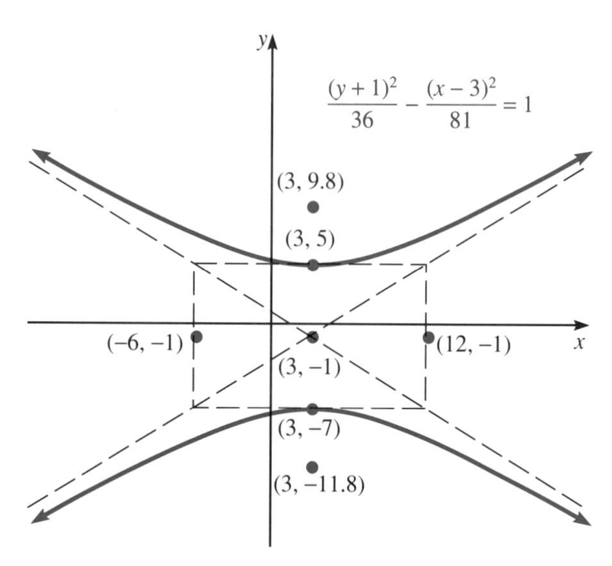

the center to the foci is given by $c = \sqrt{36 + 81} = \sqrt{117} \approx 10.8$. Figure 15–29 shows the graph, and the results are summarized as follows:

$$\begin{aligned} \text{center:} &\quad (3, -1) \\ \text{vertices:} &\quad (3, 5) \text{ and } (3, -7) \\ \text{endpoints of conjugate axis:} &\quad (12, -1) \text{ and } (-6, -1) \\ \text{foci:} &\quad (3, 9.8) \text{ and } (3, -11.8) \\ \text{slope of asymptotes:} &\quad m = \pm\frac{2}{3} \end{aligned}$$

■

EXAMPLE 3 Determine the foci, center, vertices, endpoints of the conjugate axis, and equations of the asymptotes of the hyperbola given by

$$4y^2 - 24y - x^2 + 6x = 37$$

Also sketch the graph.

Solution First, we must arrange the equation in standard form by completing the square.

$$\begin{aligned} 4y^2 - 24y - x^2 + 6x &= 37 \\ 4(y^2 - 6y) - (x^2 - 6x) &= 37 \\ 4(y^2 - 6y + 9) - (x^2 - 6x + 9) &= 37 + 36 - 9 \\ 4(y - 3)^2 - (x - 3)^2 &= 64 \\ \frac{(y - 3)^2}{16} - \frac{(x - 3)^2}{64} &= 1 \end{aligned}$$

From the equation in standard form, we know that the center is (3, 3), $a = 4$, $b = 8$, and the transverse axis is vertical. Figure 15–30 shows the graph, and the results are summarized as follows:

center:	(3, 3)
vertices:	(3, 7) and (3, −1)
endpoints of conjugate axis:	(11, 3) and (−5, 3)
foci:	(3, 11.9) and (3, −5.9)
	$c = \sqrt{a^2 + b^2} \approx 8.9$
slopes of asymptotes:	$m = \pm\frac{1}{2}$

■

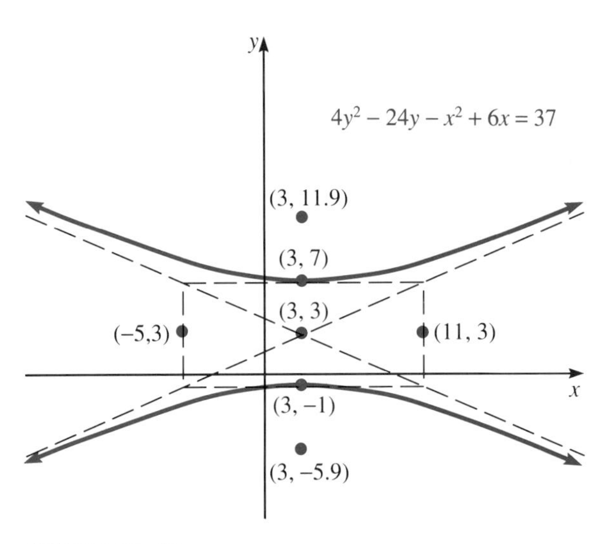

FIGURE 15–30

y
Vertex
(−3, 0)
Center
Focus
(5, 0)
x

FIGURE 15–31

Finding the Equation of a Hyperbola

EXAMPLE 4 Find the equation of the hyperbola with its center at the origin, one focus at (5, 0), and one vertex at (−3, 0). This information is shown in Figure 15–31.

Solution In order to find the equation of the hyperbola, we must fill in values for h, k, a^2, and b^2. From the information given, we know that $h = 0$ and $k = 0$. We also know that $a = 3$ and $c = 5$. From this information we can determine b^2.

$$c^2 = a^2 + b^2$$
$$b^2 = c^2 - a^2$$
$$b^2 = 25 - 9$$
$$b^2 = 16$$

From the location of the vertex and focus, we can also determine that the transverse axis is horizontal (the x term is positive). The equation is

$$\frac{x^2}{9} - \frac{y^2}{16} = 1$$

■

EXAMPLE 5 Determine the equation of the hyperbola given that its center is at $(-4, 2)$, its vertex is at $(-4, 7)$, and the slope of an asymptote is 5/2.

Solution To determine the equation, we must find h, k, a^2, and b^2. Since the center is $(-4, 2)$, $h = -4$ and $k = 2$. Since the vertex is located at $(-4, 7)$, the transverse axis is a vertical line (the y term in the equation is positive). From the equation of the asymptote, we know that $b = 2$ and $a = 5$. The equation is

$$\frac{(y - 2)^2}{25} + \frac{(x + 4)^2}{4} = 1$$

■

Application

EXAMPLE 6 The impedance of an ac circuit is given by

$$Z^2 = R^2 + (X_L - X_C)^2$$

where R = resistance, X_L = inductive reactance, and X_C = capacitive reactance. Graph the relationship between Z and R if $(X_L - X_C) = 20\ \Omega$. Let Z be the independent variable.

Solution Since $(X_L - X_C) = 20$, the equation becomes

$$Z^2 = R^2 + 400$$

$$Z^2 - R^2 = 400$$

$$\frac{Z^2}{400} - \frac{R^2}{400} = 1$$

Since $a^2 = 400$, $a = 20$, $b^2 = 400$, $b = 20$, $c = \sqrt{a^2 + b^2}$, $c \approx 28.3$, and because Z is the independent variable, the transverse axis is horizontal. The graph is shown in Figure 15–32. ■

Hyperbolas of the Form $xy = k$

The hyperbolas discussed thus far are quadratic in form. However, there is another type of equation that produces a hyperbola. An equation of the form $xy = k$ where k is some nonzero constant produces a hyperbola. The x and y axes serve as the asymptotes. Also, if k is positive, the hyperbola lies in quadrants I and III, and if k is negative, the hyperbola lies in quadrants II and IV. The vertices are given by $(\sqrt{|k|}, \sqrt{|k|})$ placed in the appropriate quadrant with appropriate signs. The foci are given by $(\sqrt{|2k|}, \sqrt{|2k|})$ placed in the appropriate quadrant with appropriate signs. The transverse axis is $y = x$ if $k > 0$, and $y = -x$ if $k < 0$.

FIGURE 15–32

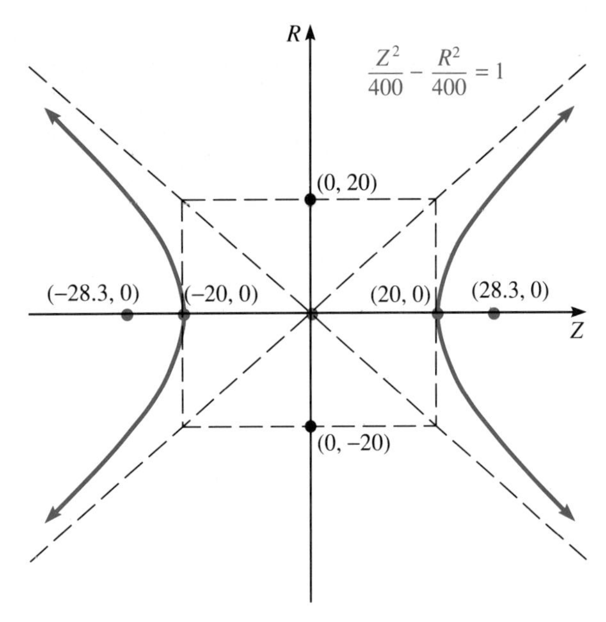

EXAMPLE 7 Graph $xy = -4$.

Solution Since $k < 0$, the hyperbola is located in quadrants II and IV. The transverse axis is the line $y = -x$. The vertices are located at $(-2, 2)$ and $(2, -2)$, and the foci are located at $(-2.8, 2.8)$ and $(2.8, -2.8)$. The graph is shown in Figure 15–33.

FIGURE 15–33

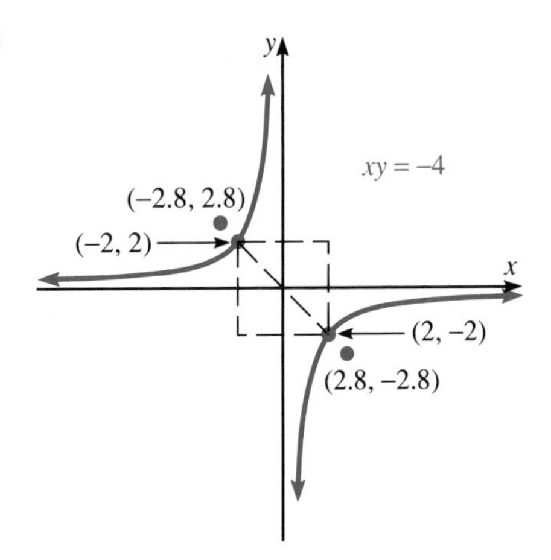

■

15–5 EXERCISES

For the following hyperbolas, determine the coordinates of the center, vertices, foci, and endpoints of the conjugate axis. Find the slopes of the asymptotes, and sketch the graph. Round to tenths if necessary.

1. $\dfrac{x^2}{24} - \dfrac{y^2}{4} = 1$

2. $\dfrac{x^2}{36} - \dfrac{y^2}{9} = 1$

3. $\dfrac{y^2}{49} - \dfrac{x^2}{16} = 1$

4. $\dfrac{y^2}{81} - \dfrac{x^2}{64} = 1$

5. $\dfrac{x^2}{64} - \dfrac{y^2}{25} = 1$

6. $\dfrac{y^2}{9} - \dfrac{x^2}{49} = 1$

7. $\dfrac{(y + 1)^2}{16} - \dfrac{(x - 2)^2}{81} = 1$

8. $\dfrac{(x + 4)^2}{49} - \dfrac{(y + 3)^2}{36} = 1$

9. $\dfrac{(x + 2)^2}{100} - \dfrac{(y - 3)^2}{49} = 1$

10. $\dfrac{(y - 6)^2}{36} - \dfrac{(x - 4)^2}{81} = 1$

11. $\dfrac{(y - 7)^2}{20} - \dfrac{(x - 3)^2}{16} = 1$

12. $\dfrac{(x + 9)^2}{36} - \dfrac{(y - 5)^2}{81} = 1$

13. $3x^2 - 12x - y^2 + 10y = 22$

14. $x^2 + 12x - 4y^2 - 8y = 8$

15. $2y^2 - x^2 + 12y - 14x = 25$

16. $y^2 - 5x^2 + 10x + 18y = -12$

17. $y^2 - 4x^2 + 16y - 24x = 8$

18. $3x^2 - 18x - 2y^2 + 16y = 1$

19. $5x^2 - 20x - 3y^2 + 12y = 7$

20. $2x^2 - 7y^2 + 16x - 14y = 67$

From the given information, find the equation for the hyperbola in standard form.

21. Center (0, 0); focus (−5, 0); and slope of asymptote 3/4.

22. Center (0, 0); vertex (−8, 0); and focus (−11, 0).

23. Center (5, 3); slope of asymptote 7/3; and focus (10, 3).

24. Center (−2, 6); vertex (−2, 3); and endpoint of conjugate axis (3, 6).

25. Vertices (0, 4) and (6, 4); slope of asymptote 8/3.

26. Foci (−6, −4) and (2, −4); endpoint of conjugate axis (−2, −1).

27. Center (−1, −5); vertex (−1, −2); and length of conjugate axis is 10.

28. Center (5, 2); focus (0, 2); and length of conjugate axis is 6.

E 29. Sketch the relationship between impedance and resistance if $X_L - X_C = 14\ \Omega$ (see Example 6).

30. Boyle's law states that for a constant temperature, the product of the pressure P and volume V of an ideal gas is a constant k ($PV = k$). Graph the relationship between P and V if $k = 1{,}800$. (V is the independent variable.)

31. The area of a triangle is $A = bh/2$. Sketch the relationship between the base and height of a triangle whose area is 250 ft^2. (Let b be the independent variable.)

32. The volume V of a conduit 2.5 m long is given by $V = 2.5\pi(R^2 - r^2)$ where R = radius of the outside of the conduit, and r = inside radius. Sketch a graph of the relationship between the radii when the volume is fixed at 60π m^3. (Let R be the independent variable.)

15–6

THE SECOND-DEGREE EQUATION

Thus far in this chapter we have discussed the equations for the circle, parabola, ellipse, and hyperbola. The equation for these conic sections are special cases of the more general second-degree equation. The general form of a second-degree equation is given by

$$Ax^2 + Bxy + Cy^2 + Dx + Ey + F = 0$$

The type of conic section is determined by coefficients A, B, and C of this equation. These results are summarized below.

The Second-Degree Equation

$$Ax^2 + Bxy + Cy^2 + Dx + Ey + F = 0$$

1. Parabola:

$B = 0$; $A = 0$ or $C = 0$ (quadratic in only x or y)

2. Circle:

$B = 0$, $A = C$ (quadratic in both x and y, and the coefficients of the quadratic terms are the same)

3. Ellipse:

$B = 0$, $A \neq C$ but have the same sign (quadratic in both x and y; the coefficients of the quadratic terms are different, but they have the same sign)

4. Hyperbola:

$B = 0$, A and C have opposite signs (quadratic in both x and y, but they have different signs)

5. Hyperbola:

$A = C = D = E = 0$ but $B \neq 0$ (not a quadratic equation but has the xy term)

EXAMPLE 1 Identify the curve represented by the following equation:

$$3x^2 - 6x - y^2 + 8y = 1$$

Determine the appropriate information for the figure and sketch its graph.

Solution To determine the type of figure, let us arrange the equation in the form of the general second-degree equation.

$$\underset{\uparrow\atop A}{3x^2} \underset{\uparrow\atop C}{-\,y^2} - 6x + 8y - 1 = 0$$

Because $A = 3$, $B = 0$, and $C = -1$, this equation represents a hyperbola. To determine the pertinent information and sketch the graph, we arrange the equation in standard form.

$$\begin{aligned}
3x^2 - 6x - y^2 + 8y &= 1 \\
3(x^2 - 2x) - (y^2 - 8y) &= 1 \\
3(x^2 - 2x + 1) - (y^2 - 8y + 16) &= 1 + 3 - 16 \\
3(x - 1)^2 - (y - 4)^2 &= -12 \\
-\frac{(x - 1)^2}{4} + \frac{(y - 4)^2}{12} &= 1 \\
\frac{(y - 4)^2}{12} - \frac{(x - 1)^2}{4} &= 1
\end{aligned}$$

$$\begin{aligned}
a &= \sqrt{12} \approx 3.5 \\
c^2 &= a^2 + b^2 \\
c^2 &= 12 + 4 \\
c &= 4
\end{aligned}$$

center: $(1, 4)$

vertices: $(1, 7.5)$ and $(1, 0.5)$

endpoints of conjugate axis: $(-1, 4)$ and $(3, 4)$

foci: $(1, 8)$ and $(1, 0)$

slope of asymptotes: $m = \pm\dfrac{a}{b} = \pm\dfrac{\sqrt{12}}{2} = \pm\sqrt{3} \approx \pm 1.7$

The graph is shown in Figure 15–34.

FIGURE 15–34

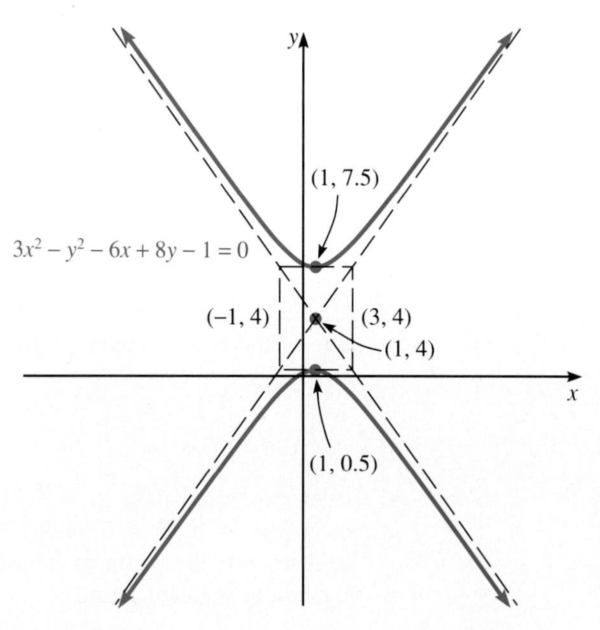

■

EXAMPLE 2 Identify the curve represented by the equation

$$x^2 - 3y - 8x = 2$$

Determine the appropriate information for the figure and sketch its graph.

Solution First, we arrange the equation in general second-degree form.

$$x^2 - 8x - 3y - 2 = 0$$

Since $A = 1$, $B = 0$, and $C = 0$, this equation represents a parabola. Second, we arrange the equation in standard form.

$$(x^2 - 8x) = 3y + 2$$
$$(x^2 - 8x + 16) = 3y + 2 + 16$$
$$(x - 4)^2 = 3(y + 6)$$

$$\text{vertex: } (4, -6)$$
$$\text{focus: } (4, -5.25)$$
$$\text{directrix: } y = -6.75$$

The graph is given in Figure 15–35.

FIGURE 15–35

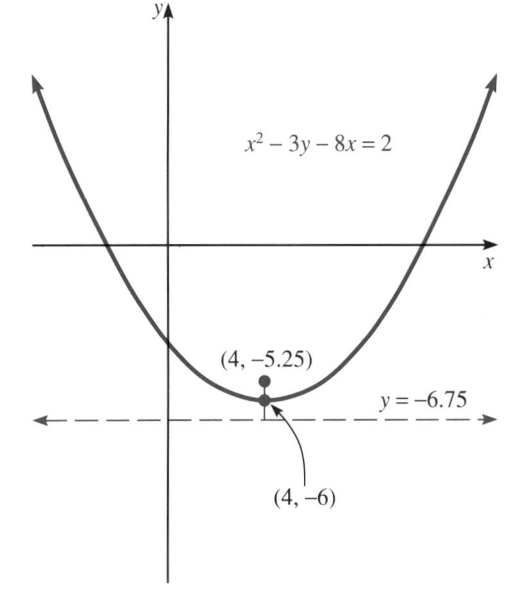

■

EXAMPLE 3 Identify the curve represented by the following equation:

$$4x^2 + 8x - 24y + 4y^2 = 5$$

Determine the appropriate information for the figure and sketch its graph.

Solution Arranging the equation in general second-degree form gives

$$\underset{\substack{\uparrow\\ A}}{4x^2} + \underset{\substack{\uparrow\\ C}}{4y^2} + 8x - 24y - 5 = 0$$

Since $B = 0$ and $A = C$, the equation represents a circle. To determine the pertinent information, we arrange the equation in standard form.

$$\begin{aligned}
4x^2 + 4y^2 + 8x - 24y - 5 &= 0\\
4x^2 + 8x + 4y^2 - 24y &= 5\\
4(x^2 + 2x) + 4(y^2 - 6y) &= 5\\
4(x^2 + 2x + 1) + 4(y^2 - 6y + 9) &= 5 + 36 + 4\\
4(x + 1)^2 + 4(y - 3)^2 &= 45\\
(x + 1)^2 + (y - 3)^2 &= 11.25
\end{aligned}$$

The center is $(-1, 3)$ and the radius is 3.4. Figure 15–36 shows the graph of this circle.

FIGURE 15–36

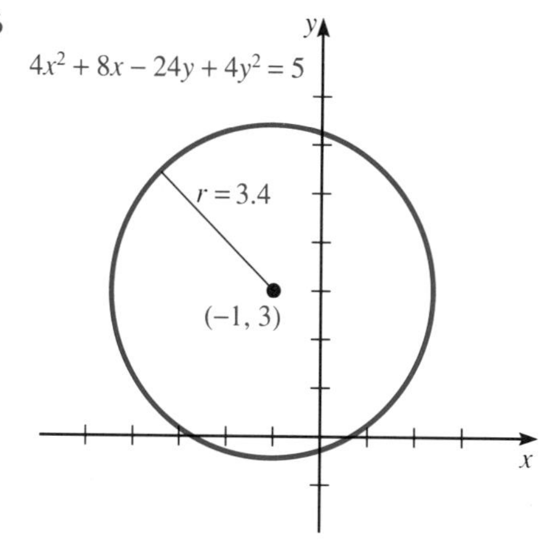

■

EXAMPLE 4 Identify the curve represented by the following equation:

$$9x^2 + 36x + 4y^2 - 16y = 20$$

Determine the appropriate information for the figure and sketch its graph.

Solution Arranging the equation in general second-degree form gives

$$9x^2 + 4y^2 + 36x - 16y - 20 = 0$$

Since $B = 0$ and $A \neq C$ but both have the same sign, this equation represents an ellipse. Next, we arrange the equation in standard form for an ellipse.

$$
\begin{aligned}
9x^2 + 36x + 4y^2 - 16y &= 20 \\
9(x^2 + 4x) + 4(y^2 - 4y) &= 20 \\
9(x^2 + 4x + 4) + 4(y^2 - 4y + 4) &= 20 + 36 + 16 \\
9(x + 2)^2 + 4(y - 2)^2 &= 72 \\
\frac{(x + 2)^2}{8} + \frac{(y - 2)^2}{18} &= 1
\end{aligned}
$$

$$
\begin{aligned}
a &= \sqrt{18} \approx 4.2 \\
b &= \sqrt{8} \approx 2.8 \\
c &= \sqrt{a^2 - b^2} = \sqrt{10} \approx 3.2
\end{aligned}
$$

center: $(-2, 2)$
vertices: $(-2, 6.2)$ and $(-2, -2.2)$
endpoints of minor axis: $(-4.8, 2)$ and $(0.8, 2)$
foci: $(-2, 5.2)$ and $(-2, -1.2)$

Figure 15–37 shows the graph.

FIGURE 15–37

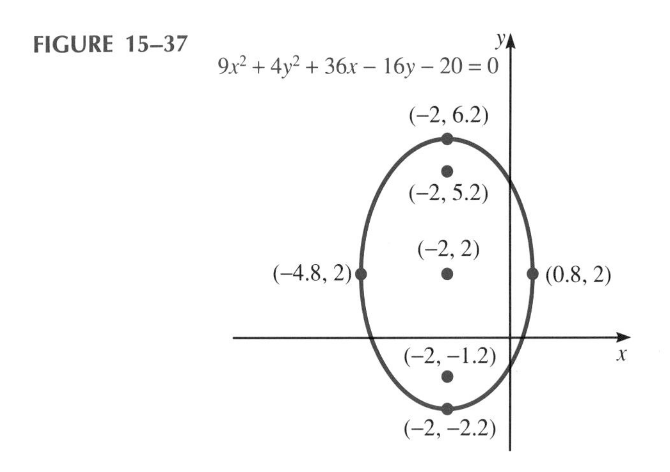

■

15–6 EXERCISES

Identify the following equations as representing a circle, parabola, ellipse, or hyperbola.

1. $3x^2 - 4y + 6x = 12$

2. $8x + 3y^2 - 6y + 4x^2 = 20$

3. $4x^2 + 10x - y^2 + 6y - 13 = 0$

4. $x^2 - y^2 + 3x - 12y + 10 = 0$

5. $6x + 2y^2 - 4y + 2x^2 = 6$

6. $3x(x - 1) = 4y + 3x - 7$
7. $4x + 14 = 3y^2 + 15y$
8. $6y^2 + 8x - 18y + 6x^2 = 0$
9. $5y^2 + 8 - 10x + 6y = 2 - 4x^2$
10. $4x + 18 - y^2 + 6y = -1$
11. $3 - 7x^2 + 21y + y^2 - 14x = 0$
12. $3x - x^2 + 7y - 4y^2 + 18 = 0$
13. $4x - 3y + x^2 - 10 + y^2 = 7$
14. $5x^2 + 5y^2 = 1$
15. $9x^2 + 4y^2 - 36x + 12y = -3$
16. $xy + 8 = 0$

Identify the curve represented by each of the given equations. Identify the information pertinent to the curve and sketch its graph.

17. $x^2 + 2y - 6x = -17$
18. $x^2 + 6x + y^2 - 10y = 43$
19. $4x^2 + 6y^2 - 8x + 36y = -26$
20. $y^2 - 6x + 8y + 58 = 0$
21. $8x + y^2 + 2y + x^2 + 13 = 0$
22. $49y^2 - 16x^2 + 196y - 160x = 988$
23. $3 - xy = 0$
24. $4x^2 + 25y^2 - 24x + 150y + 161 = 0$
25. $25x^2 - 54y - 9y^2 = 306$
26. $2x^2 + 2y^2 - 2x + 20y + 18 = 0$
27. The length of a room is 9 ft more than the width. Express the volume of the room as a function of the length if the room is 8 ft high. Identify the curve represented by this equation.
28. The height of a triangle is three more than five times the base. Express the area of the triangle as a function of the base, and identify the resulting curve.
29. The two legs of a right triangle are represented by x and y, and the hypotenuse is $y + 1$. Identify the curve represented by the equation relating x and y.

CHAPTER SUMMARY

Summary of Terms

axis (p. 540)
center (pp. 534, 547, 554)
circle (p. 534)
conjugate axis (p. 554)
directrix (p. 540)
ellipse (p. 547)
endpoints of minor axis (p. 547)
focus/foci (pp. 540, 547, 554)
general form of a line (p. 526)
general form of a circle (p. 535)
hyperbola (p. 554)
major axis (p. 547)
minor axis (p. 547)
parabola (p. 540)
point–slope form (p. 529)
radius (p. 534)
slope–intercept form (p. 526)
transverse axis (p. 554)
vertex/vertices (pp. 540, 547, 554)

Summary of Formulas

$y = mx + b$ slope–intercept form

$y - y_1 = m(x - x_1)$ point–slope form

$(x - h)^2 + (y - k)^2 = r^2$ circle

$$\left.\begin{aligned}(x - h)^2 &= 4p(y - k)\\ (y - k)^2 &= 4p(x - h)\end{aligned}\right\}\ \text{parabola}$$

$$\left.\begin{aligned}\frac{x - h)^2}{a^2} + \frac{(y - \mathrm{k})^2}{b^2} &= 1\\ \frac{(x - h)^2}{b^2} + \frac{(y - k)^2}{a^2} &= 1\end{aligned}\right\}\ \text{ellipse}$$

$$\left.\begin{array}{l}\dfrac{(x-h)^2}{a^2} - \dfrac{(y-k)^2}{b^2} = 1 \\ \dfrac{(y-k)^2}{a^2} - \dfrac{(x-h)^2}{b^2} = 1\end{array}\right\} \text{hyperbola}$$

$Ax^2 + Bxy + Cy^2 + Dx + Ey + F = 0$ second-degree equation

CHAPTER REVIEW

Section 15–1

Using the information given, find the equation of the line. Place your answer in general form, $Ax + By + C = 0$.

1. Slope $= -3/4$; passes through $(-1/2, 6)$.

2. Slope $= -4$; x intercept $= 5$.

3. Passing through $(9, -5)$ and $(-4, -3)$.

4. Vertical line; passes through $(-8, -7)$.

5. Undefined slope; passes through $(-3, 4)$.

6. Passing through $(6, -2)$ and $(1, 5)$.

7. Horizontal line through $(6, -8)$.

8. Perpendicular to $y = 3x + 8$; passing through $(1, 4)$.

9. Parallel to $3x - 2y = 6$; passing through $(-3, -5)$.

10. Slope $= 0$; y intercept $= -2$.

11. Perpendicular to $x - 5y = 10$; passing through $(7, -4)$.

12. Parallel to $4x - y = 5$; passing through $(6, -1)$.

Find the slope, x intercept, and y intercept of the following lines. Use this information to graph the line.

13. $7x - 4y = 10$

14. $2x + y = 10$

15. $2x + 3y + 6 = 0$

16. $3y - 4x = 6$

17. $x - 3y = 9$

18. $5x - 2y + 15 = 0$

E 19. The resistance of a resistor is found to increase 0.02 Ω for each 1°C increase in temperature. Write an equation to relate resistance and temperature if the resistance at 0°C is 2.3 Ω.

20. The pressure in a lake is measured to be 28.5 lb/in.^{2} at a depth of 32 ft and 39 lb/in.^{2} at 56 ft. Find the linear equation for pressure as a function of depth. (Let depth be the independent variable.)

21. The acceleration of an object is the change in velocity divided by the change in time. Find the linear equation relating velocity and time if the acceleration is 10 m/s^2 and the velocity is 3.5 m/s when $t = 0$ s.

Section 15–2

Find the radius and center of the circle and sketch its graph.

22. $(x - 3)^2 + y^2 = 16$

23. $(y + 3)^2 + (x + 4)^2 = 4$

24. $(x - 1)^2 + (y + 2)^2 = \sqrt{7}$

25. $(x + 1)^2 + (y - 6)^2 = \sqrt{6}$

26. $2x^2 + 2y^2 - 4y - 6 = 0$

27. $3x^2 + 3y^2 - 9x + 15y = 18$

Using the given information, find the equation, in standard form, of the circle.

28. Center at $(4, -2)$; radius of 2.

29. Center at $(-2, -4)$; radius of $\sqrt{3}$.

30. Center at $(4, 0)$; passing through $(6, 5)$.

31. Center at $(6, -5)$; tangent to the x axis.

32. Diameter is a segment from $(-4, 6)$ to $(2, -4)$.

33. Diameter is a segment from $(-7, 4)$ to $(3, -6)$.

34. The legs of a right triangle are represented by x and y, and the hypotenuse is 9 ft. Write an equation relating x and y, and sketch the graph of the equation.

E 35. The impedance Z in an ac circuit is given by

$$Z^2 = R^2 + (X_L - X_C)^2$$

where R = resistance, X_L = inductive reactance, and X_C = capacitive reactance. Graph the relationship between resistance and inductive reactance when $Z = 25\ \Omega$ and $X_C = 40\ \Omega$.

Section 15–3

For the given equations of parabolas, find the coordinates of the vertex and focus, determine the equation of the directrix, and sketch the graph.

36. $3(y + 1) = x^2$ **37.** $y^2 = -2(x - 3)$

38. $y^2 + x + 4y = -12$

39. $x^2 + 3y - 4x + 13 = 0$

40. $x^2 - x - 8y = 3$

41. $x^2 + 8x + 2y + 10 = 0$

Find the equation of the parabola from its given geometric properties.

42. Vertex at $(-1, 4)$; focus $(-1, 7)$.

43. Vertex at $(-2, 0)$; focus $(1, 0)$.

44. Directrix $x = 2$; focus $(4, -6)$.

45. Directrix $y = -3$; focus $(4, 0)$.

46. Vertex at $(1, 3)$; passes through $(4, 6)$; and axis parallel to the y axis.

C 47. A cable suspended between two bridge supports that are 200 ft apart approximates the shape of a parabola. The vertical distance from where the cable is attached to its lowest point is 35 ft. Assuming that the origin is at the lowest point, find the equation representing the shape of the cable.

E 48. Heat H in a resistor of resistance R and current i develops at a rate given by $H = Ri^2$. Graph H as a function of i if $R = 25\ \Omega$.

Section 15–4

Determine the coordinates for the center, vertices, endpoints of the minor axis, and foci of the given ellipse. Sketch the graph.

49. $\dfrac{(x + 1)^2}{36} + \dfrac{(y - 3)^2}{49} = 1$

50. $\dfrac{(x - 2)^2}{16} + \dfrac{(y - 5)^2}{9} = 1$

51. $81x^2 + 25y^2 - 300y = 1{,}800$

52. $20x^2 + 14y^2 - 40x - 112y = 46$

53. $9y^2 + 24x + 18y + 4x^2 + 44 = 0$

54. $16y^2 + 9x^2 - 18x = 135$

Determine the equation of the ellipse from the given geometric properties.

55. Center at $(5, 1)$; focus at $(5, -4)$; and endpoint of minor axis at $(8, 1)$.

56. Endpoints of minor axis at $(3, -4)$ and $(3, 2)$; focus at $(7, -1)$.

57. Vertices at $(-4, 2)$ and $(4, 2)$; length of minor axis is 6.

58. Center at $(5, 3)$; foci at $(2, 3)$ and $(8, 3)$; and length of minor axis is 4.

59. A satellite orbits the earth in an elliptical orbit. The maximum height is 400 mi, the minimum height is 175 mi, and the earth's center is at one focus. Determine the equation of the satellite's path. (The radius of the earth is approximately 3955 mi.)

C 60. A covered bridge in the shape of a semiellipse spans 150 ft and is 50 ft high at its maximum. What is the height of the bridge 60 ft from the center?

Section 15–5

Determine the coordinates of the center, vertices, endpoints of the conjugate axis, and foci for each hyperbola. Sketch the graph.

61. $\dfrac{(x + 3)^2}{16} - \dfrac{(y - 1)^2}{4} = 1$

62. $\dfrac{(x - 2)^2}{49} - \dfrac{(y - 3)^2}{9} = 1$

63. $\dfrac{(y + 4)^2}{25} - \dfrac{(x + 3)^2}{36} = 1$

64. $\dfrac{(y - 8)^2}{81} - \dfrac{(x + 5)^2}{36} = 1$

65. $5y^2 + 40 = 4x^2 + 40y$

66. $25x^2 + 325 = 36y^2 + 350x$

Determine the equation of the hyperbola from its geometric properties.

67. Center $(0, 0)$; vertex $(0, 3)$; and focus $(0, -5)$.

68. Center $(3, -4)$; focus $(3, 2)$, and length of conjugate axis $= 8$.

69. Vertices $(5, 7)$ and $(-1, 7)$; length of conjugate axis $= 10$.

70. Foci $(-1, -3)$ and $(-7, -3)$; length of transverse axis $= 4$.

71. Boyle's law states that for a constant temperature, the product of pressure P and volume V of an ideal gas is a constant k ($PV = k$). Graph the relationship between P and V if $k = 400$.

72. The area of a rectangle is $A = LW$. Sketch the relationship between length L and width W if the area is fixed at 200 m^2.

Section 15–6

Identify the curve represented by the following equations. Identify the information pertinent to the curve and sketch its graph.

73. $2x^2 - 4x + 2y^2 + 24y = -69$

74. $4y^2 + 9x^2 + 144x - 36y + 576 = 0$

75. $x^2 = -4y + 8x - 44$

76. $\dfrac{(y + 7)^2}{81} - \dfrac{(x - 1)^2}{64} = 1$

77. $3(x - 3)^2 + (y + 1)^2 = 18$

78. $3y^2 - 30y + 3x^2 = -67$

79. $\dfrac{(x - 3)^2}{8} - \dfrac{(y + 1)^2}{25} = 1$

80. $3y^2 - 6x + 6y + 3 = 0$

81. $16y^2 + 32y + 25x^2 + 150x = 159$

82. $xy = -7$

CHAPTER TEST

The number in parentheses refers to the appropriate learning objective given at the beginning of the chapter.

1. Determine the slope, y intercept, and x intercept of $2x - 10y = 9$. Sketch the graph. **(2)**

2. Find the equation of a circle with center at $(2, -5)$ and passing through $(3, -1)$. **(4)**

3. Determine the center, foci, vertices, endpoints of conjugate axis, and slopes of the asymptotes of the following hyperbola: **(9)**

$$\frac{(x - 1)^2}{16} - \frac{(y + 3)^2}{4} = 1$$

Also sketch the graph.

4. The impedance Z in an ac circuit is given by **(12)**

$$Z^2 = R^2 + (X_L - X_C)^2$$

where R = resistance, X_L = inductive reactance, and X_C = capacitive reactance. Graph the relationship between resistance and inductive reactance when $Z = 36\ \Omega$ and $X_C = 20\ \Omega$.

5. Find the center and radius of the circle given by **(3)**

$$(x - 4)^2 + (y + 8)^2 = 25$$

6. Determine the vertex, focus, and directrix of the following parabola: **(5)**

$$(x + 3)^2 = -8(y + 6)$$

Also sketch its graph.

7. A satellite orbits the earth in an elliptical orbit. The maximum height is 200 mi, the minimum height is 110 mi, and the earth's center is at one focus. Find the equation of the satellite's path. (Assume radius of earth = 3,960 mi.) **(12)**

8. Find the equation of the ellipse with its center at $(-7, 4)$, its vertex at $(-7, 1)$, and its minor axis of length 4. **(8)**

9. Determine the equation of a parabola whose focus is $(4, -5)$ and whose directrix is $x = -2$. **(6)**

10. The pressure in a lake is measured to be 22.5 lb/in.^{2} at a depth of 18 ft and 36.33 lb/in.^{2} at 50 ft. If the relationship between pressure and depth is linear, find the equation relating the variables. **(12)**

11. Determine the equation of the straight line passing through $(-7, 6)$ with a slope of -3. **(1)**

12. Identify the conic section represented by the following equation: **(11)**

$$4x^2 + 9y^2 - 32x + 144y + 631 = 0$$

13. Determine the coordinates for the center, vertices, foci, and endpoints of the minor axis of the ellipse represented by **(7)**

$$16y^2 + 9x^2 - 64y + 54x + 1 = 0$$

Also sketch the graph.

14. Find the equation of a hyperbola with its center at $(6, -3)$, its vertex at $(2, -3)$, and the length of its conjugate axis 12. **(10)**

SOLUTION TO CHAPTER INTRODUCTION

Figure 15–38 illustrates this problem. For convenience, we placed the center of the semiellipse at the origin. We know that the height of the bridge is less than its length. Therefore, the equation of the ellipse is

$$\frac{x^2}{a^2} + \frac{y^2}{b^2} = 1$$

FIGURE 15–38

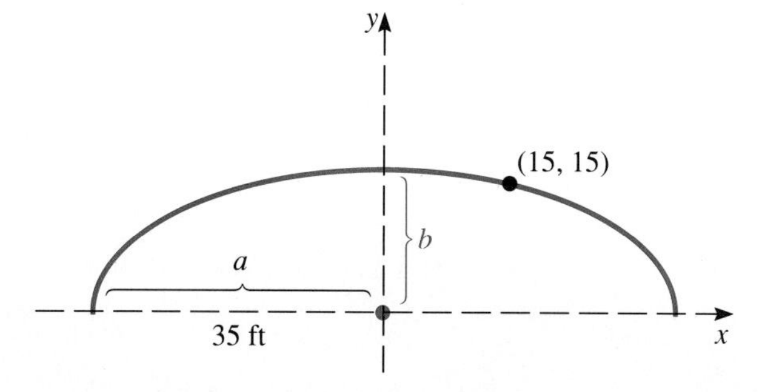

We also know that the point (15, 15) lies on the ellipse and that $a = 35$. Then we calculate b to determine the height of the bridge as follows:

$$\begin{aligned}
\frac{(15)^2}{(35)^2} + \frac{(15)^2}{b^2} &= 1 \\
\frac{225}{1{,}225} + \frac{225}{b^2} &= 1 \\
225b^2 + 275{,}625 &= 1{,}225b^2 \\
275{,}625 &= 1{,}000b^2 \\
275.625 &= b^2 \\
17 &\approx b
\end{aligned}$$

The bridge is 17 ft high.

As a biologist, you measure the number n of bacteria in a petri dish as a function of time t. You know that the function is defined by an equation of the form $n = me^{0.2t} + b$. Using the following data, you are to predict the number of bacteria at the end of 18 hours:

t (h)	0	1	2	3	4	5
n (thousands)	19.2	21	23	25	28	32

(The solution to this problem is given at the end of the chapter.)

Problems of this type involve a statistical method called *least squares* whereby experimental data are used to generate an equation that best fits the curve representing the relationship between the variables. In turn, you can use the equation to approximate other values. In this chapter we will discuss methods of displaying empirical data, measures of central tendency, measures of dispersion, and the least squares method of curve fitting.

Learning Objectives

After completing this chapter, you should be able to

1. Organize data into a frequency distribution and give a graphical representation using a histogram or a frequency polygon (Section 16–1).
2. Calculate the arithmetic mean, median, and mode from empirical data (Section 16–2).
3. Calculate the range and standard deviation from empirical data (Section 16–3).
4. Use the mean and standard deviation to calculate the variation of data values that should fall within one or two standard deviations of the mean (Section 16–3).
5. Use the least squares method to fit empirical data to an equation (Section 16–4).

Chapter 16

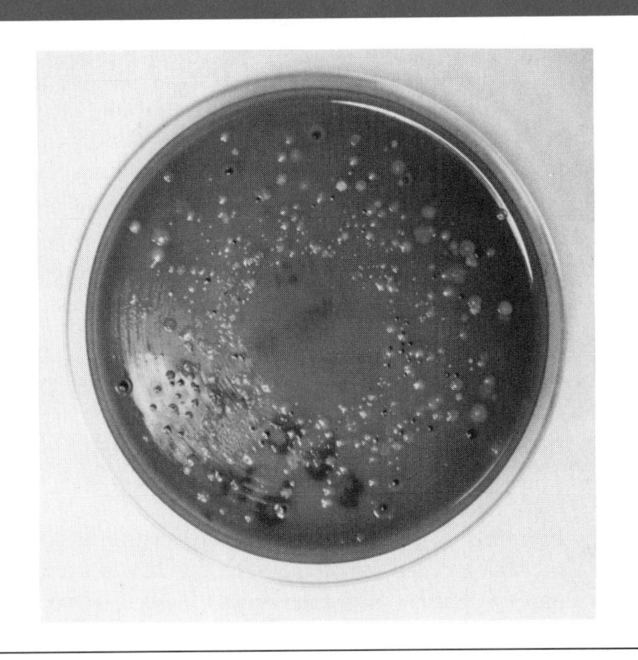

Introduction to Statistics and Empirical Curve Fitting

16–1

FREQUENCY DISTRIBUTIONS

In this section we will discuss methods of organizing and displaying statistical data. First, we will discuss the frequency distribution as a method of organizing the data. Then we will discuss the histogram and the frequency polygon as means to graphically represent the data.

Frequency Distribution

When a statistical experiment is conducted, repeated readings of measurements, or **empirical data,** are taken. Since the number of measurements can be quite large, we need some method of organizing the data. One such organizational technique, called a **frequency distribution,** involves a *table used to tabulate the number of occurrences* of a particular measurement. The next example illustrates how to create a frequency distribution.

EXAMPLE 1 The following data concern the number of cars passing through a chosen intersection in one-minute intervals for a 15-minute period:

6	8	15	12	9
15	14	10	6	6
8	13	7	10	9

Construct a frequency distribution for these data.

Solution To construct a frequency distribution, we arrange the data values in ascending order and tabulate the number of times each value occurs. The frequency distribution for these data is given in the following table:

Number of cars	6	7	8	9	10	11	12	13	14	15
Frequency	3	1	2	2	2	0	1	1	1	2

■

When data contain a large number of observations with a wide range of values, it is often helpful to organize the data into equal intervals or groups. There are several rules that you must follow when grouping data:

- All intervals must be of equal width and cannot overlap.
- A data value can belong to only one interval.

If we chose different interval widths, the frequency distribution would look different. There is no single correct rule for choosing interval widths for grouped data. However, if the interval width is too small or too large, it may be difficult to determine trends or patterns in the data. As a general rule, construct between 5 and 12 intervals. To determine an approximate width of the intervals, subtract the smallest data value from the largest, and divide the result by the number of intervals. This process is illustrated in the next example.

EXAMPLE 2 A quality control engineer conducted a survey of faulty integrated circuits. She randomly selected groups of 100 integrated circuits and recorded the number that were faulty in each group as follows:

9	21	16	12	31	27	46	18	7
36	17	29	41	6	26	35	39	11
23	38	45	10	30	32	8		

Construct a frequency distribution using grouped data.

Solution First, we divide the range of values into nine equal intervals. The approximate width of each interval is $(46 - 6)/9 \approx 4$. Then we determine the frequency distribution of each interval. This information is summarized in the following table:

Interval	4–8	9–13	14–18	19–23	24–28	29–33	34–38	39–43	44–48
Frequency	3	4	3	2	2	4	3	2	2

■

Graphic Representation

The graph of a frequency distribution makes recognition of a pattern or trend in the data easier to detect. One such graphical representation is the **histogram,** *a bar graph in which the data values are represented on the horizontal axis and the frequency is represented on the vertical axis.*

EXAMPLE 3 The following grades were recorded on a mathematics test:

84	67	73	81	94	57	78	71
81	53	62	91	77	64	89	43
98	76	78	69	93	80	51	70

Draw a histogram to depict the distribution of the grades.

Solution First, we will divide the grades into equal intervals. The approximate interval width for six intervals is $(98 - 43)/6 \approx 9$. The frequency distribution is given in the following table:

Interval	40–49	50–59	60–69	70–79	80–89	90–99
Frequency	1	3	4	7	5	4

Next, we construct a histogram using rectangles whose width represents the length of the data interval and whose height represents the frequency of that interval. In the case of grouped data, generally the horizontal axis is labeled with the midpoint of the interval. Recall from a previous chapter that the midpoint of the interval is half the difference in interval endpoints. The histogram for this data set is shown in Figure 16–1. ■

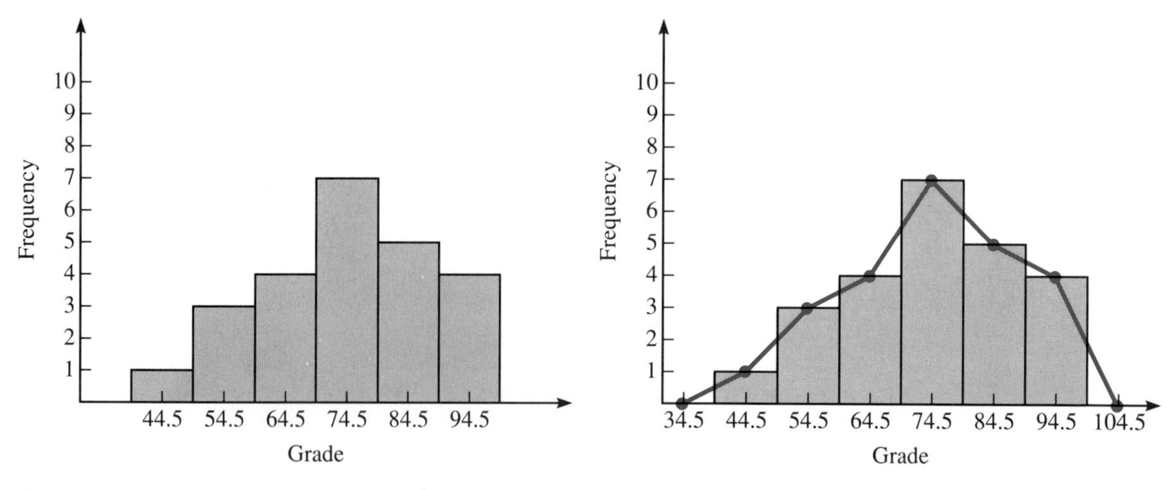

FIGURE 16–1

FIGURE 16–2

A second method of graphically representing a frequency distribution is called a **frequency polygon,** *a broken-line graph with the data value on the horizontal axis and the frequency of its occurrence on the vertical axis*. To construct a frequency polygon, place a point above the data group at the appropriate frequency. Then connect successive points with straight line segments. Since a polygon is a closed figure, always add an additional data group with a frequency of zero on each end of the horizontal axis.

EXAMPLE 4 Draw a frequency polygon for the data given in Example 3.

Solution You can construct a frequency polygon from a histogram by joining the midpoints of the tops of the rectangles with line segments. The frequency polygon for this histogram is shown in Figure 16–2. ■

NOTE ✦ As shown in the next example, you do not have to draw a histogram before drawing a frequency polygon.

EXAMPLE 5 In an effort to monitor fuel consumption in compact cars, an auto manufacturer asked 150 car owners to report their mpg results. Using the following grouped data, construct a frequency polygon.

Miles per gallon	25.0–25.4	25.5–25.9	26.0–26.4	26.5–26.9	27.0–27.4	27.5–27.9
Frequency	6	9	10	16	15	17
Miles per gallon	28.0–28.4	28.5–28.9	29.0–29.4	29.5–29.9	30.0–30.4	
Frequency	26	20	18	8	5	

Solution To construct the frequency polygon, add the two additional data groups of 24.5–24.9 and 30.5–30.9, each with a frequency of zero. Then label the midpoint of each data group on the horizontal axis, and place a dot at the corresponding frequency. Connect successive points with a line segment as shown in Figure 16–3.

FIGURE 16–3

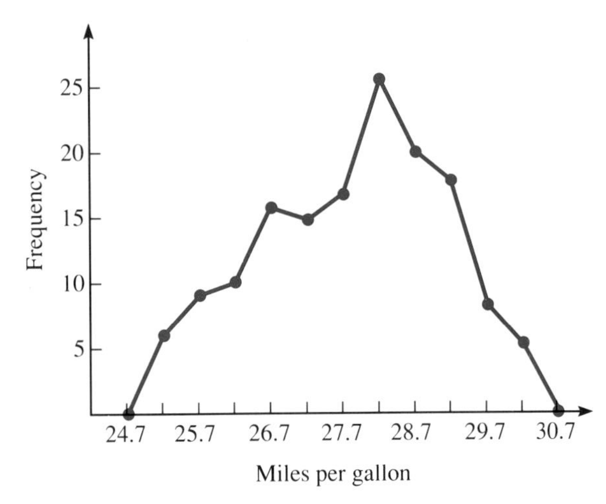

■

Often, it is necessary to interpret data organized in a histogram. This process is illustrated in the next example.

EXAMPLE 6 A group of steel shafts were weighed (in ounces). Figure 16–4 is a histogram of these data. Using the histogram, answer the following:

(a) How many shafts were measured?
(b) How many steel shafts weighed 1.85 oz?
(c) How many steel shafts weighed more than 1.95 oz?

(Exact values are used on the horizontal axis rather than the midpoint of an interval.)

Solution

(a) From the histogram, we can determine the number of steel shafts measured by totaling the frequency.

$$\begin{aligned} \text{number measured} &= 6 + 8 + 13 + 17 + 16 + 12 + 10 + 6 \\ &= 88 \end{aligned}$$

Eighty-eight steel shafts were measured.

FIGURE 16–4

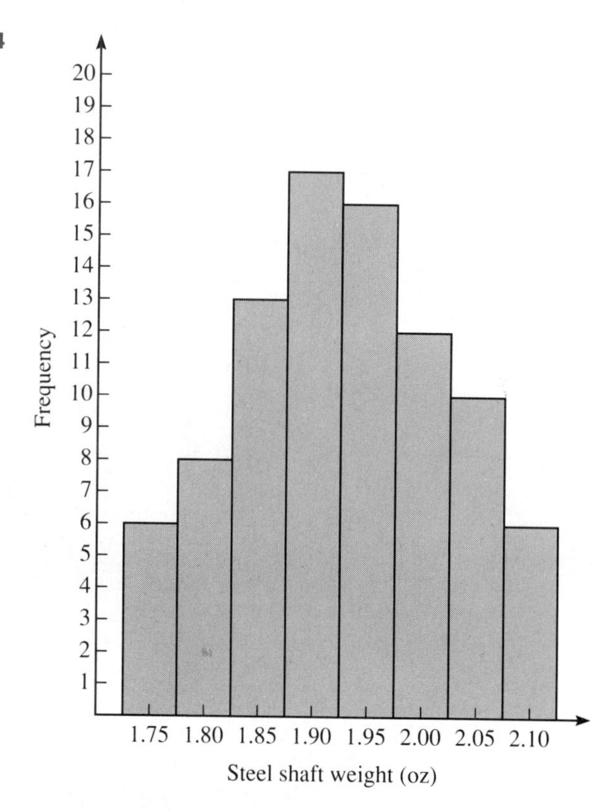

(b) To determine the number of shafts that weighed 1.85 oz, we find 1.85 on the horizontal axis; then we move vertically to the top of the rectangle and horizontally to read the frequency. There are 13 steel shafts that weigh 1.85 oz.

(c) The number of shafts that weighed more than 1.95 oz is the total of the frequencies in groups 2.00, 2.05, and 2.10 oz: 12 + 10 + 6. Twenty-eight of the shafts weigh more than 1.95 oz. ■

16–1 EXERCISES

Use data sets *A–F* for Problems 1–18.

A Grades on a math test: 86, 65, 74, 93, 88, 72, 86, 81, 74, 88, 72, 65, 69, 74, 97

B Number of defective diodes in groups of 100: 3, 10, 8, 6, 14, 10, 13, 8, 5, 18, 8, 7, 12, 3, 14, 7, 12, 6, 8, 15

C Number of hours worked in a week by technicians at C Corporation: 43, 35, 41, 56, 50, 48, 42, 40, 38, 46, 40, 35, 50, 46, 40, 48, 43

D Age of employees at M Corporation: 30, 45, 56, 47, 25, 47, 30, 32, 27, 33, 63, 42, 35, 39, 25, 35, 56, 42, 33, 42, 45, 39

E The IQ of math students at Little Rock School: 86, 110, 95, 134, 100, 89, 97, 110, 98, 80, 100, 83, 147, 110, 84, 98, 100, 120, 134, 80

F Starting salary for electrical engineers (in thousands): 18, 23, 15, 17, 20, 18, 21, 17, 22, 18, 23, 19, 21, 18, 20, 23, 19, 17

Construct a frequency distribution for the indicated data set.

1. Set *A* **2.** Set *B*

3. Set *C* **4.** Set *D*

5. Set *E* **6.** Set *F*

Construct a histogram for the indicated data set.

7. Set *A* **8.** Set *B*

9. Set *C* **10.** Set *D*

11. Set *E* **12.** Set *F*

Construct a frequency polygon for the indicated data set.

13. Set *A* **14.** Set *B*

15. Set *C* **16.** Set *D*

17. Set *E* **18.** Set *F*

19. Construct a frequency distribution from the histogram shown in Figure 16–5. (Exact values are used on the horizontal axis rather than the midpoint of the interval because the data are not grouped.)

20. From the histogram in Figure 16–5, how many factories were surveyed?

21. From the histogram in Figure 16–5, how many employees were counted?

22. From the histogram in Figure 16–5, how many factories employed 800 people?

23. Construct a frequency distribution from the histogram in Figure 16–6.

24. From the histogram in Figure 16–6, how many students took the test?

25. From the histogram in Figure 16–6, how many students scored in the interval for 80 or above on the test?

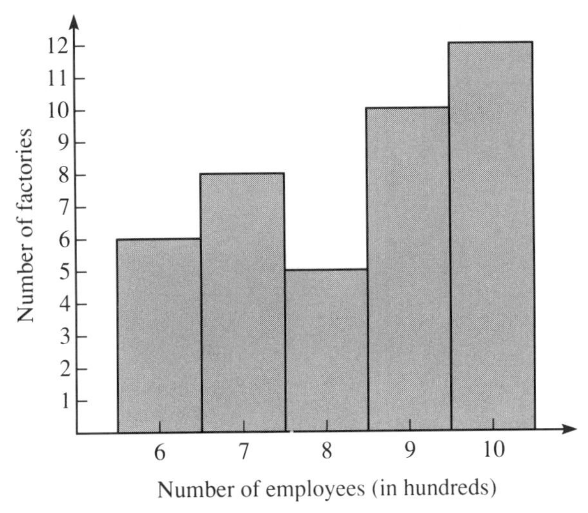

FIGURE 16–5

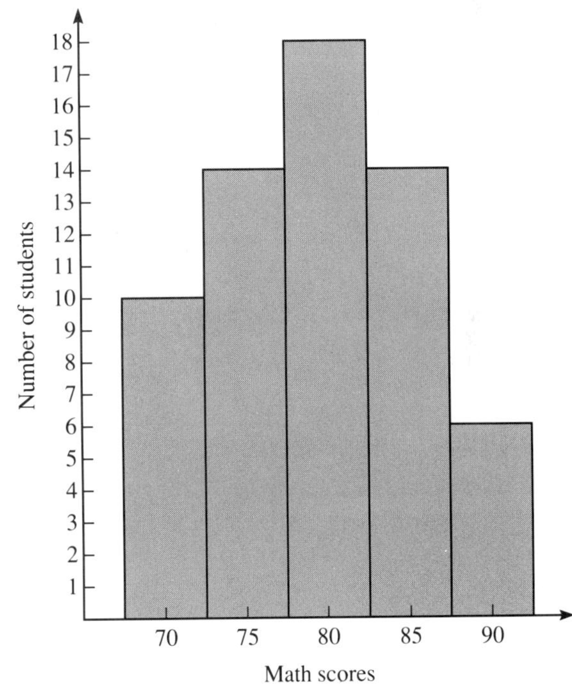

FIGURE 16–6

16–2

MEASURES OF CENTRAL TENDENCY

Although frequency distributions and graphs provide useful information about data, it is often helpful to obtain numbers that are representative of the distribution of the data. These numbers are called **measures of central tendency.** In this section we will discuss three measures of central tendency: mean, median, and mode.

Mean

The most frequently used measure of central tendency is the **arithmetic mean,** often called the **average.** To calculate the arithmetic mean $\bar{x}$ of a group of data, *add the data values and divide this total by the number of values.* (Some textbooks use the Greek letter mu, μ, to denote the arithmetic mean.) The following formula summarizes this calculation.

Arithmetic Mean

Given data values $x_1, x_2, x_3, \ldots, x_n$, the arithmetic mean $\bar{x}$ is given by

$$\bar{x} = \frac{x_1 + x_2 + x_3 + \cdots + x_n}{n} = \frac{\sum_{i=1}^{n} x_i}{n}$$

where the Greek letter Σ (uppercase sigma) denotes a sum. The notation $\sum_{i=1}^{n}$ means to add $x_1, x_2, \ldots$ through x_n. The variable n equals the number of values in the data set.

EXAMPLE 1 The monthly electrical energy consumption (in kilowatt-hours) of 8 families with the same size house was found to be

618 597 715 674 703 637 657 586

Find the average monthly energy consumption of these families.

Solution The average or arithmetic mean is given by

$$\bar{x} = \frac{\sum_{i=1}^{n} x_i}{n} = \frac{\sum_{i=1}^{8} x_i}{8} = \frac{618 + 597 + 715 + 674 + 703 + 637 + 657 + 586}{8}$$

$$\bar{x} = \frac{5{,}187}{8}$$

$$\bar{x} \approx 648$$

■

We can also calculate the arithmetic mean given a frequency distribution. The formula is given below, and its application is illustrated in the next example.

Arithmetic Mean from a Frequency Distribution

$$\bar{x} = \frac{\sum_{i=1}^{n} f_i x_i}{\sum_{i=1}^{n} f_i}$$

where x_i = a data value and f_i = frequency of that value.

EXAMPLE 2 A quality control engineer counted the number of defective light bulbs in lots of 100. From the following frequency distribution, determine the average number of defective bulbs per 100:

Number of defective bulbs	3	5	6	8	4
Frequency	6	2	1	2	4

Solution Substituting into the formula for the arithmetic mean from a frequency distribution gives

$$\bar{x} = \frac{\sum_{i=1}^{n} f_i x_i}{\sum_{i=1}^{n} f_i}$$

$$\bar{x} = \frac{3(6) + 5(2) + 6(1) + 8(2) + 4(4)}{6 + 2 + 1 + 2 + 4}$$

$$\bar{x} = 4.4$$

On average, there are approximately 4 defective bulbs per hundred. ■

If the frequency distribution is given in terms of intervals rather than exact values, substitute the midpoint of each interval for x_i into the mean formula.

EXAMPLE 3 Find the mean for the following grouped data:

x	10–14	15–19	20–24	25–29	30–34
Frequency	6	8	12	7	4

Solution To find the mean, we use a frequency distribution to substitute into the mean formula. Since the data are grouped, use the midpoint of each interval.

$$\bar{x} = \frac{\sum_{i=1}^{n} f_i x_i}{\sum_{i=1}^{n} f_i}$$

midpoint 10–14 → 12; midpoint 15–19 → 17; midpoint 20–24 → 22; midpoint 25–29 → 27; midpoint 30–34 → 32

$$\bar{x} = \frac{12(6) + 17(8) + 22(12) + 27(7) + 32(4)}{6 + 8 + 12 + 7 + 4}$$

$$\bar{x} \approx 21$$

■

EXAMPLE 4 Using the information in Example 5 of Section 16–1, find the average fuel consumption of the cars.

Solution Since the data are grouped, we use the midpoint of each interval as the data value and substitute into the formula for the arithmetic mean from a frequency distribution.

$$\bar{x} = (25.2)(6) + (25.7)(9) + (26.2)(10) + (26.7)(16) + (27.2)(15)$$
$$+ (27.7)(17) + (28.2)(26) + (28.7)(20) + (29.2)(18) + (29.7)(8)$$
$$+ (30.2)(5)$$

$$\bar{x} = 6 + 9 + 10 + 16 + 15 + 17 + 26 + 20 + 18 + 8 + 5$$

$$\bar{x} \approx 27.8$$

The average fuel consumption of these compact cars is 27.8 mpg. ■

Median

A second measure of central tendency is the **median,** which is *the middle number when the data are arranged in numerical order in an array*. If the number N of data items is odd, count $N/2$ (rounded up) units from either end of the array to reach the median.

EXAMPLE 5 Find the median of the following beginning salaries of nine technicians:

$19,160 $20,283 $20,476 $20,537 $21,571
$21,963 $22,419 $22,849 $23,473

Solution To find the median salary, count $9/2 \approx 5$ from either end of the array. The median salary is \$21,571 since there are four salaries higher and four salaries lower than it.

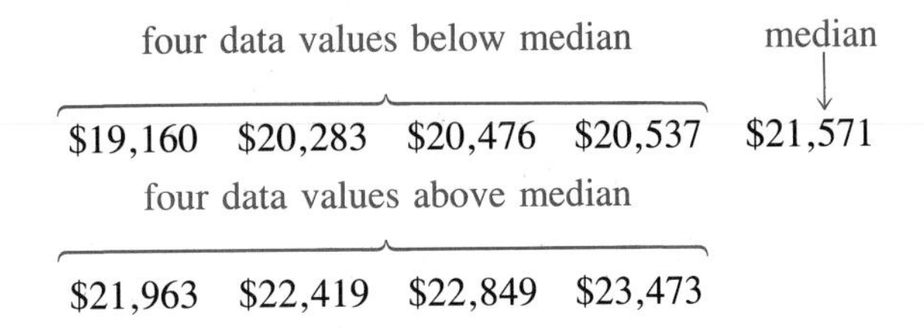

■

The median of an even number of data values is the mean of the middle two values. Count $N/2$ units from each end of the array of numbers, and find the mean of the two values. The next example illustrates how to calculate the median in this case.

EXAMPLE 6 The following numbers represent the age of Mr. Carter's employees:

37 21 41 32 18 40 28 47

Find the median of these data values.

Solution First, arrange the data values in ascending order. To find the median value, count $8/2 = 4$ values from each end and take the mean of the two values. The median is the mean of the fourth and fifth values.

median = mean of these values

$$\overbrace{18 \quad 21 \quad 28 \quad 32}^{4} \quad \overbrace{37 \quad 40 \quad 41 \quad 47}^{4}$$

$$\text{median} = \frac{32 + 37}{2} = \frac{69}{2} = 34.5$$

The median age of these employees is 34.5. ■

Mode

A third measure of central tendency, rarely used in technical applications, is the **mode,** *the data value that occurs most frequently*. There may be more than one mode for a given data set if several data values have the same maximum frequency. The mode is used primarily in quality control as an inspection method for determining central tendency. It gives a quick and approximate measure of the central tendency and is used to describe the most typical value of a distribution.

EXAMPLE 7 The following data result from a motion-time study on the time duration of a task:

6 8 10 7 6 11 7 6 12
17 9 8 6 14 11 2 8 10

Find the mode for this data set.

Solution To determine the mode, we construct the following frequency distribution:

Data	2	6	7	8	9	10	11	12	14	17
Frequency	1	4	2	3	1	2	2	1	1	1

From the frequency distribution, the mode is 6 because it occurs most frequently, four times. ■

16–2 EXERCISES

For the data sets in Problems 1–14, find the mean, median, and mode.

1. 16, 21, 18, 15, 26, 40, 13, 29, 16, 14, 20, 19, 27
2. 3, 14, 8, 7, 2, 5, 6, 8, 3, 4, 9
3. 73, 86, 93, 60, 73, 84, 82, 97
4. 205, 310, 846, 350
5. 3, 3, 3, 6, 8, 8, 10, 10, 10,12, 15
6. 750, 310, 946, 1150, 1365, 310
7. 75, 95, 80, 60, 65, 75, 90, 75, 85
8. 4, 4, 4, 9, 9, 10, 10, 10, 13, 13, 18, 20
9. 1, 360, 2108, 50, 1840, 780
10. 1, 16, 8, 12, 10, 9, 24, 2, 6, 10, 8
11. Number of defective integrated circuits per lot of 100: 6, 10, 14, 5, 6, 8, 10, 6
12. IQ of math students: 120, 86, 95, 100, 89, 115, 108, 90, 80, 100, 110, 130, 85, 99, 106, 95, 100, 103, 108, 96
13. Test scores for a math class: 73, 60, 85, 78, 85, 94, 73, 85, 77, 87, 82, 72, 85, 70, 82
14. Salary (in thousands) of civil engineers: 18, 21, 19, 24, 16, 28, 21
15. A math student needs an average of 75 on nine math exams to get a grade of C in the course. In his first eight tests, Fred made 86, 64, 70, 77, 75, 69, 76, and 83. What must he make on the last test to make a grade of C in the course?
16. During the last five months a photo lab used the following amounts of a certain chemical: 56 lb, 68 lb, 32 lb, 47 lb, and 60 lb. Based on the average monthly consumption, how much chemical is needed for the next 12 months?
17. A technician must estimate the labor cost of repairing a computer. She knows that five similar jobs required 6.1 h, 3.3 h, 4.2 h, 5.1 h, and 7.2 h. Using the arithmetic mean of the previous jobs and an hourly rate of $35 per hour, determine how much the technician should estimate for labor cost.
18. A company has 46 employees. Fifteen earn $5.50 per hour, 12 earn $10.00 per hour, and 19 earn $14.50 per hour. Determine the mean earnings per hour.
19. The monthly utility bills for XYZ Corporation are as follows:

    $96.25 $108.10 $74.12 $60.58 $87.34
    $74.12 $136.45 $103.15 $94.83 $81.50
    $96.25 $72.51

    Construct a frequency distribution and histogram for the data, and find the mean, median, and mode of this data set.

16–3

MEASURES OF DISPERSION

In the previous section we discussed measures of central tendency, which are used to determine data values representative of a collection of data. However, it is also important to have some indication of the spread of the data values, known as **measures of dispersion** or **variation.** In this section we will discuss two measures of dispersion: range and standard deviation.

Range

Range is the measure of dispersion that is least used but easiest to calculate. Because it is not sensitive to variation within a data group or to variation from the mean, it is used infrequently. The **range** is defined as *the difference between the two extreme values in a data set.*

EXAMPLE 1 A machine is tooling a part that should be 26 cm long. A quality control engineer randomly selects 8 parts and finds their lengths (in centimeters) to be as follows:

24.7 26.8 23.2 27.3 25.8 24.9 26.8 26.2

Determine the range of these values.

Solution From the list of lengths, we determine that the largest value is 27.3 cm and the smallest value is 23.2 cm. Therefore,

$$\text{range} = \text{highest value} - \text{lowest value}$$

$$\text{range} = 27.3 - 23.2$$

$$\text{range} = 4.1$$

The range of this data set is 4.1 cm. ■

Standard Deviation

The most widely used measure of dispersion is the **standard deviation,** which is a measure of how much the data relate to the arithmetic mean. The standard deviation is given by the following formula.

Standard Deviation

$$s = \sqrt{\frac{\sum_{i=1}^{n} (x_i - \bar{x})^2}{n}}$$

where $\bar{x}$ = mean of the set of n values $x_1, x_2, \ldots, x_n$. (Some textbooks use the Greek letter σ (lowercase sigma) to denote standard deviation.)

Using the formula above, we can calculate standard deviation by computing the following values successively:

- ☐ Calculate the arithmetic mean $\bar{x}$ of the data values.
- ☐ Subtract the mean from each data value.
- ☐ Square each of these differences.
- ☐ Calculate the arithmetic mean of the squared differences.
- ☐ Calculate the square root of this arithmetic mean.

It is usually best to organize these calculations into a table, as illustrated in the next example.

EXAMPLE 2 The following values represent the number of defective diodes in lots of 100:

1 6 5 2 9 6 4 8 7 5

Find the standard deviation of these data values.

Solution First, we use the information in Table 16–1 to calculate the mean of the data values.

$$\bar{x} = \frac{\Sigma\, x}{n} = \frac{53}{10} = 5.3$$

TABLE 16–1

x	$x - \bar{x}$	$(x - \bar{x})^2$
1	−4.3	18.49
6	0.7	0.49
5	−0.3	0.09
2	−3.3	10.89
9	3.7	13.69
6	0.7	0.49
4	−1.3	1.69
8	2.7	7.29
7	1.7	2.89
5	−0.3	0.09
Σ 53		Σ 56.1

Second, we calculate the difference between the mean and the data values as shown in the second column of Table 16–1. Third, we square each entry in the second column and place the result in the third column of Table 16–1. Fourth, we find the arithmetic mean of these squared differences.

$$\frac{\Sigma\,(x - \bar{x})^2}{n} = \frac{56.1}{10} = 5.61$$

Last, the standard deviation is the square root of this mean.

$$s = \sqrt{5.61} \approx 2.4$$

The average number of defective diodes per 100 is 5 with a standard deviation of 2. ■

Standard Deviation from Grouped Data

To calculate the standard deviation from grouped data, use the following formula.

Standard Deviation from a Frequency Distribution

$$s = \sqrt{\frac{\sum_{i=1}^{n} f_i(x_i - \bar{x})^2}{\sum_{i=1}^{n} f_i}}$$

where $\bar{x}$ = mean of the data set $x_1, x_2, \ldots, x_n$ and f_i = frequency of data value x_i.

EXAMPLE 3 The first two columns in Table 16–2 show the test scores for a mathematics class. Using the frequency distribution, find the standard deviation.

Solution First, fill in the $f \cdot x$ column in Table 16–2 by multiplying the columns labeled x and f. Then calculate the mean as follows:

$$\bar{x} = \frac{\Sigma f \cdot x}{f} = \frac{1{,}968}{25} = 78.72$$

TABLE 16–2

x	f	$f \cdot x$	$x - \bar{x}$	$(x - \bar{x})^2$	$f(x - \bar{x})^2$
70	4	280	−8.72	76.04	304.16
73	6	438	−5.72	32.72	196.32
78	3	234	−0.72	0.52	1.56
80	5	400	1.28	1.64	8.20
88	7	616	9.28	86.12	602.84
	Σ 25	Σ 1,968			Σ 1,113.08

Next, fill in the remainder of the table. Last, use the equation and the information in Table 16–2 to calculate the standard deviation.

$$s = \sqrt{\frac{1{,}113.08}{25}} = 6.67$$

The average grade on the test is 79 with a standard deviation of 7 points. ■

Alternate Formula for Standard Deviation

If you use a calculator to determine standard deviation, the following formula is easier.

Alternate Formula for Standard Deviation

$$s = \sqrt{\frac{\sum_{i=1}^{n} (x_i)^2}{n} - \left(\frac{\sum_{i=1}^{n} x_i}{n}\right)^2}$$

mean of the squares (first term); square of the mean (second term)

EXAMPLE 4 The following measurements (in millimeters) represent the variation of a metal part from its specified length:

1.83 4.31 3.42 4.08 2.67 3.95

Find the standard deviation of these measurements.

Solution First, calculate the square of the mean, $\left(\frac{\Sigma x}{n}\right)^2$.

1.83 [+] 4.31 [+] 3.42 [+] 4.08 [+] 2.67 [+]
3.95 [=] [÷] 6 [=] [x^2] [STO] → 11.40187778

Second, calculate the mean of the squares, $\frac{\Sigma x^2}{n}$.

1.83 [x^2] [+] 4.31 [x^2] [+] 3.42 [x^2] [+] 4.08
[x^2] [+] 2.67 [x^2] [+] 3.95 [x^2] [=] [÷] 6 [=] → 12.16653333

Third, subtract these values.

[−] [RCL] [=] → 0.764655550

Last, take the square root of the resulting value.

[$\sqrt{\ }$] → 0.874445853

The standard deviation is approximately 0.87 mm. ■

NOTE ✦ Some calculators have special keys to aid in statistical calculations, such as $\boxed{\Sigma x}$, $\boxed{\Sigma x^2}$, and $\boxed{s}$ (standard deviation). Consult your owner's manual for instructions.

Significance of Standard Deviation

Thus far in this section, we have seen that the standard deviation is a measure of the dispersion or variation of a data set from the mean, and we have discussed how to calculate it. Now, we will discuss the significance of the standard deviation.

If we take enough experimental measurements, the curve of the data set usually approaches a normal, bell-shaped distribution as shown in Figure 16–7. Notice that for the normal distribution, approximately 68% of the measurements fall within one standard deviation of the mean, or within $\bar{x} - s$ and $\bar{x} + s$. Also, roughly 95% of the measurements fall within the two standard deviations of the mean, or within $\bar{x} - 2s$ and $\bar{x} + 2s$.

FIGURE 16–7

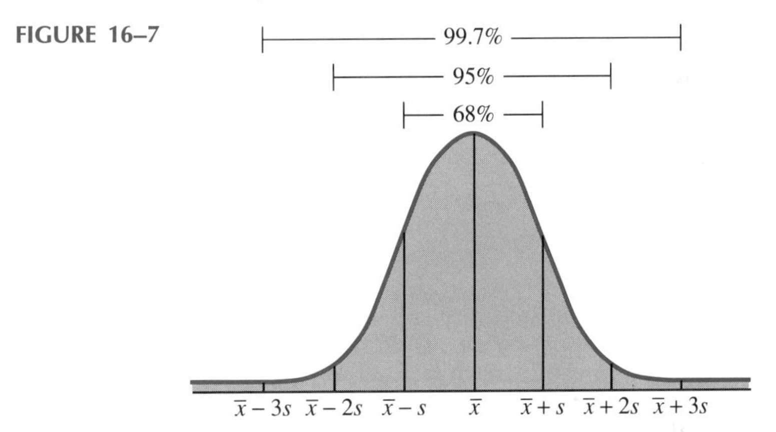

EXAMPLE 5 For the data set in Example 3, calculate the interval of values within one standard deviation of the mean. Compare this result with the actual percentage of the data set that falls within this range.

Solution To find the interval of values within one standard deviation of the mean, we calculate $\bar{x} - s$ and $\bar{x} + s$. From the calculations in Example 3, $\bar{x} = 78.72$ and $s = 6.67$. Thus,

$$\bar{x} - s = 78.72 - 6.67 = 72.05$$
$$\bar{x} + s = 78.72 + 6.67 = 85.39$$

The interval of values within one standard deviation of the mean is 72.05 to 85.39. From the data set itself, we find that 14 of the 25 values, or 56%, fall within this interval. ■

16–3 EXERCISES

Use data sets *A–E* for Problems 1–10.

A Average daily temperature (°F) for February of major cities: 36°, 72°, 65°, 19°, 18°, 25°, 60°, 58°, 30°, 15°, 48°

B Age of employees at ABC Stores: 18, 26, 43, 32, 19, 54, 39, 29, 22, 47, 39, 21, 60, 34, 23, 25, 58, 48, 18, 53

C Test scores: 63, 87, 73, 98, 60, 78, 85, 81, 70, 98, 93, 90, 72, 93, 83, 68, 88, 81, 72, 79, 61, 65, 75, 63

D Yearly income of technicians (in thousands): 18, 28, 21, 29, 31, 23, 25, 19, 20, 28, 18, 30, 20, 26, 18, 27, 23

E IQ of math students: 86, 140, 100, 96, 103, 110, 115, 97, 90, 108, 89, 97, 100, 105, 120, 118, 136, 102, 109

Find the range and standard deviation of the indicated data set.

1. Set *A* **2.** Set *B*

3. Set *C* **4.** Set *D*

5. Set *E*

Using the indicated data set, determine the interval of values one standard deviation from the mean. Compare this value to the actual percentage of data values in this range.

6. Set *A* **7.** Set *B*

8. Set *C* **9.** Set *D*

10. Set *E*

11. Using the frequency distribution in Table 16–3, construct a histogram and find the mean, median, mode, and standard deviation.

12. Using the frequency distribution in Table 16–4, construct a frequency polygon and calculate the mean, median, mode, and standard deviation.

TABLE 16–3

Grade	Frequency
60	2
65	1
70	3
80	4
85	6
90	3
95	2
100	1

TABLE 16–4

Employee's Age	Frequency
18	2
25	5
28	2
36	6
39	4
42	3
48	3
56	1

13. Calculate the standard deviation for Problem 4 of Exercises 16–1.

14. Calculate the standard deviation for Problem 12 of Exercises 16–2.

15. Calculate the standard deviation for Problem 13 of Exercises 16–2.

16. Calculate the standard deviation for Problem 14 of Exercises 16–2.

16–4

EMPIRICAL CURVE FITTING

In this chapter we have discussed statistical methods as they relate to a single variable. We have discussed methods of tabulating and displaying data and of calculating measures of central tendency and dispersion for one variable. However, we know from our study of functions in Chapter 2 that a relationship may exist between two variables. In this section we will discuss a method of determining the relationship between two variables using empirical data, or measurements.

Many of the relationships between two variables can be derived mathematically and stated in a formula, such as $A = \pi r^2$ for the area of a circle. However, in technology we are often given the relationship between two variables in the form of measurements. This relationship is activated by systematically varying one variable while measuring the second variable. Due to measurement errors, these relationships do not give exact formulas. The method of finding the curve that best fits the data is called **empirical curve fitting.**

Before attempting to fit a curve, we must first determine the type of curve that best fits the data. We can determine the form of the curve by using prior knowledge of the relationship between the variables or by graphing the data set as ordered pairs on the Cartesian coordinate plane.

EXAMPLE 1 Jack repeatedly clocks the velocity of a moving car as a function of time as follows:

t	1	2	3	4	5	6
v	28.5	35	46	50	60	66

Plot the data points on the Cartesian plane and determine the apparent relationship between the variables.

Solution The graph of the data points is given in Figure 16–8. From the graph, the relationship between the variables appears to be linear. From physics, we know that the velocity v of an object as a function of time is given by $v = v_0 + at$, where v_0 = initial velocity, a = acceleration, and t = time. Both from the graph of the data and from prior knowledge, we should fit the data to a linear equation.

FIGURE 16–8

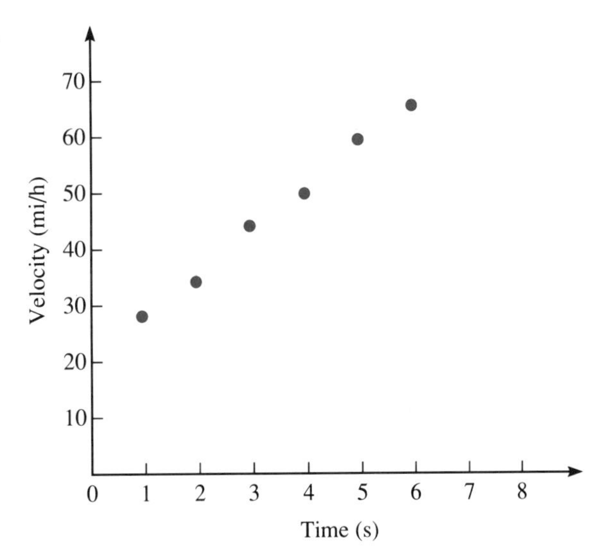

■

Least Squares Method

The data points from Example 1 and the equation representing that particular relationship, $v = 20 + 8t$ (we will discuss how to determine this equation later), are plotted on the same Cartesian plane in Figure 16–9. Note that the data points do not all fall on the graph. The difference between the y value of the data point and the y value of the curve is called the **deviation,** as illustrated in Figure 16–9.

The most commonly used technique of curve fitting is called the **least squares method.** The method is used to find the equation of the curve that minimizes the deviation.

FIGURE 16–9

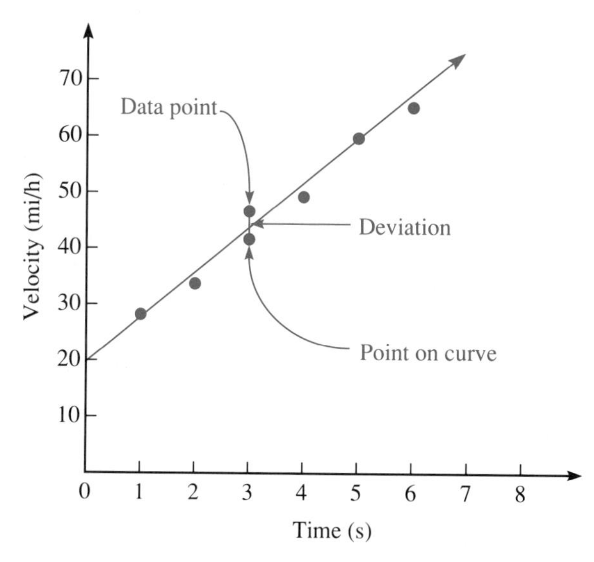

Linear Curve Fitting

We will begin our discussion of curve fitting with linear equations because they are easier. Also, fitting a nonlinear curve is based on the methods used in linear curve fitting.

The equation of a straight line can be expressed in slope–intercept form as $y = mx + b$ where m represents the slope of the line and b represents the y coordinate of the y intercept. Using advanced mathematics, we can derive the following formula to calculate the value of m and b.

Curve Fitting a Straight Line

$$y = mx + b$$

where

$$m = \frac{n \sum xy - \sum x \sum y}{n \sum x^2 - (\sum x)^2}$$

$$b = \frac{\sum x^2 \sum y - \sum x \sum xy}{n \sum x^2 - (\sum x)^2}$$

(x, y) represents the data points, and n is the number of data points.

EXAMPLE 2 A teacher wants to be able to predict a student's grade on a test based on the number of hours of study. Assuming that the relationship between the variables is linear, find the least squares equation of the line that fits the data, and use it to predict the test score of a student who studies 4 hours.

x (hours)	8	6	5	7	8.3	7.5	6.3	8.7
y (grade)	91	72	62	81	94	86	76	98

Solution Normally, we would plot the data to determine the nature of the relationship between the variables, but it is given to be linear. To find the equation of the line, we must calculate m and b using the formulas above. Table 16–5 provides the necessary information.

TABLE 16–5

x (hours)	y (grade)	xy	x^2
8	91	728	64
6	72	432	36
5	62	310	25
7	81	567	49
8.3	94	780.2	68.89
7.5	86	645	56.25
6.3	76	478.8	39.69
8.7	98	852.6	75.69
Σ56.8	Σ660	Σ4,793.6	Σ414.52

Using the formulas to calculate m and b gives

$$m = \frac{n \Sigma xy - \Sigma x \Sigma y}{n \Sigma x^2 - (\Sigma x)^2}$$

$$m = \frac{8(4{,}793.6) - 56.8(660)}{8(414.52) - (56.8)^2} \approx 9.57$$

$\boxed{(}$ 8 $\boxed{\times}$ 4793.6 $\boxed{-}$ 56.8 $\boxed{\times}$ 660 $\boxed{)}$ $\boxed{\div}$

$\boxed{(}$ 8 $\boxed{\times}$ 414.52 $\boxed{-}$ 56.8 $\boxed{x^2}$ $\boxed{)}$ $\boxed{=}$ $\rightarrow$ 9.57

$$b = \frac{\Sigma x^2 \ \Sigma y - \Sigma x \ \Sigma xy}{n \Sigma x^2 - (\Sigma x)^2}$$

$$b = \frac{414.52(660) - 56.8(4{,}793.6)}{8(414.52) - (56.8)^2} \approx 14.5$$

$\boxed{(}$ 414.52 $\boxed{\times}$ 660 $\boxed{-}$ 4793.6 $\boxed{\times}$ 56.8 $\boxed{)}$ $\boxed{\div}$

$\boxed{(}$ 8 $\boxed{\times}$ 414.52 $\boxed{-}$ 56.8 $\boxed{x^2}$ $\boxed{)}$ $\boxed{=}$ $\rightarrow$ 14.5

The least squares equation of the straight line that best fits the data is

$$y = 9.57x + 14.5$$

Using this equation to predict the grade of a student who studies 4 hours gives

$$y = 9.57(4) + 14.5$$
$$y \approx 53$$

The graph in Figure 16–10 shows the least squares line and the data points. ■

CAUTION ✦ You should always graph the data points and the least squares curve on the same axes to point out calculation errors.

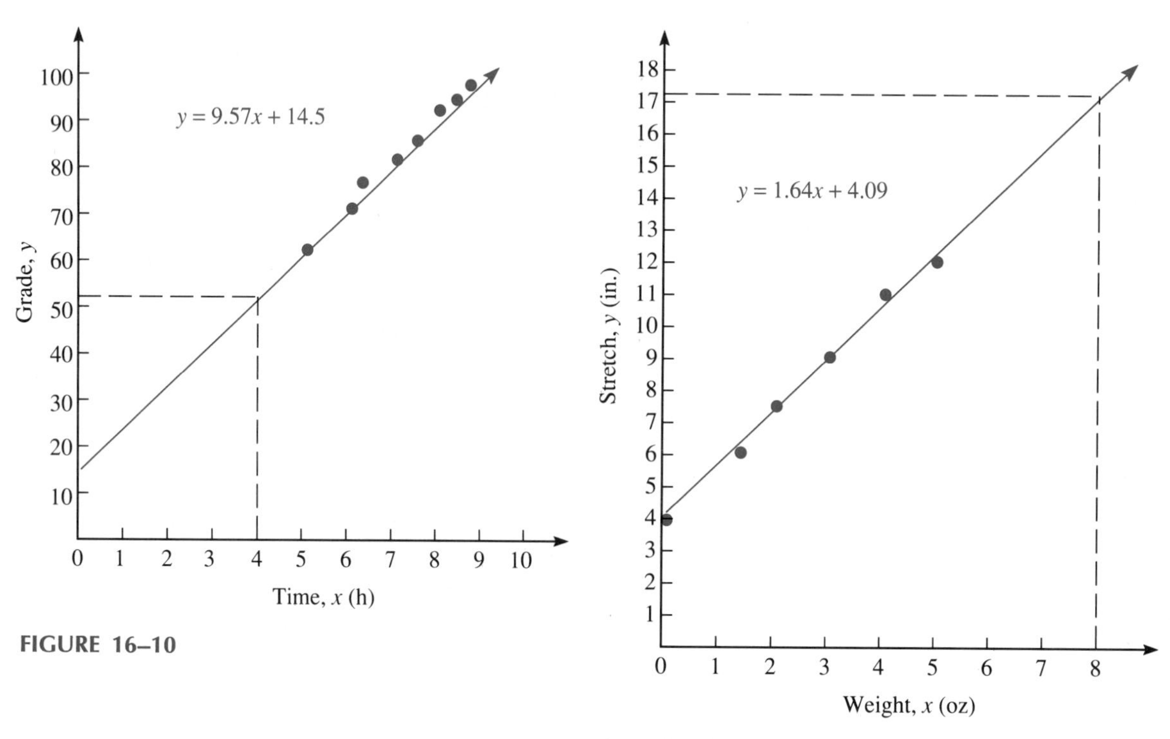

FIGURE 16–10

FIGURE 16–11

EXAMPLE 3 The following data resulted from attaching a weight to a spring and measuring the length of the spring:

(0, 4) (2, 7.5) (4, 11) (1.2, 6) (3, 9) (5, 12)

Determine a linear least squares line that fits the data, and use it to determine the length of the spring when 8 oz is attached to the spring.

Solution To determine the linear equation that best fits the data, we fill in Table 16–6.

TABLE 16–6

x (oz)	y (in.)	xy	x^2
0	4	0	0
1.2	6	7.2	1.44
2	7.5	15	4
3	9	27	9
4	11	44	16
5	12	60	25
Σ15.2	Σ49.5	Σ153.2	Σ55.44

Next, we substitute the required values from the table into the formulas for m and b.

$$m = \frac{n \Sigma xy - \Sigma x \Sigma y}{n \Sigma x^2 - (\Sigma x)^2}$$

$$m = \frac{6(153.2) - 15.2(49.5)}{6(55.44) - (15.2)^2}$$

$$m = 1.64$$

$$b = \frac{\Sigma x^2 \Sigma y - \Sigma x \Sigma xy}{n \Sigma x^2 - (\Sigma x)^2}$$

$$b = \frac{55.44(49.5) - 15.2(153.2)}{6(55.44) - (15.2)^2}$$

$$b = 4.09$$

The linear equation that best fits the data is

$$y = 1.64x + 4.09$$

To determine the length of the spring when an 8-oz object is attached, we substitute $x = 8$ into the equation.

$$y = 1.64(8) + 4.09$$
$$y = 17.21 \text{ in.}$$

Figure 16–11 shows the least squares equation and the data points. ■

Nonlinear Curve Fitting

So far in this section, we have discussed only a linear relationship between two variables. However, from previous chapters we know that variables can also be related by nonlinear functions. If the graph of the data points indicates a nonlinear relationship $f(x)$, then we can express the least squares equation by rewriting the slope–intercept form in the following manner. Since nonlinear curves are harder to determine, you will be given the form of the equation for the least squares curve.

Nonlinear Least Squares Curve

$$y = m[f(x)] + b$$

where

$$m = \frac{n \sum ([f(x)]y) - \sum f(x) \sum y}{n \sum [f(x)]^2 - [\sum f(x)]^2}$$

$$b = \frac{\sum [f(x)]^2 \sum y - \sum f(x) \sum ([f(x)]y)}{n \sum [f(x)]^2 - [\sum f(x)]^2}$$

$f(x)$ is a function, m is the slope of the line, and b is the y coordinate of the y intercept.

EXAMPLE 4 Find the least square equation of the form $y = mx^2 + b$ for the following data:

x (s)	0	1	2	3	4	5
y (ft)	8.7	9.8	13.7	19.6	28	38.6

Then use the equation to find y when $x = 8$ s.

Solution Using $f(x) = x^2$, we fill in Table 16–7.

TABLE 16–7

x	y	$f(x) = x^2$	$f(x)y$	$[f(x)]^2$
0	8.7	0	0	0
1	9.8	1	9.8	1
2	13.7	4	54.8	16
3	19.6	9	176.4	81
4	28	16	448	256
5	38.6	25	965	625
	Σ118.4	Σ55	Σ1,654.0	Σ979

Next, we substitute into the formulas to calculate m and b.

$$m = \frac{n \Sigma ([f(x)]y) - \Sigma f(x) \Sigma y}{n \Sigma [f(x)]^2 - [\Sigma f(x)]^2}$$

$$m = \frac{6(1{,}654.0) - 55(118.4)}{6(979) - (55)^2}$$

$$m = 1.2$$

$$b = \frac{\Sigma [f(x)]^2 \Sigma y - \Sigma f(x) \Sigma ([f(x)]y)}{n \Sigma [f(x)]^2 - [\Sigma f(x)]^2}$$

$$b = \frac{979(118.4) - 55(1{,}654.0)}{6(979) - (55)^2}$$

$$b = 8.8$$

The least squares equation that best fits the data is given by

$$y = 1.2x^2 + 8.8$$

When $x = 8$ s,

$$y = 1.2(8^2) + 8.8$$

$$y \approx 85.6 \text{ ft}$$

Figure 16–12 shows the graph of the curve and the data points.

FIGURE 16–12

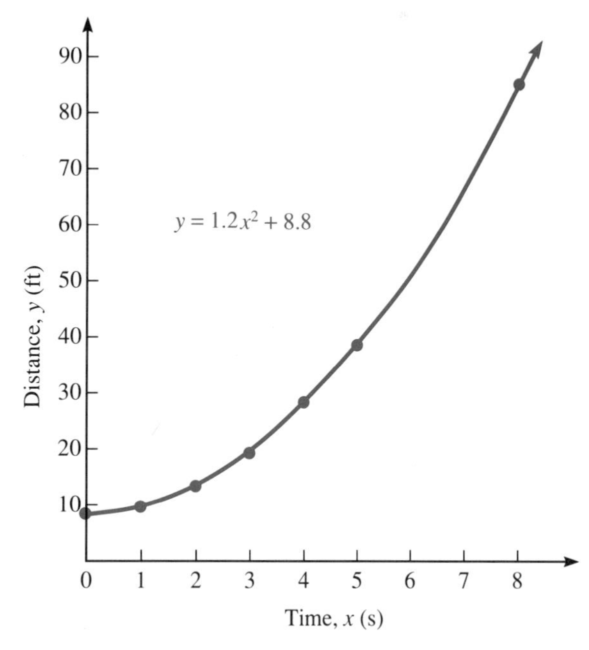

■

EXAMPLE 5 The temperature of a metal pipe was measured as it was heated, and the following data were obtained:

x (min)	1	2	3	4	5
y (°C)	77	90	107	131	165

Fit a curve to the data if the equation is of the form $y = me^{0.3x} + b$. Find y when the time is 8 min.

Solution Using $f(x) = e^{0.3x}$, we set up Table 16–8.

TABLE 16–8

x	y	$f(x) = e^{0.3x}$	$f(x)y$	$[f(x)]^2$
1	77	1.35	103.94	1.82
2	90	1.82	163.99	3.32
3	107	2.46	263.18	6.05
4	131	3.32	434.94	11.02
5	165	4.48	739.48	20.09
	Σ570	Σ13.43	Σ1,705.53	Σ42.30

From Table 16–8, we substitute the necessary information into the formulas for m and b.

$$m = \frac{n \sum ([f(x)]y) - \sum f(x) \sum y}{n \sum [f(x)]^2 - [\sum f(x)]^2}$$

$$m = \frac{5(1{,}705.53) - 13.43(570)}{5(42.30) - (13.43)^2}$$

$$m = 28.0$$

$$b = \frac{\sum [f(x)]^2 \sum y - \sum f(x) \sum ([f(x)]y)}{n \sum [f(x)]^2 - [\sum f(x)]^2}$$

$$b = \frac{42.30(570) - 13.43(1{,}705.53)}{5(42.30) - (13.43)^2}$$

$$b = 38.7$$

The least squares equation that best fits the data is

$$y = 28.0e^{0.3x} + 38.7$$

When $x = 8$ min, the temperature is

$$y = 28.0e^{0.3(8)} + 38.7 = 347.3°$$

Figure 16–13 shows a plot of the least squares curve and the data points.

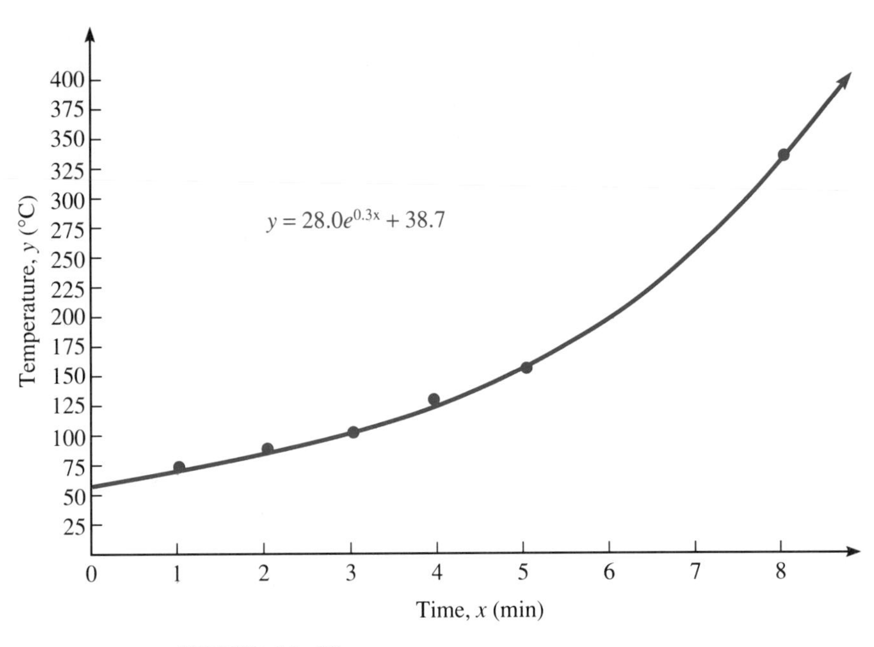

FIGURE 16–13

■

16–4 EXERCISES

Using the least squares method, find the equation of the straight line that best fits each data set. Plot the data points and the least squares equation on the same Cartesian plane.

1.

x	1	2	3	4	5	6
y	8	11	14	18	20	24

2.

x	1	2	4	6	9	11
y	4	10	22	35	53	66

3.

x	2	7	9	12	16	20
y	10	24	29	37	48	59

4.

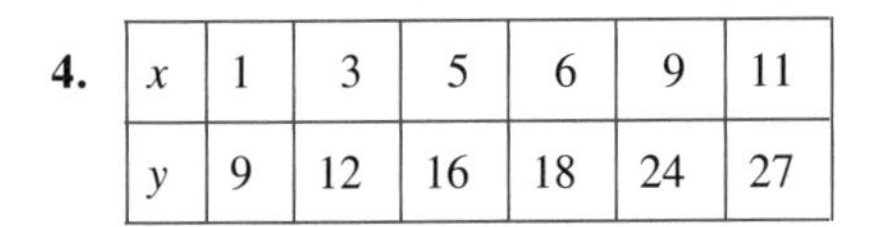

x	1	3	5	6	9	11
y	9	12	16	18	24	27

Using the least squares method, find the best-fitting curve of the given form for each data set. Plot the data points and the curve on the same graph.

5.

x	1	2	3	4	5
y	2	13	32	57	91

$y = mx^2 + b$

6.

x	4	8	12	20	30
y	20	24	29	36	42

$y = m\sqrt{x} + b$

7.

x	0.1	0.2	0.3	0.4	0.5
y	10	11	12	13	15

$y = m(10^x) + b$

8.

x	1	3	5	8
y	16	85	599	11,930

$y = m(e^x) + b$

9. The displacement s of a pendulum bob from its equilibrium position as a function of time t was measured as follows:

t (s)	0	5.4	6.8	7.9	9.3
s (cm)	14.3	15.2	15.5	15.7	15.8

Find the least squares equation of the form $s = mt + b$, and determine the displacement when $t = 12$ s.

10. The growth n of bacteria in a petri dish as a function of time t is given in the following table:

t (h)	0.75	1.0	1.5	1.75	2.0	2.25
n (thousands)	18	21	27	33	38	46

Using the least squares method, fit the data to a curve of the form $n = m(e^t) + b$. How many bacteria will be in the dish at $t = 3.5$ h?

11. The displacement s of an object moving at a constant velocity as a function of time t was measured as follows:

t (s)	3	4	7	9	10	15	20
s (ft)	131	165	280	352	390	573	760

Find the least squares equation of the form $s = mt + b$, and find the displacement when $t = 12$ s.

12. The pressure P and volume V of a gas at a constant temperature were measured. Using the following data, find the least squares equation of the form

$$P = m\frac{1}{V} + b$$

V (cm^3)	20	30	35	40	50
P (kPa)	70	45	40	35	27

13. The resistance R of a wire at various temperatures T was as follows:

T (°F)	32	46	60	72	85	93
R (Ω)	620	867	1,120	1,333	1,570	1,710

Use the least squares method to find the line of the form $R = mT + b$. Use the equation to evaluate the resistance at 80°F.

14. The distance d an object rolls down an inclined plane as a function of time t was measured and is summarized in the following table:

t (s)	1	2	3.4	5.6	7.2
d (cm)	13.9	23	48	110	170

Find the least squares equation of the form $d = mt^2 + b$, and find the distance when $t = 9$ s.

CHAPTER SUMMARY

Summary of Terms

arithmetic mean (p. 582)
average (p. 582)
deviation (p. 594)
empirical curve fitting (p. 593)
empirical data (p. 576)
frequency distribution (p. 576)
frequency polygon (p. 578)
histogram (p. 577)
least squares method (p. 594)
measures of central tendency (p. 582)
measures of dispersion (p. 587)
median (p. 584)
mode (p. 585)
range (p. 587)
standard deviation (p. 587)
variation (p. 587)

Summary of Formulas

$$\bar{x} = \frac{x_1 + x_2 + x_3 + \cdots + x_n}{n} = \frac{\sum_{i=1}^{n} x_i}{n} \quad \text{arithmetic mean}$$

$$\bar{x} = \frac{\sum_{i=1}^{n} f_i x_i}{\sum_{i=1}^{n} f_i} \quad \text{arithmetic mean from a frequency distribution}$$

$$s = \sqrt{\frac{\sum_{i=1}^{n} (x_i - \bar{x})^2}{n}} \quad \text{standard deviation}$$

$$s = \sqrt{\frac{\sum_{i=1}^{n} f_i(x_i - \bar{x})^2}{\sum_{i=1}^{n} f_i}} \quad \text{standard deviation from a frequency distribution}$$

$$s = \sqrt{\left(\frac{\sum_{i=1}^{n} (x_i)^2}{n}\right) - \left(\frac{\sum_{i=1}^{n} x_i}{n}\right)^2} \quad \text{standard deviation (alternate formula)}$$

$y = mx + b$ linear least squares

where

$$m = \frac{n \Sigma xy - \Sigma x \Sigma y}{n \Sigma x^2 - (\Sigma x)^2}$$

$$b = \frac{\Sigma x^2 \Sigma y - \Sigma x \Sigma xy}{n \Sigma x^2 - (\Sigma x)^2}$$

$y = m\,[f(x)] + b$ nonlinear least squares

where

$$m = \frac{n \Sigma ([f(x)]y) - \Sigma f(x) \Sigma y}{n \Sigma [f(x)]^2 - [\Sigma f(x)]^2}$$

$$b = \frac{\Sigma [f(x)]^2 \Sigma y - \Sigma f(x) \Sigma ([f(x)]y)}{n \Sigma [f(x)]^2 - [\Sigma f(x)]^2}$$

CHAPTER REVIEW

Use data sets *A–D* as required.

A Age of employees at ABC Corporation: 27, 35, 18, 45, 27, 36, 18, 27, 57, 27, 18, 35, 63, 27

B Grades on algebra test: 68, 94, 83, 75, 71, 88, 62, 91, 83, 77, 64, 75, 94, 83, 81, 79, 86, 96

C Salary of civil engineers (in thousands): 19, 24, 30, 21, 24, 18, 30, 19, 27, 23, 21, 23, 24, 18, 32

D Number of defective diodes per package of 200: 6, 20, 14, 18, 7, 10, 11, 18, 6, 12, 24, 14, 18, 10, 13, 14, 18

Section 16–1

Construct a frequency distribution, histogram, and frequency polygon for the indicated data set.

1. Set *A*
2. Set *B*
3. Set *C*
4. Set *D*

Section 16–2

Determine the mean, median, and mode for the indicated data set.

5. Set *A*
6. Set *B*
7. Set *C*
8. Set *D*

9. Fred must estimate the cost of repairing a television. He knows that the parts will cost $36.50, but he must calculate the cost of labor. He knows that four similar jobs required 3.1 h, 4.7 h, 2.9 h, and 4.3 h. Using the arithmetic mean of the previous jobs and an hourly rate of $26 per hour, determine how much Fred should estimate for the cost of repairing the television.

10. Carlos scored 86, 94, 90, 81, and 85 on his first five algebra tests. What must he score on his last test (the sixth) to earn an average of 90?

11. The monthly electric bills during the last year for the Smith family are as follows:

$50.27 $42.78 $67.34 $86.73 $94.86

$150.78 $126.81 $105.73 $93.70 $73.85

$62.38 $56.71

If Mrs. Smith is attempting to make a budget for the coming year, based on the monthly average, what should she budget for the coming year?

12. A company has 35 employees: 5 are 26 years old, 3 are 30 years old, 7 are 33 years old, 7 are 46 years old, 5 are 50 years old, and 8 are 56 years old. Determine the mean age of the employees.

Section 16–3

Determine the range and standard deviation for the indicated data set.

13. Set *A*
14. Set *B*
15. Set *C*
16. Set *D*

Using the indicated data set, determine the range of values one standard deviation from the mean. Compare this value to the actual percentage of the data values within this range.

17. Set *A*
18. Set *B*
19. Set *C*
20. Set *D*

21. Using the frequency distribution in Table 16–9, calculate the standard deviation.

TABLE 16–9

Employee's Salary (thousands)	Frequency
18	2
23	8
27	10
29	7
30	4
32	1

22. Using the frequency distribution in Table 16–10, calculate the standard deviation.

TABLE 16–10

Grade	Frequency
61	1
64	1
68	2
74	3
77	10
81	6
83	8
88	5
90	3
94	2
96	2
98	1

Section 16–4

Using the least squares method, find the equation of the straight line that best fits each data set. Plot the data points and the least squares equation on the same Cartesian plane.

23.

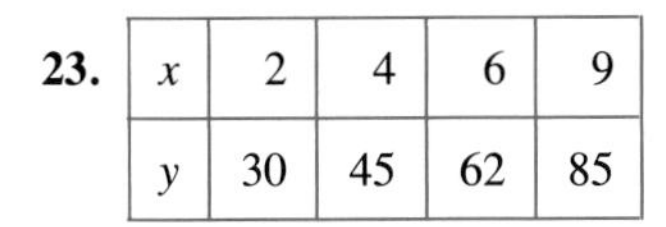

x	2	4	6	9
y	30	45	62	85

24.

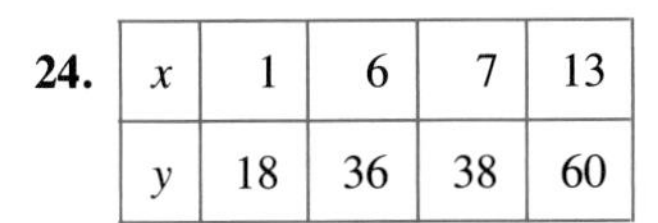

x	1	6	7	13
y	18	36	38	60

25.

x	2	5	9	12	15
y	22	36	50	60	74

26.

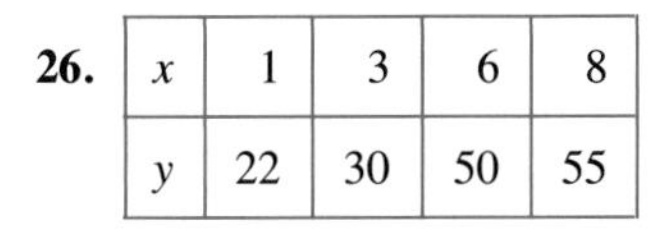

x	1	3	6	8
y	22	30	50	55

Using the least squares method, find the best-fitting curve of the given form for each data set. Plot the data points and the curve on the same graph.

27.

x	1	4	6	9
y	11	45	85	179

$y = mx^2 + b$

28.

x	1	2	3	4	5	6
y	18	40	77	130	190	274

$y = mx^2 + b$

29.

x	0.1	0.4	0.8	1.0	1.3	1.8	2.0
y	14	18	29	43	70	205	320

$y = m10^x + b$

30.

x	0.3	0.7	0.9	1.6	2.1
y	15	18	20	30	45

$y = me^x + b$

CHAPTER TEST

The number in parentheses refers to the appropriate learning objective given at the beginning of the chapter.

1. Calculate the mean, median, and mode for the following data set: **(2)**

18 23 16 27 18 36 21 23 18 21 25 16 27

2. Draw a frequency polygon for the following data set: **(1)**

250 220 286 200 250 197 200
190 250 200 186 186 250 197

3. Calculate the standard deviation for the following data set: **(3)**

25 16 29 16 34 10 15 19 24 21 28 32 12 16 24

4. Using the least squares method, determine the straight line that best fits the following data set: **(5)**

x	1	2	3	4	5
y	10	16	23	30	36

Using the equation, calculate y when $x = 8$.

5. Organize the following data set into a frequency distribution, and draw a histogram. **(1)**

2 6 3 8 11 3 14 7 8 2 3 14 8 2 9 18 7

6. Using the mean and standard deviation for the following data set, determine the range of values within one standard deviation of the mean: **(4)**

16 28 40 10 18 25 36 14 28 41 26 27 19 25 38

7. KLZ Corporation employs 47 people. Ten earn $24,000 annually, 7 earn $28,500 annually, 9 earn $32,000 annually, 11 earn $36,000 annually, 6 earn $42,000 annually, and 4 earn $50,000 annually. Calculate the mean salary and the standard deviation. **(2, 3)**

8. Using the least squares method, determine the equation of the curve of the form $y = mx^2 + b$ that best fits the following data set: **(5)**

x	1	3	4	6	9
y	15	35	55	100	215

SOLUTION TO CHAPTER INTRODUCTION

To determine the equation for the curve, we must calculate m and b. First, we fill in Table 16–11.

TABLE 16–11

t	n	$f(t) = e^{0.2t}$	$[f(t)n]$	$[f(t)]^2$
0	19.2	1.00	19.20	1.00
1	21	1.22	25.65	1.49
2	23	1.49	34.31	2.23
3	25	1.82	45.55	3.32
4	28	2.23	62.32	4.95
5	32	2.72	86.99	7.39
	Σ 148.2	Σ 10.48	Σ 274.02	Σ 20.38

Next, we substitute into the formulas to calculate m and b.

$$m = \frac{N\,\Sigma\,([f(t)]n) - \Sigma\,f(t)\,\Sigma\,n}{N\,\Sigma\,[f(t)]^2 - [\Sigma\,f(t)]^2}$$

$$m = \frac{6(274.02) - 10.48(148.2)}{6(20.38) - (10.48)^2}$$

$$m = 7.31$$

$$b = \frac{\Sigma\,[f(t)]^2\,\Sigma\,n - \Sigma\,f(t)\,\Sigma\,([f(t)]n)}{N\,\Sigma\,[f(t)]^2 - [\Sigma\,f(t)]^2}$$

$$b = \frac{20.38(148.2) - 10.48(274.02)}{6(20.38) - (10.48)^2}$$

$$b = 11.9$$

The least squares equation for the growth of bacteria is

$$n = 7.31e^{0.2t} + 11.9$$

To predict the number of bacteria at 18 hours, we substitute 18 into the equation.

$$n = 7.31e^{0.2(18)} + 11.9$$

$$n = 279$$

At the end of 18 hours, there are 279,000 bacteria in the petri dish.

You are offered a job as a technician at XYZ Corporation for a yearly salary of \$23,480 with a promise of yearly raises of 8% for eight years. You are also offered a job with ABC Company for \$26,700 with a promise of yearly raises of \$1,500 for eight years. Which job should you accept if your decision is based solely on your salary at the end of eight years? (The answer to this problem is given at the end of the chapter.)

To determine your salaries at the end of eight years, you must calculate the *n*th term of a geometric progression and of an arithmetic progression, both topics discussed in this chapter. We will also discuss finding the sum of an infinite geometric series and the binomial formula.

Learning Objectives

After completing this chapter, you should be able to

1. Find the *n*th term of an arithmetic progression from given information (Section 17–1).

2. Determine the sum of the first *n* terms of an arithmetic progression (Section 17–1).

3. Find the *n*th term of a geometric progression from given information (Section 17–2).

4. Determine the sum of the first *n* terms of a geometric progression (Section 17–2).

5. Find the sum of an infinite geometric progression (Section 17–3).

6. Represent a repeating decimal in fractional form (Section 17–3).

7. Apply the binomial formula to raise a binomial to a given power (Section 17–4).

8. Apply Pascal's triangle to expand a binomial to a given power (Section 17–4).

9. Find a specified term in a binomial expansion (Section 17–4).

Chapter 17

Sequences, Series, and the Binomial Formula

17–1

ARITHMETIC PROGRESSIONS

A **sequence** is a collection of numbers in a certain order. For example, both of the following represent a sequence:

$$7, 8, 9 \quad \text{and} \quad 1, 7, 13, 19, \ldots$$

Definitions

Each member of the sequence is called a **term.** *If the sequence has a last term,* it is called a **finite sequence,** as exemplified by the first sequence above. *Any sequence that is not finite* is called an **infinite sequence,** as exemplified by the second sequence above. The notation a_n denotes the nth term of the sequence; thus, a_3 in the infinite sequence above is 13, or $a_3 = 13$.

Arithmetic Progression

An **arithmetic progression** is a sequence of numbers in which the difference between successive terms is a constant, called the **common difference.** For example, the sequence of numbers 4, 11, 18, 25, . . . represents an arithmetic progression with a common difference of 7.

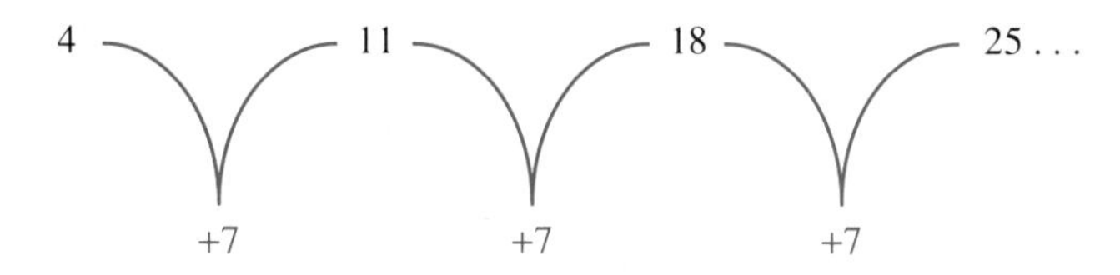

In general, if a_1 is the first term and d is the common difference of the sequence, you can generate an arithmetic progression as follows:

$$\begin{array}{ccccc} a_1 & a_1 + d & a_1 + 2d & a_1 + 3d & a_1 + 4d \\ \downarrow & \downarrow & \downarrow & \downarrow & \downarrow \\ a_1 & a_2 & a_3 & a_4 & a_5 \end{array}$$

Using this information, we can generate the following formula to find the nth term of an arithmetic progression.

---

nth Term of an Arithmetic Progression

The nth term of an arithmetic progression is given by

$$a_n = a_1 + (n - 1)d$$

where a_1 = the first term of the progression and d = the common difference.

EXAMPLE 1 Find the eighth term of the following arithmetic progression:

$$6, 10, 14, \ldots$$

Solution From the list of terms in the progression, we know that the first term $a_1 = 6$. We can determine the common difference d by subtracting any two consecutive terms. The common difference d is

$$d = 10 - 6 = 4$$

Furthermore, since we are asked to find the eighth term of the progression, $n = 8$. Substituting $a_1 = 6$, $d = 4$, and $n = 8$ into the formula gives

$$a_n = a_1 + (n - 1)d$$

$$a_8 = 6 + (8 - 1)4$$

$$a_8 = 6 + 7(4)$$

$$a_8 = 34$$

The eighth term of this arithmetic progression is 34. ■

EXAMPLE 2 Determine how many numbers between 7 and 800 are divisible by 3.

Solution First, we must find the smallest number, a_1, and the largest number, a_n, that are in the range from 7 to 800 and that are divisible by 3. A number is divisible by 3 if the sum of its digits is divisible by 3. Therefore, we start with 7 and count upwards, and then we start with 800 and count downwards to find the smallest and largest numbers in this range divisible by 3. The smallest a_1 is 9, and the largest a_n is 798. Next, we substitute $a_n = 798$, $a_1 = 9$, and $d = 3$ into the formula and solve for n.

$$a_n = a_1 + (n - 1)d$$

$$798 = 9 + (n - 1)3$$

$$798 = 9 + 3n - 3$$

$$792 = 3n$$

$$264 = n$$

There are 264 numbers between 7 and 800 that are divisible by 3. ■

EXAMPLE 3 An object moving in a line is given an initial velocity of 4.3 m/s and a constant acceleration of 1.2 m/s^2. How long will it take the object to reach a velocity of 14.8 m/s?

Solution We are given that $a_1 = 4.3$, $a_n = 14.8$, and $d = 1.2$, and we are asked to determine n. Substituting these values into the formula gives

$$a_n = a_1 + (n - 1)d$$

$$14.8 = 4.3 + (n - 1)1.2$$

$$14.8 = 4.3 - 1.2 + 1.2n$$

$$n \approx 9.8$$

After 9.8 s, the object is traveling 14.8 m/s.

14.8 [−] 4.3 [+] 1.2 [=] [÷] 1.2 [=] → 9.75 ■

Sum of an Arithmetic Progression

You can calculate the sum S_n of the first n terms of an arithmetic progression manually by adding the terms. However, to develop a formula, write an expression for the sum of the first n terms forwards and backwards, and add the resulting two equations.

$$\begin{aligned} S_n &= a_1 + (a_1 + d) + \cdots + (a_n - d) + a_n \\ S_n &= a_n + (a_n - d) + \cdots + (a_1 + d) + a_1 \\ \hline 2S_n &= (a_1 + a_n) + (a_1 + a_n) + \cdots + (a_1 + a_n) + (a_1 + a_n) \end{aligned}$$

Since there are n terms of $(a_1 + a_n)$,

$$2S_n = n(a_1 + a_n)$$

Solving the equation for S_n gives the following formula.

Sum of First n Terms of an Arithmetic Progression

The sum of the first n terms of an arithmetic progression is given by

$$S_n = \frac{n(a_1 + a_n)}{2}$$

EXAMPLE 4 Find the sum of the positive even integers up to and including 250.

Solution From the information given, we know that $a_1 = 2$, $d = 2$, and $a_n = 250$. First, we must use the previous formula to determine the number n of even integers in this range.

$$\begin{aligned} a_n &= a_1 + (n - 1)d \\ 250 &= 2 + (n - 1)2 \\ 250 &= 2 + 2n - 2 \\ n &= 125 \end{aligned}$$

Substituting $a_1 = 2$, $a_n = 250$, and $n = 125$ into the sum formula gives

$$\begin{aligned} S_n &= \frac{n(a_1 + a_n)}{2} \\ S_{125} &= \frac{125(2 + 250)}{2} \\ S_{125} &= 15{,}750 \end{aligned}$$

The sum of the positive, even integers through 250 is 15,750. ■

Application

EXAMPLE 5 In an integrated circuit with an initial current of 960 mA, the temperature in the components decreased from 25% to 22% to 19%. Assuming that each temperature decrease is caused by a decrease in the initial current, what is the value of the current at the fifth measurement?

Solution We know that $n = 5$, $a_1 = 25\%$, and $d = -3\%$. First, we must calculate a_5 before we can determine S_5.

$$a_n = a_1 + (n - 1)d$$

$$a_5 = 25\% + 4(-3\%)$$

$$a_5 = 13\%$$

Then we can calculate S_5.

$$S_n = \frac{n(a_1 + a_n)}{2}$$

$$S_5 = \frac{5(25 + 13)}{2}$$

$$S_5 = 95\%$$

Last, we calculate the value of the current at this point. Since 95% of the current is lost in heat, 5% of the 960-mA current remains.

$$\text{current} = (5\%)(960 \text{ mA}) = 48 \text{ mA}$$

■

Under certain circumstances we may need to solve for two unknowns in an arithmetic progression. Such a problem requires a system of two equations: one resulting from the a_n formula and the second from the S_n formula. The next example illustrates this technique.

EXAMPLE 6 A technician accepts a job to pay off his $12,000 college loan. If he pays $457.50 toward the loan the first month and increases his payment by $15 each month, how long will it take him to pay off the loan?

Solution We are given that $a_1 = \$457.5$, $S_n = \$12{,}000$, and $d = \$15$, and we must calculate n, which represents months. Therefore,

$$a_n = a_1 + (n - 1)d$$

$$a_n = 457.5 + (n - 1)15$$

$$a_n = 442.5 + 15n$$

Also,

$$S_n = \frac{n(a_1 + a_n)}{2}$$

$$12{,}000 = \frac{n(457.5 + a_n)}{2}$$

Then we solve the system of two equations by substitution.

$$
\begin{aligned}
12{,}000 &= \frac{n[457.5 + (442.5 + 15n)]}{2} \\
24{,}000 &= n(900 + 15n) \\
24{,}000 &= 15n^2 + 900n \\
15n^2 + 900n - 24{,}000 &= 0 \\
n^2 + 60n - 1{,}600 &= 0 \\
(n + 80)(n - 20) &= 0 \\
n = -80 \text{ and } n &= 20
\end{aligned}
$$

Since n must be positive, the technician will repay his loan in 20 months. ■

NOTE ✦ If you had difficulty with Example 6, review Chapter 5 on solving quadratic equations and Chapter 6 on solving systems of equations.

17–1 EXERCISES

Find the indicated term of the following arithmetic progressions.

1. The 5th term of 3, 9, 15, . . .

2. The 8th term of −5, −2, 1, . . .

3. The 10th term of 8, $\frac{15}{2}$, 7, . . .

4. The 13th term of 13, 7, 1, . . .

5. The 25th term of −3, 4, 11, . . .

6. The 15th term of −2, −4, −6, . . .

7. The 9th term where $a_1 = 7$ and $d = -3$

8. The 18th term where $a_1 = -4$ and $d = 6$

9. The 23rd term where $a_1 = -15$ and $d = 4$

10. The 30th term where $a_1 = 10$ and $d = 0.5$

Find the sum of the first n terms of each arithmetic progression.

11. $n = 18$, $a_1 = -3$, and $a_{18} = 31$

12. $n = 9$, $a_1 = 6$, and $a_9 = -26$

13. $n = 20$, $a_1 = 4$, and $a_{20} = 14$

14. $n = 33$, $a_1 - 18$, and $a_{33} = -10$

15. $n = 15$, $a_1 = -6$, and $d = 3$

16. $n = 27$, $a_1 = 38$, and $d = -2$

Use the information given in the following problems to find the requested quantities.

17. If $a_1 = 3$, $d = -2$, and $a_n = -11$, find n and S_n.

18. If $a_1 = -18$, $d = 1/2$, and $a_n = -13$, find n and S_n.

19. If $a_1 = 5$ and $a_{18} = 56$, find d and S_{18}.

20. If $a_1 = -0.5$ and $a_{12} = -33.5$, find d and S_{12}.

21. If $a_1 = -5$, $n = 10$, and $d = 4$, find a_{10} and S_{10}.

22. If $a_1 = 16$, $n = 30$, and $d = -2$, find a_{30} and S_{30}.

23. If $a_1 = 10$, $a_n = 31$, and $S_n = 164$, find d and n.

24. If $a_1 = -8$, $a_n = 1$, and $S_n = -45.5$, find d and n.

Solve the following problems.

25. A technician accepts a job at a starting salary of \$23,186 and receives a \$1,500 increase each year for six years. Find the technician's salary during the sixth year and the total earnings for the six-year period.

I 26. A computer depreciates \$350 per year. If the computer is currently worth \$7,860, what will its value be after seven years?

27. A taxi driver charges \$1.50 for the first mile and \$0.60 for each additional mile. How much does the driver charge for a 15-mile trip?

28. An object falls 16 ft during the first second, 48 ft during the second second, and 80 ft during the third second. Find the distance the object falls in the sixth second and the total distance fallen in the first six seconds.

29. The population of Glenwood increases arithmetically from 75,230 to 125,280 in eight years. Assuming that the growth rate is constant, find the yearly increase in population.

30. A computer manufacturer reported sales of 1,860 units at the end of January. If sales are to increase 46 units per month, in what month will sales reach 2,320 units?

17–2

GEOMETRIC PROGRESSIONS

Definitions

A **geometric progression** is a sequence in which each successive term is a *constant multiple,* called the **common ratio** r, of the previous term. For example, the sequence 1, 3, 9, 27, . . . is a geometric progression with a common ratio of 3.

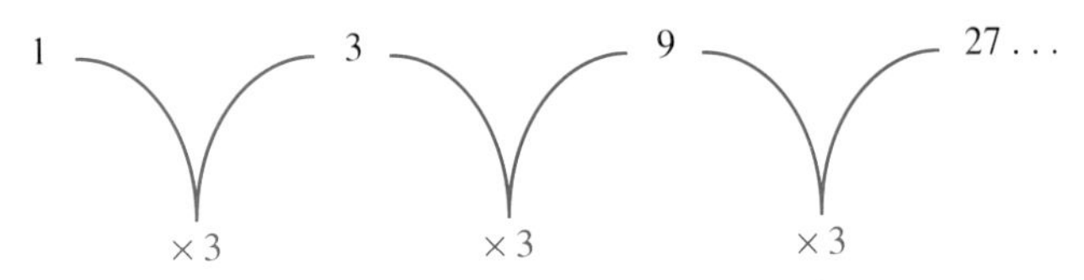

In general, if a_1 is the first term of a geometric progression with common ratio r, the terms are given by

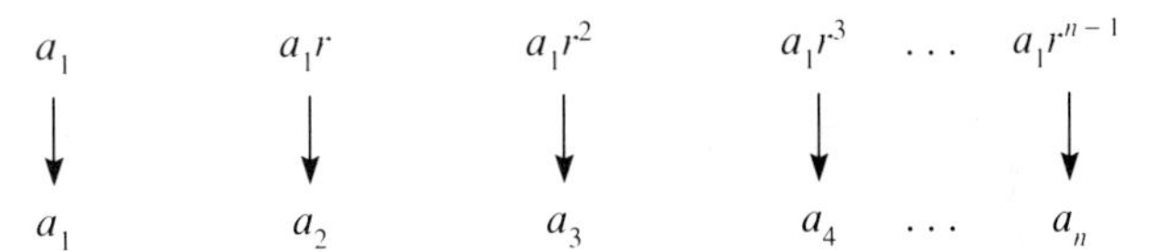

The nth term of a geometric progression is given by the following formula.

---

nth Term of a Geometric Progression

The nth term of a geometric progression is given by

$$a_n = a_1r^{n-1}$$

where a_1 = the first term of the progression and r = the common ratio.

EXAMPLE 1 Find the sixth term of the following geometric progression:

$$4, 20, 100, . . .$$

Solution To find the sixth term a_6, we determine a_1, r, and n.

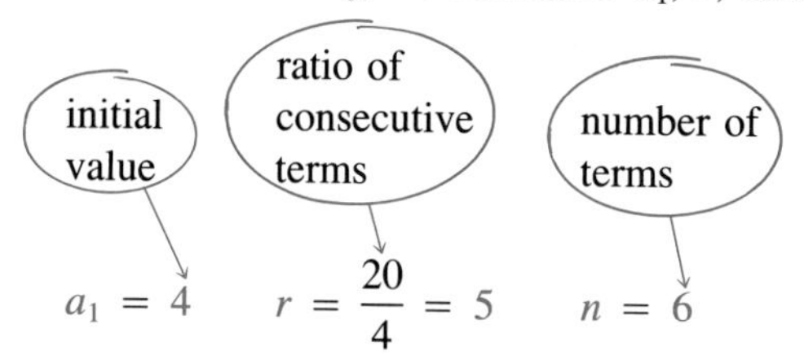

Substituting $a_1 = 4$, $r = 5$, and $n = 6$ into the formula gives

$$a_n = a_1 r^{n-1}$$
$$a_6 = 4(5)^{6-1}$$
$$a_6 = 12{,}500$$

4 [×] 5 [y^x] [(] 6 [−] 1 [)] [=] → 12,500 ■

EXAMPLE 2 Each layer of 3-in. insulation reduces fuel consumption (in kilowatts) by 8%. How many layers would reduce consumption from 948 kW to 894 kW?

Solution From the information given, we know that

$$a_1 = 948 \qquad a_n = 894 \qquad r = 92\%$$

and we must solve for n. The common ratio between successive values is 92% since each layer reduces consumption by 8%. Substituting these values into the formula gives

$$a_n = a_1 r^{n-1}$$
$$894 = 948(0.92)^{n-1}$$

To solve for the exponent n, we use logarithms.

$$\log 894 = \log [948(0.92)^{n-1}]$$
$$\log 894 = \log 948 + (n - 1)\log 0.92$$
$$\log 894 = \log 948 + n \log 0.92 - \log 0.92$$
$$\frac{\log 894 - \log 948 + \log 0.92}{\log 0.92} = n$$
$$n = 1.7$$

Approximately two layers, 6 in. (1.7 × 3 in.), of insulation are needed.

894 [log] [−] 948 [log] [+] 0.92 [log] [=]
[÷] 0.92 [log] [=] → 1.7034 ■

NOTE ✦ If you have difficulty solving the logarithmic equation in Example 2, review Chapter 8 on exponential and logarithmic functions.

EXAMPLE 3 A radioactive product has a half-life of 5 years. If the radioactivity level is 56.0 microcuries after 23 years, determine the original level of radioactivity.

Solution We are given that $a_n = 56$ and $r = 0.5$. To determine a_1, we must calculate n, the number of half-lives elapsed.

$$n = \frac{23 \text{ years}}{5 \text{ years}} = 4.6$$

Therefore, including a_1, the number of terms n is 5.6. Substituting into the formula gives

$$a_n = a_1 r^{n-1}$$

$$56.0 = a_1(0.5)^{5.6-1}$$

$$a_1 = 1{,}360$$

56 [÷] [(] 0.5 [y^x] 4.6 [)] [=] → 1358.1 ■

Sum of a Geometric Progression

To derive the formula for the sum of the first n terms of a geometric progression, multiply the equation representing the sum by r, subtract this result from the original sum, and solve for S_n.

$$S_n = a_1 + a_1 r + a_1 r^2 + \cdots + a_1 r^{n-2} + a_1 r^{n-1} \quad \text{sum of } n \text{ terms}$$

$$-rS_n = -(a_1 r + a_1 r^2 + \cdots + a_1 r^{n-2} + a_1 r^{n-1} + a_1 r^n) \quad \text{multiply by } r$$

$$S_n - rS_n = a_1 + 0 + 0 + 0 + \cdots + 0 + 0 - a_1 r^n$$

$$S_n(1 - r) = a_1 - a_1 r^n$$

$$S_n = \frac{a_1(1 - r^n)}{1 - r}$$

Sum of the First n Terms of a Geometric Progression

The sum of the first n terms of an arithmetic progression is given by

$$S_n = \frac{a_1(1 - r^n)}{1 - r}$$

where a_1 = the first term, r = the common ratio, and n = the number of terms in the progression.

EXAMPLE 4 Find the sum of the first seven terms of the geometric progression

$$9, 6, 4, \ldots$$

Solution From the given information, we know that $a_1 = 9$, $r = \frac{6}{9} = \frac{2}{3}$, and $n = 7$. Substituting these values into the formula gives

$$S_n = \frac{a_1(1 - r^n)}{1 - r}$$

$$S_7 = \frac{9\left(1 - \frac{2^7}{3}\right)}{1 - \frac{2}{3}}$$

$$S_7 \approx 25.4$$

1 [−] [(] 2 [÷] 3 [)] [y^x] 7 [=] [×] 9 [÷] [(] 1 [−] 2 [÷] 3 [)] [=] → 25.4198 ■

CAUTION ✦ In calculating an expression such as $1 - 0.5^6$, raise 0.5 to the sixth power before subtracting the result from 1.

Application

EXAMPLE 5 Determine the total worth of a yearly \$500 investment after 15 years if the interest rate is 5.5% compounded annually.

Solution At the end of the first year, the total worth is

principal ↘ \$500 + \$500(0.055) ↙ interest

$$500(1 + 0.055)$$

$$500(1.055)$$

At the end of the second year, the total worth is

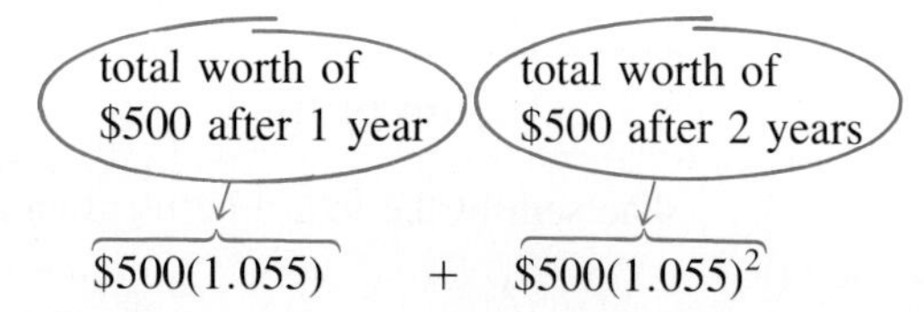

$$\$500(1.055) + \$500(1.055)^2$$

Continuing this pattern gives the following geometric progression:

$$500(1.055) + 500(1.055)^2 + 500(1.055)^3 + \cdots + 500(1.055)^{15}$$

or

$$500[(1.055) + (1.055)^2 + 1.055)^3 + \cdots + (1.055)^{15}]$$

Since the expression in brackets represents a geometric progression being added, we will substitute the following values into the formula:

$$a_1 = 1.055 \qquad r = \frac{(1.055)^2}{1.055} = 1.055 \qquad n = 15$$

$$S_n = \frac{a_1(1 - r^n)}{1 - r}$$

Substituting gives

$$S_{15} = \frac{1.055\,[1 - (1.055)^{15}]}{1 - 1.055}$$

$$S_{15} \approx 23.64$$

The total value of the investment is

$$\$500(23.64) = \$11{,}820.57$$ ■

17–2 EXERCISES

Find the indicated term of each geometric progression.

1. The 8th term of 1, 4, 16, . . .
2. The 5th term of 24, 6, 1.5, 0.375, . . .
3. The 9th term of 8, -4, 2, . . .
4. The 10th term of -9, 7.5, -6.25, . . .
5. The 7th term of 54, 9, 1.5, . . .
6. The 4th term where $a_1 = 8$ and $r = \frac{1}{2}$
7. The 6th term where $a_1 = -10$ and $r = -4.1$
8. The 8th term where $a_1 = 72$ and $r = -3$
9. The 11th term where $a_1 = 86$ and $r = 2.5$
10. The 7th term where $a_1 = 108$ and $r = -\frac{1}{3}$

Find the sum of the first n terms of each geometric progression.

11. $-6, -3, -1.5, \ldots; n = 8$
12. $4, 12, 36, \ldots; n = 7$
13. $36, -12, 4, \ldots; n = 6$
14. $n = 5$, $a_1 = 7$, and $r = 3$
15. $n = 8$, $a_1 = -4$, and $r = -0.5$
16. $n = 6$, $a_1 = 60$, and $a_6 = 14{,}580$
17. $n = 7$, $a_1 = -0.5$, and $a_7 = -2{,}048$

Use the information given in the following problems to find the requested quantities for each geometric progression.

18. If $a_1 = 6$, $r = 3$, and $a_n = 486$, find n and S_n.
19. If $a_1 = \frac{1}{8}$ and $r = \frac{1}{2}$, find a_{10} and S_{10}.
20. If $r = 2$ and $S_5 = 558$, find a_1 and a_5.
21. If $a_1 = 9$, $r = \frac{1}{3}$, and $S_n = 13\frac{1}{3}$, find n and a_n.
22. If the 7th term is 8,748 and the common ratio is 3, find the first term and the sum of the first seven terms.
23. If the first term is 8, the common ratio is 3, and the nth term is 5,832, find n and the sum of the first n terms.
24. If the sum of the first five terms is -93 and the common ratio is 2, find the first term and the third term.
25. If the second term is $\frac{1}{2}$ and the seventh term is $\frac{1}{64}$, find the tenth term.
26. If the 4th term is 1,296 and the seventh term is 279,936, find the second term.

Solve the following problems.

27. An engineer accepts a job with a starting salary of \$20,130 with an 8% raise each year for seven years. What is the technician's salary during the seventh year?

28. The number of bacteria in a culture increases by 30% each hour. If the culture contains 25,000 bacteria at 1:00 P.M., how many bacteria are present at 7:00 P.M.?

I 29. Joe invests \$1,500 per year in a savings account for nine years. Find the total value of the account if the interest rate of 6% is compounded annually.

I 30. A technician receives a job offer for \$23,500 per year. If he receives a 5% raise each year, how many years will pass before his annual salary is \$34,720.20?

17–3

INFINITE GEOMETRIC SERIES

In the previous section we found the sum of a finite number of terms in a geometric progression. In this section we will develop a formula for the sum of all the terms of an infinite geometric series.

Consider the following geometric progressions:

$$3, 6, 12, 24, \ldots \quad \text{where } r = 2$$

and

$$36, 18, 9, 4.5, \ldots \quad \text{where } r = \frac{1}{2}$$

The sum of the terms in the first geometric progression increases without bound because each successive term of the progression is larger. However, the sum of the terms of the second geometric progression approaches some limiting value because successive terms are smaller. From Table 17–1, you can see that as n, the number of terms, increases, the sum S_n for the second progression approaches 72.

TABLE 17–1

n	Series	S_n
3	$36 + 18 + 9$	63
4	$36 + 18 + 9 + 4.5$	67.5
5	$36 + 18 + 9 + 4.5 + 2.25$	69.75
6	$36 + 18 + 9 + 4.5 + 2.25 + 1.125$	70.875
11	$36 + 18 + 9 + 4.5 + \cdots + 0.0352$	71.965
15	$36 + 18 + 9 + \cdots + 0.0022$	71.9978
20	$36 + 18 + 9 + \cdots + 6.8665 \times 10^{-5}$	71.9999
30	$36 + 18 + 9 + \cdots + 6.7055 \times 10^{-8}$	71.9999999

If we compare the two geometric progressions given earlier, when $|r| > 1$, the sum of the terms increases without bound; but when $|r| < 1$, the sum of the terms approaches some limiting value. If we consider the formula for the sum of the first n terms of a geometric progression,

$$S_n = \frac{a_1(1 - r^n)}{1 - r}$$

we see that for $|r| < 1$, the expression r^n becomes smaller as the value of n becomes larger. In calculus, we say that

$$\lim_{n \to \infty} r^n = 0 \quad \text{for } |r| < 1$$

This notation is read, "the limit of r^n as n increases without bound is zero." If we apply this knowledge about r^n for $|r| < 1$ to the sum formula, we obtain the following equation.

Sum of an Infinite Geometric Progression

The sum of an infinite geometric progression is given by

$$S = \frac{a_1}{1 - r} \quad \text{for } |r| < 1$$

where a_1 = the first term of the sequence and r = the common ratio.

EXAMPLE 1 Find the sum of the following infinite geometric progression:

$$18, 6, 2, \ldots$$

Solution From the progression, we know that $a_1 = 18$ and $r = \frac{6}{18} = \frac{1}{3}$.

The sum of the geometric progression is

$$S = \frac{a_1}{1 - r}$$

$$S = \frac{18}{1 - 1/3}$$

$$S = 27$$

■

EXAMPLE 2 Find the sum of the following infinite geometric progression:

$$-8, 3, -1.125, \ldots$$

Solution We know that $a_1 = -8$ and $r = \frac{3}{-8} = -\frac{3}{8}$. The sum is

$$S = \frac{-8}{1 - (-3/8)}$$

$$S = -\frac{64}{11}$$

■

Repeating Decimals

You can use a geometric progression to find the fractional equivalent of a repeating decimal. The next two examples illustrate the procedure.

EXAMPLE 3 Find the fractional equivalent to the repeating decimal

$$0.\overline{39} = 0.393939\ldots$$

Solution The repeating decimal $0.\overline{39}$ can be written as a series in the following manner:

$$0.\overline{39} = 0.39 + 0.0039 + 0.000039 + \cdots$$

The terms in this series form a geometric progression with

$$a_1 = 0.39 \quad \text{and} \quad r = \frac{0.0039}{0.39} = 0.01$$

Using the formula for the sum of an infinite geometric progression gives

$$S = \frac{0.39}{1 - 0.01} = \frac{0.39}{0.99} = \frac{39}{99} = \frac{13}{33}$$

The fractional equivalent of $0.\overline{39}$ is $\frac{13}{33}$.

■

EXAMPLE 4 Find the fractional equivalent of $0.6\overline{813}$.

Solution First, write the repeating decimal as a series in the following manner:

$$0.6\overline{813} = 0.6 + 0.0813 + 0.0000813 + \cdots$$

Second, determine a_1 and r for the terms that comprise the geometric progression.

$$a_1 = 0.0813 \quad \text{and} \quad r = 0.001$$

Third, find the sum of the infinite geometric progression.

$$S = \frac{0.0813}{1 - 0.001} = \frac{0.0813}{0.999} = \frac{813}{9{,}990}$$

The fractional form of the repeating decimal is

$$0.6 + S = \frac{3}{5} + \frac{813}{9{,}990} = \frac{6{,}807}{9{,}990} = \frac{2{,}269}{3{,}330}$$ ■

Application

EXAMPLE 5 A ball attached to the end of an elastic band oscillates up and down. The total distance traveled during its initial oscillation is 7.8 cm, and the distance traveled during each successive oscillation is 68% of the previous distance. Find the total vertical distance the ball traveled before coming to rest.

Solution We know that $a_1 = 7.8$ and $r = 0.68$. The total vertical distance is given by

$$S = \frac{a_1}{1 - r}$$

$$S = \frac{7.8}{1 - 0.68}$$

$$S = 24 \text{ cm}$$

7.8 [÷] [(] 1 [−] 0.68 [)] [=] → 24.375 ■

17–3 EXERCISES

Find the sum of each infinite geometric progression.

1. $-8, 4, -2, \ldots$
2. $15, -10, \frac{20}{3}, \ldots$
3. $12, 4, \frac{4}{3}, \ldots$
4. $4.32, 3.6, 3.0, \ldots$
5. $42, 36, \frac{216}{7}, \ldots$
6. $54, 18, 6, \ldots$
7. $69.12, 28.8, 12, \ldots$
8. $18, 12, 8, \ldots$
9. $36, 24, 16, \ldots$
10. $63, 21, 7, \ldots$
11. $64, 16, 4, \ldots$
12. $25a^3, 5a, \frac{1}{a}, \ldots$
13. $14, -7\sqrt{2}, 7, \ldots$
14. $15, 5\sqrt{3}, 5, \ldots$
15. $6x^2, 3x, 1.5, \ldots$

Find the fractional equivalent for each repeating decimal.

16. $0.\overline{6}$
17. $0.\overline{9}$
18. $0.\overline{17}$
19. $0.\overline{83}$
20. $0.1\overline{35}$
21. $0.4\overline{23}$
22. $0.3\overline{15}$
23. $3.\overline{738}$
24. $8.\overline{217}$
25. A pendulum swings through an arc of 36 in. On each successive swing, the pendulum covers an arc equal to 90% of the previous swing. Find the length of the arc on the sixth swing and the total distance the pendulum travels before coming to rest.
26. An object suspended from a spring oscillates up and down. The first oscillation is 25 cm, and each successive oscillation is 80% of the preceding one. Find the total distance the object travels before coming to rest.
27. A rubber ball rebounds 3/5 of its height. If it is initially 30 ft high, what total vertical distance does it travel before coming to rest?
28. An object decelerates such that it travels 45 m during the first second, 15 m during the second, and 5 m during the third second. Determine the total distance the object travels before coming to rest.

17–4

THE BINOMIAL FORMULA

In previous algebraic operations, we have raised a binomial to the second or third power, but rarely to any higher power. However, in more advanced mathematics such multiplication will be necessary. In this section we will develop a formula for raising a binomial to any power.

Expanding $(x + y)^n$

Let us expand the expression $(x + y)^n$ where $n = 0, 1, 2, 3, 4$, and 5.

$$(x + y)^0 = 1$$
$$(x + y)^1 = x + y$$
$$(x + y)^2 = x^2 + 2xy + y^2$$
$$(x + y)^3 = x^3 + 3x^2y + 3xy^2 + y^3$$
$$(x + y)^4 = x^4 + 4x^3y + 6x^2y^2 + 4xy^3 + y^4$$
$$(x + y)^5 = x^5 + 5x^4y + 10x^3y^2 + 10x^2y^3 + 5xy^4 + y^5$$

The process of multiplying $(x + y)$ times itself five times is quite lengthy. However, the multiplications above illustrate certain patterns that will prove helpful in defining a formula.

☐ Each expansion of $(x + y)^n$ contains $n + 1$ terms. For example $(x + y)^4$ contains $n + 1 = 5$ terms.

☐ The first term is always x^n, and the last term is always y^n. For example, when $n = 3$,

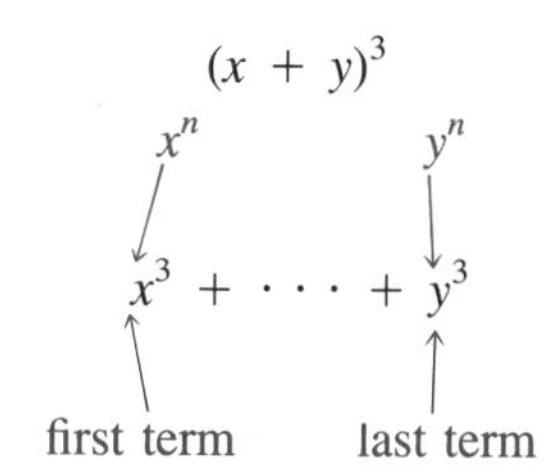

☐ Proceeding from left to right, the powers of x decrease by 1 in each term, and the powers of y increase by 1. Moreover, the sum of the exponents in each term equals n. For example,

powers of x	4	3	2	1	0
$(x + y)^4 =$	x^4 +	$4x^3y$ +	$6x^2y^2$ +	$4xy^3$ +	y^4
powers of y	0	1	2	3	4
sum of exponents	4	4	4	4	4

☐ The coefficients of the terms form a symmetrical, triangular pattern, called **Pascal's triangle.**

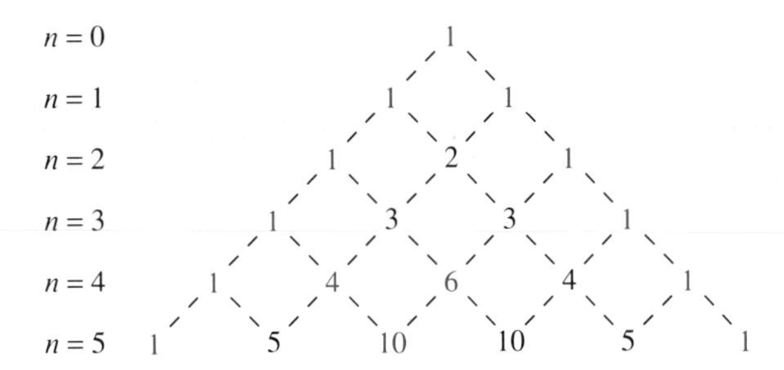

Notice that each row begins and ends with 1, and each coefficient is the sum of the two numbers above it on each side. For example,

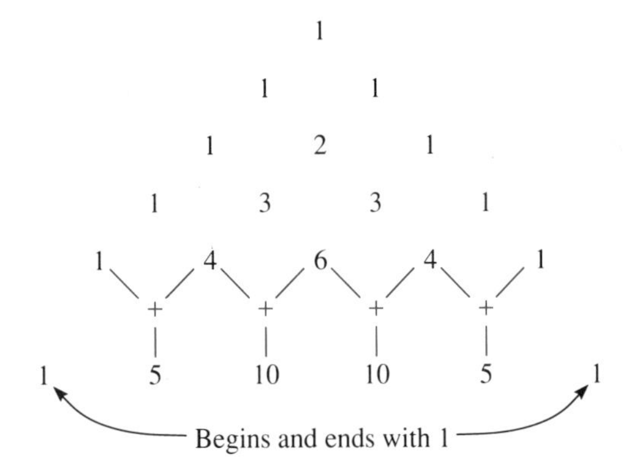

EXAMPLE 1 Using Pascal's triangle, find $(a + b)^6$.

Solution First, we use Pascal's triangle to determine the coefficients. From our previous expansion, we know the coefficient for $n = 5$.

$n = 5$: 1 5 10 10 5 1

1 + 5, 5 + 10, 10 + 10, 10 + 5, 5 + 1

$n = 6$: 1 6 15 20 15 6 1

Second, we determine the powers of a and b for each term.

descending powers of a

6 5 4 3 2 1 0

$$1a^6 + 6a^5b + 15a^4b^2 + 20a^3b^3 + 15^2b^4 + 6ab^5 + 1b^6$$

0 1 2 3 4 5 6

ascending powers of b

The result is

$$(a + b)^6 = a^6 + 6a^5b + 15a^4b^2 + 20a^3b^3 + 15a^2b^4 + 6ab^5 + b^6 \quad \blacksquare$$

EXAMPLE 2 Expand $(4a - 3b)^4$.

Solution From Pascal's triangle, the coefficients for $n = 4$ are

$$1 \quad 4 \quad 6 \quad 4 \quad 1$$

However, the terms that are substituted in descending and ascending order for x and y also contain a coefficient. To determine the expanded form, we *substitute 4a for x in descending powers and −3b for y in ascending powers in each term*. The expanded form is

$$\begin{aligned}(4a - 3b)^4 &= 1(4a)^4(-3b)^0 + 4(4a)^3(-3b)^1 + 6(4a)^2(-3b)^2 \\ &\quad + 4(4a)^1(-3b)^3 + 1(4a)^0(-3b)^4 \\ &= 256a^4 - 768a^3b + 864a^2b^2 - 432ab^3 + 81b^4 \quad \blacksquare\end{aligned}$$

CAUTION ✦ When the binomial you are expanding contains coefficients, you must be very careful in applying Pascal's triangle.

The Factorial Function

Pascal's triangle is helpful in expanding smaller powers of a binomial, but it can be difficult for large powers. A formula has been developed to do this, but it uses the **factorial function,** defined in the following box. *The number n! is the product of the first n positive integers.*

The Factorial Function

The factorial function is given by

$$n! = n(n - 1)(n - 2)(n - 3) \ldots 3 \cdot 2 \cdot 1$$

where $n =$ any positive integer. By definition,

$$0! = 1$$
$$1! = 1$$

EXAMPLE 3 Calculate the following:

(a) $6!$ (b) $15!$

Solution

(a) $6! = 6 \cdot 5 \cdot 4 \cdot 3 \cdot 2 \cdot 1 = 720$
(b) $15! = 1.31 \times 10^{12}$

15 [x!] → 1.307674368 12 ■

The Binomial Formula

The expansion of $(a + b)^n$, where n is any positive integer, can be written using factorial notation.

The Binomial Formula

The binomial formula is given by

$$(a + b)^n = a^n + na^{n-1}b + \frac{n(n-1)}{2!}a^{n-2}b^2 + \frac{n(n-1)(n-2)}{3!}a^{n-3}b^3 + \cdots + b^n$$

where $n =$ any positive integer.

EXAMPLE 4 Use the binomial formula to expand $(x + y)^8$.

Solution We substitute $a = x$, $b = y$, and $n = 8$ into the binomial formula.

$$(x + y)^8 = x^8 + 8x^7y + \frac{8\cdot 7}{2\cdot 1}x^6y^2 + \frac{8\cdot 7\cdot 6}{3\cdot 2\cdot 1}x^5y^3 + \frac{8\cdot 7\cdot 6\cdot 5}{4\cdot 3\cdot 2\cdot 1}x^4y^4 + \frac{8\cdot 7\cdot 6\cdot 5\cdot 4}{5\cdot 4\cdot 3\cdot 2\cdot 1}x^3y^5 + \frac{8\cdot 7\cdot 6\cdot 5\cdot 4\cdot 3}{6\cdot 5\cdot 4\cdot 3\cdot 2\cdot 1}x^2y^6 + \frac{8\cdot 7\cdot 6\cdot 5\cdot 4\cdot 3\cdot 2}{7\cdot 6\cdot 5\cdot 4\cdot 3\cdot 2\cdot 1}xy^7 + \frac{8\cdot 7\cdot 6\cdot 5\cdot 4\cdot 3\cdot 2\cdot 1}{8\cdot 7\cdot 6\cdot 5\cdot 4\cdot 3\cdot 2\cdot 1}y^8$$

$$(x + y)^8 = x^8 + 8x^7y + 28x^6y^2 + 56x^5y^3 + 70x^4y^4 + 56x^3y^5 + 28x^2y^6 + 8xy^7 + y^8$$

■

EXAMPLE 5 Use the binomial formula to expand $(5x - 2y)^4$.

Solution Substitute $a = 5x$, $b = -2y$, and $n = 4$ into the binomial formula.

$$(5x - 2y)^4 = (5x)^4 + 4(5x)^3(-2y) + \frac{4\cdot 3}{2\cdot 1}(5x)^2(-2y)^2 + \frac{4\cdot 3\cdot 2}{3\cdot 2\cdot 1}(5x)(-2y)^3 + (-2y)^4$$

$$(5x - 2y)^4 = 625x^4 - 1{,}000x^3y + 600x^2y^2 - 160\,xy^3 + 16y^4$$

■

Finding a Given Term

We can use the binomial formula to find a specific term of a binomial expansion without finding all the terms. The $(k + 1)$ term of the expansion $(x + y)^n$ is given by

$$\frac{n(n-1)(n-2)\cdots(n-k+1)}{k!}x^{(n-k)}y^k$$

EXAMPLE 6 Find the ninth term of expansion of $(x + y)^{12}$.

Solution To find the ninth term of the expansion, we compute $k + 1 = 9$ or $k = 8$ and $n = 12$. The ninth term is

$$\frac{12 \cdot 11 \cdot 10 \cdot 9 \cdot 8 \cdot 7 \cdot 6 \cdot 5}{8 \cdot 7 \cdot 6 \cdot 5 \cdot 4 \cdot 3 \cdot 2 \cdot 1} x^4 y^8$$

$$495x^4y^8$$

■

Expanding $(1 + y)^n$

The binomial formula is valid only for positive integral powers of n. However, if $x = 1$ and $|y| < 1$, the binomial formula is valid for any real value of n. Under these conditions, the expansion is given by

$$(1 + y)^n = 1 + ny + \frac{n(n - 1)}{2!}y^2 + \frac{n(n - 1)(n - 2)}{3!}y^3 + \cdots$$

For negative values of n, the expansion becomes an infinite series; however, the first few terms do provide an approximation for the expansion.

EXAMPLE 7 Write the first five terms in the expansion of $(1 + m)^{-4}$, where $|m| < 1$.

Solution Using the formula gives

$$(1 + m)^{-4} = 1 - 4m + \frac{-4 \cdot -5}{2 \cdot 1}m^2 + \frac{-4 \cdot -5 \cdot -6}{3 \cdot 2 \cdot 1}m^3 + \frac{-4 \cdot -5 \cdot -6 \cdot -7}{4 \cdot 3 \cdot 2 \cdot 1}m^4 + \cdots$$

$$(1 + m)^{-4} = 1 - 4m + 10m^2 - 20m^3 + 35m^4 + \cdots$$

■

17–4 EXERCISES

Using Pascal's triangles, write the expansion of each binomial.

1. $(a + 5)^4$
2. $(x - 7)^5$
3. $(m - 1)^7$
4. $(y - 3)^6$
5. $(2x - 5y)^3$
6. $(2b - 3)^4$
7. $(3y + x)^6$
8. $(4m + n)^5$

Use the binomial formula to write the expansion of each binomial.

9. $(m + 3)^5$
10. $(d - 7)^4$
11. $(b - 6)^7$
12. $(y + 8)^5$
13. $(3x - 1)^4$
14. $(2m + 5)^5$
15. $(5m - 4)^6$
16. $(8x - 3y)^4$
17. $(x - 4y)^7$
18. $(7x + 2y)^3$
19. $(2a^2 - 3b)^4$
20. $(4x + 5y^2)^3$

Write the first four terms in the binomial expansion of each expression for $|x| < 1$.

21. $(1 - x)^8$
22. $(1 + x)^7$
23. $(1 + x)^{-3}$
24. $(1 - x)^{-2}$

25. Find the 7th term of the expansion of $(y - 8)^{10}$.

26. Find the 4th term of the expansion of $(x + 7)^6$.

27. Find the x^3 term of the expansion of $(x - 2)^8$.

28. Find the term containing y^5 of the expansion of $(m - y)^7$.

CHAPTER SUMMARY

Summary of Terms

arithmetic progression (p. 610)
common difference (p. 610)
common ratio (p. 615)
factorial function (p. 626)
finite sequence (p. 610)
geometric progression (p. 615)
infinite sequence (p. 610)
Pascal's triangle (p. 624)
sequence (p. 610)
term (p. 610)

Summary of Formulas

$a_n = a_1 + (n - 1)d$ nth term of an arithmetic progression

$S_n = \dfrac{n(a_1 + a_n)}{2}$ sum of the first n terms of an arithmetic progression

$a_n = a_1r^{n-1}$ nth term of a geometric progression

$S_n = \dfrac{a_1(1 - r^n)}{1 - r}$ sum of the first n terms of a geometric progression

$S = \dfrac{a_1}{1 - r}$ sum of an infinite geometric progression for $|r| < 1$

$n! = n(n - 1)(n - 2)(n - 3) \ldots 3 \cdot 2 \cdot 1$ factorial function

$(a + b)^n = a^n + na^{n-1}b + \dfrac{n(n - 1)}{2!}a^{n-2}b^2 + \dfrac{n(n - 1)(n - 2)}{3!}a^{n-3}b^3 + \cdots + b^n$ binomial formula

$\dfrac{n(n - 1)(n - 2) \cdots (n - k + 1)}{k!}x^{(n-k)}y^k$; $(k + 1)$ term of $(x + y)^n$

CHAPTER REVIEW

Section 17–1

Find the indicated term of the arithmetic progression.

1. The 8th term of $-4, 2, 8, \ldots$

2. The 15th term of $-12, -9, -6, \ldots$

3. The 23rd term of $48, 42, 36, \ldots$

4. The 18th term of $-1, -5, -9, \ldots$

Find the sum of the first n terms of each arithmetic progression.

5. $6, 9, 12, \ldots; n = 9$

6. $-10, -2, 6, 14, \ldots; n = 7$

7. $n = 10$, $a_1 = 3$, and $d = -2$

8. $n = 15$, $a_1 = -5$, and $d = 6$

9. The population of Orange Grove increased arithmetically from 25,000 to 40,000 in 12 years. What was each year's population growth?

10. Pablo accepts a job earning $28,250 per year. If he receives a $945 raise each year, what will his salary be at the end of six years?

Section 17–2

Find the indicated term of the geometric progression.

11. The 5th term of $-3, 9, -27, \ldots$

12. The 6th term of $64, -16, 4, \ldots$

13. The 9th term of 625, 125, 25, . . .

14. The 13th term of 6, 12, 24, . . .

Find the sum of the first n terms of the geometric progression.

15. 2, 6, 18, . . .; $n = 8$

16. -3, 12, -48, . . .; $n = 5$

17. $n = 10$, $a_1 = -8$, and $r = -2$

18. $n = 12$, $a_1 = 4$, and $r = 0.5$

19. Rick accepts a job earning $18,500 per year. Each year for three years, he receives an 8% pay raise. How much does he earn during the third year?

I 20. Manufacturing equipment currently worth $60,000 depreciates 15% each year. Determine the value of the equipment after eight years.

Section 17–3

Find the sum of each infinite geometric progression.

21. $\frac{1}{3}, \frac{1}{12}, \frac{1}{48}, \ldots$

22. 24, 6, 1.5, . . .

23. 18, 6, 2, . . .

24. $48x^3, 12x^2, 3x, \ldots$

Find the fractional equivalent for each repeating decimal.

25. $0.\overline{34}$

26. $0.\overline{67}$

27. $3.\overline{124}$

28. $7.\overline{318}$

29. A rubber ball rebounds 45% of its height on successive bounces. If it is initially 50 ft high, what total distance does it travel before coming to rest?

30. An object suspended from an elastic band oscillates up and down. The first oscillation is 26 in., and each successive oscillation is 65% of the preceding one. Find the total distance the object travels.

Section 17–4

Using Pascal's triangle, write the expansion of each binomial.

31. $(x - z)^3$

32. $(m - n)^4$

33. $(2x - 3y)^4$

34. $(5a + b)^3$

Use the binomial formula to write the expansion of each binomial.

35. $(a + 2b)^4$

36. $(3x - y)^5$

37. $(y^2 - 5z)^3$

38. $(m + 4n)^4$

39. Find the 5th term of the expansion of $(x - 6)^8$.

40. Find the 4th term of the expansion of $(y + 5)^9$.

41. Find the x^4 term of the expansion of $(x + 3)^7$.

42. Find the y^5 term of the expansion of $(y - 3)^9$.

CHAPTER TEST

The number in parentheses refers to the appropriate learning objective given at the beginning of the chapter.

1. Find the sum of the following infinite geometric progression: **(5)**

80, 40, 20, . . .

2. Find the 34th term of the following arithmetic progression: **(1)**

7, 13, 19, . . .

3. Paul accepts a job earning $20,000 a year. If he receives a 7% raise each year for six years, what will his salary be during the sixth year? **(3)**

4. Find the fractional form of the repeating decimal $0.\overline{281}$. **(6)**

5. Apply the binomial formula to expand $(2x - y)^4$. **(7)**

6. Determine the sum of the first 12 terms of the following geometric progression: **(4)**

3, 9, 27, . . .

7. Using Pascal's triangle, expand $(m - n)^4$. **(8)**

8. Find the tenth term of the following geometric progression: **(3)**

$$81, 27, 9, \ldots$$

9. Determine the sum of the first 26 terms of the following arithmetic progression: **(2)**

$$9, 16, 23, \ldots$$

10. Find the fractional equivalent of $3.\overline{27}$. **(6)**

11. Determine the sum of the first 19 terms of the following arithmetic progression: **(2)**

$$2, 11, 20, \ldots$$

12. Find the sum of the following infinite geometric progression: **(5)**

$$128, 16, 2, \ldots$$

13. Find the x^4 term in the expansion of $(x + 3)^8$. **(9)**

14. The population of Palm Coast increases 4% each year. If the current population is 46,280, determine what the population will be in 20 years. **(3)**

15. Find the sum of the first eleven terms of the following geometric progression: **(4)**

$$-1, 3, -9, 27, \ldots$$

16. Apply the binomial formula to expand $(2x - y)^5$. **(7)**

17. Find the fifth term of the expansion of $(3y + 8)^8$. **(9)**

18. Find the ninth term of the following geometric progression: **(3)**

$$32, 8, 2, \ldots$$

SOLUTION TO CHAPTER INTRODUCTION

Your salary at XYZ Corporation represents a geometric progression. To determine your salary at the end of the eighth year, substitute $a_1 = 23{,}480$, $r = 1.08$, and $n = 8$ into the formula.

$$a_n = a_1 r^{n-1}$$
$$a_8 = (23{,}480)(1.08)^7$$
$$a_8 = \$40{,}240.59$$

To determine your salary at ABC Company, substitute $a_1 = 26{,}700$, $d = 1{,}500$, and $n = 8$ into the formula for an arithmetic progression.

$$a_n = a_1 + (n - 1)d$$
$$a_8 = 26{,}700 + 7(1{,}500)$$
$$a_8 = \$37{,}200$$

From the above calculation, you should accept the job at XYZ Corporation.

You are an industrial designer responsible for constructing an open cardboard box from a 12-in.-by-16-in. piece of cardboard by cutting a square from each corner and folding up the resulting tabs. How large should you make the square to obtain the maximum volume of the box? (The solution to this problem is given at the end of the chapter.)

We have been able to solve a wide variety of problems by using the techniques of algebra and trigonometry. However, many technical problems arise that you cannot solve by using these techniques. In this chapter we define and discuss the derivative, which deals with the rate of change of one variable with respect to another.

Learning Objectives

After completing this chapter, you should be able to

1. Determine whether a given function is continuous (Section 18–1).
2. Determine the limit of a given function (Section 18–1).
3. Use the delta process to find the derivative of a function and evaluate the result at a specified value (Section 18–2).
4. Find the derivative of a polynomial function by using the constant, constant multiplier, sum, or difference rules (Section 18–3).
5. Find the derivative of functions that are products, quotients, or powers of polynomial functions (Section 18–4).
6. Use implicit differentiation to find the derivative of a function (Section 18–5).
7. Apply the concepts of differentiation to technical problems (Sections 18–3, 18–4, and 18–5).

Chapter 18

Differentiation with Applications

18–1

LIMITS AND CONTINUITY

Many technical problems cannot be solved by either algebra or trigonometry. Therefore, we need to develop additional methods for solving technical problems. In this chapter we discuss differential calculus, which we can use to solve problems involving the rate of change of one variable with respect to another. For example, velocity is the rate of change of distance with respect to time. We can use integral calculus to solve problems where the rate of change is known. For example, electrical current is the time rate of change of electrical charge. Integral calculus will be discussed in Chapter 19.

Before beginning our study of differential calculus, we will discuss the concepts of limits and continuity of a function. Although the treatment of these concepts is not mathematically rigorous, you should develop an intuitive understanding necessary for calculus.

Limit of a Function

In everyday life, a *limit* refers to a boundary or an endpoint. This meaning extends to calculus. In previous chapters where we discussed asymptotes of a graph, we said that the graph ''approaches'' the asymptote but never touches it. That is, the graph gets closer and closer to that line but never touches it. The idea of the limit of a function is very similar. The **limit of a function** is the value (number) that the function approaches as x approaches a given value, denoted a. Mathematically, this definition can be written as follows:

function

$$\lim_{x \to a} f(x) = L$$

x approaches a limit

Let us examine the limit of the function

$$f(x) = 3 + \frac{1}{x}$$

as x becomes larger and larger. We know that x cannot equal zero because that would lead to division by zero. However, we know that x can take on values very close to zero. A table of values for this function follows.

x	-8	-6	-3	-2	-1	-0.5	-0.25	0	0.7	1	3	4	6	8	10
$f(x)$	2.9	2.8	2.7	2.5	2	1	-1	undef.	4.4	4	$3.\overline{3}$	3.3	3.2	3.1	3.1

FIGURE 18–1

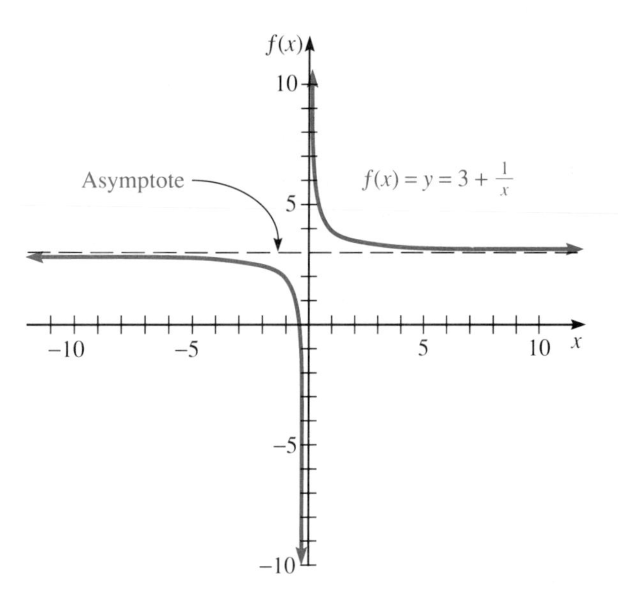

From the table of values and the graph shown in Figure 18–1, we can see that as x gets larger and larger, the function approaches the line $f(x) = 3$. As the absolute value of x gets extremely large, the limit of $f(x)$ is 3. Using the limit notation given earlier, this statement becomes

$$\lim_{x\to\infty} f(x) = 3 \qquad \text{or} \qquad \lim_{x\to\infty}\left(3 + \frac{1}{x}\right) = 3$$

↑ as x gets larger and larger ↑ limit (asymptote)

EXAMPLE 1 What is the limit of the function $f(x) = 3x - 4$ as x approaches 3 from the left and right?

Solution We fill in a table of values for this function, choosing values less than 3 and greater than 3 as shown in the following tables:

For $x < 3$

x	0	1	2	2.5	2.6	2.7	2.8	2.9	2.95	2.98	2.99
$f(x)$	−4	−1	2	3.5	3.8	4.1	4.4	4.7	4.85	4.94	4.97

For $x > 3$

x	3.01	3.08	3.1	3.4	3.8	3.9	4.0	4.5	5.0
$f(x)$	5.03	5.24	5.3	6.2	7.4	7.7	8.0	9.5	11.0

FIGURE 18–2

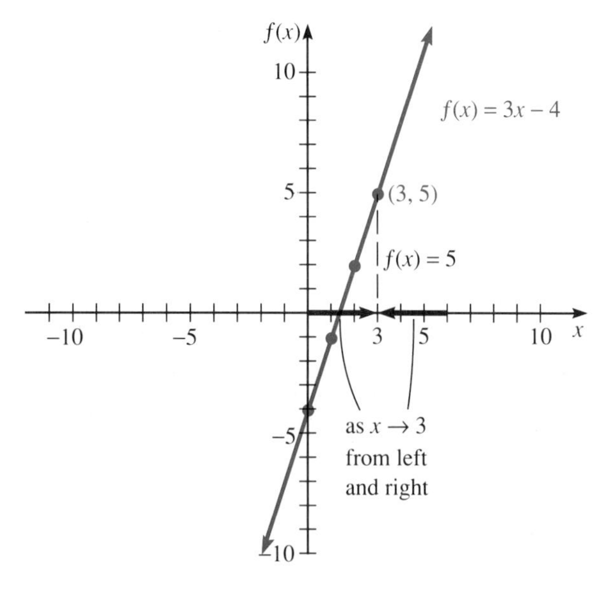

From the tables of values, we can see that $f(x)$ approaches 5 as x approaches 3 from the left and from the right. The graph shown in Figure 18–2 confirms this fact. Using the limit notation, we write the equation as follows:

$$\lim_{x \to 3}(3x - 4) = 5$$

■

The Definition of a Limit

The limit of a function $f(x)$ equals the number L that the function approaches as x approaches some value a.

$$\underbrace{\lim_{x \to a} f(x)}_{x \text{ approaches } a} = \underbrace{\lim_{x \to a^+} f(x)}_{\substack{x \text{ approaches } a \\ \text{from the right}}} = \underbrace{\lim_{x \to a^-} f(x)}_{\substack{x \text{ approaches } a \\ \text{from the left}}} = L$$

A function has a limit L at a only if both the left and right limits exist and are equal.

LEARNING HINT ✦ If a function is defined as $x \to a^+$ and $x \to a^-$, you may evaluate the limit of the function by calculating $f(a)$. However, you should use this technique cautiously.

EXAMPLE 2 For two resistors r and R connected in series in a dc circuit of voltage V, the current is given by

$$I = \frac{V}{r + R}$$

Find $\lim_{r\to 8} I$ if $V = 9$ V and $R = 5\ \Omega$.

Solution Since this function is defined at $r = 8$, we substitute $r = 8$ into the function.

$$\lim_{r\to 8}\left(\frac{V}{r + R}\right) = \frac{9}{8 + 5} \approx 0.7$$

■

EXAMPLE 3 Find $\lim_{x\to 4}\left(\frac{x^2 - x - 12}{x - 4}\right)$.

Solution We cannot just calculate $f(4)$ to find the limit because this function is not defined at $x = 4$. We will use a table of values to determine the limit.

x	3.8	3.9	3.95	3.99	4	4.09	4.10	4.15	4.2	4.25
$f(x)$	6.8	6.9	6.95	6.99	undef.	7.09	7.1	7.15	7.2	7.25

From the table of values, $f(x) \to 7$ as $x \to 4$. Therefore,

$$\lim_{x\to 4}\left(\frac{x^2 - x - 12}{x - 4}\right) = 7$$

The limit of the function does exist as $x \to 4$ even though the function does not exist at $x = 4$.

■

LEARNING HINT ✦ In general, when finding the limit of a function that does not exist at that value, first try to simplify the function using algebraic techniques. If the simplified function exists at the value, calculate $f(a)$. The next example illustrates this technique.

EXAMPLE 4 Find $\lim_{x\to -1}\left(\frac{2x^2 - x - 3}{x + 1}\right)$.

Solution Since $x = -1$ results in division by zero, this function is not defined at $x = -1$. However, let us try to simplify the function using the algebraic technique of factoring.

$$\lim_{x \to -1} \left(\frac{2x^2 - x - 3}{x + 1} \right) = \lim_{x \to -1} \left[\frac{(2x - 3)(x + 1)}{(x + 1)} \right] = \lim_{x \to -1} (2x - 3)$$

Since the simplified function does exist as $x \to -1$, we find the limit by calculating $f(-1)$.

$$\lim_{x \to -1} (2x - 3) = 2(-1) - 3 = -5$$

$$\lim_{x \to -1} \left(\frac{2x^2 - x - 3}{x + 1} \right) = -5$$

■

EXAMPLE 5 Find $\lim_{x \to -2} \sqrt{2x + 3}$.

Solution This function is not defined for $x < -1.5$ because these numbers lead to imaginary values for $\sqrt{2x + 3}$. Since this function does not exist as $x \to -2$, and since we cannot simplify the expression using algebra, the limit of the function does not exist. ■

Limits at Infinity

We have briefly discussed the limit of a function as x becomes larger and larger without bound, written as $x \to \infty$ where the symbol ∞ represents infinity. Now we will discuss this topic in more detail in the next two examples.

EXAMPLE 6 Find $\lim_{x \to \infty} \left(6 + \frac{1}{x} \right)$.

Solution The symbol ∞ does not represent a number that we can just substitute into the function. Thus, we will fill in a table of values by choosing increasingly larger values of x. A more efficient method of evaluating limits as $x \to \infty$ will be developed later.

x	1	10	50	100	175	300	1,000	2,000	4,000	8,000	100,000
$f(x)$	7	6.1	6.02	6.01	6.0057	6.0033	6.001	6.0005	6.00025	6.000125	6.00001

From the table of values, we can easily see that as $x \to \infty$, $f(x) \to 6$. Therefore,

$$\lim_{x \to \infty} \left(6 + \frac{1}{x} \right) = 6$$

■

LEARNING HINT ✦ Notice from Example 6 that as $x \to \infty$, $1/x$ gets smaller and smaller, and, in fact, $1/x \to 0$. This is also true of $1/x^2$, $1/x^3$, $1/x^4$, $1/x^n$, and c/x^n where c is a constant. This fact is very helpful in finding limits at infinity.

EXAMPLE 7 Find $\lim_{x\to\infty}\left(-1 + \frac{1}{x} + \frac{1}{x^2}\right)$.

Solution From Example 6, we know that $1/x$ and $1/x^2$ approach zero as $x \to \infty$. Therefore,

$$\lim_{x\to\infty}\left(-1 + \overset{\overset{0}{\uparrow}}{\frac{1}{x}} + \overset{\overset{0}{\uparrow}}{\frac{1}{x^2}}\right) = -1$$

■

EXAMPLE 8 Find $\lim_{x\to\infty}\left(\frac{3x^2 - 4}{2x^2 + x + 1}\right)$.

Solution We know that as $x \to \infty$, both the numerator and the denominator increase without bound. To aid us in determining the limit of this function, we fill in the following table of values:

x	1	10	50	100	500	1,000
$f(x)$	−0.25	1.403	1.484	1.492	1.498	1.499

From the table of values,

$$\lim_{x\to\infty}\left(\frac{3x^2 - 4}{2x^2 + x + 1}\right) = 1.5$$

We can also determine this limit by using algebraic techniques. Let us divide the numerator and denominator by the largest power of x in the entire fraction, x^2.

$$\frac{\dfrac{3x^2 - 4}{x^2}}{\dfrac{2x^2 + x + 1}{x^2}} = \frac{\dfrac{3x^2}{x^2} - \dfrac{4}{x^2}}{\dfrac{2x^2}{x^2} + \dfrac{x}{x^2} + \dfrac{1}{x^2}} = \frac{3 - \dfrac{4}{x^2}}{2 + \dfrac{1}{x} + \dfrac{1}{x^2}}$$

From our previous discussion, $4/x^2$, $1/x$, and $1/x^2$ all approach zero as $x \to \infty$.

approaches 0

$$\lim_{x\to\infty}\frac{3 - \dfrac{4}{x^2}}{2 + \dfrac{1}{x} + \dfrac{1}{x^2}} = \frac{3 - 0}{2 + 0 + 0} = \frac{3}{2} = 1.5$$

approaches 0

■

EXAMPLE 9 Find the limit of the following expression:

$$\lim_{x\to\infty}\left(\frac{4x^5 + 6x^2 - 7}{x^5 + 7x^3 + 3}\right)$$

Solution We will use the algebraic technique developed in Example 8 to evaluate this limit. We divide the numerator and denominator of the fraction by x^5.

$$\lim_{x\to\infty}\left(\frac{4x^5 + 6x^2 - 7}{x^5 + 7x^3 + 3}\right) = \lim_{x\to\infty}\left(\frac{4 + \overset{\to 0}{\frac{6}{x^3}} - \overset{\to 0}{\frac{7}{x^5}}}{1 + \underset{\to 0}{\frac{7}{x^2}} + \underset{\to 0}{\frac{3}{x^5}}}\right) = 4$$

■

The Limits of a Function

- ☐ If the function exists at $x = a$, find the limit by calculating $f(a)$ (Examples 1 and 2).
- ☐ If the function does not exist at $x = a$, simplify the function by using algebraic techniques, and then calculate $f(a)$ (Example 4).
- ☐ For limits where $x \to \infty$, divide the numerator and denominator by the largest power of x in the entire fraction. Then use the fact that $1/x$, $1/x^2$, . . ., $1/x^n$ and c/x^n for any constant c approach zero as $x \to \infty$ (Examples 7, 8, and 9).

Because they will be useful later, the following theorems on limits are given without proof:

If $\lim_{x\to a} f(x) = F$ and $\lim_{x\to a} g(x) = G$, then

- ☐ $\lim_{x\to a}[f(x) \pm g(x)] = F \pm G$

  (The limit of a sum equals the sum of the limits.)
- ☐ $\lim_{x\to a} f(x)\, g(x) = FG$

  (The limit of a product equals the product of the limits.)
- ☐ $\lim_{x\to a} \dfrac{f(x)}{g(x)} = \dfrac{F}{G}$

  (The limit of a quotient equals the quotient of the limits.)
- ☐ $\lim_{x\to a} kf(x) = kF$

  (The limit of the product of a constant and a function is the product of the constant and the limit of the function.)

Continuity

Mathematically, a function is said to be **continuous at a point** if the following conditions are satisfied:

- ☐ $f(x)$ is defined at $x = a$
- ☐ $\lim_{x \to a} f(x)$ exists
- ☐ $\lim_{x \to a} f(x) = f(a)$

A function is said to be **continuous over an interval** if it is continuous at every point in that interval.

Intuitively, a function is said to be continuous over an interval if you can trace the graph over that interval without lifting your pencil. In other words, there are no breaks in the graph. For example, the graph of $f(x) = 2x - 1$ shown in Figure 18–3 is continuous over the arbitrarily chosen interval from -3 to 2 because we can trace the graph from $x = -3$ to $x = 2$ without lifting the pencil.

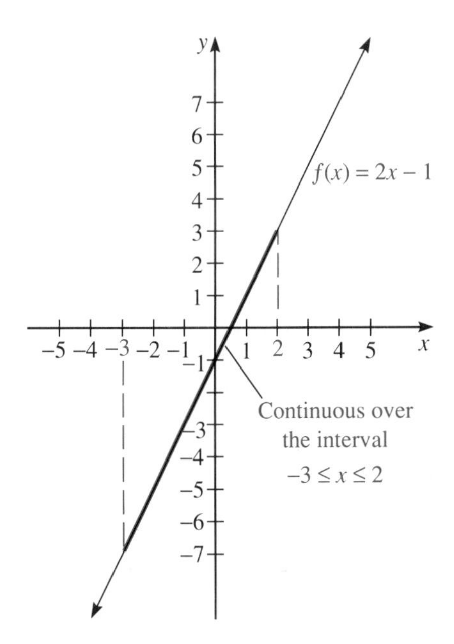

FIGURE 18–3

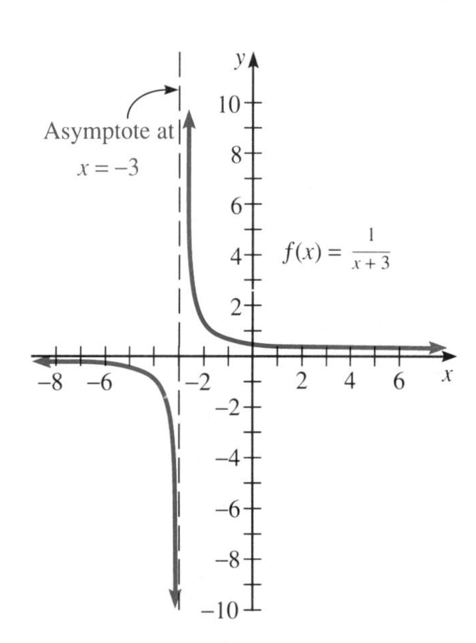

FIGURE 18–4

On the other hand, the graph of

$$f(x) = \frac{1}{x + 3}$$

shown in Figure 18–4 is not continuous at $x = -3$ because of the asymptote. Remember that the asymptote results when the function is not defined at $x = -3$ because it results in division by zero. We could not trace the graph of this function without lifting our pencil at $x = -3$.

EXAMPLE 10 Determine whether the graph shown in Figure 18–5 is continuous over the interval $x = -4$ to $x = 4$.

Solution Since there are no breaks in the graph from $x = -4$ to $x = 4$, the function is continuous over the interval. ■

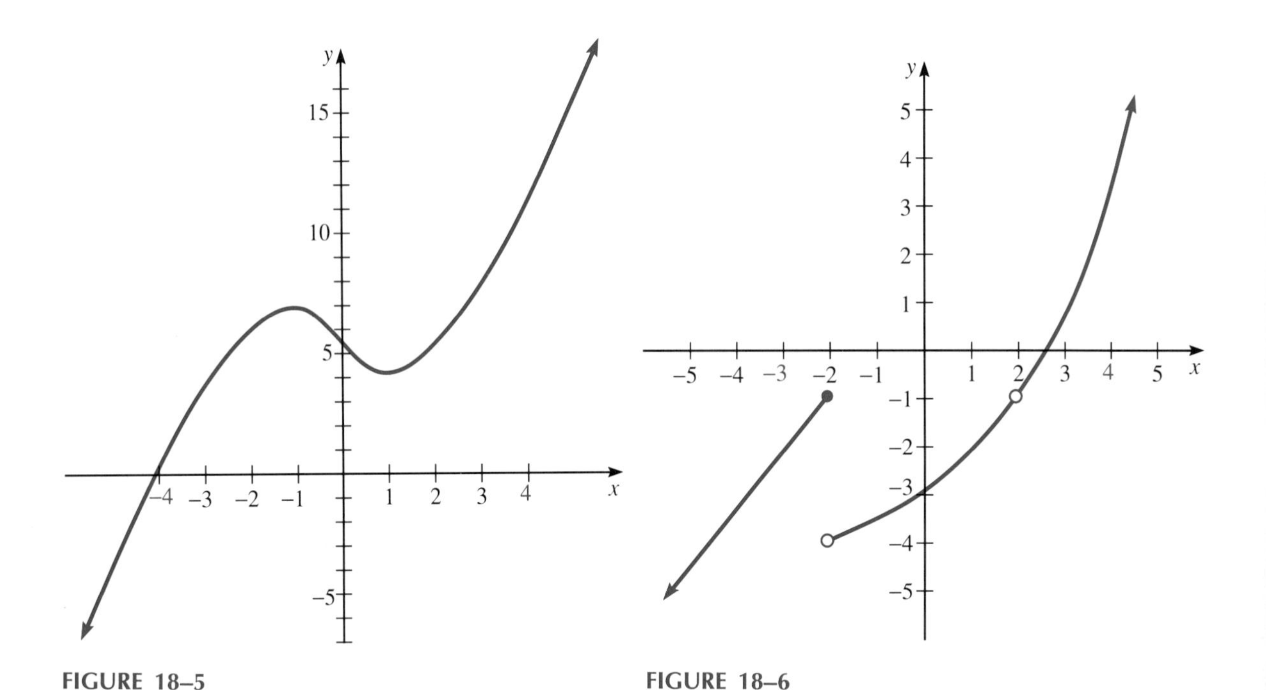

FIGURE 18–5

FIGURE 18–6

EXAMPLE 11 Determine whether the graph shown in Figure 18–6 is continuous over the interval $x = -3$ to $x = 4$.

Solution The function is not continuous at $x = 2$, as denoted by the small open circle. Remember, the open circle means that the point is not part of the solution. At $x = -2$ the function is defined, as denoted by the closed circle at $f(x) = -1$. However, a small change in x results in a jump in $f(x)$. Since the function is not defined at $x = 2$, and since the function changes abruptly at $x = -2$, this function is not continuous over the interval $x = -3$ to $x = 4$. ■

EXAMPLE 12 Determine whether the function $f(x) = 3x^2 + x - 1$ is continuous at $x = 1$.

Solution To show that this function is continuous at $x = 1$, we must show that it satisfies all three of the conditions given earlier. The function is defined at $x = 1$ because $f(1) = 3(1)^2 + (1) - 1 = 3$. Furthermore,

$$\lim_{x \to 1} f(x) = 3 \quad \text{and} \quad \lim_{x \to 1} f(x) = f(1)$$

The function $f(x) = 3x^2 + x - 1$ is continuous at $x = 1$. ■

18–1 EXERCISES

Using a table of values, evaluate the limit of the following functions.

1. $\lim_{x \to 3}(5x - 1)$
2. $\lim_{x \to -2}(3x + 7)$
3. $\lim_{x \to -1}\left(\dfrac{2x^2 + 3x + 1}{x + 1}\right)$
4. $\lim_{x \to -4}\left(\dfrac{x^2 - 16}{x + 4}\right)$
5. $\lim_{x \to 2}\left(\dfrac{x^3 - 7x^2 + 11x - 2}{x - 2}\right)$
6. $\lim_{x \to \infty}\left(8 - \dfrac{3}{x^2}\right)$
7. $\lim_{x \to \infty}\left(4 + \dfrac{7}{x}\right)$
8. $\lim_{x \to \infty}\left(\dfrac{2x^3 + x^2 - 4}{5x + x^3}\right)$
9. $\lim_{x \to \infty}\left(\dfrac{3 + 5x^2 - 7x}{x^2 + 5x}\right)$

Use any convenient method to find the indicated limits, if they exist.

10. $\lim_{x \to 1}(3x^3 + 6x - 8)$
11. $\lim_{x \to -1}(x^3 - 4x^2 + 10)$
12. $\lim_{x \to 0}(1 - x^2)^2$
13. $\lim_{x \to 2}\left(\dfrac{x^2 - 6x + 15}{x + 3}\right)$
14. $\lim_{x \to 2}\left(\dfrac{x^2 + x + 8}{x - 3}\right)$
15. $\lim_{x \to 1}\left(\dfrac{2x^2 + 3x - 5}{x - 1}\right)$
16. $\lim_{x \to 0}\left(\dfrac{x^3 + 6x}{x}\right)$
17. $\lim_{x \to 3}\left(\dfrac{2x^2 - 7x + 3}{2x^2 - x - 15}\right)$
18. $\lim_{x \to -2}\left(\dfrac{3x^2 + 5x - 2}{x + 2}\right)$
19. $\lim_{x \to 2}\sqrt{x - 9}$
20. $\lim_{x \to -1}\sqrt{2x - 3}$
21. $\lim_{x \to 0}\left(\dfrac{(3 - x)^2 - 9}{x}\right)$
22. $\lim_{x \to 0}\sqrt{x^2 + 3}$
23. $\lim_{x \to 3}\sqrt{4x - 1}$
24. $\lim_{x \to \infty}\left(\dfrac{2x^2 + 7}{5 - 3x + x^2}\right)$
25. $\lim_{x \to \infty}\left(\dfrac{4x^3 + 4x + 6}{x^2 - 5 + 3x^3}\right)$
26. $\lim_{x \to \infty}\left(\dfrac{x^2 + 3x - 1}{4x^2 - 5}\right)$
27. $\lim_{x \to \infty}\left(\dfrac{x^5 - 2x^3}{6 + x^2 + 2x^5}\right)$
28. $\lim_{x \to \infty}\left(\dfrac{x + 7x^3 - 1}{x^4 - 2x + 5}\right)$
29. $\lim_{x \to \infty}\left(\dfrac{x^2 - x - 12}{x^2 + 4x - 5}\right)$

Determine whether the following functions are continuous. If not, state why.

30. $f(x) = 2x + 9$
31. $f(x) = 2 + \dfrac{5}{x}$
32. $f(x) = \dfrac{1}{2x + 1}$
33. $f(x) = x^2$
34. $f(x) = \sqrt{3x + 9}$
35. $f(x) = \dfrac{\sqrt{x + 1}}{x}$

From the figure given, determine whether the function is continuous. If not, state why.

FIGURE 18–7

36.

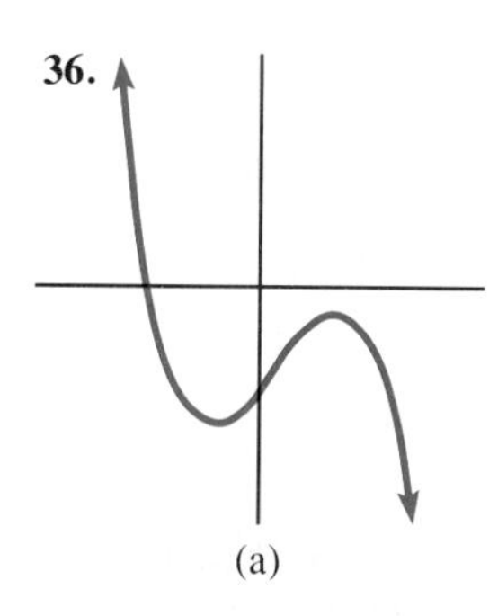

(a)

37.

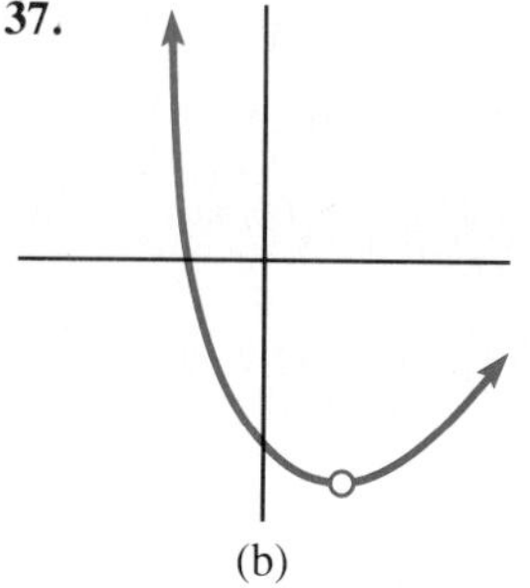

(b)

38.

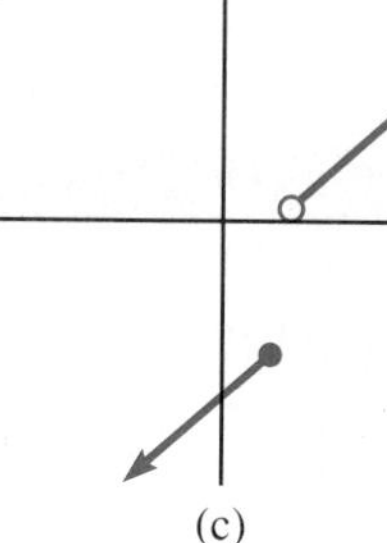

(c)

FIGURE 18–8

39.

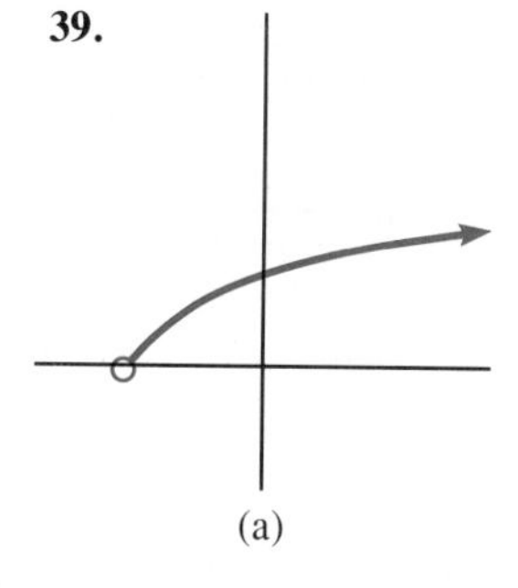

(a)

40.

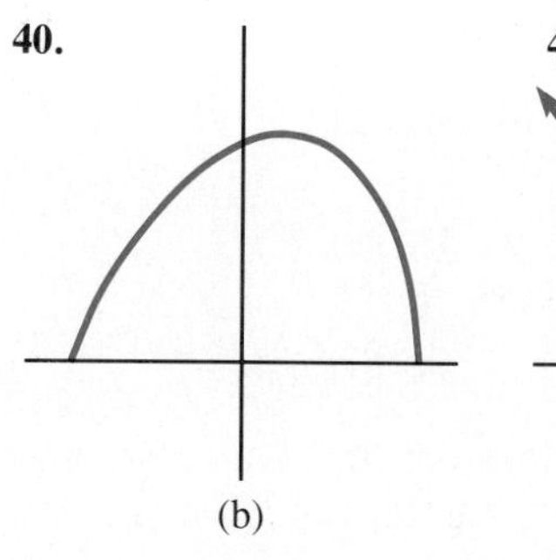

(b)

41.

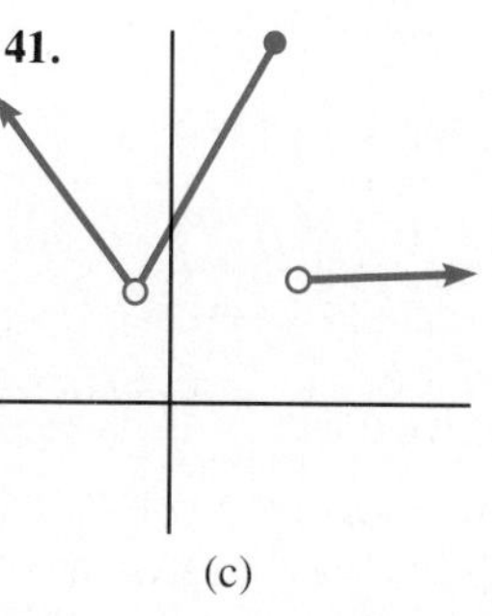

(c)

42. The displacement s of an object moving through a viscous solution is described by

$$s = (2 - t)e^{-0.5t}$$

What value does s approach as $t \to 3$?

43. The population of bacteria in a petri dish is given by

$$N = 10{,}000 + \frac{18{,}000}{(n + 3)}$$

Find the long-range population trend; that is, find $\lim_{n\to\infty} N$.

44. The total effective capacities for two capacitors in series is

$$C_e = \frac{C_1C_2}{C_1 + C_2}$$

Find $\lim_{C_1\to 4} C_e$ if $C_2 = 2\ \mu\text{F}$.

45. In surveying, the average photographic scale is given by

$$S = \frac{f}{H - h}$$

where f = focal length of the camera, H = flying height of the airplane, and h = average height of terrain. Find $\lim_{H\to 10{,}000} S$ if $f = 0.5$ ft and $h =$ 2,500 ft.

46. The mass m of a particle as a function of velocity v is given by

$$m = \frac{m_0}{\sqrt{1 - \frac{v^2}{c^2}}}$$

where m_0 = mass at rest and c = velocity of light. Find $\lim_{v\to c}$ m.

18–2

THE DERIVATIVE

In the last section we developed the concepts of continuity and limits of a function by using a table of values. In this section we will discuss a graphical interpretation of the rate of change of a function. This interpretation deals with the slope of a line tangent to the curve of a function.

Slope of Tangent to a Curve

To illustrate this idea, let us examine the graph of the function shown in Figure 18–9(a). The straight line that touches the graph at point P is called a **tangent line** to the graph, and we want to find the slope of that tangent line. This problem does not sound very important, but many problems in science and technology are related to the slope of the tangent line to a curve.

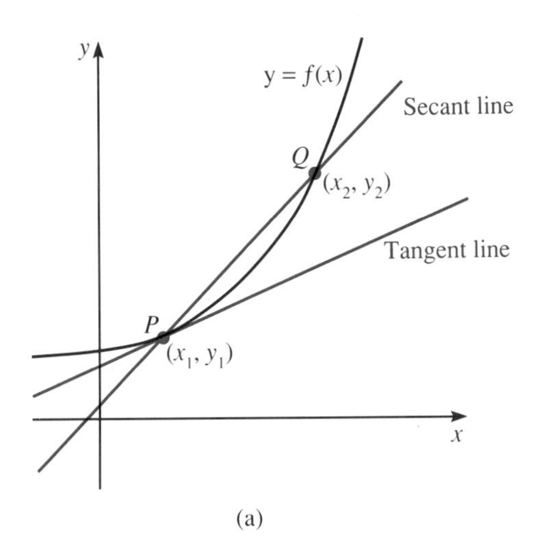

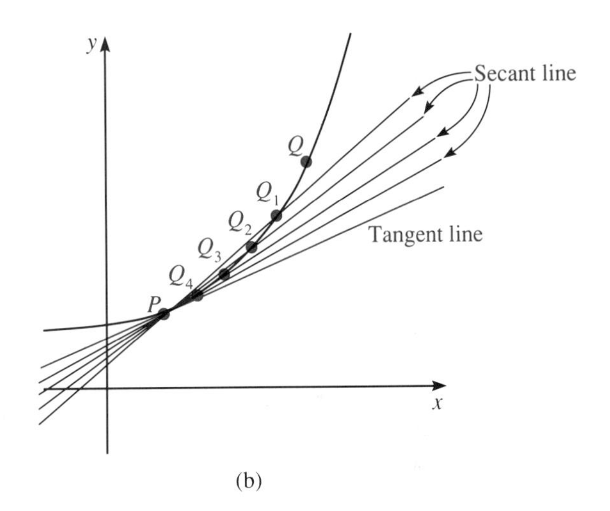

FIGURE 18–9

First, we must define a secant line. Let us choose a second point, Q, on the curve in Figure 18–9(a). The line PQ that intersects the curve twice is called the **secant line.** From our discussion of slope of a line in Chapter 2, we know that the slope of line PQ is given by

$$m_{\text{sec}} = \frac{y_2 - y_1}{x_2 - x_1}$$

To determine the slope of the tangent line at P, let us move point Q along the curve toward P. This process is shown in Figure 18–9(b) by the lines through P and points Q_1, Q_2, Q_3, and Q_4, respectively. As you can see from this figure, as point Q gets closer and closer to point P, the secant line gets closer and closer to the tangent line. Thus, as Q approaches P, the slope of the secant line approaches the slope of the tangent line. Using the limit notation from Section 18–1, this statement can be written as follows:

$$m_{tan} = \overbrace{\lim_{Q \to P} m_{sec}}^{\text{limit of slope of secant}} = \lim_{x_2 \to x_1} \left(\frac{y_2 - y_1}{x_2 - x_1} \right)$$

(m_{tan}: slope of tangent)

Since $Q \to P$ implies that $x_2 \to x_1$, the slope of the tangent line can be as given in the following box.

The Slope of a Tangent Line

The slope of a line tangent to the curve at point (x_1, y_1) is given by

$$m_{tan} = \lim_{x_2 \to x_1} \left(\frac{y_2 - y_1}{x_2 - x_1} \right)$$

EXAMPLE 1 Find the slope of the line tangent to the graph of $y = x^2 + 8x - 5$ at the point $P = (1, 4)$.

Solution As point Q approaches point P along the curve, $x_2 \to x_1$ and $(x_1, y_1) = (1, 4)$, the point given to us. Therefore, substituting into the formula for the slope of a tangent line gives the following:

$$m_{tan} = \lim_{x_2 \to x_1} \left(\frac{y_2 - y_1}{x_2 - x_1} \right)$$

$y_2 = f(x_2)$ $\qquad$ $y_1 = f(x_1) = 4$ $\qquad$ $x_1 = 1$

$$= \lim_{x_2 \to 1} \left[\frac{(x_2^2 + 8x_2 - 5) - 4}{x_2 - 1} \right] \qquad \text{substitute}$$

$$= \lim_{x_2 \to 1} \left(\frac{x_2^2 + 8x_2 - 9}{x_2 - 1} \right) \qquad \text{simplify}$$

$$= \lim_{x_2 \to 1} \left[\frac{(x_2 + 9)(x_2 - 1)}{(x_2 - 1)} \right] \qquad \text{factor}$$

$$= \lim_{x_2 \to 1} (x_2 + 9) \qquad \text{reduce fraction}$$

$$m_{tan} = 10 \qquad \text{evaluate limit}$$

FIGURE 18–10

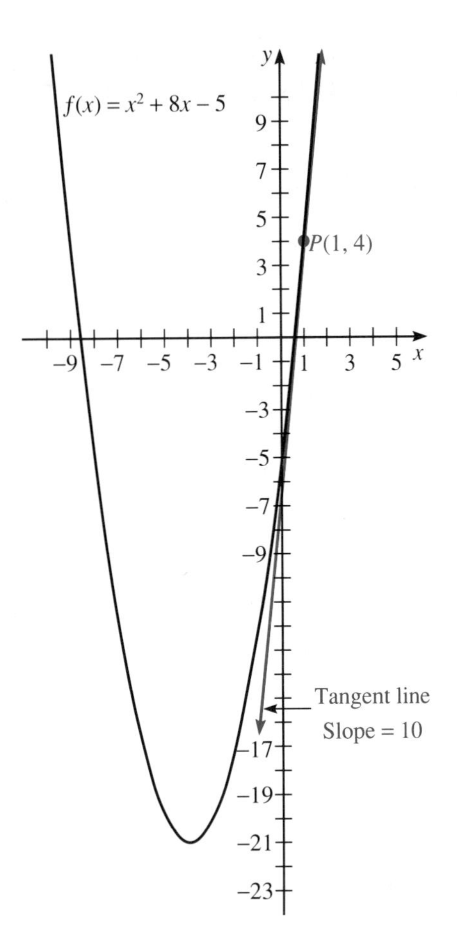

The slope of the line tangent to $y = x^2 + 8x - 5$ at the point (1, 4) is 10 as shown in Figure 18–10. ■

Instantaneous Velocity

When we discussed the slope of the tangent line, we saw that when point Q approached point P, the slope of the secant line approached the slope of the tangent line. This concept can also be viewed as the rate of change in y with respect to x. One application of this principle is the velocity of an object.

The ratio of displacement s of an object to the elapsed time t is called the **average velocity,** expressed in the following formula:

$$v_{av} = \frac{s_2 - s_1}{t_2 - t_1}$$

If the time interval becomes smaller and smaller, the average velocity approximates the **instantaneous velocity** of the object at some particular instant in time. Therefore, as $t_2 \to t_1$, the average velocity approaches the instantaneous velocity. This relationship can be written as

$$v = \lim_{t_2 \to t_1} \left(\frac{s_2 - s_1}{t_2 - t_1} \right)$$

EXAMPLE 2 A ball is thrown vertically upward with an initial velocity of 40 ft/s. Find the velocity of the ball at $t_1 = 1$ s given that the displacement s as a function of time is given by $s = 40t - 16t^2$.

Solution Using the formula for the instantaneous velocity at $t_1 = 1$ s gives the following:

$$v = \lim_{t_2 \to t_1} \left(\frac{s_2 - s_1}{t_2 - t_1} \right)$$

$s_2 = f(t_2)$ $\quad$ $s_1 = f(t_1) = 40(1) - 16(1)^2$

$$= \lim_{t_2 \to 1} \left[\frac{(40t_2 - 16t_2^2) - (24)}{t_2 - 1} \right]$$

$t_1 = 1$

$$= \lim_{t_2 \to 1} \left(\frac{-16t_2^2 + 40t_2 - 24}{t_2 - 1} \right)$$

$$= \lim_{t_2 \to 1} \left[\frac{-8(2t_2^2 - 5t_2 + 3)}{t_2 - 1} \right]$$

$$= \lim_{t_2 \to 1} \left[\frac{-8(2t_2 - 3)(t_2 - 1)}{(t_2 - 1)} \right]$$

$$= \lim_{t_2 \to 1} [-8(2t_2 - 3)]$$

$$v = 8$$

The ball's velocity is 8 ft/s. Remember from Chapter 11 that velocity is a vector. Therefore, the positive sign in our answer means that the ball is moving upward. ■

Delta (Δ) Notation

In the previous discussion in this section on the slope of the tangent line, we have used $y_2 - y_1$ and $x_2 - x_1$ in the calculations. We can interpret $y_2 - y_1$ to be the change or difference in y values and $x_2 - x_1$ to be the change in x values. We can also use **delta notation** to represent these quantities. The change in x is represented as Δx (read as "delta x"), and the change in y is represented as Δy. Therefore,

$$\Delta x = x_2 - x_1 \quad \text{and} \quad \Delta y = y_2 - y_1$$

or

$$x_2 = x_1 + \Delta x \quad \text{and} \quad y_2 = y_1 + \Delta y$$

CAUTION ✦ The notation Δx or Δy does not mean delta times x or y. It represents the change in x or the change in y.

Definition of the Derivative

If we take the definition of the slope of the tangent line given earlier and use the delta notation, we obtain the definition of the **derivative** whose derivation follows.

FIGURE 18–11

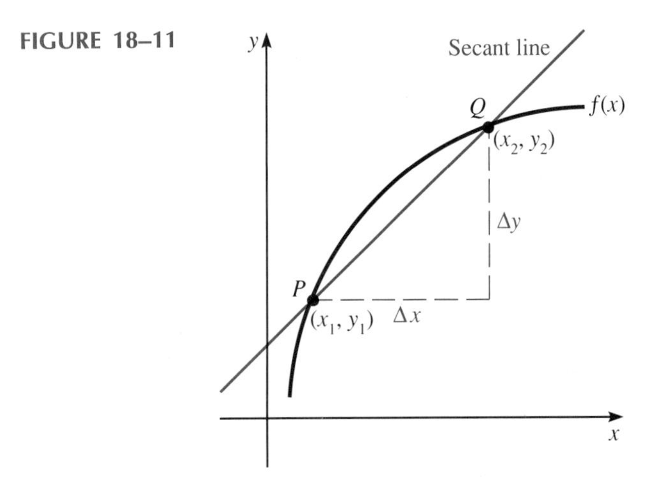

From Figure 18–11, you can see that as $x_2 \to x_1$, then $\Delta x \to 0$ (the change in x approaches zero). Using the definition of delta notation, $\Delta y = y_2 - y_1$ and $\Delta x = x_2 - x_1$, we can write the slope of the tangent line as follows:

$$m_{\tan} = \lim_{\Delta x \to 0} \left(\frac{\Delta y}{\Delta x} \right)$$

In functional notation, if $y = f(x)$, then the following statements are true:

$$y_1 = f(x_1)$$
$$y_2 = f(x_2)$$
$$x_2 = x_1 + \Delta x$$
$$y_2 = f(x_1 + \Delta x)$$

Substituting for y_2 and y_1 gives

$$\Delta y = y_2 - y_1$$
$$\Delta y = f(x_1 + \Delta x) - f(x_1)$$

Substituting the delta notation for the slope of the tangent line gives

$$\lim_{\Delta x \to 0} \frac{\Delta y}{\Delta x} = \lim_{\Delta x \to 0} \left[\frac{\overbrace{f(x_1 + \Delta x)}^{y_2} - \overbrace{f(x_1)}^{y_1}}{\underbrace{\Delta x}_{x_2 - x_1}} \right]$$

Since the formula $(y_2 - y_1)/(x_2 - x_1)$ is called the **difference quotient,** the formula for the slope of the tangent to a curve is called the **derivative.** The derivative of function $f(x)$ is a new function $f'(x)$ defined below. The general formula for the derivative of a function commonly denoted $f'(x)$, y', or dy/dx, is stated below.

The Derivative of a Function

$$f'(x) = y' = \frac{dy}{dx} = \lim_{\Delta x \to 0} \left[\frac{f(x + \Delta x) - f(x)}{\Delta x} \right]$$

A function is said to be **differentiable** at a point only if the limit exists at the point. The process of finding the derivative of a function is called **differentiation.**

NOTE ✦ The derivative can be interpreted as the instantaneous rate of change of the dependent variable with respect to the independent variable because $\dfrac{f(x + \Delta x) - f(x)}{\Delta x}$ represents the average rate of change.

EXAMPLE 3 Find the derivative of the function $y = x^2 - 3x$.

Solution We use the definition of the derivative given above. First, we replace x with $x + \Delta x$ and y with $y + \Delta y$ in the function.

replace y with $y + \Delta y$ replace x with $(x + \Delta x)$

$$y + \Delta y = (x + \Delta x)^2 - 3(x + \Delta x)$$

Second, we solve for Δy, substitute for y, and simplify the result.

$$\Delta y = (x + \Delta x)^2 - 3(x + \Delta x) - y$$

substitute $y = x^2 - 3x$

$$= (x + \Delta x)^2 - 3(x + \Delta x) - (x^2 - 3x)$$
$$= x^2 + 2x\,\Delta x + (\Delta x)^2 - 3x - 3\,\Delta x - x^2 + 3x$$
$$\Delta y = 2x\,\Delta x + (\Delta x)^2 - 3\,\Delta x$$

parentheses used for clarity

Third, we divide each side of the equation by Δx.

$$\frac{\Delta y}{\Delta x} = \frac{2x\,\Delta x + (\Delta x)^2 - 3\,\Delta x}{\Delta x}$$
$$\frac{\Delta y}{\Delta x} = \frac{\Delta x\,(2x + \Delta x - 3)}{\Delta x}$$
$$\frac{\Delta y}{\Delta x} = 2x + \Delta x - 3$$

Fourth, we find the limit of the expression as $\Delta x \to 0$.

$$\lim_{\Delta x \to 0} \frac{\Delta y}{\Delta x} = \lim_{\Delta x \to 0} (2x + \Delta x - 3) = 2x - 3$$

The derivative of $x^2 - 3x$ is $2x - 3$. Thus, $y' = 2x - 3$. This expression represents the slope of the tangent at any point on the graph of this function. ■

The method used to find the derivative in Example 3 is called the **delta (Δ) process.** The steps are summarized in the following box.

The Derivative Using the Delta Process

1. Replace x with $x + \Delta x$, and replace y with $y + \Delta y$, in the function.
2. Solve the result for Δy, substitute for y, and simplify the result. The resulting equation should contain only Δy, x, and Δx.
3. Divide each side of the equation by Δx.
4. Find the limit of the resulting expression as $\Delta x \to 0$.

EXAMPLE 4 Find $f'(x)$ for the function $y = x^3$, and determine the slope of the line tangent to the curve at $x = 2$.

Solution We will use the delta process to find the derivative of the function.

$$y = x^3$$

$$y + \Delta y = (x + \Delta x)^3$$

replace y with $y + \Delta y$ replace x with $x + \Delta x$

$$\Delta y = (x + \Delta x)^3 - y \qquad \text{solve for } \Delta y$$

$$\Delta y = x^3 + 3x^2\,\Delta x + 3x(\Delta x)^2 + (\Delta x)^3 - x^3 \qquad \text{substitute for } y$$

$$\Delta y = 3x^2\,\Delta x + 3x(\Delta x)^2 + (\Delta x)^3 \qquad \text{simplify}$$

$$\frac{\Delta y}{\Delta x} = \frac{3x^2\,\Delta x + 3x(\Delta x)^2 + (\Delta x)^3}{\Delta x} \qquad \text{divide by } \Delta x$$

$$\frac{\Delta y}{\Delta x} = 3x^2 + 3x\,\Delta x + (\Delta x)^2 \qquad \text{simplify}$$

$$\begin{aligned}\lim_{\Delta x\to 0}\frac{\Delta y}{\Delta x} &= \lim_{\Delta x\to 0}(3x^2 + 3x\,\Delta x + (\Delta x)^2)\\ &= \lim_{\Delta x\to 0}3x^2 + \lim_{\Delta x\to 0}3x\,\Delta x + \lim_{\Delta x\to 0}(\Delta x)^2\\ &= 3x^2 + 0 + 0\end{aligned}$$

$$\lim_{\Delta x\to 0}\frac{\Delta y}{\Delta x} = 3x^2 \qquad \text{take the limit of both sides}$$

$$f'(x) = 3x^2$$

This expression represents the slope of the line tangent to the curve $y = x^3$. To find the slope of the line at $x = 2$, we evaluate the derivative when $x = 2$, denoted as follows:

$$f'(x)\big|_{x=2} \quad \text{or} \quad \left.\frac{dy}{dx}\right|_{x=2}$$

Evaluating the derivative at $x = 2$ gives

$$f'(x)\big|_{x=2} = 3(2)^2 = 12$$

The slope of the line tangent to the curve $y = x^3$ at $x = 2$ is 12, as illustrated in Figure 18–12. ■

EXAMPLE 5 Find dy/dx for the function $y = (x + 2)/x$ and evaluate it when $x = 3$.

Solution First, replace x with $x + \Delta x$ and y with $y + \Delta y$.

$$y + \Delta y = \frac{x + \Delta x + 2}{x + \Delta x}$$

Second, solve for Δy, substitute for y, and simplify.

FIGURE 18–12

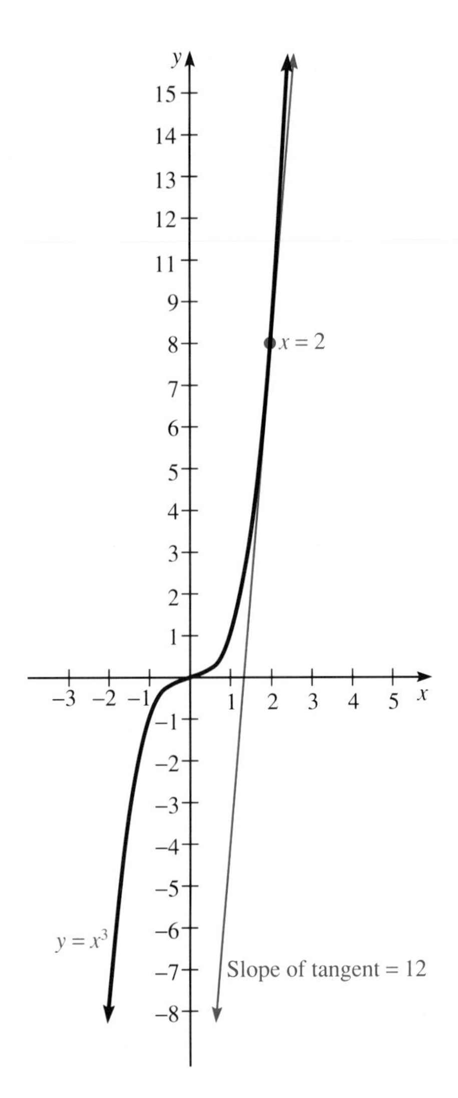

$$\Delta y = \frac{x + \Delta x + 2}{x + \Delta x} - y \qquad \text{solve for } \Delta y$$

$$= \frac{x + \Delta x + 2}{x + \Delta x} - \frac{x + 2}{x} \qquad \text{substitute}$$

$$= \frac{x(x + \Delta x + 2) - (x + 2)(x + \Delta x)}{x(x + \Delta x)} \qquad \text{combine fractions}$$

$$= \frac{x^2 + x\,\Delta x + 2x - x^2 - 2x - x\,\Delta x - 2\,\Delta x}{x(x + \Delta x)} \qquad \text{multiply numerator}$$

$$\Delta y = \frac{-2\,\Delta x}{x(x + \Delta x)} \qquad \text{combine terms}$$

Third, divide both sides of the equation by Δx.

$$\frac{\Delta y}{\Delta x} = \frac{\dfrac{-2\,\Delta x}{x(x + \Delta x)}}{\Delta x}$$

$$= \frac{\dfrac{-2\,\Delta x}{\Delta x}}{x^2 + x\,\Delta x}$$

$$\frac{\Delta y}{\Delta x} = \frac{-2}{x^2 + x\,\Delta x}$$

Fourth, find the limit as $\Delta x \to 0$.

$$\lim_{\Delta x \to 0} \frac{\Delta y}{\Delta x} = \lim_{\Delta x \to 0} \left(\frac{-2}{x^2 + x\Delta x} \right)$$

$$\frac{dy}{dx} = \frac{-2}{x^2}$$

Last, evaluate the derivative when $x = 3$.

$$\left.\frac{dy}{dx}\right|_{x=3} = \frac{-2}{(3)^2} = -\frac{2}{9}$$

The slope of the line tangent to $y = (x + 2)/x$ at $x = 3$ is $-2/9$. ■

Higher Derivatives

If $f(x)$ is a differentiable function whose derivative exists, then its derivative is also a function. The first derivative is denoted dy/dx, $f'(x)$, or y'. If dy/dx is also a differentiable function, its derivative is called the *second derivative,* denoted d^2y/dx^2 or y''. This process can continue to yield the third, fourth, . . . , or n^{th} derivative, denoted by d^3y/dx^3, $f'''(x)$ or y'''; d^4y/dx^4, $f^{(4)}(x)$ or y''''; . . . , d^ny/dx^n, $f^{(n)}(x)$ or $y^{(n)}$. In physics, acceleration is the second derivative of displacement, and voltage is $-L$ times the second derivative of charge.

Acceleration

Acceleration is defined as the time rate of change of velocity. If the velocity of an object is known as a function of time, then the acceleration is the derivative of velocity with respect to time. Remember that the derivative may be considered to be the rate of change of a variable. Therefore, the acceleration is given as follows:

$$a = \frac{dv}{dt} \qquad \text{where } v = \text{velocity} \quad \text{and} \quad t = \text{time}$$

EXAMPLE 6 A rocket is traveling in a straight line so that its position in space is given by the equation $s = t^3 + 2t^2 + 5$ where s = displacement in meters and t = time in seconds. Find expressions for the velocity and the acceleration of the satellite, and evaluate both the velocity and the acceleration when $t = 2$ s.

Solution From our earlier discussion in this section, we know that velocity is the time rate of change of displacement and can be expressed as

$$v = \frac{ds}{dt}$$

To find the velocity of the satellite, we use the delta process to take the derivative of the displacement with respect to time.

$$\begin{aligned}
s + \Delta s &= (t + \Delta t)^3 + 2(t + \Delta t)^2 + 5 \\
\Delta s &= t^3 + 3t^2\,\Delta t + 3t(\Delta t)^2 + (\Delta t)^3 + 2t^2 + 4t\,\Delta t \\
&\quad + 2(\Delta t)^2 + 5 - t^3 - 2t^2 - 5 \\
\Delta s &= 3t^2\,\Delta t + 3t(\Delta t)^2 + (\Delta t)^3 + 4t\,\Delta t + 2(\Delta t)^2 \\
\frac{\Delta s}{\Delta t} &= 3t^2 + 3t\,\Delta t + (\Delta t)^2 + 4t + 2\,\Delta t \\
v = \frac{ds}{dt} &= \lim_{\Delta t \to 0} \frac{\Delta s}{\Delta t} = 3t^2 + 4t
\end{aligned}$$

Next, we evaluate the derivative when $t = 2$.

$$\left.\frac{ds}{dt}\right|_{t=2} = 3(2)^2 + 4(2) = 20$$

The instantaneous velocity of the satellite when $t = 2$ is 20 m/s. The acceleration of the satellite is the time rate of change of velocity ($a = dv/dt$). Therefore, we take the derivative of the velocity with respect to time.

$$\begin{aligned}
v &= 3t^2 + 4t \\
v + \Delta v &= 3(t + \Delta t)^2 + 4(t + \Delta t) \\
\Delta v &= 3t^2 + 6t\,\Delta t + 3(\Delta t)^2 + 4t + 4\,\Delta t - 3t^2 - 4t \\
\Delta v &= 6t\,\Delta t + 3(\Delta t)^2 + 4\,\Delta t \\
\frac{\Delta v}{\Delta t} &= 6t + 3\,\Delta t + 4 \\
a &= \frac{dv}{dt} = 6t + 4
\end{aligned}$$

Then we evaluate this derivative when $t = 2$.

$$\left.\frac{dv}{dt}\right|_{t=2} = 6(2) + 4 = 16$$

The velocity of the satellite at $t = 2$ s is 20 m/s, and the acceleration is 16 m/s^2. ■

From Example 6, we can see that acceleration is the derivative of the derivative of displacement. In other words, acceleration is the second derivative of displacement and can be written as follows:

denotes second derivative

$$a = \frac{d^2s}{dt^2} \quad \text{or} \quad a = f''(t) \quad \text{or} \quad a = v' = s''$$

Current

The electrical current i in a circuit is the rate of transferring electrical charge past a point in the circuit. If $q(t)$ denotes the electrical charge as a function of time, then

$$i = \frac{dq}{dt}$$

where i is measured in amperes (A), q in coulombs (C), and t in seconds (s).

EXAMPLE 7 Find the current when $t = 2$ s if the charge transferred is given by

$$q = \frac{1}{t+2}$$

Solution First, we find an expression for the current by using the delta process to find the derivative of q with respect to t.

$$q = \frac{1}{t+2}$$

$$q + \Delta q = \frac{1}{t + \Delta t + 2} \qquad \text{replace } q \text{ with } q + \Delta q \text{ and } t \text{ with } t + \Delta t$$

$$\Delta q = \frac{1}{t + \Delta t + 2} - \frac{1}{t+2} \qquad \text{solve for } \Delta q \text{ and substitute for } q$$

$$\Delta q = \frac{t + 2 - t - \Delta t - 2}{(t+2)(t + \Delta t + 2)} \qquad \text{combine fractions}$$

$$\Delta q = \frac{-\Delta t}{(t+2)(t + \Delta t + 2)} \qquad \text{combine like terms}$$

$$\frac{\Delta q}{\Delta t} = \frac{-1}{(t+2)(t + \Delta t + 2)} \qquad \text{divide by } \Delta t$$

$(\Delta t \to 0)$

$$i = \frac{dq}{dt} = \frac{-1}{(t+2)^2} \qquad \text{take } \lim_{\Delta t \to 0} \frac{\Delta q}{\Delta t}$$

The current is represented by the expression $-1/(t + 2)^2$. To determine the current when $t = 2$ s, we evaluate the derivative.

$$\left.\frac{dq}{dt}\right|_{t=2} = \frac{-1}{(2 + 2)^2} = -0.0625 \text{ A}$$

The current at 2 seconds is 0.0625 A. The negative sign denotes the current flows in the direction opposite to that chosen in the circuit. ■

EXAMPLE 8 The value (in thousands of dollars) of an investment is given by $V = 8t^2 - 5t + 2000$ where $t =$ time in years. Find a general expression for the rate of change of the value V with respect to time t, and evaluate this expression when $t = 2$ years.

Solution We must first find the derivative of V with respect to t, denoted dV/dt.

$$V = 8t^2 - 5t + 2000$$

$$V + \Delta V = 8(t + \Delta t)^2 - 5(t + \Delta t) + 2000$$

$$\Delta V = 8t^2 + 16t\,\Delta t + 8(\Delta t)^2 - 5t - 5\,\Delta t + 2000 - 8t^2 + 5t - 2000$$

$$\Delta V = 16t\,\Delta t + 8(\Delta t)^2 - 5\,\Delta t$$

$$\frac{\Delta V}{\Delta t} = 16t + 8\,\Delta t - 5$$

$$\frac{dV}{dt} = \lim_{\Delta t \to 0}(16t + 8\,\Delta t - 5) = 16t - 5$$

Next, we evaluate the derivative when $t = 2$.

$$\left.\frac{dV}{dt}\right|_{t=2} = 16(2) - 5 = 27$$

After two years the investment is worth \$27,000. ■

18–2 EXERCISES

Using the delta process, find the derivative of each function.

1. $y = 6x + 8$
2. $y = 4x - 5$
3. $y = 9 - 5x$
4. $y = 12 + 8x$
5. $y = x^2 - 3$
6. $a = v^2 + 9$
7. $v = 4t^2 - 6t$
8. $v = 6t^2 + 5t$
9. $a = 2v^3 - 5v + 8$
10. $y = 5x^3 + 9x - 10$
11. $y = 6 + 7x - x^2$
12. $y = 9 - 8x + 2x^2$
13. $y = x + 3x^3$
14. $y = x^3 + 8x$
15. $V = -8t + 5t^2$
16. $s = 16t - 4t^2$
17. $y = 3 + \dfrac{1}{x}$
18. $y = x^2 - \dfrac{1}{x}$
19. $y = \dfrac{8}{x - 4}$
20. $y = \dfrac{1}{x} + \dfrac{3}{x^2}$
21. $y = \dfrac{4}{x^2}$
22. $y = \dfrac{x + 1}{x}$
23. $y = \dfrac{x - 3}{x}$
24. $y = \dfrac{6}{x + 5}$

Use the delta process to find dy/dx for each function, and evaluate the derivative at the point given.

25. $y = x^2 - 4x$; $(1, -3)$

26. $y = 4x + x^2$; $(-2, -4)$

27. $y = x + 3x^2$; $(-1, 2)$

28. $y = 4x - 5x^2$; $(2, -12)$

29. $y = \frac{1}{x} - 3$; $(1, -2)$

30. $y = x - \frac{1}{x}$; $(1, 0)$

31. $y = \frac{4}{x - 1}$; $(2, 4)$

32. $y = \frac{3}{x + 2}$; $(1, 1)$

33. The equation for the displacement of a particle along a straight line is $s = 2t^2 + 5t$ ft. What is the velocity of the particle when $t = 4$ s?

34. Find an expression for the rate of change of the volume of a sphere with respect to its radius ($V = 4\pi r^3/3$).

35. Find the instantaneous velocity when $t = 3$ s of a particle whose displacement is given by $s = 8t^2 + 5t$.

36. Find a formula for the current in a circuit if the charge transferred as a function of time is given by $q = 18 - 3t$ (see Example 7).

37. Find a general expression for the instantaneous rate of change of the area of a circle with respect to its radius. Then evaluate the result when $r = 4$ cm. (Note: $A = \pi r^2$.)

38. What is the equation for the acceleration of an object if its velocity as a function of time is $v = 2t^2 - 3t + 5$ ft/s?

39. The displacement of a satellite in space as a function of time is given by $s = 3t^3 + 8t - 7$ km. Find an expression for the velocity and acceleration of the satellite, and evaluate each when $t = 3$ s.

40. An 8-Ω resistor and a variable resistor are connected in parallel. The total resistance R_T is given by

$$R_T = \frac{8R}{8 + R}$$

Determine the instantaneous rate of change of the total resistance with respect to R, and then evaluate this expression when $R = 6\ \Omega$.

18–3

THE DERIVATIVE OF A POLYNOMIAL

The delta process for finding the derivative of a function can be quite lengthy and difficult. In this section we will use the delta process to develop formulas to find the derivative of polynomial functions of the form

$$f(x) = a_0x^n + a_1x^{n-1} + \cdots + a_n$$

These formulas will save time in finding the derivative of a polynomial function.

Derivative of a Constant

First, we will develop the formula for the derivative of a constant function of the form $y = c$. From the graph of $y = c$ shown in Figure 18–13, we know

FIGURE 18–13

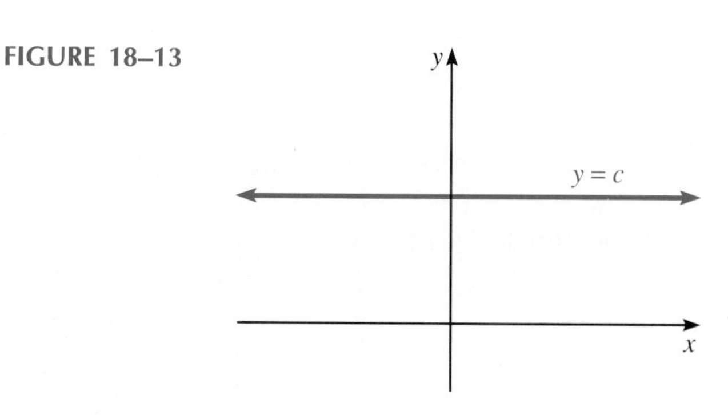

that the slope of any horizontal line is zero. Therefore, the slope of a tangent line is zero ($dc/dx = 0$) as verified by the following delta process:

$$\begin{aligned} y &= c \\ y + \Delta y &= c \\ \Delta y &= c - y \\ \Delta y &= c - c \\ \frac{\Delta y}{\Delta x} &= 0 \\ \frac{dy}{dx} &= \lim_{\Delta x \to 0} 0 = 0 \end{aligned}$$

The Constant Rule

If $y = c$, where c is a constant, then

$$\frac{dy}{dx} = \frac{dc}{dx} = 0$$

EXAMPLE 1 Find the derivative of the following functions:

(a) $y = -10$ (b) $y = 3$

Solution

(a) Because -10 is a constant, we can use the constant rule to determine $dy/dx = 0$.
(b) Similarly, 3 is a constant. Therefore, $dy/dx = 0$. ■

Derivative of a Power

Next, we shall develop a formula for the derivative of a power function of the form $y = x^n$. To find the general formula for a function of this type, we must use the binomial formula discussed in Chapter 16. Remember that for n, a positive integer,

$$(a + b)^n = a^n + na^{n-1}b + \frac{n(n-1)}{2!}a^{n-2}b^2 + \cdots + b^n$$

Using the delta process on $y = x^n$ gives

$$\begin{aligned} y + \Delta y &= (x + \Delta x)^n \\ \Delta y &= (x + \Delta x)^n - y \\ \Delta y &= (x + \Delta x)^n - x^n \end{aligned}$$

Using the binomial formula to expand $(x + \Delta x)^n$ gives

$$\Delta y = x^n + nx^{n-1}\,\Delta x + \frac{n(n-1)}{2!}x^{n-2}(\Delta x)^2 + \cdots + (\Delta x)^n - x^n$$

$$\Delta y = nx^{n-1}\,\Delta x + \frac{n(n-1)}{2!}x^{n-2}(\Delta x)^2 + \cdots + (\Delta x)^n$$

$$\frac{\Delta y}{\Delta x} = \frac{nx^{n-1}\,\Delta x + \frac{n(n-1)}{2!}x^{n-2}(\Delta x)^2 + \cdots + (\Delta x)^n}{\Delta x}$$

$$\frac{\Delta y}{\Delta x} = nx^{n-1} + \frac{n(n-1)}{2}x^{n-2}\,\Delta x + \cdots + (\Delta x)^{n-1}$$

As $\Delta x \to 0$, every term containing Δx approaches zero.

$$\frac{dy}{dx} = \lim_{\Delta x \to 0} \frac{\Delta y}{\Delta x} = nx^{n-1}$$

The Power Formula

If $y = x^n$, then the derivative is given by

decrease exponent by 1

$$\frac{dy}{dx} = nx^{n-1}$$

original exponent becomes coefficient

EXAMPLE 2 Find the derivative of the following functions:

(a) $y = x^5$ (b) $y = x^8$ (c) $y = x^{-4}$

Solution

(a) If $y = x^5$, then

decrease exponent by 1

$$\frac{dy}{dx} = 5x^{5-1} = 5x^4$$

original exponent becomes coefficient

(b) If $y = x^8$, then

$$\frac{dy}{dx} = 8x^{8-1} = 8x^7$$

(c) The power formula also applies to negative exponents.

$$\frac{dy}{dx} = -4x^{-4-1} = -4x^{-5}$$

■

Derivative of cx^n

When x^n is multiplied by a constant, the constant can be moved outside the derivative symbol as follows:

$$\frac{d(cx^n)}{dx} = c\,\frac{dx^n}{dx}$$

The Constant Multiplier Rule

If $y = cx^n$, then

constant times derivative ↓

$$\frac{dy}{dx} = c\,\frac{dx^n}{dy} = cnx^{n-1}$$

EXAMPLE 3 Find the derivative of the following functions:

(a) $y = 3x^7$ (b) $y = -2x^3$ (c) $y = 6x$ (d) $y = 8x^{3/4}$

Solution

move constant outside derivative ↓; result of power rule ↘; ↑ constant times derivative

(a) $$\frac{dy}{dx} = \frac{d(3x^7)}{dx} = 3\,\frac{d(x^7)}{dx} = 3(7x^6) = 21x^6$$

(b) $$\frac{dy}{dx} = \frac{d(-2x^3)}{dx} = -2\,\frac{d(x^3)}{dx} = -2(3x^2) = -6x^2$$

(c) $$\frac{dy}{dx} = \frac{d(6x)}{dx} = 6\,\frac{d(x)}{dx} = 6(1x^0) = 6$$

(d) The constant multiplier rule and the power formula also apply to an expression with a fractional exponent.

$$\frac{dy}{dx} = \underset{\substack{\uparrow \\ \text{constant multiplier}}}{8}\,\frac{d(x^{3/4})}{dx} = 8\left(\underset{\substack{\uparrow \\ \text{exponent becomes coefficient}}}{\frac{3}{4}}x^{-1/4}\right) = 6x^{-1/4}$$

(decrease exponent by 1) ■

CAUTION ✦ Do not confuse the derivative of a constant times a function such as $8x$ and the derivative of a constant 8.

Derivative of a Sum

Next, we will discuss the derivative of a polynomial function with more than one term. Let $y = u + v$, where $u = f(x)$ and $v = g(x)$, represent such a function. Applying the delta process to find the derivative gives

$$y = u + v$$

$$y + \Delta y = (u + \Delta u) + (v + \Delta v) \quad \text{substitute } y + \Delta y \text{ for } y,\ u + \Delta u \text{ for } u, \text{ and } v + \Delta v \text{ for } v$$

$$\Delta y = u + \Delta u + v - \overbrace{u - v}^{y} \quad \text{solve for } \Delta y \text{ and substitute for } y$$

$$\Delta y = \Delta u + \Delta v \quad \text{combine like terms}$$

$$\frac{\Delta y}{\Delta x} = \frac{\Delta u}{\Delta x} + \frac{\Delta v}{\Delta x} \quad \text{divide by } \Delta x$$

$$\lim_{\Delta x \to 0} \frac{\Delta y}{\Delta x} = \lim_{\Delta x \to 0} \frac{\Delta u}{\Delta x} + \lim_{\Delta x \to 0} \frac{\Delta v}{\Delta x} \quad \text{find } \lim_{\Delta x \to 0} \frac{\Delta y}{\Delta x}$$

$$\frac{dy}{dx} = \frac{du}{dx} + \frac{dv}{dx} \quad \text{simplify (apply the definition)}$$

The Sum Rule

If $y = u + v$, where u and v are both functions of x, then

$$\frac{dy}{dx} = \frac{du}{dx} + \frac{dv}{dx}$$

The derivative of the sum of functions is the sum of the derivative of each function.

EXAMPLE 4 Find the derivative of each of the following functions:

(a) $y = x^3 + x^7$ (b) $y = 3x^5 + 2x^9$

Solution

(a) To find the derivative of this sum of functions, find the derivative of each function.

$$\frac{dy}{dx} = \frac{d(x^3)}{dx} + \frac{d(x^7)}{dx} = 3x^2 + 7x^6$$

(b)

$$\frac{dy}{dx} = \frac{d(3x^5)}{dx} + \frac{d(2x^9)}{dx} \qquad \text{sum rule}$$

$$= 3\frac{d(x^5)}{dx} + 2\frac{d(x^9)}{dx} \qquad \text{constant multiplier rule}$$

$$= 3(5x^4) + 2(9x^8) \qquad \text{power rule}$$

$$\frac{dy}{dx} = 15x^4 + 18x^8$$

■

Derivative of a Difference

The sum rule may be extended to include the difference of two functions since $y = u - v$ can be written as $y = u + (-v)$.

The Difference Rule

If $y = u - v$, where u and v are both functions of x, then

$$\frac{dy}{dx} = \frac{du}{dx} - \frac{dv}{dx}$$

The derivative of the difference of two functions is the difference of the derivative of each function.

EXAMPLE 5 Find the derivative of the following function:

$$y = 7x^5 + 5x^3 - 8x^2 - 9x + 15$$

Solution

$$\frac{dy}{dx} = 7\frac{d(x^5)}{dx} + 5\frac{d(x^3)}{dx} - 8\frac{d(x^2)}{dx} - 9\frac{d(x)}{dx} + \frac{d(15)}{dx}$$

$$= 7(5x^4) + 5(3x^2) - 8(2x^1) - 9(1x^0) + (0)$$

$$\frac{dy}{dx} = 35x^4 + 15x^2 - 16x - 9$$

■

EXAMPLE 6 Find the slope of the line tangent to the following function:

$$y = 3x^3 - 6x^2 + 8x + 10 \quad \text{at } x = 2$$

Solution The slope of the tangent line is represented by the derivative of the function.

$$\frac{dy}{dx} = 3\frac{d(x^3)}{dx} - 6\frac{d(x^2)}{dx} + 8\frac{d(x)}{dx} + \frac{d(10)}{dx} \quad \text{constant multiplier rule}$$

$$= 3(3x^2) - 6(2x) + 8(1x^0) + 0 \quad \text{power formula and constant rule}$$

$$\frac{dy}{dx} = 9x^2 - 12x + 8$$

We find the slope of the tangent line at $x = 2$ by evaluating the derivative at $x = 2$.

$$\left.\frac{dy}{dx}\right|_{x=2} = 9(2)^2 - 12(2) + 8 = 20$$

The slope of the line tangent to the curve $y = 3x^3 - 6x^2 + 8x + 10$ at $x = 2$ is 20. ■

EXAMPLE 7 Find the first, second, and third derivatives of the following function:

$$y = 8x^4 + 6x^2 - x + 10$$

Solution Using the techniques developed in this section, the first derivative is

$$\frac{dy}{dx} = 32x^3 + 12x - 1$$

To find the second derivative of the function, take the derivative of the first derivative.

$$\frac{d^2y}{dx^2} = 96x^2 + 12$$

Similarly, the third derivative equals the derivative of the second derivative.

$$\frac{d^3y}{dx^3} = 192x$$

■

Applications

Power is defined as the time rate of change of doing work, given by

$$p = \frac{dw}{dt}$$

where p = power in ft-lb/s, w = work in ft-lb, and t = time in seconds.

EXAMPLE 8 The work done by an electrical motor as a function of time is given by $w = 18 - 3t^2 + 2t^4$. Find an expression for the power.

Solution The expression for power is given by the derivative of work with respect to time.

$$p = \frac{dw}{dt} = \frac{d(18)}{dt} - \frac{d(3t^2)}{dt} + \frac{d(2t^4)}{dt}$$

$$p = 0 - 6t + 8t^3$$

The power equation is $p = 8t^3 - 6t$ ft-lb/s. ■

When an electrical current in a coil opposes the source of the current, the induced voltage is given by

$$V = -L\frac{di}{dt} \text{ volts}$$

where L = inductance in henries (H), i = current in amperes (A) through the coil, and t = time in seconds (s).

EXAMPLE 9 Find the equation for the voltage in a coil if $L = 6$ H and the current is given by $i = 3t^3$ A. Also find the voltage when $t = 1.5$ s.

Solution The voltage is given by the expression

$$\begin{aligned} V &= -L\frac{di}{dt} \\ &= -6\frac{d(3t^3)}{dt} \\ &= -6(9t^2) \\ V &= -54t^2 \end{aligned}$$

At $t = 1.5$ s, the voltage is

$$V = -54(1.5)^2 = -121.5 \text{ V}$$ ■

An important application of calculus involves the maximum or minimum value of a quantity. The next two examples illustrate this application.

EXAMPLE 10 The voltage drop V in millivolts (mV) across a circuit is given by the relationship $V = 2t^3 - 17t^2 + 20t$ where t = time in seconds. Find the maximum value of the voltage.

Solution Graphically, a maximum or minimum of a function denotes a change in the sign of the slope of the tangent line: from negative to positive for a minimum, or from positive to negative for a maximum. Figure 18–14 illustrates this concept.

FIGURE 18–14

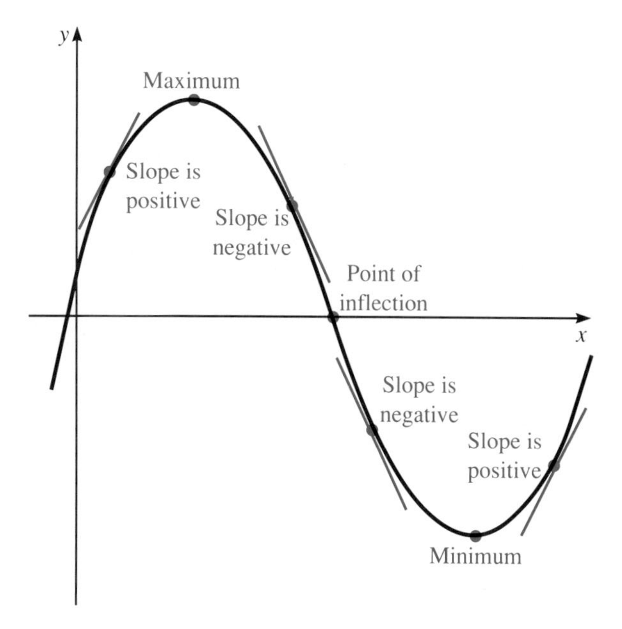

The value(s) where the derivative of a function changes signs is called the **critical value(s).** To find the critical value, set the first derivative equal to zero and solve the resulting equation as follows:

$$V = 2t^3 - 17t^2 + 20t$$

$$\frac{dV}{dt} = 6t^2 - 34t + 20$$

Setting the derivative equal to zero and solving for t gives

$$6t^2 - 34t + 20 = 0$$

$$3t^2 - 17t + 10 = 0$$

$$(3t - 2)(t - 5) = 0$$

$$t = \frac{2}{3} \quad \text{and} \quad t = 5$$

The critical values are $t = 2/3$ s and $t = 5$ s. To determine the maximum voltage, use the first derivative test. Choose a number slightly less than the critical value and a second number slightly greater. Then evaluate the first derivative at each of these values. If the sign of the derivative changes from positive to negative, the critical value is a local maximum. If the sign of the derivative changes from negative to positive, the critical value is a local minimum. For $t = 2/3$, choose $t = 0$ and $t = 1$ to determine the sign of the first derivative.

$$\frac{dV}{dt} = 6(0)^2 - 34(0) + 20 = 20 \Rightarrow \text{positive}$$

$$\frac{dV}{dt} = 6(1)^2 - 34(1) + 20 = -8 \Rightarrow \text{negative}$$

For $t = 5$, choose $t = 4.5$ and $t = 5.5$.

$$\frac{dV}{dt} = 6(4.5)^2 - 34(4.5) + 20 = -11.5 \Rightarrow \text{negative}$$

$$\frac{dV}{dt} = 6(5.5)^2 - 34(5.5) + 20 = 14.5 \Rightarrow \text{positive}$$

Since the first derivative changes from positive to negative, $t = 2/3$ is a local maximum. Calculating the maximum voltage gives the following result:

$$V\left(\frac{2}{3}\right) = 2\left(\frac{2}{3}\right)^3 - 17\left(\frac{2}{3}\right)^2 + 20\left(\frac{2}{3}\right)$$

$$\approx 6.4 \text{ mV}$$

The maximum voltage is 6.4 mV. ■

Another way to determine whether a value is a maximum or a minimum is called the *second derivative test*. The *point of inflection* is the point where the concavity of the function changes, as shown in Figure 18–14. To find the point of inflection, set the second derivative equal to zero and solve. The second derivative test states the following:

- ☐ If $f''(x) > 0$ when $f'(x) = 0$, then $f(x)$ is a minimum value.
- ☐ If $f''(x) < 0$ when $f'(x) = 0$, then $f(x)$ is a maximum value.
- ☐ If $f''(x) = 0$ when $f'(x) = 0$, the second derivative test fails, and the first derivative test must be used.

If we apply the second derivative test at $t = 2/3$ in Example 9, we find

$$\frac{d^2V}{dt^2} = 12t - 34 = 12\left(\frac{2}{3}\right) - 34 = -26$$

Since the second derivative is negative, $t = 2/3$ is a maximum.

Both the first and second derivative tests allow you to determine whether a critical value is a local maximum or a local minimum. Use the second derivative test if the second derivative is easy to find; otherwise, use the first derivative test.

In most technical problems, the function whose maximum or minimum is to be found is not explicitly given but must be determined from the problem statement, as in the following example.

EXAMPLE 11 An industrial designer wants to construct a cylindrical oil storage tank that will have a volume of 8,000 ft^3. In order to minimize the cost of producing the tank, he wants to use as little material as possible in its construction. Find the dimensions of the storage tank having the minimum surface area for the specified volume. Note that we are neglecting the waste of materials in cutting these figures.

Solution We must begin solving this problem by finding an equation relating the variables. The volume V of the cylinder is given by

$$\begin{aligned} V &= \pi r^2 h \\ 8{,}000 &= \pi r^2 h \end{aligned}$$

and the surface area A is given by

$$A = 2\pi r^2 + 2\pi rh$$

Since we want to minimize the surface area, we must determine an equation containing A and either r or h. Let us solve for h because solving for r would result in an expression involving a radical. We solve the volume equation for h and substitute it into the surface area equation.

$$\begin{aligned} 8{,}000 &= \pi r^2 h \\ \frac{8{,}000}{\pi r^2} &= h \end{aligned}$$

$$\begin{aligned} A &= 2\pi r^2 + 2\pi r\left(\frac{8{,}000}{\pi r^2}\right) \\ A &= 2\pi r^2 + \frac{16{,}000}{r} \\ A &= 2\pi r^2 + 16{,}000 r^{-1} \end{aligned}$$

Next, we find the critical values by taking the first derivative, setting it equal to zero, and solving for r.

$$\begin{aligned} \frac{dA}{dr} &= 4\pi r - 16{,}000 r^{-2} \\ 4\pi r - 16{,}000 r^{-2} &= 0 \\ 4\pi r^3 - 16{,}000 &= 0 \qquad \text{multiply by } r^2 \\ r^3 &= \frac{16{,}000}{4\pi} \\ r &\approx 10.8 \text{ ft} \end{aligned}$$

To check whether this is a minimum, we use the first derivative test and substitute values of 10 and 11 into the first derivative as follows:

$$\frac{dA}{dr} = 4\pi r - 16{,}000r^{-2}$$

$$\frac{dA}{dr} = 4\pi(10) - 16{,}000(10)^{-2} \approx -34.3$$

$$\frac{dA}{dr} = 4\pi(11) - 16{,}000(11)^{-2} \approx 6.0$$

When $r = 10.8$, the first derivative changes sign from negative to positive; thus, $r = 10.8$ results in a minimum area. To find the height, substitute this value for r.

$$h = \frac{8{,}000}{\pi r^2} = \frac{8{,}000}{\pi(10.8)^2} \approx 21.8 \text{ ft}$$

The minimum area of 2,212.18 ft^2 results from a radius of 10.8 ft and a height of 21.8 ft. ■

18–3 EXERCISES

Find the derivative of each function.

1. $y = -8$

2. $y = 4$

3. $y = x^8$

4. $y = x^3$

5. $y = -6x^{-2}$

6. $y = 5x^{-3}$

7. $y = 3x^4$

8. $y = x^3 - x$

9. $y = 5x^2 - 9x$

10. $y = 3x^4 + 6x$

11. $y = 2x^3 - 7x + 15$

12. $y = x^4 - 6x^2 - 8$

13. $y = \frac{1}{3}x^2$

14. $y = \frac{x^3}{4}$

15. $y = 3x^4 + x^2 + 8x$

16. $y = x^9 - 7x^3 + 10x$

17. $y = \frac{1}{2}x^4 + \frac{2}{3}x^3 + \frac{1}{6}$

18. $y = \frac{1}{3}x^6 + \frac{1}{4}x^8 + \frac{1}{2}$

19. $y = 7x^{-3} - 6x$

20. $y = -2x + 8x^{-4} + 7$

21. $y = x^5 - 4x^3 + 15$

Find the slope of the tangent line to the curve at the given value of the variable.

22. $y = 6x^2 - 8x + 6$ at $x = -1$

23. $y = x^3 - 2x - 5$ at $x = 2$

24. $s = 18t - 16t^2$ at $t = 2$

25. $a = 2t^2 + 8t - 10$ at $t = 1$

26. $v = t^2 - 2t + 6$ at $t = 3$

27. $s = 50t - 5t^2$ at $t = 4$

28. The linear displacement of a missile as a function of time t is given by $s = 16t^2 - 7t + 8$. Find the expression for the velocity of the missile, and show that the acceleration of the missile is constant.

29. A cannon is fired vertically upward. The displacement of the cannonball as a function of time is given by $s = 150 + 480t - 16t^2$ ft. What is the maximum altitude of the cannonball? (*Hint:* The velocity of the ball is zero when it reaches its maximum height.)

30. The displacement for a charged particle in an electrical field as a function of time is given by $s = 0.63t^{1.2} + 2.8t$ cm. Find the velocity and acceleration of the particle when $t = 1.75$ s.

31. Find the current in a resistor when $t = 0.3$ s if the amount of charge through it is given by the expression $q = 6t^{2/3}$ C. (*Remember:* $i = dq/dt$.)

32. A capacitor of 3 farads has a voltage given by $V = 30 + t^2$ V applied to its terminal. Find an expression for the current in the capacitor. (*Note:* $i = C dv/dt$.)

33. The charge transferred through a coil with an inductance of 1.5 H is given by $q = 1.3t^{0.2} + 0.75$ C. What is the induced voltage when $t = 0.1$ s.? (*Note:* $i = dq/dt$ and $V = L\, di/dt$.)

34. The strength of a rectangular beam varies directly with the width and square of the height. What is the strongest beam that can be cut from a log 10 inches in diameter?

35. Find the dimensions of an open (having no top) cylindrical tank having a minimum surface area and a volume of 75 ft^3.

36. An open box with a square base is to be constructed from 432 in^2 of cardboard. What is the maximum volume of this box?

37. A rectangular box is to be made from a 6-in.-by-9-in. piece of cardboard by cutting a square from each corner, folding up the resulting tabs, and taping the corners. Find the dimensions of the box that yield a maximum volume.

38. Find the third derivative of the function $y = 2x^4 + 7x^3 + 8x - 5$.

39. Find d^2y/dx^2 for the function $y = 5x^3 - 9x^2 + 4x + 2$.

18–4

THE DERIVATIVES OF PRODUCTS, QUOTIENTS, AND POWERS OF FUNCTIONS

The formulas developed in Section 18–3 allow us to find the derivative of polynomial functions. In this section we will develop formulas for finding the derivative of functions that are products, quotients, or powers of polynomial functions. Let us begin the discussion with the derivative of the product of functions.

Product of Functions

To find the derivative of a function such as

$$y = (3x^3 + 8x - 7)(5 + 6x^5 - 4x^4)$$

we could multiply the factors and apply the sum and difference rules to the result. However, it is easier to differentiate the product as given. Now we will derive a formula for such functions.

Let u and v both represent functions of x, and let $y = u \cdot v$. Applying the delta process to this function gives

$$y + \Delta y = (u + \Delta u)(v + \Delta v)$$

$$\Delta y = (u + \Delta u)(v + \Delta v) - u \cdot v$$

$$\Delta y = u\,\Delta v + v\,\Delta u + \Delta u\,\Delta v$$

$$\frac{\Delta y}{\Delta x} = u\frac{\Delta v}{\Delta x} + v\frac{\Delta u}{\Delta x} + \Delta u\frac{\Delta v}{\Delta x}$$

$$\lim_{\Delta x \to 0}\frac{\Delta y}{\Delta x} = u\lim_{\Delta x \to 0}\frac{\Delta v}{\Delta x} + v\lim_{\Delta x \to 0}\frac{\Delta u}{\Delta x} + \lim_{\Delta x \to 0}\left(\Delta u\frac{\Delta v}{\Delta x}\right)$$

As $\Delta x \to 0$, $\Delta v/\Delta x \to dv/dx$, but $\Delta u \to 0$.

$$\frac{dy}{dx} = u\frac{dv}{dx} + v\frac{du}{dx}$$

The Product Rule

If $y = u \cdot v$, where u and v are both functions of x, then

$$\frac{dy}{dx} = \frac{d(u \cdot v)}{dx} = u\frac{dv}{dx} + v\frac{du}{dx}$$

The derivative of the product of two functions is the first function times the derivative of the second plus the second function times the derivative of the first.

EXAMPLE 1 Find the derivative of each of the following functions:

(a) $y = (2x^2 + 3)(x^3 + 7x)$ (b) $y = (3x^3 + 8x - 7)(5 + 6x^5 - 4x^4)$

Solution

(a) First, let

$$u = 2x^2 + 3 \quad \text{and} \quad v = x^3 + 7x$$

Then

$$\frac{du}{dx} = 4x \quad \text{and} \quad \frac{dv}{dx} = 3x^2 + 7$$

Applying the product rule gives

$$\begin{aligned}\frac{dy}{dx} &= \underbrace{u}\;\underbrace{\frac{dv}{dx}} + \underbrace{v}\;\underbrace{\frac{du}{dx}}\\ &= (2x^2 + 3)(3x^2 + 7) + (x^3 + 7x)(4x)\\ &= 6x^4 + 23x^2 + 21 + 4x^4 + 28x^2\\ \frac{dy}{dx} &= 10x^4 + 51x^2 + 21\end{aligned}$$

We can check this derivative by multiplying the binomials first and then taking the derivative as follows:

$$\begin{aligned}y &= (2x^2 + 3)(x^3 + 7x)\\ &= 2x^5 + 17x^3 + 21x\\ \frac{dy}{dx} &= 10x^4 + 51x^2 + 21\end{aligned}$$

(b) If $u = 3x^3 + 8x - 7$ and $v = 5 + 6x^5 - 4x^4$, then

$$\frac{du}{dx} = 9x^2 + 8 \quad \text{and} \quad \frac{dv}{dx} = 30x^4 - 16x^3$$

Applying the product rule gives the following result for the derivative:

$$\begin{aligned}\frac{dy}{dx} &= \overbrace{u}^{}\overbrace{\frac{dv}{dx}}^{} + \overbrace{v}^{}\overbrace{\frac{du}{dx}}^{} \\ &= (3x^3 + 8x - 7)(30x^4 - 16x^3) + (5 + 6x^5 - 4x^4)(9x^2 + 8) \\ &= 90x^7 + 240x^5 - 210x^4 - 48x^6 - 128x^4 + 112x^3 + 45x^2 \\ &\quad + 54x^7 - 36x^6 + 40 + 48x^5 - 32x^4 \\ \frac{dy}{dx} &= 144x^7 - 84x^6 + 288x^5 - 370x^4 + 112x^3 + 45x^2 + 40\end{aligned}$$

Checking this derivative by multiplying the trinomials first and then taking the derivative is more difficult than in Part (a), thus justifying the use of the product rule. ■

EXAMPLE 2 Use the product rule to find dy/dx for the following function:

$$y = (x^5 - 6x^4)(8 + 9x^3 - 8x^4)$$

Solution First, let

$$u = x^5 - 6x^4 \quad \text{and} \quad v = 8 + 9x^3 - 8x^4$$

Applying the rules for the derivative gives

$$\frac{du}{dx} = 5x^4 - 24x^3 \quad \text{and} \quad \frac{dv}{dx} = 27x^2 - 32x^3$$

Applying the product rule gives

$$\begin{aligned}\frac{dy}{dx} &= u\frac{dv}{dx} + v\frac{du}{dx} \\ &= (x^5 - 6x^4)(27x^2 - 32x^3) + (8 + 9x^3 - 8x^4)(5x^4 - 24x^3) \\ &= 27x^7 - 32x^8 - 162x^6 + 192x^7 + 40x^4 - 192x^3 + 45x^7 \\ &\quad - 216x^6 - 40x^8 + 192x^7 \\ \frac{dy}{dx} &= -72x^8 + 456x^7 - 378x^6 + 40x^4 - 192x^3\end{aligned}$$

■

Derivative of a Quotient

In the previous two examples, we could have multiplied the functions and then taken its derivative. However, when the function is expressed as a quotient of polynomials, there is usually no simple way to find the derivative. Next, we will use the delta process to derive a formula to find the derivative of the quotient of two functions.

Let $y = u/v$ where u and v both represent differentiable functions of x. Then applying the delta process gives

$$y + \Delta y = \frac{u + \Delta u}{v + \Delta v}$$

$$\Delta y = \frac{u + \Delta u}{v + \Delta v} - \frac{u}{v}$$

$$\Delta y = \frac{uv + v\,\Delta u - uv - u\,\Delta v}{v(v + \Delta v)}$$

$$\frac{\Delta y}{\Delta x} = \frac{v\dfrac{\Delta u}{\Delta x} - u\dfrac{\Delta v}{\Delta x}}{v^2 + v\,\Delta v}$$

$$\lim_{\Delta x \to 0} \frac{\Delta y}{\Delta x} = \frac{v \displaystyle\lim_{\Delta x \to 0} \frac{\Delta u}{\Delta x} - u \lim_{\Delta x \to 0} \frac{\Delta v}{\Delta x}}{v^2 + \displaystyle\lim_{\Delta x \to 0} v\,\Delta v}$$

$$\frac{dy}{dx} = \frac{v\dfrac{du}{dx} - u\dfrac{dv}{dx}}{v^2}$$

The Quotient Rule

If $y = u/v$, where u and v represent functions of x, then

$$\frac{dy}{dx} = \frac{v\dfrac{du}{dx} - u\dfrac{dv}{dx}}{v^2}$$

The derivative of the quotient of two functions is the denominator times the derivative of the numerator minus the numerator times the derivative of the denominator, all divided by the square of the denominator.

EXAMPLE 3 Find the derivative of the following function:

$$y = \frac{6 + 3x^2}{2x^2 + 5}$$

Solution First, let

$$u = 6 + 3x^2 \quad \text{and} \quad v = 2x^2 + 5$$

Applying the rules for the derivatives gives

$$\frac{du}{dx} = 6x \quad \text{and} \quad \frac{dv}{dx} = 4x$$

Applying the quotient rule gives

$$\frac{dy}{dx} = \frac{v\dfrac{du}{dx} - u\dfrac{dv}{dx}}{v^2}$$

$$= \frac{\overbrace{(2x^2 + 5)}^{v}\overbrace{(6x)}^{du/dx} - \overbrace{(6 + 3x^2)}^{u}\overbrace{(4x)}^{dv/dx}}{\underbrace{(2x^2 + 5)^2}_{v^2}}$$

$$= \frac{12x^3 + 30x - 24x - 12x^3}{(2x^2 + 5)^2}$$

$$\frac{dy}{dx} = \frac{6x}{(2x^2 + 5)^2}$$

■

EXAMPLE 4 Find the derivative of the following function:

$$y = \frac{(2x - 5)(x^2 + 3)}{(x - 1)}$$

Solution Notice that this function is a quotient where the numerator consists of a product.

$$\frac{dy}{dx} = \frac{\overbrace{(x - 1)}^{v}\overbrace{[(2x - 5)(2x) + (x^2 + 3)(2)]}^{du/dx} - \overbrace{(2x - 5)(x^2 + 3)}^{u}\overbrace{(1)}^{dv/dx}}{\underbrace{(x - 1)^2}_{v^2}}$$

$$= \frac{(x - 1)(6x^2 - 10x + 6) - (2x^3 - 5x^2 + 6x - 15)}{(x - 1)^2}$$

$$= \frac{6x^3 - 16x^2 + 16x - 6 - 2x^3 + 5x^2 - 6x + 15}{(x - 1)^2}$$

$$\frac{dy}{dx} = \frac{4x^3 - 11x^2 + 10x + 9}{(x - 1)^2}$$

■

Powers of a Function: The Chain Rule

If $y = v^2$ where $v = 3x^3 + 6$, then y is called a **composite function** because y is a function of x. Next, we will derive a formula for finding the derivative of composite functions.

Let $y = f(u)$ where $u = g(x)$. If Δx represents the change in x, then Δy and Δu represent the corresponding changes in y and u, respectively. Thus,

$$\frac{\Delta y}{\Delta x} = \frac{\Delta y}{\Delta u} \cdot \frac{\Delta u}{\Delta x}$$

As $\Delta x \to 0$, Δu and Δy both approach zero since each is a function of x. Therefore, we can write

$$\lim_{\Delta x \to 0} \frac{\Delta y}{\Delta x} = \lim_{\Delta x \to 0} \left(\frac{\Delta y}{\Delta u} \cdot \frac{\Delta u}{\Delta x} \right)$$

Applying the rule discussed earlier in this chapter that the limit of a product is the product of the limits gives the following result:

$$\lim_{\Delta x \to 0} \frac{\Delta y}{\Delta x} = \lim_{\Delta x \to 0} \frac{\Delta y}{\Delta u} \cdot \lim_{\Delta x \to 0} \frac{\Delta u}{\Delta x}$$

$$\frac{dy}{dx} = \frac{dy}{du} \cdot \frac{du}{dx}$$

The Chain Rule

If $y = f(u)$ and $u = g(x)$, then

$$\frac{dy}{dx} = \frac{dy}{du} \cdot \frac{du}{dx}$$

EXAMPLE 5 Find the derivative of the following functions:

(a) $y = (6x^3 - 7x)^4$ (b) $y = \sqrt[3]{7x^4 + x^3}$

Solution

(a) The function $y = (6x^3 - 7x)^4$ can be written as $y = u^4$ where $u = 6x^3 - 7x$. Then

power rule ↓

$$\frac{dy}{du} = 4u^3 \quad \text{and} \quad \frac{du}{dx} = 18x^2 - 7$$

Applying the chain rule gives the derivative of this function as follows:

$$\frac{dy}{dx} = \frac{dy}{du} \cdot \frac{du}{dx}$$
$$= 4u^3(18x^2 - 7)$$
$$\frac{dy}{dx} = \underbrace{4(6x^3 - 7x)^3}_{\text{substitute for } u}(18x^2 - 7)$$

(b) First, we use fractional exponents to write this radical.

$$y = (7x^4 + x^3)^{1/3}$$

Second, the function can be written as $y = u^{1/3}$ where $u = 7x^4 + x^3$. Then

$$\frac{dy}{du} = \frac{1}{3}u^{-2/3} \quad \text{and} \quad \frac{du}{dx} = 28x^3 + 3x^2$$

Applying the chain rule gives the following result for the derivative:

$$\frac{dy}{dx} = \frac{dy}{du} \cdot \frac{du}{dx}$$
$$= \frac{1}{3}u^{-2/3} \cdot (28x^3 + 3x^2)$$
$$\frac{dy}{dx} = \frac{1}{3}(7x^4 + x^3)^{-2/3}(28x^3 + 3x^2)$$

■

Extended Power Rule

If the function y has the form $y = [u(x)]^n$, then the chain rule gives

$$\frac{dy}{dx} = \frac{d(u)^n}{du} \cdot \frac{du}{dx}$$

Applying the power rule derived in the previous section gives

$$\frac{dy}{dx} = nu^{n-1}\frac{du}{dx}$$

The Extended Power Rule

If $y = [u(x)]^n$, then

$$\frac{dy}{dx} = nu^{n-1}\frac{du}{dx}$$

EXAMPLE 6 Find the derivative of each of the following:

(a) $y = (3x^3 - 6x + 8)^4$ (b) $y = (2x + 6)(x^6 - 2x + 7)^3$

Solution

(a) Write the function $y = (3x^3 - 6x + 8)^4$ as $y = u^4$ where $u = 3x^3 - 6x + 8$ and $du/dx = 9x^2 - 6$. Applying the extended power rule gives

$$\begin{aligned}\frac{dy}{dx} &= nu^{n-1}\left(\frac{du}{dx}\right)\\ &= 4u^{4-1}(9x^2 - 6)\\ &= 4(3x^3 - 6x + 8)^3(9x^2 - 6)\\ \frac{dy}{dx} &= (36x^2 - 24)(3x^3 - 6x + 8)^3\end{aligned}$$

(b) To find the derivative of $y = (2x + 6)(x^6 - 2x + 7)^3$, we must apply the product rule.

$$\frac{dy}{dx} = f(u)\frac{dg}{dx} + g(v)\frac{df}{dx}$$

where $f(u) = 2x + 6$ and $g(v) = (x^6 - 2x + 7)^3$. To find dg/dx, we use the extended power rule. Since $g(v) = v^3$ where $v = x^6 - 2x + 7$, then $dv/dx = 6x^5 - 2$ and

$$\frac{dg}{dx} = 3v^2\frac{dv}{dx} = 3(x^6 - 2x + 7)^2(6x^5 - 2)$$

Substituting into the product rule gives the following results:

$$\begin{aligned}\frac{dy}{dx} &= f(u) \cdot \frac{dg}{dx} + g(v) \cdot \frac{df}{dx}\\ &= (2x + 6) \cdot [3(x^6 - 2x + 7)^2(6x^5 - 2)] + (x^6 - 2x + 7)^3 \cdot 2\\ \frac{dy}{dx} &= (x^6 - 2x + 7)^2(38x^6 + 108x^5 - 16x - 22)\end{aligned}$$

■

CAUTION ✦ The most frequent error in applying the extended power rule is neglecting to include du/dx in the derivative.

Applications

EXAMPLE 7 The curve of a racing track is banked according to the equation $y = 9/x$ where $x =$ the horizontal distance from the outer edge $(1 < x \leq 10)$

and y = the vertical height of the track (see Figure 18–15). Find the equation of the line normal to the car 5 ft horizontally from the outer edge of the track.

FIGURE 18–15

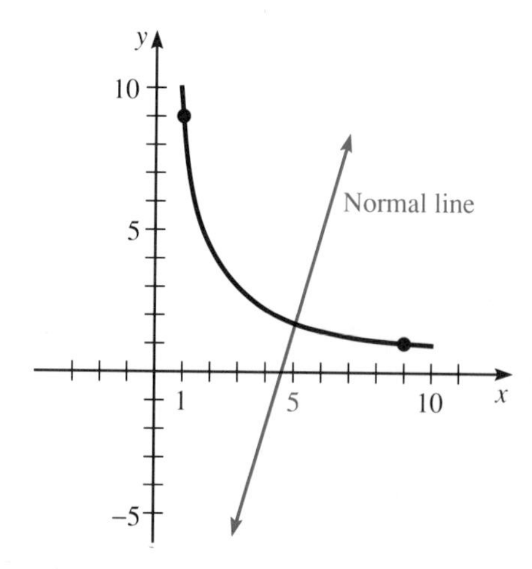

Solution To find the equation of the normal (perpendicular to the tangent line at any point on the curve) line, we start by determining its slope. The slope of the equation of the track is given by

$$\frac{dy}{dx} = \frac{x(0) - 9}{x^2}$$

$$\frac{dy}{dx} = \frac{-9}{x^2}$$

The slope of the tangent line at $x = 5$ is

$$\left.\frac{dy}{dx}\right|_{x=5} = -\frac{9}{25}$$

and the slope of the normal line is the negative reciprocal of the tangent line. Thus, the slope of the normal line is 25/9. At $x = 5$, $y = 9/5$. Substituting the slope and the point into the point–slope formula gives

$$y - \frac{9}{5} = \frac{25}{9}(x - 5)$$

$$45y - 81 = 125x - 625$$

$$125x - 45y - 544 = 0$$

The equation of the normal line is $125x - 45y - 544 = 0$. ■

18–4 EXERCISES

Find the derivative of the following functions.

1. $y = 3(x^4 - 3x^3)$
2. $y = -4(6x + x^3)$
3. $y = x^3(7x^2 - 5)$
4. $y = x^2(8x^3 + x)$
5. $y = (1 + x^2)(3x + 7)$
6. $y = (3 - x^3)(5x - 1)$
7. $y = (x^2 - 4)(7x^2 + 6x)$
8. $y = (2x^2 + x)(2x + 1)$
9. $y = (x + 3)(x^2 + 7x - 9)$
10. $y = (x - 1)(2x^2 - 3x + 1)$
11. $y = (x^2 - 3x)(7x^2 + x)$
12. $y = (9x^3 - x)(3x^2 + 4x)$
13. $y = (x^2 + x - 5)(x^2 - 3x + 7)$
14. $y = (x^2 + x - 2)(x^2 + 5x - 8)$
15. $y = (9x^2 - 7x)(2x^3 + 8x^2 - x)$
16. $y = (x^3 + x^2)(4x^2 + 7x - 5)$
17. $y = \dfrac{2x}{x - 3}$
18. $y = \dfrac{1}{x^2}$
19. $y = \dfrac{3 + x^2}{7x}$
20. $y = \dfrac{5 + 2x^3}{x}$
21. $y = \dfrac{2x + 9}{3x - 5}$
22. $y = \dfrac{4x - 7}{2x + 1}$
23. $y = \dfrac{x^2}{3x^2 + 5x}$
24. $y = \dfrac{x^3}{7x + x^2}$
25. $y = \dfrac{6x}{2x^2 + 3x - 7}$
26. $y = \dfrac{9x}{x^2 - 5x + 8}$
27. $y = \dfrac{4x^2 - 3x}{7x^3 + x^2}$
28. $y = \dfrac{6 + 7x^3}{x^3 - 3x}$
29. $y = \dfrac{(x + 1)(x^2 - 3x)}{2x + 5}$
30. $y = \dfrac{(2x - 3)(x + 6)}{x - 1}$
31. $y = \dfrac{(x^3 - x^2)(4x^2 + x)}{x^3}$
32. $y = \dfrac{(x^2 + x)(3x^3 - x^2)}{x^4}$
33. $y = (3 + 7x)^2$
34. $y = (4x - 7)^2$
35. $y = (4x^3 + x^2)^3$
36. $y = (8x^3 + 3x)^4$
37. $y = \sqrt[3]{6x^3 - 5x^2}$ [*Hint:* $\sqrt[n]{a} = (a)^{1/n}$.]
38. $y = \sqrt[4]{x^3 - 7x^2}$
39. $y = x^3(2x^3 - 4x)^4$
40. $y = 2x^3(x^4 + 6x)^3$
41. $y = \dfrac{(3x^2 + 7)^3}{4x^3 - x}$
42. $y = \dfrac{(2x - 5)^2}{8x^3 + x^2}$
43. $y = \left(\dfrac{3x}{1 - x}\right)^2$
44. $y = \left(\dfrac{x + 4}{x - 3}\right)^3$
45. The charge transferred through a certain inductance coil is given by $q = 0.45(3t + 1)^{1.4}$ C. Find an expression for the current.
46. The displacement s (in meters) of a car as a function of time (in seconds) is given by $s = \sqrt{4t^2 + 3t}$. Find the velocity of the car after 6 s.
47. The energy output E of an electrical heater as a function of time t is given by $E = 8(1 + t^2)^3$. Find the power (in kilowatts) generated by the heater at $t = 2.5$ s. (*Note*: $P = dE/dt$.)
48. The kinetic energy of an object is given by $E = \frac{1}{2}mv^2$. Find dE/dt when $t = 3.5$ s if $m = 20$ kg and $v = 1 - 3t^2 + t^3$.
49. The resistance R, frequency f, and voltage V are held constant, and the function is given by

$$I = \frac{V}{\sqrt{R^2 + (2\pi fL)^2}}$$

Find an expression for the rate of change of current I with respect to inductance L.

50. The electrical power P (in watts) produced by a certain source is given by

$$P = \frac{121r}{(r + 0.5)^2}$$

where r = resistance. Find the value of r that results in maximum power.

51. If two identical electrolytic cells with internal resistance r and electromotive force V are connected in parallel with a resistance R, the power dissipated is

$$P = \frac{V^2(r + 2R)^2}{Rr^2}$$

For what value of R will the power dissipated be a maximum? (Assume that V and r are constants.)

52. Find the velocity at $t = 4$ s of an object whose displacement is given by

$$S = \frac{t}{\sqrt{t^2 + 9}} \text{ ft}$$

53. Find the equation of the line normal to the curve $y = \sqrt{1 + 3x^2}$ at $x = 2$.

54. Find the equation of the line normal to the curve $y = (1 + x^2)^3$ at $x = 1$.

55. Find the equation of the line tangent to the curve $y = (4x - x^3)(7x + 1)^4$ at $x = 1$.

56. Find the equation of the line tangent to the curve $y = (8x^3 - x^2 + 5)^4$ at $x = -1$.

18–5

IMPLICIT DIFFERENTIATION

In the previous sections of this chapter, we have differentiated functions of the form $y = f(x)$ where the functional equation is solved for y. Thus, the variable y appears only on one side of the equation. A function of this type is called an **explicit function.** However, in this section we will discuss how to differentiate **implicit equations,** that is, equations in which the function is not explicitly solved for y but is expressed as a function of x and y. For example, both of the following equations are explicit functions:

$$y = 3x^2 - 5x + 8 \quad \text{and} \quad y = \sqrt{2x - 6}$$

On the other hand, the following equations contain y as an implicit function of x:

$$xy = 7 \quad \text{and} \quad x^2y + 3y^2 = 9$$

An equation in x and y may define more than one explicit function. For example, $y^2 + x = 2$ can be solved explicitly for y as follows:

$$y^2 + x = 2$$
$$y^2 = 2 - x$$
$$y = \pm\sqrt{2 - x}$$

Although this equation does not meet the definition of a function given in Chapter 2, we can define two functions $y = \sqrt{2 - x}$ and $y = -\sqrt{2 - x}$ from this equation. As shown here, we can solve some implicit functions for y and differentiate them by using the techniques given in the previous sections of this chapter. However, many implicit functions are difficult or impossible to solve for y. In these cases we use **implicit differentiation** to find dy/dx. We do so by differentiating each term with respect to x, regarding y as a function of x. Then we solve the resulting equation for dy/dx.

Implicit Differentiation

1. Differentiate each term of the equation with respect to x (the independent variable). Treat y as a function of x.
2. Solve the resulting equation for dy/dx.

EXAMPLE 1 Using implicit differentiation, find dy/dx for $xy + 3x^3 = 5$.

Solution First, we take the derivative with respect to x of each term in the equation.

$$xy + 3x^3 = 5$$
$$\frac{d(xy)}{dx} + \frac{d(3x^3)}{dx} = \frac{d(5)}{dx}$$

Notice that $d(xy)/dx$ is the product of two functions and will require us to use the product rule. To find $d(3x^3)/dx$, we use the constant multiplier and power rules; and to find $d(5)/dx$, we use the constant rule.

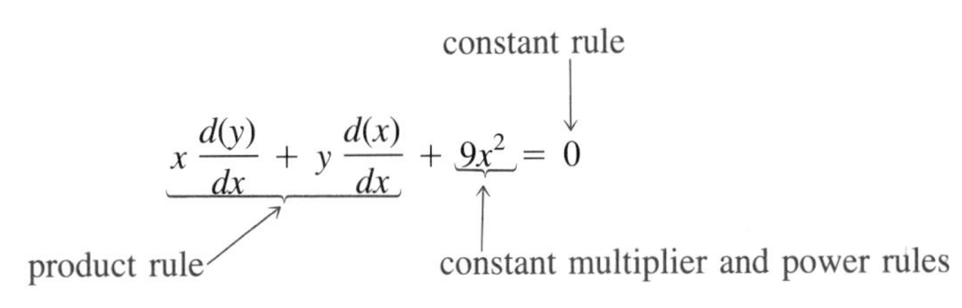

$$y + x\frac{dy}{dx} + 9x^2 = 0$$

$$\frac{dx}{dx} = 1$$

Second, we solve the resulting equation for dy/dx.

$$x\frac{dy}{dx} = -9x^2 - y$$
$$\frac{dy}{dx} = \frac{-9x^2 - y}{x}$$

■

LEARNING HINT ✦ When you are using implicit differentiation, any term containing a y will have a dy/dx term in the result. On the other hand, any term containing only a constant or x will not have dy/dx in its derivative.

EXAMPLE 2 Find dy/dx for $2x^3 - 3xy + y^2 = 5x - 8$, and evaluate the result at the point $(1, -2)$.

Solution First, we find the derivative with respect to x for each term in the equation.

$$\frac{d(2x^3)}{dx} - \frac{d(3xy)}{dx} + \frac{d(y^2)}{dx} = \frac{d(5x)}{dx} - \frac{d(8)}{dx}$$

$$6x^2 - 3\left(x\frac{dy}{dx} + y\right) + 2y\frac{dy}{dx} = 5 - 0$$

$$6x^2 - 3x\frac{dy}{dx} - 3y + 2y\frac{dy}{dx} = 5$$

Second, we solve the resulting equation for dy/dx.

$$-3x\frac{dy}{dx} + 2y\frac{dy}{dx} = 5 + 3y - 6x^2$$

$$(2y - 3x)\frac{dy}{dx} = 5 + 3y - 6x^2$$

$$\frac{dy}{dx} = \frac{5 + 3y - 6x^2}{2y - 3x}$$

Last, we evaluate the derivative at the point $(1, -2)$.

$$\left.\frac{dy}{dx}\right|_{(1,-2)} = \frac{5 + 3(-2) - 6(1)^2}{2(-2) - 3(1)} = \frac{-7}{-7} = 1$$

■

EXAMPLE 3 Find dy/dx for $(4y + 7)^3 + (y - 5)^2 = 7x$.

Solution First, we find dy/dx for each term. The derivative for terms $(4y + x)^3$ and $(y - 5)^2$ will each require the use of the chain rule.

$$\frac{d(4y + 7)^3}{dx} + \frac{d(y - 5)^2}{dx} = \frac{d(7x)}{dx}$$

$$3(4y + 7)^2 \cdot 4\frac{dy}{dx} + 2(y - 5)\left(\frac{dy}{dx}\right) = 7$$

Next, we solve for dy/dx.

$$[12(4y + 7)^2 + 2(y - 5)]\frac{dy}{dx} = 7$$

$$\frac{dy}{dx} = \frac{7}{12(4y + 7)^2 + 2(y - 5)}$$

■

EXAMPLE 4 Find dy/dx for the function $y = \dfrac{3x}{x^2 + y}$.

Solution First, we take the derivative of each term. The term $\dfrac{3x}{x^2 + y}$ requires us to use the quotient rule.

$$\frac{dy}{dx} = \frac{(x^2 + y)(3) - 3x\left(2x + \dfrac{dy}{dx}\right)}{(x^2 + y)^2}$$

Then we solve the resulting equation for dy/dx.

$$(x^2 + y)^2 \frac{dy}{dx} = 3x^2 + 3y - 6x^2 - 3x \frac{dy}{dx}$$

$$(x^2 + y)^2 \frac{dy}{dx} + 3x \frac{dy}{dx} = 3x^2 + 3y - 6x^2$$

$$[(x^2 + y)^2 + 3x] \frac{dy}{dx} = 3x^2 + 3y - 6x^2$$

$$\frac{dy}{dx} = \frac{3x^2 + 3y - 6x^2}{(x^2 + y)^2 + 3x}$$

■

Applications

EXAMPLE 5 Find the equation of the line tangent to the curve of $3y^2 + xy - 4 = 0$ at the point $(-4, 2)$.

Solution First, let us determine the slope of the tangent line by finding dy/dx and evaluating it at $(-4, 2)$. Using implicit differentiation gives

$$6y \frac{dy}{dx} + x \frac{dy}{dx} + y = 0$$

$$6y \frac{dy}{dx} + x \frac{dy}{dx} = -y$$

$$(6y + x) \frac{dy}{dx} = -y$$

$$\frac{dy}{dx} = \frac{-y}{6y + x}$$

$$\left.\frac{dy}{dx}\right|_{(-4,2)} = \frac{-(2)}{6(2) + (-4)} = -\frac{1}{4}$$

The slope of the tangent line is $-1/4$. In addition to the slope, we also know that the tangent line passes through the point $(-4, 2)$. Substituting into point–slope form of the line and simplifying gives the following result:

$$(y - y_1) = m(x - x_1)$$

$$(y - 2) = -\frac{1}{4}[x - (-4)]$$

$$4y - 8 = -x - 4$$

$$x + 4y - 4 = 0$$

■

EXAMPLE 6 The displacement (in meters) and time (in seconds) of a moving particle are given by the relationship $s^3 + t^2 = 28$. Find the velocity of the particle ds/dt when $t = 3$ s.

Solution The velocity of the particle is given by ds/dt, which we will find using implicit differentiation.

$$\frac{d(s^3)}{dt} + \frac{d(t^2)}{dt} = \frac{d(28)}{dt}$$

$$3s^2\frac{ds}{dt} + 2t = 0$$

$$3s^2\frac{ds}{dt} = -2t$$

$$\frac{ds}{dt} = \frac{-2t}{3s^2}$$

From the original equation, when $t = 3$, $s \approx 2.7$. Then the velocity at $t =$ 3 s is

$$\frac{ds}{dt} = \frac{-2(3)}{3(2.7)^2} = -0.27 \text{ m/s}$$

■

EXAMPLE 7 The frequency f of a pendulum of length L is given by

$$f = \frac{1}{2\pi}\cdot\sqrt{\frac{g}{L}}$$

where g = the acceleration due to gravity (a constant). Find an expression for the rate of change of frequency with respect to the pendulum length.

Solution The rate of change of frequency with respect to length is given by df/dL. Differentiating implicitly gives

$$f = \frac{\sqrt{g}}{2\pi}\cdot L^{-1/2} \qquad \text{rearrange equation}$$

$$\frac{df}{dL} = \frac{\sqrt{g}}{2\pi}\cdot\frac{d(L^{-1/2})}{dL} \qquad \text{constant multiplier rule}$$

$$\frac{df}{dL} = \frac{\sqrt{g}}{2\pi}\left(-\frac{1}{2}L^{-3/2}\right) \qquad \text{power rule}$$

$$\frac{df}{dL} = \frac{\sqrt{g}}{2\pi}\left(-\frac{1}{2(\sqrt{L})^3}\right) \qquad \text{convert to radical form}$$

$$\frac{df}{dL} = \frac{-1}{4\pi}\cdot\sqrt{\frac{g}{L^3}} \qquad \text{simplify}$$

$$\frac{df}{dL} = -\frac{1}{4\pi L} \cdot \sqrt{\frac{g}{L}} \qquad \text{simplify radical}$$

■

EXAMPLE 8 The radius of a circular oil spill in the Atlantic Ocean increases at a rate of 1.5 m/s. How fast is the area of the spill increasing when the radius is 240 m?

Solution The area of the oil spill is given by the equation $A = \pi r^2$. The area A and the radius r both vary with time. We must determine dA/dt when $r = 240$ m and $dr/dt = 1.5$ m/s. Differentiating and substituting known values gives

$$\frac{dA}{dt} = \pi\left(2r\frac{dr}{dt}\right)$$

$$= \pi(2)(240 \text{ m})(1.5 \text{ m/s})$$

$$\frac{dA}{dt} \approx 2300 \text{ m}^2/\text{s}$$

The area of the spill is increasing at about 2300 m^2/s. ■

In Section 18–2, we discussed the derivative as a method of describing the rate of change of one variable with respect to another. In the remainder of this section we will discuss the rate of change of several variables with respect to time, each of which is a function of time. Problems of this type are called **related rate.** The procedure for solving these problems is given in the following box.

Related Rate Problems

1. Determine an equation relating the variables, if necessary.
2. Determine the known and unknown quantities.
3. Differentiate both sides of the equation with respect to time.
4. Solve the resulting equation for the unknown and substitute in the known quantities.

EXAMPLE 9 The resistance R in ohms of a strip of metal wire as a function of temperature T is given by $R = 21.7 + 1.38T + 0.63T^2$. Find the rate at which the resistance is changing if $T = 31°\text{C}$ and the temperature is changing at 0.35°C/s.

Solution The resistance and temperature both vary with time. The change in resistance is given by dR/dt, and the change in temperature is dT/dt. We are asked to find dR/dt when $T = 31°\text{C}$ and $dT/dt = 0.35°\text{C/s}$. Taking the time derivative of the equation relating resistance and temperature gives

$$\frac{dR}{dt} = 0 + 1.38\frac{dT}{dt} + 1.26T\frac{dT}{dt}$$

Substituting the known values gives

$$\frac{dR}{dt} = 1.38(0.35) + 1.26(31)(0.35)$$

$$\frac{dR}{dt} \approx 14\ \Omega/\text{s}$$

The resistance is changing at a rate of approximately 14 Ω/s. ■

EXAMPLE 10 Two planes leave an airport at 3 P.M. Plane A travels west at 350 mi/h, and plane B travels north at 425 mi/h. At what rate is the distance between the two planes changing at 5 P.M?

Solution Figure 18–16 shows a drawing of this problem. Let x represent the distance traveled by plane A, y represent the distance traveled by plane B, and z represent the distance between the planes.

FIGURE 18–16

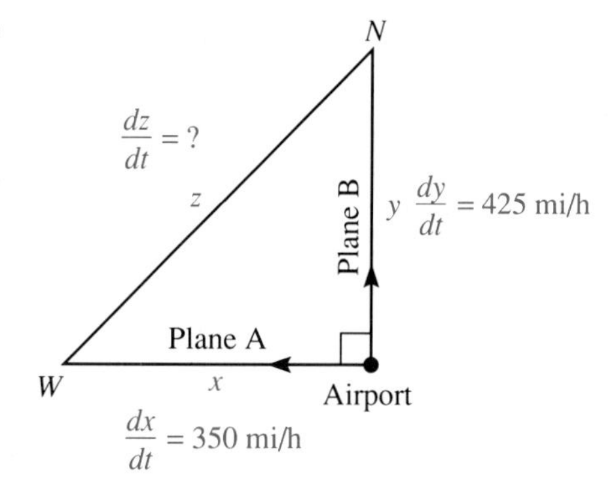

First, we must find an equation to represent the relationship between these variables. From the Pythagorean theorem,

$$z^2 = x^2 + y^2$$

We are asked to find dz/dt. Differentiating both sides of this equation with respect to t gives

$$2z\frac{dz}{dt} = 2x\frac{dx}{dt} + 2y\frac{dy}{dt}$$

$$z\frac{dz}{dt} = x\frac{dx}{dt} + y\frac{dy}{dt}$$

Then we solve this equation for dz/dt.

$$\frac{dz}{dt} = \frac{x\frac{dx}{dt} + y\frac{dy}{dt}}{z}$$

Last, we substitute known values for x, dx/dt, y, dy/dt, and z. Notice that at 5 P.M. the planes have traveled for two hours and $x = 700$ mi and $y = 850$ mi. We also know that $dx/dt = 350$ mi/h and $dy/dt = 425$ mi/h. However, we do not have a value for z, but we can calculate it as follows:

$$z^2 = x^2 + y^2$$
$$z = \sqrt{(700)^2 + (850)^2}$$
$$z \approx 1{,}101 \text{ mi}$$

Therefore,

$$\frac{dz}{dt} = \frac{(700)(350) + (850)(425)}{1{,}101}$$
$$\frac{dz}{dt} \approx 551 \text{ mi/h}$$

■

18–5 EXERCISES

Using implicit differentiation, find dy/dx for each function.

1. $3x + y = 6$
2. $2x - 9y = 15$
3. $2x^3 - y + 8 = 0$
4. $5x^2 - 6xy + y^3 = 0$
5. $x^2 + 7y^2x - 4y = 8 + y^2$
6. $3y^3 = 9x - 5x^3y$
7. $8x^3 + 9y^2x = x^4 - 6y$
8. $(3x^3 + 2y)^4 = 8$
9. $(x + 7y^2 - x^3)^2 = 0$
10. $2x^3 - \frac{x}{y^2} = 2y^3 + 7$
11. $\frac{1}{x} = \frac{1}{y} + x^2$
12. $(4y^2 - 3x)(5x^2 + y) + x^4 = 0$
13. $(2x - y)(3x^2 + y^3) = 0$
14. $y^2 = \frac{4x}{x^2 - y}$
15. $y^3 = \frac{x^2}{3x - y}$
16. $xy = -6$
17. $x^2 + xy + y^3 = 0$
18. $(4x^2 - y)^3 + 8y^2 + 3x = 0$
19. $(3y^2 + 8)^3 + (x^2 - 3y)^2 = 0$
20. $(x - 4)^2 + (3y + 7)^3 = 6$
21. $4(y^2 - 3)^2 - (2x^2 + 7)^3 = 1$

Find dy/dx for each function and evaluate it at the given point.

22. $4x^3y + 7x = 6y$ at $\left(-1, -\frac{7}{10}\right)$
23. $-5xy + y^3 = 3x - 7y$ at $\left(\frac{22}{13}, 2\right)$
24. $y = \frac{x}{y + 1}$ at $(12, 3)$
25. $x^2 = -3y - y^3$ at $(2, -1)$
26. $\frac{1}{x^2} + \frac{1}{y^2} = 10$ at $\left(-1, \frac{1}{3}\right)$
27. $\frac{x}{y} + 3x = 6$ at $(1.5, 1)$

28. The relationship between displacement s and time t of a moving particle is given by $2s^3 + 7t^3 = 48$. Find the particle's velocity when $t = 3$ s.

29. Find the current in a circuit when $t = 3$ s if the charge q as a function of time t is given by $q^3t^2 + t = 9$.

30. Find the equation of the line tangent to the curve of $4 - y^2 = x^2$ at the point (2, 0).

31. Find the equation of the line tangent to the curve of $\frac{x^2}{3} + \frac{y^2}{9} = 1$ at the point $(0, -3)$.

32. A 32-ft ladder leans against a vertical wall. If the top of the ladder slides down the wall at the rate of 3 ft/s, how fast is the bottom of the ladder moving away from the wall when it is 12 ft from the wall?

33. Two cars start from an intersection and travel on roads perpendicular to each other. One car travels north at 45 mi/h, and the other car travels east at 30 mi/h. Find the rate at which the distance between them is changing 2 h after they start.

34. A conical water tank (point down) is being filled with water at the rate of 20 ft^3/s. At what rate is the fill level changing when the water is 10 ft deep if the radius is 15 ft and the height is 30 ft? (*Note:* volume of cone is $\pi r^2 h/3$.)

35. A 12-ft ladder rests against a wall. If the ladder begins to slip so the bottom moves 2.1 ft/s along the floor, at what speed is the top of the ladder moving down the wall when the bottom is 3.5 ft. from the wall?

36. A weather balloon is released at a point 1,500 ft east of a tracking station. If the balloon is rising vertically at a rate of 45 ft/s, how fast is the distance between the station and the balloon changing 1.5 min after the balloon is released?

37. The electrical resistance R of a wire as a function of temperature T is given by $R = 65 - 0.5T + 0.01T^2$ Ω. Find the rate of change of resistance when $T = 215°$ if the temperature is changing at a rate of 10°/min.

CHAPTER SUMMARY

Summary of Terms

acceleration (p. 654)
average velocity (p. 647)
composite function (p. 675)
continuous at a point (p. 641)
continuous over an interval (p. 641)
critical values (p. 666)
delta (Δ) notation (p. 648)
delta (Δ) process (p. 651)
derivative (p. 649–50)
difference quotient (p. 650)
differentiable (p. 650)
differentiation (p. 650)
explicit function (p. 680)
implicit equations (p. 680)
implicit differentiation (p. 680)
instantaneous velocity (p. 647)
limit of a function (p. 634)
related rate (p. 685)
secant line (p. 645)
tangent line (p. 645)

Summary of Formulas

$m_{\tan} = \lim_{x_2 \to x_1} \left(\frac{y_2 - y_1}{x_2 - x_1}\right)$ slope of tangent line

$f'(x) = \lim_{\Delta x \to 0} \left[\frac{f(x + \Delta x) - f(x)}{\Delta x}\right]$ derivative of a function

$\frac{dy}{dx} = 0$ where y = a constant constant rule

$\frac{dy}{dx} = nx^{n-1}$ where $y = x^n$ power formula

$\frac{dy}{dx} = cnx^{n-1}$ where $y = cx^n$ constant multiplier rule

$\frac{dy}{dx} = \frac{du}{dx} + \frac{dv}{dx}$ where u and v are functions of x and $y = u + v$ sum rule

$\frac{dy}{dx} = \frac{du}{dx} - \frac{dv}{dx}$ where u and v are functions of x and $y = u - v$ difference rule

$\frac{dy}{dx} = u\frac{dv}{dx} + v\frac{du}{dx}$ where $y = u \cdot v$ product rule

$\frac{dy}{dx} = \frac{v\frac{du}{dx} - u\frac{dv}{dx}}{v^2}$ where $y = \frac{u}{v}$ quotient rule

$\frac{dy}{dx} = \frac{dy}{du} \cdot \frac{du}{dx}$ where $y = f(u)$ and $u = g(x)$ chain rule

$\frac{dy}{dx} = nu^{n-1}\frac{du}{dx}$ where $y = [u(x)]^n$ extended power rule

CHAPTER REVIEW

Section 18–1

Determine whether the following functions are continuous. If not, state why.

1. $f(x) = 3x - 6$

2. $f(x) = 3 - \frac{1}{x}$

3. $f(x) = \frac{2}{3x - 1}$

4. $f(x) = x^3$

5. $f(x) = \sqrt{x + 1}$

6. $f(x) = 2x^2$

Evaluate the limit of the following functions.

7. $\lim_{x \to 1}(2x^2 - 5x + 1)$

8. $\lim_{x \to -2}(3x + 7)$

9. $\lim_{x \to 0}\left(\frac{4x^3 - x}{x}\right)$

10. $\lim_{x \to -1}\sqrt{3x - 1}$

11. $\lim_{x \to 2}\sqrt{x - 4}$

12. $\lim_{x \to 0}\left(\frac{6x^2 + 16x}{16x^2 - 36x}\right)$

13. $\lim_{x \to 0}\left(\frac{3x^2 + 15x}{3x^2 + 14x - 5}\right)$

14. $\lim_{x \to -1}\left(\frac{x^2 + x - 20}{2x^2 - 7x + 4}\right)$

15. $\lim_{x \to 2}\left(\frac{2x^2 + x - 3}{x^2 + 3x - 4}\right)$

16. $\lim_{x \to \infty}\left(\frac{2 + 3x + x^2}{4x^2 - 2x + 1}\right)$

17. $\lim_{x \to \infty}\left(\frac{5 + 3x^2 + x^3}{3 + 4x - 7x^3}\right)$

Section 18–2

Using the delta process, find the derivative of each function.

18. $y = 7x - 4$

19. $y = 3 + 8x$

20. $y = \frac{1}{x}$

21. $v = 8s^2 + 16s - 5$

22. $v = 4s^3 + 7s$

23. $y = 5 + \frac{1}{x}$

24. $y = 4 - \frac{1}{x}$

25. $a = 7v^3 - 9v^2 + 5$

26. $a = 8v^3 + 3v - 2$

27. $y = \frac{6}{x - 2}$

28. $y = \frac{7}{x^2}$

29. $y = \frac{x - 4}{x}$

30. Find a general expression for the rate of change of the volume of a cube with respect to side length ($v = s^3$), and evaluate the result when $s = 4$ in.

31. The displacement s in miles of a comet as a function of time t is given as $s = t^3 + 3t^2 + 6$. Use the delta process to find the velocity of the comet when $t = 20$ h.

Section 18–3

Find the derivative of the following functions.

32. $y = -12$

33. $y = 16$

34. $y = x^3$

35. $y = x^5$

36. $y = 7x^{-3}$

37. $y = 4x^{-2}$

38. $y = 3x^4 - 6x^2 + 7$

39. $y = 7x^5 + 4x^2 - x + 10$

40. $y = 4x^{-3} + 7x + 9$ **41.** $y = x^{1/2} + 7x^3 - x$

42. If the displacement s of a bullet as a function of time t is given by $s = 7.8t^2 + 3t^{4.1}$, find an expression for its velocity.

43. Find the dimensions of a cylinder whose surface area is a minimum if its volume is 60 ft³.

44. If a rock is thrown vertically upward with an initial velocity of 16 m/s, its displacement s, neglecting wind resistance, is given by $s = 16t - 4.9t^2$. Find the time at which the displacement is a maximum.

Section 18–4

Find the derivative of the following functions.

45. $y = 7x^3(x^2 + 4x^4)$ **46.** $y = x^4(3x^2 - x^3)$

47. $y = (3x^2 + x^5)(x^3 - 4x^2)$

48. $y = (x^4 - 3x^2)(7 + x^3)$

49. $y = \dfrac{3x}{x^2 + 7}$ **50.** $y = \dfrac{x}{x^2 - 4}$

51. $y = \dfrac{x^2 + 3x}{x^3 - 8x^2}$ **52.** $y = \dfrac{x^4 - x^3}{3x + x^2}$

53. $y = (3x + 8)^3$ **54.** $y = (x^2 + 4x)^2$

55. $y = (x - 2)(x + 1)^3$ **56.** $y = (2x + 3)^2(x + 1)$

57. $y = \dfrac{(x - 1)^2}{x^2 + 6}$ **58.** $y = \dfrac{(x + 3)^3}{x^3}$

59. The distance-time equation of a bulldozer is given by

$$s = \sqrt{3t^3 + t^2} \text{ ft}$$

Find the velocity in ft/s of the bulldozer after 12 s.

60. The charge q transferred in a certain circuit as a function of time t is given by

$$q = \sqrt{\frac{t^2 + 3}{t^3 + 8}}$$

Find an expression for the current. ($i = dq/dt$)

Section 18–5

Find the derivative of the following equations by using implicit differentiation.

61. $2x^2 + 3y = 10$ **62.** $x^3 + 7xy^2 - 4y = 0$

63. $(3x + y)(x + 5y) = x$

64. $(6x^2 + y)^3 = 4$

65. $y^2 = \dfrac{x}{(x^2 - y^2)}$ **66.** $y^3 = \dfrac{x + y}{3y^2}$

67. $(3y^2 + 8x)^2 = 10$

68. $(2x^3 + y)(3y^2 + 4x) = 7$

69. $8x^3 - 7xy^2 + x^2 = y^3$

70. A weather balloon is released at a point 2,500 ft west of a tracking station. If the balloon rises vertically at a rate of 5.6 ft/s, how fast is the distance between the station and the balloon changing 70 s after its release?

CHAPTER TEST

The number in parentheses refers to the appropriate learning objective given at the beginning of the chapter.

1. Use the delta process to find the derivative of $y = 7x^2 + x - 8$, and evaluate the result when $x = -1$. (3)

2. The current i in a certain electrical circuit varies with time according to the relationship (7)

$$i = \frac{6(t - 3)}{t^2 + 9}$$

At what time is the current a maximum?

3. Find the following limit: **(2)**

$$\lim_{x \to -2} \frac{3x^2 + x - 10}{x + 2}$$

4. The displacement s of an electrically charged particle as a function of time t is $s = 7.8t^2 + t^{1.2}$ meters. Find the velocity of the particle when $t = 1.5$ s. **(4, 7)**

5. Find the derivative of the function $y = 7x^{-3} + 4x^2 - 9$. **(4)**

6. Find dy/dx for $y = (x^3 + 3x^2)(x^4 + 7x - 10)$. **(5)**

7. Use the delta process to find the velocity of a drag racer when $t = 5$ s if its displacement is given by $s = 11t^2 + 7t + 9$ ft. **(3)**

8. Find the derivative of the following function: **(5)**

$$y = (x^2 + 3x)(x^3 + 4x^2 - 6)^3$$

9. Find the current i in a resistor when $t = 0.5$ s if the charge q is given by $q = 7.3t^{1.7}$ C. *(Note:* $i = dq/dt$.*)* **(4, 7)**

10. Find the induced voltage in a circuit with a 2-H inductance at $t = 1.2$ s if the current is given by $i = (t^2 + t)^3$ A. $\left(V = -L\dfrac{di}{dt}\right)$ **(7)**

11. Determine whether the following function is continuous, and explain why or why not: **(1)**

$$f(x) = 4 + \frac{1}{x}$$

12. Find the derivative of the function $6x^2y + 4y^3 = 7xy + 8$. **(6)**

13. Find the derivative of $y = \dfrac{3x^2 + 5}{x + 4}$. **(5)**

14. Find the limit of the following function as $x \to 3$: **(2)**

$$f(x) = 2x^2 - 5x + 8$$

15. Find the derivative of the following function: **(6)**

$$(7x^2 + y)(4xy + 9y^3) = 3$$

16. Find the derivative of $y = 9x^5 - 4x^3 + 7x - 8$. **(4)**

17. A 12-ft ladder leans against a vertical wall. If the ladder begins to slip so that it is moving down the wall at the rate of 5 ft/s, at what rate is the bottom of the ladder moving away from the wall when the ladder is 3 ft from the wall? **(7)**

18. Find the derivative of the following function: **(5)**

$$y = \frac{3x(1 + x)^2}{x^3}$$

SOLUTION TO CHAPTER INTRODUCTION

To find the dimension of the square that results in a box with maximum volume, we start by writing a formula for the volume of the box.

$$V = \underbrace{(16 - 2x)}_{\text{length}} \cdot \underbrace{(12 - 2x)}_{\text{width}} \cdot \underbrace{x}_{\text{height}}$$

$$V = 192x - 56x^2 + 4x^3$$

Then we find dV/dx, set the resulting equation equal to zero, and find the critical values.

$$\frac{dV}{dx} = 192 - 112x + 12x^2$$

$$12x^2 - 112x + 192 = 0$$

$$3x^2 - 28x + 48 = 0$$

$$x = \frac{28 \pm \sqrt{(-28)^2 - 4(3)(48)}}{6}$$

$$x = \frac{28 \pm \sqrt{208}}{6}$$

$$x \approx 7.1 \quad \text{or} \quad x \approx 2.3$$

The first solution is not acceptable since it would result in a negative number for the width of the box.

Next, we use the first derivative test to determine if $x \approx 2.3$ will produce a maximum volume. We choose to substitute $x = 2$ and $x = 3$ into the first derivative.

$$\frac{dV}{dx} = 192 - 112(2) + 12(2)^2 = 16$$

$$\frac{dV}{dx} = 192 - 112(3) + 12(3)^2 = -36$$

Since the first derivative changes from positive to negative, the function has a maximum at $x \approx 2.3$ in. The approximate dimensions of the box are 11.4 in. by 7.4 in. by 2.3 in.

As shown on the opposite page, a 6-in. parabolic mirror has the shape resulting from rotating about the y axis the region bounded by the parabola $y = x^2/9 + 1$, $x = 0$, $y = 0$, and $x = 3$. Find the volume of glass used in the mirror. (The solution to this problem is given at the end of the chapter.)

In the previous chapter we discussed differential calculus where we were given a function and we found its derivative. However, in technical applications it is quite common to have information about the rate of change of a variable. Such a problem requires that we reverse the process of differentiation—a process called *integration*.

In this chapter we will begin our study of integral calculus. First we will discuss differentials, which tie together the concept of differentiation and integration. Then we will discuss indefinite and definite integrals and their applications.

Learning Objectives

After completing this chapter, you should be able to

1. Find the differential of a given function (Section 19–1).

2. Calculate absolute and relative error of a variable (Section 19–1).

3. Integrate a given polynomial function (Section 19–2).

4. Find the area under a curve by using the limit method (Section 19–3).

5. Evaluate a given definite integral (Section 19–4).

6. Use the trapezoidal rule to find the area under a curve (Section 19–5).

7. Apply integration to technical problems (Sections 19–2, 19–3, and 19–4).

Chapter 19

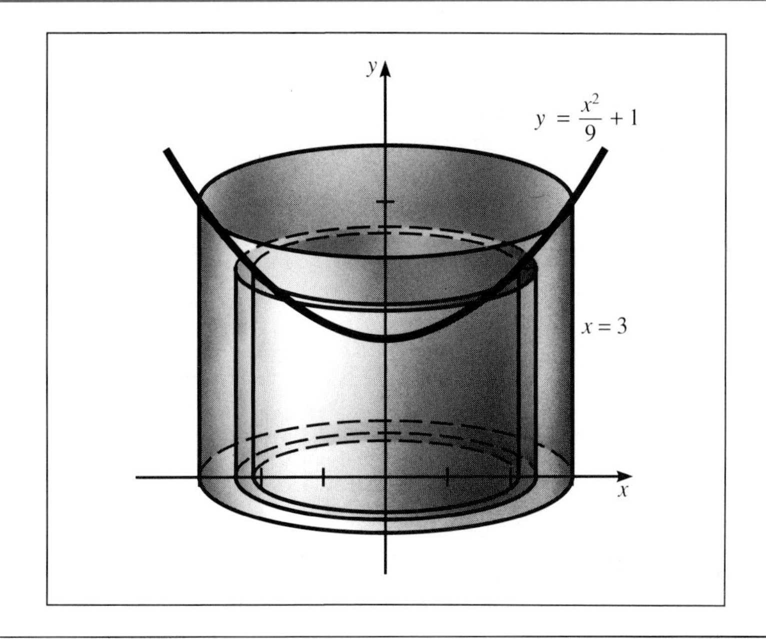

Integration with Applications

19–1

DIFFERENTIALS

In this chapter we will discuss integration, the inverse operation of differentiation. First, however, we will discuss the concept of differentials.

Concept of a Differential

If $y = f(x)$ is a function that can be differentiated, and the independent variable x is changed by an amount Δx, then the differentials dx and dy are *defined* as follows:

$$dx = \Delta x$$
$$dy = f'(x)\, dx$$

Graphically, Δx and Δy are the change in x and y from point P to point Q, as shown in Figure 19–1. The differential dy is defined above with respect to the derivative, the slope of the tangent line at point P. Graphically, dy is the change in y with respect to the tangent line at P as denoted in Figure 19–1. For very small changes in x, Δy is very closely approximated by dy.

FIGURE 19–1

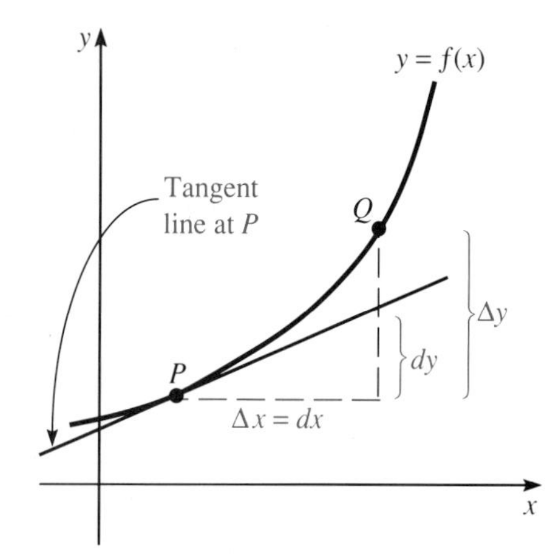

The slope of the line tangent to the curve at P is given by

$$\text{slope} = \frac{dy}{\Delta x} = \frac{dy}{dx} = f'(x)$$

From this information, $dy = f'(x)\, dx$, which is the definition given previously of the differential dy. In this way, the derivative can be interpreted as the ratio of the differential of y to x, respectively. This interpretation allows us to regard the derivative as a fraction.

EXAMPLE 1 Find the differential dy of each of the following:

(a) $y = 4x^3 - 6x$ (b) $y = (2x^3 - 4)^4$ (c) $y = \dfrac{x^3}{3x - 4}$

Solution Using the definition of the differential gives the following:

(a) $dy = f'(x)\,dx$ power rule

$dy = (12x^2 - 6)\,dx$

(b) $dy = 4(2x^3 - 4)^3(6x^2)\,dx$ chain rule

$dy = 24x^2(2x^3 - 4)^3\,dx$

(c) $dy = \dfrac{(3x - 4)(3x^2) - x^3(3)}{(3x - 4)^2}\,dx$ quotient rule

$= \dfrac{9x^3 - 12x^2 - 3x^3}{(3x - 4)^2}\,dx$

$dy = \dfrac{6x^3 - 12x^2}{(3x - 4)^2}\,dx$

■

EXAMPLE 2 A rocket is launched so that it follows a path given by the equation $s = 350x - 6x^2$. Find the following:

(a) the increment in displacement Δs at $x = 100$ km for $\Delta x = 0.1$ km
(b) the differential ds at $x = 100$ km for $\Delta x = 0.1$ km

Solution

(a) We use the delta process to solve for Δs.

$$
\begin{aligned}
s &= 350x - 6x^2 \\
s + \Delta s &= 350(x + \Delta x) - 6(x + \Delta x)^2 \\
\Delta s &= 350(x + \Delta x) - 6(x + \Delta x)^2 - s \\
\Delta s &= 350x + 350\,\Delta x - 6x^2 - 12x\,\Delta x - 6(\Delta x)^2 - 350x + 6x^2 \\
\Delta s &= 350\,\Delta x - 12x\,\Delta x - 6(\Delta x)^2
\end{aligned}
$$

Substituting $x = 100$ and $\Delta x = 0.1$ into the equation gives

$$
\begin{aligned}
\Delta s &= 350(0.1) - 12(100)(0.1) - 6(0.1)^2 \\
\Delta s &= -85.06 \text{ km}
\end{aligned}
$$

(b) Using the definition for the differential ds gives

$$
\begin{aligned}
ds &= f'(x)\,dx \\
&= f'(350x - 6x^2)\,dx \\
ds &= (350 - 12x)\,dx
\end{aligned}
$$

Substituting the values given ($dx = \Delta x$) gives

$$
\begin{aligned}
ds &= [350 - 12(100)](0.1) \\
ds &= -85.00 \text{ km}
\end{aligned}
$$

FIGURE 19–2

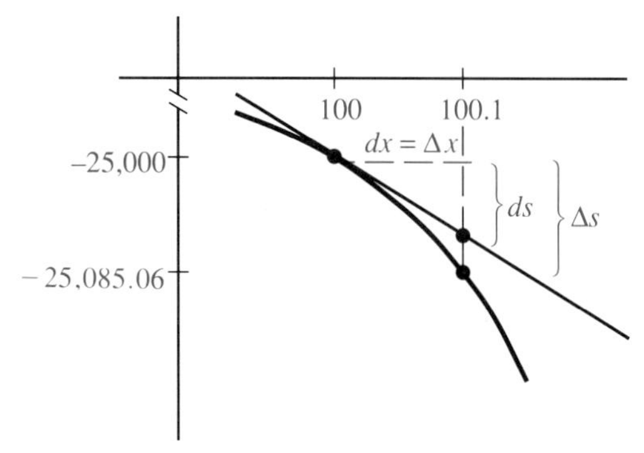

The estimated change in displacement as x changes from 100 to 100.1 km is -85.00 km. Notice that the values of Δs and ds are very close. Figure 19–2 gives a graphical representation of Δs and ds. ■

Error

We can consider Δx as the error in the measurement of the variable x. Then Δy is the resulting **absolute error** in the calculated value of the variable y, and dy is an approximation to the calculated error.

EXAMPLE 3 In calculating the power dissipated in a 5-Ω resistor, an engineer measured the current through the resistor as 3.5 A. Later, a second engineer found that the current measurement was 0.15 A too low. Find the error in the calculated value of the power.

Solution The power in the resistor is given by the equation

$$P = I^2R$$

Finding the differential gives

$$dP = f'(P)\,dI$$
$$dP = 2IR\,dI$$

Note in the differential above that R is a constant, while the power and current are variables. The quantity dP will represent the error in the measurement of the power given that $I = 3.5$ A, $R = 5\ \Omega$ and $dI = 0.15$ A. Calculating dP gives

$$dP = 2(3.5)(5)(0.15)$$
$$dP = 5.25 \text{ W}$$

The error in the calculated value of the power is 5.25 W. ■

Often the actual magnitude of an error is less important than its ratio to the measurement itself. This ratio, commonly expressed as a percentage, is called **relative error.** The calculation for relative error is given by

$$\text{relative error} = \left(\frac{\Delta y}{y} \times 100\right)\%$$

EXAMPLE 4 An exactly measured 8-Ω resistor is connected in parallel with a variable resistor R. The total resistance R_T is given by

$$R_T = \frac{8R}{8 + R}$$

What is the relative error in total resistance if R is measured to be 15 Ω with a possible error of 0.75 Ω?

Solution The error in measured resistance, $\Delta R = 0.75$, produces an error ΔR_T in the total resistance. However, for small values of ΔR, $\Delta R_T \approx dR_T$. The differential is given by

$$dR_T = f'(R_T)\, dR$$

$$= \frac{(8 + R)(8) - 8R(1)}{(8 + R)^2}\, dR$$

$$dR_T = \frac{64 + 8R - 8R}{(8 + R)^2}\, dR$$

$$dR_T = \frac{64}{(8 + R)^2}\, dR$$

To find the error in the calculated value of R_T, substitute $R = 15$ and $dR = 0.75$ into the differential.

$$dR_T = \frac{64}{(8 + 15)^2}(0.75)$$

$$dR_T \approx 0.091$$

The relative error is given by

$$\frac{dR_T}{R_T} \times 100 = \frac{0.091}{5.217} \times 100 = 1.7\%$$

where $R_T = 8(15)/(8 + 15) \approx 5.217$. The relative error in the calculated value of the total resistance is approximately 1.7%. ■

EXAMPLE 5 The taxable horsepower P of an engine is given by $P = nD^2$ where n = the number of cylinders and D = the diameter of each cylinder. Find the approximate increase in power of a four-cylinder engine if the cylinders are rebored from 3.50 in. to 3.54 in.

Solution We can find an *approximation* of the increase in power by using the differential

$$dP = f'(P)\,dD$$
$$dP = 2nD\,dD$$

From the given information, $n = 4$, $D = 3.50$ in., and $dD = 0.04$ in. Substituting these values into the equation gives

$$dP = 2(4)(3.50)(0.04)$$
$$dP \approx 1.12$$

The increase in power is approximately 1.12 hp. ■

EXAMPLE 6 A cylindrical cask used to age wine is secured using a continuous steel band around its circumference. If the temperature of the band increases 20°F during a day, the length of the band increases by 0.01 ft. Find the amount by which the radius of the band increases.

Solution The length of the band is represented by the circumference of a circle.

$$C = 2\pi r$$

We must find the change in radius given the change in length, circumference. Finding the differential gives

$$dC = f'(C)\,dr$$
$$dC = 2\pi\,dr$$

Substituting $dC = 0.01$ ft and solving for dr gives

$$0.01 = 2\pi\,dr$$
$$0.002 \approx dr$$

The radius of the band increases approximately 0.002 ft. ■

19–1 EXERCISES

In Exercises 1–12, find the differential of each function.

1. $y = 2x^3 - x^2$

2. $y = x + 8x^3$

3. $y = (x^2 - 5)^3$

4. $y = (2x + 7)^3$

5. $y = 4x^3(9 + 3x^2)$

6. $y = x^3(2x^3 + x)$

7. $y = (6 + 3x)^{1/3}$

8. $y = \sqrt{4x^2 - 3x}$

9. $y = \dfrac{x}{5x^2 + 4}$

10. $y = \dfrac{2x - 1}{7 - 3x^2}$

11. $x^3y + x^2 = y$

12. $4x^3 - 6xy + y^2 = 8$

In Exercises 13–18, find the values of dy and Δy for the given values of x and Δx.

13. $y = 4x - 3x^2$; $x = 3$; $\Delta x = 0.1$

14. $y = 5 + 8x - x^3$; $x = -1$; $\Delta x = 0.06$

15. $y = (1 + 8x^3)^2$; $x = 1$; $\Delta x = 0.02$

16. $y = (x - 2x^3)^4$; $x = 2$; $\Delta x = 0.1$

17. $y = 3x\sqrt{1 + x^2}$; $x = -3$, $\Delta x = 0.8$

18. $y = x^2(3x + 5x^3)$; $x = -2$; $\Delta x = 0.01$

In Exercises 19–24, find the approximate change in the indicated variable by calculating the differential.

19. $V = s^3$; $s = 16$ cm; $\Delta s = 0.1$ cm; $\Delta V = ?$

20. $V = 8\pi r^2$; $r = 2$ m, $\Delta r = 0.04$ m; $\Delta V = ?$

21. $A = \pi r^2$; $r = 5$ ft; $\Delta r = 0.5$ ft; $\Delta A = ?$

22. $V = \frac{4}{3}\pi r^3$; $r = 16$ ft; $\Delta r = 0.6$ ft; $\Delta V = ?$

23. $A = 4\pi r^2$; $r = 18$ in.; $\Delta r = 0.01$ in.; $\Delta A = ?$

24. $A = \frac{\pi d^2}{4}$; $d = 3.6$ yd; $\Delta d = 0.1$ yd; $\Delta A = ?$

25. A precisely measured 5-Ω resistor is wired in parallel with a variable resistor R. The total resistance is given by

$$R_T = \frac{5R}{5 + R}$$

Find the relative error in the total resistance if $R = 20\ \Omega$ with a possible error of 3.2 Ω.

26. A circular watch battery is to be made with a radius of 6 mm. If the maximum permissible relative error for quality control is 0.02 mm, find the maximum permissible relative error in the area of the circular top of the battery.

27. A 30-cm square plate is to be made so that the maximum error in area is 0.19 cm^2. What is the approximate error allowable in the length of the side?

28. A washer has an inside radius of 9 mm and an outside radius of 19 mm. What change in the inside radius will decrease the surface area of one side of the washer by 2 mm^2?

29. A circular plate 1 inch thick was originally 12 inches in diameter, but 0.5 inch was trimmed from the edge. Estimate the volume of wasted material.

30. Through repeated measurements, the diameter of a ball bearing is found to be 3.2 ± 0.0002 mm. Find the percent error in its volume.

19–2

THE INDEFINITE INTEGRAL

In Chapter 18, we discussed the derivative $f'(x)$ of a given function $f(x)$. In this section we will discuss the inverse of this process by finding a function $f(x)$ whose derivative is $f'(x)$. This reverse process is known as **antidifferentiation.**

EXAMPLE 1 Find the antiderivative of the function $f(x) = 8x^3$.

Solution From the power rule for differentiation, we know that the exponent is decreased by 1 and the original exponent is multiplied by the coefficient. Then we can guess that the antiderivative of $8x^3$ is of the form ax^4 where a is some constant. If we take the derivative of ax^4 and compare the result with $8x^3$, we can solve for the constant a.

$$\frac{d(ax^4)}{dx} = (4a)x^3$$

Therefore,

$$4a = 8$$
$$a = 2$$

The antiderivative of $8x^3$ is $2x^4$. We can verify this by taking the derivative of $2x^4$.

$$\frac{d(2x^4)}{dx} = 4(2x^{4-1}) = 8x^3$$ ■

Notice that each of the functions $2x^4 + 3$, $2x^4 - 8$, and $2x^4 - 1$ has $8x^3$ as its derivative. For each function $f(x)$, the antiderivative $g(x)$ is not a unique function but a series of functions that differ by an additive constant. This fact is due to the constant rule of differentiation discussed in the previous chapter. Figure 19–3 shows that the slope of a tangent line at $x = 1$ is the same for each of these curves because the tangent lines are parallel (parallel lines have the same slope).

FIGURE 19–3

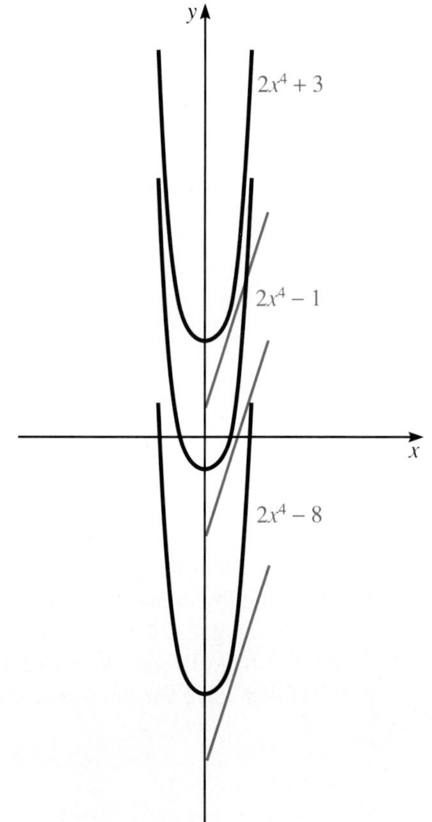

Integration

The process of antidifferentiation is commonly called **integration.** The notation of integration is given in the following box.

The Indefinite Integral

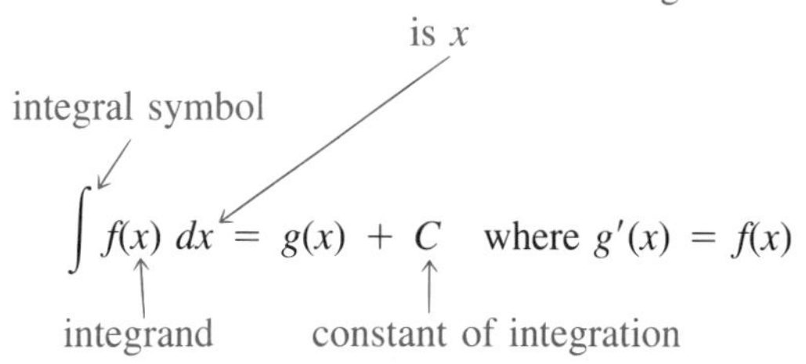

$$\int f(x)\,dx = g(x) + C \quad \text{where } g'(x) = f(x)$$

The quantity $\int f(x)\,dx$ is the **indefinite integral** of the function $f(x)$, and the process of finding the function $g(x)$ is called **integrating** $f(x)$.

We will use the rules of differentiation developed in the last chapter to develop formulas for integration of polynomial functions.

Integration of a Constant

$$\int k\,dx = k\int dx = kx + C$$

where k = a constant and $C = kc$. The integral of a constant equals the constant times the integral. The integral of dx equals $x + c$.

EXAMPLE 2 Find $\int 3\,dy$ and check the result by using the derivative.

Solution Using the constant rule, we can write the constant outside the integral.

$$\int 3\,dy = 3\int dy$$

The integral of dy is y. Therefore,

$$\int 3\,dy = 3y + C$$

We would need additional information to determine the value of the constant of integration C. We can check this result by taking the derivative with respect to y of $3y + C$, which is 3. ■

Recall from the previous chapter that to find the derivative of the power of a variable, we decrease the exponent by 1 and multiply by the original power. We reverse this process to integrate the power of a variable. Therefore, we increase the power by 1 and divide by this new power as summarized in the following box.

The Power Rule of Integration

$$\int x^n\,dx = \frac{x^{n+1}}{n+1} + C$$

where $n \neq -1$.

EXAMPLE 3 Integrate the following:

(a) $\int x^5\,dx$ (b) $\int -6y^4\,dy$

Solution

(a) Using the power rule to integrate this function, we increase the power by 1 and divide by the increased power.

increase power by 1

$$\int x^5\,dx = \frac{x^{5+1}}{5+1} + C$$

divide by new power

$$\int x^5\,dx = \frac{x^6}{6} + C$$

We check this result by differentiation.

$$\frac{d\left(\frac{x^6}{6} + C\right)}{dx} = \frac{6x^5}{6} + 0 = x^5$$

(b) To integrate this function, we place the constant before the integral sign.

$$\int -6y^4\,dy = -6\int y^4\,dy$$

Second, we use the power rule to integrate $y^4\,dy$.

$$\int -6y^4\,dy = -6\left(\frac{y^5}{5}\right) + C = -\frac{6y^5}{5} + C$$

This result is correct since the derivative with respect to y of $-\frac{6y^5}{5} + C$ is $-6y^4$. ■

The sum and difference rules for differentiation also directly apply to integration. The sum rule of integration follows.

The Sum Rule of Integration

$$\int [f(x) + g(x)]\, dx = \int f(x)\, dx + \int g(x)\, dx$$

The integral of a sum of terms equals the sum of the integrals of the terms.

EXAMPLE 4 Integrate the following:

$$\int (10x^4 + 6x^2 - x + 6)\, dx$$

Solution First, we use the sum rule to write the integral in the following form:

$$\begin{aligned}
\int (10x^4 + 6x^2 - x + 6)\, dx \\
&= \int 10x^4\, dx + \int 6x^2\, dx - \int x\, dx + \int 6\, dx \\
&= 10 \int x^4\, dx + 6 \int x^2\, dx - \int x\, dx + 6 \int dx && \text{constant rule} \\
&= 10\left(\frac{x^5}{5}\right) + 6\left(\frac{x^3}{3}\right) - \frac{x^2}{2} + 6x + C && \text{power rule of integration} \\
&= 2x^5 + 2x^3 - \frac{x^2}{2} + 6x + C
\end{aligned}$$

The result checks since its derivative with respect to x is $10x^4 + 6x^2 - x + 6$. ■

EXAMPLE 5 Integrate the following:

$$\int \left(\sqrt{x} + \frac{1}{x^2}\right) dx$$

Solution First, we write the integrand as follows:

$$\int (x^{1/2} + x^{-2})\, dx$$

Second, we apply the sum rule of integration.

$$\int x^{1/2}\,dx + \int x^{-2}\,dx$$

Third, we apply the power rule to integrate.

$$\frac{x^{3/2}}{\frac{3}{2}} + \frac{x^{-1}}{-1} + C$$

$$\frac{2x^{3/2}}{3} - \frac{1}{x} + C$$

■

CAUTION ✦ The quotient and product rules of differentiation do not have corresponding rules for integration.

Generalized Power Rule

The previous examples of integration involved a power of x, the variable. However, the function to be integrated may be more complicated than this, such as $\int 2x(3 + x^2)^4\,dx$. We will use a method of substitution to transform a function of this form into a power of the variable so that we can use existing formulas to integrate. The generalized power rule is given below, and its application is illustrated in the next examples.

The Generalized Power Rule

If u is a function of x, where $u = f(x)$ and $du = g(x)\,dx$, then

$$\int [f(x)^n g(x)]\,dx = \int u^n\,du = \frac{u^{n+1}}{n+1} + C$$

where $n \neq -1$.

EXAMPLE 6 Integrate the following:

$$\int 2x(3 + x^2)^4\,dx$$

Solution It would be difficult to integrate this function by multiplying and expanding $(3 + x^2)^4$. Let $u = 3 + x^2$. Then $du = 2x\,dx$. The integral can be written as follows:

$$\int \underbrace{(3 + x^2)^4}_{u}\underbrace{(2x\,dx)}_{du} = \int u^4\,du$$

Then we can integrate the function by using the generalized power rule as follows:

$$\int (3 + x^2)^4(2x\,dx) = \frac{u^5}{5} + C$$

Substituting the value of u gives the integral of this function as

$$\frac{(3 + x^2)^5}{5} + C$$

■

CAUTION ✦ When integrating by u substitution, you may multiply or divide the integral only by a constant. A variable must never be introduced or factored out and shown as a multiplier in front of the integral.

EXAMPLE 7 Integrate the following:

$$\int (4x^3 - 9)^3 x^2\,dx$$

Solution If we let $u = 4x^3 - 9$, then $du = 12x^2\,dx$. If we rearrange the integral as follows, notice that $x^2\,dx$ is not equal to du, which is $12x^2\,dx$.

$$\int \underbrace{(4x^3 - 9)^3}_{u}\underbrace{(x^2\,dx)}_{\text{not equal to } du}$$

However, we are missing only a numerical factor of 12. To obtain the correct value of du, we multiply and divide the integral by 12 as follows:

divide by 12 multiply by 12

$$\frac{1}{12}\int \underbrace{(4x^3 - 9)^3}_{u} \cdot \underbrace{(12x^2\,dx)}_{du}$$

Next, we integrate this function by using the generalized power rule, with the following result:

$$\frac{1}{12}\int u^3\,du = \frac{1}{12}\left(\frac{u^4}{4}\right) + C = \frac{(4x^3 - 9)^4}{48} + C$$

■

Applications

EXAMPLE 8 Find the equation of the curve passing through the point $(0, -2)$ and having a slope given by

$$f'(x) = x\sqrt{3x^2 + 4}$$

Solution The function $f(x)$ is given by

$$f(x) = \int f'(x)\,dx = \int x\sqrt{3x^2 + 4}\,dx$$

We integrate this function as follows:

$$u = 3x^2 + 4 \quad \text{and} \quad du = 6x\,dx$$

multiply and divide by 6

$$f(x) = \frac{1}{6}\int \underbrace{(3x^2 + 4)}_{u}{}^{1/2}\underbrace{(6x\,dx)}_{du}$$

$$= \frac{1}{6}\int u^{1/2}\,du$$

$$= \frac{1}{6}\left(\frac{u^{3/2}}{\frac{3}{2}}\right) + C$$

$$= \frac{1}{9}u^{3/2} + C$$

$$= \frac{1}{9}(3x^2 + 4)^{3/2} + C$$

substitute for u

From the problem statement, we are given that $f(x) = -2$ when $x = 0$. We can substitute these values into the equation and calculate the value of C.

$$f(x) = \frac{1}{9}(3x^2 + 4)^{3/2} + C$$

$$-2 = \frac{1}{9}[3(0)^2 + 4]^{3/2} + C$$

$$-2 = \frac{8}{9} + C$$

$$-\frac{26}{9} = C$$

The function $f(x)$ is given by

$$\frac{1}{9}(3x^2 + 4)^{3/2} - \frac{26}{9}$$

■

EXAMPLE 9 A bullet fired vertically upward from the roof of a 100-ft building has an initial velocity of 450 ft/s.

(a) Find the velocity of the bullet after 8 s.
(b) When will the bullet reach its maximum height?

Solution

(a) The bullet undergoes acceleration due to gravity pulling it down toward the earth. Thus, the acceleration-time equation is $a = -32$. The velocity of the bullet is given by

$$v = \int a\,dt = \int -32\,dt = -32t + C$$

We can calculate the constant of integration using the fact that $t = 0$ and $v = 450$.

$$450 = -32(0) + C$$
$$450 = C$$

The velocity when $t = 8$ s is

$$v = -32t + 450$$
$$= -32(8) + 450$$
$$v = 194 \text{ ft/s}$$

(b) The maximum height occurs when the velocity equals zero. Solving for t when $v = 0$ gives

$$v = -32t + 450$$
$$0 = -32t + 450$$
$$14 \text{ s} \approx t$$

■

19–2 EXERCISES

Integrate each of the following expressions.

1. $\int 8\,dx$
2. $\int -4\,dx$
3. $\int -2x\,dx$
4. $\int 6t^3\,dt$
5. $\int 4y^5\,dy$
6. $\int x^{-8}\,dx$
7. $\int x^{-3}\,dx$
8. $\int y^{2/3}\,dy$
9. $\int x^{4/3}\,dx$
10. $\int (3x^2 - 2)\,dx$
11. $\int (4x^3 + 8)\,dx$
12. $\int (x^2 - 3x + 10)\,dx$
13. $\int (4y^2 + y - 7)\,dy$
14. $\int \left(\sqrt{x} - \frac{1}{x^3}\right)dx$
15. $\int \left(\sqrt[4]{y} + \frac{5}{y^2}\right)dy$
16. $\int (3x^2 + 7)^3\,6x\,dx$
17. $\int (y^2 - 8)^4\,2y\,dy$
18. $\int (6x^2 + 8)^4\,x\,dx$
19. $\int (x^4 - 3x^2 + 5)^3\,(2x^3 - 3x)\,dx$
20. $\int (5x^4 - 12x)\,(x^5 - 6x^2)^4\,dx$
21. $\int 2x^3(3x^4 - 5)\,dx$
22. $\int \frac{x + 3x^2}{(x^2 + 2x^3)^4}\,dx$
23. $\int \frac{x^3 - 3x}{(x^4 - 6x^2)^3}\,dx$
24. $\int \frac{1}{\sqrt{x}}(1 - \sqrt{x})^3\,dx$

25. $\int \sqrt{6x - 8}\, dx$

26. $\int \sqrt[3]{4x + 7}\, dx$

27. $\int \frac{3x^2 - 2x}{(x^3 - x^2)^4}\, dx$

28. $\int \frac{2x + 3}{(x^2 + 3x - 1)^3}\, dx$

29. $\int \frac{2y}{\sqrt{y^2 + 6}}\, dy$

30. $\int \frac{-3x^2}{\sqrt{4 - x^3}}\, dx$

31. $\int (2x^4 - x^2)^5 (4x^3 - x)\, dx$

32. $\int (x^3 + 8x + 9)^4 (3x^2 + 8)\, dx$

33. Find the equation of the function whose graph has a slope of $3x^2 - 8$ and whose graph passes through the point (0, 3).

34. Given that the graph of a function has a slope of $x^3 - 3x^2 + 10$, find the function if its graph passes through the point (2, 1).

35. In an electrical circuit the current as a function of time is given by $i = 3t^2 + t$. What is the equation for the electrical charge q if $q = 0$ when $t = 0$? (*Note:* $i = dq/dt$.)

36. During each cycle a machine develops power given by $P = 12t^2 + 8t - 11$. Find the equation for the work done if $W = 0$ when $t = 0$. (*Note:* $P = dW/dt$.)

37. The acceleration of an object that starts from rest is given by $a = (4t^2 - t + 9)$ m/s^2. What is the velocity of the object at $t = 5$ s?

38. The acceleration of an object that starts from rest is given by $a = (3t^2 + 4t + 8)$ ft/s^2. What is the velocity of the object at $t = 3$ s?

39. A bullet fired vertically upward from the top of a 100-ft high building has an initial velocity of 600 ft/s. What is the velocity of the bullet after 5 s? How long does it take for it to reach its maximum height?

40. The power output of a mechanical system as a function of time is given by

$$P = \frac{4t}{\sqrt{t^2 + 5}}$$

Find an expression for the work done if $W = 0$ when $t = 0$. (*Note:* $P = dW/dt$.)

19–3

THE AREA UNDER A CURVE

The area under a curve can also be found by integration. In geometry, we have formulas to find the area of figures such as a triangle, circle, rectangle, or trapezoid. Using calculus, we can find the area between curves for which we have equations. The next example illustrates this concept.

EXAMPLE 1 Approximate the area bounded by the curve $y = x^2 + 3$, the x axis, $x = 1$, and $x = 3$.

Solution Figure 19–4(a) shows the graph of $y = x^2 + 3$. We must calculate the area of the shaded region. We can approximate this area by dividing it into two rectangles and adding their area, as shown in Figure 19–4(b). Both rectangles have a width of 1, and the left rectangle has a height of 4, while the right rectangle has a height of 7. The area of the two rectangles is

$$A = 1(4) + 1(7) = 11$$

Even though we know that the total area of the rectangles is 11, notice from Figure 19–4(b) that a considerable area bounded by the curve $y = x^2 + 3$ is not included in either rectangle. We can find a much better approximation to the area under the curve by increasing the number of rectangles to four, as

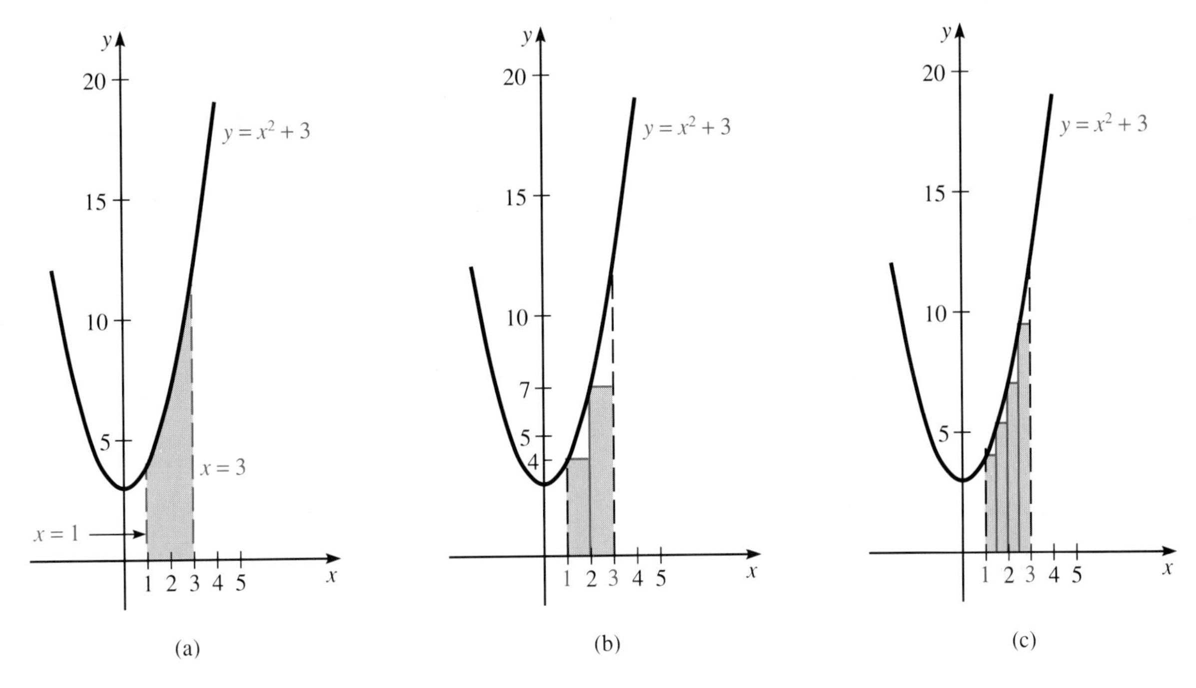

FIGURE 19–4

shown in Figure 19–4(c). The width of each resulting rectangle is 1/2, and the heights of the rectangles are 4, 5.25, 7, and 9.25, respectively. The area under the curve is approximated by the sum of the area of these four rectangles.

$$A = \frac{1}{2}(4 + 5.25 + 7 + 9.25)$$

$$A = 12.75$$

As you can see from Figure 19–4(c), portions of the area bounded by the curve $y = x^2 + 3$ are omitted from these rectangles. However, by increasing the number of rectangles, we do more closely approximate this area. The greater the number of rectangles, the more closely the sum of their areas equals the area bounded by the curve. In this section we will develop a method of integration to determine that the exact area under this curve is 14.66. ■

To generalize this method, we will find the area under the curve $y = f(x)$ from $x = a$ to $x = b$. As in the previous example, we divide the interval $a \leq x \leq b$ into n smaller intervals by choosing points $x_1 = a, x_2, x_3, \ldots, x_n = b$. The width of each interval is given by Δx, which is calculated as follows:

$$\Delta x = \frac{b - a}{n}$$

FIGURE 19–5

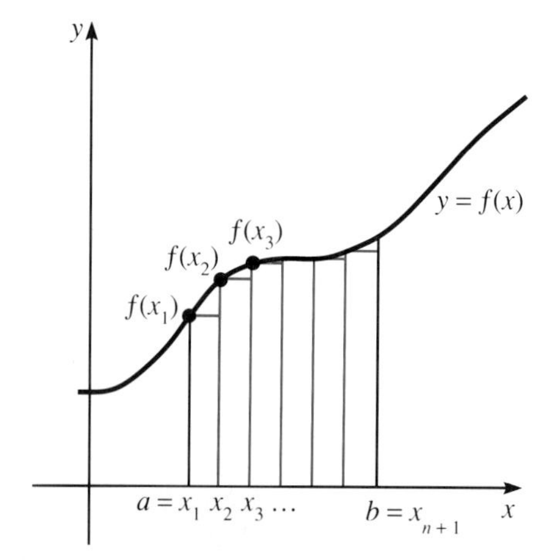

Then we construct rectangles whose width is Δx and whose height is given by $y_i = f(x_i)$, as Figure 19–5 illustrates. We approximate the area bounded by the curve by adding the area of these rectangles. From the previous example, we know that as we increase the number of rectangles, as Δx approaches zero, the approximation approaches the area bounded by the curve more closely as expressed in the following formula.

The Area under a Curve

$$A = \lim_{n \to \infty} \sum_{i=1}^{n} f(x_i)\ \Delta x$$

EXAMPLE 2 Find the area under the curve $y = x^2 + 1$ from $x = 1$ to $x = 3$ when

(a) $n = 2$ (b) $n = 8$

Solution

(a) First, we sketch the curve shown in Figure 19–6(a) and calculate Δx given that $a = 1$, $b = 3$, and $n = 2$.

$$\Delta x = \frac{3 - 1}{2} = 1$$

Second, we calculate the initial point of each interval.

$$x_1 = a = 1$$

$$x_2 = a + \Delta x = 1 + 1 = 2$$

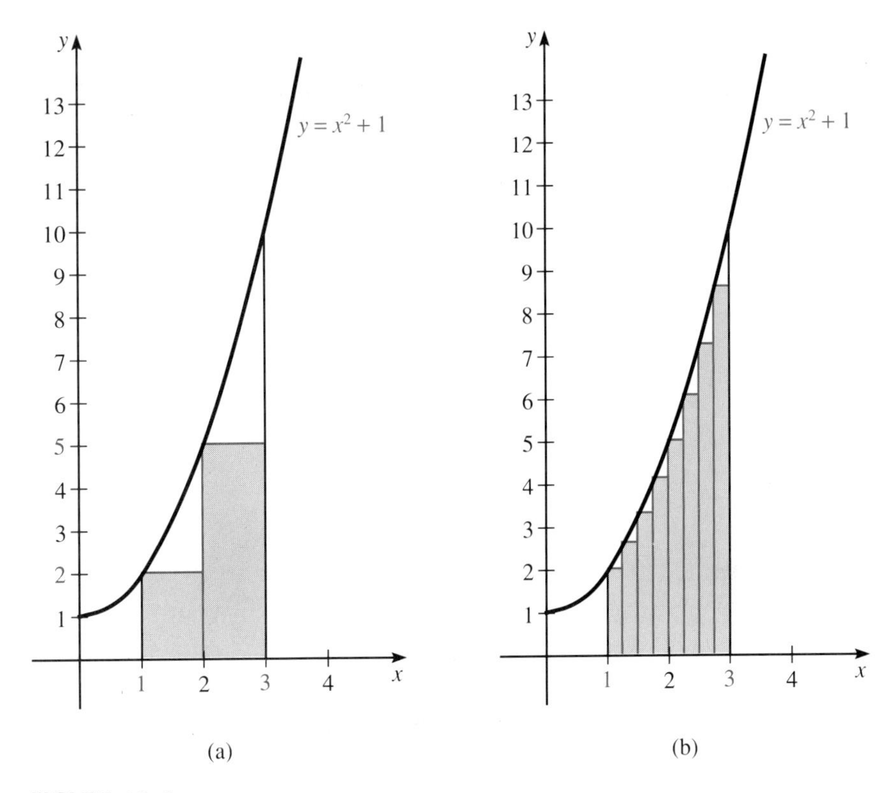

FIGURE 19–6

Third, we calculate the height of each rectangle.

$$y_1 = f(x_1) = f(1) = (1)^2 + 1 = 2$$
$$y_2 = f(x_2) = f(2) = (2)^2 + 1 = 5$$

Fourth, we calculate the area under the curve by adding the area of the two rectangles.

$$\begin{aligned} A &= f(x_1)\,\Delta x + f(x_2)\,\Delta x \\ &= 2(1) + 5(1) \\ A &= 7 \end{aligned}$$

The total area of the rectangles is 7 square units.

(b) For $n = 8$, as shown in Figure 19–6(b),

$$\Delta x = \frac{3 - 1}{8} = \frac{1}{4}$$

We calculate the initial point of each interval.

$$x_1 = a = 1$$

$$x_2 = a + \Delta x = 1 + \frac{1}{4} = 1\frac{1}{4}$$

$$x_3 = a + 2\,\Delta x = 1 + 2\left(\frac{1}{4}\right) = 1\frac{1}{2}$$

$$x_4 = a + 3\,\Delta x = 1 + 3\left(\frac{1}{4}\right) = 1\frac{3}{4}$$

$$x_5 = a + 4\,\Delta x = 1 + 4\left(\frac{1}{4}\right) = 2$$

$$x_6 = a + 5\,\Delta x = 2\frac{1}{4}$$

$$x_7 = a + 6\,\Delta x = 2\frac{1}{2}$$

$$x_8 = a + 7\,\Delta x = 2\frac{3}{4}$$

Third, we calculate the height of each rectangle.

$$y_1 = f(x_1) = f(1) = 2$$

$$y_2 = f(x_2) = f\left(1\frac{1}{4}\right) = 2.5625$$

$$y_3 = f(x_3) = f\left(1\frac{1}{2}\right) = 3.25$$

$$y_4 = f(x_4) = f\left(1\frac{3}{4}\right) = 4.0625$$

$$y_5 = f(x_5) = f(2) = 5$$

$$y_6 = f(x_6) = f\left(2\frac{1}{4}\right) = 6.0625$$

$$y_7 = f(x_7) = f\left(2\frac{1}{2}\right) = 7.25$$

$$y_8 = f(x_8) = f\left(2\frac{3}{4}\right) = 8.5625$$

Fourth, we calculate the area bounded by the curve by adding the areas of the eight rectangles.

$$A = \sum_{i=1}^{n} f(x_i)\,\Delta x$$

$$= \frac{1}{4}(2 + 2.5625 + 3.25 + 4.0625 + 5 + 6.0625 + 7.25 + 8.5625)$$

$$A = 9.6875$$

■

To find the exact area bounded by the curve using the limit formula given earlier, we must develop a formula for the total area of n rectangles and take the limit of this expression as the number of rectangles increases without bound. However, we need the following formulas to do so:

$$\sum_{i=1}^{n} c = nc \quad \text{where } c = \text{a constant}$$

$$\sum_{i=1}^{n} i = \frac{1}{2}n(n + 1)$$

$$\sum_{i=1}^{n} i^2 = \frac{1}{6}n(n + 1)(2n + 1)$$

$$\sum_{i=1}^{n} i^3 = \frac{1}{4}n^2(n + 1)^2$$

The next example shows how to use the limit formula to find the exact area under a curve.

EXAMPLE 3 Find the exact area bounded by the curves $y = x^2 + 2$ and $y = 0$ between $x = 2$ and $x = 3$.

Solution Let $\Delta x = \dfrac{3 - 2}{n} = \dfrac{1}{n}$. Then

$$x_1 = 2$$

$$x_2 = 2 + \Delta x = 2 + \frac{1}{n}$$

$$x_3 = 2 + 2\,\Delta x = 2 + \frac{2}{n}$$

$$\vdots$$

$$x_i = 2 + (i - 1)\,\Delta x = 2 + \frac{(i - 1)}{n}$$

The height of the ith rectangle at $x = x_i$ is given by

$$y_i = (x_i)^2 + 2 = \left(2 + \frac{(i-1)}{n}\right)^2 + 2$$

$$y_i = 6 + \frac{4(i-1)}{n} + \frac{(i-1)^2}{n^2}$$

The area of the ith rectangle is given by

$$A = y_i\,\Delta x = \left(6 + \frac{4(i-1)}{n} + \frac{(i-1)^2}{n}\right)\frac{1}{n}$$

$$A = \frac{6}{n} + \frac{4(i-1)}{n^2} + \frac{(i-1)^2}{n^3}$$

The area bounded by the curve is the sum of the areas of the n rectangles, given by

$$A_n = \sum_{i=1}^{n} A_i = \sum_{i=1}^{n}\left(\frac{6}{n} + \frac{4(i-1)}{n^2} + \frac{(i-1)^2}{n^3}\right)$$

Using the formulas for summation given earlier yields

$$A_n = \frac{1}{n}\sum_{i=1}^{n} 6 + \frac{4}{n^2}\sum_{i=1}^{n}(i-1) + \frac{1}{n^3}\sum_{i=1}^{n}(i-1)^2$$

$$= \frac{1}{n}(6n) + \frac{4}{n^2}\left[\frac{1}{2}n(n-1)\right] + \frac{1}{n^3}\left[\frac{1}{6}n(n-1)(2n-1)\right]$$

$$= 6 + 2 - \frac{2}{n} + \frac{1}{3} - \frac{1}{2n} + \frac{1}{6n^2}$$

$$A_n = 8\frac{1}{3} - \frac{2}{n} - \frac{1}{2n} + \frac{1}{6n^2}$$

The area bounded by the curve is given by

$$A = \lim_{n\to\infty}\left(8\frac{1}{3} - \frac{2}{n} - \frac{1}{2n} + \frac{1}{6n^2}\right)$$

$$A = 8\frac{1}{3}$$

Remember from Chapter 18 that $\frac{1}{n}, \frac{1}{2n}$, and $\frac{1}{6n^2}$ approach zero as n increases without bound. ■

EXAMPLE 4 Use the limit method to calculate the area under the curve $y = x^3$ from $x = 0$ to $x = 2$.

Solution If

$$\Delta x = \frac{2-0}{n} = \frac{2}{n}$$

then

$$x_1 = 0$$

$$x_2 = 0 + \Delta x = \frac{2}{n}$$

$$x_3 = 0 + 2\,\Delta x = \frac{4}{n}$$

$$\vdots$$

$$x_i = 0 + (i-1)\,\Delta x = \frac{2(i-1)}{n}$$

$$y_i = (x_i)^3 = \left[\frac{(2i-2)}{n}\right]^3 = \frac{8i^3 - 24i^2 + 24i - 8}{n^3}$$

The area of the ith rectangle is

$$A_i = y_i\,\Delta x = \left(\frac{8i^3 - 24i^2 + 24i - 8}{n^3}\right)\frac{2}{n}$$

The total area of the rectangles is

$$\begin{aligned}
A_n &= \sum_{i=1}^{n} A_i = \sum_{i=1}^{n} \left(\frac{8i^3 - 24i^2 + 24i - 8}{n^3}\right)\frac{2}{n} \\
&= \sum_{i=1}^{n} \left(\frac{16i^3 - 48i^2 + 48i - 16}{n^4}\right) \\
&= \frac{16}{n^4}\sum_{i=1}^{n} i^3 - \frac{48}{n^4}\sum_{i=1}^{n} i^2 + \frac{48}{n^4}\sum_{i=1}^{n} i - \frac{1}{n^4}\sum_{i=1}^{n} 16 \\
&= \frac{16}{n^4}\cdot\frac{1}{4}\cdot n^2(n+1)^2 - \frac{48}{n^4}\cdot\frac{1}{6}n(n+1)(2n+1) \\
&\quad + \frac{48}{n^4}\cdot\frac{1}{2}n(n+1) - \frac{1}{n^4}\cdot 16n \\
&= 4 + \frac{8}{n} + \frac{4}{n^2} - \frac{16}{n} - \frac{24}{n^2} - \frac{8}{n^3} + \frac{24}{n^2} + \frac{24}{n^3} - \frac{16}{n^3} \\
A_n &= 4 - \frac{8}{n} + \frac{4}{n^2}
\end{aligned}$$

The area bounded by the curve is given by

$$A = \lim_{n\to\infty}\left(4 - \underbrace{\frac{8}{n} + \frac{4}{n^2}}_{\text{approach zero as } n \to \infty}\right)$$

$$A = 4$$

The area under the curve is 4. ■

19–3 EXERCISES

Use the rectangle sum method to approximate the area under the curve and above the x-axis for the given values of n.

1. $y = 2x$ from $x = 0$ to $x = 3$ for $n = 5$ and $n = 10$
2. $y = \frac{2}{3}x$ from $x = 0$ to $x = 3$ for $n = 4$ and $n = 10$
3. $y = x^2$ from $x = 2$ to $x = 4$ for $n = 6$ and $n = 10$
4. $y = 3x^2$ from $x = 0$ to $x = 6$ for $n = 6$ and $n = 10$
5. $y = x^2 + 4$ from $x = 1$ to $x = 2$ for $n = 4$ and $n = 10$
6. $y = x^2 + 6$ from $x = 0$ to $x = 4$ for $n = 2$ and $n = 10$
7. $y = 3x - x^2$ from $x = 0$ to $x = 2$ for $n = 6$ and $n = 10$
8. $y = x + x^2$ from $x = 1$ to $x = 4$ for $n = 3$ and $n = 10$
9. $y = \frac{1}{x^2}$ from $x = 1$ to $x = 2$ for $n = 4$ and $n = 10$
10. $y = \frac{x^3}{3}$ from $x = 0$ to $x = 3$ for $n = 6$ and $n = 10$
11. $y = x^2 + x + 1$ from $x = 0$ to $x = 2$ for $n = 4$ and $n = 10$
12. $y = x^2 + x + 3$ from $x = 0$ to $x = 5$ for $n = 5$ and $n = 10$
13. $y = \frac{x^3}{2}$ from $x = 1$ to $x = 2$ for $n = 8$ and $n = 12$
14. $y = 1 + \sqrt{x}$ from $x = 4$ to $x = 9$ for $n = 5$ and $n = 10$
15. $y = \sqrt{x}$ from $x = 0$ to $x = 3$ for $n = 2$ and $n = 10$

Using the limit method, find the exact area under the curve and above the x-axis.

16. $y = 4x$ from $x = 0$ to $x = 3$
17. $y = \frac{x}{4}$ from $x = 0$ to $x = 4$
18. $y = x^2$ from $x = 1$ to $x = 3$
19. $y = x^2 - 1$ from $x = 3$ to $x = 4$
20. $y = x^2 - 2$ from $x = 2$ to $x = 5$
21. $y = 4x - x^2$ from $x = 1$ to $x = 3$
22. $y = x^3$ from $x = 2$ to $x = 4$
23. $y = 2x^3$ from $x = 0$ to $x = 3$

19–4

THE DEFINITE INTEGRAL

The limit process discussed in the previous section allows us to find the exact area under a curve; however, the calculations can be tedious and difficult. In this section we will show how to use integration in a simpler way to find the area under a curve.

FIGURE 19–7

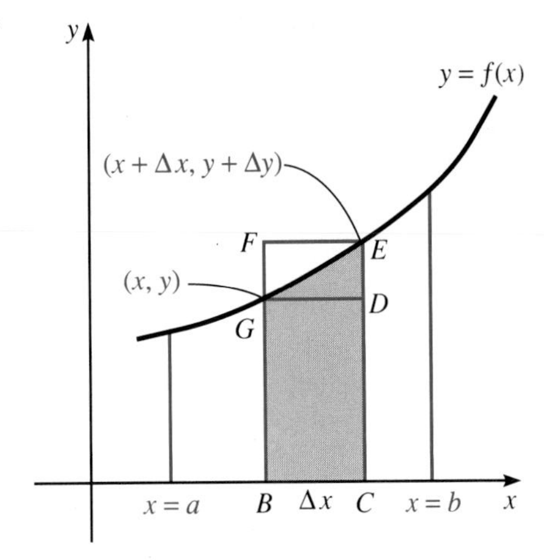

Consider the curve shown in Figure 19–7. Let ΔA represent the area *BCEG* under the curve, the shaded area. The following inequality is true:

$$A_{BCDG} < \Delta A < A_{BCEF}$$

If we let point *G* be represented as (x, y) and point *E* as $(x + \Delta x, y + \Delta y)$, then

$$y\,\Delta x < \Delta A < (y + \Delta y)\,\Delta x$$

Dividing this inequality by Δx gives

$$y < \frac{\Delta A}{\Delta x} < y + \Delta y$$

If we take the limit of this function as $\Delta x \to 0$ (Δy would also approach zero), then

$$y < \lim_{\Delta x \to 0} \frac{\Delta A}{\Delta x} < y$$

$$\frac{dA}{dx} = y \quad \text{or} \quad dA = y\,dx$$

To find the function $A(x)$, we take the indefinite integral of both sides of the equation.

$$\int dA = \int y\,dx$$

$$A(x) = \int y\,dx$$

From the definition of the indefinite integral,

$$A(x) = g(x) + C$$

where $g(x)$ = the antiderivative of y. Therefore, when $x = a$,

$$\begin{aligned} A(a) &= g(a) + C \\ 0 &= g(a) + C \\ -g(a) &= C \end{aligned}$$

In general, the area under the curve from $x = a$ to $x = b$ is given by

$$A(b) = g(b) - g(a)$$

We have related the problem of finding the area under the curve $f(x)$ to evaluating its antiderivative $g(x)$, or integral. The **definite integral** of a function is defined as follows.

The Definite Integral

$$\int_a^b f(x)\,dx = \lim_{n \to \infty} \sum_{i=1}^{n} f(x_i)\,\Delta x$$

where $f(x)$ is defined over the interval $a \leq x \leq b$.

The real numbers a and b are called the **limits of integration.** The number b is the **upper limit,** and the number a is the **lower limit.** Combining these two equations yields the Fundamental Theorem of Calculus, given in the following box.

The Fundamental Theorem of Calculus

$$\int_a^b f(x)\,dx = g(b) - g(a)$$

where $f(x)$ is continuous over the interval $a \leq x \leq b$, and $g(x)$ is the antiderivative of $f(x)$.

The difference $g(b) - g(a)$ is denoted in the equation as

$$\int_a^b f(x)\,dx = g(b) - g(a) = g(x)\Big|_a^b$$

EXAMPLE 1 Find the area under the curve $y = x^3$ from $x = 1$ to $x = 3$.

Solution The area is given by the definite integral

$$\int_1^3 x^3\,dx$$

Integration gives

$$\left.\frac{x^4}{4}\right|_1^3$$

Last, we evaluate this function at the upper limit of 3 and at the lower limit of 1.

evaluate at upper limit (points to $(3)^4$); evaluate at lower limit (points to $(1)^4$); subtract (points to the minus sign)

$$\left.\frac{x^4}{4}\right|_1^3 = \frac{(3)^4}{4} - \frac{(1)^4}{4} = \frac{81}{4} - \frac{1}{4} = \frac{80}{4} = 20 \text{ square units}$$

■

EXAMPLE 2 Integrate the following:

$$\int_1^4 (x^{-3} + 4)\,dx$$

Solution The sum and constant rules hold true for both the definite and indefinite integrals. Applying these rules gives

sum rule (points to $+$); constant rule (points to 4)

$$\int_1^4 x^{-3}\,dx + 4\int_1^4 dx$$

Integrating gives

$$\left.\left(\frac{x^{-2}}{-2} + 4x\right)\right|_1^4$$

Evaluating the function at its limits gives the following results:

$$\left(\frac{(4)^{-2}}{-2} + 4(4)\right) - \left(\frac{(1)^{-2}}{-2} + 4(1)\right)$$

$$-\frac{1}{32} + 16 + \frac{1}{2} - 4$$

$$12\frac{15}{32}$$

■

EXAMPLE 3 Integrate the following:

$$\int_0^1 x(1 - x^2)^{1/4}\, dx$$

Solution To integrate this function, we apply u substitution. Let $u = 1 - x^2$; then $du = -2x\, dx$. Thus,

$$-\frac{1}{2}\int_0^1 (1 - x^2)^{1/4}(-2x\, dx)$$

Integrating this function gives the following results:

$$-\frac{\frac{1}{2}(1 - x^2)^{5/4}}{5/4} = -\frac{2}{5}(1 - x^2)^{5/4}\Bigg|_0^1$$

Evaluating this integral at the limits of integration gives

$$-\frac{2}{5}([1 - (1)^2]^{5/4} - [1 - (0)^2]^{5/4}) = 0.4$$

■

EXAMPLE 4 Integrate the following:

$$\int_0^3 \frac{x + 1}{(x^2 + 2x - 4)^3}\, dx$$

Solution We use the u substitution method again. Let $u = x^2 + 2x - 4$; then $du = (2x + 2)\, dx = 2(x + 1)\, dx$. Note that we must multiply and divide the integral by 2 to obtain the required du.

$$\frac{1}{2}\int_0^3 (x^2 + 2x - 4)^{-3}2(x + 1)\,dx = \left.\frac{\frac{1}{2}(x^2 + 2x - 4)^{-2}}{-2}\right|_0^3$$

$$= -\frac{1}{4}(x^2 + 2x - 4)^{-2}\Big|_0^3$$

Evaluating the limits of integration gives

$$-\frac{1}{4}[(9 + 6 - 4)^{-2} - (0 + 0 - 4)^{-2}]$$

$$= -\frac{1}{4}\left(\frac{1}{121} - \frac{1}{16}\right)$$

$$\approx 0.0136$$

■

EXAMPLE 5 Find the area between the curve $y = x^2 - 1$ and the x axis from $x = 0$ to $x = 2$.

Solution Figure 19–8 shows a graph of $y = x^2 - 1$. Notice that the area is

FIGURE 19–8

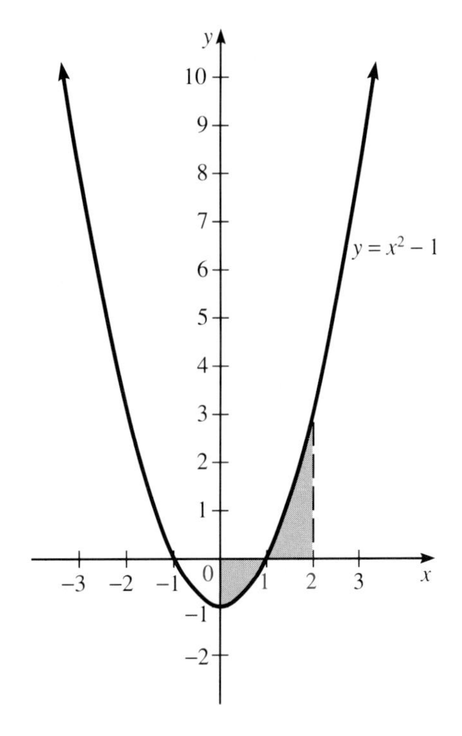

both above and below the x axis. In previous problems, we did not have to be concerned with a change in the sign of $f(x)$ because we were not finding the area over a curve. To find the area under the curve when $f(x)$ changes sign, we divide the limits of integration based upon where the function intersects the x axis. Since the area bounded by the curve is always positive whether it is above or below the x axis, we use the absolute value symbols.

note the limits

$$A = \underbrace{\left| \int_0^1 (x^2 - 1)\, dx \right|}_{\text{area below the } x \text{ axis}} + \underbrace{\left| \int_1^2 (x^2 - 1)\, dx \right|}_{\text{area above the } x \text{ axis}}$$

Next we integrate and evaluate the results at the respective upper and lower limits.

$$A = \left| \left(\frac{x^3}{3} - x \right) \Bigg|_0^1 \right| + \left| \left(\frac{x^3}{3} - x \right) \Bigg|_1^2 \right|$$

$$= \left| \left(\frac{1}{3} - 1 \right) - 0 \right| + \left| \left(\frac{8}{3} - 2 \right) - \left(\frac{1}{3} - 1 \right) \right|$$

$$= \left| -\frac{2}{3} \right| + \left| \frac{4}{3} \right|$$

$$= \frac{2}{3} + \frac{4}{3}$$

$$A = 2 \text{ square units}$$

■

Applications

One application of the definite integral involves finding the area bounded by two curves. This application utilizes the integral as a summation interpretation, as Example 6 illustrates.

To find the area between two curves, first we choose a representative rectangle in the area, as shown in Figure 19–9. The top of the rectangle falls on the curve $f(x)$, and the bottom of the rectangle falls on the curve $g(x)$. The area ΔA of the rectangle is given by

$$\Delta A = \text{length} \cdot \text{width}$$

$$\Delta A = [f(x) - g(x)]\, \Delta x$$

Then we find the area between the curves by summing all the rectangles in the area over the interval $a \leq x \leq b$. This area is given by

$$A = \int_a^b [f(x) - g(x)]\, dx$$

FIGURE 19–9

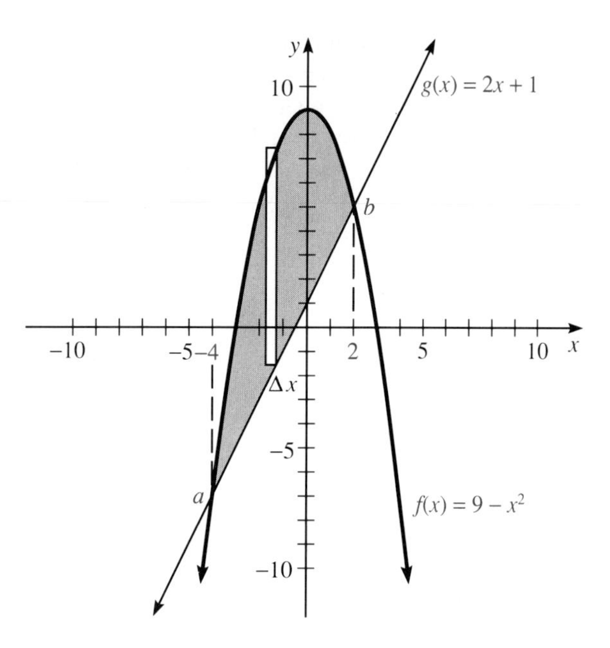

Example 6 illustrates how to find the area between two curves using a vertical representative rectangle.

EXAMPLE 6 Find the area bounded by the curves $y = 9 - x^2$ and $y = 2x + 1$.

Solution Figure 19–9 shows a sketch of these curves with the desired area shaded. We choose a representative vertical rectangular strip in the shaded area, shown in the figure.

Next, we use algebra to find the x coordinates of the points of intersection of the two curves.

$$\begin{aligned} 2x + 1 &= 9 - x^2 \\ x^2 + 2x - 8 &= 0 \\ (x + 4)(x - 2) &= 0 \\ x &= -4 \quad \text{and} \quad x = 2 \end{aligned}$$

These values of x become the limits of integration, and the upper curve throughout is $y = 9 - x^2$. Last, we set up the integral and integrate by using the power rule.

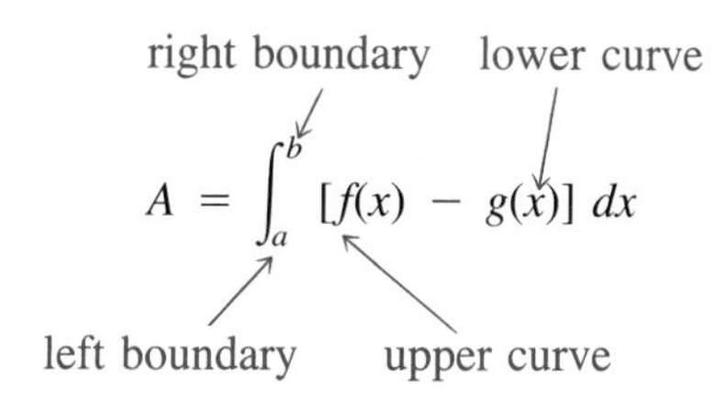

$$A = \int_{-4}^{2} [(9 - x^2) - (2x + 1)]\, dx$$

$$= \left(9x - \frac{x^3}{3}\right)\Bigg|_{-4}^{2} - (x^2 + x)\Bigg|_{-4}^{2}$$

$$= 18 - \frac{8}{3} + 36 - \frac{64}{3} - 6 + 12$$

$$A = 36$$

The area between the two curves is 36 square units. ■

We can also find the area between two curves by using a horizontal representative rectangular strip. Figure 19–10 shows two curves where a horizontal strip is advantageous. The area of the representative rectangle is given by

$$\Delta A = \text{length} \cdot \text{width}$$
$$\Delta A = [p(y) - q(y)]\, \Delta y$$

FIGURE 19–10

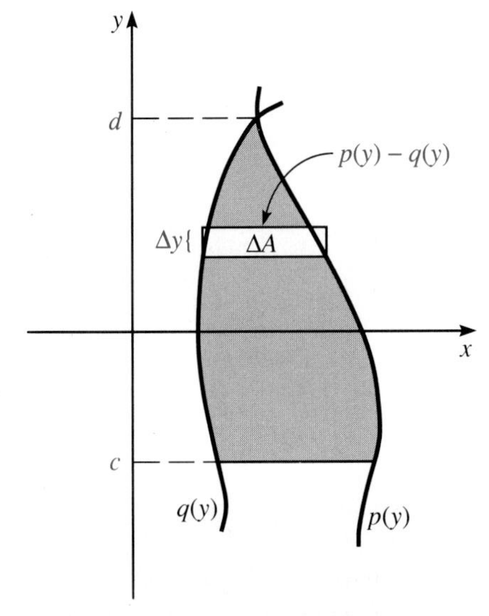

Summing all the rectangles in the area over the interval $c \leq y \leq d$ gives the following result:

$$A = \int_{c}^{d} [p(y) - q(y)]\, dy$$

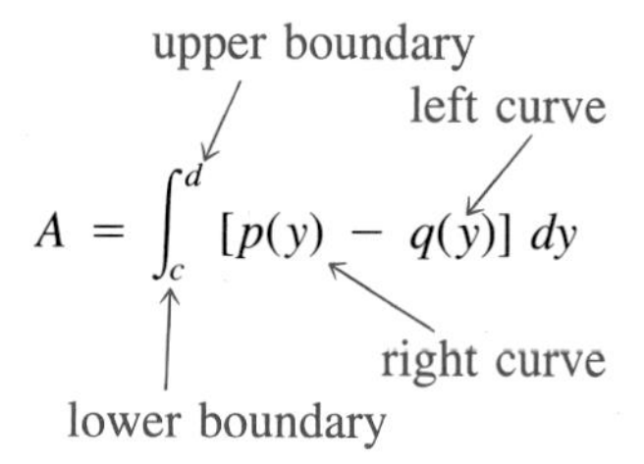

$$A = \int_c^d [p(y) - q(y)]\, dy$$

NOTE ✦ If you choose a vertical representative rectangle to find the area between two curves, sum them from left to right. If you choose a horizontal representative rectangle, sum them from bottom to top.

The center of gravity or balance point of an object is the point where the weight of the object is concentrated. If we consider a homogeneous flat surface area, the center of gravity is called the *centroid* of the area. The process of finding the coordinates of the centroid of a flat surface is similar to finding the area bounded by several curves. The method of finding the centroid of an area is based on taking a representative rectangle whose centroid is at its geometric center. The moment of each rectangle is the product of its area times the distance from the centroid of the rectangle to the axis. The moment of the total area is the limit of the sum of these rectangle moments. Using the vertical rectangle shown in Figure 19–11, the x coordinate of the centroid is given by

$$\bar{x} = \frac{M_y}{A}$$

where M_y = the moment of the area from the y axis, and A = the area. We know from our previous discussion how to calculate the area. To calculate M_y, the moment of each rectangle is the product of the area of the rectangle $(y_2 - y_1)\, dx$, and the distance of the rectangle from the y axis, x. Then we sum all rectangles over the area. The formula for the x coordinate of the centroid is

$$\bar{x} = \frac{\overbrace{\int_a^b x(y_2 - y_1)\, dx}^{M_y}}{\underbrace{\int_a^b (y_2 - y_1)\, dx}_{A}}$$

To find the y coordinate of the centroid, choose a horizontal rectangle. The y

coordinate of the centroid is given by

$$\bar{y} = \frac{M_x}{A}$$

where M_x = the moment of the area from the x axis. The calculation of M_x is the product of the rectangle, $(x_2 - x_1)\,dy$, and the distance of the rectangle from the x axis, y. Summing over all the rectangles gives the following result for the y coordinate of the centroid:

$$\bar{y} = \frac{\overbrace{\int_c^d y(x_2 - x_1)\,dy}^{M_x}}{\underbrace{\int_c^d (x_2 - x_1)\,dy}_{A}}$$

EXAMPLE 7 Find the coordinates of the centroid of the area bounded by $y = 4 - x$, $x = 0$, and $y = 0$.

FIGURE 19–11

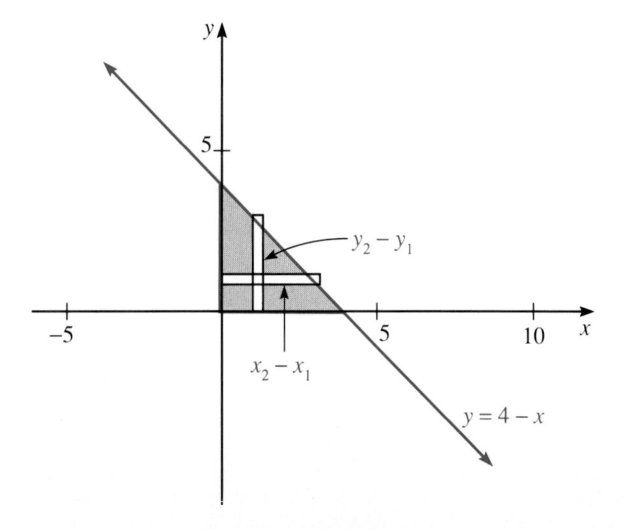

Solution Figure 19–11 shows the graph of these curves and the area of interest. We choose a representative vertical rectangle to calculate the x coordinate. The limits of integration are 0 to 4 because the area includes these x values. The value of y_2 is given by the equation of the upper curve bounding the area, $y = 4 - x$; and y_1 is given by the lower curve, $y = 0$. Substituting these values gives the x coordinate as follows:

$$\bar{x} = \frac{\int_0^4 x[(4 - x) - 0]\,dx}{\int_0^4 [(4 - x) - 0]\,dx}$$

$$= \frac{\int_0^4 (4x - x^2)\,dx}{\int_0^4 (4 - x)\,dx}$$

$$= \frac{\left(2x^2 - \frac{x^3}{3}\right)\Big|_0^4}{\left(4x - \frac{x^2}{2}\right)\Big|_0^4}$$

$$\bar{x} = \frac{4}{3}$$

Then we choose a representative horizontal rectangle for the y coordinate. The limits of integration are 0 to 4 because the area includes these y values. The value of x_2 is the right curve bounding the area, $y = 4 - x$ or $x = 4 - y$, and x_1 is the farthest left curve, $x = 0$. Substituting these values into the formula for the y coordinate gives

$$\bar{y} = \frac{\int_0^4 y[(4 - y - 0)]\,dy}{\int_0^4 [(4 - y) - 0]\,dy}$$

$$= \frac{\int_0^4 (4y - y^2)\,dy}{\int_0^4 (4 - y)\,dy}$$

$$= \frac{\left(2y^2 - \frac{y^3}{3}\right)\Big|_0^4}{\left(4y - \frac{y^2}{2}\right)\Big|_0^4}$$

$$\bar{y} = \frac{4}{3}$$

The coordinates for the centroid of this area are $\left(\frac{4}{3}, \frac{4}{3}\right)$. ■

A third application of integration is to find volumes. If we have an area and its representative rectangle rotated about either the x axis or the y axis, a volume is generated. Depending on the position of the representative rectangle to the axis of rotation, either a circular disk or a cylindrical shell is generated. We will do an example of each.

If we take the representative rectangle perpendicular to the axis of rotation, a circular disk is generated. The volume is the product of the area of the circle and the thickness of the disk, represented by

$$V = \pi(\text{radius})^2(\text{thickness})$$

If we sum all the circular disks from $x = a$ to $x = b$ and take the limit of the sum, the volume is given by

$$V = \int_a^b \pi[f(x)]^2\, dx$$

EXAMPLE 8 Find the volume of the solid of revolution generated by revolving the area bounded by $y = x^2 + 2$, $x = 0$, $x = 2$, and $y = 0$ about the x axis.

Solution Figure 19–12 shows the graph of these curves and the area of interest. If we take a representative rectangle perpendicular to the axis of rota-

FIGURE 19–12

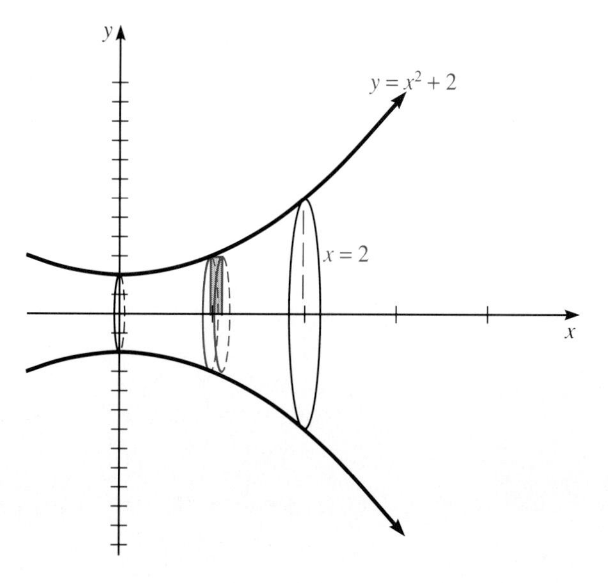

tion, the volume generated in a disk, the volume is given by the definite integral

$$V = \pi \int_a^b [f(x)]^2 \, dx$$

Substituting the information for this problem gives

$$V = \pi \int_0^2 (x^2 + 2)^2 \, dx$$

Integrating gives

$$V = \pi \int_0^2 (x^4 + 4x^2 + 4) \, dx$$

$$= \pi\left(\frac{x^5}{5} + \frac{4x^3}{3} + 4x\right)\Bigg|_0^2$$

$$= \pi\left(\frac{32}{5} + \frac{32}{3} + 8 - 0\right)$$

$$V \approx 78.7 \text{ cubic units}$$

■

An alternate method of calculating volumes of revolution is called the *cylindrical shell method*. If a representative rectangle is taken parallel to the axis of rotation, the solid formed by rotation is a cylinder whose volume is the product of the cross-sectional area and the height. This relationship is expressed as follows:

$$V = 2\pi(\text{radius})(\text{height})(\text{thickness})$$

Summing the volumes of all the cylinders generated by the representative rectangles from $x = a$ to $x = b$ and taking the limit of the sum results in the following integral:

$$V = \int_a^b 2\pi x [f(x) - g(x)] \, dx$$

where $f(x)$ is the upper curve and $g(x)$ is the lower curve. The cylindrical shell method is illustrated in the next example.

EXAMPLE 9 Find the volume of the solid of revolution generated by revolving the area bounded by $y = x^2$, $x = 4$, and $y = 0$ about the y axis.

FIGURE 19–13

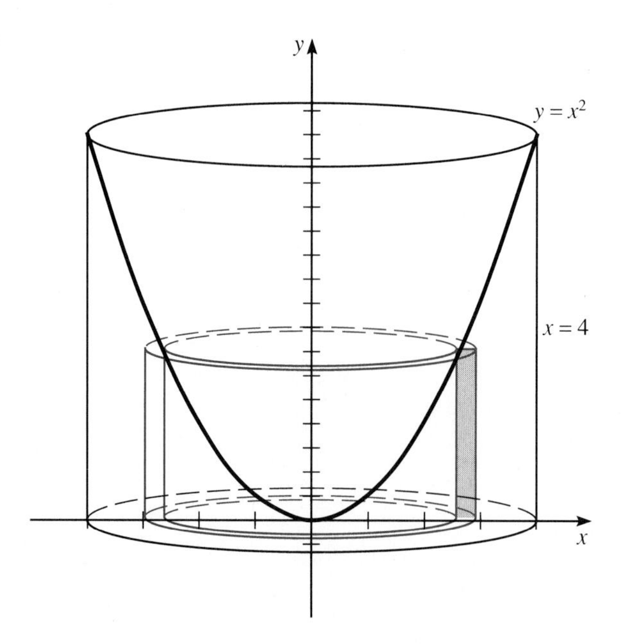

Solution Figure 19–13 shows the region bounded by these curves. If we choose a vertical representative rectangle, a cylindrical shell is generated. The volume is given by the formula

$$V = 2\pi \int_a^b x[f(x) - g(x)]\, dx$$

$$= 2\pi \int_0^4 x(x^2 - 0)\, dx$$

upper curve (x^2); lower curve (0)

$$= 2\pi \int_0^4 x^3\, dx$$

$$= 2\pi\left(\frac{x^4}{4}\right)\Bigg|_0^4$$

$$= 2\pi(64)$$

$$V \approx 402.1 \text{ cubic units}$$ ■

Another application of integration is finding the average value of a function. The arithmetic mean is quite useful in finding the average of discrete functions. However, the integral is more useful in calculating the average value of a continuous function. The average value of a continuous function is given by

$$M = \frac{1}{b - a} \int_a^b f(x)\, dx$$

EXAMPLE 10 Find the average value of the current 5 s after the switch is thrown if $i = 12t - 3t^2$.

Solution The limits of integration are 0 to 5. The average value is given by

$$M = \frac{1}{5 - 0} \int_0^5 (12t - 3t^2)\, dt$$

$$= \frac{1}{5}[6t^2 - t^3] \Big|_0^5$$

$$= \frac{1}{5}(150 - 125)$$

$$M = 5$$

The average current is 5 amperes. ■

19–4 EXERCISES

Integrate the following:

1. $\int_0^3 3x\, dx$
2. $\int_0^4 -6x\, dx$
3. $\int_{-1}^4 6x^3\, dx$
4. $\int_6^{10} 2x^4\, dx$
5. $\int_{-2}^2 x^{2/3}\, dx$
6. $\int_0^3 x^{1/2}\, dx$
7. $\int_0^3 x^{-2/3}\, dx$
8. $\int_{-1}^1 (2x^2 + 3x - 8)\, dx$
9. $\int_{-2}^3 (x^3 + 4x^2 - 6)\, dx$
10. $\int_2^4 3x(x^2 - 6)^3\, dx$
11. $\int_{-3}^1 (x^2 - 2)(x^3 - 6x + 8)^2\, dx$
12. $\int_4^7 \frac{3x^2}{\sqrt{x^3 - 4}}\, dx$
13. $\int_1^3 \frac{x}{\sqrt{x^2 + 3}}\, dx$
14. $\int_0^2 x\sqrt{3x^2 + 4}\, dx$
15. $\int_0^2 x^2\sqrt{1 - x^3}\, dx$
16. $\int_{-2}^0 x^2(2 + 3x^3)\, dx$
17. $\int_{-1}^0 x(1 + 3x^2)^3\, dx$
18. $\int_2^5 \frac{2x^3 + x}{(x^4 + x^2)^3}\, dx$
19. $\int_1^2 \frac{x^2 + 1}{(2x^3 + 6x)^2}\, dx$
20. Find the area between the curves $y = 2x^2$, $y = 0$, and $x = 2$.
21. Find the area bounded by the curves $y = x^3 - x$, $y = 0$, $x = 1$, and $x = 0$.
22. Find the area bounded by the curves $y = x^2$ and $y = x + 1$.
23. Find the area between the curves $y = 9 - x^2$ and $y = 0$.

24. Find the area between the curves $y = 4x^2 - 1$, $y = 0$, $x = 1$, and $x = 4$.

25. Find the coordinates for the centroid of the region bounded by $y = 6 - x$, $x = 0$, $y = 0$.

26. Find the coordinates for the centroid of the region bounded by $y = x^2$ and $y = 4$.

27. Find the coordinates for the centroid of the region bounded by $y = x^3$, $y = 0$, and $x = 2$.

28. Find the coordinates for the centroid of the region bounded by $x = y$, $x = 0$, and $3x + y = 4$.

29. Find the coordinates for the centroid of the region bounded by $y = 16 - x^2$ and $y = 0$.

30. Find the volume generated by revolving the area bounded by $y = x^3$, $x = 2$, and $y = 0$ about the x axis. Use the disk method.

31. Find the volume generated by revolving the area bounded by $y = x + 3$, $x = 0$, $y = 0$, and $x = 4$ about the x axis. Use the disk method.

32. Find the volume generated by revolving the area bounded by $y^2 = x$, $y = 3$, and $x = 0$ about the y axis. Use the disk method.

33. Find the volume generated by revolving the area bounded by $y = x^3$, $y = 2$, and $x = 0$ about the y axis. Use the disk method.

34. Find the volume generated by revolving the area bounded by $y = 4x$, $y = 3$, and $x = 0$ about the x axis. Use the shell method.

35. Find the volume generated by revolving the area bounded by $y = 4 - 2x$, $x = 0$, and $y = 0$ about the x axis. Use the shell method.

36. Find the volume generated by revolving the area bounded by $y = x$, $y = 0$, and $x = 3$ about the y axis. Use the shell method.

37. Find the volume generated by revolving the area bounded by $y = x^2$, $x = 3$, and $y = 0$ about the y axis. Use the shell method.

38. The flow rate through a certain valve is given by the equation $f = 4.3 - 2.3t^2$. What is the average rate of flow in m^3/s from $t = 1$ s to $t = 3$ s?

39. An object moves along the x axis according to the function $x = t^3 - t$. Find the average velocity of the object from $t = 0$ to $t = 4$ s. ($V = dx/dt$)

19–5

NUMERICAL INTEGRATION

In Section 19–4, we used the definite integral to find the area under a curve. However, there are some functions whose antiderivatives are difficult or impossible to find. The Fundamental Theorem of Calculus cannot be used to find the area under the curve of such a function. Therefore, we must use some numerical method, such as the rectangle method discussed in Section 19–3. In this section we will approximate the area under a curve using the trapezoid rather than the rectangle.

Trapezoidal Rule

To find the integral of a function by approximating the area under the curve, we begin by dividing the interval from a to b into n intervals of equal width Δx as shown in Figure 19–14. As before, the value of Δx is given by the formula

$$\Delta x = \frac{b - a}{n}$$

Remember from geometry that the area of a trapezoid is one-half the sum of the bases times the height. The height of each trapezoid is represented by

FIGURE 19–14

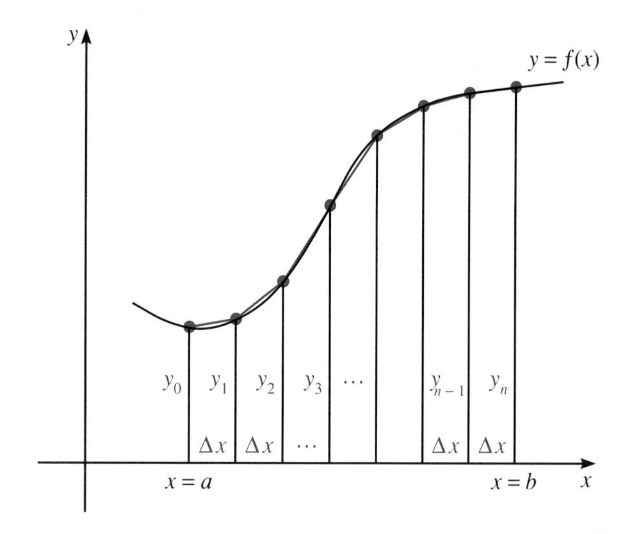

Δx, and the length of the bases equals the y coordinate of the respective points on the curve, represented by $y_0, y_1, y_2, \ldots, y_n$. The integral of this function or area under the curve is approximated by the sum of the areas of the trapezoids as follows:

$$A = \left(\frac{1}{2}y_0 + y_1 + \cdots + \frac{1}{2}y_n\right) \Delta x$$

The trapezoidal rule is summarized in the following box.

The Trapezoidal Rule

$$A = \int_a^b f(x)\,dx \approx \left(\frac{1}{2}y_0 + y_1 + y_2 + \cdots + \frac{1}{2}y_n\right) \Delta x$$

where $\Delta x = \dfrac{b - a}{n}$

EXAMPLE 1 Use the trapezoidal rule to approximate

$$\int_2^4 x^2\,dx$$

Let $n = 4$.

Solution For $n = 4$,

$$\Delta x = \frac{4 - 2}{4} = \frac{1}{2}$$

$$y_0 = f(a) = f(2) = 4$$

$$y_1 = f(2.5) = 6.25$$

$$y_2 = f(3) = 9$$

$$y_3 = f(3.5) = 12.25$$

$$y_4 = f(b) = f(4) = 16$$

The area under the curve from $x = 2$ to $x = 4$ is

$$A \approx \left[\frac{1}{2}(4) + 6.25 + 9 + 12.25 + \frac{1}{2}(16)\right]\frac{1}{2}$$

$$A \approx 18.75$$

Using the trapezoidal rule,

$$\int_2^4 x^2\,dx \approx 18.75$$

If we integrate this function directly, the exact value of the integral is $18\frac{2}{3}$. As in the case of the rectangle rule, increasing the number of intervals results in a closer approximation to the exact value of the integral. ■

EXAMPLE 2 Use the trapezoidal rule to approximate

$$\int_0^2 \sqrt{x^2 + 4}\,dx$$

Let $n = 10$.

Solution The graph of this function is shown in Figure 19–15. For $n = 10$,

$$\Delta x = \frac{2 - 0}{10} = 0.2$$

$$y_0 = f(0) = 2$$

$$y_1 = f(0.2) = 2.0099751$$

$$y_2 = f(0.4) = 2.0396078$$

$$y_3 = f(0.6) = 2.0880613$$

$$y_4 = f(0.8) = 2.1540659$$

FIGURE 19–15

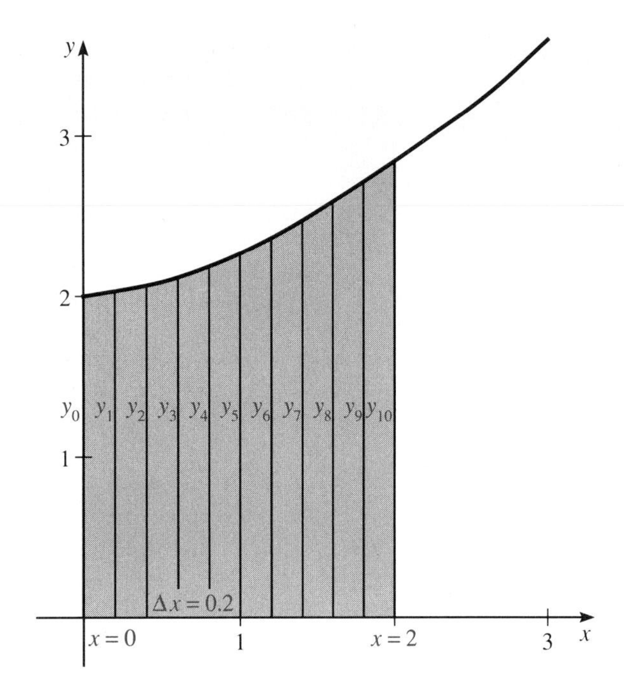

$$y_5 = f(1) = 2.236068$$
$$y_6 = f(1.2) = 2.3323808$$
$$y_7 = f(1.4) = 2.441311$$
$$y_8 = f(1.6) = 2.5612497$$
$$y_9 = f(1.8) = 2.6907248$$
$$y_{10} = f(2) = 2.8284271$$

The area under the curve is

$$A \approx \left[\frac{1}{2}(2) + 2.0099751 + 2.0396078 + \cdots + 2.6907248 + \frac{1}{2}(2.8284271)\right](0.2)$$

$$A \approx 4.5935316$$

Therefore,

$$\int_0^2 \sqrt{x^2 + 4}\, dx \approx 4.5935$$

Note that this integral could not be evaluated using any of the methods explained thus far. ■

Empirical Data

We can also use the trapezoidal rule to approximate an integral when the function is not defined by a rule or an equation. The next two examples illustrate the application of the trapezoidal rule to empirical data.

EXAMPLE 3 The work W done by a miniature airplane is

$$W = \int_{v_0}^{v_n} P\,dv$$

where v = volume in ft^3, and P = pressure in lb/ft^2. Using the data in Table 19–1, what is the work done when the volume increases from 0.25 ft^3 to 0.50 ft^3?

TABLE 19–1

Volume V (ft^3)	Pressure, P (lb/ft^2)
0.25	4.8
0.30	5.2
0.35	6.1
0.40	7.8
0.45	10.0
0.50	13.8

Solution

$$\Delta v = (0.30 - 0.25) = 0.05 \qquad \text{or} \qquad \Delta v = \frac{0.50 - 0.25}{5} = \frac{0.25}{5} = 0.05$$

$$\begin{aligned}
v_0 &= 0.25 & P(v_0) &= 4.8 \\
v_1 &= 0.30 & P(v_1) &= 5.2 \\
v_2 &= 0.35 & P(v_2) &= 6.1 \\
v_3 &= 0.40 & P(v_3) &= 7.8 \\
v_4 &= 0.45 & P(v_4) &= 10.0 \\
v_5 &= 0.50 & P(v_5) &= 13.8
\end{aligned}$$

Using the trapezoidal rule, the work done is given by the following expression:

$$W \approx \left[\frac{1}{2}(4.8) + 5.2 + 6.1 + 7.8 + 10.0 + \frac{1}{2}(13.8)\right](0.05)$$

$$W \approx 1.92$$

The engine does 1.92 ft-lb of work. ■

EXAMPLE 4 Using the data shown in Table 19–2, find the velocity of a car given its acceleration as a function of time.

TABLE 19–2

Time, t (s)	Acceleration, a (m/s^2)
1	24
2	28
3	32
4	36

Solution Recall from previous discussions that $v = \int a\, dt$. Applying the trapezoidal rule gives

$$\Delta t = \frac{4 - 1}{3} = \frac{3}{3} = 1$$

$$\begin{aligned} t_0 &= 1 & a(t_0) &= 24 \\ t_1 &= 2 & a(t_1) &= 28 \\ t_2 &= 3 & a(t_2) &= 32 \\ t_3 &= 4 & a(t_3) &= 36 \end{aligned}$$

$$v = \left[\frac{1}{2}(24) + 28 + 32 + \frac{1}{2}(36)\right](1)$$

$$v = 90$$

The velocity of the car is 90 m/s. ■

19–5 EXERCISES

Use the trapezoidal rule to approximate the value of each integral. Check Exercises 1–4 by direct integration. Round to two decimal places where necessary.

1. $\int_1^3 3x^2\, dx;\ n = 4$
2. $\int_0^2 (3 + x^2)\, dx;\ n = 6$
3. $\int_2^4 \sqrt{3x - 1}\, dx;\ n = 4$
4. $\int_1^2 x^{-3}\, dx;\ n = 4$
5. $\int_1^3 \sqrt{x^2 + 1}\, dx;\ n = 6$
6. $\int_1^6 x\sqrt{x - 1}\, dx;\ n = 5$
7. $\int_0^2 (1 + x^2)^3\, dx;\ n = 6$
8. $\int_1^3 \frac{1}{x(x + 2)}\, dx;\ n = 8$
9. $\int_0^1 \frac{1}{x^2 + x + 1}\, dx;\ n = 4$
10. $\int_0^2 \frac{1}{x^2 + 2}\, dx;\ n = 8$
11. $\int_0^2 2^x\, dx;\ n = 10$
12. $\int_0^1 e^{2x}\, dx;\ n = 10$

Use the given set of points to find the approximate value of each integral. Round to two decimal places.

13. Find $\int_1^6 y\,dx$ for the following data:

x	1	2	3	4	5	6
y	8	9.6	11	12.5	15	18.3

14. Find $\int_2^4 y\,dx$ for the following data:

x	2	2.5	3	3.5	4
y	0.7	1.2	2	2.8	3.7

15. Find $\int_0^3 y\,dx$ for the following data:

x	0	1	2	3
y	1.3	2.1	3.7	4.2

16. Find $\int_{0.1}^{0.4} y\,dx$ for the following data:

x	0.1	0.2	0.3	0.4
y	4.3	5.2	6.3	7.8

17. The displacement s (in ft) of an object is given by $s = \int v\,dt$. Using the following empirical data, find the displacement:

t (s)	1	2	3	4	5
v (ft/s)	58	76	94	112	130

18. The electrical charge q transferred past a point is given by $q = \int i\,dt$. Find the electrical charge for the following empirical data:

t (s)	0.1	0.3	0.5	0.7	0.9
i (A)	3.03	3.09	3.16	3.22	3.28

19. The current I required to cause a voltage drop V is given by

$$I = \int \frac{V}{L}\,dt$$

where L = inductance. Using the following empirical data, find the current in a circuit:

t	0.10	0.15	0.20	0.25	0.30
V/L	0.285	0.301	0.316	0.331	0.345

20. The work done in moving an object is given by $W = \int F(x)\,dx$. Using the following empirical data, find the work done in moving an object:

x (lb)	10	15	20	25	30	35
$F(x)$ (ft)	19	24	29	34	39	44

CHAPTER SUMMARY

Summary of Terms

absolute error (p. 698)

antidifferentiation (p. 701)

definite integral (p. 720)

indefinite integral (p. 703)

integrating (p. 703)

integration (p. 703)

limits of integration (p. 720)

lower limit (p. 720)

relative error (p. 699)

upper limit (p. 720)

Summary of Formulas

$$\int k\,dx = k\int dx = kx + C \quad \text{integration of a constant}$$

$$\int x^n\,dx = \frac{x^{n+1}}{n+1} + C \quad \text{power rule of integration}$$

$$\int [f(x) + g(x)]\,dx = \int f(x)\,dx + \int g(x)\,dx \quad \text{sum rule of integration}$$

$$\int [f(x)^n g(x)]\,dx = \int u^n\,du = \frac{u^{n+1}}{n+1} + C \quad \text{generalized power rule}$$

$$A = \lim_{n\to\infty} \sum_{i=1}^{n} f(x_i)\,\Delta x \quad \text{area under a curve}$$

$$\int_a^b f(x)\,dx = g(b) - g(a) \quad \text{where } g(x) = \text{the antiderivative of } f(x) \quad \text{fundamental theorem of calculus}$$

$$A = \int_a^b f(x)\,dx \approx \left(\frac{1}{2}y_0 + y_1 + y_2 + \cdots + \frac{1}{2}y_n\right)\Delta x \quad \text{trapezoidal rule}$$

CHAPTER REVIEW

Section 19–1

Find the differential of the following functions.

1. $y = 7x^3 - 8x^2 + 6$

2. $y = 7 + 9x - 3x^4$

3. $y = (3x^2 + 7)(x - 1)$

4. $y = (2x - 5)(9x^2 - 2)$

5. $y = \dfrac{8x^2 + x}{7x + 3}$

6. $y = \dfrac{7 + 5x^2}{3 + x}$

Find the approximate change in the indicated variable by calculating the differential.

7. $V = s^3$; $s = 10$ cm; $\Delta s = 0.6$ cm; $\Delta V = ?$

8. $A = \pi r^2$; $r = 16$ in.; $\Delta r = 0.2$ in.; $\Delta A = ?$

9. $V = \dfrac{4}{3}\pi r^3$; $r = 16$ in.; $\Delta r = 0.2$ in.; $\Delta V = ?$

10. $A = \dfrac{\pi d^2}{4}$; $d = 2.7$ yd; $\Delta d = 0.3$ yd; $\Delta A = ?$

11. A circular watch battery is 8 mm in radius. If the maximum allowable error is 0.03 mm, find the maximum allowable relative error in the top surface area of the battery.

12. A square plate, 15 in. on a side, is to be made so that the maximum error in area is 0.18 in^2. What is the approximate allowable error in the length of the side?

Section 19–2

Integrate the following expressions.

13. $\int 6x^3\,dx$

14. $\int 7x^{1/2}\,dx$

15. $\int (\sqrt{x} + x^{-1/3} + 9x^3)\,dx$

16. $\int (x^{-4} + 3\sqrt{x} - 7x^2)\,dx$

17. $\int \sqrt[4]{3x + 15}\,dx$

18. $\int \sqrt[3]{7x - 1}\,dx$

19. $\int 4x^3(x^4 + 9)^2\,dx$

20. $\int 2x(x^2 + 10)^3\,dx$

21. $\int x^2(x^3 + 9)^4\,dx$

22. $\int x(4x^2 + 5)^2\,dx$

23. A bullet shot vertically upward from the roof of an 80-ft building has an initial velocity of 575 ft/s. Find the velocity of the bullet after 10 s. When will it reach its maximum height?

24. Find the equation of the function whose graph has a slope of $4x^2 + x - 3$ and passes through the point (2, 6).

Section 19–3

Using the limit method, find the exact area under the curve.

25. $y = 2x^2 - 1$ from $x = 1$ to $x = 3$

26. $y = x^2$ from $x = 0$ to $x = 2$

27. $y = x^2 + 2$ from $x = 1$ to $x = 4$

28. $y = x^3 + 1$ from $x = 0$ to $x = 1$

29. $y = x^3 - 1$ from $x = 2$ to $x = 3$

Section 19–4

Integrate the following functions.

30. $\int_0^2 (x^3 + 5x - 8)\, dx$

31. $\int_1^3 (5x^3 - 9x + 7)\, dx$

32. $\int_{-1}^3 x(2x^2 - 3)^3\, dx$

33. $\int_{-1}^1 x^3(x^4 + 10)^2\, dx$

34. $\int_2^4 x\sqrt{10 + x^2}\, dx$

35. $\int_{-3}^0 x^2(x^3 + 9)^{1/3}\, dx$

36. $\int_0^2 (x^4 + 3x^2 + 5)\, dx$

37. $\int_0^2 (2x^5 + 8x - 15x^2)\, dx$

38. $\int_{-1}^0 \frac{2x + 1}{(x^2 + x - 8)^2}\, dx$

39. $\int_2^4 \frac{3x^2 + 1}{(x^3 + x)^2}\, dx$

40. Find the area under the curve $y = x^2$ from $x = 1$ to $x = 3$.

41. Find the area bounded by $y = x^2 + 1$, $y = 0$, $x = 0$, and $x = 3$.

42. Find the coordinates of the centroid of the region bounded by $y = 4 - x^2$ and $y = 0$.

43. Find the coordinates of the centroid of the region bounded by $y = 3x^2$, $y = 0$, and $x = 2$.

44. Find the volume generated by revolving the area bounded by $y = x^2$, $y = 0$, and $x = 4$ about the y axis. Use the shell method.

45. Find the volume generated by rotating the area bounded by $y = 4x - x^2$ and $y = 0$ about the x axis. Use the disk method.

Section 19–5

Use the trapezoidal rule to approximate the value of each integral. Round to two decimal places where necessary.

46. $\int_0^3 x^2\, dx;\ n = 3$

47. $\int_1^2 (x^2 + x)\, dx;\ n = 4$

48. $\int_0^1 \sqrt{x^3 + 1}\, dx;\ n = 5$

49. $\int_2^3 x\sqrt{x + 2}\, dx;\ n = 3$

50. $\int_1^3 e^{x^2}\, dx;\ n = 4$

51. $\int_0^2 \frac{1}{x^2 + 3}\, dx;\ n = 6$

Using the empirical data, approximate the value of each integral. Round to two decimal places.

52. $\int_1^6 y\, dx$

x	1	2	3	4	5	6
y	3	4.1	5	6.3	7	8.5

53. $\int_2^4 y\, dx$

x	2	2.5	3	3.5	4
y	0.6	1.4	1.9	2.3	2.7

CHAPTER TEST

The number in parentheses refers to the appropriate learning objective given at the beginning of the chapter.

1. Integrate the following: (3)

$$\int (8x^3 - 6x^2 + x - 10)\, dx$$

2. A small machine has a 4-s cycle, and it develops power (in ft-lb) according to the equation $P = 3t^2 - t$ where t = time in seconds. Find the work done in one cycle. (7)

3. Integrate the following: (3)

$$\int x^2(7x^3 + 9)^4\, dx$$

4. Find the area bounded by the curves $y = x^2 + 2$, $y = 0$, $x = 0$, and $x = 3$. (7)

5. Integrate the following: (5)

$$\int_0^2 \frac{x}{\sqrt[3]{x^2 - 1}}\, dx$$

6. Find the coordinates of the centroid of the region bounded by $y = 3x^2$, $y = 0$, and $x = 2$. (7)

7. Find the differential for $y = (3x^2 - x)(5x + 7)$. (1)

8. Use the limit method to find the area under the curve $y = 2x^2$ from $x = 1$ to $x = 3$. (4)

9. Find the volume generated by revolving the area bounded by $y = x^2 + 2$, $x = 0$, $y = 0$, and $x = 2$ about the y axis. Use the shell method. (7)

10. Use the trapezoidal rule to approximate the value of (6)

$$\int_0^1 (1 + x^2)^2\, dx$$

Let $n = 8$.

11. The diameter of a ball bearing through repeated measurements is found to be 4.7 ± 0.003 mm. Find the percent error in its volume. (2)

12. Integrate the following: (5)

$$\int_{-1}^4 (x^2 + 4x)(x^3 + 6x^2 - 8)\, dx$$

13. Find the average power in a resistor if the power is given by $P = 1.7i^2 + 10$ if the current ranges from 0.15 A to 0.45 A. (7)

14. The electrical charge q transferred past a point is given by $q = \int i\, dt$. Use the trapezoidal rule to approximate the electrical charge. (6)

$t(x)$	0.1	0.3	0.5	0.7	0.9
i(A)	2.2	2.24	2.41	2.65	2.78

15. The acceleration of a moving particle is given by (7)

$$a = 3 + t + \frac{3}{(t + 2)^2}$$

Find an expression for the velocity of $V = 0$ when $t = 0$ given that $V = \int a\, dt$.

SOLUTION TO CHAPTER INTRODUCTION

FIGURE 19–16

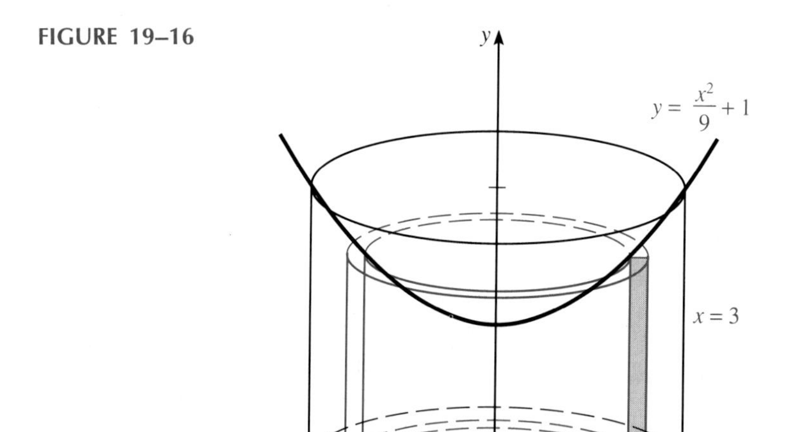

Figure 19–16 shows the region bounded by the given curves and the representative rectangle. A cylindrical shell results from revolving this rectangle about the y axis. The volume is given by the formula

$$V = 2\pi \int_a^b x[f(x) - g(x)]\, dx$$

$$V = 2\pi \int_0^3 x\left(\frac{x^2}{9} + 1\right) dx$$

Integrating and evaluating the limits gives

$$V = 2\pi \cdot \frac{9}{2} \int_0^3 \overbrace{\left(\frac{x^2}{9} + 1\right)}^{u} \overbrace{\left(\frac{2x}{9}\, dx\right)}^{du}$$

$$= 9\pi\left(\frac{u^2}{2}\right)\Bigg|_0^3 = \frac{9\pi}{2}\left(\frac{x^2}{9} + 1\right)^2\Bigg|_0^3$$

$$= \frac{9\pi}{2}[4 - 1]$$

$$V = \frac{27\pi}{2} \text{ in.}^3$$

For the circuit shown on the opposite page, you must find the current flowing 0.003 s after the switch is closed given that $V = 240$ V, $R = 2600\ \Omega$, and $C = 2 \times 10^{-6}$ F. (The solution to this problem is given at the end of the chapter.)

In Chapter 18, we found the derivative of algebraic functions, including polynomials. Any function that is not algebraic is called *transcendental*. In this chapter we will discuss the derivative of three types of transcendental functions: the trigonometric functions and their inverses; the logarithmic function; and the exponential function, which will be used to solve the problem above.

Learning Objectives

After completing this chapter, you should be able to

1. Find the derivative of functions involving the sine or cosine function (Section 20–1).
2. Find the derivative of functions involving the tangent, cotangent, secant, or cosecant functions (Section 20–2).
3. Find the derivative of the inverse trigonometric functions (Section 20–3).
4. Find the derivative of the logarithmic function (Section 20–4).
5. Find the derivative of exponential functions (Section 20–5).
6. Apply the concept of the derivative of transcendental functions to technical problems (Sections 20–1, 20–2, 20–3, 20–4, and 20–5).

Chapter 20

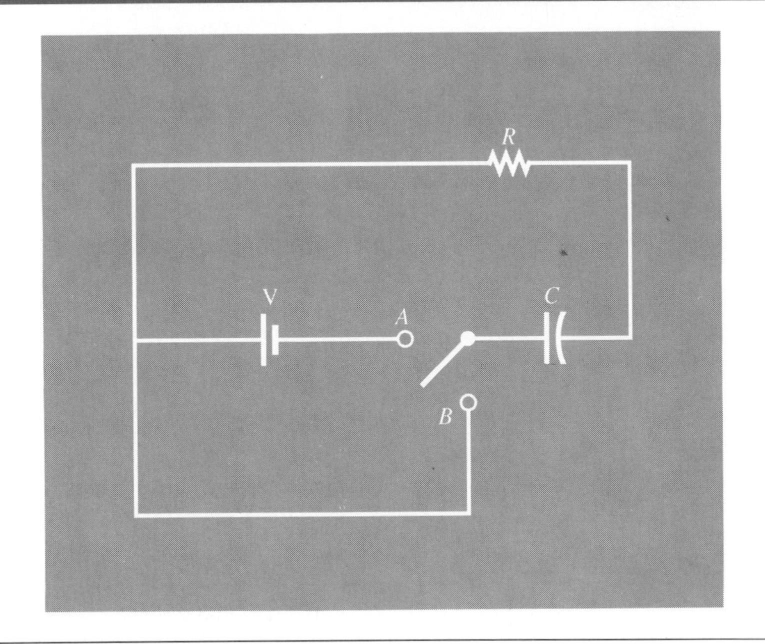

Derivatives of Transcendental Functions

20–1

DERIVATIVES OF THE SINE AND COSINE FUNCTIONS

We will begin the discussion about derivatives of the trigonometric functions by deriving the derivative of the sine function, because we can use its derivative to find the derivative of the remaining trigonometric functions and their inverses. We can find the derivative of the sine function by using the delta process; however, we need two preliminary results which we now derive.

First, we must prove

$$\lim_{\theta \to 0} \frac{\sin \theta}{\theta} = 1$$

where θ is measured in radians. Since the numerator and denominator approach zero as $\theta \to 0$, we cannot just substitute zero for θ and evaluate the results. We fill in the table of values shown as Table 20–1 as we did in Chapter 18.

TABLE 20–1

θ (radians)	$\sin \theta$	$\frac{\sin \theta}{\theta}$
0.5	0.4794255	0.9588511
0.1	0.0998334	0.9983342
0.05	0.0499792	0.9995834
0.01	0.0099998	0.9999833
0.005	0.00499998	0.9999958
0.001	0.0009999	0.9999998

From Table 20–1, notice that as $\theta \to 0$ $(\sin \theta)/\theta$ approaches 1. To prove this, we use the geometric approach using Figure 20–1. From this figure, we can see the following inequality is true:

area of triangle OBD < area of sector OBD < area of triangle OBC

Substituting for these areas gives the following:

$$\frac{1}{2}\overset{\text{base}}{\overset{\downarrow}{OB}} \cdot \underset{\text{height}}{\underset{\uparrow}{DA}} < \overset{\text{area of sector}}{\overbrace{\frac{1}{2}r^2\theta}} < \frac{1}{2}\overset{\text{base}}{\overset{\downarrow}{OB}} \cdot \underset{\text{height}}{\underset{\uparrow}{CB}}$$

But $OB = r$, the radius of the circle; $DA = r \sin \theta$ (right triangle trigonometry); and $CB = r \tan \theta$ (right triangle trigonometry). Substituting these values

FIGURE 20–1

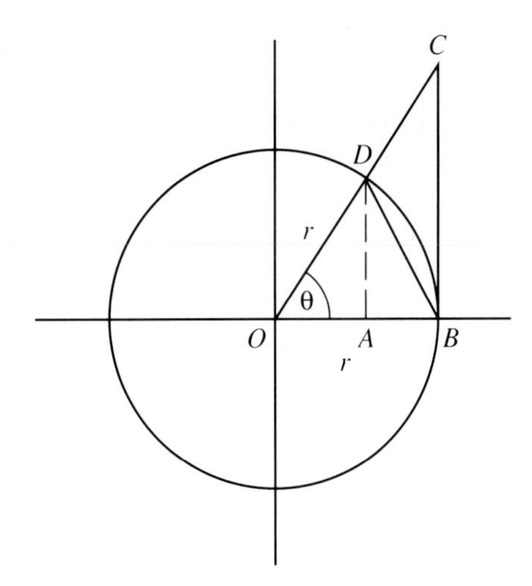

into the inequality gives the following result:

$$\frac{1}{2}r(r \sin \theta) < \frac{1}{2}r^2\theta < \frac{1}{2}r(r \tan \theta)$$

$$\frac{1}{2}r^2 \sin \theta < \frac{1}{2}r^2\theta < \frac{1}{2}r^2 \tan \theta$$

From trigonometry we also know that $\tan \theta = \sin \theta/\cos \theta$. Thus,

$$\frac{1}{2}r^2 \sin \theta < \frac{1}{2}r^2\theta < \frac{1}{2}r^2 \frac{\sin \theta}{\cos \theta}$$

Multiplying each term of the inequality by $2/(r^2 \sin \theta)$ gives the following result:

$$1 < \frac{\theta}{\sin \theta} < \frac{1}{\cos \theta}$$

Taking the reciprocal of each term and reversing the inequality symbols gives

$$1 > \frac{\sin \theta}{\theta} > \cos \theta$$

Then, if we take the limit as $\theta \rightarrow 0$, we have

$$\lim_{\theta \to 0} 1 > \lim_{\theta \to 0} \frac{\sin \theta}{\theta} > \lim_{\theta \to 0} \cos \theta$$

Since $\cos 0 = 1$, $\lim\limits_{\theta \to 0} \cos \theta = 1$.

$$1 > \lim_{\theta \to 0} \frac{\sin \theta}{\theta} > 1$$

The pinching theorem allows us to find the limit of a function by "squeezing" the function between two other simple functions whose limits are known. Applying the pinching theorem to this function gives

$$\lim_{\theta \to 0} \frac{\sin \theta}{\theta} = 1$$

The second preliminary result that we must prove is

$$\lim_{\theta \to 0} \frac{\cos \theta - 1}{\theta} = 0$$

Table 20–2 shows a few values for this function.

TABLE 20–2

θ (radians)	$\cos \theta$	$\cos \theta - 1$	$\frac{\cos \theta - 1}{\theta}$
0.5	0.877583	−0.122417	−0.244835
0.1	0.995004	−0.004996	−0.049958
0.05	0.998750	−0.001250	−0.024995
0.01	0.999950	−0.000050	−0.005000
0.005	0.999988	−0.000013	−0.002500
0.001	0.9999995	−0.0000005	−0.000500

From Table 20–2, the value of $\frac{\cos \theta - 1}{\theta}$ approaches zero as θ approaches zero. The proof is given below.

$$\begin{aligned} \frac{\cos \theta - 1}{\theta} &= \frac{\cos \theta - 1}{\theta} \cdot \frac{\cos \theta + 1}{\cos \theta + 1} \\ &= \frac{\cos^2\theta - 1}{\theta(\cos \theta + 1)} \\ &= \frac{-\sin^2\theta}{\theta(\cos \theta + 1)} \\ &= -\frac{\sin \theta}{\theta} \cdot \frac{\sin \theta}{\cos \theta + 1} \end{aligned}$$

Taking the limit as $\theta \to 0$ gives

$$\lim_{\theta \to 0} \left(\frac{\cos \theta - 1}{\theta} \right) = \lim_{\theta \to 0} \left(\frac{-\sin \theta}{\theta} \right) \cdot \lim_{\theta \to 0} \frac{\sin \theta}{\cos \theta + 1}$$
$$= (-1)(0) = 0$$

Therefore,

$$\lim_{\theta \to 0} \frac{\cos \theta - 1}{\theta} = 0$$

Derivative of the Sine Function

Now, we can use the delta process to find the derivative of the sine function. If $y = \sin u$ for u in radians, then

$$y + \Delta y = \sin(u + \Delta u) \qquad \text{substitute}$$
$$\Delta y = \sin(u + \Delta u) - \sin u \qquad \text{solve for } \Delta y$$
$$\frac{\Delta y}{\Delta u} = \frac{\sin(u + \Delta u) - \sin u}{\Delta u} \qquad \text{divide by } \Delta u$$

From Chapter 13, we know that $\sin(A + B) = \sin A \cos B + \cos A \sin B$. Substituting this identity gives the following equation:

$$\frac{\Delta y}{\Delta u} = \frac{\sin u \cos \Delta u + \cos u \sin \Delta u - \sin u}{\Delta u}$$
$$= \frac{\cos u \sin \Delta u + \sin u(\cos \Delta u - 1)}{\Delta u} \qquad \text{factor}$$
$$= \cos u \left(\frac{\sin \Delta u}{\Delta u} \right) + \sin u \left(\frac{\cos \Delta u - 1}{\Delta u} \right)$$
$$\frac{dy}{du} = \cos u \underbrace{\lim_{\Delta u \to 0} \left(\frac{\sin \Delta u}{\Delta u} \right)}_{\text{equals } 1} + \sin u \underbrace{\lim_{\Delta u \to 0} \left(\frac{\cos \Delta u - 1}{\Delta u} \right)}_{\text{equals } 0} \qquad \text{take the limit}$$
$$= \cos u$$

Last, if u is a function of x, we can find the derivative using the chain rule.

$$\frac{dy}{dx} = \frac{dy}{du} \cdot \frac{du}{dx} = \cos u \frac{du}{dx}$$

Derivative of the Sine Function

$$\frac{d(\sin u)}{dx} = \cos u \frac{du}{dx}$$

where u = a function of x

EXAMPLE 1 Find the derivative of the following functions:

(a) $y = \sin 3x$ (b) $y = \sin(x^3)$

Solution

(a) Let $u = 3x$; then $du/dx = 3$.

$$\frac{d(\sin 3x)}{dx} = \cos \underset{u}{3x} \cdot \underset{du/dx}{3} = 3 \cos 3x$$

(b) Let $u = x^3$; then $du/dx = 3x^2$.

$$\frac{d[\sin(x^3)]}{dx} = \cos(x^3) \cdot 3x^2 = 3x^2 \cos(x^3)$$

■

CAUTION ✦ Be very careful in taking the derivative of $\sin u$. Remember to include du/dx in the answer.

EXAMPLE 2 Find the derivative of the following functions:

(a) $y = \sin^3 x$ (b) $y = 4 \sin^2(2x^3)$

Solution

(a) Remember that $\sin^3 x$ equals $(\sin x)^3$. Therefore, we must use the power rule developed for polynomial functions combined with the derivative of the sine function to find the derivative of this function. Let $u = \sin x$; then $du/dx = d(\sin x)/dx = \cos x$. Then the function can be written as $y = u^n$.

$$\frac{d(\sin x)^3}{dx} = \overbrace{3(\sin x)^2}^{\frac{du^n}{dx}} \cdot \overbrace{\cos x}^{\frac{du}{dx}}$$

$$= 3 \sin^2 x \cos x$$

(b) $$\frac{d[4 \sin^2(2x^3)]}{dx} = 4 \frac{d[\sin^2(2x^3)]}{dx}$$

Let $u = \sin(2x^3)$. Then the function can be written as $y = u^2$, and the derivative, using the power rule, is

$$\begin{aligned}\frac{dy}{dx} &= 4 \cdot 2u \frac{du}{dx} \\ &= 4 \cdot 2 \sin(2x^3) \frac{du}{dx} \\ &= 8 \sin(2x^3) \frac{du}{dx}\end{aligned}$$

Then we must find du/dx when $u = \sin(2x^3)$.

$$\begin{aligned}\frac{du}{dx} &= \cos(2x^3) \cdot \frac{d(2x^3)}{dx} \\ &= \cos(2x^3) \cdot (6x^2)\end{aligned}$$

The derivative is

$$\begin{aligned}\frac{dy}{dx} &= 8 \sin(2x^3) \cdot (6x^2)\cos(2x^3) \\ &= 48x^2\sin(2x^3)\cos(2x^3)\end{aligned}$$

■

EXAMPLE 3 Find dy/dx if

(a) $y = x^2\sin(2x^2)$ (b) $y = \dfrac{\sin^2 x}{x^3}$

Solution

(a) We use the product rule.

$$\frac{dy}{dx} = x^2 \frac{d[\sin(2x^2)]}{dx} + \sin 2x^2 \frac{d(x^2)}{dx}$$

$$u = 2x^2$$

$$\frac{d(\sin u)}{dx} = \cos u \frac{du}{dx}$$

$$\frac{d[\sin(2x^2)]}{dx} = \cos 2x^2 \cdot 4x$$

$$\begin{aligned}\frac{dy}{dx} &= x^2[4x \cos(2x^2)] + \sin(2x^2) \cdot (2x) \\ &= 4x^3\cos(2x^2) + 2x \sin(2x^2)\end{aligned}$$

(b) We use the quotient rule.

$$\frac{dy}{dx} = \frac{x^3 \dfrac{d(\sin^2 x)}{dx} - \sin^2 x \dfrac{d(x^3)}{dx}}{(x^3)^2}$$

$u = \sin x$ and $y = u^2$

$2u \dfrac{du}{dx}$

$2 \sin x \cos x$

$$= \frac{2x^3 \sin x \cos x - 3x^2 \sin^2 x}{x^6}$$

$$\frac{dy}{dx} = \frac{2x \sin x \cos x - 3 \sin^2 x}{x^4}$$

■

EXAMPLE 4 Find dy/dx if $y^3 = x^2y + \sin(4x^3)$.

Solution We must use implicit differentiation to find dy/dx. Taking the derivative of both sides of the equation gives

product rule

$$3y^2 \frac{dy}{dx} = \overbrace{x^2 \frac{dy}{dx} + 2xy}^{\text{product rule}} + (12x^2)\cos(4x^3)$$

$$3y^2 \frac{dy}{dx} - x^2 \frac{dy}{dx} = 2xy + 12x^2\cos(4x^3)$$

$$\frac{dy}{dx}(3y^2 - x^2) = 2xy + 12x^2\cos(4x^3)$$

$$\frac{dy}{dx} = \frac{2xy + 12x^2\cos(4x^3)}{3y^2 - x^2}$$

■

Derivative of the Cosine Function

To derive the formula for the derivative of the cosine function, we use the delta process. Let $y = \cos u$. Then

$$y + \Delta y = \cos(u + \Delta u)$$

$$\Delta y = \cos(u + \Delta u) - \cos u$$

$$\frac{\Delta y}{\Delta u} = \frac{\cos(u + \Delta u) - \cos u}{\Delta u}$$

Then we use the following trigonometric identity:

$$\cos(A + B) = \cos A \cos B - \sin A \sin B$$

$$\frac{\Delta y}{\Delta u} = \frac{\cos u \cos \Delta u - \sin u \sin \Delta u - \cos u}{\Delta u}$$

$$= \frac{\cos u(\cos \Delta u - 1) - \sin u \sin \Delta u}{\Delta u}$$

$$\frac{\Delta y}{\Delta u} = \cos u \frac{\cos \Delta u - 1}{\Delta u} - \sin u \frac{\sin \Delta u}{\Delta u}$$

Evaluating the limit as $\Delta u \to 0$ gives

$$\frac{dy}{du} = \cos u \underbrace{\left(\frac{\cos \Delta u - 1}{\Delta u}\right)}_{\text{equals } 0} - \sin u \underbrace{\left(\frac{\sin \Delta u}{\Delta u}\right)}_{\text{equals } 1}$$

$$\frac{dy}{du} = -\sin u$$

Finally, if u is a function of x, we use the chain rule to find $\frac{d(\cos u)}{dx}$.

Derivative of the Cosine Function

$$\frac{d(\cos u)}{dx} = -\sin u \frac{du}{dx}$$

where u = a function of x

EXAMPLE 5 Find dy/dx for the following functions:

(a) $y = \cos(x^2 + 2x)$ (b) $y = 2 \cos^2 8x$

Solution

(a) Let $u = x^2 + 2x$; then $du/dx = 2x + 2$. The derivative of this function follows.

$$\frac{d[\cos(x^2 + 2x)]}{dx} = -\sin\overbrace{(x^2 + 2x)}^{u} \cdot \overbrace{(2x + 2)}^{du/dx}$$

$$= -(2x + 2)\sin(x^2 + 2x)$$

(b) $$\frac{d(2 \cos^2 8x)}{dx} = 2\frac{d(\cos^2 8x)}{dx}$$

constant multiplier rule

Let $u = \cos 8x$; then $y = 2u^2$.

$$\frac{dy}{dx} = 2\left(2u\,\frac{du}{dx}\right)$$
$$= 2\left(2\cos 8x\,\frac{du}{dx}\right)$$

We must find du/dx when $u = \cos 8x$.

$$\frac{du}{dx} = -\sin 8x \cdot \frac{d(8x)}{dx}$$
$$= -8\sin 8x$$

The derivative of $y = 2\cos^2 8x$ is

$$\frac{dy}{dx} = 4\cos 8x \cdot -8\sin 8x$$
$$= -32\cos 8x\sin 8x$$

■

EXAMPLE 6 Find dy/dx for

(a) $y = \cos(x^2)\sin^2 x$ (b) $y = \dfrac{\cos^2 x}{3x}$

Solution

(a) We must use the product rule to find the derivative.

$$\frac{dy}{dx} = \cos(x^2)\underbrace{\frac{d(\sin^2 x)}{dx}}_{2\sin x\cos x} + \sin^2 x\underbrace{\frac{d[\cos(x^2)]}{dx}}_{-\sin(x^2)\cdot 2x}$$

$$= 2\cos(x^2)\sin x\cos x - 2x\sin^2 x\sin(x^2)$$

(b) We must use the quotient rule to find the derivative.

$$\frac{dy}{dx} = \frac{3x\overbrace{\frac{d(\cos^2 x)}{dx}}^{2\cos x\cdot -\sin x} - \cos^2 x\overbrace{\frac{d(3x)}{dx}}^{3}}{(3x)^2}$$

$$= \frac{-6x\cos x\sin x - 3\cos^2 x}{9x^2}$$

$$= \frac{-2x\cos x\sin x - \cos^2 x}{3x^2}$$

■

EXAMPLE 7 Find the following:

(a) the value(s) where the tangent to the curve $y = 3 \cos x^2 + 2x^2$ is horizontal
(b) the slope of this curve at $x = 3$ rad

Solution

(a) For the tangent line to be horizontal, the derivative must equal zero (the slope of a horizontal line is zero). Therefore, we take the derivative of this function, set it equal to zero, and solve for the value(s) of x.

$$\frac{dy}{dx} = 3\frac{d[\cos(x^2)]}{dx} + 2\frac{d(x^2)}{dx}$$

$$= 3[-\sin(x^2) \cdot 2x] + 2(2x)$$

$$\frac{dy}{dx} = -6x \sin(x^2) + 4x$$

Then we set the derivative equal to zero and solve for x.

$$-6x \sin(x^2) + 4x = 0$$

$$2x[-3 \sin(x^2) + 2] = 0$$

$$2x = 0 \qquad -3 \sin(x^2) + 2 = 0$$

$$x = 0 \qquad \sin(x^2) = \frac{2}{3}$$

The tangent line is horizontal at $x = 0$ and $x \approx \pm 0.85$.

(b) To find the slope of the curve when $x = 3$, substitute $x = 3$ into the derivative found in Part (a).

$$\left.\frac{dy}{dx}\right|_{x=3} = -6x \sin(x^2) + 4x$$

$$= -6(3) \sin(3^2) + 4(3)$$

The slope is approximately 4.58. ■

Applications

EXAMPLE 8 Under certain forces, the shear stress on an inclined plane at angle θ is given by $s = 54.7 \sin \theta + 17.8 \cos 2\theta$. Find an expression for the rate of change of shear stress with respect to the angle of the plane.

Solution We are required to take the derivative of this function with respect to θ.

$$\frac{ds}{d\theta} = 54.7\frac{d(\sin\theta)}{d\theta} + 17.8\frac{d(\cos 2\theta)}{d\theta}$$

$$= 54.7\cos\theta + 17.8(-\sin 2\theta)\cdot 2$$

$$\frac{ds}{d\theta} = 54.7\cos\theta - 35.6\sin 2\theta$$

■

EXAMPLE 9 Find the velocity, after 3 seconds, of a particle whose displacement in feet is given by $s = 3\sin 4t + t^2\cos t$.

Solution To find the velocity, find ds/dt and evaluate the result when $t = 3$ s.

$$v = \frac{ds}{dt} = 3\frac{d(\sin 4t)}{dt} + \frac{d(t^2\cos t)}{\text{dt}}$$

product rule

$$= 3(\cos 4t)\cdot 4 + \overbrace{t^2\frac{d(\cos t)}{dt} + \cos t\frac{d(t^2)}{dt}}$$

$$= 12\cos 4t + t^2(-\sin t) + 2t\cos t$$

$$v = 12\cos 4t - t^2\sin t + 2t\cos t$$

Evaluating this expression when $t = 3$ gives

$$v = 12\cos 12 - 9\sin 3 + 6\cos 3$$

$$v \approx 2.9$$

The velocity of the particle is 2.9 ft/s. ■

20–1 EXERCISES

Find the derivative of each of the following functions.

1. $y = \sin 5x$

2. $y = \cos 7x$

3. $y = 2\cos 6x$

4. $y = 5\sin 8t$

5. $y = \sin 2t^2$

6. $y = \cos 3x^4$

7. $y = -7\cos 2x^3$

8. $y = 12\sin 7x^5$

9. $y = \sin(t^2 + 8)$

10. $y = \cos(3t^2 + 1)$

11. $y = \cos^3 x$

12. $y = \sin^4 x$

13. $y = \sin^2(4x^3)$

14. $y = \cos^2(3t^2)$

15. $y = \cos^3(4t^2)$

16. $y = (\sin 3x)^{1/3}$

17. $y = \sqrt{\cos x^2}$

18. $y = \cos\sqrt{3t + 1}$

19. $y = \sin\sqrt{4x - 3}$

20. $y = \sin(2x^2 + 5)^2$

21. $y = \cos(8x^3 - 3x)^2$

22. $y = t^3\sin t - 4\cos(t^2)$

23. $y = x\sin x - x^3\cos(2x)$

24. $y = \cos 3x + x^3 \sin x$

25. $y = \cos^3 3x \sin^2(7x)$

26. $y = \sin(2t) \cos t$

27. $y = 3x^3 \cos x$

28. $y = t^2 \sin t$

29. $y = 6t^2 - 4 \cos t$

30. $y = t^2 + 3 \sin t$

31. $y = 9x \cos(x^2)$

32. $y = 3x \sin^2 x$

33. $y = \dfrac{7x^3}{\cos^2(x - 5)}$

34. $y = \dfrac{4x^2}{3 \sin(x^2 + 1)}$

35. $y = \dfrac{\sin^2 4x}{\cos^3 x}$

36. $y = \dfrac{\cos(3x^2)}{\sin x}$

37. $y = -2 \sin^2 x \cos(x^3)$

38. $y = 3 \cos x^3 \sin x$

39. $y = 6 \sin(x^2) \cos 3x$

40. $y = 3 \sin x \cos(2x)$

41. $y = \cos^3(2x) + 8x^2$

42. $y = \sin^2 x + 3x^4$

43. Find the slope of a line tangent to the curve $y = 3 \cos(x^2)$ at $x = \pi/2$.

44. Find the slope of a line tangent to the curve $y = -2 \sin^2 x$ at $x = \pi/4$.

45. Find the value(s) of x where the tangent to the curve $y = 5 \sin(2x)$ is horizontal.

46. Find the value(s) of x where the tangent to the curve $y = 3 \sin x + x$ is horizontal.

47. Find the velocity, after 2 seconds, of a particle whose displacement is given by $s = 3 \cos^2 t + t \sin(2t)$.

48. The current i in a circuit containing a coil of inductance L is given by

$$i = 10 \sin(120\pi t + \pi)$$

Find the expression for the induced voltage if $L = 5$ H. $\left(\textit{Note: } V = L\dfrac{di}{dt}.\right)$

49. The vertical displacement of an oscillator is given by $y = \sin 100\pi t$. Find the vertical component of the velocity of the oscillation at $t = 1.5$ s.

50. Find the acceleration of an object when $t = 5$ s if the velocity of the object is given by $v = 3 \sin(2t) + 5 \cos t$.

51. Find an expression for the power P if the expression for the work is given by $W = 3t \cos t + t^2$. *(Note: $P = dW/dt$.)*

52. The true period T in seconds of a pendulum of constant period P for an arc θ is given by

$$T = \frac{P}{1 + \frac{1}{4} \sin^2 \frac{\theta}{4}}$$

Find an expression for the rate of change of the period with respect to the arc.

53. The displacement for a harmonic oscillator is given by

$$y = 7.1 \cos(0.23t + 2.3)$$

Find the speed of the object when $t = 3.1$ s.

20 – 2

DERIVATIVES OF THE OTHER TRIGONOMETRIC FUNCTIONS

We will find the derivatives of the remaining trigonometric functions by using the derivative of the sine and cosine functions and trigonometric identities developed earlier.

Derivative of Tangent Function

Remember that $\tan u = \dfrac{\sin u}{\cos u}$. We take the derivative of both sides of the equation, using the quotient rule on the right side.

$$\frac{d(\tan u)}{dx} = \frac{\cos u \dfrac{d(\sin u)}{dx} - \sin u \dfrac{d(\cos u)}{dx}}{(\cos u)^2}$$

$$= \frac{\cos u\left(\cos u \dfrac{du}{dx}\right) - \sin u\left(-\sin u \dfrac{du}{dx}\right)}{\cos^2 u}$$

$$= \frac{\cos^2 u + \sin^2 u}{\cos^2 u} \cdot \frac{du}{dx}$$

$$= \frac{1}{\cos^2 u} \cdot \frac{du}{dx}$$

$$\frac{d(\tan u)}{dx} = \sec^2 u \frac{du}{dx}$$

Derivative of the Tangent Function

$$\frac{d(\tan u)}{dx} = \sec^2 u \frac{du}{dx}$$

where u = a function of x

EXAMPLE 1 Find dy/dx for the following functions:

(a) $y = \tan(3x)$ (b) $y = \tan^2(4x^3)$ (c) $y = x^3\tan(8x^2)$

Solution

(a) Let $u = 3x$; then $du/dx = 3$.

$$\frac{d[\tan(3x)]}{dx} = \sec^2 u \frac{du}{dx}$$
$$= \sec^2(3x) \cdot 3$$
$$= 3\sec^2(3x)$$

(b) Let $u = \tan(4x^3)$; then $y = u^2$. To find the derivative of this function, we must use the power rule.

$$\frac{d[\tan(4x^3)]^2}{dx} = 2u \frac{du}{dx}$$

$$= 2\underbrace{[\tan(4x^3)]}_{u} \underbrace{\frac{d[\tan(4x^3)]}{dx}}_{\sec^2(4x^3)\cdot 12x^2}$$

$$= 24x^2\tan(4x^3)\sec^2(4x^3)$$

(c) We use the product rule to find this derivative.

$$\frac{d[x^3\tan(8x^2)]}{dx} = x^3 \underbrace{\frac{d[\tan(8x^2)]}{dx}}_{\sec^2(8x^2)\,\cdot\,16x} + \tan(8x^2)\underbrace{\frac{d(x^3)}{dx}}_{3x^2}$$

$$= 16x^4\sec^2(8x^2) + 3x^2\tan(8x^2)$$

■

Derivative of Cotangent Function

Since the cotangent is the reciprocal of the tangent, we can use the identity $\cot u = \dfrac{\cos u}{\sin u}$ and proceed as we did in deriving the derivative for the tangent.

$$\frac{d(\cot u)}{dx} = \frac{\sin u \dfrac{d(\cos u)}{dx} - \cos u \dfrac{d(\sin u)}{dx}}{(\sin u)^2}$$

$$= \frac{\sin u\left(-\sin u \dfrac{du}{dx}\right) - \cos u\left(\cos u \dfrac{du}{dx}\right)}{\sin^2 u}$$

$$= \frac{-\sin^2 u - \cos^2 u}{\sin^2 u} \cdot \frac{du}{dx}$$

$$= -\frac{1}{\sin^2 u} \cdot \frac{du}{dx}$$

$$\frac{d(\cot u)}{dx} = -\csc^2 u \frac{du}{dx}$$

Derivative of the Cotangent Function

$$\frac{d(\cot u)}{dx} = -\csc^2 u \frac{du}{dx}$$

where u = a function of x

EXAMPLE 2 Find the derivative for the following functions:

(a) $y = \cot(5x)$ (b) $y = 2\cot^3 x$ (c) $y = x^2\cot(4x^2)$

Solution

(a) Let $u = 5x$; then $du/dx = 5$.

$$\frac{d(\cot 5x)}{dx} = -\csc^2(5x)\,\frac{d(5x)}{dx}$$
$$= -\csc^2(5x) \cdot 5$$
$$= -5\csc^2(5x)$$

(b) Let $u = \cot x$; then $y = 2u^3$.

$$\frac{d(2\cot^3 x)}{dx} = 2\,\frac{d(\cot^3 x)}{dx}$$
$$= 2 \cdot 3(\cot^2 x) \cdot \underbrace{\frac{d(\cot x)}{dx}}_{-\csc^2 x \cdot \frac{d(x)}{dx}}$$
$$= -6\cot^2 x \csc^2 x$$

(c) We use the product rule to find the derivative of this function.

$$\frac{d[x^2\cot(4x^2)]}{dx} = x^2\,\underbrace{\frac{d[\cot(4x^2)]}{dx}}_{-\csc^2(4x^2)\cdot\frac{d(4x^2)}{dx}} + \cot(4x^2)\,\underbrace{\frac{d(x^2)}{dx}}_{2x}$$
$$= x^2[-\csc^2(4x^2) \cdot 8x] + \cot(4x^2) \cdot 2x$$
$$= -8x^3\csc^2(4x^2) + 2x\cot(4x^2)$$

■

Derivative of Cosecant Function

We can use the identity $\csc u = \dfrac{1}{\sin u} = (\sin u)^{-1}$ to find the derivation of the cosecant function.

$$\frac{d(\csc u)}{dx} = \frac{d(\sin u)^{-1}}{dx}$$
$$= -1(\sin u)^{-2} \cdot \Big(\underbrace{\cos u \frac{du}{dx}}_{\frac{d(\sin u)}{dx}}\Big)$$
$$= \frac{-\cos u}{\sin^2 u} \cdot \frac{du}{dx}$$
$$= -\frac{1}{\sin u} \cdot \frac{\cos u}{\sin u} \cdot \frac{du}{dx}$$
$$\frac{d(\csc u)}{dx} = -\csc u \cot u\,\frac{du}{dx}$$

Derivative of the Cosecant Function

$$\frac{d(\csc u)}{dx} = -\csc u \cot u \frac{du}{dx}$$

where u = a function of x

EXAMPLE 3 Find dy/dx for each of the following functions:

(a) $y = \csc(3x)$ (b) $y = \csc^3(5x^2)$ (c) $y = x^3\csc^2(4x)$

Solution

(a) Let $u = 3x$; then $du/dx = 3$.

$$\frac{d[\csc(3x)]}{dx} = -\csc(3x)\cot(3x) \cdot \underbrace{\frac{d(3x)}{dx}}_{3}$$

$$= -3\csc(3x)\cot(3x)$$

(b) Let $u = \csc 5x^2$; then $y = u^3$. We use the power rule to find the derivative of this function.

$$\frac{d[\csc^3(5x^2)]}{dx} = 3[\csc(5x^2)]^2 \underbrace{\frac{d[\csc(5x^2)]}{dx}}_{-\csc(5x^2)\cot(5x^2) \cdot \frac{d(5x^2)}{dx}}$$

$$= 3[\csc^2(5x^2)] \cdot [-\csc(5x^2) \cdot \cot(5x^2) \cdot 10x]$$

$$= -30x \csc^3(5x^2)\cot(5x^2)$$

(c) Once again, we use the product rule to find the derivative of this function.

$$\frac{d[x^3\csc^2(4x)]}{dx} = x^3 \underbrace{\frac{d[\csc^2(4x)]}{dx}}_{2\csc(4x)[-\csc(4x)\cot(4x) \cdot 4]} + \csc^2(4x) \underbrace{\frac{d(x^3)}{dx}}_{3x^2}$$

$$= x^3[-8\csc^2(4x)\cot(4x)] + \csc^2(4x)(3x^2)$$

$$= -8x^3\csc^2(4x)\cot(4x) + 3x^2\csc^2(4x)$$

$$= x^2\csc^2(4x)[-8x\cot(4x) + 3]$$

■

Derivative of Secant Function

The derivation of the secant function is very similar to that of the cosecant. Let us use the identity $\sec u = (\cos u)^{-1}$.

$$\frac{d(\sec u)}{dx} = \frac{d(\cos u)^{-1}}{dx}$$

$$\frac{d(\cos u)}{dx}$$

$$= -(\cos u)^{-2} \cdot \left(-\sin u \frac{du}{dx}\right)$$

$$= \frac{\sin u}{\cos^2 u} \cdot \frac{du}{dx}$$

$$= \frac{1}{\cos u} \cdot \frac{\sin u}{\cos u} \cdot \frac{du}{dx}$$

$$\frac{d(\sec u)}{dx} = \sec u \tan u \frac{du}{dx}$$

Derivative of the Secant Function

$$\frac{d(\sec u)}{dx} = \sec u \tan u \frac{du}{dx}$$

where u = a function of x

EXAMPLE 4 Find the derivative of each of the following functions:

(a) $y = \sec^2(3x)$ (b) $y = \dfrac{\sec(x^2)}{4x^3}$

Solution

(a) Let $u = \sec 3x$; then $y = u^2$.

$$\frac{d[\sec^2(3x)]}{dx} = 2u \frac{du}{dx}$$

$$= 2 \sec 3x \frac{d[\sec (3x)]}{dx}$$

$$\sec 3x \tan 3x \frac{d(3x)}{dx}$$

$$= 2 \sec(3x) \cdot [3 \sec(3x)\tan(3x)]$$

$$= 6 \sec^2(3x)\tan(3x)$$

(b) We use the quotient rule to find the derivative of this function.

$$\frac{dy}{dx} = \frac{4x^3 \dfrac{d[\sec(x^2)]}{dx} - \sec(x^2)\dfrac{d(4x^3)}{dx}}{(4x^3)^2}$$

$$= \frac{4x^3[\sec(x^2)\tan(x^2) \cdot 2x] - 12x^2\sec(x^2)}{16x^6}$$

$$= \frac{8x^4\sec(x^2)\tan(x^2) - 12x^2\sec(x^2)}{16x^6}$$

$$= \frac{4x^2\sec(x^2) \cdot [2x^2\tan(x^2) - 3]}{16x^6}$$

$$= \frac{\sec(x^2) \cdot [2x^2\tan(x^2) - 3]}{4x^4}$$

■

Applications

EXAMPLE 5 A searchlight rotates at a rate of 0.08 rad/s, illuminating a prison wall 615 ft from the searchlight. How fast is the light beam moving along the wall at a point 675 ft from the searchlight?

Solution Figure 20–2 illustrates this problem, which is one of related rate, requiring us to find dx/dt.

FIGURE 20–2

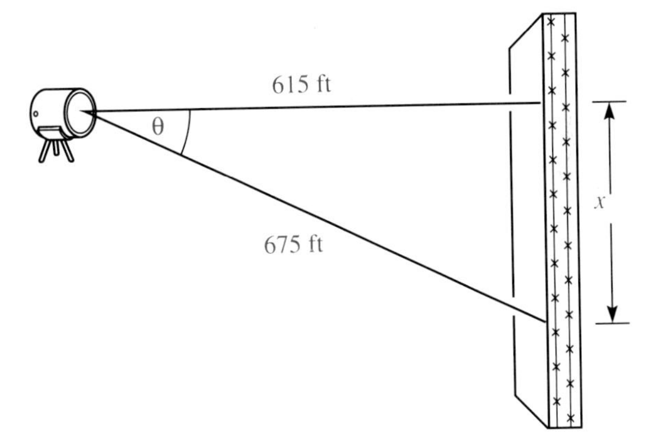

The equation relating this information is

$$\tan\theta = \frac{x}{615}$$

We differentiate this equation implicitly with respect to time, t.

$$\sec^2\theta \frac{d\theta}{dt} = \frac{1}{615} \cdot \frac{dx}{dt}$$

Then we solve for dx/dt and substitute known values.

$$\frac{dx}{dt} = 615 \sec^2\theta \frac{d\theta}{dt}$$

$$= 615\overbrace{\left(\frac{675}{615}\right)^2}^{\sec\theta} \cdot \overbrace{(0.08)}^{\frac{d\theta}{dt}}$$

$$\approx 59$$

The searchlight is scanning the wall at a rate of 59 ft/s when it is 675 ft from the light. ■

EXAMPLE 6 The displacement of an object in feet is given by $s = t^2 + \sec 2t$ $(0 \leq t < \pi/2)$. Find the acceleration when $t = 0.5$ s.

Solution The acceleration of an object is given by the derivative of its velocity, which is given by the derivative of its displacement. Therefore, we want the second derivative of the function for displacement.

$$a = \frac{d^2s}{dt^2}$$

$$= \frac{d}{dt}\overbrace{(2t + 2\sec 2t \tan 2t)}^{\frac{ds}{dt} = v}$$

$$= 2 + 2[\sec 2t(2\sec^2 2t) + \tan 2t(2\sec 2t \tan 2t)]$$

$$a = 2 + 4\sec^3 2t + 4\sec 2t \tan^2 2t$$

Last, we must evaluate this derivative when $t = 0.5$ s.

$$a = 2 + 4\sec 2t(\sec^2 2t + \tan^2 2t)$$

$$a \approx 45.3 \text{ ft/s}^2$$

The object is accelerating at a rate of approximately 45.3 ft/s². ■

20–2 EXERCISES

Find the derivative of each of the following functions.

1. $y = \tan 7x$
2. $y = \sec 8x$
3. $y = \csc(x^2)$
4. $y = \cot(3x^4)$
5. $y = \sec(x^2 + 3x)$
6. $y = \sec^3(3x)$
7. $y = 5\cot(6t^2)$
8. $y = \tan(t^2 - 5)$
9. $y = \csc(x^3 - 2)^{1/2}$
10. $y = \sec^2(4x - 5)^{2/3}$

11. $y = \cot^2(2\theta)$

12. $y = \sin(\tan 3\theta)$

13. $y = \sec(\cot 5x)$

14. $y = \sqrt{\cot 7t}$

15. $y = \sqrt{\sec 3x}$

16. $y = \csc^3(8\theta^2)$

17. $y = \tan^4(2x)$

18. $y = \cot^3(4x^3)$

19. $y = \sin 3x - 4\tan(x^2)$

20. $y = \sec^2 x + x \sin 3x$

21. $y = \cot^3(x^2) + 4 \sec 8x$

22. $y = \tan\theta \cos 2\theta$

23. $y = 4x \csc(x^2)$

24. $y = t^3 \cot^2 t$

25. $y = \dfrac{\tan 8x}{x^2}$

26. $y = \dfrac{\sec^2 t}{3t}$

27. $y = \cos(4x^2)\tan^2(5x)$

28. $y = \sin^2\theta \cot(\theta^3)$

29. $y = \dfrac{1 - \sin^2 x}{\csc^2 x}$

30. $y = \dfrac{\cot x}{3 - 4\sin x}$

31. $y = x^3 - 3\tan x + \csc^2 x$

32. $y = 3x^2 \sec^2 x$

33. $y = x^2 \cot x$

34. $y = \csc x(3 + \sec^2 x)$

35. A searchlight is scanning a wall 275 ft away at a rate of 0.09 rad/s. How fast is the light beam moving along the wall at a point 340 ft from the searchlight?

36. A radar antenna is on a ship which is 2.8 mi straight from shore. If the antenna rotates at 0.25 rad/s, how fast is the radar beam moving along the shore at a point 3.5 mi from the ship?

37. In an ac circuit containing an inductor and a variable resistor, the current is given by $I = I_0 \cot\theta$ where I_0 is the inductor current and θ is the phase angle. Find dI/dt if I_0 is a constant.

38. Find the current in a circuit when $t = 0.15$ s given that the electrical charge is $q = 0.10 \tan 0.7t$ C. *(Note:* $I = dq/dt$.*)*

39. Find the acceleration, when $t = 0.15$ s, of an object whose displacement (in meters) is given by $s = 3 \sin t \cot 2t$. $(0 < t < \pi/2)$

40. Find the velocity, when $t = 1.25$ s, of an object whose displacement is given by $s = 3t^2 + 8\sec^2 t$. $(0 \le t < \pi/2)$

20–3

DERIVATIVES OF THE INVERSE TRIGONOMETRIC FUNCTIONS

We discussed the inverse trigonometric functions in Chapter 13. Using the graph of $y = \text{Arcsin } x$ shown in Figure 20–3, we can make some intuitive statements about the derivative of $y = \text{Arcsin } x$. For example, the derivative of $y = \text{Arcsin } x$ does not exist at $x = 1$ or at $x = -1$ because the tangent line at those points is vertical, and its slope is undefined. Also, since the slope of the tangent line is positive, the derivative of $y = \text{Arcsin } x$ is positive. Now, we will develop a formula for the derivative of the inverse trigonometric functions.

FIGURE 20–3

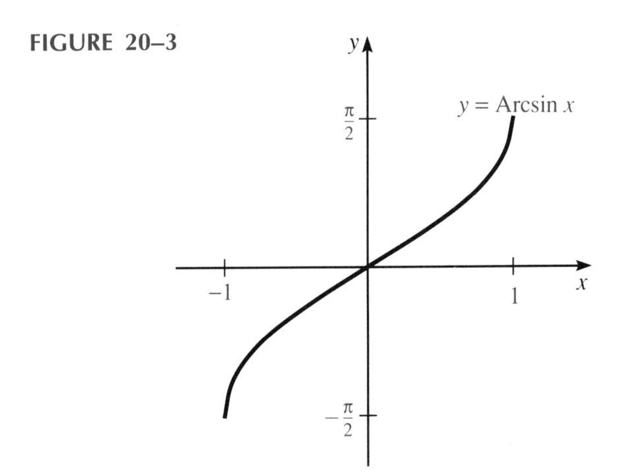

Derivative of Arcsin *u*

From the definition of the Arcsin function, we know that $y = \text{Arcsin } u$. Therefore, we write

$$u = \sin y \quad \text{for } \frac{-\pi}{2} \le y \le \frac{\pi}{2}$$

Differentiating both sides of the equation gives

$$\frac{du}{dx} = \cos y \frac{dy}{dx}$$

Solving for dy/dx gives

$$\frac{dy}{dx} = \frac{1}{\cos y} \cdot \frac{du}{dx}$$

If we solve the trigonometric identity $\sin^2 y + \cos^2 y = 1$ for $\cos y$, we have

$$\cos y = \sqrt{1 - \sin^2 y} \quad \text{where } \cos y > 0 \text{ for } \frac{-\pi}{2} < y < \frac{\pi}{2}$$

Substituting yields

$$\frac{dy}{dx} = \frac{1}{\sqrt{1 - \sin^2 y}} \cdot \frac{du}{dx}$$

Last, we substitute $u = \sin y$.

$$\frac{dy}{dx} = \frac{1}{\sqrt{1 - u^2}} \cdot \frac{du}{dx}$$

Derivative of the Arcsine Function

$$\frac{d(\text{Arcsin } u)}{dx} = \frac{1}{\sqrt{1 - u^2}} \cdot \frac{du}{dx}$$

where $-1 < u < 1$

EXAMPLE 1 Find dy/dx for each of the following:

(a) $y = \text{Arcsin } 6x$ (b) $y = \text{Arcsin } 4x^2$

Solution

(a) Let $u = 6x$; then $du/dx = 6$.

$$\frac{d(\text{Arcsin } u)}{dx} = \frac{1}{\sqrt{1 - u^2}} \cdot \frac{du}{dx}$$

$$= \frac{1}{\sqrt{1 - \underbrace{36x^2}_{u^2 = (6x)^2}}} \cdot \overbrace{6}^{du/dx}$$

$$= \frac{6}{\sqrt{1 - 36x^2}}$$

(b) Let $u = 4x^2$; then $du/dx = 8x$.

$$\frac{d(\text{Arcsin } 4x^2)}{dx} = \frac{1}{\sqrt{1 - \underbrace{(4x^2)^2}_{u^2}}} \cdot \overbrace{8x}^{du/dx}$$

$$= \frac{8x}{\sqrt{1 - 16x^4}}$$

■

Derivative of Arccos u

Similarly, we can find the derivative of the inverse cosine function. We know that $y = \text{Arccos } u$ is equivalent to

$$u = \cos y \quad \text{for } 0 \le y \le \pi$$

Taking the derivative of both sides of the equation gives

$$\frac{du}{dx} = -\sin y \frac{dy}{dx}$$

Next, we solve the equation for dy/dx.

$$\frac{dy}{dx} = -\frac{1}{\sin y} \cdot \frac{du}{dx} \quad \text{where } \sin y > 0 \text{ for } 0 < y < \pi$$

Last, we substitute $\sin y = \sqrt{1 - \cos^2 y}$ and $u = \cos y$ to give the final result.

$$\frac{dy}{dx} = -\frac{1}{\sqrt{1 - \cos^2 y}} \cdot \frac{du}{dx}$$

$$\frac{dy}{dx} = -\frac{1}{\sqrt{1 - u^2}} \cdot \frac{du}{dx}$$

Derivative of the Arccos Function

$$\frac{d(\text{Arccos } u)}{dx} = -\frac{1}{\sqrt{1 - u^2}} \cdot \frac{du}{dx}$$

where $-1 < u < 1$

EXAMPLE 2 Find the derivative of the following functions:

(a) $y = \text{Arccos}(7x^3)$ (b) $y = 4x \text{ Arccos}(x^3)$

Solution

(a) Let $u = 7x^3$; then $du/dx = 21x^2$.

$$\frac{d[\text{Arccos}(7x^3)]}{dx} = -\frac{1}{\sqrt{1 - \underbrace{(7x^3)^2}_{u^2}}} \cdot \overbrace{21x^2}^{\frac{du}{dx}}$$

$$= -\frac{21x^2}{\sqrt{1 - 49x^6}}$$

(b) We must apply the product rule to find the derivative of this function.

$$\frac{d(4x \text{ Arccos } x^3)}{dx} = 4x \underbrace{\frac{d(\text{Arccos } x^3)}{dx}}_{u = x^3,\ \frac{du}{dx} = 3x^2} + \text{Arccos } x^3 \underbrace{\frac{d(4x)}{dx}}_{4}$$

$$= 4x \cdot \left(-\frac{3x^2}{\sqrt{1 - (x^3)^2}}\right) + 4 \text{ Arccos } x^3$$

$$= -\frac{12x^3}{\sqrt{1 - x^6}} + 4 \text{ Arccos } x^3$$

■

Derivative of Arctan *u*

From the definition of the Arctangent function, if $y = \text{Arctan } u$, then

$$u = \tan y \quad \text{for } -\frac{\pi}{2} < y < \frac{\pi}{2}$$

Taking the derivative of both sides of the equation and solving for dy/dx gives

$$\frac{du}{dx} = \sec^2 y \frac{dy}{dx} \quad \text{where } \sec y > 0 \text{ for } -\frac{\pi}{2} < y < \frac{\pi}{2}$$

$$\frac{dy}{dx} = \frac{1}{\sec^2 y} \cdot \frac{du}{dx}$$

Since $\sec^2 y = 1 + \tan^2 y$ and $u = \tan y$,

$$\frac{dy}{dx} = \frac{1}{1 + u^2} \cdot \frac{du}{dx}$$

Derivative of the Arctangent Function

$$\frac{d(\text{Arctan } u)}{dx} = \frac{1}{1 + u^2} \cdot \frac{du}{dx}$$

EXAMPLE 3 Find dy/dx if

(a) $y = \text{Arctan}^2(4x^3)$ (b) $y = \dfrac{\text{Arctan } 2x}{1 - 3x^2}$

Solution

(a) Let $u = \text{Arctan}(4x^3)$; then $y = u^2$.

$$\frac{dy}{dx} = 2u \frac{du}{dx}$$

$$= 2 \text{ Arctan}(4x^3) \frac{du}{dx}$$

$$\frac{du}{dx} = \frac{1}{[1 + (4x^3)^2]} \cdot 12x^2$$

$$= \frac{24x^2}{1 + 16x^6} \text{Arctan}(4x^3)$$

(b) We use the quotient rule to find the derivative of this function.

$$\frac{dy}{dx} = \frac{(1 - 3x^2)\overbrace{\frac{d(\text{Arctan } 2x)}{dx}}^{\frac{1}{(1+4x^2)}\cdot 2} - \text{Arctan } 2x \overbrace{\frac{d(1 - 3x^2)}{dx}}^{-6x}}{(1 - 3x^2)^2}$$

$$= \frac{\frac{2}{(1 + 4x^2)} \cdot (1 - 3x^2) + 6x \text{ Arctan } 2x}{(1 - 3x^2)^2}$$

$$= \frac{2(1 - 3x^2) + 6x \text{ Arctan } 2x(1 + 4x^2)}{(1 - 3x^2)^2(1 + 4x^2)}$$

■

Applications

EXAMPLE 4 Fred sights a ship off the coast of Florida. The ship is traveling north, parallel to the coastline, at 22 knots (nautical miles per hour). The ship is 12 nautical miles offshore, approximately northeast of Fred's position. The angle between the coastline and a line from his position to the ship is 0.7 radian. Figure 20–4 is a drawing of the problem statement. How fast is this angle changing?

FIGURE 20–4

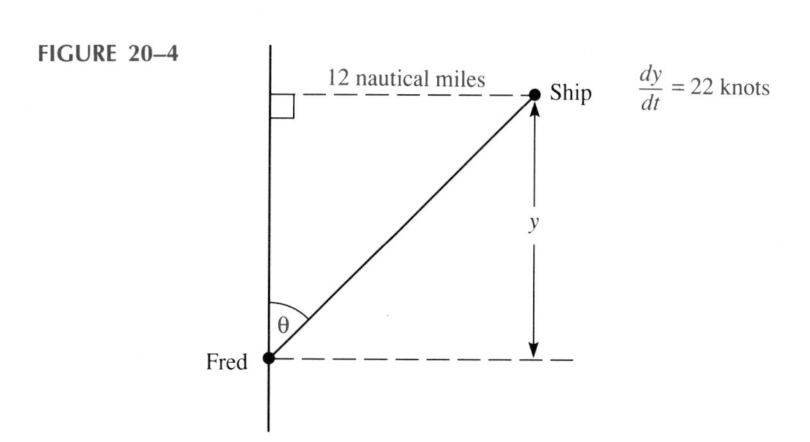

Solution This is a change of rate problem requiring us to find $d\theta/dt$. Using trigonometry, we have

$$\tan \theta = \frac{12}{y}$$

$$\theta = \text{Arctan } \frac{12}{y}$$

Taking the derivative of both sides of the equation gives

$$\frac{d\theta}{dt} = \frac{1}{1 + \left(\frac{12}{y}\right)^2}\left[-1(12y^{-2}) \cdot \frac{dy}{dt}\right]$$

$$= \frac{-12}{\not{y}^2\left(\frac{y^2 + 144}{\not{y}^2}\right)} \cdot \frac{dy}{dt}$$

$$= \frac{-12}{y^2 + 144} \cdot \frac{dy}{dt} \quad \text{where } y = \frac{12}{\tan \theta}$$

$$\frac{d\theta}{dt} = \frac{-12}{\left(\frac{12}{\tan \theta}\right)^2 + 144} \cdot \frac{dy}{dt}$$

Evaluating this expression for $\theta = 0.7$ and $dy/dt = 22$ gives

$$\frac{d\theta}{dt} = \frac{-12}{202.97 + 144} \cdot 22$$

$$\frac{d\theta}{dt} = -0.761$$

The angle θ is decreasing at a rate of 0.761 rad/h. ■

FIGURE 20–5

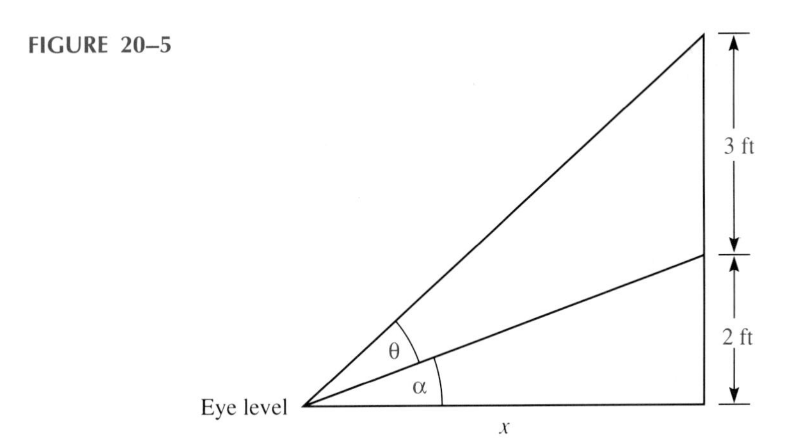

EXAMPLE 5 A travel poster 3 ft tall is placed so that the lower edge is 2 ft above eye level, as illustrated in Figure 20–5. How far from the poster should an observer stand so that the viewing angle is a maximum?

Solution From Figure 20–5,

$$\tan(\theta + \alpha) = \frac{5}{x} \quad \text{or} \quad \theta + \alpha = \text{Arctan}\left(\frac{5}{x}\right)$$

and

$$\tan \alpha = \frac{2}{x} \quad \text{or} \quad \alpha = \text{Arctan}\left(\frac{2}{x}\right)$$

Since $\theta = (\theta + \alpha) - \alpha$, then

$$\theta = \text{Arctan}\left(\frac{5}{x}\right) - \text{Arctan}\left(\frac{2}{x}\right)$$

Next, we differentiate this equation.

$$\frac{d\theta}{dx} = \frac{1}{\left(1 + \frac{25}{x^2}\right)} \cdot \left(-\frac{5}{x^2}\right) - \frac{1}{\left(1 + \frac{4}{x^2}\right)} \cdot \left(-\frac{2}{x^2}\right)$$

Using algebraic simplification, we obtain

$$\frac{d\theta}{dx} = \frac{-5}{x^2 + 25} + \frac{2}{x^2 + 4}$$

The maximum of θ occurs when $d\theta/dx = 0$. Thus,

$$\begin{aligned} \frac{-5}{x^2 + 25} + \frac{2}{x^2 + 4} &= 0 \\ -5(x^2 + 4) + 2(x^2 + 25) &= 0 \\ -3x^2 + 30 &= 0 \\ x^2 &= 10 \\ x &= \pm\sqrt{10} \\ x &\approx 3.2 \end{aligned}$$

The second derivative is more difficult to calculate; therefore, we use the first derivative test to determine the maximum. Since the first derivative of this function changes sign from positive to negative, this is a maximum. The observer should stand approximately 3 ft from the poster. ■

20–3 EXERCISES

Find the derivative of each function.

1. $y = \text{Arccos}(x^2)$

2. $y = 3 \text{ Arctan } 5x$

3. $y = 8 \text{ Arcsin}(4x^3)$

4. $y = \text{Arcsin}(\tan 4x)$

5. $y = \text{Arctan}(\sec x^2)$

6. $y = 4 \text{ Arccos}(8x + 5)$

7. $y = \text{Arctan}\left(\dfrac{3x + 5}{x + 1}\right)$

8. $y = \text{Arcsin}\left(\dfrac{x - 1}{2x}\right)$

9. $y = \text{Arccos}\sqrt{4x - 1}$

10. $y = (\text{Arcsin } 3x^2)^2$

11. $y = \dfrac{1}{\text{Arctan } 5x}$

12. $y = \sqrt{\text{Arccos } x}$

13. $y = (3x - 4)\text{ Arctan } x$

14. $y = (x + 1)\text{Arcsin } x$

15. $y = x\text{ Arccos } x$

16. $y = 2x\text{ Arcsin}(x^2)$

17. $y = \text{Arctan}^3(x)$

18. $y = \text{Arccos}^2(4x)$

19. $y = \dfrac{\text{Arctan } 4x}{3x + 7}$

20. $y = \dfrac{\text{Arccos } x}{x^2 - 1}$

21. $y = \dfrac{1}{x} + \text{Arcsin}^2(x)$

22. $y = 8x^3 + \text{Arccos}(5x^3)$

23. $y = 7x\text{ Arcsin}(x^3)$

24. $y = 4x^3\text{Arctan}(3x^5)$

25. $y = x^3\text{Arccos}(2x)$

26. $y = \dfrac{2x - 5}{\text{Arcsin } 9x}$

27. $y = \dfrac{3x}{\text{Arccos}(x^3)}$

28. A ship traveling north at 20 knots is sighted off the coast. The ship is 18 nautical miles offshore, approximately northeast of your position. The angle between the coastline and a line from your position to the ship is 0.65 radian. How fast is this angle changing? (See Example 4.)

29. As shown in Figure 20–6, a winch, 8 ft from the ground on a loading dock, is used to move a large container along the ground at a rate of 2.3 ft/s. At what rate is the angle between the winch and the ground changing when the container is 12 ft away?

30. A plane flying at an altitude of 29,000 ft and a speed of 450 ft/s passes over a control tower. At what rate is the angle of elevation of the plane from the control tower changing 15 s later?

31. Find the slope of the line tangent to $y = x^2\text{Arctan } 2x$ when $x = 2.5$.

32. Find the slope of the line tangent to $y = 3\text{ Arccos } x^2$ when $x = 0.5$.

33. A 20-ft ladder is leaning against a wall at a point 10 ft from the floor. The bottom of the ladder starts to slip away from the wall at a rate of 2.7 ft/s. Find the rate at which the angle between the top of the ladder and the wall is changing.

34. A jogger 6 ft tall runs at 8 ft/s toward a building 60 ft tall. What is the rate of change of the angle of elevation to the top of the building when the jogger is 480 ft from the building?

35. A billboard 16 ft tall is located so that its lower edge is 7 ft above eye level. How far from the billboard should an observer stand so that the viewing angle is a maximum? (See Example 5.)

36. A painting 10 ft tall is hung so that its lower edge is 3 ft above eye level. How far should an observer stand from the painting so that the viewing angle is a maximum?

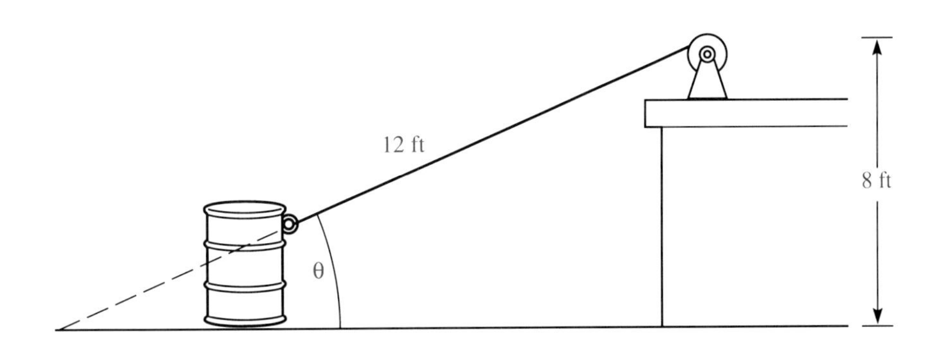

FIGURE 20–6

20 – 4

DERIVATIVE OF THE LOGARITHMIC FUNCTION

A second type of transcendental function is the logarithmic function, which we discussed in Chapter 8. We can use the delta process to derive a formula for the derivative of the logarithmic function. If $y = \log_b u$, then

$$y + \Delta y = \log_b(u + \Delta u)$$
$$\Delta y = \log_b(u + \Delta u) - \log_b u$$

Using the property for the logarithm of a quotient, we can write

$$\Delta y = \log_b\left(\frac{u + \Delta u}{u}\right)$$
$$\Delta y = \log_b\left(1 + \frac{\Delta u}{u}\right)$$

Then we divide the equation by Δu.

$$\frac{\Delta y}{\Delta u} = \frac{1}{\Delta u} \cdot \log_b\left(1 + \frac{\Delta u}{u}\right)$$

To evaluate the limit, we multiply the right side of the equation by $\frac{u}{u}$.

$$\frac{\Delta y}{\Delta u} = \frac{1}{u} \cdot \frac{u}{\Delta u} \cdot \log_b\left(1 + \frac{\Delta u}{u}\right)$$

Using the property for logarithms of a power gives

$$\frac{\Delta y}{\Delta u} = \frac{1}{u} \log_b\left(1 + \frac{\Delta u}{u}\right)^{u/\Delta u}$$

Taking the limit as $\Delta u \to 0$ gives

$$\frac{dy}{du} = \lim_{\Delta u \to 0}\left[\frac{1}{u} \log_b\left(1 + \frac{\Delta u}{u}\right)^{u/\Delta u}\right]$$
$$\frac{dy}{du} = \frac{1}{u} \lim_{\Delta u \to 0}\left[\log_b\left(1 + \frac{\Delta u}{u}\right)^{u/\Delta u}\right]$$

To evaluate this limit, we must know the value of $\lim_{x \to 0} (1 + x)^{1/x}$. We can approximate the value of this limit by using a table of values as we did in Chapter 18. The calculations in Table 20–3 show that the value of the function approaches 2.718 as $x \to 0$. The limiting value of this function is the irrational number e, also discussed previously in Chapter 8. Using this limit in the derivative gives

$$\frac{dy}{du} = \frac{1}{u} \cdot \log_b e$$

Then we apply the chain rule to differentiate this function with respect to x.

$$\frac{dy}{dx} = \frac{1}{u} \cdot \log_b e \cdot \frac{du}{dx}$$

TABLE 20–3

x	$(1 + x)^{1/x}$
0.1	2.593742
0.01	2.704814
0.001	2.716924
0.0001	2.718146
0.00001	2.718268

Derivative of the Logarithmic Function

$$\frac{d(\log_b u)}{dx} = \frac{1}{u} \log_b e \frac{du}{dx}$$

Derivative of Natural Logarithms

If the base of the logarithm is e, then

$$\begin{aligned} \log_b e &= \log_e e \\ &= \ln e \\ \log_b e &= 1 \end{aligned}$$

and the derivative for natural logarithms can be written as follows.

Derivative of the Natural Logarithmic Function

$$\frac{d(\ln u)}{dx} = \frac{1}{u} \cdot \frac{du}{dx}$$

CAUTION ✦ Remember from Chapter 8 that $\log_b u$ is defined only for positive values of u. When the base of the logarithm is 10, the base of the logarithmic expression does not need to be written.

EXAMPLE 1 Find the derivative of the following functions:

(a) $y = \log 6x$ (b) $y = \ln(3x^2 + x)$ (c) $y = (\log 3x^2)^4$

Solution

(a) Let $u = 6x$; then $du/dx = 6$.

$$\frac{d(\log 6x)}{dx} = \underbrace{\frac{1}{6x}}_{u} \cdot \log e \cdot \underbrace{6}_{du/dx}$$

$$= \frac{\log e}{x}$$

(b) Let $u = 3x^2 + x$; then $du/dx = 6x + 1$.

$$\frac{d[\ln(3x^2 + x)]}{dx} = \frac{1}{\underbrace{3x^2 + x}_{u}} \cdot \underbrace{(6x + 1)}_{du/dx}$$

$$= \frac{6x + 1}{3x^2 + x}$$

(c) If $u = \log 3x^2$, then $y = u^4$. We use the power rule to differentiate this function.

$$\frac{dy}{dx} = 4u^3 \cdot \frac{du}{dx}$$

$$= 4(\log 3x^2)^3 \cdot \underbrace{\frac{du}{dx}}_{\frac{1}{3x^2} \log e \cdot 6x}$$

$$= \frac{8 \log e}{x}[\log (3x^2)^3]$$

■

EXAMPLE 2 Find the derivative of

$$y = \ln \frac{x^2}{x^3 + 2}$$

Solution First, we will use the properties of logarithms to write the function in the following form:

$$y = \ln x^2 - \ln(x^3 + 2)$$

Then we can apply the sum rule to give the following derivative:

$$\begin{aligned}\frac{dy}{dx} &= \frac{d(\ln x^2)}{dx} - \frac{d[\ln (x^3 + 2)]}{dx}\\ &= \frac{1}{x^2} \cdot 2x - \frac{1}{x^3 + 2} \cdot 3x^2\\ &= \frac{2}{x} - \frac{3x^2}{x^3 + 2}\\ &= \frac{2x^3 + 4 - 3x^3}{x(x^3 + 2)}\\ &= \frac{4 - x^3}{x(x^3 + 2)}\end{aligned}$$

■

EXAMPLE 3 Find dy/dx for $y = \log(\cos x^2)$

Solution Let $u = \cos x^2$; then

$$\frac{d}{dx}[\log(\cos x^2)] = \frac{1}{\cos x^2} \log e \frac{du}{dx}$$

$-2x \sin x^2$

$$= \frac{-2x \sin x^2}{\cos x^2} \log e$$

$\tan x^2 = \dfrac{\sin x^2}{\cos x^2}$

$$= -2x \tan x^2 \log e$$

■

EXAMPLE 4 Find dy/dx for $y = \ln[\tan x(1 + x^2)^3]$.

Solution The multiplication and power properties of logarithms allow us to write this function as follows:

$$y = \ln(\tan x) + \ln(1 + x^2)^3$$
$$y = \ln(\tan x) + 3 \ln(1 + x^2)$$

Then the derivative is

$$\frac{dy}{dx} = \frac{d}{dx}\ln(\tan x) + 3\frac{d}{dx}\ln(1 + x^2)$$
$$= \frac{1}{\tan x}\underbrace{\frac{du}{dx}}_{\sec^2 x} + 3\frac{1}{1 + x^2}\underbrace{\frac{du}{dx}}_{2x}$$
$$= \frac{1}{\tan x} \cdot \sec^2 x + 3 \cdot \frac{1}{1 + x^2} \cdot 2x$$
$$= \frac{\sec^2 x}{\tan x} + \frac{6x}{1 + x^2}$$
$$= \frac{1}{\sin x \cos x} + \frac{6x}{1 + x^2}$$

■

Applications

EXAMPLE 5 A coaxial cable consists of an inner cylindrical conductor of radius r inside a conducting tube of radius R. The inductance in a certain coaxial cable is given by $L = 80 \ln(R/r)$. Find an expression for the rate of change of inductance with respect to the inner radius. (R is a constant.)

Solution Let $u = R/r$; then the derivative is

$$L = 80 \ln \frac{R}{r}$$
$$\frac{dL}{dr} = 80\left[\frac{1}{\left(\frac{R}{r}\right)}\right]\frac{du}{dr}$$
$$= \frac{80r}{R} \cdot \overbrace{\left(-\frac{R}{r^2}\right)}^{\frac{du}{dr}}$$
$$= -\frac{80}{r}$$

■

EXAMPLE 6 The current in a 5-H inductor is given by $I = \ln(2t - 1)$. Find the voltage when $t = 1.75$ s.

Solution Recall from previous chapters that $V = dI/dt$. Therefore, we must calculate dI/dt and evaluate the result when $t = 1.75$. Thus,

$$V = \frac{dI}{dt} = \frac{d[\ln(2t - 1)]}{dt}$$

Let $u = 2t - 1$; then $du/dt = 2$.

$$\frac{dI}{dt} = \frac{1}{2t - 1} \cdot 2$$

$$= \frac{2}{2t - 1}$$

Last, we evaluate this result when $t = 1.75$ s.

$$V = \frac{2}{2(1.75) - 1}$$

$$= \frac{dI}{dt} = 0.8 \text{ V}$$

20–4 EXERCISES

Find the derivative of each of the following functions.

1. $y = \log 7x$

2. $y = \ln 4x$

3. $y = 3 \ln(2x + 5)$

4. $y = 6 \log(5x - 1)$

5. $y = \log 3x^2$

6. $y = \ln(8x^3)$

7. $y = \ln^2(x^3)$

8. $y = \log^2(4x^2)$

9. $y = \log(\sin x)$

10. $y = \ln(\tan x)$

11. $y = \ln(\sec^2 x)$

12. $y = \log(\cot^2 x)$

13. $y = \log(x \cos 2x)$

14. $y = \ln(3x \sec x)$

15. $y = \ln(2x + \cos x)$

16. $y = \log(x^2 + \tan 3x)$

17. $y = (\log 4x^5)^3$

18. $y = (\ln 2x^3)^4$

19. $y = \ln\left(\frac{3x + 1}{2x - 3}\right)$

20. $y = \log\left(\frac{x^2 - 3}{8x + 1}\right)$

21. $y = \log\frac{(2x - x^2)}{2x}$

22. $y = x^2 \ln 3x^4$

23. $y = x \log 7x^2$

24. $y = \sin 2x \log 3x^2$

25. $y = 4x \ln(3x^2 + 4x)$

26. $y = \frac{\ln 8x^3}{2x^2 + 1}$

27. $y = \frac{(\log 6x^2)^3}{\cos^2 x}$

28. Find the slope of the line tangent to $y = \ln(x^2 - 3)$ at $x = 4$.

29. Find the slope of the line tangent to $y = x \ln 3x^2$ at $x = 1.8$.

30. The loudness of a sound in decibels is given by $L = 10 \log(I/I_0)$. Find the rate of change in loudness when $I = 7.0$ w/m^2 if I_0 is fixed at 3×10^{-11} w/m^2.

31. The capacitance in a capacitor is given by

$$C = \frac{kL}{\ln(R/r)}$$

where R = the outer radius, r = the inner radius, L = the length, and k = a constant. Find dC/dr when $R = 2.0$ cm, $r = 1.3$ cm, $L = 3.5$ cm, and $k = 5.4 \times 10^{-11}$ cm.

20–5

DERIVATIVE OF THE EXPONENTIAL FUNCTION

In this section we will discuss the derivative of another type of transcendental function, the exponential function. Recall from our discussion of exponential functions in Chapter 8 that an exponential function contains a variable in the exponent. Therefore, the exponential function is of the form $y = a^u$ where u is some function of the variable x, and a is a positive constant other than 1.

Derivative of $y = a^u$

To find the derivative of this exponential function, we take the natural logarithm of both sides of the equation and use implicit differentiation to differentiate the result.

$$y = a^u$$

$$\ln y = \ln a^u$$

$$\ln y = u \ln a$$

$$\frac{d(\ln y)}{dx} = \frac{d(u \ln a)}{dx}$$

$$\frac{1}{y} \cdot \frac{dy}{dx} = \underbrace{\ln a}_{\ln a \text{ is a constant}} \frac{du}{dx}$$

$$\frac{dy}{dx} = y \ln a \frac{du}{dx}$$

$$\frac{dy}{dx} = \underbrace{a^u}_{\text{substitute for } y} \ln a \frac{du}{dx}$$

Derivative of a^u

$$\frac{d(a^u)}{dx} = a^u \ln a \frac{du}{dx}$$

where u = a function of x

Derivative of e^u

For the special case where the base of the exponential function is e.

$$\frac{dy}{dx} = e^u \underbrace{\ln e}_{\ln e = 1} \frac{du}{dx}$$

$$\frac{dy}{dx} = e^u \frac{du}{dx}$$

Derivative of e^u

$$\frac{d(e^u)}{dx} = e^u \frac{du}{dx}$$

where u = a function of x

EXAMPLE 1 Find dy/dx if

(a) $y = 6^{4x}$ (b) $y = e^{2x^3}$ (c) $y = 3^{3x^2-x}$

Solution

(a) Using the formula for the derivative of a^u, let $a = 6$ and $u = 4x$; then $du/dx = 4$.

$$\begin{aligned}\frac{d(6^{4x})}{dx} &= 6^{4x} \ln 6 \cdot 4 \\ &= 4(6^{4x} \ln 6)\end{aligned}$$

(b) Using the formula for the derivative of e^u, let $u = 2x^3$; then $du/dx = 6x^2$.

$$\begin{aligned}\frac{d(e^{2x^3})}{dx} &= e^{2x^3} \cdot 6x^2 \\ &= 6x^2 \cdot e^{2x^3}\end{aligned}$$

(c) If $a = 3$, then $u = 3x^2 - x$ and $du/dx = 6x - 1$.

$$\begin{aligned}\frac{d(3^{3x^2-x})}{dx} &= 3^{3x^2-x} \ln 3(6x - 1) \\ &= (6x - 1)(3^{3x^2-x})(\ln 3)\end{aligned}$$

■

EXAMPLE 2 Find the derivative of the following functions:

(a) $y = x^3 e^{\sin 3x}$ (b) $y = \dfrac{e^{-4x}}{\sin x^2}$ (c) $y = 2 \operatorname{Arctan}(e^{x^2})$

Solution

(a) We use the product rule to find the derivative of this function.

$$\begin{aligned}\frac{dy}{dx} &= x^3 \frac{d(e^{\sin 3x})}{dx} + e^{\sin 3x} \frac{d(x^3)}{dx} \\ &= x^3(e^{\sin 3x} \cdot \cos 3x \cdot 3) + e^{\sin 3x} \cdot 3x^2 \\ &= 3x^3 \cdot \cos 3x \cdot e^{\sin 3x} + 3x^2 e^{\sin 3x} \\ &= 3x^2 e^{\sin 3x}(x \cos 3x + 1)\end{aligned}$$

(b) We use the quotient rule to find this derivative.

$$\frac{dy}{dx} = \left[\sin x^2 \frac{d(e^{-4x})}{dx} - e^{-4x}\frac{d(\sin x^2)}{dx}\right] \div (\sin x^2)^2$$

$$= \frac{\sin x^2(e^{-4x}) \cdot -4 - e^{-4x} \cdot \cos x^2 \cdot 2x}{\sin^2 x^2}$$

$$= \frac{-4e^{-4x} \cdot \sin x^2 - 2xe^{-4x} \cdot \cos x^2}{\sin^2 x^2}$$

(c) Let $u = e^{x^2}$; then $du/dx = e^{x^2} \cdot 2x$.

$$\frac{dy}{dx} = 2\frac{d(\text{Arctan } e^{x^2})}{dx}$$

$$= 2\frac{1}{1 + (e^{x^2})^2} \cdot e^{x^2} \cdot 2x$$

$$= \frac{4xe^{x^2}}{1 + e^{2x^2}}$$

■

Applications

EXAMPLE 3 For damped harmonic motion the amplitude of motion is given by

$$y = Ae^{-bt}\cos(kt + \theta)$$

where A = the maximum amplitude; t = time; and b, k, and θ = constants. Find the vertical speed dy/dt at $t = 1.1$ s when $A = 8$ cm, $b = 2$, $k = 4$, and $\theta = \pi/8$.

Solution First, we substitute the values given in the problem statement.

$$y = 8e^{-2t}\cos\left(4t + \frac{\pi}{8}\right)$$

Next, we use the product rule to find dy/dt.

$$\frac{dy}{dt} = 8e^{-2t}\frac{d}{dt}\left[\cos\left(4t + \frac{\pi}{8}\right)\right] + 8\cos\left(4t + \frac{\pi}{8}\right) \cdot \frac{d}{dt}(e^{-2t})$$

$$= 8e^{-2t}\left[-\sin\left(4t + \frac{\pi}{8}\right)\right] \cdot 4 + 8\cos\left(4t + \frac{\pi}{8}\right)[e^{-2t} \cdot -2]$$

$$= -32e^{-2t}\sin\left(4t + \frac{\pi}{8}\right) - 16e^{-2t}\cos\left(4t + \frac{\pi}{8}\right)$$

$$\frac{dy}{dt} = -16e^{-2t}\left[2\sin\left(4t + \frac{\pi}{8}\right) + \cos\left(4t + \frac{\pi}{8}\right)\right]$$

Last, we evaluate this expression when $t = 1.1$ s.

$$\frac{dy}{dt} \approx 3.4 \text{ cm/s}$$

■

EXAMPLE 4 Exponential growth is approximated by the equation $N = N_0e^{kt}$ where N_0 = the initial amount, k = the growth rate, and t = time in years. Lakeland has a population of 210,000 in 1987, and the growth rate is 3% annually. Find the yearly rate at which the population is increasing in 1993.

Solution First, we substitute the values given in the problem statement.

$$N = 210{,}000e^{0.03t}$$

Next, we take the derivative of this function with respect to t.

$$\frac{dN}{dt} = 210{,}000e^{0.03t} \cdot 0.03$$

$$= 6{,}300e^{0.03t}$$

Last, we evaluate the derivative when $t = 1993 - 1987 = 6$.

$$\frac{dN}{dt} \approx 7{,}542$$

The population is increasing 7,542 per year in 1993. ■

EXAMPLE 5 When a heated object is allowed to cool, the temperature of the object at time t is approximated by $T = T_0 + Ce^{-kt}$ where T_0 = the temperature of the cooling medium, C = the difference in temperature between the object and the cooling medium, t = time in minutes, and k = a constant of proportionality. An overheated computer at 58°C is allowed to cool in a room whose temperature is 13°C. If the constant of proportionality is 0.2, find the rate at which the temperature is falling after 5 min.

Solution First, we substitute the values given in the problem.

$$T = 13 + 45e^{-0.2t}$$

$$C = (58 - 13)^\circ$$

Next, we differentiate this function with respect to t.

$$\frac{dT}{dt} = 45(e^{-0.2t} \cdot -0.2)$$

$$= -9.0e^{-0.2t}$$

Last, we evaluate the derivative when $t = 5$ min.

$$\frac{dT}{dt} \approx -3.3$$

The temperature of the computer is falling 3.3°C/min. ■

20–5 EXERCISES

Find the derivative of each function.

1. $y = 6^{2x}$

2. $y = 8^{x/3}$

3. $y = 7^{x^2}$

4. $y = 2^{x+x^3}$

5. $y = 5^{-2x}$

6. $y = e^{3x}$

7. $y = e^{\sqrt{x}}$

8. $y = (3 + e^x)$

9. $y = (2 - 3^x)^3$

10. $y = x(1 + 4^{5x})$

11. $y = \ln(\sin e^x)$

12. $y = \ln(\cos e^{-x})$

13. $y = 3^x + 4^{2x}$

14. $y = 5(1 - e^x)$

15. $y = \dfrac{1}{x^2} - e^{-x}$

16. $y = 8x^3 - e^{-4x}$

17. $y = x^2 + e^{2x}$

18. $y = \dfrac{9^x}{\cos x}$

19. $y = \dfrac{e^x}{x^2}$

20. $y = x^2(7^x)$

21. $y = xe^x$

22. $y = 4^{\tan 3x}$

23. $y = \log(1 - 5^x)$

24. $y = \ln(1 + e^{7x})$

25. $y = 2^{\sin x}$

26. $y = e^{x^2}\log 2x$

27. $y = e^x \ln x^3$

28. $y = \sin(e^{4x})$

29. $y = \dfrac{e^{2x} + e^x}{x}$

30. $y = e^{2x} - e^{x^2}$

31. $y = \cos e^{2x^2}$

32. $y = 4 \text{ Arcsin } e^{3x}$

33. $y = 8 \text{ Arctan}(e^x)$

34. Find the rate at which the temperature of a steel pin is cooling after 1.5 min if its initial temperature is 340°F and the air is 42°F. (See Example 5.)

35. The growth of current in an inductance coil connected in series is given by

$$i = \frac{V}{R}(1 - e^{-Rt/L})$$

Find the rate at which the current is changing when $t = 1.3 \times 10^{-3}$ s if $V = 9$ V, $R = 15\ \Omega$, and $L = 0.1$ H.

36. Find the velocity of an object whose x and y components of displacement are given by $x = e^t \sin^2 t$ and $y = e^t \cos 2t$ when $t = \pi/8$.

CHAPTER SUMMARY

Summary of Formulas

$$\frac{d(\sin u)}{dx} = \cos u \frac{du}{dx}$$ derivative of sine function

$$\frac{d(\cos u)}{dx} = -\sin u \frac{du}{dx}$$ derivative of cosine function

$$\frac{d(\tan u)}{dx} = \sec^2 u \frac{du}{dx}$$ derivative of tangent function

$$\frac{d(\cot u)}{dx} = -\csc^2 u \frac{du}{dx}$$ derivative of cotangent function

$$\frac{d(\csc u)}{dx} = -\csc u \cot u \frac{du}{dx}$$ derivative of cosecant function

$$\frac{d(\sec u)}{dx} = \sec u \tan u \frac{du}{dx}$$ derivative of secant function

$$\frac{d(\text{Arcsin } u)}{dx} = \frac{1}{\sqrt{1 - u^2}} \cdot \frac{du}{dx}$$ derivative of Arcsin function

$$\frac{d(\text{Arccos } u)}{dx} = -\frac{1}{\sqrt{1 - u^2}} \cdot \frac{du}{dx}$$ derivative of Arccos function

$$\frac{d(\text{Arctan } u)}{dx} = \frac{1}{1 + u^2} \cdot \frac{du}{dx} \quad \text{derivative of Arctangent function}$$

$$\frac{d(\log_b u)}{dx} = \frac{1}{u} \log_b e \frac{du}{dx} \quad \text{derivative of logarithmic function}$$

$$\frac{d(\ln u)}{dx} = \frac{1}{u} \cdot \frac{du}{dx} \quad \text{derivative of natural logarithmic function}$$

$$\frac{d(a^u)}{dx} = a^u \ln a \frac{du}{dx} \quad \text{derivative of } a^u$$

$$\frac{d(e^u)}{dx} = e^u \frac{du}{dx} \quad \text{derivative of } e^u$$

CHAPTER REVIEW

Section 20–1

Find the derivative of the following functions.

1. $y = \sin^2 4x$

2. $y = 3 \cos 4t^2$

3. $y = (\cos 3x)^{1/4}$

4. $y = \sin(2t - 3)^2$

5. $y = \cos(t^2 - 4)^3$

6. $y = \cos^4(3x^2)$

7. $y = x^3 \cos(5x^2)$

8. $y = \dfrac{\sin^3 x}{\cos 6x^3}$

9. $y = 4 \sin^3 x + x \cos(3x)$

10. Find the slope of a line tangent to the curve $y = x \sin x^2$ at $x = 1$.

11. The current i in a certain circuit is given by $i = 8 \sin(120\pi t + 1.5\pi)$. Find the expression for the induced voltage. (*Note:* $V = di/dt$.)

12. Find the acceleration of an object when $t = 2.5$ s if the velocity of the object is given by $y = t^2 \cos 3t$.

Section 20–2

Find dy/dx for each function.

13. $y = \tan^3(7x)$

14. $y = 4 \sec^2(3x^4)$

15. $y = \csc^2(4x^3)$

16. $y = x^3 \cot^2(x)$

17. $y = \dfrac{\csc^2 x}{3x}$

18. $y = \dfrac{x \tan 2x}{\cos x}$

19. $y = \tan(2x)\csc(x^2)$

20. $y = \cot(x^2)\sec^3(x)$

21. $y = \sin^2[4x(1 - \tan^2 x)]$

22. A searchlight rotating at a rate of 0.10 rad/s is scanning a wall 215 ft away. How fast is the light beam scanning the wall at a point 295 ft from the searchlight?

23. Find the slope of the line tangent to the curve $y = \tan(3x + \pi)$ at $x = \pi/4$.

24. Find the acceleration, when $t = 0.15$ s, of an object whose displacement (in feet) is given by $s = 3 \sin^2 t + \cos 2t$.

25. What is the current in a circuit when $t = 0.1$ s if the charge is given by $q = 0.04 \tan 0.7t$ C?

Section 20–3

Find the derivative of each function.

26. $y = 7 \text{ Arcsin } 6x^3$

27. $y = (\text{Arccos } x^2)^2$

28. $y = \text{Arctan}\left(\dfrac{x}{x - 1}\right)$

29. $y = 3x^2 \text{Arcsin}(x^3)$

30. $y = x \text{ Arctan}(1 + x)$

31. $y = \dfrac{\text{Arccos } 3x^2}{x + 8}$

32. $y = 3 \text{ Arcsin } x - 6x \text{ Arccos}(x^2)$

33. $y = \dfrac{x}{\text{Arctan } x}$

34. $y = \dfrac{1}{\text{Arcsin}(3x^2)}$

35. Find the slope of the line tangent to $y = x \text{ Arctan } 3x^2$ when $x = \pi/4$.

36. Find the slope of the line tangent to $y = 4x \text{ Arctan}(3x + \pi)$ when $x = \pi/8$.

37. An airplane flying at 35,000 ft and a velocity of 450 ft/s passes over a radar station. At what rate is the angle of elevation of the radar changing 18 seconds later?

38. Fred approaches a building 70 ft tall at a rate of 12 ft/s. At what rate is the angle of elevation changing when he is 275 ft from the building? (Neglect Fred's height.)

Section 20–4

Find the derivative of each function.

39. $y = \ln(3x + 1)^2$

40. $y = \log(8x + 9)^2$

41. $y = (\log x^3)^2$

42. $y = \log \sqrt{5x}$

43. $y = \ln\left(\dfrac{x^3 + 1}{x^2}\right)$

44. $y = \log(4x^3) \sec x^2$

45. $y = \ln x^2\cos(3x)$

46. $y = \dfrac{\ln 2x}{3x^2}$

47. $y = \dfrac{\log 7x^2}{\csc x}$

48. Find the slope of the line tangent to the curve $y = \ln \sin x^2$ at $x = \pi/4$.

49. Find the slope of the line tangent to the curve $y = \tan(\ln x)$ when $x = \pi/8$.

50. Find the change in capacitance C with respect to inner radius r when $R = 3.4$ cm, $r = 2.7$ cm, $L = 4.1$ cm, and $k = 5.4 \times 10^{-11}$ cm when $C = \dfrac{kL}{\ln(R/r)}$.

51. The energy dissipated in a resistor is given by $E = \ln(t + 1) - \dfrac{t}{2} + 10$ J. Find dE/dt when $t = 3$ s.

Section 20–5

Find the derivative of each function.

52. $y = 5e^{-2x}$

53. $y = 9^{(x+1)}$

54. $y = 4 \text{ Arcsin}(e^{5x})$

55. $y = e^{2x}\ln(3x - 1)$

56. $y = \dfrac{e^x}{\sec^2 x}$

57. $y = \dfrac{e^{x^2}}{3x^2}$

58. $y = \text{Arctan}(e^{6x})$

59. $y = 5^{2x}\cos(x^2)$

60. $y = e^x\sin(x + \pi)$

61. Find the rate at which an overheated computer cools after 20 min if its initial temperature is 110°F and the air is 68°F.

62. The growth of current in an inductance coil is given by

$$i = \frac{V}{R}(1 - e^{-Rt/L})$$

where V, R, and L are constants. Find di/dt.

CHAPTER TEST

The number in parentheses refers to the appropriate learning objective given at the beginning of the chapter.

1. Find the derivative of $y = \text{Arcsin}(e^{2x})$. **(3, 5)**

2. A searchlight rotates at a rate of 0.08 rad/s, illuminating a wall 300 ft from the searchlight. How fast is the light beam moving over the wall at a point 415 ft from the searchlight? **(6)**

3. Find dy/dx for $y = (3x^2 + 5)\sec(x^2)$. **(2)**

4. Under certain conditions, the energy dissipated in a resistor is given by **(6)**

$$E = \ln(t + 1) - \frac{t}{2} + 10 \text{ J}$$

When is the energy output a maximum?

5. Find the slope of the line tangent to $y = 3 \sin 4x^2$ at $x = \pi/3$. **(1)**

6. Find $\left.\frac{dy}{dx}\right|_{x=2}$ for $y = \ln(\cos 5x^2)$. **(1, 4)**

7. Find the derivative of the function $y = \log(t^2 + 5)$. **(4)**

8. Find dy/dx for $y = \frac{\tan x}{1 + e^x}$. **(2,5)**

9. A missile moves in the path of a parabola with the x and y components given by $x = t \sin t$ and $y = \cos^2(2t)$. Find the formula for the x and y components of the velocity. **(6)**

10. Find the derivative of the function $y = e^x \text{Arctan}(3x^2)$. **(3, 5)**

11. A 30-ft ladder leans against a wall at a point 18 ft from the floor. If the bottom of the ladder starts slipping away from the wall at a rate of 2 ft/s, find the rate at which the angle between the top of the ladder and the wall is changing. **(6)**

12. Find the derivative of $y = \frac{\sin^2(5x)}{5^{3x}}$. **(1, 5)**

13. Find the slope of the line tangent to the curve $y = e^{3t^2}\cot(t^2)$ at $t = 2$. **(2, 5)**

14. Find the yearly rate by which the population of Greenbriar is increasing in 1995 if there are 2,586 people in 1988 and the yearly rate of increase is 1.5%. **(6)**

15. Find the derivative of $y = \frac{x^2 \sec^2 x}{e^{\ln x^2}}$. **(2, 4, 5)**

SOLUTION TO CHAPTER INTRODUCTION

The equations for this problem are

$$q = VC(1 - e^{-t/(RC)}) \quad \text{and} \quad i = \frac{dq}{dt}$$

Therefore, we substitute the given values into the equation, take the derivative with respect to t, and substitute $t = 0.003$ into the result.

$$q = 240(2 \times 10^{-6})[1 - e^{-t(2,600(2 \times 10^{-6}))}]$$
$$q = 4.8 \times 10^{-4}[1 - e^{-t/0.0052}]$$
$$q = 4.8 \times 10^{-4} - 4.8 \times 10^{-4}e^{-t/0.0052}$$
$$i = \frac{dq}{dt} = -4.8 \times 10^{-4}e^{-t/0.0052} \times -\frac{1}{0.0052}$$
$$\approx 0.0923e^{-t/0.0052}$$
$$\left.\frac{dq}{dt}\right|_{t=0.003} \approx 0.0923e^{-0.003/0.0052}$$
$$\approx 0.052$$

You want to find the average force exerted on a piston from $t = 0$ to $t = 1.75$ s if the force is given by $F = 12 - e^{-2t}$ lb. (The solution to this problem is given at the end of the chapter.)

The calculation for the average force on the piston requires that we integrate the exponential function representing the force. Now that we have discussed the derivatives of transcendental functions, we are prepared to discuss their integrals. In this chapter we will discuss integration of the exponential and logarithmic functions, trigonometric functions, and inverse trigonometric functions. In addition, we will discuss algebraic techniques, trigonometric substitution, and the use of tables to integrate a function.

Learning Objectives

After you complete this chapter, you should be able to

1. Apply the general power rule to integrate transcendental functions (Section 21–1).
2. Find the integral of the exponential and logarithmic functions (Section 21–2).
3. Integrate the basic trigonometric functions (Section 21–3).
4. Integrate the product and power of trigonometric functions (Section 21–4).
5. Integrate inverse trigonometric functions (Section 21–5).
6. Use trigonometric substitutions to integrate (Section 21–5).
7. Perform integration by parts (Section 21–6).
8. Use a table of integration to integrate a function (Section 21–7).
9. Apply the techniques of integration to technical problems (Sections 21–1 through 21–7).

Chapter 21

Techniques of Integration

21–1

THE GENERAL POWER RULE

In this section we will extend the power rule for integration discussed in Chapter 19 to include the transcendental functions. Recall from our previous discussion that

$$\int u^n \, du = \frac{u^{n+1}}{n+1} + C \quad \text{where } n \neq -1$$

When you apply the general power rule, it is extremely important to choose u, n, and du correctly. To do so, you must be familiar with the derivatives developed in the previous chapter, and you must use a certain amount of trial and error. The examples in this section illustrate this process.

EXAMPLE 1 Integrate the following:

$$\int \sin^5 x \cos x \, dx$$

Solution Since $d(\sin x)$ is $\cos x$, let $u = \sin x$, $n = 5$, and $du = \cos x \, dx$. Then the integral is of the form $\int u^5 \, du$. Therefore,

$$\int \sin^5 x \cos x \, dx = \int \overbrace{(\sin x)}^{u}{}^{\overset{n}{5}} \underbrace{(\cos x \, dx)}_{du}$$

$$= \int u^5 \frac{du}{dx}$$

$$= \frac{u^6}{6} + C$$

$$= \frac{\sin^6 x}{6} + C$$

■

NOTE ✦ Remember to include the constant of integration for an indefinite integral.

LEARNING HINT ✦ In trying to determine u and du correctly, first, let u equal the function raised to a power, and then determine whether du is present. Technically, you should change the limits of integration to reflect the value of u. However, you can avoid this process if you substitute for u before evaluating the limits. As a result, the form of the integral will be easier, and you will not have to find new limits of integration.

EXAMPLE 2 Integrate the following:

(a) $\int -\frac{\text{Arccos } 2x}{\sqrt{1 - 4x^2}}\, dx$

(b) $\int (3 \sin x + \sin^2 x)\cos x \, dx$

Solution

(a) Let $u = \text{Arccos } 2x$; then

$$du = -\frac{2}{\sqrt{1 - 4x^2}}\, dx$$

Therefore, we can write the integral as follows:

$$\int -\frac{\text{Arccos } 2x}{\sqrt{1 - 4x^2}}\, dx = \frac{1}{2}\int \text{Arccos } 2x\left(-\frac{2}{\sqrt{1 - 4x^2}}\, dx\right)$$

Substituting in terms of u and integrating give

$$= \frac{1}{2}\int u^1 \, du$$

$$= \frac{1}{2} \cdot \frac{u^2}{2} + C$$

Substituting $u = \text{Arccos } 2x$ gives the integral.

$$= \frac{1}{2}\left(\frac{1}{2} \text{Arccos}^2 2x\right) + C$$

$$= \frac{1}{4} \text{Arccos}^2 2x + C$$

(b) Let us write this integral as

$$\int (3 \sin x + \sin^2 x)\cos x \, dx = 3\int \sin x \cos x \, dx + \int \sin^2 x \cos x \, dx$$

Let $u = \sin x$ and $du = \cos x \, dx$. Then this integral can be written as

$$= 3\int u \, du + \int u^2 \, du$$

Integrating gives the following results:

$$= 3\frac{u^2}{2} + \frac{u^3}{3} + C$$

Substituting for u gives the result from integration.

$$= \frac{3}{2} \sin^2 x + \frac{1}{3} \sin^3 x + C$$

■

EXAMPLE 3 Integrate the following:

$$\int \frac{e^x}{\sqrt[3]{1 + e^x}}\, dx$$

Solution Let $u = 1 + e^x$; then $du = e^x\, dx$. This integral can be written as follows:

$$\begin{aligned}\int \frac{e^x}{\sqrt[3]{1 + e^x}}\, dx &= \int (1 + e^x)^{-1/3}(e^x\, dx)\\ &= \int u^{-1/3}\, du\\ &= \frac{3}{2}u^{2/3} + C\\ &= \frac{3}{2}(1 + e^x)^{2/3} + C\end{aligned}$$

■

EXAMPLE 4 Integrate the following:

(a) $\displaystyle\int_1^2 \frac{(\ln x)^3}{x}\, dx$

(b) $\displaystyle\int \frac{1 + t^2}{(3t + t^3)^4}\, dt$

Solution

(a) Let $u = \ln x$; then $du = \frac{1}{x}\, dx$. We write the integral in terms of u and integrate. Notice that the limits of integration are omitted to avoid confusion.

$$\begin{aligned}\int \frac{(\ln x)^3}{x}\, dx &= \int u^3\, du\\ &= \frac{u^4}{4}\end{aligned}$$

Substituting $\ln x$ for u and evaluating the result at the original limits give

$$\begin{aligned}\int_1^2 \frac{(\ln x)^3}{x}\, dx &= \left.\frac{(\ln x)^4}{4}\right|_1^2\\ &\approx 0.06\end{aligned}$$

(b) Let $u = 3t + t^3$; then $du = (3 + 3t^2)\, dx$. Since du needs a factor of 3, we multiply and divide the integral by 3.

$$\int \frac{1 + t^2}{(3t + t^3)^4} dt = \frac{1}{3} \int \frac{3(1 + t^2)}{(3t + t^3)^4} dt$$
$$= \frac{1}{3} \int u^{-4} du$$
$$= \frac{1}{3} \cdot \frac{u^{-3}}{-3} + C$$
$$= -\frac{1}{9}(3t + t^3)^{-3} + C$$
$$= -\frac{1}{9(3t + t^3)^3} + C$$

■

21–1 EXERCISES

Integrate the following functions. Round to two decimal places where necessary.

1. $\int \frac{(1 + e^{-x})^3}{e^x} dx$
2. $\int \frac{(1 + \ln 3x^2)^4}{6x} dx$
3. $\int -\cos^5 x \sin x \, dx$
4. $\int -\cot^4 x \csc^2 x \, dx$
5. $\int \sec^5 x(\sec x \tan x \, dx)$
6. $\int_0^4 \sin^2 3x \cos 3x \, dx$
7. $\int \frac{\text{Arctan}^2 3x}{1 + 9x^2} dx$
8. $\int \tan^3 4x \sec^2 4x \, dx$
9. $\int \frac{\text{Arcsin } 2x \, dx}{\sqrt{1 - 4x^2}}$
10. $\int (2 \sin x + \sin^3 x)\cos x \, dx$
11. $\int \frac{e^{3x}}{(1 + e^{3x})^2} dx$
12. $\int \frac{\sec^2 t}{(1 + \tan t)^2} dt$
13. $\int_{1.5}^2 \frac{\ln(6x - 7)}{6x - 7} dx$
14. $\int_0^2 \frac{3e^t}{\sqrt{1 + 6e^t}} dt$
15. $\int \frac{(1 - 4e^{-x})^3}{e^x} dx$
16. $\int \frac{\cos \theta}{\sqrt{1 - \sin \theta}} d\theta$
17. $\int_{-1}^1 e^t(e^t - 4)^2 \, dt$
18. $\int \frac{(8 - 4 \ln 2x)^2}{x} dx$
19. $\int \frac{\cot x}{\sin^2 x} dx$
20. $\int_0^{1/2} \frac{(\text{Arccos } 2x)^2}{\sqrt{1 - 4x^2}} dx$
21. $\int (3 \cos x - \cos^2 x) \sin x \, dx$
22. $\int (1 + \csc^2 x)^3(\csc^2 x \cot x \, dx)$
23. $\int_0^\pi \cos 2\theta \sqrt{\sin 2\theta} \, d\theta$
24. $\int \sin^{1/3}(2x)\cos(2x) \, dx$
25. $\int [\ln(x - 2)]^3 \frac{dx}{x - 2}$
26. $\int_{\pi/3}^{\pi/4} \frac{\csc^2 \theta}{(1 + \cot \theta)^3} d\theta$
27. $\int \frac{2 + t}{(4t + t^2)^3} dt$
28. $\int (4 - 5e^{3t})e^{3t} \, dt$
29. $\int \sqrt{1 + \cot x} \csc^2 x \, dx$
30. $\int \frac{\text{Arcsin } 3x}{\sqrt{1 - 9x^2}} dx$
31. Find the area under the curve $y = \sin x \cos x$ from $x = 0$ to $x = \pi/4$.
32. The slope of a certain curve is given by $(\ln 3x)^2/x$. Find the equation of the curve if the curve passes through $(2, 0)$.
33. The charge q (in coulombs) stored in a capacitive circuit decreases at a rate given by $dq/dt = e^{-t}(e^{-t} - 2)^2$. How much charge is dissipated from $t = 0.25$ s to $t = 1.3$ s?
34. Determine the average value of the function $y = \cos x \sin^3 x \, dx$ over the interval $x = 0$ to $x = \pi/2$.
35. Find the area of the region bounded by the curve $y = \frac{(\ln x)^2}{x}$, $y = 0$, $x = 1$, and $x = 3$.

36. The charge (in coulombs) stored in a capacitor decreases at a rate given by

$$\frac{dq}{dt} = e^{-t}(e^{-t} - 1)^2$$

What is the charge dissipated from $t = 0.4$ s to $t = 1.75$ s?

37. The voltage across a capacitor is given by

$$V = 10^6 \int \frac{e^t}{\sqrt{2 + e^t}}\, dt$$

Determine the function of V in terms of time t by integrating.

21–2

INTEGRATING EXPONENTIAL AND LOGARITHMIC FUNCTIONS

The Reciprocal Function

The general power rule discussed in the previous section is valid for integration of power functions for all exponents n other than -1. In this section we will develop a formula for $n = -1$. A function of this type, called the **reciprocal function,** takes the form du/u. When we discussed the derivative of the logarithmic function, we found that

$$\frac{d(\ln u)}{dx} = \frac{1}{u} \cdot \frac{du}{dx}$$

Reversing this process, we find that $\int (du/u) = \ln u + C$. Since we cannot take the logarithm of a negative number or zero, we restrict u to always being positive by using $|u|$. The formula for the integral of the reciprocal function is given in the following box.

Integral of the Reciprocal Function

$$\int \frac{du}{u} = \ln|u| + C$$

where $u \neq 0$

EXAMPLE 1 Integrate the following function:

$$\int \frac{3\, dx}{3x - 4}$$

Solution Let $u = 3x - 4$; then $du = 3\, dx$. Integrating gives the following result:

$$\int \frac{3\, dx}{3x - 4} = \int \frac{du}{u} = \ln|u| + C$$
$$= \ln|3x - 4| + C$$

■

EXAMPLE 2 Integrate the following functions:

(a) $\displaystyle\int \frac{\sec^2 x}{1 + \tan x}\, dx$

(b) $\displaystyle\int_1^2 \frac{e^{-3x}}{1 - e^{-3x}}\, dx$

Solution

(a) Let $u = 1 + \tan x$; then $du = \sec^2 x\, dx$. Since this result fits the formula for integrating the reciprocal function, the integral is as follows:

$$\int \frac{\sec^2 x}{1 + \tan x}\, dx = \int \frac{du}{u} = \ln|u| + C$$
$$= \ln|1 + \tan x| + C$$

(b) Let $u = 1 - e^{-3x}$; then $du = -e^{-3x} \cdot -3\, dx = 3e^{-3x}\, dx$. The value for du requires that we multiply and divide the integral by 3. Integrating gives

multiply and divide by 3

$$\int_1^2 \frac{e^{-3x}}{1 - e^{-3x}}\, dx = \frac{1}{3}\int_1^2 \frac{3e^{-3x}\, dx}{1 - e^{-3x}} = \frac{1}{3}\ln|1 - e^{-3x}|$$

Integrating and evaluating the limits of integration give

$$= \frac{1}{3}\, ln|1 - e^{-3x}| \Big|_1^2$$

$$\approx \frac{1}{3}[-0.002481829 + 0.05106918]$$

$$\approx 0.016$$ ■

The Exponential Function

When we differentiate the exponential function, we find that

$$\frac{d(e^u)}{dx} = e^u \cdot \frac{du}{dx}$$

Reversing this process to find the integral of the exponential function gives the following result.

Integral of the Exponential Function

$$\int e^u\, du = e^u + C$$

EXAMPLE 3 Integrate the following:

$$\int e^{3x}\,dx$$

Solution Since $u = 3x$, then $du = 3\,dx$. Thus, we must multiply and divide the integral by 3 in order to obtain the required du.

$$\int e^{3x}\,dx = \frac{1}{3}\int e^{\overbrace{3x}^{u}} \overbrace{(3\,dx)}^{du}$$

$$= \frac{1}{3}e^{3x} + C$$

■

EXAMPLE 4 Integrate the following:

$$\int_0^1 xe^{3x^2}\,dx$$

Solution Since $u = 3x^2$, $du = 6x\,dx$. We multiply and divide the integral by 6 to obtain du.

$$\int_0^1 xe^{3x^2}\,dx = \frac{1}{6}\int_0^1 e^{\overbrace{3x^2}^{u}}\overbrace{(6x\,dx)}^{du}$$

Evaluating the integral at the limits of integration gives

$$= \frac{1}{6}e^{3x^2}\Big|_0^1$$

$$\approx \frac{1}{6}[20.086 - 1]$$

$$\approx 3.18$$

■

EXAMPLE 5 Integrate the following:

(a) $\displaystyle\int_0^{\pi/6} \sec^2 2x\, e^{\tan 2x}\,dx$

(b) $\displaystyle\int \frac{e^{\sqrt{x-2}}}{\sqrt{x-2}}\,dx$

Solution

(a) Let $u = \tan 2x$; then $du = \sec^2 2x \cdot 2\,dx = 2\sec^2 2x\,dx$. Therefore, we multiply and divide the integral by 2.

$$\int_0^{\pi/6} \sec^2 2x e^{\tan 2x}\,dx = \frac{1}{2}\int_0^{\pi/6} \overbrace{e^{\tan 2x}}^{u}\overbrace{(2\sec^2 2x\,dx)}^{du} = \frac{1}{2}e^{\tan 2x}$$

Evaluating the integral at 0 and $\pi/6$ gives the following result:

$$= \frac{1}{2}e^{\tan 2x}\Big|_0^{\pi/6}$$

$$\approx \frac{1}{2}[5.652 - 1]$$

$$\approx 2.33$$

(b) Let $u = \sqrt{x-2} = (x-2)^{1/2}$; then

$$du = \frac{1}{2}(x-2)^{-1/2}\,dx = \frac{dx}{2\sqrt{x-2}}$$

To obtain the required du, we must multiply and divide the integral by 2. The integral is written as follows:

$$\int \frac{e^{\sqrt{x-2}}}{\sqrt{x-2}}\,dx = 2\int e^{\sqrt{x-2}}\left(\frac{dx}{2\sqrt{x-2}}\right)$$

$$= 2e^{\sqrt{x-2}} + C$$

■

Integrating a^u

From the previous chapter recall that

$$\frac{d(a^u)}{dx} = a^u \ln a \frac{du}{dx}$$

Reversing the differentiation formula gives the following formula for integration.

Integral of a^u

$$\int a^u\,du = \frac{a^u}{\ln a} + C$$

where $a > 0$, and $a \neq 1$

EXAMPLE 6 Integrate the following:

(a) $\int 8^{x+1}\, dx$ (b) $\int_0^1 3^{4x}\, dx$

Solution

(a) If $u = x + 1$, then $du = 1\, dx$.

$$\int \underset{\substack{\uparrow \\ a}}{8}^{\overbrace{x+1}^{u}}\, dx = \frac{\overbrace{8^{x+1}}^{a^u}}{\underbrace{\ln 8}_{\ln a}} + C$$

(b) If $u = 4x$, then $du = 4\, dx$, and we must multiply and divide the integral by 4.

$$\int_0^1 3^{4x}\, dx = \frac{1}{4}\int_0^1 3^{4x}(4\, dx) = \frac{3^{4x}}{4 \ln 3}$$

Evaluating the limits of integration gives

$$= \frac{1}{4} \cdot \frac{3^{4x}}{\ln 3}\Bigg|_0^1$$

$$\approx \frac{1}{4}(73.729 - 0.910)$$

$$\approx 18.20$$

■

Applications

EXAMPLE 7 The current (in amperes) for a certain circuit is given by

$$i = \frac{t}{(1 + 3t^2)}$$

Find the charge that passes a given point in the first 1.5 s.

Solution Remember from previous discussions that the charge is the integral of the current with respect to time. Therefore, this problem can be expressed as follows:

$$q = \int_0^{1.5} \frac{t}{(1 + 3t^2)}\, dt$$

Let $u = 1 + 3t^2$; then $du = 6t\, dt$. We multiply and divide the integral by 6 to obtain the necessary value of du. Integrating gives

$$q = \frac{1}{6}\int_0^{1.5} \frac{6t\,dt}{1 + 3t^2}$$

$$= \frac{1}{6}\ln|1 + 3t^2|\Big|_0^{1.5}$$

Evaluating the integral at its limits gives

$$q \approx \frac{1}{6}(2.048 - 0)$$

$$\approx 0.34$$

The charge in the circuit is approximately 0.34 coulombs. ■

EXAMPLE 8 The current in a circuit varies with time according to the equation $i = 5e^{2t}$. Find the charge flowing through the circuit from $t = 0$ to $t = 1.75$ s.

Solution The problem statement can be written as the following integral:

$$q = \int_0^{1.75} 5e^{2t}\,dt$$

$$= 5\int_0^{1.75} e^{2t}\,dt$$

If we let $u = 2t$, then $du = 2\,dt$. To obtain the required value of du, we multiply and divide the integral by 2.

$$q = 5 \cdot \frac{1}{2}\int_0^{1.75} e^{2t}(2\,dt)$$

$$= \frac{5}{2}e^{2t}\Big|_0^{1.75}$$

$$\approx \frac{5}{2}(33.115 - 1)$$

$$\approx 80.29$$

Approximately 80.29 coulombs flow through the circuit in the first 1.75 seconds. ■

EXAMPLE 9 Find the volume generated by revolving the area bounded by $y = e^{2x}$, $x = 0$, $y = 0$, and $x = 1$ about the x axis.

FIGURE 21–1

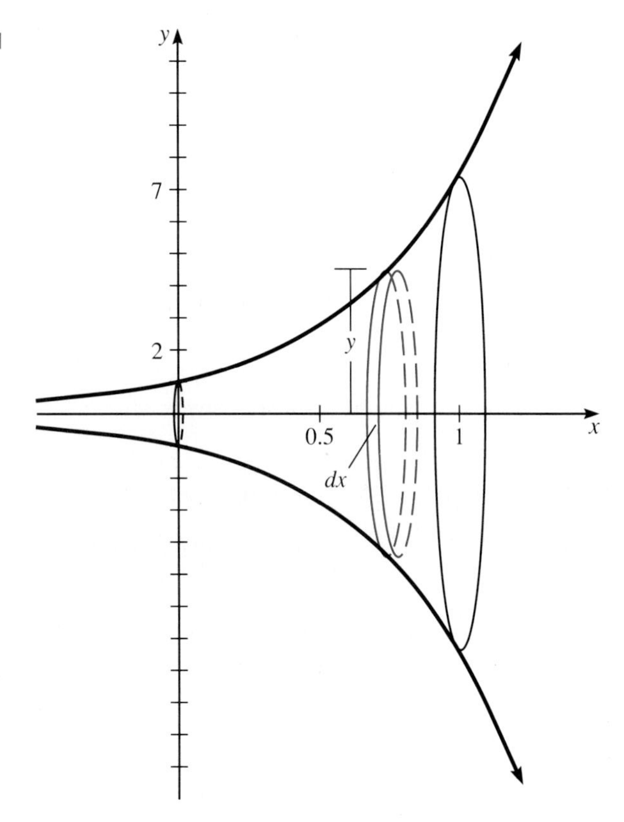

Solution Figure 21–1 shows the area bounded by these curves. If we choose a vertical representative rectangle and revolve it about the x axis, a disk is formed. Therefore, we will use the volume formula for the disk method.

$$V = \pi \int_a^b y^2\, dx$$

Substituting the information for this problem gives the following integral:

$$\begin{aligned} V &= \pi \int_0^1 (e^{2x})^2\, dx \\ &= \pi \int_0^1 e^{4x}\, dx \\ &= \pi \cdot \frac{1}{4} \int_0^1 e^{4x}\,(4\, dx) \end{aligned}$$

Integrating gives the following result:

$$V = \frac{\pi}{4} e^{4x} \Big|_0^1$$

Evaluating the limits of integration gives the volume as

$$V = \frac{\pi}{4}[e^4 - e^0]$$

$$\approx \frac{\pi}{4}(54.6 - 1)$$

$$\approx 42.1 \text{ cubic units}$$

■

EXAMPLE 10 A particle moves such that its velocity is given by

$$v = (\cos 4t)e^{\sin 4t} \text{ m/s}$$

Find its displacement after 2 seconds.

Solution Remember that $s = \int v\, dt$. Therefore, we can find the displacement by integrating the expression for the velocity.

$$s = \int_0^2 (\cos 4t)e^{\sin 4t}\, dt$$

If we let $u = \sin 4t$, then $du = 4 \cos 4t\, dt$, and we must multiply and divide the integral by 4.

$$s = \frac{1}{4}\int_0^2 e^{\sin 4t}(4 \cos 4t\, dt)$$

$$= \frac{1}{4}\left(e^{\sin 4t} \Big|_0^2\right)$$

Evaluating the limits of integration gives

$$s \approx \frac{1}{4}[2.69 - 1]$$

$$\approx 0.42$$

The particle has moved approximately 0.42 m.

■

21–2 EXERCISES

Integrate. Round to two decimal places when necessary.

1. $\int \frac{dt}{1 - 6t}$
2. $\int \frac{dt}{8t - 3}$
3. $\int \frac{x\, dx}{4x^2 - 5}$
4. $\int \frac{x^2}{7x^3 - 3}\, dx$
5. $\int \frac{\cos 3x}{\sin 3x}\, dx$
6. $\int \frac{\sec x \tan x}{\sec x}\, dx$
7. $\int \frac{\sin 4t}{\cos 4t - 7}\, dt$
8. $\int \frac{-e^{-x} + 1}{e^{-x} + x}\, dx$
9. $\int_1^2 \frac{6x + 1}{3x^2 + x}\, dx$
10. $\int_2^3 \frac{dx}{x \ln x}$
11. $\int (e^x + e^{2x})\, dx$
12. $\int \frac{e^x}{(1 + e^x)}\, dx$

13. $\int (6 + e^{3x})\, dx$

14. $\int_{\pi/4}^{\pi/3} \frac{\cos 2t}{\sin 2t}\, dt$

15. $\int_1^3 \frac{dx}{e^x}$

16. $\int_0^1 e^{2x}(e^{2x} - 3)^3\, dx$

17. $\int \frac{\sqrt{3 - e^{-x}}}{e^x}\, dx$

18. $\int x \sqrt[3]{e^{x^2}}\, dx$

19. $\int_2^3 \frac{x\, dx}{3x^2 - 4}$

20. $\int 2^{x^2}(x\, dx)$

21. $\int 8^{3x-1}\, dx$

22. $\int_3^5 9^{4x}\, dx$

23. $\int_0^1 7^{2x^3} x^2\, dx$

24. $\int_0^1 \frac{x}{9 + x^2}\, dx$

25. $\int \frac{x\, dx}{8 - 3x^2}$

26. $\int (2x - 1)e^{(x^2 - x)}\, dx$

27. $\int_1^2 (3 - e^x)^2(-e^x)\, dx$

28. $\int_1^2 \frac{e^{1/x}}{x^2}\, dx$

29. $\int \frac{e^{\text{Arctan } x}}{1 + x^2}\, dx$

30. $\int_1^3 \frac{dx}{1 - 7x}$

31. $\int \frac{e^{2x}}{5 - e^{2x}}\, dx$

32. $\int x^2 e^{x^3}\, dx$

33. $\int_0^{\pi} (\cos x)e^{\sin x}\, dx$

34. $\int_0^2 \frac{e^{\sin 4x}}{\sec 4x}\, dx$

35. Find the volume generated by rotating the area bounded by $y = e^x$, $y = 0$, $x = 0$, and $x = 3$ about the x axis.

36. Find the area under the curve $y = e^{x/4}$ from $x = 0$ to $x = 8$.

37. Find the average force exerted on a piston from $t = 0$ to $t = 3$ s if the force is given by $F = 12 - e^{-t}$.

38. Find the centroid of the region bounded by $y = \frac{1}{2x + 3}$, $x = 0$, $y = 0$, and $x = 2$.

39. Find the velocity of a charged particle if its acceleration is given by

$$a = \frac{2}{12t + 1} \text{ m/s}^2$$

after 3 s.

40. The current in a capacitor as a function of time is given by

$$i = \frac{t}{2t^2 + 1} \text{ A}$$

Find the charge transferred to the capacitor in the first 3 seconds.

41. The current in a circuit is given by $i = 12e^{-t/2}$. Find the charge flowing through the circuit from $t = 0$ to $t = 1.3$ s.

42. Find the work done in moving an object 8 ft if the force as a function of displacement is given by

$$F = \frac{1}{(3s + 4)}$$

(*Note:* $W = \int F\, ds$.)

21–3

INTEGRATING BASIC TRIGONOMETRIC FUNCTIONS

In this section we will discuss integrating the six basic trigonometric functions. Other trigonometric forms will be discussed later. Since antidifferentiation is the inverse of differentiation, every formula for the derivative of a trigonometric function has a corresponding integration formula. The formulas for the integral of some trigonometric functions are as summarized in the following box.

Integral of Trigonometric Functions

$$\int \sin u\, du = -\cos u + C$$

$$\int \cos u\, du = \sin u + C$$

$$\int \sec^2 u\, du = \tan u + C$$

$$\int \csc^2 u \, du = -\cot u + C$$
$$\int \sec u \tan u \, du = \sec u + C$$
$$\int \csc u \cot u \, du = -\csc u + C$$

where u represents some differentiable function of x

EXAMPLE 1 Integrate the following:

$$\int \cos 5x \, dx$$

Solution From the given integral, let $u = 5x$; then $du = 5\,dx$. Therefore, we must multiply and divide the integral by 5 to obtain the required value of du.

$$\int \cos 5x \, dx = \frac{1}{5} \int \overbrace{\cos 5x}^{\cos u} \overbrace{(5\,dx)}^{du}$$
$$= \frac{1}{5} \sin 5x + C$$

■

EXAMPLE 2 Integrate the following:

$$\int x \csc^2(3x^2) \, dx$$

Solution If $u = 3x^2$, then $du = 6x\,dx$, and we must multiply and divide the integral by 6.

$$\int x \csc^2(3x^2) \, dx = \frac{1}{6} \int \overbrace{\csc^2(3x^2)}^{\csc^2 u} \overbrace{(6x\,dx)}^{du}$$
$$= -\frac{1}{6} \cot 3x^2 + C$$

■

EXAMPLE 3 Integrate the following:

$$\int \frac{\cot(x + 3)}{\sin(x + 3)} \, dx$$

Solution We use the trigonometric identity $\csc x = 1/\sin x$ to transform this integral.

$$\int \frac{\cot(x + 3)}{\sin(x + 3)}\,dx = \int \csc(x + 3)\cot(x + 3)\,dx$$

If $u = x + 3$, then $du = dx$.

$$\int \underbrace{\csc(x + 3)}_{\csc u}\,\underbrace{\cot(x + 3)}_{\cot u}\,\underbrace{dx}_{du} = -\csc(x + 3) + C$$

Note that you could also integrate this function by substituting $\cot u = \cos u/\sin u$ and simplifying. ■

Integrating tan u

We use the trigonometric identity $\tan x = \sin x/\cos x$ to develop a formula for the integral of $\tan u$.

$$\begin{aligned} \int \tan u &= -\int \frac{-\sin u}{\cos u}\,du \\ &= -\ln|\cos u| + C \end{aligned}$$

Integral of Tangent Function

$$\int \tan u\,du = -\ln|\cos u| + C$$

where $\cos u \neq 0$

EXAMPLE 4 Integrate the following:

$$\int_0^{\pi/4} x \tan 2x^2\,dx$$

Solution If $u = 2x^2$, then $du = 4x\,dx$, and we must multiply and divide the integral by 4. Integrating gives

$$\int_0^{\pi/4} \tan 2x^2(x\,dx) = \frac{1}{4}\int_0^{\pi/4} \underbrace{\tan 2x^2}_{\tan u}\,\underbrace{(4x\,dx)}_{du}$$

Evaluating the result at the limits of integration gives

$$= -\frac{1}{4}\left(\ln|\cos 2x^2| \Big|_0^{\pi/4}\right)$$

$$\approx -\frac{1}{4}(-1.106 - 0)$$

$$\approx 0.28$$

■

Integrating cot u

We develop a formula for $\int \cot u\, du$ using the same method used for $\int \tan u\, du$. We use the trigonometric identity $\cot x = \cos x/\sin x$.

$$\int \cot u\, du = \int \frac{\cos u}{\sin u}\, du$$

Once again, using the formula to integrate a reciprocal function, we have

$$\int \cot u\, du = \ln|\sin u| + C$$

Integral of Cotangent Function

$$\int \cot u\, du = \ln|\sin u| + C$$

where $\sin u \neq 0$

EXAMPLE 5 Integrate the following:

$$\int \cot \frac{3x}{4}\, dx$$

Solution If $u = \frac{3}{4}x$, then $du = \frac{3}{4}\, dx$, and we must multiply and divide the integral by $\frac{3}{4}$.

$$\int \cot \frac{3x}{4}\, dx = \frac{4}{3}\int \cot \frac{3x}{4}\left(\frac{3}{4}\, dx\right)$$

$$= \frac{4}{3} \ln\left|\sin \frac{3x}{4}\right| + C$$

■

Integrating sec u

We must perform algebraic manipulations to obtain an integration formula for $\int \sec u\, du$.

$$\int \sec u\, du = \int \sec u \cdot \overbrace{\frac{\sec u + \tan u}{\sec u + \tan u}}^{\text{multiply by 1}} \cdot du$$

$$= \int \frac{\sec u \tan u + \sec^2 u}{\sec u + \tan u}\, du$$

$$= \int \underbrace{\frac{1}{\sec u + \tan u}}_{\text{reciprocal function}} \cdot \underbrace{([\sec u \tan u + \sec^2 u]\, du)}_{\text{derivative of denominator}}$$

$$= \ln|\sec u + \tan u| + C$$

Integral of Secant Function

$$\int \sec u\, du = \ln|\sec u + \tan u| + C$$

where $\sec u + \tan u \neq 0$

EXAMPLE 6 Integrate the following:

$$\int e^{2x} \sec e^{2x}\, dx$$

Solution If $u = e^{2x}$, then $du = 2e^{2x}\, dx$, and we must multiply and divide the integral by 2.

$$\int e^{2x} \sec e^{2x}\, dx = \frac{1}{2} \int \overbrace{\sec e^{2x}}^{\sec u}\, \overbrace{(2e^{2x}\, dx)}^{du}$$

$$= \frac{1}{2} \ln|\sec e^{2x} + \tan e^{2x}| + C$$

■

Integrating csc u

We use the same method to derive a formula for the cosecant integral as we did for the secant.

$$\int \csc u\, du = \int \csc u \cdot \underbrace{\frac{\csc u - \cot u}{\csc u - \cot u}}_{\text{multiply by 1}} \cdot du$$

$$= \int \underbrace{\frac{1}{\csc u - \cot u}}_{\text{reciprocal function}} \cdot \underbrace{(-\csc u \cot u + \csc^2 u)\, du}_{\text{derivative of denominator}}$$

$$= \ln|\csc u - \cot u| + C$$

Integral of Cosecant Function

$$\int \csc u \, du = \ln|\csc u - \cot u| + C$$

where $\csc u - \cot u \neq 0$

EXAMPLE 7 Integrate the following:

$$\int (x + \csc 3x) \, dx$$

Solution First, we apply the sum rule for integrals. Then we integrate each of the resulting integrals.

$$\begin{aligned}\int (x + \csc 3x) \, dx &= \int x \, dx + \int \csc 3x \, dx \\ &= \int x \, dx + \frac{1}{3} \int \csc \underbrace{3x}_{u} \underbrace{(3 \, dx)}_{du} \\ &= \frac{x^2}{2} + \frac{1}{3} \ln|\csc 3x - \cot 3x| + C\end{aligned}$$

■

EXAMPLE 8 Integrate the following:

$$\int_{\pi/4}^{\pi/3} \frac{\sin t + 1}{\cos t} \, dt$$

Solution We use trigonometric identities to simplify this integral.

$$\begin{aligned}\int_{\pi/4}^{\pi/3} \frac{\sin t + 1}{\cos t} \, dt &= \int_{\pi/4}^{\pi/3} \frac{\sin t}{\cos t} + \frac{1}{\cos t} \, dt \\ &= \int_{\pi/4}^{\pi/3} (\tan t + \sec t) \, dt \\ &= \int_{\pi/4}^{\pi/3} \tan t \, dt + \int_{\pi/4}^{\pi/3} \sec t \, dt \\ &= (-\ln|\cos t| + \ln|\sec t + \tan t|) \Big|_{\pi/4}^{\pi/3} \\ &\approx (0.693 + 1.317) - (0.347 + 0.881) \\ &\approx 0.78\end{aligned}$$

■

EXAMPLE 9 Integrate the following:

$$\int_0^{\pi/4} \frac{\sin 2x}{\cos^2 x}\,dx$$

Solution To integrate this function, we must make trigonometric substitutions and simplify the result.

$$\begin{aligned}
\int_0^{\pi/4} \frac{\sin 2x}{\cos^2 x}\,dx &= \int_0^{\pi/4} \frac{2 \sin x \cos x}{\cos^2 x}\,dx \\
&= \int_0^{\pi/4} 2 \cdot \frac{\sin x}{\cos x} \cdot dx \\
&= \int_0^{\pi/4} 2 \tan x\,dx \\
&= -2 \ln|\cos x| \Big|_0^{\pi/4} \\
&\approx -2(-0.347 - 0) \\
&\approx 0.69
\end{aligned}$$

■

Applications

EXAMPLE 10 A sinusoidal current in a circuit is given by

$$i = 12 \sin(60\pi t + \pi)$$

Find the average current passing through the circuit in the first 0.4 s.

Solution The formula for the average value of a function is

$$M = \frac{1}{b - a} \int_a^b f(x)\,dx$$

The integral for the average current is as follows:

$$\frac{1}{0.4 - 0} \int_0^{0.4} 12 \sin(60\pi t + \pi)\,dt$$

Integrating gives

$$\frac{12}{0.4} \cdot \frac{1}{60\pi} \int_0^{0.4} \sin(60\pi t + \pi)(60\pi\,dt)$$

$$\frac{1}{2\pi}\left([-\cos(60\pi t + \pi)] \Big|_0^{0.4} \right)$$

Evaluating the limits of integration gives

$$\frac{1}{2\pi}[1 - 1]$$

The average current through the circuit is zero. ■

EXAMPLE 11 The current in an RC circuit is given by

$$i = 2.25 \sin 55\pi t$$

Find the voltage across a 750-μF capacitor after 0.25 s if $V = 0$ at $t = 0$.

Solution The voltage across the capacitor is given by

$$V = \frac{1}{C}\int_0^{0.25} i\, dt$$

Substituting the pertinent information gives the voltage as

$$\begin{aligned} V &= \frac{1}{7.50 \times 10^{-4}}\int_0^{0.25} 2.25 \sin 55\pi t\, dt \\ &= \frac{2.25}{7.50 \times 10^{-4}} \cdot \frac{1}{55\pi}\int_0^{0.25} \sin 55\pi t(55\pi\, dt) \\ &\approx 17.36\left[-\cos 55\pi t \Big|_0^{0.25}\right] \\ &\approx 17.36(-0.7071 + 1) \\ &\approx 5.085 \end{aligned}$$

The voltage across the capacitor is approximately 5 V. ■

21–3 EXERCISES

Integrate the given functions. Round answers to two decimal places on the definite integrals.

1. $\int \cos 4x\, dx$
2. $\int \sin(3x - 2)\, dx$
3. $\int 3 \csc^2 7t\, dt$
4. $\int x \csc x^2 \cot x^2\, dx$
5. $\int x \sin 4x^2\, dx$
6. $\int_0^{\pi/4} 2 \tan \frac{x}{3}\, dx$
7. $\int_0^1 \frac{\sin e^{-t}}{e^t}\, dt$
8. $\int 3x \sec x^2 \tan x^2\, dx$
9. $\int (3x + 2)\sin(3x^2 + 4x)\, dx$
10. $\int \csc 5t \cot 5t\, dt$
11. $\int x^2 \tan 3x^3\, dx$
12. $\int \frac{\sin y}{\cos^2 y}\, dy$
13. $\int_0^{\pi/4} \frac{dx}{\cos 5x}$
14. $\int \frac{dt}{\sin^2 3t}$
15. $\int (\cot 4x + \csc 4x)\, dx$
16. $\int \frac{\cot y}{\sin y}\, dy$
17. $\int_0^{\pi/6} \frac{\sin x}{\cos^2 x}\, dx$
18. $\int \frac{\csc 3x}{\tan 3x}\, dx$
19. $\int \frac{\cos(\ln x)}{x}\, dx$
20. $\int_0^2 \frac{\cos y + 1}{\cos^2 y}\, dy$
21. $\int_0^{\pi/4} (1 + \sec t)^2\, dt$
22. $\int \cot(3t + 1)\, dt$
23. $\int \frac{1 + \cot^2 x}{\cos^2 x}\, dx$
24. $\int_0^2 (\cos x)e^{\sin x}\, dx$
25. Find the average value of the function $y = 2 \tan \pi x$ on the interval $x = 0$ to $x = \pi/4$.
26. Find the average value of a sinusoidal current given by $i = 3.1 \sin 6\pi t$ during the first 1.8 s.
27. Find the volume generated by revolving the area bounded by $y = \sec x$ from $x = 0$ to $x = \pi/4$ about the x axis.

28. Find the volume generated by revolving the area bounded by $y = \cot x$ from $x = \pi/4$ to $x = \pi/2$ about the x axis.

29. Find the area bounded by the curves $y = \sec(2x - \pi)$, $y = 0$, $x = 0$, and $x = \pi/8$.

30. Find the area bounded by the curves $y = -3\cos\left(x + \frac{\pi}{2}\right)$, $y = 0$, and $x = \pi/4$.

31. Find an expression for the velocity of a particle starting from rest if its acceleration is $a = \sin(3t + 1)$.

32. Find the current through a 3.0-H inductor in an *RL* circuit if

$$V = 8\cos\left(120\,\pi t + \frac{\pi}{4}\right)$$

and $i = 0$ when $t = 0$. (*Note:* $i = 1/L \int V\,dt$.)

33. The velocity of an object is given by $V = 2\cos 4t$. Determine an expression for its displacement if $V = 0$ when $t = 0$.

21–4

INTEGRATING OTHER TRIGONOMETRIC FORMS

In this section we will discuss integrating the product and powers of the trigonometric functions. We will need to use several trigonometric identities discussed in Chapter 13. They are summarized in the following box.

Trigonometric Identities

$$\cos^2 x + \sin^2 x = 1$$

$$\tan^2 x + 1 = \sec^2 x$$

$$\cot^2 x + 1 = \csc^2 x$$

$$\cos^2 x = \frac{1}{2}(1 + \cos 2x)$$

$$\sin^2 x = \frac{1}{2}(1 - \cos 2x)$$

The integrals that we will discuss will be one of the following three general forms:

$$\int \sin^n u \cos^m u\,du$$

$$\int \tan^n u \cot^m u\,du$$

$$\int \sec^n u \csc^n u\,du$$

Integrating $\sin^n u \cos^m u\,du$

The identity we use to integrate is determined by the exponents of the trigonometric functions.

LEARNING HINT ✦ When integrating a function of the form $\sin^n u \cos^m u$ where either n or m is odd, use the trigonometric identity

$$\cos^2 x + \sin^2 x = 1$$

Example 1 illustrates this procedure.

EXAMPLE 1 Integrate the following function:

$$\int \sin^2 x \cos^3 x \, dx$$

Solution To integrate this function, choose the trigonometric function with an odd exponent, factor out a power of this function, and group it with the differential. Then make a trigonometric substitution and use the power rule to integrate the result.

$$\int \sin^2 x \cos^3 x \, dx = \int \cos^2 x \underbrace{(\cos x \sin^2 x \, dx)}$$

factor out cos x, and group with dx

$$= \int \underbrace{(1 - \sin^2 x)}(\cos x \sin^2 x \, dx)$$

trigonometric substitution

$$= \int \cos x \sin^2 x \, dx - \int \sin^4 x \cos x \, dx$$

Last, use the power rule to integrate.

$$= \int \underbrace{(\sin x)^2}_{u^2} \underbrace{(\cos x \, dx)}_{du} - \int \underbrace{(\sin x)^4}_{u^4} \underbrace{(\cos x \, dx)}_{du}$$

$$= \frac{u^3}{3} - \frac{u^5}{5} + C$$

$$= \frac{1}{3} \sin^3 x - \frac{1}{5} \sin^5 x + C$$

■

LEARNING HINT ✦ If the integral contains an odd exponent of either the sine or cosine function but not the other function at all, use the same technique to integrate as used to integrate

$$\int \sin^n u \cos^m u \, du \quad \text{where } n \text{ or } m \text{ is odd}$$

Example 2 shows this procedure.

EXAMPLE 2 Integrate the following function:

$$\int \sin^3 5x \, dx$$

Solution The form of this trigonometric function is different from that of the previous example because it contains only the sine function with an odd exponent. However, we use the same technique and trigonometric substitution as used in Example 1.

$$\begin{aligned}
\int \sin^3 5x\, dx &= \int \sin^2 5x \sin 5x\, dx \\
&= \int (1 - \cos^2 5x) \sin 5x\, dx \\
&= \int (\sin 5x\, dx - \cos^2 5x \sin 5x\, dx) \\
&= \int \sin 5x\, dx - \int \cos^2 5x \sin 5x\, dx \\
&= \frac{1}{5}\int \sin \underbrace{5x}_{u} \underbrace{(5\, dx)}_{du} + \frac{1}{5}\int \underbrace{(\cos 5x)^2}_{u^2}(-5 \sin \underbrace{5x\, dx}_{du}) \\
&= -\frac{1}{5}\cos 5x + \frac{1}{5} \cdot \frac{\cos^3 5x}{3} + C \\
&= -\frac{1}{5}\cos 5x + \frac{1}{15}\cos^3 5x + C
\end{aligned}$$

■

LEARNING HINT ✦ To integrate an even power of one or both the sine and cosine function, write as a power of $\sin^2 x$ or $\cos^2 x$ and use one of the following trigonometric identities:

$$\sin^2 x = \frac{1}{2}(1 - \cos 2x) \quad \text{or} \quad \cos^2 x = \frac{1}{2}(1 + \cos 2x)$$

This process is shown in Examples 3 and 4.

EXAMPLE 3 Integrate the following function:

$$\int \cos^4 x\, dx$$

Solution To integrate this function of the cosine with an even exponent, write the function as a power of $\cos^2 x$. Then substitute the trigonometric identity $\cos^2 x = (1/2)(1 + \cos 2x)$ and simplify the result to yield a function that we can integrate.

$$\int \cos^4 x\, dx = \int (\cos^2 x)^2\, dx$$

$$= \int \underbrace{\left[\frac{1}{2}(1 + \cos 2x)\right]}_{\text{trigonometric substitution}}^2 dx$$

$$= \frac{1}{4}\int (1 + 2\cos 2x + \cos^2 2x)\, dx$$

$$= \frac{1}{4}\int \left[1 + 2\cos 2x + \underbrace{\frac{1}{2}(1 + \cos 4x)}_{\text{trigonometric substitution}}\right] dx$$

$$= \frac{1}{4}\int \left(1 + 2\cos 2x + \frac{1}{2} + \frac{1}{2}\cos 4x\right) dx$$

$$= \frac{1}{4}\int \left(\frac{3}{2} + 2\cos 2x + \frac{1}{2}\cos 4x\right) dx$$

$$= \frac{1}{4}\int \frac{3}{2}\, dx + \frac{1}{4}\int 2\cos 2x\, dx + \frac{1}{4}\int \frac{1}{2}\cos 4x\, dx$$

$$= \frac{3}{8}\int dx + \frac{1}{4}\cdot\int \cos \underbrace{2x}_{u}\, \underbrace{(2\, dx)}_{du} + \frac{1}{4}\cdot\frac{1}{8}\int \cos \underbrace{4x}_{u}\, \underbrace{(4\, dx)}_{du}$$

$$= \frac{3}{8}x + \frac{1}{4}\sin 2x + \frac{1}{32}\sin 4x + C$$

■

EXAMPLE 4 Integrate the following function:

$$\int \sin^2 x \cos^2 x\, dx$$

Solution In this case, we will substitute for both $\sin^2 x$ and $\cos^2 x$ in terms of the $\cos 2x$. Then we will proceed as in the previous examples.

$$\int \sin^2 x \cos^2 x \, dx = \int \frac{1}{2}(1 - \cos 2x)\frac{1}{2}(1 + \cos 2x)\, dx$$

$$= \int \frac{1}{4}(1 - \cos^2 2x)\, dx$$

$$= \int \left(\frac{1}{4} - \frac{1}{4}\cos^2 2x\right) dx$$

$$\cos^2 2x = \frac{1}{2}(1 + \cos 4x)$$

$$= \int \left[\frac{1}{4} - \frac{1}{8}(1 + \cos 4x)\right] dx$$

$$= \int \left(\frac{1}{4} - \frac{1}{8} - \frac{1}{8}\cos 4x\right) dx$$

$$= \int \frac{1}{8}\, dx - \frac{1}{8}\int \cos 4x \, dx$$

$$= \frac{1}{8}\int dx - \frac{1}{8}\cdot\frac{1}{4}\int \cos \underbrace{4x}_{u}\ \underbrace{(4\, dx)}_{du}$$

$$= \frac{1}{8}x - \frac{1}{32}\sin 4x + C$$ ■

Integrating $\tan^m u \sec^n u \, du$ or $\cot^m u \csc^n u \, du$

LEARNING HINT ✦ In integrals of the form $\int \tan^m u \sec^n u \, du$ or $\int \cot^m u \csc^n u \, du$, if m is an odd number, then separate out a factor of ($\sec u \tan u \, du$) or ($\csc u \cot u \, du$) and substitute

$$\tan^2 u = \sec^2 u - 1 \quad \text{or} \quad \cot^2 u = \csc^2 u - 1$$

respectively. This technique is illustrated in Example 5.

EXAMPLE 5 Integrate the following function:

$$\int \tan^3 x \sec^2 x \, dx$$

Solution First, we will factor out ($\tan x \sec x \, dx$) as follows:

$$\int \tan^2 x \sec x(\tan x \sec x\, dx)$$

Next, we substitute $(\sec^2 x - 1)$ for $\tan^2 x$ and simplify the result.

$$\int (\sec^2 x - 1) \sec x(\tan x \sec x\, dx)$$

$$\int (\sec^3 x - \sec x)(\tan x \sec x\, dx)$$

$$\int \underbrace{\sec^3 x}_{u^3}\underbrace{(\tan x \sec x\, dx)}_{du} - \int \underbrace{\sec x}_{u}\underbrace{(\tan x \sec x\, dx)}_{du}$$

$$\frac{1}{4} \sec^4 x - \frac{1}{2} \sec^2 x + C$$

■

LEARNING HINT ✦ If n is a positive, even exponent, separate out a factor of $\sec^2 u\, du$ or $\csc^2 u\, du$. Then substitute

$$\tan^2 u + 1 = \sec^2 u \quad \text{or} \quad \cot^2 u + 1 = \csc^2 u$$

respectively. This process is shown in Example 6.

EXAMPLE 6 Integrate the following function:

$$\int \cot^3 x \csc^4 x\, dx$$

Solution First, we factor out $\csc^2 x\, dx$.

$$\int \cot^3 x \csc^2 x(\csc^2 x\, dx)$$

Next, we substitute $\cot^2 x + 1$ for the first $\csc^2 x$.

$$\int \cot^3 x(\cot^2 x + 1)(\csc^2 x\, dx)$$

$$\int (\cot^5 x + \cot^3 x)(\csc^2 x\, dx)$$

$$-\int \cot^5 x(-\csc^2 x\, dx) - \int \cot^3 x(-\csc^2 x\, dx)$$

Last, we apply the power rule and integrate as follows:

$$-\frac{1}{6} \cot^6 x - \frac{1}{4} \cot^4 x + C$$

■

LEARNING HINT ✦ If the integral contains only $\tan^m x$ or $\cot^m x$, factor out $\tan x$ or $\cot x$, respectively, for m an odd, positive integer. Factor out $\tan^2 x$ or $\cot^2 x$, respectively, for m an even, positive integer. Example 7 illustrates this procedure.

EXAMPLE 7 Integrate the following functions:
(a) $\int \tan^3 x\, dx$ (b) $\int \cot^4 2x\, dx$

Solution

(a) First, we factor out $\tan x$.

$$\int \tan^2 x\, (\tan x\, dx)$$

Next, substitute $\sec^2 x - 1$ for $\tan^2 x$, simplify, and integrate.

$$\int (\sec^2 x - 1) \tan x\, dx$$

$$\int \sec^2 x \tan x\, dx - \int \tan x\, dx$$

$$\int \underbrace{\tan x}_{u}\underbrace{(\sec^2 x\, dx)}_{du} - \int \tan \underset{u}{x}\, \underset{du}{dx}$$

$$\frac{1}{2}\tan^2 x + \ln|\cos x| + C$$

(b) First, factor out $\cot^2 2x$.

$$\int \cot^2 2x\, (\cot^2 2x\, dx)$$

Next, substitute $\csc^2 2x - 1$ for one of the factors $\cot^2 2x$, simplify and integrate.

$$\int (\csc^2 2x - 1)(\cot^2 2x\, dx)$$

$$\int (\csc^2 2x \cot^2 2x - \cot^2 2x)\, dx$$

$$\int [\csc^2 2x \cot^2 2x - \underbrace{(\csc^2 2x - 1)}_{\text{substitute for } \cot^2 2x}]\, dx$$

$$\int (\csc^2 2x \cot^2 2x - \csc^2 2x + 1)\, dx$$

$$\int \csc^2 2x \cot^2 2x\, dx - \int \csc^2 2x\, dx + \int dx$$

We use the technique from Example 6 to integrate the first integral.

$$\frac{1}{2}\int \underbrace{\cot^2 2x}_{u}\,\underbrace{(\csc^2 2x \cdot 2\,dx)}_{du} - \frac{1}{2}\int \csc^2 \underbrace{2x}_{u}\underbrace{(2\,dx)}_{du} + \int dx$$

$$-\frac{1}{2}\cdot\frac{1}{3}\cot^3 2x + \frac{1}{2}\cot 2x + x + C$$

$$-\frac{1}{6}\cot^3 2x + \frac{1}{2}\cot 2x + x + C$$ ■

LEARNING HINT ✦ If you integrate $\sec^n u\ du$ or $\csc^n u\ du$ where n is an even, positive integer, substitute in terms of the tangent or cotangent functions. (We will discuss integrating the secant and cosecant when n is an odd, positive integer in Section 21–6, Integration by Parts.) This procedure is shown in Example 8.

EXAMPLE 8 Integrate the following function:

$$\int \csc^4 x\ dx$$

Solution First, we substitute for $\csc^4 x$ in terms of the cotangent function.

$$\int (\csc^2 x)^2\ dx$$

$$\int (\cot^2 x + 1)^2\ dx$$

Next, we simplify by squaring the binomial.

$$\int (\cot^4 x + 2\cot^2 x + 1)\ dx$$

Last, we use the techniques in the previous example to integrate.

$$\int \cot^4 x\ dx + 2\int \cot^2 x\ dx + \int dx$$

$$\int \cot^2 x \cot^2 x\ dx + 2\int \cot^2 x\ dx + \int dx$$

$$\int \cot^2 x(\csc^2 x - 1)dx + 2\int (\csc^2 x - 1)\ dx + \int dx$$

$$\int \csc^2 x \cot^2 x\ dx - \int \cot^2 x\ dx + \int 2\csc^2 x\ dx - 2\int dx + \int dx$$

$$\int \csc^2 x \cot^2 x\ dx - \int (\csc^2 x - 1)\ dx + \int 2\csc^2 x\ dx - \int dx$$

$$-\int \cot^2 x(-\csc^2 x\ dx) + \int \csc^2 x\ dx$$

$$-\frac{1}{3}\cot^3 x - \cot x + C$$ ■

21–4 EXERCISES

Integrate the following functions. Round to two decimal places when necessary.

1. $\int \cos^2 x \sin^3 x\, dx$
2. $\int \sin^4 x \cos^3 x\, dx$
3. $\int \sin^2 2x \cos^5 2x\, dx$
4. $\int_{\pi/4}^{\pi/3} \cos^2 t \sin^4 t\, dt$
5. $\int \sin^3 x \cos^3 x\, dx$
6. $\int \cos t \sin^2 t\, dt$
7. $\int \sin^2 x\, dx$
8. $\int_0^{\pi/3} \cos^3 3t\, dt$
9. $\int_0^{\pi} \sin^5 x\, dx$
10. $\int \cos^4(x - 2)\, dx$
11. $\int \sin^3 5x\, dx$
12. $\int \cot^3 2x\, dx$
13. $\int_{\pi/6}^{\pi/3} \tan^2 t\, dt$
14. $\int \tan^5 3x\, dx$
15. $\int \cot^2 x\, dx$
16. $\int \sec^4 x\, dx$
17. $\int_{\pi/4}^{\pi/2} \csc^6 t\, dt$
18. $\int \tan^3 x \sec x\, dx$
19. $\int \cot^2 3x \csc^2 3x\, dx$
20. $\int \tan^2 x \sec^2 x\, dx$
21. $\int \cot^3 2x \csc^3 2x\, dx$
22. $\int \tan^3 2x \sec^3 2x\, dx$
23. $\int_0^{\pi} \sin^2 3x\, dx$
24. $\int \cot^3 \pi t\, dt$
25. $\int \tan^3 x\, dx$
26. $\int \cos^2 5t\, dt$
27. $\int \tan^7 x\, dx$
28. $\int \frac{dx}{\cot^3 x}$
29. $\int \frac{1 - \cot x}{\sin^4 x}\, dx$
30. Under certain conditions, the rate of radiation by an accelerated charge is given by

$$\int \cos^3\theta\, d\theta$$

Find an expression for the rate of radiation.

31. Calculate the volume generated by revolving the region bounded by $y = \cos^2 x$, $y = 0$, and $x = \pm\pi/2$ about the x axis.
32. Find the root mean square average of the power P dissipated in an ac circuit if $P = 30 \cos\left(50\pi t + \frac{\pi}{2}\right)$ in the first 1/60 s. $\left(\textit{Note:}\ \text{root mean square} = \sqrt{\frac{\int_a^b P(t)^2\, dt}{b - a}}.\right)$
33. Find the root mean square value of the current of $i = 5 \sin t$ for $t = 0$ to $t = 2\pi$.
34. Find the volume generated by revolving the region bounded by $y = \sin x$, $y = 0$, and $x = \pi$ about the x axis.

21–5

INVERSE TRIGONOMETRIC FORMS AND TRIGONOMETRIC SUBSTITUTION

Several integrals containing expressions such as $a^2 + u^2$ result directly from the derivative of the inverse trigonometric functions. In this section we will discuss integrating functions that lead to inverse trigonometric functions.

Integrals Leading to the Arcsine Function

Recall from our earlier discussion that

$$\frac{d}{dx}(\text{Arcsin } u) = \frac{1}{\sqrt{1 - u^2}}\frac{du}{dx} \quad \text{for } -1 < u < 1$$

We can also write this equation in the following general form:

$$\frac{d}{dx}\left(\text{Arcsin}\,\frac{u}{a}\right) = \frac{1}{\sqrt{1 - \left(\frac{u}{a}\right)^2}}\,\frac{d}{dx}\left(\frac{u}{a}\right)$$

$$= \left(\frac{1}{a}\right)\frac{1}{\sqrt{\frac{a^2 - u^2}{a^2}}}\,\frac{du}{dx}$$

$$= \frac{1}{\sqrt{a^2 - u^2}}\,\frac{du}{dx}$$

Integrals Leading to the Arcsine Function

$$\int \frac{1}{\sqrt{a^2 - u^2}}\,du = \text{Arcsin}\,\frac{u}{a} + C$$

where $a > 0$ and $-1 < \frac{u}{a} < 1$

EXAMPLE 1 Integrate the following function:

$$\int \frac{1}{16 - x^2}\,dx$$

Solution This integral fits the formula that results in the Arcsine function.

$$\int \frac{1}{\sqrt{16 - x^2}}\,dx = \text{Arcsin}\,\frac{x}{4} + C$$

a^2
$a = 4$

u^2
$u = x$
$du = dx$

■

Integrals Leading to the Arctangent Function

From Chapter 20, we know that

$$\frac{d}{dx}(\text{Arctan}\,u) = \frac{1}{1 + u^2}\,\frac{du}{dx}$$

which leads to the following integral formula.

Integrals Leading to the Arctangent Function

$$\int \frac{du}{a^2 + u^2} = \frac{1}{a} \text{Arctan} \frac{u}{a} + C$$

where $a \neq 0$

EXAMPLE 2 Integrate the following:

$$\int \frac{dx}{x^2 + 9}$$

Solution This integral is of the form that leads to the Arctangent function where $u = x$ and $a = 3$. Remember that $x^2 + 9 = 9 + x^2$.

$$\int \frac{dx}{x^2 + 9} = \int \frac{dx}{9 + x^2} = \frac{1}{3} \text{Arctan} \frac{x}{3} + C$$

■

EXAMPLE 3 Integrate the following:

$$\int_0^1 \frac{dx}{25 + 9x^2}$$

Solution This integral is of the form that leads to the Arctangent function. In this integral, $a^2 = 25$, $a = \pm 5$, $u^2 = 9x^2$, $u = \pm 3x$, and $du = 3\ dx$. Therefore, we must multiply and divide the integral by 3 to obtain the required du.

$$\int_0^1 \frac{dx}{25 + 9x^2} = \frac{1}{3} \int_0^1 \frac{3\ dx}{25 + 9x^2}$$

$$= \frac{1}{3}\left(\underset{\uparrow a}{\frac{1}{5}} \text{Arctan} \frac{\overset{u}{\overbrace{3x}}}{\underset{\uparrow a}{5}} \Bigg|_0^1\right)$$

Evaluating the limits of integration gives the following result:

$$= \frac{1}{15}\left(\text{Arctan} \frac{3x}{5} \Bigg|_0^1\right)$$

$$\approx 0.04$$

■

EXAMPLE 4 Integrate the following:

$$\int \frac{dx}{x^2 + 8x + 25}$$

Solution This integral does not appear to be of the form to give the Arcsine or Arctangent functions. However, by completing the square, we can transform the denominator into a required form. We write the denominator as $x^2 + 8x + 25 = (x^2 + 8x + 16) + 9 = (x + 4)^2 + 3^2$. Thus,

$$\int \frac{dx}{x^2 + 8x + 25} = \int \frac{dx}{(x + 4)^2 + 3^2}$$

u^2, $u = x + 4$ a^2, $a = 3$

$$= \frac{1}{3} \text{Arctan}\left(\frac{x + 4}{3}\right) + C$$

■

CAUTION ✦ Be very careful in integrating functions that lead to the inverse trigonometric functions. You must be careful to distinguish them from the integral of the general power rule or logarithmic form. This distinction is illustrated in the following example.

EXAMPLE 5 Integrate the following:

$$\int \frac{2x + 3}{x^2 + 4}\, dx$$

Solution This integral is not of the form to give the Arctangent function because of the factor $2x + 3$ in the numerator. It also does not fit logarithmic form or the general power rule because $2x + 3$ is not the differential of $x^2 + 4$. We can write the fraction as

$$\frac{2x}{x^2 + 4} + \frac{3}{x^2 + 4}$$

and integrate the result as the sum of two integrals. Thus,

$$\int \frac{2x + 3}{x^2 + 4}\, dx = \int \frac{2x}{x^2 + 4}\, dx + \int \frac{3}{x^2 + 4}\, dx$$

The first integral is of logarithmic form with $u = x^2 + 4$ and $du = 2x\, dx$. The second integral is of the form leading to the Arctangent function with $u^2 = x^2$ and $a^2 = 4$. Integrating gives

$$\int \frac{2x + 3}{x^2 + 4}\, dx = \int \frac{(2x\, dx)}{x^2 + 4} + 3 \int \frac{dx}{x^2 + 4}$$

$$= \ln|x^2 + 4| + \frac{3}{2} \text{Arctan}\, \frac{x}{2} + C$$

■

EXAMPLE 6 Integrate the following:

(a) $\displaystyle\int \frac{dx}{\sqrt{9 - x^2}}$ (b) $\displaystyle\int \frac{2x}{\sqrt{9 - x^2}}\, dx$

Solution

(a) This integral is of the form that leads to the Arcsine function with $a = 3$ and $u = x$. Integration gives the following result:

$$\int \frac{dx}{\sqrt{9 - x^2}} = \text{Arcsin}\,\frac{x}{3} + C$$

(b) This integral is not of the form for the Arcsine function because of the $2x$ factor in the numerator. We can integrate this function using the general power rule with $u = 9 - x^2$, $n = -\frac{1}{2}$, and $du = -2x\,dx$. We must multiply and divide the integral by -1 to obtain the required value of du. Integrating gives

$$\int \frac{2x}{\sqrt{9 - x^2}}\, dx = -\int \underbrace{(9 - x^2)^{-1/2}}_{u^n} \underbrace{(-2x\, dx)}_{du}$$

$$= -2(9 - x^2)^{1/2} + C$$

$$= -2\sqrt{9 - x^2} + C$$

■

Summary

The following integrals are of the form indicated:

$\displaystyle\int \frac{dx}{\sqrt{1 - x^2}}$ Arcsine

$\displaystyle\int \frac{dx}{1 + x^2}$ Arctangent

$\displaystyle\int \frac{x^{n-1}\, dx}{1 + x^n}$ Logarithm

$\displaystyle\int \frac{x\, dx}{\sqrt{1 - x^2}}$ General power

Trigonometric Substitution

We can use trigonometric identities to transform algebraic integrals into a form that can be integrated. This process is illustrated in the following examples. The trigonometric substitutions are summarized in the following box.

Trigonometric Substitution

- ☐ If the integrand is of the form $\sqrt{a^2 - u^2}$, where $a > 0$, then substitute $u = a \sin \theta$ with $du = a \cos \theta \, d\theta$, and the integrand becomes $a \cos \theta$.
- ☐ If the integrand contains $\sqrt{a^2 + u^2}$, where $a > 0$, then substitute $u = a \tan \theta$ with $du = a \sec^2\theta \, d\theta$, and the integrand becomes $a \sec \theta$.
- ☐ If the integrand contains $\sqrt{u^2 - a^2}$, where $a > 0$, then substitute $u = a \sec \theta$ with $du = a \sec \theta \tan \theta \, d\theta$, and the integrand becomes $a|\tan \theta|$.

EXAMPLE 7 Integrate the following:

$$\int \frac{dx}{x^2\sqrt{4 - x^2}}$$

Solution This integrand contains an expression of the form $\sqrt{a^2 - u^2}$ where $a^2 = 4$ or $a = 2$ and $u^2 = x^2$ or $u = x$. Therefore, we substitute $u = 2 \sin \theta$ and $du = dx = 2 \cos \theta \, d\theta$. Making these substitutions into the integral gives

$$\int \frac{dx}{x^2\sqrt{4 - x^2}} = \int \frac{\overbrace{2 \cos \theta \, d\theta}^{du}}{\underbrace{(2 \sin \theta)^2}_{x^2}\sqrt{4 - \underbrace{(2 \sin \theta)^2}_{x^2}}}$$

$$= \int \frac{2 \cos \theta \, d\theta}{4 \sin^2\theta\sqrt{4 - 4 \sin^2\theta}}$$

$$= \int \frac{2 \cos \theta \, d\theta}{4 \sin^2\theta\sqrt{4(1 - \sin^2\theta)}}$$

$$= \int \frac{2 \cos \theta \, d\theta}{4 \sin^2\theta \; 2 \cos \theta}$$

$$= \int \frac{d\theta}{4 \sin^2\theta}$$

$$= \int \frac{1}{4} \underbrace{\csc^2\theta}_{\csc^2\theta \,=\, 1/\sin^2\theta} \, d\theta$$

$$= \int \frac{1}{4} \csc^2\theta \, d\theta$$

$$= -\frac{1}{4} \cot \theta + C$$

FIGURE 21–2

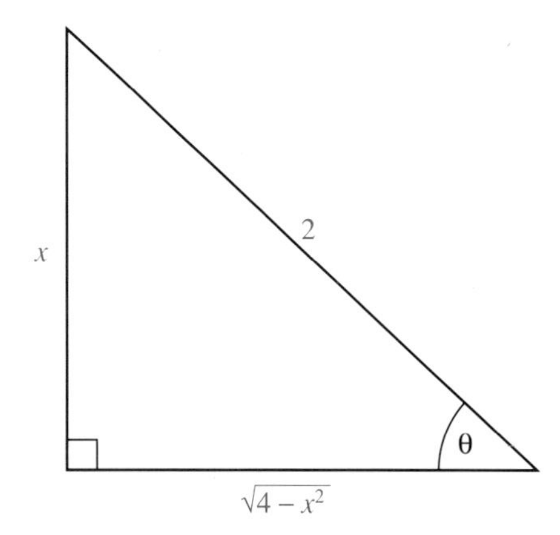

Last, we must make the substitution so that the result is expressed in terms of the original variable x. Figure 21–2 shows the right triangle formed by an angle θ with sides 2 and x such that $x = a \sin\theta$ or $x = 2\sin\theta$. The remaining side of the triangle is $\sqrt{4 - x^2}$. Then we can find an expression in terms of x for $\tan\theta$ and substitute it into our answer.

$$\int \frac{dx}{x^2\sqrt{4 - x^2}} = -\frac{1}{4} \cdot \underbrace{\frac{\sqrt{4 - x^2}}{x}}_{\cot\theta} + C$$

■

EXAMPLE 8 The current in a circuit (in amperes) is given by

$$I = \frac{1}{\left(\sqrt{1 + t^2}\right)^3}$$

Find the charge q (in coulombs) transmitted from $t = 0$ to $t = 0.4$ s.

Solution Remember from previous discussions that the charge q is the integral of the current. Therefore,

$$q = \int I\,dt = \int_0^{0.4} \frac{1}{\left(\sqrt{1 + t^2}\right)^3}\,dt$$

This integrand contains an expression of the form $\sqrt{a^2 + u^2}$ where $a^2 = 1$ or $a = 1$ and $u^2 = t^2$ or $u = t$. Therefore, we use the trigonometric substitutions $u = a\tan\theta = \tan\theta$ and $du = \sec^2\theta\,d\theta$. Substituting into the integral gives

$$\int_0^{0.4} \frac{1}{(\sqrt{1 + t^2})^3}\, dt = \int \frac{\sec^2\theta\, d\theta}{(\sqrt{1 + \tan^2\theta})^3}$$

$$= \int \frac{\sec^2\theta\, d\theta}{\left(\sqrt{\sec^2\theta}\right)^3}$$

$$= \int \frac{\sec^2 d\theta}{\sec^3\theta}$$

$$= \int \frac{d\theta}{\sec\theta}$$

$$= \int \cos\theta\, d\theta$$

$$= \sin\theta$$

Notice that the limits of integration have been omitted to this point due to the change in variable. Figure 21–3 shows a reference triangle for angle θ whose $\tan\theta = t$. Using the Pythagorean theorem, we find the hypotenuse and substitute for $\sin\theta$ in the integral above. Last, we evaluate the limits of integration.

$$q = \int_0^{0.4} I\, dt = \overbrace{\frac{t}{\sqrt{1 + t^2}}}^{\sin\theta}\Bigg|_0^{0.4}$$

$$= 0.37 \text{ C}$$

The charge in the first 0.4 s is 0.37 C.

FIGURE 21–3

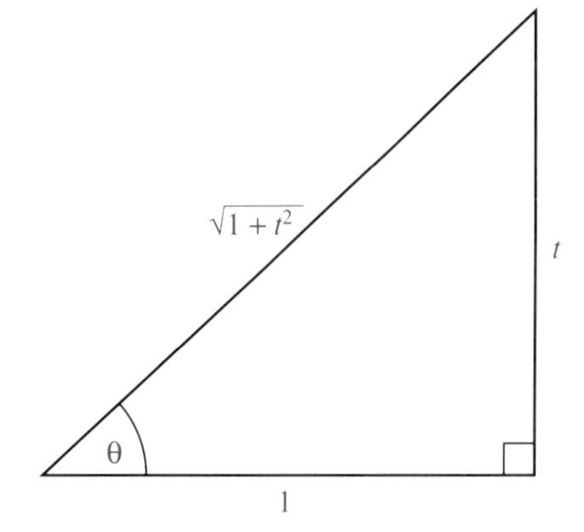

■

EXAMPLE 9 Integrate the following:

$$\int \frac{\sqrt{16x^2 - 1}}{x}\, dx$$

Solution The integrand contains an expression of the form $\sqrt{u^2 - a^2}$ where $u^2 = 16x^2$ or $u = 4x$ and $a^2 = 1$ or $a = 1$. We simplify this integral by substituting $x = \dfrac{1}{4}\sec\theta$ and $dx = \dfrac{1}{4}\sec\theta\tan\theta\, d\theta$. Substituting into the integral and integrating the simplified result gives

$$\begin{aligned}\int \frac{\sqrt{16x^2 - 1}}{x}\,dx &= \int \frac{\sqrt{\sec^2\theta - 1}}{\frac{1}{4}\sec\theta} \cdot \frac{1}{4}\sec\theta\tan\theta\,d\theta \\ &= \int \frac{4|\tan\theta|}{\sec\theta} \cdot \frac{1}{4}\sec\theta\tan\theta\,d\theta \\ &= \int \tan^2\theta\,d\theta \\ &= \int (\sec^2\theta - 1)\,d\theta \\ &= \tan\theta - \theta + C\end{aligned}$$

Figure 21–4 gives a reference triangle for angle θ. Using this figure, we find that $\tan\theta = \sqrt{16x^2 - 1}$ and $\sec\theta = 4x$ or $\theta = \text{Arcsec } 4x$. Therefore,

$$\int \frac{\sqrt{16x^2 - 1}}{x}\,dx = \sqrt{16x^2 - 1} - \text{Arcsec } 4x + C$$

FIGURE 21–4

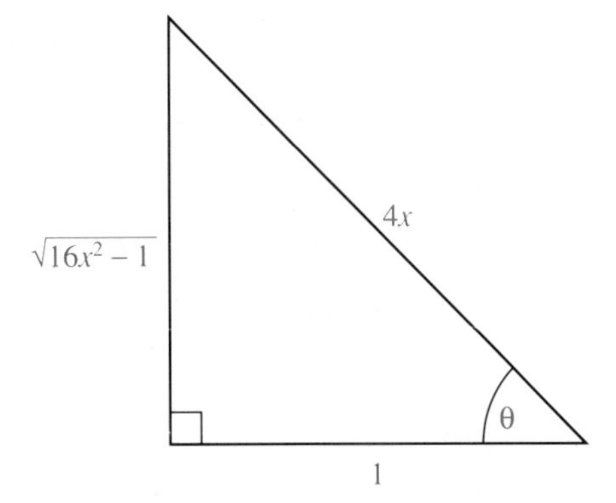

■

21–5 EXERCISES

Integrate the following functions. Round to hundredths where necessary.

1. $\displaystyle\int \frac{dx}{\sqrt{4 - x^2}}$

2. $\displaystyle\int \frac{dy}{\sqrt{36 - y^2}}$

3. $\displaystyle\int \frac{dt}{\sqrt{49 - t^2}}$

4. $\displaystyle\int \frac{dy}{\sqrt{81 - y^2}}$

5. $\displaystyle\int_0^1 \frac{dx}{9 + x^2}$

6. $\displaystyle\int_{-1}^1 \frac{dx}{36 + x^2}$

7. $\int \frac{dt}{4t^2 + 49}$

8. $\int \frac{dy}{9y^2 + 16}$

9. $\int \frac{dx}{\sqrt{1 - 8x^2}}$

10. $\int t^3\sqrt{16 + t^2}\, dt$

11. $\int \frac{2 - x}{\sqrt{9 - 4x^2}}\, dx$

12. $\int \frac{e^{2x}}{\sqrt{1 - e^{2x}}}\, dx$

13. $\int \frac{e^{2t}}{\sqrt{1 - e^{4t}}}\, dt$

14. $\int \frac{dx}{\sqrt{2 - x^2}}$

15. $\int \frac{dx}{25 + x^2}$

16. $\int_0^{\pi/4} \frac{\sin 2x}{1 + \cos^2 2x}\, dx$

17. $\int_0^{\pi} \frac{\cos 2x}{1 + \sin^2 2x}\, dx$

18. $\int \frac{\csc^2 x}{\sqrt{1 - \cot^2 x}}\, dx$

19. $\int \frac{\sec^2 x}{\sqrt{1 - \tan^2 x}}\, dx$

20. $\int \frac{3x}{\sqrt{16 - 9x^2}}\, dx$

21. $\int \frac{2x\, dx}{\sqrt{4 - x^4}}$

22. $\int y^3\sqrt{y^4 - 1}\, dy$

23. $\int \frac{3}{12 + 16t^2}\, dt$

24. $\int \frac{dx}{16 + x^2}$

25. The current in a transformer is given by

$$I = \int \frac{\sqrt{t^2 + 1}}{4t^2}\, dt$$

Find an expression for the current.

26. The radial force (in newtons) exerted on an electrically charged particle moving in an electrical field is given by

$$F(r) = \frac{\frac{1}{4\pi e_0} \cdot qq'}{r^2}$$

where $4\pi e_0$, q, and q' are constants. Find an expression in terms of r for the work done by the force.

21–6

INTEGRATION BY PARTS

We have discussed several methods of transforming an integrand into a form that can be integrated. In this section we will discuss integration by parts. Recall that in Chapter 18 we discussed differentiation of the product of two functions. If u and v represent differentiable functions of x, then

$$\frac{d(uv)}{dx} = u\frac{dv}{dx} + v\frac{du}{dx}$$

which, in differential form, can be written as

$$d(uv) = u\, dv + v\, du$$

or

$$u\, dv = d(uv) - v\, du$$

Taking the integral of each term gives the formula for integration by parts.

Integration by Parts

$$\int u\, dv = uv - \int v\, du$$

The constant of integration is added after the final integral is found.

In the formula for integration by parts, $\int u\,dv$ indicates that the integrand should be separated into two parts: a function of u and a differential of the function dv. The key to integration by parts is choosing u and dv correctly. Some trial and error is involved in this process, but general guidelines are as follows:

- ☐ You must be able to integrate dv.
- ☐ Choose u so that du/dx is a simpler form than u.
- ☐ Let $u = x$ if there is an extra factor of x.
- ☐ Let $u = \ln x$ if $\ln x$ is one of the factors.

EXAMPLE 1 Integrate the following:

$$\int xe^{-3x}\,dx$$

Solution This integral does not fit any of the methods of integration discussed previously because neither factor is the differential of the other. Therefore, we use integration by parts and separate the integrand into factors u and dv. Let us choose $u = x$ and $dv = e^{-3x}\,dx$. Then $du = dx$, and

$$\begin{aligned} v &= \int dv = \int e^{-3x}\,dx \\ &= -\frac{1}{3}\int e^{-3x}(-3\,dx) = -\frac{1}{3}e^{-3x} \end{aligned}$$

Integrating by parts gives

$$\int \overset{\overset{u}{\downarrow}}{x}\underbrace{(e^{-3x}\,dx)}_{\overset{\uparrow}{dv}} = uv - \int v\,du$$

$$= \underset{\underset{u}{\uparrow}}{x}\underbrace{\left(-\frac{1}{3}e^{-3x}\right)}_{\overset{\uparrow}{v}} - \int \underbrace{-\frac{1}{3}e^{-3x}}_{\overset{\uparrow}{v}}\,\underset{\underset{du}{\uparrow}}{dx}$$

$$= -\frac{1}{3}xe^{-3x} + \frac{1}{3}\int e^{-3x}\,dx$$

Integrating gives

$$\begin{aligned} &= -\frac{1}{3}xe^{-3x} + \frac{1}{3}\left[-\frac{1}{3}\int e^{-3x}(-3\,dx)\right] \\ &= -\frac{1}{3}xe^{-3x} - \frac{1}{9}e^{-3x} + C \end{aligned}$$

■

EXAMPLE 2 Integrate the following:

$$\int \text{Arctan } x \, dx$$

Solution Let $u = \text{Arctan } x$ and $dv = dx$. Then

$$du = \frac{dx}{1 + x^2} \quad \text{and} \quad v = \int dx = x$$

Integrating by parts gives

$$\int \text{Arctan } x \, dx = \underset{v}{x} \; \underset{u}{\text{Arctan } x} - \int \underset{v}{x} \, \underset{du}{\frac{dx}{1 + x^2}}$$

$$= x \text{ Arctan } x - \frac{1}{2} \int \frac{2x \, dx}{1 + x^2}$$

$$= x \text{ Arctan } x - \frac{1}{2} \ln|1 + x^2| + C$$

■

EXAMPLE 3 Integrate the following:

$$\int x^2 \ln x \, dx$$

Solution Let $u = \ln x$ and $dv = x^2 \, dx$. Then

$$u = \ln x \quad \text{and} \quad du = \frac{1}{x} \, dx$$

$$dv = x^2 \, dx \quad \text{and} \quad v = \int x^2 \, dx = \frac{x^3}{3}$$

Integrating by parts gives

$$\int x^2 \ln x \, dx = \underset{v}{\frac{x^3}{3}} \; \underset{u}{\ln x} - \int \underset{v}{\frac{x^3}{3}} \; \underset{du}{\frac{1}{x} \, dx}$$

$$= \frac{x^3}{3} \ln x - \frac{1}{3} \int x^2 \, dx$$

$$= \frac{x^3}{3} \ln x - \frac{1}{3} \cdot \frac{x^3}{3} + C$$

$$= \frac{x^3}{3} \ln x - \frac{x^3}{9} + C$$

■

Sometimes it is necessary to integrate by parts more than once, as shown in the next example.

EXAMPLE 4 Integrate the following:

$$\int x^2 \sin x \, dx$$

Solution Let $u = x^2$ and $dv = \sin x \, dx$. Then

$$du = 2x \, dx \quad \text{and} \quad v = \int \sin x \, dx = -\cos x$$

Integrating by parts gives

$$\begin{aligned} \int x^2 \sin x \, dx &= -x^2 \cos x - \int (-\cos x)(2x \, dx) \\ &= -x^2 \cos x + 2 \int x \cos x \, dx \end{aligned}$$

To evaluate the remaining integral, we must use integration by parts a second time. Let $u = x$ and $dv = \cos x \, dx$. Then

$$du = dx \quad \text{and} \quad v = \int \cos x \, dx = \sin x$$

$$\int x^2 \sin x \, dx = -x^2 \cos x + 2\left(x \sin x - \int \sin x \, dx\right)$$

Integrating the final time gives

$$\int x^2 \sin x \, dx = -x^2 \cos x + 2x \sin x + 2 \cos x + C$$ ■

The next example shows an integral where two applications of integration by parts lead back to the original integral. We must use algebraic techniques to integrate.

EXAMPLE 5 Integrate the following:

$$\int e^x \cos x \, dx$$

Solution Let $u = e^x$ and $dv = \cos x \, dx$. Then

$$u = e^x \quad \text{and} \quad du = e^x \, dx$$

$$dv = \cos x \, dx \quad \text{and} \quad v = \int \cos x \, dx = \sin x$$

Integrating by parts gives

$$\int e^x \cos x \, dx = e^x \sin x - \int e^x \sin x \, dx$$

To simplify this result, we must integrate $e^x \sin x\, dx$ using integration by parts. Let $u = e^x$ and $dv = \sin x\, dx$. Then

$$u = e^x \quad \text{and} \quad du = e^x\, dx$$

$$dv = \sin x\, dx \quad \text{and} \quad v = \int \sin x\, dx = -\cos x$$

Integrating gives

$$\int e^x \cos x\, dx = e^x \sin x - \Big[\underbrace{e^x}_{u}\underbrace{(-\cos x)}_{v} - \int \underbrace{-\cos x}_{v}\underbrace{(e^x\, dx)}_{du}\Big]$$

$$= e^x \sin x + e^x \cos x - \int e^x \cos x\, dx$$

Notice that we have returned to the original integral on the right side of the equation. We add this integral to both sides of the equation and simplify the result.

$$2\int e^x \cos x\, dx = e^x \sin x + e^x \cos x$$

$$\int e^x \cos x\, dx = \frac{e^x}{2}(\sin x + \cos x) + C \qquad \blacksquare$$

EXAMPLE 6 Integrate the following:

$$\int \csc^3 x\, dx$$

Solution Let $u = \csc x$ and $dv = \csc^2 x\, dx$. Then

$$u = \csc x \quad \text{and} \quad du = -\csc x \cot x\, dx$$

$$dv = \csc^2 x\, dx \quad \text{and} \quad v = \int \csc^2 x\, dx = -\cot x$$

$$\int \csc^3 dx = \csc x \cdot (-\cot x) - \int -\cot x \cdot (-\csc x \cot x)\, dx$$

$$= -\cot x \csc x - \int (\csc^2 x - 1)\csc x\, dx$$

$$= -\cot x \csc x - \int \csc^3 x\, dx + \int \csc x\, dx$$

$$= -\cot x \csc x - \int \csc^3 x\, dx + \ln|\csc x - \cot x|$$

$$2\int \csc^3 x\, dx = -\cot x \csc x + \ln|\csc x - \cot x|$$

$$\int \csc^3 x\, dx = \frac{1}{2}[-\cot x \csc x + \ln|\csc x - \cot x|] + C \qquad \blacksquare$$

21–6 EXERCISES

Integrate the following functions.

1. $\int x^2 \cos x\, dx$
2. $\int 3x \sin x\, dx$
3. $\int x\, e^{3x} dx$
4. $\int t^2 \ln t\, dt$
5. $\int_0^{\pi/4} \cos^2 y \sin y\, dy$
6. $\int \text{Arcsin}\, x\, dx$
7. $\int (\ln x)^3\, dx$
8. $\int x^3 \ln x\, dx$
9. $\int x^2 \cos x\, dx$
10. $\int e^x \sin x\, dx$
11. $\int e^t \cos 2t\, dt$
12. $\int x^3 e^{-x}\, dx$
13. $\int \text{Arccos}\, 3x\, dx$
14. $\int \frac{x}{\sqrt{1 + x}}\, dx$
15. $\int y \cot^2 y\, dy$
16. $\int_0^2 t^3 \sqrt{1 + t^2}\, dt$
17. $\int y(3 - y)^3\, dy$
18. $\int \sec^3 x\, dx$
19. $\int x(3 + x^3)^2\, dx$
20. $\int x^2 \sqrt{1 - x}\, dx$
21. $\int y(y + 8)^2\, dy$
22. $\int \csc^5 t\, dt$
23. $\int x^3 e^x\, dx$
24. Find the x coordinate of the centroid of the area bounded by $y = \cos x$, $y = 0$, $x = 0$, and $x = \pi/2$.
25. Find the x coordinate of the centroid of the area bounded by $y = \sin x$, $y = 0$, $x = 0$, and $x = \pi$.
26. Find the volume generated by rotating the area bounded by $y = \ln x$, $y = 0$, and $x = 2$ about the x axis.
27. Find the displacement s of a particle whose velocity is given by $V = t\sqrt{1 + t}$ if $s = 0$ when $t = 0$.

21–7

INTEGRATION USING TABLES

In this chapter we have expanded the number of integrals that we can integrate. We have used algebraic manipulation, trigonometric identities, and integration by parts to rewrite integrals to an easier form. However, we can use a computer or table of integrals to integrate integrals whose transformations are difficult or tricky.

In this section we will discuss integration using the brief table of integrals given in Appendix C. Being able to correctly use a table of integrals depends on being able to recognize the applicable form. Recognition of the correct integral from the table is easier if you rewrite the given integral to a form matching the one in the table. The following examples illustrate the use of a table to integrate.

EXAMPLE 1 Integrate the following:

$$\int \frac{x}{\sqrt{4 + 3x}}\, dx$$

Solution This integral fits the form of Formula 36 in Appendix C. This formula is

$$\int \frac{u\, du}{\sqrt{a + bu}} = \frac{2(bu - 2a)}{3b^2} \cdot \sqrt{a + bu} + C$$

From the given integral, we see that $u = x$, $a = 4$, and $b = 3$. Substituting these values into this formula gives the following result for the integral:

$$\int \frac{x}{\sqrt{4 + 3x}}\, dx = \frac{2(3x - 8)}{27} \cdot \sqrt{4 + 3x} + C$$ ■

EXAMPLE 2 Integrate the following:

$$\int \frac{x^2\, dx}{\sqrt{9 - x^2}}$$

Solution This integral fits the form of Formula 49 in Appendix C.

$$\int \frac{u^2\, du}{\sqrt{a^2 - u^2}} = -\frac{u}{2}\sqrt{a^2 - u^2} + \frac{a^2}{2}\,\text{Arcsin}\left(\frac{u}{a}\right) + C$$

where $u^2 = x^2$ or $u = x$ and $a^2 = 9$ or $a = 3$. Substituting these values into this formula gives the following result:

$$\int \frac{x^2\, dx}{\sqrt{9 - x^2}} = -\frac{x}{2}\sqrt{9 - x^2} + \frac{9}{2}\,\text{Arcsin}\left(\frac{x}{3}\right) + C$$ ■

EXAMPLE 3 Integrate the following:

$$\int \csc^3 x\, dx$$

Solution We integrated this function in Example 6 in Section 21–6 by using integration by parts. In this section we integrate using a table of integrals, which is easier in this case.This integral fits the form of Formula 78 in Appendix C.

$$\int \csc^n u\, du = -\frac{\csc^{n-2} u \cot u}{n - 1} + \frac{n - 2}{n - 1}\int \csc^{n-2} u\, du$$

where $n = 3$ and $u = x$. Substituting into the formula gives

$$\int \csc^3 x\, dx = \frac{-\csc x \cot x}{2} + \frac{1}{2}\int \csc x\, dx$$

Integrating once again gives the final result.

$$= \frac{-\csc x \cot x}{2} + \frac{1}{2}\ln|\csc x - \cot x| + C$$ ■

EXAMPLE 4 Integrate the following:

$$\int_1^2 x^3 \ln x\, dx$$

Solution This integral fits the form of Formula 100 in Appendix C.

$$\int u^n \ln u \, du = \frac{u^{n+1} \ln u}{n+1} - \frac{u^{n+1}}{(n+1)^2}$$

where $u = x$ and $n = 3$. Substituting these values into the formula gives

$$\int_1^2 x^3 \ln x \, dx = \left[\frac{x^4 \ln x}{4} - \frac{x^4}{16}\right]\Bigg|_1^2$$

Evaluating the limits of integration gives

$$\int_1^2 x^3 \ln x \, dx \approx (2.77 - 1) - (0 - 0.0625)$$
$$\approx 1.84$$

■

21–7 EXERCISES

Use the table of integrals in Appendix C to integrate the following functions.

1. $\int_0^{\pi} x \cos x \, dx$
2. $\int_0^{\pi/2} \sin^2\theta \, d\theta$
3. $\int x \, e^{4x} \, dx$
4. $\int \frac{dx}{x^2\sqrt{16 - x^2}}$
5. $\int \frac{dt}{t(5 + 3t)}$
6. $\int_0^{\pi} x^2 \sin 3x \, dx$
7. $\int x^2\sqrt{8 + 3x} \, dx$
8. $\int \frac{dx}{\sqrt{7 + 5x}}$
9. $\int_1^3 \frac{dx}{4 + 3x}$
10. $\int \tan^3 x \, dx$
11. $\int \sin 3x \sin x \, dx$
12. $\int_0^1 \frac{\sqrt{x^2 - 9}}{x} \, dx$
13. $\int \frac{x^2}{1 + 5x} \, dx$
14. $\int \sin 5x \cos 2x \, dx$
15. $\int_2^4 \frac{dt}{t^2(1 + 4t)}$
16. $\int_{-1}^1 xe^{4x} \, dx$
17. $\int \frac{d\theta}{\theta \ln \theta}$
18. $\int x \operatorname{Arcsin} x \, dx$
19. $\int_{-1}^1 x^2 e^{2x} \, dx$
20. $\int \frac{dx}{x(1 + 2x)^2}$
21. $\int x^3\sqrt{1 - x^2} \, dx$
22. $\int \frac{x^2 dx}{\sqrt{4 - x^2}}$

CHAPTER SUMMARY

Summary of Terms

reciprocal function (p. 796)

Summary of Formulas

$$\int \frac{du}{u} = \ln|u| + C \quad \text{where } u \neq 0 \quad \text{integral of reciprocal function}$$

$$\int e^u \, du = e^u + C \quad \text{integral of exponential function}$$

$$\int a^u\,du = \frac{a^u}{\ln a} + C \quad \text{where } a > 0 \text{ and } a \neq 1 \quad \text{integral of } a^u$$

$$\left.\begin{aligned}
\int \sin u\,du &= -\cos u + C \\
\int \cos u\,du &= \sin u + C \\
\int \sec^2 u\,du &= \tan u + C \\
\int \csc^2 u\,du &= -\cot u + C \\
\int \sec u \tan u\,du &= \sec u + C \\
\int \csc u \cot u\,du &= -\csc u + C
\end{aligned}\right\} \quad \text{integral of trigonometric functions}$$

$$\int \tan u\,du = -\ln|\cos u| + C \quad \text{where } \cos u \neq 0 \quad \text{integral of tangent function}$$

$$\int \cot u\,du = \ln|\sin u| + C \quad \text{where } \sin u \neq 0 \quad \text{integral of cotangent function}$$

$$\int \sec u\,du = \ln|\sec u + \tan u| + C \quad \text{where } \sec u + \tan u \neq 0 \quad \text{integral of secant function}$$

$$\int \csc u\,du = \ln|\csc u - \cot u| + C \quad \text{where } \csc u - \cot u \neq 0 \quad \text{integral of cosecant function}$$

$$\int \frac{du}{\sqrt{a^2 - u^2}} = \text{Arcsin}\,\frac{u}{a} + C \quad \text{where } a > 0 \text{ and } -1 < \frac{u}{a} < 1 \quad \text{integrals leading to Arcsin}$$

$$\int \frac{du}{a^2 + u^2} = \frac{1}{a}\,\text{Arctan}\,\frac{u}{a} + C \quad \text{where } a \neq 0 \quad \text{integrals leading to Arctan}$$

$$\int u\,dv = uv - \int v\,du \quad \text{integration by parts}$$

CHAPTER REVIEW

Section 21–1

Integrate the following functions. Round to two decimal places where necessary.

1. $\displaystyle\int_0^{\pi} (\sin x - \sin 3x)\,dx$

2. $\displaystyle\int \frac{e^{3t}}{(1 + e^{3t})^2}\,dt$

3. $\displaystyle\int \frac{(\text{Arcsin } 2x)^3}{\sqrt{1 - 4x^2}}\,dx$

4. $\displaystyle\int_1^2 (\ln x)^4\,\frac{dx}{x}$

5. $\displaystyle\int_0^1 (6 + 7e^{3t})^2 e^{3t}\,dt$

6. $\displaystyle\int \frac{\sin x\,dx}{(1 + \cos x)^2}$

7. $\int \tan^2 x \sec^2 x\, dx$

8. $\int_0^{\pi/4} \frac{(\text{Arctan } 2x)^4}{1 + 4x^2}\, dx$

Section 21–2

Integrate each of the given functions.

9. $\int \frac{e^{\text{Arcsin } x}}{\sqrt{1 - x^2}}\, dx$

10. $\int \frac{e^{\cos x}}{\sec x}\, dx$

11. $\int_0^1 \frac{e^{4x}\, dx}{2 + e^{4x}}$

12. $\int_{-1}^2 \frac{x\, dx}{3 - 4x^2}$

13. $\int_0^{\pi/2} \sin x e^{\cos x}\, dx$

14. $\int 8^{7x}\, dx$

15. $\int 11^{2x+1}\, dx$

16. $\int_0^1 x^2 e^{2x^3}\, dx$

17. Find the average force exerted on a piston from $t = 0$ to $t = 1.75$ s if the force is given by $F = 8 - e^{-t}$.

18. Find the velocity of a charged particle after 3 s if its acceleration is given by $V = 8(10t + 3)$ m/s^2.

Section 21–3

Integrate the given functions.

19. $\int_0^1 3x \cos 2x^2\, dx$

20. $\int \frac{\csc 4x}{\tan 4x}\, dx$

21. $\int \csc 3\theta \cot 3\theta\, d\theta$

22. $\int_0^{\pi/2} \cos \theta \sin \theta + \sin 3\theta\, d\theta$

23. $\int_0^{\pi/4} e^x \sin e^x\, dx$

24. $\int_0^{\pi/2} \sec(3x - 1)\, dx$

25. $\int x \sec 3x^2 \tan 3x^2\, dx$

26. $\int \frac{1 + \tan^2 x}{\sin^2 x}\, dx$

27. Find the volume generated by revolving the area bounded by $y = \tan x$, $y = 0$, and $x = \pi/4$ about the x axis.

28. Find the displacement of an object whose velocity is given by $V = 3 \sin 5t$ if $s = 0$ when $t = 0$.

Section 21–4

Integrate the following functions.

29. $\int_0^{\pi} \sin^2 x \cos^4 x\, dx$

30. $\int \csc^2 x\, dx$

31. $\int \cot^3 x\, dx$

32. $\int_0^{\pi/4} \tan x \sec^4 x\, dx$

33. $\int \frac{\sin x}{\cos^2 x}\, dx$

34. $\int \cos^5 \theta\, d\theta$

35. $\int \sec^6 t\, dt$

36. $\int \cot^2 x \csc^2 x\, dx$

Section 21–5

Integrate the following functions.

37. $\int \frac{dx}{\sqrt{9 - x^2}}$

38. $\int \frac{dx}{4 + x^2}$

39. $\int t^3 \sqrt{9 + t^2}\, dt$

40. $\int \frac{3x\, dx}{\sqrt{16 - 9x^4}}$

41. $\int \frac{x\, dx}{\sqrt{16x^2 - 25}}$

42. $\int \frac{1 + t}{1 + t^2}\, dt$

43. $\int \frac{dx}{25 + x^2}$

44. $\int \frac{\csc^2 x}{\sqrt{1 - \cot^2 x}}\, dx$

Section 21–6

Integrate the following functions.

45. $\int_0^1 x^2 e^x\, dx$

46. $\int_0^{\pi/2} x \sin x\, dx$

47. $\int x(\ln x)^4\, dx$

48. $\int x(x - 1)^4\, dx$

49. $\int_0^{\pi} x^2 \cos x\, dx$

50. $\int y\sqrt{y + 4}\, dy$

51. $\int x^2 \sqrt{3 - x^3}\, dx$

52. $\int_{-1}^1 x^3 e^x\, dx$

Section 21–7

Use the table of integrals in Appendix C to integrate the following functions.

53. $\int \cos^5 x\, dx$

54. $\int \frac{dx}{(16 - x^2)^{3/2}}$

55. $\int_1^3 \frac{x\, dx}{9 - 4x}$

56. $\int_0^2 \frac{t\, dt}{(1 + 3t)^2}$

57. $\int x^2 e^{4x}\, dx$

58. $\int \cot^2 x\, dx$

59. $\int \tan^2 t\, dt$

60. $\int \sin 2x \sin 5x\, dx$

CHAPTER TEST

The number in parentheses refers to the appropriate learning objective at the beginning of the chapter.

1. Integrate the following: (1)

$$\int e^x(e^x - 4)^2\, dx$$

2. Find the work done in moving an object 12 ft if the force is given by (9)

$$F = \frac{1}{5s - 1} \quad \text{where } W = \int F\, ds$$

3. Integrate the following: (1)

$$\int (1 + e^{3x})^2 e^{3x}\, dx$$

4. Use the table of integrals in Appendix C to integrate the following: (8)

$$\int_0^1 xe^{3x}\, dx$$

5. Integrate the following: (3)

$$\int_{\pi/8}^{\pi/4} \frac{\sin \ln x}{x}\, dx$$

6. Find the x coordinate of the centroid of the area bounded by $y = e^x$, $y = 0$, $x = 0$, and $x = 2$. (9)

7. Integrate the following: (2)

$$\int_{-1}^{1} \frac{x\, dx}{2 + 3x^2}$$

8. Integrate the following: (3)

$$\int \sin x(\cos x - 3\cos^2 x)\, dx$$

9. Integrate the following: (5)

$$\int \frac{dx}{x^2 + 16}$$

10. Integrate the following by parts: (7)

$$\int_0^{\pi} x \sin^2 4x\, dx$$

11. Find the displacement s, after 2 s, of a particle whose velocity is given by $v = 4x^2e^{8x}$ m/s. (9)

12. Integrate the following: (1)

$$\int \frac{(\text{Arccos } 3x)^2}{\sqrt{1 - 9x^2}}\, dx$$

13. Use the table of integrals in Appendix C to integrate the following: (8)

$$\int \frac{\sqrt{x^2 - 4}}{x}\, dx$$

14. Integrate the following: (4)

$$\int \cot^3 x \csc x\, dx$$

15. Integrate the following by parts: (7)

$$\int y(y + 5)^3\, dy$$

16. Integrate the following: (6)

$$\int \frac{dx}{x^2\sqrt{9 - x^2}}$$

SOLUTION TO CHAPTER INTRODUCTION

The average value of a function is given by the formula

$$M = \frac{1}{b - a} \int_a^b f(x)\, dx$$

Relating this formula to the average force on the piston gives the following integral:

$$M = \frac{1}{1.75 - 0} \int_0^{1.75} (12 - e^{-2t})\, dt$$

Performing the integration and evaluating the limits of integration gives

$$\begin{aligned} M &= \frac{1}{1.75}\left[\int_0^{1.75} 12\, dt - \int_0^{1.75} e^{-2t}\, dt\right] \\ &= \frac{1}{1.75}\left[\int_0^{1.75} 12\, dt + \frac{1}{2}\int_0^{1.75} e^{-2t}(-2\, dt)\right] \\ &\approx \frac{1}{1.75}\left(12t\Big|_0^{1.75} + \frac{1}{2}e^{-2t}\Big|_0^{1.75}\right) \\ &\approx \frac{1}{1.75}\left[(21 - 0) + \frac{1}{2}(0.03 - 1)\right] \\ &= 11.72 \end{aligned}$$

The average force on the piston is 11.72 lb.

You are an engineer responsible for programming a computer chip to calculate cosh x (hyperbolic cosine function). Find the Maclaurin series so that you can use the algebraic expression to program the calculation, which is given by the following equation:

$$\cosh x = \frac{1}{2}(e^x + e^{-x})$$

(The solution to this problem is given at the end of the chapter.)

In this chapter we will discuss how to represent a nonalgebraic function in terms of a polynomial. Once we have this representation, we will use it to evaluate the function to any desired degree of accuracy. We can also use this representation to further our methods of integration.

Learning Objectives

After completing this chapter, you should be able to

1. Find the Maclaurin series expansion for a given function (Section 22–1).
2. Use Maclaurin series expansions to find the Maclaurin series for other transcendental functions (Section 22–2).
3. Use the Maclaurin series expansion to find the derivative or integral of a given function (Section 22–2).
4. Use Maclaurin series expansions to calculate specific functional values (Section 22–3).
5. Find the Taylor series expansion of a function about $x = a$ (Section 22–4).
6. Find the Fourier series expansion of a periodic function (Section 22–5).
7. Apply power series to technical problems (Sections 22–1 through 22–5).

Chapter 22

Series Expansion of Functions

22–1

MACLAURIN SERIES

If we divide $1 + x$ into 1 using long division, we get

$$\frac{1}{1 + x} = 1 - x + x^2 - x^3 + \cdots + x^{2n} - \cdots + x^n - \cdots$$

Since x represents a real number, the infinite series on the right side of the equation is a geometric progression. Recall from our discussion in Chapter 16 that the first term of this series is $a_1 = 1$ and the common ratio is $r = -x$. Also recall from Chapter 16 that if $|r| < 1$, the sum of the infinite geometric progression is given by

$$S = \frac{a}{1 - r} = \frac{1}{1 + x}$$

which equals the left side of the equation. Therefore, for $|r| < 1$, the geometric series above represents the function $1/(1 + x)$.

Power Series

If $a_0, a_1, a_2, \ldots, a_n$ are constants and x is a variable, then a series of the form $a_0 + a_1x + a_2x^2 + \cdots + a_nx^n + \cdots$ is called a **power series expansion** of a function $f(x)$.

Convergence

Recall from our discussion of infinite series in Chapter 16 that the sum of the first n terms in the series is given by $S_n = a_1 + a_2 + \cdots + a_n$. This series **converges** if $\lim_{n\to\infty} S_n = S$ where S represents the sum of the series. In other words, if the sum of the terms approaches some limiting value as $n \to \infty$, the series converges. If the series does not approach such a limit, the series is said to **diverge.** For example, the series $\frac{1}{2} + \frac{1}{4} + \frac{1}{8} + \cdots$ converges, while the series $2 + 4 + 8 + \cdots$ diverges. The interval of values for x where the series converges is called the **interval of convergence.** Outside this interval, the series diverges. For example, $1 - x + x^2 - x^3 + \cdots$ converges on the interval $-1 < x < 1$. If the series converges, then for a larger and larger number of terms in the series, the difference in the series expansion and the function approaches zero. Therefore, the function $f(x)$ is represented by the power series expansion on the interval of convergence. Not all functions can be represented by a power series, but a large class of functions called *analytic functions,* which we will discuss later in this chapter, can be represented by a power series.

Maclaurin Series

We will assume, unless noted otherwise, that the functions discussed can be represented by a power series expansion. To specify a power series exactly, we must evaluate the coefficients $a_0, a_1, a_2, \ldots, a_n$. We can do so by finding successive derivatives of the power series and evaluating them at $x = 0$ as follows:

$$\begin{aligned}
f(x) &= a_0 + a_1x + a_2x^2 + a_3x^3 + a_4x^4 + \cdots & f(0) &= a_0 \\
f'(x) &= a_1 + 2a_2x + 3a_3x^2 + 4a_4x^3 + \cdots & f'(0) &= a_1 \\
f''(x) &= 2a_2 + 2 \cdot 3a_3x + 3 \cdot 4a_4x^2 + \cdots & f''(0) &= 2a_2 \\
f'''(x) &= 2 \cdot 3a_3 + 2 \cdot 3 \cdot 4a_4x + \cdots & f'''(0) &= 2 \cdot 3a_3
\end{aligned}$$

From the calculations above, we can note the following pattern for the coefficients a_n:

$$\begin{aligned}
a_0 &= f(0) \\
a_1 &= f'(0) \\
a_2 &= \frac{f''(0)}{2!} \\
a_3 &= \frac{f'''(0)}{3!}
\end{aligned}$$

If we continue this pattern, then

$$a_n = \frac{f^{(n)}(0)}{n!}$$

Substituting these values for the coefficients into the power series gives

$$f(x) = f(0) + f'(0)x + \frac{f''(0)x^2}{2!} + \frac{f'''(0)x^3}{3!} + \cdots + \frac{f^{(n)}(0)x^n}{n!} + \cdots$$

This equation is called the **Maclaurin series expansion** of a function. For a function to be represented by a Maclaurin series expansion, the function and its derivatives must exist at $x = 0$.

Maclaurin Series Expansion

$$f(x) = f(0) + f'(0)x + \frac{f''(0)x^2}{2!} + \frac{f'''(0)x^3}{3!} + \cdots + \frac{f^{(n)}(0)x^n}{n!} + \cdots$$

The Maclaurin series is named after Colin Maclaurin, the Scottish mathematician who first published his results. Johann Bernoulli actually first used a series expansion of this type, but his work was not published until well after Maclaurin's name was associated with this power series expansion.

EXAMPLE 1 Find the Maclaurin series expansion of $f(x) = e^{2x}$.

Solution First, we successively differentiate this function and evaluate the resulting derivatives at $x = 0$.

$$
\begin{aligned}
f(x) &= e^{2x} & f(0) &= e^{2(0)} = 1\\
f'(x) &= 2e^{2x} & f'(0) &= 2e^{2(0)} = 2\\
f''(x) &= 4e^{2x} & f''(0) &= 4e^{2(0)} = 4\\
f'''(x) &= 8e^{2x} & f'''(0) &= 8e^{2(0)} = 8
\end{aligned}
$$

Second, we substitute these values into the formula for the Maclaurin series.

$$
\begin{aligned}
f(x) &= f(0) + f'(0)x + \frac{f''(0)x^2}{2!} + \frac{f'''(0)x^3}{3!} + \cdots\\
&= 1 + 2x + \frac{4x^2}{2!} + \frac{8x^3}{3!} + \cdots\\
&= 1 + 2x + 2x^2 + \frac{4}{3}x^3 + \cdots
\end{aligned}
$$

The Maclaurin series expansion represents the function $f(x) = e^{2x}$ for $-\infty < x < \infty$, the interval of convergence. Figure 22–1 shows the graph of $f(x) = e^{2x}$ and the approximation to this curve by addition of successive terms of its Maclaurin expansion.

FIGURE 22–1

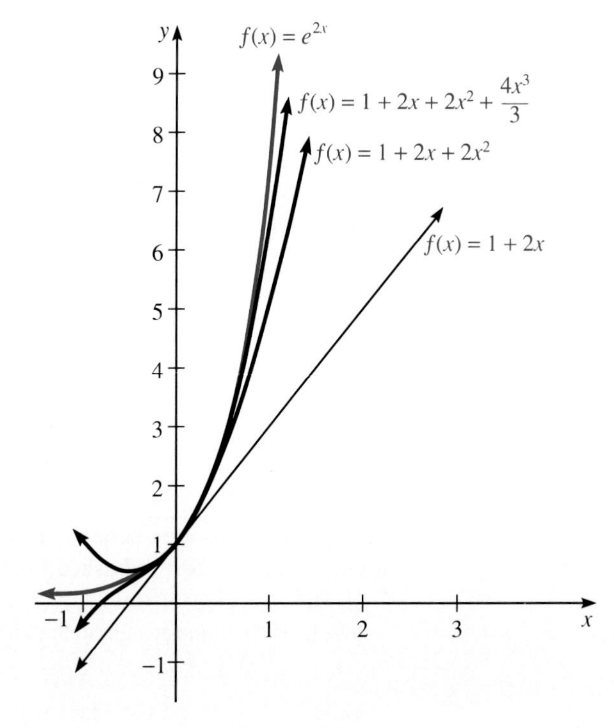

■

EXAMPLE 2 Find the first three nonzero terms in the Maclaurin expansion of $f(x) = \cos x$.

Solution First, we find successive derivatives and evaluate them at $x = 0$.

$$\begin{aligned} f(x) &= \cos x & f(0) &= 1 \\ f'(x) &= -\sin x & f'(0) &= -\sin 0 = 0 \\ f''(x) &= -\cos x & f''(0) &= -\cos 0 = -1 \\ f'''(x) &= \sin x & f'''(0) &= \sin 0 = 0 \\ f^{\text{IV}}(x) = f^{(4)}(x) = f''''(x) &= \cos x & f''''(0) &= \cos 0 = 1 \end{aligned}$$

Next, we substitute these values for the coefficients into the formula for the Maclaurin series.

$$f(x) = f(0) + f'(0)x + \frac{f''(0)x^2}{2!} + \frac{f'''(0)x^3}{3!} + \frac{f''''(0)x^4}{4!} + \cdots$$

$$\cos x = 1 + 0 + \frac{(-1)x^2}{2} + 0 + \frac{(1)x^4}{4!} + \cdots$$

$$\cos x = 1 - \frac{x^2}{2} + \frac{x^4}{24} - \cdots$$

Note that the Maclaurin series expansion for $\cos x$ is an **alternating series** because the signs of the terms alternate. The interval of convergence for $f(x) = \cos x$ is $-\infty < x < \infty$. Figure 22–2 shows the graph of $f(x) = \cos x$ and the effect that addition of successive terms of the Maclaurin series expansion has on the graph.

FIGURE 22–2

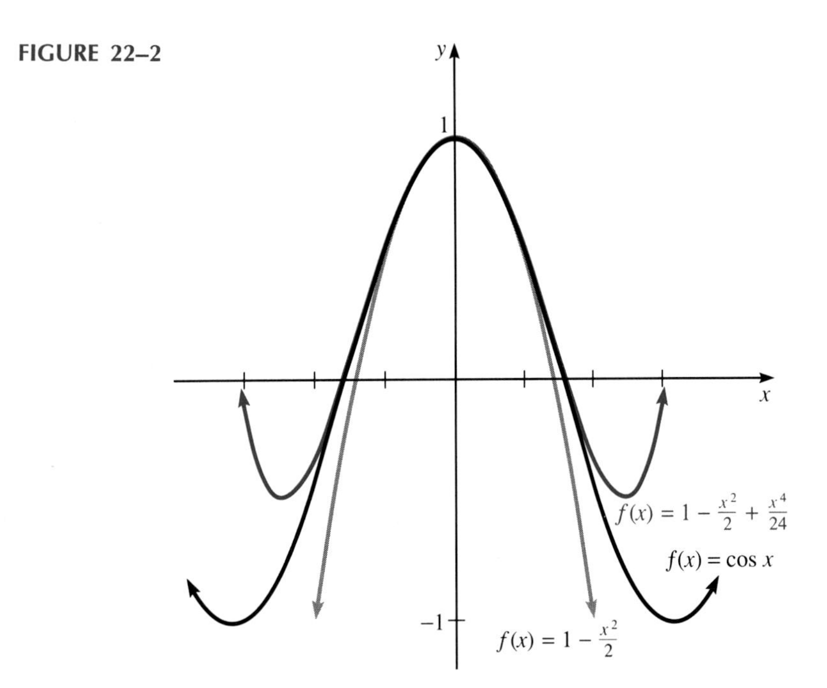

■

EXAMPLE 3 Find the first two nonzero terms of the Maclaurin expansion for $f(x) = \ln(\cos x)$.

Solution First, we take the required derivatives and evaluate them at $x = 0$.

$$\begin{aligned} f(x) &= \ln(\cos x) & f(0) &= 0 \\ f'(x) &= \frac{1}{\cos x} \cdot -\sin x = -\tan x & f'(0) &= 0 \\ f''(x) &= -\sec^2 x & f''(0) &= -1 \\ f'''(x) &= -2\sec^2 x \tan x & f'''(0) &= 0 \\ f''''(x) &= -2\sec^4 x - 4\tan^2 x \sec^2 x & f''''(0) &= -2 \end{aligned}$$

Second, we substitute the coefficients into the Maclaurin series expansion.

$$\begin{aligned} \ln(\cos x) &= 0 + 0x + \frac{(-1)x^2}{2!} + \frac{0x^3}{3!} + \frac{(-2)x^4}{4!} \\ &= -\frac{x^2}{2} - \frac{x^4}{12} \end{aligned}$$

■

EXAMPLE 4 Find the first three nonzero terms of the Maclaurin expansion of $f(x) = x^2e^x$.

Solution First, we differentiate the function and evaluate each derivative at $x = 0$.

$$\begin{aligned} f(x) &= x^2e^x & f(0) &= 0 \\ f'(x) &= e^x(x^2 + 2x) & f'(0) &= 0 \\ f''(x) &= e^x(x^2 + 4x + 2) & f''(0) &= 2 \\ f'''(x) &= e^x(x^2 + 6x + 6) & f'''(0) &= 6 \\ f''''(x) &= e^x(x^2 + 8x + 12) & f''''(0) &= 12 \end{aligned}$$

Next, we substitute the coefficients into the Maclaurin series expansion.

$$\begin{aligned} x^2e^x &= 0 + 0x + \frac{2x^2}{2!} + \frac{6x^3}{3!} + \frac{12x^4}{4!} \\ &= x^2 + x^3 + \frac{x^4}{2} \end{aligned}$$

■

EXAMPLE 5 Find the first three nonzero terms of the Maclaurin series expansion of $y = \text{Arccos } 2x$.

Solution First, we take the derivatives of the function and evaluate them at $x = 0$.

$$f(x) = \text{Arccos } 2x \qquad f(0) = \frac{\pi}{2}$$

$$f'(x) = \frac{-2}{\sqrt{1 - 4x^2}} \qquad f'(0) = -2$$

$$f''(x) = -8x(1 - 4x^2)^{-3/2} \qquad f''(0) = 0$$

$$f'''(x) = -96x^2(1 - 4x^2)^{-5/2} - 8(1 - 4x^2)^{-3/2} \qquad f'''(0) = -8$$

Next, we substitute the coefficients into the Maclaurin series expansion.

$$\text{Arccos } 2x = \frac{\pi}{2} - 2x + \frac{0x^2}{2!} + \frac{-8x^3}{3!}$$

$$= \frac{\pi}{2} - 2x - \frac{4x^3}{3}$$

■

22–1 EXERCISES

Find the first three nonzero terms of the Maclaurin series expansion for each of the following functions.

1. $f(x) = \sin x$
2. $f(x) = e^x$
3. $f(t) = \cos 3t$
4. $f(x) = \ln(2x + 3)$
5. $f(x) = 3e^{-x}$
6. $f(x) = x \cdot e^x$
7. $f(x) = \sqrt{1 + 3x}$
8. $f(x) = \dfrac{1}{(1 + x)^2}$
9. $f(t) = \dfrac{1}{3t + 1}$
10. $f(x) = \dfrac{x + 3}{x - 3}$
11. $f(x) = e^x \sin x$
12. $f(x) = x^2 \sin x$
13. $f(x) = \ln(1 - 3x)$
14. $f(t) = \sin^2 t$
15. $f(x) = 3x^3 + 5x - 8$
16. $f(x) = (3x + 4)^2$
17. $f(x) = \cos \frac{1}{2}x$
18. $f(t) = \text{Arcsin } t$

Find the first two nonzero terms of the Maclaurin series expansion of each function.

19. $f(x) = \ln(2 + 5x)$
20. $f(x) = e^x \cos x$
21. $f(x) = \text{Arctan}(x + 1)$
22. $f(x) = \tan 3x$
23. $f(x) = \cos e^x$
24. $f(x) = e^{\sin x}$
25. $f(x) = \cos^2 x$
26. $f(x) = \ln(1 - \sin x)$
27. The hyperbolic sine is given by

$$\sinh x = \frac{1}{2}(e^x - e^{-x})$$

Find the first three nonzero terms in the Maclaurin series expansion for this function.

28. If $f(x) = x^4 + 8x$, show that the Maclaurin series expansion of this function is $x^4 + 8x$.
29. The function $f(x) = \sec x$ is used in the study of fluid flow. Find the first three nonzero terms in the Maclaurin series expansion of this function.
30. Under certain circumstances, the displacement of an object oscillating at the end of a spring is given by

$$s(x) = e^{-x}(\cos x - \sin x)$$

Find the first two nonzero terms of the Maclaurin series expansion of this function.

22–2

OPERATIONS WITH SERIES

Table 22–1 summarizes the Maclaurin series expansion and interval of convergence of the most important functions. You can use them to find the Maclaurin series of other transcendental functions and other functions without deriving the series directly.

TABLE 22–1

Function	Maclaurin Series Expansion	Convergence
e^x	$1 + x + \frac{x^2}{2!} + \frac{x^3}{3!} + \cdots$	$-\infty < x < \infty$
$\sin x$	$x - \frac{x^3}{3!} + \frac{x^5}{5!} - \frac{x^7}{7!} + \cdots$	$-\infty < x < \infty$
$\cos x$	$1 - \frac{x^2}{2!} + \frac{x^4}{4!} - \frac{x^6}{6!} + \cdots$	$-\infty < x < \infty$
$\ln(1 + x)$	$x - \frac{x^2}{2} + \frac{x^3}{3} - \frac{x^4}{4} + \cdots$	$-1 < x \leq 1$
$\frac{1}{1 + x}$	$1 - x^1 + x^2 - x^3 + \cdots$	$-1 < x < 1$
$(1 + x)^n$	$1 + nx + \frac{n(n - 1)x^2}{2!} + \frac{n(n - 1)(n - 2)x^3}{3!} + \cdots$	$-1 < x < 1$ for any real number n

One very important property of power series is that many of the basic algebraic operations with polynomials also apply to them. The operations of substitution of variables, addition, subtraction, multiplication, and division may be performed with Maclaurin power series. In addition, term-by-term differentiation and integration also apply. We discuss each of these operations in this section.

Variable Substitution

In Chapter 2, we discussed such functions as $f(x)$, $f(3x)$, or $f(x^2)$. Remember that if $f(x) = 8x - 4$, then $f(3)$ is found by substituting 3 for x to give $f(3) = 8(3) - 4$. Using functional notation, we can find the Maclaurin series expansion of numerous functions without the use of direct expansion. The next example illustrates this procedure.

EXAMPLE 1 Find the first four nonzero terms of the Maclaurin series expansion of each of the following:

(a) e^{-4x}
(b) $\cos x^2$

Solution

(a) Instead of deriving the Maclaurin series expansion of this function directly as in the last section, we will substitute $-4x$ for x in the series expansion of e^x. The expansion of e^x is as follows:

$$f(x) = e^x = 1 + x + \frac{x^2}{2!} + \frac{x^3}{3!} + \cdots$$

To determine the Maclaurin series expansion of e^{-4x}, we find $f(-4x)$ as follows:

$$f(-4x) = e^{-4x} = 1 + \underbrace{(-4x)} + \frac{\underbrace{(-4x)}^2}{2!} + \frac{\underbrace{(-4x)}^3}{3!} + \cdots \quad \text{(substitute } -4x \text{ for } x\text{)}$$

$$= 1 - 4x + 8x^2 - \frac{32x^3}{3} + \cdots$$

(b) To find $\cos x^2$, we substitute x^2 for x in the series expansion of $\cos x$. The series for $\cos x$ is

$$f(x) = \cos x = 1 - \frac{x^2}{2!} + \frac{x^4}{4!} - \frac{x^6}{6!} + \cdots$$

Substituting gives

$$f(x^2) = \cos x^2 = 1 - \frac{(x^2)^2}{2!} + \frac{(x^2)^4}{4!} - \frac{(x^2)^6}{6!} + \cdots$$

$$= 1 - \frac{x^4}{2} + \frac{x^8}{24} - \frac{x^{12}}{720} + \cdots$$

■

Multiplication of Series

EXAMPLE 2 Find the first four nonzero terms of the Maclaurin series expansion for $e^{-x}\cos x$.

Solution Rather than calculating this series directly, we multiply the series for e^{-x} and $\cos x$. First, we must find the series expansion for e^{-x}. We do so using the same technique as in the previous example and substitute $-x$ for x in the Maclaurin series expansion for e^x. Thus,

$$e^{-x} = 1 + (-x) + \frac{(-x)^2}{2!} + \frac{(-x)^3}{3!} + \cdots$$

Then we determine the series for $\cos x$.

$$\cos x = 1 - \frac{x^2}{2} + \frac{x^4}{24} - \frac{x^6}{720} + \cdots$$

Last, we determine the Maclaurin series expansion for $e^{-x}\cos x$ by multiplying these two series term-by-term.

$$
\begin{array}{l}
1 - x + \frac{x^2}{2} - \frac{x^3}{6} + \cdots \\
1 - \frac{x^2}{2} + \frac{x^4}{24} - \frac{x^6}{720} + \cdots \\
\hline
1 - x + \frac{x^2}{2} - \frac{x^3}{6} + \frac{x^4}{24} + \cdots \\
\qquad - \frac{x^2}{2} + \frac{x^3}{2} - \frac{x^4}{4} + \frac{x^5}{12} + \cdots \\
\qquad\qquad\qquad + \frac{x^4}{24} + \cdots \\
\hline
1 - x + 0 + \frac{x^3}{3} - \frac{x^4}{6} + \cdots
\end{array}
$$

Therefore,

$$e^{-x}\cos x = 1 - x + \frac{x^3}{3} - \frac{x^4}{6} + \cdots \quad \text{for all } x$$

In multiplying these two series term-by-term, it is necessary to multiply only enough terms to obtain the desired number of terms in the product. ■

NOTE ✦ The interval of convergence for the product of two Maclaurin series is the interval common to both series. For example, if the interval of convergence for two series is $-\infty$ to ∞ and -1 to 1, then the interval of convergence for the product is -1 to 1.

Division of Series

EXAMPLE 3 Find the first three nonzero terms of the Maclaurin series expansion of $\frac{\sin x}{1 + x}$.

Solution Rather than determine this series directly, we will divide the Maclaurin series for $\sin x$ by the series for $1 + x$. Note that the Maclaurin series expansion of the polynomial $1 + x$ is $1 + x$, which can be written as follows:

$$\frac{\sin x}{1 + x} = \frac{x - \frac{x^3}{3!} + \frac{x^5}{5!} - \frac{x^7}{7!} + \cdots}{1 + x}$$

Then we use polynomial long division to obtain the desired series.

$$
\begin{array}{r l}
 & x - x^2 + \frac{5}{6}x^3 - \\
1 + x \,\big) & x + 0x^2 - \frac{x^3}{6} + 0x^4 \\
 & \underline{x + x^2} \\
 & \quad -x^2 - \frac{x^3}{6} \\
 & \quad \underline{-x^2 - x^3} \\
 & \qquad \frac{5x^3}{6} + 0x^4 + \cdots \\
 & \qquad \underline{\frac{5x^3}{6} + \frac{5x^4}{6}} + \cdots
\end{array}
$$

Therefore, from the division,

$$\frac{\sin x}{1 + x} = x - x^2 + \frac{5}{6}x^3 - \cdots$$ ■

Differentiation

A power series can be differentiated term-by-term within its interval of convergence, excluding the endpoints. This technique allows us to differentiate functions that would be difficult to differentiate directly. The next example illustrates this procedure.

EXAMPLE 4 The displacement (in meters) of an object oscillating at the end of a spring is given by $s = t \sin 2t$. Use a Maclaurin series to find the velocity of this object when $t = 1.25$ s.

Solution The velocity of the object is determined as the derivative of its displacement. First, we will develop a Maclaurin series for $t \sin 2t$.

Maclaurin series for $\sin 2t$

$$
\begin{aligned}
t \sin 2t &= t\left(2t - \frac{(2t)^3}{3!} + \frac{(2t)^5}{5!} - \frac{(2t)^7}{7!} + \frac{(2t)^9}{9!} - \cdots\right) \\
&= t\left(2t - \frac{8t^3}{6} + \frac{32t^5}{120} - \frac{128t^7}{5{,}040} + \frac{512t^9}{362{,}880} - \cdots\right) \\
&= 2t^2 - \frac{4t^4}{3} + \frac{4t^6}{15} - \frac{8t^8}{315} + \frac{4t^{10}}{2{,}835} - \cdots
\end{aligned}
$$

Next, we find the velocity by differentiating the Maclaurin series term-by-term and evaluating the result when $t = 1.25$ s.

$$v = 4t - \frac{16t^3}{3} + \frac{24t^5}{15} - \frac{64t^7}{315} + \frac{40t^9}{2{,}835}\Bigg|_{t=1.25}$$
$$\approx -1.40 \text{ m/s}$$

The velocity of the object is approximately 1.40 m/s. The negative sign denotes. that the velocity is in the opposite direction to motion. Direct differentiation and evaluation of $s = t \sin 2t$ at $t = 1.25$ give the following result:

$$\frac{ds}{dt} = t(\cos 2t)(2) + \sin 2t(1) = 2t \cos 2t + \sin 2t\big|_{t=1.25}$$
$$\approx -1.40$$

■

Integration

We can also integrate a power series term-by-term, thereby obtaining an additional method of integration. This procedure is illustrated in the next example.

EXAMPLE 5 Use the first 5 nonzero terms of a Maclaurin series expansion to find the integral of

$$\int e^x \cos x \, dx$$

Solution First, we must find the Maclaurin series expansion of $e^x \cos x$ by multiplying term-by-term the power series for e^x and $\cos x$.

$$e^x \cos x = \left(1 + x + \frac{x^2}{2} + \frac{x^3}{6} + \frac{x^4}{24} + \frac{x^5}{120} + \cdots\right)\left(1 - \frac{x^2}{2} + \frac{x^4}{24} - \frac{x^6}{720} + \cdots\right)$$
$$= 1 + x - \frac{x^3}{3} - \frac{x^4}{6} - \frac{x^5}{30} - \cdots$$

Next, we find the integral by integrating the power series term-by-term.

$$\int e^x \cos x \, dx = \int dx + \int x \, dx - \int \frac{x^3}{3} dx - \int \frac{x^4}{6} dx - \int \frac{x^5}{30} dx$$
$$= x + \frac{x^2}{2} - \frac{x^4}{12} - \frac{x^5}{30} - \frac{x^6}{180} + C$$

■

EXAMPLE 6 The charge (in microcoulombs) on a capacitor is given by

$$Q = \int_0^{0.5} (1.0 - e^{0.2t}) \, dt$$

Use the first four terms of the Maclaurin series to evaluate this integral.

Solution First, we will find a series for $e^{0.2t}$ by using substitution.

$$\begin{aligned} e^{0.2t} &= 1 + (0.2t) + \frac{(0.2t)^2}{2!} + \frac{(0.2t)^3}{3!} + \cdots \\ &= 1 + 0.2t + \frac{0.04t^2}{2} + \frac{0.008t^3}{6} + \cdots \\ &\approx 1 + 0.2t + 0.02t^2 + 0.00133t^3 + \cdots \end{aligned}$$

Substituting this series into the integral and simplifying the result give

$$\begin{aligned} Q &= \int_0^{0.5} [1 - (1 + 0.2t + 0.02t^2 + 0.00133t^3 + \cdots)]\, dt \\ &= \int_0^{0.5} (-0.2t - 0.02t^2 - 0.00133t^3 - \cdots)\, dt \end{aligned}$$

Integrating and evaluating the result give

$$\begin{aligned} Q &= -\frac{0.2t^2}{2} - \frac{0.02t^3}{3} - \frac{0.00133t^4}{4} \Bigg|_0^{0.5} \\ &\approx (-0.025 - 0.0008\overline{3} - 0.00002) - 0 \\ &\approx -0.025854 \ldots \end{aligned}$$

The charge is approximately 0.03 μC in the opposite direction. ■

Euler's Identity

In chapter 14, we defined the exponential form of a complex number as

$$e^{j\theta} = \cos\theta + j\sin\theta \quad \text{where } j^2 = -1$$

Using the Maclaurin series expansion for e^x, $\sin x$, and $\cos x$, we can now prove this formula. Using the Maclaurin series for e^x, we find the series for $e^{j\theta}$ by substituting $j\theta$ for x.

$$\begin{aligned} e^{j\theta} &= 1 + j\theta + \frac{(j\theta)^2}{2!} + \frac{(j\theta)^3}{3!} + \frac{(j\theta)^4}{4!} + \frac{(j\theta)^5}{5!} + \cdots \\ &= 1 + j\theta - \frac{\theta^2}{2!} - \frac{j\theta^3}{3!} + \frac{\theta^4}{4!} + \frac{j\theta^5}{5!} + \cdots \end{aligned}$$

Separating the imaginary and real terms of the complex number gives

$$e^{j\theta} = \left(1 - \frac{\theta^2}{2!} + \frac{\theta^4}{4!} - \cdots\right) + j\left(\theta - \frac{\theta^3}{3!} + \frac{\theta^5}{5!} - \cdots\right)$$

But

$$\cos\theta = 1 - \frac{\theta^2}{2!} + \frac{\theta^4}{4!} - \cdots \quad \text{and} \quad \sin\theta = \theta - \frac{\theta^3}{3!} + \frac{\theta^5}{5!} - \cdots$$

Therefore,

$$e^{j\theta} = \cos\theta + j\sin\theta$$

The **Euler identity** results from evaluating this expression when $\theta = \pi$.

$$e^{j\pi} = \cos\pi + j\sin\pi = -1$$

This equation is called the Euler identity because it relates e, π, j, and -1, symbols used extensively by the Swiss mathematician Leonhard Euler.

22–2 EXERCISES

Find the first four nonzero terms of the Maclaurin series expansion of each function by using the series expansion given in Table 22–1.

1. $f(x) = e^{-3x}$
2. $f(x) = e^{x^2}$
3. $f(x) = \ln(1 + x^2)$
4. $f(t) = \ln(1 - 3t)$
5. $f(t) = \sin\left(\frac{5t}{3}\right)$
6. $f(\theta) = \cos 3\theta^2$
7. $f(x) = x^2e^{-x}$
8. $f(x) = \frac{e^x}{x}$
9. $f(x) = \frac{\sin x}{1 - x}$
10. $f(x) = \frac{\ln(1 + x)}{x^2}$
11. $f(\theta) = \sin\theta\cos\theta$
12. $f(x) = e^x(1 + x)^2$
13. $f(x) = \frac{3}{1 + x^2}$
14. $f(t) = \frac{\cos t}{e^t}$

Evaluate the given integrals by using the first three terms of the Maclaurin series expansion in Table 22–1.

15. $\int \cos x^2\,dx$
16. $\int \frac{\ln(1 + x)}{x^2}\,dx$
17. $\int e^{\theta}\sin\theta\,d\theta$
18. $\int \sin\sqrt{x}\,dx$
19. $\int_2^3 \frac{\ln(1 - x^2)}{x}\,dx$
20. $\int_0^1 t^2\ln(1 - 3t)\,dt$

Differentiate the given functions by using the first three terms of the Maclaurin series expansion in Table 22–1.

21. $f(x) = \ln(1 + x^2)$
22. $f(x) = (1 + 3x)^4$
23. $f(x) = x^2\cos x$
24. $f(x) = \sin x\ln(1 + x)$
25. $f(x) = \frac{e^x}{\sin x}$
26. $f(x) = \frac{\cos 2x}{x}$

27. Maclaurin series are particularly useful in working with digital computers. Using a Maclaurin series, a programmer can express a trigonometric function as an algebraic expression that the computer can evaluate. Write the first three terms of the Maclaurin series for $f(x) = \sin(\sin x)$.

28. The displacement of a simple harmonic oscillator as a function of time is given by $s = \sin 2\pi t$. Find the first three terms of the Maclaurin series expansion.

29. Find the area of the region bounded by $y = x\sin x$, $x = \pi$, and $y = 0$ by using the first three terms of the Maclaurin series.

30. Find the volume generated by revolving the area bounded by $y = \cos x$, $x = 0$, $y = 0$, and $x = \pi$ about the x axis. Use the first three terms of the Maclaurin series.

22–3

COMPUTATIONS WITH SERIES

We can use the Maclaurin series expansion of a function to compute numerical values near zero of transcendental functions. By including enough terms from the Maclaurin series, we can calculate the value of the function to any desired degree of accuracy. Tables for e^x, logarithms, and trigonometric functions may be generated using this method. The next three examples will illustrate this procedure.

EXAMPLE 1 Use the first four terms of a Maclaurin series to approximate $e^{0.3}$.

Solution From Table 22–1, we know that the following series represents e^x:

$$e^x = 1 + x + \frac{x^2}{2!} + \frac{x^3}{3!} + \cdots$$

Next, we substitute 0.3 for x in the Maclaurin series and calculate the resulting value.

$$\begin{aligned} e^{0.3} &\approx 1 + (0.3) + \frac{(0.3)^2}{2!} + \frac{(0.3)^3}{3!} \\ &\approx 1.3495 \end{aligned}$$

■

EXAMPLE 2 Calculate the value of cos 5° by using the first four terms of the Maclaurin series.

Solution From Table 22–1, we know that the Maclaurin series for $\cos x$ is given by

$$\cos x = 1 - \frac{x^2}{2!} + \frac{x^4}{4!} - \frac{x^6}{6!} + \cdots$$

Before substituting, we convert 5° to radians as follows:

$$\begin{aligned} \frac{5°}{x} &= \frac{180°}{\pi} \\ x &= \frac{\pi}{36} \end{aligned}$$

Substituting $\pi/36$ for x and performing the calculation give the following result:

$$\begin{aligned} \cos 5° &\approx 1 - \frac{(\pi/36)^2}{2!} + \frac{(\pi/36)^4}{4!} - \frac{(\pi/36)^6}{6!} \\ &\approx 0.9962 \end{aligned}$$

Use your calculator to check this result. ■

EXAMPLE 3 Calculate the value of ln 1.4 by using the first four terms of the Maclaurin series.

Solution The Maclaurin series for $\ln(1 + x)$ is

$$\ln(1 + x) = x - \frac{x^2}{2} + \frac{x^3}{3} - \frac{x^4}{4} + \cdots$$

To calculate ln 1.4, we write ln(1 + 0.4). Therefore, we substitute $x = 0.4$ into the formula and perform the calculations.

$$\ln 1.4 \approx 0.4 - \frac{(0.4)^2}{2} + \frac{(0.4)^3}{3} - \frac{(0.4)^4}{4}$$
$$\approx 0.3349$$

■

Error

Recall that in Chapter 19 we discussed the differential as a measurement of errors in calculated values due to errors in measured values. When a function is approximated by a truncated Maclaurin series, an error results as given by the following formula.

Error in Truncated Maclaurin Series

$$E = \left|\frac{f^{(n+1)}(c)x^{n+1}}{(n + 1)!}\right|$$

where c = a number between 0 and x, and x^n = the last term in the series approximation

EXAMPLE 4 Compute $e^{0.7}$ by using the first four terms of a Maclaurin series, and compute the maximum error of this approximation.

Solution First, we find the approximation for $e^{0.7}$. From Table 22–1,

$$e^x = 1 + x + \frac{x^2}{2!} + \frac{x^3}{3!} + \frac{x^4}{4!} + \cdots$$

Substituting 0.7 for x gives

$$e^{0.7} \approx 1 + (0.7) + \frac{(0.7)^2}{2!} + \frac{(0.7)^3}{3!}$$
$$\approx 2.002166667$$

To find the upper bound on the error in this approximation, use the formula

$$E = \left|\frac{f^{(n+1)}(c)x^{(n+1)}}{(n + 1)!}\right|$$

where $x = 0.7$; $n + 1 = 4$, the first term not included in the series approximation; and $c = 0.7$. Thus,

$$E = \left|\frac{f^{(4)}(0.7)(0.7)^4}{4!}\right|$$

where $f^{(4)}(0.7)$ = the fourth derivative of e^x evaluated at 0.7. Let us use $e \approx 3$. Then

$$E = \left|\frac{3^{0.7}(0.7)^4}{4!}\right|$$
$$\approx 0.021585683$$

Since $e^{0.7}$ is an increasing function, its value is less than 2.002166667 + 0.021585683 = 2.023752350. Therefore, to four decimal places, $e^{0.7} \approx 2.0238$. ■

For the decreasing function, use $c = 0$ to find the maximum error. For an alternating series, the error of the approximation is less than the absolute value of the first term not included in the Maclaurin series approximation.

The Fresnel integrals given by

$$\int_0^z \cos\left(\frac{\pi}{2}t^2\right) dt \quad \text{or} \quad \int_0^z \sin\left(\frac{\pi}{2}t^2\right) dt$$

are used to describe the lengthwise displacement of a beam subjected to a periodic force.

EXAMPLE 5 Use the first two terms in a Maclaurin series to approximate

$$\int_0^{0.3} \sin\left(\frac{\pi}{2}t^2\right) dt$$

and find the maximum error in this approximation.

Solution First, we find the Maclaurin series for $\sin\left(\frac{\pi}{2}t^2\right)$. From Table 22–1,

$$\sin x = x - \frac{x^3}{3!} + \cdots$$

$$\sin\left(\frac{\pi}{2}t^2\right) \approx \frac{\pi}{2}t^2 - \frac{\left(\frac{\pi}{2}t^2\right)^3}{6}$$
$$\approx \frac{\pi}{2}t^2 - \frac{\pi^3 t^6}{48}$$

Therefore,

$$\int_0^{0.3} \sin\left(\frac{\pi}{2}t^2\right) dt \approx \int_0^{0.3} \left(\frac{\pi}{2}t^2 - \frac{\pi^3 t^6}{48}\right) dt$$

$$\approx \left.\frac{\pi t^3}{6} - \frac{\pi^3 t^7}{336}\right|_0^{0.3}$$

$$\approx 0.014116985$$

Since the sine function is an alternating series, the maximum error is less than the next term in the sum. The third term in the series for $\sin\left(\frac{\pi}{2}t^2\right)$ is

$$\frac{\left(\frac{\pi}{2}t^2\right)^5}{5!} = \frac{\pi^5 t^{10}}{3{,}840}$$

Therefore, the next term in the integral is

$$\frac{\pi^5 t^{11}}{11 \cdot 3{,}840} = \frac{\pi^5 t^{11}}{42{,}240}$$

and the error is

$$\frac{\pi^5 (0.3)^{11}}{42{,}240} \approx 0.0000000128339$$

The approximation is accurate to six decimal places, and the value of the integral is 0.0141170. ■

22–3 EXERCISES

Calculate the value of each function by using the first three terms of a Maclaurin series.

1. $e^{0.03}$

2. $e^{-0.25}$

3. $\sin 0.3$

4. $\cos 0.75$

5. $\cos 4°$

6. $\cos 14°$

7. $\sqrt{e}$

8. $\sqrt[4]{e}$

9. $\ln(0.5)$

10. $\ln 1.2$

11. $\sqrt[3]{1.1}$

12. $\sqrt{0.5}$

13. $\frac{1}{e}$

14. $\frac{1}{1.4}$

15. $(1.5)^3$

16. $\ln 1.7$

17. $\sqrt[3]{1.176}$

18. $\sqrt[3]{1.2375}$

19. $\sqrt{1.43}$

20. $\sqrt{1.18}$

21. The displacement of a simple harmonic oscillator is given by $s = \sin 2\pi t$. Use the first three terms in the Maclaurin series to approximate the velocity of this object ($v = ds/dt$) when $t = 2.25$ s, and estimate the maximum error in this approximation.

22. The voltage in a circuit is given by $V = 0.7e^{0.7t}$. Use the Maclaurin series to approximate the voltage when $t = 0.55$ s.

23. The charge q (in microcoulombs) on a capacitor is given by

$$q = \int_0^{1.25} (1.0 - e^{0.3t^2})\,dt$$

Use a Maclaurin series to approximate this integral.

24. The accumulation of a charge on a certain capacitor from $t = 1$ s to $t = 3$ s is found by evaluating the integral

$$\int_1^3 e^{0.5t^2}\,dt$$

Find the accumulation of charge by using a Maclaurin series, and estimate the maximum error from this approximation.

25. The image distance as a function of object distance for a particular lens is given by

$$q = \frac{10p}{p - 10}$$

Use a Maclaurin series to approximate q when $p = 8$ cm.

22–4

TAYLOR SERIES

The Maclaurin series was defined so that it converges rapidly for values of x close to zero. However, for values of x within the convergence interval that are not close to zero, we must have a large number of terms of a Maclaurin expansion to obtain a good approximation to the function. A generalization of the Maclaurin series used to provide a good approximation to the function centered at $x = a$ is called the **Taylor series.**

We can formulate an expression for a Taylor series expansion by using the same technique used with the Maclaurin series expansion; namely, determine successive derivatives, but evaluate them at $x = a$ instead of zero. This method results in the $(x - a)$ factor in the binomial expansion.

$$f(x) = a_0 + a_1(x - a) + a_2(x - a)^2 + a_3(x - a)^3 + \cdots \qquad f(a) = a_0$$
$$f'(x) = a_1 + 2a_2(x - a) + 3a_3(x - a)^2 + \cdots \qquad f'(a) = a_1$$
$$f''(x) = 2a_2 + 2 \cdot 3a_3(x - a) + \cdots \qquad f''(a) = 2a_2$$

From the pattern above for the coefficients,

$$f^{(n)}(a) = n!a_n$$

If we substitute for the coefficients in the power series, the resulting series, called the Taylor series about a, is as given in the following box.

Taylor Series Expansion about a

$$f(x) = f(a) + f'(a)(x - a) + \frac{f''(a)(x - a)^2}{2!} + \cdots + \frac{f^{(n)}(a)(x - a)^n}{n!} + \cdots$$

EXAMPLE 1 Find the Taylor series expansion of e^x about $x = 2$.

Solution First, we find the successive derivatives of e^x and evaluate them at $x = 2$.

$$\begin{aligned} f(x) &= e^x & f(2) &= e^2 \\ f'(x) &= e^x & f'(2) &= e^2 \\ f''(x) &= e^x & f''(2) &= e^2 \\ f'''(x) &= e^x & f'''(2) &= e^2 \end{aligned}$$

Next, we substitute the coefficients into the Taylor series expansion.

$$\begin{aligned} e^x &= e^2 + e^2(x - 2) + \frac{e^2(x - 2)^2}{2!} + \frac{e^2(x - 2)^3}{3!} + \cdots \\ &= e^2\left[1 + (x - 2) + \frac{(x - 2)^2}{2} + \frac{(x - 2)^3}{6} + \cdots\right] \end{aligned}$$

For values of x near 2, this series converges more rapidly than the Maclaurin series for e^x. ■

EXAMPLE 2 Find the Taylor series expansion for $\sqrt[4]{x}$ about $x = 1$.

Solution First, we take successive derivatives of $x^{1/4}$ and evaluate them at $x = 1$.

$$\begin{aligned} f(x) &= x^{1/4} & f(1) &= 1 \\ f'(x) &= \frac{1}{4}x^{-3/4} & f'(1) &= \frac{1}{4} \\ f''(x) &= -\frac{3}{16}x^{-7/4} & f''(1) &= -\frac{3}{16} \\ f'''(x) &= \frac{21}{64}x^{-11/4} & f'''(1) &= \frac{21}{64} \end{aligned}$$

Next, we substitute these coefficients into the Taylor series expansion.

$$\begin{aligned} \sqrt[4]{x} &= 1 + \frac{(x - 1)}{4} - \frac{3(x - 1)^2}{16 \cdot 2!} + \frac{21(x - 1)^3}{64 \cdot 3!} - \cdots \\ &= 1 + \frac{(x - 1)}{4} - \frac{3(x - 1)^2}{32} + \frac{7(x - 1)^3}{128} - \cdots \end{aligned}$$

■

The Taylor series is very useful in evaluating functions for which tables or calculator keystrokes are not available.

EXAMPLE 3 Expand $\sin x$ in a Taylor series about $a = \pi/2 \approx 1.57$, and use the first three nonzero terms in this series to approximate $\sin 1.62$.

Solution First, we find the successive derivatives of $\sin x$ and evaluate them at $\pi/2$.

$$\begin{aligned} f(x) &= \sin x & f(\pi/2) &= 1 \\ f'(x) &= \cos x & f'(\pi/2) &= 0 \\ f''(x) &= -\sin x & f''(\pi/2) &= -1 \\ f'''(x) &= -\cos x & f'''(\pi/2) &= 0 \\ f^{(4)}(x) &= \sin x & f^{(4)}(\pi/2) &= 1 \\ f^{(5)}(x) &= \cos x & f^{(5)}(\pi/2) &= 0 \end{aligned}$$

Next, we substitute the coefficients into the Taylor series.

$$\begin{aligned} \sin x &= 1 + 0(x - \pi/2) + \frac{(-1)(x - \pi/2)^2}{2!} + \frac{0(x - \pi/2)^3}{3!} \\ &\quad + \frac{(1)(x - \pi/2)^4}{4!} + \frac{(0)(x - \pi/2)^5}{5!} + \cdots \\ &= 1 - \frac{(x - \pi/2)^2}{2!} + \frac{(x - \pi/2)^4}{4!} + \cdots \end{aligned}$$

Then, for $x = 1.62$, $(x - \pi/2) \approx 0.05$, and we can use the first three terms of the series to approximate $\sin 1.62$.

$$\begin{aligned} \sin 1.62 &\approx 1 - \frac{(0.05)^2}{2!} + \frac{(0.05)^4}{4!} \\ &\approx 0.9988 \end{aligned}$$

■

EXAMPLE 4 Use a Taylor series expansion to find the value of $\cos 31°$.

Solution First, we must choose a convenient value for a. Let us choose $a = 30° = \pi/6$ since the sine and cosine values can be calculated directly. Therefore, we will use

$$\sin\frac{\pi}{6} = \frac{1}{2} \quad \text{and} \quad \cos\frac{\pi}{6} = \frac{\sqrt{3}}{2}$$

Second, we take successive derivatives of $\cos x$ and evaluate them at $x = a = \pi/6$.

$$\begin{aligned} f(x) &= \cos x & f(a) &= \frac{\sqrt{3}}{2} \\ f'(x) &= -\sin x & f'(a) &= -\frac{1}{2} \\ f''(x) &= -\cos x & f''(a) &= -\frac{\sqrt{3}}{2} \\ f'''(x) &= \sin x & f'''(a) &= \frac{1}{2} \end{aligned}$$

Third, we find the Taylor series expansion about $a = \pi/6$.

$$\cos\frac{\pi}{6} = \frac{\sqrt{3}}{2} + \left(-\frac{1}{2}\right)\left(x - \frac{\pi}{6}\right) + \left(-\frac{\sqrt{3}}{2}\right)\left[\frac{\left(x - \frac{\pi}{6}\right)^2}{2!}\right]$$

$$+ \frac{1}{2}\left[\frac{\left(x - \frac{\pi}{6}\right)^3}{3!}\right] + \cdots$$

Last, we use the Taylor series to find cos 31°. For $x = 31°$,

$$x - a = (31° - 30°) \approx (0.5411 - 0.5236) \text{ radians} \approx 0.017$$

Therefore, substituting 0.017 for $x - a$ in the Taylor series gives the following value for cos 31°:

$$\cos 31° \approx \frac{\sqrt{3}}{2} - \frac{0.017}{2} - \frac{\sqrt{3}\cdot(0.017)^2}{4} + \frac{(0.017)^3}{12}$$

$$\approx 0.8572$$

■

22–4 EXERCISES

Find the first three nonzero terms in a Taylor series about the given value of a.

1. $\cos x$; $a = \dfrac{\pi}{4}$
2. $\sin x$; $a = 60°$
3. e^x; $a = 3$
4. e^{-2x}; $a = 1$
5. $\tan x$; $a = 30°$
6. $\ln x$; $a = 2$
7. $\dfrac{1}{x+1}$; $a = 2$
8. $\sqrt[3]{x}$; $a = 8$
9. $\dfrac{1}{x^2}$; $a = 1$
10. x^3; $a = 3$

Evaluate each function by using the first three nonzero terms of a Taylor series.

11. e^{π}; $a = 3$
12. $\ln \pi$; $a = 3$
13. $\sin 43°$; $a = 45°$
14. $\cos 93°$; $a = \dfrac{\pi}{2}$
15. $\ln(0.98)$; $a = 1$
16. $e^{0.97}$; $a = 1$
17. $\sqrt{8.7}$; $a = 9$
18. $\sqrt[3]{7.7}$; $a = 8$
19. The damped current in an electronic circuit has the form

$$i = \frac{\sin \pi t}{t}$$

Expand this function as a Taylor series about $t = 2$.

22–5

FOURIER SERIES

The Maclaurin and Taylor series provide good approximations to a function near some point $x = a$. As values of x vary from $x = a$, the series representation of the function gets worse. In this section we will discuss the Fourier series, which applies to periodic functions.

Periodic Functions

Recall that trigonometric functions are periodic. A function is said to be **periodic** if, for all real values of x, there is some number P such that $f(x + P) = f(x)$. The number P is called the **period** of the function. For example, the period of $\cos x$ is 2π. That is, $\cos x = \cos(x + 2\pi) = \cos(x + 4\pi)$. . . . In general, the periodic function $f(x)$ with period P can be written $f(x) = f(x + nP)$ where $n =$ any integer.

The sine and cosine functions are the simplest periodic functions. Therefore, we will use these functions to represent more complex periodic functions that occur frequently in ac voltages and oscillating mechanical systems.

Fourier Series

If $a_0, a_1, a_2, \ldots$ and $b_1, b_2, b_3, \ldots$ are constants and x is a variable, a trigonometric series is given by

$$a_0 + a_1\cos x + a_2\cos 2x + a_3\cos 3x + \cdots + a_n\cos nx + \cdots$$
$$+ b_1\sin x + b_2\sin 2x + b_3\sin 3x + \cdots + b_n\sin nx + \cdots$$

Since both the sine and cosine functions have a period of 2π, the series expansion will have a period of 2π. If a function $f(x)$ is defined over the interval $-\pi \leq x < \pi$, the **Fourier series** is a periodic approximation to this function and is defined as follows.

Fourier Series

$$f(x) = a_0 + a_1\cos x + a_2\cos 2x + \cdots + a_n\cos nx + \cdots$$
$$+ b_1\sin x + b_2\sin 2x + \cdots + b_n\sin nx + \cdots$$

where

$$a_0 = \frac{1}{2\pi}\int_{-\pi}^{\pi} f(x)\,dx$$

$$a_n = \frac{1}{\pi}\int_{-\pi}^{\pi} f(x)\cos nx\,dx \qquad n = 1, 2, 3, \ldots$$

$$b_n = \frac{1}{\pi}\int_{-\pi}^{\pi} f(x)\sin nx\,dx \qquad n = 1, 2, 3, \ldots$$

The derivation of the coefficients for the Fourier series is omitted for the sake of brevity. The Fourier series is named for Jean Fourier, who used the series in his work in heat conduction and published his results in 1812.

EXAMPLE 1 In digital circuits, the square-wave function is used as a timing signal. The function is defined as

$$f(x) = \begin{cases} 0 & \text{for } -\pi \le x < 0 \\ 1 & \text{for } 0 \le x < \pi \end{cases}$$

Find the Fourier series for this function.

FIGURE 22–3

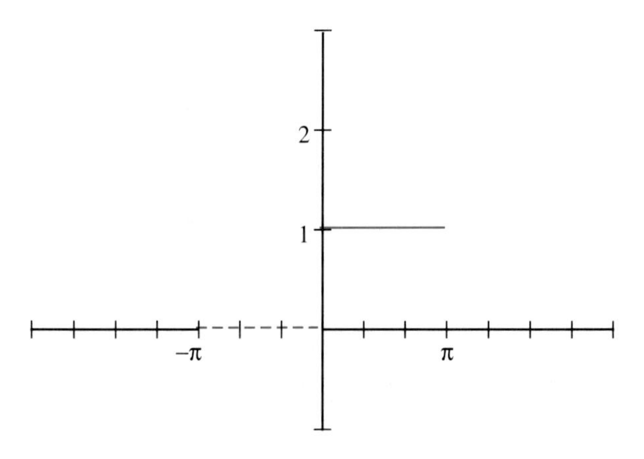

Solution Figure 22–3 shows a graph of this function for $-\pi \le x < \pi$. We must calculate the coefficients for the Fourier series. Let us begin by calculating a_0. Since the function is defined by two ranges for x, we use two integrals, one for each range.

$$a_0 = \frac{1}{2\pi}\int_{-\pi}^{0} f(x)\,dx + \frac{1}{2\pi}\int_{0}^{\pi} f(x)\,dx$$

$$= \frac{1}{2\pi}\int_{-\pi}^{0} 0\,dx + \frac{1}{2\pi}\int_{0}^{\pi} 1\,dx$$

$f(x) = 0$ for $-\pi \le x < 0$

$f(x) = 1$ for $0 \le x < \pi$

$$= \left[0\right]\Big|_{-\pi}^{0} + \left[\frac{1}{2\pi}(x)\right]\Big|_{0}^{\pi}$$

$$= \frac{1}{2} - 0$$

$$= \frac{1}{2}$$

Second, we use the same techniques to calculate the values of a_n.

$$\begin{aligned}
a_n &= \frac{1}{\pi}\int_{-\pi}^{0} f(x)\cos nx\,dx + \frac{1}{\pi}\int_{0}^{\pi} f(x)\cos nx\,dx \\
&= \frac{1}{\pi}\int_{-\pi}^{0} (0)\cos nx\,dx + \frac{1}{\pi}\int_{0}^{\pi} (1)\cos nx\,dx \\
&= \frac{1}{\pi}\int_{0}^{\pi} \cos nx\,dx \\
&= \left[\frac{1}{n\pi}\sin nx\right]\Bigg|_{0}^{\pi} \\
&= \frac{1}{n\pi}[\sin n\pi - \sin 0] \\
&= 0
\end{aligned}$$

Third, we calculate b_n.

$$\begin{aligned}
b_n &= \frac{1}{\pi}\int_{-\pi}^{0} f(x)\sin nx\,dx + \frac{1}{\pi}\int_{0}^{\pi} f(x)\sin nx\,dx \\
&= \frac{1}{\pi}\int_{-\pi}^{0} (0)\sin nx\,dx + \frac{1}{\pi}\int_{0}^{\pi} (1)\sin nx\,dx \\
&= \frac{1}{\pi}\int_{0}^{\pi} \sin nx\,dx \\
&= -\frac{1}{n\pi}\cos nx\Bigg|_{0}^{\pi} \\
&= -\frac{1}{n\pi}[\cos n\pi - \cos 0] \\
&= -\frac{1}{n\pi}(\cos n\pi - 1)
\end{aligned}$$

$$b_1 = -\frac{1}{\pi}(\cos \pi - 1) = \frac{2}{\pi} \qquad \text{for } n = 1$$

$$b_2 = -\frac{1}{2\pi}(\cos 2\pi - 1) = 0 \qquad \text{for } n = 2$$

$$b_3 = -\frac{1}{3\pi}(\cos 3\pi - 1) = \frac{2}{3\pi} \qquad \text{for } n = 3$$

From the calculated values of b_n, we see a pattern for the coefficients. If n is even, $b_n = 0$; and if n is odd, $b_n = 2/(n\pi)$. Last, we substitute the coefficients into the Fourier series.

$$f(x) = \frac{1}{2} + \frac{2}{\pi}\sin x + \frac{2}{3\pi}\sin 3x + \cdots$$

Figure 22–4 shows the graph of $f(x)$ using the first three terms. The dotted line represents the graph of the original square-wave function.

FIGURE 22–4

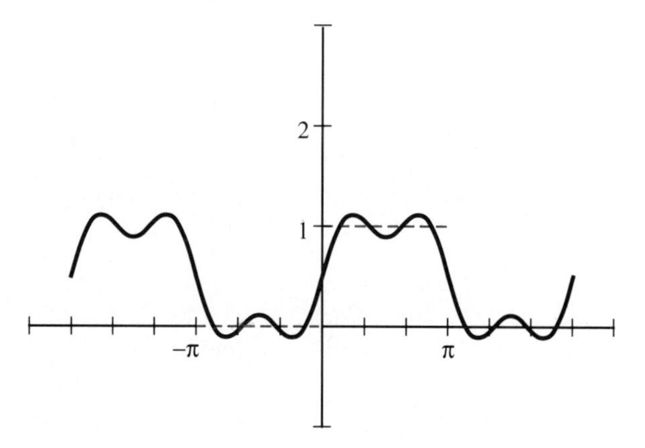

■

A half-wave rectifier circuit converts ac input to pulsating dc on every positive half-cycle. When an alternating current is applied, the current exists for only half the cycle, and all negative portions of the input signal are deleted. Figure 22–5(a) shows the initial sinusoidal voltage, and Figure 22–5(b) shows the resulting output voltage.

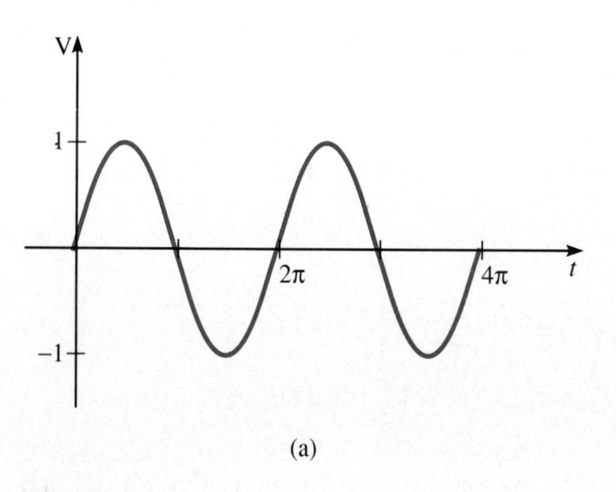

(a)

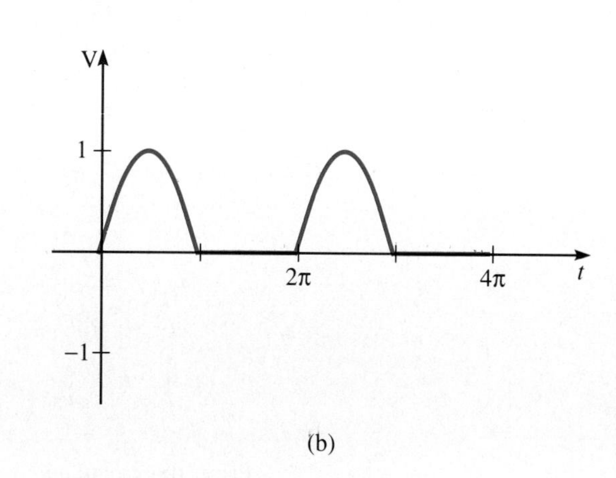

(b)

FIGURE 22–5

EXAMPLE 2 Find the Fourier series for the half-wave rectifier circuit defined by

$$V(t) = \begin{cases} \sin t & \text{for } 0 \le t < \pi \\ 0 & \text{for } \pi \le t < 2\pi \end{cases}$$

Solution First, we must calculate the coefficients.

$$\begin{aligned} a_0 &= \frac{1}{2\pi}\int_0^{\pi} f(t)\,dt + \frac{1}{2\pi}\int_{\pi}^{2\pi} f(t)\,dt \\ &= \frac{1}{2\pi}\int_0^{\pi} \sin t\,dt + \frac{1}{2\pi}\int_{\pi}^{2\pi} 0\,dt \\ &= \left[\frac{1}{2\pi}(-\cos t)\right]\Bigg|_0^{\pi} \\ &= \frac{1}{2\pi}(1 + 1) \\ &= \frac{1}{\pi} \end{aligned}$$

$$\begin{aligned} a_n &= \frac{1}{\pi}\int_0^{\pi} f(t)\cos nt\,dt + \frac{1}{\pi}\int_{\pi}^{2\pi} f(t)\cos nt\,dt \\ &= \frac{1}{\pi}\int_0^{\pi} \sin t\cos nt\,dt + \frac{1}{\pi}\int_{\pi}^{2\pi} 0\cos nt\,dt \\ &= \frac{1}{\pi}\int_0^{\pi} \sin t\cos nt\,dt \end{aligned}$$

We integrate this function by using Formula 88 from Appendix C, where $a = 1$ and $b = n$.

$$\begin{aligned} a_n &= \frac{1}{\pi}\left(-\frac{\cos[(1+n)t]}{2(1+n)} - \frac{\cos[(1-n)t]}{2(1-n)}\right)\Bigg|_0^{\pi} \\ &= \frac{1}{\pi}\left(-\frac{[\cos(1+n)\pi]}{2(1+n)} - \frac{\cos[(1-n)\pi]}{2(1-n)} + \frac{\cos[(1+n)(0)]}{2(1+n)} + \frac{\cos[(1-n)(0)]}{2(1-n)}\right) \\ &= \frac{1}{\pi}\left(-\frac{\cos[(1+n)\pi]}{2(1+n)} - \frac{\cos[(1-n)\pi]}{2(1-n)} + \frac{1}{2(1+n)} + \frac{1}{2(1-n)}\right) \end{aligned}$$

This formula is valid for all values of n except $n = 1$.

$$a_1 = \frac{1}{\pi}\int_0^{\pi} \sin t \cos t \, dt = \frac{1}{2\pi} \sin^2 t \Big|_0^{\pi} = 0$$

$$a_2 = \frac{1}{2\pi}\left(\frac{1}{3} - 1 + \frac{1}{3} - 1\right) = -\frac{2}{3\pi}$$

$$a_3 = \frac{1}{2\pi}\left(-\frac{1}{4} + \frac{1}{2} + \frac{1}{4} - \frac{1}{2}\right) = 0$$

$$a_4 = \frac{1}{2\pi}\left(\frac{1}{5} - \frac{1}{3} + \frac{1}{5} - \frac{1}{3}\right) = -\frac{2}{15\pi}$$

$$b_n = \frac{1}{\pi}\int_0^{\pi} f(t)\sin nt \, dt + \frac{1}{\pi}\int_{\pi}^{2\pi} f(t)\sin nt \, dt$$

$$= \frac{1}{\pi}\int_0^{\pi} \sin t \sin nt \, dt + \frac{1}{\pi}\int_{\pi}^{2\pi} 0 \sin nt \, dt$$

Using Formula 86 from Appendix C, where $a = 1$ and $b = n$, gives the following integral:

$$b_n = \frac{1}{\pi}\left(-\frac{\sin[(1 + n)t]}{2(1 + n)} + \frac{\sin[(1 - n)t]}{2(1 - n)}\right)\Bigg|_0^{\pi}$$

$$= \frac{1}{2\pi}\left(-\frac{\sin[(1 + n)\pi]}{1 + n} + \frac{\sin[(1 - n)\pi]}{1 - n}\right)$$

This formula is valid for all n except $n = 1$. Thus,

$$b_1 = \frac{1}{\pi}\int_0^{\pi} \sin t \sin t \, dt$$

$$= \frac{1}{\pi}\int_0^{\pi} \sin^2 t \, dt$$

Using Formula 69 from Appendix C, where $u = t$, gives the following result:

$$b_1 = \frac{1}{\pi}\left(\frac{1}{2}t - \frac{1}{4}\sin 2t\right)\Bigg|_0^{\pi}$$

$$= \frac{1}{\pi}\left[\frac{1}{2}t - \frac{1}{4}(2 \sin t \cos t)\right]\Bigg|_0^{\pi}$$

$$= \frac{1}{\pi}\left(\frac{1}{2}t - \frac{1}{2}\sin t \cos t\right)\Bigg|_0^{\pi}$$

$$= \frac{1}{2\pi}(t - \sin t \cos t)\Bigg|_0^{\pi}$$

$$= \frac{1}{2}$$

For $n > 1$, $b_n = 0$ because the sine of multiples of π equals zero. Since $V(t) = 0$ for the other half of the wave, the Fourier series for the half-wave rectifier is

$$f(t) = \frac{1}{\pi} - \frac{2}{3\pi}\cos 2t - \frac{2}{15\pi}\cos 4t + \cdots + \frac{1}{2}\sin t$$ ■

Functions with Other Periods

The Fourier series we have just discussed is the standard form, where the function has a period of 2π. However, many applied problems deal with functions of time that are periodic over a different interval. The Fourier series can also be used to give the mathematical equation of these time-dependent periodic functions. If the period of the function is $2P$, then substitute $\pi t/P$ for x in the equation for the Fourier series. Thus,

$$f(t) = a_0 + a_1\cos\frac{\pi t}{P} + a_2\cos\frac{2\pi t}{P} + \cdots + a_n\cos\frac{n\pi t}{P} + \cdots$$
$$+ b_1\sin\frac{\pi t}{P} + b_2\sin\frac{2\pi t}{P} + \cdots + b_n\sin\frac{n\pi t}{P} + \cdots$$

where

$$a_0 = \frac{1}{2P}\int_0^{2P} f(t)\,dt$$
$$a_n = \frac{1}{P}\int_0^{2P} f(t)\cos\frac{n\pi t}{P}\,dt \quad n = 1, 2, 3, \ldots$$
$$b_n = \frac{1}{P}\int_0^{2P} f(t)\sin\frac{n\pi t}{P}\,dt \quad n = 1, 2, 3, \ldots$$

The electron beam that forms the picture on your television screen is controlled basically by sawtooth voltages and currents. Figure 22–6 shows the graph of a sawtooth function.

FIGURE 22–6

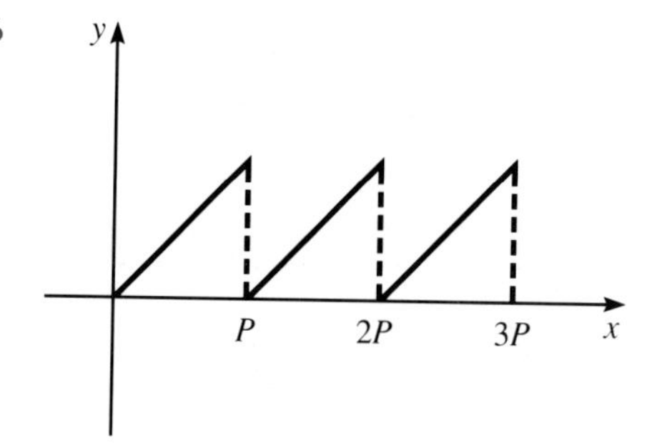

EXAMPLE 3 In a certain electrical circuit, the current takes the form of a sawtooth wave given by

$$i = f(t) = t \quad \text{for } 0 \le t < 2$$

Find the Fourier series of the current.

Solution Applying the formula for periods other than 2π where $2P = 2$ gives

$$\begin{aligned} a_0 &= \frac{1}{2P}\int_0^{2P} f(t)\,dt \\ &= \frac{1}{2}\int_0^2 t\,dt \\ &= \frac{1}{2}\cdot\frac{t^2}{2}\Bigg|_0^2 \\ &= 1 \\ a_n &= \frac{1}{P}\int_0^{2P} f(t)\cos\frac{n\pi t}{P}\,dt \\ &= \int_0^2 t\cos n\pi t\,dt \end{aligned}$$

Integrating by parts gives

$$\begin{aligned} a_n &= \frac{1}{n^2\pi^2}\left(\cos n\pi t + n\pi t\sin n\pi t\Bigg|_0^2\right) \\ &= \frac{1}{n^2\pi^2}[(\cos 2n\pi + 2n\pi\sin 2n\pi) - (\cos 0 + 0\sin 0)] \\ &= 0 \end{aligned}$$

$$\begin{aligned} b_n &= \frac{1}{P}\int_0^{2P} f(t)\sin\frac{n\pi t}{P}\,dt \\ &= \int_0^2 t\sin n\pi t\,dt \end{aligned}$$

Integrating by parts gives

$$\begin{aligned} b_n &= \frac{1}{n^2\pi^2}\left(\sin n\pi t - n\pi t\cos n\pi t\Bigg|_0^2\right) \\ &= \frac{1}{n^2\pi^2}[(\sin 2n\pi - 2n\pi\cos 2n\pi) - (\sin 0 - 0\cos 0)] \\ &= -\frac{2}{n\pi} \end{aligned}$$

$$b_1 = -\frac{2}{\pi} \quad \text{for } n = 1$$

$$b_2 = -\frac{1}{\pi} \quad \text{for } n = 2$$

$$b_3 = -\frac{2}{3\pi} \quad \text{for } n = 3$$

The Fourier series for $f(t)$ is

$$i = f(t) = 1 - \frac{2}{\pi}\sin \pi t - \frac{1}{\pi}\sin 2\pi t - \frac{2}{3\pi}\sin 3\pi t + \cdots$$

■

22–5 EXERCISES

Find the Fourier series representation for each function.

1. $f(x) = \begin{cases} 0 & \text{for } -2 \le x < 0 \\ 1 & \text{for } 0 \le x < 2 \end{cases}$

2. $f(x) = 2x \quad \text{for } -1 \le x < 1$

3. $f(t) = \begin{cases} 0 & \text{for } -\pi \le t < 0 \\ t & \text{for } 0 \le t < \pi \end{cases}$

4. $f(x) = \begin{cases} -1 & \text{for } -\pi \le x < 0 \\ 1 & \text{for } 0 \le x < \pi \end{cases}$

5. $f(x) = \begin{cases} 1 & \text{for } -\pi \le x < 0 \\ 2 & \text{for } 0 \le x < \pi \end{cases}$

6. $f(x) = \begin{cases} 1 & \text{for } -\pi \le x < 0 \\ 0 & \text{for } 0 \le x < \pi \end{cases}$

7. $f(t) = \begin{cases} -1 & \text{for } -\pi \le t < 0 \\ 3 & \text{for } 0 \le t < \pi \end{cases}$

8. $f(t) = t + \pi \quad \text{for } -\frac{\pi}{2} \le t < \frac{\pi}{2}$

9. $f(x) = \begin{cases} 1 & \text{for } -1 \le x < 1 \\ 2x & \text{for } 1 \le x < 2 \end{cases}$

10. $f(x) = \begin{cases} 0 & \text{for } -1 \le x < 0 \\ 1 & \text{for } 0 \le x < 1 \\ 0 & \text{for } 1 \le x < 2 \end{cases}$

11. $f(x) = \begin{cases} 0 & \text{for } -\pi \le x < 0 \\ x & \text{for } 0 \le x < \pi \\ 0 & \text{for } \pi \le x < 2\pi \end{cases}$

12. Find the Fourier expansion of an electronic circuit that provides a full-wave rectification of a sinusoidal voltage input. The function is given by

$$v(t) = \begin{cases} \sin t & \text{for } 0 \le t < \pi \\ -\sin t & \text{for } \pi \le t < 2\pi \end{cases}$$

13. Find the Fourier series of an exponential wave given by $y = e^t$, $0 \le t < \pi$, that is used in the frequency control circuit of a radar unit.

CHAPTER SUMMARY

Summary of Terms

Summary of Formulas

$$f(x) = f(0) + f'(0)x + \frac{f''(0)x^2}{2!} + \frac{f'''(0)x^3}{3!} + \cdots + \frac{f^{(n)}(0)x^n}{n!} + \cdots \quad \text{Maclaurin series expansion}$$

$$E = \left| \frac{f^{(n+1)}(c)x^{n+1}}{(n+1)!} \right| \quad \text{error in truncated Maclaurin series}$$

$$f(x) = f(a) + f'(a)(x-a) + \frac{f''(a)(x-a)^2}{2!} + \cdots + \frac{f^{(n)}(a)(x-a)^n}{n!} + \cdots \quad \text{Taylor series expansion about } a$$

Fourier series expansion:

$$f(x) = a_0 + a_1\cos x + a_2\cos 2x + \cdots + a_n\cos nx + \cdots + b_1\sin x + b_2\sin 2x + \cdots + b_n\sin nx + \cdots$$

where

$$a_0 = \frac{1}{2\pi}\int_{-\pi}^{\pi} f(x)\,dx$$

$$a_n = \frac{1}{\pi}\int_{-\pi}^{\pi} f(x)\cos nx\,dx \quad n = 1, 2, 3, \cdots$$

$$b_n = \frac{1}{\pi}\int_{-\pi}^{\pi} f(x)\sin nx\,dx \quad n = 1, 2, 3, \cdots$$

CHAPTER REVIEW

Section 22–1

Find the first three nonzero terms of the Maclaurin series expansion for the following functions.

1. $f(x) = \cos 2x$
2. $f(x) = e^{\pi x}$
3. $f(x) = \ln(1 + 2x)$
4. $f(t) = \sin 4t$
5. $f(t) = e^{-3t}$
6. $f(x) = x^2\sin x$
7. $f(x) = e^x\cos x$

Section 22–2

Find the first three nonzero terms of the Maclaurin series expansion of each function by using the series given in Table 22–1.

8. $f(x) = \sin x^2$
9. $f(x) = e^{-4x}$
10. $f(x) = \ln(2x + 8)$
11. $f(x) = \cos 2x$
12. $f(x) = e^{5x}\cos x$
13. $f(x) = \dfrac{e^x}{\sin x}$
14. $f(x) = \dfrac{\sin x}{\cos x}$

Use the first three terms of the Maclaurin series expansion in Table 22–1 to evaluate the following integrals.

15. $\int \ln \cos x\,dx$
16. $\int xe^x\,dx$
17. $\int_0^1 \tan x^2 dx$
18. $\int_0^2 e^{5x^2}\,dx$

Differentiate the given functions by using the first three terms of the Maclaurin series expansion in Table 22–1.

19. $f(x) = x^2\sin x$
20. $f(x) = e^x\cos x$
21. $f(x) = \ln e^x$
22. Find the average value of the following function from $x = 0.4$ to $x = 1.2$:

$$f(x) = \frac{\sin x}{x}$$

Use the first three terms of a Maclaurin series.

23. The electrical potential near a charged place is given by

$$V = V_0 + \frac{ae^r}{r}$$

Use a Maclaurin series to approximate the electrical field E. (*Note:* $E = dV/dr$.)

Section 22–3

Use the first three terms of a Maclaurin series to calculate the value of each function.

24. $e^{1.4}$

25. $\sin 0.18$

26. $\dfrac{1}{1.76}$

27. $\ln 1.86$

28. $\cos 2°$

29. $\sqrt[3]{1.67}$

Section 22–4

Find the first three nonzero terms in a Taylor series about the given value of a.

30. e^x; $a = 3$

31. e^{-x}; $a = 2$

32. $\ln x$; $a = 1$

33. $\cos x$; $a = \pi$

34. $\sin x$; $a = \pi/2$

35. $\sqrt[3]{x}$; $a = 1$

36. $\tan x$; $a = \pi/4$

37. x^2; $a = 4$

Evaluate each function by using the first three terms of an appropriate Taylor series.

38. $\sin 42°$

39. $\cos 62°$

40. $\tan 44°$

41. $\sin 31°$

42. $\sqrt[3]{8.1}$

43. $\sqrt{26}$

Section 22–5

Find the Fourier series of each function.

44. $f(t) = \begin{cases} \pi & \text{for } 0 \le t < \pi \\ 0 & \text{for } \pi \le t < 2\pi \end{cases}$

45. $f(t) = 2t$ for $0 \le t \le 2\pi$

46. $f(x) = \begin{cases} 2 & \text{for } 0 \le x < \pi \\ -1 & \text{for } \pi \le x < 2\pi \end{cases}$

CHAPTER TEST

The number in parentheses refers to the appropriate learning objective at the beginning of the chapter.

1. Find the Maclaurin series expansion for $f(x) = xe^{x^2}$ by using the Maclaurin series for $f_1(x) = x$ and $f_2(x) = e^{x^2}$. (2)
2. Find the first three terms in the Taylor series for $\sin 2x$ about $a = \pi/3$. (5)
3. Find the Fourier series for (6)

$$f(t) = \begin{cases} 0 & \text{for } -\pi \le t < 0 \\ t^2 & \text{for } 0 \le t < \pi \end{cases}$$

4. Find the first three terms in the Maclaurin series expansion of $f(x) = e^x \sin x$. (2)
5. Use the first three terms of the Maclaurin series expansion to find the derivative of $f(x) = x^2 \ln(1 + x)$ at $x = 1$. (3)
6. Use the first three terms of a Maclaurin series to calculate the value of $\sqrt[3]{1.89}$. (4)
7. Use the first three terms in an appropriate Taylor series to calculate $e^{1.1}$ where $e^1 = 2.718$. (5)
8. The current at time t in an RL circuit with voltage V is given by (7)

$$i = \frac{V}{R}(1 - e^{-Rt/L})$$

Represent this expression by a Maclaurin series.

9. Find the first two nonzero terms of the Maclaurin series expansion for $f(x) = \text{Arcsin } x$. (1)
10. When heat travels along a metal bar, the change in temperature between two (7)

points on the bar is given by

$$T = \int_a^b e^{-x^2} \cos 2x \, dx$$

Use a Maclaurin series to approximate the change in temperature when $a = 4$ and $b = 7$.

11. Find the Fourier series for **(6)**

$$f(t) = \begin{cases} 2 & \text{for } 0 \le t < \pi \\ 0 & \text{for } \pi \le t < 2\pi \end{cases}$$

SOLUTION TO CHAPTER INTRODUCTION

To calculate the Maclaurin series for the hyperbolic cosine, we must find the Maclaurin series for e^x and e^{-x}. From Table 22–1, we know that

$$e^x = 1 + x + \frac{x^2}{2!} + \frac{x^3}{3!} + \frac{x^4}{4!} + \cdots$$

To derive the Maclaurin series for e^{-x}, we substitute $-x$ for x in the series for e^x. Thus,

$$\begin{aligned} e^{-x} &= 1 + (-x) + \frac{(-x)^2}{2!} + \frac{(-x)^3}{3!} + \frac{(-x)^4}{4!} + \cdots \\ &= 1 - x + \frac{x^2}{2!} - \frac{x^3}{3!} + \frac{x^4}{4!} + \cdots \end{aligned}$$

Next, we will find a Maclaurin series for $(e^x + e^{-x})$ by adding the respective series.

$$e^x + e^{-x} = \left(1 + x + \frac{x^2}{2!} + \frac{x^3}{3!} + \frac{x^4}{4!} + \cdots\right) + \left(1 - x + \frac{x^2}{2!} - \frac{x^3}{3!} + \frac{x^4}{4!} - \cdots\right)$$

$$= 2 + x^2 + \frac{x^4}{12} + \cdots$$

Last,

$$\begin{aligned} \cosh x &= \frac{1}{2}(e^x + e^{-x}) \\ &= \frac{1}{2}\left(2 + x^2 + \frac{x^4}{12} + \cdots\right) \\ &= 1 + \frac{x^2}{2} + \frac{x^4}{24} + \cdots \end{aligned}$$

The computer would calculate the hyperbolic cosine by calculating the algebraic expression $1 + \frac{x^2}{2} + \frac{x^4}{24} + \cdots$.

You are in airport operations awaiting takeoff of an airplane at Dulles International Airport. You know that ice forms on the wings at a rate given by $dT/dt = 2T^2$ where T represents the thickness of the ice. If $T = 0.1$ cm when $t = 0$ min, how long can you allow the pilot to sit on the runway if he cannot take off after the ice becomes 1.2 cm thick? (The answer to this problem is given at the end of the chapter.)

In physics and engineering, the rate of change of one variable with respect to another variable is used to describe a dynamic relationship. Equations that involve a derivative are called *differential* equations. In this chapter we will discuss several techniques used to solve differential equations. In addition, we will discuss applications of differential equations to technology.

Learning Objectives

After completing this chapter, you should be able to

1. Identify the order and degree of a differential equation (Section 23–1).
2. Solve a differential equation by direct integration (Section 23–1).
3. Find the general or particular solution to a differential equation by using separation of variables (Section 23–1).
4. Use an integrable combination to solve a differential equation (Section 23–2).
5. Solve a linear, first-order differential equation by using an integrating factor (Section 23–2).
6. Apply first-order differential equations to technical problems (Section 23–3).
7. Solve higher-order differential equations by direct integration (Section 23–4).
8. Solve homogeneous differential equations with real, complex, or repeated roots (Section 23–4).
9. Solve nonhomogeneous differential equations with undetermined coefficients (Section 23–5).
10. Solve differential equations by using the Laplace transform (Section 23–6).

Chapter 23

Introduction to Differential Equations

23–1

SOLVING DIFFERENTIAL EQUATIONS

A **differential equation** is an equation that contains a derivative or a differential. For example, the following equations are both differential equations:

$$\frac{dy}{dx} = 3x^3 + 8x - 6 \quad \text{and} \quad x\,dy = 4xy + y^2\,dx$$

Order

The **order** of a differential is the highest derivative in the equation. If the equation contains only a first derivative, the equation is called a **first-order differential equation.** The equations $y' + x^3 - x = 4$ and $dy/dx = xy^2 + 3x^3$ are both first-order differential equations. If the highest derivative is two, the equation is called a **second-order differential equation.** For example, both of the following equations are second-order:

$$xy' + y^2y'' = 8 \quad \text{and} \quad \frac{d^2y}{dx^2} + 8x^2 - 3x = 0$$

Degree

The **degree** of a differential equation is the power of the highest-order derivative. For example,

$$x\frac{d^2y}{dx^2} + y^2\frac{dy}{dx} - 3y = 6$$

is a first-degree, second-order differential equation because the degree of the highest derivative is one, and

$$\left(\frac{d^3y}{dx^3}\right)^2 + 3\frac{dy}{dx} = 0$$

is a second-degree, third-order differential equation.

EXAMPLE 1 Determine the order and degree of the following differential equations:

(a) $xy''' + (3y')^2 = 6x - 8$

(b) $3x\left(\frac{d^2y}{dx^2}\right)^2 - y^2\frac{dy}{dx} = 0$

(c) $xy' + 8(y'')^3 = 3x - 5y$

Solution

(a) The highest derivative is three, and its degree is one. This is a third-order, first-degree differential equation.
(b) The order is two, and the degree is two.
(c) The order is two, and the degree is three. ■

The Solution

A **solution** to a differential equation is a relationship between the variables and differentials that satisfies the equation. Thus, when $f(x)$ is substituted for y in the differential equation, an identity results.

EXAMPLE 2 Show that each equation is a solution to the given differential equation:

(a) $y = 6x^3 + 2x + 8$ is a solution to $y' = 18x^2 + 2$.

(b) $y = e^x - 3$ is a solution to $\dfrac{dy}{dx} = 3 + y$.

(c) $y = \cos 2x$ is a solution to $\dfrac{d^2y}{dx^2} + 4y = 0$.

Solution

(a) If $y = 6x^3 + 2x + 8$, then $y' = 18x^2 + 2$. Substituting this quantity into the differential equation gives

$$\begin{aligned} y' &= 18x^2 + 2 \\ 18x^2 + 2 &= 18x^2 + 2 \end{aligned}$$

(b) If $y = e^x - 3$, then $dy/dx = e^x$. Substituting into the differential equation gives

$$\begin{aligned} \frac{dy}{dx} &= 3 + y \\ e^x &= 3 + (e^x - 3) \\ e^x &= e^x \end{aligned}$$

(c) If $y = \cos 2x$, then

$$\frac{dy}{dx} = -2 \sin 2x \quad \text{and} \quad \frac{d^2y}{dx^2} = -4 \cos 2x$$

Substituting into the differential equation gives

$$\begin{aligned} \frac{d^2y}{dx^2} + 4y &= 0 \\ -4 \cos 2x + 4(\cos 2x) &= 0 \\ 0 &= 0 \end{aligned}$$

■

General and Particular Solutions

Each of the solutions in Example 2 is only one of an infinite number of solutions of the given differential equation. For example, $y' = 18x^2 + 2$ also has solutions $y = 6x^3 + 2x - 10$ or $y = 6x^3 + 2x + 3$. In general, a differential equation has an infinite family of solutions. The **general solution** to an nth-order differential equation contains n arbitrary constants. When additional information is given to determine at least one of these constants, the solution is called a **particular solution.**

Direct Integration

Some differential equations can be solved by integration. If a first-order differential equation can be represented as $y' = f(x)$, where $f(x)$ = an integrable function, then we can find the solution by integrating both sides of the equation. The next two examples illustrate this procedure.

EXAMPLE 3 Solve the differential equation

$$y' = 8x^3 + 6x^2 + 4$$

Solution First, we write the equation in differential form and solve by integrating both sides of the equation.

$$\frac{dy}{dx} = 8x^3 + 6x^2 + 4$$

$$dy = (8x^3 + 6x^2 + 4)\,dx$$

$$\int dy = \int (8x^3 + 6x^2 + 4)\,dx$$

$$y = 2x^4 + 2x^3 + 4x + C$$

↑ first-order differential equation has one arbitrary constant

The general solution to the differential equation is $y = 2x^4 + 2x^3 + 4x + C$. We can check this solution by differentiating and substituting into the original differential equation.

$$y' = \frac{d(2x^4 + 2x^3 + 4x + C)}{dx} = 8x^3 + 6x^2 + 4$$

Substituting this into the original equation gives

$$y' = 8x^3 + 6x^2 + 4$$

$$8x^3 + 6x^2 + 4 = 8x^3 + 6x^2 + 4$$

■

EXAMPLE 4 A particular moves along a straight line under an applied force. The motion of the particle can be described by the equation $dv/dt = t \cos t$ where v = the velocity of the particle. If the initial velocity of the particle at $t = 0$ is $v = 3$ m/s, find the equation to represent the velocity of the particle.

Solution The differential equation that we solve by direct integration is

$$\frac{dv}{dt} = t \cos t$$

In differential notation, the equation becomes

$$dv = (t \cos t)\, dt$$
$$\int dv = \int (t \cos t)\, dt$$

We can evaluate the integral on the right side of the equation by using integration by parts where $u = t$ and $dw = \cos t\, dt$. Therefore,

$$du = dt \quad \text{and} \quad w = \int \cos t\, dt = \sin t$$

$$v = \underset{u}{t}\ \underset{w}{\sin t} - \int \underset{w}{\sin t}\ \underset{du}{dt}$$

$$v = t \sin t + \cos t + C$$

From the initial condition, $v = 3$ when $t = 0$, we can determine the particular solution by calculating the value of C.

$$3 = 0 \sin 0 + \cos 0 + C$$
$$2. = C$$

The velocity of the particle is given by the equation

$$v = t \sin t + \cos t + 2$$

■

Separation of Variables

A first-order, first-degree differential equation $dy/dx = f(x, y)$ is called **separable** if it can be written in the form

$$\frac{dy}{dx} = \frac{A(x)}{B(y)} \quad \text{or} \quad A(x)\, dx = B(y)\, dy$$

In this equation, $A(x)$ is a function solely of x, and $B(y)$ is a function solely of y. In other words, the variables have been separated from each other. We find the solution to the separated equation by integrating both sides.

EXAMPLE 5 Solve the differential equation

$$3x^2y\, dx + dy = 0$$

Solution First, we must arrange this equation in a form such that the dx term is a function of only x, and the dy term is a function of only y. We do so by dividing each term by y as follows:

$$\frac{3x^2y}{y}\, dx = -\frac{1}{y}\, dy$$
$$3x^2\, dx = -\frac{dy}{y}$$

Second, we integrate both sides of the equation.

$$\int 3x^2\,dx = -\int \frac{dy}{y}$$
$$x^3 + C_1 = -\ln|y| + C_2$$
$$x^3 + \ln|y| = C$$

The constant of integration $C = C_2 - C_1$ becomes the arbitrary constant of the solution. ■

EXAMPLE 6 Solve the following differential equation:

$$e^{4x}\cos y\,dx + e^x\sec y\,dy = 0$$

Solution First, we separate the variables by dividing each term by $e^x\cos y$.

$$\frac{e^{4x}\cos y}{e^x\cos y}\,dx = -\frac{e^x\sec y}{e^x\cos y}\,dy$$
$$e^{3x}\,dx = -\sec^2 y\,dy$$

Second, we integrate each side of the equation.

$$\int e^{3x}\,dx = \int -\sec^2 y\,dy$$
$$\frac{1}{3}e^{3x} + C_1 = -\tan y + C_2$$
$$\frac{1}{3}e^{3x} + \tan y = C \qquad (C = C_2 - C_1)$$

■

EXAMPLE 7 Solve the differential equation

$$4x^3y\,dx + y^3\,dy - y\,dx = 0$$

subject to the condition $x = 1$ when $y = 3$.

Solution First, we separate the variables.

$$4x^3y\,dx + y^3\,dy - y\,dx = 0$$
$$(4x^3y - y)\,dx + y^3\,dy = 0$$
$$y(4x^3 - 1)\,dx + y^3\,dy = 0$$
$$\frac{y(4x^3 - 1)\,dx}{y} + \frac{y^3\,dy}{y} = \frac{0}{y}$$
$$(4x^3 - 1)\,dx = -y^2\,dy$$

Second, we integrate both sides of the equation.

$$\int (4x^3 - 1)\,dx = -y^2\,dy$$

$$\int 4x^3\,dx - \int 1\,dx = \int -y^2\,dy$$

$$x^4 - x + C_1 = -\frac{1}{3}y^3 + C_2$$

$$x^4 - x + \frac{1}{3}y^3 = C$$

Last, we find the particular solution by substituting $x = 1$ and $y = 3$ in order to solve for C.

$$(1)^4 - (1) + \frac{1}{3}(3)^3 = C$$

$$1 - 1 + 9 = C$$

$$9 = C$$

The particular solution is

$$x^4 - x + \frac{1}{3}y^3 = 9$$

■

23–1 EXERCISES

Find the order and degree of the following differential equations.

1. $x\dfrac{dy}{dx} = 3xy + \dfrac{d^2y}{dx^2}$

2. $x\,dy + \dfrac{1}{y+1}\,dx = 3$

3. $dy - x\,dx + \cos x\,dx = 0$

4. $y' = 3xy + (y'')^3$

5. $\dfrac{d^2y}{dx^2} + y = 0$

6. $y''' + 3(y'')^2 = xy^3$

7. $1 - xy'' + y' = 2xy^3$

8. $(y'')^2 + y' - 8y = x$

9. $e^xy' + \sin x = xe^x$

Determine whether the first equation is a solution to the corresponding differential equation.

10. $y = x^2 + 4;\ dy - 2x\,dx = 0$

11. $y = 2\sin 4t + 3\cos 4t;\ \dfrac{d^2y}{dt^2} + 16y = 0$

12. $y = 4 + 3e^{-3x};\ \dfrac{d^2y}{dx^2} + 3\dfrac{dy}{dx} = 0$

13. $y = x + 7x^2;\ x\dfrac{dy}{dx} - 2y = -x$

14. $y = 3e^{4x} + 2e^{-3x};\ y'' - y' - 12y = 0$

15. $y = \dfrac{1}{x} + x;\ xy' + y = 2$

Solve the following differential equations.

16. $4xy + (x^2 + 1)y' = 0$

17. $(y + 2)\,dx + y(x + 4)\,dy = 0$

18. $y' = \sec^2 x$

19. $x\sin y\,dx + (x^2 + 1)\cos y\,dy = 0$

20. $\csc y\,dx + \sec x\,dy = 0$

21. $\dfrac{dy}{dx} = \sin x - x$

22. $\dfrac{dy}{dx} = \cos 3x$

23. $\dfrac{dy}{dx} = x \cos 2x$

24. $4\,dy = 6^x\,dx$

25. $dy = dx + e^x\,dx$

26. $(x^2 + 1)y\dfrac{dy}{dx} = 1$

27. $y' + 3y \sin x = 0$

28. $xy\dfrac{dy}{dx} = 2(y + 3)$

29. $(x^2 + 1)y' + y^2 + 1 = 0$

30. $(y^2 + 1)y' = 2 \cos x$

31. $y^2\dfrac{dy}{dx} - \cos^2 x = 0$

32. $(x^2 - x + 1)y' + 4(1 - 2xy) = 0$

33. $y' + y \tan x = 0$

34. $(x^2 + 1)yy' = 1$

35. Find the curve through the point $(3, -4)$ if the slope of the curve at any point is $2y/x$.

36. Find the curve through the point $(1, 0)$ if the slope of the curve at any point is $-x/y$.

37. In an electric circuit containing a constant resistance R, a constant inductance L, and voltage V, the current satisfies the equation

$$L\frac{di}{dt} + iR = V$$

Solve this equation for i if $i = 0$ when $t = 0$.

38. The mathematical formulation of Newton's law of cooling for a certain temperature is

$$\frac{dT}{dt} = k(T - 20)$$

where k = a negative constant. Solve this equation for T.

23–2

INTEGRABLE COMBINATIONS

Separation of variables may be difficult or impossible to use on many differential equations. Therefore, we must develop other methods of solving such equations. One method is based on combining terms of the equation as the differential of an expression and integrating directly. The following differentials suggest some of these combinations.

Integrable Combinations

$$d(xy) = x\,dy + y\,dx \tag{23–1}$$

$$d(x^2 + y^2) = 2(x\,dx + y\,dy) \tag{23–2}$$

$$d\left(\frac{y}{x}\right) = \frac{x\,dy - y\,dx}{x^2} \tag{23–3}$$

$$d\left(\frac{x}{y}\right) = \frac{y\,dx - x\,dy}{y^2} \tag{23–4}$$

These integrable combinations are based on the rules for differentiation. For example, Equation (23–3) results from the quotient rule as follows:

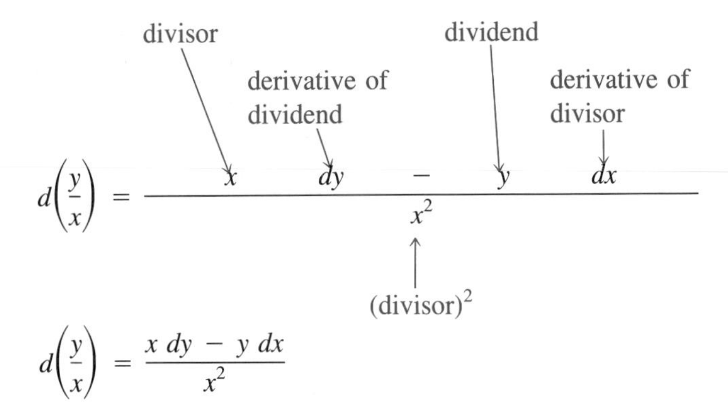

$$d\left(\frac{y}{x}\right) = \frac{x\,dy - y\,dx}{x^2}$$

EXAMPLE 1 Solve the following differential equation:

$$x\,dy - y\,dx - x^4\,dx = 0$$

Solution First, we try to group differential terms to yield one of the combinable forms.

$$x\,dy - y\,dx = x^4\,dx$$

The left side of the equation fits the form of the numerator for Equation (23–3). To complete this integrable form, we divide both sides of the equation by x^2.

$$\frac{x\,dy - y\,dx}{x^2} = \frac{x^4}{x^2}\,dx$$

Then we substitute the differential for the integrable form and integrate both sides of the resulting equation.

$$d\left(\frac{y}{x}\right) = x^2\,dx$$

$$\underbrace{\int d\left(\frac{y}{x}\right)} = \int x^2\,dx$$

↑ integral of a differential

$$\frac{y}{x} = \frac{x^3}{3} + C$$

$$y = \frac{x^4}{3} + Cx$$

■

EXAMPLE 2 Solve the differential equation

$$3x^2y^2 + 2x^3yy' = 4$$

subject to the condition that $y = 2$ when $x = 1$.

Solution First, we write this equation in differential form.

$$3x^2y^2 + 2x^3y\frac{dy}{dx} = 4$$

$$3x^2y^2\,dx + 2x^3y\,dy = 4\,dx$$

Second, we arrange the equation to fit one of the integrable forms. Notice that the expression on the left side of the equation is $d(uv)$ where $u = y^2$ and $v = x^3$, which fits Equation (23–1).

$$\underset{u}{\underset{\uparrow}{y^2}}\;\underset{dv}{\underset{\uparrow}{\underbrace{(3x^2\,dx)}}} + \underset{v}{\underset{\uparrow}{x^3}}\;\underset{du}{\underset{\uparrow}{\underbrace{(2y\,dy)}}}$$

Therefore, we write the equation as

$$d(uv) = 4\,dx$$

Next, we integrate both sides of the equation.

$$\int d(uv) = \int 4\,dx$$

$$uv = 4x + C$$

$$x^3y^2 = 4x + C \quad \text{substitute for } u \text{ and } v$$

Last, we solve for C by substituting $x = 1$ and $y = 2$.

$$(1)^3(2)^2 = 4(1) + C$$

$$0 = C$$

The particular solution is $x^3y^2 = 4x$. ■

NOTE ✦ As a general rule, try solving a first-order differential equation by separation of variables before attempting any other method.

EXAMPLE 3 Solve the differential equation

$$2y\sin x\,dy - y^2\cos x\,dx = y^3\sin^2 x\,dy$$

subject to the condition that $y = 1$ when $x = \pi/2$.

Solution First, we would attempt to solve this equation using separation of variables. However, this equation is not separable. We will try to arrange this equation to fit one of the integrable forms. Notice that the expression on the left side of the equation has the form $u\,dv - v\,du$ where $u = \sin x$ and $v = y^2$.

$$\underbrace{\sin x}_{u}\underbrace{(2y\,dy)}_{dv} - \underbrace{y^2}_{v}\underbrace{(\cos x\,dx)}_{du} = \underbrace{\sin^2 x}_{u^2}(y^3\,dy)$$

As a result of dividing by u^2, we can write this equation in the form of Equation (23–3).

$$\frac{u\,dv - v\,du}{u^2} = y^3\,dy$$

Substituting the integrable form and integrating give

$$d\left(\frac{v}{u}\right) = y^3\,dy$$

$$\int d\left(\frac{v}{u}\right) = \int y^3\,dy$$

$$\frac{v}{u} = \frac{y^4}{4} + C$$

$$\frac{y^2}{\sin x} = \frac{y^4}{4} + C \quad \text{substitute for } u \text{ and } v$$

Last, we find the particular solution.

$$\frac{(1)^2}{\sin \pi/2} = \frac{(1)^4}{4} + C$$

$$1 = \frac{1}{4} + C$$

$$\frac{3}{4} = C$$

The particular solution is

$$\frac{y^2}{\sin x} = \frac{y^4}{4} + \frac{3}{4}$$

■

Linear Differential Equations

Arranging a differential equation in integrable form can be quite difficult. In the remainder of this section, we will discuss a general procedure to solve a first-order, linear differential equation.

A first-order differential equation is said to be **linear** if it can be written in the form

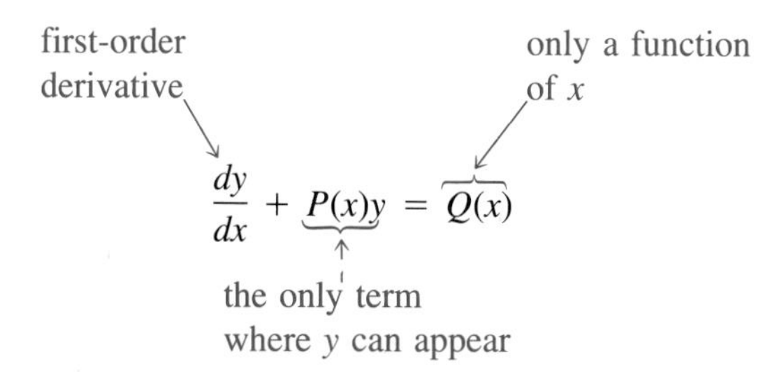

The differential form of this equation is $dy + P(x)y\,dx = Q(x)\,dx$, also called the **standard form** of a first-order linear differential equation. If each term of this equation is multiplied by $e^{\int P(x)\,dx}$, the equation becomes integrable.

$$e^{\int P(x)\,dx}\,dy + e^{\int P(x)\,dx}P(x)y\,dx = e^{\int P(x)\,dx}Q(x)\,dx$$

We simplify the left side of this equation by recognizing that if

$$u = ye^{\int P(x)\,dx}$$

then

$$du = e^{\int P(x)\,dx}\,dy + e^{\int P(x)\,dx}\,P(x)y\,dx$$

Therefore, we can write the left side of the equation as $d(ye^{\int P(x)\,dx})$. Thus,

$$d(ye^{\int P(x)\,dx}) = e^{\int P(x)\,dx}\,Q(x)\,dx$$

Integrating both sides of this equation gives the solution to this type of equation.

$$\int d(ye^{\int P(x)\,dx}) = \int e^{\int P(x)\,dx}Q(x)\,dx$$

$$ye^{\int P(x)\,dx} = \int Q(x)e^{\int P(x)\,dx}\,dx + C$$

First-Order Linear Differential Equations

If a first-order differential equation can be written as

$$\frac{dy}{dx} + P(x)y = Q(x)$$

its solution is

$$y = \frac{1}{e^{\int P(x)\,dx}}\int Q(x)e^{\int P(x)\,dx}\,dx$$

EXAMPLE 4 Solve the following differential equation:

$$\frac{dy}{dx} = y + \sin x$$

Solution First, we must arrange this equation in the required form and identify $P(x)$ and $Q(x)$.

$$\frac{dy}{dx} \underbrace{-\ 1}_{P(x)}y = \underbrace{\sin x}_{Q(x)}$$

Second, we must find $e^{\int P(x)\,dx}$, which is $e^{\int -1\,dx} = e^{-x}$. The solution is

$$\begin{aligned} y &= \frac{1}{e^{-x}} \int \sin x(e^{-x})\,dx \\ &= \frac{1}{e^{-x}} \int e^{-x} \sin x\,dx \\ &= e^x \int e^{-x} \sin x\,dx \end{aligned}$$

Using Formula 94 from Appendix C, where $a = -1$ and $b = 1$, gives the following result:

$$\begin{aligned} y &= e^x\left(\frac{e^{-x}(-\sin x - \cos x)}{2} + C\right) \\ &= \frac{-\sin x - \cos x}{2} + e^x C \end{aligned}$$

■

LEARNING HINT ✦ The key to solving this type of equation is correctly identifying $P(x)$ and $Q(x)$. Remember that the term containing $P(x)$ is the only term that contains y, while $Q(x)$ contains only x.

EXAMPLE 5 Solve the differential equation

$$(x^2 + 1)\,dy + 4xy\,dx = x\,dx$$

subject to the initial condition $y = 1$ when $x = 2$.

Solution First, we arrange the equation in the required form and identify $P(x)$ and $Q(x)$.

$$(x^2 + 1)\,dy + 4xy\,dx = x\,dx$$

$$(x^2 + 1)\frac{dy}{dx} + 4xy = x \qquad \text{divide by } dx$$

$$\frac{dy}{dx} + \frac{4xy}{(x^2 + 1)} = \frac{x}{(x^2 + 1)} \qquad \text{divide by } x^2 + 1$$

$$\frac{dy}{dx} + \underbrace{\frac{4x}{x^2 + 1}}_{\substack{\uparrow \\ P(x)}} y = \underbrace{\frac{x}{x^2 + 1}}_{\substack{\uparrow \\ Q(x)}}$$

Next, we determine the integrating factor $e^{\int P(x)\,dx}$.

$$e^{\left(\int \frac{4x}{x^2 + 1}\,dx\right)} = e^{\ln (x^2 + 1)^2} = (x^2 + 1)^2$$

The general solution is

$$\begin{aligned}
y &= \frac{1}{(x^2 + 1)^2} \int \left[\frac{x}{x^2 + 1} \cdot (x^2 + 1)^2\right] dx \\
&= \frac{1}{(x^2 + 1)^2} \int x(x^2 + 1)\,dx \\
&= \frac{1}{(x^2 + 1)^2} \int (x^3 + x)\,dx \\
&= \frac{1}{(x^2 + 1)^2} \cdot \left(\frac{x^4}{4} + \frac{x^2}{2} + C\right)
\end{aligned}$$

The particular solution is

$$\begin{aligned}
1 &= \frac{1}{(4 + 1)^2}\left(\frac{16}{4} + \frac{4}{2} + C\right) \\
1 &= \frac{1}{25}(6 + C) \\
25 &= 6 + C \\
19 &= C \\
y &= \frac{1}{(x^2 + 1)^2}\left(\frac{x^4}{4} + \frac{x^2}{2} + 19\right)
\end{aligned}$$

■

NOTE ✦ The coefficient of dy/dx must be 1.

EXAMPLE 6 Find the particular solution to the differential equation

$$y' - 3y = e^x \quad \text{if } x = 0 \text{ when } y = 2$$

Solution Writing the equation in the required form and identifying $P(x)$ and $Q(x)$ give

$$\frac{dy}{dx} - 3y = e^x$$

$$\uparrow \quad \uparrow$$

$$P(x) \quad Q(x)$$

Then we find $e^{\int P(x)\,dx}$

$$e^{\int -3\,dx} = e^{-3x}$$

The solution is

$$\begin{aligned} y &= \frac{1}{e^{-3x}} \int e^{-3x} e^x\,dx \\ &= e^{3x} \int e^{-2x}\,dx \\ &= e^{3x}\left[-\frac{1}{2}\int e^{-2x}(-2\,dx)\right] \\ &= e^{3x}\left(-\frac{1}{2}e^{-2x} + C\right) \\ &= -\frac{1}{2}e^x + e^{3x}C \end{aligned}$$

The particular solution is

$$2 = -\frac{1}{2} + C$$

$$\frac{5}{2} = C$$

$$y = -\frac{1}{2}e^x + \frac{5}{2}e^{3x}$$

■

23–2 EXERCISES

Solve the following differential equations by using an integrable combination.

1. $x\,dx + y\,dy - dx = 0$

2. $x\,dx + y\,dy - x\,dx = 0$

3. $x\,dx - y\,dy = 5\,dy$

4. $y\,dx + 2y^3\,dy = dy$

5. $xy^2(x\,dy + y\,dx) = dy$

6. $x\,dx - y\,dy - (y^2 - x^2)\,dx = 0$

7. $x\,dy - y\,dx = y^2 \cos x\,dx$

8. $y\,dx - x\,dy = xy(2xy + y)\,dx$

9. $x^2\,dy + (xy + 2)\,dx = 0$

Use an integrating factor to solve the following.

10. $dy - y\,dx = e^x\,dx$

11. $\dfrac{dy}{dx} + \dfrac{3y}{x} = 6x^2$

12. $\dfrac{dy}{dx} + 3x^2y = x^2$

13. $2x(y + 1)\,dx - (x^2 + 1)\,dy = 0$

14. $\dfrac{ds}{dt} - s = \sin 2t$

15. $\dfrac{dy}{dx} = x - 2xy$

16. $\dfrac{dy}{dx} - x^2y = x^2$

17. $\dfrac{dy}{dx} - \dfrac{2y}{x} = 1$

18. $dy + y \tan x \, dx = \cos x \, dx$

19. $y' + y - e^{-x} = 0$

20. $x \, dy = 4x \, dx - y \, dx$

21. $dy = (x^2 e^{-3x} - 3y) \, dx$

22. $x \dfrac{dy}{dx} = y - 2x^3$

23. $4y - y' = xe^{2x}$

24. $x \, dy - x \, dx + y \, dx = 0$

Find the particular solution for each differential equation.

25. $\dfrac{dy}{dx} - \dfrac{4}{x}y - x^2 - 1 = 0$ when $x = 1, y = 1$

26. $xy' - y = x$; $y = 2$ when $x = 1$

27. $xy' + y = \sec^2 x$; $y = 0$ when $x = \pi/4$

28. $y' + 20y = \sin (60 \, x)$; $x = 0$ when $y = 0$

29. $y' - y = e^x$; $y = 0$ when $x = -1$

30. $x \dfrac{dy}{dx} + (1 + x)y = e^{-x}$; $y = 0$ when $x = 1$

22–3

APPLICATIONS OF FIRST-ORDER DIFFERENTIAL EQUATIONS

The first-order, first-degree differential equations discussed in the previous sections of this chapter have numerous applications in geometry and technology. In this section we will discuss several of these applications.

EXAMPLE 1 Find the equation of the curve in the Cartesian plane that passes through the point (2, 1) and whose tangent line at any point has the slope $2x^3/y$.

Solution First, we write a differential equation describing the curve.

$$\text{slope} = \frac{dy}{dx} = \frac{2x^3}{y}$$

Second, we solve this differential equation by separation of variables.

$$dy = \frac{2x^3}{y} \, dx$$

$$y \, dy = 2x^3 \, dx$$

$$\int y \, dy = \int 2x^3 \, dx$$

$$\frac{1}{2}y^2 = \frac{2x^4}{4} + C$$

$$y^2 = x^4 + 2C$$

Next, we solve for C by using the boundary condition that $y = 1$ when $x = 2$.

$$(1)^2 = (2)^4 + 2C$$

$$-\frac{15}{2} = C$$

The equation is given by

$$y^2 = x^4 - 15$$

We can check the solution by substituting (2, 1) into the equation.

$$y^2 = x^4 + 2C$$
$$(1)^2 = (2)^4 + 2(-7.5)$$
$$1 = 1$$

We must also determine the slope of this equation at any point. Differentiating gives

$$2y\frac{dy}{dx} = 4x^3$$
$$\frac{dy}{dx} = \frac{2x^3}{y}$$

■

Orthogonal Trajectory

A curve that intersects all members of a family of curves at right angles is called an **orthogonal trajectory** of the family. A family of curves is a specified set of curves that satisfy the equation $F(x, y, k) = 0$ for each real value of the parameter k. Systems of orthogonal curves are important in the study of gravitational, electrostatic, and magnetic fields; heat conduction; and fluid flow.

EXAMPLE 2 Find the equations of the orthogonal trajectories of the hyperbolas $xy = k$.

Solution First, we find an expression for the slope of the family of hyperbolas.

$$y = \frac{k}{x} \qquad \text{solve for } y$$

$$y' = -\frac{k}{x^2} \qquad \text{differentiate}$$

This equation contains the constant k, which depends on the point (x, y). Therefore, we eliminate k from this equation by substituting $k = xy$.

$$y' = -\frac{xy}{x^2}$$
$$= -\frac{y}{x}$$

We now have an expression for the slope of the family of hyperbolas. The slope of a curve perpendicular to these hyperbolas must be the negative reciprocal of this expression. Therefore, the slope of the orthogonal trajectories is

$$y' = \frac{x}{y}$$

Solving this differential equation using separation of variables gives the following result:

$$\frac{dy}{dx} = \frac{x}{y}$$

$$dy = \frac{x}{y}\,dx$$

$$y\,dy = x\,dx$$

$$\frac{y^2}{2} = \frac{x^2}{2} + C_1$$

$$y^2 = x^2 + 2C_1$$

$$y^2 = x^2 + C \qquad \text{where } 2C_1 = C$$

The equation of the orthogonal trajectories is the family of hyperbolas given by the equation $y^2 - x^2 = C$. This relationship is shown in Figure 23–1. ■

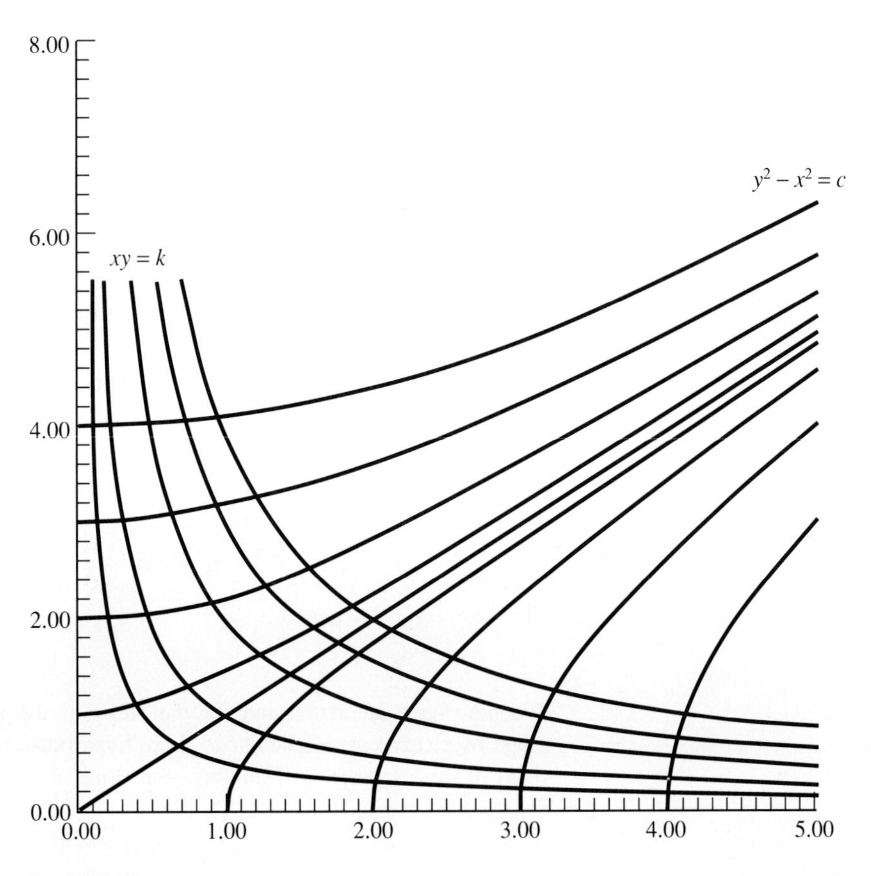

FIGURE 23–1

Exponential Growth and Decay

A quantity undergoes **exponential growth** or **exponential decay** if the rate of increase or decrease at any time is directly proportional to the amount present. This relationship is represented by the following equation:

$$\underbrace{\frac{dN}{dt}}_{\substack{\uparrow \\ \text{rate of change of} \\ N \text{ at time } t}} = k\overbrace{N}^{\substack{\text{amount at} \\ \text{time } t \\ \downarrow}}$$

EXAMPLE 3 A particular isotope of uranium decays to half its original amount in 25 days. Find the equation relating the amount present as a function of time, and find the percentage remaining after 45 days.

Solution The equation relating the rate of decay as a function of time is

$$\frac{dN}{dt} = kN$$

We solve this differential equation by separation of variables.

$$dN = kN\,dt$$

$$\frac{dN}{N} = k\,dt$$

$$\int \frac{dN}{N} = k \int dt$$

$$\ln N = kt + \ln C$$

Using the initial conditions that $N = N_0$ when $t = 0$, we can evaluate the constant.

$$\ln N_0 = k(0) + \ln C$$

$$N_0 = C$$

The solution to the equation is

$$\ln N = kt + \ln N_0$$

$$\ln N - \ln N_0 = kt$$

$$\ln\left(\frac{N}{N_0}\right) = kt$$

$$N = N_0 e^{kt}$$

This equation should look familiar since we used it in Chapter 8 to represent exponential growth and decay. Since half the isotope decays in 25 days, then $N = N_0/2$ when $t = 25$ days. Thus,

$$\frac{N_0}{2} = N_0 e^{25k}$$

$$\frac{1}{2} = (e^k)^{25}$$

$$\left(\frac{1}{2}\right)^{1/25} = e^k$$

Substituting into the equation relating N and t gives

$$N = N_0\left(\frac{1}{2}\right)^{t/25}$$

To find the percentage remaining after 45 days, we evaluate the equation when $t = 45$.

$$N = N_0\left(\frac{1}{2}\right)^{45/25} \approx 0.287N_0$$

Roughly 29% of the original isotope remains after 45 days. ■

RL Circuits

The general equation relating current i, voltage V, inductance L, capacitance C, and resistance R of an electrical circuit is

$$L\frac{di}{dt} + iR + \frac{q}{C} = V$$

where q = the charge on the capacitor. If a series electrical circuit contains an inductance, a resistance, and a voltage source, the differential equation becomes

$$L\frac{di}{dt} + iR = V$$

EXAMPLE 4 Find the value of the current after 2 s if $R = 10\ \Omega$, $L = 75$ H, and $V = 220$ V if $i = 0$ when $t = 0$.

Solution We must solve the differential equation

$$L\frac{di}{dt} + iR = V$$

for i by using separation of variables.

$$\frac{di}{dt} + \frac{iR}{L} = \frac{V}{L}$$

$$\frac{di}{dt} = \frac{V}{L} - \frac{iR}{L}$$

$$= \frac{R}{L}\left(\frac{V}{R} - i\right)$$

$$di = \frac{R}{L}\left(\frac{V}{R} - i\right) dt$$

$$\frac{di}{\left(\frac{V}{R} - i\right)} = \frac{R}{L}\, dt$$

$$\int \frac{di}{\left(\frac{V}{R} - i\right)} = \int \frac{R}{L}\, dt$$

$$-\ln\left(\frac{V}{R} - i\right) = \frac{Rt}{L} + C_1 \quad \text{if } R,\ V, \text{ and } L \text{ are constants}$$

Then we solve this equation for i.

$$\ln\left(\frac{V}{R} - i\right) = -\frac{Rt}{L} + \ln C \quad \text{where } C_1 = -\ln C$$

$$\ln\left(\frac{\frac{V}{R} - i}{C}\right) = -\frac{Rt}{L}$$

$$e^{-Rt/L} = \left(\frac{\frac{V}{R} - i}{C}\right)$$

$$Ce^{-Rt/L} = \frac{V}{R} - i$$

Substituting the initial conditions $i = 0$ when $t = 0$ gives

$$Ce^{0} = \frac{V}{R} - 0$$

$$C = \frac{V}{R}$$

and the equation becomes

$$\frac{V}{R}e^{-Rt/L} = \frac{V}{R} - i$$

$$i = \frac{V}{R}(1 - e^{-Rt/L})$$

Substituting the known values gives

$$i = \frac{220}{10}(1 - e^{-[10(2)]/75})$$

$$i \approx 5 \text{ A}$$

■

Related Rates

EXAMPLE 5 A tank contains a brine solution made by dissolving 30 lb of salt in 90 gallons of water. A more concentrated solution of 2 lb of salt per gallon of water is pumped into the tank at a rate of 2 gal/min. The solution is kept uniform by stirring and is pumped out at a rate of 2 gal/min.

(a) Find an equation relating the amount of salt in the tank as a function of time.

(b) Find the amount of salt in the tank when $t = 20$ min.

Solution

(a) Let x represent the amount of salt in the tank at time t. Then dx/dt represents the change in the amount of salt, and

$$\frac{dx}{dt} = \text{rate salt is added} \quad - \text{rate salt is removed}$$

$$= (2 \text{ lb/gal})(2 \text{ gal/min}) - \left(\frac{x}{90} \text{ lb/gal}\right)(2 \text{ gal/min})$$

$$= 4 - \frac{x}{45}$$

$$= \frac{180 - x}{45}$$

We solve this equation by using separation of variables.

$$dx = \frac{180 - x}{45}\,dt$$

$$\frac{dx}{180 - x} = \frac{dt}{45}$$

$$-\ln(180 - x) \approx 0.022t + C_1$$

$$\ln(180 - x) = -0.022t + \ln C \quad \text{where } \ln C = -C_1$$

$$\ln\left(\frac{180 - x}{C}\right) = -0.022t$$

$$\frac{180 - x}{C} = e^{-0.022t}$$

$$x = 180 - Ce^{-0.022t}$$

Using the initial condition that $x = 30$ lb at $t = 0$ gives the constant of integration as

$$30 = 180 - Ce^{\circ}$$

$$C = 150$$

$$x = 180 - 150e^{-0.022t}$$

(b) When $t = 20$ min,

$$x = 180 - 150e^{-0.022(20)}$$

$$x \approx 83 \text{ lb}$$

■

Motion with a Resisting Force

An object moving through a resisting medium experiences a retarding force directly proportional to the velocity. This relationship can be expressed as follows:

$$m\frac{dv}{dt} = F - kv$$

(where m = mass, F = applied force, kv = retarding force)

EXAMPLE 6 An object whose mass is 6 kg is falling under the influence of earth's gravity. Find its velocity after 3 s if it starts from rest and experiences a retarding force equal to 0.2 of its velocity.

Solution The differential equation that we must solve for v is

$$m\frac{dv}{dt} = mg - kv$$

force = weight

We can solve this equation using separation of variables.

$$\frac{dv}{dt} = g - \frac{kv}{m}$$

$$\frac{dv}{dt} = \frac{k}{m}\left(\frac{mg}{k} - v\right)$$

$$\frac{dV}{\frac{mg}{k} - v} = \frac{k}{m}\,dt$$

$$\int \frac{dv}{\frac{mg}{k} - v} = \int \frac{k}{m}\,dt$$

$$-\ln\left(\frac{mg}{k} - v\right) = \frac{k}{m}t + C_1$$

$$\ln\left(\frac{mg}{k} - v\right) = -\frac{kt}{m} + \ln C \text{ (where } \ln C = -C_1\text{)}$$

$$\ln\left(\frac{\frac{mg}{k} - v}{C}\right) = -\frac{kt}{m}$$

$$e^{-kt/m} = \left(\frac{\frac{mg}{k} - v}{C}\right)$$

$$Ce^{-kt/m} = \frac{mg}{k} - v$$

$$\frac{mg}{k} - Ce^{-kt/m} = v$$

Since $v = 0$ when $t = 0$, then we calculate C to be

$$\frac{mg}{k} - Ce^0 = 0$$

$$C = \frac{mg}{k}$$

The equation becomes

$$\frac{mg}{k} - \frac{mg}{k}e^{-kt/m} = v$$

$$\frac{mg}{k}\left(1 - e^{-kt/m}\right) = v$$

Then we substitute the known values: $m = 6$ kg, $g = 9.8$ m/s^2, $k = 0.2$ kg/s, and $t = 3$ s.

$$v = \frac{6(9.8)}{0.2}\left(1 - e^{[-0.2(3)]/6}\right)$$
$$\approx 28 \text{ m/s}$$

■

23–3 EXERCISES

1. Find the equation for the current for a series *RL* circuit given that the initial current is zero, the resistance is 18 Ω, the inductance is 0.25 H, and the voltage is 9 V. (Refer to Example 4.)

2. Find the equation for the current in an *RL* circuit given that the initial current is zero, $R = 5\ \Omega$, $L = 0.75$ H, and V is given by e^t. (Refer to Example 4 but use an integrating factor.)

3. Find an expression for the velocity of an object starting from rest if its mass is 12 slugs, the applied force is 175 lb, and the retarding force k is 1. (See Example 6.)

4. Find the equation of the curve that passes through (1, 4) if its slope at any point is given by $y' = y/2x$.

5. Find the equation of the curve that passes through (0, 2) if the slope of the curve at any point is given by $y' + y = 1 + x$.

6. Find the orthogonal trajectories of the family $y = ce^{-x}$.

7. Find the orthogonal trajectories of the family $cx^2 + y^2 = 1$.

8. Find the orthogonal trajectories of the family of parabolas $y = cx^2$.

9. Find the orthogonal trajectories of the family $y = cx^3$.

10. A tank initially contains 60 gal of pure water. Starting at $t = 0$, a brine containing 2 lb of salt/gal flows into the tank at the rate of 3 gal/min. The mixture is kept uniform by stirring. If the brine flows out of the tank at a rate of 3 gal/min, how much salt is present after 30 min?

11. A tank initially contains 100 gal of pure water. Starting at $t = 0$, a brine containing 4 lb of salt/gal flows into the tank at the rate of 5 gal/min. The mixture is kept uniform by stirring. If the brine flows out of the tank at a rate of 5 gal/min, how much salt is present after 45 min?

12. A large tank initially contains 100 gal of brine in which 15 lb of salt is dissolved. Starting at $t = 0$, pure water flows into the tank at a rate of 2 gal/min. The mixture is kept uniform by stirring. If brine flows from the tank at a rate of 2 gal/min, how much salt is in the tank after 20 min?

13. A tank initially contains 200 gal of brine in which 30 lb of salt is dissolved. Starting at $t = 0$, brine containing 4 lb of salt/gal flows into the tank at a rate of 5 gal/min. The mixture is kept uniform by stirring. If brine flows from the tank at a rate of 5 gal/min, how much salt is in the tank after 45 min?

14. A chemical reaction converts one chemical into a second chemical. The rate at which the first chemical is converted is proportional to the amount of this chemical present at any time. At the end of one hour, 50 grams of the first chemical remain, while at the end of 3 hours, only 25 grams remain. How many grams of the first chemical were present initially? How many grams of the first chemical will remain at the end of 5 hours?

15. A chemical reaction converts one chemical into a second chemical at a rate proportional to the amount of the first chemical present at any time. If 20% of the original amount of the first chemical has been converted in 30 min, what percent of the first chemical will have been converted in 45 min?

16. Newton's law of cooling is given by the differential equation

$$\frac{dT}{dt} = -k(T - T_0)$$

where $k > 0$ and T_0 = the temperature of the surrounding medium. An object is heated to 80°F and then placed in a room maintained at 68°F. At the end of 5 min, the object has cooled to 70°F. What is the temperature of the object after 15 min?

17. A body cools from 50°C to 40°C in 15 min in air which is maintained at 30°C. What is the temperature of the object after 40 min? (See Exercise 16.)

18. A culture of bacteria grows at a rate proportional to the number of bacteria present. If 40,000 bacteria are present at the start of an experiment, and if the number of bacteria doubles in 3 hours, how many bacteria will be present after 5 hours?

19. If 10,000 bacteria are present at the end of 2 hours and 40,000 at the end of 5 hours, how many were there initially? (See Exercise 18.)

20. A certain radioactive substance decays at a rate equal to 30% of the mass of the substance. Find a formula for the mass of the substance as a function of time if $m = 150$ g when $t = 0$.

21. What is the half-life of a radioactive substance if 25% disappears in 50 years?

22. A radioactive substance decays at a rate such that half of the original amount disappears in 750 years. What part of the original amount disappears in 200 years?

23. After 3 hours, 2% of a radioactive material has decayed. Find the half-life of the substance.

24. If interest on a bank account is compounded continuously, the amount grows at a rate proportional to the amount present. Determine the amount in an account after two years if $1,500 is placed in a 6% account, compounded continuously.

25. A radioactive element with a half-life of 30 min is used to find leaks in a hydraulic system. How much time is required for the element to decay to 10% of its initial amount?

26. A mass of 3 slugs is falling under the influence of gravity. Find its velocity after 3 s if it starts from rest and experiences a retarding force equal to 0.2 times its velocity.

27. The acceleration of a solenoid-activated plunger is proportional to its velocity. If the initial velocity of the plunger is 24 in./s, and if 0.2 s later it is 300 in./s, what is the velocity when $t = 0.1$ s?

28. When a boat moves through the water, the retarding force is proportional to the square of its velocity ($m\, dv/dt = -kv^2$). If a boat whose mass is 32 slugs is traveling 45 ft/s and its power is cut, it slows to 8 ft/s in 12 s. What is the value of the drag coefficient k?

29. A parachutist is falling at a rate of 150 ft/s when the parachute opens. If air resistance is $0.6v^2$, find an expression for the velocity as a function of time if the mass of the parachutist and equipment is 6.25 slugs.

30. An RL circuit with $R = 20\ \Omega$ and $L = 0.5$ H has an applied voltage of 60 V. Find an equation relating current as a function of time if $i = 0$ when $t = 0$.

31. In an RL circuit with $R = 125\ \Omega$, $L = 25$ H, and $V(t) = 2 \cos t$, find an expression for current if $i = 0$ when $t = 0$.

23–4

HIGHER-ORDER DIFFERENTIAL EQUATIONS

So far in this chapter we have discussed first-order differential equations. In this section we will discuss second-order differential equations. A second-order, linear differential equation is of the form

$$P_0(x)\frac{d^2y}{dx^2} + P_1(x)\frac{dy}{dx} + P_2(x)y = Q(x)$$

In certain cases, we can solve a second-order, linear differential equation by using direct integration.

Direct Integration

A second-order, linear differential equation of the form

$$P_0(x)\frac{d^2y}{dx^2} = Q(x)$$

can be solved as follows:

$$\frac{d^2y}{dx^2} = \frac{Q(x)}{P_0(x)}$$

$$\frac{d}{dx}\left(\frac{dy}{dx}\right) = \frac{Q(x)}{P_0(x)}$$

$$d\left(\frac{dy}{dx}\right) = \frac{Q(x)}{P_0(x)}\,dx$$

$$\int d\left(\frac{dy}{dx}\right) = \int \frac{Q(x)}{P_0(x)}\,dx$$

$$\frac{dy}{dx} = \int \frac{Q(x)}{P_0(x)}\,dx + C_1$$

We repeat this procedure until y appears on the left side of the equation.

EXAMPLE 1 Solve the following differential equation:

$$\frac{d^2y}{dx^2} = 3x - 4$$

Solution First, we write the left side of the equation as $\frac{d}{dx}\left(\frac{dy}{dx}\right)$. Thus,

$$\frac{d}{dx}\left(\frac{dy}{dx}\right) = 3x - 4$$

Second, we multiply the equation by dx.

$$dx\,\frac{d}{dx}\left(\frac{dy}{dx}\right) = dx\,[3x - 4]$$

$$d\left(\frac{dy}{dx}\right) = (3x - 4)\,dx$$

Next, we integrate both sides of the equation.

$$\int d\left(\frac{dy}{dx}\right) = \int (3x - 4)\,dx$$

$$\frac{dy}{dx} = \frac{3x^2}{2} - 4x + C_1$$

Last, we solve this equation by separation of variables.

$$dy = \left(\frac{3x^2}{2} - 4x + C_1\right) dx$$

$$\int dy = \int \left(\frac{3x^2}{2} - 4x + C_1\right) dx$$

$$y = \frac{x^3}{2} - 2x^2 + C_1x + C_2$$

You can check the solution by differentiating and substituting into the original equation. If

$$y = \frac{x^3}{2} - 2x^2 + C_1x + C_2$$

then

$$\frac{dy}{dx} = \frac{3x^2}{2} - 4x + C_1 \quad \text{and} \quad \frac{d^2y}{dx^2} = 3x - 4$$

Substituting gives

$$\frac{d^2y}{dx^2} = 3x - 4$$

$$3x - 4 = 3x - 4$$

■

If the variable y does not appear explicitly in the equation, substitute $u = dy/dx$ to make a second-order equation into a first-order differential equation. Then solve this equation by using the integrating factor method discussed in Section 23–2. Last, substitute for u and integrate to yield the solution.

EXAMPLE 2 Solve the following:

$$x\frac{d^2y}{dx^2} = \frac{dy}{dx} - 2x$$

Solution First, let $u = dy/dx$; then $d^2y/dx^2 = du/dx$. The equation can be written as follows:

$$x\frac{du}{dx} = u - 2x$$

$$x\frac{du}{dx} - u = -2x$$

$$\frac{du}{dx} - \frac{u}{x} = -2$$

Second, we solve this equation by using the integrating factor

$$e^{\int P(x)\, dx} = e^{\int -dx/x} = e^{-\ln x} = x^{-1}$$

The solution is

$$\begin{aligned} u &= \frac{1}{x^{-1}} \int \frac{-2\, dx}{x} \\ &= x \int \frac{-2\, dx}{x} \\ &= -2x \ln x + C_1 \end{aligned}$$

Third, we substitute $u = dy/dx$ into the equation.

$$\begin{aligned} \frac{dy}{dx} &= -2x \ln x + C_1 \\ dy &= (-2x \ln x + C_1)\, dx \end{aligned}$$

Then we solve by integrating both sides of the equation.

$$\begin{aligned} \int dy &= \int (-2x \ln x + C_1)\, dx \\ y &= -x^2 \ln x + \frac{x^2}{2} + C_1 x + C_2 \end{aligned}$$

■

Homogeneous Equations

Using the earlier notation in this section for a second-order differential equation, an equation is called **homogeneous** if $Q(x) = 0$. If $Q(x) \neq 0$, the equation is said to be **nonhomogeneous,** which we will discuss in the next section. We will now discuss homogeneous differential equations where the functions $P_i(x)$ are constants. Let us write the general form of the linear homogeneous differential equation using the operator notation where $Dy = dy/dx$ and $D^2y = d^2y/dx^2$ as follows:

$$\begin{aligned} a_0D^2y + a_1Dy + a_2y &= 0 \\ (a_0D^2 + a_1D + a_2)y &= 0 \end{aligned}$$

where a_0, a_1, and a_2 are constants. The D operator is not an algebraic quantity, but many algebraic operations with it are valid. We solve this quadratic equation by factoring. If m_1 and m_2 are the roots of the equation, it can be written as

$$(D - m_1)(D - m_2)y = 0$$

The Zero Factor Property gives the equations

$$(D - m_1)y_1 = 0 \quad \text{and} \quad (D - m_2)y_2 = 0$$

These equations translate to the differential equations

$$(D - m_1)y_1 = 0 \quad \text{and} \quad (D - m_2)y_2 = 0$$
$$\frac{dy}{dx} - m_1y_1 = 0 \qquad \frac{dy}{dx} - m_2y_2 = 0$$

These linear differential equations can be solved by separation of variables or an integrating factor. Thus,

$$y_1 = C_1e^{m_1x} \quad \text{and} \quad y_2 = C_2e^{m_2x}$$

Using the sum rule, we combine these solutions to obtain

$$y = y_1 + y_2$$
$$y = C_1e^{m_1x} + C_2e^{m_2x}$$

In general, a homogeneous linear differential equation with constant coefficients of the form

$$a_0\frac{d^ny}{dx^n} + a_1\frac{d^{n-1}y}{dx^{n-1}} + \cdots + a_{n-1}\frac{dy}{dx} + a_ny = 0$$

can be written as follows:

$$a_0m^n + a_1m^{n-1} + \cdots + (a_{n-1})m + a_n = 0$$

called the *auxiliary equation*. If $m_1, m_2, \ldots, m_n$ are the distinct roots of this equation, then the solution of the differential equation is

$$y = C_1e^{m_1x} + C_2e^{m_2x} + \cdots + C_ne^{m_nx}$$

EXAMPLE 3 Solve the following differential equation:

$$3\frac{d^2y}{dx^2} + 2\frac{dy}{dx} - y = 0$$

Solution Before proceeding, we must check to see whether this equation is a linear homogeneous differential equation with constant coefficients. Since the equation satisfies the definition, we write the equation using the D operator notation.

$$3D^2y + 2Dy - y = 0$$
$$(3D^2 + 2D - 1)y = 0$$

Next, we write the auxiliary equation by substituting m for D.

$$3m^2 + 2m - 1 = 0$$

Then we solve this quadratic equation for m by factoring.

$$(3m - 1)(m + 1) = 0$$

$$m_1 = \frac{1}{3} \quad \text{and} \quad m_2 = -1$$

Last, we write the general solution to the differential equation.

$$y = C_1e^{x/3} + C_2e^{-x}$$

Check the solution by substituting into the original differential equation. ■

EXAMPLE 4 Solve the following differential equation:

$$y'' - 9y' = 0$$

Solution First, we write the equation using the D operator.

$$(D^2 - 9D)y = 0$$

The auxiliary equation is

$$m^2 - 9m = 0$$

Next, we solve this quadratic equation by factoring.

$$m(m - 9) = 0$$

$$m_1 = 0 \quad \text{and} \quad m_2 = 9$$

The solution is

$$\begin{aligned} y &= C_1e^{0x} + C_2e^{9x} \\ &= C_1 + C_2e^{9x} \end{aligned}$$

■

EXAMPLE 5 Solve the following differential equation:

$$3y'' + 5y' - 4y = 0$$

Solution The equation written in D operator notation is

$$(3D^2 + 5D - 4)y = 0$$

The auxiliary equation is

$$3m^2 + 5m - 4 = 0$$

Using the quadratic formula to solve this quadratic equation gives

$$\begin{aligned} m &= \frac{-5 \pm \sqrt{25 - 4(3)(-4)}}{2(3)} \\ &= \frac{-5 \pm \sqrt{73}}{6} \end{aligned}$$

The general solution is

$$y = C_1 e^{\left(\frac{-5+\sqrt{73}}{6}x\right)} + C_2 e^{\left(\frac{-5-\sqrt{73}}{6}x\right)}$$ ■

Repeated Roots

The solution developed for homogeneous equations is valid only for real and unequal roots. Now we will develop a formula for roots that are equal, or repeated.

If the roots of a homogeneous differential equation are equal ($m_1 = m_2$), then the D operator form can be written $(D - m_1)(D - m_1)y = 0$. If we let $(D - m_1)y = u$, then the auxiliary equation can be written as $(D - m_1)u = 0$, whose solution is $u = C_1 e^{m_1 x}$. The auxiliary equation is

$$(D - m_1)y = C_1 e^{m_1 x}$$

$$\frac{dy}{dx} - m_1 y = C_1 e^{m_1 x}$$

We can solve this differential equation by using the integrating factor

$$e^{-\int m_1\, dx} = e^{-m_1 x}$$

The solution is

$$y = \frac{1}{e^{-m_1 x}} \int e^{-m_1 x} C_1 e^{m_1 x}\, dx$$

$$y = e^{m_1 x}(C_1 x + C_2)$$

Repeated Roots

If the auxiliary equation has n real roots, $m_1 = \cdots = m_n = m$, the solution to the differential equation is

$$y = C_1 e^{mx} + C_2 x e^{mx} + C_3 x^2 e^{mx} + \cdots + C_n x^{n-1} e^{mx}$$

EXAMPLE 6 Solve the following differential equation:

$$y'' + 6y' + 9y = 0$$

Solution

$(D^2 + 6D + 9)y = 0$	D operator form
$m^2 + 6m + 9 = 0$	auxiliary equation
$(m + 3)^2 = 0$	factor
$m = -3$	solve for m

The general solution to the equation is

$$y = C_1 e^{-3x} + C_2 x e^{-3x}$$

$$= e^{-3x}(C_1 + C_2 x)$$ ■

Complex Roots

If the auxiliary equation has a pair of complex conjugate roots ($m_1 = a + bj$ and $m_2 = a - bj$), then the solution to the differential equation is

$$y = C_1e^{(a+bj)x} + C_2e^{(a-bj)x}$$
$$= C_1e^{ax}e^{bjx} + C_2e^{ax}e^{-bjx}$$
$$y = e^{ax}(C_1e^{bjx} + C_2e^{-bjx})$$

However, we can simplify this equation by using the exponential form of the complex numbers.

$$e^{j\theta} = \cos\theta + j\sin\theta \quad \text{and} \quad e^{-j\theta} = \cos\theta - j\sin\theta$$

Using these identities gives

$$y = e^{ax}[C_1(\cos bx + j\sin bx) + C_2(\cos bx - j\sin bx)]$$
$$= e^{ax}[(C_1 + C_2)\cos bx + j(C_1 - C_2)\sin bx]$$
$$y = e^{ax}(C_3\cos bx + C_4\sin bx)$$

where $C_3 = C_1 + C_2$ and $C_4 = (C_1 - C_2)j$.

Complex Roots

If the auxiliary equation has complex roots, the general solution to the differential equation is

$$y = e^{ax}(C_1\cos bx + C_2\sin bx)$$

EXAMPLE 7 Solve the following differential equation:

$$y'' + 4y' + 8y = 0$$

Solution The D operator form of this equation is

$$(D^2 + 4D + 8)y = 0$$

The auxiliary equation that we then solve is

$$m^2 + 4m + 8 = 0$$
$$m = \frac{-4 \pm \sqrt{16 - 4(1)(8)}}{2}$$
$$= \frac{-4 \pm 4j}{2}$$
$$= \underset{\uparrow a}{-2} \pm \overset{b \downarrow}{2j}$$

The general solution is

$$y = e^{-2x}(C_1 \cos 2x + C_2 \sin 2x)$$ ■

23–4 EXERCISES

Solve by direct integration.

1. $y'' - x^3 = 0$
2. $y'' - 3x^2 = 0$
3. $y'' = e^x$
4. $y'' = \cos x$
5. $\frac{d^2y}{dx^2} - \frac{1}{x} = 0$
6. $\frac{d^2y}{dx^2} = 4$

Solve the following homogeneous linear differential equations.

7. $\frac{d^2y}{dx^2} - 3\frac{dy}{dx} + 2y = 0$
8. $\frac{d^2y}{dx^2} - 5\frac{dy}{dx} + 6y = 0$
9. $y'' - 2y' - 3y = 0$
10. $4y'' - 12y' + 5y = 0$
11. $3D^2y - 14Dy = 5y$
12. $D^2y = 8Dy - 16y$
13. $D^2y - Dy - 12y = 0$
14. $9D^2y = 6Dy - y$
15. $\frac{d^2y}{dx^2} + 9y = 0$
16. $4\frac{d^2y}{dx^2} + y = 0$
17. $\frac{d^2y}{dx^2} - y = 0$
18. $\frac{d^2y}{dx^2} - 4\frac{dy}{dx} + 13y = 0$
19. $D^2y = 4Dy - 29y$
20. $4D^2y + 4Dy + 37y = 0$
21. $y'' - 2y' + 5y = 0$
22. $y'' + 16y = 0$
23. $4y'' - 4y' + y = 0$
24. $y'' + 8y' + 16y = 0$
25. The radioactive decay of an element is given by the differential equation

$$\frac{dm}{dt} - km = 0$$

where m = mass, t = time, and k = constant rate of decay. Solve for m in terms of t.

23–5

NON-HOMOGENEOUS DIFFERENTIAL EQUATIONS

In this section we will solve nonhomogeneous differential equations. Recall from our discussion in the previous section that a nonhomogeneous equation is of the form

$$a_0 \frac{d^2y}{dx^2} + a_1 \frac{dy}{dx} + a_2 y = Q(x)$$

or, in operator notation,

$$(a_0 D^2 + a_1 D + a_2)y = Q(x)$$

When the solution to this type of equation is substituted into the left side of the equation, the result is the right side. Solutions to homogeneous equations will yield zero when substituted, but these solutions contain arbitrary constants. If we can find a particular solution that yields the right side of the equation when substituted, then we can add this solution to the solution containing the arbitrary constants. The solution of nonhomogeneous differential equations is

$$y = y_c + y_p$$

where y_c, called the **complementary solution,** is the solution to the corresponding homogeneous equation, and y_p is a particular solution of the nonhomogeneous equation.

Undetermined Coefficients

We can use the methods of the previous section to find y_c. The **method of undetermined coefficients** is used to find y_p.

EXAMPLE 1 Solve the following equation:

$$\frac{d^2y}{dx^2} + 2\frac{dy}{dx} - 3y = 6x$$

Solution First, we solve the corresponding homogeneous equation.

$$\begin{aligned} (D^2 + 2D - 3)y &= 0 \quad D \text{ operator} \\ m^2 + 2m - 3 &= 0 \quad \text{auxiliary equation} \\ (m + 3)(m - 1) &= 0 \quad \text{factor to solve} \\ m_1 = -3 \quad \text{and} \quad m_2 &= 1 \quad \text{solve for } m \end{aligned}$$

The complementary solution is $y_c = C_1e^{-3x} + C_2e^x$.

Second, we determine a general form of the particular solution. Its form should include an x term and a constant, which is the derivative of the x term. Let us use $y_p = A + Bx$ where A and B are unspecified constants. To find the numerical value of A and B, we substitute into the differential equation and equate like terms. Since $y_p = A + Bx$, then

$$y'_p = B \quad \text{and} \quad y''_p = 0$$

$$\begin{aligned} y'' + 2y' - 3y &= 6x \\ 0 + 2B - 3(A + Bx) &= 6x \\ 2B - 3A - 3Bx &= 6x \end{aligned}$$

We solve this equation for A and B by equating like powers of x.

$$\begin{aligned} (2B - 3A)x^0 - 3Bx &= 0x^0 + 6x \\ 2B - 3A &= 0 \\ -3Bx &= 6x \quad \text{or} \quad B = -2 \end{aligned}$$

Substituting $B = -2$ into the equation $2B - 3A = 0$ gives

$$\begin{aligned} 2(-2) - 3A &= 0 \\ 3A &= -4 \\ A &= -\frac{4}{3} \end{aligned}$$

Therefore, the particular solution is

$$y_p = -\frac{4}{3} - 2x$$

and the general solution of the original equation is

$$y = y_c + y_p$$

$$y = C_1 e^{-3x} + C_2 e^x - \frac{4}{3} - 2x$$

To check this solution, substitute into the original differential equation. ■

Table 23–1 provides some guidance on choosing the general form of the particular solution to a nonhomogeneous differential equation.

TABLE 23–1

$Q(x)$	General Form of y_p
x^n ax^n $ax^n + Bx^{n-1} + \cdots$	$A + Bx + Cx^2 + \cdots + kx^n$
ae^{bx}	Ae^{bx}
axe^{bx}	$Ae^{bx} + Bxe^{bx}$
$a \cos bx$ $a \sin bx$	$A \sin bx + B \cos bx$
$ax \cos bx$ $ax \sin bx$	$A \sin bx + B \cos bx + Cx \cos bx + Ex \sin bx$

EXAMPLE 2 Solve the differential equation

$$D^2y - Dy - 12y = 3e^x$$

Solution First, we solve the corresponding homogeneous equation.

$$(D^2 - D - 12)y = 0$$

$$m^2 - m - 12 = 0$$

$$(m - 4)(m + 3) = 0$$

$$m_1 = 4 \quad \text{and} \quad m_2 = -3$$

The complementary solution to the homogeneous differential equation is

$$y_c = C_1e^{4x} + C_2e^{-3x}$$

Second, we find the particular solution y_p. From Table 23–1, we assume that the general form of this solution should be $y_p = Ae^x$ since $b = 1$. Then we solve for A by substituting into the original differential equation.

$$\underset{\substack{\uparrow \\ Ae^x}}{y''_p} - \underset{\substack{\uparrow \\ Ae^x}}{y'_p} - 12y_p = 3e^x$$

$$Ae^x - Ae^x - 12Ae^x = 3e^x$$

$$-12Ae^x = 3e^x$$

$$A = -\frac{1}{4}$$

The particular solution is $y_p = -\frac{1}{4}e^x$, and the general solution of the nonhomogeneous differential equation is

$$y = y_c + y_p$$

$$= C_1e^{4x} + C_2e^{-3x} - \frac{1}{4}e^x$$

Check this solution. ■

EXAMPLE 3 Solve the differential equation

$$\frac{d^2y}{dx^2} - 9y = 4 \sin 2x$$

Solution First, we solve the corresponding homogeneous equation.

$$y'' - 9y = 0$$

$$(D^2 - 9)y = 0$$

$$m^2 - 9 = 0$$

$$(m + 3)(m - 3) = 0$$

$$m_1 = -3 \quad \text{and} \quad m_2 = 3$$

The complementary solution is

$$y_c = C_1e^{-3x} + C_2e^{3x}$$

Second, we find the particular solution. From Table 23–1, the general form of the particular solution is $A \sin 2x + B \cos 2x$. We solve for A and B.

$$y'_p = 2A \cos 2x - 2B \sin 2x$$

$$y''_p = -4A \sin 2x - 4B \cos 2x$$

$$y'' - 9y_p = 4 \sin 2x$$

$$\underbrace{(-4A \sin 2x - 4B \cos 2x)}_{y''_p} - 9\underbrace{(A \sin 2x + B \cos 2x)}_{y_p} = 4 \sin 2x$$

$$-13A \sin 2x - 13B \cos 2x = 4 \sin 2x$$

$$-13A \sin 2x = 4 \sin 2x \rightarrow -13A = 4 \quad A = -\frac{4}{13}$$

$$-13B \cos 2x = 0 \rightarrow B = 0$$

The particular solution is

$$y_p = -\frac{4}{13} \sin 2x$$

and the general solution to the homogeneous differential equation is

$$y = C_1 e^{-3x} + C_2 e^{3x} - \frac{4}{13} \sin 2x$$

■

EXAMPLE 4 Solve the differential equation

$$D^2y - 2Dy + 2y = x + 3xe^{2x}$$

subject to the initial conditions $y(0) = 0$ and $y'(0) = 1$.

Solution For the corresponding homogeneous equation $y'' - 2y' + 2y = 0$, the auxiliary equation $m^2 - 2m + 2 = 0$ yields the solutions

$$m_1 = 1 + j \quad \text{and} \quad m_2 = 1 - j$$

Therefore, using the formula for complex roots, we find that the complementary solution is

$$e^x(C_1 \cos x + C_2 \sin x)$$

From Table 23–1, the particular solution is of the form

$$y_p = A + Bx + Ce^{2x} + Fxe^{2x}.$$

Since $Q(x)$ is a sum of two expressions in Table 23–1, y_p is the sum of the corresponding general forms in the table. Solving for A, B, C, and F gives

$$y'_p = B + 2Ce^{2x} + 2Fxe^{2x} + Fe^{2x} = B + e^{2x}(2C + F + 2Fx)$$
$$y''_p = 2Fe^{2x} + 2e^{2x}(2C + F + 2Fx)$$

Substituting into the original differential equation gives

$$2Fe^{2x} + 2e^{2x}(2C + F + 2Fx) - 2[B + e^{2x}(2C + F + 2Fx)] + 2(A + Bx + Ce^{2x} + Fxe^{2x}) = x + 3xe^{2x}$$
$$2(Fe^{2x} - B + A + Bx + Ce^{2x} + Fxe^{2x}) = x + 3xe^{2x}$$
$$(-2B + 2A) + 2Bx + 2e^{2x}(F + C) + 2Fxe^{2x} = x + 3xe^{2x}$$

$$-2B + 2A = 0 \to A = \frac{1}{2}$$

$$B = \frac{1}{2} \quad \text{and} \quad F = \frac{3}{2}$$

$$F + C = 0 \to C = -\frac{3}{2}$$

Therefore, the particular solution is

$$y_p = \frac{1}{2} + \frac{1}{2}x - \frac{3}{2}e^{2x} + \frac{3}{2}xe^{2x}$$

and the general solution to the differential equation is

$$y = e^x(C_1\cos x + C_2\sin x) + \frac{1}{2} + \frac{1}{2}x - \frac{3}{2}e^{2x} + \frac{3}{2}xe^{2x}$$

$$y(0) = 0 \to C_1 = 1 \quad \text{and} \quad y'(0) = 1 \to C_2 = 1$$

Therefore, the particular solution is

$$y = e^x(\cos x + \sin x) + \frac{1}{2} + \frac{1}{2}x - \frac{3}{2}e^{2x} + \frac{3}{2}xe^{2x}$$

■

Applications

One application of nonhomogeneous differential equations is **simple harmonic motion,** which is defined as motion in a straight line for which the acceleration is proportional to the displacement and whose direction is opposite to the displacement. The oscillation of a weight attached to a spring, the movement of a clock pendulum, and the oscillation of an electrical charge in a circuit are all examples of simple harmonic motion.

Remember from earlier chapters that acceleration is the second derivative of displacement. Therefore, if x represents displacement, then d^2x/dt^2 represents acceleration. From the definition of simple harmonic motion,

$$m\frac{d^2x}{dt^2} = -kx$$

where k = the constant of proportion. Solving this differential equation gives

$$mD^2x + kx = 0$$

$$x = C_1\sin\left(t\sqrt{\frac{k}{m}}\right) + C_2\cos\left(t\sqrt{\frac{k}{m}}\right)$$

We can see the initial condition to solve for C_1 and C_2. When $t = 0$, the object is motionless with a displacement of $x = A$. Therefore,

$$x = A = C_1\sin\left(t\sqrt{\frac{k}{m}}\right) + C_2\cos\left(t\sqrt{\frac{k}{m}}\right) \quad \text{for } t = 0$$

$$A = C_1\sin 0 + C_2\cos 0$$

$$A = C_2$$

and $dx/dt = 0$ since the object is motionless. Note that for ease in writing this expression, $k_1 = \sqrt{\frac{k}{m}}$,

$$\frac{dx}{dt} = C_1k_1\cos\left(t\sqrt{\frac{k}{m}}\right) - C_2k_1\sin\left(t\sqrt{\frac{k}{m}}\right)$$

$$0 = C_1k_1\cos\left(t\sqrt{\frac{k}{m}}\right) - C_2k_1\sin\left(t\sqrt{\frac{k}{m}}\right) \quad \text{for } t = 0$$

$$0 = C_1k_1\cos 0 - C_2k_1\sin 0$$

$$0 = C_1k_1 \quad \text{since } k_1 \neq 0$$

$$0 = C_1$$

and the solution is

$$x = A\cos\left(t\sqrt{\frac{k}{m}}\right)$$

Since the period of the cosine function is 2π, the period of this function is $2\pi\sqrt{m/k}$ and the frequency is

$$f = \frac{1}{P} = \frac{1}{2\pi}\sqrt{\frac{k}{m}}$$

EXAMPLE 5 When a 5-kg block is attached to a spring at equilibrium, the spring stretches 0.05 m. If the block is moved 0.2 m and released, find the following:

(a) the spring constant,
(b) the equation relating displacement and time,
(c) the frequency of oscillation of the block, and
(d) the maximum speed of the block.

Solution

(a) The spring constant is represented by k. Thus,

$$k = \frac{\text{weight}}{\text{displacement}} = \frac{\text{mass} \cdot 9.8}{\text{displacement}} = \frac{5 \cdot 9.8}{0.05}$$
$$= 980 \text{ N/m}$$

(b) Since the block moves in simple harmonic motion, its motion is given by

$$x = A \cos\left(t\sqrt{\frac{k}{m}}\right)$$

(c) The frequency of oscillation is

$$f = \frac{1}{2\pi}\sqrt{\frac{980}{5}} = \frac{7}{\pi} \approx 2.2 \text{ Hz}$$

(d) The equation for displacement is

$$x = 0.2 \cos 14t$$

and the speed at any time is given by

$$v = \frac{dx}{dt}$$

$$\frac{dx}{dt} = -2.8 \sin 14t$$

The maximum speed is 2.8 m/s. ■

In practice, an object is subject to frictional forces when it undergoes simple harmonic motion. The frictional force or damping is proportional to the speed of the object. The differential equation of motion becomes

$$m\frac{d^2x}{dt^2} = -a\frac{dx}{dt} - kx$$

(with $-a\frac{dx}{dt}$ labelled: friction or damping force)

The solution to this second-order homogeneous equation with constant coefficients is

$$m\frac{d^2x}{dt^2} + a\frac{dx}{dt} + kx = 0$$

$$(mD^2 + aD + k)x = 0$$

$$\text{roots} = \frac{-a \pm \sqrt{a^2 - 4mk}}{2m}$$

The solution depends on the values of a and k. If $a^2 - 4mk < 0$, the roots are complex conjugate numbers, and this case is called **underdamped.** If $a^2 - 4mk > 0$, the roots are unequal, negative real numbers, and this case is called **overdamped.** If $a^2 - 4mk = 0$, the roots are equal, negative real numbers, and this case is called **critically damped.** Figure 23–2 illustrates these three cases.

FIGURE 23–2

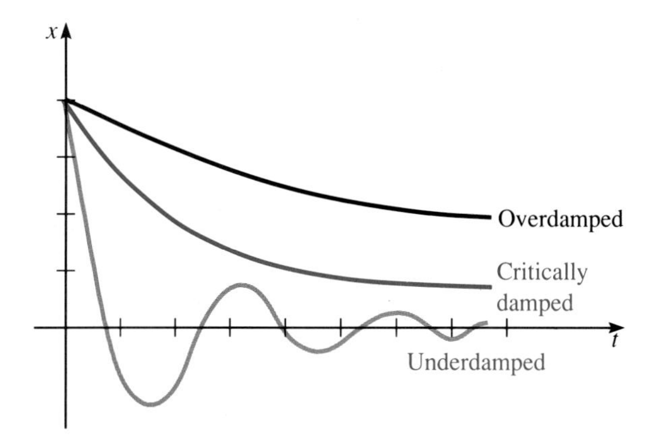

RCL Circuits

Consider an electrical circuit that contains a resistance R (in ohms), a capacitance C (in farads), an inductance L (in henries), and a voltage V (in volts). The voltage drop across the inductor is $L\,di/dt$, across the resistor iR, and across the capacitor q/C where q = the charge on the capacitor. Since $i = dq/dt$, the charge on the capacitor at any time t is given by

$$L\frac{d^2q}{dt^2} + R\frac{dq}{dt} + \frac{q}{C} = V$$

EXAMPLE 6 Find an equation for the charge on a capacitor at time t in an RCL circuit where $L = 1$ H, $R = 7\ \Omega$, $C = 0.1$ F, and $V = 100 \sin 60t$.

Solution The differential equation is

$$\frac{d^2q}{dt^2} + 7\frac{dq}{dt} + 10q = 100 \sin 60t$$

which is a second-order, nonhomogeneous linear differential equation. To solve this equation, first, we solve the corresponding homogeneous equation.

$$\frac{d^2q}{dt^2} + 7\frac{dq}{dt} + 10q = 0$$

$$m^2 + 7m + 10 = 0 \quad \text{auxiliary equation}$$

$$(m + 5)(m + 2) = 0 \quad \text{factor to solve for } m$$

$$m_1 = -5 \quad \text{and} \quad m_2 = -2 \quad \text{solve for } m$$

The complementary solution is

$$q_c = C_1e^{-5t} + C_2e^{-2t}$$

To find the particular solution, we assume that

$$q_p = A \sin 60t + B \cos 60t$$

Then

$$q'_p = 60A \cos 60t - 60B \sin 60t$$

$$q''_p = -3{,}600A \sin 60t - 3{,}600B \cos 60t$$

Substituting into the original differential equation gives

$$\begin{aligned}
&(-3{,}600A \sin 60t - 3{,}600B \cos 60t) \\
&\quad +7(60A \cos 60t - 60B \sin 60t) \\
&\quad +10(A \sin 60t + B \cos 60t) = 100 \sin 60t \\
&\sin 60t(-3{,}600A - 420B + 10A) \\
&\quad +\cos 60t(-3{,}600B + 420A + 10B) = 100 \sin 60t
\end{aligned}$$

Therefore,

$$-3{,}590A - 420B = 100$$

$$-3{,}590B + 420A = 0$$

Solving this system of equations for A and B gives

$$A \approx -0.0275 \quad \text{and} \quad B \approx -0.00321$$

The general solution to the differential equation is

$$\begin{aligned}
q &= q_c + q_p \\
&= C_1e^{-5t} + C_2e^{-2t} - 0.0275 \sin 60t - 0.00321 \cos 60t
\end{aligned}$$ ■

Notice that the solution in Example 6 contains both an exponential term and a sinusoidal term. After a short time, the exponential term in the complementary solution becomes negligible. For this reason, it is referred to as the **transient term,** and the particular solution is called the **steady-state solution.**

Thus, to find the steady-state solution, we must find only the particular solution to the differential equation.

EXAMPLE 7 Find the steady-state solution for the current in a circuit composed of a 0.1-F capacitor, a 1-H inductor, an 8-Ω resistor, and a voltage source of 120 sin $50t$.

Solution The differential equation we must solve is

$$\frac{d^2q}{dt^2} + 8\frac{dq}{dt} + 10q = 120 \sin 50t$$

and then we may find the current by taking the derivative of the resulting equation ($i = dq/dt$).

$$q_p = A \sin 50t + B \cos 50t$$

and

$$q'_p = 50A \cos 50t - 50B \sin 50t$$
$$q''_p = -2{,}500A \sin 50t - 2{,}500B \cos 50t$$

Substituting into the original differential equation and solving for A and B give

$$(-2{,}500A \sin 50t - 2{,}500B \cos 50t) + 8(50A \cos 50t - 50B \sin 50t) + 10(A \sin 50t + B \cos 50t) = 120 \sin 50t$$
$$-2{,}490A - 400B = 120$$
$$-2{,}490B + 400A = 0 \rightarrow A = 6.225B$$

Solving this system of equations gives

$$A \approx -0.046980393 \quad \text{and} \quad B \approx -0.007547051$$

Therefore,

$$q_p = -0.04698 \sin 50t - 0.007547 \cos 50t$$

and the current is

$$i = \frac{dq}{dt} = -2.349 \cos 50t + 0.37735 \sin 50t$$

■

23–5 EXERCISES

Solve each differential equation.

1. $y'' - 3y' + 2y = 3x^2$
2. $y'' - 4y' + 3y = 9x^2 + 4$
3. $y'' - 4y' = 6$
4. $y'' + y' = 2$
5. $y'' - 2y' - 6y = -12(1 + x)x^2 - 4$
6. $y'' + 3y' - 2y = -2x^2 + 11$
7. $y'' + 2y' + 4y = \cos 4x$
8. $y'' + 2y' + 2y = 2 \sin 4x$

9. $y'' - 2y' - 3y = 4 \sin 2x - 7 \cos 2x$

10. $y'' + y = -6 \sin 2x$

11. $y'' - 2y' - 3y = e^x - \sin x$

12. $y'' + 2y' + 5y = \sin 2x + 7 \cos 2x$

13. $y'' - y = 2e^x$

14. $y'' + 4y = 5e^{-x}$

15. $y'' + y = x^2 + 2;\ y(0) = 0,\ y'(0) = 2$

16. $y'' - y = 2 - x^2;\ y(0) = 2,\ y'(0) = 0$

17. $y'' + y = -9 \cos 2x;\ y(0) = 2,\ y'(0) = 1$

18. $y'' + y = -60 \sin 4x;\ y(0) = 8,\ y'(0) = 14$

19. $y'' + 4y' + 13y = 3 \sin 2x$

20. $y'' - 4y = 12e^{2x};\ y(0) = 1,\ y'(0) = 2$

21. $y'' + y = x^2 - e^{-x};\ y(0) = 0,\ y'(0) = 0$

22. A 10-lb weight is placed on the lower end of a coil spring suspended from the ceiling. The weight comes to rest in its equilibrium position, thereby stretching the spring 1.6 inches. The weight is then pulled down 2 inches below its equilibrium position and released from rest at $t = 0$. Find the spring constant, the displacement as a function of time, the period and frequency of oscillation, and the maximum speed.

23. In simple harmonic motion, the force and acceleration are proportional to the displacement s, as given by the differential equation

$$\frac{d^2s}{dt^2} = -\omega^2 s$$

where ω = angular velocity. Solve this equation for s in terms of t.

24. The vertical displacement of an atomic particle is given by

$$\frac{dy}{dt} + gt = v_0 \sin \alpha$$

where t = time, g = gravitational force, V_0 = initial velocity, and α = angle between the x axis and the direction of motion. Solve the equation for y as a function of t if $y = 0$ when $t = 0$.

25. For a given electric circuit, $L = 0.2$ H, $R = 0$, $C = 50\ \mu$F, and $E = 110$ V. Find the equation relating charge on the capacitor and time if $q = 0$ and $i = 0$ when $t = 0$.

26. Find the relationship relating current and time for the circuit in Exercise 25.

27. Find the steady-state current for the circuit in Exercise 25.

23–6 THE LAPLACE TRANSFORM

The Laplace transform is a powerful method of solving a differential equation because it reduces the problem of solving a differential equation to an algebraic problem. The Laplace transform allows you to obtain a particular solution without first finding the general solution. It also allows you to obtain the solution to a nonhomogeneous equation without first solving the homogeneous equation. In this section we will first discuss the Laplace transform in general and then apply it to solving differential equations. This discussion is intended only as an introduction to Laplace transforms.

Definition

Let $f(t)$ be a function defined for all positive values of t. We multiply $f(t)$ by e^{-st} and then integrate with respect to t from zero to infinity, obtaining the function $F(s)$ as follows:

$$F(s) = \int_0^{\infty} e^{-st} f(t)\, dt$$

The function $F(s)$ is called the **Laplace transform** of the original function $f(t)$, denoted $L(f)$.

The Laplace Transform

$$L(f) = F(s) = \int_0^{\infty} e^{-st} f(t)\, dt$$

The integral contained in the Laplace transform is called an *improper* integral because one of the limits of integration is infinity. To evaluate such an integral, we substitute C for t and determine the limit as $C \to \infty$.

EXAMPLE 1 Find the Laplace transform for the function $f(t) = \sin t$.

Solution From the definition of the Laplace transform, we have

$$L(f) = L(\sin t) = \int_0^{\infty} e^{-st} \sin t\, dt$$

Using Formula 94 from Appendix C, where $a = -s$ and $b = 1$, gives the following result:

$$\begin{aligned}
L(\sin t) &= \int_0^{\infty} e^{-st} \sin t\, dt \\
&= \lim_{C\to\infty} \left(\int_0^{C} e^{-st} \sin t\, dt \right) \\
&= \lim_{C\to\infty} \left[\frac{e^{-st}(-s \sin t - \cos t)}{s^2 + 1} \right] \Bigg|_0^C \\
&= \lim_{C\to\infty} \left(\frac{e^{-sC}(-s \sin C - \cos C)}{s^2 + 1} \right) - \left(-\frac{1}{s^2 + 1} \right) \\
&= 0 + \frac{1}{s^2 + 1} \\
&= \frac{1}{s^2 + 1}
\end{aligned}$$

The Laplace transform of $\sin t$ is $\dfrac{1}{s^2 + 1}$. ■

Linearity Property

The Laplace transform is a linear operator; that is,

$$L[af(t) + bg(t)] = aL(f) + bL(g)$$

where a and b are constants. The linearity property is based on the definition of the Laplace transform and the sum rule of integrals, as shown in the following equations:

$$L[af(t) + bg(t)] = \int_0^\infty e^{-st}[af(t) + bg(t)]\,dt$$

$$= a\int_0^\infty e^{-st}f(t)\,dt + b\int_0^\infty e^{-st}g(t)\,dt$$

$$= aL(f) + bL(g)$$

The **linearity property** states that the Laplace transform of a sum of functions is the sum of Laplace transforms. A short table of Laplace transforms is given in Table 23–2.

TABLE 23–2
Laplace transforms

	$f(t) = L^{-1}(F)$	$L(f) = F(s)$		$f(t) = L^{-1}(F)$	$L(f) = F(s)$
1.	1	$\frac{1}{s}$	11.	te^{-at}	$\frac{1}{(s+a)^2}$
2.	$\frac{t^{n-1}}{(n-1)!}$	$\frac{1}{s^n}$ (n = 1, 2, 3, . . .)	12.	$t^{n-1}e^{-at}$	$\frac{(n-1)!}{(s+a)^n}$
3.	e^{at}	$\frac{1}{s-a}$	13.	$e^{-at}(1-at)$	$\frac{s}{(s+a)^2}$
4.	$1 - e^{-at}$	$\frac{a}{s(s+a)}$	14.	$[(b-a)t + 1]e^{-at}$	$\frac{s+b}{(s+a)^2}$
5.	$\cos at$	$\frac{s}{s^2+a^2}$	15.	$\sin at - at\cos at$	$\frac{2a^3}{(s^2+a^2)^2}$
6.	$\sin at$	$\frac{a}{s^2+a^2}$	16.	$t\sin at$	$\frac{2as}{(s^2+a^2)^2}$
7.	$1 - \cos at$	$\frac{a^2}{s(s^2+a^2)}$	17.	$\sin at + at\cos at$	$\frac{2as^2}{(s^2+a^2)^2}$
8.	$at - \sin at$	$\frac{a^3}{s^2(s^2+a^2)}$	18.	$t\cos at$	$\frac{s^2-a^2}{(s^2+a^2)^2}$
9.	$e^{-at} - e^{-bt}$	$\frac{b-a}{(s+a)(s+b)}$	19.	$e^{-at}\sin bt$	$\frac{b}{(s+a)^2+b^2}$
10.	$ae^{-at} - be^{-bt}$	$\frac{s(a-b)}{(s+a)(s+b)}$	20.	$e^{-at}\cos bt$	$\frac{s+a}{(s+a)^2+b^2}$

Laplace Transform of Derivatives

Another Laplace transform important to the solution of differential equations is the transform of a derivative. Applying the definition of the Laplace transform to the function $f'(t)$ gives

$$L(f') = \int_0^\infty e^{-st} f'(t)\, dt$$

We integrate by using integration by parts where $u = e^{-st}$ and $dv = f'(t)\, dt$. Thus,

$$du = -se^{-st}\, dt \quad \text{and} \quad v = \int f'(t)\, dt = f(t)$$

Therefore,

$$\begin{aligned} L(f') &= e^{-st} f(t) \Big|_0^\infty - \int -se^{-st} f(t)\, dt \\ &= 0 - f(0) + s \int e^{-st} f(t)\, dt \\ &= -f(0) + sL(f) \\ &= sL(f) - f(0) \end{aligned}$$

Applying the same analysis to the Laplace transform of the second derivative gives the following result:

$$L(f'') = s^2 L(f) - sf(0) - f'(0)$$

Laplace Transform of Derivatives
$L(f') = sL(f) - f(0)$
$L(f'') = s^2 L(f) - sf(0) - f'(0)$

EXAMPLE 2 If $f(0) = 0$ and $f'(0) = 2$, express the transform of $f''(t) - 5f'(t)$ in terms of s and the transform of $f(t)$.

Solution Applying the linearity property gives

$$L[f''(t) - 5f'(t)] = L(f'') - 5L(f')$$

Using the formula for the transform of a derivative gives

$$\begin{aligned} L[f''(t) - 5f'(t)] &= [s^2 L(f) - sf(0) - f'(0)] - 5[sL(f) - f(0)] \\ &= s^2 L(f) - 0 - 2 - 5sL(f) - 0 \\ &= (s^2 - 5s)L(f) - 2 \end{aligned}$$

■

Inverse Laplace Transform

If the Laplace transform of a function is known, it is possible to find the original function by finding the inverse transform, denoted L^{-1}.

$$L^{-1}(F) = f(t)$$

The original function will be denoted by a lowercase letter, and its transform by the same letter in uppercase.

EXAMPLE 3 Find the inverse Laplace transform of

$$\frac{1}{(s+2)^2}$$

Solution From Transform (11) in Table 23–2, we see that $a = 2$. Therefore,

$$L^{-1}(F) = L^{-1}\left(\frac{1}{(s+2)^2}\right) = te^{-2t}$$

or

$$f(t) = te^{-2t}$$ ■

EXAMPLE 4 Find the inverse transform of the function

$$F(s) = \frac{s+5}{s^2+4s+13}$$

Solution Since this formula does not fit any of the formulas in Table 23–2, we must rearrange its form. We can write $s^2 + 4s + 13$ as

$$(s^2 + 4s + 4) + 9 = (s+2)^2 + 3^2$$

In turn, the numerator can be written as $(s + 2) + 3$. Thus,

$$\begin{aligned} F(s) &= \frac{s+5}{s^2+4s+13} \\ &= \frac{(s+2)+3}{(s+2)^2+3^2} \\ &= \frac{s+2}{(s+2)^2+3^2} + \frac{3}{(s+2)^2+3^2} \end{aligned}$$

Taking the inverse Laplace transform of each term gives

$$\begin{aligned} L^{-1}(F) &= e^{-2t}\cos 3t + e^{-2t}\sin 3t \\ &= e^{-2t}(\cos 3t + \sin 3t) \end{aligned}$$ ■

Differential Equations

Now we will show how to solve ordinary linear differential equations with constant coefficients by using Laplace transforms. The Laplace transform provides an algebraic method by finding the particular solution to such equations.

EXAMPLE 5 Solve the following differential equation by using the Laplace transform:

$$y' + 4y = 0 \quad \text{where } y(0) = 2$$

Solution Using the formula for the Laplace transform of a derivative, we can write the equation as follows:

$$\begin{aligned} L(y') + L(4y) &= L(0) \\ L(y') + 4L(y) &= 0 \\ [sL(y) - y(0)] + 4L(y) &= 0 \\ sL(y) - y(0) + 4L(y) &= 0 \\ sL(y) - 2 + 4L(y) &= 0 \end{aligned}$$

Solving for $L(y)$ gives

$$\begin{aligned} (s + 4)L(y) &= 2 \\ L(y) &= \frac{2}{s + 4} \\ L(y) &= 2\left(\frac{1}{s + 4}\right) \end{aligned}$$

We find the inverse transform from Table 23–2 by using Transform (3) where $a = -4$.

$$y = 2e^{-4t}$$

■

EXAMPLE 6 A small body of mass 2 kg is attached to a spring whose spring constant is 10. Find the displacement y of the body, starting from the initial position $y(0) = 2$ with the initial velocity $y'(0) = -4$. The differential equation is $y'' + 2y' + 5y = 0$.

Solution Solving the equation using the Laplace transform gives

$$\begin{aligned} y'' + 2y' + 5y &= 0 \\ L(y'') + 2L(y') + 5L(y) &= L(0) \\ [s^2L(y) - sy(0) - y'(0)] + 2[sL(y) - y(0)] + 5L(y) &= 0 \\ s^2L(y) - 2s + 4 + 2sL(y) - 4 + 5L(y) &= 0 \\ (s^2 + 2s + 5)L(y) &= 2s \\ L(y) &= \frac{2s}{s^2 + 2s + 5} \end{aligned}$$

It appears that this function does not fit any of the formulas in Table 23–2 for the inverse Laplace transform. However,

$$s^2 + 2s + 5 = (s^2 + 2s + 1) + 4 = (s + 1)^2 + 2^2$$

Therefore, the equation can be written as follows:

$$\begin{aligned} L(y) &= \frac{2s}{s^2 + 2s + 5} \\ &= \frac{2(s + 1) - 2}{(s + 1)^2 + 2^2} \\ &= 2\frac{s + 1}{(s + 1)^2 + 2^2} - \frac{2}{(s + 1)^2 + 2^2} \end{aligned}$$

Using Transforms (20) and (19) from Table 23–2 gives the following result:

$$L^{-1}(y) = e^{-t}2 \cos 2t - e^{-t}\sin 2t$$

or

$$y = e^{-t}(2 \cos 2t - \sin 2t)$$

■

EXAMPLE 7 An electric circuit contains a 1-H inductor, a 40-Ω resistor, and a 6-V battery. Find the current as a function of time if the initial current is zero.

Solution The differential equation to solve is

$$\frac{di}{dt} + 40i = 6$$

Following the procedure given in Example 6 gives

$$\begin{aligned} L\left(\frac{di}{dt}\right) + L(40i) &= L(6) \\ [sL(i) - 0] + 40L(i) &= \frac{6}{s} \\ (s + 40)L(i) &= \frac{6}{s} \\ L(i) &= \frac{6}{s(s + 40)} \\ L(i) &= 0.15\left(\frac{40}{s(s + 40)}\right) \\ i &= 0.15(1 - e^{-40t}) \end{aligned}$$

■

23–6 EXERCISES

Use Table 23–2 to find the Laplace transform of the given functions.

1. $f(t) = 1 - e^{-4t}$

2. $f(t) = \cos 2t + \sin 3t$

3. $f(t) = 1 - \cos 2t + te^{-t}$

4. $f(t) = t \sin 2t + 1$

5. $f(t) = e^{-t} - e^{-2t}$

6. $f(t) = e^{-2t}\sin 4t$

7. $f(t) = (3t + 1)e^{-t}$

8. $f(t) = t^2e^{-2t} + t \sin 3t$

9. $f(t) = 3e^{-3t} - 2e^{-2t}$

10. $f(t) = \sin 2t + 2t \cos 2t$

Find the inverse Laplace transform of the given function.

11. $F(s) = \dfrac{1}{s(s + 1)}$

12. $F(s) = \dfrac{2s - 6}{s^2 - 1}$

13. $F(s) = \dfrac{6}{(s + 5)^2 + 36}$

14. $F(s) = \dfrac{s}{s^2 + 16}$

15. $F(s) = \dfrac{8}{s^4 + 4s^2}$

16. $F(s) = \dfrac{2}{s^2 + 8s + 15}$

17. $F(s) = \dfrac{s^2 - 10s - 25}{s^3 - 25s}$

18. $F(s) = \dfrac{s^2 - 6s + 4}{s^3 - 3s^2 + 2s}$

Solve the following differential equations by using the Laplace transform.

19. $y'' - 4y' + 3y = 0;\ y(0) = 3,\ y'(0) = 7$

20. $y'' - y = 0;\ y(0) = 0,\ y'(0) = 1$

21. $y'' - 5y' + 6y = 0;\ y(0) = 0,\ y'(0) = 1$

22. $y' + 3y = 1;\ y(0) = 0$

23. $y'' + 4y = 0;\ y(0) = 0,\ y'(0) = 1$

24. $y'' - 2y' - 3y = 4;\ y(0) = 1,\ y'(0) = -1$

25. $y'' + 2y' + y = 0;\ y(0) = 0,\ y'(0) = 1$

26. $y' + 2y = te^{-t};\ y(0) = 0$

27. $y' + 4y = 5;\ y(0) = 3$

28. $y'' - 4y' + 3y = e^{2t};\ y(0) = 0,\ y'(0) = 0$

29. A 100-Ω resistor, a 200-μF capacitor, and a 6-V battery are connected in series. Find the charge as a function of time if the initial charge on the capacitor is zero.

30. A 3-H inductor, a 75-Ω resistor, and a 9-V battery are connected in series. Find the current in the circuit as a function of time if the initial current is zero.

31. A 10-H inductor, an 80-μF capacitor, and a 12-V battery are connected in series. Find the current as a function of time if the initial current is zero and the initial charge on the capacitor is zero ($q(0) = 0$).

32. An object moves with simple harmonic motion according to the differential equation $y'' + 200y = 0$. Find the displacement y as a function of time if $y(0) = 0$ and $y'(0) = \sqrt{200}$.

33. An object attached to a vibrating spring undergoes a displacement given by $y'' + 8y = 3 \sin 2t$. Find its displacement as a function of time if $y(0) = 4$ and $y'(0) = 0$.

CHAPTER SUMMARY

Summary of Terms

complementary solution (p. 913)
critically damped (p. 920)
degree (p. 880)
differential equation (p. 880)
exponential decay (p. 897)
exponential growth (p. 897)
first-order differential equation (p. 880)
general solution (p. 881)
homogeneous (p. 907)
Laplace transform (p. 923)
linear (p. 890)
linearity property (p. 925)
method of undetermined coefficients (p. 913)
nonhomogeneous (pp. 907 and 912)
order (p. 880)
orthogonal trajectory (p. 895)
overdamped (p. 920)
particular solution (p. 882)
second-order differential equation (p. 880)

Summary of Formulas

$$\left.\begin{aligned} d(xy) &= x\,dy + y\,dx \\ d(x^2 + y^2) &= 2(x\,dx + y\,dy) \\ d\left(\frac{y}{x}\right) &= \frac{x\,dy - y\,dx}{x^2} \\ d\left(\frac{x}{y}\right) &= \frac{y\,dx - x\,dy}{y^2} \end{aligned}\right\} \text{integrable combinations}$$

$$\left.\begin{aligned} &\frac{dy}{dx} + P(x)y = Q(x) \text{ whose solution is} \\ &y = \frac{1}{e^{\int P(x)\,dx}} \int Q(x)e^{\int P(x)\,dx}\,dx \end{aligned}\right\} \text{first-order linear}$$

$$y = C_1e^{m_1x} + C_2xe^{m_2x} + C_3x^2e^{m_3x} + \cdots + C_nx^{n-1}e^{mx} \quad \text{homogeneous; repeated roots}$$

$$y = e^{ax}(C_1\cos bx + C_2\sin bx) \quad \text{homogeneous; complex roots}$$

$$L(f) = F(s) = \int_0^\infty e^{-st}f(t)\,dt \quad \text{Laplace transform}$$

$$\left.\begin{aligned} L(f') &= sL(f) - f(0) \\ L(f'') &= s^2L(f) - sf(0) - f'(0) \end{aligned}\right\} \text{Laplace transform of derivatives}$$

CHAPTER REVIEW

Section 23–1

Solve the following differential equations by direct integration or separation of variables.

1. $y\,dx - 4x\,dy = 0$
2. $\dfrac{dI}{dt} = 2\sin t - 5\cos t$
3. $\dfrac{dI}{dt} = e^t\cos t$
4. $\dfrac{dy}{dx} = \dfrac{\cos x}{\sin y}$; $x = \dfrac{\pi}{2}$ when $y = 0$
5. $ye^x\,dx + dy = 0$
6. $(x - 1)\,dx + e^{x+y}\,dy = 0$
7. $xy' + 3x^2y = y$; $x = 0$, $y = 1$
8. $y' = \dfrac{y}{x}$; $x = 1$, $y = 2$
9. $x^2y\,dx + y^2\,dy = 0$; $x = 1$, $y = 1$
10. $(2 + x^2)y' = 2y$
11. The slope of a curve is given by $3x^2y$. Find the equation of the curve if it passes through the point (1, 1).
12. If a resistance R and a capacitance C are connected in parallel, the voltage across the capacitor V_c, is given by

$$V_c + iR = 0 \quad \text{where } i = C\frac{dV_c}{dt}$$

Solve for V in terms of t if $V = 0$ when $t = 0$.

Section 23–2

Solve the following differential equations by using an integrable combination or integrating factor.

13. $x^2\,dx + y\,dy = -y^2\,dx - x\,dx$

14. $(y^2 - y)\,dx + x\,dy = 0$

15. $xy' - y = 7x^2$

16. $dy + y\,dx = e^x\,dx$

17. $\dfrac{dy}{dx} = x^3 - 4x^3y$

18. $\dfrac{dy}{dx} = \sin x - y$

19. $x\dfrac{dy}{dx} + y = xe^x$

20. $\dfrac{dy}{dx} + xy - 2x = 0$

Section 23–3

21. A tank initially contains 100 gal of brine in which there is dissolved 25 lb of salt. Starting at time $t = 0$, brine containing 4 lb/gal flows into the tank at a rate of 5 gal/min. The mixture is kept uniform by stirring. How much salt is in the tank after 20 min if the brine flows out of the tank at a rate of 5 gal/min?

22. Find the orthogonal trajectories of the family of curves $y = e^{cx}$.

23. A culture of bacteria grows at a rate proportional to the number present. If 25,000 bacteria are present at the start of an experiment, and if the number of bacteria doubles in 2 hours, how many bacteria will be present after 3 hours?

24. The intensity of light I emitted by the phosphor of a television tube decreases at a rate proportional to the intensity. Find the expression for the intensity as a function of time if the initial intensity is 30 and the intensity at $t = 1$ is 25.

25. A radioactive isotope decays in such a way that half the original amount decays in 7 hours. Find the amount present after 10 hours.

26. A mass of 5 slugs is falling under the influence of gravity. Find its velocity after 4 s if it starts from rest and experiences a retarding force equal to 0.1 times its velocity.

27. An RL circuit with $R = 15\ \Omega$ and $L = 0.2$ H has an applied voltage of 50 V. Find an equation relating current as a function of time if $I = 0$ when $t = 0$.

Section 23–4

Solve the following homogeneous differential equations.

28. $y'' + y' - 2y = 0$

29. $y'' + 7y = 0$

30. $D^2y + 4Dy + 13y = 0$

31. $D^2y - 4Dy = -4y$

32. $\dfrac{d^2y}{dx^2} + 3\dfrac{dy}{dx} + 2y = 0$; $y(0) = 1$, $y'(0) = 0$

33. $\dfrac{d^2x}{dt^2} + x = 0$; $x(0) = 1$, $x(\pi/2) = 1$

34. $2y'' + y' - y = 0$

35. $y'' + 4y = 0$; $y(0) = 2$, $y'(0) = -8$

36. $y'' - 4y' + 4y = 0$; $y(0) = -4$, $y'(0) = -6$

37. $y'' = 2y' + y = 0$

Section 23–5

Solve the following nonhomogeneous differential equations.

38. $y'' + 2y' + 5y = \sin x$

39. $y'' - y' - 12y = e^{3x}$

40. $D^2y - 2Dy = 3\sin 2x + \cos 2x$

41. $D^2y - y = \cos 2x$

42. $\dfrac{d^2y}{dx^2} - 4\dfrac{dy}{dx} = 12e^{2t}$; $y(0) = 1$, $y'(0) = 2$

43. $y'' + y = x^2 + 2$; $y(0) = 0$, $y'(0) = 2$

44. $y'' - 4y' + 3y = 4e^{3x}$

45. $y'' + 4y' + 4y = 4\cos x + 3\sin x$

Section 23–6

Use Table 23–2 to find the Laplace transform of the given function.

46. $f(t) = t\cos 3t + e^{-2t}$

47. $f(t) = \sin 3t + 3t\cos 3t$

48. $f(t) = 1 + t^4e^{-3t}$

49. $f(t) = e^{-4t}\cos 5t$

Find the inverse Laplace transform of the given function.

50. $F(s) = \dfrac{1}{(s + 2)^2}$

51. $F(s) = \dfrac{s + 4}{(s + 3)^2}$

52. $F(s) = \dfrac{4 - s}{s(s^2 + s - 2)}$

53. $F(s) = \dfrac{s + 13}{s^2 + s - 6}$

Solve the following differential equations by using Laplace transforms.

54. $y'' + 4y = \sin 2t;\ y(0) = 1,\ y'(0) = 1$

55. $y'' + y' - 2y = 3\cos 3t - 11\sin 3t;\ y(0) = 0,$ $y'(0) = 6$

56. $y'' - 9y = e^t;\ y(0) = 0,\ y'(0) = 1$

57. $y'' - 9y = \sin t;\ y(0) = 1,\ y'(0) = 0$

58. $y'' - y' - 12y = -24e^t - 12;$ $y(0) = 5,\ y'(0) = 3$

59. $y'' - 4y' + 5y = 0;\ y(0) = 0,\ y'(0) = 1$

CHAPTER TEST

The number in parentheses refers to the appropriate learning objective given at the beginning of the chapter.

1. Solve the following by direct integration: **(2)**

$$2x\frac{dy}{dx} + x^2 = 3x$$

2. Solve the following linear differential equation by using an integrating factor: **(5)**

$$y'' + y' + 4y = 0$$

3. Solve the following homogeneous differential equation: **(8)**

$$D^2y - 8Dy + 16y = 0$$

4. A tank initially contains 100 gal of brine in which 20 lb of salt is dissolved. Starting at $t = 0$, pure water flows into the tank at a rate of 10 gal/min. The mixture is kept uniform by stirring. How much salt is in the tank at the end of 10 min if the brine flows out of the tank at a rate of 10 gal/min? **(6)**

5. Find the order and degree of the following differential equation: **(1)**

$$4(y')^2 - 3x(y'')^3 + 6y'' = \sin x + y$$

6. Solve the following by separation of variables: **(3)**

$$y' = \frac{xy}{x^2 + 2} + 3y$$

7. Solve the following linear differential equation by using an integrating factor: **(5)**

$$\frac{dy}{dx} + \frac{y}{x} = \sin x$$

8. Solve the following by using an integrable combination: **(4)**

$$x(1 - 3x)\,dx = -y\,dy$$

9. A generator having voltage given by $V = 20 \sin 8t$ volts is connected in series with a 7-Ω resistor and an inductor of 3 H. If $i = 0$ at $t = 0$, find the current as a function of time. **(6)**

10. Solve the differential equation **(10)**

$$y'' + 2y = \sin t \quad \text{where } y(0) = 0 \text{ and } y'(0) = 1$$

by using the Laplace transform.

11. Solve the following nonhomogeneous equation: **(9)**

$$y'' - 2y' - 3y = 2e^x - 10 \sin x$$

12. Solve the following homogeneous differential equation: **(8)**

$$\frac{d^2y}{dx^2} - 4\frac{dy}{dx} + 13y = 0$$

13. Find an equation for the charge on a capacitor at time t in an RCL circuit where $L = 1$ H, $R = 9\ \Omega$, $C = 0.2$ F, and $V = 15 \sin 60t$. **(6, 9)**

14. Solve the following differential equation by using direct integration: **(7)**

$$\frac{d^2y}{dx^2} = 2x + 7$$

15. A 6-H inductor and 20-Ω resistor are connected in series. Find the current as a function of time if the initial current is 10 A. Use Laplace transforms. **(10)**

SOLUTION TO CHAPTER INTRODUCTION

The differential equation that we must solve is

$$\frac{dT}{dt} = 2T^2$$

We solve this equation by separation of variables.

$$dT = 2T^2\,dt$$

$$\frac{dT}{T^2} = 2\,dt$$

Then we integrate both sides of the equation and obtain

$$-\frac{1}{T} = 2t + C$$

Next, we solve for the constant of integration by substituting $T = 0.1$ and $t = 0$. Thus,

$$-\frac{1}{0.1} = 2(0) + C \rightarrow C = -10$$

$$-\frac{1}{T} = 2t - 10$$

Last, we solve for t when $T = 1.2$ cm.

$$-\frac{1}{1.2} = 2t - 10$$
$$4.58 \approx t$$

The pilot can sit on the runway about 4.5 min before the wings have too much ice on them for takeoff.

Appendix A

A Review of Basic Algebraic Concepts

A basic understanding of algebra is important to every student entering a technical field. This appendix reviews the algebraic concepts found in most elementary algebra courses. Topics include the following:

- ☐ Real numbers
- ☐ Laws and operations of real numbers
- ☐ Zero and the order of operations
- ☐ Exponents
- ☐ Scientific notation
- ☐ Roots and radicals
- ☐ Addition and subtraction of algebraic expressions
- ☐ Multiplying polynomials
- ☐ Polynomial division
- ☐ Percentages

We will begin the review with a discussion of real numbers.

A–1 REAL NUMBERS

Imagine all the computations a technician must make daily in order to solve problems or finalize reports. Such calculations require a thorough understanding and working knowledge of the operations, rules, and symbols of real numbers.

Integers

Historically, the different types of numbers evolved from the need for a number to express a real-life situation. For example, counting the number of desks in a classroom or expressing how many items are contained in a shipment requires the use of numbers called **integers.** The counting numbers such as 1, 2, 3 are called **positive integers.** The **negative integers,** such as -1, -2, -3 are often used to show a decrease in a quantity. **Zero** is also an integer, but it is neither positive nor negative. The group of numbers designated as integers includes the positive and negative integers and zero.

To represent part of a group or a part of a whole requires the use of a rational number. A **rational number** is any number in the form a/b (ratio) where a is an integer and b is nonzero integer. The numbers 3/5, $-1/2$, and $-15/4$ are examples of rational numbers. Since any integer can also be expressed in the form a/b where b equals 1, all integers are also rational numbers.

Rational Numbers

A rational number may be expressed as a decimal by dividing the denominator into the numerator. The result is either a terminating or repeating decimal.

EXAMPLE 1 Express the following rational numbers in decimal form:

(a) $\frac{4}{5}$ (b) $\frac{1}{3}$

Solution

(a) Dividing 5 into 4 gives 0.8, a terminating decimal.

(b) $\frac{1}{3} = 0.333\ldots$ is a repeating decimal because the digit 3 repeats. ■

When the length of the diagonal of a unit square (See Figure A–1) is computed, the result is the number $\sqrt{2}$, read "square root of two." Such numbers are called irrational numbers.

FIGURE A–1

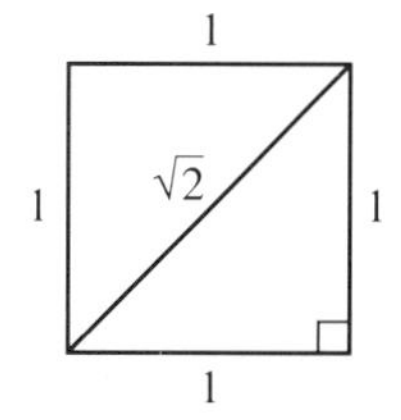

Irrational Numbers

Irrational numbers are numbers that cannot be expressed in the form of a rational number (a/b where a is an integer and b is nonzero integer). In decimal form, irrational numbers neither terminate nor repeat. For instance, the $\sqrt{2}$ expressed in decimal form equals 1.4142 . . . . The decimal part of this number will never form a repetitive pattern nor terminate. For this reason, when decimal forms of irrational numbers are used in calculations, it is necessary to round to some approximate value.

EXAMPLE 2 Using a calculator, express the following irrational numbers in decimal form:

(a) $\sqrt{8}$ (b) $\sqrt[3]{12}$

Solution

(a) $\sqrt{8} = 2.8284271\ldots$ is a nonrepeating decimal.
(b) $\sqrt[3]{12} = 2.2894285\ldots$ is a nonrepeating decimal. ■

Irrational numbers are often encountered when finding the square root, or higher root, of certain numbers. The square root of any number that is not a perfect squared number is an irrational number. Two quantities often found in technical formulas, π and e, are both irrational numbers because they cannot be written as either terminating or repeating decimals.

EXAMPLE 3 Using a calculator, express the following irrational numbers in decimal form:

(a) π (b) e

Solution

(a) Pressing the π key on your calculator gives $3.1415926\ldots$, which is a nonrepeating decimal.
(b) Pressing 1 and e^x on your calculator gives $2.7182818\ldots$, which is a nonrepeating decimal. ■

The majority of technical applications involve only rational and irrational numbers. The rational and irrational numbers are combined to make the real numbers. In this appendix when we speak of numbers, we mean real numbers unless stated otherwise.

EXAMPLE 4 Classify each of the following numbers as integer, rational, or irrational. A given number may be in more than one category.

(a) $\frac{2}{3}$ (b) -8 (c) $-\sqrt{4}$ (d) 0.5

(e) $\sqrt{3}$ (f) $-\frac{\sqrt{5}}{8}$ (g) $\frac{\pi}{4}$ (h) $\frac{e}{2}$

Solution

(a) $\frac{2}{3}$ is a rational number because it is represented as a ratio.
(b) -8 is both an integer and a rational number.
(c) $-\sqrt{4} = -2$ is both an integer and a rational number.
(d) 0.5 is a rational number because it is expressed as a terminating decimal.
(e) $\sqrt{3}$ is an irrational number because $\sqrt{3} = 1.7320508\ldots$, a nonrepeating, nonterminating decimal.
(f) $-\frac{\sqrt{5}}{8}$ is an irrational number because it is a nonrepeating, nonterminating decimal.
(g) $\frac{\pi}{4}$ is an irrational number.
(h) $\frac{e}{2}$ is an irrational number. ■

Complex Numbers

Some applications in physics and in the field of electronics involve complex numbers. **Complex numbers** are numbers in the form $a + bj$ where a and b are real numbers and $j = \sqrt{-1}$. When working with quadratic equations, you will encounter the square root of a negative number, such as $\sqrt{-36}$. Such a number is called an **imaginary number** and is equivalent to $6j$ or the imaginary part of a complex number. Complex numbers are explained in greater detail in Chapters 4 and 5.

Real Number Line

The **real number line** is a pictorial representation of the real number system. Each real number can be associated with one and only one point on the line; conversely, each point on the line can be assigned only one real number.

The point associated with the real number zero is called the **origin.** It is customary to locate positive values to the right of zero and negative values to the left as shown in Figure A–2.

FIGURE A–2

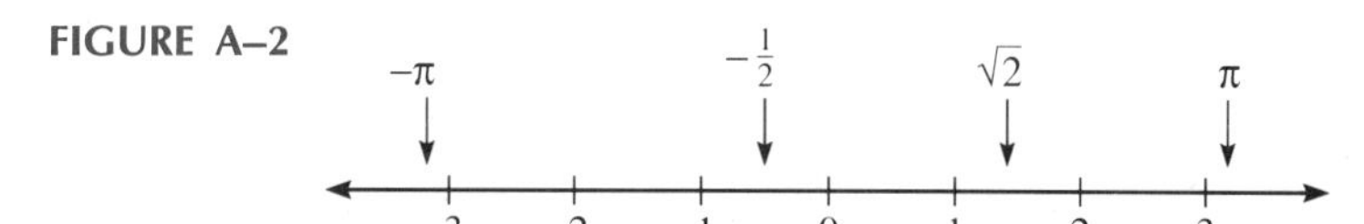

The relative position of two real numbers on the number line can be clarified using the equality symbol, = (equal to), and the two inequality symbols, $<$ (less than) and $>$ (greater than).

EXAMPLE 5 For each of the following, place the appropriate symbol ($=$, $<$, or $>$) between the two numbers to make the statement true.

(a) $\frac{10}{2} \square 5$ (b) $-1.5 \square 0$ (c) $3 \square -2$

Solution

(a) $\frac{10}{2} = 5$ (Read 10 divided by 2 is equal to 5.) Since these values are equal, they would occupy the same position on the number line.
(b) $-1.5 < 0$ (Read -1.5 is less than 0.) Figure A–3 shows the position of these numbers on the number line.
(c) $3 > -2$ (Read 3 is greater than -2.) Figure A–4 shows the relative position of -2 and 3 on the number line.

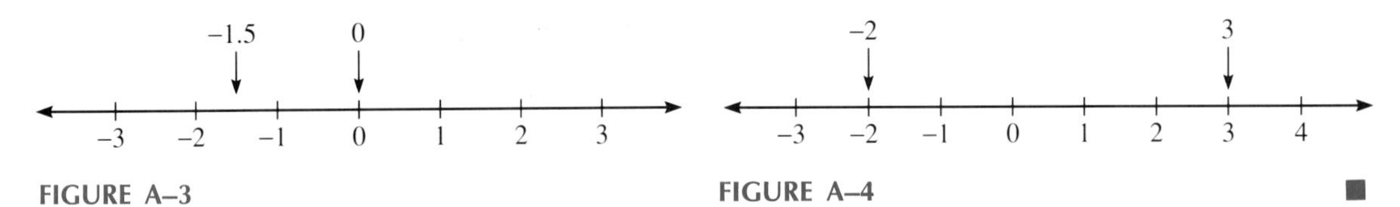

FIGURE A–3

FIGURE A–4

■

The pictorial or graphical representation of the numbers in parts (b) and (c) in Example 5 demonstrates the following fact: Any number to the left of another number on the number line is less than that number; any number to the right of another number on the number line is greater than that number.

EXAMPLE 6 Decide if the following statements are true or false.

(a) $-5 < -2$ (b) $\sqrt{3} > 2$ (c) $\frac{1}{3} > -\frac{3}{5}$ (d) $-\frac{\pi}{2} < -0.5$

Solution

(a) True, because -5 is to the left of -2 (Figure A–5).
(b) False, because $\sqrt{3} \approx 1.7$ is not to the right of 2 (Figure A–6).
(c) True (see Figure A–7).
(d) True (see Figure A–8).

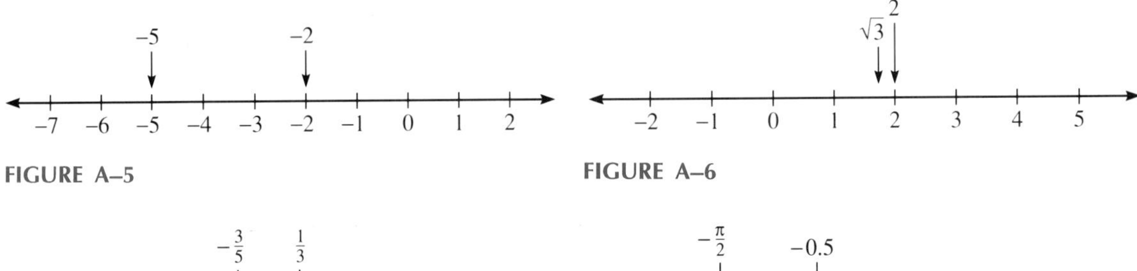

FIGURE A–5

FIGURE A–6

FIGURE A–7

FIGURE A–8

■

Absolute Value

When applying the rules for adding or subtracting real numbers, you need to know the meaning of absolute value. The **absolute value** of a number is the distance from the origin to the number. Since distance is always positive or zero, the absolute value of a number is always positive or zero. Two vertical bars $|\ \ |$ placed around a number is the mathematical symbol used to indicate the absolute value of that number.

EXAMPLE 7 Find the absolute value of the following numbers:

(a) $|-3|$ (b) $\left|\frac{1}{2}\right|$ (c) $|-\sqrt{30}|$

Solution

(a) As shown in Figure A–9, the distance from the origin to -3 equals 3. Therefore, $|-3| = 3$.

(b) $\left|\frac{1}{2}\right| = \frac{1}{2}$

(c) $|-\sqrt{30}| = \sqrt{30}$ ■

FIGURE A–9

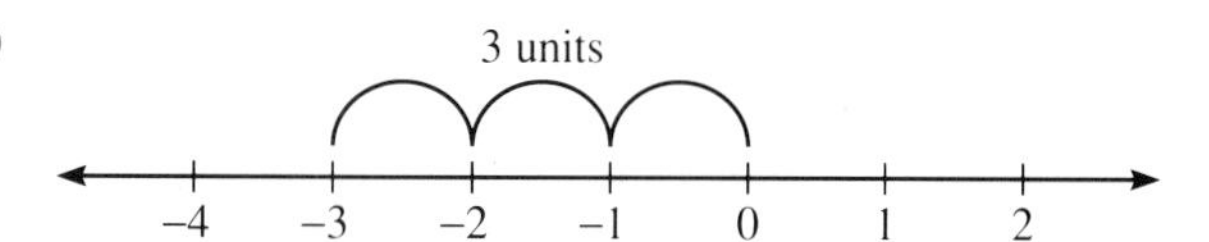

A–1 EXERCISES

Classify each of the following numbers as integer, rational number, or irrational number. A given number may be in more than one category.

1. $\frac{1}{3}$

2. $-\sqrt{4}$

3. 0.575

4. $\sqrt{12}$

5. -5

6. $\frac{\pi}{2}$

7. $-\sqrt{144}$

8. 0

9. $15\frac{1}{4}$

10. $-\frac{1}{8}$

Decide if the following are true or false.

11. $-\frac{1}{2} < -1$

12. $\pi > 3$

13. $-5 < -3$

14. $\frac{1}{8} = 0.128$

15. $\frac{5}{13} > \frac{4}{15}$

16. $0 > |-0.5|$

17. $13\frac{2}{3} > \sqrt{271}$

18. $-3.2 < -3.5$

19. $-|5| < 2$

20. $|-8| < |-4|$

21. Locate the following numbers on the same number line:

$$-\frac{1}{2}, \quad 1.3, \quad \sqrt{5}, \quad -3, \quad 4.5$$

22. Locate the following numbers on the same number line:

$$\sqrt{20}, \quad -2\frac{2}{3}, \quad \left|\frac{\pi}{3}\right|, \quad -1.45, \quad 0.33$$

23. Locate the following numbers on the number line.

a. The temperature when it drops to 10 degrees below zero.

b. The length of a structural beam found to be $\sqrt{40}$ m.

c. The circumference of a circle found to be π cm.

d. A current source of 5 negative units.

A–2

LAWS AND OPERATIONS OF REAL NUMBERS

One main characteristic of algebra that distinguishes it from arithmetic is its use of literal numbers. Literal numbers are used extensively in algebra, and the laws pertaining to literal numbers must be defined explicitly. A **literal number** is a letter or other symbol used to represent a real number. Using literal numbers allows you to write a single expression to represent several different situations. For example, the perimeter of any square is the sum of all its sides. The formula $P = 4s$ describes the perimeter of any square no matter what the length of the sides.

When adding or multiplying literal numbers, understanding the laws of operations is essential.

Laws of Operations

Commutative Law

Addition:	$a + b = b + a$
Multiplication:	$a \cdot b = b \cdot a$

Associative Law

Addition:	$(a + b) + c = a + (b + c)$
Multiplication:	$(a \cdot b) \cdot c = a \cdot (b \cdot c)$

Distributive Law

$$a(b + c) = a \cdot b + a \cdot c$$

$$(b + c)a = b \cdot a + c \cdot a$$

Commutative Law

The commutative law states that the order in which you add or multiply two real numbers does not affect the sum or product. For example, the commutative law means that $5 + 3 = 3 + 5$ and $2 \cdot 8 = 8 \cdot 2$. However, the commutative law does not apply to the operations of subtraction or division.

Associative Law

Another important law is the associative law. The associative law states that the grouping of real numbers in addition or multiplication does not alter the results. For example, $(5 + 3) + 2 = 5 + (3 + 2)$ and $(2 \cdot 6) \cdot 8 = 2 \cdot (6 \cdot 8)$. Again, the associative law does not apply to the operations of subtraction or division.

Distributive Law

Many times in formulas and applications you need to multiply a number by the sum of two other numbers. The distributive law allows the expression of this product. Applying the distributive law gives $5(3 + 4) = 5 \cdot 3 + 5 \cdot 4$ and $(8 + 2)6 = 8 \cdot 6 + 2 \cdot 6$. The distributive law is used extensively in evaluating and solving equations and formulas.

EXAMPLE 1 For each of the following equations, indicate whether the commutative, associative, or distributive law is involved.

(a) $3.5 + 2.4 = 2.4 + 3.5$
(b) $1.002 \cdot 2 = 2 \cdot 1.002$
(c) $(6 + 24) + 8 = 6 + (24 + 8)$
(d) $(3 \cdot 5) \cdot 10 = 3 \cdot (5 \cdot 10)$
(e) $9(2 + 4) = 9 \cdot 2 + 9 \cdot 4$

Solution

(a) commutative because the order of the numbers is changed
(b) commutative because the order of the numbers is changed
(c) associative because the grouping of the numbers is changed
(d) associative because the grouping is changed
(e) distributive because multiplication is distributed over addition ■

Addition and Subtraction

The laws stated thus far are true for all real numbers and help to simplify operations with algebraic expressions. To clarify the four fundamental operations of addition, subtraction, multiplication, and division with signed numbers more rules are needed. The following two rules are given for the operation of addition.

Adding Real Numbers

Rule 1 To add two real numbers with like signs, find the sum of their absolute values and attach their common sign to the sum.

Rule 2 To add two real numbers with unlike signs, find the difference of their absolute values and attach the sign of the number with the larger absolute value.

EXAMPLE 2 Perform the operation of addition.

(a) $3 + 8$ (b) $(-4) + (-5)$ (c) $(-10) + 3$
(d) $12 + (-8)$ (e) $(-2.5) + (5.4)$

Solution

(a) $3 + 8 = 11$; since the numbers have the same sign, we add their absolute values and use their common sign, +.
(b) $(-4) + (-5) = -9$; both numbers have the same sign. Therefore, we add the numbers and use their common sign.
(c) $(-10) + (3) = -7$; the numbers have unlike signs. Therefore, we subtract their absolute values and use the sign of the larger.
(d) $12 + (-8) = 4$; the sum is positive because 12 is larger in absolute value.
(e) $(-2.5) + (5.4) = 2.9$; the sum is positive because 5.4 is larger in absolute value. ■

The operation of subtraction is not difficult if the following rule is applied carefully.

Subtracting Real Numbers

Rule 3 To subtract two real numbers, change the subtraction sign to an addition sign and change the sign of the subtrahend (the number to be subtracted). Then apply the rules of addition.

EXAMPLE 3 Perform the operation of subtraction.

(a) $18 - 20$ (b) $(-7) - 2$ (c) $21 - (-3)$
(d) $(-3.8) - (-12.9)$ (e) $(-25) - 10.5$

Solution

(a) $18 - 20 = 18 + (-20) = -2$; change to addition, change sign of subtrahend, apply rules of addition.
(b) $(-7) - 2 = -7 + (-2) = -9$
(c) $21 - (-3) = 21 + (+3) = 24$
(d) $(-3.8) - (-12.9) = (-3.8) + (+12.9) = 9.1$
(e) $(-25) - (10.5) = (-25) + (-10.5) = -35.5$ ■

Multiplication and Division

The next rule applies for multiplication or division of signed numbers.

Multiplication or Division of Real Numbers

Rule 4 The product or quotient of two real numbers with like signs is positive. The product or quotient of two real numbers with unlike signs is negative.

In other words, multiplying or dividing an even number of negative signs results in a positive answer; multiplying or dividing an odd number of negative signs results in a negative answer.

EXAMPLE 4 Perform the indicated operations.

(a) $5 + -4$ (b) $(-3)(-4)$
(c) $45 \div (-5)$ (d) $12 - (-4)$
(e) $(-9) + (-4)$ (f) $10 - 15$
(g) $(-2.5)(-1.8)(-3.6)$ (h) $-8.2 \div (-2)$

Solution

(a) 1 (b) 12 (c) −9 (d) 16
(e) −13 (f) −5 (g) −16.2 (h) 4.1 ■

A–2 EXERCISES

Perform the indicated operations.

1. $(-12) + 6$
2. $7 + 2$
3. $(-3) + (-14)$
4. $8 + (-12)$
5. $13 - (-5)$
6. $24 - 45$
7. $(-6) - 8$
8. $3(-4)$
9. $(-9)(6)$
10. $15 \div (-5)$
11. $(-28) \div (-7)$
12. $(-2.5) + 1$
13. $(-4.5)(3.1)$
14. $-100 \div 10$
15. $-10.5 - 2.5$
16. $(-9)(3)(-4)$
17. $10 - (-3) + 6$
18. $(-134.6) + 12.5 - 28.7$
19. $-5.3 + (-7.8) - (-1.3)$
20. $6 + (-8) - (-4) - 10$

For each of the following equations, identify the operational law being illustrated.

21. $5(x + y) = 5x + 5y$
22. $45(3 \cdot 8) = (45 \cdot 3)8$
23. $4 + 5 = 5 + 4$
24. $(9 + b)a = 9a + ba$

Solve the following problems.

25. In Denver, Colorado the temperature at 3 P.M. was 12°C. By 10 P.M. it had dropped to −5°C. What is the degree difference between the two temperatures?

26. Temperature on the Celsius (C) scale is related to the Fahrenheit (F) scale by the formula $C = 5/9(F - 32)$. Convert each of the following Fahrenheit temperatures to Celsius temperatures.

a. $F = -22$, find C. b. $F = 77$, find C.
c. $F = -12.2$, find C. d. $F = 48.9$, find C.

27. When a body is moved by a force in the line of action of the force, work is done. The formula for work (w) is w = force (f) × distance (d).

a. f = 100 lb and d = 3 ft, find w.
b. f = −60 lb and d = −6 ft, find w.
c. f = −22 lb and d = −2 ft, find w.
d. f = 34 lb and d = 12.5 ft, find w.

28. Using the formula $F = 9/5\ (C + 32)$, convert the following Celsius (C) temperatures to Fahrenheit (F) temperatures.

a. C = 57, find F. b. C = −2, find F.
c. C = −263, find F. d. C = 145, find F.

FIGURE A–10

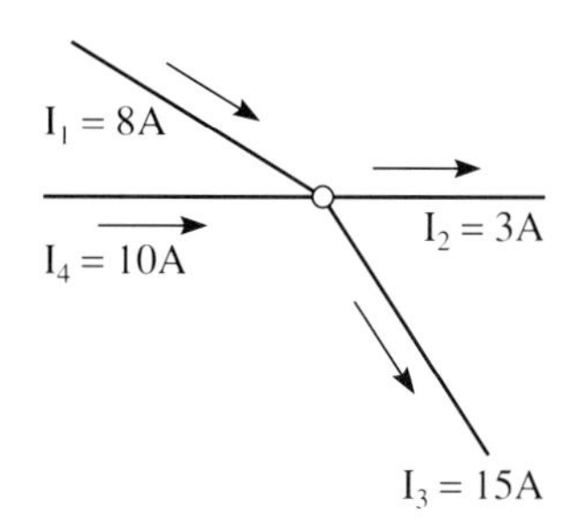

29. In many physics applications the sum of component vectors must be found. Find the sum of each of the following:

a. $\mathbf{A}_x = -24.0$, $\mathbf{A}_y = 15.4$
b. $\mathbf{B}_x = -283$, $\mathbf{B}_y = -246$
c. $\mathbf{A}_x = 35.6$, $\mathbf{A}_y = -10.8$
d. $\mathbf{B}_x = -102.7$, $\mathbf{B}_y = -20.3$

30. The algebraic sum of all currents at a point in a circuit must be zero. Currents toward a point are positive, currents away from it are negative. Find the sum of the currents for the circuit diagrams in Figures A–10 and A–11.

FIGURE A–11

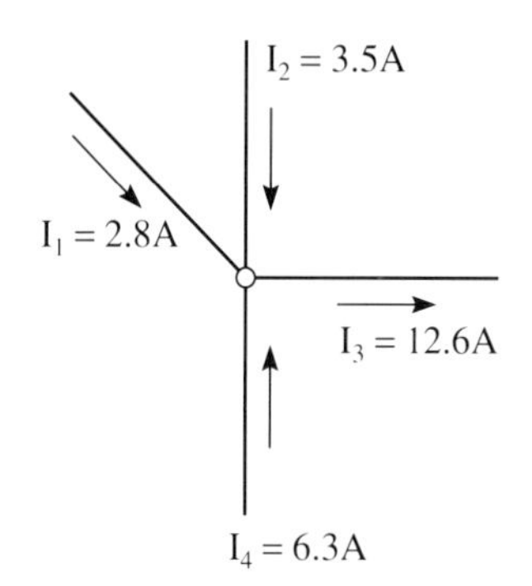

A–3

ZERO AND THE ORDER OF OPERATIONS

Zero is an important real number that is encountered quite often in algebraic and arithmetic operations. Also, the order of performing operations is an important topic in algebra and arithmetic. This section will present the rules and guidelines for both of these topics.

Operations with Zero

The four basic operations of addition, subtraction, multiplication, and division should be performed with zero as summarized below.

Operations with Zero

For any real number, the following statements are true:

Addition	$a + 0 = a$ and $0 + a = a$
Subtraction	$a - 0 = a$ and $0 - a = -a$
Multiplication	$a \cdot 0 = 0$ and $0 \cdot a = 0$
Division	$0/a = 0$ and $a/0$ = undefined

As shown in the box, division by zero is undefined. The reason for this fact is given as follows: $6 \div 3 = 2$ because $6 = 3 \cdot 2$. Similarly,

$0 \div 8 = 0$ because $0 = 8 \cdot 0$. However, applying this same reasoning to division by zero gives $3 \div 0 = x$, but we can find no real number x such that $0 \cdot x = 3$. Therefore, division by zero is undefined.

EXAMPLE 1 Evaluate each of the following:

(a) $0 - 5$ (b) $(-8) + 0$ (c) $4(0)$ (d) $\frac{0}{7}$ (e) $\frac{3}{0}$

Solution

(a) -5 by the subtraction rule
(b) -8 by the addition rule
(c) 0 by the multiplication rule
(d) 0 by the division rule
(e) undefined by the division rule ■

Order of Operations

Often in algebraic problems an expression or formula is encountered that involves a variety of operations and/or grouping symbols. The rules for order of operations are as follows.

Rules for Order of Operations

1. Simplify within grouping symbols.
2. Perform all multiplication and division, proceeding from left to right.
3. Perform all addition and subtraction, proceeding from left to right.

The first rule refers to grouping symbols. The most common symbols of grouping are parentheses ( ), brackets [ ], braces { }, and the fraction bar /. Use the rules to calculate the following example.

EXAMPLE 2 Calculate $10 - 3(5 + 1) \div 2 + 5$.

Solution First, simplify within the parentheses.

$$10 - 3(5 + 1) \div 2 + 5 = 10 - 3\textcircled{6} \div 2 + 5$$

Second, multiply and divide from left to right.

$$= 10 - \textcircled{18} \div 2 + 5 \quad \text{multiply}$$
$$= 10 - \textcircled{9} + 5 \quad \text{divide}$$

Third, add and subtract from left to right.

$$= \textcircled{1} + 5 \quad \text{subtract}$$
$$= 6 \quad \text{add}$$

■

Within a given step from the previous box, remember to simplify from left to right. For example, in the expression $8 \div 2 \cdot 3 \div 6$, multiplication and division are done at the same time moving from left to right as shown.

$$\begin{aligned} 8 \div 2 \cdot 3 \div 6 &= 4 \cdot 3 \div 6 \\ &= 12 \div 6 \\ &= 2 \end{aligned}$$

EXAMPLE 3 Calculate $12 \div 4 + 6(10 - 3) - 2$.

Solution First, simplify within the parentheses.

$$12 \div 4 + 6(10 - 3) - 2 = 12 \div 4 + 6\textcircled{7} - 2$$

Second, multiply and divide from left to right.

$$\begin{aligned} &= \textcircled{3} + 6(7) - 2 && \text{divide} \\ &= 3 + \textcircled{42} - 2 && \text{multiply} \end{aligned}$$

Third, add and subtract from left to right.

$$\begin{aligned} &= \textcircled{45} - 2 && \text{add} \\ &= 43 && \text{subtract} \end{aligned}$$

■

Although the fraction bar is a grouping symbol, it is used to separate a numerator from a denominator denoting division. The next example illustrates the use of the fraction bar as well as the previous rules for order of operations.

EXAMPLE 4 Calculate $\dfrac{5 \cdot 2 - 3 \cdot 8}{10 - 3(4 - 2)}$.

Solution Simplify the numerator and then the denominator according to the rules for the order of operations.

$$\begin{aligned} \frac{5 \cdot 2 - 3 \cdot 8}{10 - 3(4 - 2)} &= \frac{10 - 24}{10 - 3(4 - 2)} && \text{multiply} \\ &= \frac{-14}{10 - 3(4 - 2)} && \text{subtract} \\ &= \frac{-14}{10 - 3(2)} && \text{parentheses} \\ &= \frac{-14}{10 - 6} && \text{multiply} \\ &= \frac{-14}{4} && \text{subtract} \\ &= -3.5 && \text{divide} \end{aligned}$$

■

Try the problem in Example 4 with your calculator. Follow these steps.

[(] 5 [×] 2 [−] 3 [×] 8 [)] [÷] [(] 10 [−] 3 [×] [(] 4 [−] 2 [)] [)] [=]

CAUTION ✦ When performing these operations on your calculator, remember to enter the parentheses.

EXAMPLE 5 Perform the following operations:

(a) $12 + 5(6 - 3 \cdot 4) - 7$ (b) $\dfrac{56 - 4(5.2 - 3.4)}{12.4 + 3(-1.2)}$

Solution

(a)

$$\begin{aligned} 12 + 5(6 - 3 \cdot 4) - 7 &= 12 + 5(6 - 12) - 7 && \text{multiply within parentheses} \\ &= 12 + 5(-6) - 7 && \text{subtract within parentheses} \\ &= 12 - 30 - 7 && \text{multiply} \\ &= -25 && \text{subtract} \end{aligned}$$

(b)

$$\begin{aligned} \frac{56 - 4(5.2 - 3.4)}{12.4 + 3(-1.2)} &= \frac{56 - 4(1.8)}{12.4 + 3(-1.2)} && \text{parentheses} \\ &= \frac{56 - 7.2}{12.4 - 3.6} && \text{multiply} \\ &= \frac{48.8}{8.8} && \text{subtract} \\ &= 5.5454\ldots && \text{fraction bar} \end{aligned}$$

■

A–3 EXERCISES

Perform the indicated operations.

1. $23 + 0$

2. $3(0)$

3. $0 - 12$

4. $6 - 0$

5. $0 - (-8)$

6. $\dfrac{0}{10}$

7. $\dfrac{-16}{0}$

8. $23 - 0$

9. $8 - 3(4)$

10. $3(0) + 2(-4)$

11. $12 \div 2 + 8$

12. $\dfrac{3(-6)}{2} - 3(-4)$

13. $\dfrac{-44}{11} + \dfrac{10 - 2}{0 - 4}$

14. $10 - (36 \div 6) + 3(-5)$

15. $\dfrac{(-12)(-2) - 7(3)}{2(5) + 9}$

16. $1.3 - 2(4.5) + (6.8)$

17. $\dfrac{0(-7) + 45 \div 9}{(-3.5)2 - 3(1.5)}$

18. $\dfrac{6.7 - 3.4(2.2) + 5.4}{(-45.1)(-3) - 23.4(-2)}$

19. $\dfrac{34.5}{(-.5)} + 12.4$

20. $34.5(-100) - 67(0) + 96 - (-4)$

Solve the following problems.

21. The perimeter of a rectangle with dimensions L and W can be found when using the formula $P = 2L + 2W$.
a. Find P, when $L = 40$ ft and $W = 8$ ft.
b. Find P, when $L = 100$ m and $W = 25$ m.

22. The formula $v_2 = v_1 + a(t_2 - t_1)$ is often used in physics. Find v_2 when $a = 60$ ft/s^2, $t_1 = 2$ s, $t_2 = 4$ s, and $v_1 = 30$ ft/s.

23. Using the same formula given in Exercise 22, find v_2 for $a = 12$ ft/s^2, $t_1 = 9$ s, and $t_2 = 12$ s and $v_1 = 25$ ft/s.

24. The area of a geometric figure called a trapezoid is found using the formula, $A = .5h(B + b)$. Find A if $h = 4.5$ in., $B = 5$ in., and $b = 3$ in.

25. Using the formula in Exercise 24, find A, if $h = 12.6$ ft, $B = 10.5$ ft, and $b = 8.9$ ft.

A– 4

EXPONENTS

One important process in algebra involves a way of expressing repeated multiplication by using exponents. This section will present the definitions and rules for this essential process.

Laws of Exponents

When two numbers are multiplied, each number is called a **factor.** For example, in the expression $2 \cdot 5 = 10$, 2 and 5 are both factors of the product 10. However, if one factor is multiplied repeatedly, such as $2 \cdot 2 \cdot 2 \cdot 2 = 16$, we need a shortcut notation for this. Exponential notation a^n represents a as the factor being repeated and n as the number of times it is multiplied. For example, $3 \cdot 3 \cdot 3 \cdot 3 \cdot 3 = 3^5$. In the expression a^n, the number a is called the **base,** the number n is called the **exponent,** and the expression is read as the "nth power of a."

EXAMPLE 1 Write the following multiplications in exponential form:

(a) $4 \cdot 4 \cdot 4$ (b) $(-1)(-1)(-1)(-1)$ (c) $x \cdot x$

Solution

(a) $4 \cdot 4 \cdot 4 = 4^3$ read the third power of 4 or four cubed
(b) $(-1)(-1)(-1)(-1) = (-1)^4$ read the fourth power of -1
(c) $x \cdot x = x^2$ read the second power of x or x squared ■

NOTE ✦ A base written without an exponent is understood to have an exponent of 1. Thus, $x = x^1$.

EXAMPLE 2 Complete each of the following:

(a) Name the base and the exponent: $(-3)^4$ and x^5

(b) Write the following using exponents: $\frac{1}{2} \cdot \frac{1}{2} \cdot \frac{1}{2}$ and $a \cdot a$.

(c) Calculate: $\left(\frac{1}{4}\right)^2$, 5^3, $(-3)^3$, $(-3)^4$

(d) Write each value as a power of 10: 100,000 and 100

Solution

(a) -3 is the base and 4 is the exponent; x is the base and 5 is the exponent

(b) $\left(\frac{1}{2}\right)^3$; a^2

(c) $\left(\frac{1}{4}\right)\left(\frac{1}{4}\right) = \frac{1}{16}$

$5^3 = 5 \cdot 5 \cdot 5 = 125$

$(-3)^3 = (-3)(-3)(-3) = -27$

$(-3)^4 = (-3)(-3)(-3)(-3) = 81$

(d) 10^5; 10^2 ■

Rather than writing repeated factors to perform algebraic operations, the Laws of Exponents make computations involving exponents more efficient.

Laws of Exponents

For any real numbers a and b and any positive integers m and n, the following rules apply:

$$a^m \cdot a^n = a^{m+n} \qquad \text{[Rule 1–1]}$$

If the bases are the same, add exponents to multiply.

$$(ab)^m = a^m b^m \qquad \text{[Rule 1–2]}$$

The exponent applies to each factor.

$$(a^n)^m = a^{nm} \qquad \text{[Rule 1–3]}$$

Multiply exponents to raise to a power.

$$\frac{a^m}{a^n} = a^{m-n} \quad \text{where } a \neq 0 \qquad \text{[Rule 1–4]}$$

If the bases are the same, subtract exponents to divide.

$$\left(\frac{a}{b}\right)^m = \frac{a^m}{b^m} \quad \text{where } b \neq 0 \qquad \text{[Rule 1–5]}$$

The exponent applies to the numerator and denominator.

CAUTION ✦ Be careful in applying the Laws of Exponents. Rules 1–1 and 1–4 can only be used when the bases are the same.

EXAMPLE 3 Perform the indicated operations.

(a) $3^5 \cdot 3^2$ (b) $x^2 \cdot x^3 \cdot x^2$ (c) $\dfrac{2^8}{2^3}$ (d) $\dfrac{a^4}{a}$

Solution

(a) $3^5 \cdot 3^2 = 3^{5+2} = 3^7$ [Rule 1–1]
(b) $x^2 \cdot x^3 \cdot x^2 = x^{2+3+2} = x^7$ [Rule 1–1]
(c) $\dfrac{2^8}{2^3} = 2^{8-3} = 2^5$ [Rule 1–4]
(d) $\dfrac{a^4}{a} = a^{4-1} = a^3$ [Rule 1–4] ■

Rules 1–2, 1–3, and 1–5 are the power laws. The use of parentheses in expressions where these rules apply are important to the simplification process.

EXAMPLE 4 Simplify the following:

(a) $(xy)^3$ (b) $(2^2)^3$ (c) $(x^3)^4$ (d) $\left(\dfrac{2}{5}\right)^3$ (e) $\left(\dfrac{x}{y}\right)^5$

Solution

(a) $(xy)^3 = x^3y^3$ [Rule 1–2]
(b) $(2^2)^3 = 2^{2\cdot 3} = 2^6$ [Rule 1–3]
(c) $(x^3)^4 = x^{3\cdot 4} = x^{12}$ [Rule 1–3]
(d) $\left(\dfrac{2}{5}\right)^3 = \dfrac{2^3}{5^3}$ [Rule 1–5]
(e) $\left(\dfrac{x}{y}\right)^5 = \dfrac{x^5}{y^5}$ [Rule 1–5] ■

Negative Exponents

Two special laws of exponents involve numbers with negative and zero exponents. The first is the law for negative exponents. Consider applying the division law [Rule 1–4] to the following example:

$$\frac{5^3}{5^5} = \frac{5 \cdot 5 \cdot 5}{5 \cdot 5 \cdot 5 \cdot 5 \cdot 5} = \frac{1}{5^2}$$

Law of Negative Exponents

For any real number a and integer m,

$$a^{-m} = \frac{1}{a^m} \quad \text{(where } a \neq 0\text{)} \qquad \text{[Rule 1–6]}$$

EXAMPLE 5 Simplify and express each quotient with positive exponents:

(a) $\frac{5}{5^4}$ (b) $\frac{x^3}{x^7}$ (c) $\frac{t^5}{t^{12}}$

Solution

(a) $5^{1-4} = 5^{-3} = \frac{1}{5^3}$

(b) $x^{3-7} = x^{-4} = \frac{1}{x^4}$

(c) $t^{5-12} = t^{-7} = \frac{1}{t^7}$ ■

All the laws of exponents given earlier apply to negative exponents also. You must pay particular attention to the rules for signed numbers to obtain the correct answer.

EXAMPLE 6 Perform the indicated operation.

(a) $6^{-2} \cdot 6^{-3}$ (b) $\frac{x^{-1}}{x^{-2}}$ (c) $\frac{(ab)^{-5}}{(a^2b^3)^{-1}}$

Solution

(a) $6^{-2} \cdot 6^{-3} = 6^{-2+(-3)} = 6^{-5} = \frac{1}{6^5}$ [Rules 1–1, 1–6]

(b) $\frac{x^{-1}}{x^{-2}} = x^{-1-(-2)} = x^{-1+2} = x$ [Rule 1–4]

(c) $\frac{(ab)^{-5}}{(a^2b^3)^{-1}} = \frac{a^{-5} \cdot b^{-5}}{a^{-2} \cdot b^{-3}} = a^{-5-(-2)}b^{-5-(-3)}$ [Rules 1–2, 1–4]

$= a^{-3}b^{-2} = \frac{1}{a^3b^2}$ [Rule 1–6] ■

Zero Exponent

The Law of Zero Exponent also involves the division rule 1–4. Consider the following example:

$$\frac{8^2}{8^2} = 8^{2-2} = 8^0$$

Since we know that any number divided by itself, except zero, equals 1, we can conclude that any nonzero number raised to the zero power is equal to 1.

Law of Zero Exponent

For any nonzero real number a,

$$a^0 = 1 \qquad \text{[Rule 1–7]}$$

EXAMPLE 7 Evaluate the following:

(a) 10^0 (b) x^0 (c) $(-3)^0$

Solution

(a) $10^0 = 1$ [Rule 1–7]
(b) $x_0 = 1$ [Rule 1–7]
(c) $(-3)^0 = 1$ [Rule 1–7] ■

In the order of operation, exponential factors should be evaluated before the steps given previously.

EXAMPLE 8 Calculate $5(2 + 4) - 4^2$.

Solution

$$\begin{aligned} 5(2+4) - 4^2 &= 5(2+4) - \textcircled{16} && \text{exponent} \\ &= 5\textcircled{(6)} - 16 && \text{parentheses} \\ &= \textcircled{30} - 16 && \text{multiply} \\ &= 14 && \text{subtract} \end{aligned}$$

■

Before doing the exercises for this section complete the following example.

EXAMPLE 9 Identify the base in each of the following exponential expressions:

(a) $4x^2$ (b) $-a^3$ (c) $(-3)^2$ (d) -3^2

Solution

(a) $4x^2$ means $4 \cdot x^2$; x is the base.

(b) $-a^3$ means $-1 \cdot a^3$; a is the base.

(c) $(-3)^2 = -3 \cdot -3 = 9$; -3 is the base.

(d) $-3^2 = -1 \cdot 3 \cdot 3 = -9$; 3 is the base. ■

A–4 EXERCISES

Simplify using the Laws of Exponents. Express results with positive exponents only.

1. $(-2)^5$

2. $3^0 \cdot 2^3$

3. $x \cdot x^3$

4. $(x^2)^3$

5. $\dfrac{10^5}{10^4}$

6. $\dfrac{8^2 \cdot 3^4}{8 \cdot 3^2}$

7. $\left(\dfrac{2}{5}\right)^2$

8. $\dfrac{x^5}{x^3}$

9. $\dfrac{2x^2}{x^5}$

10. $(2x^3)^3$

11. $\left(\dfrac{10}{y^2}\right)^3$

12. $(-2x)^5$

13. $\dfrac{(a^0b^2)}{ab^6}$

14. $-8x^0$

15. $\dfrac{12s^0}{-6s^3}$

16. $(20x^2y)^{-1}$

17. $(-2)^4 \cdot x^{-1} \cdot x^3$

18. $\dfrac{x^{-3}}{x^4}$

19. $\dfrac{10^{-3} \cdot 10}{10^6 \cdot 10^{-4}}$

20. $\left(\dfrac{3a^2b}{2ab^3}\right)^{-2}$

Evaluate the given expressions.

21. $3(-8)^2 - (-6)$

22. $8 \div (-2) - (-1)^3$

23. $(-5)^2 + (-3)(20) \div 10$

24. $10^3 - (-2)10^4$

Solve the given problems. Round to three decimal places when necessary.

25. The formula for the total surface area of a cylinder is $2\pi r^2 + 2\pi rh$. Find the total surface area of a cylinder with a diameter of 4 m and a height of 12 m.

26. If a body in space moves from location $d_0 = 500$ ft to a location $d = 2800$ ft, with a velocity $v_0 = 245$ ft/s in time $t = 5$ s, what local gravity is operating? Use the formula $g = \dfrac{d - d_0 - v_0 t}{.5t^2}$.

27. The volume of a sphere is $V = (4\pi r^3/3$, where r is the radius. What is the volume of a sphere with a radius of 1.5 m?

28. Using the formula given in Exercise 27, find the volume of a sphere with a radius of 12 ft.

29. Prove that $(a + b)^2 \neq a^2 + b^2$ using $a = 3$ and $b = 2$.

30. The volume of a cube is $V = e^2$, where e is the length of a side. What is the volume of a cube whose sides are 7.2 cm long?

A–5 SCIENTIFIC NOTATION

Numbers such as 16,300,000,000 or 0.000000015 are often encountered in technical work. However, actually working with these numbers in this form is cumbersome or even impossible with some calculators. Scientific notation is a technique for writing such very large and very small numbers in a much more concise form.

A number is said to be in **scientific notation** when it is written as the product of a number between 1 and 10 and a power of 10. Stated formally, a number is expressed in scientific notation when it is in the form $N \times 10^k$ where $1 < N \leq 10$ and k is an integer.

To write a number in scientific notation we first determine the factor N between 1 and 10. To do this we write the original number with a decimal point after the left most nonzero digit. Second, we must find the value of k. The value of k, the exponent of 10, is equal to the number of places the decimal point was moved from its original position. If the decimal point was moved to the left, k is positive. If the decimal point was moved to the right, k is negative.

EXAMPLE 1 Write the following numbers in scientific notation:

(a) 16,300,000,000 (b) 0.000000015

Solution

(a) $16{,}300{,}000{,}000 = 1.63 \times 10^{10}$ ($N = 1.63$ and $k = 10$)

(b) $0.000000015 = 1.5 \times 10^{-8}$ ($N = 1.5$ and $k = -8$) ■

To change a number from scientific notation back to standard decimal notation, reverse the process given above.

EXAMPLE 2 Rewrite each of the following in standard form:

(a) 3.58×10^8 (b) 2.3×10^{-6}

Solution

(a) $3.58 \times 10^8 = 3.58000000$

$= 358{,}000{,}000$

(b) $2.3 \times 10^{-6} = 0000002.3$

$= 0.0000023$ ■

Multiplication and Division

Multiplying and dividing very large or very small numbers is much easier when the numbers are expressed in scientific notation because the Laws of Exponents can be used.

EXAMPLE 3 Calculate (123,000,000)(0.0000083).

Solution

Step 1: Rewrite each number in scientific notation.

$$(1.23 \times 10^8)(8.3 \times 10^{-6})$$

Step 2: Multiply the numbers 1.23×8.3.

$$1.23 \times 8.3 = 10.209$$

Step 3: Use the Laws of Exponents on the powers of 10.

$$10^8 \times 10^{-6} = 10^{8-6} = 10^2$$

Step 4: Combine the results of steps 2 and 3.

$$10.209 \times 10^2 = 1020.9$$

■

EXAMPLE 4 Calculate $\dfrac{(0.000043)(9{,}000{,}000)}{(0.0004)}$

Solution

Step 1: $\dfrac{(4.3 \times 10^{-5})(9.0 \times 10^6)}{(4.0 \times 10^{-4})}$ convert to scientific notation

Step 2: $\dfrac{4.3 \times 9.0}{4.0} = 9.675$ perform operation on numbers

Step 3: $\dfrac{10^{-5} \times 10^6}{10^{-4}} = 10^5$ perform operation on powers of 10

Step 4: $9.675 \times 10^5 = 967{,}500$ combine results of steps 2 and 3 ■

A–5 EXERCISES

Rewrite each number in scientific notation.

1. 8,956,000
2. 0.000345
3. −450,000
4. 0.000000236
5. 6,900,000,000
6. 12
7. The normal red blood cell count is 5,000,000 per mm^3.
8. The thermal conductivity of wood is 0.00024 cal/cm·s.

Rewrite each number in standard decimal notation.

9. 1.23×10^7
10. -8.9×10^{-5}
11. 5.4×10^9
12. 6.7×10^{-8}
13. 2.0×10^{-3}
14. 4.5×10^{10}
15. The measure of the temperature at which hydrogen would become an ionized plasma is given by 1.05×10^5 K.
16. The radius of a typical atom is 10^{-8} cm.

Perform the following calculations by rewriting each number in scientific notation and using the Law of Exponents to calculate the answers. Write each answer in scientific notation.

17. $\dfrac{(0.00000098)(42.4)}{2300}$

18. $(89{,}000{,}000)(1{,}200{,}000)$

19. $(200)^4$

20. $(0.50200)(0.00023)(8{,}000{,}000)$

21. $\dfrac{6{,}300{,}000{,}000}{8{,}230{,}000}$

22. $\dfrac{(450{,}000)(0.004500)}{(3400)(0.000002)}$

23. The sun exerts a force of about

40,000,000,000,000,000,000,000,000,000 N

on the earth, which travels 958,000,000,000 m in its annual orbit of the sun. To compute how much work is done by the sun on the earth during the year the following computation would be necessary: $(4.0 \times 10^{28})(9.58 \times 10^{11})$. Perform this computation and express your answer in scientific notation. (Final units are joules.)

24. An electron in a beam of a TV tube has a mass of 9.1×10^{-31} kg. Its velocity is 30,000,000 m/s. Find its kinetic energy, *KE* by performing this computation: $0.5(9.1 \times 10^{-31})(3.0 \times 10^7)(3.0 \times 10^7)$. (Final units are joules.)

A–6

ROOTS AND RADICALS

To solve many equations and to evaluate various formulas, we sometimes find it necessary in algebra to take the ''square root'' of a value or of an expression. This section will explain square roots and other roots that are often found in technical applications.

Terminology

When explaining exponents in Section A–4, we discussed repeated multiplication of a factor n times. For example, $5 \cdot 5 = 5^2 = 25$ or $2 \cdot 2 \cdot 2 = 2^3 = 8$. Now let's consider this process starting with the product and working backwards. Given the product of 25, what number multiplied twice gives 25? In other words, what is the second root or square root of 25? For the second example, we would ask what is the 3rd root or cube root of 8? This process can be generalized by the following definition: The ***n*th root** of the number a, written $\sqrt[n]{a}$, is the number which, when used as a factor n times, gives a as the product. The symbol $\sqrt{}$ is called the **radical sign,** a is the **radicand,** and n is the **index.** When the index is not written, it is assumed to be 2, indicating square root.

EXAMPLE 1 Simplify the following:

(a) $\sqrt{36}$ (b) $\sqrt[3]{27}$ (c) $\sqrt[4]{16}$

Solution

(a) $\sqrt{36} = 6$ because $6 \cdot 6 = 6^2 = 36$
(b) $\sqrt[3]{27} = 3$ because $3 \cdot 3 \cdot 3 = 3^3 = 27$
(c) $\sqrt[4]{16} = 2$ because $2 \cdot 2 \cdot 2 \cdot 2 = 2^4 = 16$ ■

Technically, $\sqrt{25}$ equals both $+5$ and -5 because $+5 \cdot +5 = 25$ and $-5 \cdot -5 = 25$. To avoid confusion with which root to use, we define $\sqrt[n]{a}$ as the **principal *n*th root** of a which is $+a$. In this section we will always take the principal root.

EXAMPLE 2 Simplify the following:

(a) $\sqrt[4]{81}$ (b) $\sqrt[3]{-27}$ (c) $\sqrt{-16}$

Solution

(a) $\sqrt[4]{81} = 3$ because $(3)^4 = 81$
(b) $\sqrt[3]{-27} = -3$ because $(-3)^3 = -27$
(c) $\sqrt{-16} =$ no real solution since no real number squared gives -16 ■

NOTE ✦ On a calculator, the $\boxed{\sqrt{x}}$ key is used for square roots and the $\boxed{\sqrt[y]{x}}$ or $\boxed{x^y}$ are usually used to find other roots. A calculator will always display the principal root, and most calculators will not accept a negative radicand, regardless of the index.

Simplifying Radicals

In order to simplify radicals, you need an important property of roots. That is,

$$\sqrt{ab} = \sqrt{a} \cdot \sqrt{b}$$

This property is usually necessary when the radicand does not have an integral root. For example, when you take the $\sqrt{8}$, there is no integral value that when squared will result in 8. The radical may be rewritten using the above property.

$$\sqrt{8} = \sqrt{(4)(2)} = \sqrt{4} \cdot \sqrt{2} = 2\sqrt{2}$$

The procedure to follow when simplifying a radical is to use the above rule to remove or "take out" the square root of any perfect square factor in the radicand. Try this as you simplify the following examples.

EXAMPLE 3 Simplify the following:

(a) $\sqrt{12}$ (b) $\sqrt{125}$ (c) $\sqrt{32}$

Solution

(a) $\sqrt{12} = \sqrt{(4)(3)} = \sqrt{4} \cdot \sqrt{3} = 2\sqrt{3}$
(b) $\sqrt{125} = \sqrt{(25)(5)} = \sqrt{25} \cdot \sqrt{5} = 5\sqrt{5}$
(c) $\sqrt{32} = \sqrt{(16)(2)} = \sqrt{16} \cdot \sqrt{2} = 4\sqrt{2}$ ■

Some radicals cannot be simplified, such as $\sqrt{2}$ and $\sqrt{13}$. These radicals have radicand values that cannot be factored so that one of the factors is a perfect square.

Simplification of radicals allows us to state an exact value of the radical. Of course, the calculator can always be used to find decimal approximations of radicals. Try taking the $\sqrt{5}$ on a calculator. As you will see, $\sqrt{5} = 2.236067977\ldots$.

NOTE ✦ Recall that such numbers are classified as irrational numbers and an exact decimal value does not exist for these numbers.

Order of Operations

There are two important facts that should be stressed concerning radicals and the order of operations. The first has to do with the radical symbol; it is a grouping symbol and all operations under the radical should be performed before the root is extracted or simplified. The second important fact pertains to the order in which radicals should be evaluated when combined with other operations. Radicals are simplified in the same step as exponents and should be done before multiplication, division, addition, or subtraction.

EXAMPLE 4 Simplify the following:

(a) $\sqrt{7 + 3^2}$ (b) $\sqrt{25} - 3(2)^2 + 8$ (c) $-\sqrt{40}$
(d) $\sqrt[3]{-8} + 3^2$

Solution

(a) $\sqrt{7 + 9} = \sqrt{16} = 4$
(b) $5 - 3(4) + 8 = 5 - 12 + 8 = -7 + 8 = 1$
(c) $-\sqrt{(4)(10)} = -\sqrt{4} \cdot \sqrt{10} = -2\sqrt{10}$
(d) $-2 + 9 = 7$ ■

A–6 EXERCISES

Simplify each expression.

1. $\sqrt{81}$
2. $\sqrt{144}$
3. $\sqrt{32}$
4. $\sqrt[4]{16}$
5. $\sqrt[3]{8}$
6. $\sqrt{\frac{4}{25}}$
7. $\sqrt{\frac{16}{25}}$
8. $\sqrt{\frac{-1}{8}}$
9. $\sqrt{\frac{1}{4}}$
10. $\sqrt[5]{-243}$
11. $\sqrt{32}$
12. 45
13. $\sqrt{2500}$
14. $\sqrt[5]{-32}$
15. $\sqrt{24}$
16. $\sqrt{\frac{9}{49}}$

Evaluate the following using the order of operations rules.

17. $\sqrt{4 + 3(7)}$
18. $\sqrt{16} - 3(2)^2$
19. $\dfrac{\sqrt{100} - 8}{2^2 + 3(2)}$
20. $\dfrac{\sqrt[3]{8}}{2}$
21. $\dfrac{\sqrt[3]{-1000} - 3(-4)^3}{(-1)^3 + \sqrt{4}}$

Use your calculator to find the following roots to the nearest hundredth.

22. $\sqrt{45}$

23. $\sqrt{\frac{1}{2}}$

24. $\sqrt{15}$

25. $\sqrt{4.8}$

26. $\sqrt{13.001}$

27. $\sqrt[3]{12}$

28. $\sqrt[5]{95}$

29. $\sqrt[3]{2400}$

30. $\sqrt{0.002}$

Solve each of the following. (Round all answers to hundredths.)

31. Find the length of the diagonal of the rectangle in Figure A–12 using the Pythagorean Theorem,

$$c = \sqrt{a^2 + b^2}$$

32. Find the length of the diagonal of a 4-meter square.

33. Find the length of a side (e) of a cube with a volume of 64 in.^{3} ($V = e^3$)

FIGURE A–12

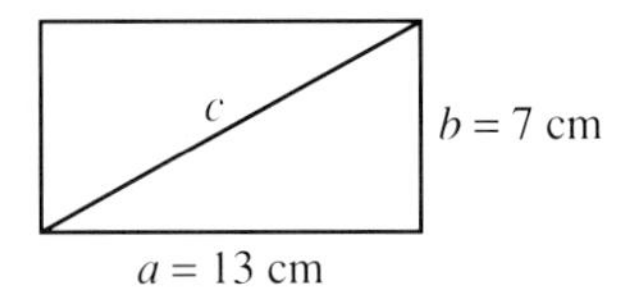

34. This calculation is necessary for finding the length of an unknown side of an oblique triangle:

$$c = \sqrt{20.4^2 + 12.3^2 - 2(20.4)(12.3)(0.5)}$$

Find the unknown side, c.

35. This calculation is necessary for finding the magnitude of a vector when its two component vectors are known:

$$\mathbf{v} + \sqrt{2.2^2 = 4.3^2}$$

Find the magnitude of vector, **v**.

A–7

ADDITION AND SUBTRACTION OF ALGEBRAIC EXPRESSIONS

Thus far we have worked mostly with the algebraic operations of addition, subtraction, multiplication, division, and taking roots. In this section we will extend our rules for working with literal numbers to include algebraic expressions. The rules discussed previously also apply to operations with algebraic expressions.

Terminology

Algebraic expressions are composed of **terms** that are literal expressions separated by plus and minus signs. Since a term may contain several factors, be careful not to confuse terms with factors because the rules are not the same.

EXAMPLE 1 Identify the terms in the following expressions:

(a) $2x + 6xy - 4y$ (b) $a - \frac{2}{b}$

Solution

(a) $2x + 6xy - 4y$ is an algebraic expression with three terms: $2x$, $6xy$ and $-4y$.

(b) $a - \frac{2}{b}$ is an algebraic expression with two terms: a and $-\frac{2}{b}$ ■

Algebraic expressions are named according to the number of terms that they contain. An expression with only one term is called a **monomial** expression, a two-term expression is called a **binomial,** and a three-term expression is called a **trinomial.** Many expressions contain more than three terms, and these expressions are called **multinomial** expressions. In fact, any expression containing two or more terms may be called a multinomial.

EXAMPLE 2 Identify the following expressions by the number of terms.

(a) $2x$ (b) $8x - y$ (c) $2xy + 3x - 5$ (d) $-3a + b - 5c + 16$

Solution

(a) $2x$ is a monomial expression with one term $2x$.
(b) $8x - y$ is a binomial expression with terms $8x$ and $-y$.
(c) $2xy + 3x - 5$ is a trinomial expression.
(d) $-3a + b - 5c + 16$ is a multinomial expression containing 4 terms. ■

As seen in Example 2, a term can be composed of both literal and numerical factors. Any set of factors in a term is the coefficient of the remaining factors. For example, the expression $3x^2y$ contains three factors. The coefficient of y is $3x^2$; while the coefficient of 3 is x^2y. A numerical factor is called the **numerical coefficient** of the term. In our previous example, 3 is the numerical coefficient. When a numerical coefficient is not written, it is understood to be 1 or -1. Terms are called **like** if they differ only in their numerical coefficients. Like terms have exactly the same variables and corresponding variables have the same exponents. For example, in the expression $3x + 5a^2 - 2xy + 4x + a^2 + 4y$, the like terms are $3x$ and $4x$, $5a^2$ and a^2, but $-2xy$ and $4y$ are not like terms.

To simplify an algebraic expression, all like terms should be combined. This is accomplished by adding or subtracting the numerical coefficients and leaving the variable unchanged.

EXAMPLE 3 Simplify the following:

(a) $2x + 5x - x$ (b) $-5a^2 + a^2$

Solution

(a) $2x + 5x - x = (2 + 5 - 1)x = 6x$ (All three terms are like, so we combine the numerical coefficients and leave the variable x unchanged.)
(b) $-5a^2 + a^2 = (-5 + 1)a^2 = -4a^2$ ■

CAUTION ✦ Be careful when performing subtraction. Group the subtrahend in parentheses in order to perform the subtraction correctly.

EXAMPLE 4 Perform the indicated operation.

(a) $(2x + 5y) + (3x - 4y)$ (b) $(5a - b) - (4a + 6b - 8c)$

Solution

(a) $(2x + 5y) + (3x - 4y) = 2x + 5y + 3x - 4y$ remove parentheses
$= (2 + 3)x + (5 - 4)y$ combine like terms
$= 5x + y$

(b) $(5a - b) - (4a + 6b - 8c)$
$= 5a - b - 4a - 6b + 8c$ remove parentheses
$= (5 - 4)a + (-1 - 6)b + 8c$ combine like terms
$= a - 7b + 8c$ ■

When expressions are added, the parentheses may just be dropped and like terms combined. However, when subtracting one expression from another, additional steps must be taken. When a grouping symbol is preceded by a subtraction sign, you must change the sign of each term inside the symbol in order to remove the grouping symbol. Review the following example for clarification of this principle.

EXAMPLE 5 Subtract $(4x - 5y + 10) - (3x + 2y - 8)$.

Solution

Step 1: Remove parentheses and change signs.

$$4x - 5y + 10 - 3x - 2y + 8$$

Step 2: Group like terms together.

$$4x - 3x - 5y - 2y + 10 + 8$$

Step 3: Combine like terms.

$$(4 - 3)x + (-5 - 2)y + (10 + 8)$$

Step 4: Simplify answer.

$$x - 7y + 18$$ ■

When you combine algebraic expressions, remember to always combine only like terms. Try the following example before starting the exercises.

EXAMPLE 6 Simplify the following:

(a) $10x - 5x + x$ (b) $x^2 + x$

(c) $(2a + 3b) + (6a + b)$ (d) $(2s - 4t) - (s + 7t)$
(e) $(5 - 3x + 8y) - (-3 + x)$

Solution

(a) $(10 - 5 + 1)x = 6x$
(b) $x^2 + x$; the two terms are not like and cannot be combined

(c) $2a + 6a + 3b + b = (2 + 6)a + (3 + 1)b$
$= 8a + 4b$

(d) $2s - 4t - (s + 7t) = 2s - s - 4t - 7t$
$= (2 - 1)s + (-4 - 7)t$
$= s - 11t$

(e) $5 - 3x + 8y + 3 - x = -3x - x + 8y + 5 + 3$
$= (-3 - 1)x + 8y + 8$
$= -4x + 8y + 8$ ■

EXERCISES A–7

Simplify the following algebraic expressions.

1. $2x + 4x - x$
2. $3y - 5y + 6y$
3. $a + 3a - 2a$
4. $-2s + 3s - 5s$
5. $12p - 3p - 9p$
6. $5x^2 - 8x^2$
7. $9xz + 4xz - 2xz$
8. $2y - 5y + 4y + 2y + 4y$
9. $2a - 3b + a - 4b + 5$
10. $3t^2 - 2t + 4t^2$

Perform the indicated operations.

11. $(5x - 3y) + (2x + 8y)$
12. $(x^2 + 2y^2) + (3x^2 + y^2)$
13. $(a + 2b) - (2a - 3b)$
14. $(3t - 5s) - (-2t + 2s)$
15. $(x - 2y + 6) + (2x - y + 6)$
16. $(2x + y - 10) + (5x + 4y + 2)$
17. $(x^2 + y^2 + x - y) - (y + x - 2y^2)$
18. $(3 - 2xy + 2x^2 + y) - (xy + 5x^2 - 3y + 8)$
19. $3(2m + 5n) - 4(6m - 2n)$
20. $-2(x - y + 5) + 5(3x + 2y - 6)$

Solve the following problems.

21. In computing the slope of a roadway, a surveyor obtains the following expression:

$$(m - 2) + m - (2m + 5) - (1 - 4m)$$

Simplify the expression.

22. The expression $2(2w - 2) + 2w$ describes the perimeter of a certain rectangle. Simplify the expression.

23. The expression $0.144x + 0.1034(2000 - x)$ represents the total interest earned from two investments. Simplify the expression.

24. Many times subscripted variables are used in expressions, such as $I_1 + 2I_2 - (3I_2 + 3)$. Simplify the expression.

25. In finding the current through an electric current circuit, an electrical engineering student is confronted with the expression:

$$(2I_1 + I_2) - (3I_1 - I_2) + (I_2 - I_1)$$

Simplify the expression.

A–8

MULTIPLYING POLYNOMIALS

The process of multiplying polynomials can be broken into three categories—multiplying by a monomial, multiplying binomials, and multiplying polynomials. Each of these categories is discussed in this section.

Multiplying by a Monomial

First, we will discuss multiplying monomials. Multiply the coefficients and then multiply variables by applying the Laws of Exponents. The next two examples illustrate this procedure.

EXAMPLE 1 Multiply $(2x^2y^3)(-6xy^4z^2)$.

Solution Multiplying the numerical coefficients gives $(2)(-6) = -12$. Next, we multiply the variables by adding exponents on like bases. This gives $x^2 \cdot x = x^3$; $y^3 \cdot y^4 = y^7$; and z^2 remains unchanged. The product is $-12x^3y^7z^2$. ■

EXAMPLE 2 Multiply $(-8m^2)(2m^3n^4)(-m^4n)$.

Solution First, we multiply the numerical coefficients.

$$(-8)(2)(-1) = 16$$

Next, we multiply the variables

$$\begin{aligned}(-8m^2)(2m^3n^4)(-m^4n) &= m^2(m^3)(m^4)\\ &= m^{2+3+4}\\ &= m^9\end{aligned}$$

$$\begin{aligned}(-8m^2)(2m^3n^4)(-m^4n) &= n^4(n)\\ &= n^{4+1}\\ &= n^5\end{aligned}$$

The product is $16m^9n^5$. ■

Next, we will multiply a polynomial by a monomial. We apply the Distributive Property to multiply the monomial and each term in the polynomial. The next two examples illustrate this procedure.

EXAMPLE 3 Multiply $2xy^2(6xy - 7x^2y^3 + 6)$.

Solution First, apply the Distributive Property by multiplying $2xy^2$ by each term of the polynomial inside the parentheses. Writing out this product gives

$$2xy^2(6xy - 7x^2y^3 + 6) \quad \text{or} \quad 2xy^2(6xy) + 2xy^2(-7x^2y^3) + 2xy^2(6)$$

Second, multiply monomial factors using the techniques given previously for monomial multiplication.

$$12x^2y^3 - 14x^3y^5 + 12xy^2$$

Since none of the terms are alike, the product is

$$12x^2y^3 - 14x^3y^5 + 12xy^2$$

■

EXAMPLE 4 Multiply $-5a^2b^3c(4ac^3 + 9a^2bc^2)$.

Solution Multiply $-5a^2b^3c$ by each term of the polynomial. Writing out this product gives

$$(-5a^2b^3c)(4ac^3) + (-5a^2b^3c)(9a^2bc^2)$$

Multiplying monomial factors gives the product of

$$-20a^3b^3c^4 - 45a^4b^4c^3$$

■

Multiplying Binomials

There are several techniques for multiplying binomials; however, the most frequently used method is called FOIL. FOIL is based on applying the Distributive Property to binomials, and the letters in FOIL represent the order in which you multiply terms (see Figure A–13). The letters in FOIL represent

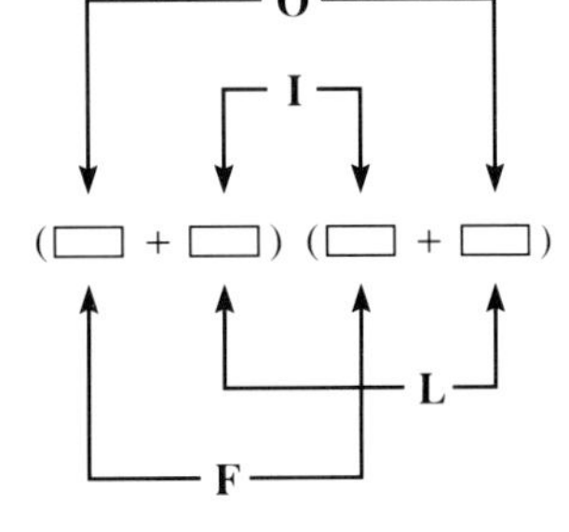

FIGURE A–13

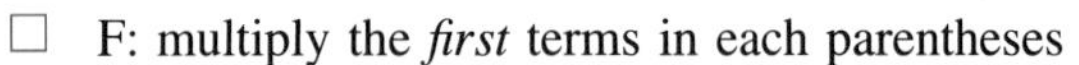

- ☐ F: multiply the *first* terms in each parentheses
- ☐ O: multiply the *outside* terms in each parentheses
- ☐ I: multiply the *inside* terms in each parentheses
- ☐ L: multiply the *last* terms in each parentheses

Multiplying binomials using FOIL is also based on multiplying monomials. When you multiply the first elements in each parentheses, you are multiplying two monomial factors.

EXAMPLE 5 Multiply $(7x - 8y)(4x - 5y)$.

Solution Figure A–14 shows the terms to multiply:

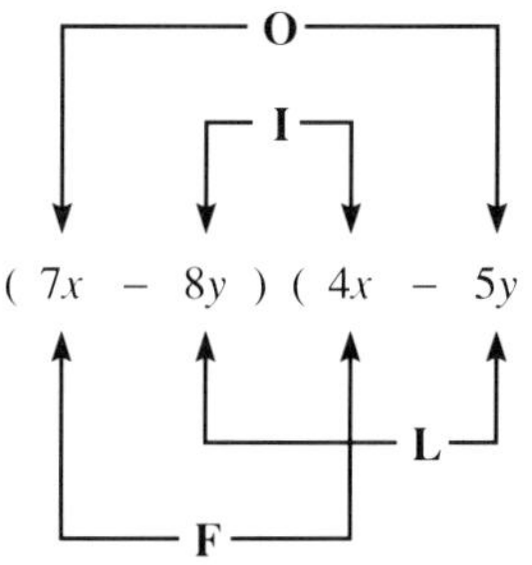

FIGURE A–14

F gives $(7x)(4x) = 28x^2$
O gives $(7x)(-5y) = -35xy$
I gives $(-8y)(4x) = -32xy$
L gives $(-8y)(-5y) = 40y^2$

FOIL gives $28x^2 - 35xy - 32xy + 40y^2$. After combining like terms the product is $28x^2 - 67xy + 40y^2$. ■

Multiplying Polynomials

Since FOIL only applies to multiplying binomials, we must use another technique when any factor is not a binomial. We will use the Distributive Property to distribute each term of one polynomial over the second polynomial. The next example illustrates this process.

EXAMPLE 6 Multiply $(2a^3 - a^2 + 5a - 1)(a + 6)$.

Solution To multiply these polynomials, we multiply each term in the first polynomial by each term in the second polynomial. Since multiplying by the binomial is easier, we will distribute $a + 6$ over the polynomial $(2a^3 - a^2 + 5a - 1)$.

$$a(2a^3 - a^2 + 5a - 1) + 6(2a^3 - a^2 + 5a - 1)$$

Multiplying gives the following result:

$$a(2a^3 - a^2 + 5a - 1) + 6(2a^3 - a^2 + 5a - 1)$$

$$2a^4 - a^3 + 5a^2 - a + 12a^3 - 6a^2 + 30a - 6$$

Last, we combine like terms to yield the following product:

$$2a^4 + 11a^3 - a^2 + 29a - 6$$

■

A–8 EXERCISES

Find each of the following products.

1. $4a(3a^2b)$
2. $(-4xy)(-x^3y^2)$
3. $(-12s^5)(2s)$
4. $(-3x)(5x^8)(x^3)$
5. $2x^2(x^2 - 3y)$
6. $-4a^3(a^2 + a - 12)$
7. $-mn(mn - m + n)$
8. $5s^4t^2(-4s^3 + 6st)$
9. $x^4(x^5 + x^3 - x)$
10. $-23xy(2xy + 4)$
11. $(x - 3)(x + 4)$
12. $(y + 5)(y + 4)$
13. $(2a - 4b)(a - b)$
14. $(3x + 6)(2x - 4)$
15. $(3s^2 + t)(5s^2 - t)$
16. $(w - 3)(w + 3)$
17. $(2xy)^2(3x)^3$
18. $(-6a^3)^2(-3)$
19. $(2x + 3y)^2$
20. $(12a - 2b)^2$
21. $(x - y)(x^2 + xy - y^2)$
22. $(2s + t)(s^3 - 2s + 4)$
23. $(x - 3)(x^2 - 2x - 2)$
24. $(2x + y)(3x - y)(4x - 1)$
25. $(6y + 2)(2y - 4)(y + 1)$
26. $(x - 2)^3$

Solve the following problems.

27. In finding the maximum power in an electric circuit, the expression $(R + 2r)^2 - r(2R + r)$ might arise. Multiply and simplify this expression.
28. In determining the area of a special rectangular plot of land, the expression $(200x + 5)(100 + 3x)$ is found. Perform the indicated multiplication.

A–9

POLYNOMIAL DIVISION

Before discussing polynomial division, we need to review the Laws of Exponents that relate to division. These laws are summarized below.

$$\frac{a^m}{a^n} = a^{m-n} \quad \text{if } m > n$$

$$\frac{a^m}{a^n} = \frac{1}{a^{n-m}} \quad \text{if } m < n$$

Since we do not want negative exponents, always subtract the smaller exponent from the larger and write the result in the part of the fraction where the larger exponent appeared. Division of polynomials can be placed into two categories—division by a monomial and division by a polynomial.

Division by a Monomial

Division by a Monomial

To divide a polynomial by a monomial,

1. Divide each term of the polynomial by the monomial.
2. Divide or reduce the numerical coefficients.
3. Divide the variables using the Laws of Exponents.

EXAMPLE 1 Divide $\dfrac{24x^2y^3 - 9x^3y^2 + 27xy}{-3xy}$.

Solution To divide a polynomial by a monomial, divide each term of the polynomial by the monomial. Writing each term of the polynomial over the monomial gives

$$\frac{24x^2y^3}{-3xy} - \frac{9x^3y^2}{-3xy} + \frac{27xy}{-3xy}$$

Dividing each term using monomial division gives the quotient as

$$-8xy^2 + 3x^2y - 9$$

■

EXAMPLE 2 Divide $\dfrac{54x^2y^4 + 6x^5y^2 - 36xy^5}{9x^3y^2}$.

Solution Divide $9x^3y^2$ into each term of the polynomial.

$$\frac{54x^2y^4}{9x^3y^2} + \frac{6x^5y^2}{9x^3y^2} - \frac{36xy^5}{9x^3y^2}$$

The quotient is

$$\frac{6y^2}{x} + \frac{2x^2}{3} - \frac{4y^3}{x^2}$$

■

Division by a Polynomial

Division by a polynomial requires a different process than the one used for division by a monomial. Division by a polynomial is very similar to long division in arithmetic. As a matter of fact, you use the same procedure in dividing by a polynomial as you use for long division in arithmetic. Compare the examples from arithmetic and algebra that follow.

$$\begin{array}{r} 2 \\ 36\overline{)785} \end{array}$$

divide 36 into 78, as a trial divisor divide 3 into 7

$$\begin{array}{r} x \\ x+2\overline{)x^2-3x-10} \end{array}$$

divide $x + 2$ into $x^2 - 3x$, as a trial divisor divide x into x^2

$$\begin{array}{r} 2 \\ 36\overline{)785} \\ 72 \end{array}$$

multiply 36 by 2

$$\begin{array}{rrrr} & & x & \\ x+2 & \overline{)x^2} & -3x & -10 \\ & x^2 & +2x & \end{array}$$

multiply $x + 2$ by x

$$\begin{array}{r} 2 \\ 36\overline{)785} \\ \underline{-72} \\ 65 \end{array}$$

subtract and bring down the next number

$$\begin{array}{rrrr} & & x & \\ x+2 & \overline{)x^2} & -3x & -10 \\ & \underline{\ominus x^2} & \underline{\overset{\ominus}{+}2x} & \\ & & -5x & -10 \end{array}$$

subtract by changing signs on the bottom poly. and add, bring down the next term

$$\begin{array}{r} 21 + \frac{29}{36} \\ 36\overline{)785} \\ \underline{-72} \\ 65 \\ \underline{-36} \\ 29 \end{array}$$

repeat steps 1, 2, & 3 until the last number has been used

$$\begin{array}{rrrr} & & x & -5 \\ x+2 & \overline{)x^2} & -3x & -10 \\ & \underline{\ominus x^2} & \underline{\overset{\ominus}{+}2x} & \\ & & -5x & -10 \\ & & \underline{\overset{\oplus}{-5x}} & \underline{\overset{\oplus}{-}10} \\ & & & 0 \end{array}$$

repeat steps 1, 2, & 3 until the last term has been used

In algebraic long division as in arithmetic, write the remainder as a fraction, writing the remainder as the numerator of the fraction and the divisor as the denominator of the fraction. Always precede the remainder with an addition sign.

Polynomial Division

In dividing by a polynomial you must perform the following steps before dividing:

1. Always arrange both polynomials in descending powers. This means place the variable with the largest exponent first, then the term with the second largest exponent next, and so on.
2. Place a zero numerical coefficient in front of any omitted variables. This acts as a place holder so you will subtract only like terms.

EXAMPLE 3 Divide $6x^2 + 13x - 5$ by $3x - 1$.

Solution Since we are dividing by a polynomial, we must use long division. Each polynomial is arranged in descending powers and all powers are present.

$$3x - 1 \overline{\smash{)}\, 6x^2 + 13x - 5} \quad \text{quotient: } 2x$$

divide $3x - 1$ into $6x^2 + 13x$

$$\begin{array}{r} 2x \phantom{{}-5} \\ 3x - 1 \overline{\smash{)}\, 6x^2 + 13x - 5} \\ \ominus \quad \oplus \\ \underline{6x^2 - 2x} \phantom{{}-5} \\ 15x - 5 \end{array}$$

multiply $3x - 1$ by $2x$, subtract and bring down the next term

$$\begin{array}{r} 2x + 5 \\ 3x - 1 \overline{\smash{)}\, 6x^2 + 13x - 5} \\ \ominus \quad \oplus \\ \underline{6x^2 - 2x} \phantom{{}-5} \\ 15x - 5 \\ \ominus \quad \oplus \\ \underline{15x - 5} \\ 0 \end{array}$$

divide $3x - 1$ into $15x - 5$ using $3x$ into $15x$ as a trial divisor; multiply; and subtract

The quotient is $2x + 5$. ■

EXAMPLE 4 Divide $2x^4 + 5x + 3x^3 - 3 - 7x^2$ by $x^2 - 3 + 2x$.

Solution Since you are dividing by a polynomial, you must use long division. Arranging the polynomials in descending powers gives

$$2x^4 + 3x^3 - 7x^2 + 5x - 3 \text{ and } x^2 + 2x - 3$$

$$\begin{array}{lr} & 2x^2 - x + 1 \\ & x^2 + 2x - 3 \overline{\smash{)}\, 2x^4 + 3x^3 - 7x^2 + 5x - 3} \\ (1) & \underline{2x^4 + 4x^3 - 6x^2} \phantom{{}+5x-3} \\ (2) & -x^3 - x^2 + 5x \phantom{{}-3} \\ & \underline{-x^3 - 2x^2 + 3x} \phantom{{}-3} \\ (3) & x^2 + 2x - 3 \\ & \underline{x^2 + 2x - 3} \\ & 0 \end{array}$$

(1) divide $x^2 + 2x - 3$ into $2x^4 + 3x^3 - 7x^2$; multiply by $2x^2$ and subtract; (2) divide $x^2 + 2x - 3$ into $-x^3 - x^2 + 5x$; multiply by $-x$ and subtract; (3) divide $x^2 + 2x - 3$ into $x^2 + 2x - 3$ into $x^2 + 2x - 3$; multiply and subtract

The quotient is $2x^2 - x + 1$. ■

EXAMPLE 5 Divide $m^4 - 2m^2 + 6$ by $m - 3$.

Solution Long division is necessary. In arranging the terms in descending powers, the m^3 and m terms are omitted. Therefore, these terms are filled in with a zero numerical coefficient.

$$
\begin{array}{rl}
 & m^3 + 3m^2 + 7m + 21 + \dfrac{69}{(m-3)} \\
m - 3 \,\big) & m^4 + 0m^3 - 2m^2 + 0m + 6 \\
 & \underline{m^4 - 3m^3} \\
 & \quad 3m^3 - 2m^2 \\
 & \quad \underline{3m^3 - 9m^2} \\
 & \qquad 7m^2 + 0m \\
 & \qquad \underline{7m^2 - 21m} \\
 & \qquad\quad 21m + 6 \\
 & \qquad\quad \underline{21m - 63} \\
 & \qquad\qquad 69
\end{array}
$$

■

A–9 EXERCISES

Find each of the following quotients.

1. $\dfrac{21a^2}{7a}$
2. $\dfrac{20x^3y^4}{-4xy^3}$
3. $\dfrac{-6a^3c^3d^2}{24a^3c}$
4. $\dfrac{5st}{-25s^3t^4}$
5. $\dfrac{15xy^6}{3xy^4}$
6. $\dfrac{p^{12}q^{10}}{p^8q^{12}}$
7. $\dfrac{(2x^2)^2}{8xy}$
8. $\dfrac{(-3x^2y)^3}{9xy^2}$
9. $\dfrac{(xy)^3(2a)^2}{-6x^3ya}$
10. $\dfrac{7x^2y - 21xy^2}{7xy}$
11. $\dfrac{12a^3b - 4ab^6}{2ab}$
12. $\dfrac{3a^2b - 9ab^2 + 6b^3}{3b}$
13. $\dfrac{32s^2t - 16st + 5}{8st^2}$
14. $\dfrac{x^2 + 11x + 30}{x + 5}$
15. $\dfrac{x^2 + 4x - 12}{x - 2}$
16. $\dfrac{4y^3 - 17y + 20}{2y + 5}$
17. $\dfrac{4x^3 + 4x^2 - 5x + 1}{2x - 1}$
18. $\dfrac{x^3 + 27}{x + 3}$
19. $\dfrac{3x^3 - 8x^2 - 9}{x - 3}$
20. $\dfrac{6x^2 + 5x - 5}{2x + 3}$
21. $\dfrac{x^3 + 11}{x + 2}$
22. $\dfrac{30x^2 - 49x + 20}{5x - 4}$
23. $\dfrac{6x^4 - 3x^2 - 63}{2x^2 - 7}$
24. $\dfrac{x^3 - 1}{x^2 - 1}$
25. $\dfrac{2x^3 - 3x^2 + 8x - 2}{x^2 - x + 2}$
26. $(3x^4 - 15x^3 - 19x^2 - 25x - 40) \div (x^2 - 5x - 8)$
27. $(15x^2 - 76x + 5) \div (x - 5)$
28. Resistance is equal to the applied voltage divided by the current. Find the resistance if the voltage is represented by $16r^2 - 22r$ and the current is represented by $2r$.
29. A free-falling body travels a distance of $-6\,gt^2 + 4t + 2$. Find an algebraic expression for one-half the distance.
30. The volume of a sphere is $4/3\pi r^3$ and its surface area is $4\pi r^2$. Find the relation between volume and the area. (Find $V \div A$.)

A–10

PERCENTAGES

The last concept we will review is percentages. Many technical applications require a thorough knowledge of this concept. In this section, applied percentage problems will be presented in depth.

The concept of percentages is difficult because there are many applications of it and each one may appear to be different. Actually, once the basics are learned all percentage problems can be solved using a procedural approach. **Percent,** denoted %, always means "per hundred" or "divided by 100." If 36% of the employees in a company received a raise, this simply is another way of saying 36 out of every 100 employees received a raise. Now, if the company employs 3000 people, how many people in all received a raise? "36 out of every 100" and 3000 divided by 100 equals 30. Thus, $30 \times 36 = 1080$ employees.

It would be time-consuming to do all percentage problems in this fashion. To simplify matters, a formula Base × Rate = Percentage is used. In the previous example the base was the total number of employees, 3000, the rate was the percent value, 36%, and the percentage was the number of employees receiving a raise, 1080. Using the formula, the problem would be set up this way.

$$\text{Base} \times \text{Rate} = \text{Percentage}$$
$$(3000)(36\%) = 1080$$

NOTE ✦ Most calculators have a percent [%] key but some do not. If the percent key is not used, then the percent value must be changed to a decimal form. Remember "36 out of every 100" is equal to the fraction 36/100 and this fraction is equal to the decimal value 0.36. A quick rule for this procedure is to drop the percent symbol and move the decimal point two places to the left.

EXAMPLE 1 Change the following percents to decimal form:

(a) 45% (b) 9% (c) 104% (d) 0.5% (e) $\frac{1}{4}\%$

Solution

(a) $\frac{45}{100} = 0.45$ or $45\% = 0.45$

(b) $\frac{9}{100} = 0.09$ or $9\% = 0.09$

(c) $\frac{104}{100} = 1.04$ or $104\% = 1.04$

(d) $\frac{.5}{100} = 0.005$ or $0.5\% = 0.005$

(e) $\frac{.25}{100} = 0.0025$ or $1/4\% = 0.25\% = 0.0025$ ■

Many times it is also necessary to rewrite a decimal value as a percent value, such as when writing the results taken from a calculator. A decimal value may be changed to a percent value by moving the decimal two places to the right and adding the percent symbol.

EXAMPLE 2 Change the following decimals to percents:

(a) 0.356 (b) 0.03 (c) 1.67 (d) 0.0075

Solution

(a) 0.356 = 35.6%
(b) 0.03 = 3%
(c) 1.67 = 167%
(d) 0.0075 = 0.75% ■

When setting up applied percentage problems, always remember to identify what quantities in the formula are given and what is the unknown quantity. Review the following examples carefully.

EXAMPLE 3 There are 45 accidents within one year for a manufacturing company with 450 employees. What is the accident rate for the company for the year?

Solution

Step 1: Identify the given quantities.
Base: 450 total
Percentage: 45 part of base
Step 2 Identify the unknown quantity.
Rate: R Number of accidents per 100 employees
Step 3 Substitute known quantities into the formula and solve for the unknown value.

$$\begin{aligned} \text{Base} \times \text{Rate} &= \text{Percentage} \\ 450R &= 45 \\ R &= \frac{45}{450} \\ &= 0.10 \\ &= 10\% \quad \text{always express in percent form} \end{aligned}$$ ■

EXAMPLE 4 A solution consists of 3 parts water and 5 parts alcohol. How many ounces of water are there in a solution of 60 oz?

Solution

Step 1: Identify the known quantities.
Base: 6 oz
Rate for water: 3/8 or .375 or 37.5%
Step 2: Identify the unknown quantity.
Percentage: P (ounces of water in total solution)
Step 3: Apply the formula and solve.

$$B \times R = P$$
$$60(37.5\%) = P$$
$$22.5 \text{ oz} = P$$

■

Example 4 is one of the cases that is often confusing. Remember to always add the parts together to obtain the denominator of the fraction that must be converted to a percent or rate. These types of percentage problems are often called ratio problems.

As stated earlier there are many ways to approach a percentage problem. Many basic problems encountered by a technician might simply be stated as

45% of 180 is what number?
68 is what percent of 88?
23 is 12% of what number?

To solve each of these, the formula Base × Rate = Percentage applies. Again, use the procedure for solving for the unknown quantity.

EXAMPLE 5 Solve each of the following:

(a) 85% of 235 is what number?
(b) 35 is what percent of 56?
(c) 78 is 25% of what number?

Solution

(a) $R = 85\%$ and $B = 235$, find P.

$$(235)(0.85) = P$$
$$P = 199.75$$

(b) $B = 56$ and $P = 35$, find R.

$$56R = 35$$
$$R = \frac{35}{56}$$
$$R = 0.625$$
$$R = 62.5\%$$

(c) $R = 25\%$ and $P = 78$, find B. Remember to convert 25% to a decimal before substituting into the formula.

$$B(0.25) = 78$$

$$B = \frac{78}{0.25}$$

$$B = 312$$

■

In summary, when solving percentage problems remember to always identify the unknown quantity in the formula, Base × Rate = Percentage ($BR = P$), and use algebra to solve for the unknown value.

A–10 EXERCISES

Change to percents.

1. 0.32

2. 0.045

3. $\frac{3}{8}$

4. $\frac{5}{1000}$

5. $\frac{1}{5}$

6. 1.34

7. 0.00625

8. 0.89

9. 0.05

10. 5.23

Change to decimals.

11. 56%

12. 130%

13. 0.25%

14. 12.3%

15. 18%

16. 12.75%

17. 255%

18. 0.01%

19. 0.03%

20. 5½%

Solve the following problems.

21. 125 is what percent of 600?

22. 346 is what percent of 890?

23. 35% of 780 is what number?

24. 150% of 456 is what number?

25. 56 is 60% of what number?

26. 904 is 28% of what number?

27. If a company employees 560 people and 23.5% of these are females, find how many females work for the company.

28. If 34% of a technician's salary is deducted for taxes and other items, find his take home pay for a salary of $34,560.

29. A company has an accident rate of 3% for a certain period. During this period there were 23 employees injured. How many employees does the company employ in all?

30. An electrical engineer for a particular company must spend 12 hours of his 40-hour work week "on call." What percent of his work week is spent "on call"?

31. A solution consists of 2 parts alcohol and 4 parts water. How many ounces of alcohol are there in 40 oz of the solution?

32. A school has 1250 males and 680 females enrolled. What percent of the student body is male? Female?

Appendix B

Geometry Review

Geometric figures, facts, and formulas are essential to many technical fields. This appendix will serve not only as a review of basic applied geometry but as a reference for the many geometric problems found in the text. For our study of geometry we will assume no formal definitions are necessary for these geometric terms: point, line, and plane.

A majority of geometric figures is made up of a combination of intersecting and nonintersecting lines or line segments. A **straight line** (Figure B–1) extends infinitely in both directions, but a **line segment** (Figure B–2) is a finite portion of a line bounded by and including two endpoints. Sometimes it is necessary to speak of a half-line or ray. A **half-line** or **ray** (Figure B–3) is a line extending infinitely to one side of a given endpoint.

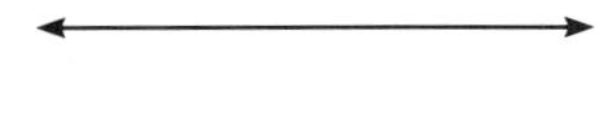

FIGURE B–1

FIGURE B–2

FIGURE B–3

Two lines that intersect are assumed to have one point in common (Figure B–4). Two lines that do not intersect and have no points in common are said to be **parallel lines** (Figure B–5).

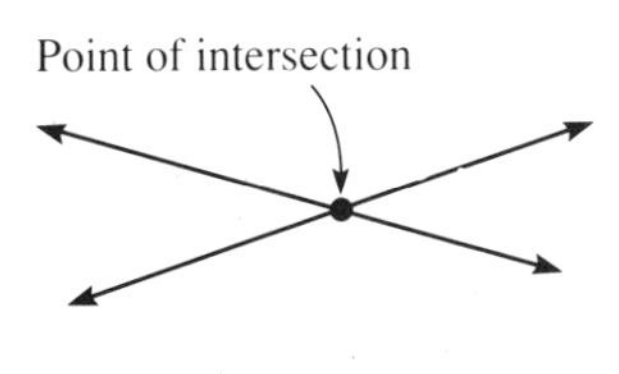

FIGURE B–4

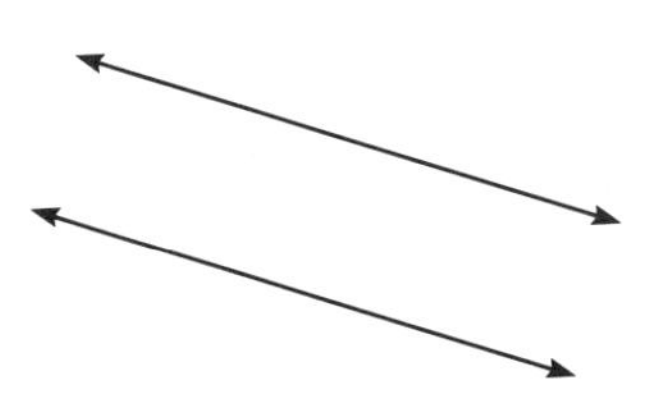

FIGURE B–5

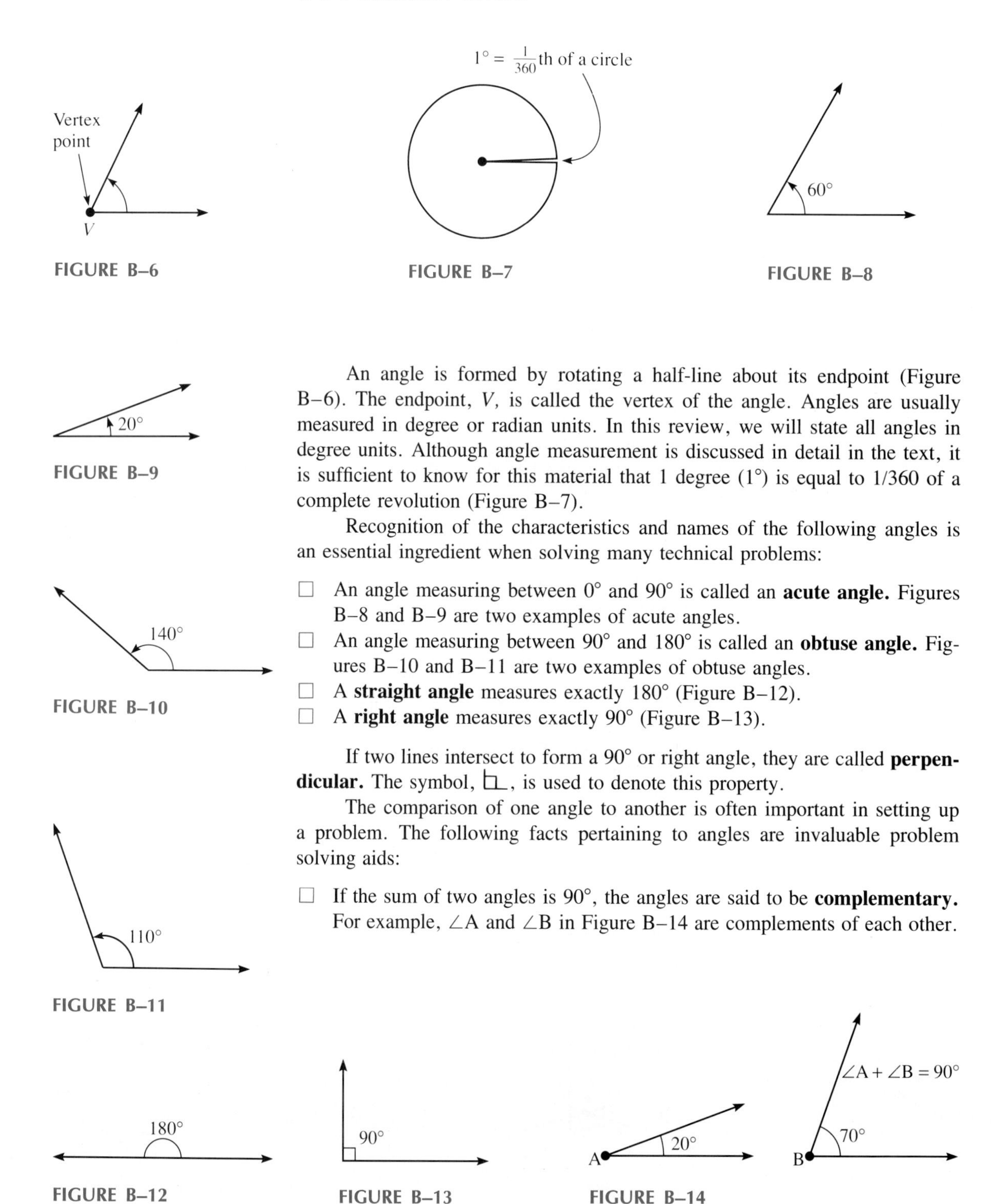

FIGURE B–6

FIGURE B–7

FIGURE B–8

FIGURE B–9

FIGURE B–10

FIGURE B–11

FIGURE B–12

FIGURE B–13

FIGURE B–14

An angle is formed by rotating a half-line about its endpoint (Figure B–6). The endpoint, *V*, is called the vertex of the angle. Angles are usually measured in degree or radian units. In this review, we will state all angles in degree units. Although angle measurement is discussed in detail in the text, it is sufficient to know for this material that 1 degree (1°) is equal to 1/360 of a complete revolution (Figure B–7).

Recognition of the characteristics and names of the following angles is an essential ingredient when solving many technical problems:

- ☐ An angle measuring between 0° and 90° is called an **acute angle.** Figures B–8 and B–9 are two examples of acute angles.
- ☐ An angle measuring between 90° and 180° is called an **obtuse angle.** Figures B–10 and B–11 are two examples of obtuse angles.
- ☐ A **straight angle** measures exactly 180° (Figure B–12).
- ☐ A **right angle** measures exactly 90° (Figure B–13).

If two lines intersect to form a 90° or right angle, they are called **perpendicular.** The symbol, ⊾, is used to denote this property.

The comparison of one angle to another is often important in setting up a problem. The following facts pertaining to angles are invaluable problem solving aids:

- ☐ If the sum of two angles is 90°, the angles are said to be **complementary.** For example, ∠A and ∠B in Figure B–14 are complements of each other.

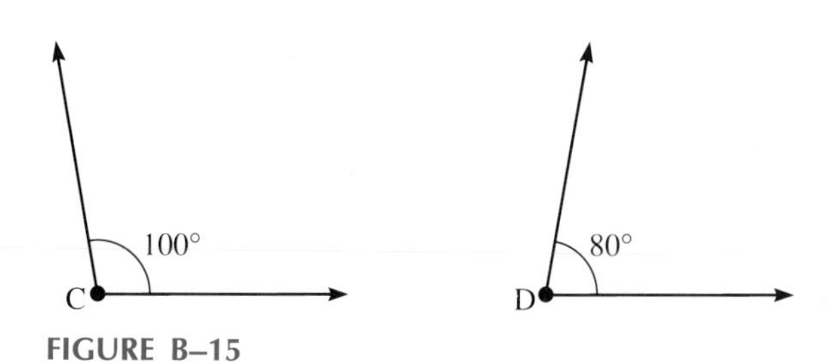

FIGURE B–15

FIGURE B–16

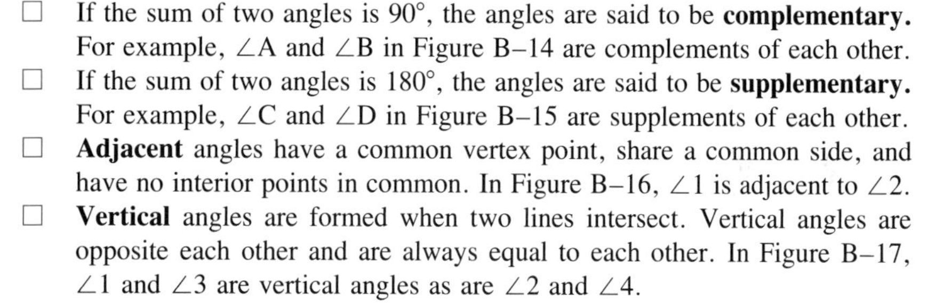

- ☐ If the sum of two angles is 90°, the angles are said to be **complementary.** For example, ∠A and ∠B in Figure B–14 are complements of each other.
- ☐ If the sum of two angles is 180°, the angles are said to be **supplementary.** For example, ∠C and ∠D in Figure B–15 are supplements of each other.
- ☐ **Adjacent** angles have a common vertex point, share a common side, and have no interior points in common. In Figure B–16, ∠1 is adjacent to ∠2.
- ☐ **Vertical** angles are formed when two lines intersect. Vertical angles are opposite each other and are always equal to each other. In Figure B–17, ∠1 and ∠3 are vertical angles as are ∠2 and ∠4.

Parallel lines intersected by a third line, called a **transversal,** form several pairs of special angles:

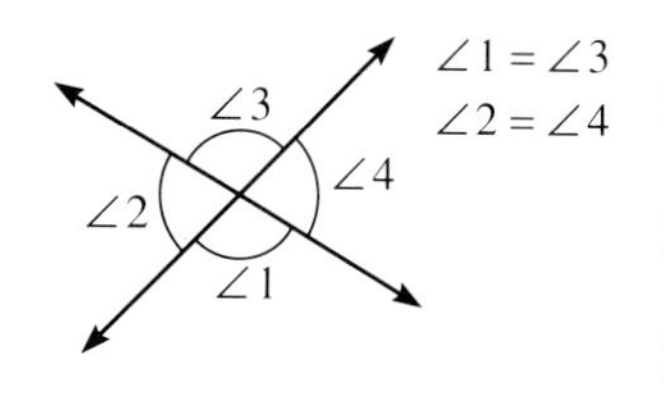

FIGURE B–17

- ☐ **Alternate interior angles** are equal to each other. In Figure B–18, ∠1 and ∠7 are alternate interior angles as are ∠4 and ∠6.
- ☐ **Alternate exterior angles** are equal to each other. In Figure B–18, ∠3 and ∠5, and ∠2 and ∠8 are two pairs of alternate exterior angles.
- ☐ **Corresponding angles** are equal to each other. The pairs of corresponding angles in Figure B–18 are ∠3 and ∠7, ∠2 and ∠6, ∠8 and ∠4, and ∠5 and ∠1.

EXAMPLE 1 Solve each of the following:

(a) A ramp makes an angle of 34° with the horizontal (See Figure B–19). What angle does it make with the vertical? Name these two angles.

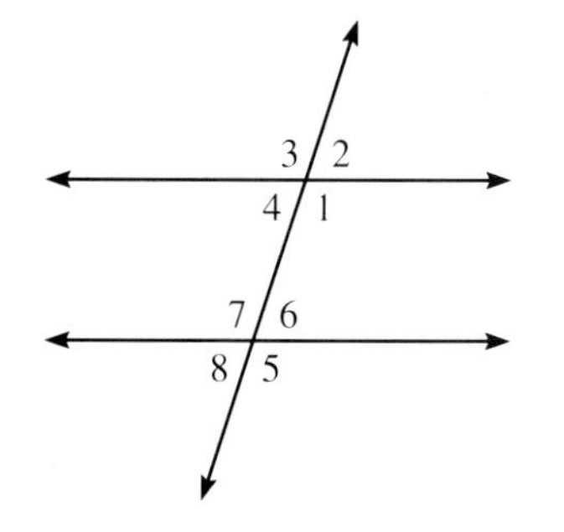

FIGURE B–18

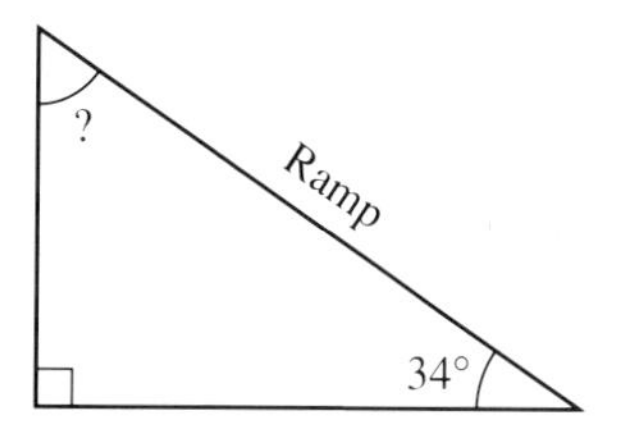

FIGURE B–19

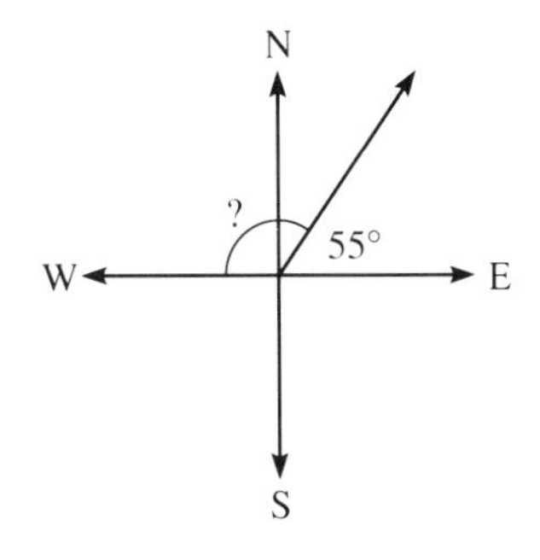

FIGURE B–20

(b) A plane takes off at an angle of 55° with the East as shown (See Figure B–20). What angle does it make with the West? Name these two angles.

Solution

(a) $90° - 34° = 56°$ complementary angles
(b) $180° - 55° = 125°$ supplementary angles ■

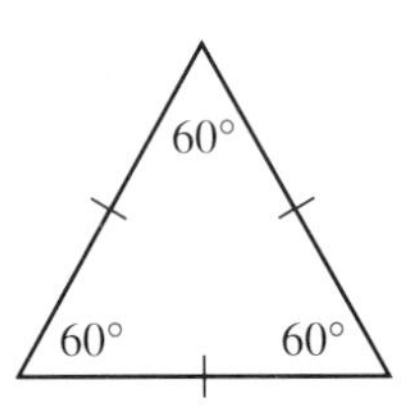

FIGURE B–21

Any closed figure in a plane with three or more straight sides is called a **polygon.** Polygons are usually named according to the number of sides they contain. The most familiar polygons are triangles, three-sided figures, and quadrilaterals, four-sided figures. A six-sided polygon is called a hexagon; a five-sided polygon is a pentagon; an eight-sided polygon is an octagon and so on.

If the sides of a polygon are all equal, and in addition, the interior angles are equal, the polygon is said to be **regular.**

An important feature of every polygon is the sum of the angles it contains. If a polygon has n sides, the angles add up to $(n - 2)(180°)$. Try this formula on the polygons we stated above. You will find that the angles of a triangle ($n = 3$) add up to 180°, a quadrilateral's ($n = 4$) to 360°, a pentagon's ($n = 5$) to 540° and a octagon's ($n = 8$) to 1080°.

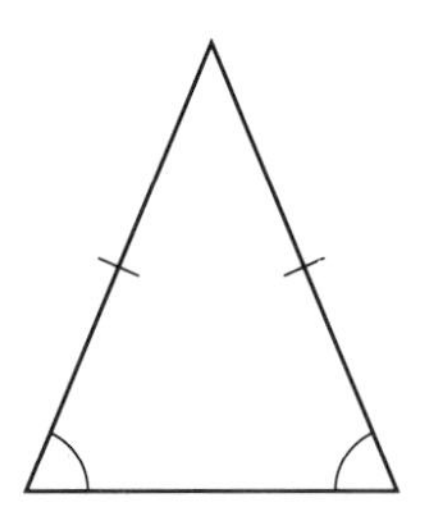
FIGURE B–22

EXAMPLE 2 What is a six-sided polygon called? Find the sum of the angles in this figure.

Solution A six-sided polygon is called a hexagon.

$$(6 - 2)(180°) = 4 \cdot 180° = 720°$$ ■

The most often used polygons in technical work, the triangle and the quadrilateral, have many features and properties that you should be able to recognize and use.

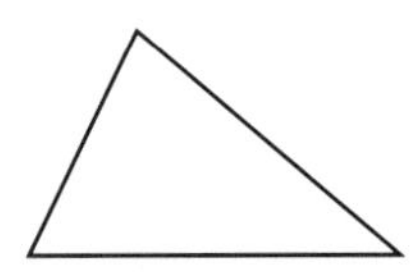
FIGURE B–23

Triangles are classified by the relations between the sides (or angles):

- ☐ If all three sides (or angles) are equal, the triangle is called **equilateral** (Figure B–21).
- ☐ If only two sides (two angles) are equal, the triangle is called **isosceles** (Figure B–22).
- ☐ If all the sides (angles) are of different size, the triangle is called **scalene** (Figure B–23).
- ☐ If the triangle has a right angle, it is called a **right** triangle (Figure B–24).

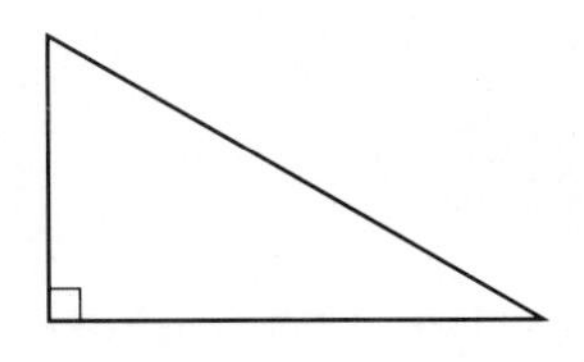
FIGURE B–24

Right triangles are of special importance because they are the basis of trigonometry. Because all right triangles have one right angle, the side opposite this angle has been given a name, the hypotenuse. The other two sides are called legs. All right triangles share a common relationship, the square of the hypotenuse is equal to the sum of the squares of the legs. This relationship is called the Pythagorean theorem and expressed as the formula

$$a^2 + b^2 = c^2$$

where a and b represent the two legs and c represents the hypotenuse of the right triangle. This formula can be used to find any missing side of a right triangle, provided two other sides are known.

EXAMPLE 3 Solve for the missing side in each of the following right triangles (Figures B–25, B–26, B–27).

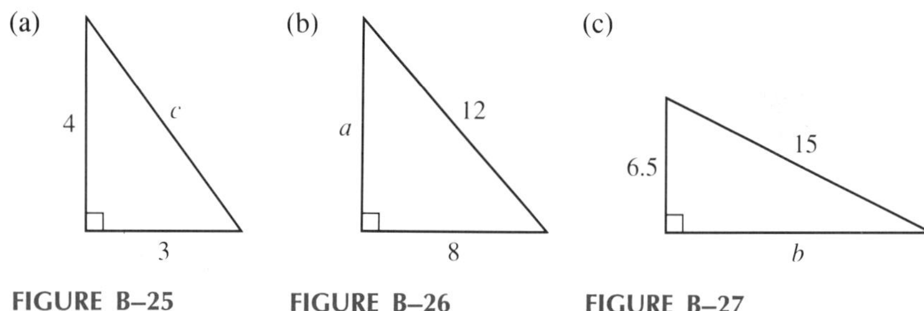

FIGURE B–25 FIGURE B–26 FIGURE B–27

Solution:

(a) $c = \sqrt{3^2 + 4^2} = \sqrt{25} = 5$

(b) $a = \sqrt{12^2 - 8^2} = \sqrt{80} = 4\sqrt{5}$

(c) $b = \sqrt{15^2 - 6.5^2} = \sqrt{182.75} = 13.5$ (rounded) ■

Similiar triangles are another classification important to the study of trigonometry. If two triangles are similar, their corresponding angles must be equal and their corresponding sides must be proportional. Figure B–28 is an example of two similiar triangles.

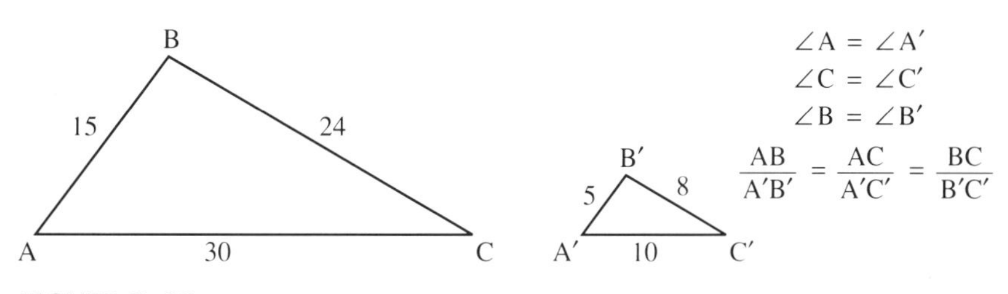

FIGURE B–28

Quadrilaterals are the second most often used polygon. There are several categories of quadrilaterals.

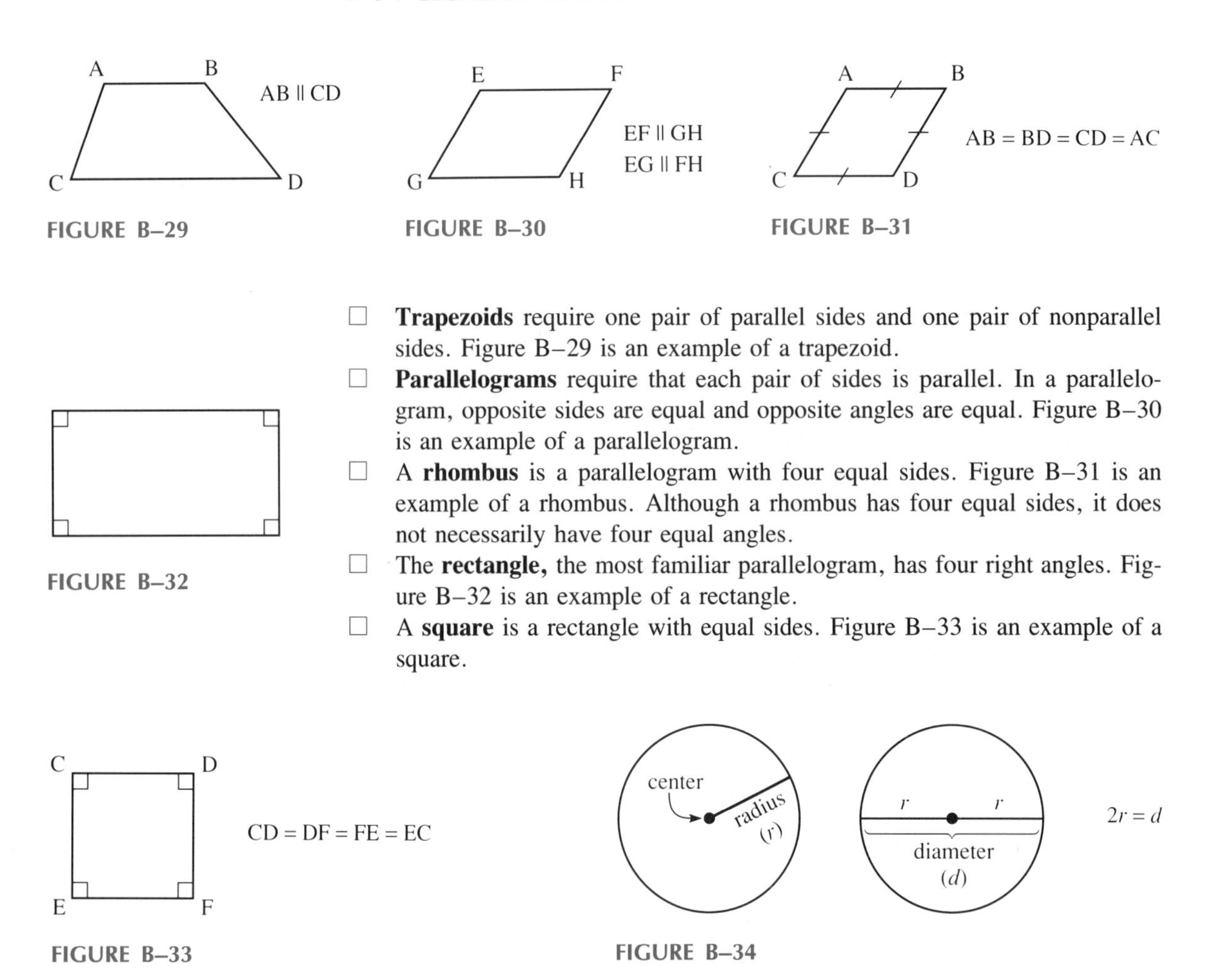

FIGURE B–29

FIGURE B–30

FIGURE B–31

FIGURE B–32

FIGURE B–33

FIGURE B–34

- ☐ **Trapezoids** require one pair of parallel sides and one pair of nonparallel sides. Figure B–29 is an example of a trapezoid.
- ☐ **Parallelograms** require that each pair of sides is parallel. In a parallelogram, opposite sides are equal and opposite angles are equal. Figure B–30 is an example of a parallelogram.
- ☐ A **rhombus** is a parallelogram with four equal sides. Figure B–31 is an example of a rhombus. Although a rhombus has four equal sides, it does not necessarily have four equal angles.
- ☐ The **rectangle,** the most familiar parallelogram, has four right angles. Figure B–32 is an example of a rectangle.
- ☐ A **square** is a rectangle with equal sides. Figure B–33 is an example of a square.

When a quadrilateral is encountered in a problem, remember to examine its characteristics carefully for identification purposes.

The last plane figure we will describe in detail is the circle. A **circle** is a closed plane figure defined to be the set (or group) of points that are the same distance from a fixed point (Figure B–34). The fixed point is called the center of the circle, and the fixed distance is called the radius, r, (plural—radii). The diameter is a line segment passing through the center of the circle and having its endpoints on the circle (Figure B–34). Thus, the diameter is twice the length of the radius. All diameters in a given circle are the same length.

Many facts and terms relating to circles are essential to solving technical problems. Following are some important ones.

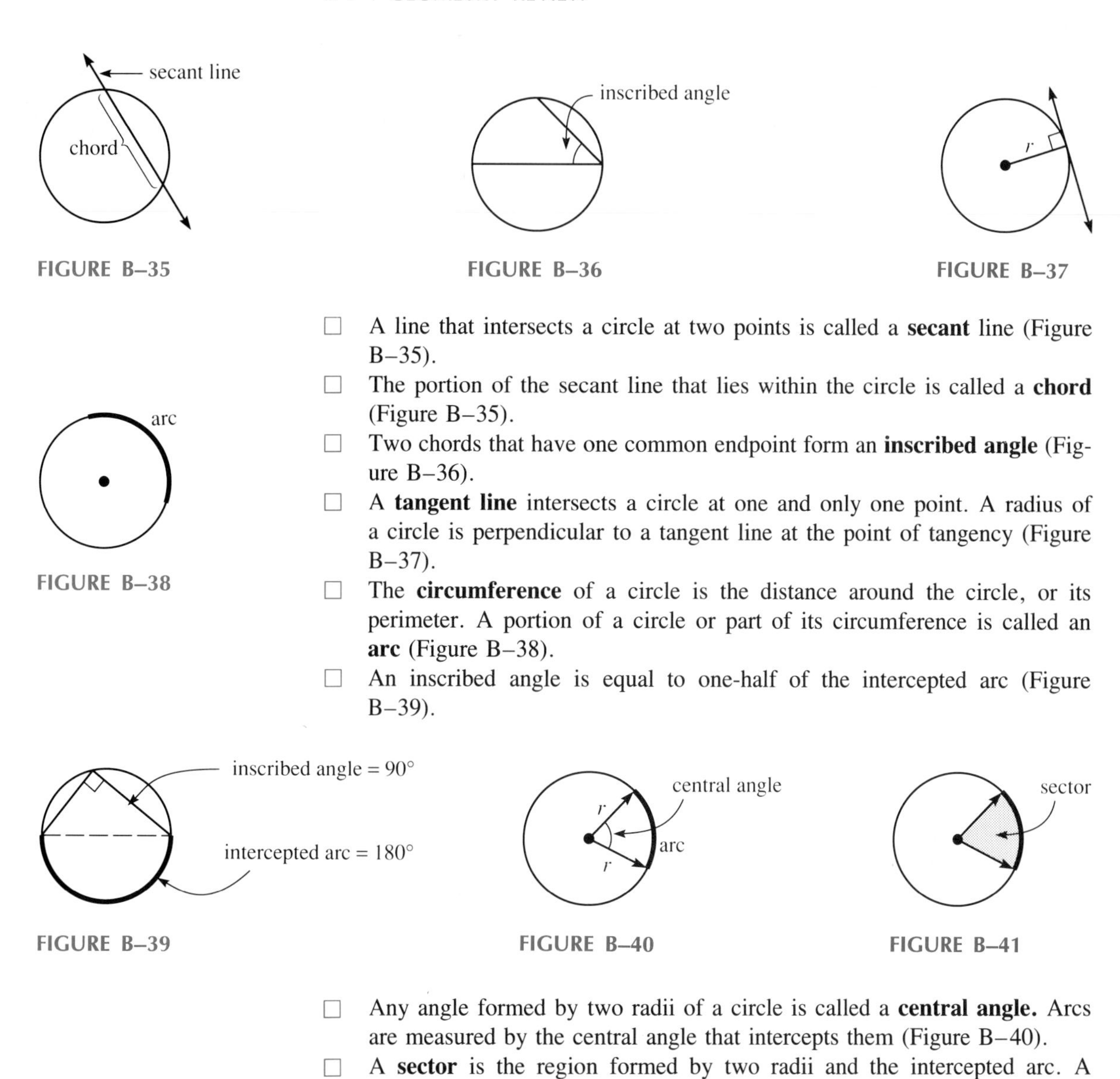

FIGURE B–35

FIGURE B–36

FIGURE B–37

FIGURE B–38

- ☐ A line that intersects a circle at two points is called a **secant** line (Figure B–35).
- ☐ The portion of the secant line that lies within the circle is called a **chord** (Figure B–35).
- ☐ Two chords that have one common endpoint form an **inscribed angle** (Figure B–36).
- ☐ A **tangent line** intersects a circle at one and only one point. A radius of a circle is perpendicular to a tangent line at the point of tangency (Figure B–37).
- ☐ The **circumference** of a circle is the distance around the circle, or its perimeter. A portion of a circle or part of its circumference is called an **arc** (Figure B–38).
- ☐ An inscribed angle is equal to one-half of the intercepted arc (Figure B–39).

FIGURE B–39

FIGURE B–40

FIGURE B–41

- ☐ Any angle formed by two radii of a circle is called a **central angle.** Arcs are measured by the central angle that intercepts them (Figure B–40).
- ☐ A **sector** is the region formed by two radii and the intercepted arc. A sector can also be identified by its central angle (Figure B–41).

The perimeter and the area can be found for each of the geometric figures we have described. The **perimeter** is the distance around the outside of a figure, and the **area** is the region bounded by the sides (or circumference) of a figure. When using the formulas stated below for computing the area and perimeter of many plane figures, it is important to understand the term **altitude.** The altitude of a figure is the line drawn from any vertex perpendicular to the opposite side (extended if necessary). The letter h is used to denote the altitude in the formulas summarized in the following box.

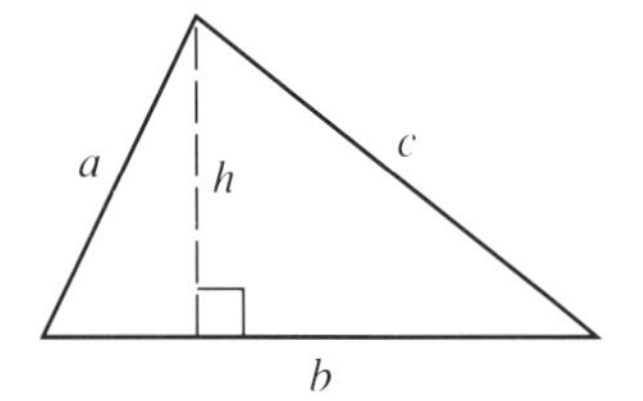

FIGURE B–42

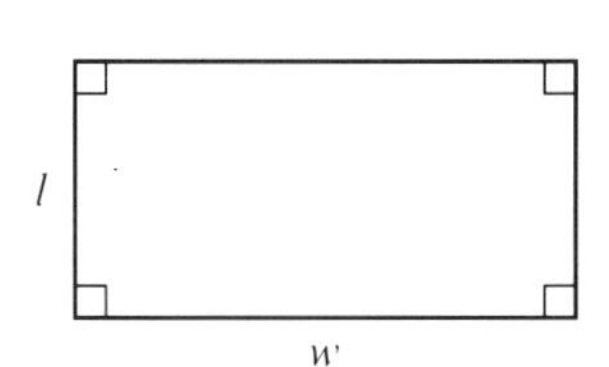

FIGURE B–43

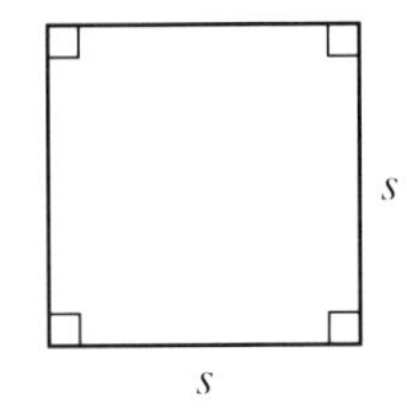

FIGURE B–44

Geometric Formulas for Area and Perimeter

Figure	*Perimeter, P*	*Area, A*
Triangle (Figure B–42)	$P = a + b + c$	$A = .5bh$
Rectangle (Figure B–43)	$P = 2l + 2w$	$A = lw$
Square (Figure B–44)	$P = 4s$	$A = s^2$
Parallelogram (Figure B–45)	$P = 2a + 2b$	$A = bh$
Trapezoid (Figure B–46)	$P = a + b + c + d$	$A = .5(a + b)h$
Circle (Figure B–47)	$C = \pi d = 2\pi r$	$A = \pi r^2$

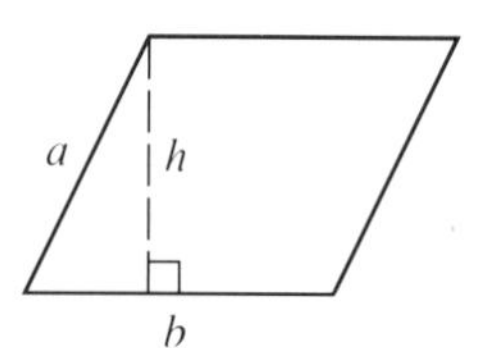

FIGURE B–45

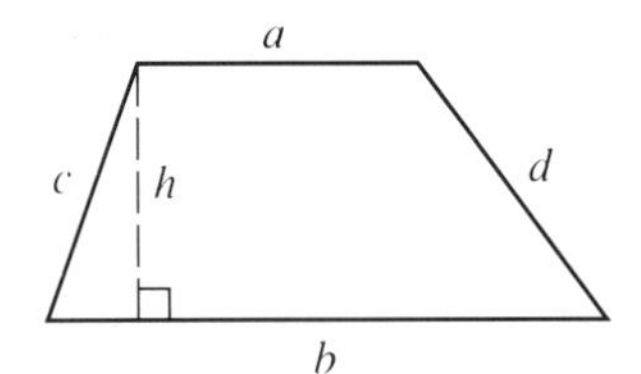

FIGURE B–46

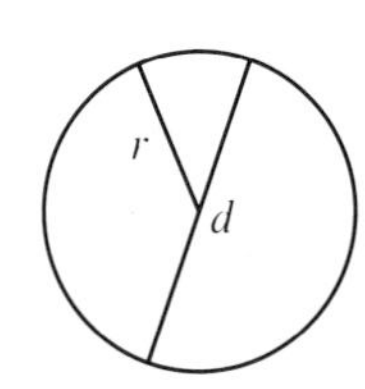

FIGURE B–47

In plane geometry (two dimensional), figures have dimensions of length and width. Technical problems also require a knowledge of solid geometry (three dimensional). In solid geometry the third dimension of height is added. Many solid figures you encounter every day, such as the sphere (ball), the cone, and the cylinder, are not part of the group of solids called polyhedrons. A *polyhedron* is a solid figure that is bounded by a finite number of polygonal regions. Thus a cube, a pyramid, and a prism are polyhedrons.

Formulas have been derived to simplify computations involving solid figures. The ones listed in the following box are for finding the volume and surface area of the most common solids. The following letters are used: B = area of base; L = lateral surface area [area of all surfaces other than the base(s)]; S = total surface area [includes area of base(s)]; V = volume.

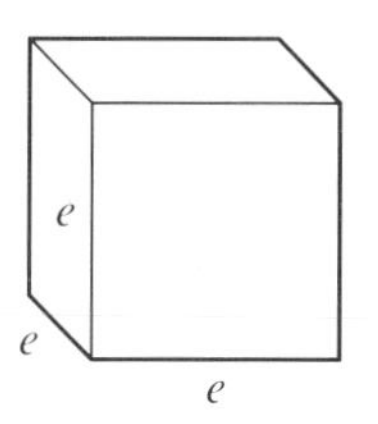

FIGURE B–48

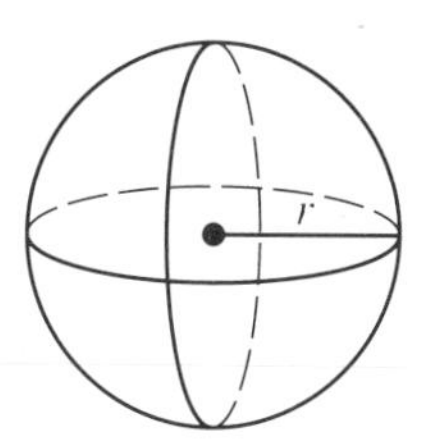

FIGURE B–49

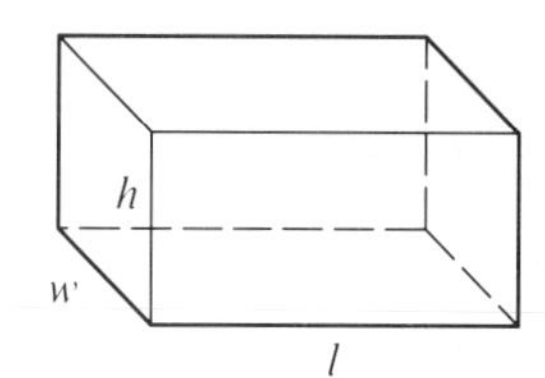

FIGURE B–50

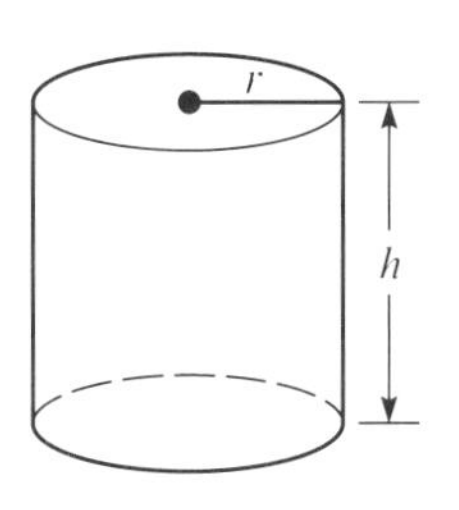

FIGURE B–51

Formulas for Surface Area and Volume

Solid	*Surface Area, S*	*Volume, V*
Cube (Figure B–48)	$S = 6e^2$	$V = e^3$
Sphere (Figure B–49)	$S = 4\pi r^2$	$V = 4/3\pi r^3$
Rectangular prism (Figure B–50)	$S = 2(lw + wh + lh)$	$V = lwh$
Right circular cylinder (Figure B–51)	$S = 2\pi rh + 2\pi r^2$	$V = \pi r^2 h$
Right circular cone (Figure B–52)	$S = \pi rs + \pi r^2$	$V = 1/3\pi r^2 h$
General cone or pyramid (Figure B–53)		$V = 1/3Bh$

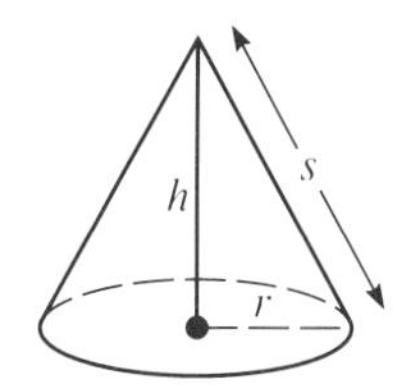

FIGURE B–52

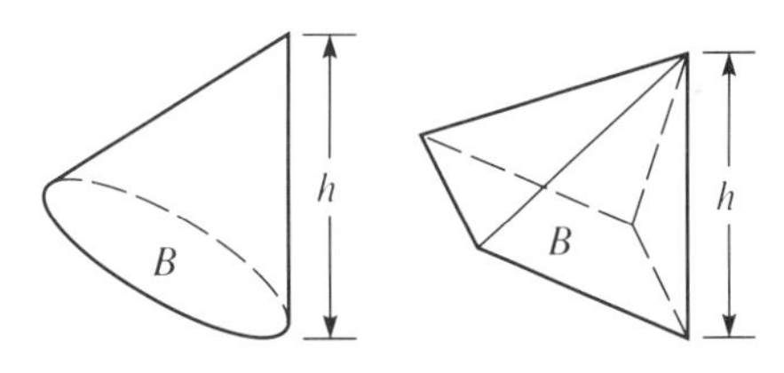

FIGURE B–53

APPENDIX B EXERCISES

Find the complement of each angle.

1. 18° **2.** 35° **3.** 56.4° **4.** 83.5°

Find the supplement of each angle.

5. 103° **6.** 65.2° **7.** 120° **8.** 95.5°

Use Figure B–54 for problems 9–13.

9. State all pairs of alternate interior angles.

10. State all pairs of alternate exterior angles.

11. State all pairs of corresponding angles.

12. Find the measurement of ∠4 and ∠8.

13. Find the measurement of ∠2 and ∠5.

65°
2 1
P
3 4
5 6
Q
7 8

$P \parallel Q$
1 = 65°

FIGURE B–54

Use Figure B–55 for problems 14–16.

14. Find $\angle x$.

15. Find side BC.

16. Name the triangle.

FIGURE B–55

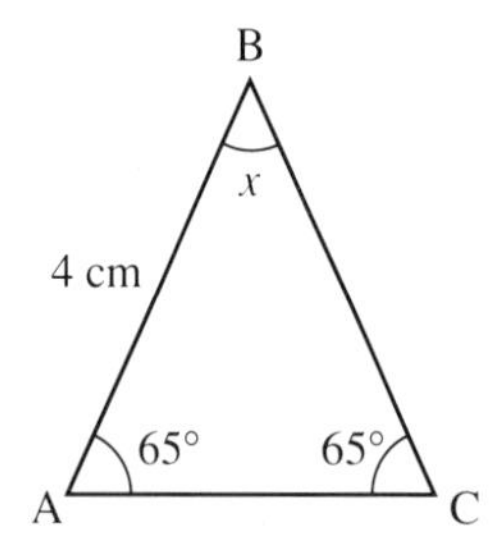

In problems 17–20, using the Pythagorean theorem and given side measurements, find the length of the missing side of the right triangle (Figure B–56). (Round to tenths place.)

17. Given a = 12 cm and b = 15 cm, find c.

18. Given b = 8 in. and c = 14 in., find a.

19. Given a = 34.6 ft and c = 54.8 ft, find b.

20. Given a = 125 m and b = 75 m, find c.

FIGURE B–56

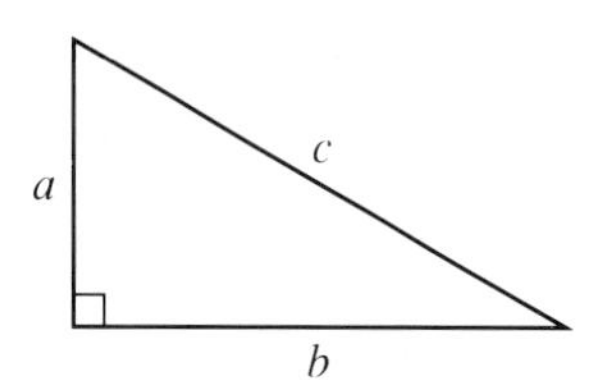

For Figures B–57 through B–63, (a) name the figure; (b) find the perimeter of the figure; and (c) find the area of the figure (Problems 21–27).

21. FIGURE B–57

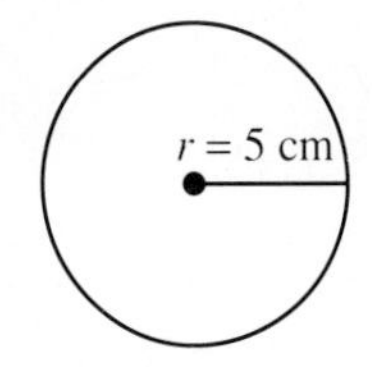

22. FIGURE B–58

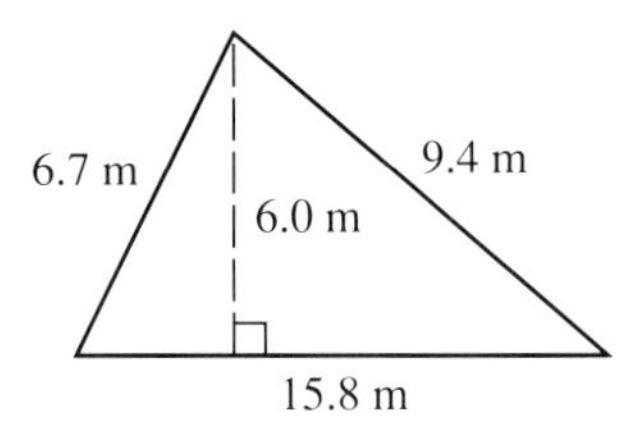

23. FIGURE B–59

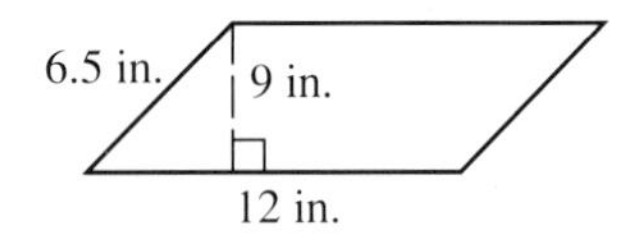

24. FIGURE B–60

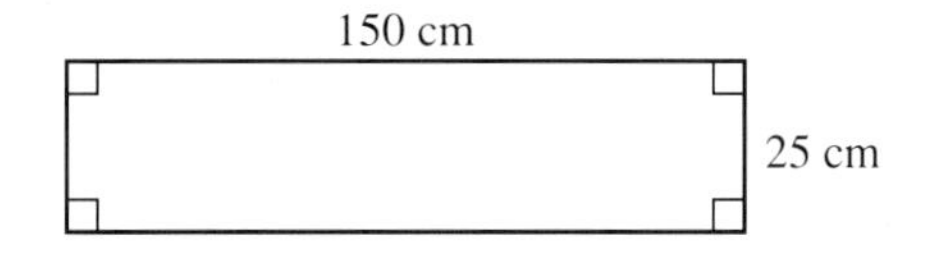

25. FIGURE B–61

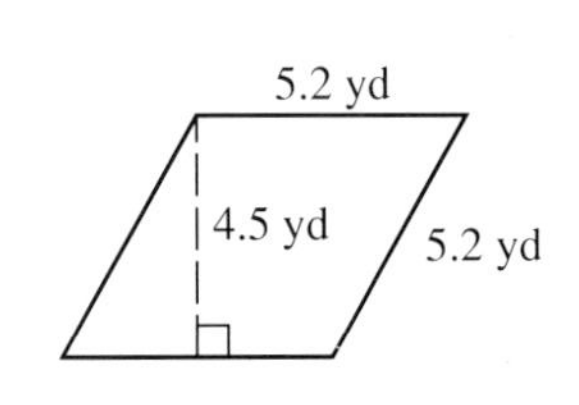

26. FIGURE B–62

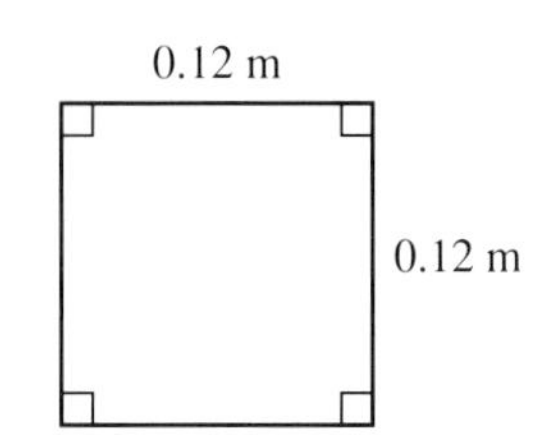

27. FIGURE B–63

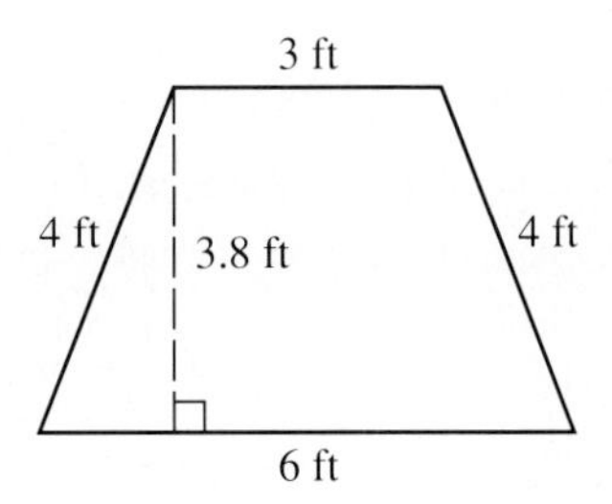

For Figures B–64 through B–66, (a) name the figure and (b) find the volume of the figure (Problems 28–30).

28. FIGURE B–64

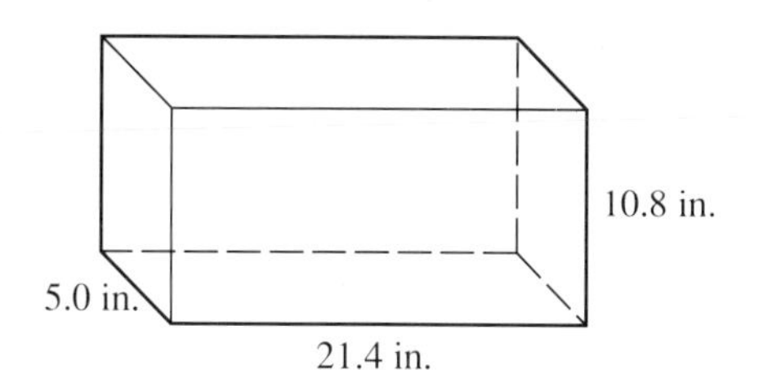

29. FIGURE B–65

30. FIGURE B–66

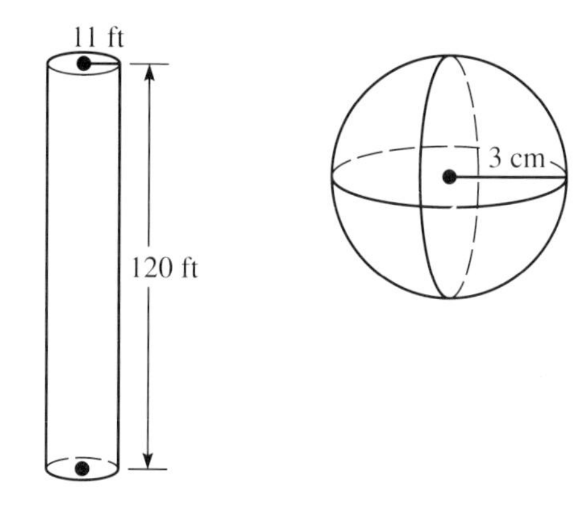

For Figures B–67 through B–69, (a) name the figure and (b) find the total surface area of the figure (Problems 31–33).

31. FIGURE B–67

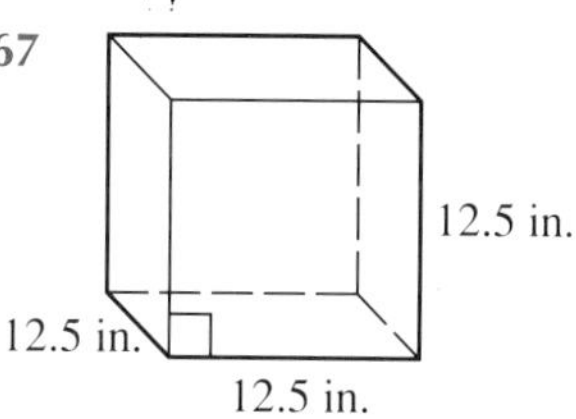

32. FIGURE B–68

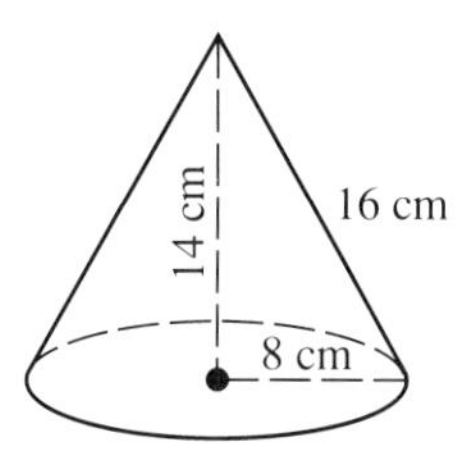

33. FIGURE B–69

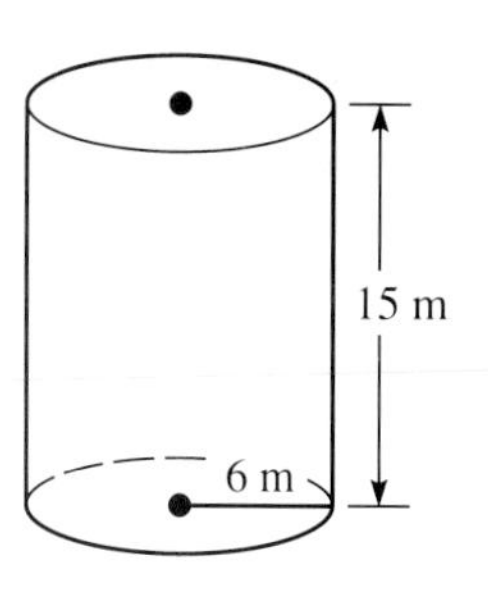

Solve the following problems.

34. A right triangle with legs 18.3 cm and 6.5 cm is rotated about the 18.3 cm leg. Find the volume of the generated cone (Figure B–70).

FIGURE B–70

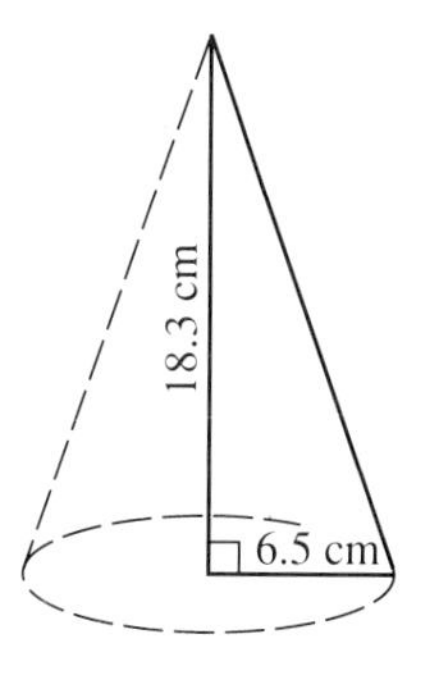

35. The triangle shown in Figure B–71 is known as an impedance triangle and is used in many ac circuit problems. The side Z is called the impedance, the side X the reactance, and the side R is the resistance. All three sides are measured in units of ohms (Ω). The angle θ is known as the phase angle. Use the impedance triangle and solve for the unknown.
(a) Given $X = 9\ \Omega$, $R = 12\ \Omega$, find Z.
(b) Given $Z = 26\ \Omega$, $R = 10\ \Omega$, find X.
(c) Given $X = 90\ \Omega$, $Z = 150\ \Omega$, find R.

FIGURE B–71

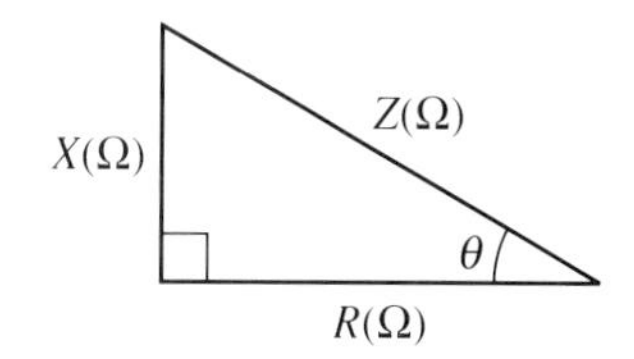

36. A rectangular tank is 12 ft long, 10 ft wide and 8 ft deep. How many gallons of water will it hold? (1 gal = 231 in.^{3})

37. A hollow pipe has an outside circumference of 10.5 in. and an inner diameter of 2.8 in. In Figure B–72, find the shaded cross–sectional area of the pipe.

FIGURE B–72

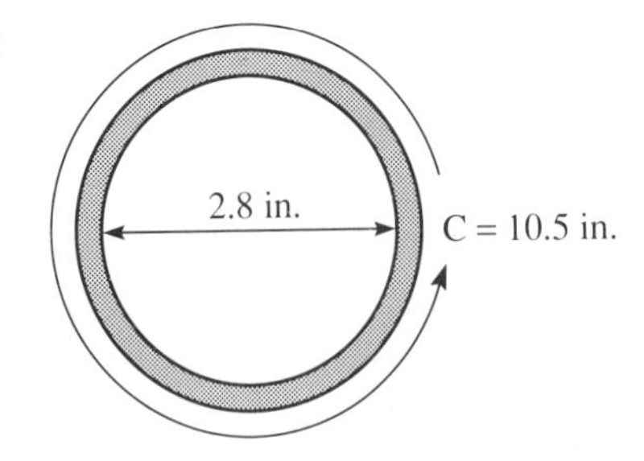

38. A wheel with a diameter of 36 cm makes 100 turns. How far does it roll?

39. An A–frame structure has a base of 18 m, while a scale drawing of the structure has a base of 20 cm (0.20 m). If the sides of the structure in the drawing are 40 cm (0.40 m) each, what are the lengths of the real sides (Figure B–73)?

FIGURE B–73

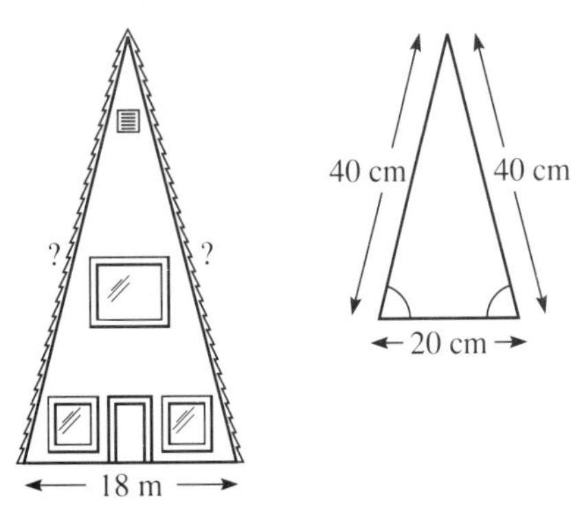

40. In the truss shown in Figure B–74, $\overleftrightarrow{BC} \| \overleftrightarrow{AD}$, $\overleftrightarrow{AB} \| \overleftrightarrow{CE}$, $\overleftrightarrow{BE} \| \overleftrightarrow{CD}$. $BAE = 45°$, $BEA = 25°$. Find angles 1, 2, 3, 4, 5, 6, and 7.

FIGURE B–74

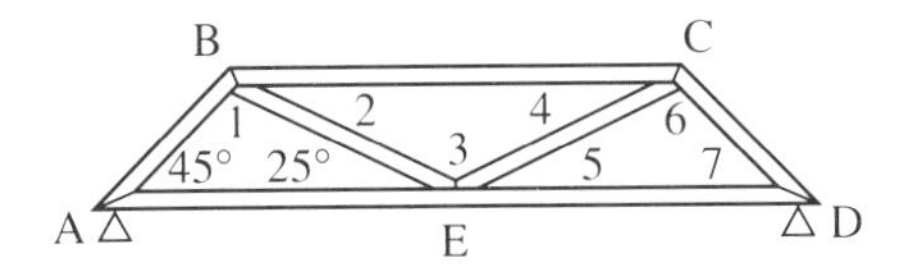

41. In Figure B–75, find the area of the shaded region.

FIGURE B–75

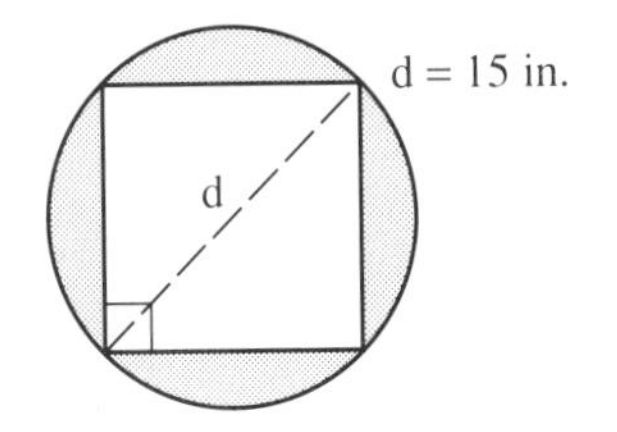

42. Find the total surface area of the I-beam shown in Figure B–76. All thicknesses are 1.5 in.

FIGURE B–76

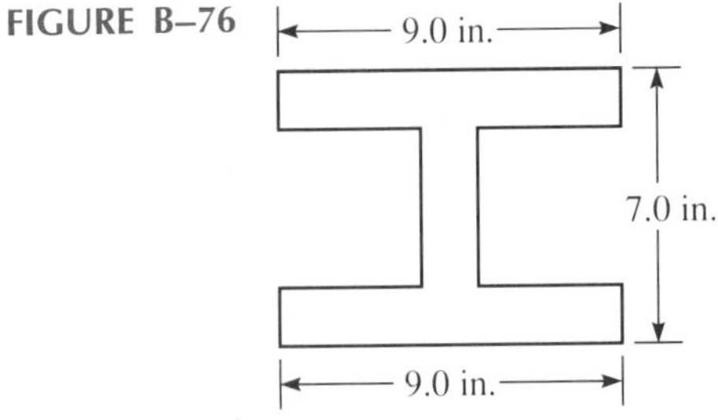

43. How many liters of gasoline can be stored in a cylindrical tank with a diameter of 5 m and a height of 12 m? (1000 L = 1 m^3)

44. Find the total surface area of the cylindrical tank described in Problem 43.

Appendix C

Tables

TABLE C–1
Table of Integrals

Basic Integrals

1. $\int u^n\,du = \dfrac{u^{n+1}}{n+1} + C \qquad [n \neq -1]$

2. $\int \dfrac{du}{u} = \ln|u| + C$

3. $\int e^u\,du = e^u + C$

4. $\int a^u\,du = \dfrac{a^u}{\ln a} + C \qquad [a > 0]$

5. $\int \sin u\,du = -\cos u + C$

6. $\int \cos u\,du = \sin u + C$

7. $\int \tan u\,du = -\ln|\cos u| + C$

8. $\int \cot u\,du = \ln|\sin u| + C$

9. $\int \sec u\,du = \ln|\sec u + \tan u| + C$

10. $\int \csc u\,du = \ln|\csc u - \cot u| + C$

11. $\int \sec^2 u\,du = \tan u + C$

12. $\int \csc^2 u\,du = -\cot u + C$

13. $\int \sec u \tan u\,du = \sec u + C$

14. $\int \csc u \cot u\,du = -\csc u + C$

15. $\displaystyle\int \frac{du}{\sqrt{a^2 - u^2}} = \text{Arcsin}\left(\frac{u}{a}\right) + C$

16. $\displaystyle\int \frac{du}{a^2 + u^2} = \frac{1}{a}\,\text{Arctan}\left(\frac{u}{a}\right) + C$

17. $\displaystyle\int \frac{du}{\sqrt{u^2 \pm a^2}} = \ln\left|u + \sqrt{u^2 \pm a^2}\right| + C$

18. $\displaystyle\int \frac{du}{u^2 - a^2} = \frac{1}{2a}\ln\left|\frac{u-a}{u+a}\right| + C$

Integrals Involving $a + bu$

19. $\displaystyle\int (a + bu)^n\,du = \frac{(a+bu)^{n+1}}{b(n+1)} + C \qquad [n \neq -1]$

20. $\displaystyle\int u\,(a + bu)^n\,du = \frac{(a+bu)^{n+1}}{b^2}\left(\frac{a+bu}{n+2} - \frac{a}{n+1}\right) + C \qquad [n \neq -1, -2]$

21. $\displaystyle\int \frac{du}{a + bu} = \frac{1}{b}\ln|a + bu| + C$

22. $\displaystyle\int \frac{du}{u(a + bu)} = -\frac{1}{a}\ln\left|\frac{a+bu}{u}\right| + C$

23. $\displaystyle\int \frac{du}{u^2(a + bu)} = -\frac{1}{au} + \frac{b}{a^2}\ln\left|\frac{a+bu}{u}\right| + C$

24. $\displaystyle\int \frac{u\,du}{a+bu} = \frac{1}{b^2}\left[a + bu - a\ln|a+bu|\right] + C$

25. $\displaystyle\int \frac{u^2\,du}{a+bu} = \frac{1}{b^3}\left[\frac{1}{2}(a+bu)^2 - 2a(a+bu) + a^2\ln|a+bu|\right] + C$

26. $\displaystyle\int \frac{du}{u(a+bu)^2} = \frac{1}{a(a+bu)} - \frac{1}{a^2}\ln\left|\frac{a+bu}{u}\right| + C$

27. $\displaystyle\int \frac{du}{u^2(a+bu)^2} = -\frac{a+2bu}{a^2u(a+bu)} + \frac{2b}{a^3}\ln\left|\frac{a+bu}{u}\right| + C$

28. $\displaystyle\int \frac{u\,du}{(a+bu)^2} = \frac{1}{b^2}\left[\ln|a+bu| + \frac{a}{a+bu}\right] + C$

Integrals Involving $\sqrt{a+bu}$

29. $\displaystyle\int \sqrt{a+bu}\,du = \frac{2}{3b}(a+bu)^{3/2} + C$

30. $\displaystyle\int (\sqrt{a+bu})^n\,du = \frac{2}{b}\frac{(\sqrt{a+bu})^{n+2}}{n+2} + C \qquad [n \neq -2]$

31. $\displaystyle\int u\sqrt{a+bu}\,du = \frac{2(3bu-2a)}{15b^2}(a+bu)^{3/2} + C$

32. $\displaystyle\int u^2\sqrt{a+bu}\,du = \frac{2(15b^2u^2 - 12\,abu + 8a^2)}{105b^3}(a+bu)^{3/2} + C$

33. $\displaystyle\int u^n\sqrt{a+bu}\,du = \frac{2u^n(a+bu)^{3/2}}{(2n+3)b} - \frac{2an}{(2n+3)b}\int u^{n-1}\sqrt{a+bu}\,du \qquad [2n \neq -3]$

34. $\displaystyle\int \frac{\sqrt{a+bu}}{u}\,du = 2\sqrt{a+bu} + a\int \frac{du}{u\sqrt{a+bu}}$

35. $\displaystyle\int \frac{du}{\sqrt{a+bu}} = \frac{2}{b}\sqrt{a+bu} + C$

36. $\displaystyle\int \frac{u\,du}{\sqrt{a+bu}} = \frac{2(bu-2a)}{3b^2}\sqrt{a+bu} + C$

37. $\displaystyle\int \frac{u^n\,du}{\sqrt{a+bu}} = \frac{2u^n\sqrt{a+bu}}{(2n+1)b} - \frac{2an}{(2n+1)b}\int \frac{u^{n-1}\,du}{\sqrt{a+bu}} \qquad [2n \neq -1]$

38. $\displaystyle\int \frac{u\,du}{\sqrt{a+bu}} = \frac{2(bu-2a)}{3b^2}\sqrt{a+bu} + C$

39. $\displaystyle\int \frac{du}{u\sqrt{a+bu}} = \frac{1}{\sqrt{a}}\ln\left|\frac{\sqrt{a+bu}-\sqrt{a}}{\sqrt{a+bu}+\sqrt{a}}\right| + C \qquad [a > 0]$

40. $\displaystyle\int \frac{du}{u\sqrt{a+bu}} = \frac{2}{\sqrt{-a}}\,\text{Arctan}\sqrt{\frac{a+bu}{-a}} + C \qquad [a < 0]$

41. $\displaystyle\int \frac{du}{u^n\sqrt{a+bu}} = -\frac{\sqrt{a+bu}}{(n-1)au^{n-1}} - \frac{(2n-3)b}{2(n-1)a}\int \frac{du}{u^{n-1}\sqrt{a+bu}} \qquad [n \neq 1]$

Integrals Involving $\sqrt{a^2 - u^2}$

42. $\displaystyle\int \sqrt{a^2-u^2}\,du = \frac{u}{2}\sqrt{a^2-u^2} + \frac{a^2}{2}\,\text{Arcsin}\left(\frac{u}{a}\right) + C$

43. $\int u\sqrt{a^2 - u^2}\,du = -\frac{1}{3}(a^2 - u^2)^{3/2} + C$

44. $\int u^n\sqrt{a^2 - u^2}\,du = -\frac{u^{n-1}(a^2 - u^2)^{3/2}}{n + 2} + \frac{(n - 1)a^2}{n + 2}\int u^{n-2}\sqrt{a^2 - u^2}\,du \qquad [n \neq -2]$

45. $\int \frac{\sqrt{a^2 - u^2}}{u}\,du = \sqrt{a^2 - u^2} - a\ln\left|\frac{a + \sqrt{a^2 - u^2}}{u}\right| + C$

46. $\int \frac{\sqrt{a^2 - u^2}}{u^2}\,du = -\frac{\sqrt{a^2 - u^2}}{u} - \text{Arcsin}\left(\frac{u}{a}\right) + C$

47. $\int \frac{du}{u\sqrt{a^2 - u^2}} = -\frac{1}{a}\ln\left|\frac{a + \sqrt{a^2 - u^2}}{u}\right| + C$

48. $\int \frac{du}{u^2\sqrt{a^2 - u^2}} = -\frac{\sqrt{a^2 - u^2}}{a^2u} + C$

49. $\int \frac{u^2\,du}{\sqrt{a^2 - u^2}} = -\frac{u}{2}\sqrt{a^2 - u^2} + \frac{a^2}{2}\text{Arcsin}\left(\frac{u}{a}\right) + C$

50. $\int \frac{u^n\,du}{\sqrt{a^2 - u^2}} = -\frac{u^{n-1}\sqrt{a^2 - u^2}}{n} + \frac{(n - 1)a^2}{n}\int \frac{u^{n-2}\,du}{\sqrt{a^2 - u^2}} \qquad [n \neq 0]$

51. $\int \frac{du}{u^n\sqrt{a^2 - u^2}} = \frac{-\sqrt{a^2 - u^2}}{(n - 1)a^2u^{n-1}} + \frac{n - 2}{(n - 1)a^2}\int \frac{du}{u^{n-2}\sqrt{a^2 - u^2}} \qquad [n \neq 1]$

52. $\int (a^2 - u^2)^{3/2}\,du = \frac{u}{4}(a^2 - u^2)^{3/2} + \frac{3a^2u}{8}\sqrt{a^2 - u^2} + \frac{3a^4}{8}\text{Arcsin}\left(\frac{u}{a}\right) + C$

53. $\int \frac{(a^2 - u^2)^{3/2}}{u}\,du = \frac{1}{3}(a^2 - u^2)^{3/2} - a^2\sqrt{a^2 - u^2} + a^3\ln\left|\frac{a + \sqrt{a^2 - u^2}}{u}\right| + C$

54. $\int \frac{du}{(a^2 - u^2)^{3/2}} = \frac{u}{a^2\sqrt{a^2 - u^2}} + C$

55. $\int \frac{u^2\,du}{(a^2 - u^2)^{3/2}} = \frac{u}{\sqrt{a^2 - u^2}} - \text{Arcsin}\left(\frac{u}{a}\right) + C$

56. $\int \frac{du}{u(a^2 - u^2)^{3/2}} = \frac{1}{a^2\sqrt{a^2 - u^2}} - \frac{1}{a^3}\ln\left|\frac{a + \sqrt{a^2 - u^2}}{u}\right| + C$

Integrals Involving $\sqrt{u^2 + a^2}$ or $\sqrt{u^2 - a^2}$

57. $\int \sqrt{u^2 \pm a^2}\,du = \frac{1}{2}\left[u\sqrt{u^2 \pm a^2} \pm a^2\ln\left|u + \sqrt{u^2 \pm a^2}\right|\right] + C$

58. $\int \frac{\sqrt{u^2 + a^2}}{u}\,du = \sqrt{u^2 + a^2} - a\ln\left|\frac{a + \sqrt{u^2 + a^2}}{u}\right| + C$

59. $\int \frac{\sqrt{u^2 - a^2}}{u}\,du = \sqrt{u^2 - a^2} - a\,\text{Arccos}\left(\frac{a}{u}\right) + C$

60. $\int \frac{\sqrt{u^2 \pm a^2}}{u^2}\,du = -\frac{\sqrt{u^2 \pm a^2}}{u} + \ln\left|u + \sqrt{u^2 \pm a^2}\right| + C$

61. $\int u^2\sqrt{u^2 \pm a^2}\,du = \frac{u}{8}(2u^2 \pm a^2)\sqrt{u^2 \pm a^2} - \frac{a^4}{8}\ln\left|u + \sqrt{u^2 \pm a^2}\right| + C$

62. $\displaystyle\int \frac{du}{u\sqrt{u^2 + a^2}} = \frac{1}{a} \ln \left| \frac{u}{a + \sqrt{u^2 + a^2}} \right| + C$

63. $\displaystyle\int \frac{du}{u\sqrt{u^2 - a^2}} = \frac{1}{a} \text{Arccos} \left(\frac{a}{u} \right) + C$

64. $\displaystyle\int \frac{u^2\, du}{\sqrt{u^2 \pm a^2}} = \frac{u}{2}\sqrt{u^2 \pm a^2} \mp \frac{a^2}{2} \ln |u + \sqrt{u^2 \pm a^2}| + C$

65. $\displaystyle\int \frac{du}{u^2\sqrt{u^2 \pm a^2}} = \frac{\mp\sqrt{u^2 \pm a^2}}{a^2 u} + C$

66. $\displaystyle\int \frac{du}{(u^2 \pm a^2)^{3/2}} = \frac{\pm u}{a^2\sqrt{u^2 \pm a^2}} + C$

67. $\displaystyle\int \frac{u^2\, du}{(u^2 \pm a^2)^{3/2}} = -\frac{u}{\sqrt{u^2 \pm a^2}} + \ln |u + \sqrt{u^2 \pm a^2}| + C$

68. $\displaystyle\int (u^2 \pm a^2)^{3/2}\, du = \frac{u}{8}(2u^2 \pm 5a^2)\sqrt{u^2 \pm a^2} + \frac{3a^4}{8} \ln |u + \sqrt{u^2 \pm a^2}| + C$

Integrals Involving Trigonometric Expressions

69. $\displaystyle\int \sin^2 u\, du = \frac{1}{2}u - \frac{1}{4} \sin 2u + C$

70. $\displaystyle\int \sin^n u\, du = -\frac{1}{n} \sin^{n-1} u \cos u + \frac{n-1}{n} \int \sin^{n-2} u\, du$

71. $\displaystyle\int \cos^2 u\, du = \frac{1}{2}u + \frac{1}{4} \sin 2u + C$

72. $\displaystyle\int \cos^n u\, du = \frac{1}{n} \cos^{n-1} u \sin u + \frac{n-1}{n} \int \cos^{n-2} u\, du$

73. $\displaystyle\int \tan^2 u\, du = \tan u - u + C$

74. $\displaystyle\int \tan^n u\, du = \frac{\tan^{n-1} u}{n-1} - \int \tan^{n-2} u\, du$

75. $\displaystyle\int \cot^2 u\, du = -\cot u - u + C$

76. $\displaystyle\int \cot^n u\, du = -\frac{\cot^{n-1} u}{n-1} - \int \cot^{n-2} u\, du$

77. $\displaystyle\int \sec^n u\, du = \frac{\sec^{n-2} u \tan u}{n-1} + \frac{n-2}{n-1} \int \sec^{n-2} u\, du$

78. $\displaystyle\int \csc^n u\, du = -\frac{\csc^{n-2} u \cot u}{n-1} + \frac{n-2}{n-1} \int \csc^{n-2} u\, du$

79. $\displaystyle\int u \sin u\, du = \sin u - u \cos u + C$

80. $\displaystyle\int u \cos u\, du = \cos u + u \sin u + C$

81. $\displaystyle\int u^n \sin au\, du = -\frac{u^n}{a} \cos au + \frac{n}{a} \int u^{n-1} \cos au\, du$ [n a positive integer]

82. $\int u^n \cos au\, du = \frac{u^n}{a}\sin au - \frac{n}{a}\int u^{n-1}\sin au\, du$ [n a positive integer]

83. $\int \frac{\sin u}{u^n}\, du = -\frac{\sin u}{(n-1)u^{n-1}} + \frac{1}{n-1}\int \frac{\cos u}{u^{n-1}}\, du$ [$n \neq 1$]

84. $\int \frac{\cos u}{u^n}\, du = -\frac{\cos u}{(n-1)u^{n-1}} - \frac{1}{n-1}\int \frac{\sin u}{u^{n-1}}\, du$ [$n \neq 1$]

85. $\int \frac{du}{\sin u \cos u} = \ln|\tan u| + C$

86. $\int \sin au \sin bu\, du = -\frac{\sin(a+b)u}{2(a+b)} + \frac{\sin(a-b)u}{2(a-b)} + C$

87. $\int \cos au \cos bu\, du = \frac{\sin(a+b)u}{2(a+b)} + \frac{\sin(a-b)u}{2(a-b)} + C$

88. $\int \sin au \cos bu\, du = -\frac{\cos(a+b)u}{2(a+b)} - \frac{\cos(a-b)u}{2(a-b)} + C$

89. $\int \sin^m u \cos^n u\, du = \frac{\sin^{m+1}u \cos^{n-1}u}{m+n} + \frac{n-1}{m+n}\int \sin^m u \cos^{n-2}u\, du$ [$m, n > 0$]

$= -\frac{\sin^{m-1}u \cos^{n+1}u}{m+n} + \frac{m-1}{m+n}\int \sin^{m-2}u \cos^n u\, du$ [$m, n > 0$]

Integrals Involving Exponential Functions

90. $\int ue^{au}\, du = \left(\frac{au-1}{a^2}\right)e^{au} + C$

91. $\int u^2 e^{au}\, du = \frac{e^{au}}{a^3}(a^2u^2 - 2au + 2) + C$

92. $\int u^n e^{au}\, du = \frac{1}{a}u^n e^{au} - \frac{n}{a}\int u^{n-1}e^{au}\, du$ [$n > 0$]

93. $\int e^{au}\ln u\, du = \frac{e^{au}\ln u}{a} - \frac{1}{a}\int \frac{e^{au}}{u}\, du$

94. $\int e^{au}\sin bu\, du = \frac{e^{au}(a\sin bu - b\cos bu)}{a^2+b^2} + C$

95. $\int e^{au}\cos bu\, du = \frac{e^{au}(a\cos bu + b\sin bu)}{a^2+b^2} + C$

96. $\int u^n a^u\, du = \frac{a^u u^n}{\ln a} - \frac{n}{\ln a}\int u^{n-1}a^u\, du$

97. $\int \frac{e^{au}}{u^n}\, du = \frac{1}{n-1}\left(-\frac{e^{au}}{u^{n-1}} + a\int \frac{e^{au}}{u^{n-1}}\, du\right)$ [n an integer > 1]

98. $\int \frac{du}{a+be^u} = \frac{u - \ln|a+be^u|}{a} + C$

Integrals Involving Logarithmic Functions

99. $\int \ln u\, du = u\ln u - u + C$

100. $\int u^n \ln u\, du = \dfrac{u^{n+1} \ln u}{n+1} - \dfrac{u^{n+1}}{(n+1)^2} + C \qquad [n \neq -1]$

101. $\int \dfrac{\ln u}{u}\, du = \dfrac{1}{2}(\ln u)^2 + C$

102. $\int \dfrac{du}{u \ln u} = \ln |\ln u| + C$

Integrals Involving Inverse Trigonometric Functions

103. $\int \text{Arcsin}\, u\, du = u\, \text{Arcsin}\, u + \sqrt{1-u^2} + C$

104. $\int \text{Arccos}\, u\, du = u\, \text{Arccos}\, u - \sqrt{1-u^2} + C$

105. $\int \text{Arctan}\, u\, du = u\, \text{Arctan}\, u - \dfrac{1}{2} \ln(1+u^2) + C$

106. $\int u^n\, \text{Arcsin}\, u\, du = \dfrac{u^{n+1}\, \text{Arcsin}\, u}{n+1} - \dfrac{1}{n+1} \int \dfrac{u^{n+1}\, du}{\sqrt{1-u^2}} \qquad [n \neq -1]$

107. $\int u^n\, \text{Arccos}\, u\, du = \dfrac{u^{n+1}\, \text{Arccos}\, u}{n+1} + \dfrac{1}{n+1} \int \dfrac{u^{n+1}\, du}{\sqrt{1-u^2}} \qquad [n \neq -1]$

108. $\int u^n\, \text{Arctan}\, u\, du = \dfrac{u^{n+1}\, \text{Arctan}\, u}{n+1} - \dfrac{1}{n+1} \int \dfrac{u^{n+1}\, du}{\sqrt{1+u^2}} \qquad [n \neq -1]$

Miscellaneous Integrals

109. $\int \dfrac{du}{au^2+bu+c} = \dfrac{2}{\sqrt{4ac-b^2}}\, \text{Arctan} \left(\dfrac{2au+b}{\sqrt{4ac-b^2}}\right) + C \qquad [4ac - b^2 > 0]$

110. $\int \dfrac{u\, du}{au^2+bu+c} = \dfrac{1}{2a} \ln |au^2+bu+c| - \dfrac{b}{a\sqrt{4ac-b^2}}\, \text{Arctan} \left(\dfrac{2au+b}{\sqrt{4ac-b^2}}\right) + C \qquad [4ac - b^2 > 0]$

111. $\int \dfrac{du}{(au^2+bu+c)^n} = \dfrac{2au+b}{(n-1)(4ac-b^2)(au^2+bu+c)^{n-1}} + \dfrac{2a(2n-3)}{(n-1)(4ac-b^2)} \int \dfrac{du}{(au^2+bu+c)^{n-1}}$

$[\text{n} \neq 1,\ 4ac - b^2 > 0]$

112. $\int \dfrac{du}{\sqrt{au^2+bu+c}} = \dfrac{1}{\sqrt{a}} \ln |2au + b + 2\sqrt{a}\sqrt{au^2+bu+c}| \qquad [a > 0]$

$= \dfrac{1}{\sqrt{-a}}\, \text{Arcsin} \left(-\dfrac{2au+b}{\sqrt{b^2-4ac}}\right) \qquad [a < 0,\ b^2 - 4ac > 0]$

113. $\int \sqrt{au^2+bu+c}\, du = \dfrac{2au+b}{4a}\sqrt{au^2+bu+c} + \dfrac{4ac-b}{8a} \int \dfrac{du}{\sqrt{au^2+bu+c}}$

114. $\int \sqrt{2au-u^2}\, du = \dfrac{u-a}{2}\sqrt{2au-u^2} + \dfrac{a^2}{2}\, \text{Arcsin} \left(\dfrac{u-a}{a}\right) + C$

115. $\int \dfrac{du}{\sqrt{2au-u^2}} = \text{Arccos} \left(\dfrac{a-u}{a}\right) + C$

TABLE C–2
Natural Logarithms

N	0	1	2	3	4	5	6	7	8	9
1.0	0000	0100	0198	0296	0392	0488	0583	0677	0770	0862
1.1	0953	1044	1133	1222	1310	1398	1484	1570	1655	1740
1.2	1823	1906	1989	2070	2151	2231	2311	2390	2469	2546
1.3	2624	2700	2776	2852	2927	3001	3075	3148	3221	3293
1.4	3365	3436	3507	3577	3646	3716	3874	3853	3920	3988
1.5	4055	4121	4187	4253	4318	4383	4447	4511	4574	4637
1.6	4700	4762	4824	4886	4947	5008	5068	5128	5188	5247
1.7	5306	5365	5423	5481	5539	5596	5653	5710	5766	5822
1.8	5878	5933	5988	6043	6098	6152	6206	6259	6313	6366
1.9	6419	6471	6523	6575	6627	6678	6729	6780	6831	6881
2.0	6931	6981	7031	7080	7129	7178	7227	7275	7324	7372
2.1	7419	7467	7514	7561	7608	7655	7701	7747	7793	7839
2.2	7885	7930	7975	8020	8065	8109	8154	8198	8242	8286
2.3	8329	8372	8416	8459	8502	8544	8587	8629	8671	8713
2.4	8755	8796	8838	8879	8920	8961	9002	9042	9083	9123
2.5	9163	9203	9243	9282	9322	9361	9400	9439	9478	9517
2.6	9555	9594	9632	9670	9708	9746	9783	9821	9858	9895
2.7	9933	9969	*0006	*0043	*0080	*0116	*0152	*0188	*0225	*0260
2.8	1.0296	0332	0367	0403	0438	0473	0508	0543	0578	0613
2.9	0647	0682	0716	0750	0784	0818	0852	0886	0919	0953
3.0	1.0986	1019	1053	1086	1119	1151	1184	1217	1249	1282
3.1	1314	1346	1378	1410	1442	1474	1506	1537	1569	1600
3.2	1632	1663	1694	1725	1756	1787	1817	1848	1878	1909
3.3	1939	1969	2000	2030	2060	2090	2119	2149	2179	2208
3.4	2238	2267	2296	2326	2355	2384	2413	2442	2470	2499
3.5	1.2528	2556	2585	2613	2641	2669	2698	2726	2754	2782
3.6	2809	2837	2865	2892	2920	2947	2975	3002	3029	3056
3.7	3083	3110	3137	3164	3191	3218	3244	3271	3297	3324
3.8	3350	3376	3403	3429	3455	3481	3507	3533	3558	3584
3.9	3610	3635	3661	3686	3712	3737	3762	3788	3813	3838
4.0	1.3863	3888	3913	3938	3962	3987	4012	4036	4061	4085
4.1	4110	4134	4159	4183	4207	4231	4255	4279	4303	4327
4.2	4351	4375	4398	4422	4446	4469	4493	4516	4540	4563
4.3	4586	4609	4633	4656	4679	4702	4725	4748	4770	4793
4.4	4816	4839	4861	4884	4907	4929	4951	4974	4996	5019
4.5	1.5041	5063	5085	5107	5129	5151	5173	5195	5217	5239
4.6	5261	5282	5304	5326	5347	5369	5390	5412	5433	5454
4.7	5476	5497	5518	5539	5560	5581	5602	5623	5644	5665
4.8	5686	5707	5728	5748	5769	5790	5810	5831	5851	5872
4.9	5892	5913	5933	5953	5974	5994	6014	6034	6054	6074
5.0	1.6094	6114	6134	6154	6174	6194	6214	6233	6253	6273
5.1	6292	6312	6332	6351	6371	6390	6409	6429	6448	6467
5.2	6487	6506	6525	6544	6563	6582	6601	6620	6639	6658
5.3	6677	6696	6715	6734	6752	6771	6790	6808	6827	6845
5.4	6864	6882	6901	6919	6938	6956	6974	6993	7011	7029
N	0	1	2	3	4	5	6	7	8	9

TABLE C–2
(Continued)

N	0	1	2	3	4	5	6	7	8	9
5.5	1.7047	7066	7084	7102	7120	7138	7156	7174	7192	7210
5.6	7228	7246	7263	7281	7299	7317	7334	7352	7370	7387
5.7	7405	7422	7440	7457	7475	7492	7509	7527	7544	7561
5.8	7579	7596	7613	7630	7647	7664	7681	7699	7716	7733
5.9	7750	7766	7783	7800	7817	7834	7851	7867	7884	7901
6.0	1.7918	7934	7951	7967	7984	8001	8017	8034	8050	8066
6.1	8083	8099	8116	8132	8148	8165	8181	8197	8213	8229
6.2	8245	8262	8278	8294	8310	8326	8342	8358	8374	8390
6.3	8405	8421	8437	8453	8469	8485	8500	8516	8532	8547
6.4	8563	8579	8594	8610	8625	8641	8656	8672	8687	8703
6.5	1.8718	8733	8749	8764	8779	8795	8810	8825	8840	8856
6.6	8871	8886	8901	8916	8931	8946	8961	8976	8991	9006
6.7	9021	9036	9051	9066	9081	9095	9110	9125	9140	9155
6.8	9169	9184	9199	9213	9228	9242	9257	9272	9286	9301
6.9	9315	9330	9344	9359	9373	9387	9402	9416	9430	9445
7.0	1.9459	9473	9488	9502	9516	9530	9544	9559	9573	9587
7.1	9601	9615	9629	9643	9657	9671	9685	9699	9713	9727
7.2	9741	9755	9769	9782	9796	9810	9824	9838	9851	9865
7.3	9879	9892	9906	9920	9933	9947	9961	9974	9988	*0001
7.4	2.0015	0028	0042	0055	0069	0082	0096	0109	0122	0136
7.5	2.0149	0162	0176	0189	0202	0215	0229	0242	0255	0268
7.6	0281	0295	0308	0321	0334	0347	0360	0373	0386	0399
7.7	0412	0425	0438	0451	0464	0477	0490	0503	0516	0528
7.8	0541	0554	0567	0580	0592	0605	0618	0630	0643	0656
7.9	0669	0681	0694	0707	0719	0732	0744	0757	0769	0782
8.0	2.0794	0807	0819	0832	0844	0857	0869	0882	0894	0906
8.1	0919	0931	0943	0956	0968	0980	0992	1005	1017	1029
8.2	1041	1054	1066	1078	1090	1102	1114	1126	1138	1150
8.3	1163	1175	1187	1199	1211	1223	1235	1247	1258	1270
8.4	1282	1294	1306	1318	1330	1342	1353	1365	1377	1389
8.5	2.1401	1412	1424	1436	1448	1459	1471	1483	1494	1506
8.6	1518	1529	1541	1552	1564	1576	1587	1599	1610	1622
8.7	1633	1645	1656	1668	1679	1691	1702	1713	1725	1736
8.8	1748	1759	1770	1782	1793	1804	1815	1827	1838	1849
8.9	1861	1872	1883	1894	1905	1917	1928	1939	1950	1961
9.0	2.1972	1983	1994	2006	2017	2028	2039	2050	2061	2072
9.1	2083	2094	2105	2116	2127	2138	2148	2159	2170	2181
9.2	2192	2203	2214	2225	2235	2246	2257	2268	2279	2289
9.3	2300	2311	2322	2332	2343	2354	2364	2375	2386	2396
9.4	2407	2418	2428	2439	2450	2460	2471	2481	2492	2502
9.5	2.2513	2523	2534	2544	2555	2565	2576	2586	2597	2607
9.6	2618	2628	2638	2649	2659	2670	2680	2690	2701	2711
9.7	2721	2732	2742	2752	2762	2773	2783	2793	2803	2814
9.8	2824	2834	2844	2854	2865	2875	2885	2895	2905	2915
9.9	2925	2935	2946	2956	2966	2976	2986	2996	3006	3016
N	0	1	2	3	4	5	6	7	8	9

TABLE C–3
Values of Trigonometric Functions

Angle	Radians	Sin	Cos	Tan	Cot	Sec	Csc		
0°00′	0.0000	0.0000	1.0000	0.0000	—	1.000	—	1.5708	90°00′
10′	0.0029	0.0029	1.0000	0.0029	343.8	1.000	343.8	1.5679	50′
20′	0.0058	0.0058	1.0000	0.0058	171.9	1.000	171.9	1.5650	40′
30′	0.0087	0.0087	1.0000	0.0087	114.6	1.000	114.6	1.5621	30′
40′	0.0116	0.0116	0.9999	0.0116	85.94	1.000	85.95	1.5592	20′
50′	0.0145	0.0145	0.9999	0.0145	68.75	1.000	68.76	1.5563	10′
1°00′	0.0175	0.0175	0.9998	0.0175	57.29	1.000	57.30	1.5533	89°00′
10′	0.0204	0.0204	0.9998	0.0204	49.10	1.000	49.11	1.5504	50′
20′	0.0233	0.0233	0.9997	0.0233	42.96	1.000	42.98	1.5475	40′
30′	0.0262	0.0262	0.9997	0.0262	38.19	1.000	38.20	1.5446	30′
40′	0.0291	0.0291	0.9996	0.0291	34.37	1.000	34.38	1.5417	20′
50′	0.0320	0.0320	0.9995	0.0320	31.24	1.001	31.26	1.5388	10′
2°00′	0.0349	0.0349	0.9994	0.0349	28.64	1.001	28.65	1.5359	88°00′
10′	0.0378	0.0378	0.9993	0.0378	26.43	1.001	26.45	1.5330	50′
20′	0.0407	0.0407	0.9992	0.0407	24.54	1.001	24.56	1.5301	40′
30′	0.0436	0.0436	0.9990	0.0437	22.90	1.001	22.93	1.5272	30′
40′	0.0465	0.0465	0.9989	0.0466	21.47	1.001	21.49	1.5243	20′
50′	0.0495	0.0494	0.9988	0.0495	20.21	1.001	20.23	1.5213	10′
3°00′	0.0524	0.0523	0.9986	0.0524	19.08	1.001	19.11	1.5184	87°00′
10′	0.0553	0.0552	0.9985	0.0553	18.07	1.002	18.10	1.5155	50′
20′	0.0582	0.0581	0.9983	0.0582	17.17	1.002	17.20	1.5126	40′
30′	0.0611	0.0610	0.9981	0.0612	16.35	1.002	16.38	1.5097	30′
40′	0.0640	0.0640	0.9980	0.0641	15.60	1.002	15.64	1.5068	20′
50′	0.0669	0.0669	0.9978	0.0670	14.92	1.002	14.96	1.5039	10′
4°00′	0.0698	0.0698	0.9976	0.0699	14.30	1.002	14.34	1.5010	86°00′
10′	0.0727	0.0727	0.9974	0.0729	13.73	1.003	13.76	1.4981	50′
20′	0.0756	0.0756	0.9971	0.0758	13.20	1.003	13.23	1.4952	40′
30′	0.0785	0.0785	0.9969	0.0787	12.71	1.003	12.75	1.4923	30′
40′	0.0814	0.0814	0.9967	0.0816	12.25	1.003	12.29	1.4893	20′
50′	0.0844	0.0843	0.9964	0.0846	11.83	1.004	11.87	1.4864	10′
5°00′	0.0873	0.0872	0.9962	0.0875	11.43	1.004	11.47	1.4835	85°00′
10′	0.0902	0.0901	0.9959	0.0904	11.06	1.004	11.10	1.4806	50′
20′	0.0931	0.0929	0.9957	0.0934	10.71	1.004	10.76	1.4777	40′
30′	0.0960	0.0958	0.9954	0.0963	10.39	1.005	10.43	1.4748	30′
40′	0.0989	0.0987	0.9951	0.0992	10.08	1.005	10.13	1.4719	20′
50′	0.1018	0.1016	0.9948	0.1022	9.788	1.005	9.839	1.4690	10′
6°00′	0.1047	0.1045	0.9945	0.1051	9.514	1.006	9.567	1.4661	84°00′
10′	0.1076	0.1074	0.9942	0.1080	9.255	1.006	9.309	1.4632	50′
20′	0.1105	0.1103	0.9939	0.1110	9.010	1.006	9.065	1.4603	40′
30′	0.1134	0.1132	0.9936	0.1139	8.777	1.006	8.834	1.4573	30′
40′	0.1164	0.1161	0.9932	0.1169	8.556	1.007	8.614	1.4544	20′
50′	0.1193	0.1190	0.9929	0.1198	8.345	1.007	8.405	1.4515	10′
7°00′	0.1222	0.1219	0.9925	0.1228	8.144	1.008	8.206	1.4486	83°00′
10′	0.1251	0.1248	0.9922	0.1257	7.953	1.008	8.016	1.4457	50′
20′	0.1280	0.1276	0.9918	0.1287	7.770	1.008	7.834	1.4428	40′
30′	0.1309	0.1305	0.9914	0.1317	7.596	1.009	7.661	1.4399	30′
40′	0.1338	0.1334	0.9911	0.1346	7.429	1.009	7.496	1.4370	20′
50′	0.1367	0.1363	0.9907	0.1376	7.269	1.009	7.337	1.4341	10′
8°00′	0.1396	0.1392	0.9903	0.1405	7.115	1.010	7.185	1.4312	82°00′
10′	0.1425	0.1421	0.9899	0.1435	6.968	1.010	7.040	1.4283	50′
20′	0.1454	0.1449	0.9894	0.1465	6.827	1.011	6.900	1.4254	40′
30′	0.1484	0.1478	0.9890	0.1495	6.691	1.011	6.765	1.4224	30′
40′	0.1513	0.1507	0.9886	0.1524	6.561	1.012	6.636	1.4195	20′
50′	0.1542	0.1536	0.9881	0.1554	6.435	1.012	6.512	1.4166	10′
9°00′	0.1571	0.1564	0.9877	0.1584	6.314	1.012	6.392	1.4137	81°00′
		Cos	Sin	Cot	Tan	Csc	Sec	Radians	Angle

TABLE C–3
(Continued)

Angle	Radians	Sin	Cos	Tan	Cot	Sec	Csc		
9°00′	0.1571	0.1564	0.9877	0.1584	6.314	1.012	6.392	1.4137	81°00′
10′	0.1600	0.1593	0.9872	0.1614	6.197	1.013	6.277	1.4108	50′
20′	0.1629	0.1622	0.9868	0.1644	6.084	1.013	6.166	1.4079	40′
30′	0.1658	0.1650	0.9863	0.1673	5.976	1.014	6.059	1.4050	30′
40′	0.1687	0.1679	0.9858	0.1703	5.871	1.014	5.955	1.4021	20′
50′	0.1716	0.1708	0.9853	0.1733	5.769	1.015	5.855	1.3992	10′
10°00′	0.1745	0.1736	0.9848	0.1763	5.671	1.015	5.759	1.3963	80°00′
10′	0.1774	0.1765	0.9843	0.1793	5.576	1.016	5.665	1.3934	50′
20′	0.1804	0.1794	0.9838	0.1823	5.485	1.016	5.575	1.3904	40′
30′	0.1833	0.1822	0.9833	0.1853	5.396	1.017	5.487	1.3875	30′
40′	0.1862	0.1851	0.9827	0.1883	5.309	1.018	5.403	1.3846	20′
50′	0.1891	0.1880	0.9822	0.1914	5.226	1.018	5.320	1.3817	10′
11°00′	0.1920	0.1908	0.9816	0.1944	5.145	1.019	5.241	1.3788	79°00′
10′	0.1949	0.1937	0.9811	0.1974	5.066	1.019	5.164	1.3759	50′
20′	0.1978	0.1965	0.9805	0.2004	4.989	1.020	5.089	1.3730	40′
30′	0.2007	0.1994	0.9799	0.2035	4.915	1.020	5.016	1.3701	30′
40′	0.2036	0.2022	0.9793	0.2065	4.843	1.021	4.945	1.3672	20′
50′	0.2065	0.2051	0.9787	0.2095	4.773	1.022	4.876	1.3643	10′
12°00′	0.2094	0.2079	0.9781	0.2126	4.705	1.022	4.810	1.3614	78°00′
10′	0.2123	0.2108	0.9775	0.2156	4.638	1.023	4.745	1.3584	50′
20′	0.2153	0.2136	0.9769	0.2186	4.574	1.024	4.682	1.3555	40′
30′	0.2182	0.2164	0.9763	0.2217	4.511	1.024	4.620	1.3526	30′
40′	0.2211	0.2193	0.9757	0.2247	4.449	1.025	4.560	1.3497	20′
50′	0.2240	0.2221	0.9750	0.2278	4.390	1.026	4.502	1.3468	10′
13°00′	0.2269	0.2250	0.9744	0.2309	4.331	1.026	4.445	1.3439	77°00′
10′	0.2298	0.2278	0.9737	0.2339	4.275	1.027	4.390	1.3410	50′
20′	0.2327	0.2306	0.9730	0.2370	4.219	1.028	4.336	1.3381	40′
30′	0.2356	0.2334	0.9724	0.2401	4.165	1.028	4.284	1.3352	30′
40′	0.2385	0.2363	0.9717	0.2432	4.113	1.029	4.232	1.3323	20′
50′	0.2414	0.2391	0.9710	0.2462	4.061	1.030	4.182	1.3294	10′
14°00′	0.2443	0.2419	0.9703	0.2493	4.011	1.031	4.134	1.3265	76°00′
10′	0.2473	0.2447	0.9696	0.2524	3.962	1.031	4.086	1.3235	50′
20′	0.2502	0.2476	0.9689	0.2555	3.914	1.032	4.039	1.3206	40′
30′	0.2531	0.2504	0.9681	0.2586	3.867	1.033	3.994	1.3177	30′
40′	0.2560	0.2532	0.9674	0.2617	3.821	1.034	3.950	1.3148	20′
50′	0.2589	0.2560	0.9667	0.2648	3.776	1.034	3.906	1.3119	10′
15°00′	0.2618	0.2588	0.9659	0.2679	3.732	1.035	3.864	1.3090	75°00′
10′	0.2647	0.2616	0.9652	0.2711	3.689	1.036	3.822	1.3061	50′
20′	0.2676	0.2644	0.9644	0.2742	3.647	1.037	3.782	1.3032	40′
30′	0.2705	0.2672	0.9636	0.2773	3.606	1.038	3.742	1.3003	30′
40′	0.2734	0.2700	0.9628	0.2805	3.566	1.039	3.703	1.2974	20′
50′	0.2763	0.2728	0.9621	0.2836	3.526	1.039	3.665	1.2945	10′
16°00′	0.2793	0.2756	0.9613	0.2867	3.487	1.040	3.628	1.2915	74°00′
10′	0.2822	0.2784	0.9605	0.2899	3.450	1.041	3.592	1.2886	50′
20′	0.2851	0.2812	0.9596	0.2931	3.412	1.042	3.556	1.2857	40′
30′	0.2880	0.2840	0.9588	0.2962	3.376	1.043	3.521	1.2828	30′
40′	0.2909	0.2868	0.9580	0.2994	3.340	1.044	3.487	1.2799	20′
50′	0.2938	0.2896	0.9572	0.3026	3.305	1.045	3.453	1.2770	10′
17°00′	0.2967	0.2924	0.9563	0.3057	3.271	1.046	3.420	1.2741	73°00′
10′	0.2996	0.2952	0.9555	0.3089	3.237	1.047	3.388	1.2712	50′
20′	0.3025	0.2979	0.9546	0.3121	3.204	1.048	3.356	1.2683	40′
30′	0.3054	0.3007	0.9537	0.3153	3.172	1.049	3.326	1.2654	30′
40′	0.3083	0.3035	0.9528	0.3185	3.140	1.049	3.295	1.2625	20′
50′	0.3113	0.3062	0.9520	0.3217	3.108	1.050	3.265	1.2595	10′
18°00′	0.3142	0.3090	0.9511	0.3249	3.078	1.051	3.236	1.2566	72°00′
		Cos	Sin	Cot	Tan	Csc	Sec	Radians	Angle

TABLE C–3
(Continued)

Angle	Radians	Sin	Cos	Tan	Cot	Sec	Csc		
18°00′	0.3142	0.3090	0.9511	0.3249	3.078	1.051	3.236	1.2566	72°00′
10′	0.3171	0.3118	0.9502	0.3281	3.047	1.052	3.207	1.2537	50′
20′	0.3200	0.3145	0.9492	0.3314	3.018	1.053	3.179	1.2508	40′
30′	0.3229	0.3173	0.9483	0.3346	2.989	1.054	3.152	1.2479	30′
40′	0.3258	0.3201	0.9474	0.3378	2.960	1.056	3.124	1.2450	20′
50′	0.3287	0.3228	0.9465	0.3411	2.932	1.057	3.098	1.2421	10′
19°00′	0.3316	0.3256	0.9455	0.3443	2.904	1.058	3.072	1.2392	71°00′
10′	0.3345	0.3283	0.9446	0.3476	2.877	1.059	3.046	1.2363	50′
20′	0.3374	0.3311	0.9436	0.3508	2.850	1.060	3.021	1.2334	40′
30′	0.3403	0.3338	0.9426	0.3541	2.824	1.061	2.996	1.2305	30′
40′	0.3432	0.3365	0.9417	0.3574	2.798	1.062	2.971	1.2275	20′
50′	0.3462	0.3393	0.9407	0.3607	2.773	1.063	2.947	1.2246	10′
20°00′	0.3491	0.3420	0.9397	0.3640	2.747	1.064	2.924	1.2217	70°00′
10′	0.3520	0.3448	0.9387	0.3673	2.723	1.065	2.901	1.2188	50′
20′	0.3549	0.3475	0.9377	0.3706	2.699	1.066	2.878	1.2159	40′
30′	0.3578	0.3502	0.9367	0.3739	2.675	1.068	2.855	1.2130	30′
40′	0.3607	0.3529	0.9356	0.3772	2.651	1.069	2.833	1.2101	20′
50′	0.3636	0.3557	0.9346	0.3805	2.628	1.070	2.812	1.2072	10′
21°00′	0.3665	0.3584	0.9336	0.3839	2.605	1.071	2.790	1.2043	69°00′
10′	0.3694	0.3611	0.9325	0.3872	2.583	1.072	2.769	1.2014	50′
20′	0.3723	0.3638	0.9315	0.3906	2.560	1.074	2.749	1.1985	40′
30′	0.3752	0.3665	0.9304	0.3939	2.539	1.075	2.729	1.1956	30′
40′	0.3782	0.3692	0.9293	0.3973	2.517	1.076	2.709	1.1926	20′
50′	0.3811	0.3719	0.9283	0.4006	2.496	1.077	2.689	1.1897	10′
22°00′	0.3840	0.3746	0.9272	0.4040	2.475	1.079	2.669	1.1868	68°00′
10′	0.3869	0.3773	0.9261	0.4074	2.455	1.080	2.650	1.1839	50′
20′	0.3898	0.3800	0.9250	0.4108	2.434	1.081	2.632	1.1810	40′
30′	0.3927	0.3827	0.9239	0.4142	2.414	1.082	2.613	1.1781	30′
40′	0.3956	0.3854	0.9228	0.4176	2.394	1.084	2.595	1.1752	20′
50′	0.3985	0.3881	0.9216	0.4210	2.375	1.085	2.577	1.1723	10′
23°00′	0.4014	0.3907	0.9205	0.4245	2.356	1.086	2.559	1.1694	67°00′
10′	0.4043	0.3934	0.9194	0.4279	2.337	1.088	2.542	1.1665	50′
20′	0.4072	0.3961	0.9182	0.4314	2.318	1.089	2.525	1.1636	40′
30′	0.4102	0.3987	0.9171	0.4348	2.300	1.090	2.508	1.1606	30′
40′	0.4131	0.4014	0.9159	0.4383	2.282	1.092	2.491	1.1577	20′
50′	0.4160	0.4041	0.9147	0.4417	2.264	1.093	2.475	1.1548	10′
24°00′	0.4189	0.4067	0.9135	0.4452	2.246	1.095	2.459	1.1519	66°00′
10′	0.4218	0.4094	0.9124	0.4487	2.229	1.096	2.443	1.1490	50′
20′	0.4247	0.4120	0.9112	0.4522	2.211	1.097	2.427	1.1461	40′
30′	0.4276	0.4147	0.9100	0.4557	2.194	1.099	2.411	1.1432	30′
40′	0.4305	0.4173	0.9088	0.4592	2.177	1.100	2.396	1.1403	20′
50′	0.4334	0.4200	0.9075	0.4628	2.161	1.102	2.381	1.1374	10′
25°00′	0.4363	0.4226	0.9063	0.4663	2.145	1.103	2.366	1.1345	65°00′
10′	0.4392	0.4253	0.9051	0.4699	2.128	1.105	2.352	1.1316	50′
20′	0.4422	0.4279	0.9038	0.4734	2.112	1.106	2.337	1.1286	40′
30′	0.4451	0.4305	0.9026	0.4770	2.097	1.108	2.323	1.1257	30′
40′	0.4480	0.4331	0.9013	0.4806	2.081	1.109	2.309	1.1228	20′
50′	0.4509	0.4358	0.9001	0.4841	2.066	1.111	2.295	1.1199	10′
26°00′	0.4538	0.4384	0.8988	0.4877	2.050	1.113	2.281	1.1170	64°00′
10′	0.4567	0.4410	0.8975	0.4913	2.035	1.114	2.268	1.1141	50′
20′	0.4596	0.4436	0.8962	0.4950	2.020	1.116	2.254	1.1112	40′
30′	0.4625	0.4462	0.8949	0.4986	2.006	1.117	2.241	1.1083	30′
40′	0.4654	0.4488	0.8936	0.5022	1.991	1.119	2.228	1.1054	20′
50′	0.4683	0.4514	0.8923	0.5059	1.977	1.121	2.215	1.1025	10′
27°00′	0.4712	0.4540	0.8910	0.5095	1.963	1.122	2.203	1.0996	63°00′
		Cos	Sin	Cot	Tan	Csc	Sec	Radians	Angle

TABLE C–3
(Continued)

Angle	Radians	Sin	Cos	Tan	Cot	Sec	Csc		
27°00′	0.47122	0.4540	0.8910	0.5095	1.963	1.122	2.203	1.0996	63°00′
10′	0.4741	0.4566	0.8897	0.5132	1.949	1.124	2.190	1.0966	50′
20′	0.4771	0.4592	0.8884	0.5169	1.935	1.126	2.178	1.0937	40′
30′	0.4800	0.4617	0.8870	0.5206	1.921	1.127	2.166	1.0908	30′
40′	0.4829	0.4643	0.8857	0.5243	1.907	1.129	2.154	1.0879	20′
50′	0.4858	0.4669	0.8843	0.5280	1.894	1.131	2.142	1.0850	10′
28°00′	0.4887	0.4695	0.8829	0.5317	1.881	1.133	2.130	1.0821	62°00′
10′	0.4916	0.4720	0.8816	0.5354	1.868	1.134	2.118	1.0792	50′
20′	0.4945	0.4746	0.8802	0.5392	1.855	1.136	2.107	1.0763	40′
30′	0.4974	0.4772	0.8788	0.5430	1.842	1.138	2.096	1.0734	30′
40′	0.5003	0.4797	0.8774	0.5467	1.829	1.140	2.085	1.0705	20′
50′	0.5032	0.4823	0.8760	0.5505	1.816	1.142	2.074	1.0676	10′
29°00′	0.5061	0.4848	0.8746	0.5543	1.804	1.143	2.063	1.0647	61°00′
10′	0.5091	0.4874	0.8732	0.5581	1.792	1.145	2.052	1.0617	50′
20′	0.5120	0.4899	0.8718	0.5619	1.780	1.147	2.041	1.0588	40′
30′	0.5149	0.4924	0.8704	0.5658	1.767	1.149	2.031	1.0559	30′
40′	0.5178	0.4950	0.8689	0.5696	1.756	1.151	2.020	1.0530	20′
50′	0.5207	0.4975	0.8675	0.5735	1.744	1.153	2.010	1.0501	10′
30°00′	0.5236	0.5000	0.8660	0.5774	1.732	1.155	2.000	1.0472	60°00′
10′	0.5265	0.5025	0.8646	0.5812	1.720	1.157	1.990	1.0443	50′
20′	0.5294	0.5050	0.8631	0.5851	1.709	1.159	1.980	1.0414	40′
30′	0.5323	0.5075	0.8616	0.5890	1.698	1.161	1.970	1.0385	30′
40′	0.5352	0.5100	0.8601	0.5930	1.686	1.163	1.961	1.0356	20′
50′	0.5381	0.5125	0.8587	0.5969	1.675	1.165	1.951	1.0327	10′
31°00′	0.5411	0.5150	0.8572	0.6009	1.664	1.167	1.942	1.0297	59°00′
10′	0.5440	0.5175	0.8557	0.6048	1.653	1.169	1.932	1.0268	50′
20′	0.5469	0.5200	0.8542	0.6088	1.643	1.171	1.923	1.0239	40′
30′	0.5498	0.5225	0.8526	0.6128	1.632	1.173	1.914	1.0210	30′
40′	0.5527	0.5250	0.8511	0.6168	1.621	1.175	1.905	1.0181	20′
50′	0.5556	0.5275	0.8496	0.6208	1.611	1.177	1.896	1.0152	10′
32°00′	0.5585	0.5299	0.8480	0.6249	1.600	1.179	1.887	1.0123	58°00′
10′	0.5614	0.5324	0.8465	0.6289	1.590	1.181	1.878	1.0094	50′
20′	0.5643	0.5348	0.8450	0.6330	1.580	1.184	1.870	1.0065	40′
30′	0.5672	0.5373	0.8434	0.6371	1.570	1.186	1.861	1.0036	30′
40′	0.5701	0.5398	0.8418	0.6412	1.560	1.188	1.853	1.0007	20′
50′	0.5730	0.5422	0.8403	0.6453	1.550	1.190	1.844	0.9977	10′
33°00′	0.5760	0.5446	0.8387	0.6494	1.540	1.192	1.836	0.9948	57°00′
10′	0.5789	0.5471	0.8371	0.6536	1.530	1.195	1.828	0.9919	50′
20′	0.5818	0.5495	0.8355	0.6577	1.520	1.197	1.820	0.9890	40′
30′	0.5847	0.5519	0.8339	0.6619	1.511	1.199	1.812	0.9861	30′
40′	0.5876	0.5544	0.8323	0.6661	1.501	1.202	1.804	0.9832	20′
50′	0.5905	0.5568	0.8307	0.6703	1.492	1.204	1.796	0.9803	10′
34°00′	0.5934	0.5592	0.8290	0.6745	1.483	1.206	1.788	0.9774	56°00′
10′	0.5963	0.5616	0.8274	0.6787	1.473	1.209	1.781	0.9745	50′
20′	0.5992	0.5640	0.8258	0.6830	1.464	1.211	1.773	0.9716	40′
30′	0.6021	0.5664	0.8241	0.6873	1.455	1.213	1.766	0.9687	30′
40′	0.6050	0.5688	0.8225	0.6916	1.446	1.216	1.758	0.9657	20′
50′	0.6080	0.5712	0.8208	0.6959	1.437	1.218	1.751	0.9628	10′
35°00′	0.6109	0.5736	0.8192	0.7002	1.428	1.221	1.743	0.9599	55°00′
10′	0.6138	0.5760	0.8175	0.7046	1.419	1.223	1.736	0.9570	50′
20′	0.6167	0.5783	0.8158	0.7089	1.411	1.226	1.729	0.9541	40′
30′	0.6196	0.5807	0.8141	0.7133	1.402	1.228	1.722	0.9512	30′
40′	0.6225	0.5831	0.8124	0.7177	1.393	1.231	1.715	0.9483	20′
50′	0.6254	0.5854	0.8107	0.7221	1.385	1.233	1.708	0.9454	10′
36°00′	0.6283	0.5878	0.8090	0.7265	1.376	1.236	1.701	0.9425	54°00′
		Cos	Sin	Cot	Tan	Csc	Sec	Radians	Angle

TABLE C–3
(Continued)

Angle	Radians	Sin	Cos	Tan	Cot	Sec	Csc		
36°00′	0.6283	0.5878	0.8090	0.7265	1.376	1.236	1.701	0.9425	54°00′
10′	0.6312	0.5901	0.8073	0.7310	1.368	1.239	1.695	0.9396	50′
20′	0.6341	0.5925	0.8056	0.7355	1.360	1.241	1.688	0.9367	40′
30′	0.6370	0.5948	0.8039	0.7400	1.351	1.244	1.681	0.9338	30′
40′	0.6400	0.5972	0.8021	0.7445	1.343	1.247	1.675	0.9308	20′
50′	0.6429	0.5995	0.8004	0.7490	1.335	1.249	1.668	0.9279	10′
37°00′	0.6458	0.6018	0.7986	0.7536	1.327	1.252	1.662	0.9250	53°00′
10′	0.6487	0.6041	0.7969	0.7581	1.319	1.255	1.655	0.9221	50′
20′	0.6516	0.6065	0.7951	0.7627	1.311	1.258	1.649	0.9192	40′
30′	0.6545	0.6088	0.7934	0.7673	1.303	1.260	1.643	0.9163	30′
40′	0.6574	0.6111	0.7916	0.7720	1.295	1.263	1.636	0.9134	20′
50′	0.6603	0.6134	0.7898	0.7766	1.288	1.266	1.630	0.9105	10′
38°00′	0.6632	0.6157	0.7880	0.7813	1.280	1.269	1.624	0.9076	52°00′
10′	0.6661	0.6180	0.7862	0.7860	1.272	1.272	1.618	0.9047	50′
20′	0.6690	0.6202	0.7844	0.7907	1.265	1.275	1.612	0.9018	40′
30′	0.6720	0.6225	0.7826	0.7954	1.257	1.278	1.606	0.8988	30′
40′	0.6749	0.6248	0.7808	0.8002	1.250	1.281	1.601	0.8959	20′
50′	0.6778	0.6271	0.7790	0.8050	1.242	1.284	1.595	0.8930	10′
39°00′	0.6807	0.6293	0.7771	0.8098	1.235	1.287	1.589	0.8901	51°00′
10′	0.6836	0.6316	0.7753	0.8146	1.228	1.290	1.583	0.8872	50′
20′	0.6865	0.6338	0.7735	0.8195	1.220	1.293	1.578	0.8843	40′
30′	0.6894	0.6361	0.7716	0.8243	1.213	1.296	1.572	0.8814	30′
40′	0.6923	0.6383	0.7698	0.8292	1.206	1.299	1.567	0.8785	20′
50′	0.6952	0.6406	0.7679	0.8342	1.199	1.302	1.561	0.8756	10′
40°00′	0.6981	0.6428	0.7660	0.8391	1.192	1.305	1.556	0.8727	50°00′
10′	0.7010	0.6450	0.7642	0.8441	1.185	1.309	1.550	0.8698	50′
20′	0.7039	0.6472	0.7623	0.8491	1.178	1.312	1.545	0.8668	40′
30′	0.7069	0.6494	0.7604	0.8541	1.171	1.315	1.540	0.8639	30′
40′	0.7098	0.6517	0.7585	0.8591	1.164	1.318	1.535	0.8610	20′
50′	0.7127	0.6539	0.7566	0.8642	1.157	1.322	1.529	0.8581	10′
41°00′	0.7156	0.6561	0.7547	0.8693	1.150	1.325	1.524	0.8552	49°00′
10′	0.7185	0.6583	0.7528	0.8744	1.144	1.328	1.519	0.8523	50′
20′	0.7214	0.6604	0.7509	0.8796	1.137	1.332	1.514	0.8494	40′
30′	0.7243	0.6626	0.7490	0.8847	1.130	1.335	1.509	0.8465	30′
40′	0.7272	0.6648	0.7470	0.8899	1.124	1.339	1.504	0.8436	20′
50′	0.7301	0.6670	0.7451	0.8952	1.117	1.342	1.499	0.8407	10′
42°00′	0.7330	0.6691	0.7431	0.9004	1.111	1.346	1.494	0.8378	48°00′
10′	0.7359	0.6713	0.7412	0.9057	1.104	1.349	1.490	0.8348	50′
20′	0.7389	0.6734	0.7392	0.9110	1.098	1.353	1.485	0.8319	40′
30′	0.7418	0.6756	0.7373	0.9163	1.091	1.356	1.480	0.8290	30′
40′	0.7447	0.6777	0.7353	0.9217	1.085	1.360	1.476	0.8261	20′
50′	0.7476	0.6799	0.7333	0.9271	1.079	1.364	1.471	0.8232	10′
43°00′	0.7505	0.6820	0.7314	0.9325	1.072	1.367	1.466	0.8203	47°00′
10′	0.7534	0.6841	0.7294	0.9380	1.066	1.371	1.462	0.8174	50′
20′	0.7563	0.6862	0.7274	0.9435	1.060	1.375	1.457	0.8145	40′
30′	0.7592	0.6884	0.7254	0.9490	1.054	1.379	1.453	0.8116	30′
40′	0.7621	0.6905	0.7234	0.9545	1.048	1.382	1.448	0.8087	20′
50′	0.7650	0.6926	0.7214	0.9601	1.042	1.386	1.444	0.8058	10′
44°00′	0.7679	0.6947	0.7193	0.9657	1.036	1.390	1.440	0.8029	46°00′
10′	0.7709	0.6967	0.7173	0.9713	1.030	1.394	1.435	0.7999	50′
20′	0.7738	0.6988	0.7153	0.9770	1.024	1.398	1.431	0.7970	40′
30′	0.7767	0.7009	0.7133	0.9827	1.018	1.402	1.427	0.7941	30′
40′	0.7796	0.7030	0.7112	0.9884	1.012	1.406	1.423	0.7912	20′
50′	0.7825	0.7050	0.7092	0.9942	1.006	1.410	1.418	0.7883	10′
45°00′	0.7854	0.7071	0.7071	1.000	1.000	1.414	1.414	0.7854	45°00′
		Cos	Sin	Cot	Tan	Csc	Sec	Radians	Angle

TABLE C–4
Powers and Roots

n	n^2	n^3	$\sqrt{n}$	$\sqrt[3]{n}$
0	0	0	0.000	0.000
1	1	1	1.000	1.000
2	4	8	1.414	1.260
3	9	27	1.732	1.442
4	16	64	2.000	1.587
5	25	125	2.236	1.710
6	36	216	2.449	1.817
7	49	343	2.646	1.913
8	64	512	2.828	2.000
9	81	729	3.000	2.080
10	100	1 000	3.162	2.154
11	121	1 331	3.317	2.224
12	144	1 728	3.464	2.289
13	169	2 197	3.606	2.351
14	196	2 744	3.742	2.410
15	225	3 375	3.873	2.466
16	256	4 096	4.000	2.520
17	289	4 913	4.123	2.571
18	324	5 832	4.243	2.621
19	361	6 859	4.359	2.668
20	400	8 000	4.472	2.714
21	441	9 261	4.583	2.759
22	484	10 648	4.690	2.802
23	529	12 167	4.796	2.844
24	576	13 824	4.899	2.884
25	625	15 625	5.000	2.924
26	676	17 576	5.099	2.962
27	729	19 683	5.196	3.000
28	784	21 952	5.292	3.037
29	841	24 389	5.385	3.072
30	900	27 000	5.477	3.107
31	961	29 791	5.568	3.141
32	1 024	32 768	5.657	3.175
33	1 089	35 937	5.745	3.208
34	1 156	39 304	5.831	3.240
35	1 225	42 875	5.916	3.271
36	1 296	46 656	6.000	3.302
37	1 369	50 653	6.083	3.332
38	1 444	54 872	6.164	3.362
39	1 521	59 319	6.245	3.391
40	1 600	64 000	6.325	3.420
41	1 681	68 921	6.403	3.448
42	1 764	74 088	6.481	3.476
43	1 849	79 507	6.557	3.503
44	1 936	85 184	6.633	3.530
45	2 025	91 125	6.708	3.557
46	2 116	97 336	6.782	3.583
47	2 209	103 823	6.856	3.609
48	2 304	110 592	6.928	3.634
49	2 401	117 649	7.000	3.659
50	2 500	125 000	7.071	3.684
51	2 601	132 651	7.141	3.708
52	2 704	140 608	7.211	3.733
53	2 809	148 877	7.280	3.756
54	2 916	157 464	7.348	3.780
55	3 025	166 375	7.416	3.803
56	3 136	175 616	7.483	3.826
57	3 249	185 193	7.550	3.849
58	3 364	195 112	7.616	3.871
59	3 481	205 379	7.681	3.893
60	3 600	216 000	7.746	3.915
61	3 721	226 981	7.810	3.936
62	3 844	238 328	7.874	3.958
63	3 969	250 047	7.937	3.979
64	4 096	262 144	8.000	4.000
65	4 225	274 625	8.062	4.021
66	4 356	287 496	8.124	4.041
67	4 489	300 763	8.185	4.062
68	4 624	314 432	8.246	4.082
69	4 761	328 509	8.307	4.102
70	4 900	343 000	8.367	4.121
71	5 041	357 911	8.426	4.141
72	5 184	373 248	8.485	4.160
73	5 329	389 017	8.544	4.179
74	5 476	405 224	8.602	4.198
75	5 625	421 875	8.660	4.217
76	5 776	438 976	8.718	4.236
77	5 929	456 533	8.775	4.254
78	6 084	474 552	8.832	4.273
79	6 241	493 039	8.888	4.291
80	6 400	512 000	8.944	4.309
81	6 561	531 441	9.000	4.327
82	6 724	551 368	9.055	4.344
83	6 889	571 787	9.110	4.362
84	7 056	592 704	9.165	4.380
85	7 225	614 125	9.220	4.397
86	7 396	636 056	9.274	4.414
87	7 569	658 503	9.327	4.431
88	7 744	681 472	9.381	4.448
89	7 921	704 969	9.434	4.465
90	8 100	729 000	9.487	4.481
91	8 281	753 571	9.539	4.498
92	8 464	778 688	9.592	4.514
93	8 649	804 357	9.644	4.531
94	8 836	830 584	9.695	4.547
95	9 025	857 375	9.747	4.563
96	9 216	884 736	9.798	4.579
97	9 409	912 673	9.849	4.595
98	9 604	941 192	9.899	4.610
99	9 801	970 299	9.950	4.626
100	10 000	1 000 000	10.000	4.642

TABLE C–5
Comparison of English and Metric Systems of Units

English	Metric		SI
	MKS	CGS	
Length:			
Yard (yd) (0.914 m)	Meter (m) (39.37 in.) (100 cm)	Centimeter (cm) (2.54 cm = 1 in.)	Meter (m)
Mass:			
Slug (14.6 kg)	Kilogram (kg) (1000 g)	Gram (g)	Kilogram (kg)
Force:			
Pound (lb) (4.45 N)	Newton (N) (100,000 dynes)	Dyne	Newton (N)
Temperature:	Celsius or	Centigrade (°C)	Kelvin (K)
Fahrenheit (°F) $\left(= \frac{9}{5}°\text{C} + 32\right)$	Centigrade (°C) $\left(= \frac{5}{9}(°\text{F} - 32)\right)$		K = 273.15 + °C
Energy:			
Foot-pound (ft-lb) (1.356 joules)	Newton-meter (N-m) or Joule (J) (0.7378 ft-lb)	Dyne-centimeter or Erg (1 joule = 10^7 ergs)	Joule (J)
Time:			
Second (s)	Second (s)	Second (s)	Second (s)

Answers to Odd-Numbered Exercises

CHAPTER 1

1–1 Exercises, p. 7

1. -17 **3.** -4 **5.** 2 **7.** 0.51 **9.** 9 **11.** 18 **13.** 16 **15.** -4.6 **17.** $\frac{-23}{6}$ **19.** 19

21. -13 **23.** 2 **25.** $\frac{42}{31}$ **27.** -20 **29.** 10 **31.** -8 **33.** 4.8 **35.** $\frac{-57}{11}$ **37.** $\frac{27}{5}$

39. $\frac{141}{68}$ **41.** 4 or -1 **43.** $\frac{6}{5}$ or $\frac{-1}{5}$ **45.** $\frac{17}{2}$ or $\frac{3}{2}$ **47.** 352°K **49.** 5 A **51.** 7.3 A **53.** 50 ft

55. 3.2 Ω **57.** 122,500 ft-lb **59.** 95°

1–2 Exercises, p. 13

1. $\frac{A}{Lw}$ **3.** $\frac{S - 2\pi r^2}{2\pi r}$ **5.** $\frac{E}{gh}$ **7.** $\frac{kLT - I}{kL}$ **9.** $\frac{V^2 - V_0^2}{2s}$ **11.** $2T + d$ **13.** $\frac{2E}{v^2}$ **15.** $\frac{L}{2\pi h}$

17. $\frac{5F - 160}{9}$ or $\frac{5}{9}(F - 32)$ **19.** $v - at$ **21.** $2v_{av} - v_0$ **23.** $\frac{L - 3.14r_2 - 2d}{3.14}$ **25.** $at + w_0$

27. $\frac{Fr^2}{kq_2}$ **29.** $\frac{y - b}{x}$ **31.** $\frac{fu + fv_s}{u}$ **33.** $\frac{2T + agd_2}{ag}$ **35.** $\frac{E - IR}{I}$ **37.** $\frac{PgJ + V_1}{V_1}$

39. $\frac{ab - ra - rb}{r}$ **41.** 1,254.4 m **43.** 7 **45.** 15 ft **47.** 3 s **49.** 12 ft

1–3 Exercises, p. 21

1. $x \geq -13$

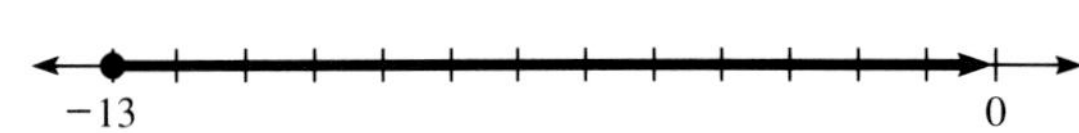

3. $x \leq 7$

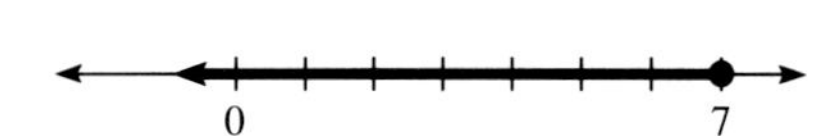

5. $x < \frac{11}{3}$

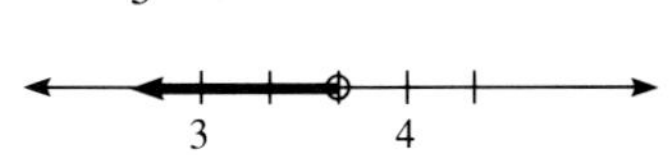

7. $x > -12$

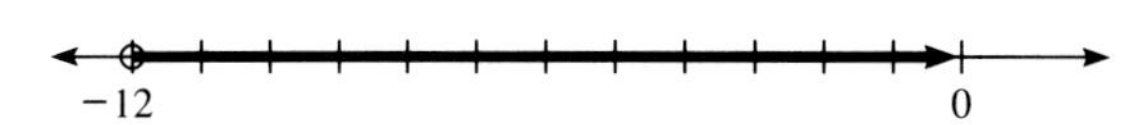

9. $x > -3$

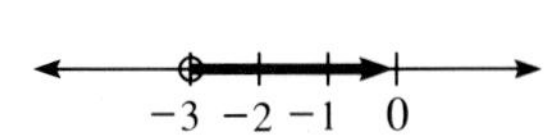

11. $x \leq -5$

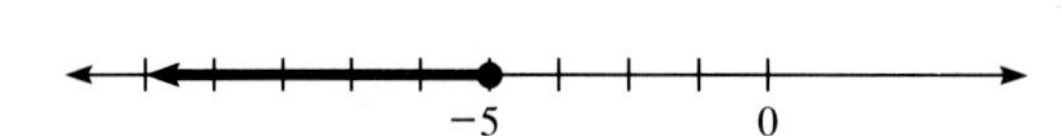

13. $x > 6$

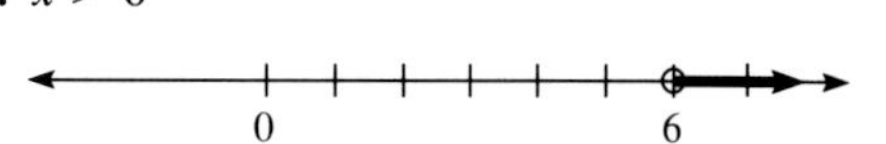

15. $x > 2$

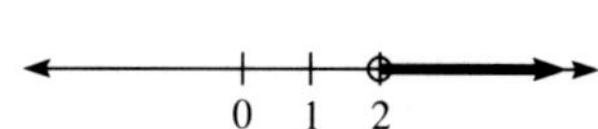

17. $x > \frac{-21}{2}$

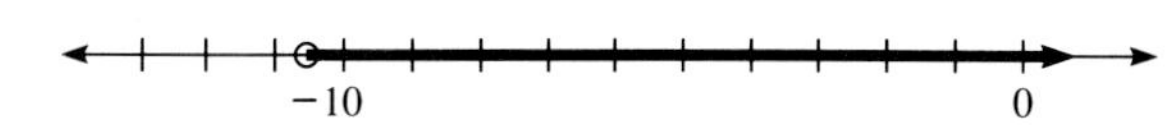

19. $x \leq 4$

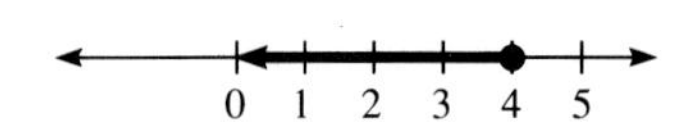

21. $x < \frac{1}{2}$

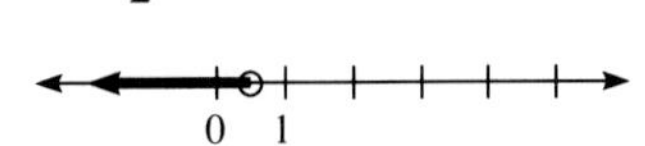

23. $x > \frac{38}{15}$

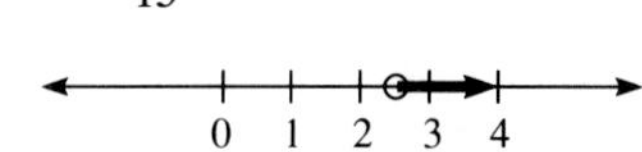

25. $x \geq \frac{21}{11}$

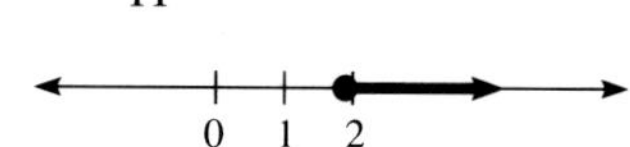

27. $x \leq \frac{1}{6}$

0 $\frac{1}{6}$ $\frac{2}{6}$

29. $x < 0$

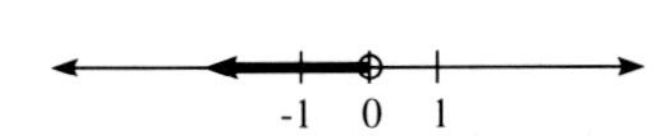

31. $x > \frac{31}{58}$

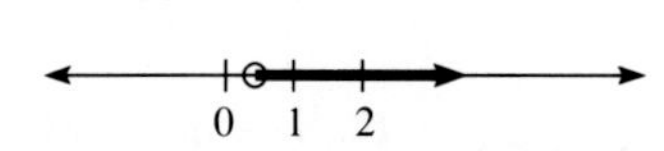

33. $x \leq 2$

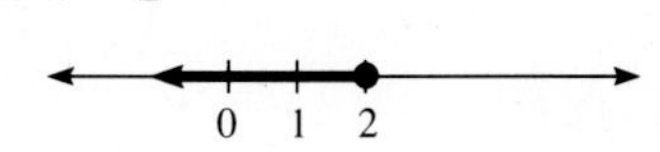

35. $x < \frac{31}{7}$

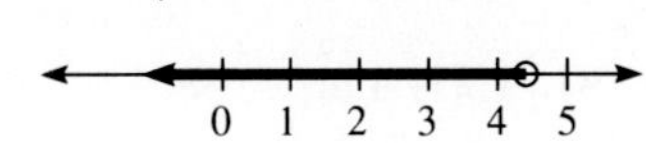

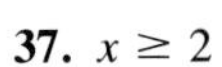

37. $x \geq 2$

39. $x > \frac{1}{2}$

41. $-4 < x < \frac{4}{3}$

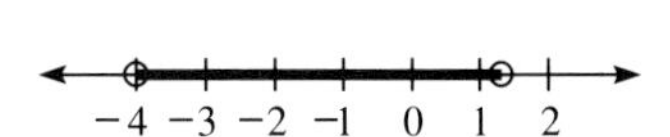

43. $x > \frac{7}{2}$ or $x < -3$

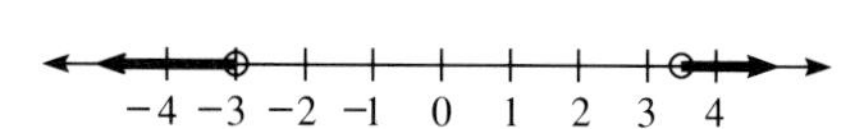

45. $\frac{3}{8} < x < \frac{45}{8}$

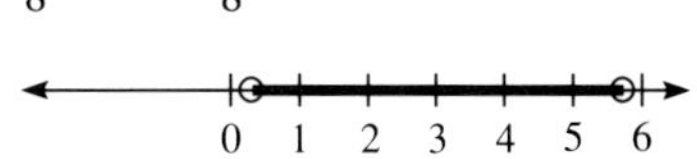

47. $x \geq 89$ **49.** $x < 18$ mi
51. 5.95 mm to 6.05 mm

1–4 Exercises, p. 26

1. $x = kz$ **3.** $a = kbc$ **5.** $x = \frac{ky}{s}$ **7.** $m = \frac{knp^3}{r}$ **9.** $s = kt^2v$ **11.** $\frac{5}{2}$ **13.** 6 **15.** 4

17. 56 **19.** $\frac{63}{2}$ **21.** 90 **23.** 2 **25.** $\frac{25}{8}$ **27.** 1.35 m **29.** 78.5 cm^2 **31.** 6,250 lb **33.** $\frac{3}{7}\ \Omega$

35. 1.7 m **37.** 2.5 s **39.** 176.4 m

1–5 Exercises, p. 34

1. 125 rods **3.** 121 ft **5.** 4 to 1 **7.** 56.56 in. **9.** 11.4 qt. milk, 8.6 qt. cream **11.** 50 mph

13. $\frac{22}{15}$ or 22 to 15 **15.** 24 yd^3 **17.** 35 l **19.** 7.5 ft/min **21.** 24, 25, 26 **23.** The bar

25. 12 hp, 9 hp, 13 hp **27.** 35 l **29.** 7 h **31.** 420 gal **33.** $0.31

Chapter Review, p. 36

1. 11 **3.** $\frac{1}{2}$ **5.** 3.7 **7.** 3 **9.** $\frac{5}{3}$ or $\frac{-2}{3}$ **11.** $at + v_0$ **13.** $\frac{y - y_0 + mx_0}{m}$ **15.** $IR + Ir$

17. $\frac{QJ}{I_2t}$ **19.** $\frac{L - a + d}{d}$ **21.** 64 N **23.** 8,000 ft^2

25. $x > \frac{3}{7}$

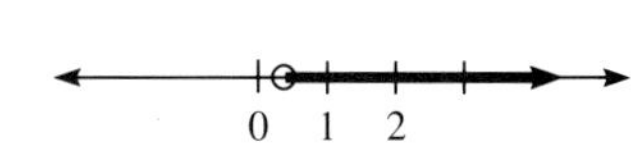

27. $x \geq \frac{90}{91}$

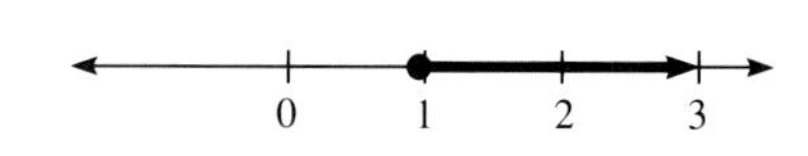

29. $x > \frac{7}{2}$ or $x < -5$

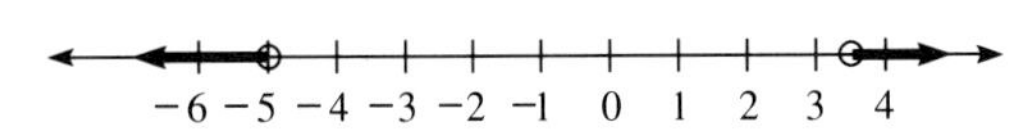

31. $x \geq \frac{16}{5}$ or $x \leq -4$

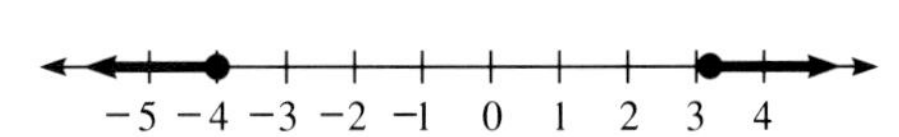

33. $x \le \frac{1}{10}$

0 .1 .2

35. $x = \frac{k\sqrt{y}}{z^2}$ **37.** $\frac{3}{2}$ **39.** 96 **41.** 11 men **43.** 8.68 h **45.** 1200 lb **47.** 15 A, 39 A
49. \$2,000, \$2,300 **51.** \$80 **53.** 8 s

Chapter Test, p. 38

1. $\frac{6}{5}$ **2.** $\frac{2s - 2v}{t^2}$ **3.** \$2200 @ 7.3%, \$3800 @ 9.5%

4. $x \ge -1$

-1 0

5. $b = 14$ ft

6. $d = \frac{kb^3}{c}$ **7.** $\frac{-7}{6}$ or $\frac{-47}{6}$

8. $x < \frac{-18}{7}$

-4 −3 −2 −1 0

9. 25 V

10. 188.7

11. $x \ge 2$ or $x \le \frac{-10}{7}$

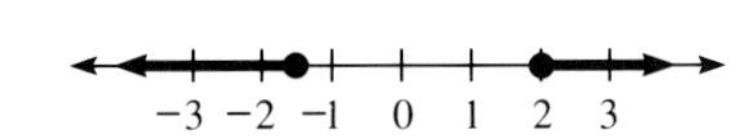

12. 3 to 1 **13.** 6 h @ 55 mph, 2 h @ 60 mph **14.** $h = \frac{V}{\pi r^2}$ **15.** $x = -14$ **16.** $1\frac{1}{2}$ gal

17. $\frac{Vk - a}{V^2}$ **18.** 1.7 **19.** 1,800 lb

20. $\frac{2}{3} < x < \frac{13}{3}$

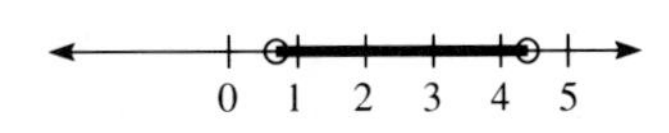

CHAPTER 2

2–1 Exercises, p. 46

1. $P(s) = 4s$ **3.** $A(w) = 5w$ **5.** $C(m) = 0.60 + 0.40m$ **7.** $I(V) = \frac{V}{10}$ **9. a.** 26 **b.** 51 **c.** 11
11. a. −19 **b.** 1 **c.** −19 **13. a.** 6.7 **b.** −7 **c.** −5.6 **15. a.** −3 **b.** 2.27 **c.** 0.98

17. a. 48 **b.** 8 **c.** 93 **19.** $3x - 38, 6s + 1$ **21.** $\frac{6n + 8}{2n - 1}, \frac{-11}{n(n - 1)}$ **23.** $30y^2 - 34y + 10$

25. All reals ≥ -7 **27.** All reals ≥ 0 except $\frac{1}{2}$ **29.** All reals except 0 and -6 **31.** 5 mi/h^2

33. 61.5% **35.** 84 m^3

2–2 Exercises, p. 50

1.

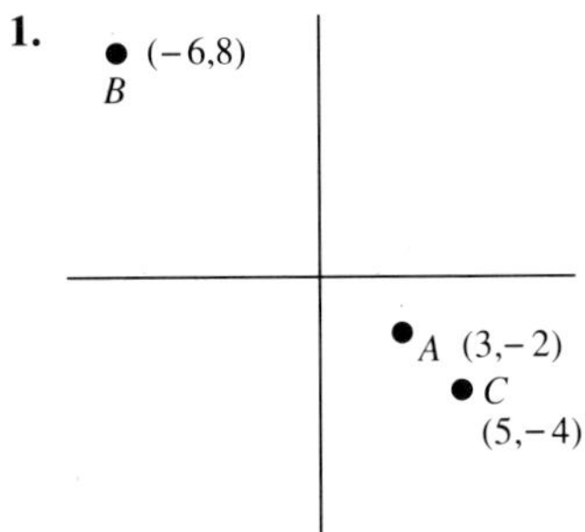

3.

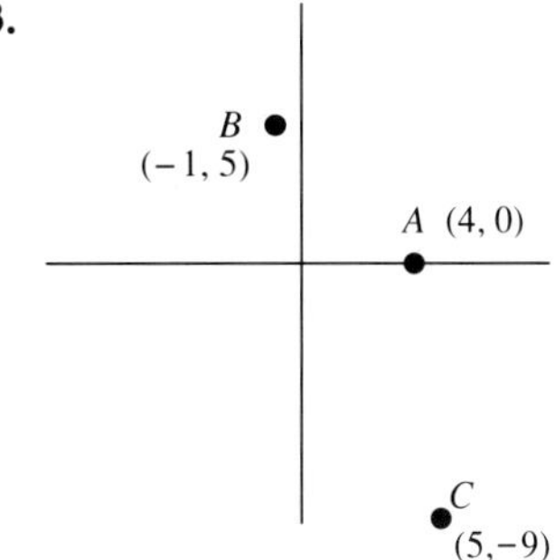

5. A: (4, 2) B: $(-1, 2)$ C: $(-4, -5)$ **7.** II **9.** III **11.** on the y axis **13.** 0 **15.** I and III

17. (0, 5) **19.** $\frac{-1}{2}$

2–3 Exercises, p. 56

1.

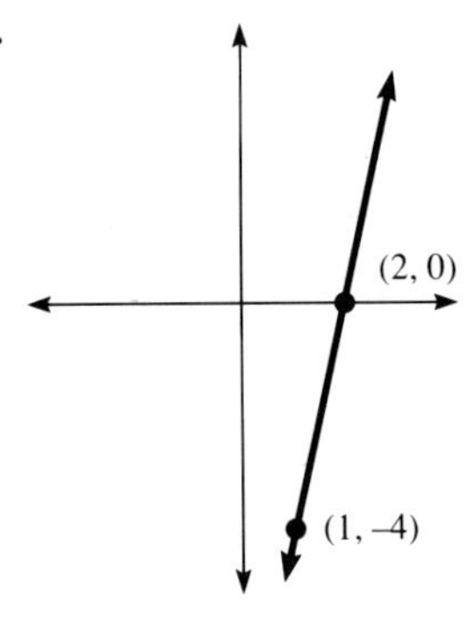

3.

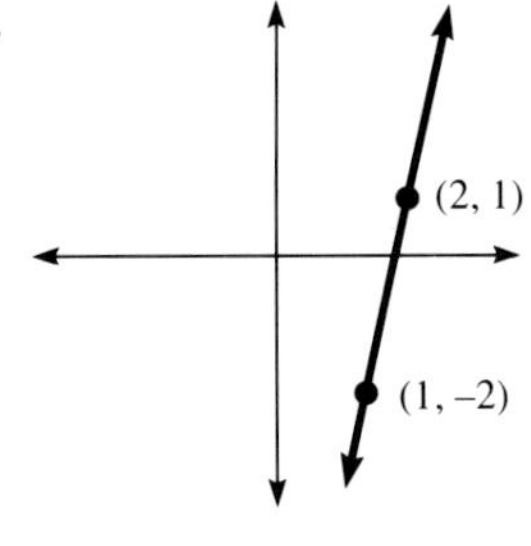

5.

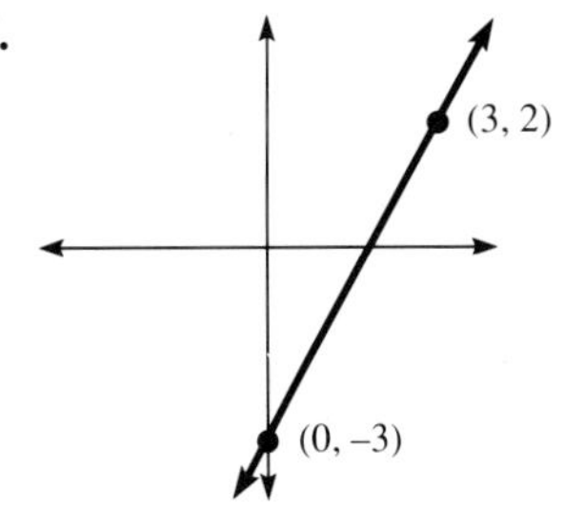

7.

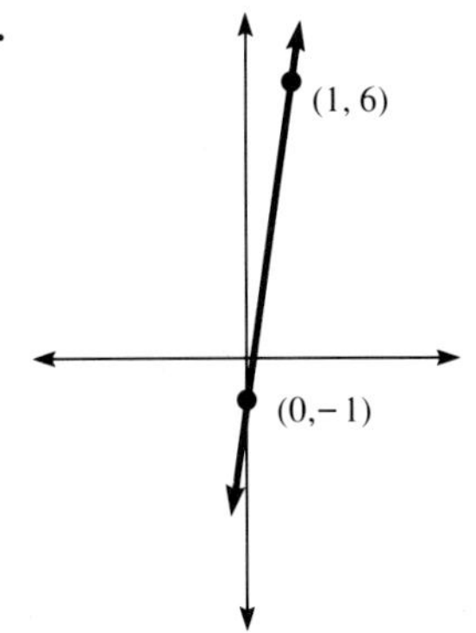

9.

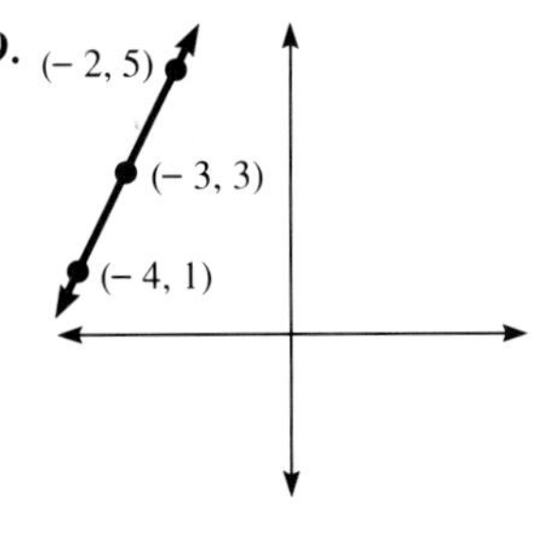

11.

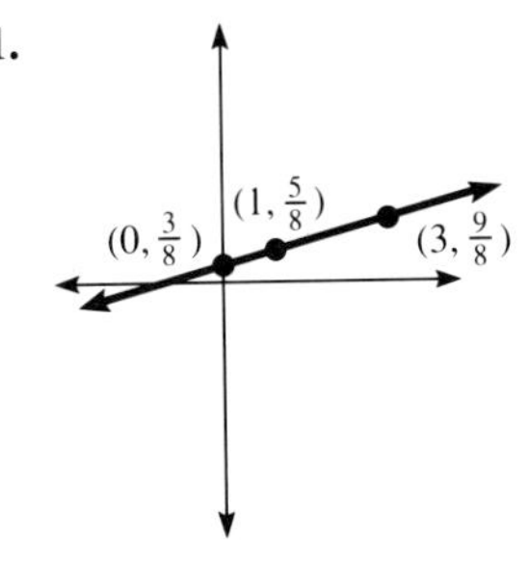

13.

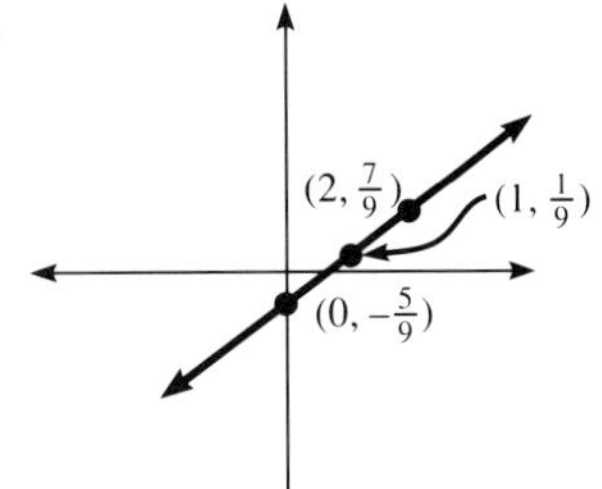

15.

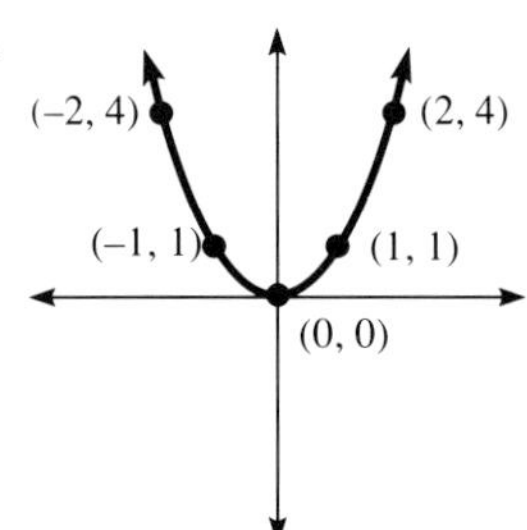

17.

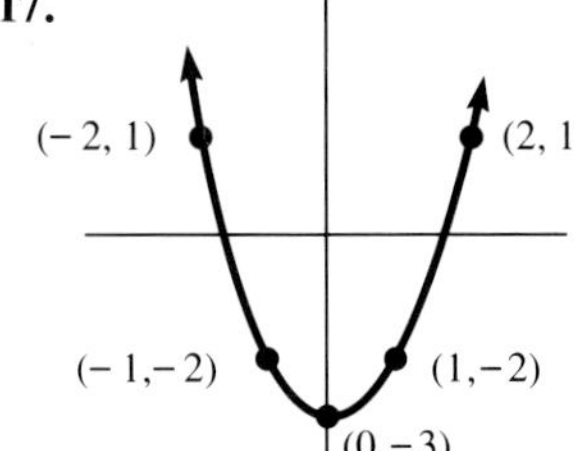

19.

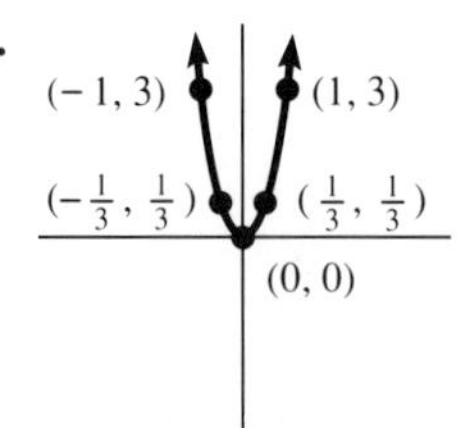

21.

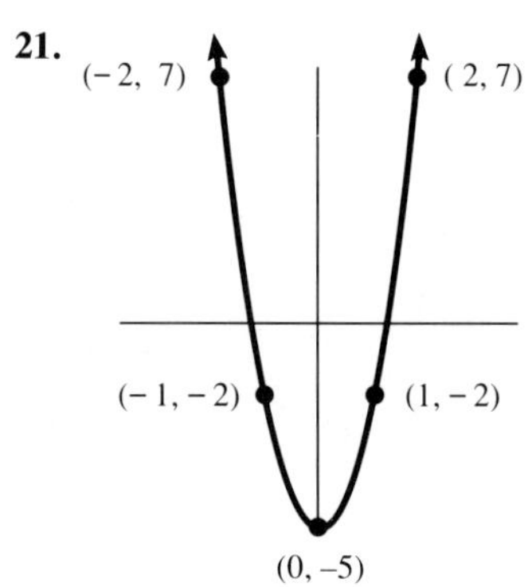

23.

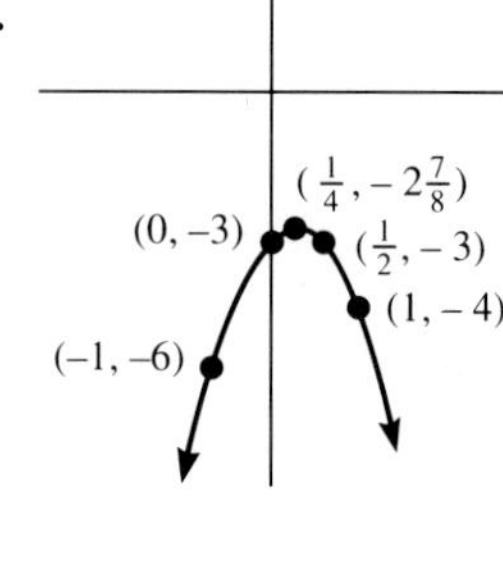

25.

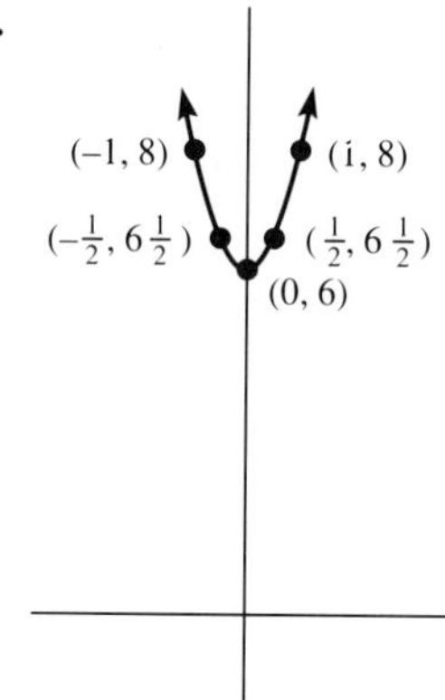

27.

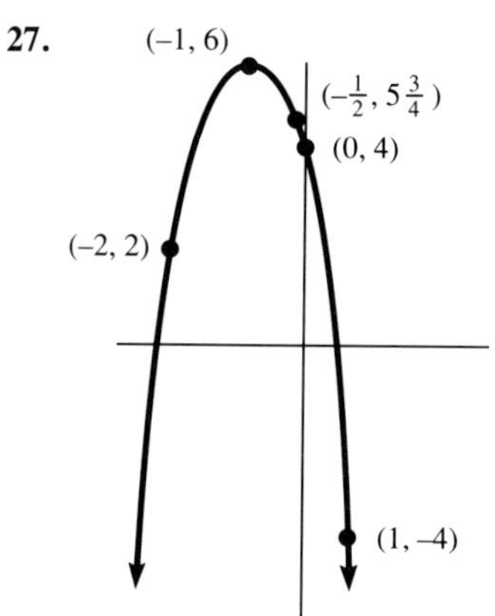

29.

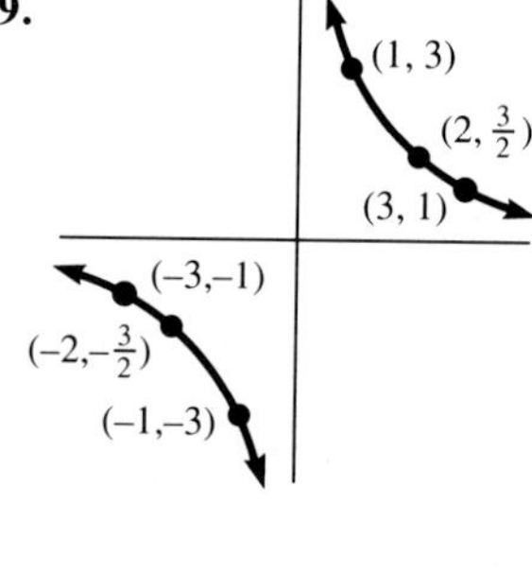

31.

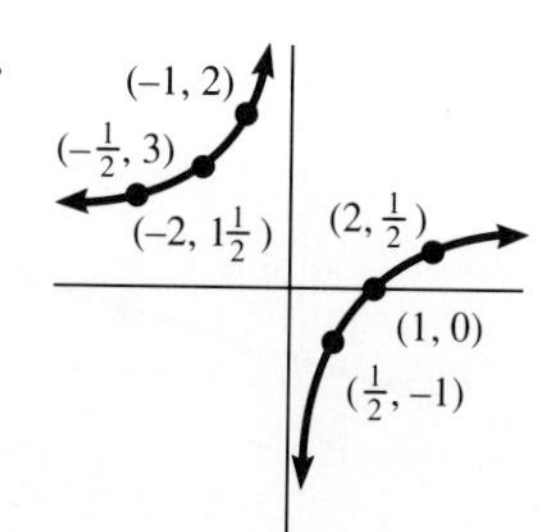

33.

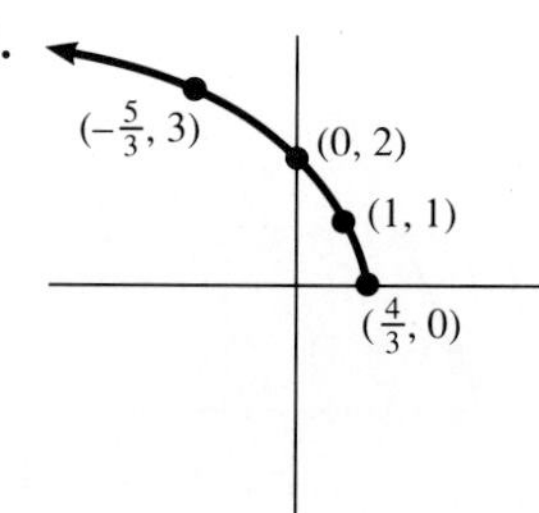

35.

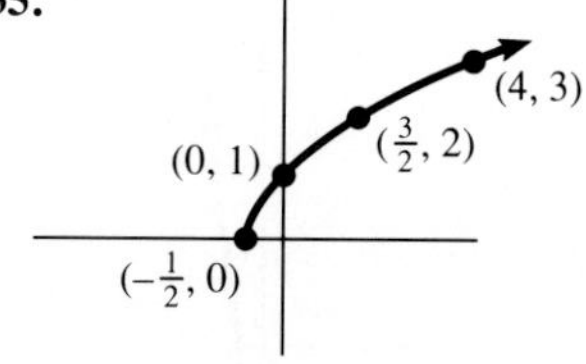

37.

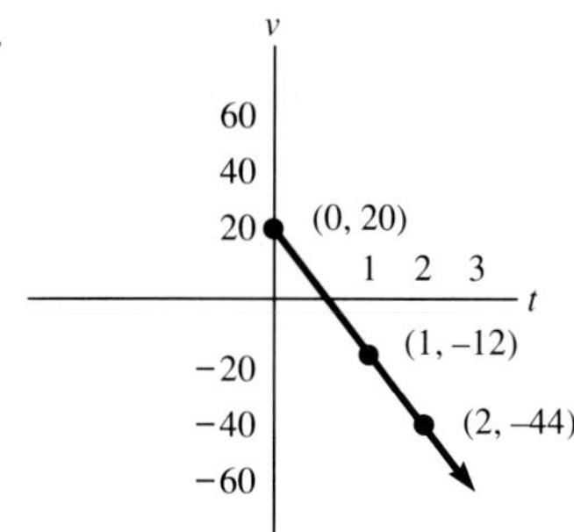

39.

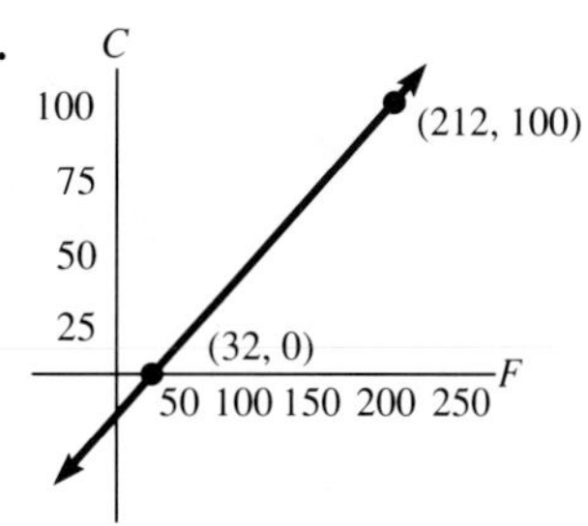

41.

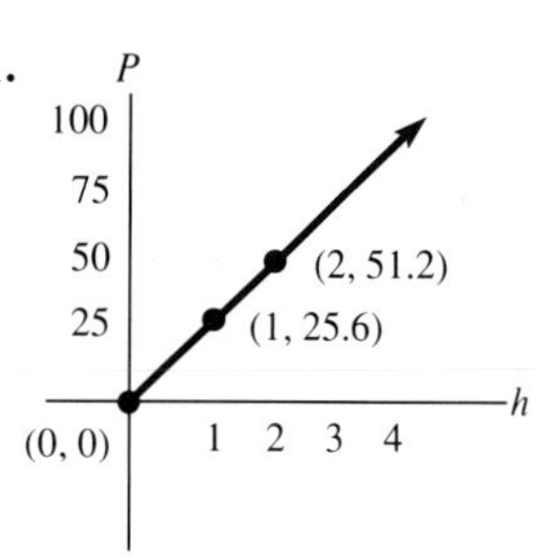

2–4 Exercises, p. 64

1. 8.6 **3.** 4.24 **5.** 3.79 **7.** 3 **9.** 6 **11.** $\left(2, \frac{-3}{2}\right)$ **13.** $\left(\frac{-5}{2}, 6\right)$ **15.** $(2.35, -3.65)$
17. $\left(\frac{11}{6}, \frac{-5}{2}\right)$ **19.** $\left(\frac{23}{6}, \frac{29}{16}\right)$ **21.** $\frac{-13}{11}$ **23.** undef **25.** $\frac{15}{2}$ **27.** 0 **29.** $\frac{-4}{7}$ **31.** neither
33. perpendicular **35.** neither **37.** $\overline{AB}$ is perpendicular to $\overline{BC}$; therefore it is a rt. △. **39.** square

2–5 Exercises, p. 72

1.

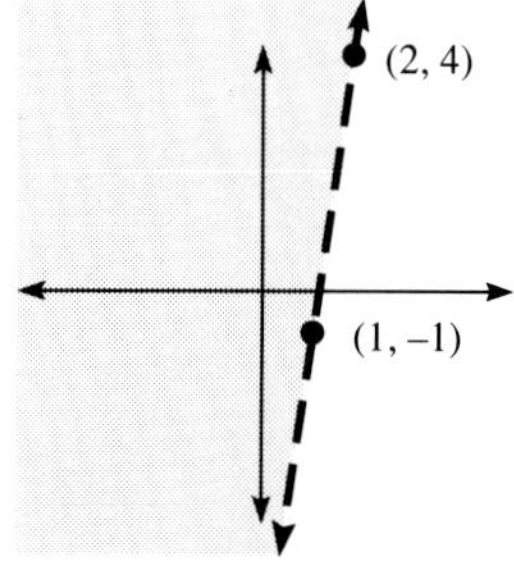

3.

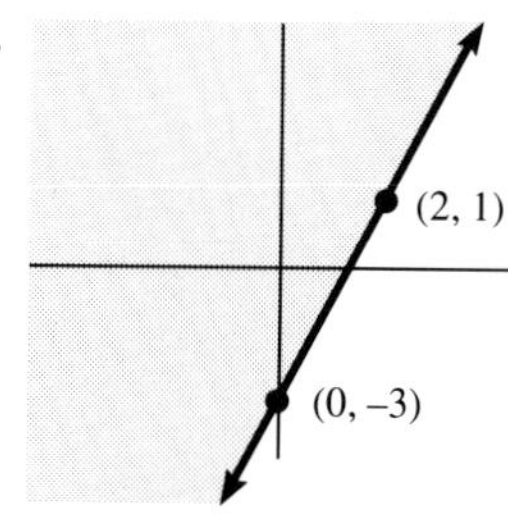

5.

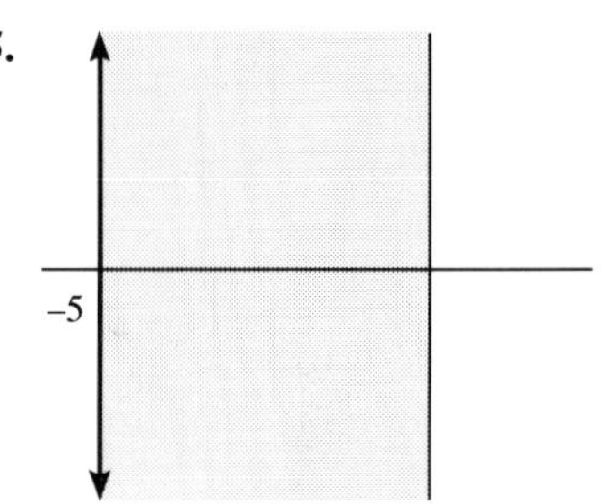

7.

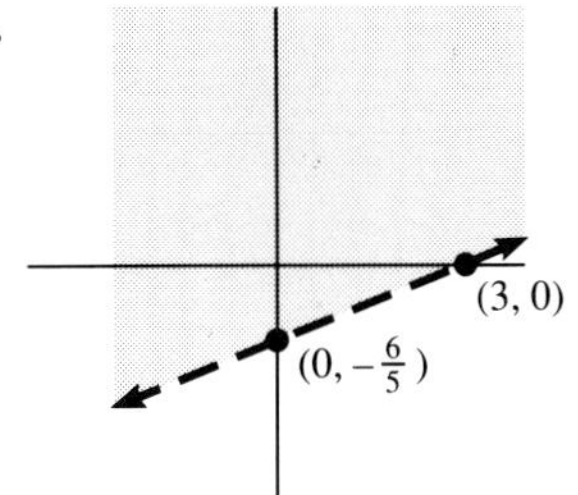

9.

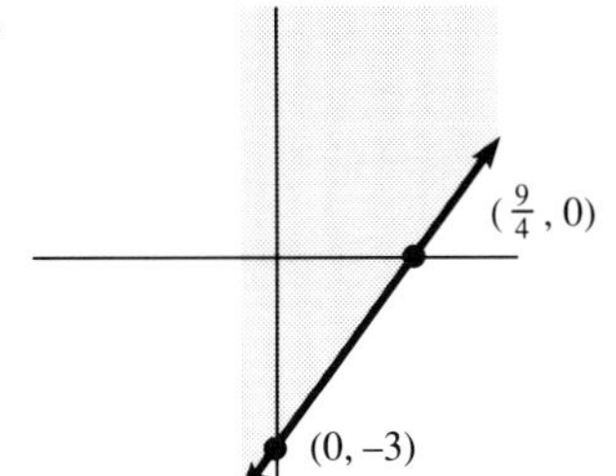

11.

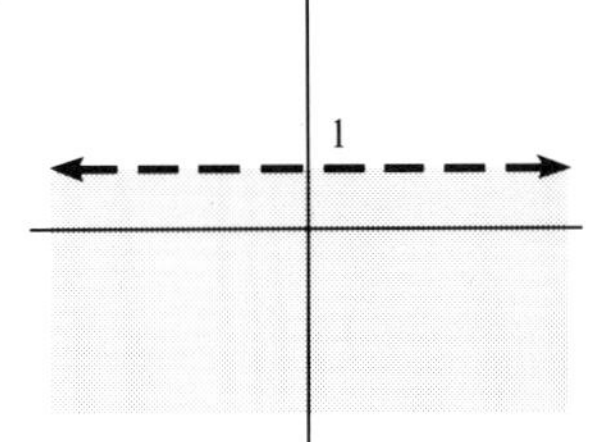

13.

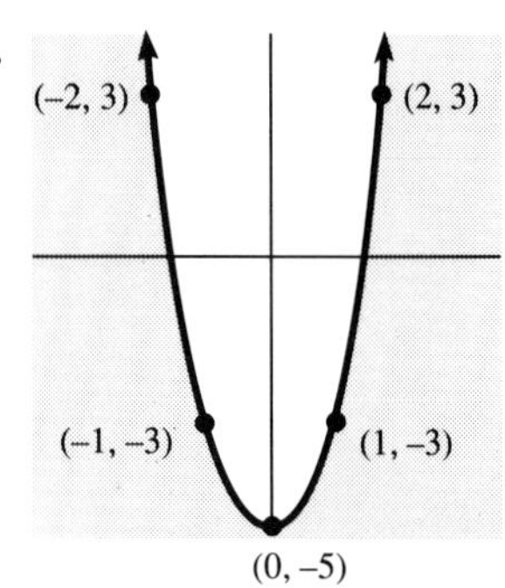

15.

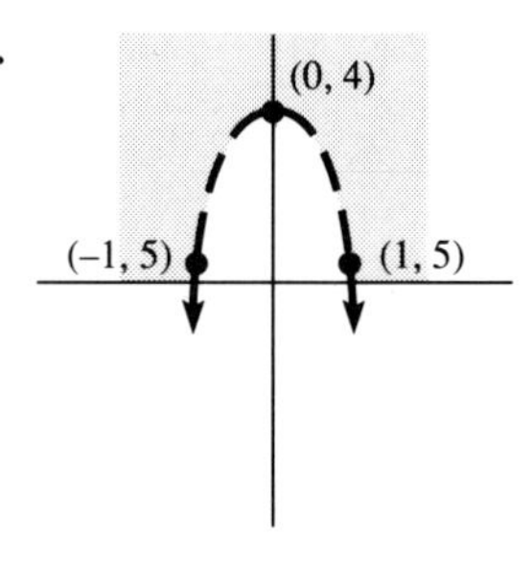

17.

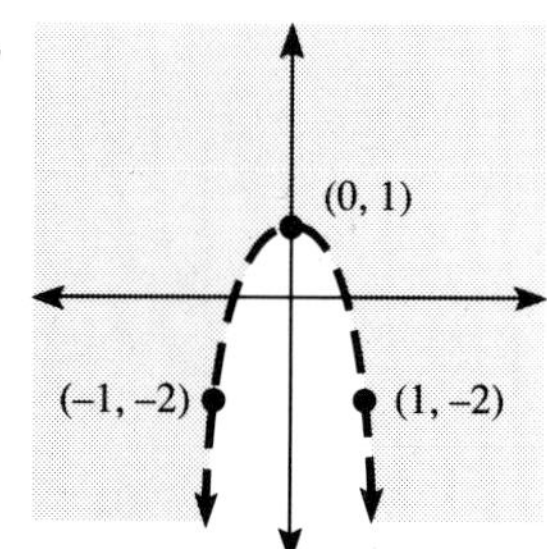

19.

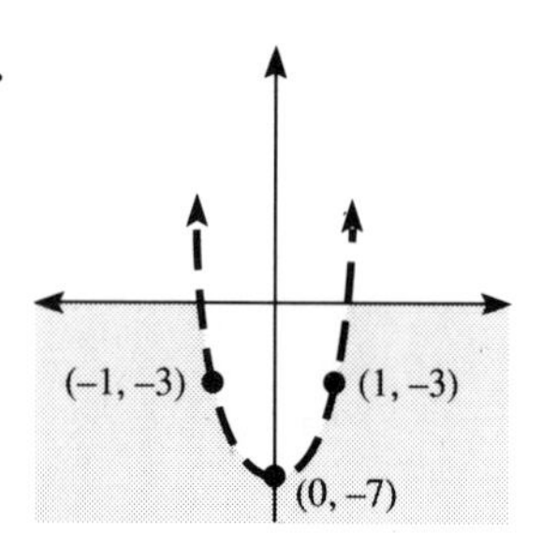

21.

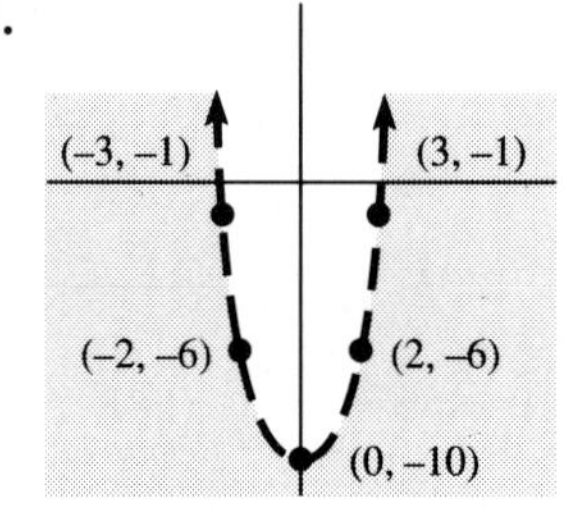

23.

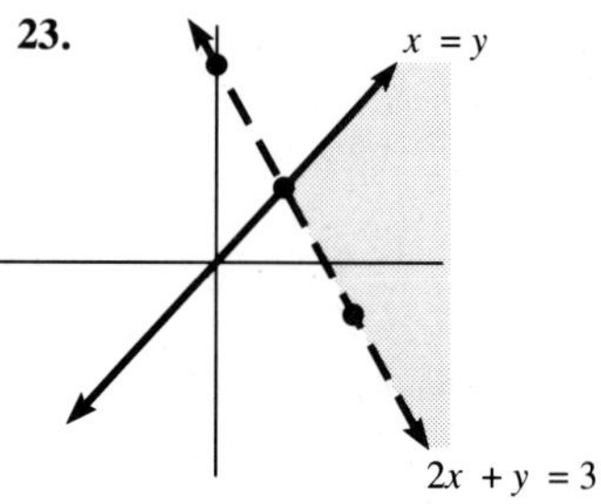

25.

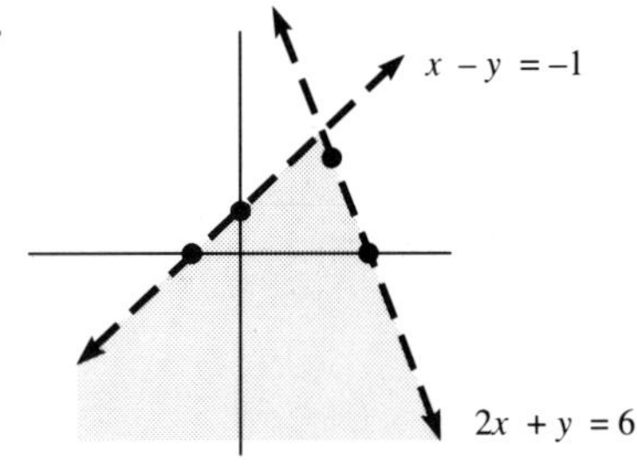

27.

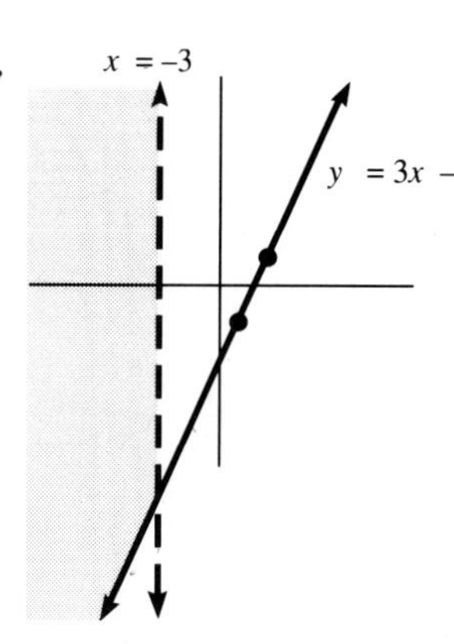

29. max profit at (7, 3.3)

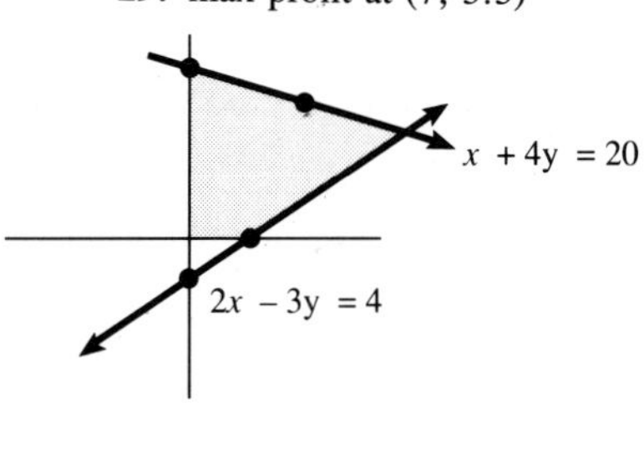

31. 0 economy and 6 deluxe 33. 270 of type one and 4,300 of type two
35. 0 of first computer and 13 of second

2–6 Exercises, p. 78

1.

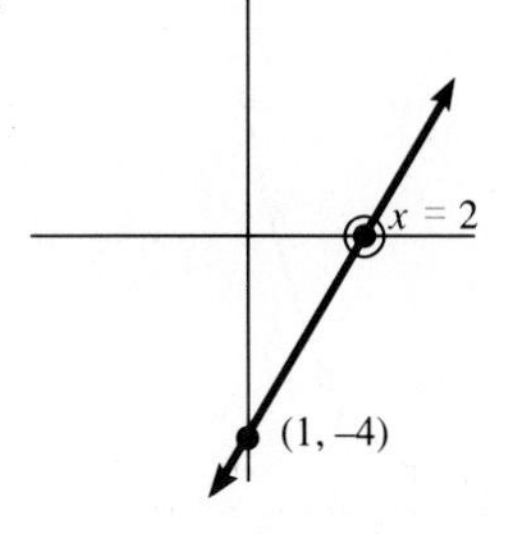

3.

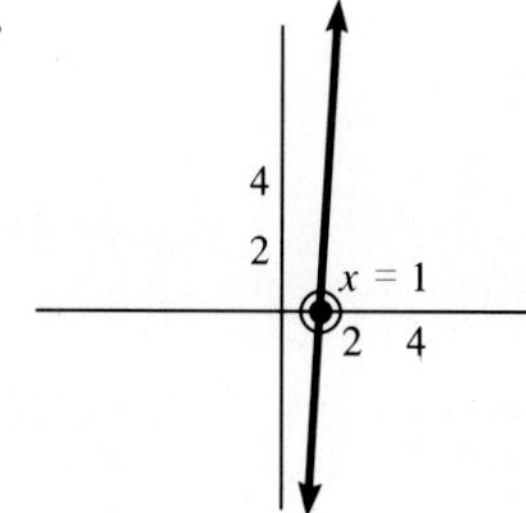

5.

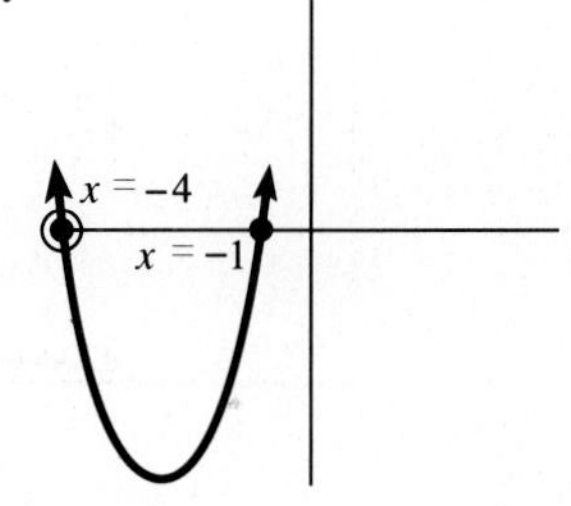

7.

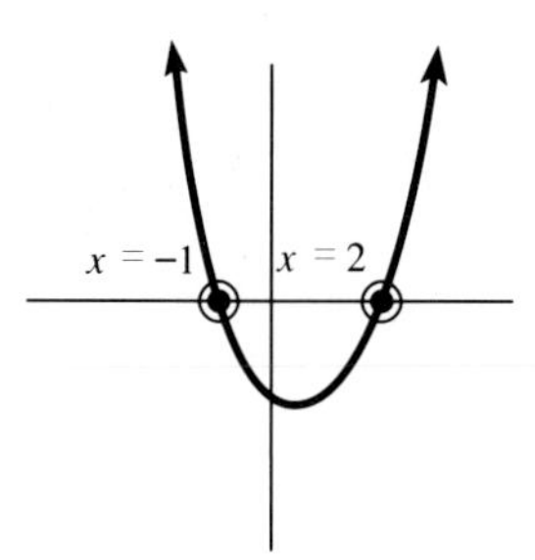

9.

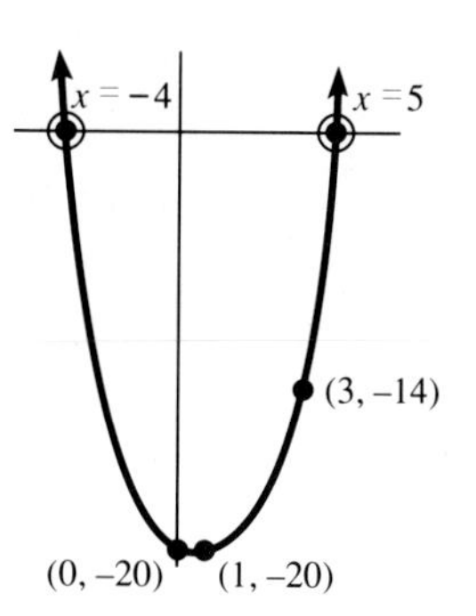

11.

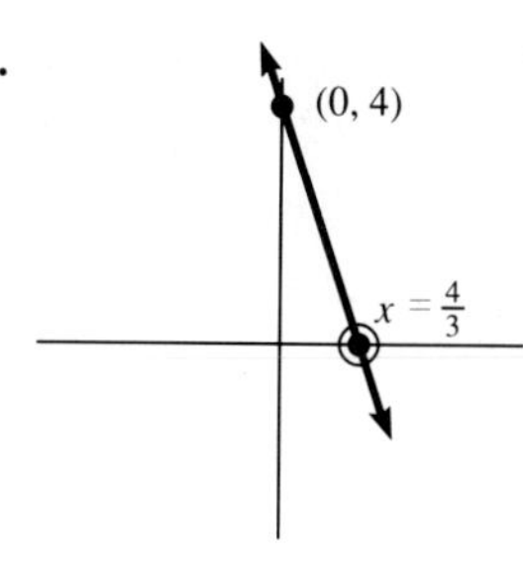

13.

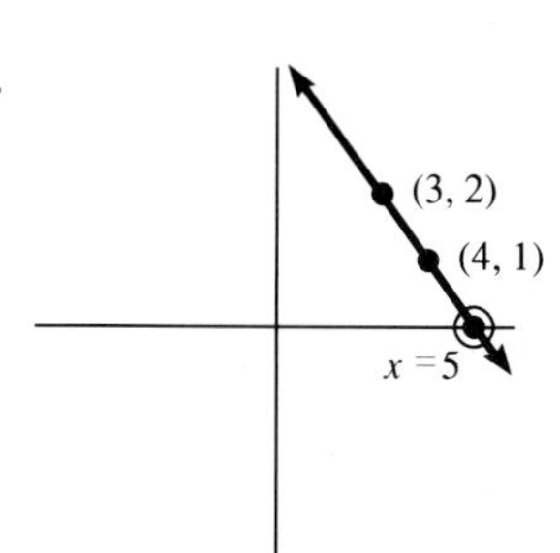

15.

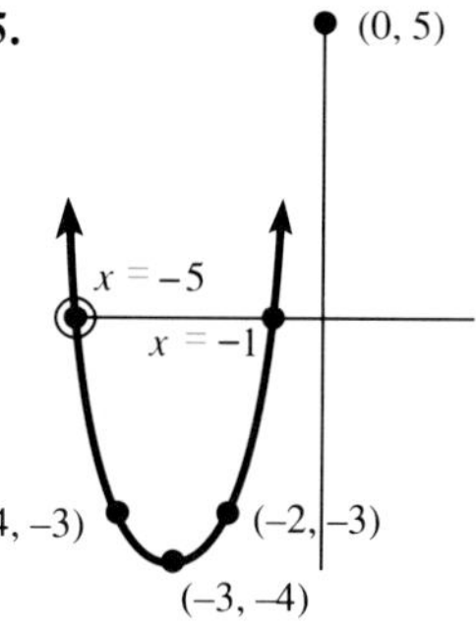

17. 4,000 units @ $12,000 **19.** 3,000 units at $42,000

21. Since he must produce 3000 units to break even, he suffers a loss at 1000 units.

23.

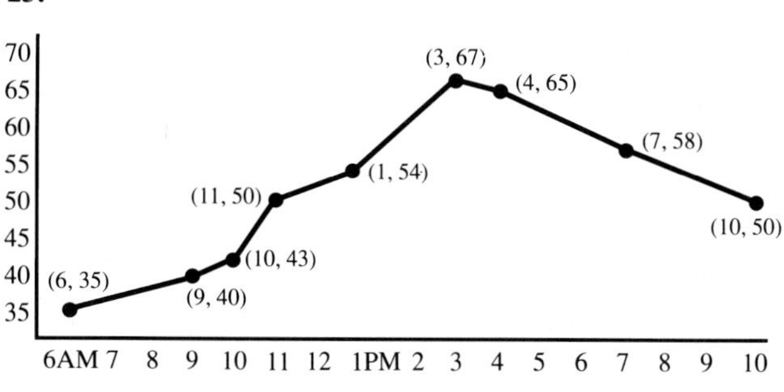

25. approx. 25 m

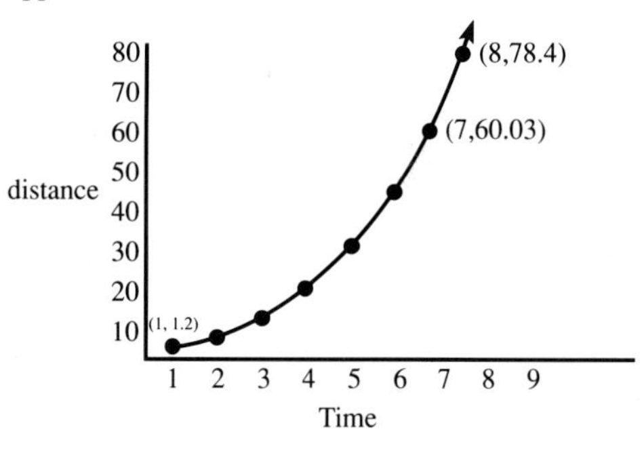

27. approx. 0.76 V

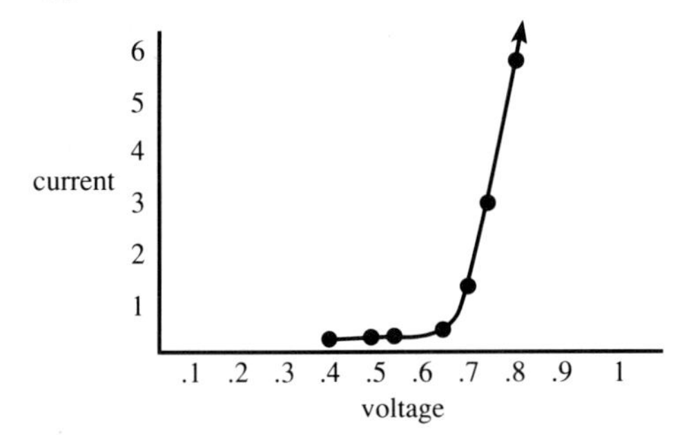

29. approx. 265 mA

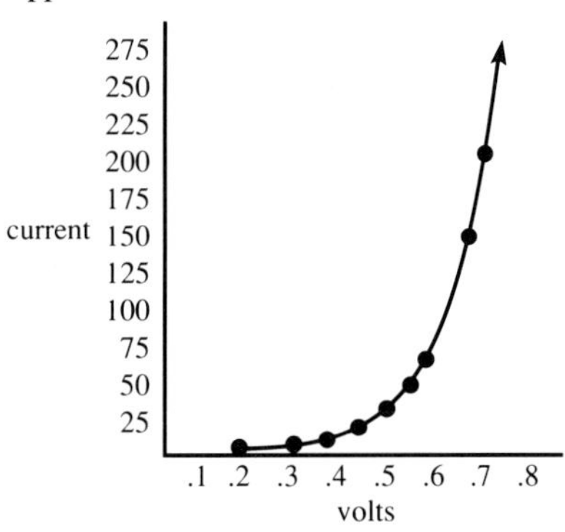

Chapter Review, p. 80

1. $P(s) = 3s$ **3.** $A(b) = \frac{1}{2}b(b + 3)$ **5.** $-6, 12$ **7.** $31, 3$ **9.** $24y - 26$ **11.** $x \geq \frac{-8}{3}$

13. $t \geq 4$ **15.** All reals except 0 and 3

17.

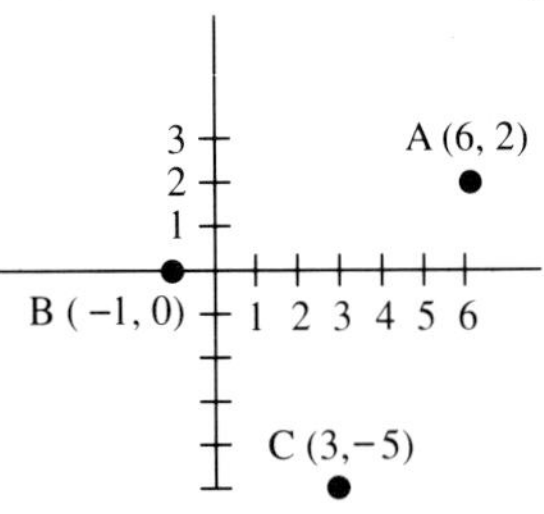

19.

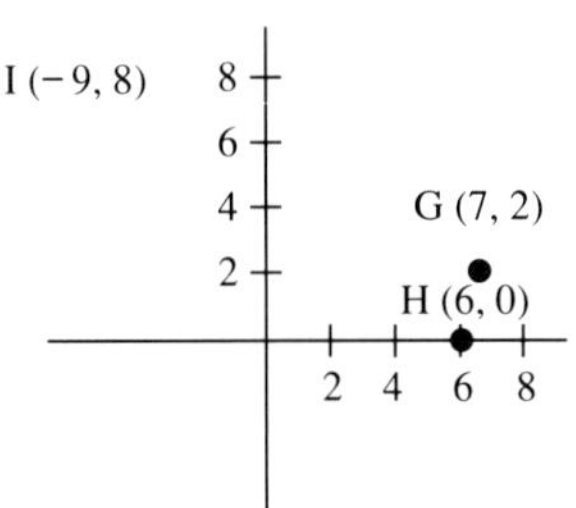

21.

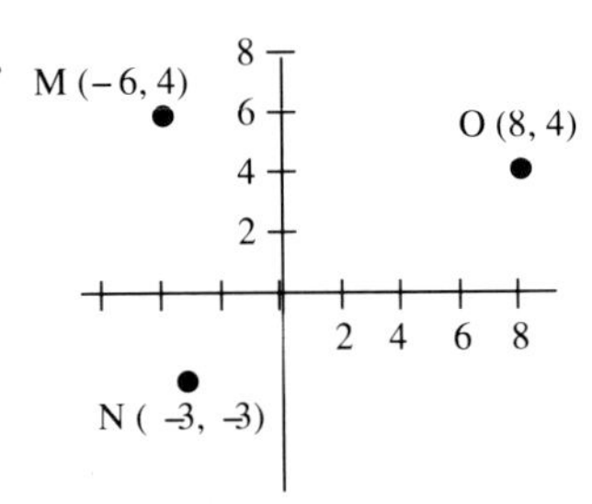

23.

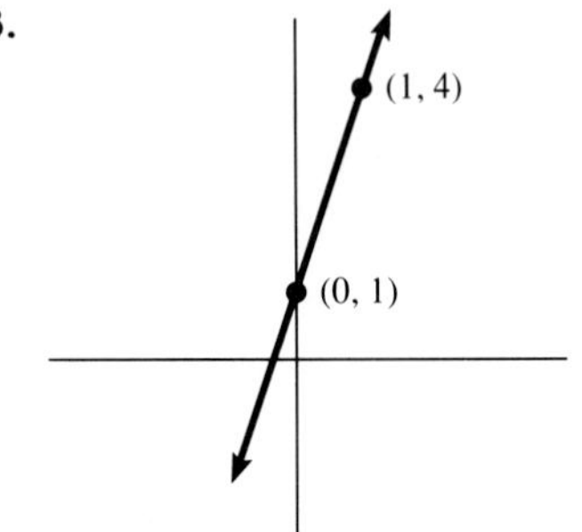

25.

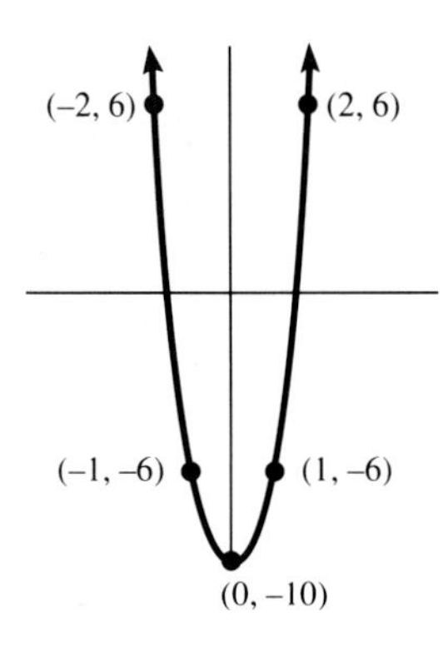

27.

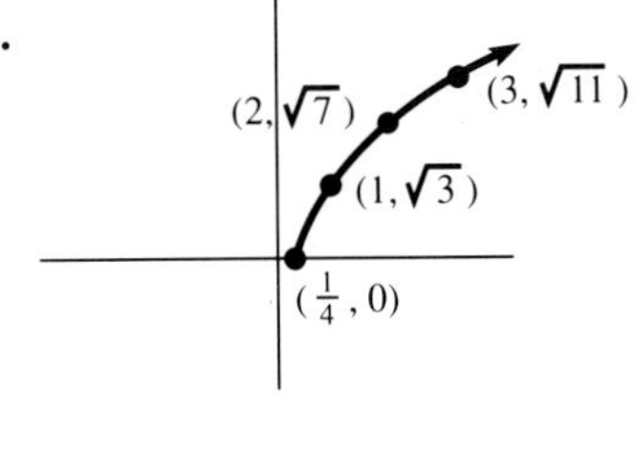

29.

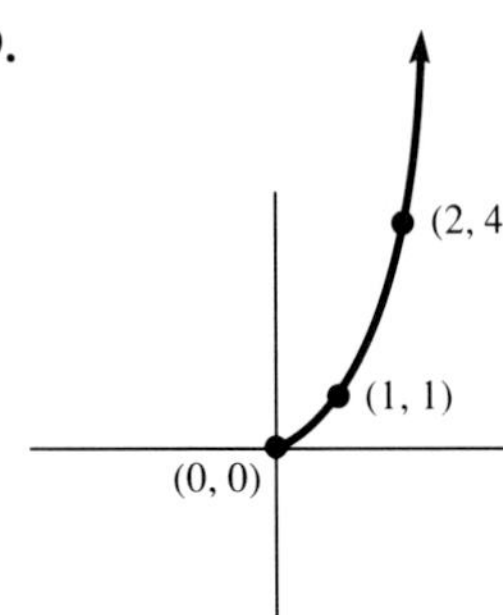

31. $11.18; \frac{-1}{2}; \left(3, \frac{7}{2}\right)$ **33.** $8.6; \frac{-5}{7}; \left(\frac{-5}{2}, \frac{-15}{2}\right)$

35. $15; 0; \left(\frac{-1}{2}, -3\right)$ **37.** $7.005; \frac{-105}{4}; \left(\frac{4}{15}, \frac{-1}{2}\right)$

39. $11.76; \frac{39}{4}; (2.3, 2.25)$

41.

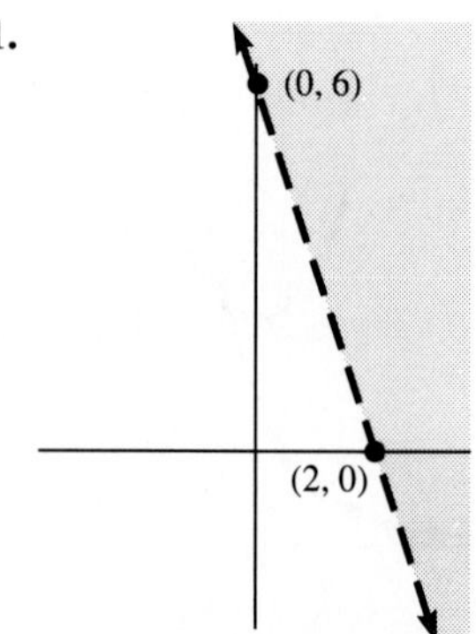

43.

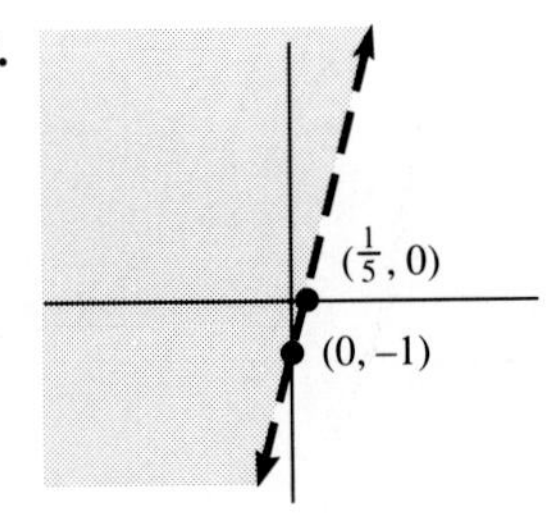

45.

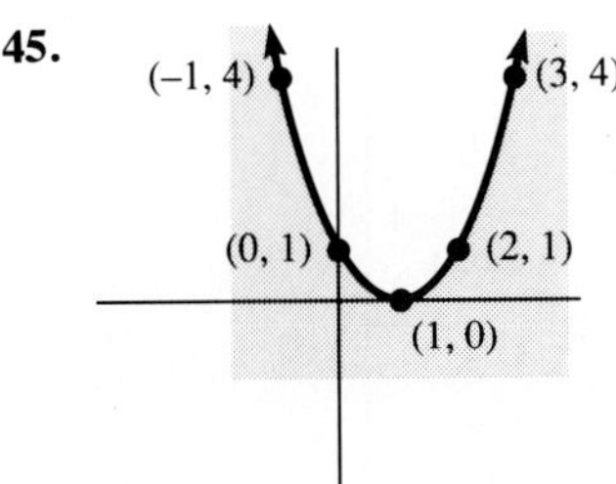

47. no solution

49.

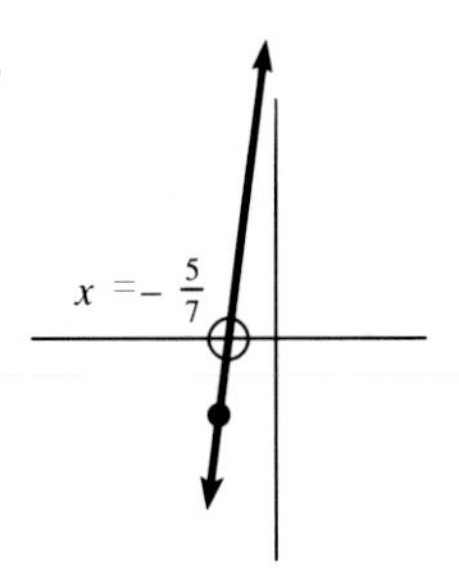

51.

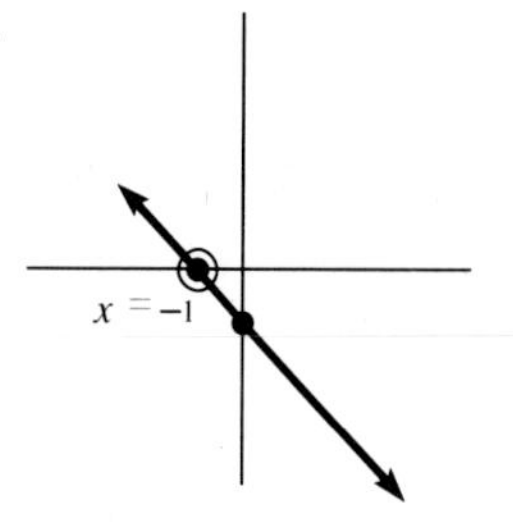

53.

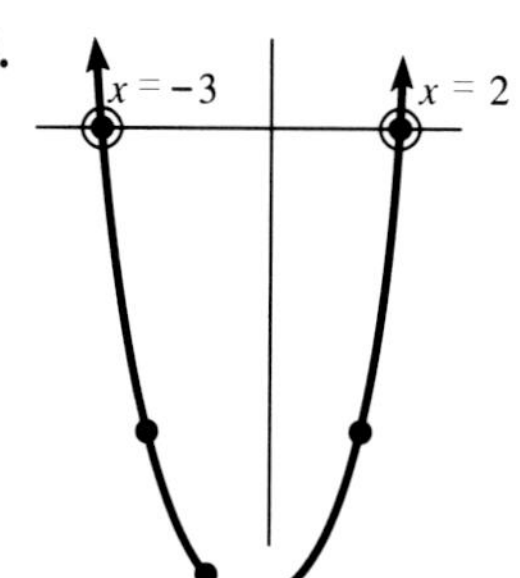

55.

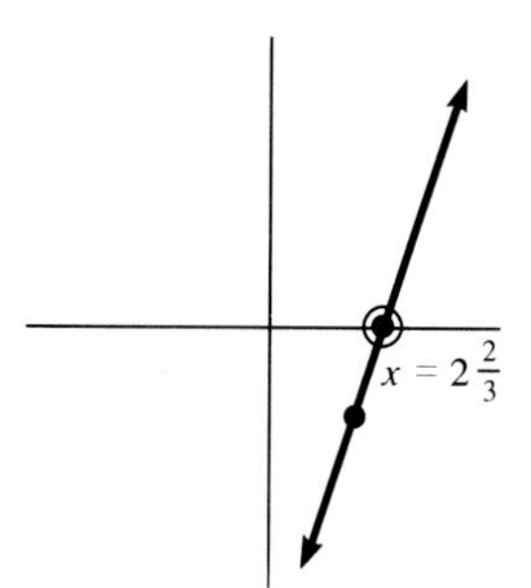

57. 2,000 units at $16,000

59. Yes

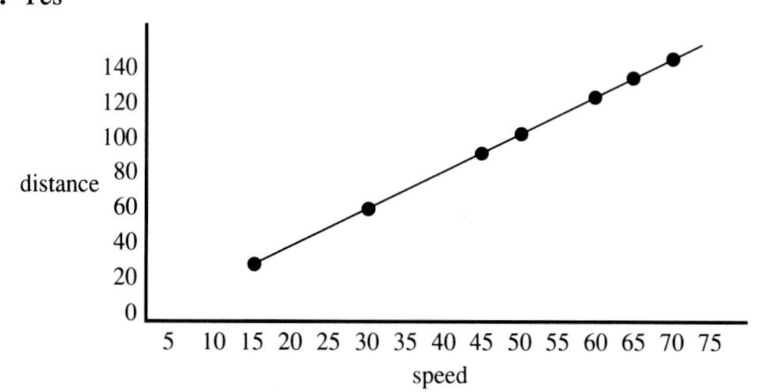

Chapter Test, p. 82

1. 14

2.

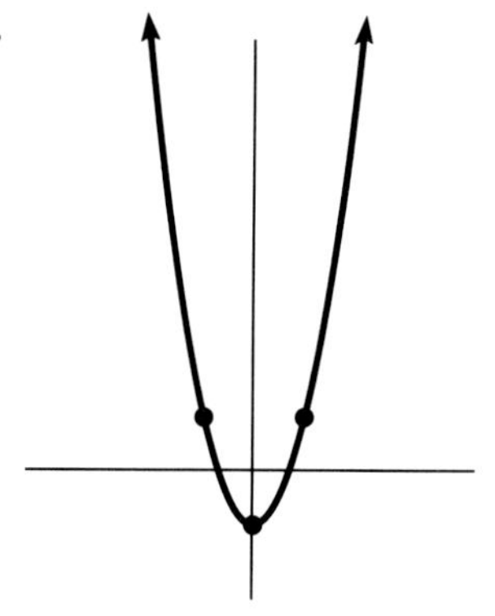

3.

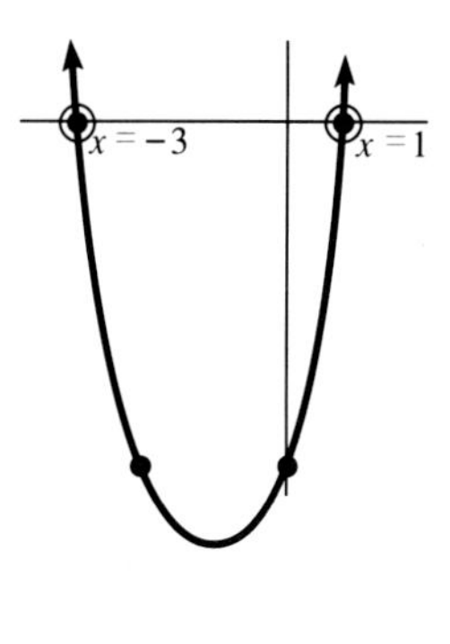

4.

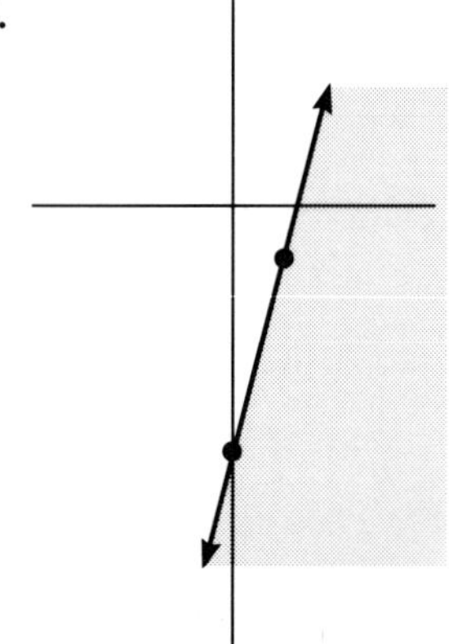

5.

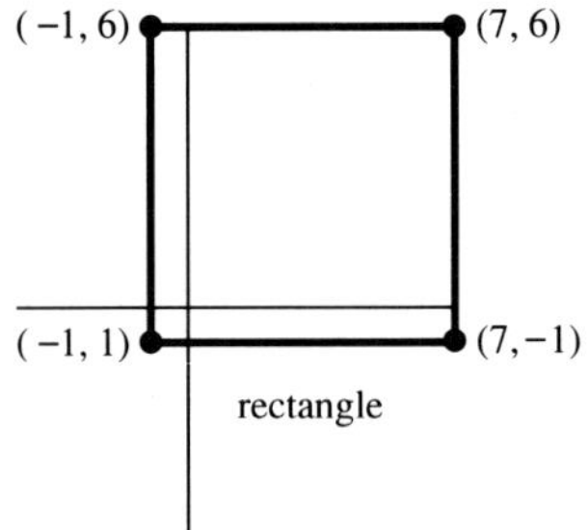

6. 3,000 units at $9,000

7.

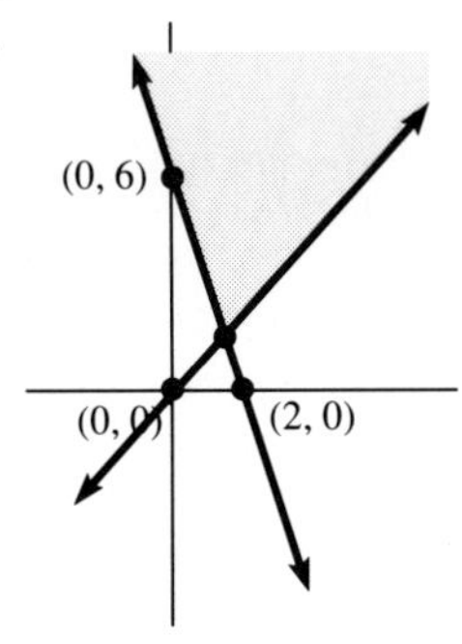

8.

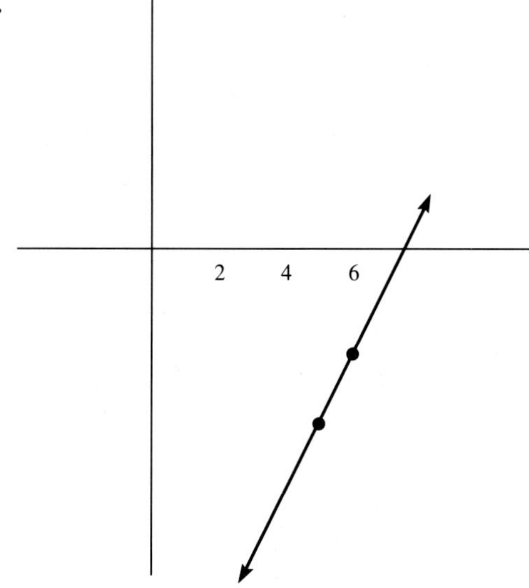

9.

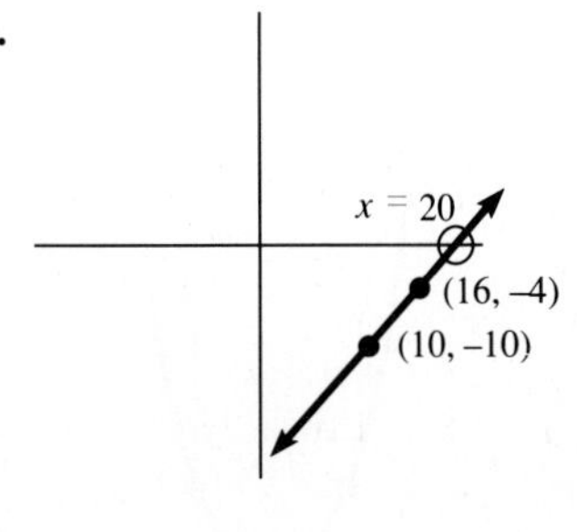

10. $\frac{4}{9}$ **11.** $x \geq 5$ **12.** $-15x^2 + 6x - 13$

13.

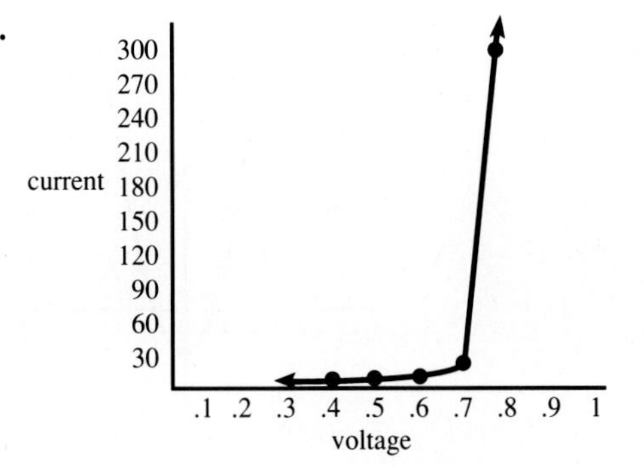

14. 19.72; $\left(-1, \frac{-3}{2}\right)$

15. No

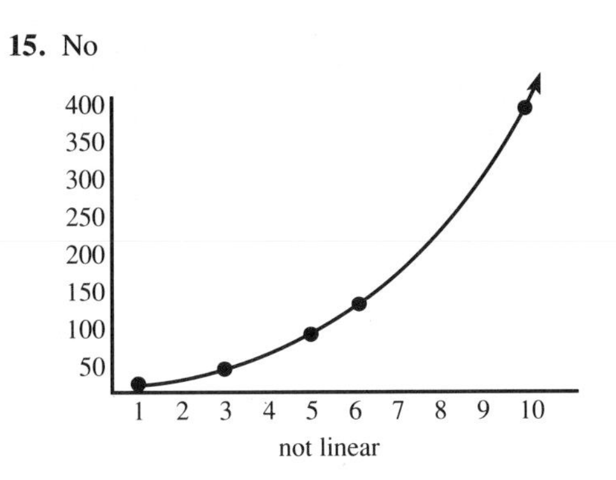

16. 16 m^2

17.

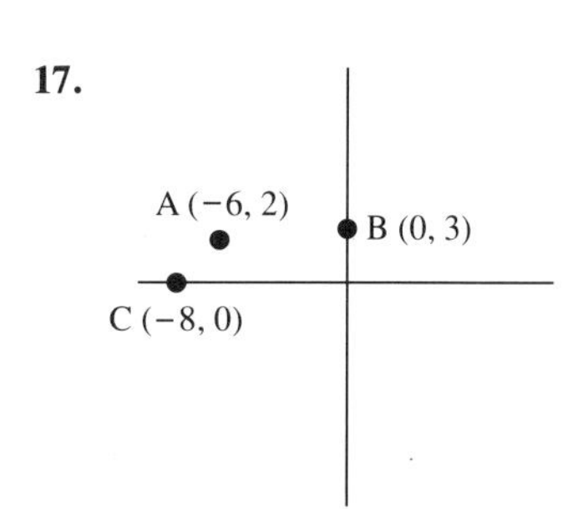

18. $r(P,t) = \frac{\text{I}}{\text{Pt}}$

CHAPTER 3

3–1 Exercises, p. 91

1. $3a^2(b + 3a)$ **3.** $7x(xy - 2 + 5x^2y^2)$ **5.** $(7x + 5y)(7x - 5y)$ **7.** $10x^2(3x^3 - 4 + 10x)$
9. $(5p + 8n)(5p - 8n)$ **11.** $y^3(9 + 7y^3 + 10yx^2)$ **13.** $(6m + 11n)(6m - 11n)$ **15.** $8st(4s - 5s^2t^2 + 1)$
17. $3m^2n^3(6mn - 3n^2 + m^2)$ **19.** $5x^2(5x^3 + 6 - 3x^4)$ **21.** $(12s + 7t)(12s - 7t)$
23. $19m^3(2mn^3 + 3n^4 - 1)$ **25.** $(4m^2 + 5n^2)(4m^2 - 5n^2)$ **27.** $16m^4(4m^4 + 2 + 3m^{11})$
29. $(13c + 6d)(13c - 6d)$ **31.** $7a^5c^{10}(9a^2b^5 - 7)$ **33.** $(2x + 7y)(2x - 7y)$ **35.** $9x^3z(6y^2 - 7xy^5z^2 + 9)$
37. $(a^2 + 6b^2)(a^2 - 6b^2)$ **39.** $23s^8t^4(2st^2 + s^4 + 3t^5)$ **41.** $(x - 3)(2x + 3)$ **43.** $(x - 5)(3 - 5x)$
45. $4\pi(r_2 + r_1)(r_2 - r_1)$ **47.** $125(p + 3)(p - 3) = 0$ **49.** $m(v_2 - v_1)$ **51.** $\pi r^2(h_1 - h_2)$
53. $P(1 + rt)$

3–2 Exercises, p. 96

1. $(a + 11b)(a - 3b)$ **3.** $(8x + 13z)^2$ **5.** $(x + 6y)(x - 5y)$ **7.** $(10m + 9n)^2$ **9.** $(11x - 3y)^2$
11. $(m - 6n)(m + 4n)$ **13.** $2(2n + 1)(n - 3)$ **15.** $(3 - 4y)(2 + 3y)$ **17.** $(12a + 7)^2$
19. $(7k + 6g)(5k - 3g)$ **21.** $(8n - 13p)^2$ **23.** $(5x^2 - 9y^2)(x^2 + 2y^2)$ **25.** $(2m^2 + 7n^2)^2$
27. $3xy(2x - y)(x + y)$ **29.** $(7x^2 - 9y^2)(3x^2 + 2y^2)$ **31.** $(8a - 3b)^2$ **33.** $(9n + 5p)^2$
35. $(6x^2 - 5y^2)^2$ **37.** $8xy(x^2y - 8x^3 - 10y^3)$ **39.** $y(7x + 5y)^2$ **41.** $z(17xy + 34x^2 + 16y^3)$
43. $9(x - 3z)^2$ **45.** $4z(2x - 9y)^2$ **47.** $5(x + 3y)$ **49.** $6(2x + 5y)(x - 3y)$ **51.** $(12x + 7y)(12x - 7y)$
53. $2(x + 6)(x - 5)$ **55.** $(2t + 3)(t - 4)$ **57.** $4(x - 6)(x - 2)$

3–3 Exercises, p. 101

1. $3xyz$ **3.** $4xyz^2$ **5.** $4a^2 + 4ab$ **7.** $18m^2$ **9.** $3xy - 12y$ **11.** $6x^2 + 3xy$ **13.** $\frac{x^4}{3y^3}$ **15.** $\frac{3a^4c^8}{b^4}$

17. $\frac{x^2}{9z^3}$ **19.** $\frac{4n^3p^2}{5m^5}$ **21.** $\frac{7a^2d^3}{c^7}$ **23.** $\frac{2b}{a}$ **25.** $\frac{7x(x - 1)}{8}$ **27.** $\frac{y - 2x}{x + 2y}$ **29.** $\frac{x + 5}{5 - x}$ **31.** $\frac{2a^3}{b^2}$

33. $\frac{x}{2}$ **35.** $\frac{a}{3b^3}$ **37.** $\frac{7x - y}{2x + 3y}$

3–4 Exercises, p. 107

1. $4x^3$ **3.** $\frac{1}{7x^4}$ **5.** $\frac{2x + 3}{2x}$ **7.** $\frac{5x(7x - 1)}{7x + 3}$ **9.** $3x^2y$ **11.** $\frac{a^4c^4}{15b}$ **13.** $\frac{m^3(a + 2b)}{a - 2b}$ **15.** $\frac{-3}{x}$

17. $\frac{3x + 2}{4x^2}$ **19.** $2(3x + 1)$ **21.** $\frac{(a - b)(2a - b)}{a^2 + b^2}$ **23.** $\frac{(3m + 8)(m - 1)}{(2m - 5)(2m + 3)}$ **25.** 1

27. $\frac{(7a - b)(5a - b)}{(a - 4b)}$ **29.** $\frac{3x + 5y}{4x + y}$ **31.** $\frac{(x - 3y)^2(2x + y)}{(x + 2y)^2(x - y)}$ **33.** $\frac{(5x + 6t)(2s - 7t)}{(s - t)^2}$ **35.** $\frac{3m + 2n}{3m - 2n}$

37. $\frac{(3 - x)(x + 4)}{x + 3}$ **39.** $\frac{wv^2}{gr}$ **41.** s **43.** 366.7 ft/s

3–5 Exercises, p. 114

1. $\frac{x^2 - 2x + 2}{4x^2}$ **3.** $\frac{10x^3 + 6x^2 - 1}{4x^2}$ **5.** $\frac{10m^2 - 15m - 6}{30m^3}$ **7.** $\frac{4x^2 - 7x - 6}{x(x + 1)}$ **9.** $\frac{4y^2 - 17y - 4}{y(7y + 2)}$

11. $\frac{2x^2 + 7x - 1)}{3x(2x - 1)}$ **13.** $\frac{-14a + 1}{3(a - 2)}$ **15.** $\frac{4x^2 - 8x + 1}{5x(x - 3)}$ **17.** $\frac{17y^2 + 2y}{(2y - 1)(3y + 2)}$ **19.** $\frac{2x - 23}{(x + 2)(2x - 5)}$

21. $\frac{7z^3 - 15z^2 + 25z + 3}{7z(z - 2)(2z + 3)}$ **23.** $\frac{14x^2 - 17x - 7}{(8x + 7)(3x - 8)}$ **25.** $\frac{b^2 - 32b - 18}{(2b + 1)(5b + 3)}$ **27.** $\frac{57x^2 + 49x - 30}{(7x - 4)(4x + 3)}$

29. $\frac{4n - 2}{n - 4}$ **31.** $\frac{9k^2 + 10k - 1}{(3k - 1)(k + 2)(2k + 1)}$ **33.** $\frac{63x + 33}{4(3x - 1)(x - 3)(x + 3)}$ **35.** $\frac{6y^2 + 21y + 19}{(y + 3)(y - 3)}$

37. $\frac{5x^2 + 8x}{(x - 4)(x + 1)(3x + 2)}$ **39.** $\frac{5x^3 + 4x^2 - 23x - 4}{(x - 3)(x + 2)(x + 4)(x - 1)}$ **41.** $\frac{x(5y - 3x^2)}{y(6 + 5x^2)}$ **43.** $\frac{4x^2 - 6x - 2}{6x^2 - 3x + 3}$

45. $\frac{-6x^2 + 4x - 7}{4x + 5}$ **47.** $\frac{3x(2x^2 + 4x + 18)}{(x - 4)^2(2x + 3)^2}$ **49.** $\frac{3a(8a^2 + 16ab - 6b^2)}{(2a - b)(4ab - 18a - 9b)}$ **51.** $\frac{2x^3 - 6x^2 + x - 4}{16x^3(x - 4)}$

53. $\frac{2x + 1}{x(x + 1)}$ **55.** $\frac{2vt + at^2}{2}$ **57.** $\frac{v_1v_2(d_1 + d_2)}{d_1v_2 + d_2v_1}$

3–6 Exercises, p. 120

1. $\frac{-21}{10}$ **3.** $\frac{16}{3}$ **5.** -12 **7.** $\frac{-16}{3}$ **9.** $\frac{2}{3}$ **11.** 6 **13.** $\frac{1}{6}$ **15.** -3 **17.** no solution

19. $\frac{-1}{4}$ **21.** $\frac{-25}{36}$ **23.** -2 **25.** 1 **27.** no solution **29.** no solution **31.** $\frac{mx}{2x - m}$

33. $M - \frac{m^2d}{2\pi^2r^2}$ **35.** $0.42m^3$ **37.** 6 A **39.** 0.735 Ω **41.** 15.625 Ω **43.** 0.023 μf **45.** 5

47. 4, 6, 8 or $-8, -6, -4$ **49.** $\frac{6}{5}$ h **51.** $\frac{10}{3}$ h

Chapter Review, p. 122

1. $x^2(7x + 38x^6 - 28)$ **3.** $5(3a^2b + 8b^2c - 5ac^3)$ **5.** $8m(4m^7 - 1 - 3m^3)$ **7.** $18x^2(xy^2 + 2z - 4x^3y^3z^2)$ **9.** $(2a + b)(7x - 6y)$ **11.** $6(m - 3n)(2x^2 + 1)$ **13.** $3r^2(-5s - 6t)$ **15.** $4r^2(6rx - 2ry + y + 4z)$ **17.** $(4x + 5y)(4x - 5y)$ **19.** prime **21.** $(13s + 11t)(13s - 11t)$ **23.** $(5c^2 + 6d^2)(5c^2 - 6d^2)$

25. $\frac{1}{3}\pi(r_1 + r_2)(r_1 - r_1)$ **27.** $8p(50 - p)$ **29.** prime **31.** $(13a + 5)^2$ **33.** $(7m - 8n)^2$

35. $(10s^2 - 13t^2)^2$ **37.** $(5x - 6y)(2x + 7y)$ **39.** $(2a + 11b)(a - 9b)$ **41.** $(m + 13)(m + 3)$ **43.** $(8s - 7t)(4s + 5t)$ **45.** $(8x + 5y)(9x - 8y)$ **47.** $(9s^2 - 5y^2)(2s^2 - 7y^2)$ **49.** $16(t^2 - 3t + 4)$

51. $\frac{m^3}{2n^2}$ **53.** $\frac{2x}{y(x^2y^2 - 3)}$ **55.** $\frac{8(x + 1)}{5x + 6}$ **57.** $-\frac{x + 1}{5 + 4x}$ **59.** $\frac{7}{12x^3y^2}$ **61.** $\frac{5a^7c^7}{9b^3}$ **63.** $\frac{7x - 3}{3x(3x - 4)}$

65. $\frac{(5x - y)(x - y)}{(3x + y)(x + 4y)}$ **67.** $\frac{54x^3}{(3x + 4y)^2}$ **69.** $\frac{28xy^3 + 30y^2 - 25x^2}{10x^2y^3}$ **71.** $\frac{3x + 7}{2(x - 4)}$

73. $\frac{10x^2 - 26x}{(x - 2)(x + 1)(x - 3)}$ **75.** $\frac{4x^2 - 24x + 31}{(3x + 4)(3x - 4)(2x - 1)}$ **77.** $\frac{15x^2 - 33x + 22}{(3x - 4)(x + 5)(x - 1)}$ **79.** $\frac{5}{12y^2}$

81. $\frac{2(3x - 1)}{3xy^2}$ **83.** $\frac{3x^2(x + 2)}{4}$ **85.** $\frac{21x^2(18x^2 + 3x + 18)}{(63x - 4)(3x - 1)(x + 2)}$ **87.** $\frac{(5x + 4)(6x - 5)(3x^2 + 10x + 3}{(x - 3)(x + 2)(17x^2 - x + 4)}$

89. -18 **91.** no solution **93.** $\frac{3}{2}$ **95.** -1 **97.** 10

Chapter Test, p. 123

1. $\frac{-3y^3}{8x}$ **2.** $9x^2y(4x - 3y^2 + 2)$ **3.** $\frac{-6x + 15}{(x - 3)(2x - 3)}$ **4.** $5x(7x - y)(2x - 3y)$ **5.** 5 **6.** $\frac{3 - x}{x - 5}$

7. $4m(2m + 3n)(m - 5n)$ **8.** $\frac{2x}{3}$ **9.** $(7x + 3y)(4x - 9y)$ **10.** $\frac{n + 1}{m(n - 1)}$

11. $e\pi(T_1^2 + T_1^2)(T_1 + T_2)(T_1 - T_2)$ **12.** $2\frac{2}{5}$ h **13.** $\frac{2}{11}$ **14.** $(5x - 2y)^2$ **15.** $8(2x + 5)$, 120 when $x = 5$

16. $\frac{1}{9}$ **17.** $6\frac{1}{4}$ ft

CHAPTER 4

4–1 Exercises, p. 133

1. b^5 **3.** $\frac{1}{m^4}$ **5.** $\frac{x^6}{27y^{12}}$ **7.** $\frac{4}{x^2y^2}$ **9.** $\frac{a^8}{36c^6}$ **11.** 1 **13.** 2 **15.** $3 + y$ **17.** 1 **19.** 1

21. $\frac{y^7z^8}{x^{10}}$ **23.** a^3c^3 **25.** $\frac{p^6}{m^3n^2}$ **27.** $\frac{a^8}{b^8c^6}$ **29.** $\frac{y^{24}}{x^{12}z^{12}}$ **31.** $\frac{8y^{12}}{x^{21}}$ **33.** $\frac{5x^2 + 37x + 68}{(2x + 5)^2} =$ $\frac{(5x + 17)(x + 4)}{(2x + 5)^2}$ **35.** $\frac{ab^2 - a + 1}{a}$ **37.** $\frac{x^2y^3 + 1}{y^3}$ **39.** $\frac{m^4n^3 + 1 + n^3}{n^3}$ **41.** $\frac{-x^2 - 2x + 48}{(x + 1)^2} =$ $-\frac{x^2 + 2x - 48}{(x + 1)^2} = -\frac{(x + 8)(x - 6)}{(x + 1)^2}$ **43.** $\frac{1 + ca^2 + 2abc + b^2c}{(a + b)^2}$ **45.** $\frac{y + x^2}{x^2y}$ **47.** $\frac{a^2}{(2 + ab)^2}$

49. $H = \frac{4mLr}{(r^2 - L^2)^2}$ **51.** $F = \frac{f_1f_2}{f_1 + f_2 - d}$ **53.** $A = \frac{y}{(1 + n)^x}$

4–2 Exercises, p. 137

1. 4 **3.** 4 **5.** 6 **7.** -2 **9.** 6 **11.** $\frac{15\sqrt[3]{3}}{4}$ **13.** $\frac{3}{32}$ **15.** $\frac{7}{2}$ **17.** 27.35

19. 0.14 **21.** 0.52 **23.** 0.78 **25.** $\frac{1}{x^{17/12}}$ **27.** $a^{41/35}$ **29.** $\frac{m^{1/2}}{n^{3/2}}$ **31.** $\frac{1}{y^{5/8}}$ **33.** $\frac{1}{b^{7/3}}$ **35.** $x^{9/8}$

37. $\frac{1}{a^{47/21}}$ **39.** $\frac{y^{7/4}}{x^{5/12}}$ **41.** 81 **43.** $\frac{1}{5}$ **45.** 6 **47.** $\frac{3x + 4}{(x - 1)^{1/4}}$ **49.** $\frac{-5y - 12}{(y + 2)^{3/5}}$ **51.** $\frac{2mL}{\sqrt{(r^2 + L^2)^3}}$

53. \$10,473.05 **55.** 0.53

4–3 Exercises, p. 143

1. $4a^2b\sqrt{5b}$ **3.** $5xy^2\sqrt{6y}$ **5.** $2xy\sqrt[3]{4x^2y^2}$ **7.** $9mn^2y^4\sqrt{2mn}$ **9.** $11b^2c^6\sqrt{3ab}$ **11.** $2xy\sqrt[5]{3y^3z^2}$

13. $\sqrt[6]{y}$ **15.** $\sqrt[6]{4x}$ **17.** $\frac{2a^3\sqrt{2a}}{b^3}$ **19.** $4xz^2\sqrt[3]{3x^2y^2z^2}$ **21.** $2mnp^3\sqrt[4]{5n^3p^3}$ **23.** $2km^2\sqrt[5]{2m^2}$

25. $\frac{\sqrt{102}}{6}$ **27.** $\frac{\sqrt{105}}{5}$ **29.** $\frac{\sqrt{138ab}}{3}$ **31.** $\frac{a\sqrt{14a}}{4}$ **33.** $\frac{9m^2\sqrt{7n}}{7n}$ **35.** $\frac{\sqrt{a}}{2}$ **37.** $\frac{\sqrt{2b}}{2}$

39. $\frac{2m\sqrt{70n}}{7n^2}$ **41.** $\frac{m\sqrt[3]{12}}{2}$ **43.** $\frac{2m\sqrt[3]{12m^2n^2}}{3n^2}$ **45.** $\frac{3\sqrt{2ab}}{2b}$ **47.** $\frac{6xy\sqrt{10yz}}{5z^2}$ **49.** $\frac{6m^3\sqrt[3]{n}}{n^2}$

51. already simplified **53.** $x + 4$ **55.** $y + \frac{5}{2}$ **57.** $\frac{(4m - 1)\sqrt{2}}{2}$ **59.** $6\sqrt{30}$ rad/s **61.** 433.01 m^2

4–4 Exercises, p. 150

1. $6\sqrt{2}$ **3.** $\sqrt{5}$ **5.** $15\sqrt{3} - 12\sqrt{5}$ **7.** $17\sqrt{2} - 6\sqrt{3}$ **9.** $5a^2\sqrt{10a} + 18a\sqrt{10}$

11. $(4ab^2 - 6a^2b)\sqrt{a}$ **13.** $-\sqrt[3]{2}$ **15.** $17\sqrt[3]{3}$ **17.** $60x\sqrt{2}$ **19.** $\frac{11\sqrt{2}}{a}$

21. $\frac{(3 - 2a^2)\sqrt{2a} + 12a\sqrt{a}}{a^2}$ **23.** $\sqrt{21} - 5\sqrt{3}$ **25.** $8\sqrt{3} - 8\sqrt{30}$ **27.** $\sqrt{6} + 2\sqrt{3} - \sqrt{15} - \sqrt{30}$

29. $\sqrt[3]{20x^2} + 7\sqrt[3]{10x} + 6\sqrt[3]{2x} + 42$ **31.** $8\sqrt{15} - 24\sqrt{35} + 2\sqrt{21} - 42$ **33.** $86 + 13\sqrt{3}$

35. $6\sqrt[4]{40} + 14\sqrt[4]{30} - 24\sqrt[4]{6} - 28\sqrt[4]{72}$ **37.** $56m + 22\sqrt{70mn} + 120n$ **39.** $240 - 80\sqrt{5}$ **41.** $157 - 84\sqrt{3}$

43. 74 **45.** $\frac{\sqrt{7}}{7}$ **47.** $\frac{2\sqrt{30}}{5}$ **49.** $\frac{2\sqrt{7} + 3\sqrt{14}}{-42}$ **51.** $\frac{12a + 8\sqrt{ab} - 3b\sqrt{a} - 2b\sqrt{b}}{16a - b^2}$

4–5 Exercises, p. 154

1. 8 **3.** 4 **5.** $\frac{17}{5}$ **7.** no solution **9.** 15 **11.** 4 **13.** -5 **15.** 5 **17.** $\frac{2}{3}$ **19.** no solution

21. 12 **23.** no solution **25.** no solution **27.** 3 **29.** $\frac{13}{5}$ **31.** $\frac{-8}{3}$ **33.** 0 **35.** no solution

37. -1 **39.** 0 **41.** $-\frac{1}{2}$ **43.** -2 **45.** no solution **47.** $\frac{v^2m}{3k}$ **49.** $\frac{4}{3}\pi r^3$ **51.** $\frac{v_0^2 - v^2}{2a}$

Chapter Review, p. 156

1. $\frac{8y^8}{x^3}$ **3.** $\frac{2n^{11}}{3m^2}$ **5.** $\frac{3p^9}{m^2n}$ **7.** $\frac{a^{12}}{b^9c^{15}}$ **9.** $\frac{y^2 + 3x}{xy^2}$ **11.** 27 **13.** $\frac{1}{2}$ **15.** $\frac{-19}{4}$ **17.** $6^{5/6}$ or 4.45

19. 2.83 **21.** $\frac{m^{11/6}}{n^{8/5}}$ **23.** $\frac{y^{2/3}z^2}{x^{13/4}}$ **25.** $6xy^2\sqrt{5y}$ **27.** $3x^2y^4\sqrt{6x}$ **29.** $\frac{2xy\sqrt{2y}}{z^3}$ **31.** $\frac{3xy^2\sqrt{3xyz}}{4z^3}$

33. $\frac{a^2b^4\sqrt{abc}}{2c}$ **35.** $\frac{(2x + 5)\sqrt{2}}{2}$ **37.** $-3\sqrt{2}$ **39.** $-9\sqrt{3} + 12\sqrt{2}$ **41.** $8 - 3\sqrt{6} + 24\sqrt{2}$

43. $3\sqrt{10} + 12\sqrt{6}$ **45.** $3 + 2\sqrt{6} - 3\sqrt{15} - 6\sqrt{10}$ **47.** -84 **49.** $\frac{2}{x}$ **51.** $\frac{3\sqrt{21}}{7}$

53. $\frac{3x - 10\sqrt{x} + 8}{x - 4}$ **55.** $\frac{2m - 3\sqrt{mn} + n}{m - n}$ **57.** 9 **59.** 11 **61.** $\frac{8}{3}$ **63.** no solution **65.** $\frac{45}{7}$

67. no solution **69.** no solution

Chapter Test, p. 157

1. $11\sqrt{3} - 14\sqrt{6}$ **2.** $3xy^2\sqrt[3]{4x^2y^2z}$ **3.** $\frac{33}{4}$ **4.** $\frac{2xy\sqrt[3]{2xz^2}}{z^2}$ **5.** $\frac{M}{LT^2}$ **6.** $x = \frac{5}{3}$ **7.** $\frac{3y^5}{5x^5z^4}$

8. $-14 - 11\sqrt{10}$ **9.** $6x^2y\sqrt{6xyz}$ **10.** $P = \$13,308.05$ **11.** $\frac{x^4z\sqrt{2xy}}{6y^3}$ **12.** $12\sqrt{6} - 54\sqrt{2}$

13. $\frac{81x^8}{4y^{14}}$ **14.** $\frac{1}{27}$ **15.** $n = 144.53$ **16.** $\frac{1}{3}$ **17.** no solution

CHAPTER 5

5–1 Exercises, p. 167

1. $-6, 3$ **3.** 5, 4 **5.** $\frac{3}{2}, -7$ **7.** $0, \frac{9}{2}$ **9.** $\frac{3}{5}, -5$ **11.** $\frac{9}{2}, -\frac{9}{2}$ **13.** $-\frac{1}{5}, -\frac{2}{3}$ **15.** $-9, 8$

17. $-\frac{3}{7}, \frac{5}{4}$ **19,** 0, 4 **21.** $\frac{5}{4}, -\frac{5}{4}$ **23.** $-\frac{5}{3}, -\frac{3}{2}$ **25.** $-3, -1$ **27.** $-1, -4$ **29.** $-6, 2$

31. $\frac{1}{6}, 1$ **33.** $6, -4$ **35.** $3, -2$ **37.** $9, -6$ **39.** 3 **41.** -2 **43.** $\frac{3}{2}$ **45.** $\frac{5}{4}$ **47.** 24 h

49. $\frac{7}{5}$ A

5–2 Exercises, p. 173

1. $2 \pm 2\sqrt{6}$ **3.** $-3 \pm \sqrt{23}$ **5.** $3, -1$ **7.** $\frac{-5 \pm \sqrt{57}}{2}$ **9.** $\frac{9 \pm \sqrt{113}}{2}$ **11.** $\frac{-4 \pm \sqrt{42}}{2}$

13. $\frac{3 \pm j\sqrt{10}}{2}$ **15.** $\frac{9 \pm \sqrt{249}}{6}$ **17.** $\frac{-10 \pm 2\sqrt{55}}{5}$ **19.** $\frac{21 \pm j\sqrt{35}}{14}$ **21.** $\frac{-5 \pm \sqrt{21}}{2}$

23. $\frac{9 \pm \sqrt{65}}{4}$ **25.** $\frac{-7 \pm j\sqrt{23}}{6}$ **27.** $\frac{17 \pm j\sqrt{31}}{8}$ **29.** $\frac{3 \pm \sqrt{107}}{7}$ **31.** 0.03 A, -0.63 A

33. 12, 16, 20 **35.** 750, 250

5–3 Exercises, p. 181

1. $\frac{3}{4}, -2$ **3.** $2 \pm j$ **5.** $\frac{5 \pm \sqrt{457}}{18}$ **7.** $\frac{9 \pm j\sqrt{47}}{8}$ **9.** $\frac{1 \pm j\sqrt{439}}{20}$ **11.** $\frac{7 \pm j\sqrt{7}}{4}$

13. $\frac{-5 \pm \sqrt{193}}{4}$ **15.** $\frac{\pm 3\sqrt{2}}{4}$ **17.** $0, \frac{15}{8}$ **19.** $\frac{3 \pm j\sqrt{31}}{5}$ **21.** $\frac{13 \pm \sqrt{145}}{4}$ **23.** $\frac{27 \pm \sqrt{985}}{16}$

25. $\frac{11 \pm \sqrt{109}}{2}$ **27.** $\frac{1 \pm \sqrt{1697}}{16}$ **29.** $\frac{-1 \pm \sqrt{6}}{3}$ **31.** $-2 \pm \sqrt{7}$ **33.** $8 \pm 5\sqrt{2}$ **35.** $\frac{9 \pm \sqrt{65}}{2}$

37. no real solution **39.** $\frac{9 - \sqrt{193}}{8}$ is only solution **41.** 2 real **43.** 2 real **45.** 2 real **47.** 2 real

49. 2 complex **51.** 2 real **53.** 2 real **55.** 2 complex **57.** $\frac{-\pi h \pm \sqrt{\pi^2 h^2 + 2\pi s}}{2\pi}$ **59.** 496.64

5–4 Exercises, p. 187

1. $x < -3$ or $x > 2$

3. $-6 \leq x \leq -\frac{1}{3}$

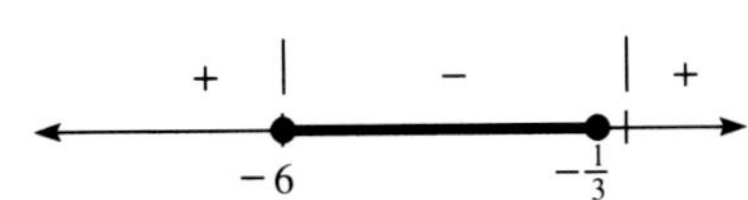

5. $x \leq 0$ or $x \geq 3$

7. $x < -\frac{5}{3}$ or $x > \frac{2}{7}$

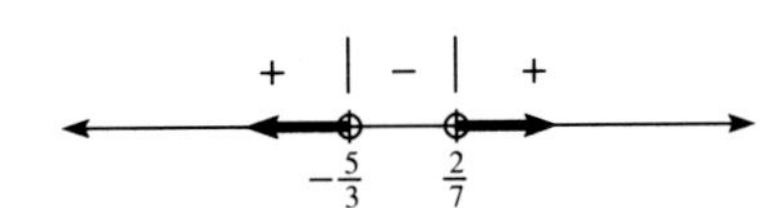

9. $x \leq -\frac{9}{7}$ or $x \geq \frac{9}{7}$

11. $-8 < x < \frac{5}{6}$

13. $-7 \leq x \leq -4$

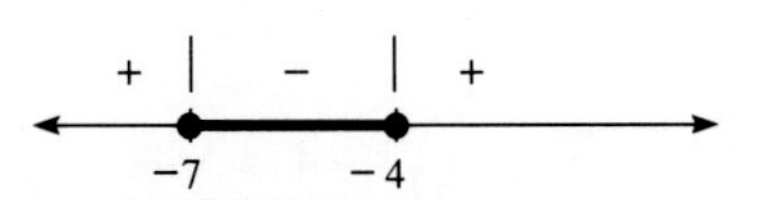

15. $-\frac{1}{2} \leq x \leq 0$ or $x \geq 5$

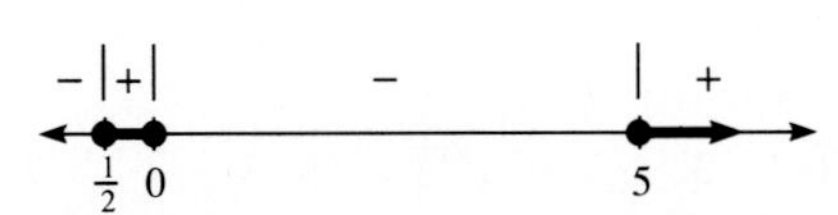

17. $x \leq -2$ or $0 \leq x \leq 2$

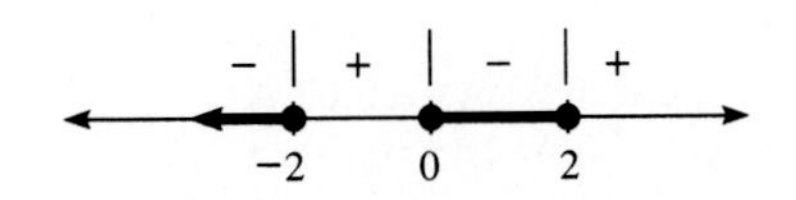

19. $x < -4$ or $x > \frac{1}{2}$

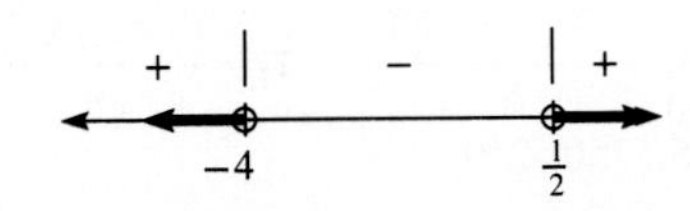

21. $-9 \le x < -3$ or $x \ge 1$

23. $x < -2$ or $0 < x < 3$

25. $x \le -6$ or $-1 < x \le \frac{4}{3}$

27. $v_0 > 149.9$

29. $l < 8$ m

5–5 Exercises, p. 194

1. $\frac{5}{2}$ s **3.** 3 ft **5.** -4 or -2 **7.** 17.64 d **9.** \$4.16 or \$4.40 **11.** 0.56 cm **13.** 1 cm

15. $-3, -1; 1, 3$ **17.** 20.84 ft **19.** 2 **21.** $6\frac{1}{4}$ units **23.** 10 and 10 **25.** 9.65 m

Chapter Review, p. 196

1. $\frac{2}{3}, -6$ **3.** $-4, 4$ **5.** $-\frac{1}{2}, -5$ **7.** $0, \frac{5}{7}$ **9.** $-\frac{5}{4}, 4$ **11.** $-10, 2$ **13.** $\frac{1}{2}$ **15.** $\frac{5}{3}, -\frac{3}{8}$

17. $-3 \pm \sqrt{21}$ **19.** 5, 0 **21.** $\frac{3 \pm \sqrt{37}}{2}$ **23.** $\frac{-7 \pm j\sqrt{155}}{6}$ **25.** $\frac{-5 \pm 7\sqrt{5}}{10}$ **27.** $\frac{-1 \pm \sqrt{3}}{3}$

29. $\frac{1 \pm j}{2}$ **31.** $\frac{\pm 2\sqrt{7}}{7}$ **33.** $\frac{6 \pm j\sqrt{29}}{5}$ **35.** $0, -\frac{8}{9}$ **37.** $\frac{-1 \pm \sqrt{229}}{6}$ **39.** $\frac{-3 \pm j\sqrt{19}}{2}$

41. $\pm 3\sqrt{2}$ **43.** $\frac{6 \pm 2j}{5}$ **45.** $\frac{3 \pm \sqrt{33}}{8}$ **47.** 2 real **49.** 2 complex **51.** 2 real **53.** 2 real

55. $x \le -1$ or $x \ge \frac{3}{4}$

57. $x < -5$ or $0 < x < \frac{3}{2}$

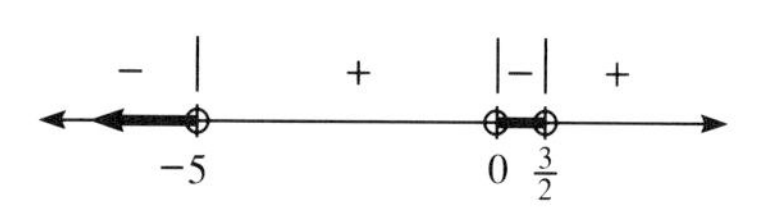

59. $-1 < x < 3$

61. $-3 \le x \le 1$ or $x > 2$

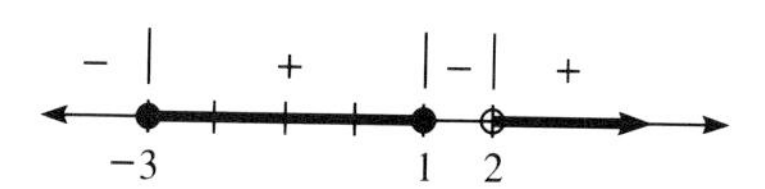

63. 18, 19 or $-19, -18$ **65.** $R_1 = 20\ \Omega$, $R_2 = 5\ \Omega$; $R_1 = 5\ \Omega$, $R_2 = 20\ \Omega$ **67.** 17.64 h, 14.64 h

Chapter Test, p. 197

1. $-\frac{4}{7}, 3$ **2.** 2 real **3.** $\frac{-13 \pm \sqrt{163}}{2}$ **4.** $-12, -10; 10, 12$ **5.** $\frac{7}{5}, \frac{3}{2}$ **6.** $\frac{-3 \pm \sqrt{35}}{2}$

7. $\dfrac{-13 \pm \sqrt{357}}{8}$ **8.** 8 ft **9.** $\dfrac{\pm\sqrt{14}}{3}$ **10.** 2 complex **11.** $\dfrac{5}{3}$, 1 **12.** 26.17 min, 22.17 min

13. $\dfrac{6}{7}, -\dfrac{4}{3}$ **14.** $-\dfrac{3}{2}, -3$ **15.** $0, \dfrac{8}{5}$ **16.** 0.9 s **17.** $\dfrac{3}{4}, -\dfrac{7}{2}$ **18.** 2 complex **19.** $-8 \pm \sqrt{82}$

20. $-3 \le x \le \dfrac{3}{2}$

CHAPTER 6

6–1 Exercises, p. 207

1. not a solution **3.** is a solution **5.** not a solution **7.** is a solution **9.** is a solution

11.

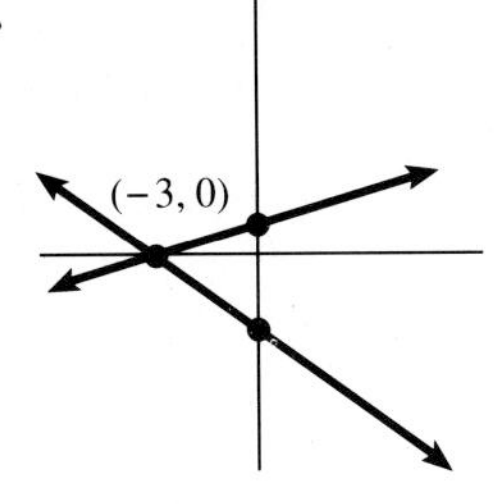

13.

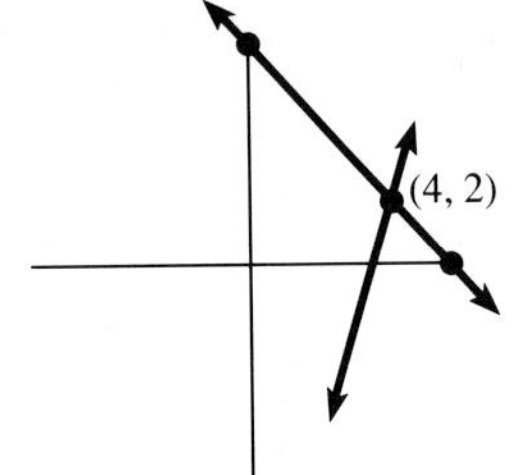

15.

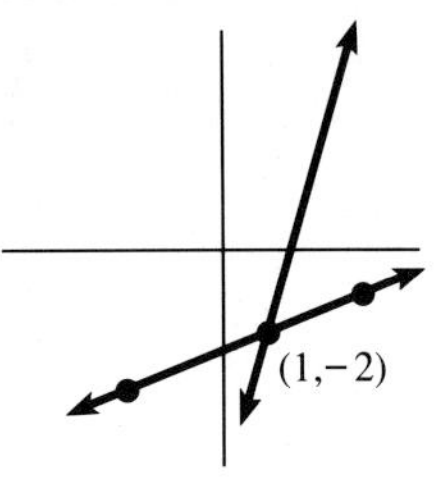

17.

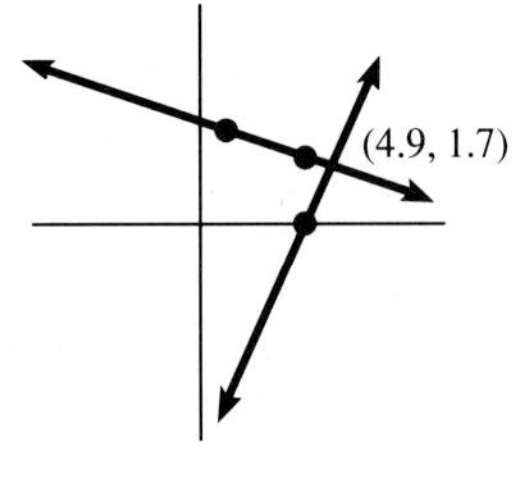

19.

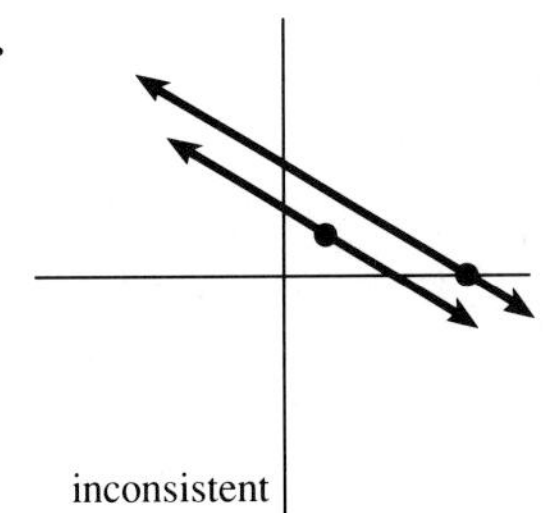

21.

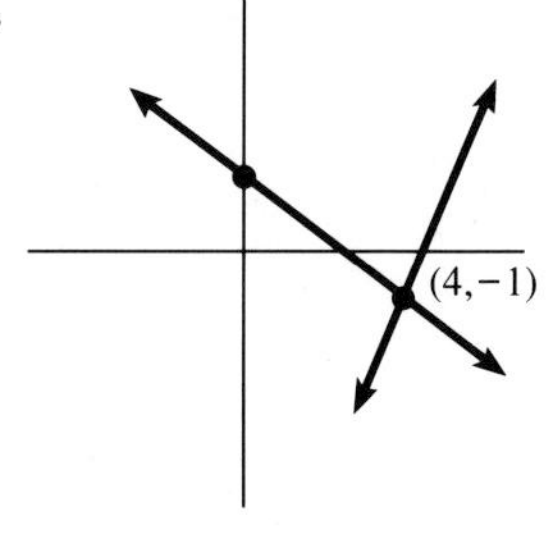

23.

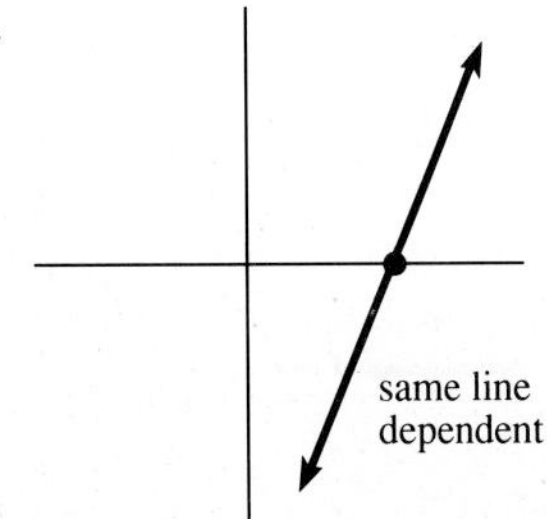

25.

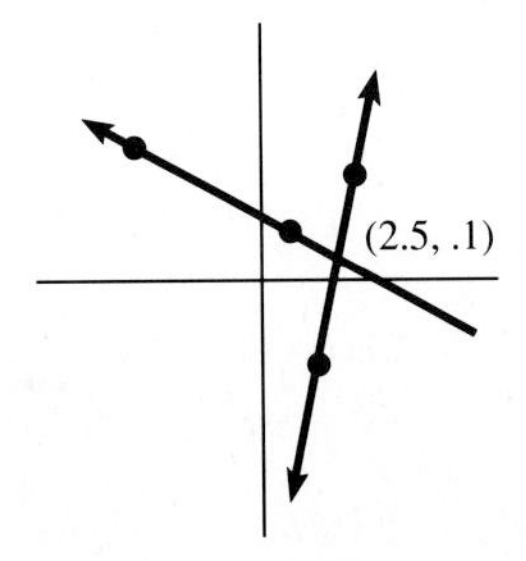

27. A = 103.5 tons, B = 113.5 tons

29.

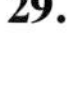

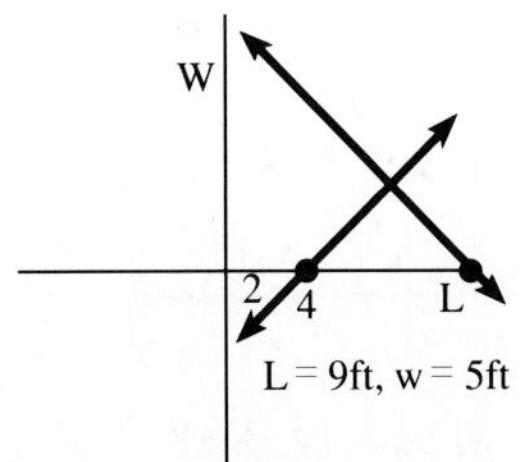

6–2 Exercises, p. 216

1. $x = 3, y = -1$ **3.** $x = -\frac{5}{2}, y = 2$ **5.** $x = -3, y = -\frac{1}{2}$ **7.** $y = -6, x = 0$ **9.** $x = -1, y = 4$ **11.** $x = -30, y = -40$ **13.** $y = -2, x = 2$ **15.** $x = -2, y = -3$ **17.** $x = \frac{38}{17}, y = \frac{14}{17}$ **19.** $y = 6, x = 2$ **21.** $x = -3, y = -7$ **23.** no solution **25.** $x = \frac{1}{2}, y = -1$ **27.** $x = 3, y = -2$ **29.** $x = \frac{100}{19}, y = \frac{18}{19}$ **31.** $2x + y = 8$ **33.** 2 lb nectarines, 3 lb apples **35.** \$6076.92 @ 7%, \$6923.08 @ 8.3%

6–3 Exercises, p. 224

1. $x = 0, y = 1, z = -2$ **3.** $x = 3, y = -2, z = 0$ **5.** $x = 2, y = 1, z = -2$ **7.** $x = 3, y = 0, z = 1$ **9.** $x = 0, y = 2, z = 0$ **11.** $x = -1, y = 0, z = -2$ **13.** $x = 2, y = 3, z = -2$ **15.** $x = 0, y = 1, z = 3$ **17.** $x = -2, y = 1, z = 2$ **19.** $x = -4, y = 1, z = -3$ **21.** $x = \frac{-9}{4}, y = \frac{9}{4}, z = \frac{-1}{2}$ **23.** $x = -2, y = -1, z = 0$ **25.** $x = \frac{1}{2}, y = -\frac{1}{3}, z = 0$ **27.** $x = 0, y = \frac{1}{5}, z = -2$ **29.** $I_1 = \frac{27}{28}$ A, $I_2 = \frac{18}{28}$ A, $I_3 = \frac{9}{28}$ A **31.** $x = 17, y = 24, z = 36$ **33.** $y = 3x^2 - 2x + 5$ **35.** Type 1 = 13, Type 2 = 22, Type 3 = 16

6–4 Exercises, p. 231

1. $x = -\frac{49}{25}, y = \frac{14}{5}; x = -1, y = -2$ **3.** no real solution **5.** $x = 1, y = \sqrt{3}; x = 1, y = -\sqrt{3}$ **7.** $x = 4, y = 2; x = -4, y = 2$ **9.** no real solution **11.** no real solution **13.** $x = \frac{\sqrt{10}}{2}, y = \frac{\sqrt{6}}{2}; x = \frac{\sqrt{10}}{2}, y = \frac{-\sqrt{6}}{2}; x = \frac{-\sqrt{10}}{2}, y = \frac{\sqrt{6}}{2}; x = \frac{-\sqrt{10}}{2}, y = \frac{-\sqrt{6}}{2}$ **15.** no solution **17.** $x = \sqrt{6}, y = \frac{\sqrt{10}}{2}; x = \sqrt{6}, y = \frac{-\sqrt{10}}{2}; x = -\sqrt{6}, y = \frac{\sqrt{10}}{2}; x = -\sqrt{6}, y = \frac{-\sqrt{10}}{2}$ **19.** no real solution **21.** $-8, -3; 8, 3$ **23.** $R_1 = 5\ \Omega, R_2 = 3\ \Omega; R_1 = 3\ \Omega, R_2 = 5\ \Omega$ **25.** $q = 18, p = \$3.00$

6–5 Exercises, p. 236

1. 10 **3.** 62 **5.** −1 **7.** −17.02 **9.** −21 **11.** −11 **13.** 14 **15.** 0 **17.** −36.162 **19.** 13 **21.** −42 **23.** 16 **25.** 8 **27.** −33 **29.** −18

6–6 Exercises, p. 243

1. $x = 0, y = -4$ **3.** $x = 2, y = -3$ **5.** $x = 6, y = 0$ **7.** $x = -9.0, y = 4.1$ **9.** $x = 0, y = 0$ **11.** $x = 3, y = 6$ **13.** $x = \frac{1}{3}, y = -\frac{1}{2}$ **15.** $x = \frac{3}{5}, y = \frac{1}{4}$ **17.** $x = \frac{22}{89}, y = \frac{305}{89}, z = -\frac{192}{89}$

19. $x = 1.69, y = -0.97, z = -1.09$ **21.** $x = 0, y = 0, z = -1$ **23.** $x = \frac{84}{19}, y = \frac{102}{19}, z = -\frac{34}{19}$ **25.** $x = 4, y = -2, z = 5$ **27.** $x = \frac{1}{2}, y = \frac{2}{3}, z = -1$ **29.** $x = 0, y = 3, z = \frac{3}{4}$ **31.** W = 23 ft, L = 37 ft **33.** 48 is largest **35.** 65 children's tickets

6–7 Exercises, p. 252

1. $x = \frac{4}{3}\text{h}, y = \frac{2}{3}\text{h}$ **3.** 160 ml of 5%, 240 ml of 30% **5.** $\frac{16x + 3}{(2x - 2)(x + 3)} = \frac{19/4}{2x - 2} + \frac{45/8}{x + 3}$ **7.** 27 yd × 34 yd **9.** $B = 117.5$ N, $A = 207.5$ N **11.** $33\frac{1}{3}$ yd^3 of 25%, $66\frac{2}{3}$ yd^3 of 40% **13.** Rate of plane = 350 mph, Wind = 50 mph **15.** fixed = \$200, variable = \$1.80 **17.** $\frac{-2x - 42}{(x + 3)(x - 1)} = \frac{9}{x + 3} - \frac{11}{x - 1}$ **19.** $I_1 = 0.58$ A, $I_2 = 0.45$ A, $I_3 = 0.13$ A **21.** $I_2 = 0.4$ A, $I_1 = 1$ A, $I_3 = 0.6$ A **23.** 12 dimes, 18 quarters

Chapter Review, p. 255

1. No, it does not satisfy the 2nd equation.

3.

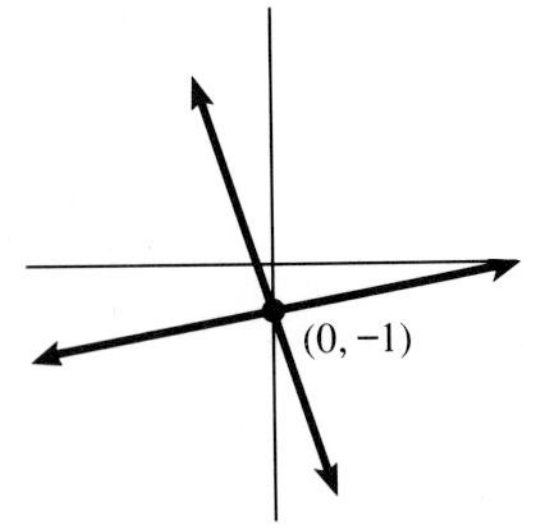

5.

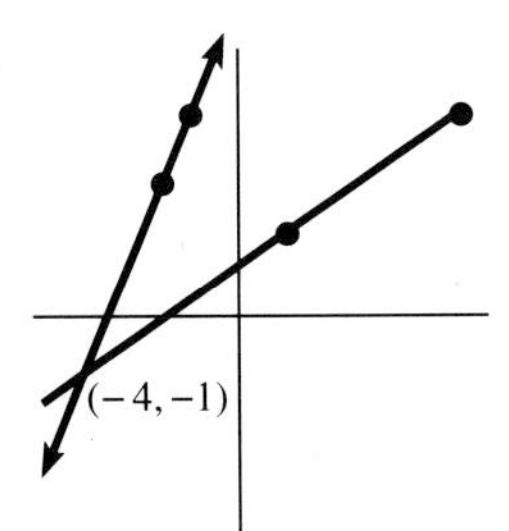

7.

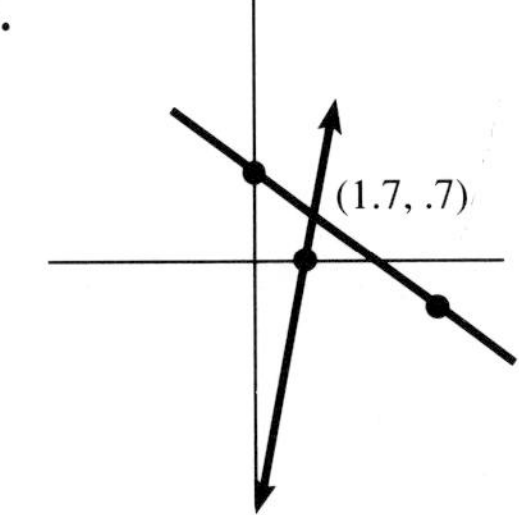

9. $x = -3, y = 6$ **11.** $x = \frac{-8}{19}, y = \frac{7}{19}$ **13.** $x = 2, y = 2$ **15.** no solution, dependent **17.** \$6,000 @ $8\frac{1}{2}$%, \$9,000 @ 9.75% **19.** $x = -2, y = 1, z = 3$ **21.** $x = 0, y = 4, z = 0$ **23.** $x = -1, y = 2, z = 3$ **25.** $x = 0, y = -3, z = 1$ **27.** 32°, 105°, 43° **29.** $x = 0, y = -8; x = 2, y = -4$ **31.** $x = 2, y = 0; x = -2, y = 0$ **33.** no solution **35.** $-2a - 3c$ **37.** 36 **39.** 8 **41.** 145 **43.** $x = 5, y = 0$ **45.** $x = -\frac{2}{3}, y = 4$ **47.** $x = 4, y = -3, z = 2$ **49.** $x = 18, y = 37$ **51.** $I_1 = 1.06$ A, $I_2 = 0.46$ A, $I_3 = 0.6$ A **53.** $\frac{34x - 18}{(4x - 3)(2x + 1)} = \frac{3}{4x - 3} + \frac{7}{2x + 1}$

Chapter Test, p. 257

1. $x = 6, y = -7$ **2.** -30 **3.** It is a solution, it satisfies both equations. **4.** $x = 40°, y = 40°, z = 100°$

5.

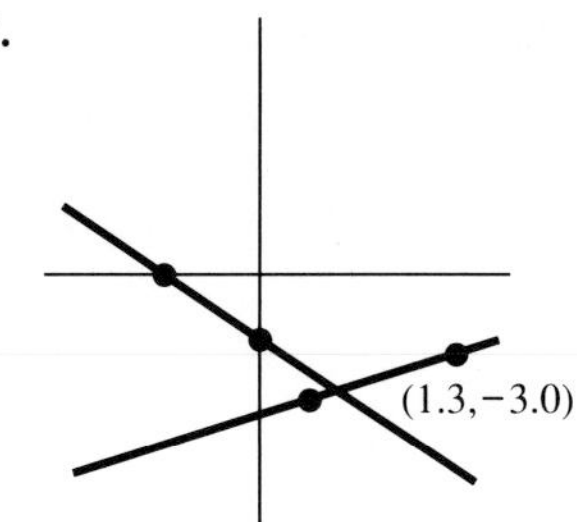

6. $x = 1, y = 3, z = 2$ **7.** $x = -1, y = 8$

8. 23 lb @ \$1.99, 27 lb @ \$3.49 **9.** 137 **10.** $x = \frac{141}{41}, y = \frac{-22}{41}$

11. $x = 0, y = 4; x = 0, y = -4$ **12.** $A = 240, B = 135$

13. no solution

14.

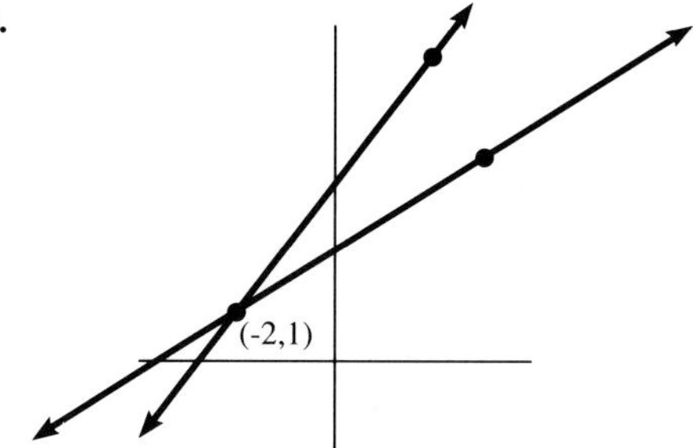

15. $x = \sqrt{13}, y = 2; x = \sqrt{13}, y = -2; x = -\sqrt{13}, y = 2; x = -\sqrt{13}, y = -2$

CHAPTER 7

7–1 Exercises, p. 264

1. $f(2) = 15$ **3.** $f(-3) = -189$ **5.** $f\left(\frac{1}{3}\right) = 17$ **7.** $f(1) = 3$ **9.** $f(-3) = 79$ **11.** $f(-2) = 71$ **13.** $f\left(-\frac{2}{3}\right) = -2$ **15.** $f(-1) = 8$ **17.** $f(2) = 8$ **19.** $f(4) = 37$ **21.** $f(-5) = -793$ **23.** $f(-2) = 7$ **25.** $f(1) = 15$ **27.** $f\left(\frac{1}{2}\right) = -\frac{57}{8}$ **29.** $f\left(\frac{2}{3}\right) = -\frac{34}{9}$ **31.** not a factor **33.** is a factor **35.** is a factor **37.** not a factor **39.** not a factor **41.** not a factor **43.** is a factor **45.** not a zero **47.** is a zero **49.** is a zero **51.** is a zero **53.** not a zero **55.** is a zero **57.** not a zero **59.** is a root **61.** is a root **63.** not a root **65.** not a root **67.** not a root **69.** not a root

7–2 Exercises, p. 270

1. $x^2 + 9x + 26 + \frac{87}{x - 3}$ **3.** $2x^2 - 7x + 10 - \frac{2}{x + 1}$ **5.** $3x^2 + 2x + 7 + \frac{12}{x - 3}$

7. $4x^2 + 3x + 9 + \frac{2}{x - 1}$ **9.** $x^2 + 5x + 10 + \frac{10}{x - 2}$ **11.** $9x^2 + 36x + 136 + \frac{555}{x - 4}$

13. $3x^3 + 13x^2 + 72x + 359 + \frac{1804}{x - 5}$ **15.** $x^3 - 2x^2 - 3x + 6 - \frac{4}{x + 2}$

17. $x^3 - 2x^2 + 3x - 7 + \frac{15}{x + 3}$ **19.** $2x^4 + 2x^3 - 4x^2 - 4x - 5$ **21.** $x^4 - 3x^3 + 12x^2 - 56x + 224 - \frac{890}{x + 4}$

23. $3x^5 + 6x^4 + 12x^3 + 21x^2 + 42x + 84 + \frac{176}{x - 2}$ **25.** is a factor **27.** is a factor **29.** not a factor **31.** is a factor **33.** not a factor **35.** not a factor **37.** is a factor **39.** not a factor **41.** is a factor **43.** not a zero **45.** is a zero **47.** is a zero **49.** not a zero

7–3 Exercises, p. 274

1. $-3 \pm j\sqrt{5}, 4$ **3.** $\frac{3 \pm j\sqrt{47}}{4}, -3$ **5.** $\frac{-3 \pm \sqrt{137}}{8}, 1$ **7.** $\frac{1 \pm j\sqrt{119}}{12}, -2$ **9.** $\frac{7 \pm \sqrt{13}}{2}, 5$ **11.** $\frac{7 \pm j\sqrt{39}}{4}, 2$ **13.** $-4, 3, -1, -2$ **15.** $4, -2, -2, -2$ **17.** $-7, -1, 3, 3$ **19.** $-2, -2, 5, 3$ **21.** $-2, 2, 3j, -3j$ **23.** $-1, -1, -1, 4$ **25.** $\frac{-3 \pm \sqrt{13}}{2}, -2, 3$ **27.** $5, 1 + j, 1 - j$ **29.** $\frac{-1 \pm \sqrt{13}}{2}, 2j, -2j$

7–4 Exercises, p. 278

1. $\pm 1, \pm 2, \pm 5, \pm 10$ **3.** $\pm 1, \pm 2, \pm 3, \pm 4, \pm 6, \pm 8, \pm 12, \pm 24$ **5.** $\pm 1, \pm 2, \pm 3, \pm 4, \pm 6, \pm 9, \pm 12, \pm 18, \pm 36$ **7.** $\pm 1, \pm 2, \pm 3, \pm 6, \pm 9, \pm 18, \pm \frac{1}{3}, \pm \frac{2}{3}$ **9.** $\pm 1, \pm \frac{7}{2}, \pm 7, \pm \frac{1}{2}, \pm 14, \pm 2$ **11.** $\pm 1, \pm 2, \pm 3, \pm 4, \pm 6, \pm 12, \pm \frac{1}{5}, \pm \frac{2}{5}, \pm \frac{3}{5}, \pm \frac{4}{5}, \pm \frac{6}{5}, \pm \frac{12}{5}$ **13.** $\pm 1, \pm 2, \pm 3, \pm 6, \pm 9, \pm 18, \pm \frac{1}{2}, \pm \frac{3}{2}, \pm \frac{9}{2}, \pm \frac{1}{3}, \pm \frac{2}{3}, \pm \frac{1}{6}$ **15.** $\pm 1, \pm 2, \pm 4, \pm 5, \pm 10, \pm 20, \pm \frac{1}{2}, \pm \frac{5}{2}, \pm \frac{1}{4}, \pm \frac{5}{4}$ **17.** 3 positive, 1 negative **19.** 1 positive, 2 negative **21.** 3 positive, 0 negative **23.** 4 positive, 3 negative **25.** 0 positive, 2 negative **27.** 3 positive, 4 negative **29.** $1, \frac{1}{2}, -3$ **31.** $-2, -\frac{2}{3}, 4$ **33.** $-3, \frac{-7 \pm \sqrt{17}}{2}$ **35.** $3, -6, 1$ **37.** $-2, -7, 1$ **39.** $1, -\frac{3}{2}, \pm j$ **41.** $4, -3, -5, 1$ **43.** $-2, 4, 1, \frac{3 \pm j\sqrt{7}}{2}$ **45.** $1, -1, 5, \frac{1 \pm j\sqrt{11}}{2}$ **47.** 5 in. × 5 in. × 8 in. or 10.4 in. × 10.4 in. × 1.8 in. **49.** 2 s **51.** 6 cm × 3 cm × 2 cm or 0.788 cm × 8.424 cm × 5.424 cm

7–5 Exercises, p. 284

1. 4.2 **3.** 2.5 and −1.2 **5.** 0.9 **7.** 2.2 **9.** −5.8, −1.6 **11.** −1.6, −5.7 **13.** 0.9, −1 **15.** 0.8, −1.2 **17.** −1.4 **19.** −1.3 **21.** increase by 3 cm **23.** $t = 5$ s

Chapter Review, p. 285

1. −198 **3.** 10 **5.** not a factor **7.** not a factor **9.** not a factor **11.** $6x^2 - 11x + 14 - \frac{15}{x + 1}$ **13.** $x^3 - 5x^2 + 10x - 13 + \frac{18}{x + 2}$ **15.** $4x^3 + 11x^2 + 35x + 105 + \frac{306}{x - 3}$ **17.** is a zero

19. not a zero **21.** is a zero **23.** $-3, \frac{1}{2}, 1$ **25.** $-1, \frac{5 \pm j\sqrt{47}}{4}$ **27.** $\frac{-1}{2}, 3, j\sqrt{3}, -j\sqrt{3}$

29. $5, 1, 1 \pm j\sqrt{7}$ **31.** $\pm 1, \pm 2, \pm\frac{1}{2}, \pm\frac{1}{4}$ **33.** $\pm 1, \pm 2, \pm 3, \pm 4, \pm 6, \pm 12, \pm\frac{1}{3}, \pm\frac{2}{3}, \pm\frac{4}{3}$ **35.** $\pm 1, \pm 3, \pm 5, \pm 15, \pm\frac{1}{5}, \pm\frac{3}{5}$ **37.** 2 positive, 1 negative **39.** 3 positive, 1 negative **41.** 2 positive, 3 negative

43. $2, \frac{-3}{2}, 4$ **45.** $9, \frac{-3}{2}, \frac{-1}{2}$ **47.** $-3, 4, -2$ **49.** $-7, -2, \frac{3}{2}, 1$ **51.** 1.1 **53.** -1.6

Chapter Test, p. 286

1. The remainder is zero, so -2 is a zero of the equation. **2.** $-1, \sqrt{2}, -\sqrt{2}$ **3.** $5, \frac{4}{3}, \frac{-1}{2}$ **4.** $8, 3 \pm j$

5. It is not a factor. **6.** $\pm 1, \pm 2, \pm 3, \pm 6, \pm\frac{1}{2}, \pm\frac{3}{2}$ **7.** $6, \frac{-1}{2}, -2$ **8.** remainder is 52

9. 3 positive, 1 negative **10.** 2 cm × 13 cm × 12 cm or 3.57 cm × 9.87 cm × 8.87 cm

11. $7x^3 - 6x^2 + 6x - 12 + \frac{34}{x + 2}$ **12.** $-7, -0.8$ **13.** 2 positive, 3 negative **14.** $-5, \frac{1}{3}, -\frac{5}{2}$

15. $x^4 + 8x^3 + 16x^2 + 25x + 50 + \frac{108}{x - 2}$ **16.** $\pm 1, \pm 2, \pm 4, \pm 8, \pm 16, \pm\frac{1}{3}, \pm\frac{2}{3}, \pm\frac{4}{3}, \pm\frac{8}{3}, \pm\frac{16}{3}$

CHAPTER 8

8–1 Exercises, p. 295

1.

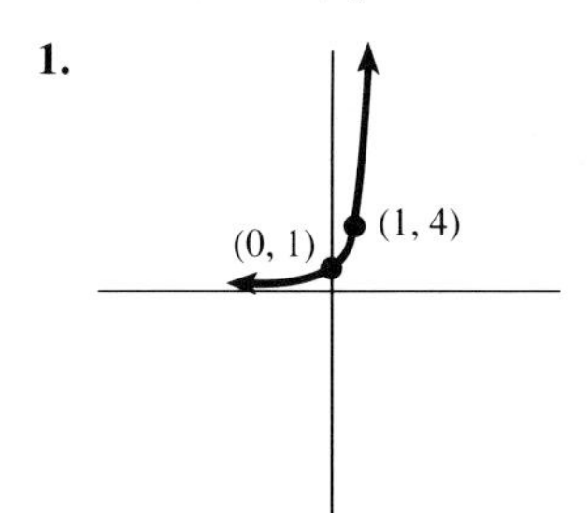

3.

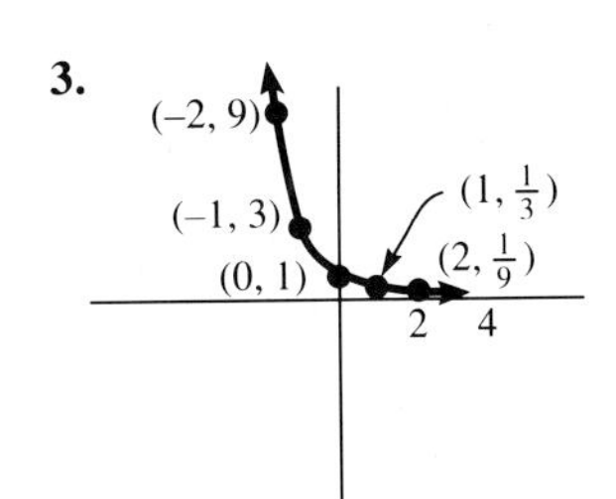

5.

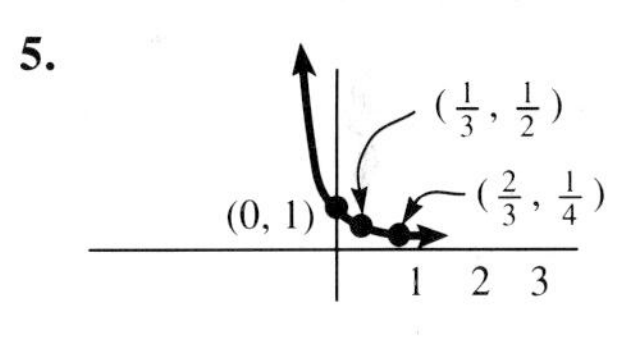

7.

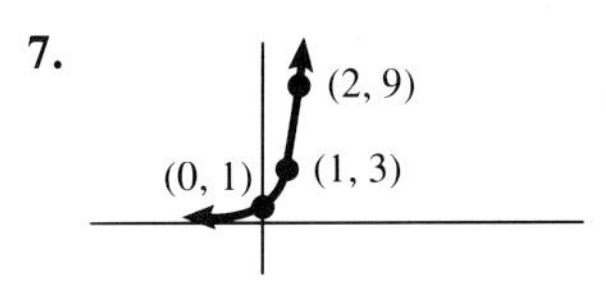

9.

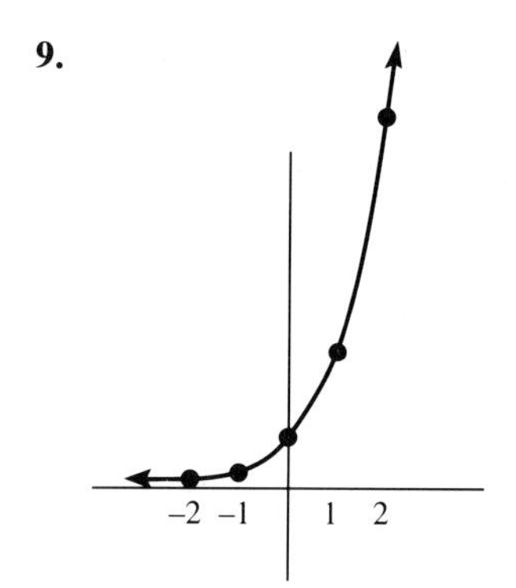

11.

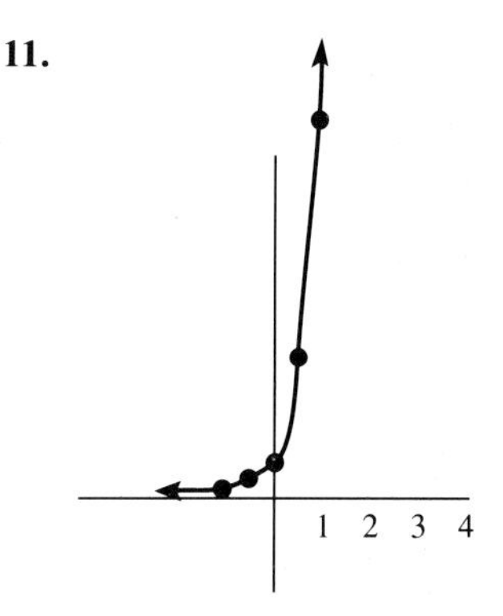

13.

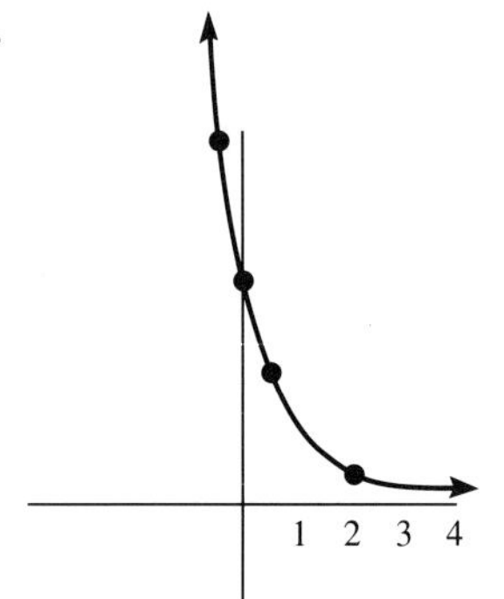

15.

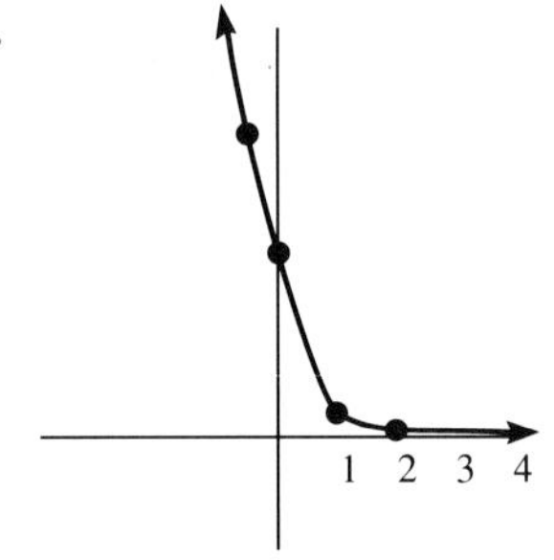

17. 8100 at 4 h

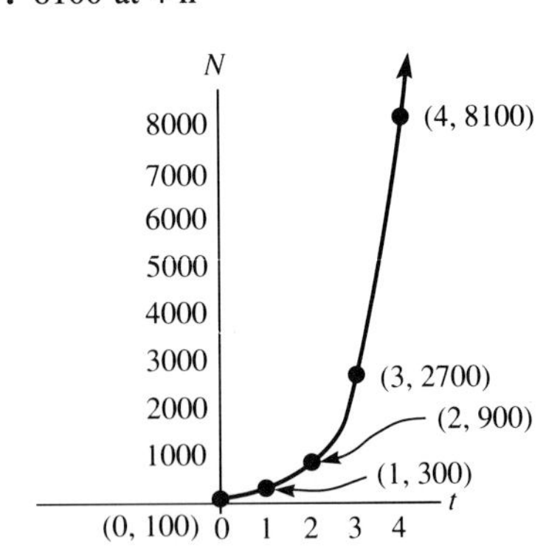

19. \$13,345.74

21. \$17,470.14

23. 4.36×10^{-26} A

25.

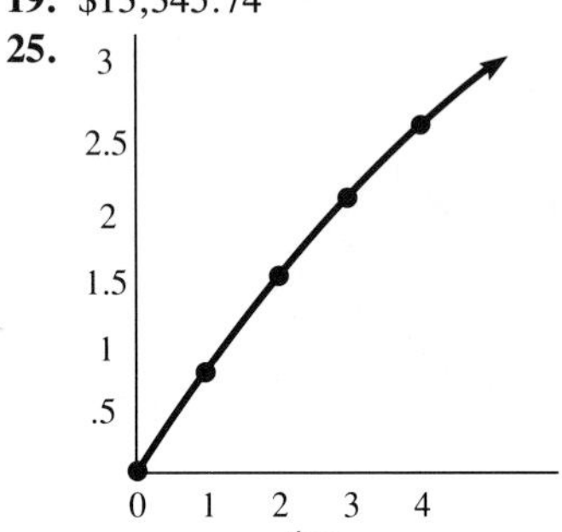

27. 1.14 A

29. \$14,902.95

8–2 Exercises, p. 301

1. $\log_6 216 = 3$ **3.** $\log_2 32 = 5$ **5.** $\log_3\left(\frac{1}{81}\right) = -4$ **7.** $\log_{1/3}\left(\frac{1}{27}\right) = 3$ **9.** $\ln 7.39 = 2$

11. $\log_{16}\left(\frac{1}{8}\right) = -\frac{3}{4}$ **13.** $\ln 0.37 = -1$ **15.** $\log_8\left(\frac{1}{4}\right) = -\frac{2}{3}$ **17.** $3^3 = 27$ **19.** $2^4 = 16$

21. $5^3 = 125$ **23.** $e^3 = 20.1$ **25.** $6^{-2} = \frac{1}{36}$ **27.** $e^2 = 7.39$ **29.** $\left(\frac{8}{27}\right)^{-2/3} = \frac{9}{4}$

31.

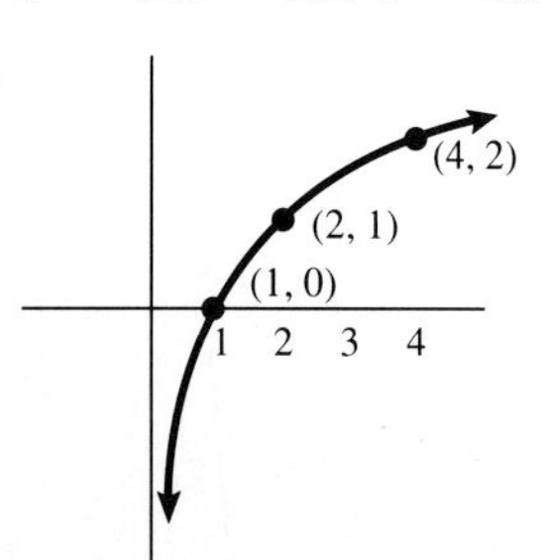

33.

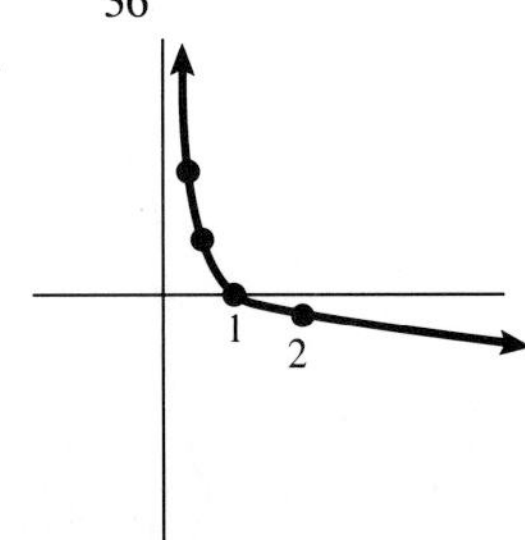

35.

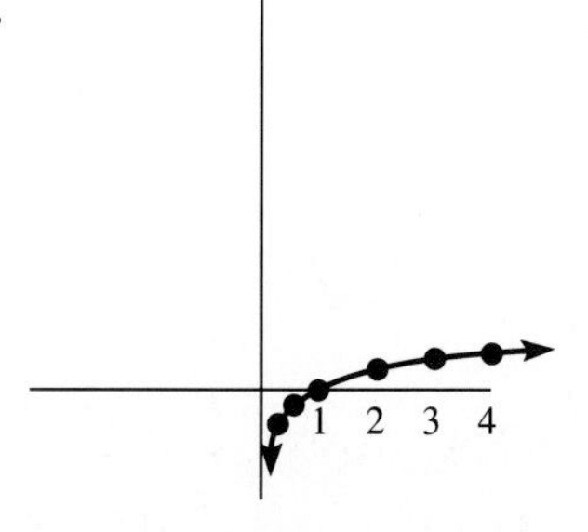

37.

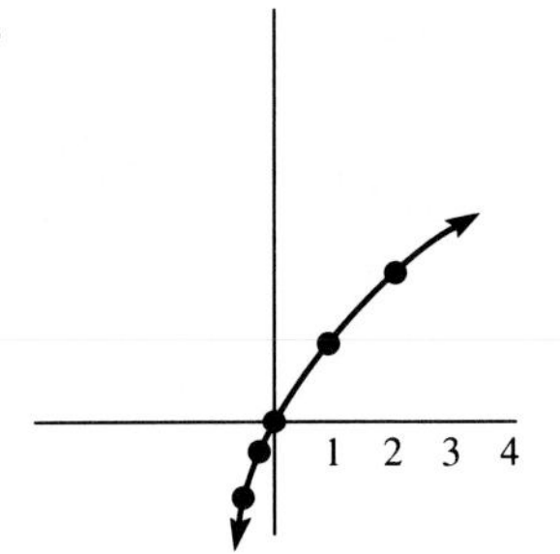

39.

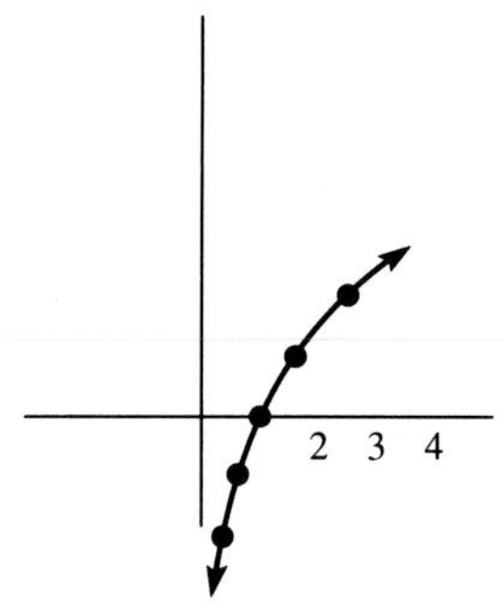

41.

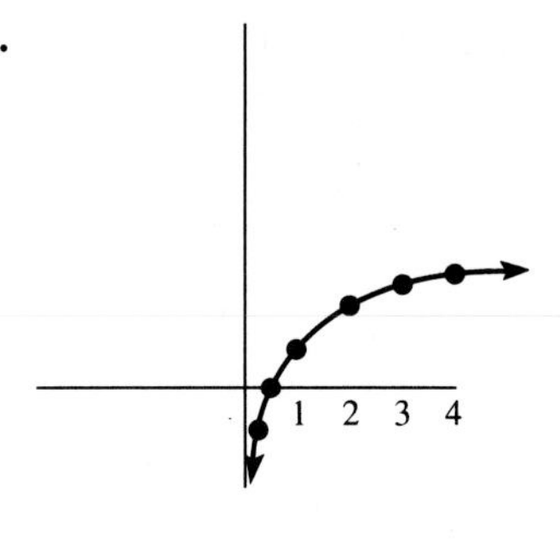

43.

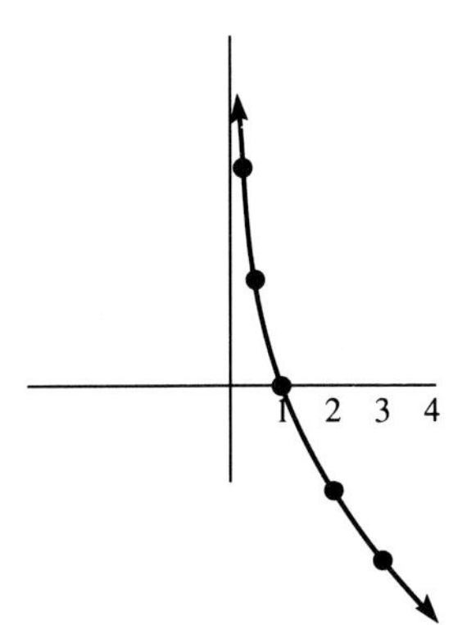

45.

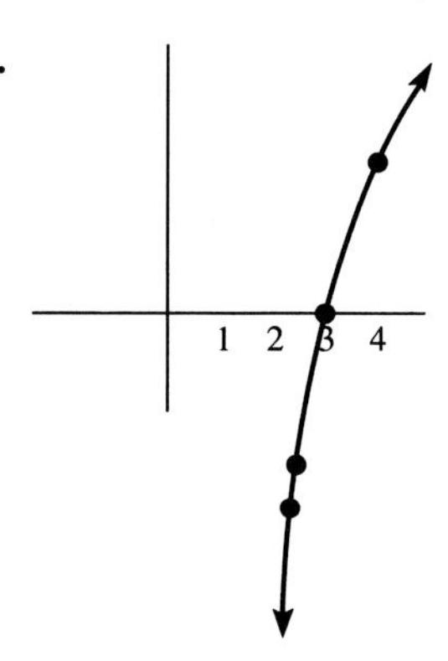

47. 6 **49.** 512 **51.** $\frac{3}{4}$

53. 8 **55.** 6 **57.** 1

59. -1 **61.** $\frac{1}{k}\ln\left(\frac{P}{P_0}\right)$

63. $-RC\ln\left(\frac{IR}{V}\right)$

65. $\frac{\ln\left(\frac{y}{100}\right)}{0.2}$ or $5\ln\left(\frac{y}{100}\right)$

8–3 Exercises, p. 307

1. 7 **3.** 8 **5.** 4 **7.** 24 **9.** 4 **11.** $\frac{2}{3}$ **13.** $\log\frac{x}{(x-3)^2}$ **15.** $\log[(x+1)^2(x-3)]$

17. $\log\left(\frac{xy^3}{\sqrt{x-5}}\right)$ **19.** $\log\left(\frac{5\sqrt[3]{2y+5}}{y}\right)$ **21.** $\log[\sqrt[3]{x^2}(x+1)]$ **23.** $\log\left[\frac{(x-1)^3}{x^9}\right]$ **25.** $\log\left(\frac{x^4}{yz^3}\right)$

27. 0.7781 **29.** 0.3980 **31.** 0.5881 **33.** 1.3801 **35.** 1.8751 **37.** 1.9542 **39.** 1.2731

41. 2.965 **43.** 5.196 **45.** 3.0566 **47.**

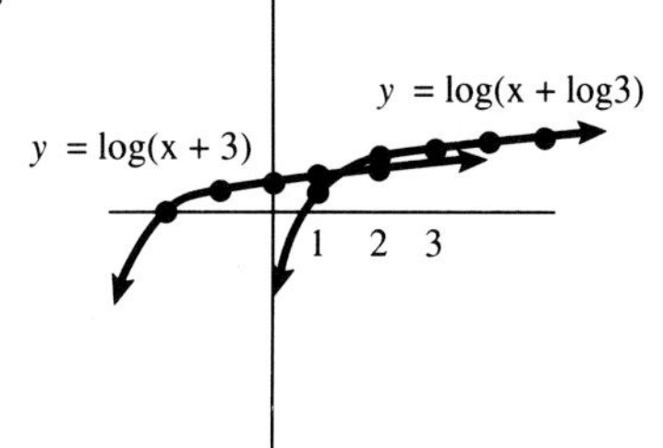

49. $14.7e^{-0.21h}$

8–4 Exercises, p. 313

1. 2.74 **3.** -2.41 **5.** 0.876 **7.** -4.75 **9.** -6.59 **11.** 0.536 **13.** 1.079 **15.** 0.924

17. $\frac{1}{17}$ **19.** 0.601 **21.** 1002 **23.** 7 **25.** no solution **27.** $\frac{1}{10}$ **29.** no solution **31.** 1.58

33. no solution **35.** $\Delta T_m = 148.26°$ **37.** 11.05 y **39.** 1.875 dB

Chapter Review, p. 314

1.

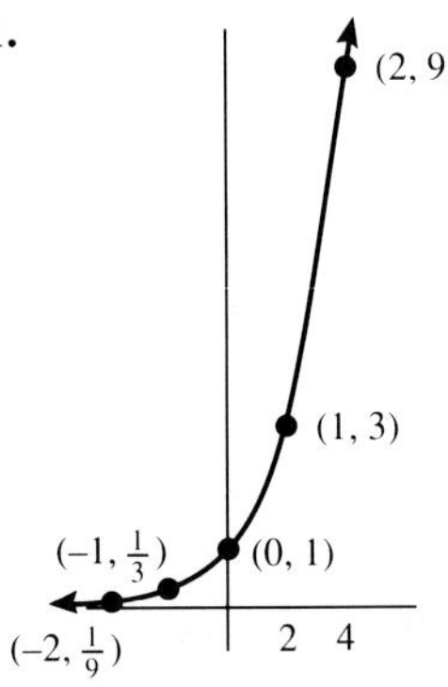

3.

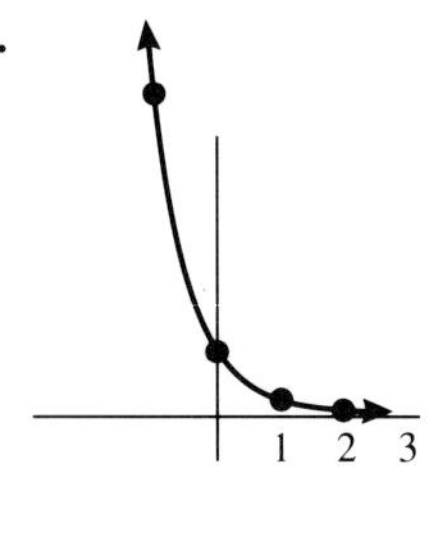

5.

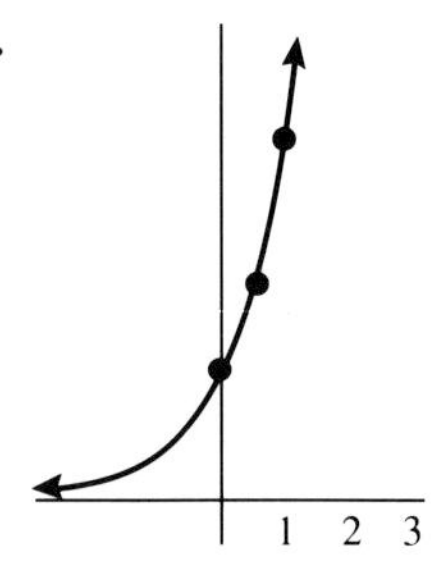

7.

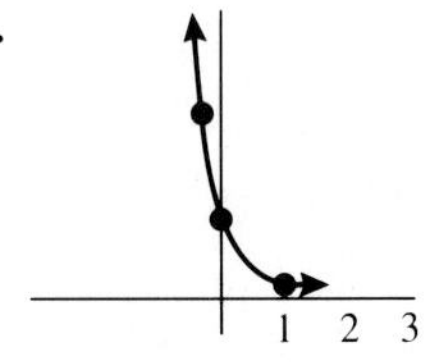

9.

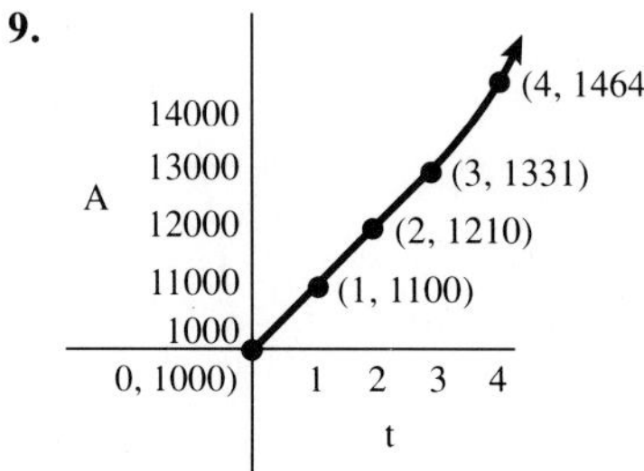

11. \$450.51

13. $\log_4\left(\frac{1}{16}\right) = -2$ **15.** $\log_{81}27 = \frac{3}{4}$ **17.** $\ln 20.1 = 3$ **19.** $\ln 0.135 = -2$ **21.** $3^2 = 9$

23. $2^5 = 32$ **25.** $e^{-1.4} \approx 0.25$ **27.** $e^3 \approx 20.1$ **29.** $\frac{1}{3}$ **31.** 6 **33.** −4 **35.** 5 **37.** 6

39. 14 **41.** $\log(x + 1)^2$ **43.** $\log[x^3(x + 1)^6]$ **45.** $\log\left(\frac{xy}{z^2}\right)$ **47.** 0.79 **49.** 1.52 **51.** 2.25

53. $-\frac{3}{7}$ **55.** 2 is the only root **57.** 0.056 **59.** 6.38 y

Chapter Test, p. 316

1.

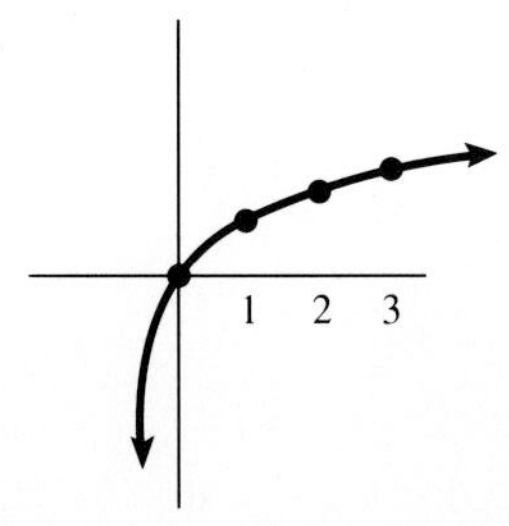

2. 32 **3.** 8.5 y

4.

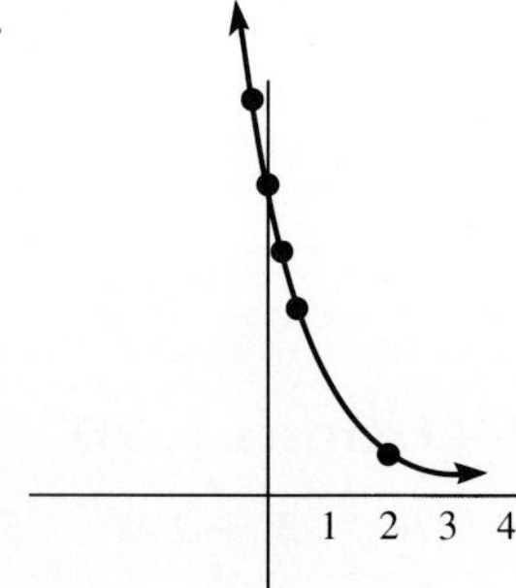

5. $C(1 - r)^n$ **6.** 2.69 **7.** $\log_{1/8}\left(\frac{1}{2}\right) = \frac{1}{3}$ **8.** The graphs are the same.

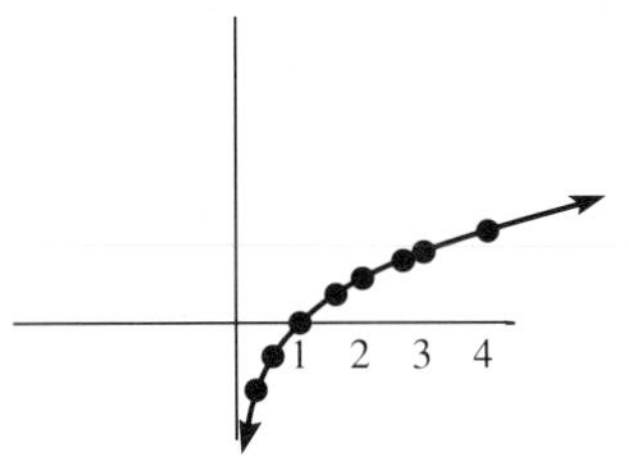

9. 136 lumens **10.** $\log\left[\frac{x^3(x + 1)}{(3x + 4)(x - 1)^2}\right]$ **11.** $\frac{3}{11}$ **12.** $e^{-1} = 0.37$ **13.** 3.7818 **14.** 1.09
15. 3 **16.** 31° **17.** 1.862×10^{-6}

CHAPTER 9

9–1 Exercises, p. 323

1. 34.77° **3.** 318.23° **5.** 148.29° **7.** 39.31° **9.** 256.43° **11.** 274°8′ **13.** 158°20′
15. 314°44′ **17.** 195°43′ **19.** 395°47′, −324°13′ **21.** 678°34′45″, −41°25′15″

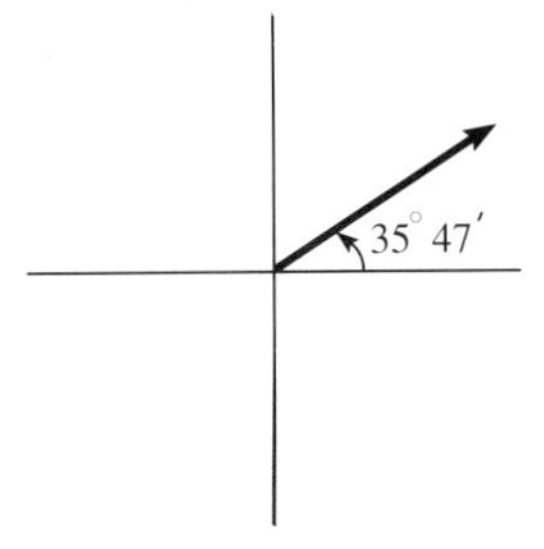

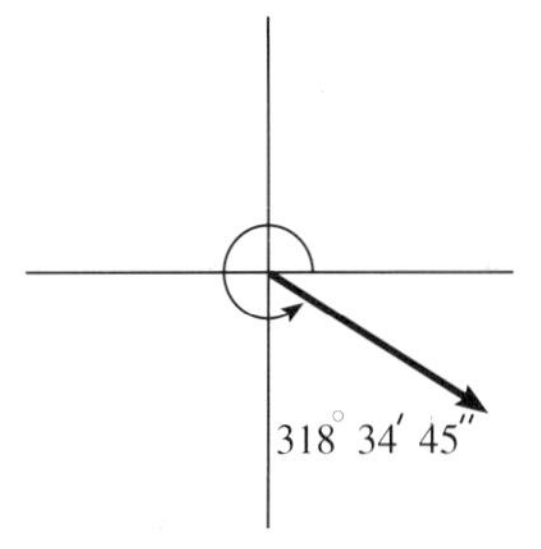

23. 416°29′41″, −303°30′19″ **25.** 378°56′, −341°4′ **27.** 613°38′45″, −106°21′15″

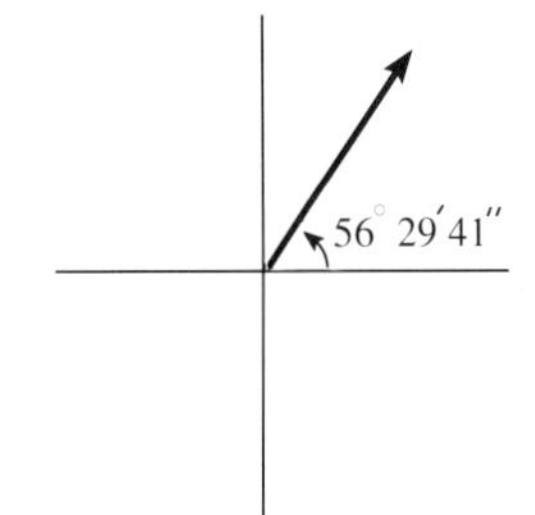

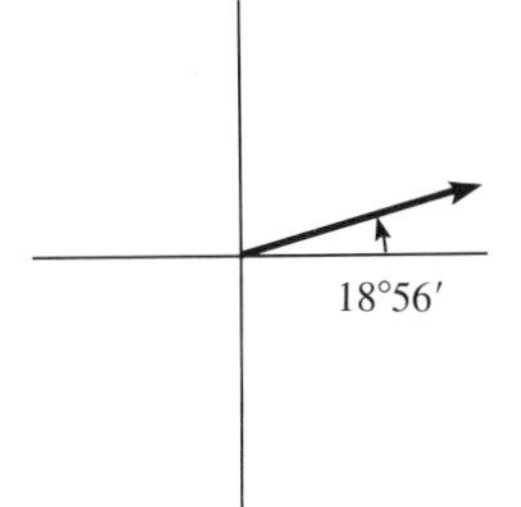

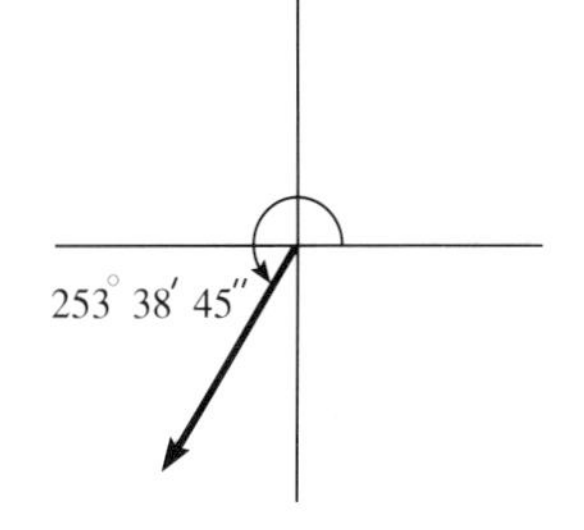

29. II

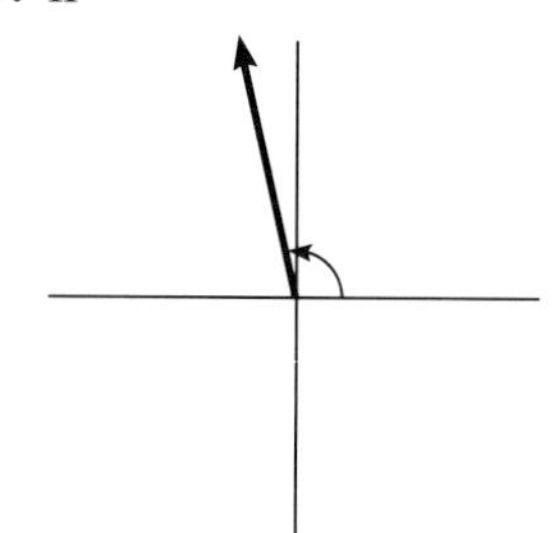

31. I

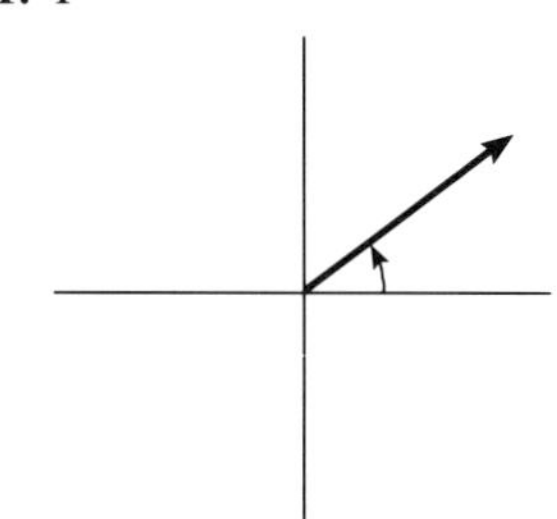

33. II

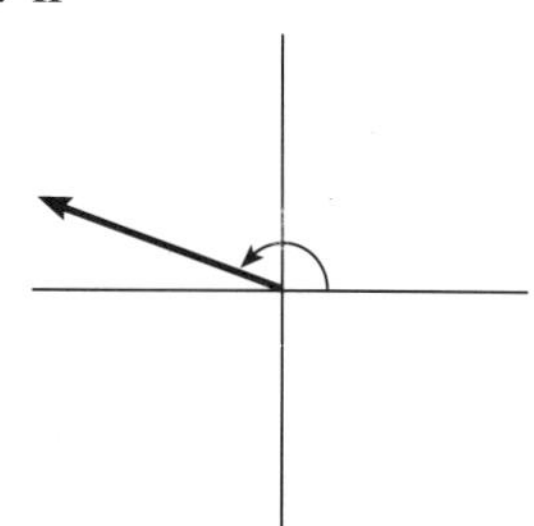

35. III

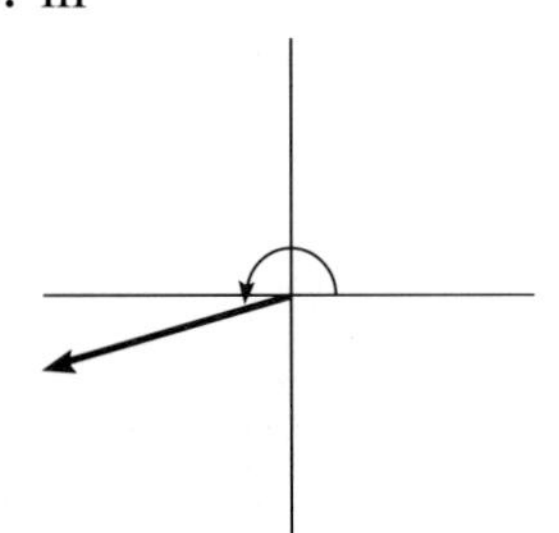

37.

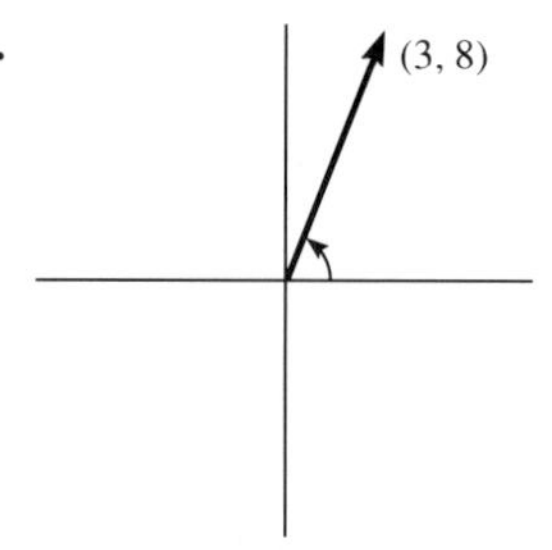

39.

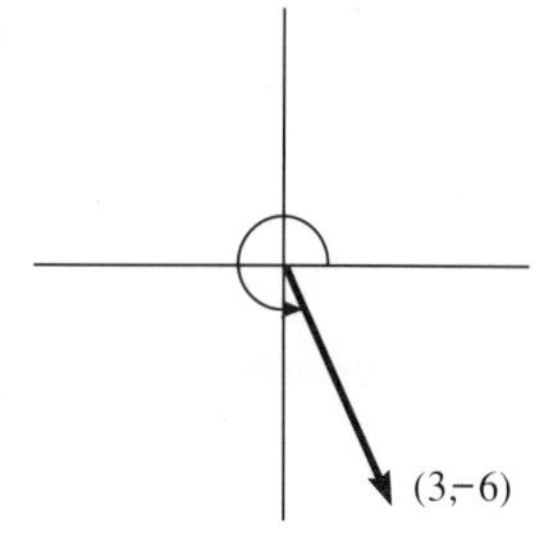

41.

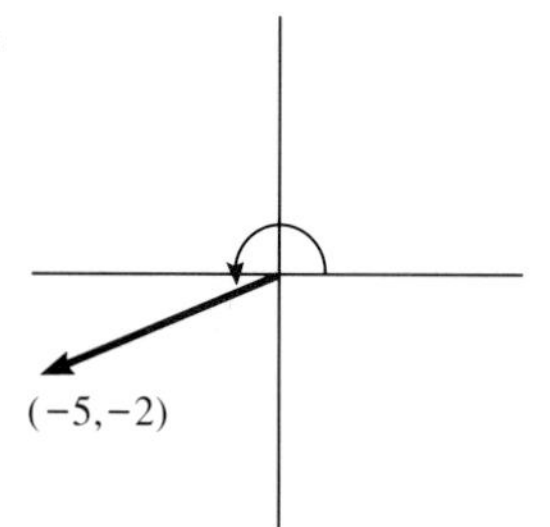

43.

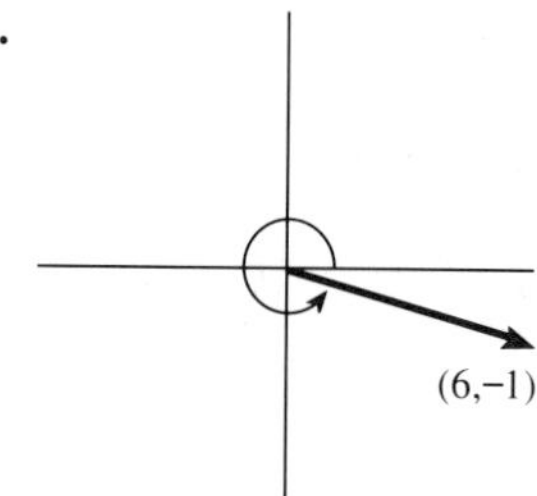

45.

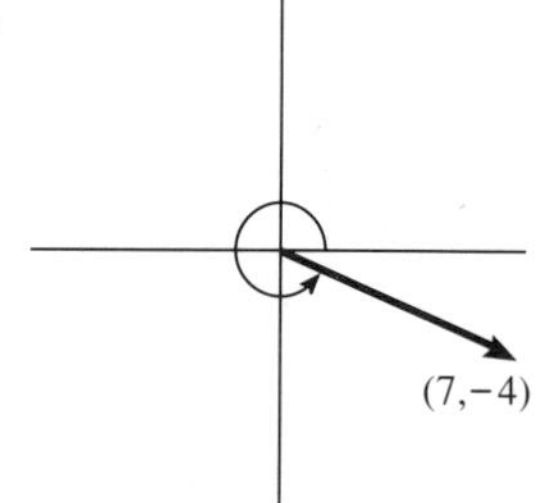

9–2 Exercises, p. 329

1. $\sin\theta = 0.5547$, $\cos\theta = 0.8321$, $\tan\theta = 0.6667$, $\csc\theta = 1.8028$, $\sec\theta = 1.2019$, $\cot\theta = 1.5000$
3. $\sin\theta = 0.9080$, $\cos\theta = 0.4191$, $\tan\theta = 2.1667$, $\csc\theta = 1.1014$, $\sec\theta = 2.3863$, $\cot\theta = 0.4615$
5. $\sin\theta = 0.6644$, $\cos\theta = 0.7474$, $\tan\theta = 0.8889$, $\csc\theta = 1.5052$, $\sec\theta = 1.3380$, $\cot\theta = 1.1250$
7. $\sin\theta = 0.5000$, $\cos\theta = 0.8660$, $\tan\theta = 0.5774$, $\csc\theta = 2.000$, $\sec\theta = 1.1547$, $\cot\theta = 1.7321$
9. $\sin\theta = 0.9042$, $\cos\theta = 0.4270$, $\tan\theta = 2.1176$, $\csc\theta = 2.3419$, $\cot\theta = 0.4722$ **11.** $\sin\theta = 0.6000$, $\cos\theta = 0.8000$, $\tan\theta = 0.7500$, $\csc\theta = 1.6667$, $\sec\theta = 1.2500$, $\cot\theta = 1.3333$ **13.** $\tan\theta = \frac{6}{\sqrt{13}} = 1.6641$, $\csc\theta = \frac{7}{6} = 1.1667$ **15.** $\sin\theta = 0.7714$, $\cot\theta = 0.8250$ **17.** $\cos\theta = 0.3162$, $\csc\theta = 1.0541$
19. $\sec\theta = 1.2534$, $\cot\theta = 1.3234$ **21.** $\cot\theta = 0.6135$, $\cos\theta = 0.5229$

9–3 Exercises, p. 336

1. 0.8368 **3.** 0.1045 **5.** 0.4429 **7.** 2.2148 **9.** 1.1736 **11.** 0.3843 **13.** 0.3134 **15.** 1.0515
17. 1.4661 **19.** 0.8785 **21.** 0.8059 **23.** 3.4570 **25.** 1.1363 **27.** 4.2273 **29.** $\theta = 49.7°$

31. $\theta = 67.8°$ **33.** $\theta = 14.9°$ **35.** $\theta = 7.9°$ **37.** $\theta = 69.6°$ **39.** $\theta = 80.0°$ **41.** $\theta = 74.5°$
43. $\theta = 12.0°$ **45.** $\theta = 81°34'$ **47.** $\theta = 13°16'$ **49.** $\theta = 49°54'$ **51.** $\theta = 53°1'$ **53.** $\theta = 46°26'$
55. $\theta = 49°17'$ **57.** 59.5 in.-lb **59.** 358.2 V

9–4 Exercises, p. 341

1. $A = 46.2°$, $b = 6.5$ m, $c = 9.4$ m **3.** $B = 53°17'$, $a = 23.6$ cm, $b = 31.7$ cm **5.** $A = 75.5°$, $b = 4.1$ m, $c = 16.5$ m **7.** $B = 24°23'$, $a = 107.29$ cm, $c = 117.79$ cm **9.** $b = 51.5$ ft, $A = 36.4°$, $B = 53.6°$ **11.** $c = 52.6$ ft, $A = 34.5°$, $B = 55.5°$ **13.** 50 ft **15.** 3.4 mi **17.** 10.5 ft **19.** 79.3 yd **21.** 1.27 km **23.** 35.87°, 13.00 cm **25.** 16° **27.** 80.2°, 320.3 ft **29.** 142 ft

Chapter Review, p. 345

1. 67.7° **3.** 274.7° **5.** 218.8° **7.** 16°47′ **9.** 318°9′ **11.** 108°26′
13. quad II, 507.8°, −212.2° **15.** quad IV, 650°18′12″, −69°41′48″ **17.** quad IV, 674.63°, −45.37°

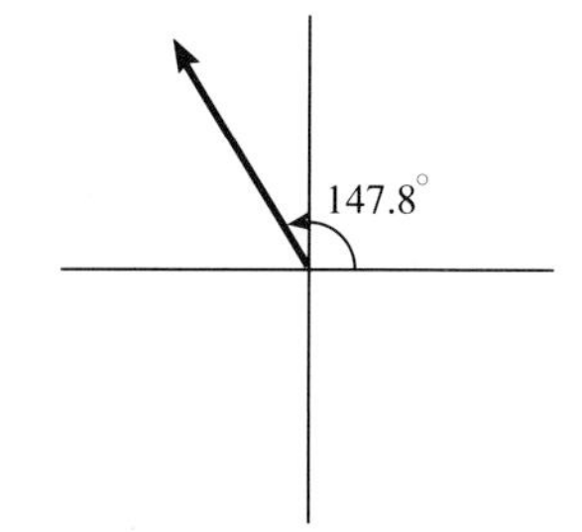

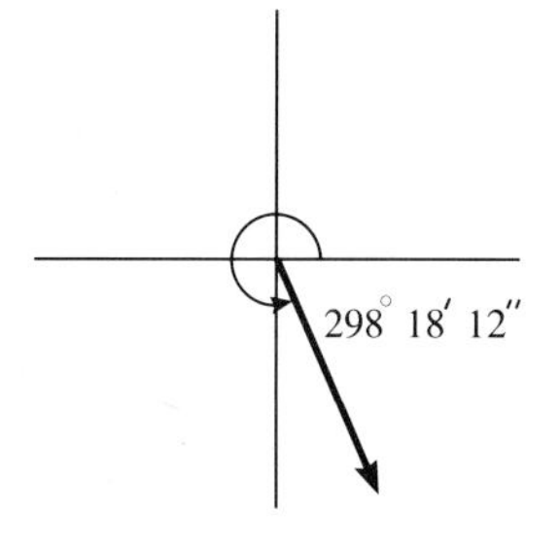

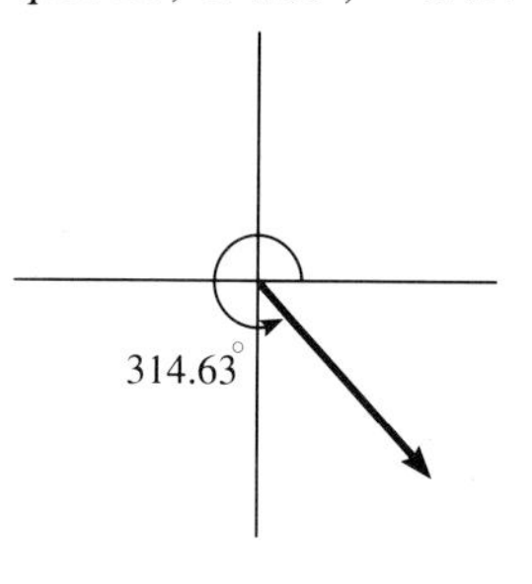

19. $\sin\theta = 0.8944$, $\cos\theta = 0.4472$, $\tan\theta = 2.0000$, $\csc\theta = 1.1180$, $\sec\theta = 2.2361$, $\cot\theta = 0.5000$
21. $\sin\theta = 0.8319$, $\cos\theta = 0.5546$, $\tan\theta = 1.5000$, $\csc\theta = 1.2020$, $\sec\theta = 1.8030$, $\cot\theta = 0.6667$
23. $\sin\theta = 0.7559$, $\cos\theta = 0.6547$, $\tan\theta = 1.1547$, $\csc\theta = 1.3229$, $\sec\theta = 1.5275$, $\cot\theta = 0.8660$
25. $\tan\theta = 2.0647$, $\sec\theta = 2.2942$ **27.** $\sin\theta = 0.9402$, $\cos\theta = 0.3406$ **29.** $\cos\theta = 0.5598$, $\tan\theta = 1.4803$ **31.** 0.957 **33.** 1.080 **35.** 3.375 **37.** $\theta = 75.2°$ **39.** $\theta = 52.2°$
41. $\theta = 21.5°$ **43.** $B = 33.3°$, $a = 11.1$ yd, $c = 13.3$ yd **45.** $a = 44.6$ m, $A = 38.4°$, $B = 51.6°$
47. $B = 36°12'$, $c = 121.8$ km, $b = 71.9$ km **49.** 66.4 ft

Chapter Test, p. 346

1. 1.0202 **2.** 52.0 **3.** $\cos A = 0.4061$, $\cot A = 0.4444$ **4.** 127.2° **5.** 23.8° **6.** 1.0415
7. 516°, −204°

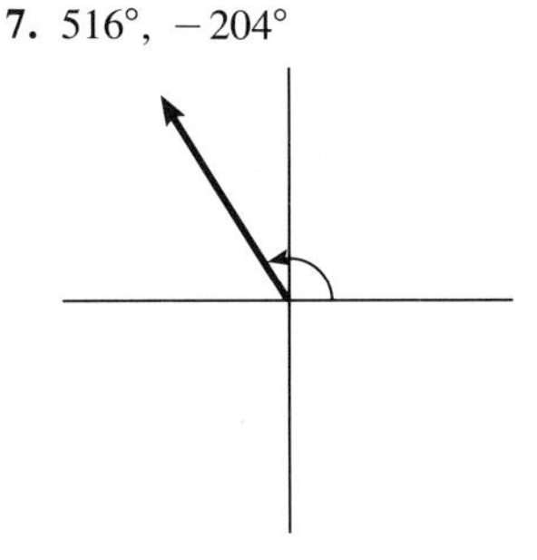

8. 11.3° **9.** $A = 1.8868$ **10.** 1.8262 **11.** 124°50′
12. 783 ft **13.** 1.0973 **14.** 117.6 **15.** 29.4° **16.** 1.8628

CHAPTER 10

10–1 Exercises, p. 355

1. − **3.** + **5.** − **7.** + **9.** + **11.** − **13.** − **15.** − **17.** − **19.** + **21.** + **23.** − **25.** − **27.** − **29.** − **31.** I and IV **33.** II and IV **35.** II and IV **37.** IV **39.** III **41.** IV **43.** I and II **45.** III **47.** $\sin\theta = 0.9138$, $\cos\theta = -0.4061$, $\tan\theta = -2.25$, $\csc\theta = 1.0943$, $\sec\theta = -2.4622$, $\cot\theta = -0.4444$ **49.** $\sin\theta = -0.8480$, $\cos\theta = -0.5300$, $\tan\theta = 1.6000$, $\csc\theta = -1.1792$, $\sec\theta = -1.8868$, $\cot\theta = -0.6250$ **51.** $\sin\theta = -0.2747$, $\cos\theta = -0.9615$, $\tan\theta = 0.2857$, $\csc\theta = -3.6401$, $\sec\theta = -1.0400$, $\cot\theta = 3.5000$ **53.** $\sin\theta = -0.4061$, $\cos\theta = -0.9138$, $\tan\theta = 0.4444$, $\csc\theta = -2.4622$, $\sec\theta = -1.0943$, $\cot\theta = 2.2500$ **55.** $\sin\theta = -0.6139$, $\cos\theta = -0.7894$, $\tan\theta = -0.7778$, $\csc\theta = -1.6288$, $\sec\theta = 1.2669$, $\cot\theta = -1.2857$ **57.** $\sin\theta = -0.5812$, $\cos\theta = 0.8137$, $\tan\theta = -0.7143$, $\csc\theta = -1.7205$, $\sec\theta = 1.2289$, $\cot\theta = -1.4000$ **59.** $\sin\theta = -0.2425$, $\cos\theta = 0.9701$, $\tan\theta = -0.2500$, $\csc\theta = -4.1231$, $\sec\theta = 1.0308$, $\cot\theta = -4.0000$

10–2 Exercises, p. 361

1. −0.233 **3.** −1.575 **5.** 0.826 **7.** 0.106 **9.** −320.775 **11.** 0.279 **13.** −0.111 **15.** 13.431 **17.** −0.967 **19.** 0.869 **21.** 0.445 **23.** 0.854 **25.** −1.788 **27.** 0.532 **29.** 66.4°, II **31.** 81.9°, IV **33.** 33.27°, IV **35.** 83.8°, I **37.** 30.9°, III **39.** 42.57°, II **41.** 64°, IV **43.** 38°13′, III **45.** 50°23′, I **47.** 80°31′, II **49.** 54.60°, 125.40° **51.** 137.90°, 317.90° **53.** 65.70°, 245.70° **55.** 81.77°, 98.23° **57.** 136.40°, 223.60° **59.** 258.39°, 281.61° **61.** 198°30′, 341°30′ **63.** 47°6′, 312°54′ **65.** 56°6′, 236°6′ **67.** 148°18′, 211°42′ **69.** −0.9622 **71.** −30.1371 **73.** 3.1189 **75.** 106.7° **77.** 336.8° **79.** 336° **81.** 308.7° **83.** 5.83 in. **85.** 40.8° **87.** 78.14 W

10–3 Exercises, p. 367

1. II **3.** III **5.** II **7.** IV **9.** $\frac{17\pi}{45}$ **11.** $\frac{53\pi}{30}$ **13.** $\frac{11\pi}{6}$ **15.** $-\frac{14\pi}{9}$ **17.** 0.82 **19.** 1.71 **21.** 3.40 **23.** −5.67 **25.** 93.97° **27.** 322.58° **29.** 420° **31.** −330° **33.** 0.278 **35.** 1.410 **37.** 0.651 **39.** 1.342 **41.** −1.161 **43.** 1 **45.** 0.383 **47.** −2 **49.** 0.58, 5.70 **51.** 3.51, 5.91 **53.** 0.41, 2.73 **55.** 2.23, 4.05 **57.** 1.91, 4.37 **59.** 2.88, 6.02 **61.** 55.82 m^2 **63.** 0; the weight is at the starting position.

10–4 Exercises, p. 372

1. 8.7 ft **3.** 0.88° **5.** 8.3 yd **7.** 1016.87 cm^2 **9.** 491.5 m **11.** 3° **13.** $r = 17.2$ in. **15.** $0.61°\left(\frac{180}{\pi}\right) = 35°$ **17.** 8.02 ft/s **19.** $1.17°\left(\frac{180}{\pi}\right) = 67°$ **21.** 1.73 ft/s **23.** 19,699 mm^2 **25.** 18.85 ft/s

Chapter Review, p. 374

1. − **3.** + **5.** + **7.** + **9.** + **11.** I **13.** II **15.** IV **17.** $\sin\theta = -0.6508$, $\cos\theta = 0.7593$, $\tan\theta = -0.8571$, $\csc\theta = -1.5365$, $\sec\theta = 1.3171$, $\cot\theta = -1.1667$ **19.** $\sin\theta = -0.5215$, $\cos\theta = 0.8533$, $\tan\theta = -0.6111$, $\csc\theta = -1.9177$, $\sec\theta = 1.1719$, $\cot\theta = -1.6364$ **21.** $\sin\theta = 0.4856$, $\cos\theta = -0.8742$, $\tan\theta = -0.5556$, $\csc\theta = 2.0591$, $\sec\theta = -1.1440$, $\cot\theta = -1.8000$ **23.** undefined **25.** 21.3557 **27.** −3.5687 **29.** 67°18′ **31.** 43.3° **33.** 65°17′ **35.** 69.5° **37.** 256.4°

39. 330° **41.** $\frac{2\pi}{3}$ **43.** $\frac{11\pi}{6}$ **45.** $\frac{31\pi}{18}$ **47.** 2.23 **49.** 5.56 **51.** 3.74 **53.** 157.9° **55.** 315° **57.** −294.9° **59.** −1.0023 **61.** −1 **63.** −0.1506 **65.** $\theta = 2.63$, $\theta = 5.77$ **67.** $\theta = 1.65$ **69.** 8.46 m **71.** 2.21° **73.** 15.4 in. **75.** 1233.7 yd^2 **77.** 451.56 m^2 **79.** 25,132.7 m/min.

Chapter Test, p. 375

1. −1.2020 **2.** $\sin\theta = \frac{-4}{\sqrt{65}} = -0.8682$, $\cot\theta = \frac{-4}{-7} = 0.5714$ **3.** 7.96 ft **4.** −2 **5.** 303.3° **6.** 68.8° **7.** 85.3° **8.** III **9.** 221.7° **10.** 66 rad/s **11.** 140 m^2 **12.** 38°34′ **13.** 0.367 or 21.05° **14.** 2.47 **15.** $\frac{7\pi}{9}$ **16.** 5.39 **17.** 230 ft^2

CHAPTER 11

11–1 Exercises, p. 386

1. 13 N, 0° **3.** 65 m, 270° **5.** 66.4 ft/s, 126.5° **7.** 97.0 lb, 217.7° **9.** 377.6 N, 144.9° **11.** 35.6 mi, 328.9° **13.** 69.9 ft/s, 127.3° **15.** 39.1 ft/s^2, 236.6° **17.** $x = 33.5$ ft/s, $y = 113.1$ ft/s **19.** $x = -40.9$ lb, $y = 39.5$ lb **21.** $x = 32.7$ yd, $y = -119.4$ yd **23.** $x = -64.4$ m, $y = -18.2$ m **25.** $x = -384.2$ mi, $y = 165.5$ mi **27.** 24.3 ft/s, 148.7° **29.** 109.2 yd, 97.0° **31.** 60.9 m/s^2, 4.2° **33.** 144.3 m, 59.4° **35.** 79.1 N, 287.9° **37.** 555.2 mi, 187.7° **39.** 932.4 N, 214.3°

11–2 Exercises, p. 391

1. 48° north of east **3.** 468 mph **5.** 231 N **7.** 14 lb **9.** 52.41° or 232.41° **11.** 51.3° **13.** $R = 77$ lb at 55.5° below horizontal **15.** 30.43 N, 22.74 N **17.** 461 mph, 328.8° **19.** 121.6 lb, 179.3 lb **21.** 432 lb

11–3 Exercises, p. 400

1. $A = 27°$, $b = 32.5$, $c = 26.4$ **3.** $b = 18.4$, $C = 31.7°$, $B = 105.3°$ **5.** $a = 16.7$, $b = 22.5$, $A = 43°$ **7.** $A = 45°$, $C = 77°$, $c = 10.3$ **9.** $A = 79.3°$, $b = 29.4$, $a = 42.6$ **11.** $B = 45°$, $a = 21.1$, $c = 13.6$ **13.** $A = 61°$, $b = 26.1$, $c = 43.0$ **15.** $B = 43.1°$ or $B = 136.9°$, $A = 106.7°$ or $A = 12.9°$, $a = 28.2$ or $a = 6.6$ **17.** $A = 62.2°$, $a = 52.1$, $c = 41.9$ **19.** no solution **21.** The pilot should fly at 41.5° north of east. **23.** 67.5 m **25.** 85.9 m **27.** 56.8 km **29.** 33.2°

11–4 Exercises, p. 406

1. $A = 40.3°$, $B = 48.2°$, $C = 91.5°$ **3.** $c = 23.2$, $A = 88.2°$, $B = 48.8°$ **5.** $a = 29.5$, $B = 33.2°$, $C = 73°$ **7.** $A = 55.4°$, $B = 75.3°$, $C = 49.3°$ **9.** $A = 111.3°$, $B = 41.3°$, $C = 27.4°$ **11.** $C = 142.5°$, $A = 21.8°$, $B = 15.7°$ **13.** $b = 91.7$, $A = 46.5°$, $C = 31.5°$ **15.** $c = 126.7$, $A = 48.1°$, $B = 56.6°$ **17.** $B = 148.9°$, $A = 20.3°$, $C = 10.8°$ **19.** $a = 106.3$, $B = 31.3°$, $C = 40.7°$ **21.** $AB = 26.1$ **23.** 80.7° and 99.3° **25.** 38.5° **27.** 9.56 knots; 12.4° S of E

Chapter Review, p. 408

1. 323.8 lb, 215.1° **3.** 40.9 m/s, 84.2° **5.** 310.7 N, 133.3° **7.** 17.3 lb, 334.3° **9.** $x = -219.6$ lb, $y = -190.91$ lb **11.** $x = 255.0$ km, $y = 414.5$ km **13.** 16 blocks 10° S of E **15.** 150.2 lb, 90.4 lb **17.** 240.6 lb, 78.2 lb **19.** 454.6 mph at 26° N of W **21.** $A = 70.7°$, $b = 78.1$, $c = 21.8$

23. $B = 26.8°$, $A = 96.2°$, $a = 30.8$ **25.** $B = 22.6°$, $C = 90.3°$, $c = 61.7$ **27.** Heading should be 62° S of W. **29.** 57.8° **31.** $b = 195.8$, $A = 55.7°$, $C = 85.3°$ **33.** $C = 87.6°$, $A = 57.5°$, $B = 34.9°$

35. $c = 196.1$, $A = 76.6°$, $B = 49.8°$ **37.** 26 ft **39.** $B = 81.6°$, $C = 98.4°$

Chapter Test, p. 409

1. 99.3 lb at 72.8° S of W **2.** 44.6 ft **3.** 7.2 **4.** 58.8° **5.** 41.6 N **6.** 39.6 lb **7.** 41.4 at 26.9° S of W **8.** 26.3 ft and 21.9 ft **9.** $c = 28.4$ **10.** $B = 79°$, $A = 45.6°$, $C = 55.4°$ **11.** 280.5 **12.** 67.6° N of W **13.** 8.7 mph at 60° relative to shore **14.** $R = 26.4$ m/s, $\theta = 324.6°$ **15.** $b = 40.3$, $C = 41°$, $c = 26.4$

CHAPTER 12

12–1 Exercises, p. 419

1.

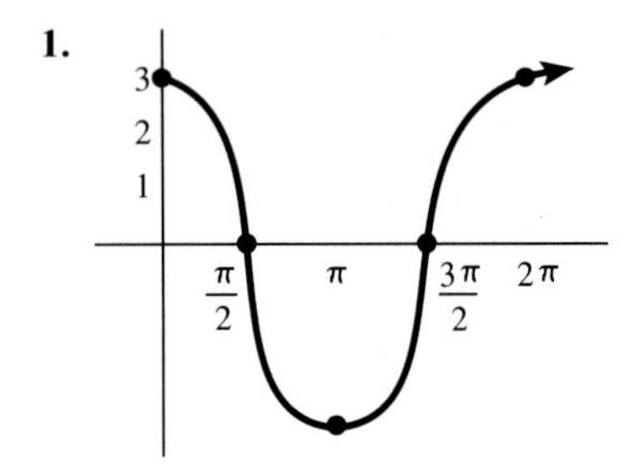

3.

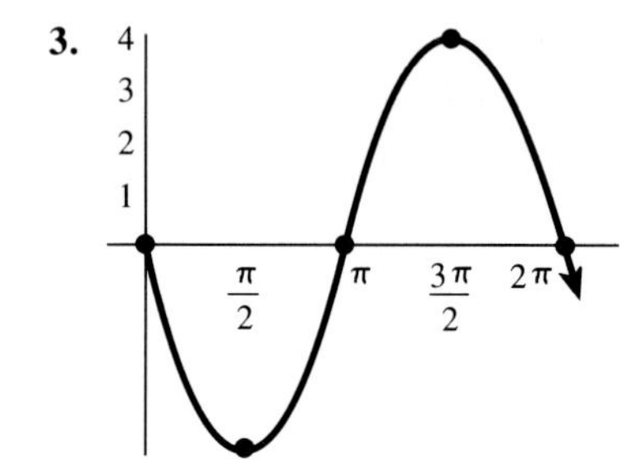

5.

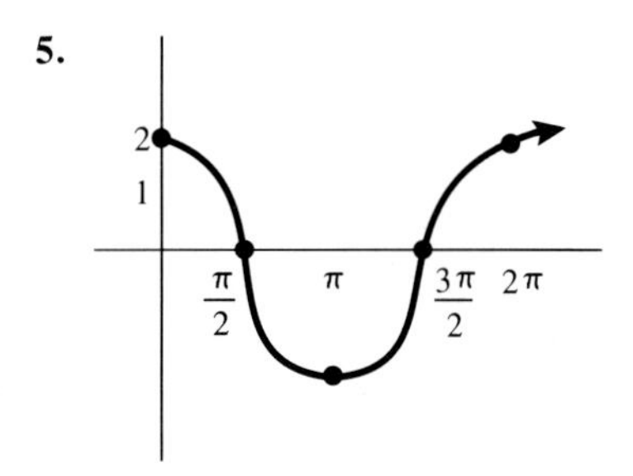

7.

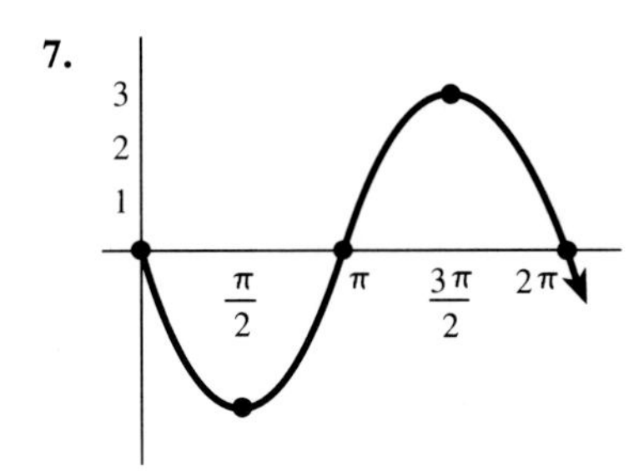

9.

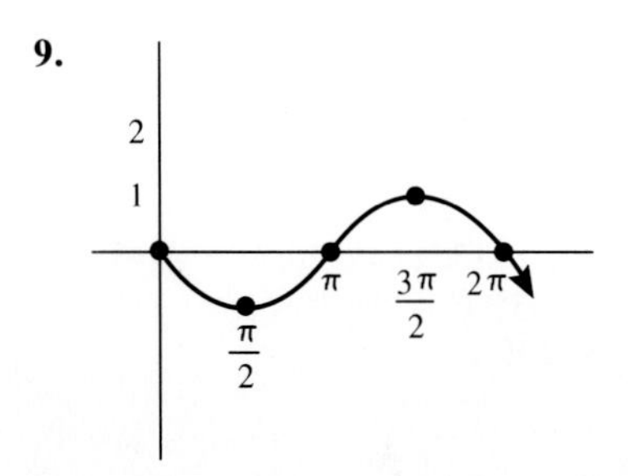

11.

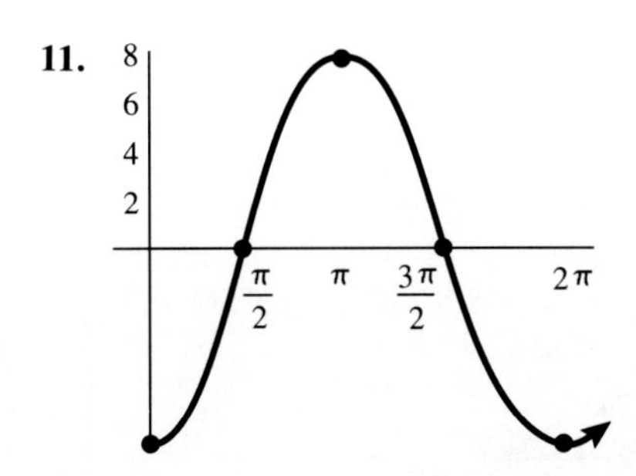

13.

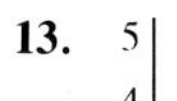

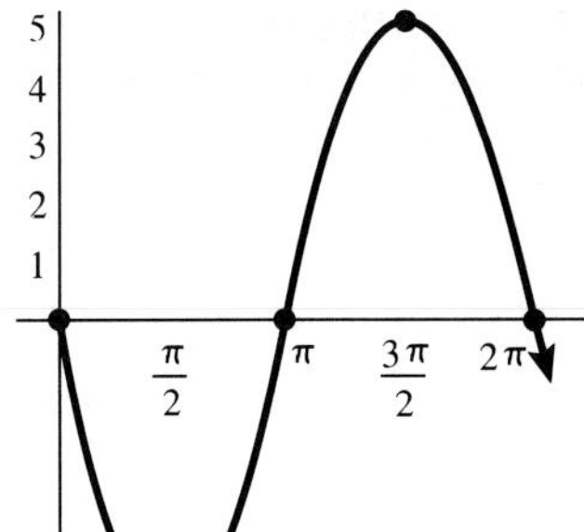

15.

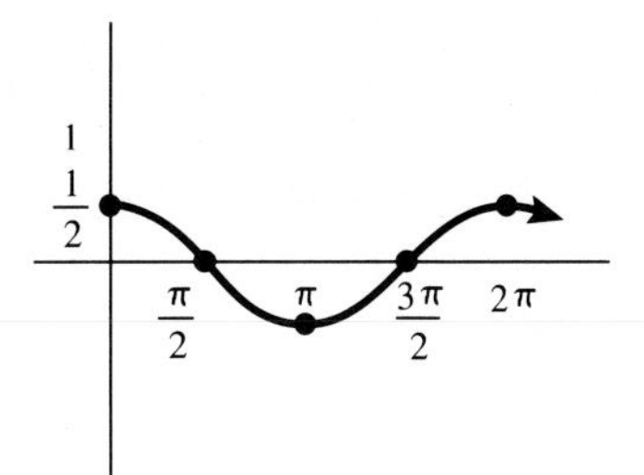

17.

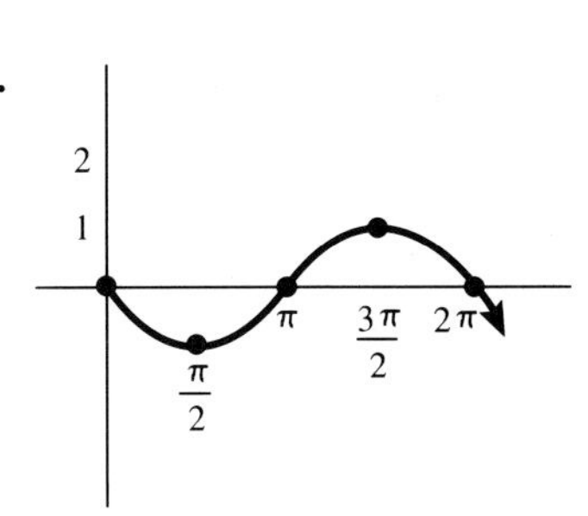

19.

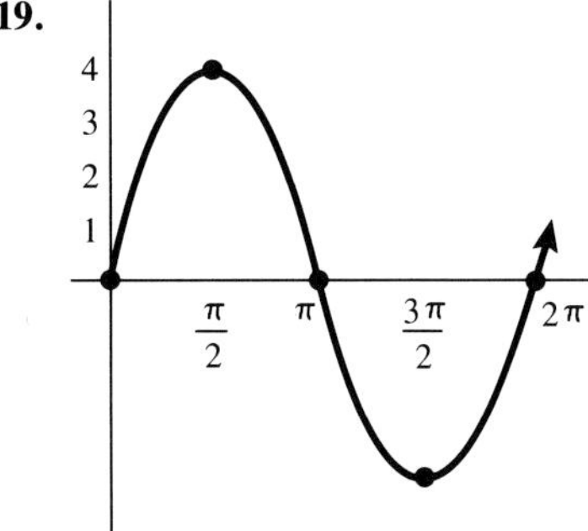

12–2 Exercises, p. 422

1. $1, \frac{\pi}{2}$

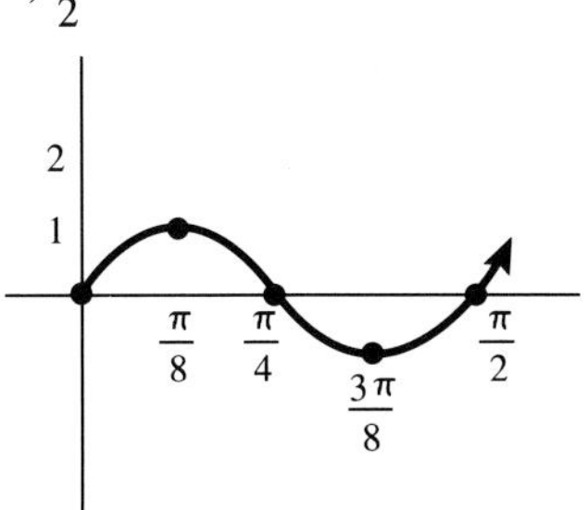

3. $1, \frac{\pi}{3}$

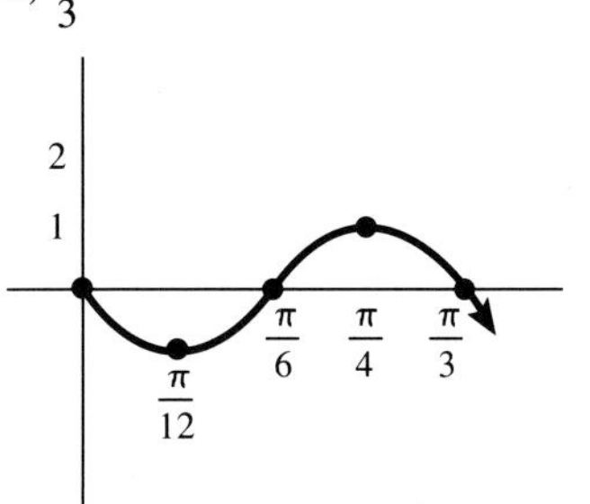

5. $1, 6\pi$

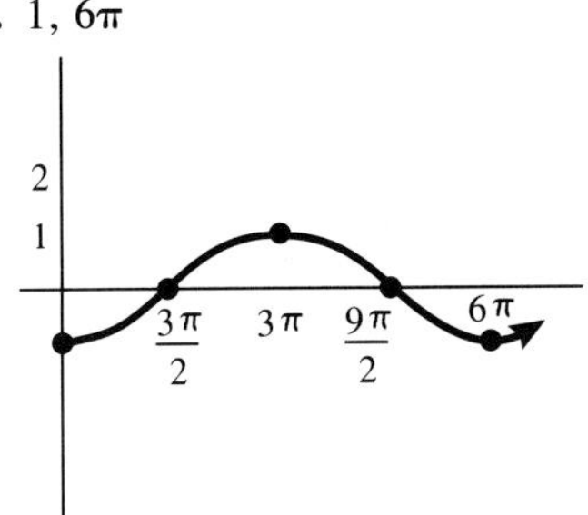

7. $3, \frac{2\pi}{3}$

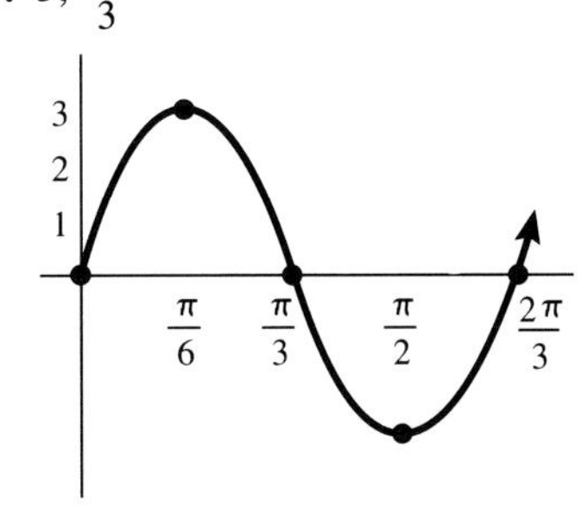

9. 2, 2

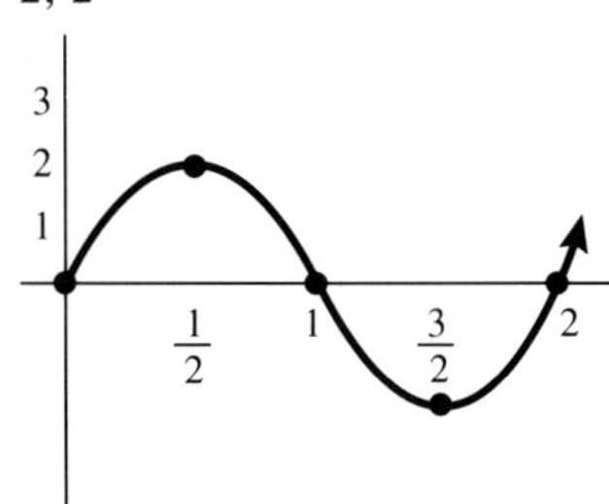

11. 2, 1

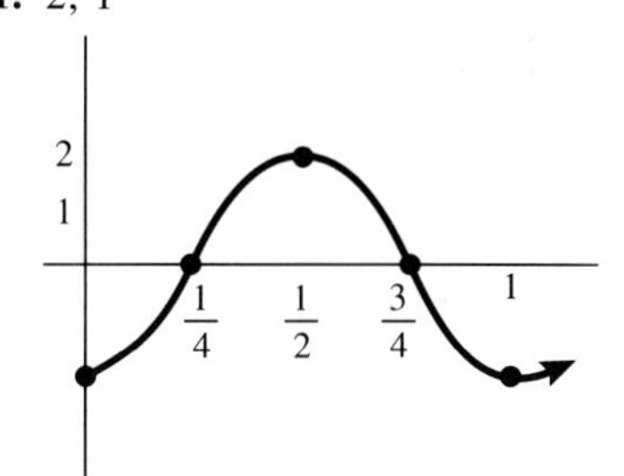

13. 6, 7π

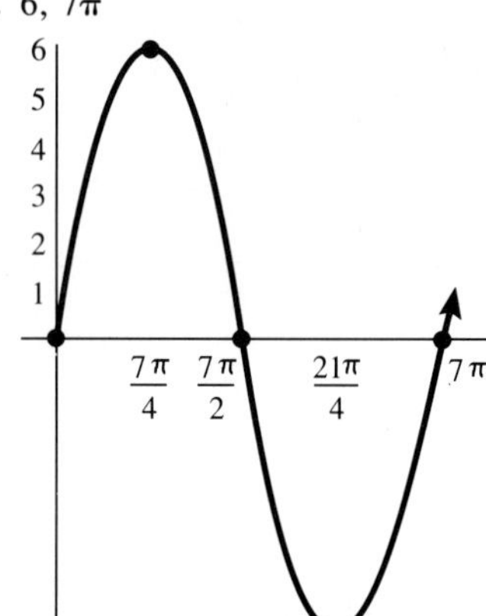

15. $\frac{1}{2}$, 4π

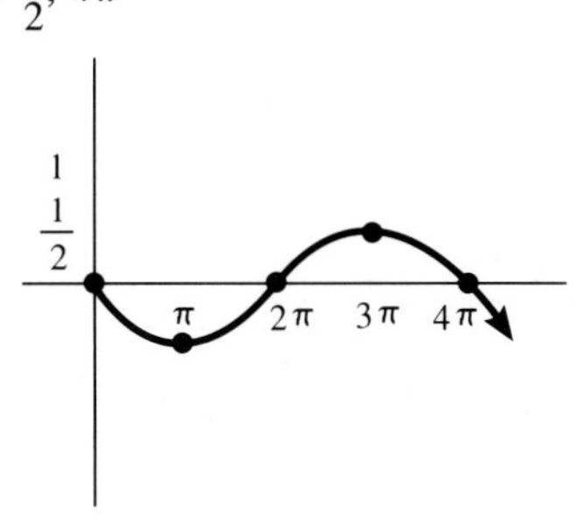

17. 4, $\frac{2}{5}$

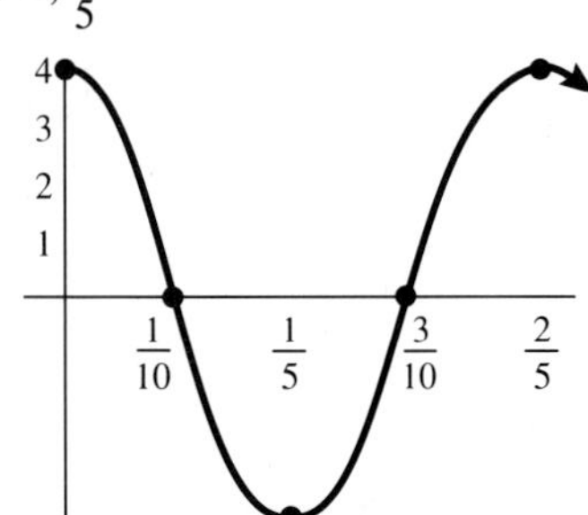

19. 3, $\frac{4\pi}{3}$

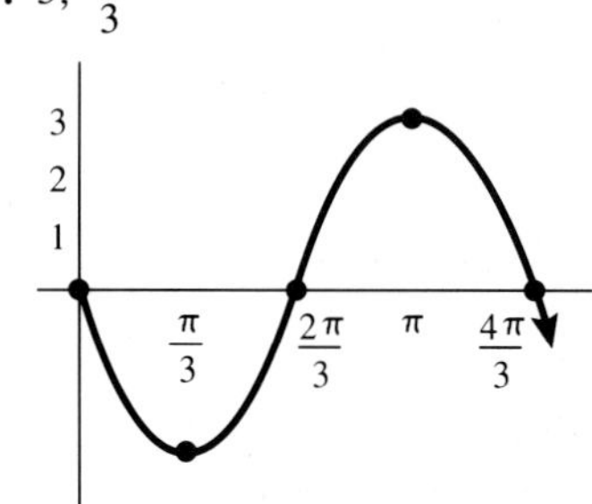

21. 3, 1

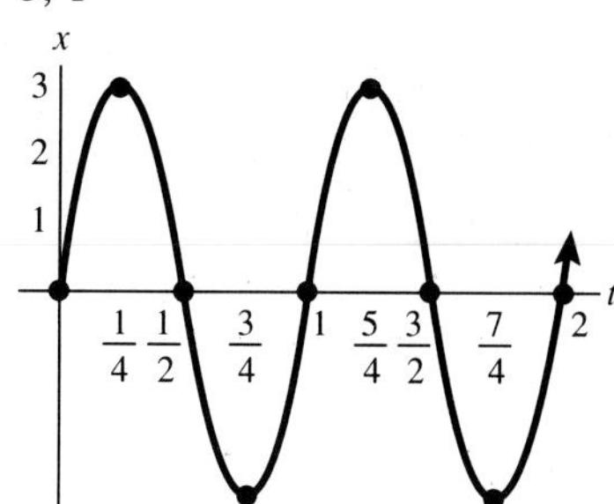

23. 8, $\frac{1}{3}$

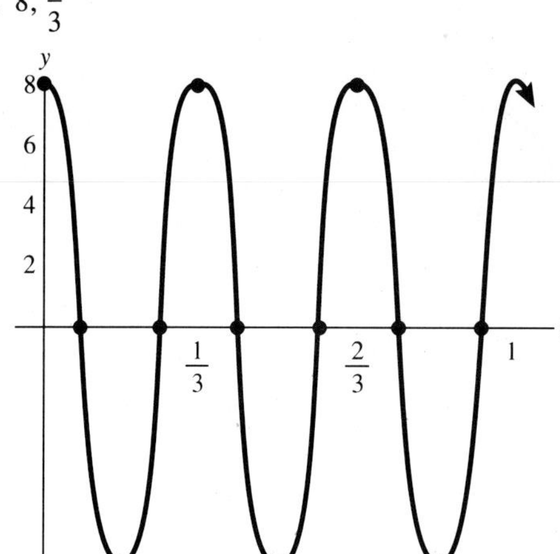

12–3 Exercises, p. 427

1. 1, 2π, $\frac{\pi}{4}$ left

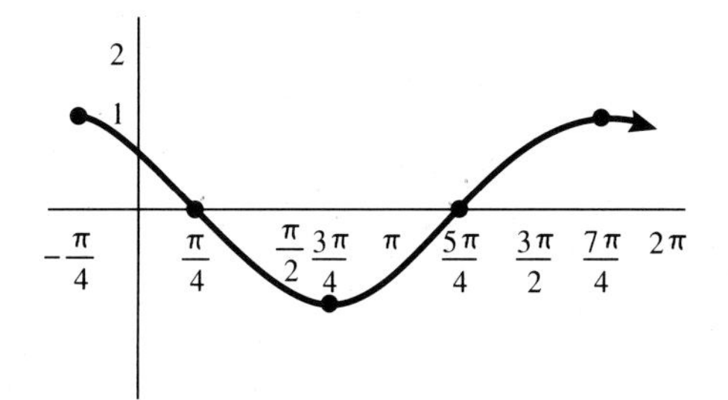

3. 1, 2π, $\frac{\pi}{3}$ right

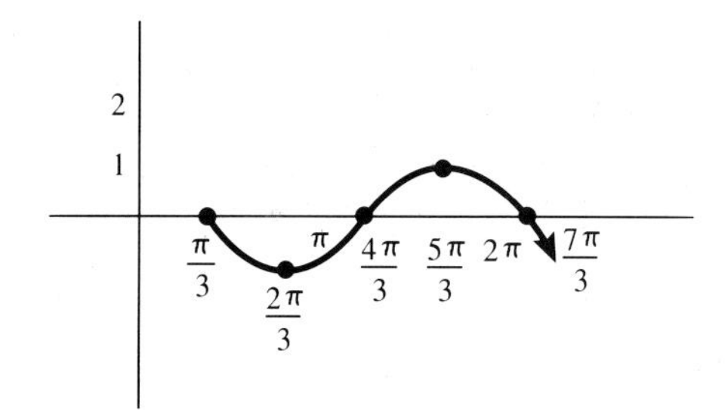

5. 2, 2π, 1 left

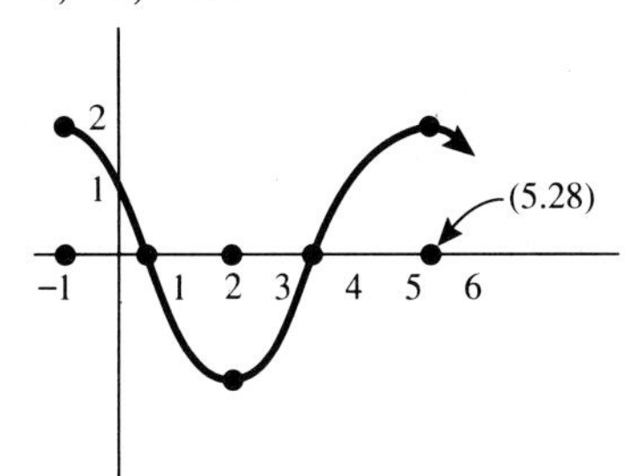

7. 1, π, $\frac{\pi}{2}$ right

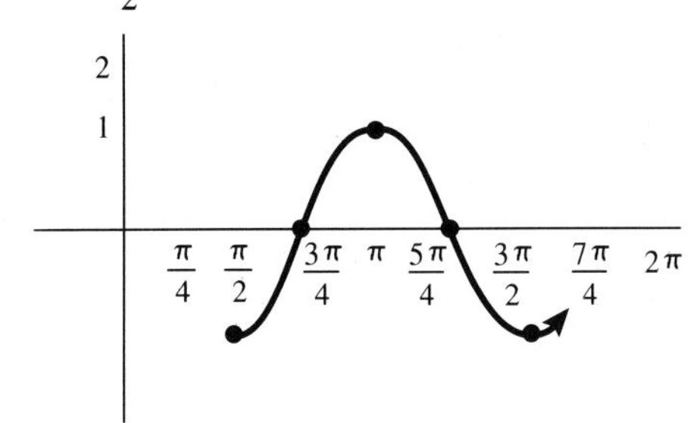

9. $1, \frac{2\pi}{3}, \frac{2\pi}{3}$ left

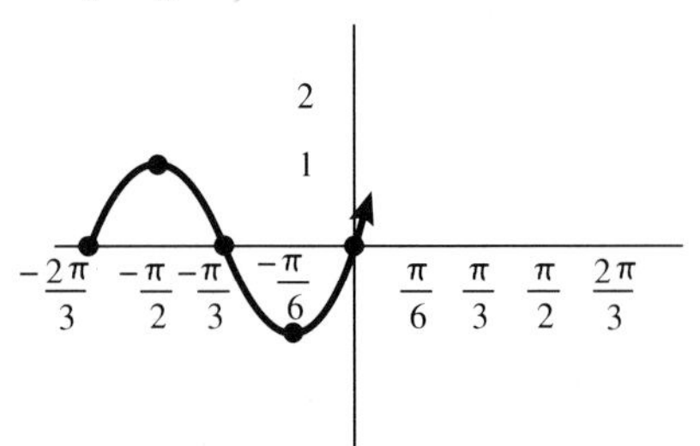

11. $2, 2\pi, \frac{\pi}{4}$ left

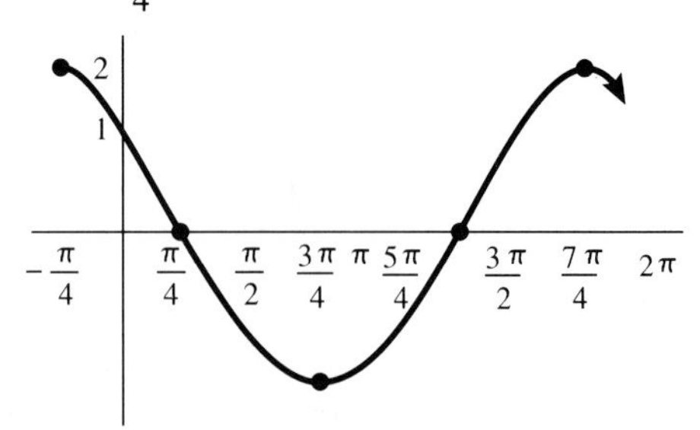

13. $3, 2\pi, 3$ left

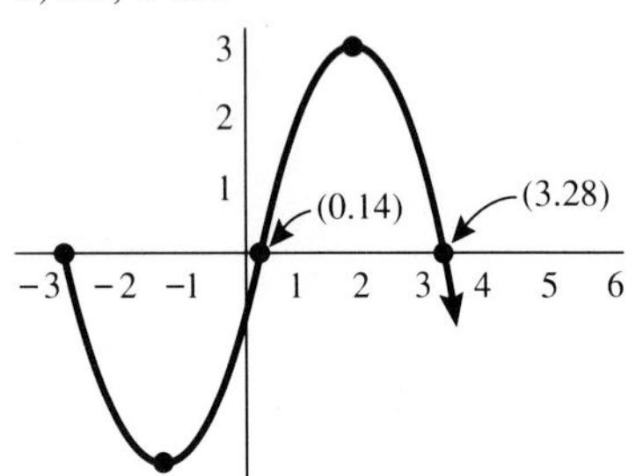

15. $2, 1, \frac{1}{2\pi}$ left (0.16)

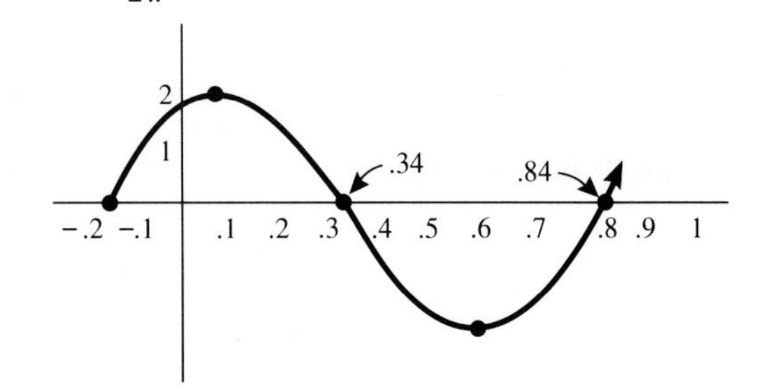

17. $1, \frac{2}{3}, \frac{2}{3\pi}$ right (0.2)

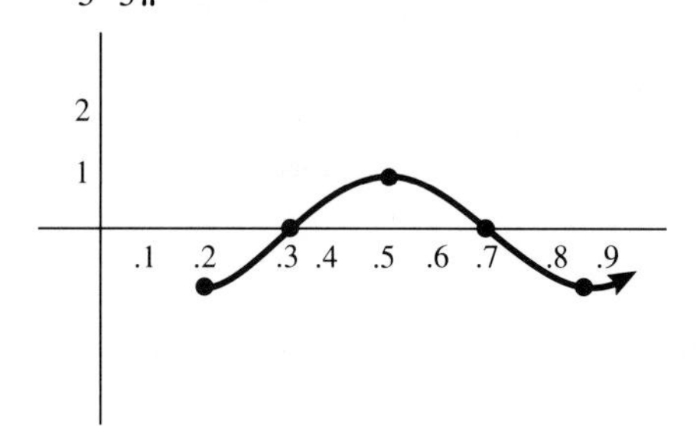

19. $4, \frac{1}{3}, \frac{1}{12\pi}$ left (0.026)

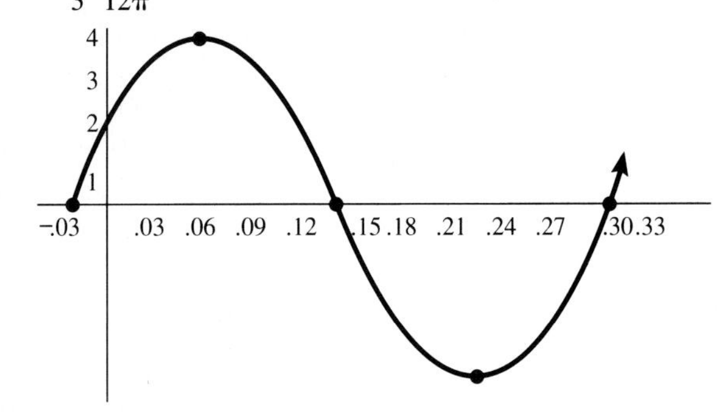

21. $2.5, \frac{2\pi}{3}, \frac{2\pi}{3}$ right

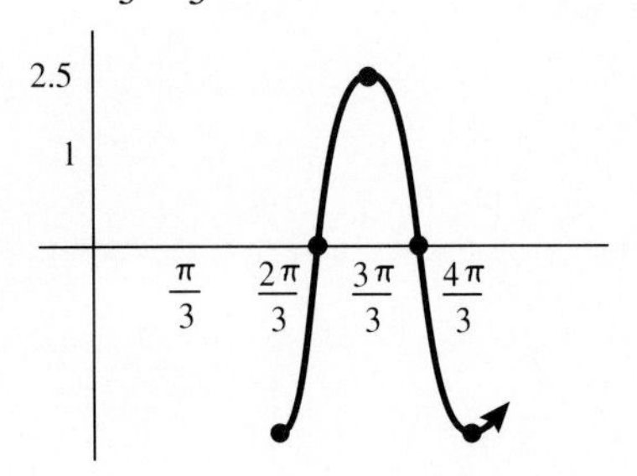

23. $1, 1, \frac{5}{2}$ right

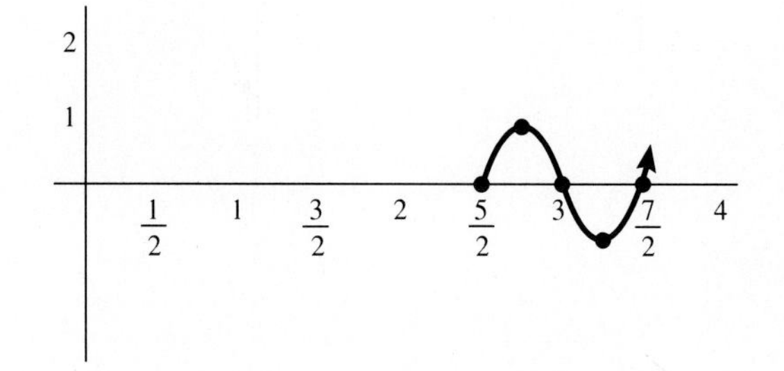

25. $y = 3\cos\left(\frac{\pi}{2}x\right)$ **27.** $y = -\pi \sin\left(\frac{\pi}{3}x + \frac{\pi^2}{9}\right)$ **29.** $y = 4\cos\left(2x - \frac{2\pi}{3}\right)$

31.

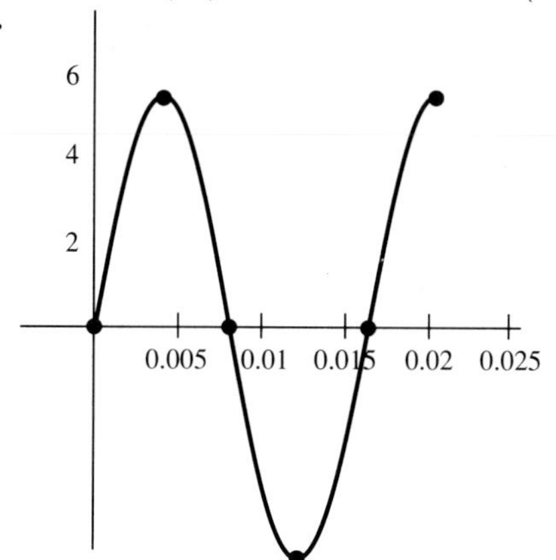

33.

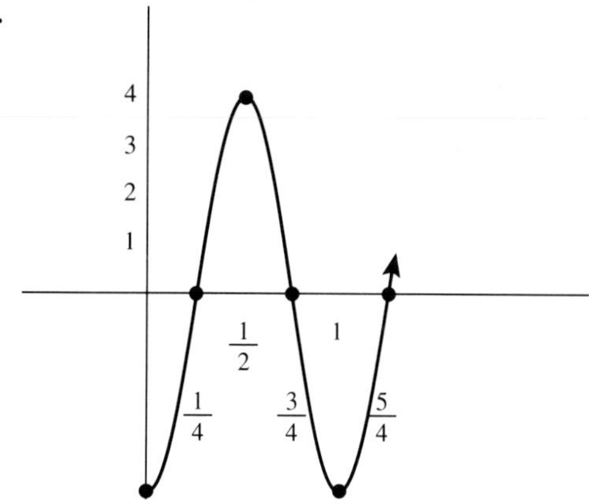

12–4 Exercises, p. 431

1.

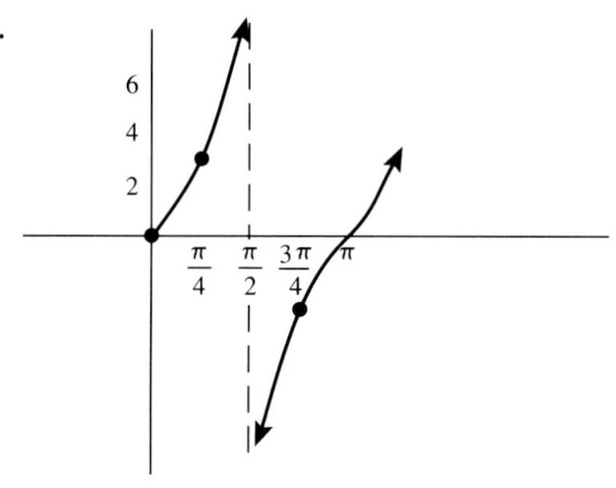

3.

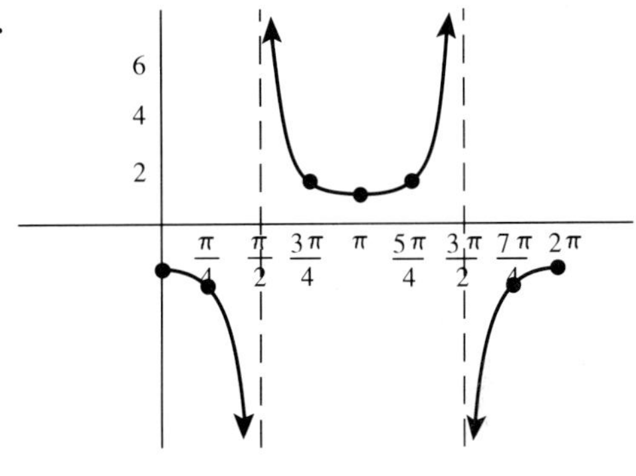

5.

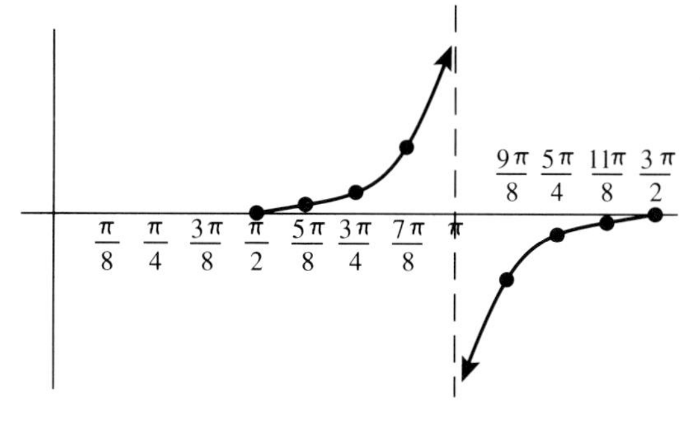

7.

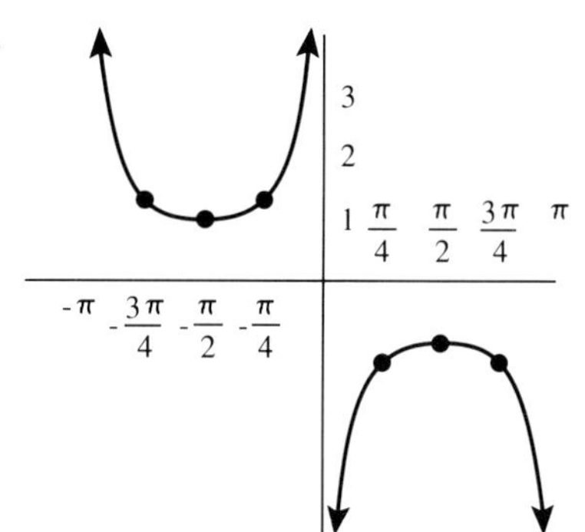

9.

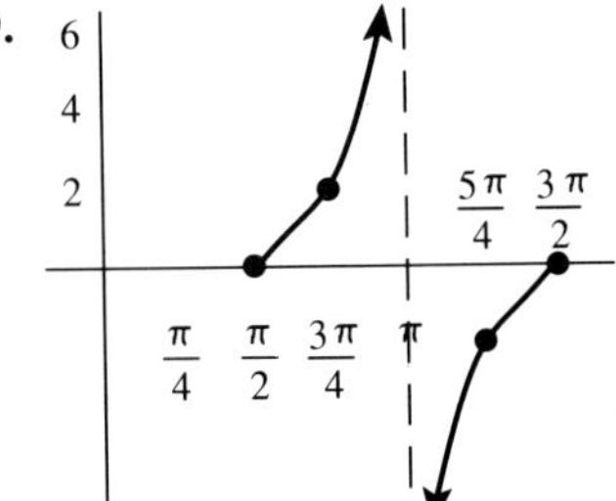

11.

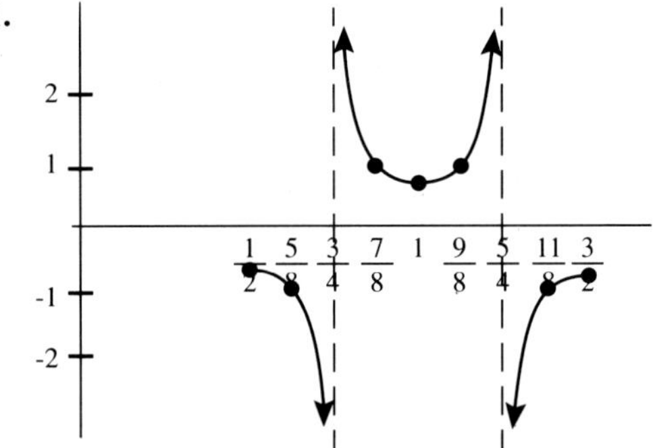

13.

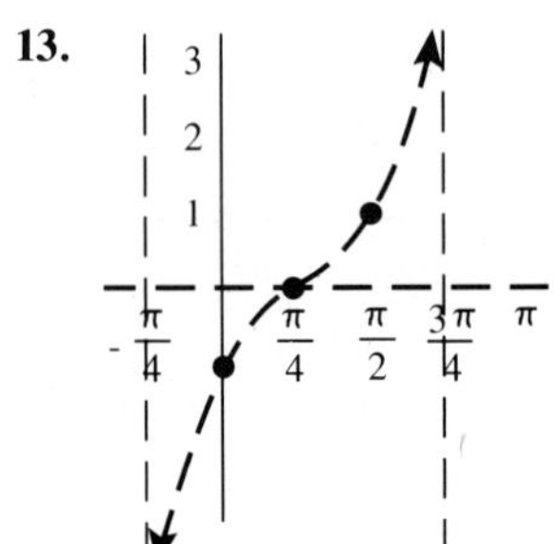

15.

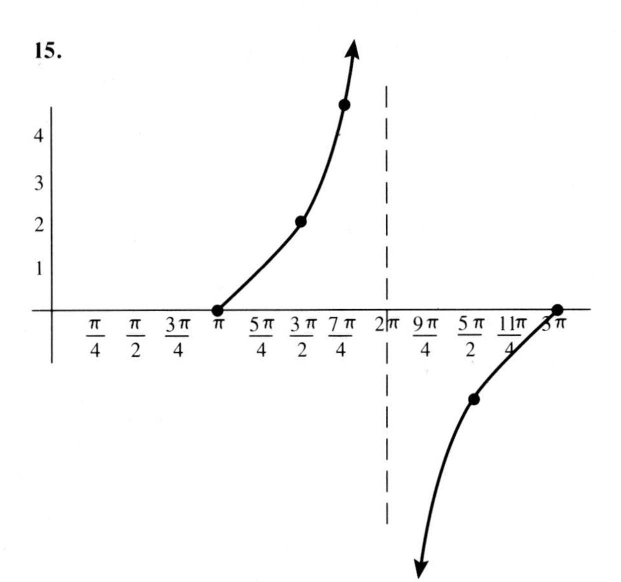

17.

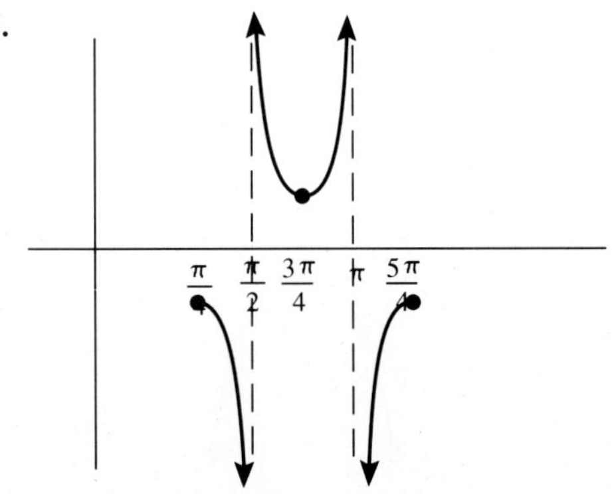

19.

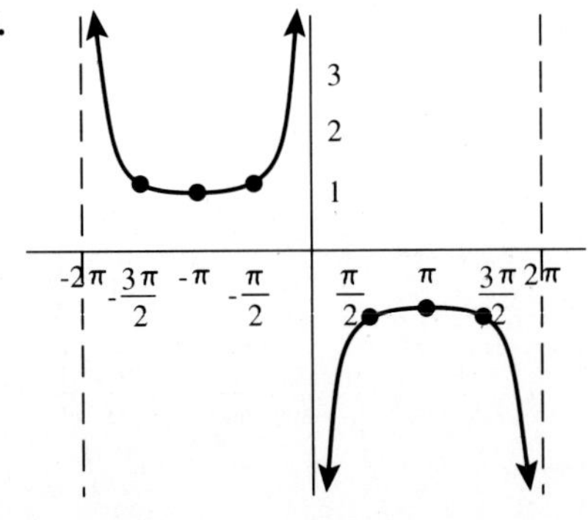

12–5 Exercises, p. 436

1. **(a)** 1.4, $\frac{1}{3}$, $\frac{-\pi}{2}$ **(b)** 1.4 at 1.5 s **(c)** 1/6 s **(d)**

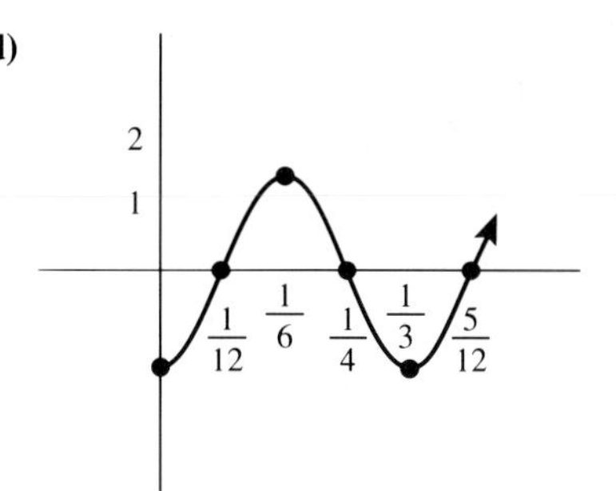

3. **(a)** 2, $\frac{2}{3}$, $\frac{\pi}{2}$ **(b)** 0 at 1.5 s **(c)** 0 s or $\frac{2}{3}$ s in 2nd cycle **(d)**

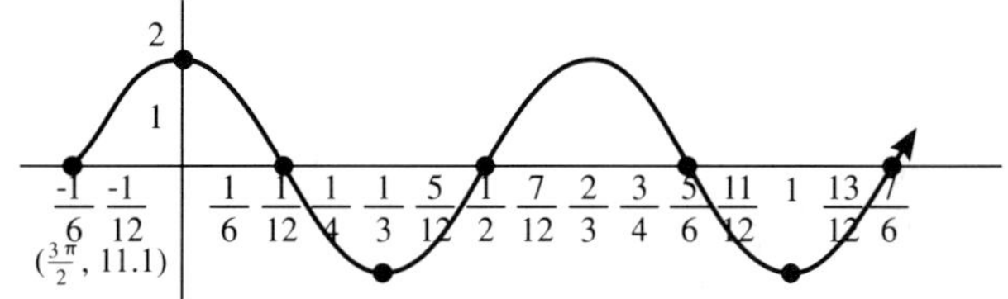

5. **(a)** 3.1, 2, $-\pi$ **(b)** 3.1 at 1.5 s **(c)** $\frac{3}{2}$ s **(d)**

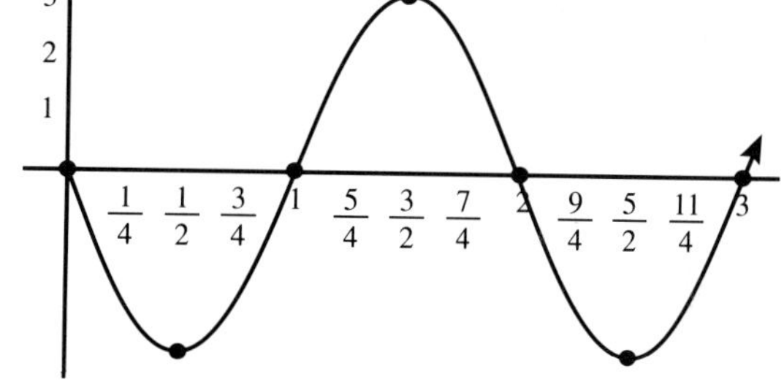

7. $y = 4\sin\left(\pi t + \frac{\pi}{3}\right)$ **9.** $y = 6\sin\left(\frac{20\pi}{13}t - \frac{\pi}{4}\right)$ **11.** $y = 2\sin\left(\frac{\pi}{2}t + \pi\right)$

13.

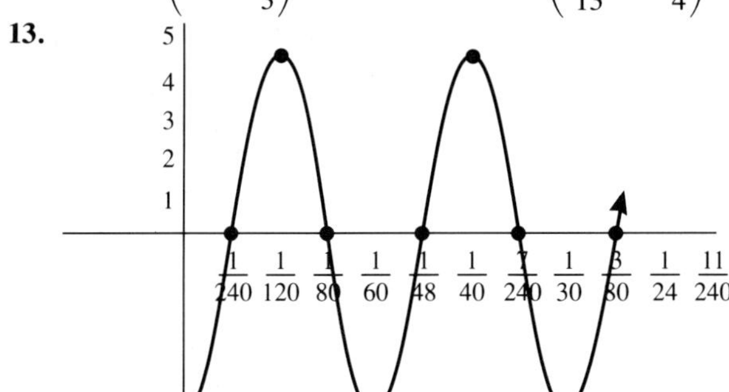

15.

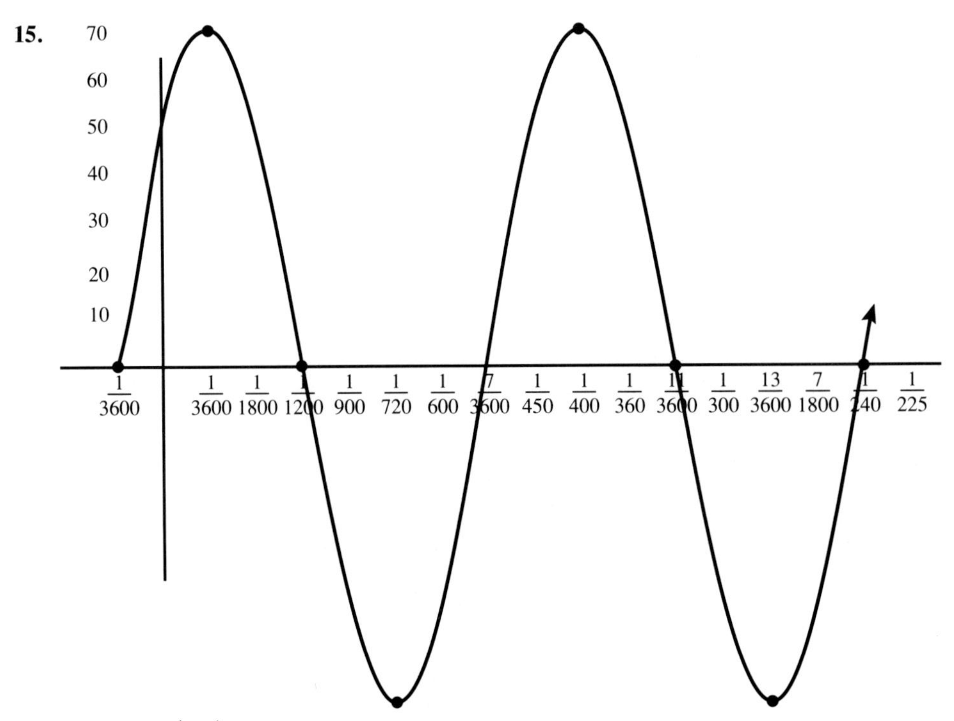

17. $y = 18 \sin\left(\frac{4\pi}{3}t\right)$, 7.32 cm at 0.65 s

19. $V = 135 \sin 100\pi t$, amp $= 135$, per $= \frac{1}{50}$

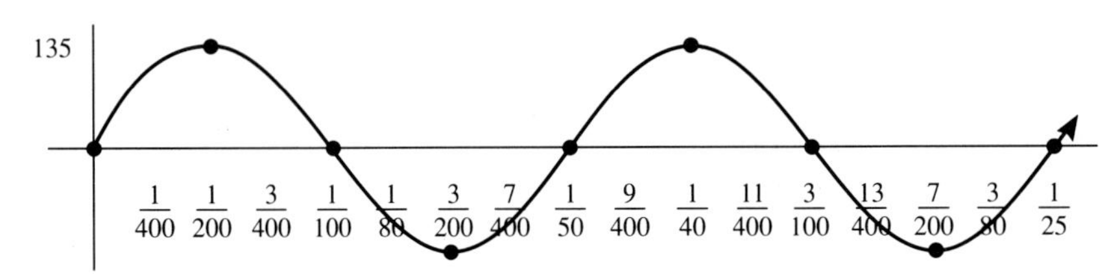

12–6 Exercises, p. 441

1.

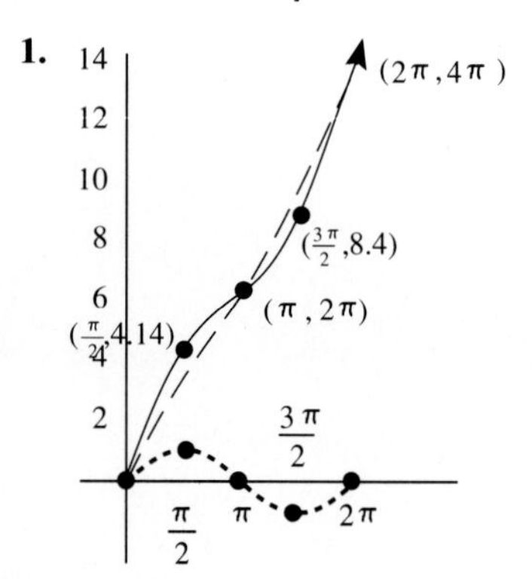

3.

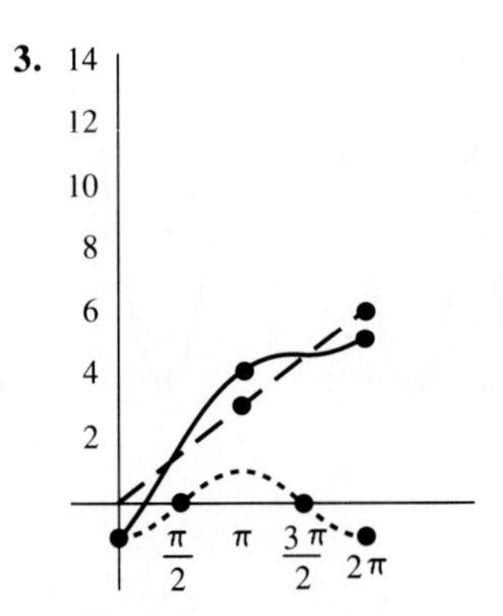

5.

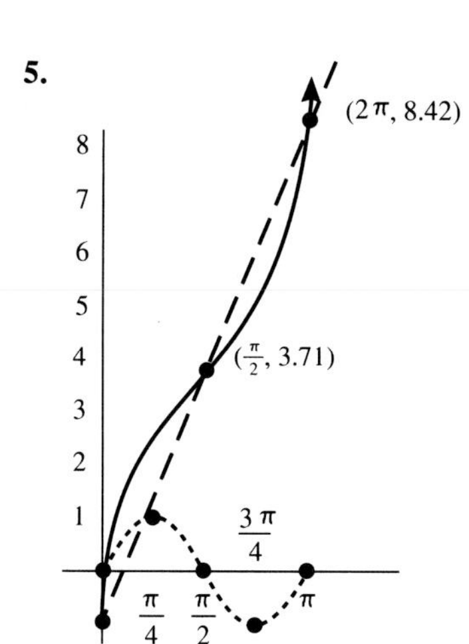

7.

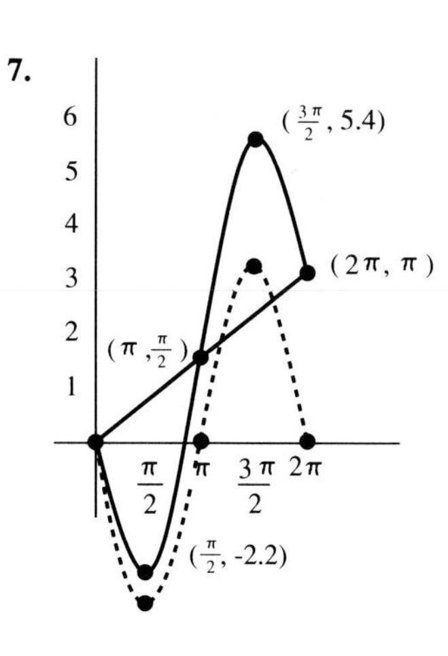

9.

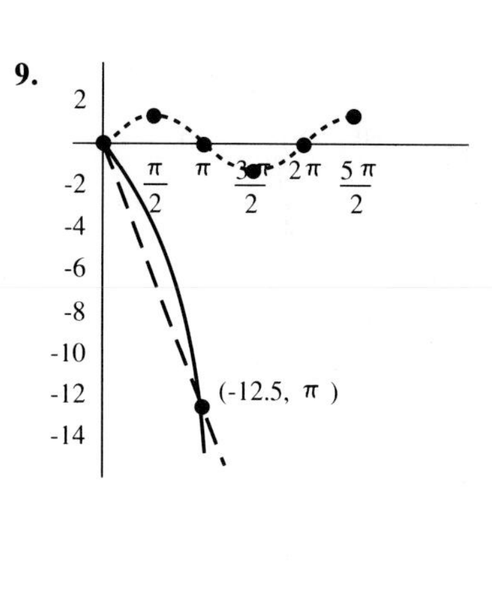

11.

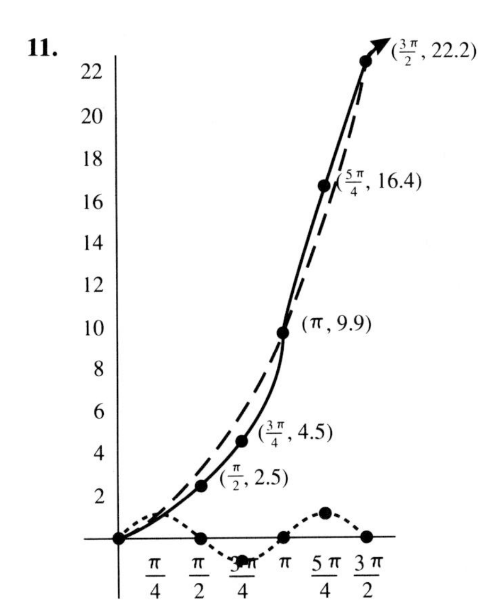

13.

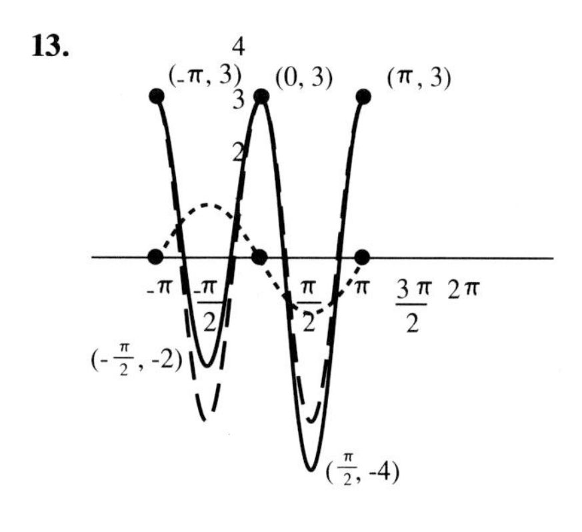

15.

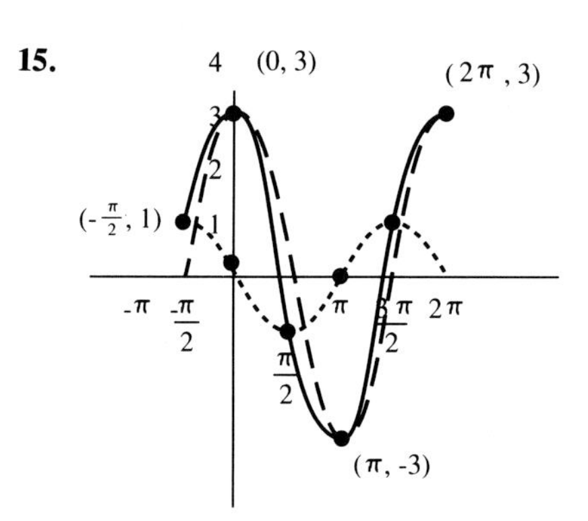

17.

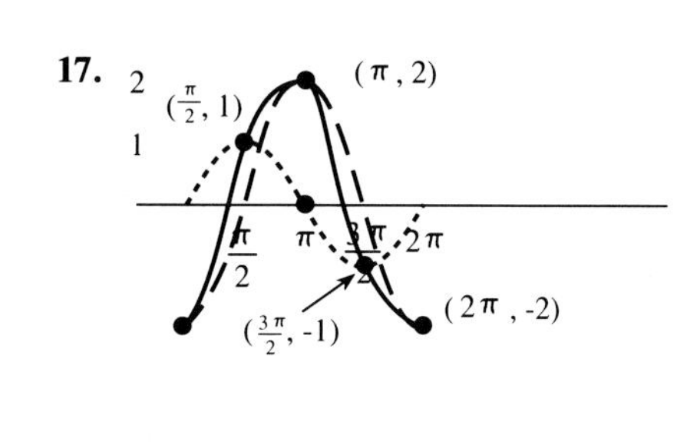

19.

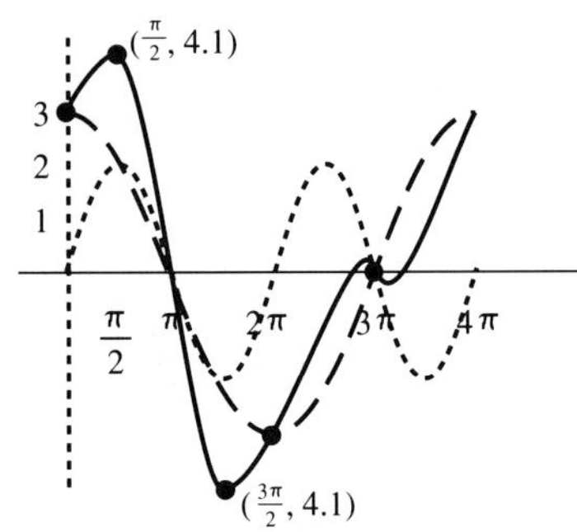

(π/2, 4.1)
3
2
1
π/2
π
2π
3π
4π
(3π/2, 4.1)

21.

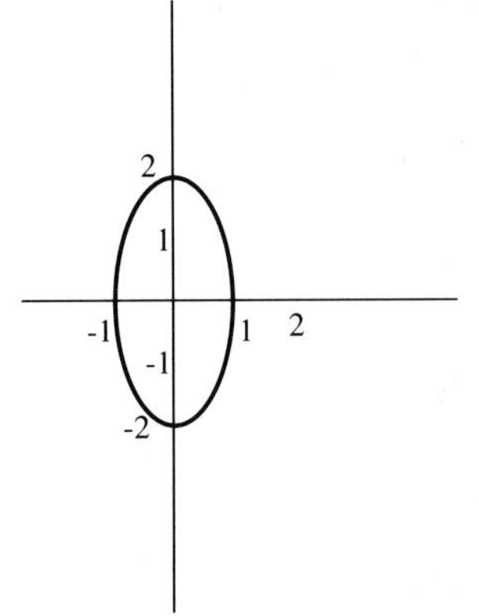

2
1
-1
1
2
-1
-2

23.

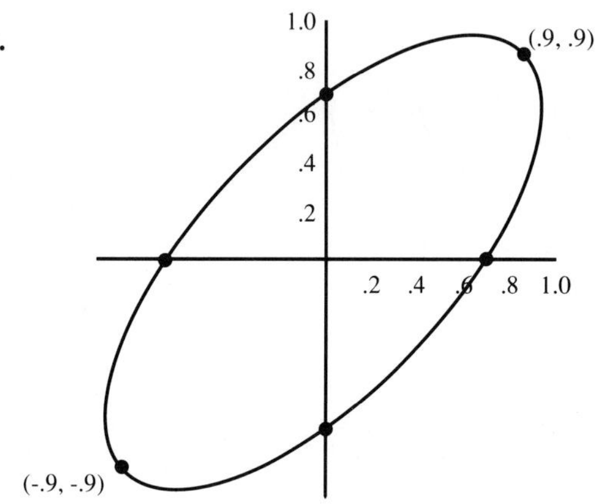

1.0
.8
.6
.4
.2
(.9, .9)
.2
.4
.6
.8
1.0
(-.9, -.9)

25.

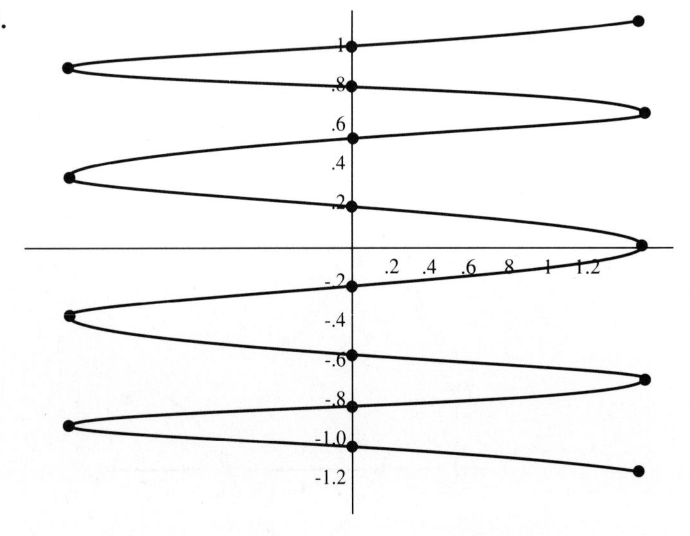

1
.8
.6
.4
.2
.2
.4
.6
.8
1
1.2
-.2
-.4
-.6
-.8
-1.0
-1.2

27.

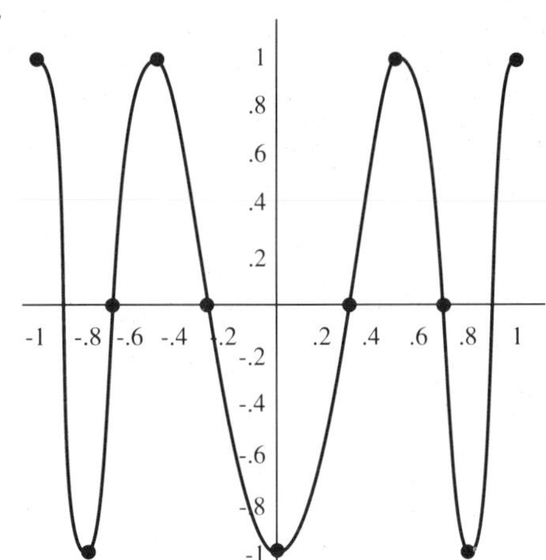

29.

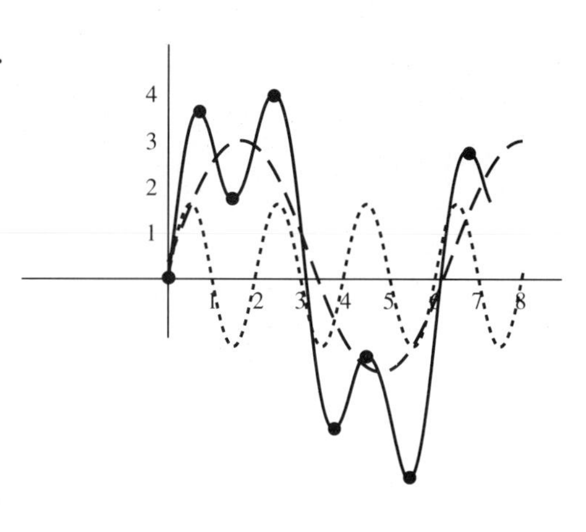

31.

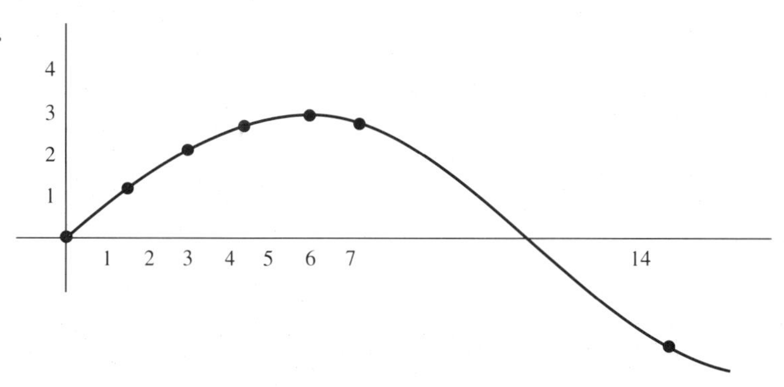

33.

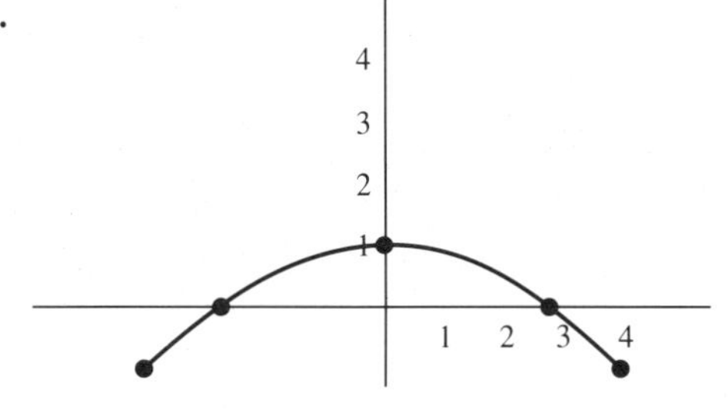

Chapter Review, p. 443

1.

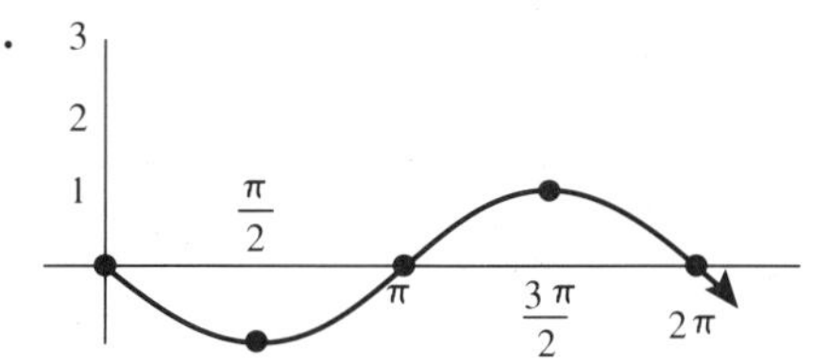

3.

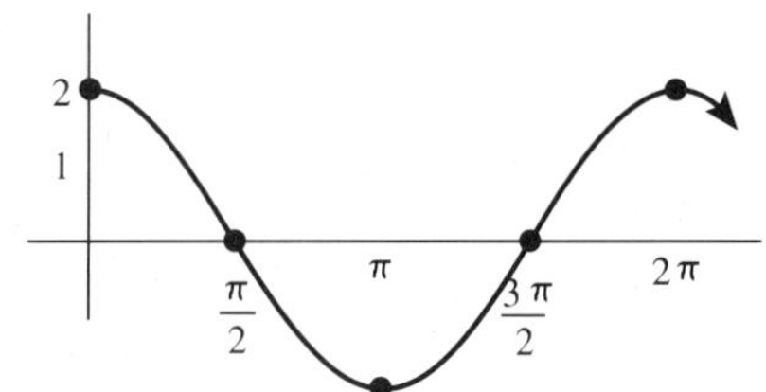

5.

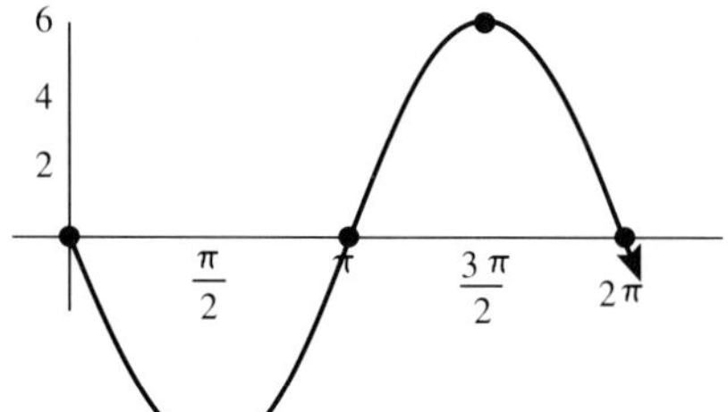

7. 1, π

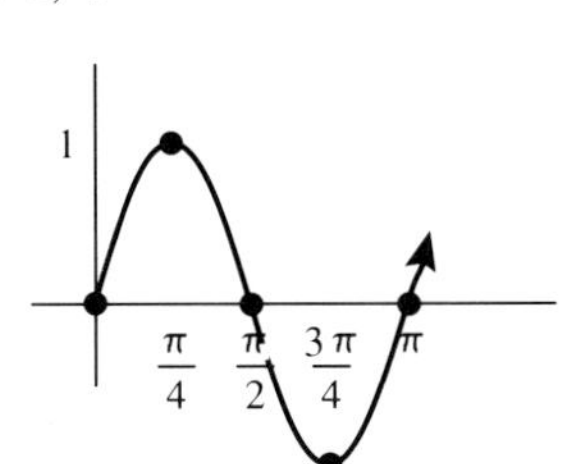

9. 3, 8π

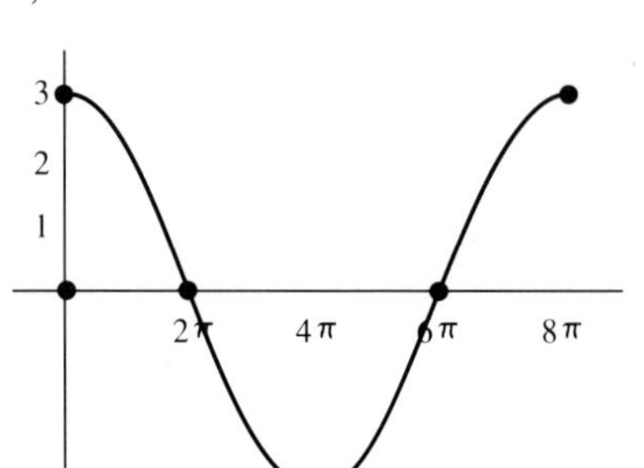

11. 6, $\frac{\pi}{2}$

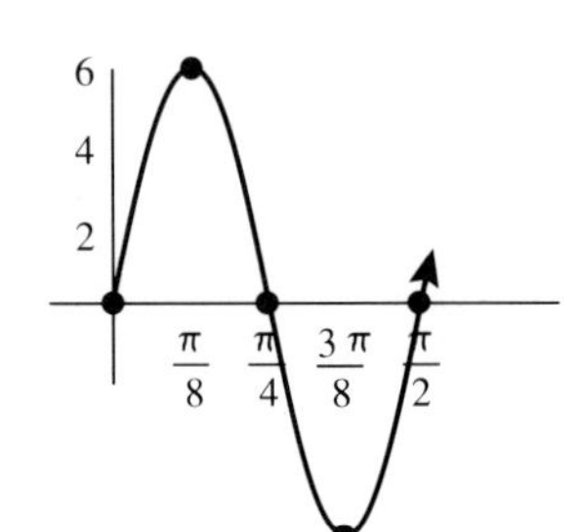

13. 1.6, $\frac{2}{3}$

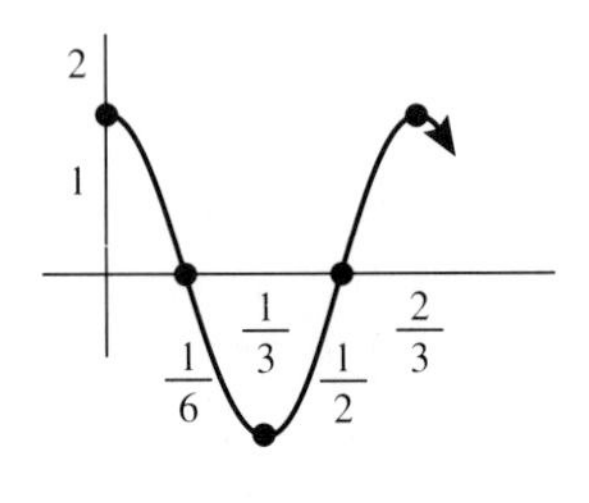

15. 4, 2

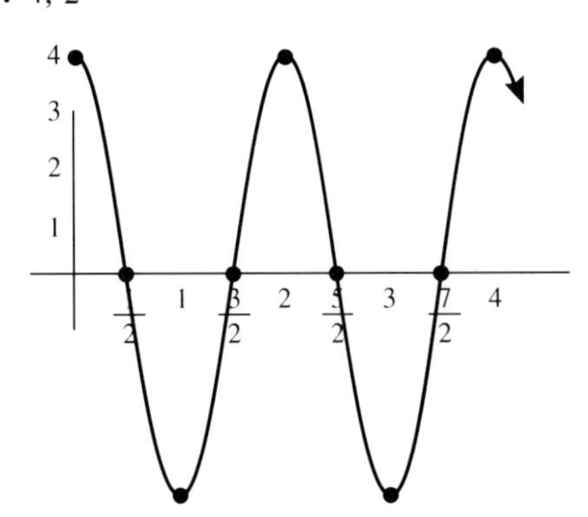

17. 2, 2π, $\frac{\pi}{4}$

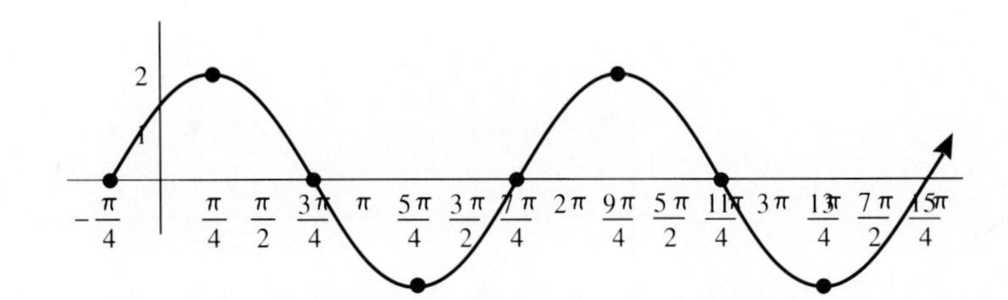

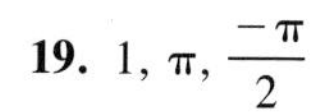

19. 1, π, $\frac{-\pi}{2}$

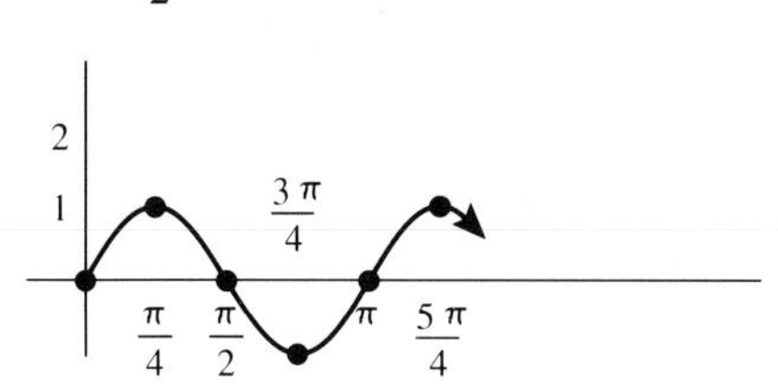

21. 3, 1, -3

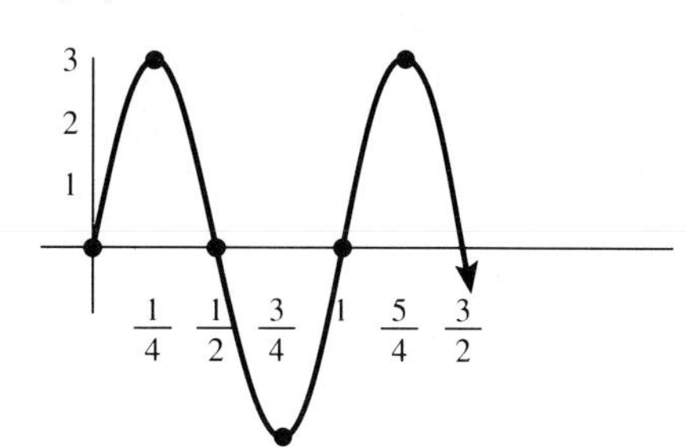

23. 4, 6π, $\frac{\pi}{2}$

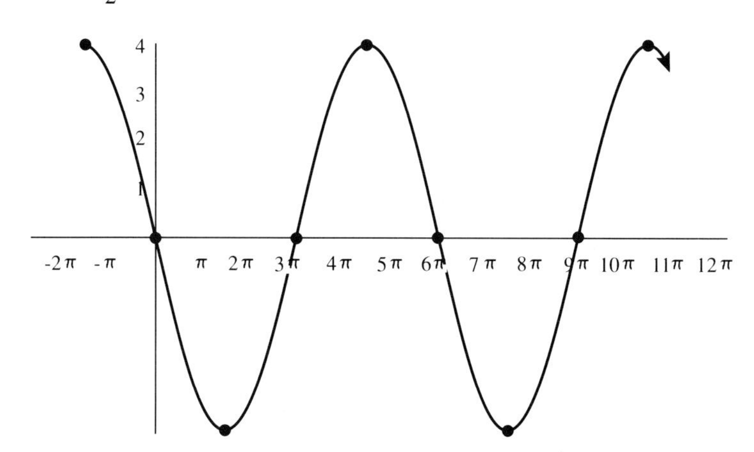

25. $y = 2\cos(2\pi^2 x - \pi^3)$

27.

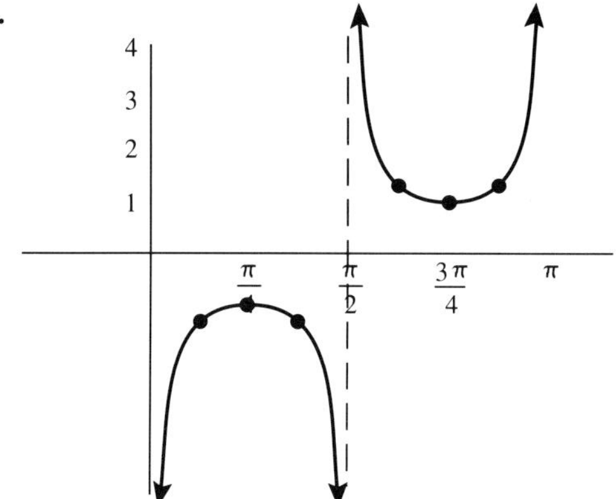

29.

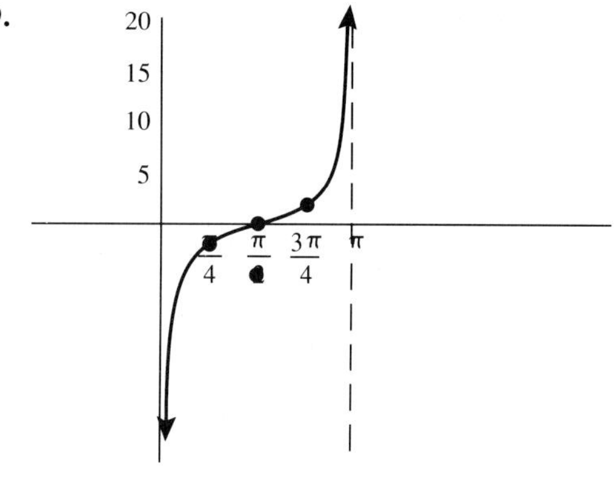

31. 

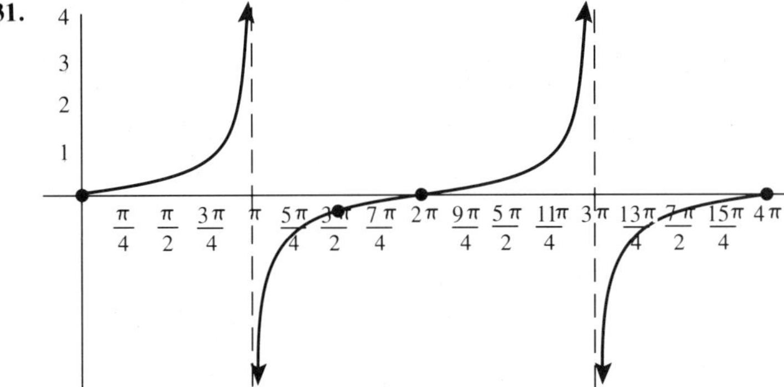

33.

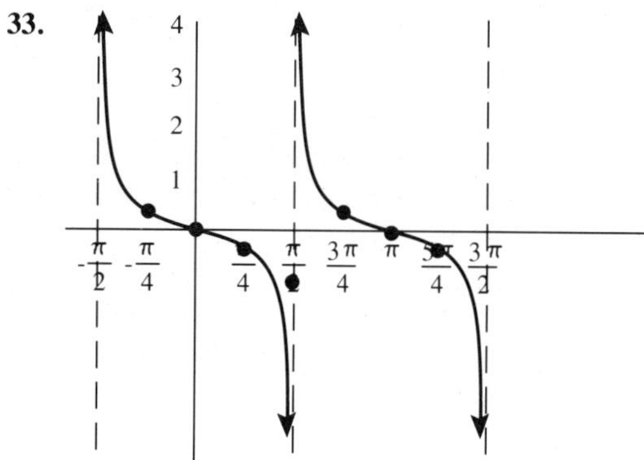

35.

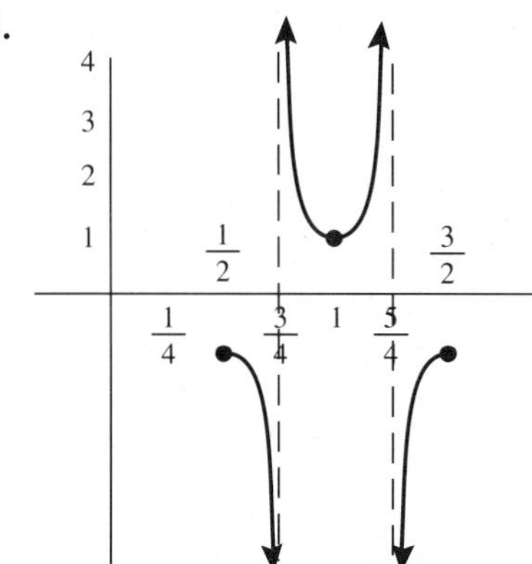

37. **(a)** 8, 2π, $\frac{\pi}{4}$ right **(b)** 8 at 2.4 s **(c)** $\frac{3\pi}{4}$

(d)

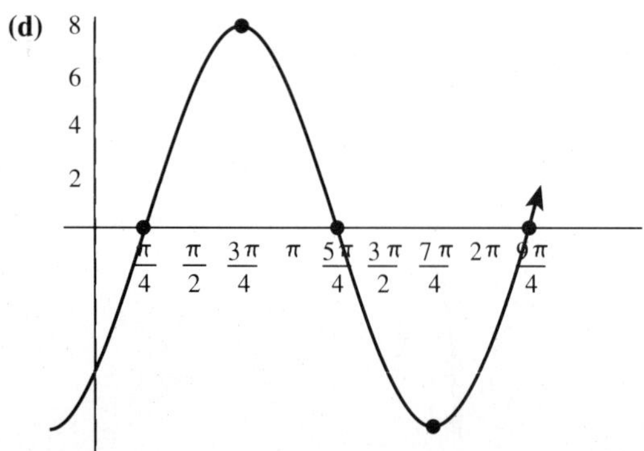

39. **(a)** 3.1, π, $\frac{\pi}{2}$ left **(b)** 3.1 at 2.4 s

(c) $\frac{-\pi}{4}$ or $\frac{3\pi}{4}$ in 2nd cycle

(d)

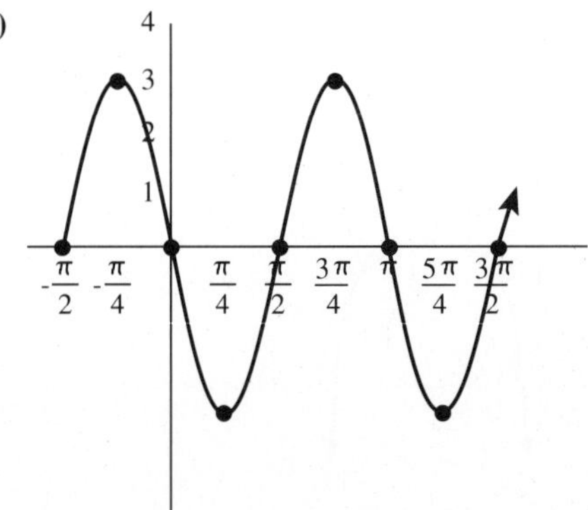

41. $60, \frac{\pi}{15}, -\pi$

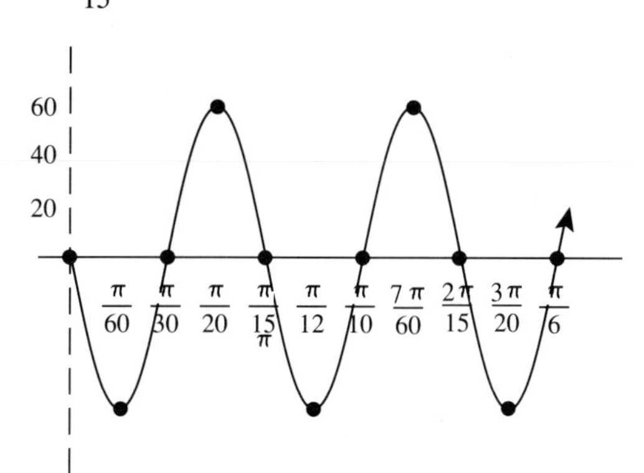

43.

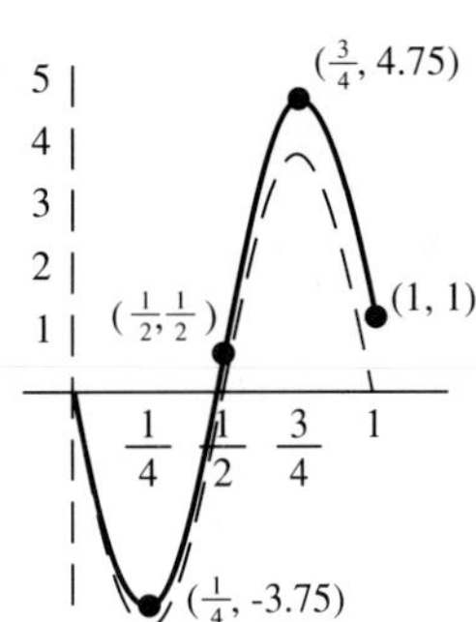

45.

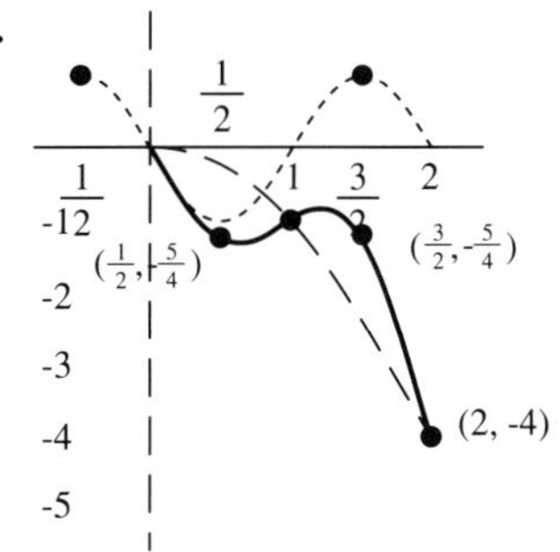

47.

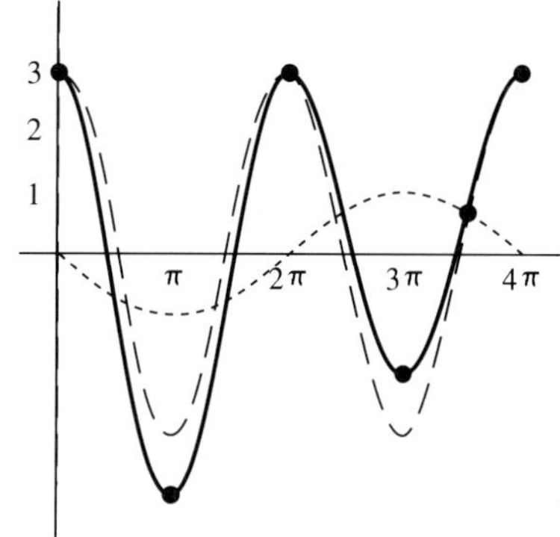

49.

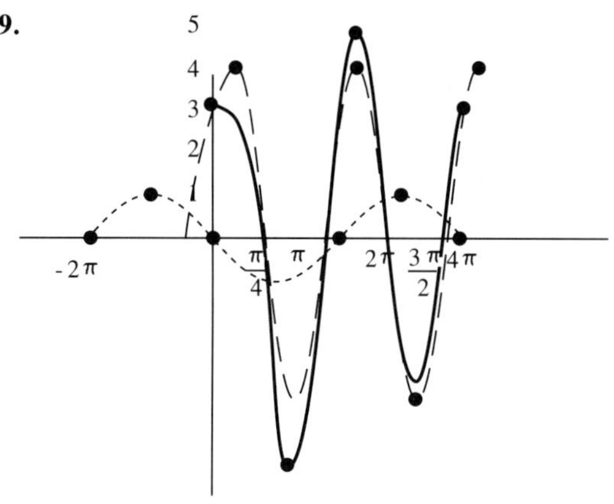

51.

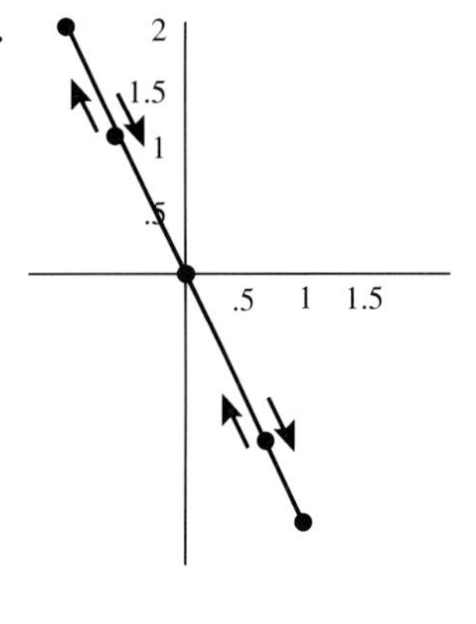

53.

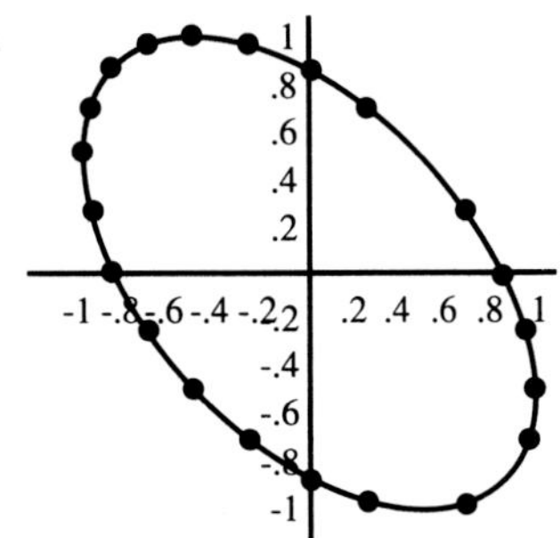

Chapter Test, p. 444

1.

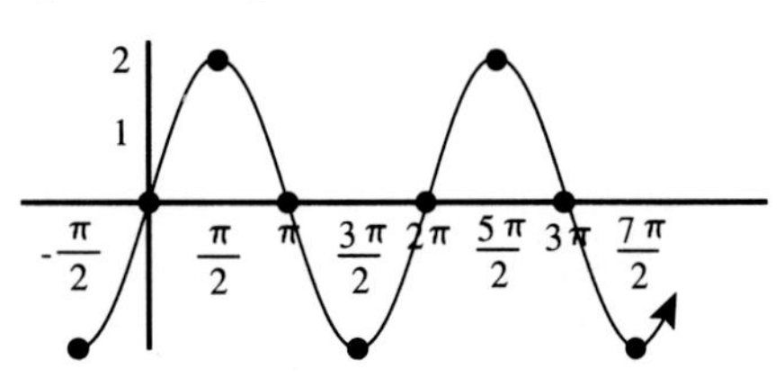

2.

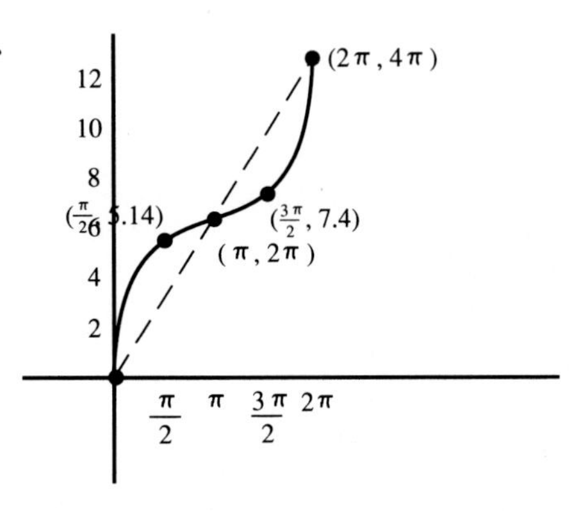

3. $y = 2\sin\left(\frac{5}{2}\pi t + \frac{\pi}{3}\right)$ **4.** $1.7, \frac{2}{3}, \frac{1}{9}$ right

5.

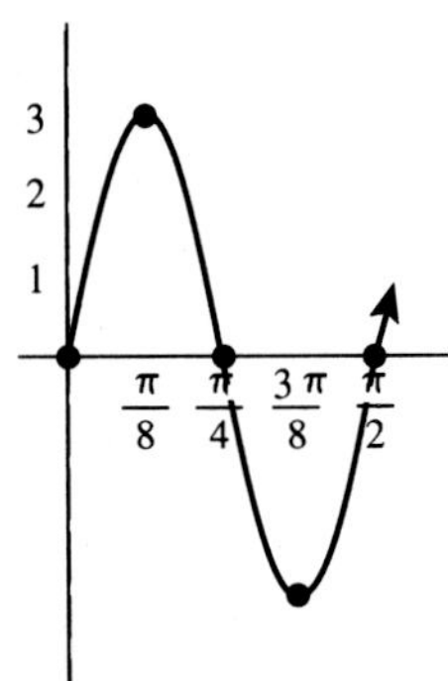

6.

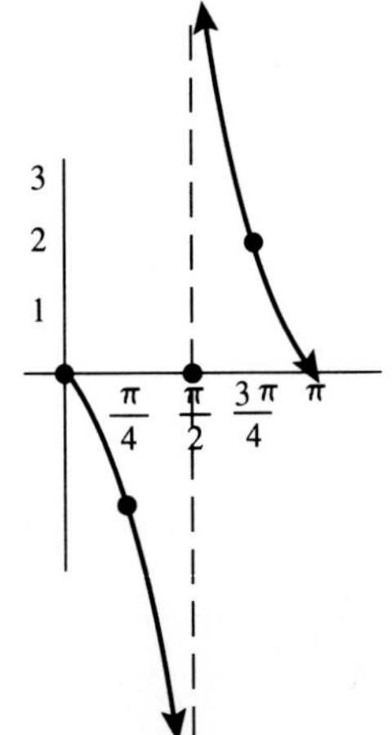

7.

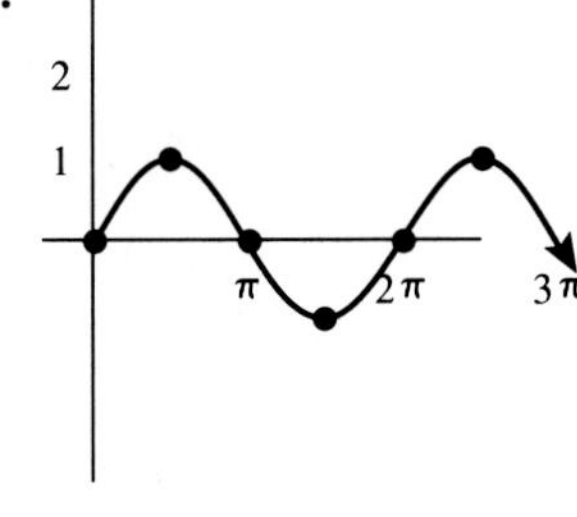

8. $I = 2.4\sin(120\pi t)$, $I = 0$ A

9. 

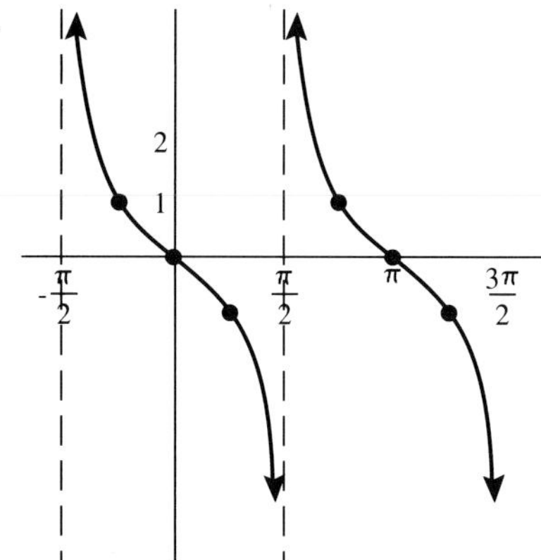

10.

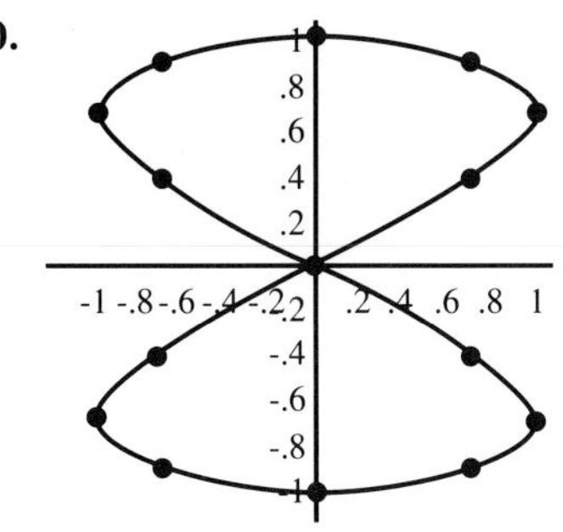

11. 3, 2, $\frac{1}{\pi}$ right

12. 

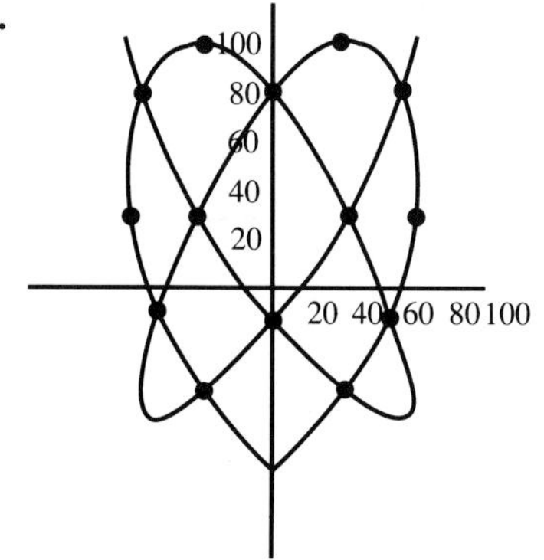

13. $y = 115\sin\left(100\pi t - \frac{\pi}{6}\right)$

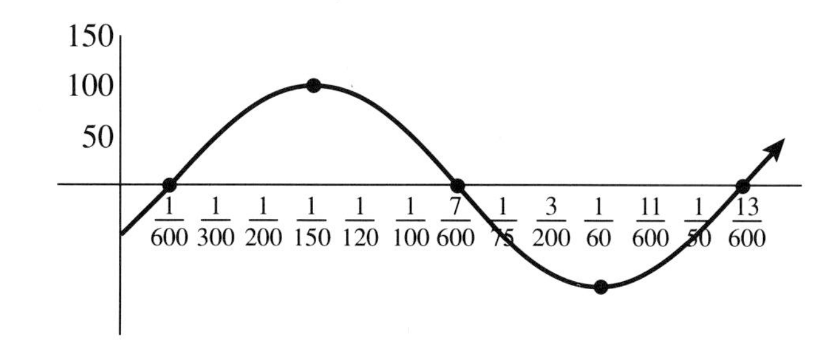

14.

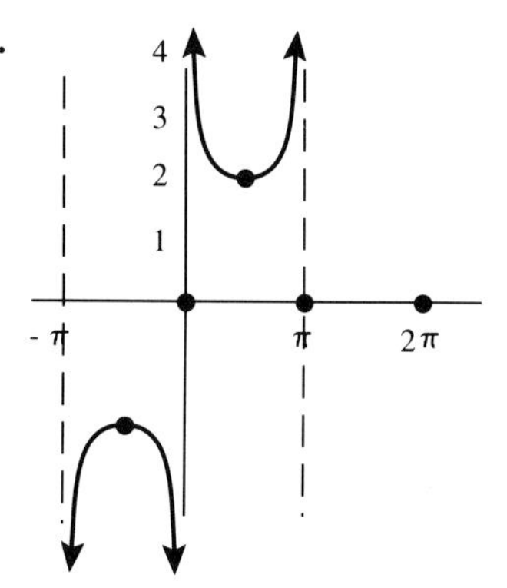

CHAPTER 13

13–1 Exercises, p. 454

The problems in this section can all be solved.

13–2 Exercises, p. 460

1. $\dfrac{\sqrt{6}+\sqrt{2}}{4}$ **3.** $\dfrac{\sqrt{6}+\sqrt{2}}{4}$ **5.** $\dfrac{-\sqrt{2}-\sqrt{6}}{4}$ **7.** $\dfrac{-\sqrt{2}-\sqrt{6}}{4}$ **9. (a)** $\dfrac{-171}{221}$ **(b)** $\dfrac{-140}{171}$ **(c)** quad II **11. (a)** $\dfrac{-171}{221}$ **(b)** $\dfrac{-140}{221}$ **(c)** quad III **13.** $\sin 35°$ **15.** $\cos 6A$ **17.** $-\cos 4A$ **19.** $\tan B$ **21–31.** These identities can be verified. **33.** $a\sin(2\pi ft + B)$ **35.** $y = 12\cos\left(\dfrac{\pi t}{3}\right)$

13–3 Exercises, p. 468

1. $\dfrac{\sqrt{2-\sqrt{3}}}{2}$ **3.** $\dfrac{-\sqrt{2-\sqrt{3}}}{2}$ **5.** $-\sqrt{3}$ **7.** $\cos 2A = \dfrac{161}{289}$, $\tan\dfrac{A}{2} = \dfrac{1}{4}$ **9.** $\tan\dfrac{A}{2} = -7$, $\sin\dfrac{A}{2} = \dfrac{7\sqrt{2}}{10}$ **11.** $2\sin 4A$ **13.** $\sin^2 2x$ **15.** $\cos 105°$ **17–31.** These problems can be proven. **33.** $n = \cos\dfrac{D}{2} + \cot\dfrac{A}{2}\sin\dfrac{D}{2}$ (others are possible)

13–4 Exercises, p. 473

1. 0.62, 2.53, $\dfrac{2\pi}{3}, \dfrac{4\pi}{3}$ **3.** $x = \dfrac{3\pi}{2}$ **5.** $\dfrac{7\pi}{6}, \dfrac{11\pi}{6}, \dfrac{\pi}{6}, \dfrac{5\pi}{6}$ **7.** 2.14, 5.28, 1.79, 4.93 **9.** $\dfrac{\pi}{4}, \dfrac{5\pi}{4}$ **11.** $\dfrac{\pi}{2}, \dfrac{3\pi}{2}, 0$ **13.** $\dfrac{\pi}{6}, \dfrac{5\pi}{6}, \dfrac{3\pi}{2}$ **15.** no solution **17.** $0, \pi, \dfrac{\pi}{6}, \dfrac{5\pi}{6}$ **19.** 1.77, 4.51, 0 **21.** $\dfrac{3\pi}{10}, \dfrac{7\pi}{10}, \dfrac{11\pi}{10}, \dfrac{15\pi}{10}, \dfrac{19\pi}{10}$ **23.** $\dfrac{\pi}{3}, \dfrac{5\pi}{3}, \pi$ **25.** $\dfrac{\pi}{9}, \dfrac{5\pi}{9}, \dfrac{7\pi}{9}, \dfrac{11\pi}{9}, \dfrac{13\pi}{9}, \dfrac{17\pi}{9}$ **27.** $\dfrac{\pi}{2}, \dfrac{3\pi}{2}, \dfrac{\pi}{6}, \dfrac{5\pi}{6}$ **29.** $\dfrac{\pi}{2}, \dfrac{3\pi}{2}, \dfrac{\pi}{4}, \dfrac{5\pi}{4}$ **31.** 20.75°, 69.25° **33.** 3.97 s, 5.43 s

13–5 Exercises, p. 481

1. $\dfrac{\pi}{3}$ **3.** $\dfrac{\pi}{3}$ **5.** 0 **7.** $\dfrac{\pi}{4}$ **9.** $\dfrac{\pi}{6}$ **11.** $-\dfrac{\pi}{2}$ **13.** 0.845 **15.** 2.305 **17.** −1.064 **19.** 0.810 **21.** 0.5 **23.** 0.5 **25.** 1 **27.** undefined **29.** $\dfrac{3}{4}$ **31.** $-\dfrac{12}{13}$ **33.** $-\dfrac{15}{17}$ **35.** $\dfrac{1}{\sqrt{1-x^2}}$ **37.** $2x\sqrt{1-x^2}$ **39.** $\dfrac{1-\sqrt{1-x^2}}{x}$ **41.** $\sqrt{1-10x^2+9x^4} - 3x^2$ **43.** $\text{Arcsin}\,\dfrac{C}{HmL}$ **45.** $4\arcsin \pm 2\sqrt{\dfrac{P-T}{T}}$

Chapter Review, p. 482

1–11. These problems can be proven. **13. (a)** $\dfrac{435}{533}$ **(b)** $\dfrac{92}{525}$ **(c)** $\dfrac{308}{533}$ **(d)** quad I **15.** $\sin(C - D + B)$ **17.** $\tan 5x$ **19.** can be proven **21.** $\sin\left(x+\dfrac{\pi}{2}\right) - \cos\left(x+\dfrac{\pi}{2}\right) = \cos x + \sin x$; $\sin x\cos\dfrac{\pi}{2} + \cos x\sin\dfrac{\pi}{2} - \left(\cos x\cos\dfrac{\pi}{2} - \sin x\sin\dfrac{\pi}{2}\right) = \cos x + \sin x$; $0 + \cos x - 0 + \sin x = \cos x + \sin x$

23. can be proven **25.** can be proven **27.** $8 \cos 2B$ **29.** $\tan \frac{5B}{2}$ **31.** $\sin \frac{B}{2} = \frac{4\sqrt{17}}{17}$, $\cos 2B = \frac{161}{289}$

33–37. These can be proven. **39.** 0.61, 2.53 **41.** 2.3, 3.98, $\frac{\pi}{3}$, $\frac{5\pi}{3}$ **43.** π, 0, $\frac{\pi}{4}$, $\frac{5\pi}{4}$

45. $\frac{\pi}{2}$ **47.** 1.15, 1.98, 4.29, 5.13 **49.** $\frac{8}{17}$ **51.** $-\frac{40}{9}$ **53.** $\frac{x}{\sqrt{1 + x^2}}$ **55.** $\sqrt{1 - x^2}$

57. $\sqrt{\frac{1 + \sqrt{1 - x^2}}{2}}$

Chapter Test, p. 483

Note: Problems 3, 6, 8, 10, 13, 14 can be proven. **1.** $\tan \frac{A}{2} = \frac{5}{4}$, $\sin 2A = -\frac{720}{1681}$ **2.** 1.11, 2.03

4. $\sqrt{\frac{1 + \sqrt{1 - 9x^2}}{2}}$ **5.** $\cos(A + B + C - D)$ **7.** $\frac{13}{5}$ **9.** 3.47, 5.94, $\frac{\pi}{6}$, $\frac{5\pi}{6}$ **11.** 39.26°

12. $\frac{1}{2} \sin \frac{A}{2}$ **15.** $\frac{\sqrt{2}}{2}A_1(\sin x + \cos x) + \frac{\sqrt{2}}{2}A_2(\cos x - \sin x)$

CHAPTER 14

14–1 Exercises, p. 492

1. $3j$ **3.** $2j\sqrt{2}$ **5.** $-0.3j$ **7.** $-2j\sqrt{10}$ **9.** $\frac{2}{7}j\sqrt{2}$ **11.** $-2\sqrt{3}$ **13.** $3j\sqrt{6}$ **15.** -54

17. $31j$ **19.** 1 **21.** $-j$ **23.** -1 **25.** $-j$ **27.** 8 **29.** $-3j$ **31.** $x = \frac{1}{2}$, $y = -7$

33. $x = 5$, $y = \frac{14}{3}$ **35.** $x = -2$, $y = -3$ **37.** $x = -8$, $y = \frac{1}{3}$ **39.** $x = -8$, $y = 11$ **41.** $8 + 5j$
43. $-7 - 9j$ **45.** $2j$ **47.** -10 **49.** $Z = 44 - 14j$

14–2 Exercises, p. 497

1. $3 - 7j$ **3.** $-14 + 16j$ **5.** $6 + 8j$ **7.** $6 + j$ **9.** $-23 + 9j$ **11.** $-14 + j$ **13.** $-8 + 3j$
15. $22 - 2j$ **17.** -33 **19.** $30 + 15j$ **21.** $35 - 6j$ **23.** $-7 + 103j$ **25.** $68 + 11j$

27. $-21 + 20j$ **29.** 53 **31.** $27 - 36j$ **33.** $-46 - 9j$ **35.** $\frac{7 - 6j}{2}$ **37.** $\frac{2 + 3j}{-3}$ **39.** $\frac{6 - 2j}{5}$

41. $-5 - 8j$ **43.** $\frac{9 - 5j}{2}$ **45.** $\frac{-51 + 3j}{29}$ **47.** $\frac{-29 - 15j}{13}$ **49.** $\frac{-53 + 22j}{37}$

51.

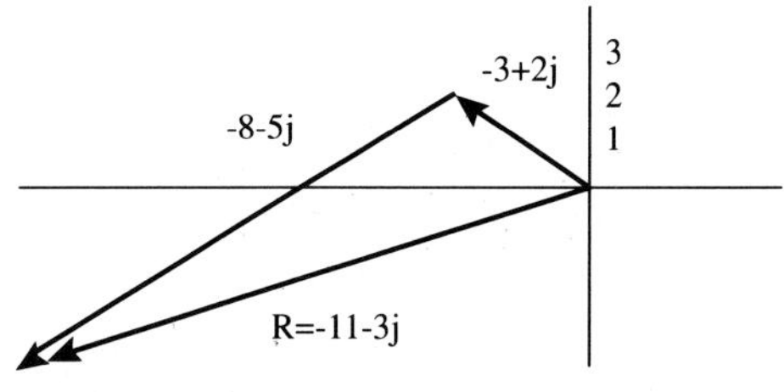

53.

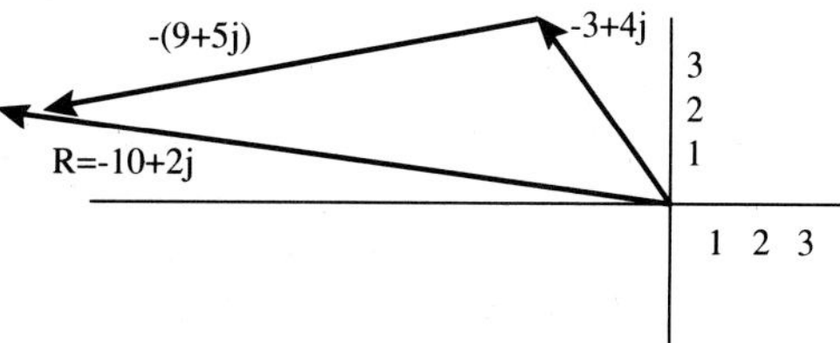

55.

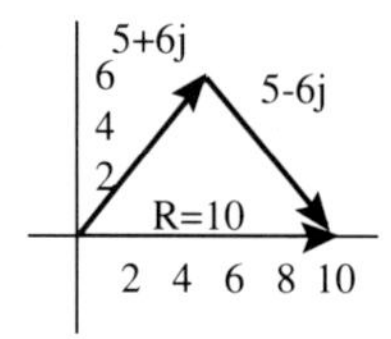

14–3 Exercises, p. 503

1. $\sqrt{13}(\cos 123.6° + j \sin 123.6°)$, $\sqrt{13}e^{2.16j}$

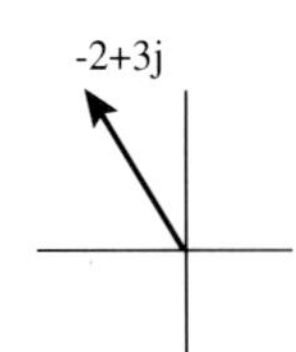

3. $\sqrt{149}(\cos 234.9° + j \sin 234.9°)$, $\sqrt{149}e^{4.1j}$

-7-10j

5. $3\sqrt{13}(\cos 56.3° + j \sin 56.3°)$, $3\sqrt{13}e^{0.98j}$

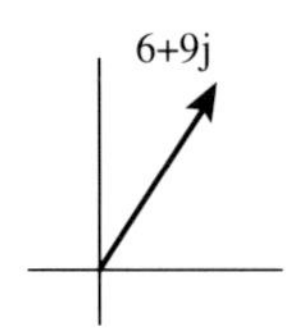

7. $\sqrt{205}(\cos 294.6° + j \sin 294.6°)$, $\sqrt{205}e^{5.14j}$

6-13j

9. $12.58(\cos 126.5° + j \sin 126.5°)$, $12.58e^{2.21j}$

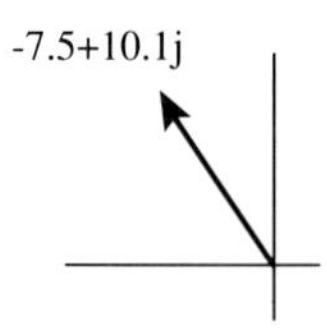

11. $10.9(\cos 238.4° + j \sin 238.4°)$, $10.9e^{4.16j}$

-5.7-9.3j

13. $8(\cos 90° + j \sin 90°)$, $8e^{1.57j}$

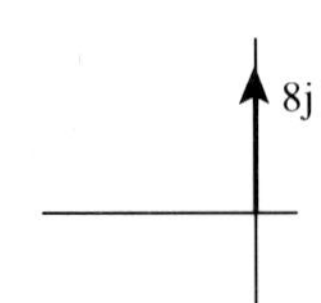

15. $15(\cos 0° + j \sin 0°)$, $15e^{0j}$

15

17. $3 - 5.2j$ **19.** $-44.8 + 254j$ **21.** $4.02 - 5.98j$ **23.** -137 **25.** $-1.59 + 6.8j$ **27.** $-139.7 + 84j$ **29.** $7.35 + 27.6j$ **31.** $0.34 + 0.66j$ **33.** $6.75(\cos 223.5° + j \sin 223.5°)$, $-4.9 - 4.6j$ **35.** $175(\cos 103.1° + j \sin 103.1°)$, $-39.7 + 170.4j$ **37.** $26.9(\cos 143.2° + j \sin 143.2°)$, $-21.5 + 16.1j$ **39.** $6.4(\cos 304° + j \sin 304°)$, $6.4e^{5.3j}$

14–4 Exercises, p. 510

1. $28e^{5j}$ **3.** $50.02e^{6.1j}$ **5.** $18.4e^{0.78j}$ **7.** $4e^{5j}$ **9.** $4.16e^{5.18j}$ **11.** $2.16e^{1.52j}$ **13.** $156e^{2.71j}$ **15.** $112(\cos 74.2° + j \sin 74.2°)$ **17.** $192(\cos 264.5° + j \sin 264.5°)$ **19.** $0.3(\cos 45° + j \sin 45°)$ **21.** $2.44(\cos 159.5° + j \sin 159.5°)$ **23.** $216(\cos 81° + j \sin 81°)$ **25.** $4220(\cos 241° + j \sin 241°)$

27. 1.12(cos 105° + j sin 105°), 1.12(cos 225° + j sin 225°), 1.12(cos 345° + j sin 345°)
29. 2.45(cos 156° + j sin 156°), 2.45(cos 336° + j sin 336°) **31.** 1(cos 22.5° + j sin 22.5°), 1(cos 112.5° + j sin 112.5°), 1(cos 202.5° + j sin 202.5°), 1(cos 292.5° + j sin 292.5°) **33.** 1(cos 0° + j sin 0°), 1(cos 60° + j sin 60°), 1(cos 120° + j sin 120°), 1(cos 180° + j sin 180°), 1(cos 240° + j sin 240°), 1(cos 300° + j sin 300°) **35.** 2.1(cos 31.7° + j sin 31.7°), 2.1(cos 211.7° + j sin 211.7°)
37. 24(cos 115° + j sin 115°) V **39.** $\frac{31}{20} + \frac{38j}{20}$

14–5 Exercises, p. 518

1. (a) 331.6 Ω **(b)** 13.57 Ω **(c)** 318.12 Ω **(d)** $\theta = -88.65°$ voltage lags current **(e)** 0.0283 A **(f)** 0.2122 V **(g)** 9.381 V **(h)** 0.384 V **3.** 393.61 Hz **5.** 8.6 Ω, −35.54° voltage lags current **7.** 49.65 V **9.** 1.06 A **11.** 265.25 Hz **13.** 2.64×10^{-12} F or 2.64 pF **15.** 0.87 W

Chapter Review, p. 520

1. $2j\sqrt{6}$ **3.** −180 **5.** $-63 + \frac{2}{5}j$ **7.** $\frac{5}{3}j$ **9.** $-j$ **11.** 1 **13.** −4 **15.** $-4j$ **17.** $14 - 22j$
19. $5 + j\sqrt{114} - 7j$ **21.** 29 **23.** $84 + 23j$ **25.** $-198 - 10j$ **27.** $7 + 10j$ **29.** $\frac{-11 + 17j}{82}$
31. 10.6(cos 131.2° + j sin 131.2°), $10.6e^{2.29j}$ **33.** 16.1(cos 209.7° + j sin 209.7°), $16.1e^{3.66j}$
35. 8(cos 90° + j sin 90°), $8e^{1.57j}$ **37.** 14(cos 0° + j sin 0°), $14e^{0j}$ **39.** $1.46 - 13.9j$ **41.** $-1.4 - 0.95j$
43. $19.2 - 12.7j$ **45.** $-7.4 + 16.2j$ **47.** $2e^{11j}$ or $2e^{4.72j}$ **49.** $15e^{2.4j}$ **51.** $3.3e^{-0.7j}$ or $3.3e^{5.6j}$
53. 60.45(cos 70° + j sin 70°) **55.** 0.53(cos 296.9° + j sin 296.9°) **57.** 60.84(cos 259.6° + j sin 259.6°)
59. 1.75(cos 97.3° + j sin 97.3°), 1.75(cos 217.3° + j sin 217.3°), 1.75(cos 337.3° + j sin 337.3°)
61. 1.22(cos 15.85° + j sin 15.85°), 1.22(cos 105.85° + j sin 105.85°), 1.22(cos 195.85° + j sin 195.85°), 1.22(cos 285.85° + j sin 285.85°) **63.** 1(cos 15° + j sin 15°), 1(cos 75° + j sin 75°), 1(cos 135° + j sin 135°), 1(cos 195° + j sin 195°), 1(cos 255° + j sin 255°), 1(cos 315° + j sin 315°) **65.** 1(cos 36° + j sin 36°), 1(cos 108° + j sin 108°), 1(cos 180° + j sin 180°), 1(cos 252° + j sin 252°), 1(cos 324° + j sin 324°) **67.** 530.5 Ω
69. 517.1 Ω **71.** 23 mA **73.** 12.20 V **75.** 127 V **77.** 41.23 pF

Chapter Test, p. 521

1. $16j$ **2.** $8.6e^{3.8j}$ **3.** 6.631×10^2 Ω **4.** −1 **5.** 1.75(cos 22.7° + j sin 22.7°), 1.75(cos 142.7° + j sin 142.7°), 1.75(cos 262.7° + j sin 262.7°) **6.** $81 - 15j$ **7.** $0.21 + 0.77j$
8. 10000(cos 147.5° + j sin 147.5°) **9.** $-12\sqrt{5}$ **10.** 6.74 Ω, −20.85° voltage lags current **11.** −1
12. 12.2(cos 305° + j sin 305°), $12.2e^{5.3j}$ **13.** 142.86 Ω **14.** $\frac{31}{4}j$ **15.** $6.46e^{3j}$ **16.** $-6.6 - j3.3$
17. 5.33 A

CHAPTER 15

15–1 Exercises, p. 533

1. $9x + 3y - 2 = 0$ **3.** $x - 2y + 2 = 0$ **5.** $3x + 5y - 14 = 0$ **7.** $4x - y + 26 = 0$
9. $x - 4 = 0$ **11.** $9x + 8y - 11 = 0$ **13.** $y + 3 = 0$ **15.** $x - 3 = 0$ **17.** $3x - y + 9 = 0$
19. $x + 2y + 9 = 0$

21. $m = \frac{3}{4}$, $b = -3$, x int $= 4$

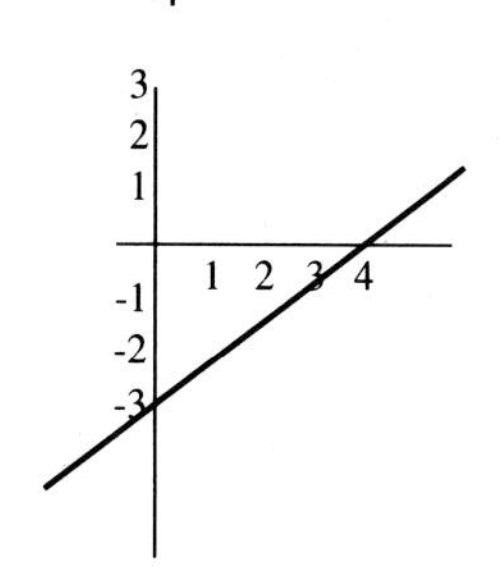

23. $m =$ undef, no y int, x int $= -8$

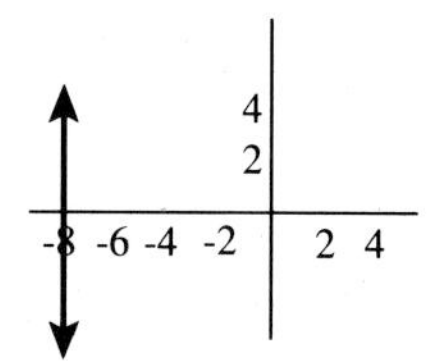

25. $m = \frac{2}{3}$, $b = \frac{5}{2}$, x int $= \frac{-15}{4}$

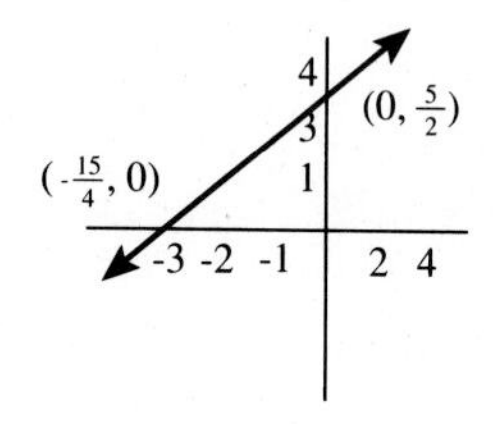

27. perpendicular **29.** neither **31.** parallel

33. $c = 3x + 800$ **35.** $P = \frac{4}{9}D + 13.6$

37. $s = 20t + 15$ **39.** $R = 0.075t + 3.5$

15–2 Exercises, p. 539

1. (0, 0), 3

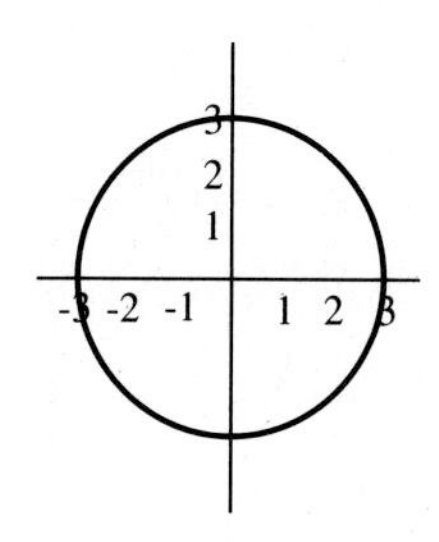

3. $(3, -1)$, $\sqrt{6}$

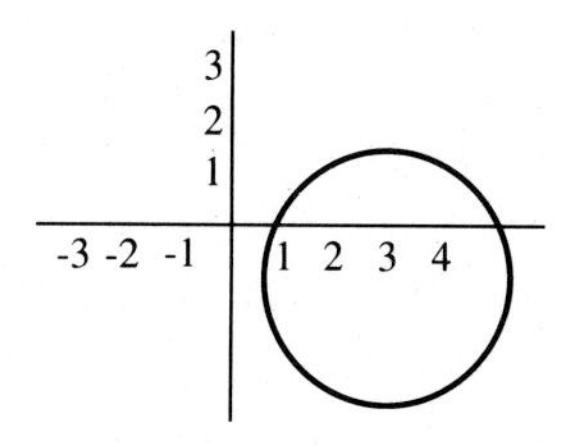

5. (1, 0), $\sqrt[4]{3}$ or 1.3

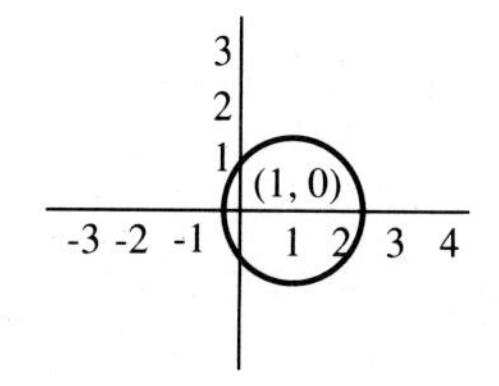

7. $(-4, -1)$, $\sqrt{22}$ or 4.7

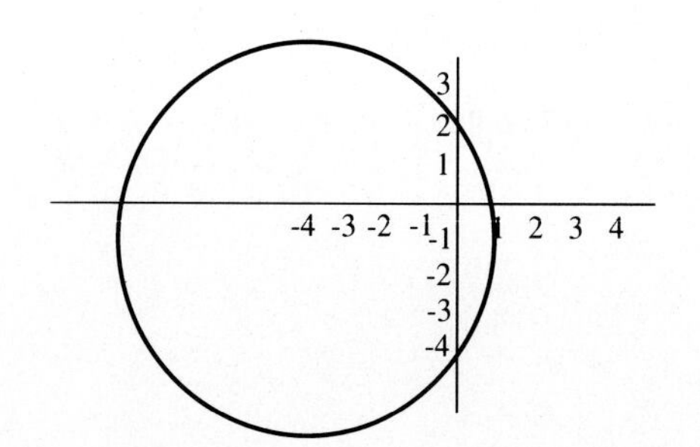

9. $(-2, 3)$, $\sqrt{15}$ or 3.8

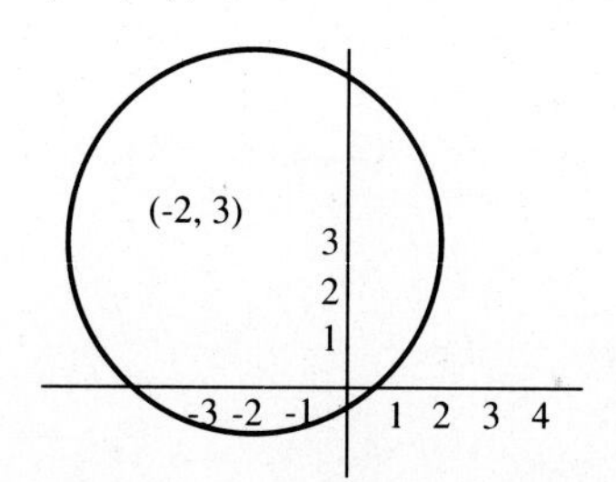

11. $\left(\frac{-5}{2}, 1\right)$, $\sqrt{10.25}$ or 3.2

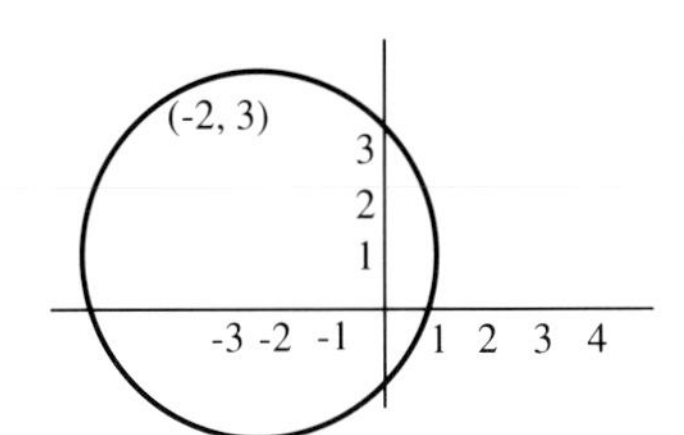

13. $\left(-1, \frac{4}{3}\right)$, $\sqrt{\frac{10}{9}}$ or 1.05

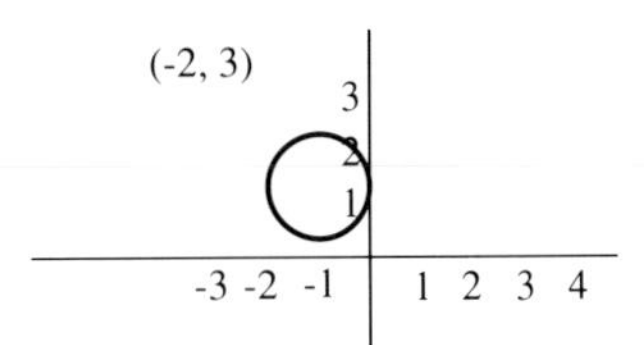

15. $x^2 + y^2 = 81$ **17.** $(x + 3)^2 + (y + 1)^2 = 9$ **19.** $x^2 + (y + 4)^2 = 6$
21. $(x + 8)^2 + (y - 5)^2 = 104$ **23.** $(x - 6)^2 + (y + 4)^2 = 45$ **25.** $(x - 3)^2 + (y - 4)^2 = 16$
27. $\left(x + \frac{3}{2}\right)^2 + \left(y - \frac{1}{2}\right)^2 = 40.5$ **29.** $(x + 4)^2 + (y - 6)^2 = 1.03 \times 10^{-11}$
31.

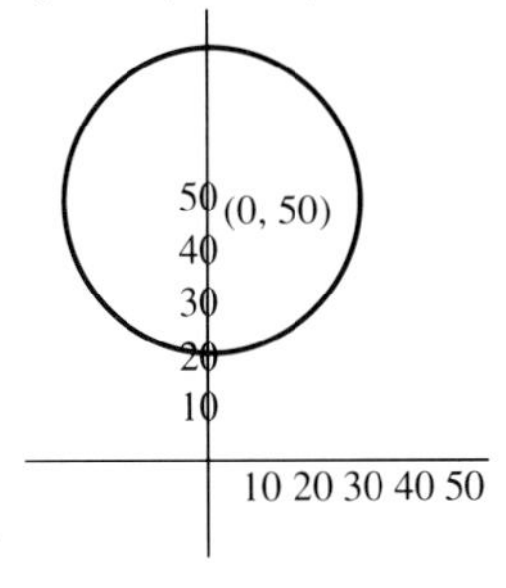

15–3 Exercises, p. 546

1. $V(0, 0)$; $F(2, 0)$; $x = -2$

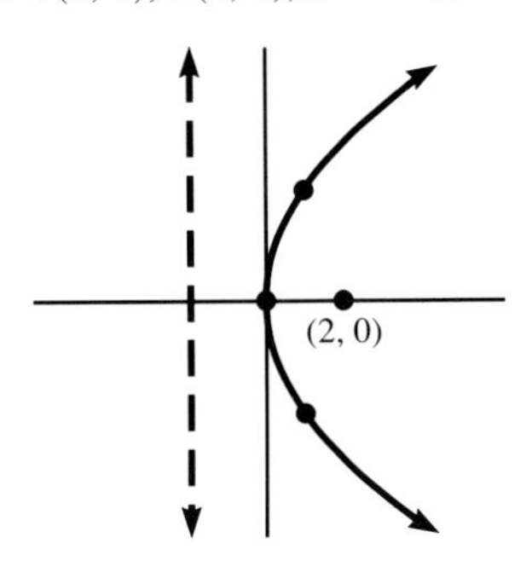

3. $V(0, 0)$; $F(0, 3)$; $y = -3$

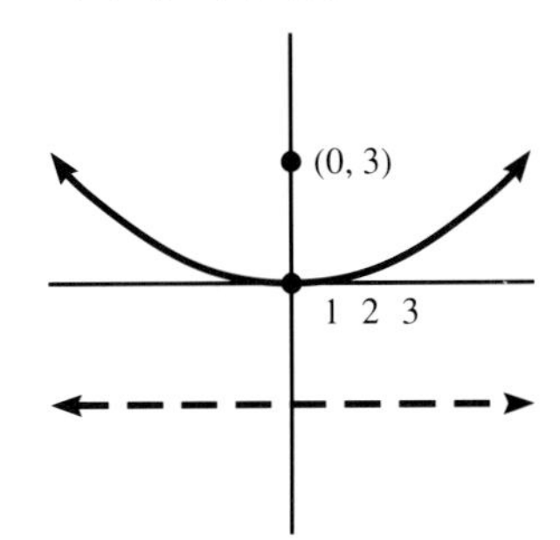

5. $V(0, 0)$; $F(-4, 0)$; $x = 4$

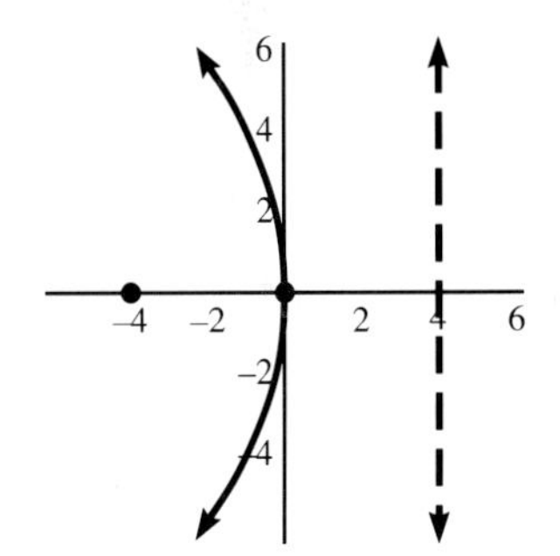

7. $V(0, 0)$; $F\left(0, \dfrac{-5}{2}\right)$; $y = \dfrac{5}{2}$

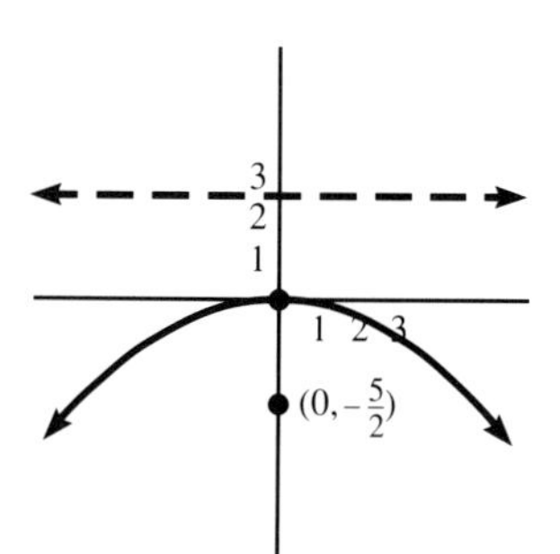

9. $V(0, 3)$; $F(-2, 3)$; $x = 2$

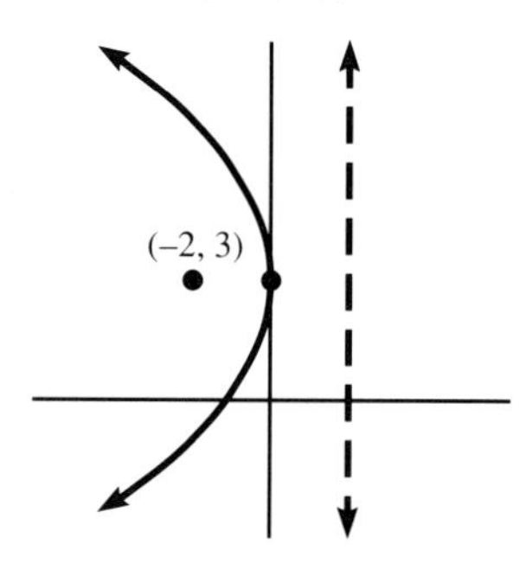

11. $V(-2, 3)$; $F(-2, 6)$; $y = 0$

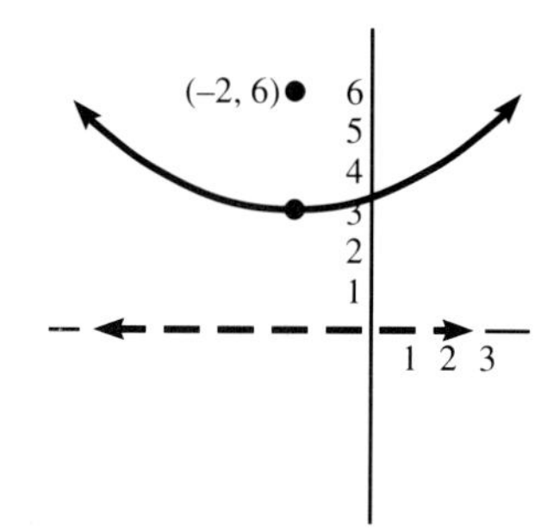

13. $V(-4, 6)$; $F(-4.5, 6)$; $x = -\dfrac{7}{2}$

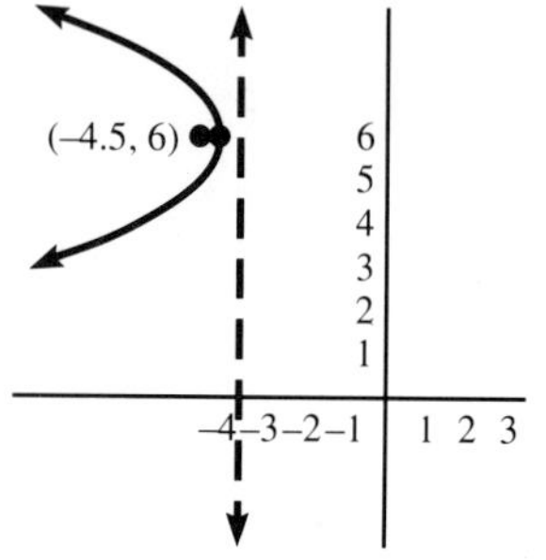

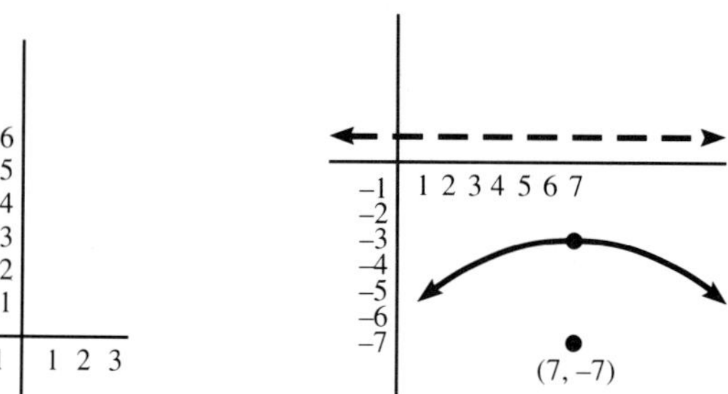

15. $V(7, -3)$; $F(7, -7)$; $y = 1$

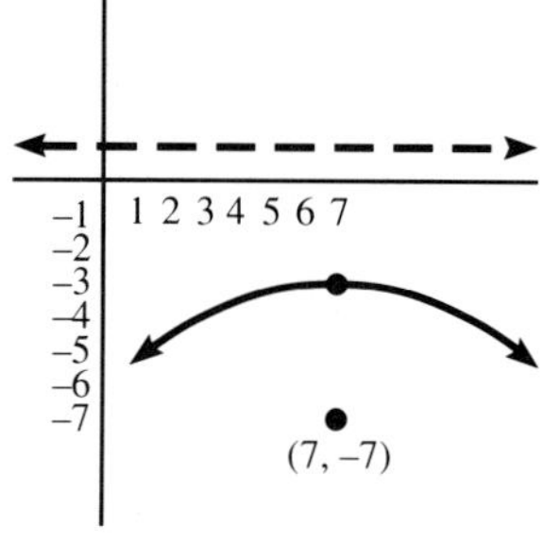

17. $V(-1, 4)$; $F\left(-1, 3\dfrac{3}{4}\right)$; $y = 4\dfrac{1}{4}$

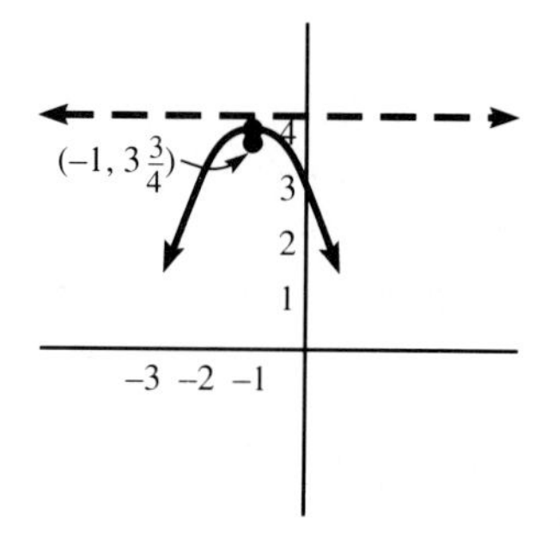

19. $V(-5, -5)$; $F(-4.5, -5)$; $x = -5.5$

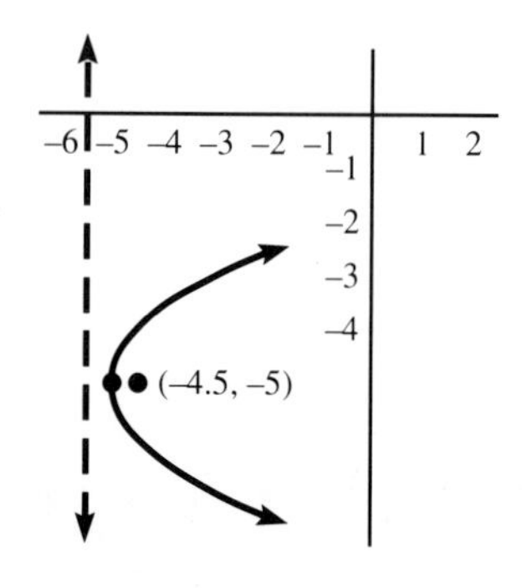

21. $V(-1, 3)$; $F\left(-1, \dfrac{10}{3}\right)$; $y = 2\dfrac{2}{3}$

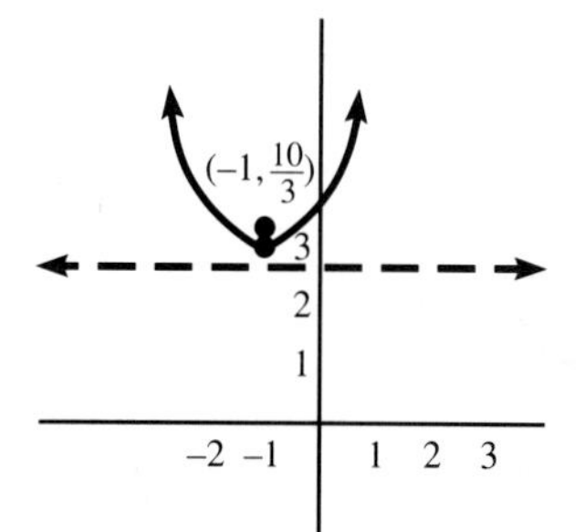

23. $x^2 = -16y$ **25.** $(x - 3)^2 = 12(y - 2)$ **27.** $(y - 4)^2 = 8(x + 1)$ **29.** $x^2 = \dfrac{8}{5}(y + 3)$

31. $(x - 3)^2 = 16(y + 6)$ **33.** $(t - 2.6)^2 = -.2(s - 31.9)$

35. $(v + 10)^2 = 20(y + 5)$

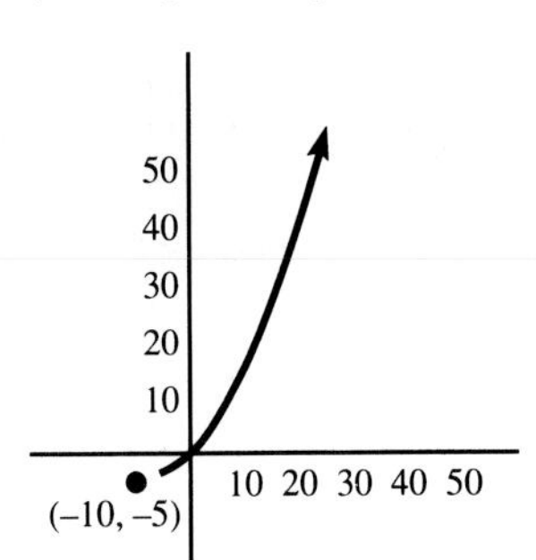

37.

H
100
90
80
70
60
50
40
30
20
10
(3, 90)
(2, 40)
(1, 10)
i
1 2 3 4 5 6

15–4 Exercises, p. 553

1. $C(0, 0)$; $V(5, 0)$ $(-5, 0)$; $MA(0, 3)$ $(0, -3)$; $F(4, 0)$ $(-4, 0)$

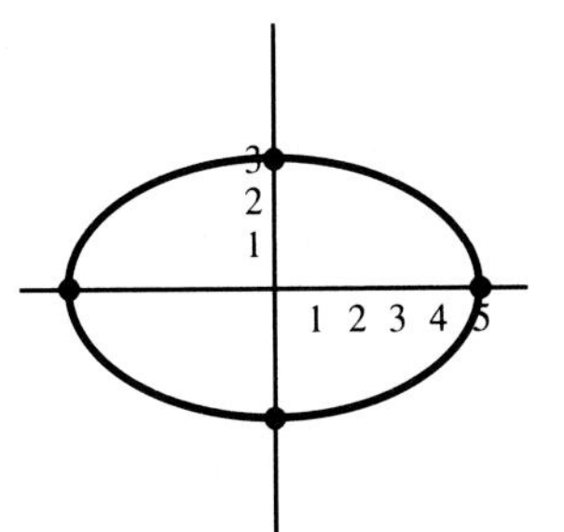

3. $C(0, 0)$; $V(0, 6)$ $(0, -6)$; $MA(4, 0)$ $(-4, 0)$; $F(0, 4.5)$ $(0, -4.5)$

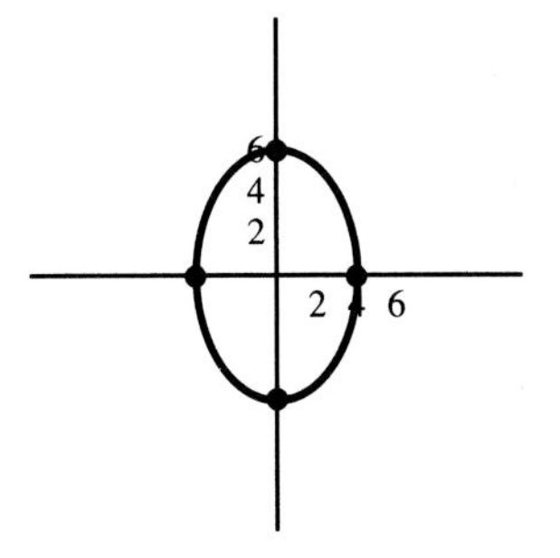

5. $C(0, 0)$; $V(3.2, 0)$ $(-3.2, 0)$; $MA(0, 2.4)$ $(0, -2.4)$; $F(2, 0)$ $(-2, 0)$

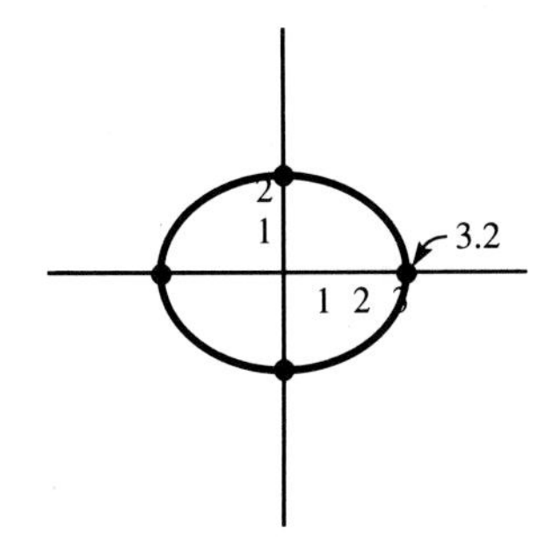

7. $C(-2, 6)$; $V(-2, 11)$ $(-2, 1)$; $MA(0, 6)$ $(-4, 6)$; $F(-2, 10.6)$ $(-2, 1.4)$

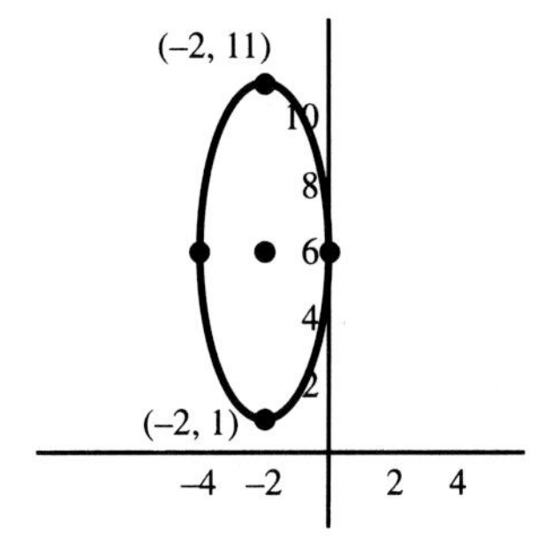

9. $C(-1, 0)$; $V(-1, 9)$ $(-1, -9)$; $MA(5, 0)$ $(-7, 0)$; $F(-1, 6.7)$ $(-1, -6.7)$

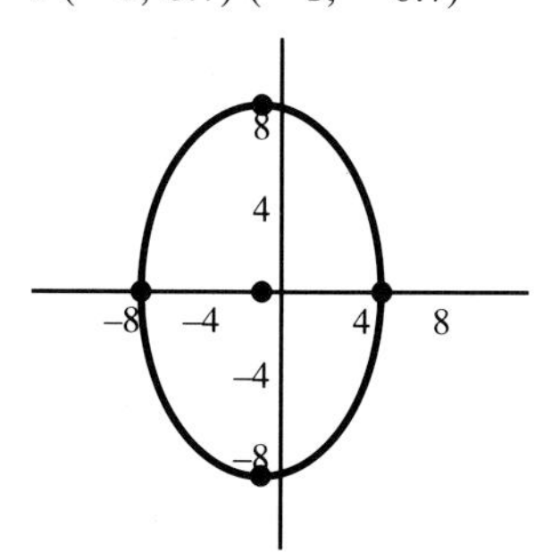

11. $C(3, 1)$; $V(6.5, 1)$ $(-0.5, 1)$; $MA(3, 3)$ $(3, -1)$; $F(5.8, 1)$ $(0.2, 1)$

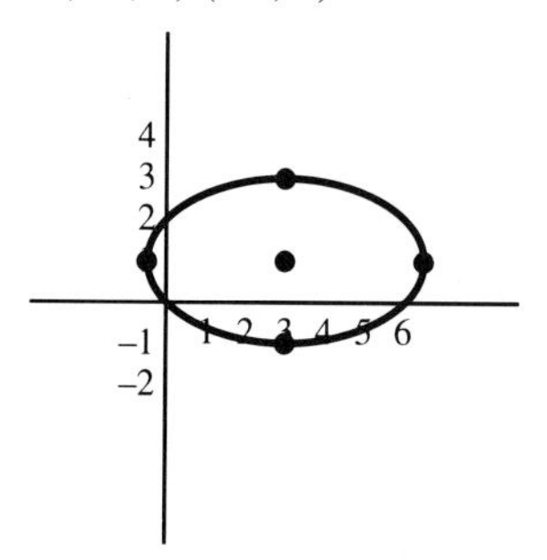

13. $C(-8, -4)$; $V(-2.5, -4)$ $(-13.5, -4)$; $MA(-8, -0.5)$ $(-8, -7.5)$; $F(-3.8, -4)$ $(-12.2, 4)$

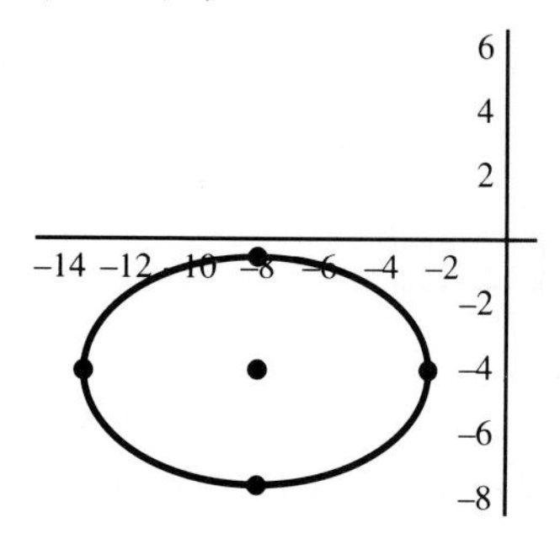

15. $C(0, 0)$; $V(3, 0)$ $(-3, 0)$; $MA(0, 2)$ $(0, -2)$; $F(2.2, 0)$ $(-2.2, 0)$

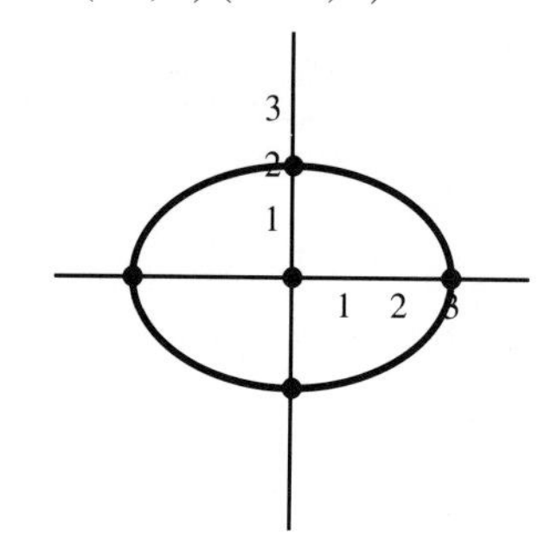

17. $C(-4, 2)$; $V(1.4, 2)$ $(-9.4, 2)$; $MA(-4, 5.1)$ $(-4, -1.1)$; $F(0.4, 2)$ $(-8.4, 2)$

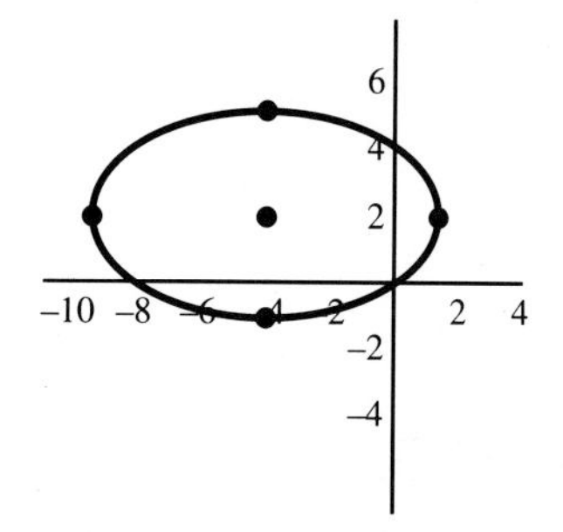

19. $C(-3, -1)$; $V(2.5, -1)$ $(-8.4, -1)$; $MA(-3, 3.1)$ $(-3, -5.1)$; $F(0.6, -1)$ $(-6.6, -1)$

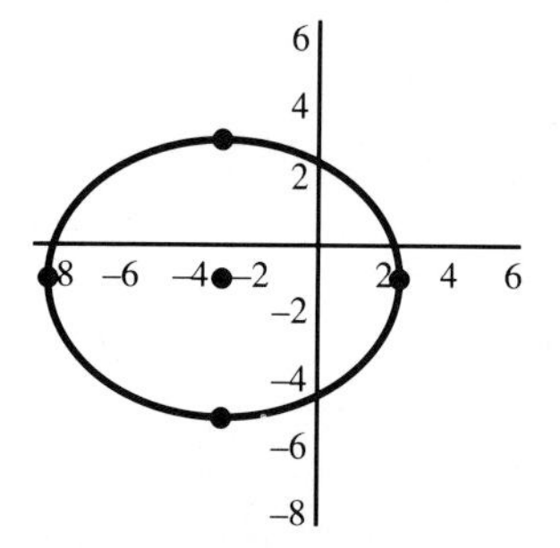

21. $C(-2, 0)$; $V(-2, 6)$ $(-2, -6)$; $MA(0.4, 0)$ $(-4.4, 0)$; $F(-2, 5.5)$ $(-2, -5.5)$

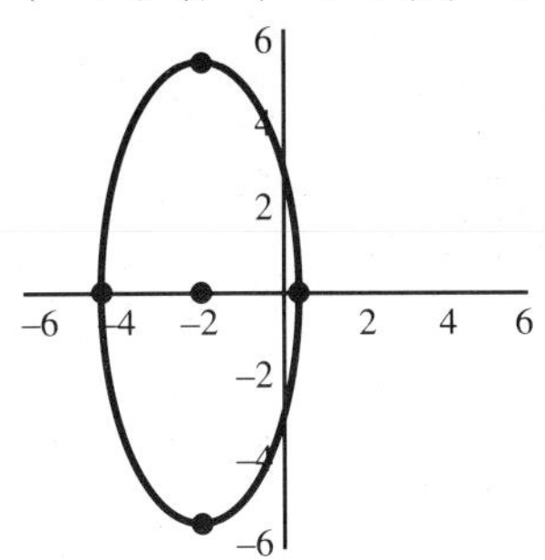

23. $\dfrac{y^2}{49} + \dfrac{x^2}{40} = 1$ **25.** $\dfrac{(x-4)^2}{5} + \dfrac{(y+3)^2}{4} = 1$ **27.** $\dfrac{(x+3)^2}{25} + \dfrac{(y+3)^2}{9} = 1$

29. $\dfrac{(y+2)^2}{36} + \dfrac{(x+3)^2}{25} = 1$ **31.** $\dfrac{x^2}{(4187.5)^2} + \dfrac{y^2}{(4186.4)^2} = 1$

15–5 Exercises, p. 562

1. $C(0, 0)$; $V(4.9, 0)$ $(-4.9, 0)$; $F(5.3, 0)$ $(-5.3, 0)$; $CA(0, 2)$ $(0, -2)$; $\pm\dfrac{\sqrt{6}}{6}$

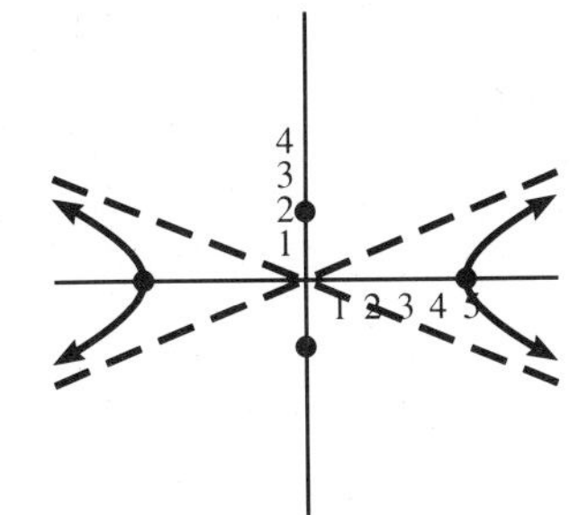

3. $C(0, 0)$; $V(0, 7)$ $(0, -7)$; $F(0, 8.1)$ $(0, -8.1)$; $CA(4, 0)$ $(-4, 0)$; $\pm\dfrac{7}{4}$

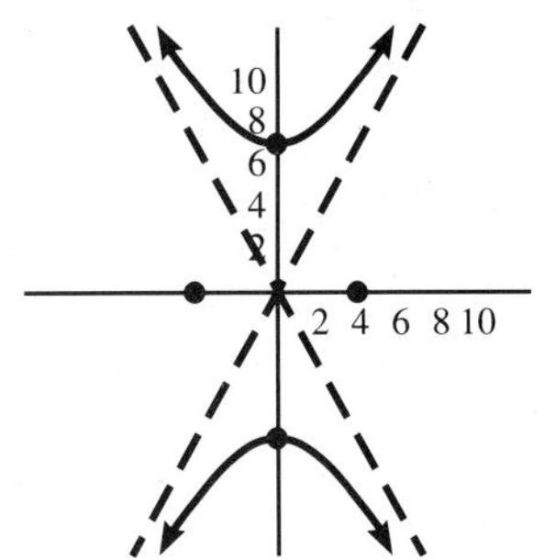

5. $C(0, 0)$; $V(8, 0)$ $(-8, 0)$; $F(9.4, 0)$ $(-9.4, 0)$; $CA(0, 5)$ $(0, -5)$; $\pm\dfrac{5}{8}$

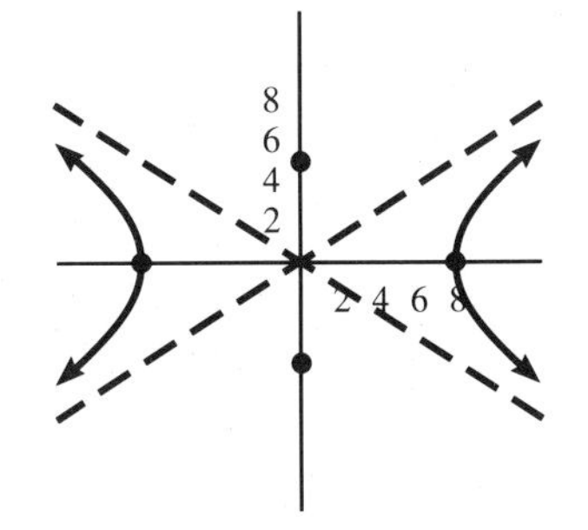

7. $C(2, -1)$; $V(2, 3)$ $(2, -5)$; $F(2, 8.8)$ $(2, -10.8)$; $CA(11, -1)$ $(-7, -1)$; $\pm\dfrac{4}{9}$

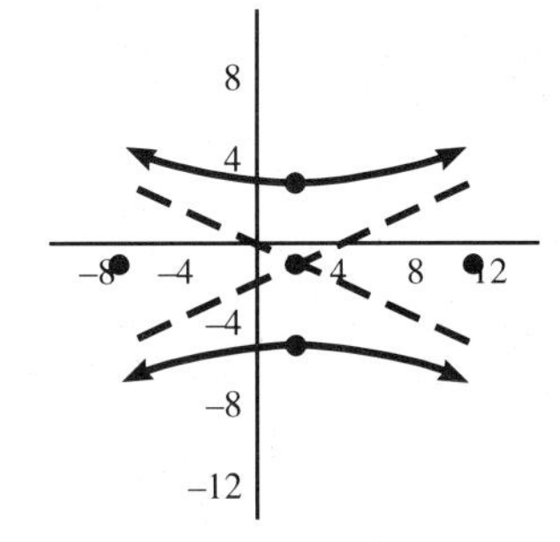

9. $C(-2, 3)$; $V(8, 3)$ $(-12, 3)$; $F(10.2, 3)$ $(-14.2, 3)$; $CA(-2, 10)$ $(-2, -4)$; $\pm\dfrac{7}{10}$

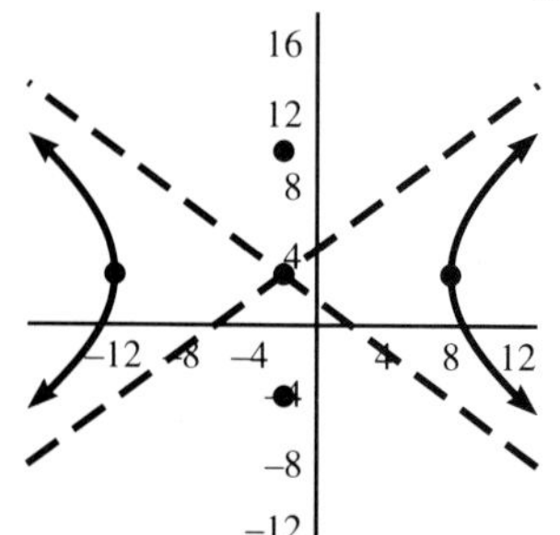

11. $C(3, 7)$; $V(3, 11.5)$ $(3, 2.5)$; $F(3, 13)$ $(3, 1)$; $CA(7, 7)$ $(-1, 7)$; $\pm\dfrac{4.5}{4}$

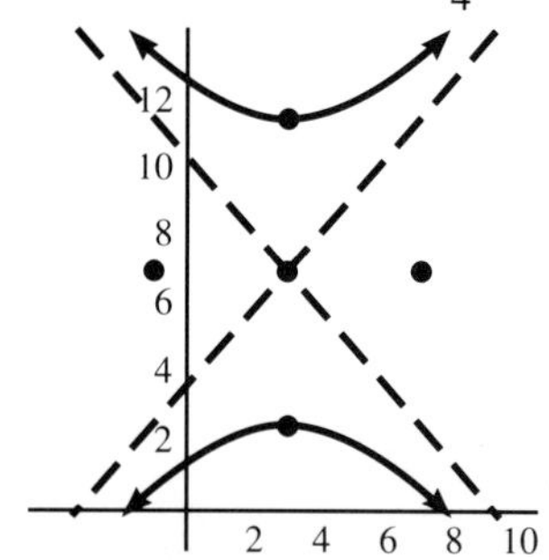

13. $C(2, 5)$; $V(0.3, 5)$ $(3.7, 5)$; $F(-1.5, 5)$ $(5.5, 5)$; $CA(2, 8)$ $(2, 2)$; $\pm\dfrac{\sqrt{3}}{1}$

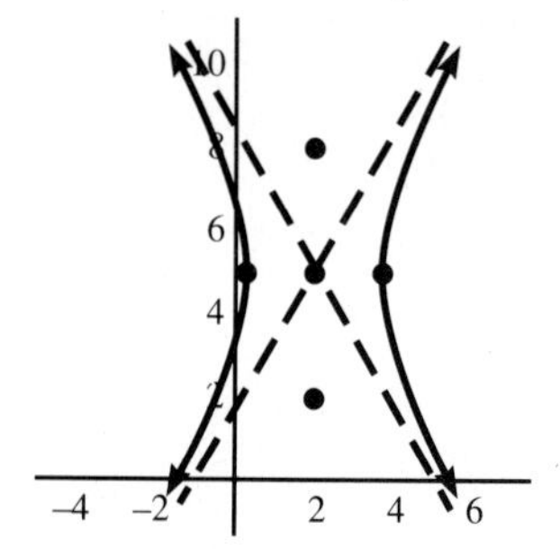

15. $C(-7, -3)$; $V(-4.6, -3)$ $(-9.4, -3)$; $F(-4, -3)$ $(-10, -3)$; $CA(-7, -1.3)$ $(-7, -4.7)$; $\pm\dfrac{\sqrt{2}}{2}$

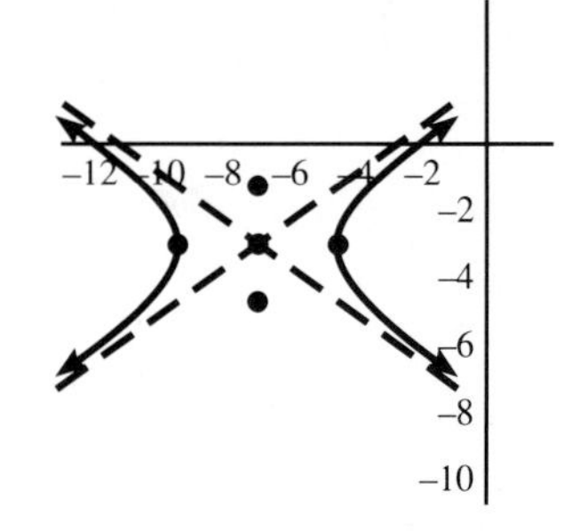

17. $C(-3, -8)$; $V(-3, -2)$ $(-3, -14)$; $F(-3, -13)$ $(-3, -14.7)$; $CA(0, -8)$ $(-6, -8)$; ± 2

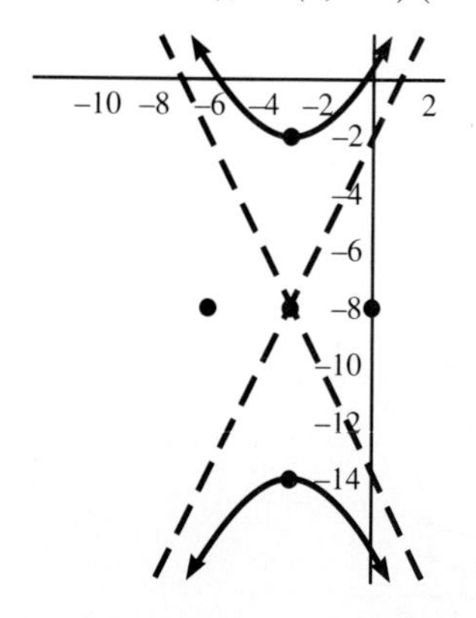

19. $C(2, 2)$; $V(0.3, 2)$ $(3.7, 2)$; $F(-0.8, 2)$ $(4.8, 2)$; $CA(2, 4.2)$ $(2, -0.2)$; $\pm\dfrac{2.2}{1.7}$

21. $\frac{x^2}{16} - \frac{y^2}{9} = 1$ **23.** $\frac{(x-5)^2}{3.9} - \frac{(y-3)^2}{21.1} = 1$ **25.** $\frac{(x-3)^2}{9} - \frac{(y-4)^2}{64} = 1$

27. $\frac{(y+5)^2}{9} - \frac{(x+1)^2}{25} = 1$

29.

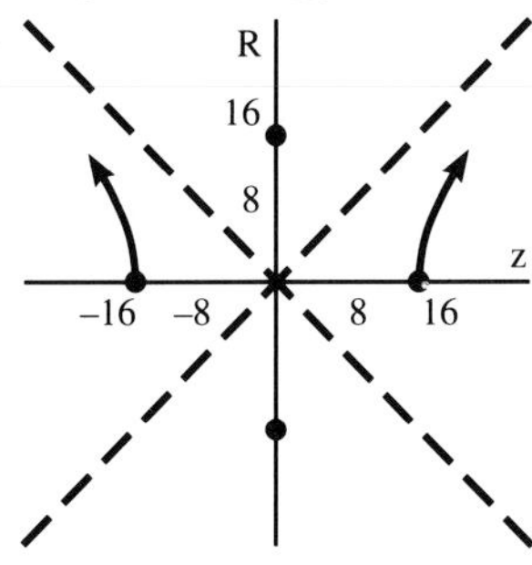

31.

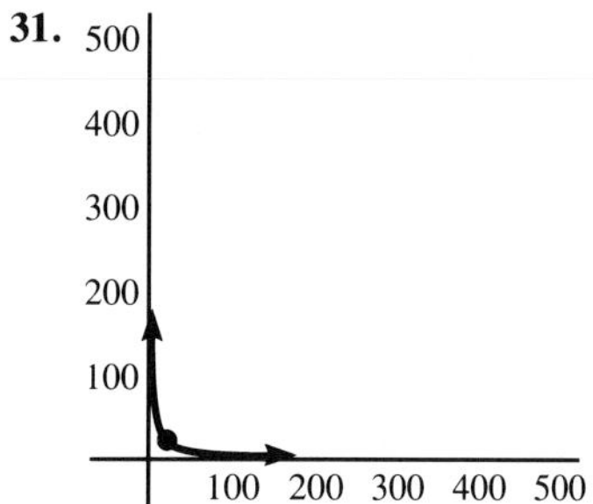

15–6 Exercises, p. 567

1. parabola **3.** hyperbola **5.** circle **7.** parabola **9.** ellipse **11.** hyperbola **13.** circle **15.** ellipse

17. parabola; $V(3, -4)$; $F(3, -4.5)$; $y = -3.5$

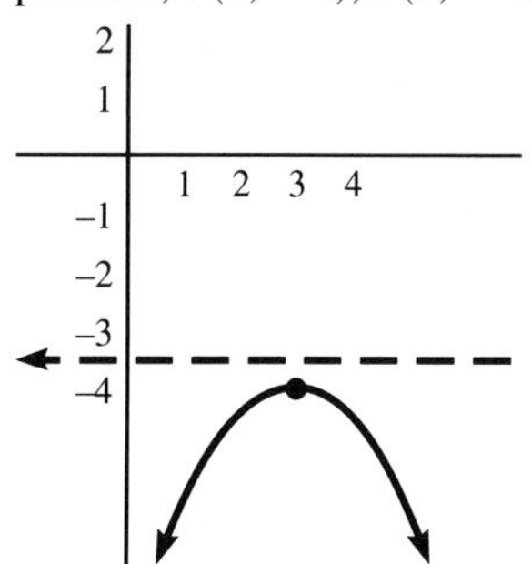

19. ellipse; $C(1, -3)$; $V(3.8\ -3)\ (-4.7, -3)$; $MA(1, -0.7)\ (1, -5.3)$; $F(2.6, -3)\ (-0.6, -3)$

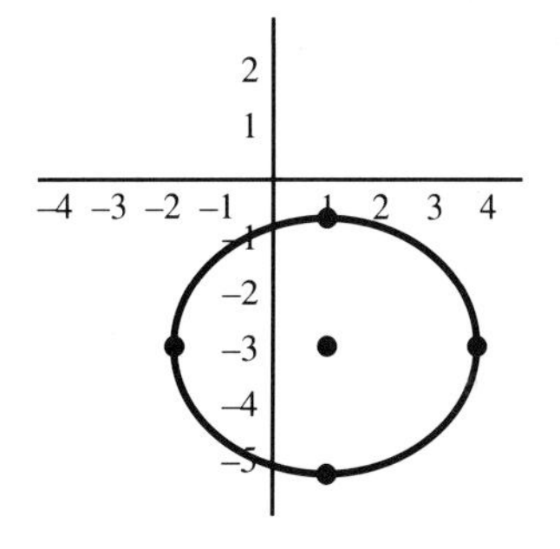

21. circle; $(-4, -1)$; $r = 2$

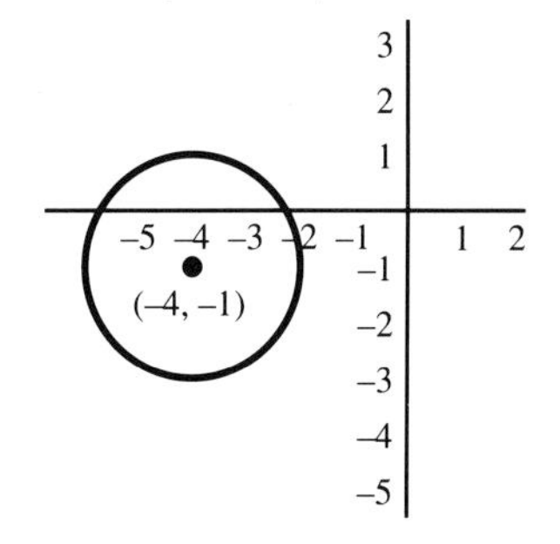

23. hyperbola; $C(0, 0)$; asymptotes are axes; $V(\sqrt{3}, \sqrt{3})\ (-\sqrt{3}, -\sqrt{3})$

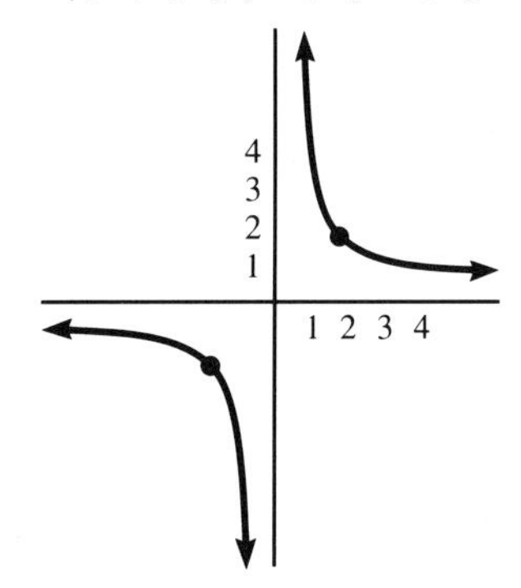

25. hyperbola; $C(0, -3)$; $V(3, -3)$ $(-3, -3)$; $CA(0, 2)$ $(0, -8)$; $F(5.8, -3)$ $(-5.8, -3)$; $\pm\frac{5}{3}$

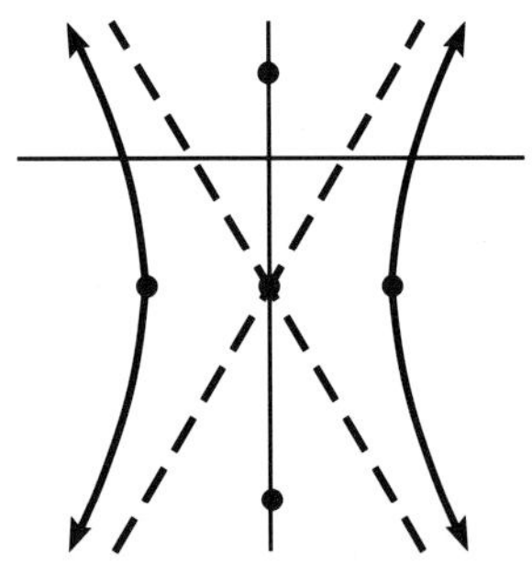

27. $v = 8L^2 - 72L$; parabola

29. $x^2 = 2y + 1$; parabola

Chapter Review, p. 569

1. $6x + 8y - 45 = 0$ **3.** $2x + 13y + 47 = 0$ **5.** $x = -3$ **7.** $y = -8$ **9.** $3x - 2y - 1 = 0$
11. $5x + y - 31 = 0$

13. $m = \frac{7}{4}$, x int $= \frac{10}{7}$, y int $= -2.5$

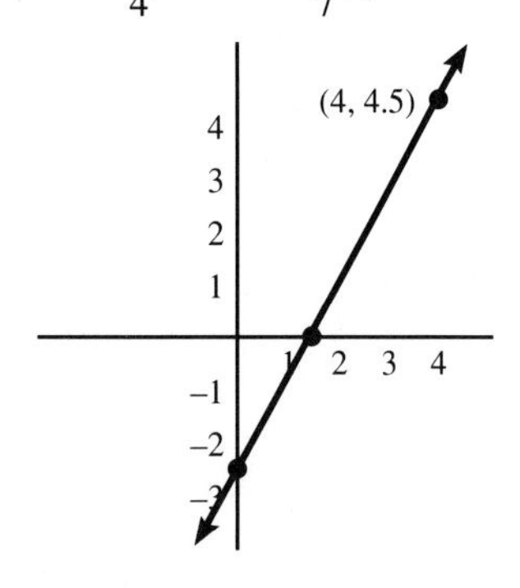

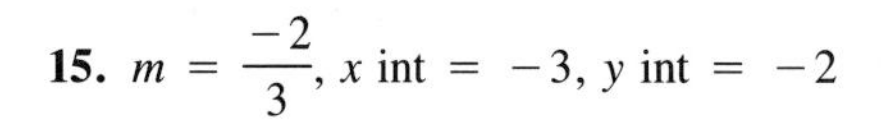

15. $m = \frac{-2}{3}$, x int $= -3$, y int $= -2$

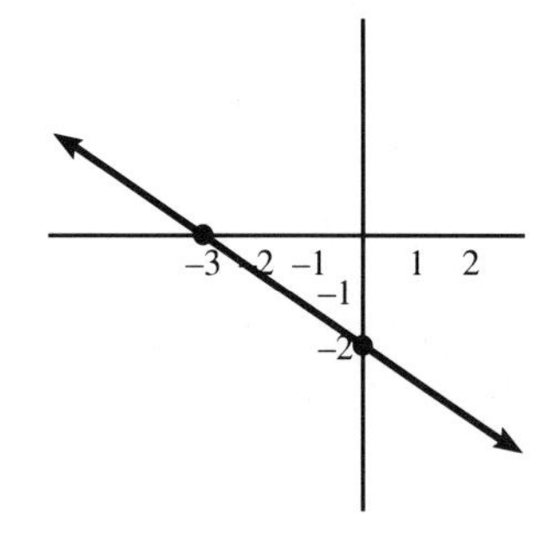

17. $m = \frac{1}{3}$, x int $= 9$, y int $= -3$

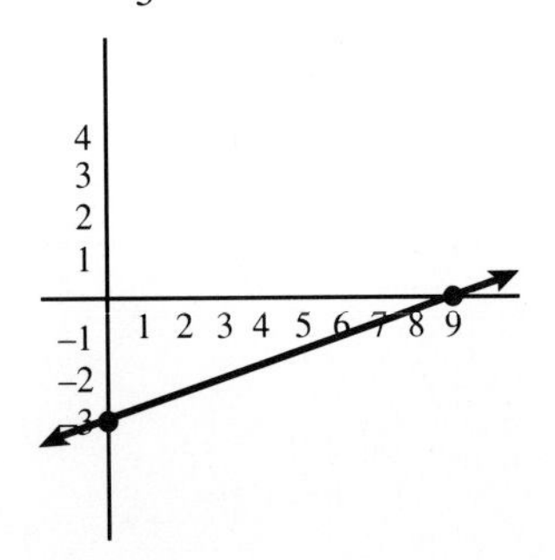

19. $2T = 100R - 230$

21. $v = 10t + 3.5$

23. $(-4, -3); r = 2$

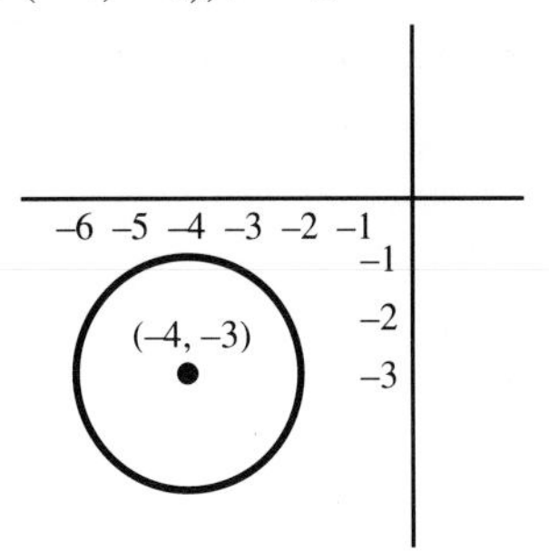

25. $(-1, 6); r = \sqrt[4]{6} = 1.6$

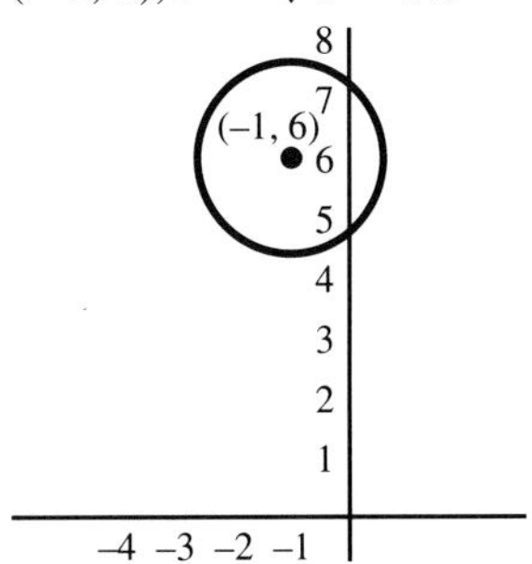

27. $\left(\frac{3}{2}, \frac{-5}{2}\right); r = \sqrt{14.5} = 3.8$

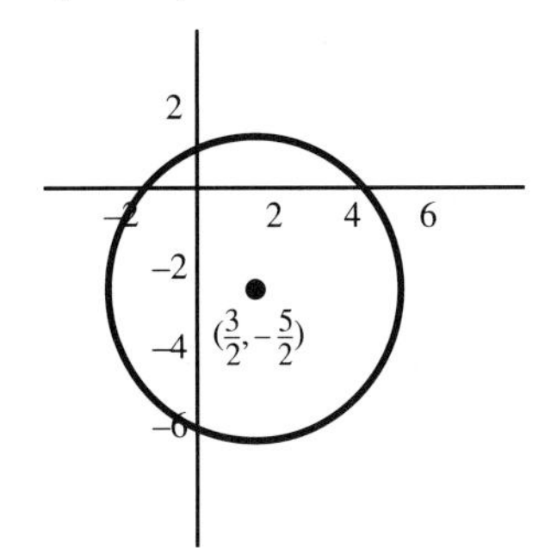

29. $(x + 2)^2 + (y + 4)^2 = 3$ **31.** $(x - 6)^2 + (y + 5)^2 = 25$ **33.** $(x + 2)^2 + (y + 1)^2 = 50$

35. $(40, 0); r = 25$

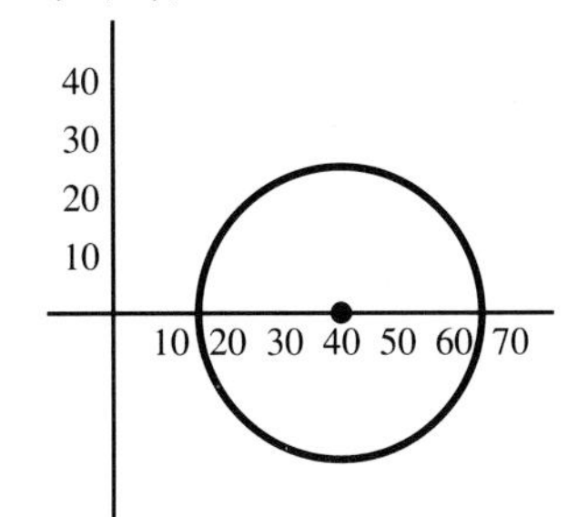

37. $V(3, 0); F(2.5, 0); x = 3.5$

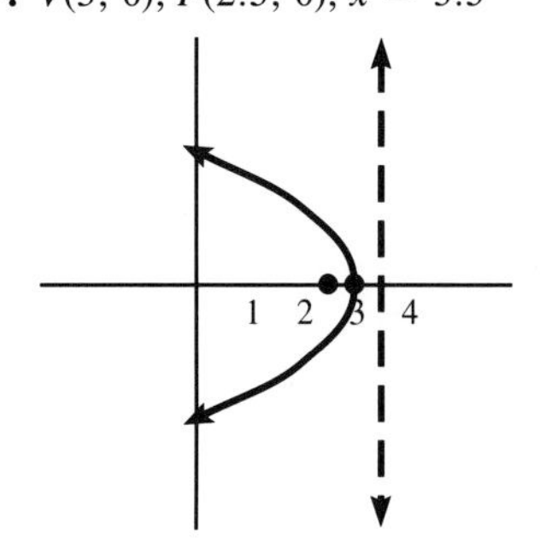

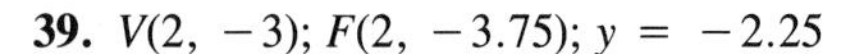

39. $V(2, -3)$; $F(2, -3.75)$; $y = -2.25$

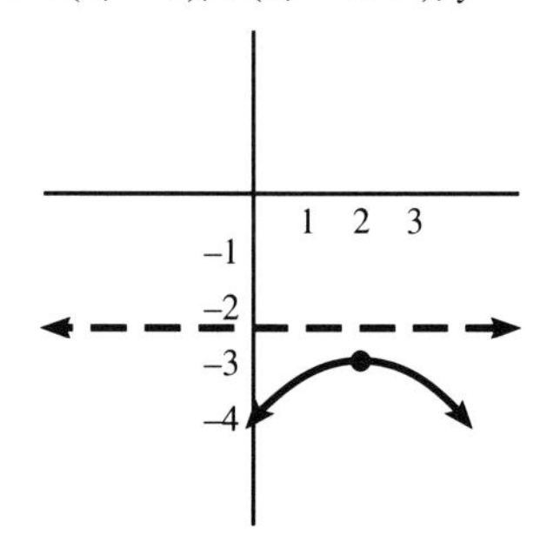

41. $V(-4, 3)$; $F(-4, 2.5)$; $y = 3.5$

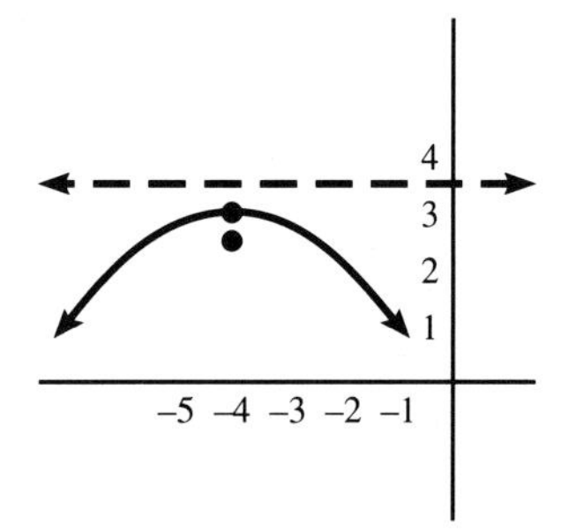

43. $y^2 = 12(x + 2)$ **45.** $(x - 4)^2 = 6(y + 1.5)$ **47.** $x^2 = 285.7y$

49. $C(-1, 3)$; $V(-1, 10)$ $(-1, -4)$; $MA(5, 3)$ $(-7, 3)$; $F(-1, 6.6)(-1, -0.6)$

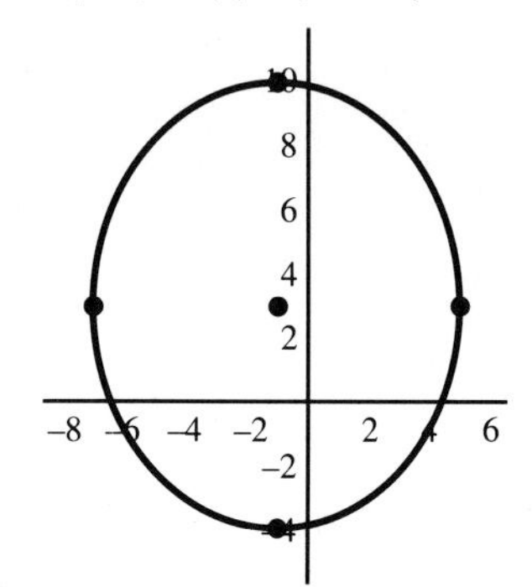

51. $C(0, 6)$; $V(0, 16.4)$ $(0, -4.4)$; $MA(5.8, 6)$ $(-5.8, 6)$; $F(0, 14.6)$ $(0, -2.6)$

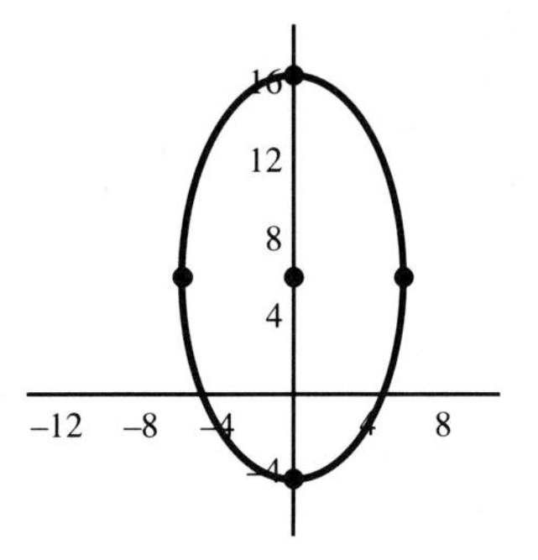

53. $C(-3, -1)$; $V(-2.5, -1)$ $(-3.5, -1)$; $MA\left(-3, \frac{-2}{3}\right)\left(-3, \frac{-4}{3}\right)$; $F(-2.6, -1)$ $(-3.4, -1)$

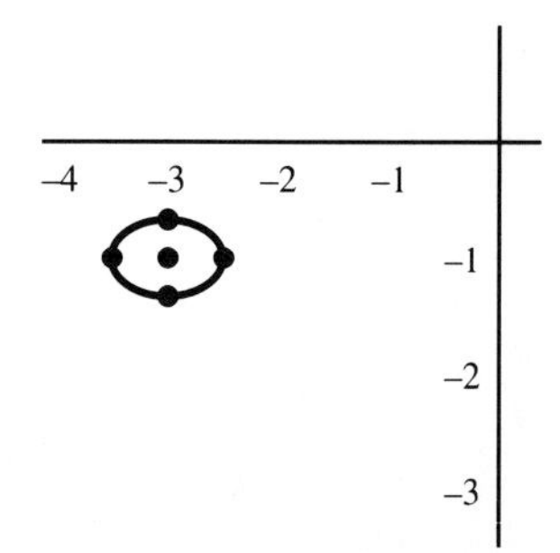

55. $\frac{(y - 1)^2}{34} + \frac{(x - 5)^2}{9} = 1$ **57.** $\frac{x^2}{16} + \frac{(y - 2)^2}{9} = 1$ **59.** $\frac{x^2}{(4242.5)^2} + \frac{(y - 112.5)^2}{(4241)^2} = 1$

61. $C(-3, 1)$; $V(1, 1)\ (-7, 1)$; $CA(-3, 3)\ (-3, -1)$; $F(1.5, 1)(-7.5, 1)$

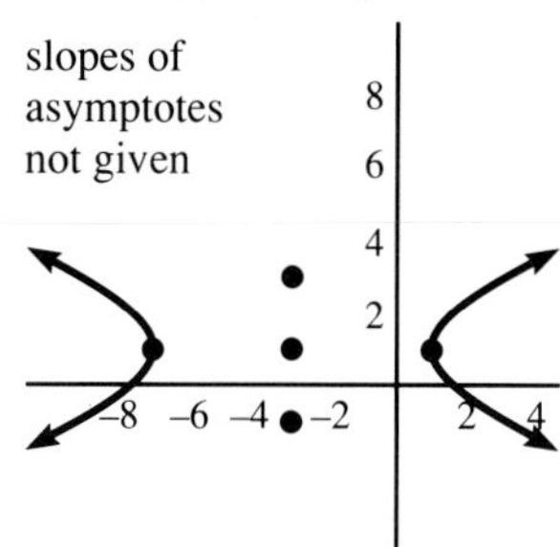

63. $C(-3, -4)$; $V(-3, -9)\ (-3, 1)$; $CA(3, -4)\ (-9, -4)$; $F(-3, -11.8)\ (-3, 3.8)$

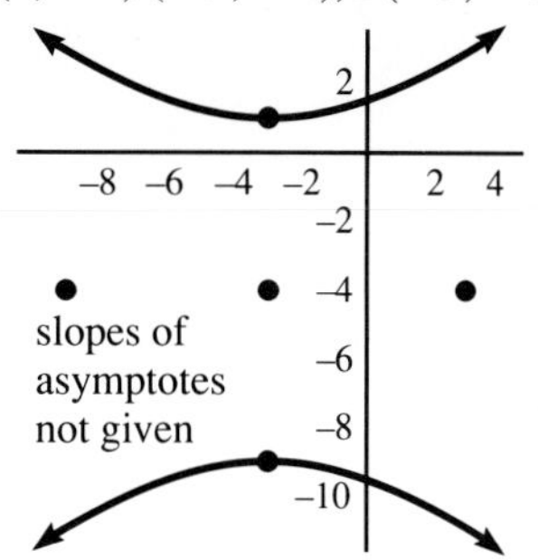

65. $C(0, 4)$; $V(0, 6.8)\ (0, 1.2)$; $CA(3.2, 4)\ (-3.2, 4)$; $F(0, 8.2)\ (0, -0.2)$

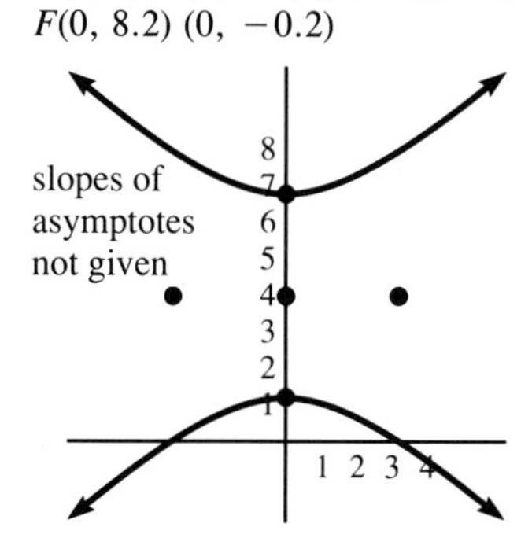

67. $\dfrac{y^2}{9} - \dfrac{x^2}{16} = 1$

69. $\dfrac{(x-2)^2}{9} - \dfrac{(y-7)^2}{25} = 1$

71.

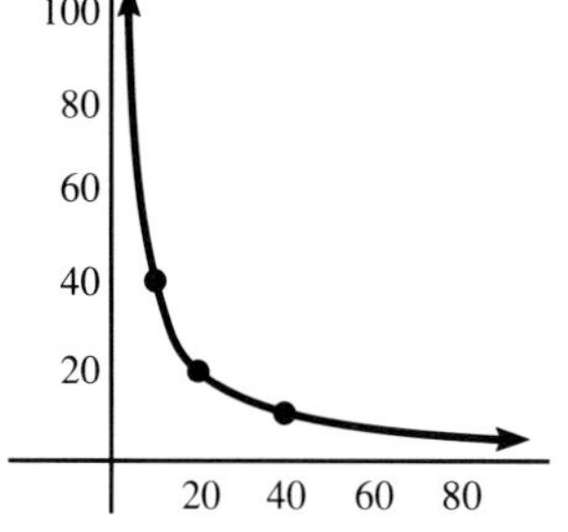

73. circle; $(1, -6)$; $r = 1.6$

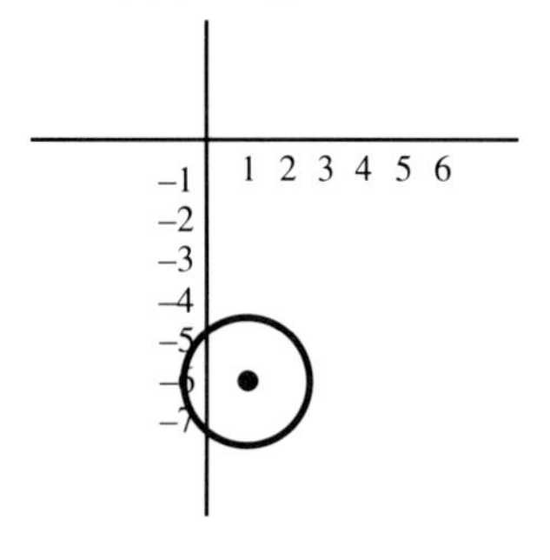

75. parabola; $V(4, -7)$; $F(4, -8)$; $y = -6$

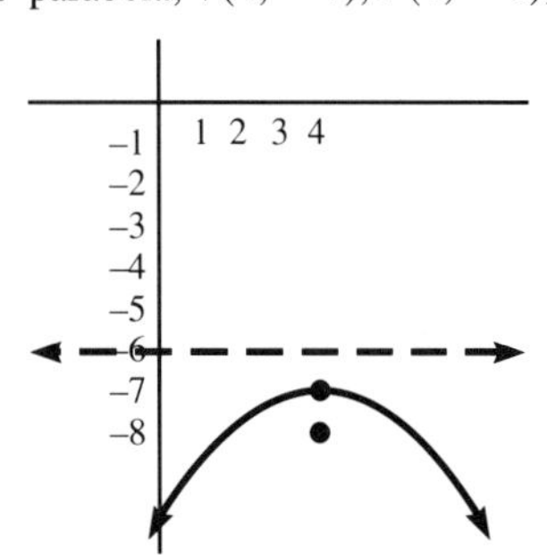

77. ellipse; $C(3, -1)$; $V(3, -5.2)$ $(3, 3.2)$; $MA(5.4, -1)$ $(0.6, -1)$; $F(3, -4.5)$ $(3, 2.5)$

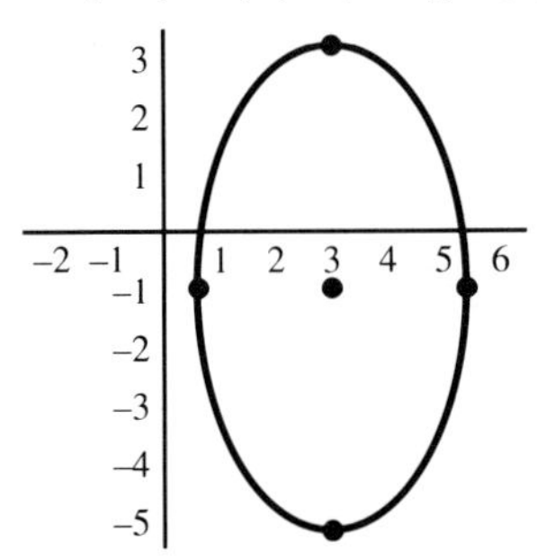

79. hyperbola; $C(3, -1)$; $V(5.8, -1)$ $(0.2, -1)$; $CA(3, 4)$ $(3, -6)$; $F(8.7, -1)$ $(-2.7, -1)$

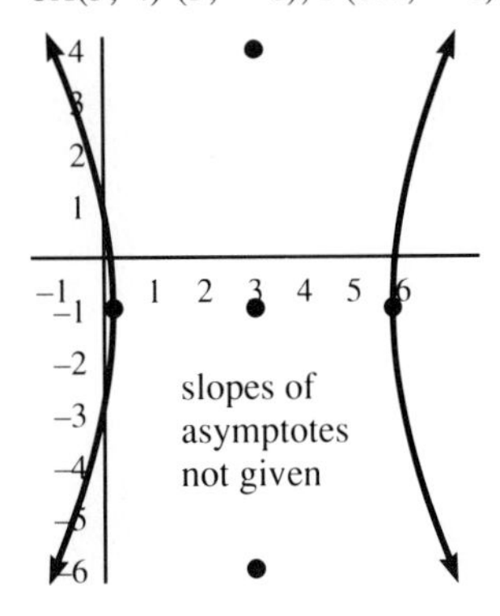

81. ellipse; $C(-3, -1)$; $V(-3, 4)$ $(-3, -6)$; $MA(1, -1)$ $(-7, -1)$; $F(-3, 2)$ $(-3, -4)$

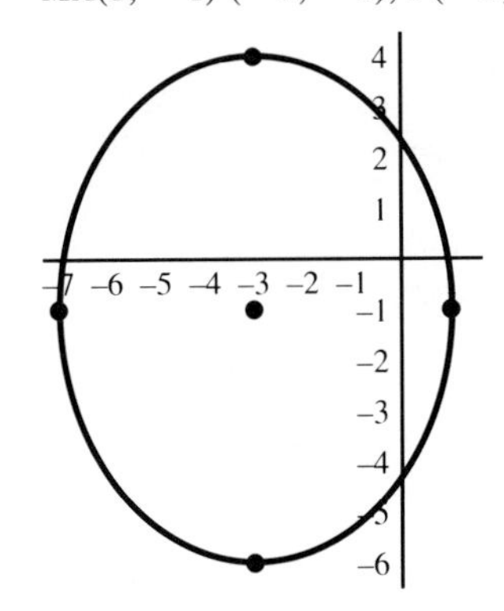

Chapter Test, p. 571

1. $m = \frac{1}{5}$, x int $= \frac{9}{2}$, y int $= -\frac{9}{10}$

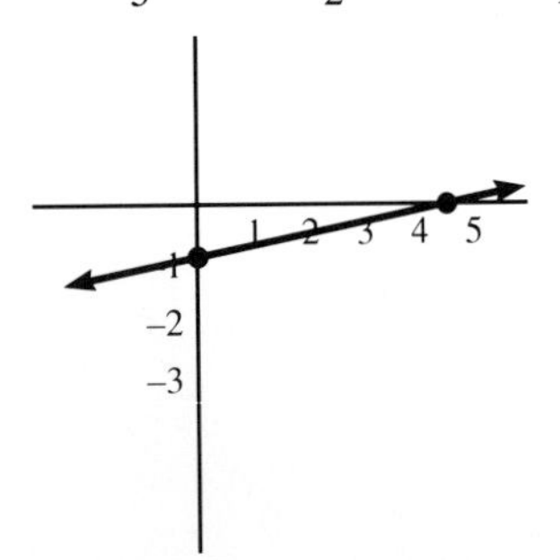

2. $(x - 2)^2 + (y + 5)^2 = 17$

3. $C(1, -3)$; $F(-3.5, -3)$ $(4.5, -3)$; $V(-3, -3)$ $(5, -3)$; $CA(1, -1)$ $(1, -5)$; $\pm\frac{1}{2}$

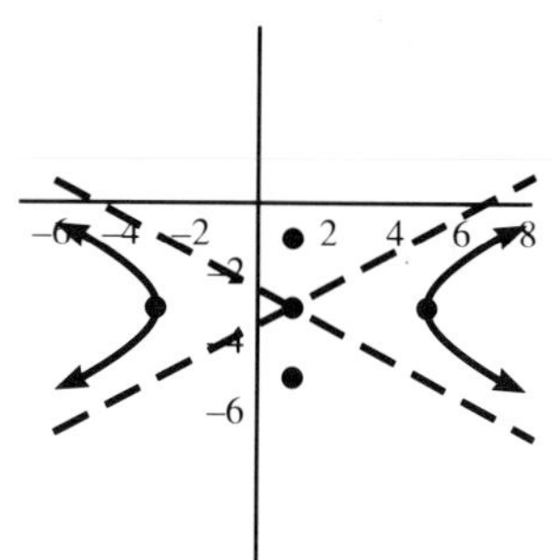

4. $(20, 0)$; $r = 36$

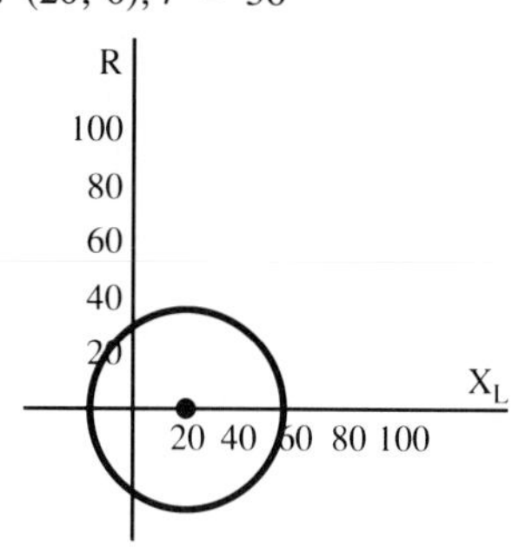

5. $(4, -8)$; $r = 5$

6. $V(-3, -6)$; $F(-3, -8)$; $y = -4$

7. $\dfrac{(x - 45)^2}{16933225} + \dfrac{y^2}{16931200} = 1$ **8.** $\dfrac{(y - 4)^2}{9} + \dfrac{(x + 7)^2}{4} = 1$ **9.** $(y + 5)^2 = 12(x - 1)$

10. $P = 0.43d + 14.7$ **11.** $3x + y + 15 = 0$ **12.** ellipse

13. $C(-3, 2)$; $V(1, 2)$ $(-7, 2)$; $F(-0.4, 2)$ $(-5.6, 2)$; $MA(-3, 5)$ $(-3, -1)$

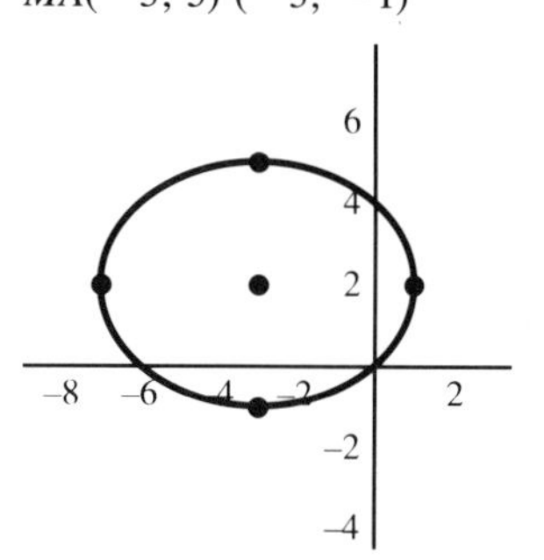

14. $\dfrac{(x - 6)^2}{16} - \dfrac{(y + 3)^2}{36} = 1$

CHAPTER 16

16–1 Exercises, p. 580

1.

grade	65	69	72	74	81	86	88	93	97
frequency	2	1	2	3	1	2	2	1	1

3.

hours	35	38	40	41	42	43	46	48	50	56
frequency	2	1	3	1	1	2	2	2	2	1

5.

IQ	80	83	84	86	89	95	97	98	100	110	120	134	147
frequency	2	1	1	1	1	1	1	2	3	3	1	2	1

7.

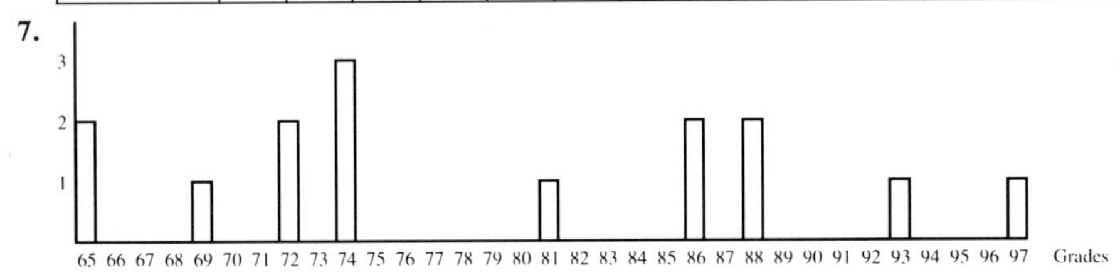

9.

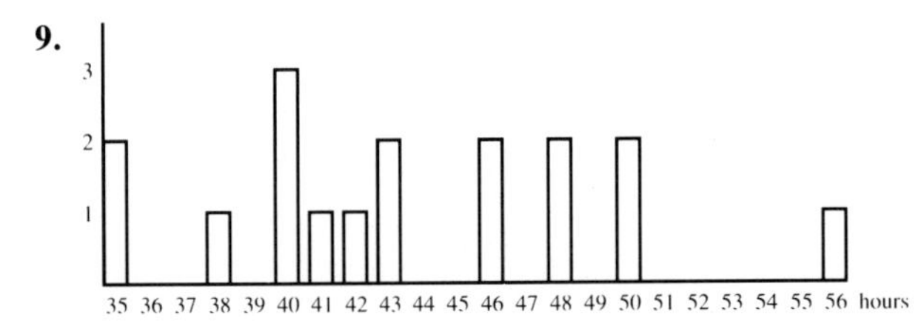

11.

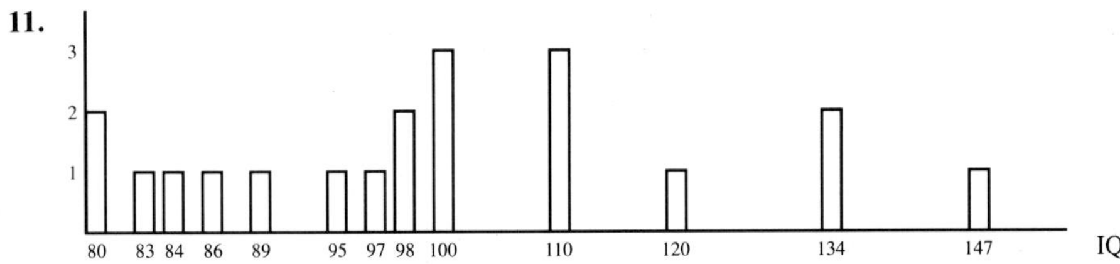

13.

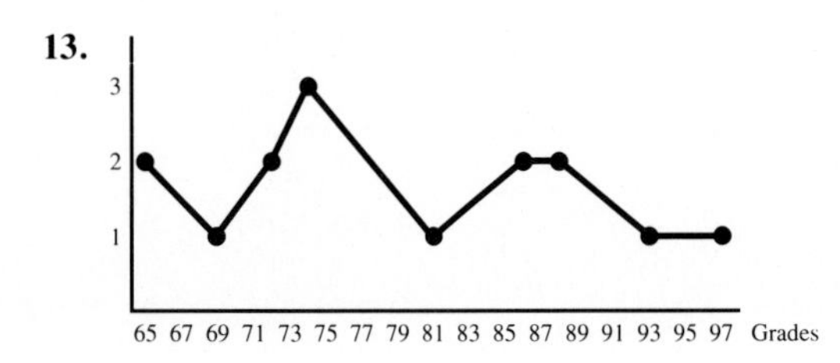

15.

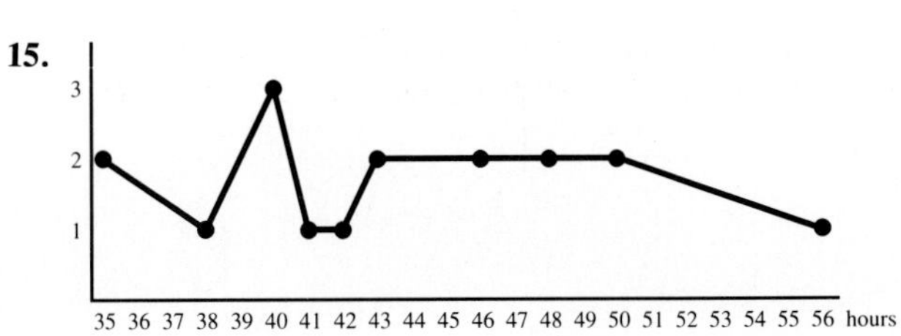

17.

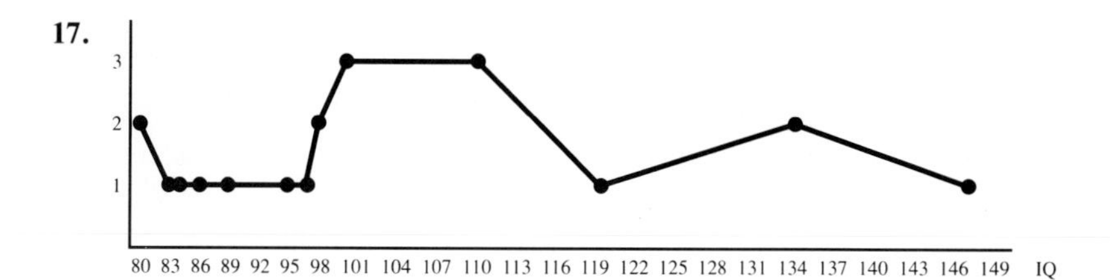

19.

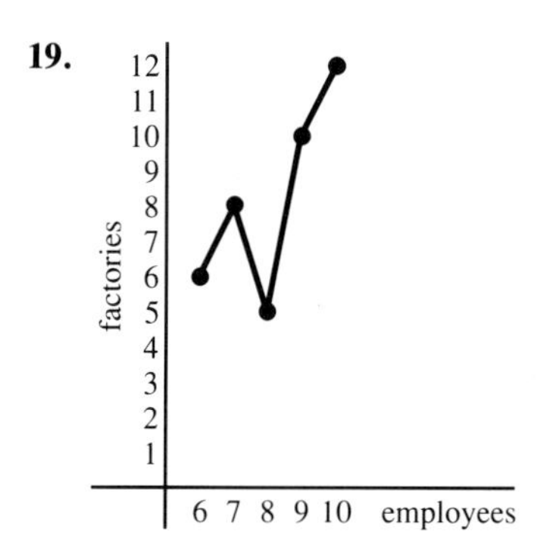

21. 34,200 employees

23.

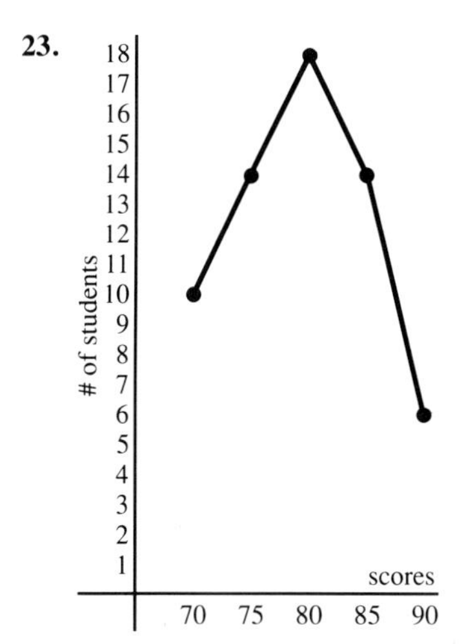

25. 38 students

16–2 Exercises, p. 586

1. 21.1, 19, 16 **3.** 81, 83, 73 **5.** 8; 8; 3, 10 **7.** 77.8, 75, 75 **9.** 856.5; 570; 1, 50, 360, 780, 1840, 2108 **11.** 8.125, 7, 6 **13.** 79.2, 82, 85 **15.** 75 **17.** $181.30

19. \$90.43; \$91.09; \$74.12, \$96.25

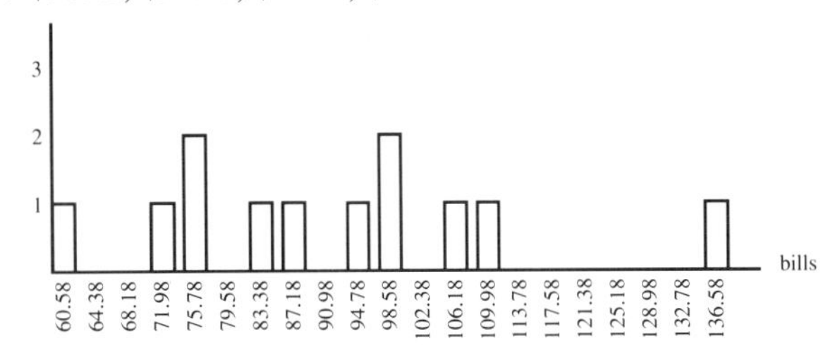

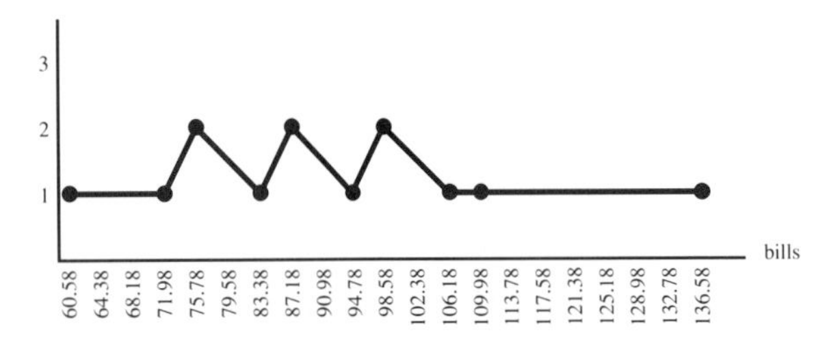

16–3 Exercises, p. 592

1. 57, 19.83 **3.** 38, 11.49 **5.** 54, 14.16 **7.** 21.6 to 49.2, 60% **9.** 19.4 to 28.2, 58.5%

11. 81.1, 85, 85, 10.9

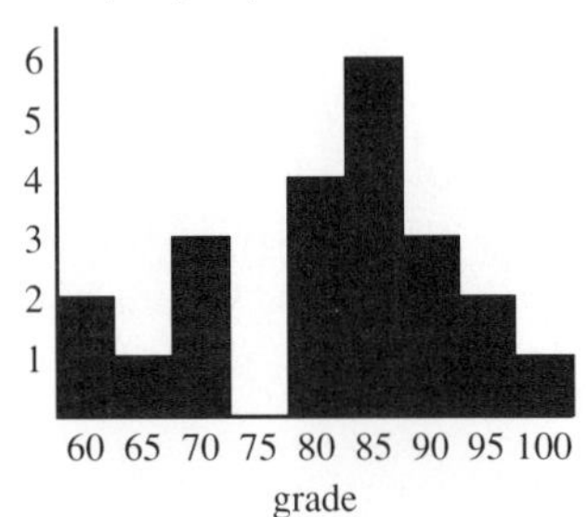

13. 101.25

15. 8.24

16–4 Exercises, p. 601

1. $y = 3.17x + 4.7$

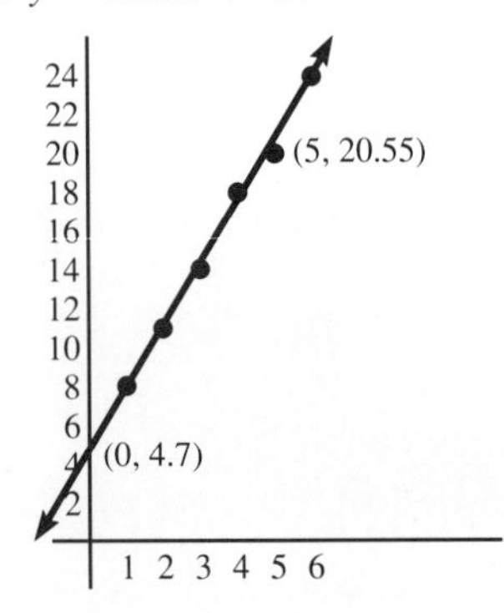

3. $y = 2.71x + 4.67$

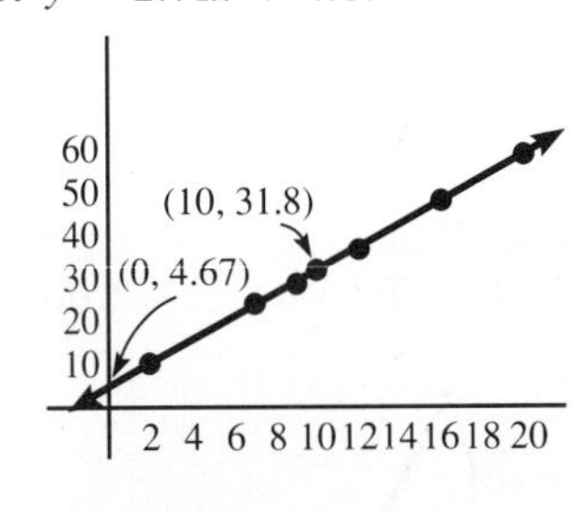

5. $y = 3.7x^2 - 1.7$

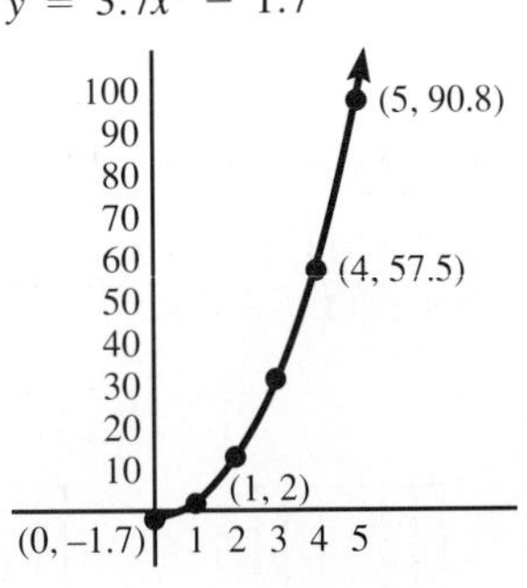

7. $y = 2.53(10)^x + 6.88$

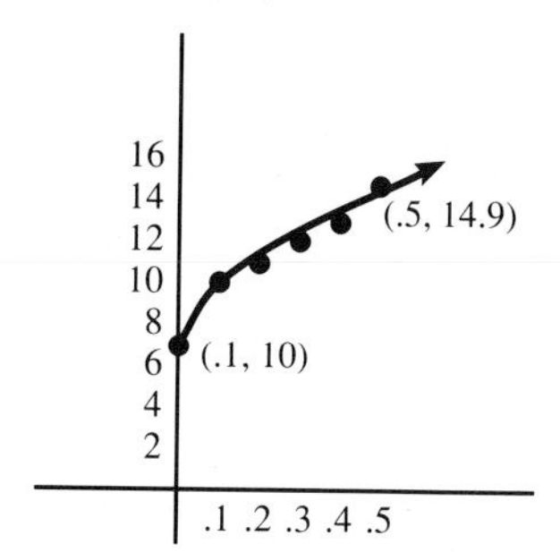

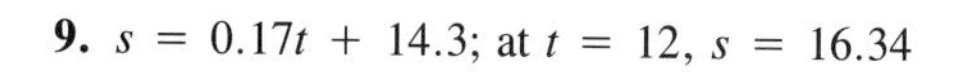

9. $s = 0.17t + 14.3$; at $t = 12$, $s = 16.34$

11. $s = 37t + 19$; at $t = 12$, $s = 463$

13. $R = 18.65\,T - 2.53$; 1489 Ω

Chapter Review, p. 604

1.

Age	18	27	35	36	45	57	63
Frequency	3	5	2	1	1	1	1

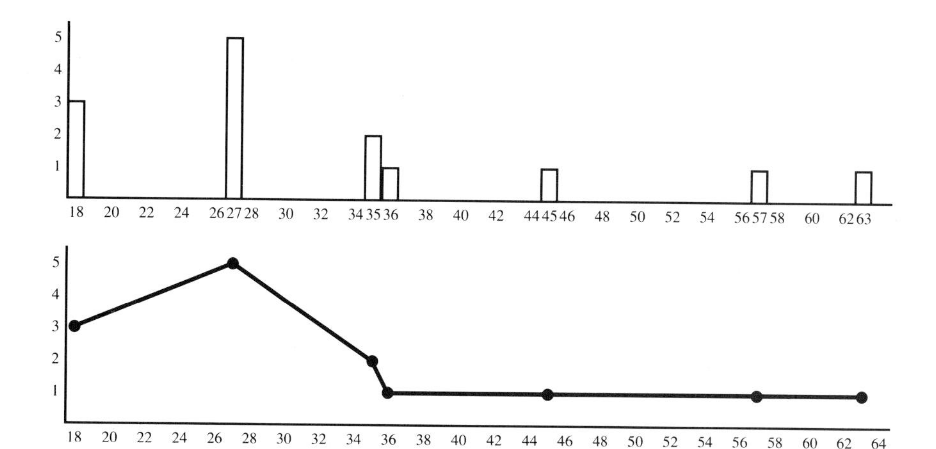

3.

salary	18	19	21	23	24	27	30	32
frequency	2	2	2	2	3	1	2	1

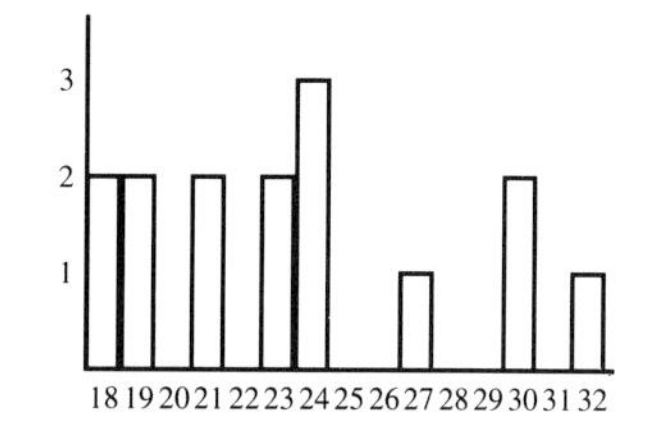

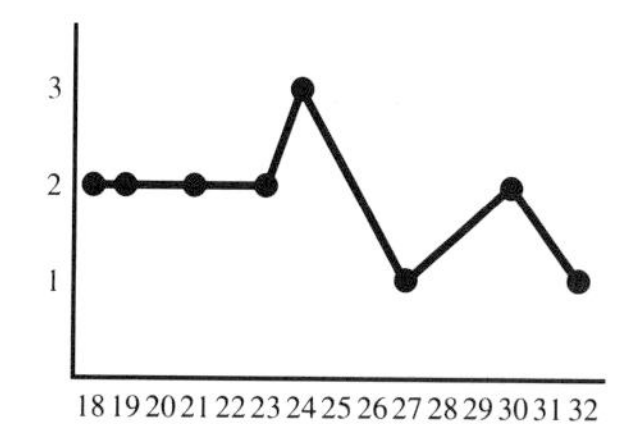

5. 32.86, 27, 27 **7.** 23.53, 23, 24 **9.** \$134.00 **11.** \$84.33 **13.** 45, 13.3 **15.** 14, 4.35
17. 19.56 to 46.16, 64.2% **19.** 19.18 to 27.88, 53.3% **21.** 3.4 (thousand)

23. $y = 7.9x + 14$

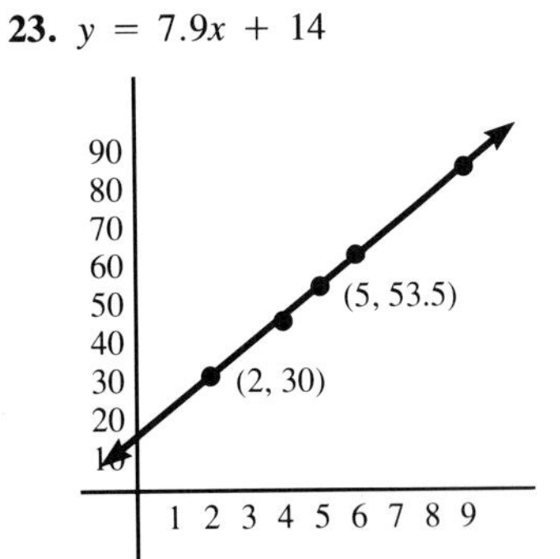

25. $y = 3.87x + 15.1$

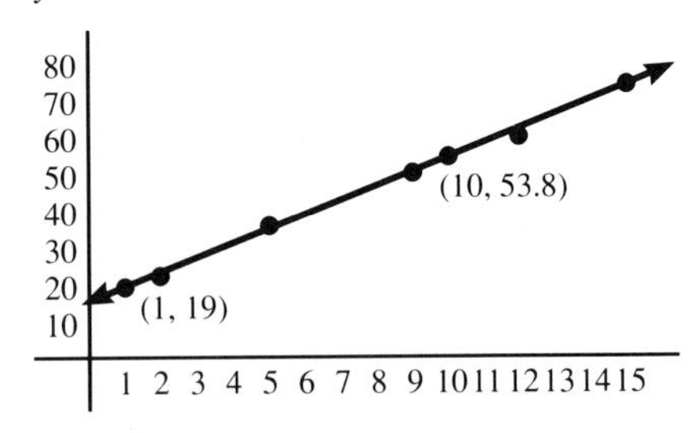

27. $y = 2.1x^2 + 10.04$

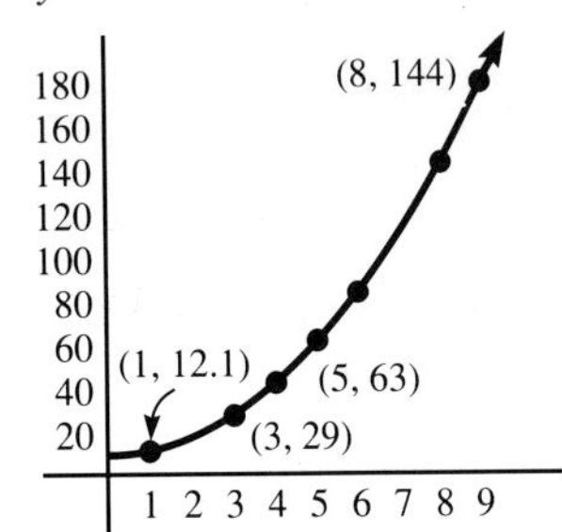

29. $y = 3.1(10)^x + 10$

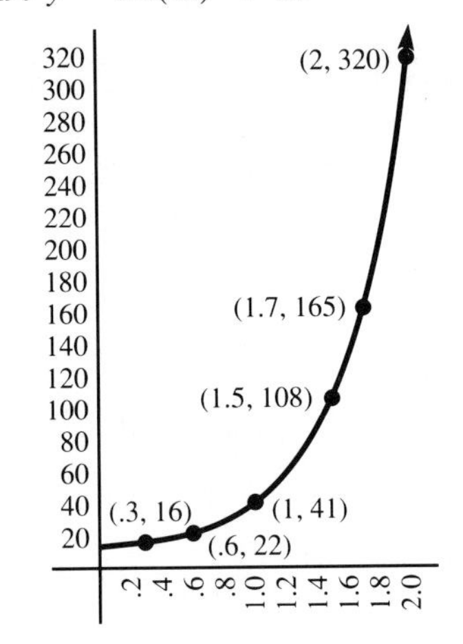

Chapter Test, p. 605

1. 22.2, 21, 18

2.

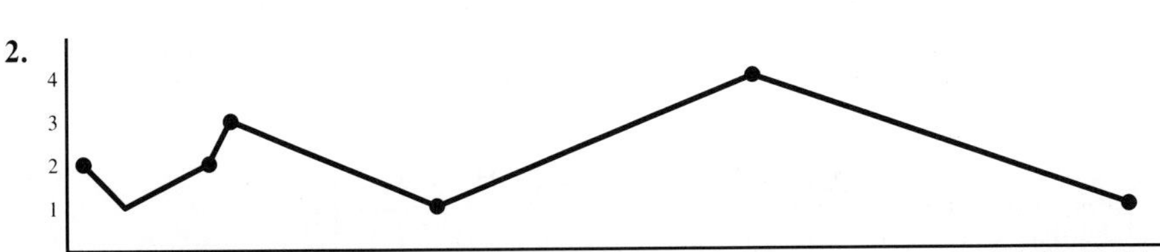

3. 7.08 **4.** $y = 6.6x + 3.2$, 56

5.

data	2	3	6	7	8	9	11	14	18
frequency	3	3	1	2	3	1	1	2	1

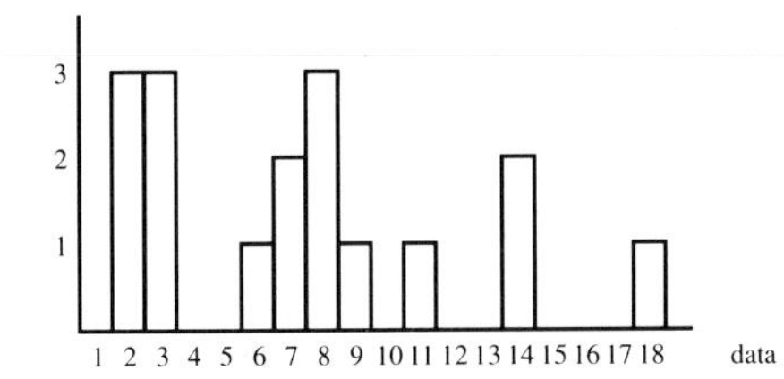

6. 9.27, 16.8 to 35.34 **7.** \$33,522; \$7,563.88 **8.** $y = 2.49x^2 + 12.8$

CHAPTER 17

17–1 Exercises, p. 614

1. 27 **3.** $\frac{7}{2}$ **5.** 165 **7.** -17 **9.** 73 **11.** 252 **13.** 180 **15.** 225 **17.** $n = 8, S_8 = -32$
19. $d = 3, S_{18} = 549$ **21.** $a_{10} = 31, S_{10} = 130$ **23.** $d = 3, n = 8$ **25.** $a_6 = \$30{,}686, S_6 = \$161{,}616$
27. $C = \$9.90$ **29.** 7,150

17–2 Exercises, p. 619

1. 16,384 **3.** 0.03125 **5.** 0.00116 **7.** 11,585.6 **9.** 820,160 **11.** -11.953 **13.** 26.963
15. -2.65625 **17.** $-2{,}730.5$ **19.** $a_{10} = 0.00024, S_{10} = 0.25$ **21.** $n = 4, a_4 = \frac{1}{3}$ **23.** $n = 7$,
$S_7 = 8{,}744$ **25.** $\frac{1}{512}$ **27.** \$31,944 **29.** \$18,271.19

17–3 Exercises, p. 623

1. $\frac{-16}{3}$ **3.** 18 **5.** 294 **7.** 118.49 **9.** 108 **11.** $\frac{256}{3}$ **13.** $14(2 - \sqrt{2})$ **15.** $\frac{12x^3}{2x - 1}$ **17.** 1
19. $\frac{83}{99}$ **21.** $\frac{419}{990}$ **23.** $\frac{415}{111}$ **25.** $a_6 = 21.25764$ in. $s = 168.68$ in **27.** $S = 75$ ft

17–4 Exercises, p. 628

1. $a^4 + 20a^3 + 150a^2 + 500a + 625$ **3.** $1m^7 - 7m^6 + 21m^5 - 35m^4 + 35m^3 - 21m^2 + 7m - 1$
5. $8x^3 - 60x^2y + 150xy^2 - 125y^3$ **7.** $729y^6 + 1{,}458xy^5 + 1{,}215x^2y^4 + 540x^3y^3 + 135x^4y^2 + 18x^5y + x^6$
9. $m^5 + 15m^4 + 90m^3 + 270m^2 + 405m + 243$
11. $b^7 - 42b^6 + 756b^5 - 7{,}560b^4 + 45{,}360b^3 - 163{,}296b^2 + 326{,}592b - 279{,}936$
13. $81x^4 - 108x^3 + 54x^2 - 12x + 1$
15. $15{,}625m^6 - 75{,}000m^5 + 150{,}000m^4 - 160{,}000m^3 + 96{,}000m^2 - 30{,}720m + 4{,}096$
17. $x^7 - 28x^6y + 336x^5y^2 - 2{,}240x^4y^3 + 8{,}960x^3y^4 - 21{,}504x^2y^5 + 28{,}672xy^6 - 16{,}384y^7$

19. $16a^8 - 96a^6b + 216a^4b^2 - 216a^2b^3 + 81b^4$
21. $1 - 8x + 28x^2 - 56x^3 \cdots$ **23.** $1 - 3x + 6x^2 - 10x^3 \cdots$ **25.** $55{,}050{,}240y^4$ **27.** $-1{,}792x^3$

Chapter Review, p. 629

1. 38 **3.** -84 **5.** 162 **7.** -60 **9.** 1,364 **11.** -243 **13.** 0.0016 **15.** 6,560 **17.** 2,728
19. \$21,578.40 **21.** $\frac{4}{9}$ **23.** 27 **25.** $\frac{34}{99}$ **27.** $\frac{3121}{999}$ **29.** $\frac{1000}{11}$ ft **31.** $x^3 - 3x^2z + 3xz^2 - z^3$
33. $16x^4 - 96x^3y + 216x^2y^2 - 216xy^3 + 81y^4$ **35.** $a^4 + 8a^3b + 24a^2b^2 + 32ab^3 + 16b^4$
37. $y^6 - 15y^4z + 75y^2z^2 - 125z^3$ **39.** $90720x^4$ **41.** $945x^4$

Chapter Test, p. 630

1. 160 **2.** 205 **3.** \$28,051.04 **4.** $\frac{31}{110}$ **5.** $16x^4 - 32x^3y + 24x^2y^2 - 8xy^3 + y^4$ **6.** 797,160
7. $m^4 - 4m^3n + 6m^2n^2 - 4mn^3 + n^4$ **8.** $\frac{1}{243}$ or 0.0041152 **9.** 2509 **10.** $\frac{36}{11}$ **11.** 1577
12. 146.286 **13.** $5{,}670x^4$ **14.** 97,505 **15.** $S_{11} = -44{,}287$
16. $32x^5 - 80x^4y + 80x^3y^2 - 40x^2y^3 + 10xy^4 - y^5$ **17.** $23{,}224{,}320y^4$ **18.** 0.000488 or $\frac{1}{2048}$

CHAPTER 18

18–1 Exercises, p. 643

1. 14

x	2	2.9	2.99	3	3.01	3.1	4
$f(x)$	9	13.5	13.95	14	14.05	14.5	19

3. -1

x	-1.5	-1.1	-1.01	-1	-0.99	-0.9	-0.5
$f(x)$	2	-1.2	-1.02	und.	-0.98	-0.8	0

5. -5

x	1.8	1.9	1.99	2	2.01	2.1	2.2
$f(x)$	-4.76	-4.89	-4.99	und.	-5.0099	-5.09	-5.16

7. 4

x	100	1,000	100,000	1,000,000
$f(x)$	4.07	4.007	4.00007	4.000007

9. 5

x	100	1,000	1,000,000
$f(x)$	4.7	4.97	4.99997

11. 5 **13.** $\frac{7}{5}$ **15.** 7 **17.** $\frac{5}{11}$ **19.** does not exist **21.** -6 **23.** $\sqrt{11}$ **25.** $\frac{4}{3}$ **27.** $\frac{1}{2}$

29. 1 **31.** discontinuous at $x = 0$ **33.** continuous **35.** function exists for $x > -1$, discontinuous at $x = 0$, because results in division by zero **37.** discontinuous at $x = 2$, point does not exist **39.** discontinuous at $x = -2$, $f(2)$ does not exist **41.** discontinuous at $x = -1$ and $x = 2$, $f(1)$ does not exist, $\lim_{x \to 2} f(x)$ does not exist

43. 10,000 **45.** 6.67×10^{-5}

18–2 Exercises, p. 657

1. 6 **3.** -5 **5.** $2x$ **7.** $8t - 6$ **9.** $6v^2 - 5$ **11.** $7 - 2x$ **13.** $1 + 9x^2$ **15.** $-8 + 10t$ **17.** $-\frac{1}{x^2}$ **19.** $\frac{-8}{(x - 4)^2}$ **21.** $-\frac{8}{x^3}$ **23.** $\frac{3}{x^2}$ **25.** -2 **27.** -5 **29.** -1 **31.** -4 **33.** 21 ft/s **35.** 53 **37.** 8π **39.** velocity at 3 s = 89 km/s, acceleration at 3 s = 54 km/s^2

18–3 Exercises, p. 669

1. 0 **3.** $8x^7$ **5.** $12x^{-3}$ **7.** $12x^3$ **9.** $10x - 9$ **11.** $6x^2 - 7$ **13.** $\frac{2}{3}x$ **15.** $12x^3 + 2x + 8$ **17.** $2x^3 + 2x^2$ **19.** $-21x^{-4} - 6$ **21.** $5x^4 - 12x^2$ **23.** 10 **25.** 12 **27.** 10 **29.** 3,750 ft **31.** 5.98 A **33.** -19.69 V **35.** $r = 2.89$ ft, $h = 2.86$ ft **37.** 6.64 in. $\times$ 3.64 in. $\times$ 1.18 in. **39.** $30x - 18$

18–4 Exercises, p. 679

1. $3(4x^3 - 9x^2)$ **3.** $35x^4 - 15x^2$ **5.** $9x^2 + 14x + 3$ **7.** $28x^3 + 18x^2 - 56x - 24$ **9.** $3x^2 + 20x + 12$ **11.** $28x^3 - 60x^2 - 6x$ **13.** $4x^3 - 6x^2 - 2x + 22$ **15.** $90x^4 + 232x^3 - 195x^2 + 14x$ **17.** $\frac{-6}{(x - 3)^2}$ **19.** $\frac{x^2 - 3}{7x^2}$ **21.** $\frac{-37}{(3x - 5)^2}$ **23.** $\frac{5}{(3x + 5)^2}$ **25.** $\frac{-12x^2 - 42}{(2x^2 + 3x - 7)^2}$ **27.** $\frac{-28x^2 + 42x + 3}{x^2(7x + 1)^2}$ **29.** $\frac{4x^3 + 11x^2 - 20x - 15}{(2x + 5)^2}$ **31.** $8x - 3$ **33.** $98x + 42$ **35.** $3(4x^3 + x^2)^2(12x^2 + 2x)$ **37.** $\frac{1}{3}(6x^3 - 5x^2)^{-2/3}(18x^3 - 10x)$ **39.** $16x^6(x^2 - 2)^3(15x^2 - 14)$ **41.** $\frac{(3x^2 + 7)^2(36x^4 - 99x^2 + 7)}{x^2(4x^2 - 1)^2}$ **43.** $\frac{18x}{(1 - x)^3}$ **45.** $1.89(3t + 1)^{0.4}$ **47.** 6,307.5 kW **49.** $-4V\pi^2 f^2 L(R^2 + 4\pi^2 f^2 L^2)^{-3/2}$ A/H **51.** $\frac{r}{2}$ **53.** $\sqrt{13}x + 6y = 8\sqrt{13}$ **55.** $47{,}104x - y = 34{,}816$

18–5 Exercises, p. 687

1. -3 **3.** $6x^2$ **5.** $\frac{-2x - 7y^2}{14xy - 4 - 2y}$ **7.** $\frac{4x^3 - 9y^2 - 24x^2}{18xy + 6}$ **9.** $\frac{4x^3 - 3x^5 - 7y^2 - x + 21x^2y^2}{14xy + 98y^3 - 14x^3y}$ **11.** $\frac{2x^3y^2 + y^2}{x^2}$ **13.** $\frac{6xy - 18x^2 - 2y^3}{6xy^2 - 4y^3 - 3x^2}$ **15.** $\frac{3x^2 - 2xy}{3y^2(3x - y)^2 - x^2}$ **17.** $\frac{-2x - y}{x + 3y^2}$ **19.** $\frac{6xy - 2x^3}{9y(3y^2 + 8)^2 + 9y - 3x^2}$ **21.** $\frac{3x(2x^2 + 7)^2}{4y(y^2 - 3)}$ **23.** $\frac{3 + 5y}{3y^2 - 5x + 7}, \frac{169}{137}$ **25.** $-\frac{4}{6}$ **27.** $\frac{y + 3y^2}{x}, \frac{8}{3}$ **29.** -0.243 A **31.** $y = -3$ **33.** 54.08 mi/h **35.** -0.64 ft/s **37.** 38 Ω/min

Chapter Review, p. 689

1. continuous **3.** discontinuous at $x = \frac{1}{3}$ **5.** continuous for $x \geq -1$ **7.** -2 **9.** -1 **11.** does not exist **13.** 0 **15.** $\frac{7}{6}$ **17.** $-\frac{1}{7}$ **19.** 8 **21.** $16s + 16$ **23.** $-\frac{1}{x^2}$ **25.** $21v^2 - 18v$

27. $\frac{-6}{(x-2)^2}$ **29.** $\frac{4}{x^2}$ **31.** 1,320 mi/h **33.** 0 **35.** $5x^4$ **37.** $-8x^{-3}$ **39.** $35x^4 + 8x - 1$

41. $\frac{1}{2}x^{-1/2} + 21x^2 - 1$ **43.** $r = 2.12$ ft, $h = 4.24$ ft **45.** $196x^6 + 35x^4$ **47.** $8x^7 - 28x^6 + 15x^4 - 48x^3$

49. $\frac{-3x^2 + 21}{(x^2+7)^2}$ **51.** $\frac{-x^2 - 6x + 24}{x^2(x-8)^2}$ **53.** $9(3x+8)^2$ **55.** $(x+1)^2(4x-5)$ **57.** $\frac{2(x-1)(x+6)}{(x^2+6)^2}$

59. 9.04 ft/s **61.** $-\frac{4x}{3}$ **63.** $\frac{1 - 6x - 16y}{16x + 10y}$ **65.** $\frac{-x^2 - y^2}{2y(x^2 - y^2)^2 - 2xy}$ **67.** $-\frac{4}{3y}$

69. $\frac{24x^2 - 7y^2 + 2x}{3y^2 + 14xy}$

Chapter Test, p. 690

1. $14x + 1$, -13 **2.** 3 s **3.** -11 **4.** 24.7 m/s **5.** $21x^{-4} + 8x$
6. $7x^6 + 18x^5 + 28x^3 + 33x^2 - 60x$ **7.** 117 m/s **8.** $(x^3 + 4x^2 - 6)^2(11x^4 + 62x^3 + 84x^2 - 12x - 18)$

9. 7.64 A **10.** 142.2 V **11.** discontinuous at $x = 0$, because $f(0)$ does not exist **12.** $\frac{7y - 12xy}{6x^2 + 12y^2 - 7x}$

13. $\frac{3x^2 + 24x - 5}{(x+4)^2}$ **14.** 11 **15.** $\frac{-4y^2 - 84x^2y - 126xy^3}{28x^3 + 189x^2y^2 + 8xy + 36y^3}$ **16.** $45x^4 - 12x^2 + 7$ **17.** 19.3 ft/s

18. $\frac{-6x - 6}{x^3}$

CHAPTER 19

19–1 Exercises, p. 700

1. $(6x^2 - 2x)\,dx$ **3.** $6x(x^2 - 5)^2\,dx$ **5.** $(60x^4 + 108x^2)\,dx$ **7.** $(6 + 3x)^{-2/3}\,dx$ **9.** $\left[\frac{-5x^2 + 4}{(5x^2+4)^2}\right]dx$

11. $\left[\frac{x(2 + x^3)}{(1 - x^3)^2}\right]dx$ **13.** $\Delta y = -1.43$, $dy = -1.40$ **15.** $\Delta y = 9.053$, $dy = 8.64$

17. $\Delta y = +12.51$, $dy = +14.41$ **19.** 76.8 cm^3 **21.** 15.71 ft^2 **23.** 4.52 in.^{2} **25.** 3.2%
27. 0.0173 cm **29.** 18.84 in.^{3}

19–2 Exercises, p. 709

1. $8x + C$ **3.** $-x^2 + C$ **5.** $\frac{2}{3}y^6 + C$ **7.** $-\frac{1}{2}x^{-2} + C$ **9.** $\frac{3}{7}x^{7/3} + C$ **11.** $x^4 + 8x + C$

13. $\frac{4}{3}y^3 + \frac{1}{2}y^2 - 7y + C$ **15.** $\frac{4}{5}y^{5/4} - \frac{5}{y} + C$ **17.** $\frac{1}{5}(y^2 - 8)^5 + C$ **19.** $\frac{1}{8}(x^4 - 3x^2 + 5)^4 + C$

21. $\frac{1}{12}(3x^4 - 5)^2 + C$ **23.** $\frac{-1}{8(x^4 - 6x^2)^2} + C$ **25.** $\frac{1}{9}(6x - 8)^{3/2} + C$ **27.** $\frac{-1}{3(x^3 - x^2)^3} + C$

29. $2\sqrt{y^2 + 6} + C$ **31.** $\frac{1}{12}(2x^4 - x^2)^6 + C$ **33.** $y = x^3 - 8x + 3$ **35.** $q = t^3 + \frac{1}{2}t^2$

37. 199.17 m/s **39.** 440 ft/s, 18.75 s

19–3 Exercises, p. 718

1. 7.2, 8.1 **3.** 16.703, 17.48 **5.** 5.97, 6.185 **7.** 2.964, 3.12 **9.** 0.602, 0.5389 **11.** 5.25, 6.08

13. 1.6625, 1.7101 **15.** 1.837, 3.1725 **17.** 2 **19.** $11\frac{1}{3}$ **21.** $\frac{22}{3}$ **23.** 40.5

19–4 Exercises, p. 733

1. $\frac{27}{2}$ **3.** $\frac{765}{2}$ **5.** 3.8098 **7.** 4.3267 **9.** 32.92 **11.** $\frac{28}{9}$ **13.** $2(\sqrt{3} - 1)$ **15.** 5.777

17. $-\frac{85}{8}$ **19.** 0.0149 **21.** $+\frac{1}{4}$ unit^2 **23.** 36 units^2 **25.** 2, 2 **27.** $\frac{8}{5}, \frac{16}{7}$ **29.** 0, 6.4

31. 330.9 units^3 **33.** 5.98 units^3 **35.** $\frac{32}{3}\pi$ units^3 **37.** $\frac{81}{2}\pi$ units^3 **39.** $\frac{30}{2}$

19–5 Exercises, p. 739

1. 26.25 **3.** 5.62 **5.** 4.51 **7.** 50.34 **9.** 0.608 **11.** 4.34 **13.** 61.25 **15.** 8.55 **17.** 376
19. 0.06315

Chapter Review, p. 741

1. $(21x^2 - 16x)\,dx$ **3.** $(9x^2 - 6x + 7)\,dx$ **5.** $\frac{56x^2 + 48x + 3}{(7x + 3)^2}\,dx$ **7.** $dV = 180$

9. $dV = 643.4$ **11.** 0.75% **13.** $\frac{3}{2}x^4 + C$ **15.** $\frac{2}{3}x^{3/2} + \frac{3}{2}x^{2/3} + \frac{9}{4}x^4 + C$

17. $\frac{4}{15}(3x + 15)^{5/4} + C$ **19.** $\frac{1}{3}(x^4 + 9)^3 + C$ **21.** $\frac{1}{15}(x^3 + 9)^5 + C$ **23.** 255 ft/s, 17.97 s **25.** $15\frac{1}{3}$

27. 27 **29.** $15\frac{1}{4}$ **31.** 78 **33.** 0 **35.** −7.11 **37.** −2.67 **39.** 0.085 **41.** 12 **43.** $\frac{3}{2}$, 3.6

45. 107.2 units^3 **47.** 3.84 **49.** 5.32 **51.** 0.49 **53.** 3.625

Chapter Test, p. 742

1. $2x^4 - 2x^3 + \frac{1}{2}x^2 - 10x + C$ **2.** 56 **3.** $\frac{1}{105}(7x^3 + 9)^5 + C$ **4.** 15 **5.** $\frac{3}{4}(\sqrt[3]{9} - 1)$ **6.** $\frac{3}{2}$, 3.6

7. $(45x^2 + 32x - 7)\,dx$ **8.** $17\frac{1}{3}$ **9.** 16π units^3 **10.** 1.877 **11.** 0.19% **12.** 1065.16 **13.** 10.17 **14.** 1.958 **15.** $3t + \frac{1}{2}t^2 - \frac{3}{t+2} + \frac{3}{2}$

CHAPTER 20

20–1 Exercises, p. 758

1. $5 \cos 5x$ **3.** $-12 \sin 6x$ **5.** $4t \cos 2t^2$ **7.** $42x^2(\sin 2x^3)$ **9.** $2t \cos(t^2 + 8)$ **11.** $-3 \sin x \cos^2 x$ **13.** $24x^2\sin(4x^3)\cos(4x^3)$ **15.** $-24t \cos^2(4t^2)\sin(4t^2)$ **17.** $-x(\cos x^2)^{-1/2}\sin x^2$ **19.** $2(4x - 3)^{-1/2}\cos \sqrt{4x - 3}$ **21.** $(-384x^5 + 192x^3 - 18x)\sin(8x^3 - 3x)^2$ **23.** $x \cos x + \sin x + 2x^3\sin(2x) - 3x^2\cos(2x)$ **25.** $14 \sin(7x)\cos(7x)\cos^3 3x - 9 \sin^2(7x)\cos^2 3x \sin 3x$ **27.** $9x^2\cos x - 3x^3\sin x$ **29.** $12t + 4 \sin t$ **31.** $9 \cos(x^2) - 18x^2\sin(x^2)$ **33.** $\dfrac{21x^2\cos(x - 5) + 14x^3\sin(x - 5)}{\cos^3(x - 5)}$ **35.** $\dfrac{8 \cos x \sin 4x + 3 \sin^2 4x \sin x}{\cos^4 x}$ **37.** $6x^2\sin^2 x\sin(x^3) - 4 \cos(x^3)\cos x \sin x$ **39.** $12x \cos 3x \cos(x^2) - 18 \sin(x^2)\sin 3x$ **41.** $16x - 6 \cos^2(2x)\sin(2x)$ **43.** -5.88 **45.** $\frac{\pi}{4}, \frac{3\pi}{4}, \frac{5\pi}{4}, \frac{7\pi}{4}, \ldots$ or $\frac{(2n - 1)\pi}{4}$, n is the set of all integers **47.** -1.1 **49.** 100π units/s **51.** $2t + 3 \cos t - 3t \sin t$ **53.** -0.209 units/s

20–2 Exercises, p. 766

1. $7 \sec^2 7x$ **3.** $-2x \csc(x^2)\cot(x^2)$ **5.** $(2x + 3)\sec(x^2 + 3x)\tan(x^2 + 3x)$ **7.** $-60t \csc^2(6t^2)$ **9.** $\dfrac{-3x^2}{2(x^3 - 2)^{1/2}} \csc(x^3 - 2)^{1/2}\cot(x^3 - 2)^{1/2}$ **11.** $-4 \cot(2\theta)\csc^2(2\theta)$ **13.** $-0.5 \sec(\cot 5x)\tan(\cot 5x)\csc^2(5x)$ **15.** $\frac{3}{2} \sec^{1/2}(3x)\tan 3x$ **17.** $8 \tan^3(2x)\sec^2(2x)$ **19.** $3 \cos 3x - 8x \sec^2(x^2)$ **21.** $-6x \cot^2(x^2)\csc^2(x^2) + 32 \sec 8x \tan 8x$ **23.** $4 \csc(x^2) - 8(x^2)\csc(x^2)\cot(x^2)$ **25.** $\dfrac{8x \sec^2 8x - 2 \tan 8x}{x^3}$ **27.** $10 \cos(4x^2)\tan(5x)\sec^2(5x) - 8x \tan^2(5x)\sin(4x^2)$ **29.** $\sin 2x \cos 2x$ **31.** $3x^2 - 3 \sec^2 x - 2 \csc^2 x \cot x$ **33.** $2x \cot x - x^2\csc^2 x$ **35.** 37.9 ft/s **37.** $I_0(-\csc^2\theta)\dfrac{d\theta}{dt}$ **39.** 130.8 m/s^2

20–3 Exercises, p. 774

1. $\dfrac{-2x}{\sqrt{1 - x^4}}$ **3.** $\dfrac{96x^2}{\sqrt{1 - 16x^6}}$ **5.** $\dfrac{2x(\sec x^2)\tan x^2}{1 + \sec^2 x^2}$ **7.** $\dfrac{-1}{5x^2 + 16x + 13}$ **9.** $\dfrac{-2}{\sqrt{2 - 4x}\sqrt{4x - 1}}$ **11.** $\dfrac{-5}{(1 + 25x^2)(\text{Arctan } 5x)^2}$ **13.** $\dfrac{3x - 4}{1 + x^2} + 3 \text{ Arctan } x$ **15.** $\dfrac{-x}{\sqrt{1 - x^2}} + \text{Arccos } x$ **17.** $\dfrac{3 \text{ Arctan}^2 x}{1 + x^2}$ **19.** $\dfrac{12x + 28 - (1 + 16x^2)(3 \text{ Arctan } 4x)}{(1 + 16x^2)(3x + 7)^2}$ **21.** $\dfrac{2 \text{ Arcsin } x}{\sqrt{1 - x^2}} - \dfrac{1}{x^2}$

23. $\dfrac{21(x^3)}{\sqrt{1 - x^6}} + 7\text{ Arcsin}(x^3)$ **25.** $\dfrac{-2x^3}{\sqrt{1 - 4x^2}} + 3x^2\text{Arccos}(2x)$

27. $\dfrac{3\sqrt{1 - x^6}\text{ Arccos}(x^3) + 9(x^3)}{\sqrt{1 - x^6}\text{ Arccos}^2(x^3)}$ **29.** 0.13 rad/s **31.** 7.38 **33.** 0.16 rad/s **35.** 12.7 ft

20–4 Exercises, p. 781

1. $\left(\dfrac{1}{x}\right)\log e$ **3.** $\dfrac{6}{2x + 5}$ **5.** $\dfrac{2 \log e}{x}$ **7.** $\dfrac{6 \ln(x^3)}{x}$ **9.** $\cot x \log e$ **11.** $2 \tan x$

13. $\dfrac{\log e(\cos 2x - 2x \sin 2x)}{x \cos 2x}$ **15.** $\dfrac{2 - \sin x}{2x + \cos x}$ **17.** $\dfrac{15(\log 4x^5)^2 \log e}{x}$ **19.** $\dfrac{-11}{(3x + 1)(2x - 3)}$

21. $\dfrac{-\log e}{2 - x}$ **23.** $2 \log e + \log 7x^2$ **25.** $\dfrac{4(6x + 4)}{3x + 4} + 4 \ln (3x^2 + 4x)$

27. $\dfrac{6 \cos x \log e (\log 6x^2)^2 + 2x \sin x(\log 6x^2)^3}{x \cos^3 x}$ **29.** 4.27 **31.** 7.83×10^{-10} farads/cm

20–5 Exercises, p. 786

1. $2(6^{2x})\ln 6$ **3.** $2x(7^{x^2})\ln 7$ **5.** $-2(5^{-2x})\ln 5$ **7.** $\dfrac{e^{\sqrt{x}}}{2\sqrt{x}}$ **9.** $-3(2 - 3^x)^2(3^x\ln 3)$ **11.** $e^x\cot e^x$

13. $3^x\ln 3 + 2(4^{2x})\ln 4$ **15.** $e^{-x} - \dfrac{2}{x^3}$ **17.** $2x + 2e^{2x}$ **19.** $\dfrac{xe^x - 2e^x}{x^3}$ **21.** $xe^x + e^x$

23. $\dfrac{-5^x\ln 5 \log e}{1 - 5^x}$ **25.** $2^{\sin x}\ln 2 \cos x$ **27.** $\dfrac{3e^x}{x} + e^x\ln x^3$ **29.** $\dfrac{2xe^{2x} + xe^x - e^{2x} - e^x}{x^2}$

31. $-4xe^{2x^2}\sin e^{2x^2}$ **33.** $\dfrac{8e^x}{1 + e^{2x}}$ **35.** 74.06 A/s

Chapter Review, p. 787

1. $8 \sin 8x$ **3.** $\dfrac{-3}{4} \sin 3x(\cos 3x)^{-3/4}$ **5.** $-6t(t^2 - 4)^2\sin(t^2 - 4)^3$ **7.** $-10x^4\sin(5x^2) + 3x^2\cos(5x^2)$

9. $12 \sin^2 x \cos x - 3x \sin (3x) + \cos(3x)$ **11.** $960\pi \cos(120\pi t + 1.5\pi)$ **13.** $21 \tan^2(7x)\sec^2(7x)$

15. $-24x^2\csc^2(4x^3)\cot(4x^3)$ **17.** $\dfrac{-2x \csc^2 x \cot x - \csc^2 x}{3x^2}$ **19.** $-2x \tan(2x)\csc(x^2)\cot(x^2) + 2 \csc(x^2)\sec^2(2x)$

21. $8(1 - \tan^2 x - 2x \tan x \sec^2 x)\sin[8x(1 - \tan^2 x)]$ **23.** 6 **25.** 0.028 A **27.** $\dfrac{-4x\text{ Arccos } x^2}{\sqrt{1 - x^4}}$

29. $\dfrac{9x^4}{\sqrt{1 - x^6}} + 6x\text{ Arcsin}(x^3)$ **31.** $\dfrac{-6x(x + 8) + \sqrt{1 - 9x^4}\text{ Arccos } 3x^2}{\sqrt{1 - 9x^4}(x + 8)^2}$ **33.** $\dfrac{(1 + x^2)\text{Arctan } x - x}{(1 + x^2)(\text{Arctan } x)^2}$

35. 62.45 **37.** 0.013 rad/s (decreasing) **39.** $\dfrac{6}{3x + 1}$ **41.** $\dfrac{6 \log x^3 \log e}{x}$ **43.** $\dfrac{x^3 - 2}{x^4 + x}$

45. $-3 \ln x^2\sin(3x) + \dfrac{2 \cos(3x)}{x}$ **47.** $\dfrac{2 \csc x \log e + x \log 7x^2\csc x \cot x}{x \csc^2 x}$ **49.** 7.22

51. $-\frac{1}{4}$ J/s **53.** $9^{(x+1)}\ln 9$ **55.** $\frac{3e^{2x}}{3x-1} + 2e^{2x}\ln(3x-1)$ **57.** $\frac{2x^2e^{x^2} - 2e^{x^2}}{3x^3}$
59. $2(5^{2x})(-x \sin x^2 + \ln 5 \cos x^2)$ **61.** -0.15°F/min

Chapter Test, p. 788

1. $\frac{2e^{2x}}{\sqrt{1-e^{4x}}}$ **2.** 45.9 ft/s **3.** $2x(3x^2+5)\sec(x^2)\tan(x^2) + 6x\sec(x^2)$ **4.** 1 s **5.** -8.05 **6.** -44.7
7. $\frac{2t \log e}{t^2+5}$ **8.** $\frac{\sec^2 x(1+e^x) - e^x \tan x}{(1+e^x)^2}$ **9.** $\frac{dx}{dt} = t\cos t - \sin t,\ \frac{dy}{dt} = -2\sin 4t$
10. $\frac{6xe^x}{1+9x^4} + e^x \text{Arctan}(3x^2)$ **11.** 0.083 rad/s **12.** $\frac{5\sin(10x) - 3\ln 5 \sin^2(5x)}{5^{3x}}$ **13.** 550183.5
14. 43.08 **15.** $2\sec^2 x \tan x$

CHAPTER 21

21–1 Exercises, p. 795

1. $-\frac{1}{4}(1+e^{-x}) + C$ **3.** $\frac{1}{6}\cos^6 x + C$ **5.** $\frac{1}{6}\sec^6 x + C$ **7.** $\frac{1}{9}\text{Arctan}^3 3x + C$ **9.** $\frac{1}{4}(\text{Arcsin } 2x)^2 + C$
11. $\frac{-1}{9(1+e^{3x})^3} + C$ **13.** 0.176 **15.** $\frac{1}{16}(1-4e^{-x})^4 + C$ **17.** 15.27 **19.** $-\frac{1}{2}\cot^2 x + C$
21. $\frac{1}{3}\cos^3 x - \frac{3}{2}\cos^2 x + C$ **23.** 0 **25.** $\frac{1}{4}[\ln(x-2)]^4 + C$ **27.** $\frac{-1}{4(4t+t^2)^2} + C$
29. $-\frac{2}{3}(1+\cot x)^{3/2} + C$ **31.** $\frac{1}{4}$ **33.** 1.11 C **35.** 0.44 **37.** $10^6[2\sqrt{2+e^t} + C]$

21–2 Exercises, p. 803

1. $-\frac{1}{6}\ln|1-6t| + C$ **3.** $\frac{1}{8}\ln|4x^2-5| + C$ **5.** $\frac{1}{3}\ln|\sin 3x| + C$ **7.** $-\frac{1}{4}\ln|\cos 4t - 7| + C$
9. 1.25 **11.** $e^x + \frac{1}{2}e^{2x} + C$ **13.** $6x + \frac{1}{3}e^{3x} + C$ **15.** 0.318 **17.** $\frac{2}{3}(3-e^{-x})^{3/2} + C$ **19.** 0.41
21. $\frac{1}{3}\left(\frac{8^{3x-1}}{\ln 8}\right) + C$ **23.** 4.11 **25.** $-\frac{1}{6}\ln|8-3x^2| + C$ **27.** -28.2 **29.** $e^{\text{Arctan } x} + C$
31. $-\frac{1}{2}\ln|5-e^{2x}| + C$ **33.** 0 **35.** 632.13 **37.** 11.68 **39.** 0.60 m/s **41.** 11.47 C

21–3 Exercises, p. 811

1. $\frac{1}{4}\sin 4x + C$ **3.** $-\frac{3}{7}\cot 7t + C$ **5.** $-\frac{1}{8}\cos 4x^2 + C$ **7.** 0.393 **9.** $-\frac{1}{2}\cos(3x^2+4x) + C$
11. $-\frac{1}{9}\ln|\cos 3x^3| + C$ **13.** -0.176 **15.** $\frac{1}{4}\ln|\sin 4x| + \frac{1}{4}\ln|\csc 4x - \cot 4x| + C$ **17.** 0.155

19. $\sin(\ln x) + C$ **21.** 3.55 **23.** $\tan x - \cot x + C$ **25.** 0.2 **27.** π **29.** -0.44

31. $-\frac{1}{3}\cos(3t + 1)$ **33.** $\frac{1}{2}\sin 4t$

21–4 Exercises, p. 820

1. $\frac{1}{5}\cos^5 x - \frac{1}{3}\cos^3 x + C$ **3.** $\frac{1}{14}\sin^7(2x) - \frac{1}{5}\sin^5(2x) + \frac{1}{6}\sin^3(2x) + C$ **5.** $\frac{1}{4}\sin^4 x - \frac{1}{6}\sin^6 x + C$

7. $\frac{1}{2}x - \frac{1}{4}\sin 2x + C$ **9.** $\frac{16}{15}$ **11.** $-\frac{1}{5}\cos 5x + \frac{1}{15}\cos^3 5x + C$ **13.** 0.632 **15.** $-\cot x - x + C$

17. $\frac{28}{15}$ **19.** $-\frac{1}{9}\cot^3 3x + C$ **21.** $-\frac{1}{10}\csc^5 2x + \frac{1}{6}\csc^3 2x + C$ **23.** $\frac{\pi}{2}$ **25.** $\frac{1}{2}\sec^2 x + \ln|\cos x| + C$

27. $\frac{1}{6}\sec^6 x - \frac{3}{4}\sec^4 x + \frac{3}{2}\sec^2 x + \ln|\cos x| + C$ **29.** $-\cot x + \frac{1}{4}\csc^4 x - \frac{1}{3}\cot^3 x + C$ **31.** $\frac{3\pi}{8}$

33. 3.53

21–5 Exercises, p. 828

1. $\text{Arcsin}\,\frac{x}{2} + C$ **3.** $\text{Arcsin}\,\frac{t}{7} + C$ **5.** 0.11 **7.** $\frac{1}{14}\text{Arctan}\,\frac{2t}{7} + C$ **9.** $\frac{1}{2\sqrt{2}}\text{Arcsin}(2\sqrt{2}x) + C$

11. $\text{Arcsin}\,\frac{2x}{3} + \frac{1}{4}\sqrt{9 - 4x^2} + C$ **13.** $\frac{1}{2}\text{Arcsin}(e^{2t}) + C$ **15.** $\frac{1}{5}\text{Arctan}\,\frac{x}{5} + C$ **17.** 0

19. $\text{Arcsin}(\tan x) + C$ **21.** $\text{Arcsin}\,\frac{x^2}{2} + C$ **23.** $\frac{\sqrt{3}}{8}\text{Arctan}\,\frac{2t}{\sqrt{3}} + C$

25. $\frac{1}{4}\ln\left|\sqrt{1 + t^2} + t\right| - \frac{\sqrt{t^2 + 1}}{t} + C$

21–6 Exercises, p. 834

1. $x^2\sin x + 2x\cos x - 2\sin x + C$ **3.** $\frac{1}{3}xe^{3x} - \frac{1}{9}e^{3x} + C$ **5.** 0.45

7. $x(\ln x)^3 - 3x(\ln x)^2 + 6x\ln x - 6x + C$ **9.** $x^2\sin x + 2x\cos x - 2\sin x + C$

11. $\frac{2}{5}e^t\sin 2t + \frac{1}{5}e^t\cos 2t + C$ **13.** $x\,\text{Arccos}\,3x - \frac{1}{3}\sqrt{1 - 9x^2} + C$ **15.** $-\frac{1}{2}y^2 - y\cot y + \ln|\sin y| + C$

17. $-\frac{1}{4}y(3 - y)^4 - \frac{1}{20}(3 - y)^5 + C$ **19.** $\frac{1}{2}x^2(3 + x^3)^2 - \frac{9}{5}x^5 - \frac{3}{8}x^8 + C$

21. $\frac{1}{3}y(y + 8)^3 - \frac{1}{12}(y + 8)^4 + C$ **23.** $x^3e^x - 3x^2e^x + 6xe^x - 6e^x + C$ **25.** $\frac{\pi}{2}$

27. $\frac{2}{3}t(1 + t)^{3/2} - \frac{4}{15}(1 + t)^{3/2} - \frac{4}{15}$

21–7 Exercises, p. 836

1. -2 **3.** $\left(\frac{4x - 1}{16}\right)e^{4x} + C$ **5.** $-\frac{1}{5}\ln\left|\frac{5 + 3t}{t}\right| + C$ **7.** $\frac{270x^2 - 576x + 1024}{2835}(8 + 3x)^{3/2} + C$

9. 0.206 **11.** $-\frac{\sin 4x}{8} + \frac{\sin 2x}{4} + C$ **13.** $\frac{1}{125}\left[\frac{1}{2}(1 + 5x)^2 - 2(1 + 5x) + \ln|1 + 5x|\right] + C$ **15.** 0.02

17. $\ln|\ln \theta| + C$ **19.** 1.67 **21.** $\frac{-x^2(1 - x^2)^{3/2}}{5} - \frac{2}{15}(1 - x^2)^{3/2} + C$

Chapter Review, p. 837

1. $\frac{4}{3}$ **3.** $\frac{1}{8}(\text{Arcsin } 2x)^4 + C$ **5.** 49,974 **7.** $\frac{1}{3}\tan^3 x + C$ **9.** $e^{\text{Arcsin } x} + C$ **11.** 0.734

13. $-1 + e$ **15.** $\frac{11^{2x+1}}{2 \ln 11} + C$ **17.** 7.53 **19.** 0.68 **21.** $-\frac{1}{3}\csc 3\theta + C$ **23.** 1.12

25. $\frac{1}{6}\sec 3x^2 + C$ **27.** 7.4 **29.** $\frac{\pi}{16}$ **31.** $-\frac{1}{2}\cot^2 x - \ln|\sin x| + C$ **33.** $\sec x + C$

35. $\frac{1}{5}\tan^5 t + \frac{2}{3}\tan^3 t + \tan t + C$ **37.** $\text{Arcsin } \frac{x}{3} + C$ **39.** $\frac{1}{3}t^2(9 + t^2)^{3/2} - \frac{2}{15}(9 + t^2)^{5/2} + C$

41. $\frac{1}{16}\sqrt{16x^2 - 25} + C$ **43.** $\frac{1}{5}\text{Arctan } \frac{x}{5} + C$ **45.** 0.718

47. $\frac{1}{2}x^2(\ln x)^4 - x^2(\ln x)^3 + \frac{3}{2}x^2(\ln x)^2 - \frac{3}{2}x^2\ln x + \frac{3}{4}x^2 + C$ **49.** -2π **51.** $-\frac{2}{9}(3 - x^3)^{3/2} + C$

53. $\frac{1}{5}\cos^4 x \sin x + \frac{4}{15}\cos^2 x \sin x + \frac{8}{15}\sin x + C$ **55.** -0.21 **57.** $\frac{e^{4x}}{64}(16x^2 - 8x + 2) + C$

59. $\tan t - t + C$

Chapter Test, p. 839

1. $\frac{1}{3}(e^x - 4)^3 + C$ **2.** 0.82 ft/lb **3.** $\frac{1}{9}(1 + e^{3x})^3 + C$ **4.** 4.57 **5.** -0.38 **6.** 1.31 **7.** 0

8. $\cos^3 x - \frac{1}{2}\cos^2 x + C$ **9.** $\frac{2}{\sqrt{3}}\text{Arctan } \frac{x}{2\sqrt{3}} + C$ **10.** $\frac{\pi^2}{4}$ **11.** $\approx 15{,}600{,}000$ m

12. $-\frac{1}{9}(\text{Arccos } 3x)^3 + C$ **13.** $\sqrt{x^2 - 4} - 2\text{ Arccos } \frac{2}{x} + C$ **14.** $\csc x - \frac{1}{3}\csc^3 x + C$

15. $\frac{1}{4}y(y + 5)^4 - \frac{1}{20}(y + 5)^5 + C$ **16.** $\frac{-\sqrt{9 - x^2}}{x} + C$

CHAPTER 22

22–1 Exercises, p. 849

1. $x - \frac{x^3}{6} + \frac{x^5}{120} - \cdots$ **3.** $1 - \frac{9t^2}{2} + \frac{27t^4}{8} - \cdots$ **5.** $3 - 3x + \frac{3}{2}x^2 + \cdots$ **7.** $1 + \frac{3}{2}x - \frac{9}{8}x^2 + \cdots$

9. $1 - 3x + 9x^2 + \cdots$ **11.** $x + x^2 + \frac{1}{3}x^3 + \cdots$ **13.** $-3x - \frac{9x^2}{2} - 9x^3 \cdots$ **15.** $-8 + 5x + 3x^3 + \cdots$

17. $1 - \frac{1}{8}x^2 + \frac{1}{384}x^4 + \cdots$ **19.** $\ln 2 + \frac{5}{2}x + \cdots$ **21.** $\frac{\pi}{4} + \frac{1}{2}x + \cdots$ **23.** $0.54 - 0.84x + \cdots$

25. $1 - x^2 + \cdots$ **27.** $x + \frac{x^3}{6} + \frac{x^5}{120} + \cdots$ **29.** $1 + \frac{1}{2}x^2 + \frac{5}{24}x^4 + \cdots$

22–2 Exercises, p. 856

1. $1 - 3x + \frac{9}{2}x^2 - \frac{9x^3}{2} + \cdots$ **3.** $x^2 - \frac{x^4}{2} + \frac{x^6}{3} - \frac{x^8}{4} + \cdots$ **5.** $\frac{5t}{3} - \frac{125}{162}t^3 + \frac{625}{5{,}832}t^5 - \frac{15{,}625}{2{,}204{,}496}t^7 + \cdots$

7. $x^2 - x^3 + \frac{x^4}{2} - \frac{x^5}{6} + \cdots$ **9.** $x + x^2 + \frac{5}{6}x^3 + \frac{5}{6}x^4 + \cdots$ **11.** $x - \frac{2}{3}x^3 + \frac{2}{15}x^5 - \frac{4}{315}x^7 + \cdots$

13. $3 - 3x^2 + 3x^4 - 3x^6 + \cdots$ **15.** $x - \frac{1}{10}x^5 + \frac{1}{216}x^9 + \cdots$ **17.** $\frac{1}{2}\theta^2 + \frac{1}{3}\theta^3 + \frac{1}{12}\theta^4 + \cdots$

19. 26.319 **21.** $2x - 2x^3 + 2x^5 + \cdots$ **23.** $2x - 2x^3 + \frac{1}{4}x^5 + \cdots$ **25.** $-\frac{1^2}{x} + \frac{2}{3} + \cdots$

27. $x - \frac{1}{3}x^3 + \frac{1}{10}x^5 + \cdots$ **29.** 3.73

22–3 Exercises, p. 860

1. 1.0345 **3.** 0.29552 **5.** 0.997551 **7.** 1.625 **9.** $-0.6666\ldots$ **11.** $1.03222\ldots$ **13.** 0.5 **15.** 3.25 **17.** 1.05522 **19.** 1.19189 **21.** $\approx 2\pi(0.01996)$, max error $= 2\pi(0.021)$ **23.** $-0.2228\ \mu\text{C}$ **25.** -19.52

22–4 Exercises, p. 864

1. $\frac{\sqrt{2}}{2} - \frac{\sqrt{2}}{2}\left(x - \frac{\pi}{4}\right) - \frac{\sqrt{2}}{4}\left(x - \frac{\pi}{4}\right)^2 + \cdots$ **3.** $e^3 + e^3(x - 3) + \frac{e^3}{2}(x - 3)^2 + \cdots$

5. $\frac{\sqrt{3}}{3} + \frac{4}{3}\left(x - \frac{\pi}{6}\right) + \frac{4\sqrt{3}}{9}\left(x - \frac{\pi}{6}\right)^2 + \cdots$ **7.** $\frac{1}{3} - \frac{1}{9}(x - 2) + \frac{1}{27}(x - 2)^2 + \cdots$

9. $1 - 2(x - 1) + 3(x - 1)^2 + \cdots$ **11.** 24.35144 **13.** 0.6819 **15.** -0.0202 **17.** 2.94958

19. $\frac{\pi}{2}(t - 2) - \frac{\pi^3}{2}(t - 2)^3 + \cdots$

22–5 Exercises, p. 874

1. $\frac{1}{2} + \frac{2}{\pi}\sin\frac{\pi x}{2} + \frac{2}{3\pi}\sin\frac{3\pi x}{2} + \frac{2}{5\pi}\sin\frac{5\pi x}{2} + \cdots$

3. $\frac{\pi}{4} - \frac{2}{9\pi}\cos 3x + \cdots + \sin x - \frac{1}{2}\sin 2x + \frac{1}{3}\sin 3x + \cdots$

5. $\frac{3}{2} + \frac{2}{\pi}\sin x + \frac{2}{3\pi}\sin 3x + \frac{2}{5\pi}\sin 5x + \cdots$ **7.** $1 + \frac{8}{\pi}\sin t + \frac{8}{3\pi}\sin 3t + \frac{8}{5\pi}\sin 5t + \cdots$

9. $\frac{5}{3} - \frac{8\sqrt{3}}{\pi}\cos\frac{2\pi x}{3} + \frac{4\sqrt{3}}{\pi}\cos\frac{4\pi x}{3} + \cdots + \left(\frac{3 - 3\sqrt{3}}{\pi^2}\right)\sin\frac{2\pi x}{3} + \left(\frac{3\sqrt{3} + 6\pi}{4\pi^2}\right)\sin\frac{4\pi x}{3}$

11. $\frac{\pi}{6} + \frac{-9 + 2\sqrt{3}\pi}{4\pi}\cos\frac{2x}{3} + \frac{-9 - 8\sqrt{3}\pi}{16\pi}\cos\frac{4x}{3} + \cdots + \frac{3\sqrt{3} + 2\pi}{4\pi}\sin\frac{2x}{3} + \frac{-3\sqrt{3} + 4\pi}{16\pi}\sin\frac{4x}{3} + \cdots$

13. $\frac{1}{\pi}(e^\pi - 1) - \frac{1}{\pi}(e^\pi + 1)\cos t + \frac{2}{5\pi}(e^\pi - 1)\cos 2t - \frac{1}{5\pi}(e^\pi + 1)\cos 3t + \cdots$

Chapter Review, p. 875

1. $1 - 2x^2 + \frac{2}{3}x^4 + \cdots$ **3.** $2x - 2x^2 - \frac{8}{3}x^3 + \cdots$ **5.** $1 - 3t + \frac{9}{2}t^2 + \cdots$ **7.** $1 + x - \frac{1}{3}x^3 + \cdots$

9. $1 - 4x + 8x^2 + \cdots$ **11.** $1 - 2x^2 + \frac{2}{3}x^4 + \cdots$ **13.** $\frac{1}{x} + 1 + \frac{2}{3}x + \cdots$

15. $\frac{-1}{6}x^3 - \frac{1}{60}x^5 - \frac{1}{336}x^7 + \cdots$ **17.** 0.39 **19.** $3x^2 + \frac{5}{6}x^4 + \frac{7}{120}x^6 + \cdots$

21. 1 **23.** $-\frac{a}{r^2} + \frac{1}{2}a + \frac{1}{3}ar + \cdots$ **25.** 0.17903 **27.** 0.7022 **29.** 1.173455 . . .

31. $e^{-2} - e^{-2}(x - 2) + \frac{e^{-2}}{2}(x - 2)^2 + \cdots$ **33.** $-1 + \frac{1}{2}(x - \pi)^2 - \frac{1}{24}(x - \pi)^4 + \cdots$

35. $1 + \frac{1}{3}(x - 1) - \frac{1}{9}(x - 1)^2 + \cdots$ **37.** $16 + 8(x - 4) + (x - 4)^2$ [no remaining terms]

39. 0.46947 **41.** 0.51504 **43.** 5.099 **45.** $2\pi - 4 \sin t - 2 \sin 2t - \frac{4}{3} \sin 3t + \cdots$

Chapter Test, p. 876

1. $x + x^3 + \frac{1}{2}x^5 + \frac{1}{6}x^7 + \cdots$ **2.** $\frac{\sqrt{3}}{2} - \left(x - \frac{\pi}{3}\right) - \sqrt{3}\left(x - \frac{\pi}{3}\right)^2 + \cdots$

3. $\frac{\pi^2}{6} + 2 \cos t - \frac{1}{2} \cos 2t + \frac{2}{9} \cos 3t + \cdots \frac{\pi^2 - 4}{\pi} \sin t - \frac{\pi}{2} \sin 2t + \cdots$ **4.** $x + x^2 + \frac{1}{3}x^3 + \cdots$ **5.** $\frac{8}{3}$

6. 1.208655 **7.** 3.0037 **8.** $\frac{V}{L}t - \frac{RV}{2L^2}t^2 + \frac{R^2V}{6L^3}t^3 + \cdots$ **9.** $x + \frac{1}{6}x^3 + \cdots$ **10.** 9,981

11. $1 + \frac{4}{\pi} \sin x + \frac{4}{3\pi} \sin 3x + \frac{4}{5\pi} \sin 5x + \cdots$

CHAPTER 23

23–1 Exercises, p. 885

1. second order, first degree **3.** first order, first degree **5.** second order, first degree **7.** second order, first degree **9.** first order, first degree **11.** is a solution **13.** is a solution **15.** is not a solution

17. $\ln|x + 4| - 2 \ln|y + 2| + y = C$ **19.** $\frac{1}{2} \ln|x^2 + 1| = \ln|\sin y| + C$ **21.** $y = -\cos x - \frac{1}{2}x^2 + C$

23. $y = \frac{1}{2}x \sin 2x + \frac{1}{4} \cos 2x + C$ **25.** $y = x + e^x + C$ **27.** $\frac{1}{3} \ln y = \cos x + C$

29. $\text{Arctan } y = -\text{Arctan } x + C$ **31.** $\frac{1}{3}y^3 = \frac{1}{4} \sin 2x + \frac{1}{2}x + C$ **33.** $-\ln|y| = -\ln|\cos x| + C$

35. $\frac{1}{2} \ln|y| = \ln|x| - 0.41$ **37.** $\frac{V}{R}\left(1 - e^{\frac{-Rt}{L}}\right)$

23–2 Exercises, p. 893

1. $\frac{1}{2}x^2 + \frac{1}{2}y^2 = x + C$ **3.** $\frac{1}{2}x^2 - \frac{1}{2}y^2 = 5y + C$ **5.** $(xy)^2 = 2\ln|y| + C$ **7.** $\frac{x}{y} + \sin x = C$

9. $xy + 2\ln|x| = C$ **11.** $y = x^3 + Cx^{-3}$ **13.** $\ln|x^2 + 1| = \ln|y + 1| + C$ **15.** $y = \frac{1}{2} + \frac{C}{2e^{x^2}}$

17. $y = -x + x^2C$ **19.** $y = \frac{x}{e^x} + \frac{C}{e^x}$ **21.** $y = \frac{x^3}{3e^{3x}} + \frac{C}{3e^{3x}}$ **23.** $y = \frac{2x + 1}{4e^{-2x}} - \frac{C}{e^{-4x}}$

25. $y = -x^3 - \frac{1}{3}x + \frac{7}{3}x^4$ **27.** $y = \frac{1}{x}\tan x - \frac{1}{x}$ **29.** $y = xe^x + e^x$

23–3 Exercises, p. 903

1. $i = \frac{1}{2}(1 - e^{-0.72t})$ **3.** $V = 0.175(1 - e^{-t/12})$ **5.** $y = x + \frac{2}{e^x}$ **7.** $\frac{1}{2}y^2 - \ln y = \frac{1}{x}x^2 + C$

9. $\frac{3}{2}y^2 = -\frac{1}{2}x^2 + C$ **11.** 357.8 lb **13.** 550 lb **15.** 28.44% **17.** 33.2° **19.** 3,970

21. 120.5 yr **23.** 102.9 yr **25.** 99.7 min **27.** 84.85 in./s **29.** $\frac{18.31(1 + e^{-3.5t-0.245})}{(1 - e^{-3.5t-0.245})}$

31. $\frac{0.08}{e^{5t}}\left[\frac{e^{5t}}{26}(5\cos t + \sin t) - 0.1925\right]$

23–4 Exercises, p. 912

1. $y = \frac{1}{20}x^5 + C_1x + C_2$ **3.** $y = e^x + C_1x + C_2$ **5.** $y = x\ln x - x + C_1x + C_2$
7. $y = C_1e^{2x} + C_2e^x$ **9.** $y = C_1e^{3x} + C_2e^{-x}$ **11.** $y = C_1e^{-1/3x} + C_2e^{5x}$ **13.** $y = C_1e^{4m} + C_2e^{-3m}$
15. $y = C_1\cos 3x + C_2\sin 3x$ **17.** $y = C_1e^{-x} + C_2e^x$ **19.** $y = e^{2x}(C_1\cos 5x + C_2\sin 5x)$
21. $y = e^x(C_1\cos 2x + C_2\sin 2x)$ **23.** $y = C_1e^{1/2x} + C_2xe^{1/2x}$ **25.** $m = e^{(t-c)}$

23–5 Exercises, p. 922

1. $y = C_1e^{2x} + C_2e^x + \frac{21}{4} + \frac{9}{2}x + \frac{3}{2}x^2$ **3.** $y = -\frac{3}{2}x + \frac{1}{4}C_1e^{4x} + C_1C_2$
5. $y = C_1e^{(1+\sqrt{7})x} + C_2e^{(1-\sqrt{7})x} + 2x + 2x^3$

7. $y = e^{-x}(C_1\cos\sqrt{3}x + C_2\sin\sqrt{3}x) + \frac{1}{26}\sin 4x - \frac{3}{52}\cos 4x$

9. $y = C_1e^{3x} + C_2e^{-x} + \cos 2x$ **11.** $y = C_1e^{3x} + C_2e^{-x} - \frac{1}{4}e^x + \frac{1}{5}\sin x - \frac{1}{10}\cos x$
13. $y = C_1e^x + C_2e^{-x} + xe^x$ **15.** $y = -2\sin x + x^2$ **17.** $y = -\cos x + \sin x + 3\cos 2x$

19. $y = e^{-2x}(C_1\cos Bx + C_2\sin 3x) + \frac{27}{145}\sin 2x - \frac{24}{145}\cos 2x$ **21.** $y = \cos x + \frac{1}{2}\sin x - 2 + x^2 - \frac{1}{2}e^x$

23. $s = C_1\sin\omega t + C_2\cos\omega t$ **25.** $q_C = (-550\cos 316.2t + 550) \times 10^{-5}$ C **27.** 0

23–6 Exercises, p. 930

1. $\dfrac{4}{s(s+4)}$ **3.** $\dfrac{4}{s(s^2+4)} + \dfrac{1}{(s+1)^2}$ **5.** $\dfrac{1}{(s+1)(s+2)}$ **7.** $\dfrac{s+4}{(s+1)^2}$ **9.** $\dfrac{s}{(s+3)(s+2)}$
11. $1 - e^{-t}$ **13.** $e^{-5t}\sin 6t$ **15.** $2t - \sin 2t$ **17.** $1 + e^{-5t} - e^{5t}$ **19.** $y = 2e^{3t} + e^t$

21. $y = e^{3t} - e^{2t}$ **23.** $y = \dfrac{1}{2}\sin 2t$ **25.** $y = te^{-t}$ **27.** $y = \dfrac{5}{4} + \dfrac{7}{4}e^{-4t}$ **29.** $0.012(1 - e^{-5t})$

31. $\dfrac{1}{1041.6}(1 - \cos 35.35t)$ **33.** $4\cos\sqrt{8}t - \dfrac{2\sqrt{8}}{3}\sin\sqrt{8}t + \dfrac{3}{4}\sin 2t$

Chapter Review, p. 931

1. $\ln y = \dfrac{1}{4}\ln 4x + C$ **3.** $I = \dfrac{e^t(\cos t + \sin t)}{2}$ **5.** $-\ln y = e^x + C$ **7.** $\ln y = \ln x - \dfrac{3}{2}x^2 + C$

9. $\dfrac{1}{3}x^3 + \dfrac{1}{2}y^2 - \dfrac{5}{6} = 0$ **11.** $\ln y = x^3 - 1$ **13.** $\ln(y^2 + x^2) + 2x = C$ **15.** $\dfrac{y}{x} = 7x + C$

17. $y = \dfrac{1}{4} + \dfrac{C}{4e^{x^4}}$ **19.** $xy = e^x(x - 1) + C$ **21.** 262 lb **23.** 70,700 **25.** 37%

27. $i = \dfrac{10}{3}(1 - e^{-75t})$ **29.** $y = C_1\cos\sqrt{7}x + C_2\sin\sqrt{7}x$ **31.** $y = C_1e^{2x} + C_2xe^{2x}$ **33.** $x = \cos t + \sin t$

35. $y = 2\cos 2x - 4\sin 2x$ **37.** $y = C_1e^x + C_2xe^x$ **39.** $y = C_1e^{4x} + C_2e^{-3x} - \dfrac{1}{6}e^{3x}$

41. $y = C_1e^x + C_2e^{-x} - \dfrac{1}{5}\cos 2x$ **43.** $y = 2\sin x + x^2$ **45.** $y = C_1e^{-2x} + C_2e^{-2x} + \sin x$

47. $\dfrac{6s^2}{(s^2+9)^2}$ **49.** $\dfrac{s+4}{(s+4)^2 + 25}$ **51.** $[t + 1]e^{-3t}$ **53.** $-2e^{-3t} + 3e^{2t}$ **55.** $y = \sin 3t - e^{-2t} + e^t$

57. $y = \dfrac{29}{60}e^{-3t} + \dfrac{31}{60}e^{3t} - \dfrac{1}{10}\sin t$ **59.** $y = e^{2t}\sin t$

Chapter Test, p. 933

1. $2y + \dfrac{1}{2}x^2 = 3x + C$ **2.** $y = e^{-x/2}\left(C_1\cos\dfrac{\sqrt{15}}{2}x + C_2\sin\dfrac{\sqrt{15}}{2}x\right)$ **3.** $y = C_1e^{4x} + C_2e^{4x}$ **4.** 7.36 lb

5. second order, third degree **6.** $\ln(y) = \dfrac{1}{2}\ln(x^2 + 2) + 3x + C$ **7.** $y = \dfrac{1}{x}\sin x - \cos x + \dfrac{c}{x}$

8. $\dfrac{1}{2}(x^2 + y^2) = x^3 + C$ **9.** $i = \dfrac{12}{125}\left(\dfrac{7}{3}\sin 8t - 8\cos 8t\right) + 0.768e^{-7t/3}$ **10.** $y = \sin t$

11. $y = C_1e^{3x} + C_2e^{-x} - \dfrac{1}{2}e^x + 2\sin x - \cos x$ **12.** $y = e^{2x}(C_1\cos 3x + C_2\sin 3x)$

13. $q = C_1e^{-6t} + C_2e^{-8.4t} - 0.004\sin 60t - 0.0006\cos 60t$ **14.** $y = \dfrac{1}{3}x^3 + \dfrac{7}{2}x^2 + C_1x + C_2$

15. $y = 10e^{-10/3t}$

APPENDIX A

A–1 Exercises, p. A–6

1. rational **3.** rational **5.** integer and rational **7.** integer and rational **9.** rational **11.** F **13.** T **15.** T **17.** F **19.** T

21.

-3 $-\frac{1}{2}$ 1.3 $\sqrt{5}$ 4.5

–6 –5 –4 –3 –2 –1 0 1 2 3 4 5 6

23.

a. –10

–10 –9 –8 –7 –6 –5 –4 –3 –2 –1 0

b. $\sqrt{40}$

–1 0 1 2 3 4 5 6 7

c. π

–1 0 1 2 3 4 5

d. –5

–6 –5 –4 –3 –2 –1 0

A–2 Exercises, p. A–10

1. -6 **3.** -17 **5.** 18 **7.** -14 **9.** -54 **11.** 4 **13.** 13.95 **15.** -13 **17.** 19 **19.** 48.28 **21.** distributive **23.** commutative **25.** 17° **27.** **(a)** 300 **(b)** 360 **(c)** 44 **(d)** 425 **29.** **(a)** -8.6 **(b)** -529 **(c)** 24.8 **(d)** -123

A–3 Exercises, p. A–14

1. 23 **3.** -12 **5.** 8 **7.** undefined **9.** -4 **11.** 14 **13.** -6 **15.** -3 **17.** $\frac{-5}{11.5}$ **19.** 5.5 **21.** **(a)** 96 ft **(b)** 250 m **23.** 61 ft/s **25.** 122.2 ft^2

A–4 Exercises, p. A–20

1. -32 **3.** x^4 **5.** 10 **7.** $\frac{4}{25}$ **9.** $\frac{2}{x^3}$ **11.** $\frac{1000}{y^6}$ **13.** $\frac{b^2}{a}$ **15.** $\frac{-2}{s^3}$ **17.** $16x^2$ **19.** $\frac{1}{10^4}$ **21.** 198 **23.** 19 **25.** 175.929 m^2 **27.** 14.137 m^3 **29.** $(3 + 2)^2 = 5^2 = 25$, $3^2 + 2^2 = 9 + 4 = 13$, $25 \neq 13$

A–5 Exercises, p. A–22

1. 8.956×10^6 **3.** -4.50×10^5 **5.** 6.9×10^9 **7.** 5.0×10^6 mm^3 **9.** 12,300,000 **11.** 5,400,000,000 **13.** 0.002 **15.** 105,000 **17.** 1.8066×10^{-8} **19.** 1.6×10^9 **21.** 7.6549×10^2 **23.** 3.76×10^{40} joules

A–6 Exercises, p. A–25

1. 9 **3.** 2 **5.** 2 **7.** $\frac{4}{5}$ **9.** $\frac{1}{2}$ **11.** $4\sqrt{2}$ **13.** 50 **15.** $2\sqrt{6}$ **17.** 5 **19.** 0.2 **21.** 182
23. 0.71 **25.** 2.19 **27.** 2.29 **29.** 13.39 **31.** 14.76 cm **33.** 4 in. **35.** 4.830

A–7 Exercises, p. A–29

1. $5x$ **3.** $2a$ **5.** 0 **7.** $11xz$ **9.** $3a - 7b + 5$ **11.** $7x + 5y$ **13.** $-a + 5b$ **15.** $3x - 3y + 12$
17. $x^2 + 3y^2 + 2x - 2y$ **19.** $-18m + 23n$ **21.** $4m - 8$ **23.** $0.0406x + 206.8$ **25.** $-2I_1 + 3I_2$

A–8 Exercises, p. A–32

1. $12a^3b$ **3.** $-24s^6$ **5.** $2x^4 - 6x^2y$ **7.** $-m^2n^2 + m^2n - mn^2$ **9.** $x^9 + x^7 - x^5$ **11.** $x^2 + x - 12$
13. $2a^2 - 6ab + 4b^2$ **15.** $15s^4 + 2s^2t - t^2$ **17.** $108x^5y^2$ **19.** $4x^2 + 12xy + 9y^2$ **21.** $x^3 - 2xy^2 + y^3$
23. $x^3 - 5x^2 + 4x + 6$ **25.** $12y^3 - 8y^2 - 28y - 8$ **27.** $R^2 + 2Rr + 3r^2$

A–9 Exercises, p. A–36

1. $3a$ **3.** $-\frac{c^2d^2}{4}$ **5.** $5y^2$ **7.** $\frac{x^3}{2y}$ **9.** $-\frac{2y^2a}{3}$ **11.** $6a^2 - 2b^5$ **13.** $\frac{4s}{t} - \frac{2}{t} + \frac{5}{8st^2}$ **15.** $x + 6$

17. $2x^2 + 3x - 1$ **19.** $3x^2 + x + 3$ **21.** $x^2 - 2x + 4 + \frac{3}{x + 2}$ **23.** $3x^2 + 9$

25. $2x - 1 + \frac{3x}{x^2 - x + 2}$ **27.** $15x - 1$ **29.** $-3gt^2 + 2t + 1$

A–10 Exercises, p. A–40

1. 32% **3.** 37.5% **5.** 20% **7.** 0.625% **9.** 5% **11.** 0.56 **13.** 0.0025 **15.** 0.18 **17.** 2.55
19. 0.0003 **21.** 20.8% **23.** 273 **25.** 93.3 **27.** 132 **29.** 767 **31.** 13.3 oz

APPENDIX B

Exercises, p. B–9

1. 72° **3.** 33.6° **5.** 77° **7.** 60° **9.** ∠3 and ∠6; ∠5 and ∠4 **11.** ∠2 and ∠5; ∠3 and ∠7; ∠4 and ∠8; ∠1 and ∠6 **13.** 115° **15.** 4 cm **17.** 19.2 cm **19.** 42.5 ft **21.** **(a)** circle **(b)** 31.4 cm **(c)** 78.5 cm^2 **23.** **(a)** parallelogram **(b)** 37 in. **(c)** 108 in.^{2} **25** **(a)** rhombus **(b)** 20.8 yd **(c)** 23.4 yd^2 **27.** **(a)** Trapezoid **(b)** 17 ft **(c)** 17.1 ft^2 **29.** **(a)** right circular cylinder **(b)** 45,592.8 ft^3 **31.** **(a)** cube **(b)** 937.5 in.^{2} **33.** **(a)** right circular cylinder **(b)** 791.3 m^2 **35.** **(a)** 15 Ω **(b)** 24 Ω **(c)** 120 Ω **37.** 2.6 in.^{2} **39.** 36 m **41.** 64.1 in.^{2} **43.** 235,500 H

Index